THE MECHANICS PROBLEM SOLVER®

REGISTERED TRADEMARK

A Complete Solution Guide to Any Textbook

Staff of Research and Education Association
Dr. M. Fogiel, Chief Editor

Research and Education Association
61 Ethel Road West
Piscataway, New Jersey 08854

THE MECHANICS
PROBLEM SOLVER®

Copyright © 2002, 1998, 1995, 1980 by Research & Education Association. All rights reserved. No part of this book may be reproduced in any form without permission of the publisher.

Printed in the United States of America

Library of Congress Control Number 2001086082

International Standard Book Number 0-87891-519-2

PROBLEM SOLVER is a registered trademark of
Research & Education Association, Piscataway, New Jersey 08854

IV-1

WHAT THIS BOOK IS FOR

Students have generally found mechanics a difficult subject to understand and learn. Despite the publication of hundreds of textbooks in this field, each one intended to provide an improvement over previous textbooks, students continue to remain perplexed as a result of the numerous conditions that must often be remembered and correlated in solving a problem. Various possible interpretations of terms used in mechanics have also contributed to much of the difficulties experienced by students.

In a study of the problem, REA found the following basic reasons underlying students' difficulties with mechanics taught in schools:

(a) No systematic rules of analysis have been developed which students may follow in a step-by-step manner to solve the usual problems encountered. This results from the fact that the numerous different conditions and principles which may be involved in a problem, lead to many possible different methods of solution. To prescribe a set of rules to be followed for each of the possible variations, would involve an enormous number of rules and steps to be searched through by students, and this task would perhaps be more burdensome than solving the problem directly with some accompanying trial and error to find the correct solution route.

(b) Textbooks currently available will usually explain a given principle in a few pages written by a professional who has an insight in the subject matter that is not shared by students. The explanations are often written in an abstract manner which leaves the students confused as to the application of the principle. The explanations given are not sufficiently detailed and extensive to make the student aware of the wide range of applications and different aspects of the principle being studied. The numerous possible variations of principles and their applications are usually not discussed, and it is left for the students to discover these for themselves while doing the exercises. Accordingly, the average student is expected to rediscover that which has been long known and practiced, but not published or explained extensively.

(c) The examples usually following the explanation of a topic are too few in number and too simple to enable the student to obtain a thorough grasp of the principles involved. The explanations do not provide sufficient basis to enable a student to solve problems that may be subsequently assigned for homework or given on examinations.

The examples are presented in abbreviated form which leaves out much material between steps, and requires that students derive the omitted material themselves. As a result, students find the examples difficult to understand--contrary to the purpose of the examples.

Examples are, furthermore, often worded in a confusing manner. They do not state the problem and then present the solution. Instead, they pass through a general discussion, never revealing what is to be solved for.

Examples, also, do not always include diagrams/graphs, wherever appropriate, and students do not obtain the training to draw diagrams or graphs to simplify and organize their thinking.

(d) Students can learn the subject only by doing the exercises themselves and reviewing them in class, to obtain experience in applying the principles with their different ramifications.

In doing the exercises by themselves, students find that they are required to devote considerably more time to mechanics than to other subjects of comparable credits, because they are uncertain with regard to the selection and application of the theorems and principles involved. It is also often necessary for students to discover those "tricks" not revealed in their texts (or review books), that make it possible to solve problems easily. Students must usually resort to methods of trial-and-error to discover these "tricks", and as a result find that they may sometimes spend several hours in solving a single problem.

(e) When reviewing the exercises in classrooms, instructors usually request students to take turns in writing solutions on the board and explaining them to the class. Students often find it difficult to explain in a manner that holds the interest of the class, and enables the remaining students to follow the material written on the board. The remaining students seated in the class are, furthermore, too occupied with copying the material from the board, to listen to the oral explanations and concentrate on the methods of solution.

This book is intended to aid students in mechanics to overcome the difficulties described, by supplying detailed illustrations of the solution methods which are usually not apparent to students. The solution methods are illustrated by problems selected from those that are most often assigned for class work and given on examinations. The problems are arranged in order of complexity to enable students to learn and understand a particular topic by reviewing the problems in sequence. The problems are illustrated with detailed step-by-step explanations, to save the students the large amount of time that is often needed to fill in the gaps that are usually found between steps of illustrations in textbooks or review/outline books.

The staff of REA considers mechanics a subject that is best learned by allowing students to view the methods of analysis and solution techniques themselves. This approach to learning the subject matter is similar to that practiced in various scientific laboratories, particularly in the medical fields.

In using this book, students may review and study the illustrated problems at their own pace; they are not limited to the time allowed for explaining problems on the board in class.

When students want to look up a particular type of problem and solution, they can readily locate it in the book by referring to the index which has been extensively prepared. It is also possible to locate a particular type of problem by glancing at just the material within the boxed portions. To facilitate rapid scanning of the problems, each problem has a heavy border around it. Furthermore, each problem is identified with a number immediately above the problem at the right-hand margin.

To obtain maximum benefit from the book, students should familiarize themselves with the section, "How To Use This Book," located in the front pages.

To meet the objectives of this book, staff members of REA have selected problems usually encountered in assignments and examinations, and have solved each problem meticulously to illustrate the steps which are usually difficult for students to comprehend. Special gratitude, for added outstanding support in this area, is due to:

Prof. R.B. Agarwal
Texas A&M University

Prof. James R. Albers
Western Washington University

Prof. John A. Brenner
Loyola Marymount University

Prof. C.S. Cheng
East Stroudsburg State College

Prof. Charles A. Eckroth
St. Cloud State University

Prof. Hubert P. Grunwald
State University of N.Y. at Brockport

Prof. Thruman R. Kremser
Albright College

Gratitude is also expressed to the many persons involved in the difficult task of typing the manuscript with its endless changes, and to the REA art staff who prepared the numerous detailed illustrations together with the layout and physical features of the book.

The difficult task of coordinating the efforts of all persons was carried out by Carl Fuchs. His conscientious work deserves much appreciation. He also trained and supervised art and production personnel in the preparation of the book for printing.

Finally, special thanks are due to Helen Kaufmann for her unique talents in rendering those difficult border-line decisions and in making constructive suggestions related to the design and organization of the book.

Max Fogiel, Ph.D.
Program Director

HOW TO USE THIS BOOK

This book can be an invaluable supplement to standard textbooks for students studying mechanics. The book is divided into 26 chapters, each dealing with a separate topic. The subject matter is developed beginning with statics and extending through friction, kinematics, impulse and momentum, systems of particles, rigid body kinetics, and vibrations. Sections on three-dimensional dynamics, moving coordinate frames, special relativity, and variational methods have also been included. An extensive number of applications have been included, since they appear to be most troublesome to students.

HOW TO LEARN AND UNDERSTAND A TOPIC THOROUGHLY

1. Refer to your class text and read the section pertaining to the topic. You should become acquainted with the principles discussed there. These principles, however, initially may not be clear to you.

2. Locate the topic you are looking for by referring to the Table of Contents in the front of the book.

3. Turn to the page where the topic begins and review the problems under each topic, in the order given. For each topic, the problems are arranged in order of complexity, from the simplest to the more difficult. Some problems may appear similar to others, but each problem has been selected to illustrate a different point or solution method.

To learn and understand a topic thoroughly and retain its contents, it will be generally necessary for students to review the problems several times. Repeated review is essential in order to gain experience in recognizing the principles that should be applied, and to select the best solution technique.

HOW TO FIND A PARTICULAR PROBLEM

To locate one or more problems related to particular subject matter, refer to the index. In using the index, be certain to note that the numbers given there refer to *problem* numbers, not page numbers. This arrangement of the index is intended to facilitate finding a problem more rapidly, since two or more problems may appear on a page.

If a particular type of problem cannot be found readily, it is recommended that the student refer to the Table of Contents, and then turn to the chapter which is applicable to the problem being sought. By scanning or glancing at the material that is boxed, it will generally be possible to find problems related to the one being sought, without consuming considerable time. After the problems have been located, the solutions can be reviewed and studied in detail.

For the purpose of locating problems rapidly, students should acquaint themselves with the organization of the book as found in the Table of Contents.

In preparing for an exam, it is useful to find the topics to be covered in the exam from the Table of Contents, and then review the problems under those topics several times. This should equip the student with what might be needed for the exam.

CONTENTS

UNITS CONVERSION FACTORS

This section includes a particularly useful and comprehensive table to aid students and teachers in converting between systems of units.

The problems and their solutions in this book use SI (International System) as well as English units. Both of these units are in extensive use throughout the world, and therefore students should develop a good facility to work with both sets of units until a single standard of units has been found acceptable internationally.

In working out or solving a problem in one system of units or the other, essentially only the numbers change. Also, the conversion from one unit system to another is easily achieved through the use of conversion factors that are given in the subsequent table. Accordingly, the units are one of the least important aspects of a problem. For these reasons, a student should not be concerned mainly with which units are used in any particular problem. Instead, a student should obtain from that problem and its solution an understanding of the underlying principles and solution techniques that are illustrated there.

To convert	To	Multiply by	For the reverse, multiply by
acres	square feet	4.356×10^{4}	2.296×10^{-5}
acres	square meters	4047	2.471×10^{-4}
ampere-hours	coulombs	3600	2.778×10^{-4}
ampere-turns	gilberts	1.257	0.7958
ampere-turns per cm.	ampere-turns per inch	2.54	0.3937
angstrom units	inches	3.937×10^{-9}	2.54×10^{8}
angstrom units	meters	10^{-10}	10^{10}
atmospheres	feet of water	33.90	0.02950
atmospheres	inch of mercury at 0°C	29.92	3.342×10^{-2}
atmospheres	kilogram per square meter	1.033×10^{4}	9.678×10^{-5}
atmospheres	millimeter of mercury at 0°C	760	1.316×10^{-3}
atmospheres	pascals	1.0133×10^{5}	0.9869×10^{-5}
atmospheres	pounds per square inch	14.70	0.06804
bars	atmospheres	9.870×10^{-7}	1.0133
bars	dynes per square cm.	10^{6}	10^{-6}
bars	pascals	10^{5}	10^{-5}
bars	pounds per square inch	14.504	6.8947×10^{-2}
Btu	ergs	1.0548×10^{10}	9.486×10^{-11}
Btu	foot-pounds	778.3	1.285×10^{-3}
Btu	joules	1054.8	9.480×10^{-4}
Btu	kilogram-calories	0.252	3.969
calories, gram	Btu	3.968×10^{-3}	252
calories, gram	foot-pounds	3.087	0.324
calories, gram	joules	4.185	0.2389
Celsius	Fahrenheit	$(°C \times 9/5) + 32 = °F$	$(°F - 32) \times 5/9 = °C$

To convert	To	Multiply	For the reverse, multiply by
Celsius	kelvin	°C + 273.1 = K	K – 273.1 = °C
centimeters	angstrom units	1×10^{8}	1×10^{-8}
centimeters	feet	0.03281	30.479
centistokes	square meters per second	1×10^{-6}	1×10^{6}
circular mils	square centimeters	5.067×10^{-6}	1.973×10^{5}
circular mils	square mils	0.7854	1.273
cubic feet	gallons (liquid U.S.)	7.481	0.1337
cubic feet	liters	28.32	3.531×10^{-2}
cubic inches	cubic centimeters	16.39	6.102×10^{-2}
cubic inches	cubic feet	5.787×10^{-4}	1728
cubic inches	cubic meters	1.639×10^{-5}	6.102×10^{4}
cubic inches	gallons (liquid U.S.)	4.329×10^{-3}	231
cubic meters	cubic feet	35.31	2.832×10^{-2}
cubic meters	cubic yards	1.308	0.7646
curies	coulombs per minute	1.1×10^{12}	0.91×10^{-12}
cycles per second	hertz	1	1
degrees (angle)	mils	17.45	5.73×10^{-2}
degrees (angle)	radians	1.745×10^{-2}	57.3
dynes	pounds	2.248×10^{-6}	4.448×10^{5}
electron volts	joules	1.602×10^{-19}	0.624×10^{18}
ergs	foot-pounds	7.376×10^{-8}	1.356×10^{7}
ergs	joules	10^{-7}	10^{7}
ergs per second	watts	10^{-7}	10^{7}
ergs per square cm.	watts per square cm.	10^{-3}	10^{3}
Fahrenheit	kelvin	(°F + 459.67)/1.8	1.8K – 459.67
Fahrenheit	Rankine	°F + 459.67 = °R	°R – 459.67 = °F
faradays	ampere-hours	26.8	3.731×10^{-2}
feet	centimeters	30.48	3.281×10^{-2}
feet	meters	0.3048	3.281
feet	mils	1.2×10^{4}	8.333×10^{-5}
fermis	meters	10^{-15}	10^{15}
foot candles	lux	10.764	0.0929
foot lamberts	candelas per square meter	3.4263	0.2918
foot-pounds	gram-centimeters	1.383×10^{4}	1.235×10^{-5}
foot-pounds	horsepower-hours	5.05×10^{-7}	1.98×10^{6}
foot-pounds	kilogram-meters	0.1383	7.233
foot-pounds	kilowatt-hours	3.766×10^{-7}	2.655×10^{6}
foot-pounds	ounce-inches	192	5.208×10^{-3}
gallons (liquid U.S.)	cubic meters	3.785×10^{-3}	264.2
gallons (liquid U.S.)	gallons (liquid British Imperial)	0.8327	1.201
gammas	teslas	10^{-9}	10^{9}
gausses	lines per square cm.	1.0	1.0
gausses	lines per square inch	6.452	0.155
gausses	teslas	10^{-4}	10^{4}
gausses	webers per square inch	6.452×10^{-8}	1.55×10^{7}
gilberts	amperes	0.7958	1.257
grads	radians	1.571×10^{-2}	63.65
grains	grams	0.06480	15.432
grains	pounds	$^{1}/_{7000}$	7000
grams	dynes	980.7	1.02×10^{-3}
grams	grains	15.43	6.481×10^{-2}

To convert	To	Multiply	For the reverse, multiply by
grams	ounces (avdp)	3.527×10^{-2}	28.35
grams	poundals	7.093×10^{-2}	14.1
hectares	acres	2.471	0.4047
horsepower	Btu per minute	42.418	2.357×10^{-2}
horsepower	foot-pounds per minute	3.3×10^{4}	3.03×10^{-5}
horsepower	foot-pounds per second	550	1.182×10^{-3}
horsepower	horsepower (metric)	1.014	0.9863
horsepower	kilowatts	0.746	1.341
inches	centimeters	2.54	0.3937
inches	feet	8.333×10^{-2}	12
inches	meters	2.54×10^{-2}	39.37
inches	miles	1.578×10^{-5}	6.336×10^{4}
inches	mils	10^{3}	10^{-3}
inches	yards	2.778×10^{-2}	36
joules	foot-pounds	0.7376	1.356
joules	watt-hours	2.778×10^{-4}	3600
kilograms	tons (long)	9.842×10^{-4}	1016
kilograms	tons (short)	1.102×10^{-3}	907.2
kilograms	pounds (avdp)	2.205	0.4536
kilometers	feet	3281	3.408×10^{-4}
kilometers	inches	3.937×10^{4}	2.54×10^{-5}
kilometers per hour	feet per minute	54.68	1.829×10^{-2}
kilowatt-hours	Btu	3413	2.93×10^{-4}
kilowatt-hours	foot-pounds	2.655×10^{6}	3.766×10^{-7}
kilowatt-hours	horsepower-hours	1.341	0.7457
kilowatt-hours	joules	3.6×10^{6}	2.778×10^{-7}
knots	feet per second	1.688	0.5925
knots	miles per hour	1.1508	0.869
lamberts	candles per square cm.	0.3183	3.142
lamberts	candles per square inch	2.054	0.4869
liters	cubic centimeters	10^{3}	10^{-3}
liters	cubic inches	61.02	1.639×10^{-2}
liters	gallons (liquid U.S.)	0.2642	3.785
liters	pints (liquid U.S.)	2.113	0.4732
lumens per square foot	foot-candles	1	1
lumens per square meter	foot-candles	0.0929	10.764
lux	foot-candles	0.0929	10.764
maxwells	kilolines	10^{-3}	10^{3}
maxwells	webers	10^{-8}	10^{8}
meters	feet	3.28	30.48×10^{-2}
meters	inches	39.37	2.54×10^{-2}
meters	miles	6.214×10^{-4}	1609.35
meters	yards	1.094	0.9144
miles (nautical)	feet	6076.1	1.646×10^{-4}
miles (nautical)	meters	1852	5.4×10^{-4}
miles (statute)	feet	5280	1.894×10^{-4}
miles (statute)	kilometers	1.609	0.6214
miles (statute)	miles (nautical)	0.869	1.1508
miles per hour	feet per second	1.467	0.6818
miles per hour	knots	0.8684	1.152
millimeters	microns	10^{3}	10^{-3}

To convert	To	Multiply	For the reverse, multiply by
mils	meters	2.54×10^{-5}	3.94×10^{4}
mils	minutes	3.438	0.2909
minutes (angle)	degrees	1.666×10^{-2}	60
minutes (angle)	radians	2.909×10^{-4}	3484
newtons	dynes	10^{5}	10^{-5}
newtons	kilograms	0.1020	9.807
newtons per sq. meter	pascals	1	1
newtons	pounds (avdp)	0.2248	4.448
oersteds	amperes per meter	7.9577×10	1.257×10^{-2}
ounces (fluid)	quarts	3.125×10^{-2}	32
ounces (avdp)	pounds	6.25×10^{-2}	16
pints	quarts (liquid U.S.)	0.50	2
poundals	dynes	1.383×10^{4}	7.233×10^{-5}
poundals	pounds (avdp)	3.108×10^{-2}	32.17
pounds	grams	453.6	2.205×10^{-3}
pounds (force)	newtons	4.4482	0.2288
pounds per square inch	dynes per square cm.	6.8946×10^{4}	1.450×10^{-5}
pounds per square inch	pascals	6.895×10^{3}	1.45×10^{-4}
quarts (U.S. liquid)	cubic centimeters	946.4	1.057×10^{-3}
radians	mils	10^{3}	10^{-3}
radians	minutes of arc	3.438×10^{3}	2.909×10^{-4}
radians	seconds of arc	2.06265×10^{5}	4.848×10^{-6}
revolutions per minute	radians per second	0.1047	9.549
roentgens	coulombs per kilogram	2.58×10^{-4}	3.876×10^{3}
slugs	kilograms	1.459	0.6854
slugs	pounds (avdp)	32.174	3.108×10^{-2}
square feet	square centimeters	929.034	1.076×10^{-3}
square feet	square inches	144	6.944×10^{-3}
square feet	square miles	3.587×10^{-8}	27.88×10^{6}
square inches	square centimeters	6.452	0.155
square kilometers	square miles	0.3861	2.59
stokes	square meter per second	10^{-4}	10^{-4}
tons (metric)	kilograms	10^{3}	10^{-3}
tons (short)	pounds	2000	5×10^{-4}
torrs	newtons per square meter	133.32	7.5×10^{-3}
watts	Btu per hour	3.413	0.293
watts	foot-pounds per minute	44.26	2.26×10^{-2}
watts	horsepower	1.341×10^{-3}	746
watt-seconds	joules	1	1
webers	maxwells	10^{8}	10^{-8}
webers per square meter	gausses	10^{4}	10^{-4}

CHAPTER 1

STATICS

REPRESENTING FORCES BY VECTORS

• PROBLEM 1-1

Express the force $\vec{f}$ as shown in the diagram in terms of the basis $\hat{i}$, $\hat{j}$ and also in terms of $\hat{e}_1$, $\hat{e}_2$.

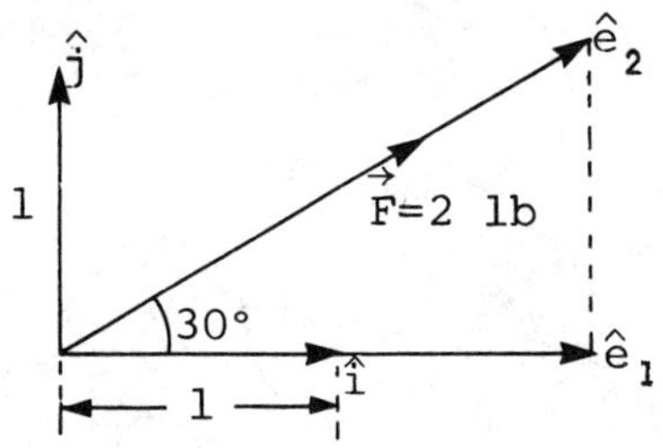

Solution: The vector $\vec{f}$ can be written

$$\vec{f} = f_1\hat{i} + f_2\,\hat{j} \tag{1}$$

where f_1 and f_2 are components of $\vec{f}$ along the coordinate axes.

Assuming that $\hat{i}$ and $\hat{j}$ are orthogonal unit vectors

$$\vec{f}_1 = \vec{f}\cdot\hat{i} = f\cos\theta \tag{2}$$

$$\vec{f}_2 = \vec{f}\cdot\hat{j} = f\sin\theta \tag{3}$$

where $f \equiv$ magnitude of f and θ is the angle between $\vec{f}$ and $\hat{i}$. ($\theta = 30°$ in this example.)

$$\therefore\ \vec{f} = (\sqrt{3}\,\hat{i} + \hat{j})\ \text{lbs.} \tag{4}$$

$\vec{f}$ can also be written in terms of the basis vectors $\hat{e}_1$ and $\hat{e}_2$. This is not an orthogonal set of unit vectors, so that simple result used above to evaluate the components cannct be used in the general case. However, since $\vec{f}$ lies alon one of the basis vectors, the result follows by inspection. Writing

$$\vec{f} = m_1 \hat{e}_1 + m_2 \hat{e}_2 , \qquad (5)$$

we see that $m_1 = 0$ and that $m_2 = \dfrac{f}{|\hat{e}_2|}$

$$\therefore \vec{f} = \frac{f}{|\hat{e}_2|} \hat{e}_2 \qquad (6)$$

Since the length of $\hat{e}_2$ is not specified, we cannot evaluate m_2 further.

• PROBLEM 1-2

A force $\vec{f}$ is expressed with respect to the basis $\hat{e}_1$, $\hat{e}_2$ by the equation $\vec{f} = 2(3\hat{e}_1 + 4\hat{e}_2)$ lb. If the directions of $\hat{e}_1$ and $\vec{f}$ are known as shown in the Figure, find the vector $\hat{e}_2$.

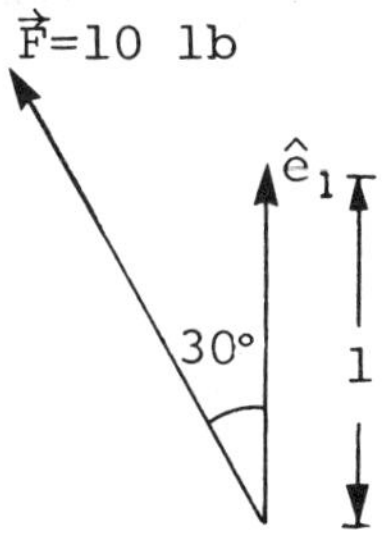

Solution: The vector $\hat{e}_2$ is shown in Figure 1.

The construction of Fig. 1 follows from the formula and the figure given in the problem. Point C is constructed so sides BC and AC of the triangle ABC intersect at right angles. In the smaller triangle (DBC) the side DC is given by (all units are lbs),

$$DC = AC - 6 = 10 \cos 30° - 6 = 2.660.$$

The side BC is given by,

$$BC = 10 \sin 30° = 5.$$

The side DB is given by,

$$DB = \sqrt{(BC)^2 + (DC)^2} = 5.664$$

Since $DB = 8\ |\hat{e}_2|$ we have,

$$|\hat{e}_2| = \frac{5.664}{8} = 0.708$$

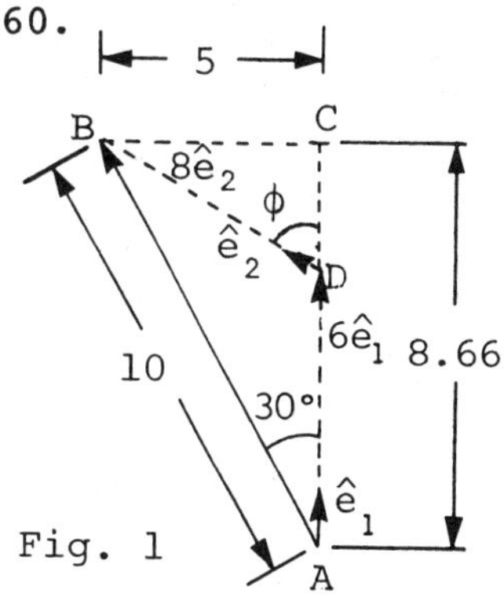

Fig. 1

We can write $\hat{e}_2$ on the $(\hat{i}, \hat{j})$ basis (defined in Figure 2) as

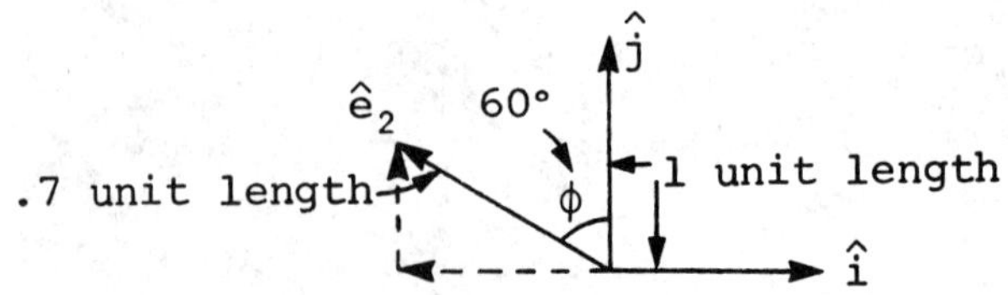

Fig. 2

$$\hat{e}_2 = m_1 \hat{i} + m_2 \hat{j}$$

$$= -|\hat{e}_2| \sin\phi\, \hat{i} + |\hat{e}_2| \cos\phi\, \hat{j}$$

$$= -(0.708)(0.883)\hat{i} + (0.708)(0.470)\hat{j}$$

$$\hat{e}_2 = -0.625\hat{i} + 0.333\hat{j}.$$

• PROBLEM 1-3

A telephone pole in a rural area supports a wire which carries a force of 50 lbs directed along the wire. The wire configuration is indicated in the figure shown below. Find the force on the pole from the wire.

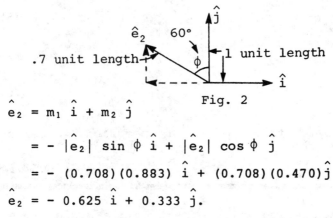

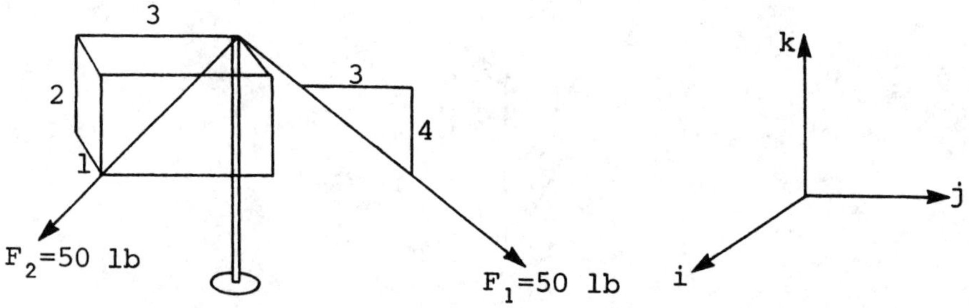

Solution: Forces F_1 and F_2 must be put into vector form using x, y, and z coordinates before they can be added to find the total force on the pole. We can write $\vec{F}_1$ directly as

$$\vec{F}_1 = 30\hat{j} - 40\hat{k}. \qquad (1)$$

F_2 has three components which can be found from vector laws as follows:

$$\vec{F}_2 = (x)\vec{i} - (3x)\vec{j} - (2x)\vec{k} \qquad (2)$$

$$|\vec{F}_2| = 50 = \sqrt{x^2 + 9x^2 + 4x^2}$$

$$x = 13.36,$$

giving $\vec{F}_2 = 13.36\vec{i} - 40.08\vec{j} - 26.72\vec{k}$ (3)

Adding $\vec{F}_1$ and $\vec{F}_2$ gives the force on the pole as

$$\vec{F} = (13.36\vec{i} - 10.08\vec{j} - 66.72\vec{k})\text{lb}. \qquad (4)$$

• PROBLEM 1-4

Find the resultant of the three forces acting at O by computing the magnitude and direction of the resultant.

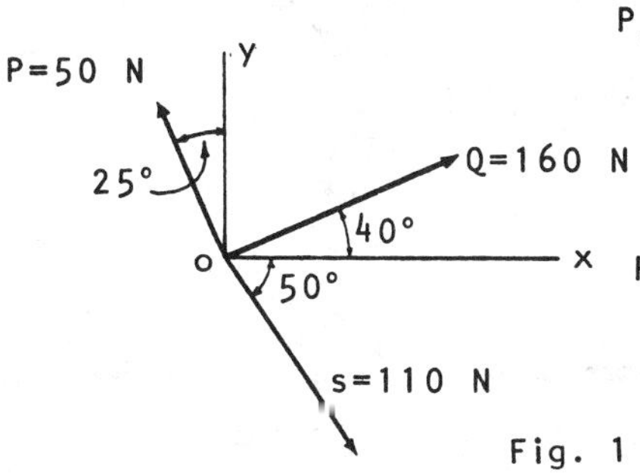

Fig. 1

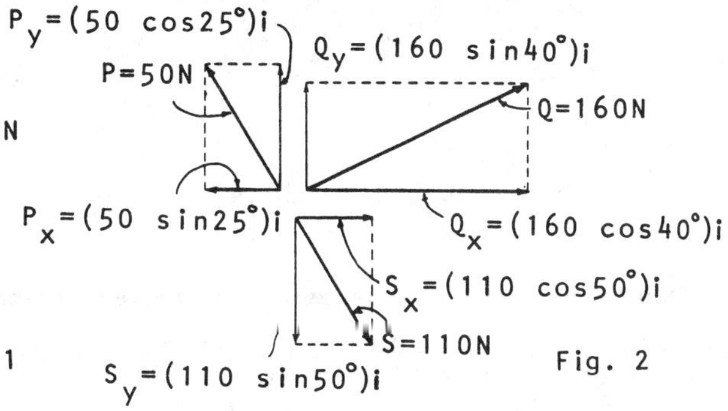

Fig. 2

Solution: Each of the three force vectors is resolved into its x and y-components using the method shown in Figure 2.

The x and y-components of force P is determined as follows:

$$P_x = -(50 \sin 25°)i = (-21.1N)i$$

$$P_y = (50 \cos 25°)j = (45.3N)j$$

For force Q; $Q_x = (160 \cos 40°)i = (122.6N)i$

and $Q_y = (160 \sin 40°)j = (102.8N)j$

For force S; $S_x = (110 \cos 50°)i = (70.7N)i$

and $S_y = -(110 \sin 50°)j = -(84.3N)j$

The total resultant is a summation of all the x and y-components of the three forces acting on A

Thus $R = R_x + R_y$

$$= (P_x + Q_x + S_x)i + (P_y + Q_y + S_y)j$$

$$= (-21.1 + 122.6 + 70.7)i + (45.3 + 102.8 - 84.3)j$$

$$= (172.2N)i + (63.8N)j$$

Notice that both the x and y-components of the resultants are represented by positive numbers. Therefore, R_x will act to the right and R_y will act upwards as shown in Figure 3.

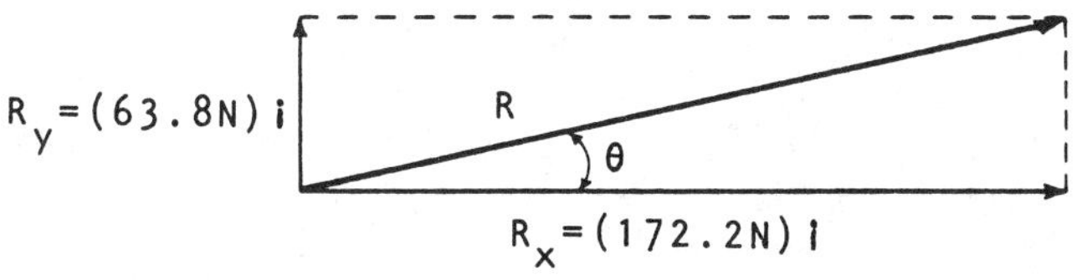

Fig. 3

The magnitude and direction of the resultant R can be determined from Figure 3.

$$\tan\theta = \frac{R_y}{R_x} = \frac{63.8N}{172.2N} = 0.371 \text{ and } \theta = 20.4°$$

$$\sin\theta = \frac{R_y}{R} \text{ and } R = \frac{R_y}{\sin\theta} = \frac{63.8N}{\sin 20.4}$$

or

$$R = 184N \quad \measuredangle\ 20.4°$$

• **PROBLEM 1-5**

One end of a rod is fixed to a point S, while the other end P, carries a 200 lb. weight. Find (a) the moment of the force about S due to the weight; (b) the smallest force at P which when applied results in the same moment about point S; (c) the magnitude of the horizontal force which when applied produces the same moment about S.

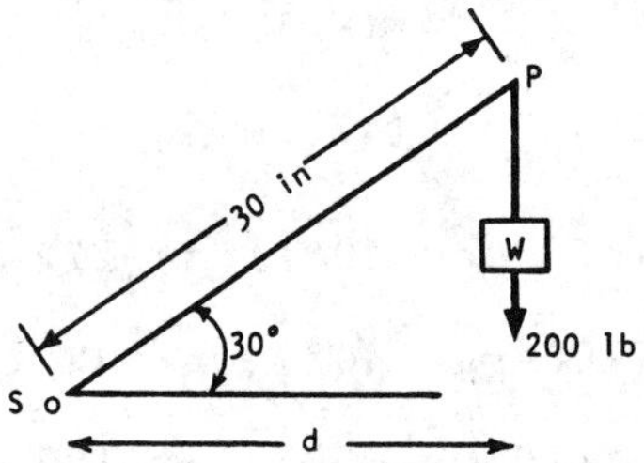

Fig. 1

Solution: (a) The moment of the force about S, M_S is given as $M_S = Fd$, where F is the force and d is the perpendicular distance from S to the line of action of the 200 lb-force.

From Figure 1, d is found from trigonometry

$$\cos 30° = \frac{d}{30\text{in}} \quad \text{and} \quad d = 30\cos 30° = 26.0\text{in}$$

Thus $M_S = (200 \text{ lb})(26.0\text{in}) = 5200 \text{ lb}\cdot\text{in}$ acting in the clockwise direction

(b) The smallest force to produce the same moment as in part(a) occurs when the perpendicular distance is a maximum.

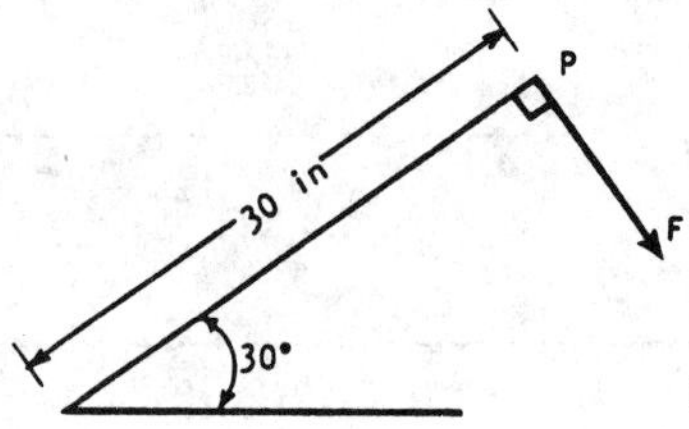

Fig. 2

Referring to Fig. 2, d = 30in

From the relation $M_S = Fd$

$$F = M_S/d = \frac{5200 \text{ lb}\cdot\text{in}}{30 \text{ in}} = 173 \text{ lb}$$

(c) To determine the magnitude of the horizontal force refer to figure 3.

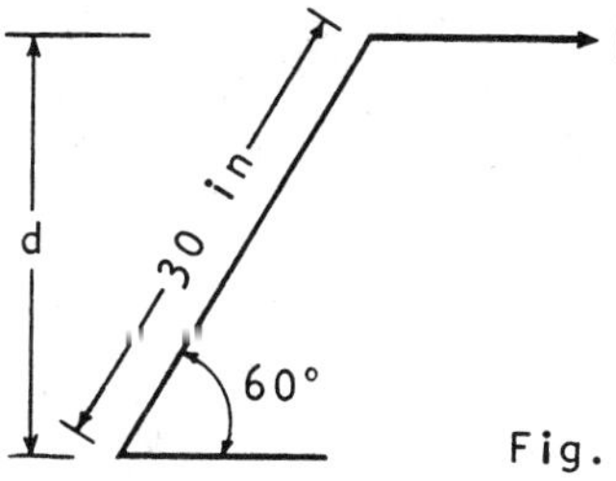

Fig. 3

In this case $M_S = 5200$ lb·in

5200 lb·in $= Fd$

From figure 3, $\sin 60° = \frac{d}{30}$ and $d = 30\sin 60°$

or d = 15in.

Thus,

$$F = \frac{5200 \text{ lb in}}{15\text{in}} = 347 \text{ lb}$$

• **PROBLEM 1-6**

Find the moment M_O about P of a 1000N force acting on point A as shown in Figure 1, using vector algebra.

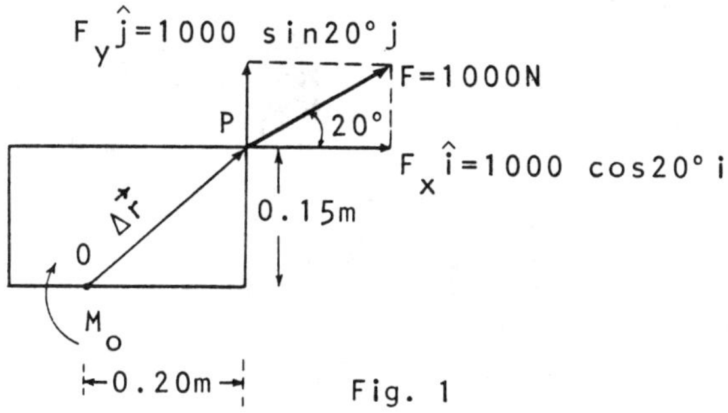

Fig. 1

Solution: The moment of the 1000N force about P is given as the vector product of $\vec{\Delta}r$ and $\vec{F}$.

Thus

$$\vec{M}_O = \vec{\Delta}r \times \vec{F} \qquad (1)$$

The position and force vectors are now resolved into rectangular x, y, and z components.

Thus

$$\vec{\Delta r} = \Delta x\hat{i} + \Delta y\hat{j} = (0.20\text{m})\hat{i} + (0.15\text{m})\hat{j}$$

and

$$\vec{F} = F_x\hat{i} + F_y\hat{j} = (1000\text{N})\cos 20°\hat{i} + (1000\text{N})\sin 20°\hat{j}$$

$$= (939.7\text{N})\hat{i} + (342\text{N})\hat{j}$$

From equation 1

$$\vec{M}_o = [(0.20\text{m})\hat{i} + (0.15\text{m})\hat{j}] \times [(939.7\text{N})\hat{i} + (342\text{N})\hat{j}]$$

$$= (68.4\text{N.m})\hat{k} - (140.9\text{N.m})\hat{k} = -(72.5\text{N.m})\hat{k}$$

According to convention the moment $\vec{M}_o$ is pointing into the plane of the paper since it has a negative value.

• **PROBLEM 1-7**

A square concrete slab acts as an anchor for two cables. These cables exert the 100 lb and 50 lb forces indicated in the Figure. Determine the moment about the corner O for these forces.

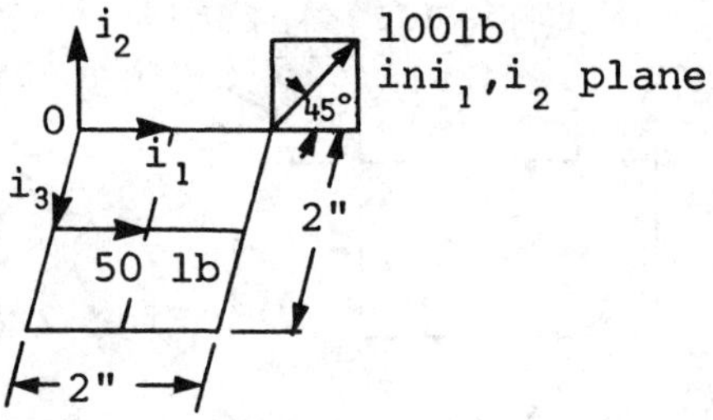

Solution: Since moments are vectors, the sum of several moment vectors acting at the same point is obtained by a vector sum. In this case,

$$\vec{m}_0 = \vec{m}_0' + \vec{m}_0''$$

where m_0' is the moment of the 100 lb force, and m_0'' is the moment of the 50 lb force (conveniently determined with the definition).

The 50 lb force acts in the $\hat{i}_1$ direction. Its moment about O is

$$1\hat{i}_3 \times 50\hat{i}_1 = 50\hat{i}_2.$$

The other moment, m_0', can be determined using the components of the 100 lb force which are in $\hat{i}_1$ direction:

$$100 \text{ lb} (\cos 45°) = 70.7 \text{ lb}$$

in $\hat{i}_2$ direction:

$$100 \text{ lb} (\sin 45°) = 70.7 \text{ lb}.$$

$$\vec{m}_0 = \begin{vmatrix} \hat{i}_1 & \hat{i}_2 & \hat{i}_3 \\ 2 & 0 & 0 \\ 70.7 & 70.7 & 0 \end{vmatrix} + 50\hat{i}_2$$

$$= (141.4\hat{i}_3 + 50\hat{i}_2) \text{in.-lb}.$$

• PROBLEM 1-8

A model iceboat is being studied to determine the effect of various rigging changes. When the boat sails with the wind coming over the left side, only the right runner and the steering runner touch the ice. During one test four of the forces acting on the boat were found to have the average values indicated on the sketch. Find the couple vector which represents their turning effect.

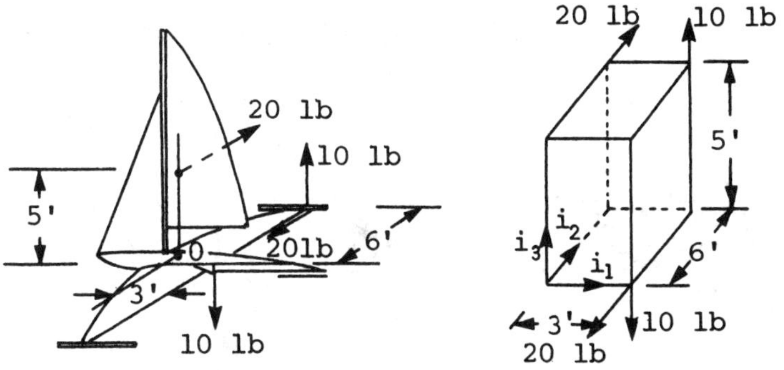

Fig. 1 Fig. 2

Solution: Choose the right-handed orthonormal basis $\hat{i}_1, \hat{i}_2, \hat{i}_3$. We select the position vectors between the 20 lb forces and between the 10 lb forces so that the position vectors will have simple component representations.

$$\vec{c} = (\vec{r} \times \vec{f})_{10 \text{lb}} + (\vec{r} \times \vec{f})_{20 \text{lb}}$$

For the 10 lb forces (from Figure 2),

$$r_1 = 0, \qquad f_1 = 0$$

$r_2 = 6$ ft, $\quad f_2 = 0$

$r_3 = 0, \quad f_3 = 10$ lb

For the 20 lb forces,

$r_1 = -3$ ft, $\quad f_1 = 0$

$r_2 = 0, \quad f_2 = 20$ lb

$r_3 = 5$ ft, $\quad f_3 = 0.$

$$\text{So} \quad \vec{c} = \begin{vmatrix} \hat{i}_1 & \hat{i}_2 & \hat{i}_3 \\ 0 & 6 & 0 \\ 0 & 0 & 10 \end{vmatrix} + \begin{vmatrix} \hat{i}_1 & \hat{i}_2 & \hat{i}_3 \\ -3 & 0 & 5 \\ 0 & 20 & 0 \end{vmatrix}$$

$$= 60\hat{i}_1 - 100\hat{i}_1 - 60\hat{i}_3$$

$$\vec{c} = (-40\hat{i}_1 - 60\hat{i}_3)\text{ft-lb.}$$

• PROBLEM 1-9

A winch-equipped bulldozer is trying to lift a pile of scrap metal from a scrap pile. The bulldozer is lifting one part of the metal with the lip of its shovel and is pulling another with the winch. If the loads from the bulldozer can be idealized as drawn in Figure 2, determine the moment of these loads about the point O.

Fig. 1

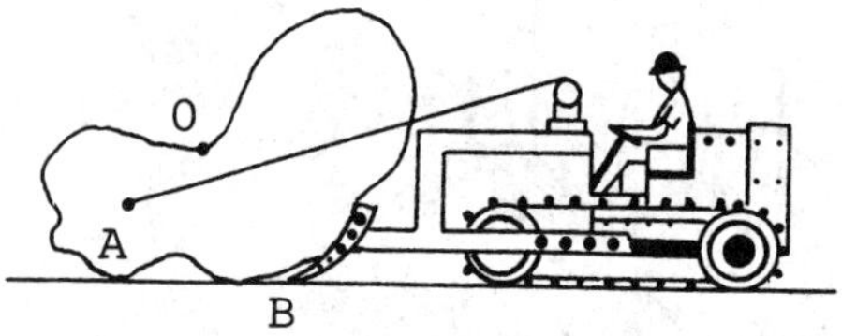

Fig. 2

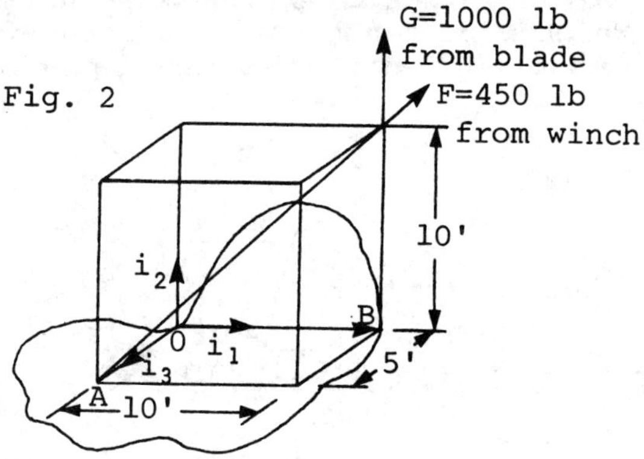

Solution:

$$\vec{m}_0 = (\vec{a} \times \vec{f}) + (\vec{b} \times \vec{G})$$

where $\vec{a}$ and $\vec{b}$ are the vectors from O to points A and B respectively.

We choose the right-handed orthonormal basis shown for the moment vector space at O. Any position vectors can be selected to the forces, but we select ones which give convenient components with respect to a basis parallel to i_1, i_2, i_3.

From Figure, 2 for the 1000 lb force:

$b_1 = 10$ ft, $\quad G_1 = 0$

$b_2 = 0$, $\quad G_2 = 1000$ lb

$b_3 = 0$, $\quad G_3 = 0$.

For the 450 lb force:

$a_1 = 0$, $\quad f_1 = 300$ lb

$a_2 = 0$, $\quad f_2 = 300$ lb

$a_3 = 5$ ft, $\quad f_3 = -150$ lb.

$$\text{So} \quad m_0 = \begin{vmatrix} i_1 & i_2 & i_3 \\ 10 & 0 & 0 \\ 0 & 1000 & 0 \end{vmatrix} + \begin{vmatrix} i_1 & i_2 & i_3 \\ 0 & 0 & 5 \\ 300 & 300 & -150 \end{vmatrix}$$

$$= i_1(0) - i_2(0) + i_3(10{,}000) + i_1(-1500) - i_2(-1500) + i_3(0)$$

$$m_0 = (-1500i_1 + 1500i_2 + 10{,}000i_3)\text{ft-lb.}$$

EQUILIBRIUM OF A PARTICLE

• PROBLEM 1-10

Determine the magnitude and direction of the resultant of forces F_1 and F_2 acting at point S in Figure 1, using a trigonometric solution.

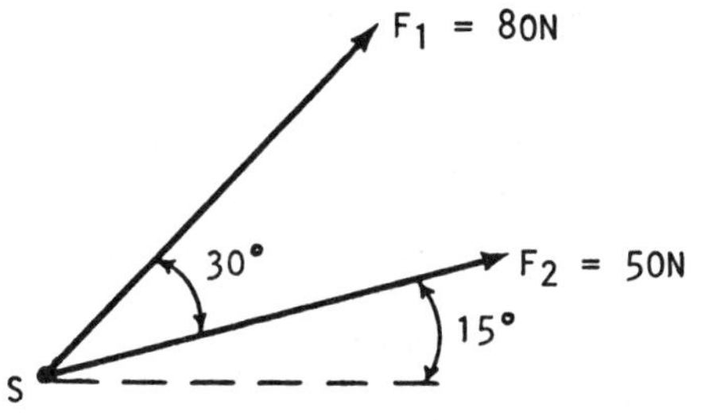

Fig. 1

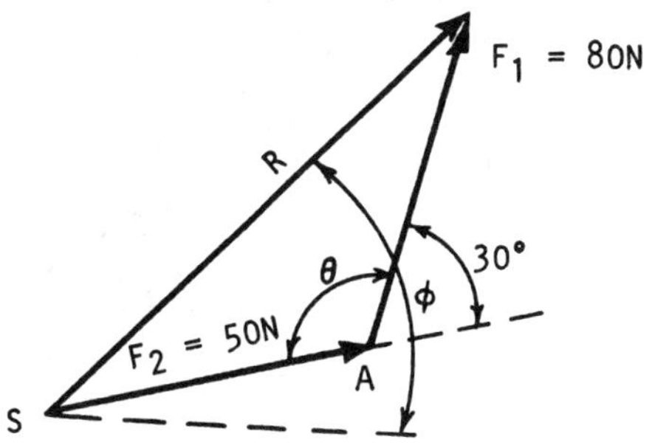

Fig. 2

Solution: The resultant of the two forces F_1 and F_2 is sketched as shown in Fig. 2 using the law of cosines. It's magnitude is determined using the relation

$$R^2 = F_1{}^2 + F_2{}^2 - 2F_1F_2 \cos\theta$$

From inspection angle = 180°-30 = 150°

$$\therefore \quad R^2 = (80N)^2 + (50N)^2 - 2(80N)(50N)\cos 150°$$

$$R = 125.8N$$

The law of sines is now used to determine the direction of the resultant

$$\frac{\sin S}{F_1} = \frac{\sin A}{R}$$

$$\frac{\sin S}{80N} = \frac{\sin 150°}{125.8N}$$

Thus,

$$\sin S = \frac{(80N)\sin 150°}{125.8N} = 0.3179$$

and

$$S = 18.5°.$$

Angle ϕ is thus 15° + S

Thus,

$$\phi = 33.5°.$$

The resultant is thus 125.8N $\measuredangle$ 33.5°

• PROBLEM 1-11

The resultant of two forces, A and B, acts vertically downard with a magnitude of 2700N. Calculate the magnitude and direction of force B if A is a 500N force.

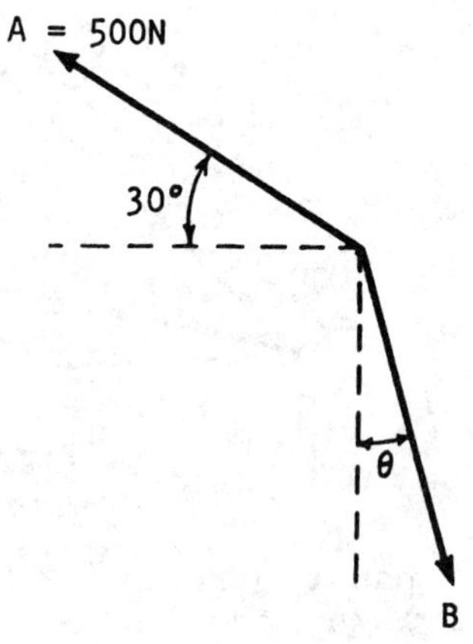

Fig. 1

Solution: In Fig. 2, forces A and B are resolved into their x and y components.

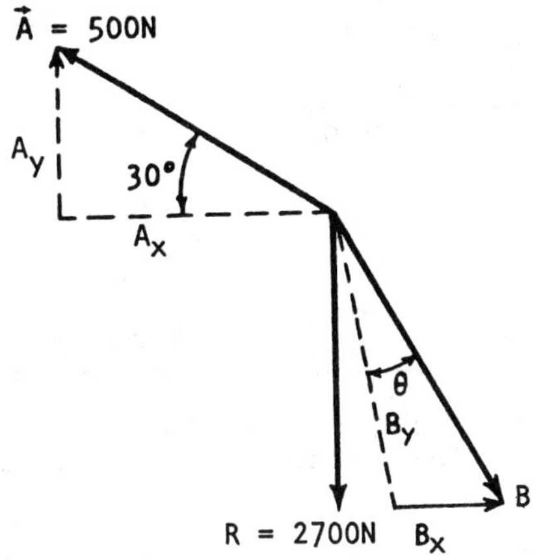

Fig. 2

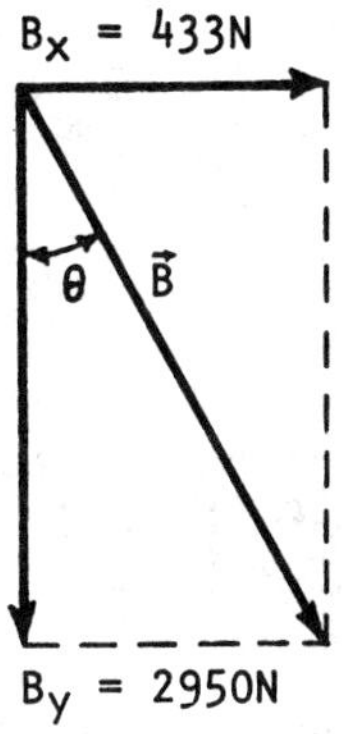

Fig. 3

$$-\vec{A}_x + \vec{B}_x = \vec{R}_x \tag{1}$$

$$\vec{A}_y - \vec{B}_y = -\vec{R}_y \tag{2}$$

Using trigonometry

$$A_x = F \cos 30° = (500N)(0.8660) = 433N$$

$$A_y = F \sin 30° = (500N)(0.50) = 250N$$

But $R_x = 0$ since it acts vertically downward, thus substituting for A_x, equation (1) becomes

$$-433N + B_x = 0$$

or

$$B_x = 433N$$

Also equation (2) becomes,

$$250N - B_y = -2700N \quad \text{or} \quad B_y = 2950N$$

Using Figure 3,

$$\tan\theta = \frac{B_x}{B_y} = \frac{433N}{2950N} = 0.1468$$

and $\theta = 8.4°$

The magnitude of $\vec{B}$ is

$$|\vec{B}| = \sqrt{B_x^{\,2} + B_y^{\,2}} = \sqrt{(433N)^2 + 2950N)^2} = 2982N$$

• PROBLEM 1-12

In Fig. 1, a 50-N tension is required to maintain the box B in equilibrium with force F. Calculate the magnitude of F given that d = 10cm and r = 5cm.

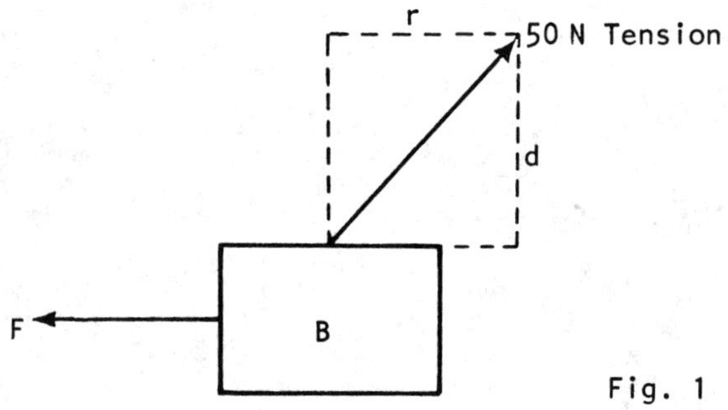

Fig. 1

Solution: The free body diagram of the box is shown in Fig. 2. It accounts for all forces acting on the box. Since only F is required, it is sufficient to consider only the x-direction. It is given that the box is in equilibrium, thus the summation of all the forces in the x-direction must be zero.

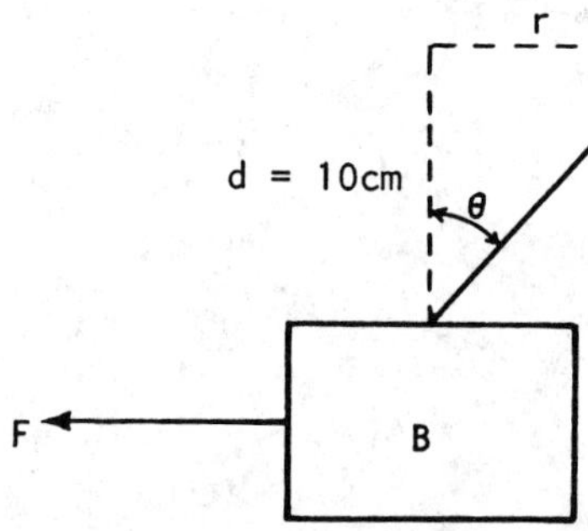

Fig. 2

$$-F + (50N)(\sin\theta) = 0 \qquad (1)$$

From trigonometry,

$$\sin\theta = \frac{r}{\sqrt{10^2 + r^2}}$$

substituting for sinθ in equation (1) gives

$$F = (50N) \frac{r}{\sqrt{10^2 + r^2}}$$

But

$$r = 5cm$$

$$F = (50N)\left(\frac{5cm}{\sqrt{10^2 + 5^2}}\right) = (50N)\left(\frac{5}{\sqrt{100 + 25}}\right)$$

or

$$F = 22.4N$$

• **PROBLEM 1-13**

An 80 lb lid to a sewage tank (Fig. 1) is held by three cables attached to a fork lift at point P. Each cable is 5 ft long, and the cables form angles of $\theta_1 = 160°$, $\theta_2 = 110°$ and $\theta_3 = 90°$ with one another. Calculate the tension in each cable, given the radius of the lid as 3 ft.

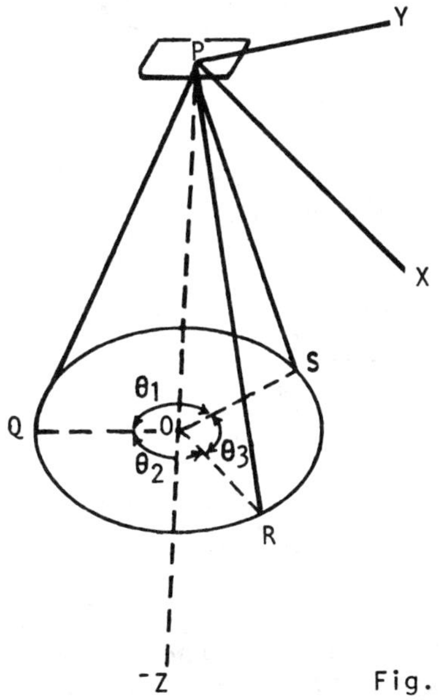

Fig. 1

Solution: This kind of 3-dimensional problem demands systematic treatment.

Consider a top view, place the x, y-axes through P parallel to OR and OS respectively, since their projections are already perpendicular.

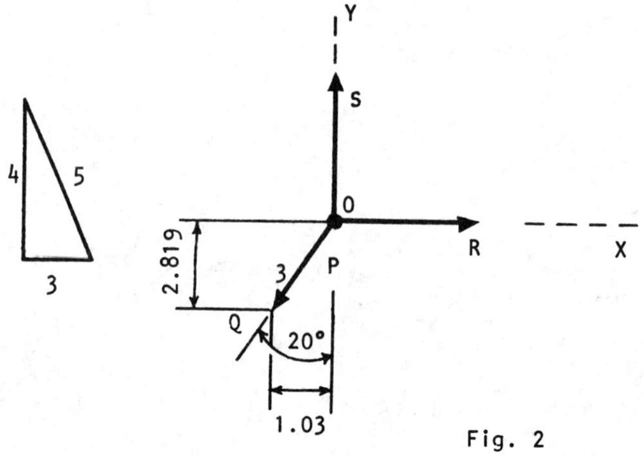

Fig. 2

Figure 2 is a free body diagram of the "knob" where the three cables join at the top. Actually it is a 2-D projection of the free-body diagram. The vertical configuration of each of the wires is the same, and is shown in figure 3.

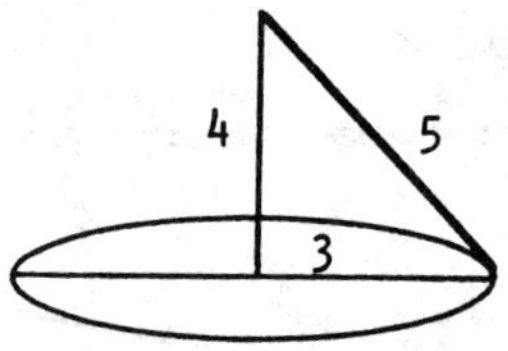

Fig. 3

Direction cosines for vectors Q, S, R are

Vector	l = x/5	m = y/5	n = z/5
Q	$-\frac{1.03}{5}$	$-\frac{2.819}{5}$	$-\frac{4}{5}$
S	0	$+\frac{3}{5}$	$-\frac{4}{5}$
R	$+\frac{3}{5}$	0	$-\frac{4}{5}$

These direction cosines are obtained from the coordinates of the ring ends of the wires relative to the overhead "knob" (origin). The same direction cosines apply to the desired tensions.

Equilibrium requires

$$\sum F_x = 0$$

$$\sum F_y = 0$$

$$\sum F_z = 0$$

since

$$A_x = A \cos[\vec{A},\hat{x}], \quad A_y = A \cos[\vec{A},\hat{y}], \text{ etc.}$$

We have,

$$Ql_Q + Sl_S + Rl_R = 0$$

$$Qm_Q + Sm_S + Rm_R = 0$$

$$Qn_Q + Sn_S + Rn_R = 80\text{lb}$$

putting in numerical values and multiplying through by the denominator, 5ft, given

$$-1.03Q - 0 + 3R = 0$$

$$-2.819Q + 3S + 0 = 0$$

$$-4Q - 4S - 4R = -(80)(5)$$

from the last of these we have

$$Q + S + R = 100$$

and, from the first two

$$Q = \frac{3}{1.03} R = 2.9R$$

$$S = \frac{2.819Q}{3} = \frac{2.819}{3} (2.9R) = 2.725R$$

substituting these last two into the immediately preceding equation gives

$$(2.9R) + (2.725R) + R = 100.$$

Therefore

$$R = \frac{100}{1 + 2.725 + 2.9} = 15.09 \text{ lb}$$

and

$$Q = 2.9R = 2.9(15.09) = 43.76 \text{ lb}$$

$$S = 2.819R = 42.54 \text{ lb}$$

• PROBLEM 1-14

A 4000-lb force is kept in equilibrium by two cables PS and RS as shown in Fig. 1. Cable PS makes a 10° angle with the vertical, while cable RS makes a 40° angle with the horizontal. Calculate the tension in the cable RS.

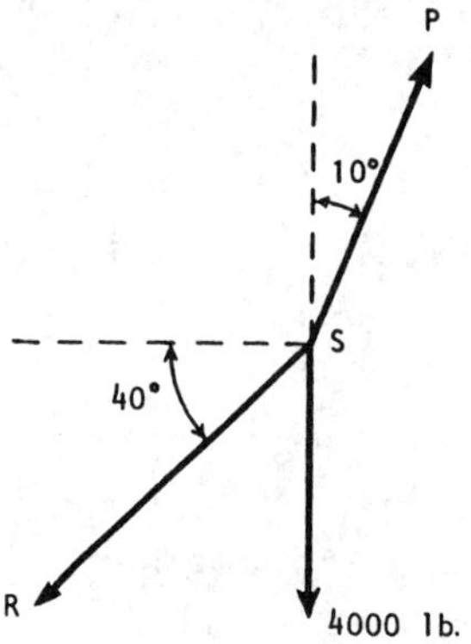

Fig. 1

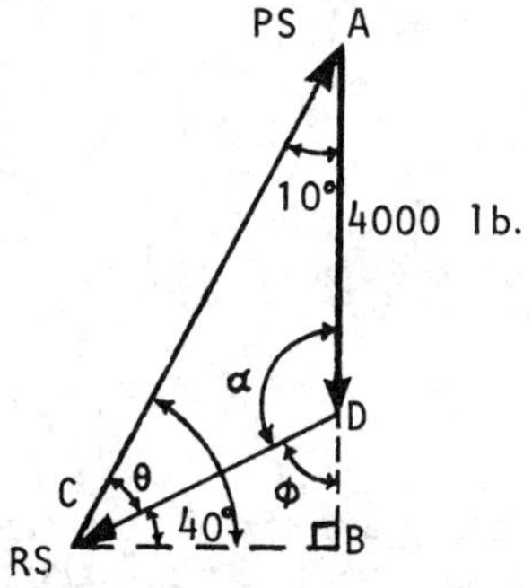

Fig. 2

Solution: The free body diagram is shown in Fig. 2.

From ΔBCD, we observe that $\phi = 50°$. Since α is the supplement of ϕ, it must equal 130°.

$$\therefore \quad \theta = 40°$$

From the law of sines

$$\frac{PS}{\sin 130°} = \frac{RS}{\sin 10°} = \frac{4000 \text{ lb}}{\sin 40°}$$

$$PS = \frac{4000 \text{ lb}(\sin 130)°}{\sin 40°} = \frac{4000 \text{ lb}(0.7660)}{0.6428} = 4766 \text{ lb}$$

$$RS = \frac{4000(\sin 10°)}{\sin 40°} = 1080 \text{ lb}$$

• PROBLEM 1-15

In Fig. 1 a 20 ft-frame PQ is supported at two points L and M, 6 ft and 4 ft respectively from the edges. If a 300 lb load is attached to edge Q, determine the range of load W that must be placed at P to keep the frame in equilibrium.

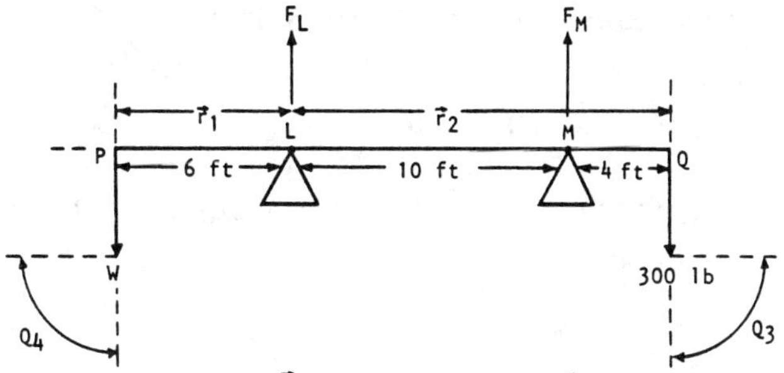

Fig. 1

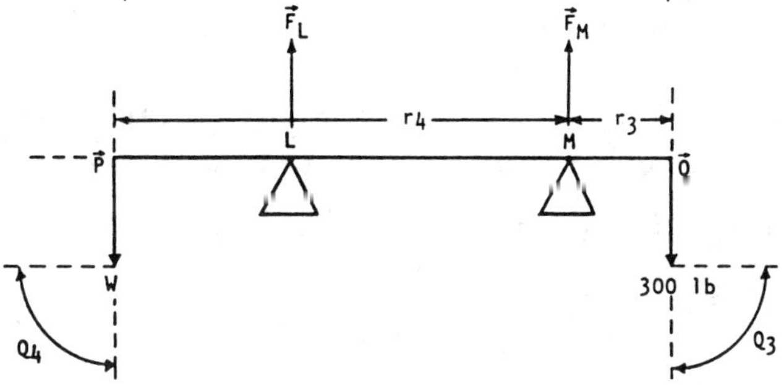

Fig. 2

Solution: The conditions for equilibrium of the beam are:

Net force acting on beam = 0

Net torque acting on beam about any axis = 0

First, identify the forces acting on the beam. (See Figure 1.) We are told to ignore the mass of the beam, therefore, there is no gravitational force acting on it.

Recall that the torque exerted by a given force lying in the plane of the paper, about a rotation axis perpendicular to the paper is given by

$$\tan = rF \sin\theta,$$

where $\vec{r}$ is the vector from the rotation axis to the line of action of the force, and θ is the angle between the force $\vec{F}$ and $\vec{r}$.

Let's arbitrarily define torques that cause a counter-clockwise rotation to be positive.

Consider two cases:

a) W is large enough to cause the beam to rotate counter-clockwise;

b) W is small enough so the beam will rotate clockwise.

a) Suppose that W is large enough to just turn the beam counter-clockwise about the axis through support point L. When W has the required magnitude the force at support M will go to zero, i.e., $\vec{F}_M \to 0$.

Now, using quantities defined in Figure 1 and equation 2) we have

$T_L \equiv$ Net torque about axis through L

$$= Wr_1 \sin\theta_1 - 300 \text{ lb } r_2 \sin\theta_2$$

(Note that the force F_L produces no torque since it acts along a line through L.)

Since $\theta_1 = \theta_2 = 90°$ we have

$$T_L = Wr_1 - (300 \text{ lb})r_2$$

If $T_L \leq 0$, the beam will not rotate counter-clockwise. This allows an upper limit to be placed on W:

$$T_L = Wr_1 - (300 \text{ lb})r_2 \leq 0$$

$$\therefore \quad W \leq 300 \text{ lb} \frac{r_2}{r_1} = 300 \text{ lb} \frac{14 \text{ ft}}{6 \text{ ft}} = 700 \text{ lb.}$$

b) Think of W decreasing in value until the beam just begins to rotate clockwise. The force exerted on the beam by the support at L will go to zero at this value of W. The net torque about an axis through support M will be:

$$T_M = r_4 W \sin\theta_4 - r_3(300 \text{ lb})\sin\theta_3$$

(refer to Figure 2)

The beam will not rotate clockwise, provided $T_M \geq 0$. We can now place a lower limit on W.

$$T_M = r\,W - r\,(300 \text{ lb}) \geq 0$$

$$\therefore \quad W \geq (300 \text{ lb}) \frac{r_3}{r_4} = (300 \text{ lb}) \frac{4 \text{ ft}}{16 \text{ ft}} = 75 \text{ lb.}$$

Therefore, the range of values for W that will leave the beam in equilibrium is

$$75 \text{ lb} \leq W \leq 700 \text{ lb.}$$

EQUIVALENT FORCE SYSTEMS

• PROBLEM 1-16

The cube shown in Fig. 1 has applied to it two couples. Represent these couples by a single equivalent couple.

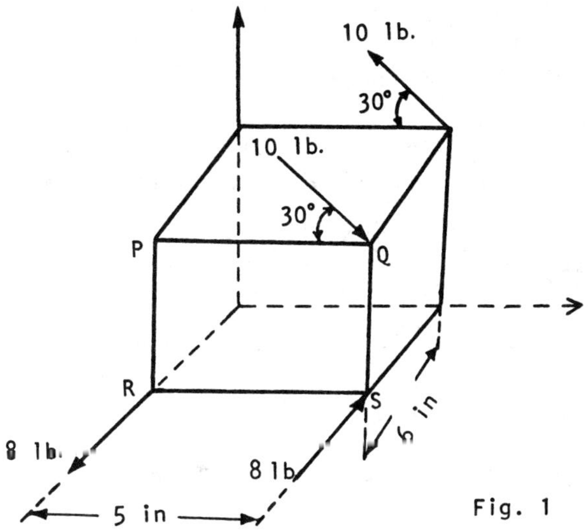

Fig. 1

Solution: A couple vector which is perpendicular to the plane of the couple is used to represent each of the two couples. The two couple vectors are shown in Fig. 2 with respect to a set of coordinate axes.

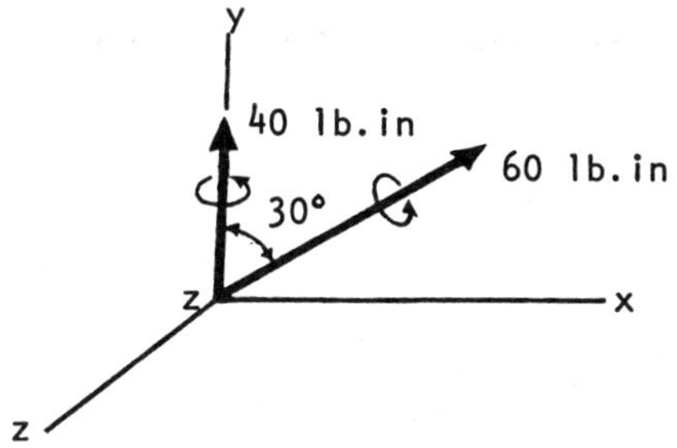

Fig. 2

The magnitude of the resultant of the two couple vectors is determined from Fig. 3 using trigonometry.

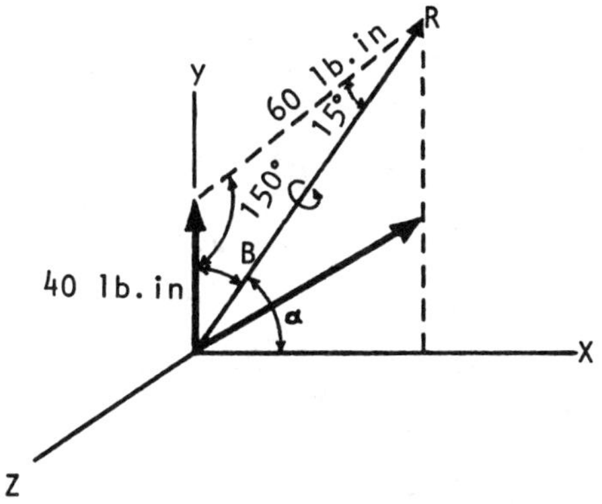

Fig. 3

$$R^2 =$$

$$(40 \text{ lb·in})^2 + (60 \text{ lb·in})^2 - (2)(40 \text{ lb·in})(60 \text{ lb·in})\cos 150°$$

$$R = 96.7 \text{ lb·in}$$

The angle β is determined also from trigonometry

$$\frac{\sin \beta}{60 \text{ lb·in}} = \frac{\sin 150°}{90.7 \text{ lb·in}}$$

$$\sin \beta = \frac{60 \sin 150°}{96.7} = 0.3102 \text{ and } \beta = 18°$$

Thus $\alpha = 90° - 18° = 72°$

The single equivalent couple vector R is represented as

$R = 96.7$ lb·in,

$\alpha = 72°$, $\beta = 18°$

• PROBLEM 1-17

In Fig. 1, a body is subjected to a couple and a single force. Replace the force-couple system with a single force applied along the line PQ.

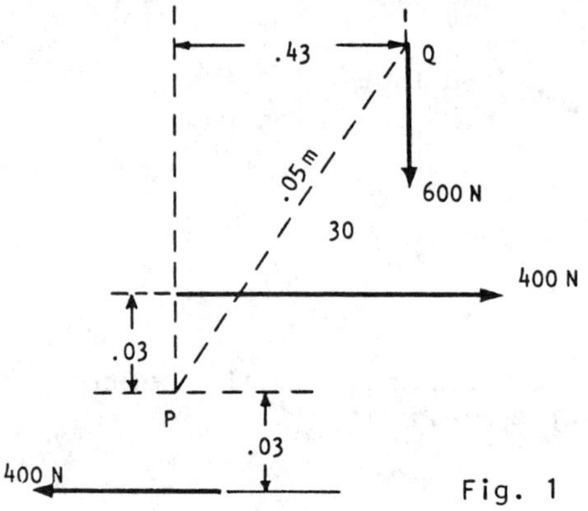

Fig. 1

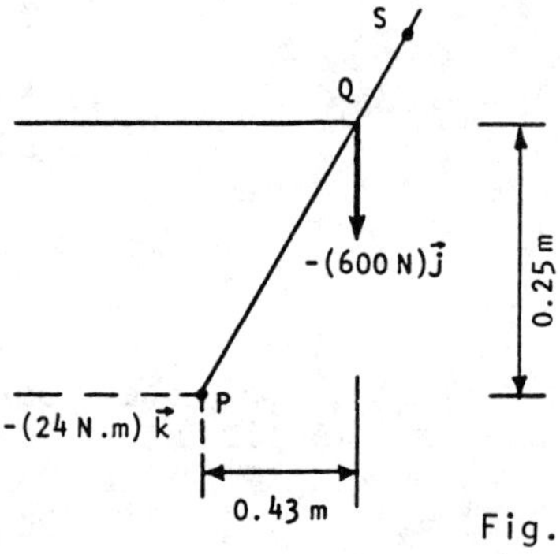

Fig. 2

Solution: The total moment about P due to the force-couple system is

$$M_P = (-.03\text{m})\vec{j} \times (-400\text{N})\vec{i} + (.03\text{m})\vec{j} \times (400\text{N})\vec{i} + 0.50\text{m}(\cos 30°\vec{i} + \sin 30°\vec{j}) \times (-600\text{N})\vec{j}$$

$$= (-284 \text{ N.m})\vec{k}$$

We must now replace our force somewhere along segment PQ such that it will produce a moment about P equivalent to M_P. If we call this point S, (see Fig. 2), then

$$\vec{r}_S \times \vec{F} = (-284 \text{ N.m})\vec{k}$$

$$PS(\cos 30°\vec{i} + \sin 30°\vec{j}) \times (-600\text{N})\vec{j} = (-284\text{N.m})\vec{k}$$

$$PS \cos 30° = \frac{-284N.m}{-600N}$$

$$PS = 0.55m$$

• PROBLEM 1-18

A loaded board hung from the ceiling is suspended by six cables. The tensions in each of the cables are indicated in Fig. 1. Determine the magnitude and point of application of a single cable to replace the six cables.

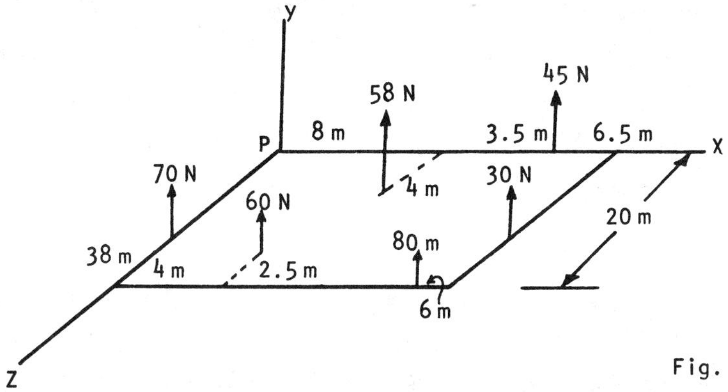

Fig. 1

Solution: By convention, any system of forces may be reduced to an equivalent force-couple system acting at a given point. In Fig. 1, these tensions can be added vectorially and replaced by a single resultant R. Their couples may also be vectorially added and replaced by a single couple M_P. Thus the equivalent force-couple system can be defined as;

$$R = \sum T \text{ and } M_P = \sum M_P = \sum (d \times T)$$

where d is the distance from the point of application of the tension to be given point P. The point of application of the six tensions and their position vectors will now be determined.

70N: $\vec{d} = 16.2\vec{k}$ $\quad \vec{T} = 70\hat{j}N;$ $\quad \vec{d} \times \vec{T} = -1134\vec{i}\ N.m$

60N: $\vec{d} = 4\hat{i} + 17.5\hat{k}$ $\quad \vec{T} = 60\hat{j}N;$ $\quad \vec{d} \times \vec{T} = (-1050\hat{i} + 240\hat{k})N.m$

80N: $\vec{d} = 12\hat{i} + 20\hat{k}$ $\quad \vec{T} = 80\hat{j}N;$ $\quad \vec{d} \times \vec{T} = (-1600\hat{i} + 960\hat{k})N.m$

30N: $\vec{d} = 18\hat{i} + 10\hat{k}$ $\quad \vec{T} = 30\hat{j}N;$ $\quad \vec{d} \times \vec{T} = (-300\hat{i} + 540\hat{k})N.m$

45N: $\vec{d} = 11.5\hat{i}$ $\quad \vec{T} = 45\hat{j}N;$ $\quad \vec{d} \times \vec{T} = 518\hat{k}\ N.m$

58N: $\vec{d} = 8\hat{i} + 4\hat{k}$ $\quad \vec{T} = 58\hat{j}N;$ $\quad \vec{d} \times \vec{T} = (-232\hat{i} + 464\hat{k})N.m$

$$R = \sum T = 343\hat{j}\ N \text{ and } \sum M_P = (-4316\hat{i} + 2722\hat{k})N.m$$

The force-couple system at P now consists of a tension R

and a couple vector M_P which are perpendicular. It could still be reduced further by moving $\vec{R}$ to a new point Q(X,O,Z) chosen such that the moment of $\vec{R}$ about Q is equal to M_P

Thus, $$\vec{d} \times \vec{R} = \vec{M}_P$$

or $$(x\hat{i} + z\hat{k}) \times Ry\hat{j} = M_{Px}\hat{i} + M_{Pz}\hat{k}$$

$$(x\hat{i} + z\hat{k}) \times (343\hat{j}) = -4316\hat{i} + 2722\hat{k}$$

$$343x\hat{k} - 343z\hat{i} = -4316\hat{i} + 2722\hat{k}$$

Thus $$343x = 2722$$

and $$x = 7.94\text{m}$$

Also $$-343z = -4316$$

and $$z = 12.58\text{m}$$

Thus the single tension to replace the six cables is

$R = 343\text{N}\uparrow$ at $x = 7.94$m, $z = 12.58$m

• PROBLEM 1-19

A beam has applied to it the forces shown in Fig. 1. Given that each force is 3KN, calculate the equivalent force-couple system at point S.

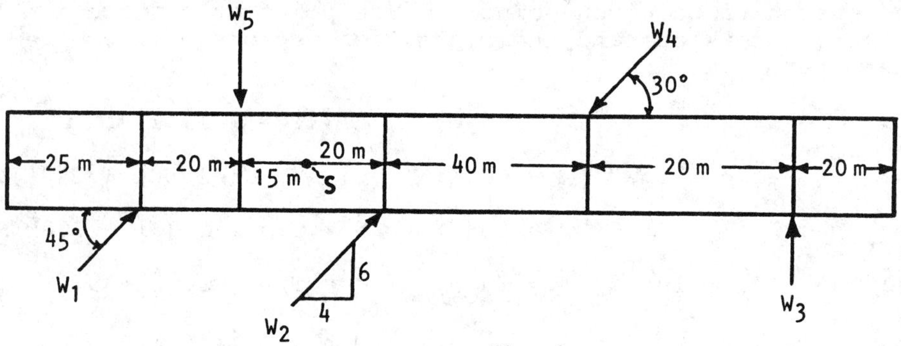

Fig. 1

Solution: The equivalent force-couple system is denoted by

$$R = \sum W$$

and $M_S = \sum(\vec{d} \times \vec{W})$

Each of the five 3KN forces are resolved into their x and y components as shown in Fig. 2.

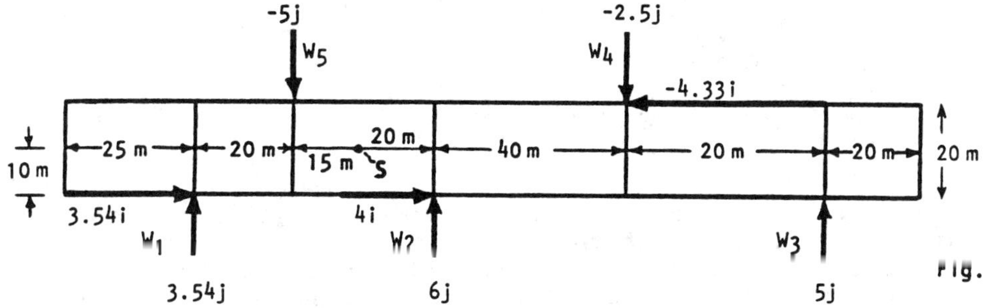

Fig. 2

$$\vec{R} = \sum\vec{W} = (3.54\hat{i} + 3.54\hat{j}) + (4\hat{i} + 6\hat{j}) + (5\hat{j}) + (-4.33\hat{i} - 2.5\hat{j}) + (-5\hat{j})$$

$$= 3.21\hat{i} + 7.04\hat{j}$$

$$\vec{M}_S = \sum(\vec{d} \times \vec{W})$$

$$= (-35\hat{i} - 10\hat{j}) \times (3.54\hat{i} + 3.54\hat{j}) + (20\hat{i} - 10\hat{j}) \times (4\hat{i} + 6\hat{j})$$

$$+ (80\hat{i} - 10\hat{j}) \times (5\hat{j}) + (60\hat{i} + 10\hat{j}) \times (-4.33\hat{i} - 2.50\hat{j})$$

$$+ (-15\hat{i} + 10\hat{j}) \times (-5\hat{j})$$

$$= (-123.9 + 35.4 + 120 + 40 + 400 - 150 + 43.3 + 75)\hat{k}$$

$$= 440\hat{k}$$

The equivalent force-couple system at point S is

$$\vec{R} = 3.21\hat{i} + 7.04\hat{j}$$

and

$$M_S = 440\hat{k}$$

• PROBLEM 1-20

(1) Reduce the system of forces acting on the cube (fig. 1) to an equivalent single force at the center of the cube, plus a couple composed of two forces acting at two adjacent corners.

(2) Reduce this system to a system of two forces, and state where the forces act.

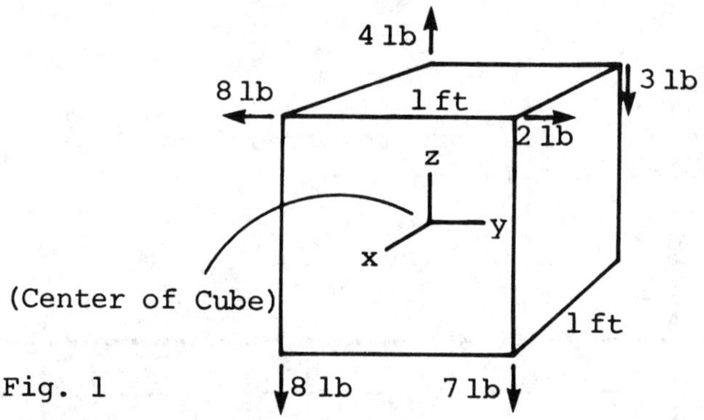

Fig. 1

Solution: We will consider first the resultant force which we will apply at the center of the cube (origin of coordinates). Next, we will consider moments about the coordinate axes and apply "couples" (equal and opposite forces at the corners which represent pure turning actions) to give the same moments.

Part 1:

Resultant force: consider the components of this force in each of the coordinate directions.(See fig.2).

$$R_x = \Sigma F_x = 0$$

(there are no forces acting on the cube which have an x-component.)

$$R_y = \Sigma F_y = 2 \text{ lb} - 8 \text{ lb} = - 6 \text{ lb}$$

$$R_z = \Sigma F_z = 4 \text{ lb} - 8 \text{ lb} - 7 \text{ lb} - 3 \text{ lb} = - 14 \text{ lb.}$$

Thus, the resultant force to be applied at the cube center is

$$\vec{R} = (R_x, R_y, R_z)$$

$$= (0, - 6 \text{ lb}, - 14 \text{ lb}).$$

Moments about each axis will be calculated using the right hand rule to assign algebraic signs.

$$M_x = \left(\frac{1}{\sqrt{2}} \text{ ft}\right) (8 \text{ lb}) - \left(\frac{1}{\sqrt{2}} \text{ ft}\right) (7 \text{ lb}) - \left(\frac{1}{\sqrt{2}} \text{ ft}\right) (4 \text{ lb}) - \left(\frac{1}{\sqrt{2}} \text{ ft}\right) (3 \text{ lb})$$

$$= \frac{-6}{\sqrt{2}} \text{ lb-ft.}$$

(This is a clockwise moment when we look down the x-axis towards the origin.)

Similarly,

$$M_y = \frac{1}{\sqrt{2}} [7 + 8 + 4 - 3] = \frac{16}{\sqrt{2}} \text{ lb-ft}$$

$$M_z = \frac{1}{\sqrt{2}} [2 - 8] = \frac{-6}{\sqrt{2}} \text{ lb-ft.}$$

These moments can be obtained in a number of ways by using couples composed of forces applied at corners of the cube. Some examples:

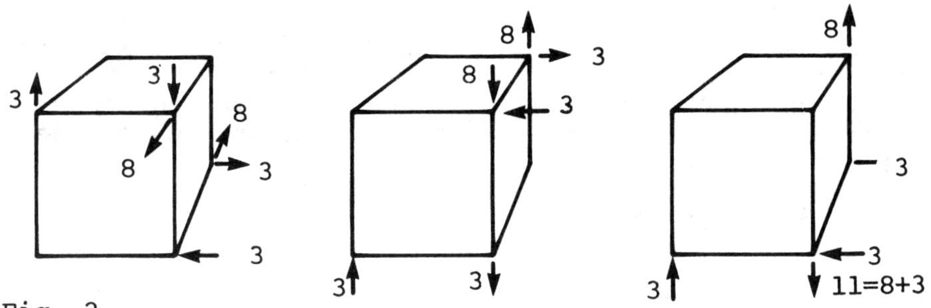

Fig. 2

Note that in each case, couples acting in different directions provide the moments separately.

If the moments are to be generated by forces applied at only two corners, then the following works:

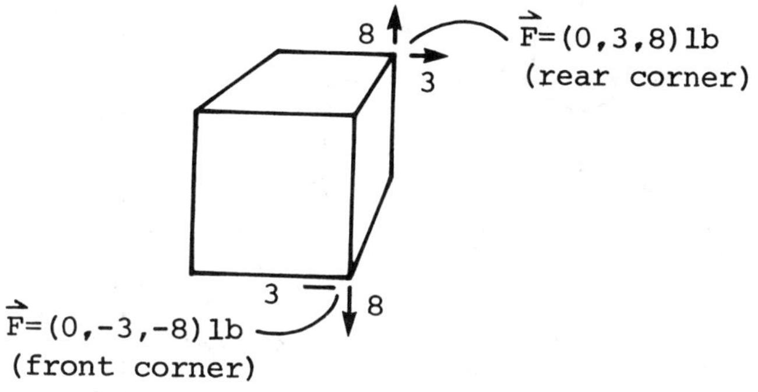

$$\vec{F}_{\text{rear corner}} = (0,\ 3,\ 8) \text{ lb}$$

$$\vec{F}_{\text{front corner}} = (0,\ -3,\ -8) \text{ lb.}$$

The couple composed of three pound forces provides M_x and M_z.

Part 2:

The solution from Part 1 is, shown in fig.3.

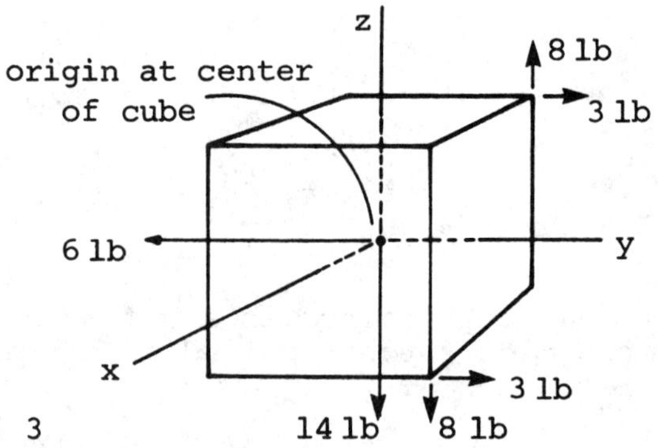

Fig. 3

To reduce this to a system of only two forces, first note the couples applied to the corners have zero resultants. Hence, our final answer will be the force at the center relocated to give the correct moments about each of the axes. Recall, these were

$M_x = -3$ lb-ft.

$M_y = 8$ lb-ft

$M_z = -3$ lb-ft.

First, view the system in the yz- plane so the x-axis is seen as a point view,(see fig. 4).

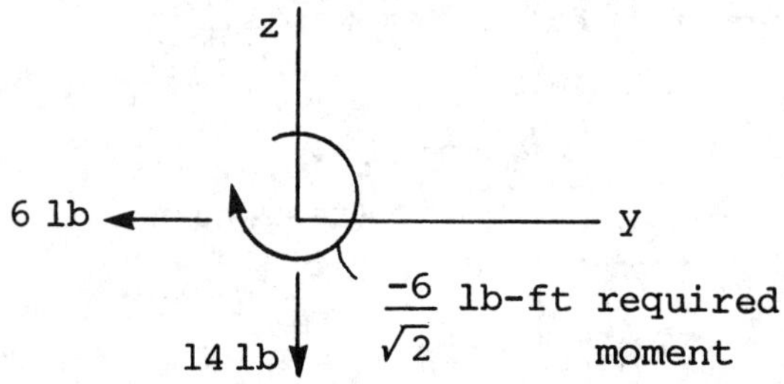

Fig. 4

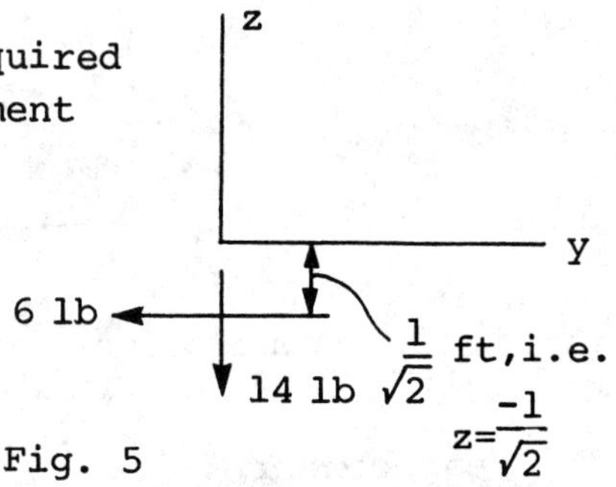

Fig. 5

By sliding the 6 lb force to a line of action $1/\sqrt{2}$ ft below and parallel to the y-axis, we have figure 5.

This gives the correct M_x and is otherwise still the same two forces.

Similarly, in the xz-plane.(see fig. 6).

And finally, in the xy-plane.(see fig. 7).

Fig. 6

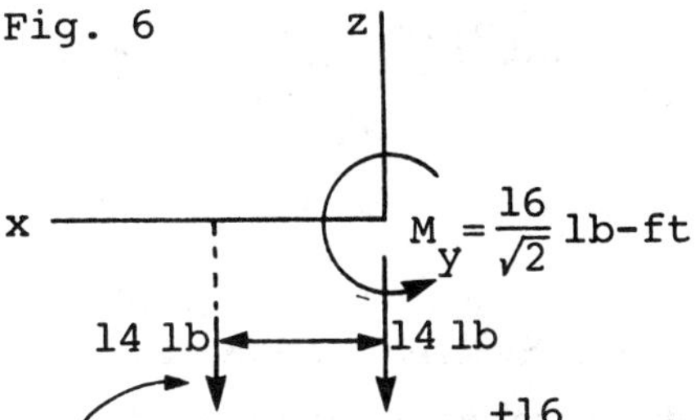

$$x = \frac{\frac{+16}{\sqrt{2}} \text{ lb-ft}}{14 \text{ lb}} = 0.808 \text{ ft}$$

Fig. 7

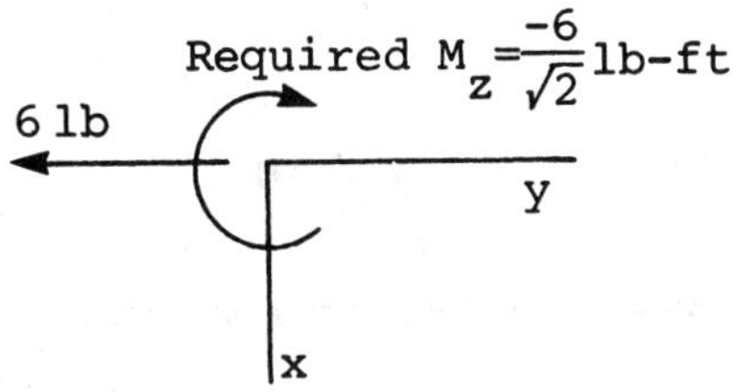

Clearly, moving the 6 lb force to a parallel line of action through

$$x = - \frac{\frac{6}{\sqrt{2}} \text{ lb-ft}}{6 \text{ lb}} = - \frac{1}{\sqrt{2}} \text{ ft}$$

will give the required moment.

Recapitulating, a solution for Part 2 is:

$$\vec{F}_1 = (0, -6, 0) \text{ lb applied at } \left(-\frac{1}{\sqrt{2}}, 0, -\frac{1}{\sqrt{2}}\right) \text{ ft}$$

and $\vec{F}_2 = (0, 0, -14)$ lb applied at (0.808, 0, 0) ft.

• PROBLEM 1-21

Replace the distributed load acting on the beam by a single rigid-body equivalent force.

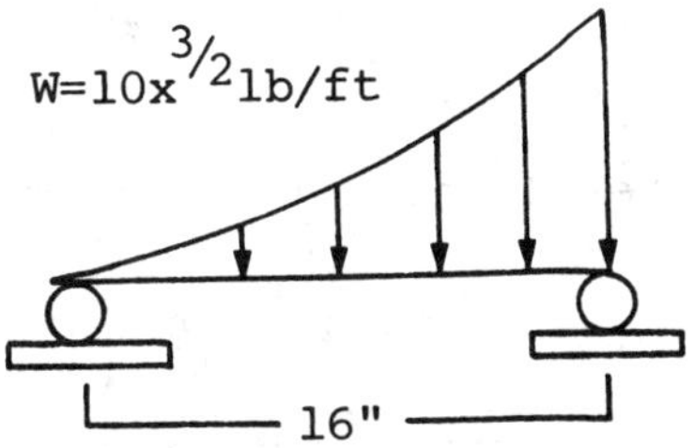

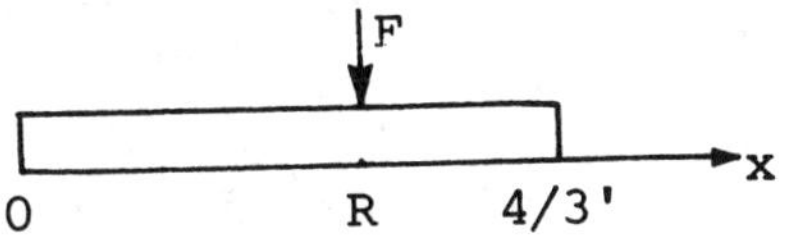

Solution: Let the equivalent force be F and its point of exertion x = R.

Then F is equal to the total load and RF is equal to the total torque with respect to the original O.

$$F = \int_0^{4/3} 10\, x^{3/2}\, dx = 10 \cdot (2/5)(4/3)^{5/2} = 8.21 \text{ lb.}$$

$$R = \frac{1}{F}\int_0^{4/3} x \cdot 10x^{3/2}\, dx = \frac{10(2/7)(4/3)^{7/2}}{8.21} = (20/21) \text{ ft}$$

$$= 11.4 \text{ in.}$$

• PROBLEM 1-22

A beam is in equilibrium under a force distribution $W = 3x^2$ lb/ft. What are the reactions at the wall and the internal resultant at the midpoint for the section to the right of the cut 3 ft from the wall?

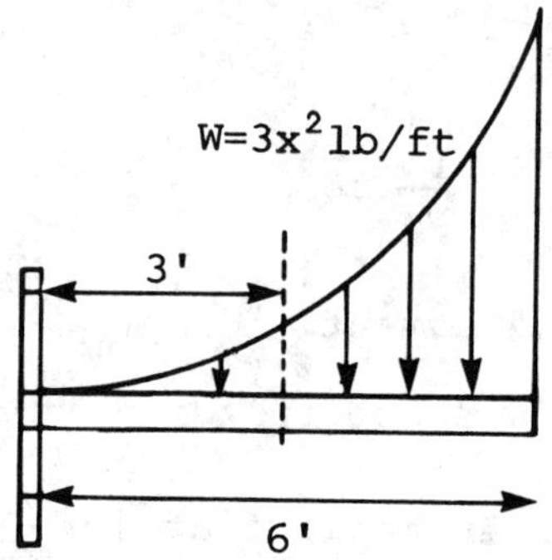

Solution: The reactions at the wall are the upward force R and the counterclockwise torque τ produced by the pair of forces (normal to the wall) as indicated. R is equal in magnitude to the total load and τ is equal to the total torque due to the load.

$$R = \int_0^6 3x^2\, dx = 216 \text{ lb.}$$

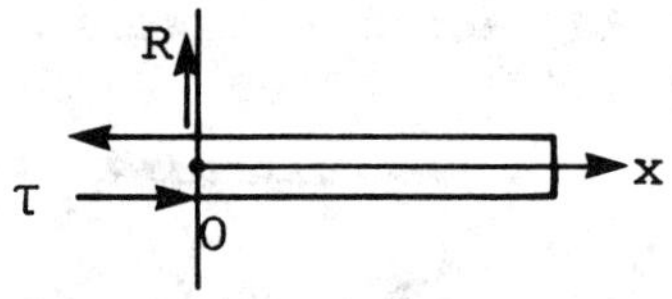

$$\tau = \int_0^6 x\, 3x^2\, dx = 972 \text{ ft-lb.}$$

Similarily, the internal resultants at the midpoint are

$$R_1 = \int_3^6 3x^2\, dx = 189 \text{ lb.}$$

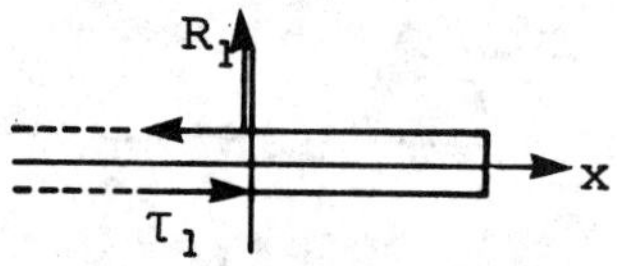

$$\tau_1 = \int_3^6 x\, 3x^2\, dx = 911.25 \text{ ft-lb.}$$

• PROBLEM 1-23

Parallel forces distributed along a straight line are frequently used to represent loads on beams. Find an expression for the resultant force and its location which is rigid-body equivalent to the distributed force system $\vec{f} = g(x)\hat{j}$ in Figure 1.

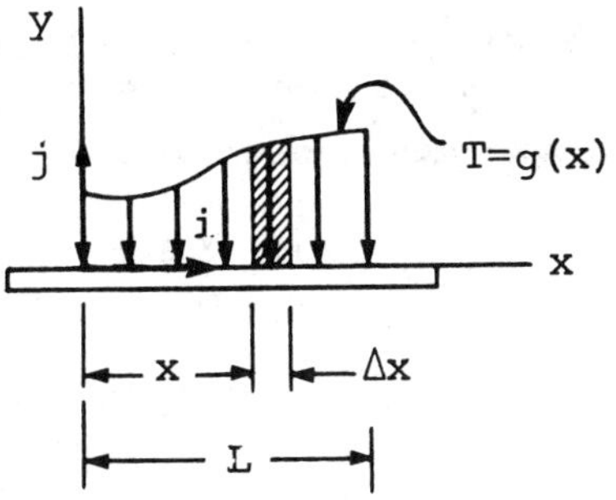

Fig. 1

Solution: Locate a basis $\hat{i}$, $\hat{j}$ at the left edge of the distributed load. The rigid-body equivalent of force $\vec{f}$ is

$$\vec{f} = \int_S \vec{T}\, ds = -\int_{x=0}^{L} g(x)dx\, \hat{j}$$

The integral can be rewritten as

$$\vec{f} = -A\hat{j}$$

Here A is the area under the force intensity curve.

The position vector $\vec{R}$ can be written in a similar manner.

$$\vec{R} \times \vec{f} = \int_S \vec{x} \times \vec{T}\, dS$$

$$R_x\hat{i} \times (-A\hat{j}) = \int_{x=0}^{L} (x\hat{i}) \times (-g(x)dx)\hat{j}$$

$$-R_x A\hat{k} = \int_{x=0}^{L} -x[g(x)dx]\hat{k}$$

or $$R_x A = \int_A x\, dA$$

where $dA = g(x)dx$. Thus

$$R_x = \frac{1}{A}\int_A x\, dA.$$

The magnitude of the resultant force equals the area A under the load curve. The line of action of the resultant force passes through the centroid of the area A.

• PROBLEM 1-24

A shaped plate is under a uniformly distributed normal

force. Where can a single force be placed to balance this load? How large must the force be?

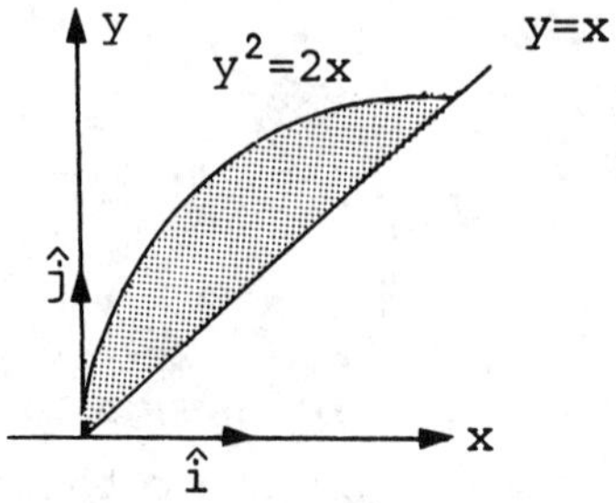

Fig. 1

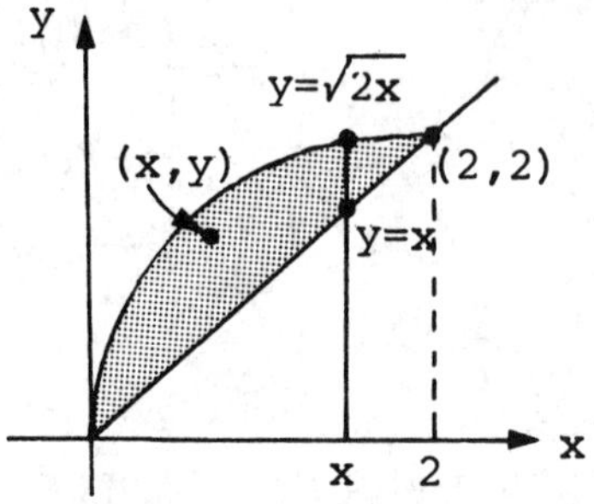

Fig. 2

Solution: Let p be the normal force on the plate per unit area (or pressure) in the direction of $-\hat{k}$. The total force acting on the plate is

$$= \iint (-p\hat{k})\,dxdy = -pA\hat{k},$$

where A is the total area of the plate. The total torque with respect to the origin is

$$\tau = \iint (x\hat{i} + y\hat{j}) \times (-p\hat{k})dx\,dy$$

$$= p\hat{j} \iint x\,dx\,dy - p\hat{i} \iint y\,dx\,dy.$$

Let $\vec{F}$ be the single force and

$$\vec{R} = X\hat{i} + Y\hat{j}$$

its point of exertion that is required to balance the load. The conditions for equilibrium are

$$\vec{F} = -\vec{f} = pA\hat{k},$$

$$\vec{R} \times \vec{F} = -\vec{\tau}\ .$$

The last equation is equivalent to

$$X = \iint x\,dx\,dy/A$$

$$Y = \iint y\,dx\,dy/A.$$

That is, the point (X, Y) is the plate's center of mass, as it might have been expected. The work reduces to finding the following integrals. See Figure 2.

$$A = \iint dx\,dy = \int_0^2 dx \int_x^{\sqrt{2}} dy = \frac{2}{3},$$

$$\iint x\,dx\,dy = \int_0^2 x\,dx \int_x^{\sqrt{2x}} dy = \frac{8}{15},$$

$$\iint y\,dx\,dy = \int_0^2 dx \int_x^{\sqrt{2x}} y\,dy = \frac{2}{3}.$$

Finally, we have

$\vec{F} = (2/3)\, p\, \hat{k}$

$\vec{R} = (4/5)\hat{i} + \hat{j}.$

• PROBLEM 1-25

> The center of pressure for a flat plate subject to a field of uniformly distributed compressive forces which act perpendicular to the plate surface is the point on the plate surface through which the line of action of the single resultant force passes. If an arbitrarily shaped flat plate of face area A is loaded in this manner, obtain the location of its center of pressure.

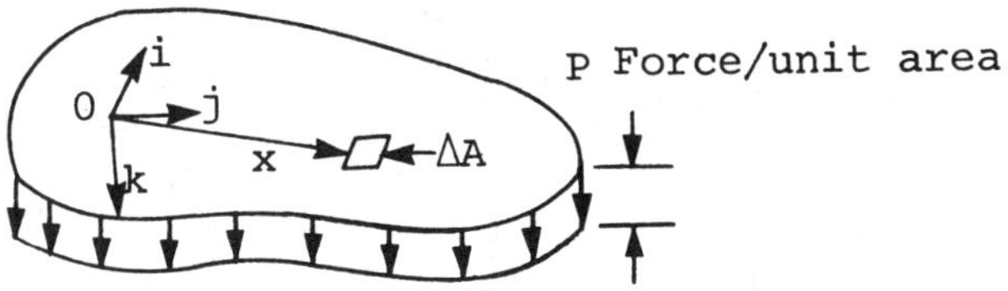

Solution: The base vectors, $\hat{i}$, $\hat{j}$, and $\hat{k}$ are chosen at some arbitrary point O on the plate.

$$\vec{f} = \int_S \vec{T}\, dS = \int_A p\, dA\, \hat{k} = pA\hat{k}$$

where p is the magnitude of the stress vector T. The resultant force equals the pressure p times the area.

Also, the resultant force couple is

$$\vec{c} = \int_A \vec{x} \times p\, dA\hat{k} = p \int_A \vec{x} dA \times \hat{k}.$$

The center of pressure is located relative to O by $\vec{R} = R_x\hat{i} + R_y\hat{j}$ where

$$\vec{R} \times \vec{f} = \vec{c}.$$

Thus $$\vec{R} \times pA\hat{k} = p \int_A \vec{x}\, dA \times \hat{k}$$

or $$\left(pA\vec{R} - p \int_A \vec{x} dA \right) \times \hat{k} = 0.$$

Let $\vec{V} = \left(pA\vec{R} - p \int_A dA \right)$. Then $\vec{V}$ is either zero or parallel to $\hat{k}$ from the definition of the cross product. Since $\vec{R}$ and $\vec{x}$ lie in the $\hat{i}$, $\hat{j}$ plane, $\vec{V}$ is not parallel to $\hat{k}$ and therefore $\vec{V} = 0$.

$$pA\vec{R} - p \int_A \vec{x} dA = 0$$

$$A\vec{R} = \int_A \vec{x} dA$$

$$A(R_x\hat{i} + R_y\hat{j}) = \int_A (x\hat{i} + y\hat{j})\, dA$$

$$AR_x\hat{i} + AR_y\hat{j} = \int_A x_{\hat{i}} dA + \int_A y_{\hat{j}} dA$$

$$AR_x = \int_A x_i\, dA$$

$$AR_y = \int_A x_y\, dA$$

Consequently,

$$R_x = \frac{\int_A x\, dA}{A}$$

$$R_y = \frac{\int_A x\, dA}{A}.$$

The center of pressure is at the centroid of area of the plane surface.

• PROBLEM 1-26

A rigid body model of a sphere loaded as in the Figure is in equilibrium. Determine the reaction forces $\vec{A}$ and $\vec{D}$.

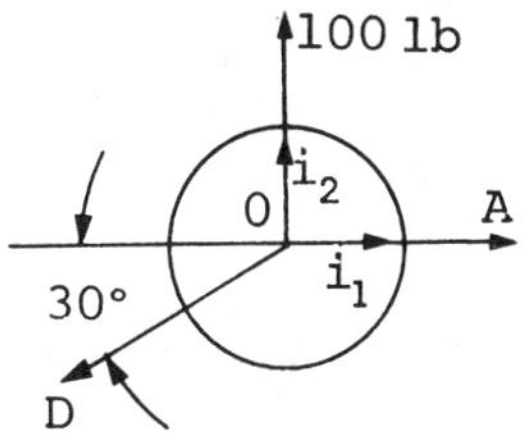

Solution: We can use the equilibrium equations.

$$\Sigma\vec{F} = 0$$

$$(A - D\cos 30°)\hat{i}_1 + (100 - D\sin 30°)\hat{i}_2 = 0$$

Therefore $D = 200$ lb

$A = 173.2$ lb

The senses and directions are shown on the figure. Note that there are two scalar equilibrium equations available for determining reactions in this 2-D force system in which all the lines of action pass through a common point (a concurrent force system).

• PROBLEM 1-27

Given that the tension in a support wire A-B is 650 N, determine the horizontal and vertical components of the reaction at pin A.

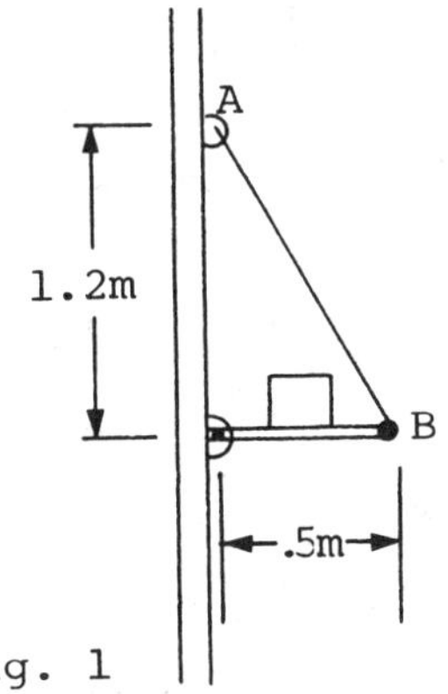

Fig. 1

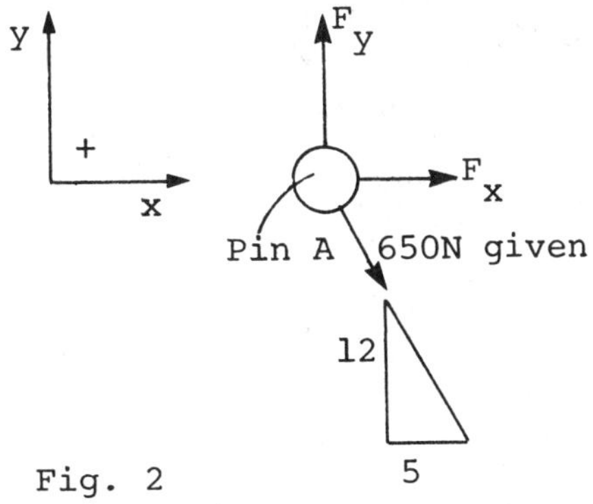

Fig. 2

Solution: A force body diagram of the pin at A is given in Fig. 2.

Note that a sign convention has been shown which is compatible with the problem formulation. (Horizontal and vertical components of the force on the pin are to be found.) Also note that F_x, F_y which are, as yet, unknown have been shown as positive in agreement with this sign convention.

Now, we apply the principle of equilibrium, working first in the x- and then in the y-direction:

$\xrightarrow[x]{+} \Sigma F_x = 0 \qquad F_x + (650\text{ N})\left(\frac{5}{\sqrt{5^2 + 12^2}}\right) = 0$

$\therefore \qquad F_x = -250\text{ N}$

$+\uparrow y \Sigma F_y = 0 \qquad F_y - (650\text{ N})\left(\frac{12}{13}\right) = 0$

$\therefore \qquad F_y = +600\text{ N.}$

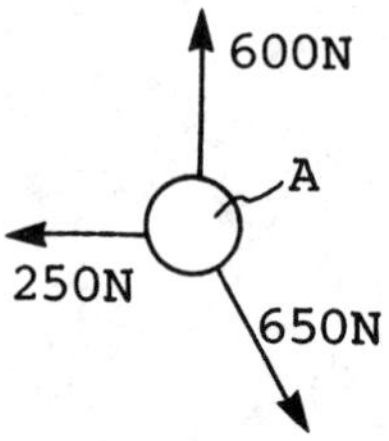

Fig. 3

Note that, in accord with our designated sign convention, the answer for F_x above means a 250 N horizontal force acting from right to left. The positive F_y acts vertically upward. See fig. 3.

• PROBLEM 1-28

A 1000lb weight is hung from the end of a pipe which is fastened to a ball and socket at the lower end and supported at the top by two cables as shown in Figure 1. Neglecting the weight of the pipe, determine the forces in each of the two cables and the reaction at point A.

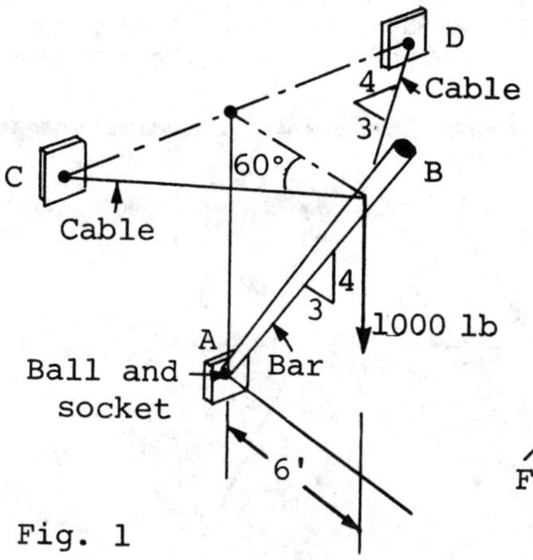

Fig. 1

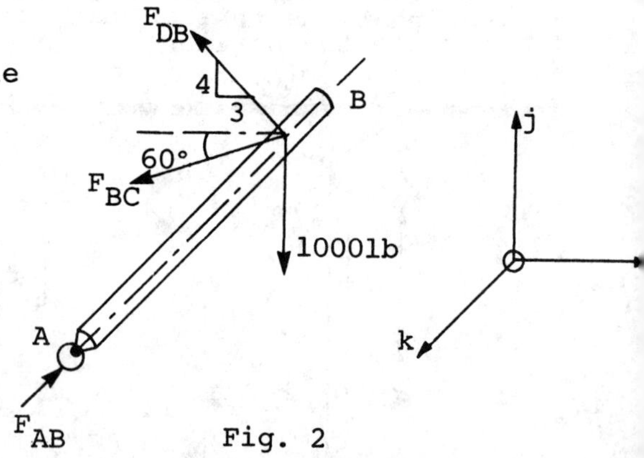

Fig. 2

<u>Solution:</u> We isolate the pipe AB and obtain the free-body diagram shown in Figure 2. Recognizing that AB is a two-force member, we obtain the following equations.

$$\vec{f}_{DB} = f_{DB}\left(-\frac{3}{5}\hat{i} - \frac{4}{5}\hat{k}\right)$$

$$\vec{f}_{CB} = f_{CB}\ (-\cos 60°\hat{i} + \sin 60°\hat{k})$$

$$\vec{f}_{AB} = f_{AB}\left(\frac{3}{5}\ \hat{i} + \frac{4}{5}\ \hat{k}\right) \qquad \text{or}$$

$$\vec{f}_{DB} = f_{DB}\ (-\ 0.6\hat{i} - 0.8\hat{k})$$

$$\vec{f}_{CB} = f_{CB}\ (-\ 0.5\hat{i} + 0.866\hat{k})$$

$$\vec{f}_{AB} = f_{AB}\ (0.6\hat{i} + 0.8\hat{j}).$$

$$\vec{W} = -\ 1000\hat{j}\ \text{lb}$$

Applying the condition $\vec{F} = 0$, we have

$$(-0.6f_{BD}-0.5f_{CB}+0.6f_{AB})\hat{i}+(0.8f_{AB}-1000)\hat{j}+(-0.8f_{BD}+0.866f_{BC})\hat{k}=0$$

The components of this equation give us the three equations needed to find the three unknowns. Solving them yields,

$$f_{AB} = 1250\ \text{lb}$$

$$f_{BD} = 705\ \text{lb}$$

$$f_{BC} = 652\ \text{lb}\ .$$

• PROBLEM 1-29

A model of an airplane standing on a runway is loaded as shown. Determine the reactions $\vec{A}$, $\vec{B}$, and $\vec{D}$ if the plane is in equilibrium.

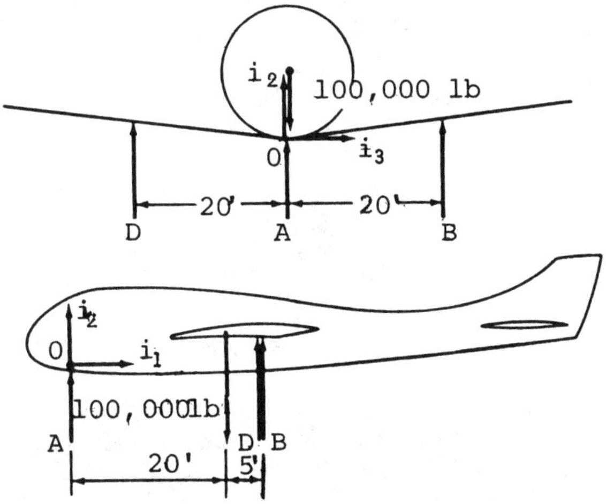

Solution: We choose the point O to locate a basis $\hat{i}_1$, $\hat{i}_2$, $\hat{i}_3$. Since the airplane is in equilibrium, we can write:

$$\sum_{i=1}^{n} \vec{f}_i = 0, \text{ implies } (A + B + D - 100{,}000)\hat{i}_2 = 0$$

$$\sum_{i=1}^{n} \vec{r}_i \times \vec{f}_i = 0, \text{ implies } (20B - 20D)\hat{i}_1 + [25(B + D) - 20(100{,}000)]\hat{i}_3 = 0,$$

The above vector equations yield the following scalar equations:

$$A + B + D - 100{,}000 = 0$$

$$20B - 20D = 0$$

$$25(B + D) - 20(100{,}000) = 0.$$

Thus,

$$A = 20{,}000 \text{ lb}$$

$$B = 40{,}000 \text{ lb}$$

$$D = 40{,}000 \text{ lb}.$$

The directions of $\vec{A}$, $\vec{B}$, and $\vec{D}$ are shown in the Figure. Note that there are three equations which can be used in solving for the reactions in this parallel force system.

• PROBLEM 1-30

A 50KN block is held on an inclined plane by a force F_1 making a 30° angle with the horizontal. The center of gravity of the block is at point O, 15m from the plane. Calculate the reactions at points P and Q and the magnitude of the restraining force F_1.

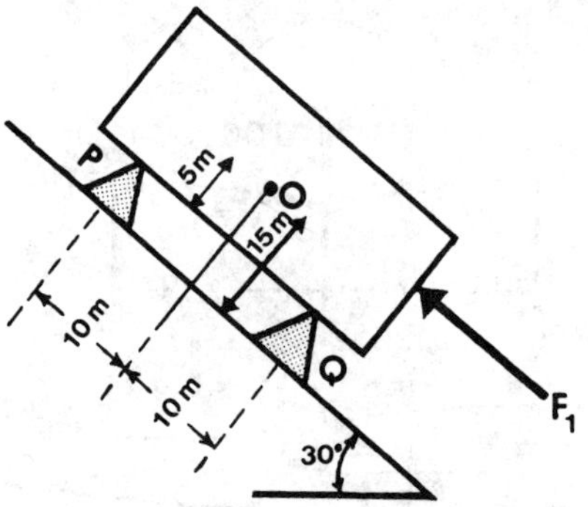

Fig. 1

Solution: In Fig. 2 all the forces acting on the block are shown. The weight of the block is resolved into its x and y-components, T_x and T_y respectively.

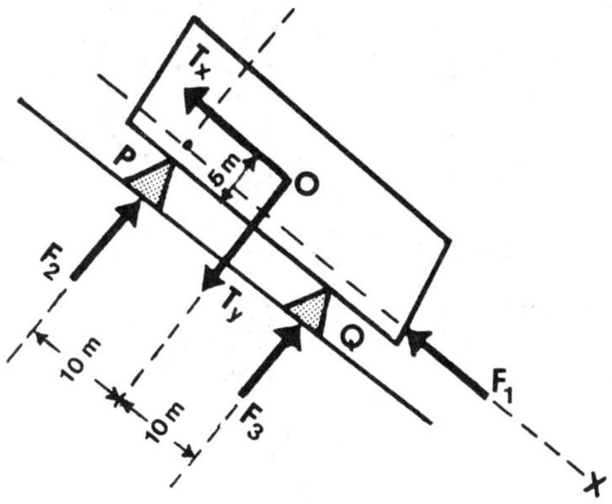

Fig. 2

$T_x = -(50KN)\cos 30° = -43.3KN$

$T_y = -(50KN)\sin 30° = -25.0KN$

Taking moments about point P to eliminate forces F_2 and F gives

$$\sum M_P = 0$$

$$= (-25KN)(10m) + (43.3KN)(5m) + F_3(20m) = 0$$

$$\therefore \quad F_3 = 1.7KN$$

Take moments about Q to eliminate F_3 and F_1 gives

$$\sum M_Q = 0$$

$$= (25KN)(10m) + (43.3KN)(5m) - F_2(20m)$$

$$F_2 = 23.3KN$$

The magnitude of the restraining force F_1 is given as

$$\sum F_{1x} = 0: \quad -43.3KN - T = 0$$

or $\quad T = -43.3KN$

• PROBLEM 1-31

> A man is standing 3 ft from the top of an 8 ft ladder which leans against a smooth wall as shown in the Figure. If the ladder is in equilibrium, determine the forces at A and D. Neglect the weight of the ladder.

Solution: This problem can be solved by using the conditions for equilibrium $\vec{F}(B) = 0$ and $\vec{C}(B) = 0$. However, an alternative procedure is to determine the line of action of $\vec{f}_D$ and then use $\vec{F}(B) = 0$. The lines of action of $\vec{f}_A$ and the 200 lb gravity force are known and intersect at point C. (See figure b).

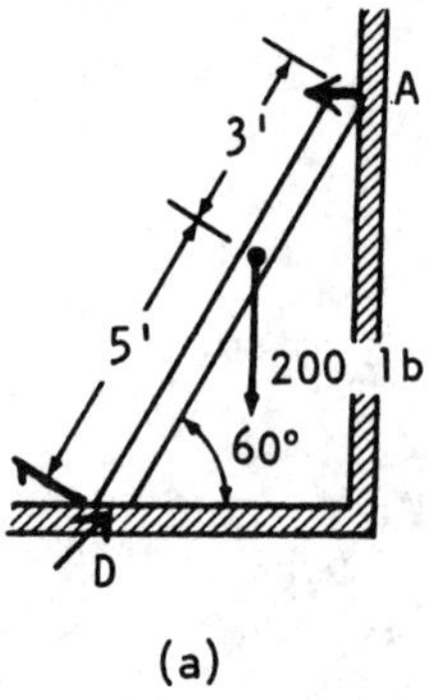

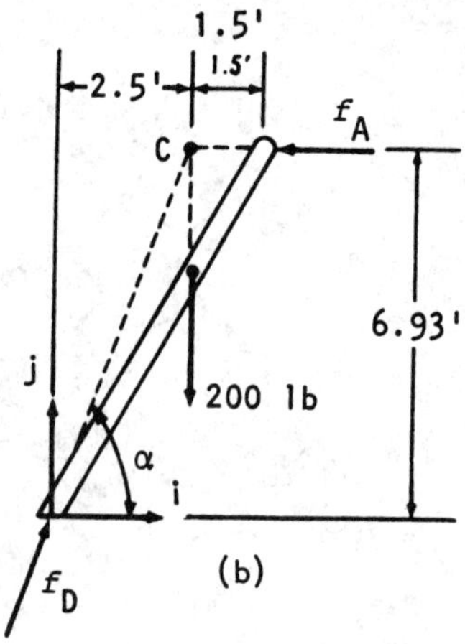

Since the ladder is a three-force member, the line of action of $\vec{f}_D$ must pass through point C. From $\vec{F} = 0$ we obtain

$$\tan \alpha = \frac{6.93}{2.5} = 2.772$$

$$\alpha = 70.16 \qquad \cos \alpha = 0.339 \quad \sin \alpha = 0.94.$$

$$(-f_A + 0.339 f_D)\hat{i} - (0.94 f_D - 200)\hat{j} = 0$$

which yields

$$-f_A = -0.339\, f_D$$

$$0.94 f_D = 200 \qquad f_D = \frac{200}{0.94}$$

$$f_A = 0.339\left(\frac{200}{0.94}\right)$$

$$f_D = 213 \text{ lb}$$

$$f_A = 72.2 \text{ lb.}$$

Since the three forces intersect at a point the condition $\vec{C} = 0$ is satisfied.

• PROBLEM 1-32

A board, weighing 30 kg and mounted on wheels at point RS and T is used to unload a truck. The wheel at T is fixed to the edge of the truck. A 100-kg box B rest on the edge of the board causing the center of gravity of the board and the box to act at point P on the ground. Calculate the reactions at R, S and T with the dimensions given.

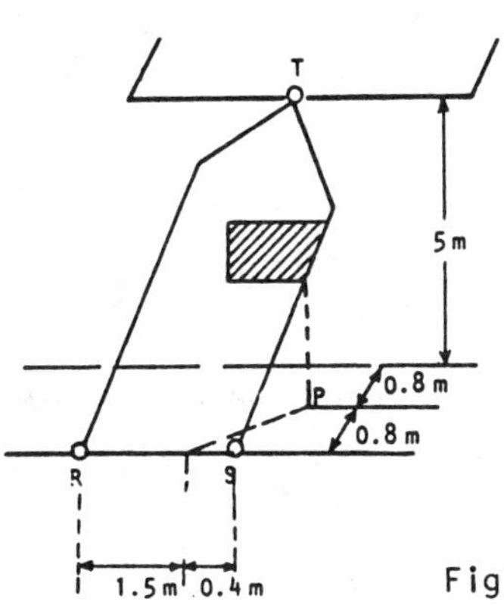

Fig. 1

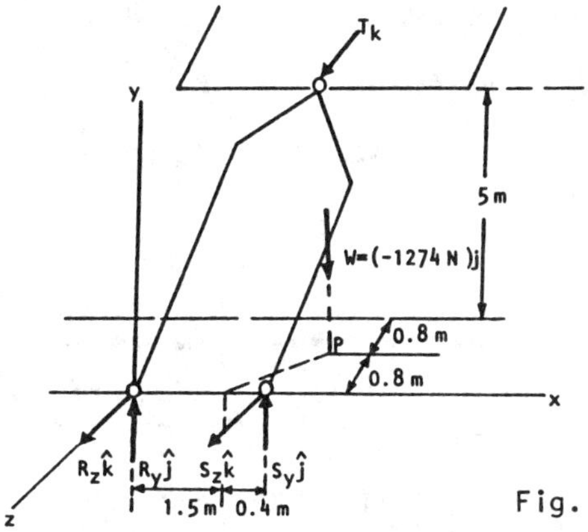

Fig. 2

Solution: All the forces acting on the board are drawn on the free-body diagram of Fig. 2. The combined weight of the box and the board acting at point P is determined as follows

$$W = -mg\hat{j} = -(100\text{ kg} + 30\text{ kg})(9.8\text{m/s}^2)\hat{j} = -(1274\text{N})\hat{j}$$

Equilibrium Equations:

Setting all the forces and moments acting on the board equal to zero.

$$\sum\vec{F} = 0: \quad R_y\vec{j} + R_z\vec{k} + S_y\vec{j} + S_z\vec{k} + T\vec{k} - (1274\text{N})\vec{j} = 0$$

$$(Ry + Sy = 1274\text{N})\vec{j} + (R_z + S_z + T)\vec{k} = 0 \qquad (1)$$

(NOTE: Ry and Sy are not subscripted. I just typed what I saw)

$$\sum\vec{M}_R = \sum(\vec{r}\times\vec{F}) = 0$$

$$1.9\vec{i} \times (Sy\vec{j} + S_z\vec{k}) + (1.5\vec{i} - 0.8\vec{k}) \times (-1274\vec{j})$$

$$+ (1.5\vec{i} + 5\vec{j} - 1.6\vec{k}) \times T\vec{k} = 0$$

Simplifying gives

$$(5T - 1019.2)\vec{i} - (1.9S_z + 1.5T)\vec{j} + (1.9S_y - 1911)\vec{k} = 0$$

From equation (2) set the coefficients of $\vec{i}$, $\vec{j}$ and $\vec{k}$ equal to zero

Thus T = 1019.2N and solving for S_z gives

$$1.9S_z + 1.5(1019.2)\text{N} = 0$$

or $\quad S_z = -804.6\text{N}$

Set the coefficients of $\hat{i}$, $\hat{j}$ and $\hat{k}$ equal to zero and determine the reactions S_y, R_y and R_z

$$S_y = 1005.8\text{N}$$

Thus, $R_y + 1005.8 - 1274 = 0$ or $R_y = 268.2N$

solving for R_z;

$$R_z - 804.6 + 1019.2 = 0$$

or $$R_z = -214.6N$$

• PROBLEM 1-33

> A 9 ft machine link rests on wheels located at points E, F and G in Fig. 1 of a steel coating machine.
>
> Determine the reactions at these three points if a 400 lb force is directed vertically downward at point D.

Fig. 1

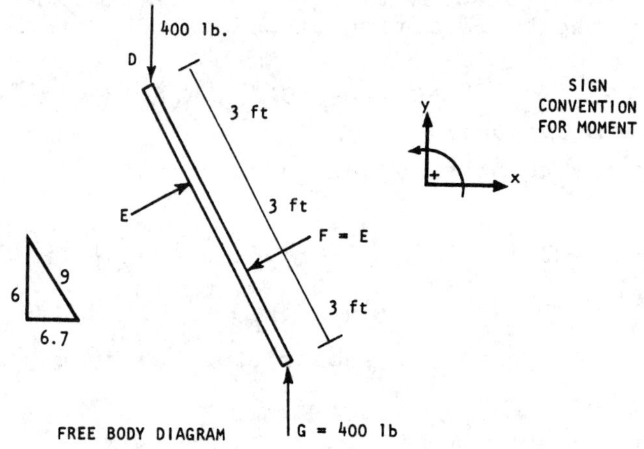

Solution: Free Body Diagram:

(a) The forces at G and D clearly form a couple of value

+ (6.7 ft) (400 lb) = 2680 lb;ft.

Thus, the couple formed by E and F must equilibrate this. Hence

$$E = \frac{2680 \text{ ft lb}}{3 \text{ ft}} = 893 \text{ lb}$$

and the directions are as shown in the free body diagram.

Summarizing:

E = 893 lb   F = 893 lb   G = 400 lb

(b) Force E comes from pressure of roller 1 on the bar; Force F from roller 4 pressing on the bar.

Thus, rollers 2, 3 can be eliminated.

• **PROBLEM 1-34**

A 2000 kg member is held in place at locations P and Q as shown in Fig. 1. Determine the reactions at P and Q if the fork lift is used to haul a 3000 kg load.

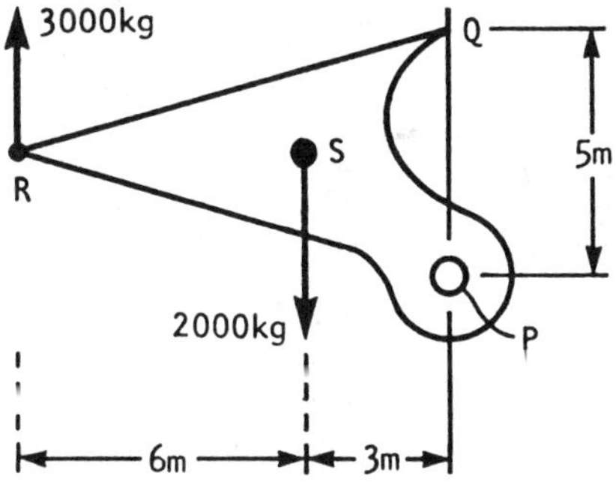

Fig 1

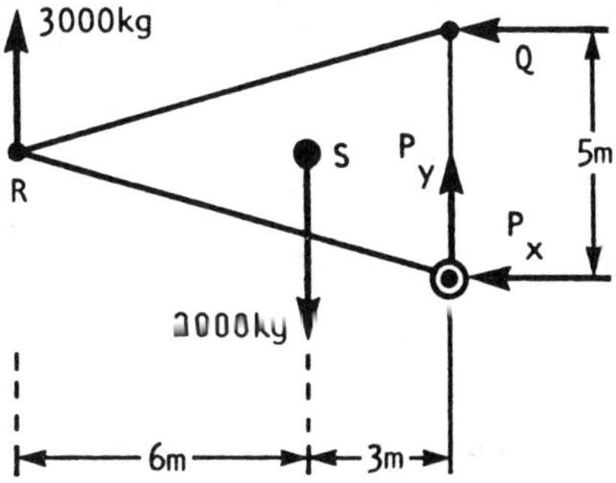

Fig 2

Solution: The weight of the member acts along the center of gravity located at point S on the free body diagram of Fig. 2. The reaction at Q has been resolved into its x and y-components. Assume the forces to act in the direction shown. The final sense of each force components will be known after the magnitudes of the forces are determined.

Taking moments about point P eliminates P_x and P_y and allows for the value of Q to be determined.

$$\Sigma M_P = 0$$

$$(3000 \text{ kg})(9\text{m}) - (2000 \text{ kg})(3\text{m}) - Q(5\text{m}) = 0$$

or $\quad Q = 4200$ kg

P_x is determined by summing the horizontal components of all external forces and setting it equal to zero.

$$\therefore \quad \Sigma F_x = 0: \quad -P_x - Q \text{ or } -P_x - 4200 \text{ kg} = 0$$

and

$$P_x = -4200 \text{ kg}$$

Since P_x is negative it means that its sense as shown in Fig. 2 should be in the opposite direction. To determine P_y, all vertical forces are summed equal to zero.

$$\therefore \quad \Sigma F_y = 0: \quad 3000 \text{ kg} - 2000 \text{ kg} + P_y = 0$$

or

$$P_y = -1000 \text{ kg}.$$

P_y is negative, thus it has an opposite sense to that shown in Fig. 2. The forces acting on the member and its direction as determined above is shown in Fig. 3.

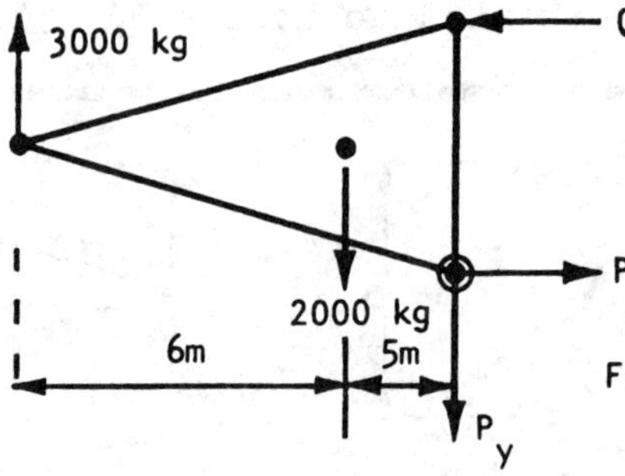

Fig. 3

• PROBLEM 1-35

A crane lifts a load of 10^4 kg mass. The boom of the crane is uniform, has a mass of 1000 kg. and a length of 10 m. Calculate the tension in the upper cable and the magnitude and direction of the force exerted on the boom by the lower pivot.

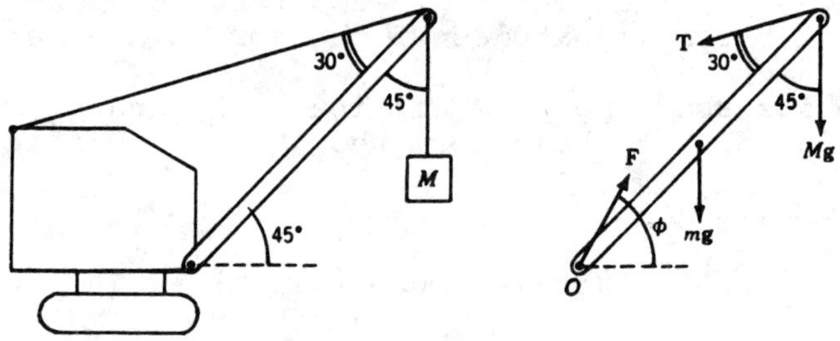

Solution: Isolate the boom analytically and indicate all forces on it as in the right-hand portion of the figure, where $\vec{T}$ is the tension in the upper cable, $\vec{F}$ is the force exerted on the boom by the lower pivot, m is the mass of the boom, and M is the mass of the load being lifted by the crane. The magnitude of $\vec{T}$ is unknown and both the magnitude and the direction of $\vec{F}$ are unknown. Set the net torque about point O equal to zero. If the length of the boom is S, this net torque is given by the equation

$$\frac{S}{2}\, mg \sin 45° + SM\, g \sin 45° - ST \sin 30° = 0$$

or
$$\frac{g(m/2 + M) \sin 45°}{\sin 30°} = T.$$

Substitute the values given above.

$$T = \frac{9.8(500 + 10{,}000)(1/\sqrt{2})}{\frac{1}{2}} = 1.46 \times 10^5 \text{ N}$$

We can find F_x and F_y, the x- and y-components of $\vec{F}$ respectively, by requiring that both the x- and y-components of the net force on the boom be equal to zero.

$$\Sigma F_x = 0. \overset{+}{\rightarrow}$$

$$F_x - T \cos 15° = 0$$

$$\Sigma F_y = 0. \; +\uparrow$$

$$F_y - T \sin 15° - mg - Mg = 0$$

$$F_x = 1.46 \times 10^5 \; (\cos 15°)$$

$$F_y = 1.46 \times 10^5 \; (\sin 15°) + 9.8(1000 + 10,000)$$

whence $\quad F_x = 1.41 \times 10^5$ N

$$F_y = 1.46 \times 10^5 \text{ N}$$

so that the magnitude of F is

$$F = \sqrt{F_x^{\,2} + F_y^{\,2}} = 2.03 \times 10^5 \text{ N.}$$

The angle ϕ which F makes with the horizontal is given by

$$\tan \phi = \frac{F_x}{F_y} = \frac{1.46}{1.41} = 1.035$$

so that $\quad \phi = 46°$.

• PROBLEM 1-36

Two rollers each weighing 200 kg with a diameter of 1.0m, have forces applied to them as shown in Fig. (a) and (b), to roll them over a 2 cm step. Determine the magnitude of the force P in each case.

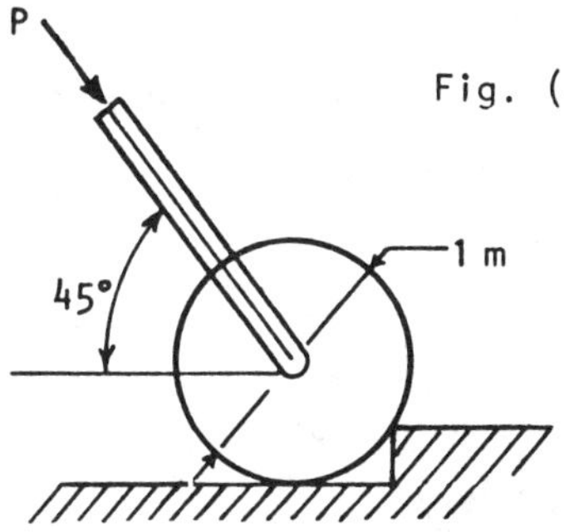

Fig. (a)

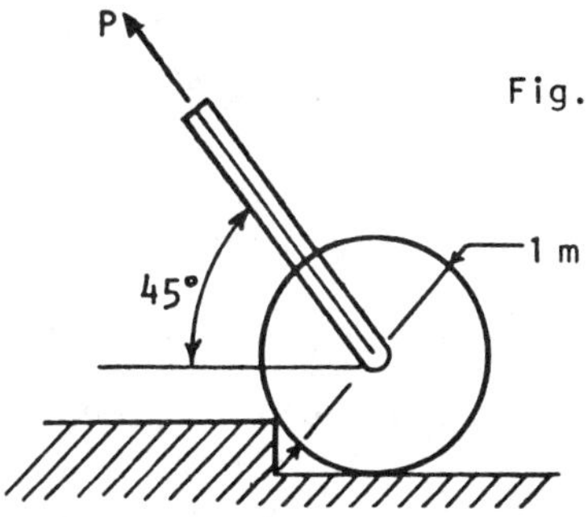

Fig. (b)

Solution: The roller is in equilibrium because the forces acting on it satisfy the conditions

Net force acting on roller = 0 (1)

Net torque about any axis acting on roller = 0 (2)

The force $\vec{P}$ can disturb this equilibrium if it is large enough. As the roller is moved over the obstruction, it will rotate about an axis through the contact point at the corner of the obstruction (A in Figure 1). The magnitude of the minimum required force P can be found by setting the torque at 0.

$$\text{Net torque about axis through A} \equiv \tau_A = P\ell_1 - W\ell_2$$

where W is the weight of the roller, ℓ_1 and ℓ_2 are the torque "lever arms" defined in Figure 1. We have used the fact that the force exerted by the ground on the roller goes to zero when the roller begins to rotate about A, and that the force exerted on the roller at the contact point with the obstruction has no torque lever arm about A and therefore produces no torque about this axis. Also note that counter clockwise torques have been defined as negative in this example.

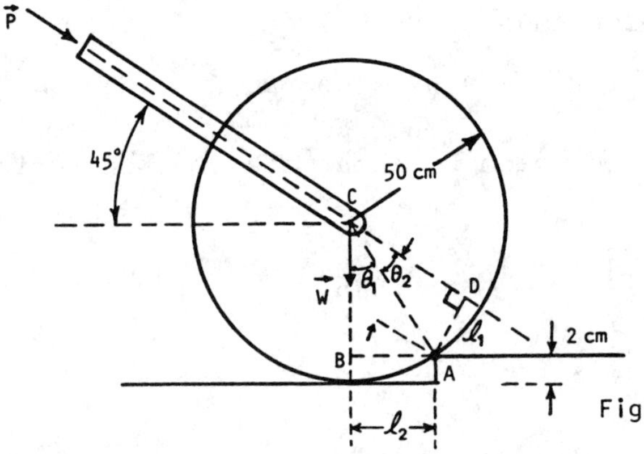

Fig. 1

To evaluate P (in equation 3) we must determine ℓ_1 and ℓ_2 . From Figure 1, note that ℓ_1 is the side AD in the right triangle ADC and ℓ_2 is the side AB in the right triangle ABC. Then we can write

4) $\ell_1 = (50\text{ cm})\sin\theta_1$

5) $\ell_2 = (50\text{ cm})\sin\theta_2$

where θ_1 is the angle ∢ ACD and θ_2 is ∢ ACB. From Figure 1 we see that

$$\cos\theta_1 = \frac{(50-2)\text{ cm}}{50\text{ cm}} = 0.96$$

Therefore, $\sin\theta_1 = \sqrt{1-\cos^2\theta_1} = 0.28$ and $\theta_1 = 16.26°$.

Also from Figure 1 we see that $\theta_2 + \theta_1 = 45°$.

Then

$$\sin \theta_2 = \sin 28.74° = 0.481$$

Putting the quantities found above into equation (3) yields

$$\tau_A = P(50 \text{ cm})(0.481) - (200 \text{ kg})(980 \text{ cm/s}^2)(50 \text{ cm})(0.28)$$

$$= P(0.5\text{m})(0.481) - (200 \text{ kg})(9.8\text{m/s}^2)(0.5\text{m})(0.28)$$

Setting $\tau_A = 0$ we have

$$P = 1141\text{N}$$

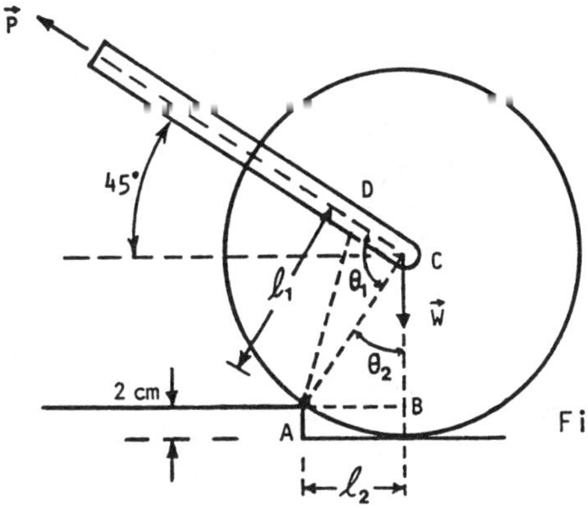

Fig. 2

b) The analysis proceeds just as for part a). To find the value of P which will just cause the roller to rotate about an axis through point A (Figure 2), we set τ_A (equation 3) equal to zero. Equations 4) and 5) are again used to determine ℓ_1 and ℓ_2, where these quantities are now defined in Figure 2. (Note that they are not the same as for part a)). We see that again

$$\cos \theta_1 = 0.96, \ \sin \theta_1 = 0.28 \text{ and } \theta_1 = 16.26°$$

But now

$$\theta_2 = 135° - \theta_1 = 118.74°$$

and

$$\sin \theta_2 = 0.877$$

Then

$$\tau_A = P(50 \text{ cm})(0.877) - (200 \text{ kg})(980 \text{ cm/S}^2)(50 \text{ cm})(0.28)$$

$$= P(0.5\text{m})(0.877) - (200 \text{ kg})(9.8\text{m/S}^2)(0.5\text{m})(0.28$$

Setting

$$\tau_A = 0$$

we have

$$P = 625.8\text{N}$$

The lifting crank is held in equilibrium by a force $\vec{T}$ applied at point T in a direction perpendicular to the plane RST. It is given that $\phi = 70°$ and the load on the crank is 900N as shown. Determine the magnitude of $\vec{T}$ and the reactions at points P and Q. The bearings at points P and Q exert no axial thrust.

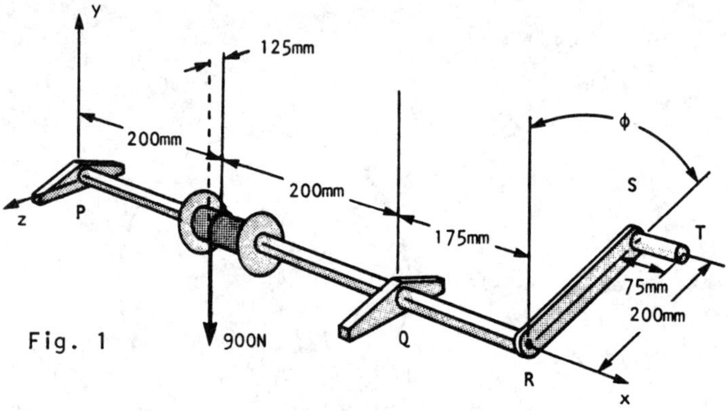

Fig. 1

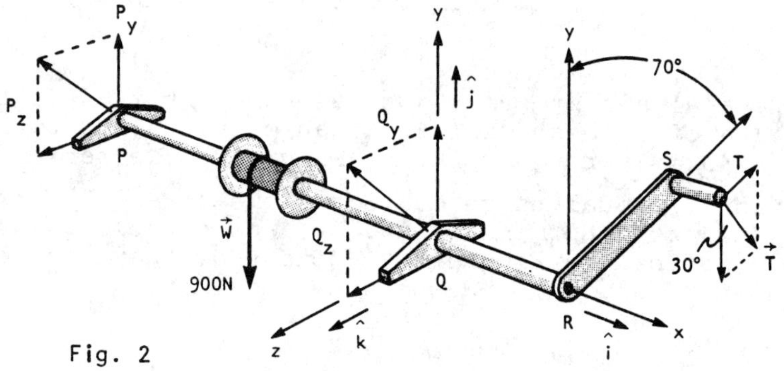

Fig. 2

Solution: Since we are not otherwise instructed, assume that the mass of the crank is zero and that there is no static friction in the system.

The conditions for equilibrium of the system are

Net Force acting on system = 0,

Net Torque about any axis acting on system = 0.

We define the "system" to be the axle, lift, and crank.

First let's consider the torque about the axle of the lift (defined as the x axis).

$$\tau_x = (900N)(125mm) - T(200mm). \tag{1}$$

In equilibrium $\tau_x = 0.$

Therefore

$$T = (900N)\frac{(125mm)}{(200mm)} = 562.5N.$$

To calculate the reactions at P and Q we apply the condition Net force = 0.

We must now identify the forces acting on the system. (See Figure 2.) There are no axial forces.

Define $\vec{P}$ = force exerted on axle by bearing P

$\vec{Q}$ = force exerted on axle by bearing Q

Then the net force on the axle is

$$\text{net force} = \vec{T} + \vec{P} + \vec{Q} + \vec{W} \quad (2)$$

$$= (562.5)(\sin 30° \,(-\hat{k}) + \cos 30°(-\hat{j}))$$

$$+ \vec{Q} + \vec{P} + (900N)(-\hat{j}).$$

At equilibrium set net force = 0. Writing components along the coordinate axes we have

$$P_x + Q_x = 0 \quad (3)$$

$$P_y + Q_y = (562.5N \cos 30°) + 900N \cong 1387.139N \quad (4)$$

$$P_z + Q_z = (562.5N \sin 30°) \cong 281.25N \quad (5)$$

To evaluate the forces $\vec{P}$ and $\vec{Q}$, we calculate the moment on the system about axes parallel to $\hat{j}$ and $\hat{k}$.

Moment about z axis $\equiv M_z$(through Q):

$$M_z = (T \cos 30°)(250mm) - (200mm)(900N) + (400mm)P_y.$$

Setting $M_z = 0$ (condition for equilibrium), we have

$$P_y = \frac{12.1785 \times 10^4 + 180000}{400m} N \cong 145.538N \quad (6)$$

From equations 4 and 6 we have

$$Q_y = 1387.139N - P_y = 1241.601N \quad (7)$$

Moment about y axis (through Q) $\equiv M_y$.

$$M_y = (T \sin 30°)(250mm) + P_z(400mm),$$

setting $M_y = 0$ yields

$$P_z \cong -175.781N \quad (8)$$

From equations 5 and 8 we have

$$Q_z = 281.25 - P_z = 457.031N \quad (9)$$

Then the forces exerted by the bearings on the axle are

$$\vec{P} = (145.538N)\hat{j} - (175.781N)\hat{k}$$

$$\vec{Q} = (1241.601N)\hat{j} + (457.031N)\hat{k}$$

• PROBLEM 1-38

A homogeneous flap of mass 20 kg is hinged at G and H and is supported horizontally by the wire JK. Determine the tension in the wire and the reactions at G and H assuming that there is no axial thrust on hinge H.

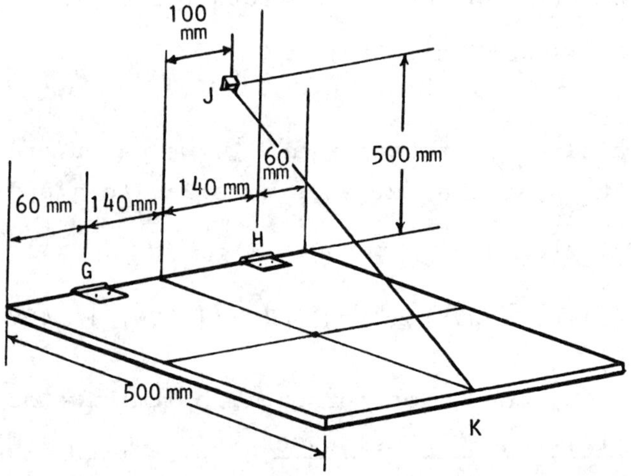

Fig. 1

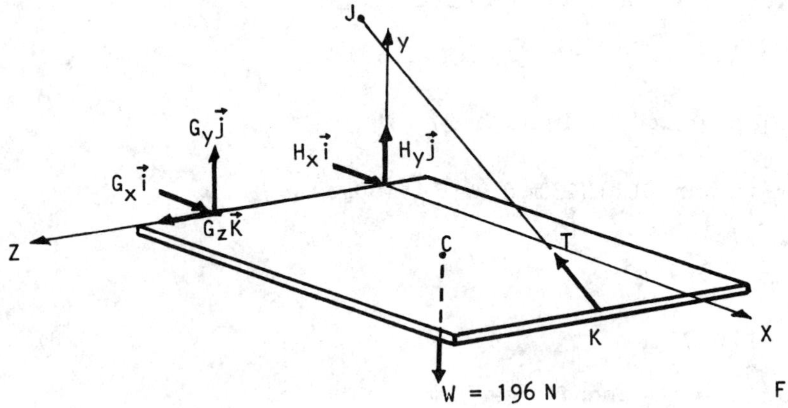

Fig. 2

Solution: Figure 2 shows the free-body diagram of the flap door with a chosen set of coordinate axes. The forces acting on the free body are the weight of the flap, the tension T, and the resolved components of the reactions at the two hinges. The weight may be found to be

$$\vec{W} = -mg\vec{j} = -(20\ \text{kg})(9.81\ \text{m/s}^2)\vec{j} = (-196\text{N})\vec{j}$$

The components of $\vec{T}$ may be expressed as the magnitude of $\vec{T}$ multiplied by the unit vector of $K\vec{J}$, denoted $K\vec{J}$ or

$$T = T\vec{K}$$

$$K\vec{J} = (-500\text{mm})\vec{i} + (500\text{mm})\vec{j} - (100\text{mm})\vec{k}$$

$$K\hat{\vec{J}} = \frac{-500\vec{i} + 500\vec{j} - 100\vec{k}}{\sqrt{(-500)^2 + (500)^2 + (-100)^2}} = \frac{-5}{\sqrt{51}}\vec{i} + \frac{5}{\sqrt{51}}\vec{j} - \frac{1}{\sqrt{51}}\vec{k}$$

EQUILIBRIUM EQUATIONS:

Setting the forces and moments acting on the flap door equal to zero, we obtain

$$\Sigma F = 0 : \left(H_x + G_x - \frac{5T}{\sqrt{51}}\vec{i}\right) + \left(H_y + G_y + \frac{5T}{\sqrt{51}} - 196\text{N}\right)\vec{j}$$

$$+ \left(G_z - \frac{T}{\sqrt{51}}\right)\vec{k} = 0 \qquad (1)$$

$$\Sigma\vec{M}_H = \Sigma(\vec{r} \times \vec{F}) = 0 : (280\ \text{mm})\vec{k} \times (G_x\vec{i} + G_y\vec{j} + G_z\vec{k})$$

$$+ (140\ \text{mm}\vec{k} + 250\ \text{mm}\vec{i}) \times (-196\text{N})\vec{j}$$

$$+ (140\ \text{mm}\vec{k} + 500\ \text{mm}\vec{i}) \times \left(\frac{-5T}{\sqrt{51}}\vec{i} + \frac{5T}{\sqrt{51}}\vec{j} - \frac{1}{\sqrt{51}}\vec{k}\right) = 0$$

Simplifying,

$$\left(27440\text{N} - 280\ G_y - \frac{700}{\sqrt{51}}T\right)\vec{i} + \left(280\ G_x + \frac{1200}{\sqrt{51}}T\right)\vec{j}$$

$$+\left(\frac{2500}{\sqrt{51}}T - 49000\text{N}\right)\vec{k} = 0 \qquad (2)$$

The coefficients of $\vec{i}$, $\vec{j}$ and $\vec{k}$ in equations (1) are set to equal zero and the reactions, G_x, and G_y and the tension T are determined.

T may be solved from $\dfrac{98\sqrt{51}\ \text{N}}{5} = 140\text{N}$

Thus,

$$G_x = \frac{\left(\frac{-1200}{\sqrt{51}}\right)\left(\frac{98\sqrt{51}}{5}\right)N}{280} = -84.0N$$

$$G_y = \frac{27440N - \left(\frac{700}{\sqrt{51}}\right)\left(\frac{98\sqrt{51}}{5}\right)N}{280} = 49.0N$$

The coefficients of equation (2) are then set equal to zero and the reactions G_z, H_y and H_x are determined.

$$G_z = \frac{98\sqrt{51}}{5\sqrt{51}} N = 19.6N$$

$$H_y = 196N - \frac{5 \times 98\sqrt{51}}{5\sqrt{51}} - 49.0N = 49N$$

$$H_x = \frac{5 \times 98\sqrt{51}}{5\sqrt{51}} N - (-84N) = 182N$$

• PROBLEM 1-39

The force intensity distribution over one end of a circular rod is

$$\vec{F}(r) = (2y\hat{i} - 2x\hat{j}) \text{ lb/in}^2$$

where $\vec{r} = x\hat{i} + y\hat{j}$. What resultant reaction is necessary to hold this torsion bar in equilibrium?

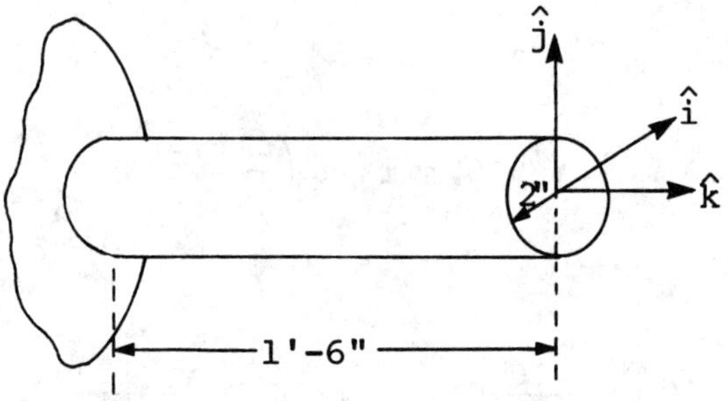

Solution: The total force and torque due to the distribution are, respectively,

$$\vec{F} = \iint (2y\hat{i} - 2x\hat{j})dA$$

$$= \int_0^a 2r^2\, dr \left[\hat{i}\int_0^{2\pi} \sin\theta\, d\theta - \hat{j}\int_0^{2\pi} \cos\theta\, d\theta\right] = 0.$$

$$\vec{N} = \iint (x\hat{i} + y\hat{j}) \times (2y\hat{i} - 2x\hat{j})\, dA$$

$$= -2\hat{k} \iint (x^2 + y^2)\, dA = -2\hat{k} \int_0^a r^2\, r\, dr \int_0^{2\pi} d\theta$$

$$= -\pi a^4 \hat{k}\ .$$

In evaluating the integrals the change of variables from (x, y) to the polar coordinates (r, θ) are used,

$$x = r \cos \theta$$

$$y = r \sin \theta$$

$$dA = r\, dr\, d\theta.$$

The resultant reactions that are necessary to hold the rod in equilibrium are

$$\vec{f} = -\vec{F} = 0, \qquad \vec{\tau} = -\vec{N} = \pi a^4 \hat{k}.$$

• PROBLEM 1-40

A plate rocker arm of mass m is attached to a frictionless bearing at O and contacts a smooth surface at P. Assuming that the plate's c.g. is at a distance d from the wall find the reactions at O and P.

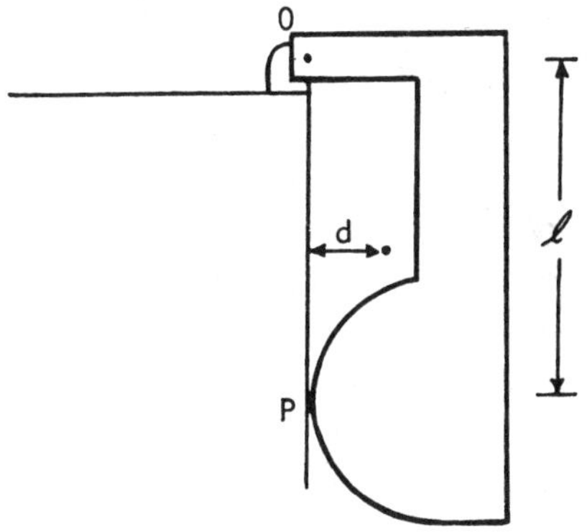

Solution: The reactions at O and P may be reduced to their horizontal and vertical components. (Since the surface at P is smooth, there will be no vertical reaction due to friction.) Figure 1 shows the free-body diagram of the mass. Applying the equations of static equilibrium,

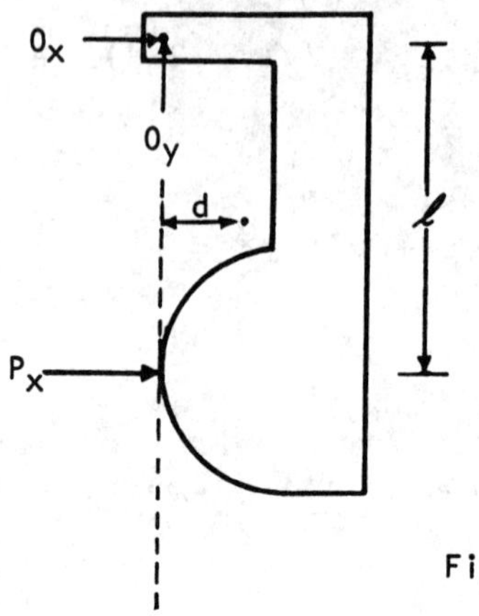

Fig. 1

$$+\circlearrowleft \Sigma M_o = 0 : \qquad P_x \ell - mgd = 0 \quad = \quad P_x = \frac{mgd}{\ell} \quad ,$$

$$\underset{\rightarrow}{+} \Sigma F_x = 0 : \qquad P_x + O_x = 0 \quad = \quad O_x = -\frac{mgd}{\ell} \quad ,$$

$$+\uparrow \Sigma F_y = 0 : \qquad O_y - mg = 0 \quad = \quad O_y = mg \; .$$

To find the total reaction at O, we must find the magnitude and direction of O. Doing so we obtain

$$|O| = \sqrt{O_x^2 + O_y^2} = mg \sqrt{1 + \left(\frac{d^2}{\ell^2}\right)}$$

$$\measuredangle\, O = \tan^{-1} \frac{O_y}{O_x} = \tan^{-1} \frac{\ell}{d} \; .$$

Thus the answers may be given as

$$O = mg\sqrt{1 + \left(\frac{d^2}{\ell^2}\right)} \measuredangle \tan^{-1} \frac{\ell}{d}$$

$$P = \frac{mgd}{\ell} \rightarrow$$

CHAPTER 2

ANALYSIS OF STRUCTURES

● PROBLEM 2-1

> The truss in Fig. 1 has forces acting as shown. Calculate the forces in members NR and LK.

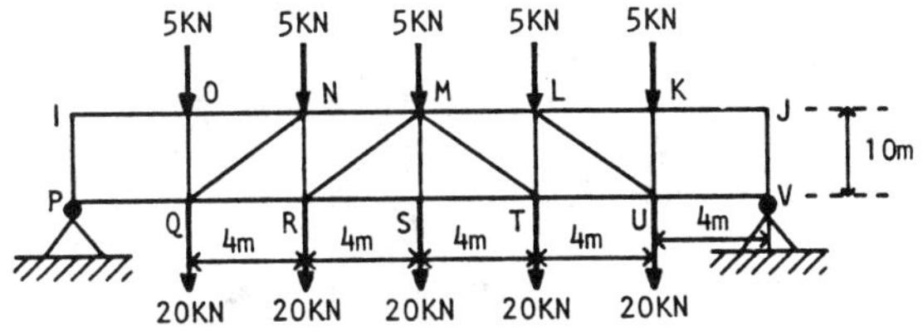

Fig. 1

Solution: The free body diagram of the whole truss and the forces acting on it is shown in Fig. 2.

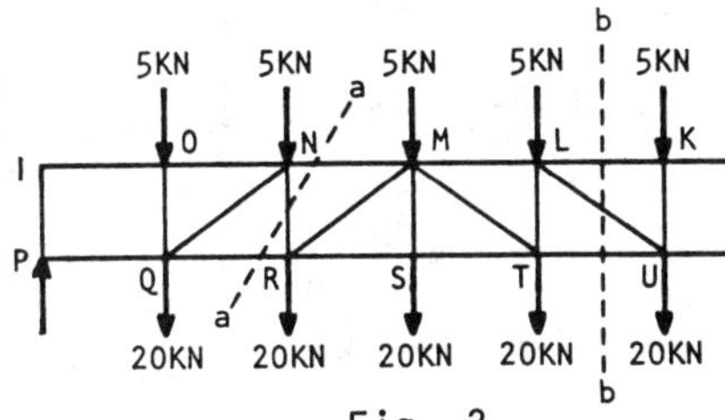

Fig. 2

The equilibrium of the entire truss is considered in order to determine the reactions at P and V $+\circlearrowright \Sigma M_V = 0$.

$$(25\ KN)(4m) + (25KN)(8m) + (25KN)(12m) + (25KN)(16m)$$

$$+ (25KN)(20m) - P(24m) - 0$$

$$\therefore \quad P = 62.5KN$$

Likewise the equation $+\circlearrowright \Sigma M_P = 0$ yields $V = 62.5KN$

To calculate the force in NR, the method of sections is used. A line aa is passed through the truss dividing it into two separate parts but not intersecting more than three members. One portion of the truss shown in Fig. 3 is now a free body, from which the force F_{NR} is determined.

The section bb intersecting member LK would be used to determine the force. A free body diagram of one portion of the truss is shown in Fig. 4.

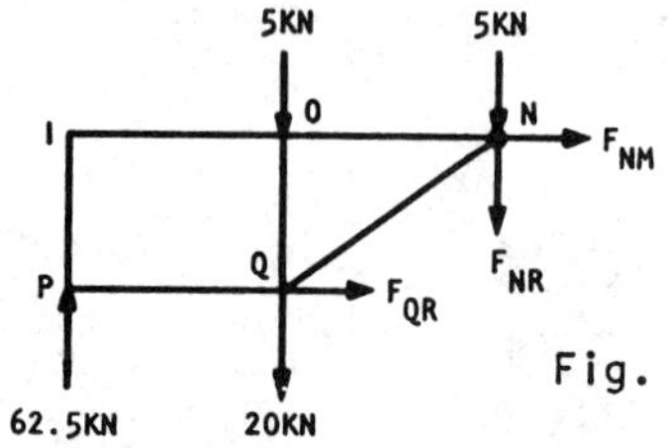

Fig. 3

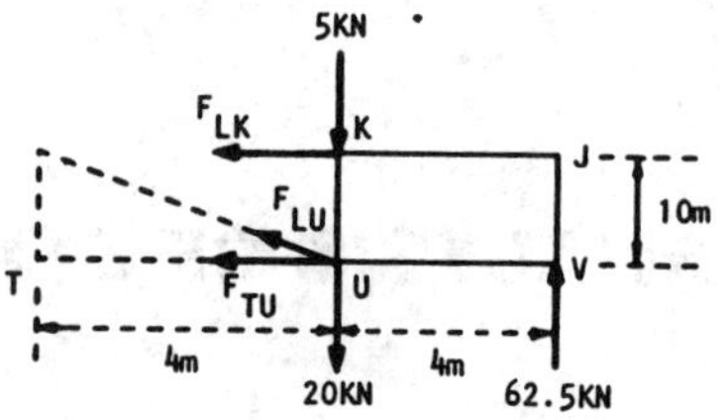

Fig. 4

$+\uparrow\Sigma F_y = 0;\ -F_{NR} - 5KN - 5KN - 20KN + 62.5KN = 0$

$F_{NR} = 32.5KN$ Tension

$$+\circlearrowright \Sigma M_U = 0;\ (F_{LK})(10m) + (62.5KN)(4m) = 0$$

or

$$F_{LK} = -25KN$$

The minus sign indicates that F_{LK} has an opposite sense to that shown in Fig. 4.

• PROBLEM 2-2

In the bridge truss of Fig. 1, calculate the forces in members UV and DE.

Fig. 1

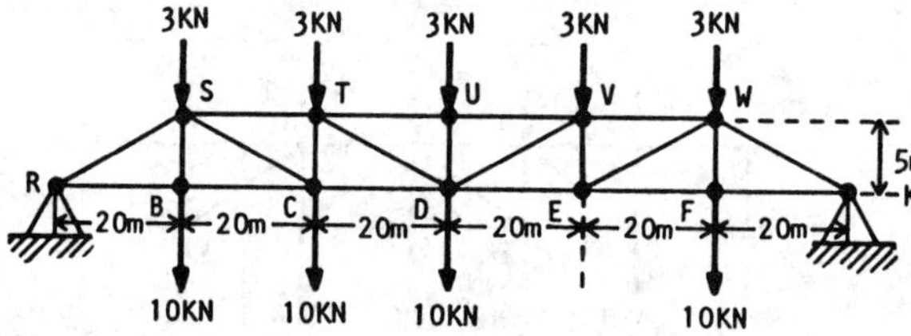

Solution: The free body diagram is shown in Fig. 2. The method of section will again be used. The section aa cuts through the truss intersecting members UV and DE. The reaction at K is determined using the equation

$$+\circlearrowright \Sigma M_R = 0$$

Fig. 2

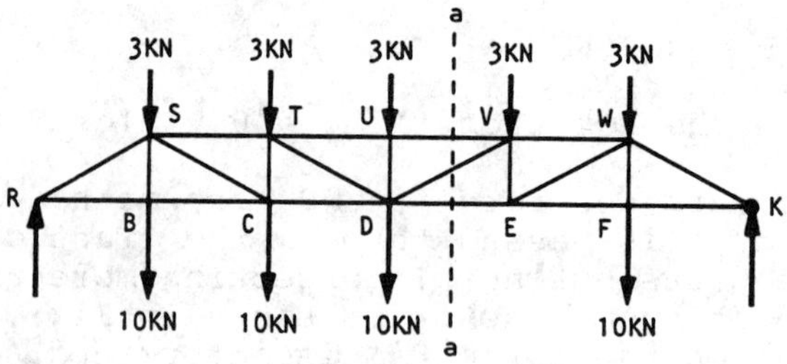

$$- (13KN)(20m) - (13KN)(40m) - (13KN)(60m) - (3KN)(80m)$$

$$- (13KN)(100m) + K(120m) = 0$$

or

$$K = 25.8KN$$

The right portion of the truss VWKE cut off by the line aa in Fig. 3 will be used as a free body to calculate the force in the three members. For force in member DE

Fig. 3

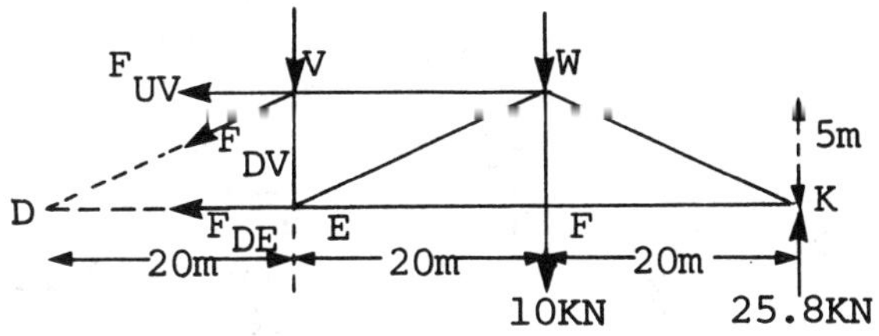

$$+\circlearrowleft \Sigma M_V = 0$$

$$(25.8KN)(40m) - (13KN)(20m) - F_{DE}(5m) = 0$$

$$\therefore \quad F_{DE} = +154KN = 154KN \text{ tension}$$

The force in member UV is determined by taking moments about joint D. From Fig. 3;

$$+\circlearrowleft \Sigma M_D = 0$$

$$(25.8KN)(60m) - (13KN)(40m) - (3KN)(20m) + (F_{UV})(5m) = 0$$

$$\therefore \quad F_{UV} = -194KN$$

The negative sign indicates F_{UV} to be in compression.

• PROBLEM 2-3

Determine the forces in the members of the truss shown in Figure 1.

Fig. 1

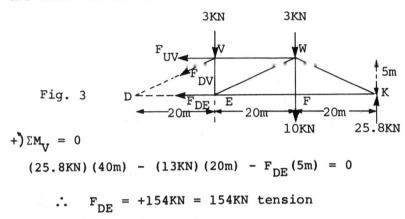

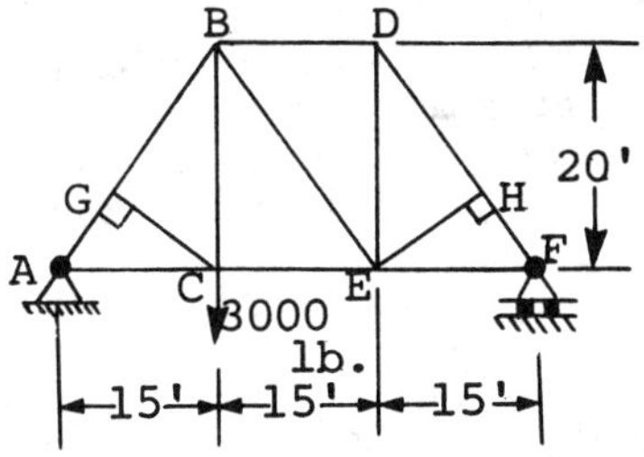

Solution: The first step in the solution is to isolate the complete truss as shown in Figure 2 and to apply the conditions for equilibrium in order to determine the external reactions.

$+\circlearrowleft \Sigma M_A = 0$:

$$F_y(45\text{ ft}) - 3000(15) = 0$$

$$F_y = \frac{45,000}{45} = 1,000\text{ lb}$$

$+\circlearrowleft \Sigma M_F = 0$:

$$-A_y(45\text{ ft}) + 3000(30\text{ ft}) = 0$$

$$A_y = \frac{90,000}{45} = 2,000\text{ lb}$$

$$\Sigma F_x = 0 \;.\; \overset{+}{\rightarrow}$$

$$A_x = 0\;.$$

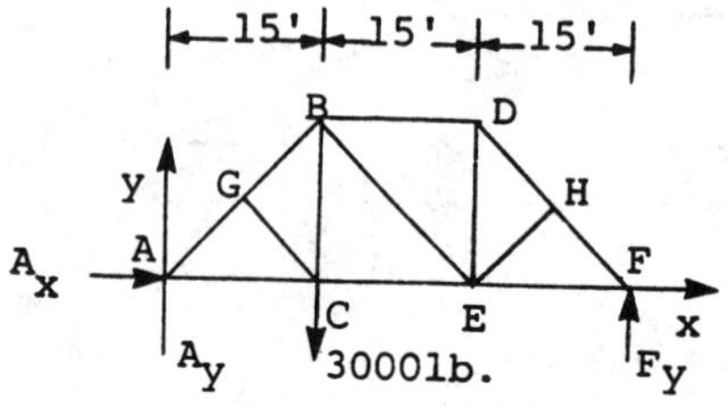

Fig. 2

Free Body Diagram of Complete Truss

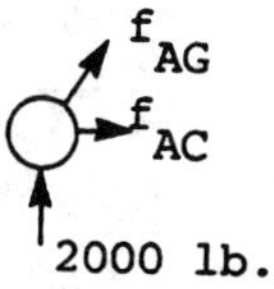

Fig. 3

Joint A

The free-body diagram for joint A is shown in Figure 3. Since the senses of the forces f_{AG} and f_{AC} are unknown, we arbitrarily assume that the bars AG and AC are in tension and show the force arrows acting away from the joint. Since only two unknown forces appear, we can completely determine the forces acting at this joint. Applying the conditions of equilibrium, we have

$$\Sigma F_y = 0:$$

$$\frac{4}{5} f_{AG} + 2000 = 0$$

$$f_{AG} = -2500\text{ lb} \qquad \text{(compression)}$$

$$\Sigma F_x = 0:$$

$$-(\tfrac{3}{5})2500 + f_{AC} = 0$$

$$f_{AC} = 1500\text{ lb} \qquad \text{(tension)}$$

Because we assumed that bar AG is in tension, the minus preceding the calculated value of f_{AG} indicates member AG is in compression.

Members GC and EG are easily recognized as zero-load members. This is clear since no member exists to balance any force that EH may exert on DF or CG may exert on AB. From the free-body diagram for joint G we find that f_{GB} = 2500 lb compression.

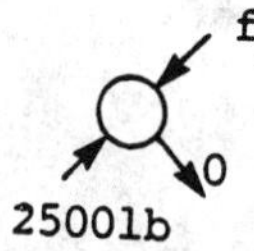

Fig. 4

Joint G

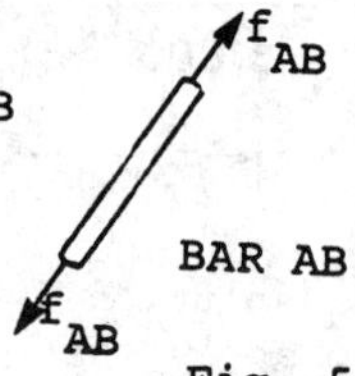

Fig. 5

Fig. 6

For joint C we obtain

$$\Sigma F_y = 0:$$
$$f_{BC} - 3000 = 0$$
$$f_{BC} = 3000 \text{ lb} \quad \text{(tension)},$$
$$\Sigma F_x = 0:$$
$$-1500 + f_{CE} = 0$$
$$f_{CE} = 1500 \text{ lb} \quad \text{(tension)} .$$

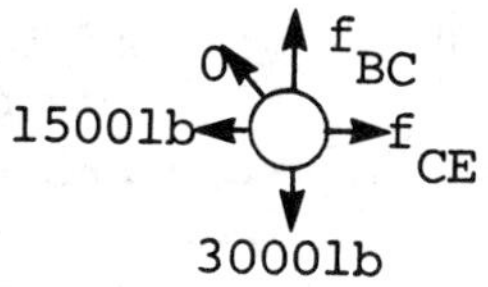

Joint C
Fig. 7

For joint B,

$$\Sigma F_y = 0:$$
$$(\tfrac{4}{5})2500 - 3000 - \tfrac{4}{5} f_{BE} = 0$$
$$f_{BE} = -1250 \text{ lb} \quad \text{(compression)},$$
$$\Sigma F_x = 0:$$
$$(\tfrac{3}{5})2500 - (\tfrac{3}{5})1250 + f_{BD} = 0$$
$$f_{BD} = -750 \text{ lb} \quad \text{(compression)} .$$

Joint B
Fig. 8

For joint D,

$$\Sigma F_x = 0:$$
$$750 + \tfrac{3}{5} f_{DH} = 0$$
$$f_{DH} = -1250 \text{ lb} \quad \text{(compression)},$$
$$\Sigma F_y = 0:$$
$$-f_{DE} + (\tfrac{4}{5})1250 = 0$$
$$f_{DE} = 1000 \text{ lb} \quad \text{(tension)}.$$

Joint D
Fig. 9

From joint H we recognize EH as a zero-load member and it follows that f_{HF} = 1250 lb compression. For joint

$$\Sigma F_x = 0:$$
$$-1500 + (\tfrac{3}{5})1250 + f_{EF} = 0$$
$$f_{EF} = 750 \text{ lb} \quad \text{(tension)}.$$

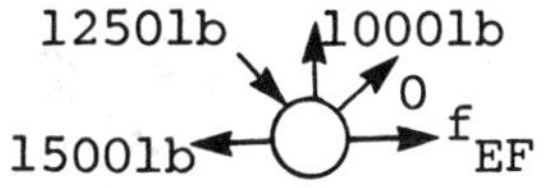

Joint E
Fig. 10

Since the force carried by each member has been determined, joint K provides us with a check on the calculations. Thus,

$$\Sigma F_y = 0:$$
$$-(\tfrac{4}{5})1250 + 1000 = 0$$
$$\Sigma F_x = 0:$$
$$-750 + (\tfrac{3}{5})1250 = 0 .$$

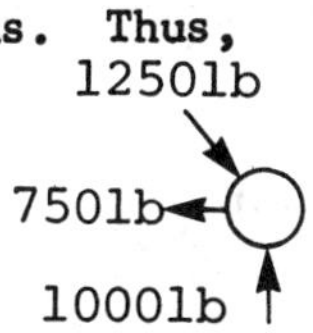

Joint F
Fig. 11

Calculate the forces in all the members of the truss shown in Fig. 1, using the method of joints.

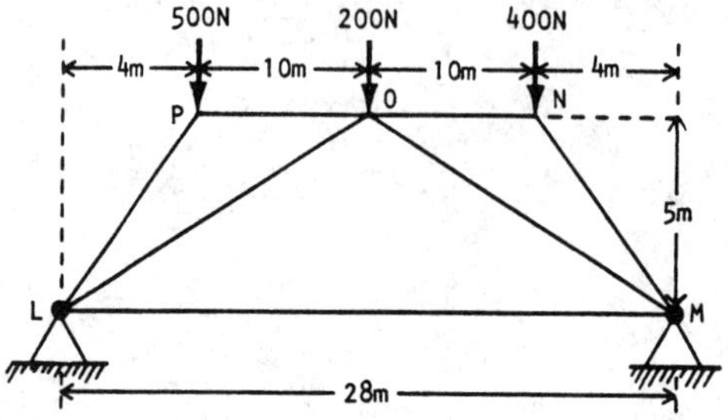

Fig. 1

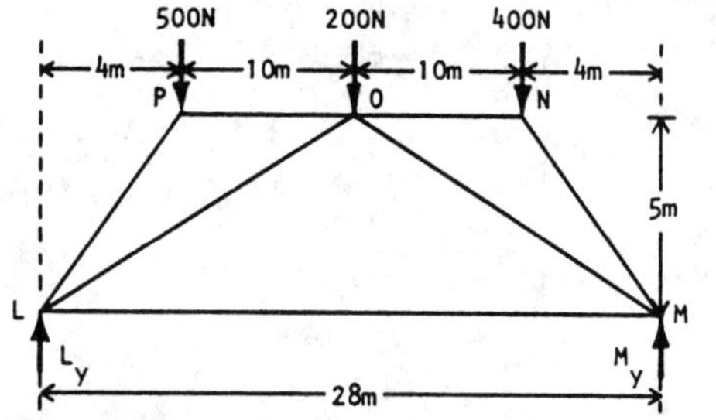

Fig. 2

Solution: A free body diagram showing all the forces acting on the truss is shown in Fig. 2.

For equilibrium,

$$+\circlearrowleft \Sigma M_L = 0; \quad -(500N)(4m) - (200N)(14m) - (400N)(24m) + M_Y(28m) = 0$$

$$M_Y = 514N \uparrow$$

$$+\uparrow \Sigma F_Y = 0; \quad -500N - 200N + 514N - 400N + L_Y = 0$$

$$L_Y = 586N \uparrow$$

Joint P: Only two unknown forces act on this joint, namely, the forces exerted by members PO and PL. A force triangle will be used to determine these two forces noting that they both push on the joint.

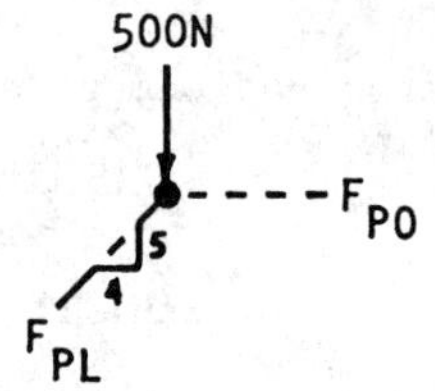

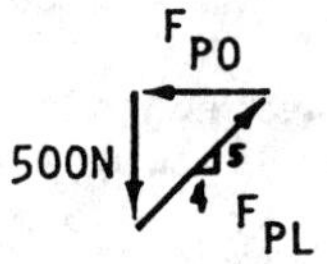

The magnitudes of the two forces are found by the proportion

$$\frac{500N}{5} = \frac{F_{PO}}{4} = \frac{F_{PL}}{\sqrt{4^2 + 5^2}}$$

$$F_{PO} = 400N \ C$$

$$F_{PL} = 640N \ C$$

Joint L: Since F_{PL} has been determined, there are only two unknown forces acting on joint L. These forces may be determined by solving the two equilibrium equation $F_x = 0$ and $F_y = 0$. Both forces may be arbitrarily assumed to act away from the joint, i.e., that the members are in tension.

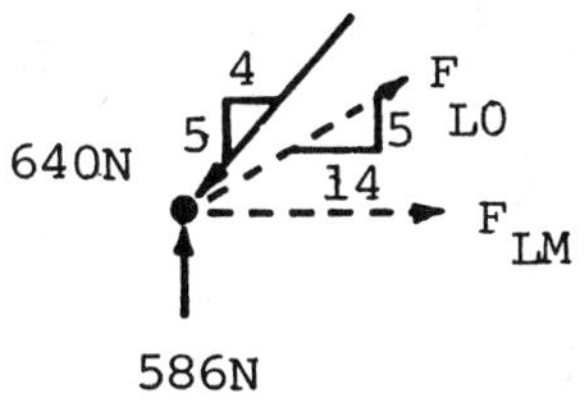

$$\stackrel{+}{\rightarrow} F_x = 0; \quad -400N + \frac{14}{\sqrt{221}} F_{LO} + F_{LM} = 0$$

$$+\uparrow \Sigma F_y = 0; \quad -500N + 586N + \frac{5}{\sqrt{221}} F_{LO} = 0$$

$$F_{LO} = -256N = 256N\ C$$

$$F_{LM} = 641N = 641N\ T$$

Joint O: Following the same procedure for Joint O as was carried out for Join L, gives

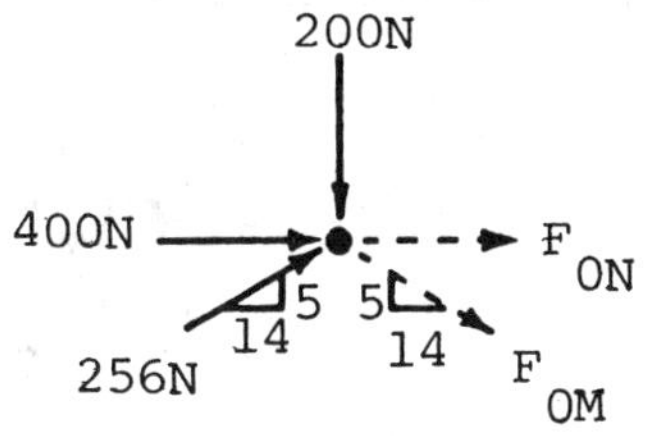

$$+\uparrow \Sigma F_y = 0; \quad -200N + \frac{5}{\sqrt{221}} (256N) - \frac{5}{\sqrt{221}} F_{OM} = 0$$

$$F_{OM} = -339N = 339N\ C$$

$$\stackrel{+}{\rightarrow} \Sigma F_x = 0; \quad 400N + \frac{14}{\sqrt{221}} (256N) + \frac{14}{\sqrt{221}} (-339N) + F_{ON} = 0$$

$$F_{ON} = -321N = 321N\ C$$

Joint N: Drawing a free-body diagram of the pin and setting equilibrium conditions yields

$\overset{+}{\rightarrow} \Sigma F_x = 0; \quad 321N + \frac{4}{\sqrt{41}} F_{NM} = 0$

$F_{NM} = -514N = 514N\ C$

As a check, $+\uparrow\Sigma F_y = 0; \quad -400N - \frac{5}{\sqrt{41}} (-514N) = 0$

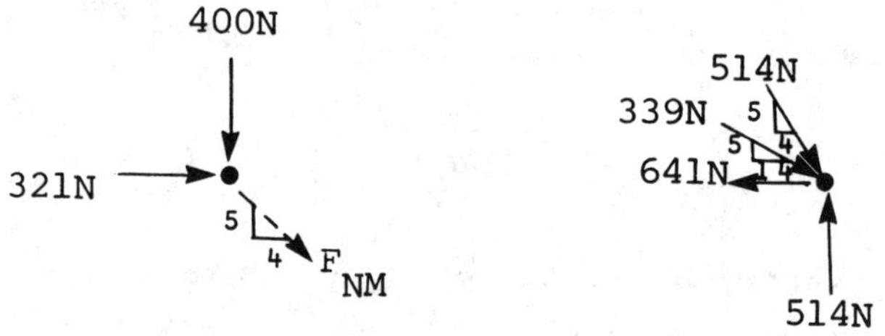

Also as a check, we may now consider the forces acting on joint M

Joint M:

$\overset{+}{\rightarrow}\Sigma F_x = 0; \quad -641N + (339N) \frac{14}{\sqrt{221}} + (514) \frac{4}{\sqrt{41}} = 0$

• PROBLEM 2-5

Determine the forces acting in each member of the frame in figure 1.

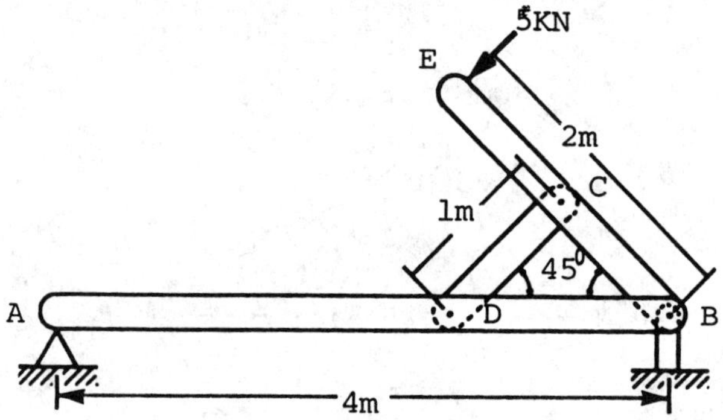

Fig. 1

Solution: First consider the entire frame in order to determine the reactions at A and B. From Figure 2,

$\overset{+}{\rightarrow}\Sigma F_x = 0; \quad -(5KN)(\sin 45°) + B_x = 0$

$B_x = 3.54\ KN\rightarrow$

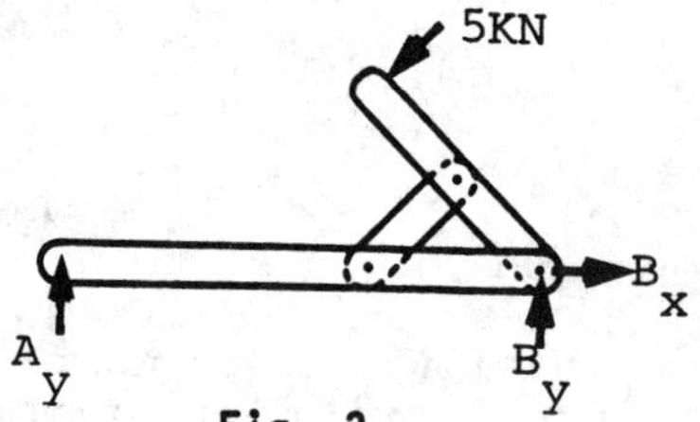

Fig. 2

$+\circlearrowleft \Sigma M_B = 0; \quad -(Ay)(4m) + (5KN)(\cos 45°)(2m)(\cos 45°) = 0$

$$A_y = 1.25KN\uparrow$$

$+\uparrow \Sigma F_y = 0; \quad A_y + B_y - (5KN)(\cos 45°) = 0$

$$B_y = 2.29 \text{ KN}\uparrow$$

Now it is possible to determine the internal reactions by taking each member as a free body. Figure 3 shows a free body diagram for each member.

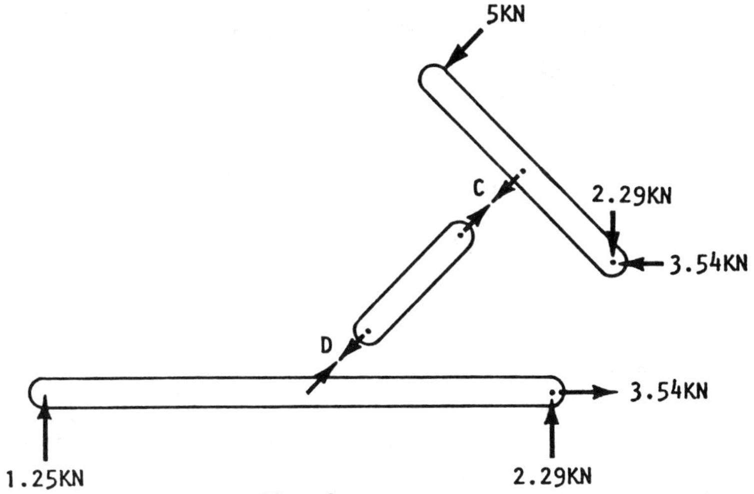

Fig. 3

$+\circlearrowleft \Sigma M_B = 0; \quad (-1.25KN)(4m) - (D \sin 45°)(\sqrt{2}\text{ m}) = 0.$

$D = 5 \text{ KN}$

$D = 5 \text{ KN} \ \nearrow 45°$

$C = 2.5 \text{ KN} \ \measuredangle 45°$

• PROBLEM 2-6

Calculate the forces acting on the two vertical members of the assembly shown in Fig. 1.

Solution: The free body diagram of Fig. 2 shows the four unknown reactions acting on the assembly.

The reaction S_y is determined by taking moment about R.

$+\circlearrowleft \Sigma M_R = 0: \quad (150)(15m) + S_y(6m) = 0$

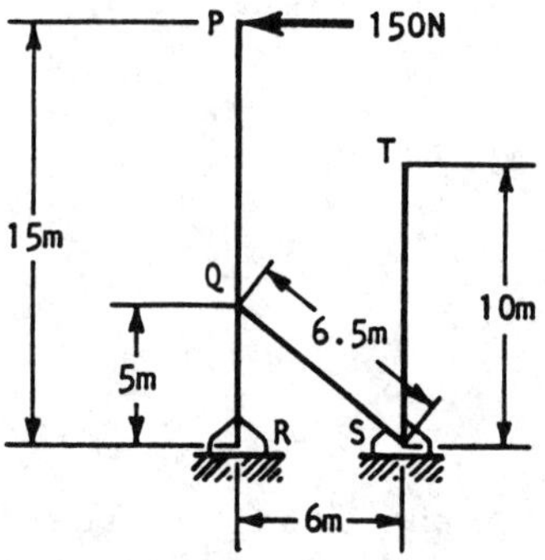

Fig. 1

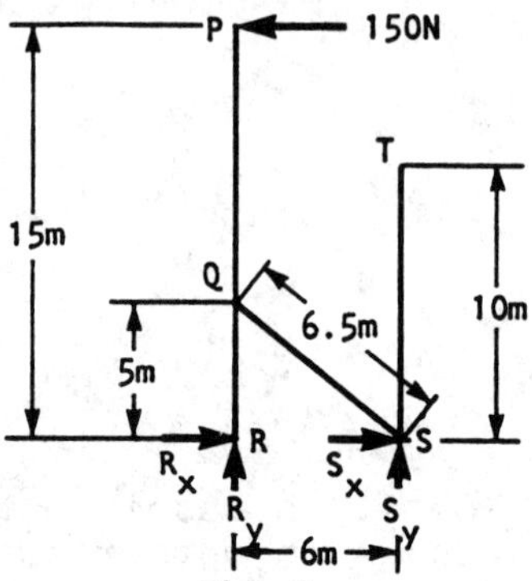

Fig. 2

or

$$S_y = -375N$$

S_y is thus acting in the opposite direction to that assumed in Fig. 2. Thus $S_y = +375N$

$+\uparrow\Sigma F_y = 0: \quad R_y + 375N = 0$

or

$$R_y = -375N$$

The force members are now dismembered in order to calculate the reactions.

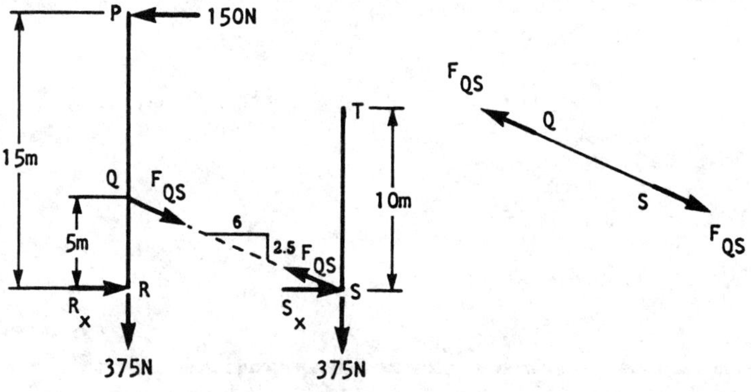

(a) (b) Fig. 3 (c)

For member PQR in Fig. 3(a)

$$+\uparrow\Sigma F_y = 0: \quad -\frac{2.5}{6.5} F_{QS} - 375N = 0$$

or

$$F_{QS} = -\ 975N$$

F_{QS} thus is acting in the opposite direction to that

assumed in Fig. 3(a).

$$\overset{+}{\rightarrow}\Sigma F_x = 0: \quad -150N + \frac{6}{7.5}(-975N) + R_x = 0$$

or

$$R_x = 930N$$

Returning to Fig. 2, the reaction S_x can now be determined since R_x is known. Thus

$$\overset{+}{\rightarrow}\Sigma F_x = 0: \quad 930N - 150N + S_x = 0$$

or

$$S_x = -780N$$

The minus sign indicates that the sense assumed for S_x in Fig. (2) was incorrect.

• PROBLEM 2-7

By drawing Maxwell's diagram, determine the force in each member of the truss shown in figure 1.

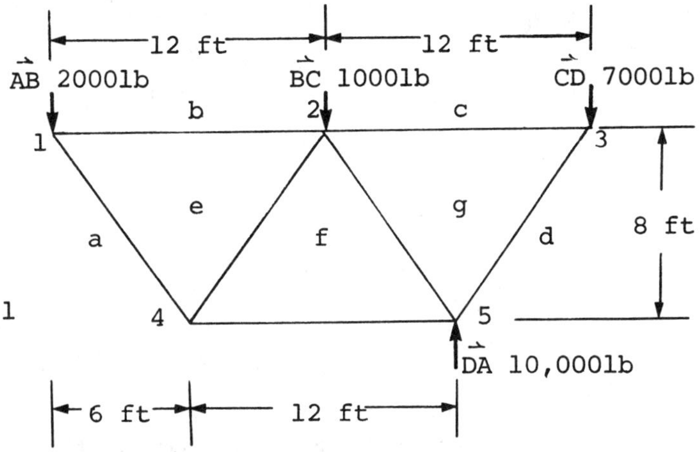

Fig. 1

Solution: Maxwell's diagram is a scale drawing of the force polygons of each joint, superimposed where appropriate. Bow's notation is used in which each member is labeled according to the region letters (a,b,c,d,e,f, and g) of each region the particular member separates. Each region bounded by an external force or truss member is given a letter designation as shown in figure 1. Here the letter a is used 2 times to make reading the notation easier. Also, the joints have been numbered to help clarify discussion.

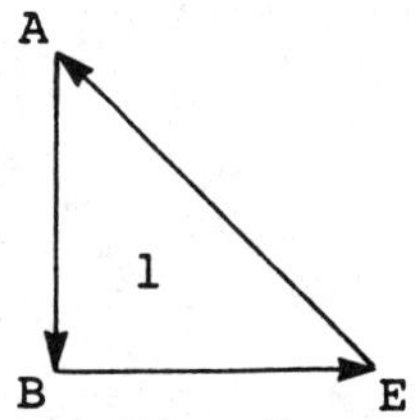

Fig. 2

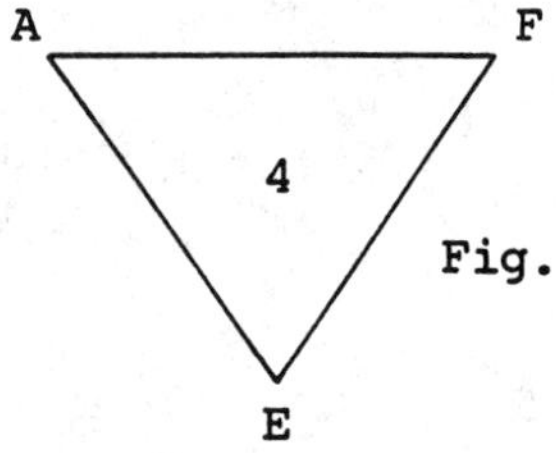

Fig. 3

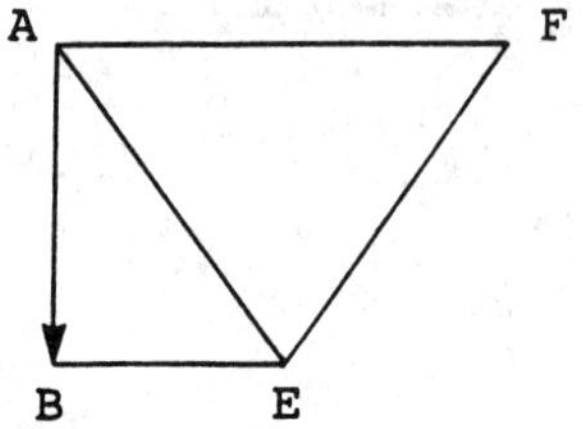

Fig. 4

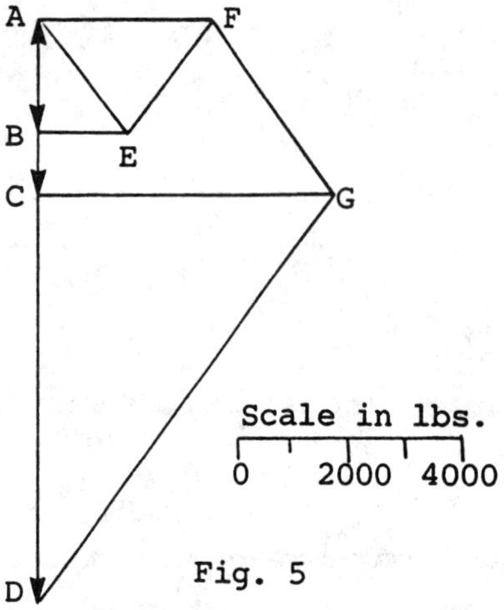

Fig. 5

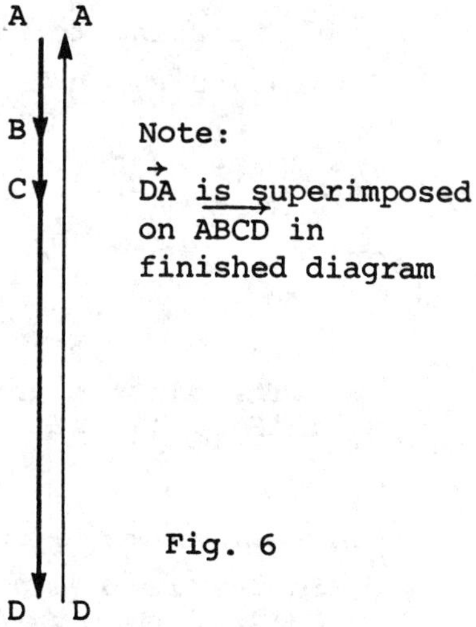

Fig. 6

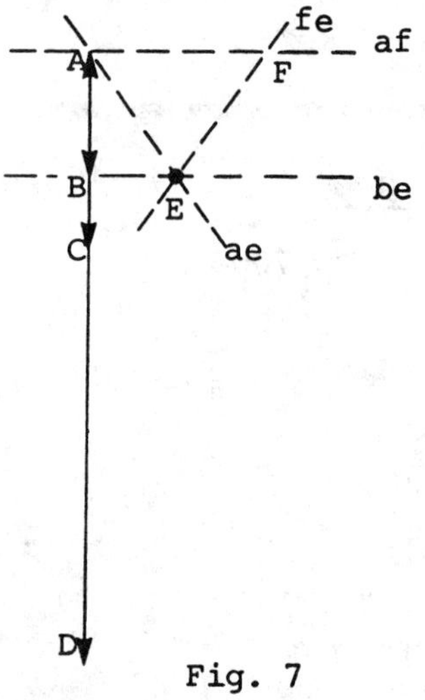

Fig. 7

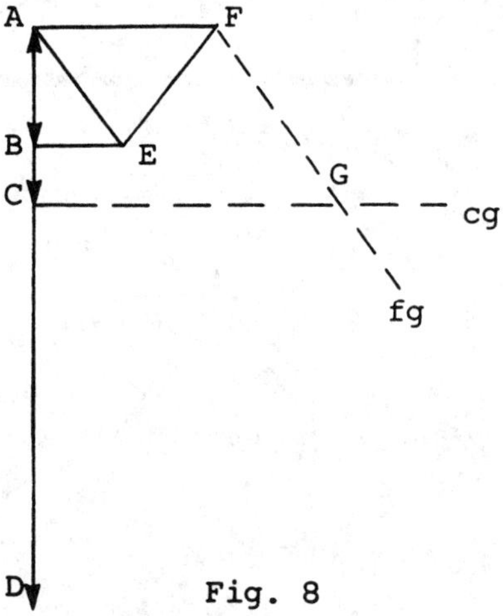

Fig. 8

Consider joint 1; the force polygon for it is shown in figure 2. Note that by the placement of the letters at the intersections of the forces, each force gets the appropriate label: AB ↓ is the external force, DE is the force in the member between b,e and AE is the force in the member between a,e. The arrowheads have been placed arbitrarily for the structure members and will be left out henceforth. The senses of the forces must be considered separately. Now examine joint 4, the force polygon for it is shown in figure 3.

The force AE is common to both joints 1 and 4 so figures 2 and 3 may be combined.

This is the procedure to be followed for the other joints.

In actually drawing the Maxwell diagram, three considerations must be made:

1) The length of the member forces is unknown.
2) The directions of the member forces must be taken from the structure.
3) Scale must be adhered to, to get accurate results.

The completed Maxwell's diagram is shown in figure 5. First, all the external forces on the truss are drawn to scale, thus locating points A,B,C and D,(figure 6). These correspond to the external labels.

The points, E,F,G (from the internal labels), are yet to be located. In doing so, since the magnitudes of the truss members are unknown, only the appropriate directions of the member force may be used. Now consider joints 1 and 4. Point E is located by drawing lines be parallel to member be and line ae parallel to member ae. Next point F is located. Lines af and fe are drawn parallel to members af and fe respectively with fe passing, appropriately, through point E. This much is shown in figure 7.

In developing figure 5 only point G remains to be located. Upon examining joint 2, it becomes clear that H lies at the intersection of lines cg and fg drawn parallel to members cg and fg and passing through the appropriate points already located. This is shown in figure 8 below. The Maxwell diagram is completed by joining the located points so that all truss members are represented. The finished diagram and scale are shown in figure 5.

The magnitude of the force in each member may now be measured, using the scale, from the Maxwell diagram. Whether the force is compression or tension is determined by a simple rule. Whether the member pushes or pulls on either of the two joints it connects, is determined by: a) selecting one of the joints on the truss (figure 1). b) reading the names of the areas adjacent to the member in clockwise order around the joint. c) reading the corresponding letters on the Maxwell diagram in the same order to get the direction. For example: chose the member connecting joints 2 and 4. Reading clockwise about point 4 gives ef. EF in figure 5 is up and to the right, so the force is a tension (pulling).

The results are given in the table below

Member	Force
be	1500 lb T
ae	2500 lb C
ef	2500 lb T
fa	3000 lb C
fg	3750 lb C
cg	5250 lb T
gd	8750 lb C

• **PROBLEM** 2-8

A 1500N-force is applied to the linkage at Q. For each member of the linkage calculate the components of the forces acting on them.

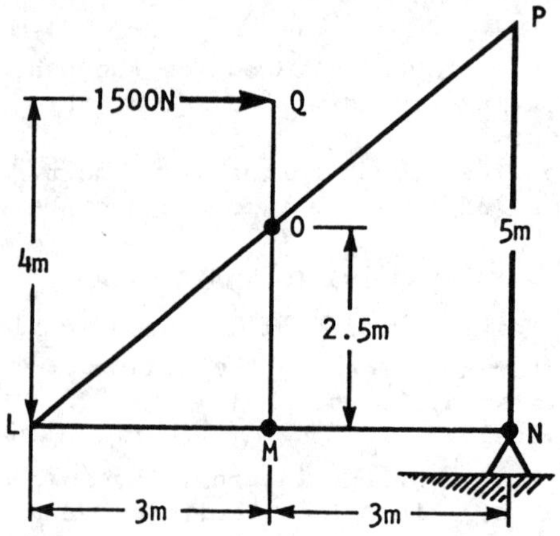

Fig. 1

Solution: The free body diagram of the whole linkage in Fig. 2 will be used to calculate the reactions at L and N.

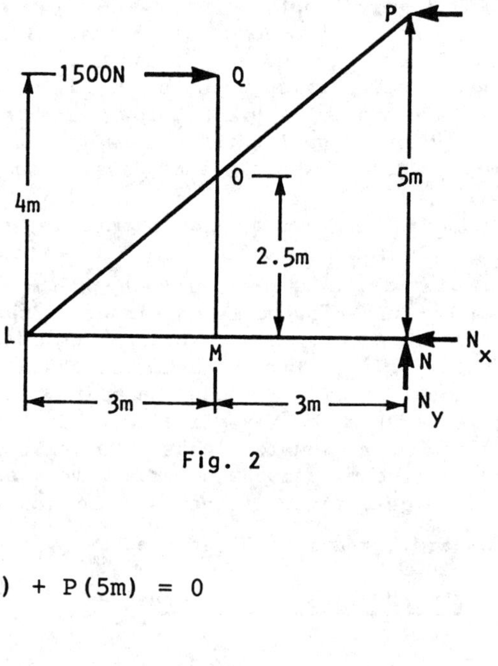

Fig. 2

$$\circlearrowright + \Sigma M_M = 0$$

$$-(1500N)(4m) + P(5m) = 0$$

or

$$P = 1200N$$

$$\overset{+}{\rightarrow}\Sigma F_x = 0: \quad 1500N - 1200N - N_x = 0$$

or

$$N_x = 300N$$

$$+\uparrow\Sigma F_y = 0: \quad N_y = 0$$

In order to calculate the components of the forces, each member of the linkage is dismembered as in Fig. 3.

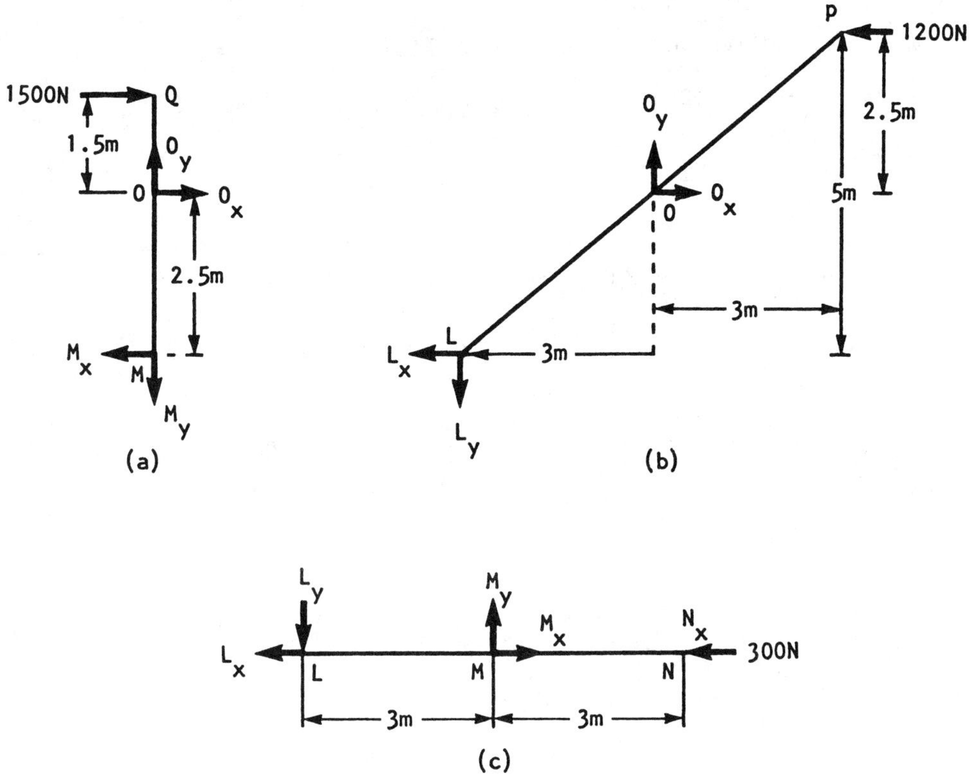

Fig. 3

For member QOM in Fig. 3(a)

$+\circlearrowleft \Sigma M_M = 0: \quad -(1500N)(4m) - O_x(2.5m) = 0$

or

$$O_x = -2400N$$

The minus sign indicates that O_x has an opposite sense to that assumed in Fig. 3(a) so

$$O_x = +2400N$$

$+\circlearrowleft \Sigma M_O = 0: \quad -(1500N)(1.5m) - M_x(2.5m) = 0$

or

$$M_x = -900N.$$

Since M_x now has a sense opposite to that shown in Fig. 3(a) because of the minus sign, the value of M_x = 900N

$+\uparrow \Sigma F_y = 0: \quad O_y - M_y = 0$

Considering member LMN in Fig. 3(c)

$\overset{+}{\curvearrowleft}\Sigma M_L = 0: \quad M_y(3m) = 0$

or

$$M_y = 0$$

$\underset{\rightarrow}{+}\Sigma F_x = 0: \quad -L_x + 900N - 300N = 0$

or

$$L_x = 600N$$

$+\uparrow\Sigma F_y = 0: \quad -L_y + M_y = 0$

or

$$L_y = 0$$

Returning to member QOM in Fig. 3(a)

$$O_y - M_y = 0$$

$$O_y - 0 = 0$$

or

$$O_y = 0$$

Considering member LOP in Fig. 3(b)

$\overset{+}{\curvearrowleft}\Sigma M_L = 0: \quad (O_y)(3m) - O_x(2.5m) + (1200N)(5m) = 0$

$$(0)(3m) - (2400N)(2.5m) + (1200N)5m) = 0$$

$$0 - 6000N + 60000N = 0$$

$$0 = 0$$

This confirms that member LOP is in equilibrium and checks the results.

• PROBLEM 2-9

A truss, simply supported at the two ends, carries two loads each of 10KN as shown in figure 1. Determine the force in each member of the truss by using the method of joints. In each case state whether the member is in tension or compression.

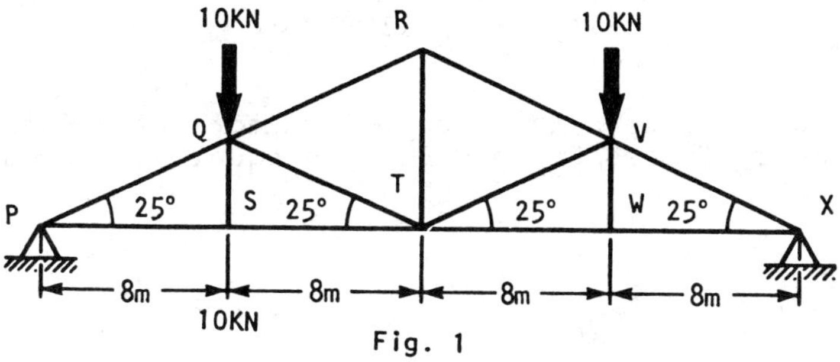

Fig. 1

Solution: Since the truss and the loads are both symmetrical with respect to RT, only the forces on one side of the truss need be calculated. Draw the free-body diagram for each joint and apply the principle of force equilibrium.

First consider the joint P whose free-body diagram is shown in Figure 2. From symmetry, the external reaction at P is 10KN. Setting the forces at P equal to zero yields

$$+\uparrow\Sigma F_y = 0; \quad 10KN - F_{PQ} \sin 25° = 0 \qquad (1)$$

$$F_{PQ} = 23.66KN \text{ C}$$

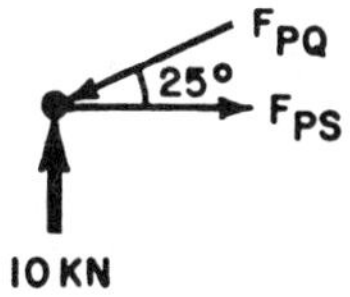

Fig. 2

$$\underset{\rightarrow}{+}\Sigma F_x = 0; \quad F_{PS} - F_{PQ} \cos 25° = 0 \qquad (2)$$

$$F_{PS} = 21.44KN \text{ T}$$

Now for joint S:

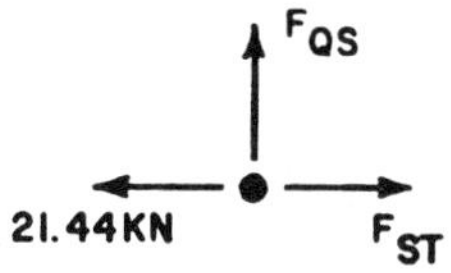

Fig. 3

$$\underset{\rightarrow}{+}\Sigma F_x = 0; \quad F_{ST} = 21.44KN \text{ T} \qquad (3)$$

$$+\uparrow\Sigma F_y = 0; \quad F_{QS} = 0 \qquad (4)$$

Looking at joint Q:

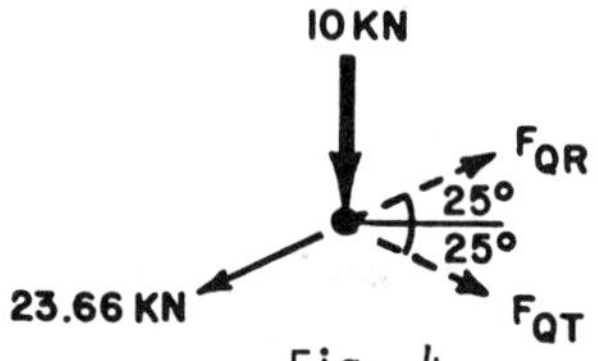

Fig. 4

$$\underset{\rightarrow}{+}\Sigma F_x = 0; \quad (23.66KN + F_{QR} + F_{QT}) \cos 25° = 0 \qquad (5)$$

$$+\uparrow\Sigma F_y = 0; \quad -10KN + (23.66KN + F_{QR} - F_{QT}) \sin 25° = 0 \qquad (6)$$

Solving equations 5 and 6 simultaneously results in

$F_{QT} = F_{QR} = -11.83KN = 11.83KN\ C$

Figure 5 shows the free-body diagram for joint T. Due to symmetry forces F_{TW} and F_{TV} are equal and opposite to forces F_{TS} and F_{TQ}. Accordingly,

$F_{TW} = 21.44KN\ T$, $F_{TV} = 11.83KN\ C$

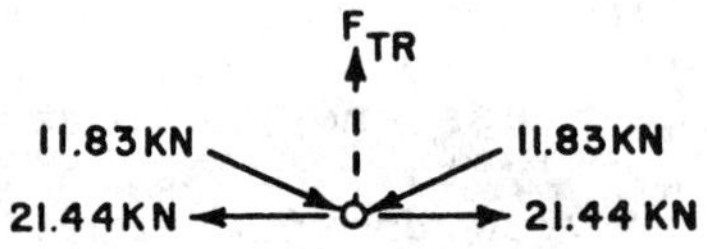

Fig. 5

Summing forces in the vertical direction yields

$+\uparrow \Sigma F_y = 0; \quad F_{TR} - 2(11.83\ KN \sin 25°) = 0$

$F_{TR} = 10\ KN\ T$

Having found all the forces in the left hand side of the truss, it may be stated due to symmetry that

$F_{RV} = 11.83\ KN\ C; \ F_{VW} = 0; \ F_{XY} = 23.66\ KN\ C;$

$F_{XW} = 21.44\ KN\ T$

To summarize,

$F_{PQ} = F_{XY} = 23.66\ KN$ Compression

$F_{PS} = F_{XW} = 21.44\ KN$ Tension

$F_{QS} = F_{VW} = 0$

$F_{QT} = F_{VT} = 11.83\ KN$ Compression

$F_{ST} = F_{WT} = 21.44\ KN$ Tension

$F_{QR} = F_{VR} = 11.83\ KN$ Compression

$F_{RT} = 10\ KN$ Tension

CHAPTER 3

FORCES IN BEAMS AND CABLES

SHEAR AND BENDING MOMENT DIAGRAMS OF BEAMS

• PROBLEM 3-1

Draw the shear and bending-moment diagrams of the rod shown in Fig. 1. A couple of magnitude M is applied to the rod at point P.

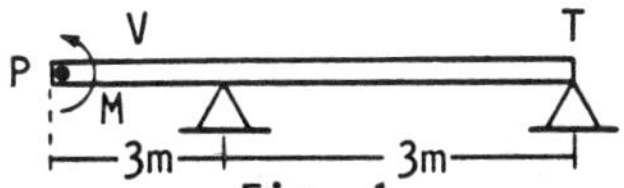

Fig. 1

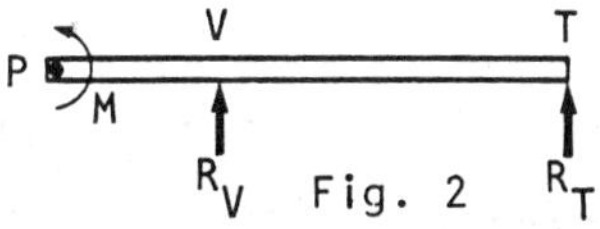

Fig. 2

Solution: The entire rod shown in Fig. 2 is considered as a free body from which the reactions $\bar{R}_V$ and $\bar{R}_T$ are determined.

$$R_V = \frac{M}{6} \uparrow \text{ and } R_T = \frac{M}{6} \downarrow$$

Figure 3 shows several free body diagrams of the rod, and the corresponding shear and bending moments at each section.

Writing equilibrium equations for each section yields:

$$S_1 = 0, \; M_1 = -M;$$

$$S_2 = 0, \; M_2 = -M;$$

$$S_3 = \frac{M}{6} \; ; \; M_3 = -M;$$

$$S_4 = \frac{M}{6} \; , \; M_4 = 0 \; .$$

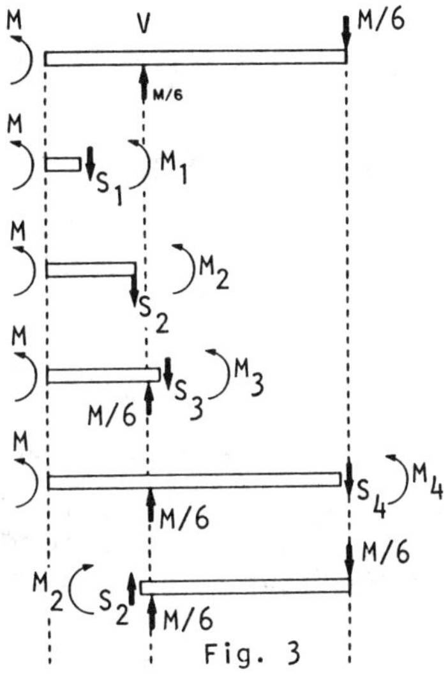

Fig. 3

Therefore the shear and bending diagrams are as follows:

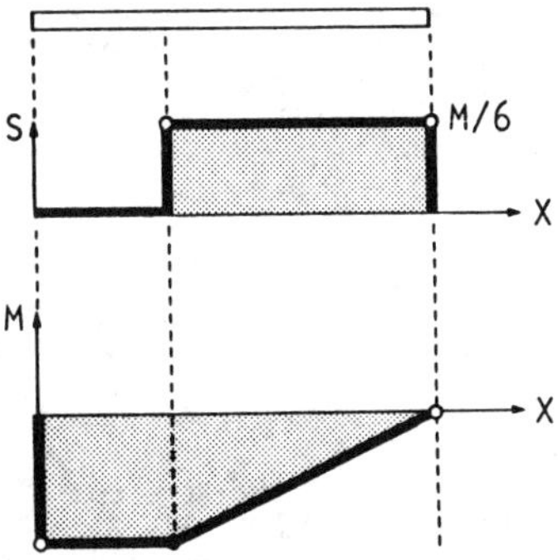

Fig. 4

Draw the shear and bending moment diagrams for the simply supported beam in Figure 1.

Solution: By considering the entire beam as a free-body (Fig. 2), the equations of equilibrium are applied to obtain

$W = \frac{\text{Force}}{\text{unit length}}$

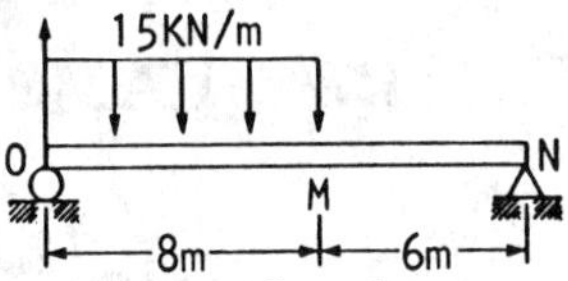

Fig. 1

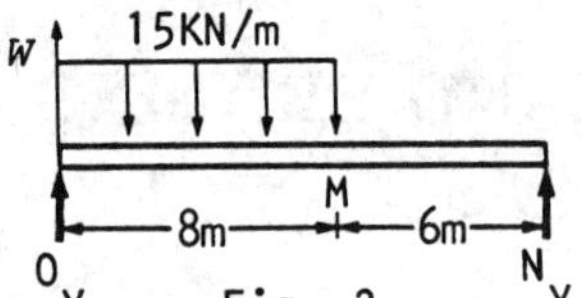

Fig. 2

$$+\uparrow\Sigma F_y = 0 \; ; \; O_y - (15 \text{ kN/m})(8\text{m}) + N_y = 0 \qquad O_y + N_y = 120$$

$$+\circlearrowright \Sigma M_N = 0 \; ; \; -O_y(14\text{m}) + 15 \frac{\text{kN}}{\text{m}} (8\text{m})(10\text{m}) = 0$$

$$O_y = 85.7 \text{ kN}$$

Therefore, $N_y = 34.3$ kN

$$\frac{dS}{dx} = -w \tag{1}$$

$$\text{or} \quad S_x - S_1 = -\int_{x_1}^{x} w\,dx = -w(x - x_1) \tag{2}$$

The shear at M is given as

$$S_2 - 85.7 \text{ kN} = -\int_0^{8\text{m}} 15 \frac{\text{kN}}{\text{m}} dx \text{ , or}$$

$$S_2 - 85.7 \text{ kN} = -15 \frac{\text{kN}}{\text{m}} (8\text{m} - 0) \qquad S_m = -34.3\text{kN}$$

Since w is a constant, the slope $\frac{dS}{dx}$ is just linear (from relation (1)) between points 0 and M. To the right of M there is no load; therefore, the shear does not change in that region. Hence

$$S_N = S_M = -34.3$$

The shear diagram would therefore be:

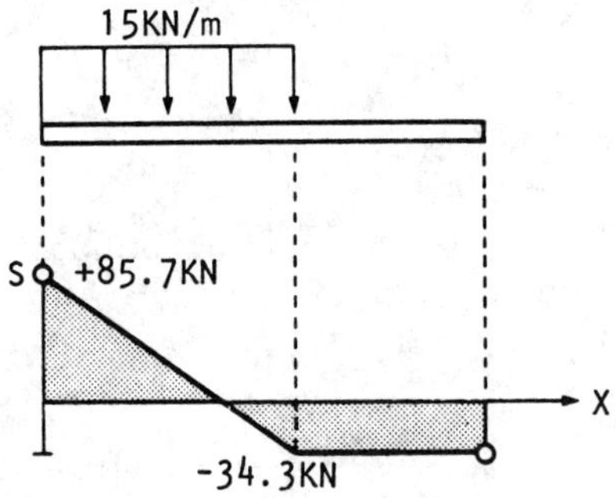

Fig. 3

BENDING MOMENT:

The bending moment is related to the shear by

$$\frac{dM}{dx} = S \tag{3}$$

Between O and M, the equation for the shear may be obtained by analyzing Fig. 3.

$$S = \frac{85.7 - (-34.3)}{0 - 8} \frac{kN}{m} x + 85.7\ kN$$

or

$$S = -15x + 85.7 \qquad 0 \le x \le 8 \tag{4}$$

It can be seen from equations (3) and (4) that the moment will be parabolic with its maximum value occurring at

$$S = \frac{dM_x}{dx} = -15x + 85.7 = 0$$

or

$$x = \frac{85.7}{15} = 5.71m$$

It is now possible to construct the bending moment diagram by looking at the shear diagram and summing areas.

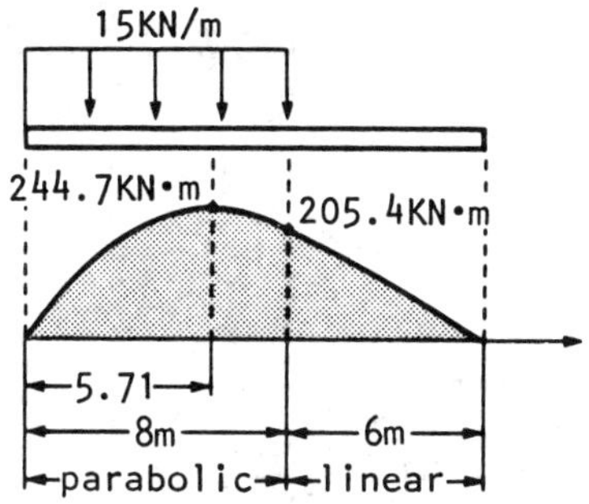

Fig. 4

$$\text{At } x = 5.71m,\ M = \frac{1}{2}(85.7 \times 5.71) = 244.7\ kN\cdot m$$

$$\text{At } x = 8m,\ M = 244.7\ kN\cdot m + \frac{1}{2}[-34.3 \times (8 - 5.71)]kN\cdot m$$

$$= 205.7\ kN\cdot m$$

As a check,

$$\text{At } x = 14m,\ M = 205.7\ kN\cdot m + [-34.3 \times 6]kN\cdot m = 0\ .$$

• PROBLEM 3-3

> Draw the shear and bending moment diagrams for the cantilever beam shown in Fig. 1.

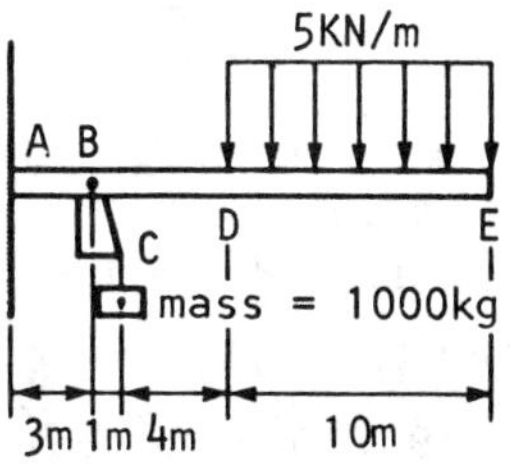

Fig. 1

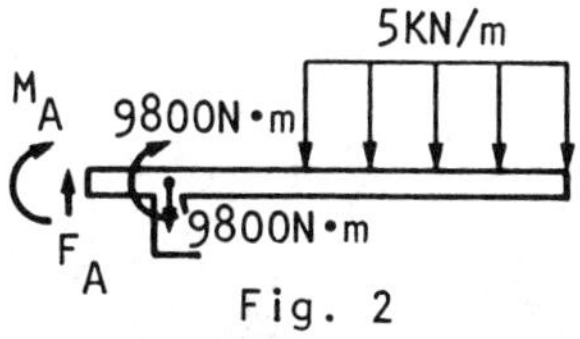

Fig. 2

<u>Solution</u>: A free-body diagram of the beam with the force at C transformed to a force-couple system at B, is shown in Figure 2.

Assuming that the weight of the beam is negligible, it is now possible to write the equations of equilibrium.

$$+\uparrow \Sigma F_y = 0 \; ; \; F_A - 9800N - (5 \text{ kN/m})(10m) = 0$$

or

$$F_A = 59.8kN$$

$$+\circlearrowleft \Sigma M_B = 0 \; ; \; -M_A - (59.8kN)(3m) - (9.8kN{\cdot}m) -$$

$$(5 \text{ kN/m})(10m)(10m) = 0$$

$$M_A = -689 \text{ kN}{\cdot}\text{m} = 689 \text{ kN}{\cdot}\text{m} \circlearrowright$$

The sections along the beam will be analyzed in order to determine the shears and moments.

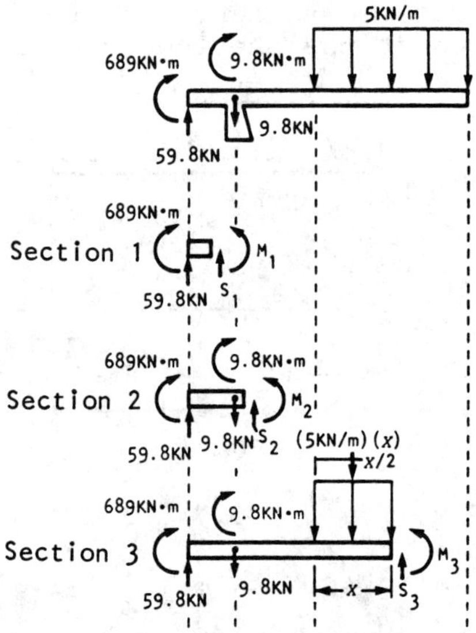

Fig. 3

Referring to Fig. 3

Section 1: $+\uparrow \Sigma F_y = 0 \; ; \; S_1 = -59.8 \text{ kN}$

$+\circlearrowleft \Sigma M_A = 0 \; ; \; M, = 689 \text{ kN}{\cdot}\text{m}$

Section 2: $+\uparrow \Sigma F_y = 0 \; ; \; 59.8 \text{ kN} - 9.8kN + y_2 = 0$

or

$$S_2 = -50kN$$

$$+\circlearrowleft \Sigma M_A = 0 \; ; \; -689 \text{ kN}{\cdot}\text{m} - (9.8kN)(3m)'$$

$$+ (-50kN)(3m) - 9.8 \text{ kN}{\cdot}\text{m} + M_2 = 0$$

$$M_2 = +878 \text{ kN}{\cdot}\text{m}$$

Section 3: $+\uparrow \Sigma F_y = 0 \; ; \; 59.8kN - 9.8kN - 5x \text{ kN} + y_3 = 0$

or $\qquad S_3 = 5x - 50 \text{ kN}$

$$+\circlearrowright \Sigma M_B = 0 \; ; \; -689 - (59.8)(3) - 9.8 - 5x\left(\frac{x}{2} + 5\right)$$

$$+ V_3(x + 5) + M_3 = 0$$

$$-879 - \frac{5}{2} x^2 - 25x + (5x - 50)(x + 5) + M_3 = 0$$

$$M_3 = 879 - \frac{5}{2} x^2 - 25x + 5x^2 - 50x + 25x - 250$$

$$= 2.5x^2 - 50x + 629$$

The shear and bending moment diagrams may now be constructed.

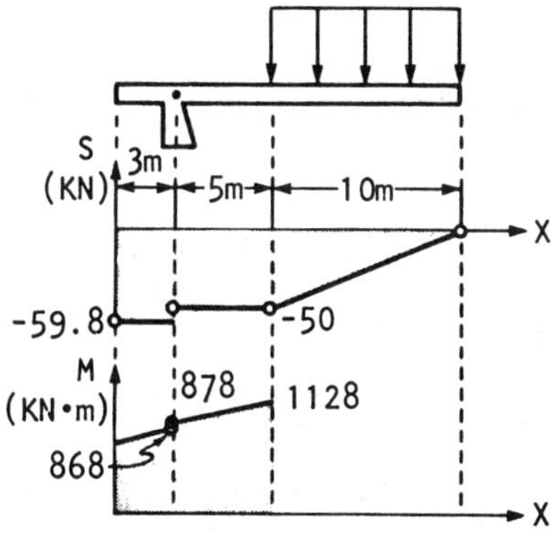

Fig. 4

• PROBLEM 3-4

Draw the shear and moment diagram for the simply supported beam shown in Figure 1.

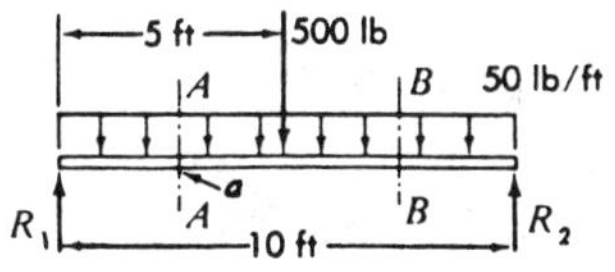

Fig. 1

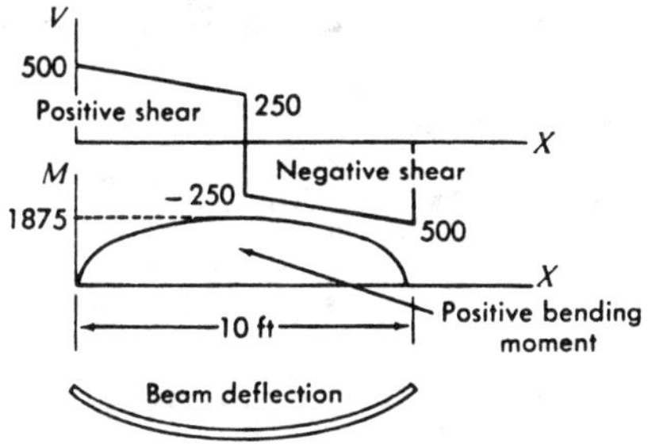

Fig. 2

Solution: Applying to the entire beam as a free body, the reactions at the end supports are calculated

$$\Sigma M_1 = 0. \; +\circlearrowright$$

$$- R_2 \text{ (10 ft)} + 500 \text{ (5 ft)} + 50 \frac{\text{lb}}{\text{ft}} \text{ (10) (5 ft)} = 0.$$

$$R_2 = \frac{2500 + 2500}{10} = 500 \text{ lbs.}$$

$\Sigma M_2 = 0.$ ↻+

$$R_1 (10) - 500 (5 \text{ ft}) - 50 (10) (5 \text{ ft}) = 0.$$

$$R_1 = 500 \text{ lbs}$$

As a check, sum up all the forces acting on the beam.

$\Sigma F_y = 0.$ + ↑ $\quad - 500 + 500 + 500 - 50 (10) = 0.$

$$0 = 0.$$

Thus, our calculated values are correct.

$\Sigma F_y = 0. \qquad R_1 - 50 (x) - V = 0.$

$$V = 500 - 50 x$$

↻+ $\Sigma M_A = 0.$

$$R_1 (x) - 50 (x) \left(\frac{x}{2}\right) - M = 0.$$

$$M = 500 x - 25 x^2$$

The above equation is applicable in the range $0 \leq x < 5$ ft. For the region $5 < x < 10$, (section B - B of Figure 1) we obtain

$\Sigma F_y = 0.$ +↑

$$R_1 - 50 (x) - 500 - V = 0.$$

$$V = 500 - 50 x - 500 = - 50 x$$

$\Sigma M = 0.$

$$R_1 (x) - 50 (x) \left(\frac{x}{2}\right) - 500 (x - 5) - M = 0$$

$$M = 500 (x) - 25 x^2 - 500 x + 2500$$

$$M = 2500 - 25 x^2$$

From the first equation for shear (section A-A), substitution of $x = 5$ ft yields $V = 250$ lb. The shear calculated at $x = 5$ ft from the second shear equation (section B-B) is $V = - 250$ lb. Because of the concentrated load, a discontinuity in the shearing force exists at $x = 5$ ft and, since V passes through zero there, the bending moment is a maximum at this point. The bending moment at $x = 5$ ft is calculated to be 1875 lb-ft from both bending-moment equations. The shear and bending-moment diagrams are plots of the above equations and are shown in Figure 2.

The rod PQ shown in Fig. 1 carries a 50kN and a 60kN loading, draw its shear and bending-moment diagram for the rod and the loadings.

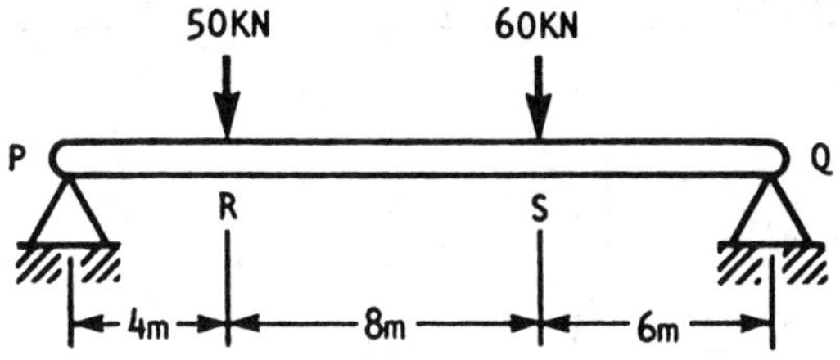

Fig 1

Solution: A free-body diagram is drawn in Fig. 2 to determine the reactions at P and Q

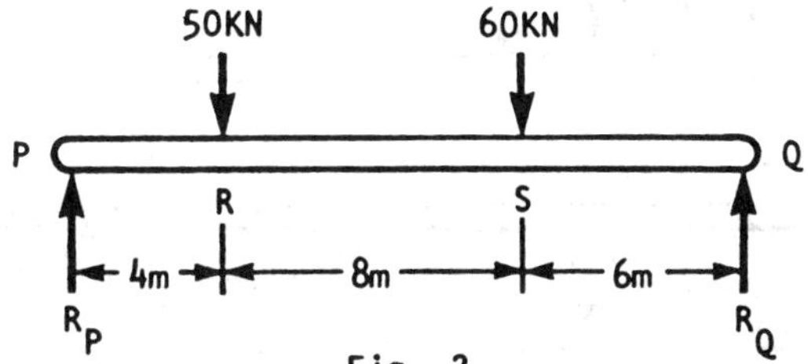

Fig. 2

$+\circlearrowleft \Sigma M_Q = 0 : \quad -R_P(18m) + (50kN)(14m) + (60kN)(6m) = 0$

$R_P = 58.8kN$

$+\uparrow \Sigma F_y = 0 : \quad 58{\cdot}8kN - 50kN - 60kN + R_Q = 0$

$R_Q = 51kN$

The internal forces acting on the rod will now be determined by dividing it into sections as shown in Fig. 3.

$+\circlearrowleft \Sigma M_1 = 0 : -(58{\cdot}8kN)(0m) + M_1 = 0$

or $\quad M_1 = 0$

$+\uparrow \Sigma F_y = 0 : 58{\cdot}8kN - S_1 = 0 ; \quad S_1 = 58.8kN$

Considering the free body #2

$+\circlearrowleft \Sigma M_2 = 0 : -(58{\cdot}8kN)(4m) + M_2 = 0 ; \quad M_2 = 235{\cdot}2kN{\cdot}m$

$+\uparrow \Sigma F_y = 0 : \quad 58{\cdot}8kN - S_2 = 0 ; \quad S_2 = 58{\cdot}8kN$

From the free body diagram #3

$+\circlearrowleft \Sigma M_3 = 0 : \quad -(58{\cdot}8kN)(4m) + (50kN)(0m) + M_3 = 0 ;$

$M_3 = 235{\cdot}2kN{\cdot}m$

$+\uparrow \Sigma F_y = 0 : \quad 58{\cdot}8kN - 50kN - S_3 = 0 ;$

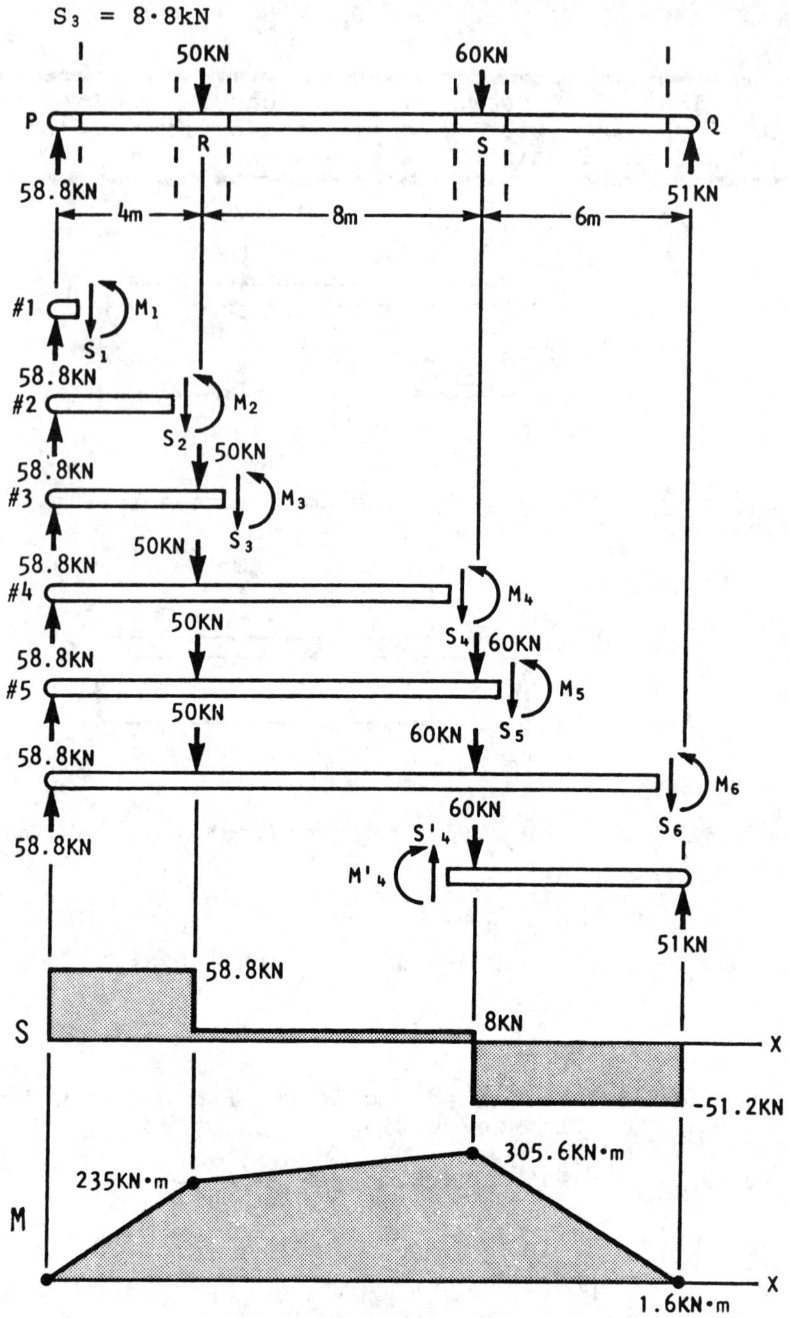

Fig. 3

Similarly, the shear and bending-moment S_4, S_5, S_6, M_4, M_5 and M_6 are determined by considering free-body diagrams #4, #5 and #6. The results are as follows:

$M_4 = 305.6\text{kN}\cdot\text{m}$ and $S_4 = 8\cdot 8\text{kN}$

$M_5 = 305.6\text{kN}\cdot\text{m}$ and $S_5 = -51\cdot 2\text{kN}$

$M_6 = 1.6\text{kN}\cdot\text{m}$ and $S_6 = -51\cdot 2\text{kN}$

Points for S and M are plotted in Fig. 3 to show the shear and bending-moment diagrams.

DISTRIBUTED LOADS ACTING ON BEAMS

• PROBLEM 3-6

A uniformly distributed load of intensity w lb per unit length acts over a length L of a beam, as shown in Figure 1. Replace the distributed load by a single rigid body equivalent force.

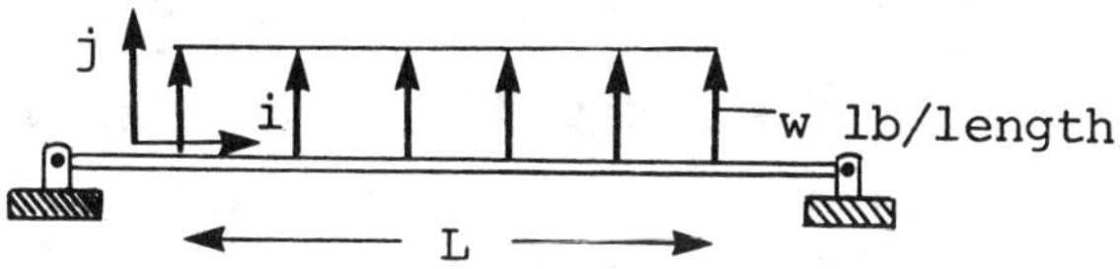

Fig. 1

Solution: The single force which would be equivalent to the distributed load must satisfy the following two conditions:

(a) The single force must be equal to the resultant of the distributed load.

(b) The moment of the single force about any point must be equal to the total moment due to the distributed load about the same point.

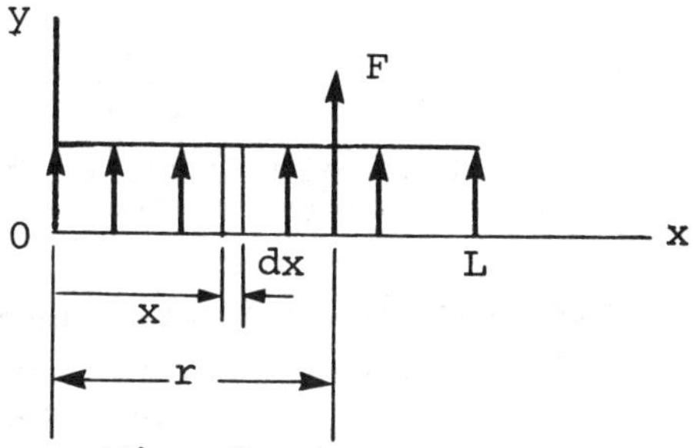

Fig. 2

Figure 2 shows the chosen coordinate system with the x axis along the beam, y axis perpendicular to it and the origin at the left end of the distributed load. Let us consider an infinitesimal portion of the distributed load at distance x from the origin and of width dx. The magnitude of this infinitesimal load is w dx, and it acts in the y direction. Let F be the single equivalent force. Then F is equal to the sum of the total distributed load over the length L.

$$F = \int_0^L w\,dx\,\hat{j} = w\,L\,\hat{j} \qquad (1)$$

Therefore F is of magnitude w L and acts in the y direction. Let its line of action be at a distance r from the origin. Let us consider moments of F and the distributed load about the origin. (This is chosen for convenience, although any other point would do). The moment of F about O is

$$M_0 = r\,\hat{i} \times F$$

$$= r\,\hat{i} \times w\,L\,\hat{j}$$

$$= r\,w\,L\,\hat{k} \qquad (2)$$

The moment of the infinitesimal portion of the distributed load about O is

$$d\,M_0 = x\,\hat{i} \times w\,dx\,\hat{j} = w\,x\,dx\,\hat{k} \qquad (3)$$

Therefore the total moment of the distributed load about O is

$$M_0 = \int_0^L w\,x\,dx\,\hat{k} = \frac{wx^2}{2}\,\hat{k}\,\Big|_0^L = \frac{wL^2}{2}\,\hat{k} \qquad (4)$$

Now we equate the right sides of equations (2) and (4) and solve for the distance r.

$$r\,w\,L = \frac{w\,L^2}{2}$$

or

$$r = \frac{L}{2} \qquad (5)$$

Thus the equivalent single force acts through the midpoint of the uniformly distributed load.

• PROBLEM 3-7

Given the bar loaded as in the figure, find a uniform load and a couple which satisfies the same relaxed boundary conditions as the triangular load.

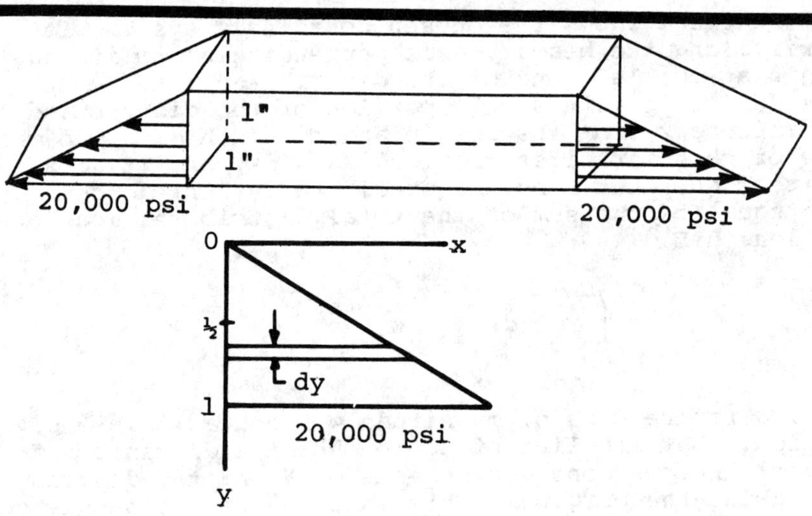

Solution: All that is required here is to find a uniform load and a couple that together are equivalent to the triangular load.

The figure above represents the triangular load. In these coordinates the loading is given by

$$p = 20000y \text{ psi.} \qquad (1)$$

On an area (1) dy this loading produces an axial force of

$$dF = p\ (1)\ dy$$
$$= 20000y\ dy. \qquad (2)$$

The total axial load, F, can be found by integrating dF over the entire bar.

$$F = \int dF$$
$$= \int_0^1 20000y\ dy$$
$$= 20000 \left. \frac{y^2}{2} \right|_0^1$$

$$F = 10000 \text{ lb.}$$

Since the bar cross-sectional area is 1 in^2, the 10000 lb distributed uniformly over the area is

$$T_i = 10000\ \hat{i} \text{ psi.}$$

The neutral surface of the bar is located at $y = \frac{1}{2}$. The moment of dF about the neutral surface is

$$dM = dF \left(y - \frac{1}{2}\right) \qquad (\text{ccw} +)$$

or, substituting from Eq. 2,

$$dM = 20000y \left(y - \frac{1}{2}\right) dy.$$

The total moment of the triangular loading about the neutral surface may be found by integrating dM over the entire cross-section

$$M = \int dM$$
$$= 20000 \int_0^1 y \left(y - \frac{1}{2}\right) dy.$$

Performing the integration leads to

$$M = 20000 \left(\frac{y^3}{3} - \frac{y^2}{2}\right) \Bigg|_0^1$$

$$= 20000 \left(\frac{1}{3} - \frac{1}{4}\right)$$

$$= 1667 \text{ lb-in.} \circlearrowleft +$$

If, in the figure, $\hat{k}$ were a unit vector out of the page, then

$$M = 1667\, \hat{k} \text{ lb-in.}$$

In summary, the axial elongation of the bar caused by the triangular load could be duplicated by a uniform load of

$$T_i = 10000\, \hat{i} \text{ psi.}$$

The bending of the bar caused by the triangular load would be identical to that caused by a couple.

$$C = 1667\, \hat{k} \text{ lb-in.}$$

• PROBLEM 3-8

A distributed load, the intensity of which increases linearly from zero to a lb per unit length over a length L, acts on a beam as shown in Figure 1. Find the single rigid body force equivalent to the distributed load.

Fig. 1

Solution: The single force which would be equivalent to the distributed load must satisfy the following two conditions:

(a) The single force must be equal to the resultant of the distributed load.

(b) The moment of the single force about any point must be equal to the total moment of the distributed load about the same point.

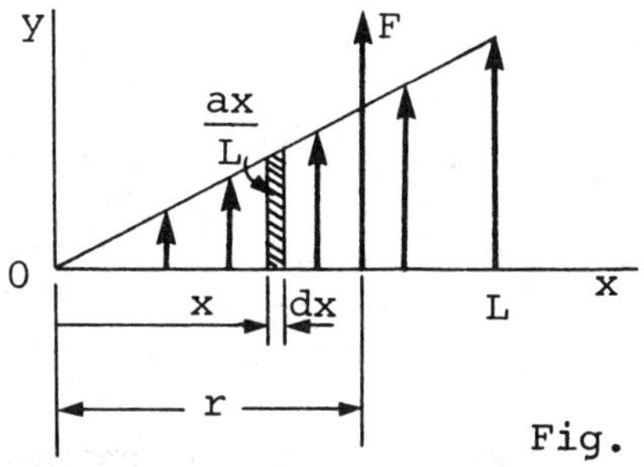

Fig. 2

Figure 2 shows the chosen coordinate system with the x axis along the beam, the y axis perpendicular to it and the origin at the left end of the distributed load. In order to find the resultant of the distributed load, let us consider an infinitesimal portion of it acting over a width dx, at distance x from the origin. By similar triangles we find that the intensity of the load at x is ax/L, acting in the y direction. The magnitude of the infinitesimal force over the width dx is axdx/L. Therefore the resultant of the distributed load, which is equal to the single force F is

$$F = \int_0^L \frac{ax}{L}\,dx\,\hat{j} = \frac{a}{L}\frac{x^2}{2}\Big|_0^L\,\hat{j} = \frac{aL}{2}\,\hat{j}. \qquad (1)$$

Thus the equivalent single force is of magnitude aL/2 and acts in the y direction. We consider the moments next, and choose the origin as the point about which moments are taken. Let r be the distance of the line of action of F from O. The moment of F about O is then

$$M_0 = r\,\hat{i} \times F = r\,\hat{i} \times \frac{aL}{2}\,\hat{j} = \frac{arL}{2}\,\hat{k}. \qquad (2)$$

The moment of the distributed load about O is found by considering the moment of the infinitesimal portion of it and then integrating over the length L. The moment of the infinitesimal load axdx/L about O is

$$dM_0 = x\,\hat{i} \times \frac{axdx}{L}\,\hat{j} = \frac{ax^2dx}{L}\,\hat{k}$$

Thus the total moment of the distributed load about O is

$$M_0 = \int_0^L \frac{ax^2}{L}\,dx\,\hat{k} = \frac{a}{L}\frac{x^3}{3}\Big|_0^L\,\hat{k} = \frac{aL^2}{3}\,\hat{k} \qquad (3)$$

We now equate the right hand sides of equations (2) and (3) and get

$$\frac{arL}{2} = \frac{aL^2}{3}$$

or $$r = 2L/3. \qquad (4)$$

Therefore the single equivalent force F acts at a distance 2L/3 from the left end of the distributed load.

• PROBLEM 3-9

The beam of length L shown in Figure 1 is loaded by a distributed load, the intensity of which increases linearly from zero at the left end to W lb per unit length at the right end. Find the shear force and bending moment at any point along the beam.

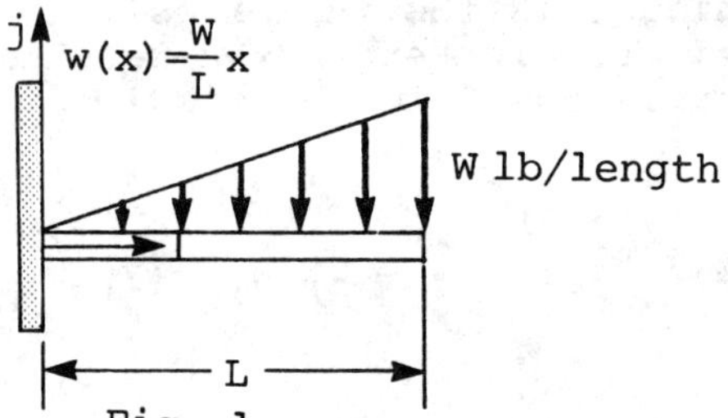

Fig. 1

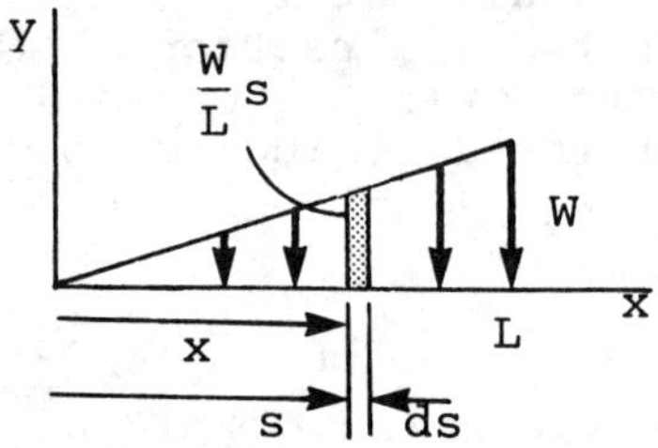

Fig. 2

Solution: The shear force at any point on the beam is the sum of all the forces acting on one (say the right) side of that point, in a direction perpendicular to the section of the beam. Figure 2 shows the coordinate system, with the x axis along the beam, the y axis perpendicular to it and the origin at the left end of the beam where the load intensity is zero. The load intensity, since it varies linearly, is given by

$$w\ (x) = \frac{W}{L}\, x. \qquad (1)$$

We wish to find the shear force at a point a distance x from the origin. Therefore, we must find the total force on the right side of this point. Consider an infinitesimal portion of the load at s, acting over a width ds. The magnitude of the infinitesimal force is $\frac{Wsds}{L}$ and it acts in the negative y direction. Let V (x) denote the shear force at x. Its value is found by integrating the infinitesimal force from s = x to s = L.

$$V\ (x) = -\int_x^L \frac{Ws}{L}\, ds = -\frac{W}{L}\frac{s^2}{2}\Bigg|_x^L$$

$$= -\frac{W}{2L}(L^2 - x^2) = \frac{Wx^2}{2L} - \frac{WL}{2}. \qquad (2)$$

The bending moment at any point on the beam is the sum of all the moments due to forces and all the couples acting on one side of that point. We find the bending moment at x by finding the moment due to the infinitesimal force at s and integrating. Let M (x) denote the bending moment at x, assumed to be positive in the counterclockwise direction. The moment due to the infinitesimal load Wsds/L about x is:

$$dM(x) = -\frac{Wsds}{L}(s - x). \qquad (3)$$

The total bending moment at x is obtained by integrating the above expression from s = x to s = L.

$$M(x) = -\int_x^L \frac{Ws(s - x)}{L}\,ds$$

$$= \frac{W}{L}\left(\frac{xs^2}{2} - \frac{s^3}{3}\right)\Bigg|_x^L$$

$$= -\frac{WL^2}{3} + \frac{WLx}{2} - \frac{Wx^3}{6L}. \qquad (4)$$

Thus, the shear force at a distance x from the left end of the beam is as given by equation (2) and the bending moment is as given by equation (4).

• PROBLEM 3-10

A rectangular bar, 2 in. by 4 in., in a structure in equilibrium is loaded in tension and bending. The internal force distribution across a cut midway along the rod is shown on Figure 1. Find the force intensities a and b.

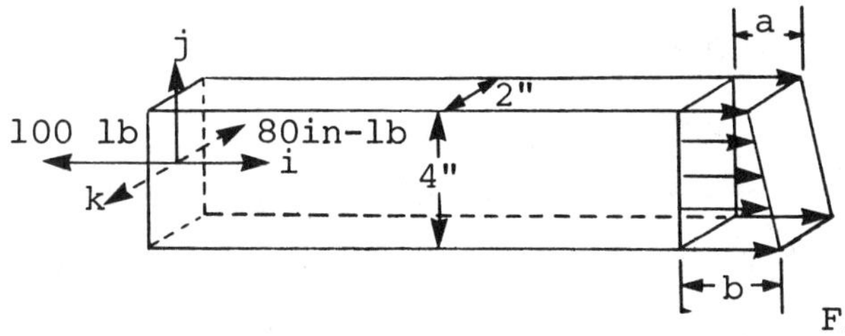

Fig. 1

Solution: Let us first find a rigid-body equivalent force f_M and couple c_M at the midpoint of the right face for the distributed load, Figure 2.

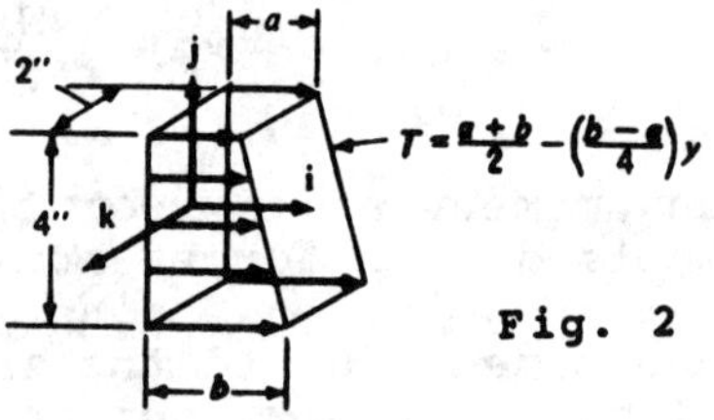

Fig. 2

The equivalent force due to tension and bending combined is $T = -\left(\frac{b - a}{4}\right) y + \frac{a + b}{2}$ where y = 2 (at the midpoint of the right face).

$$f_M = \int_{y=-2}^{2} \int_{z=-1}^{1} \left[\frac{a + b}{2} - \left(\frac{b - a}{4}\right) y\right] dz\, dy\, \hat{i}$$

$$= \int_{-2}^{2} \left[\int_{z=-1}^{1} \left(\frac{a + b}{2}\right) dz - \int_{z=-1}^{1} \left(\frac{b - a}{4}\right) y\, dz \right] dy\, \hat{i}$$

Integrating inside the bracket only,

$$\int_{-1}^{1} \frac{a + b}{2}\, dz - \int_{-1}^{1} \frac{b - a}{4}\, y\, dz$$

$$= \left(\frac{a + b}{2}\right) z \Big|_{-1}^{1} - \left[\left(\frac{b - a}{4}\right) z \Big|_{-1}^{1}\right] y$$

$$= \left(\frac{a + b}{2} - \frac{- a - b}{2} \right) - \left(\frac{b - a}{4} - \frac{- b + a}{4} \right) y$$

$$= (a + b) - \left(\frac{b - a}{2}\right) y.$$

Doing the y-integration yields

$$\int_{y=-2}^{2} (a + b)\, dy\, \hat{i} - \int_{y=-2}^{2} \frac{b - a}{2}\, y\, dy\, \hat{i}$$

$$= \left[(a + b)\, y \Big|_{-2}^{2} - \left(\frac{b - a}{2}\right) \frac{y^2}{2} \Big|_{-2}^{2} \right] \hat{i}$$

$$= \left[2\,(a + b) - (-\,2)(a + b) - \left[\left(\frac{b - a}{2}\right)\left(\frac{4}{2}\right) - \left(\frac{b - a}{2}\right)\left(\frac{4}{2}\right)\right]\right] \hat{i}$$

$$= 4\,(a + b)\, \hat{i}$$

$$c_M = \int_{y=-2}^{2} \int_{z=-1}^{1} y \left[\frac{a + b}{2} - \left(\frac{b - a}{4}\right)y\right] dz\, dy\, \hat{k}$$

$$\int_{y=-2}^{2} \left[\int_{z=-1}^{1} \frac{(a + b)}{2} y\, dz\, dy - \int_{z=-1}^{1} \frac{(b - a)}{4} y^2 dz\, dy\right] \hat{k}$$

$$\int_{y=-2}^{2} \left[\int_{-1}^{1} \frac{a + b}{2} y\, dz - \int_{-1}^{1} \frac{b - a}{4} y^2 dz\right] dy\, \hat{k}.$$

Integrating inside the bracket only,

$$\int_{-1}^{1} \frac{a + b}{2} y\, dz - \int_{-1}^{1} \frac{b - a}{4} y^2 dz$$

$$y \left(\frac{a + b}{2}\right) z \Big|_{-1}^{1} - y^2 \left(\frac{b - a}{4}\right) z \Big|_{-1}^{1}$$

$$y\left[\left(\frac{a + b}{2}\right) - (-1) \left(\frac{a + b}{2}\right)\right] - y^2\left[\left(\frac{b - a}{4}\right) - (-1) \left(\frac{b - a}{4}\right)\right]$$

$$y\, (a + b) - y^2\left(\frac{b - a}{2}\right).$$

Doing the y-integration yields

$$\int_{y=-2}^{2} (a + b)\, y\, dy\, k - \int_{y=-2}^{2} \left(\frac{b - a}{2}\right) y^2 dy\, \hat{k}$$

$$\left[(a + b) \frac{y^2}{2} \Big|_{-2}^{2} - \left(\frac{b - a}{2}\right) \frac{y^3}{3} \Big|_{-2}^{2}\right] \hat{k}$$

$$\left[(a + b) \frac{4}{2} - (a + b) \frac{4}{2} - \left(\left(\frac{b - a}{2}\right) \frac{8}{3} - \left(\frac{b - a}{2}\right)\left(\frac{-8}{3}\right)\right)\right] \hat{k}$$

$$- \left[\left(\frac{b - a}{2}\right) \frac{8}{3} + \left(\frac{b - a}{2}\right) \frac{8}{3}\right] \hat{k} = - \frac{8}{3} (b - a)\, \hat{k}$$

but this is in negative k direction so

$$- \left(- \frac{8}{3} (b - a)\right) k = \frac{8}{3} (b - a)\hat{k}.$$

$$= \frac{8}{3} (b - a)\, \hat{k}.$$

For equilibrium of the part of the body

$$4\,(a + b)\,\hat{i} - 100\,\hat{i} = 0$$

$$a + b = 25$$

$$\left[\frac{8}{3}\,(b - a) - 80\right]\hat{k} = 0$$

$$b - a = 30.$$

The values for a and b are determined from the simultaneous solution of these two equations in a and b.

$$b = 27.5 \text{ psi}$$

$$a = -\,2.5 \text{ psi}.$$

• PROBLEM 3-11

A 20 ft. long beam, simply supported at the ends, is loaded as shown in Figure 1. Concentrated loads of 1000 lb and 2000 lb act at points 5 ft from each end, and a uniformly distributed load of 100 lb/ft acts on the entire length of the beam. Find the expressions for shear force and bending moment at any point along the beam.

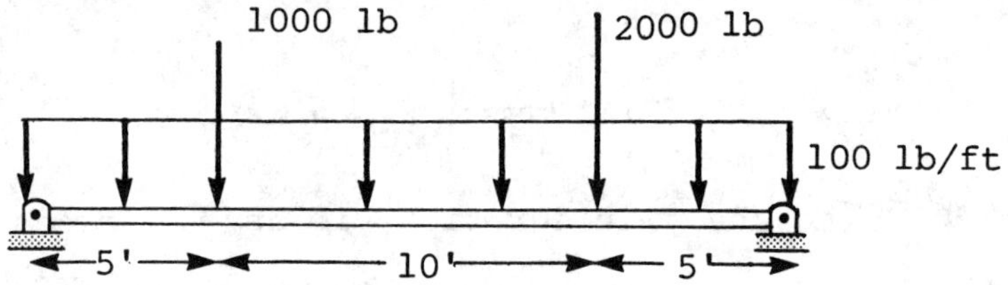

Fig. 1

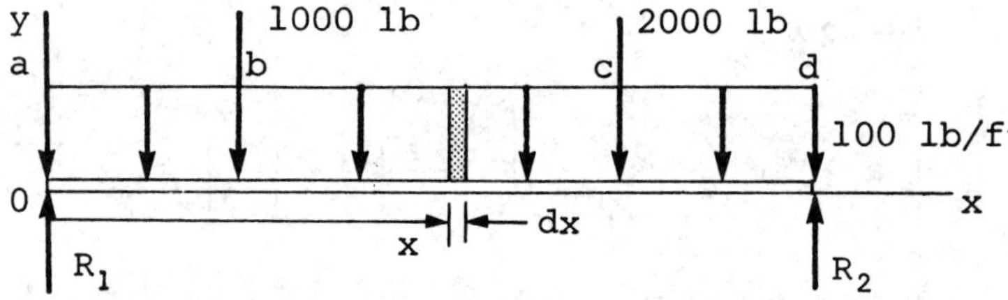

Fig. 2

Solution: Figure 2 shows the free body diagram for the beam, with the coordinate axes and the support reactions R_1 and R_2. Our first objective is to determine the magnitude of these reactions. This is accomplished by applying the two principles of static equilibrium, i.e., (a) the sum of all the forces acting on the rigid body (the beam) is zero and (b) the sum of the moments of all the forces about any point is zero. Summing the forces in the vertical direction gives us the equation

$$R_1 + R_2 - 1000 - 2000 - (100 \times 20) = 0$$

or $$R_1 + R_2 = 5000 \text{ lb.} \qquad (1)$$

We next consider the moments about O. The moment of the distributed load is found by finding the moment of the infinitesimal portion of it, at position x, and integrating over the length of the beam. Assuming the moments to be positive in the counterclockwise direction, the moment equation of equilibrium is

$$20\ R_2 - (15)(2000) - (5)(1000) - \int_0^{20} 100\ x\ dx = 0$$

or $$20\ R_2 - 30,000 - 5,000 - 20,000 = 0$$

or $$R_2 = 2,750\ \text{lb.} \qquad (2)$$

Upon substituting this value of R_2 into equation (1), R_1 is found

$$R_1 = 2,250\ \text{lb.} \qquad (3)$$

The expressions for the shear force V(x) and the bending moment M(x) must be found separately for the three sections of the beam: ab, bc and cd. The reason for this is that at the concentrated loads (at b and c) there are sudden changes (discontinuities) in the shear force and bending moment. We begin at the right end and proceed to the left.

Fig. 3

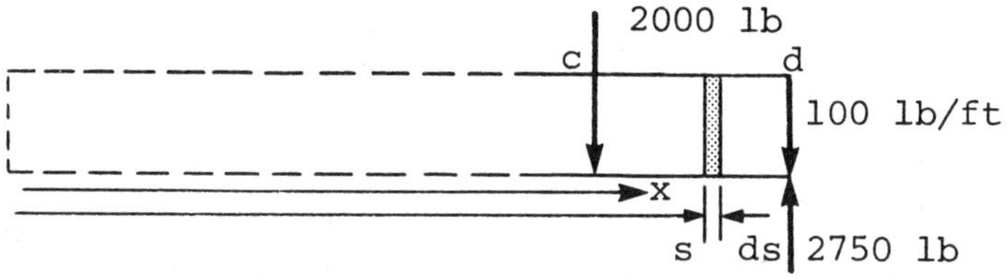

Figure 3 shows the cd portion of the beam. To find the shear force at position x we determine the total force in the vertical direction to the right of the point. The shear force is given by

$$V(x) = 2,750 - \int_x^{20} 100\ ds$$

$$= 2,750 - 100\ s \Big|_x^{20}$$

$$= 2,750 - 100\ (20-x)$$

$$= 750 + 100\ x\ \text{lb.} \qquad (4)$$

The bending moment at position x is the sum of moments of all forces to the right of this point, about this point. The moment due to the infinitesimal portion of the distributed load shown in Figure 3, about x is 100 ds (s-x). The total moment about x is, therefore

$$M(x) = (2,750)(20-x) - \int_x^{20} 100(s-x)\, ds$$

$$= 55,000 - 2750x - 100 \left(\frac{s^2}{2}\right) - xs \Big|_x^{20}$$

$$= 35,000 - 750x - 50x^2 \text{ lb-ft.} \qquad (4)$$

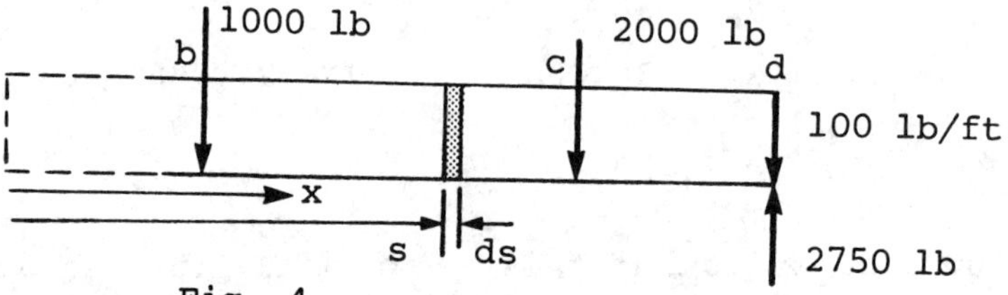

Fig. 4

Figure 4 is shown for the purpose of finding V(x) and M(x) for the bc portion. Shear force at x is the sum of all the forces to the right of that position.

$$V(x) = 2,750 - 2000 - \int_x^{20} 100\, ds$$

$$= 750 - 100\, s \Big|_x^{20}$$

$$= 750 - 100\,(20-x)$$

$$= -\,1250 + 100\,x \text{ lb.} \qquad (5)$$

To find the bending moment at position x we must consider the moments due to the 2000 lb load, the 2,750 lb reaction and the distributed load to the right of that position. The bending moment is given by

$$M(x) = (2750)(20-x) - (2000)(15-x)$$

$$- \int_x^{20} 100(s-x)\, ds$$

$$= 5,000 + 1,250x - 50\,x^2 \text{ lb-ft.} \qquad (6)$$

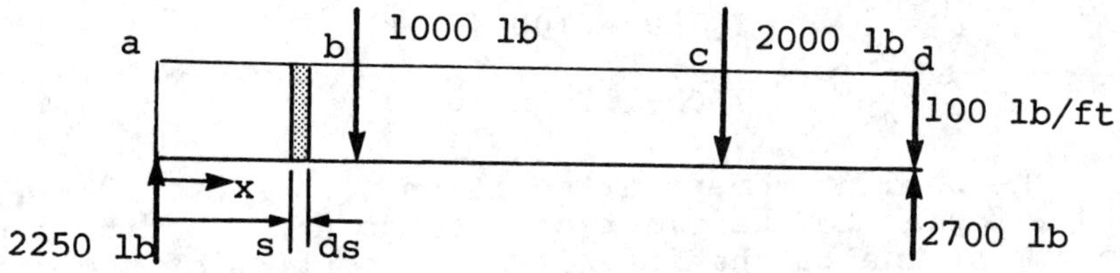

Fig. 5

Figure 5 shows the situation for the ab portion of the beam. Proceeding as before we find the shear force at position x

$$V(x) = 2,750 - 2,000 - 1,000 - \int_x^{20} 100\, ds$$

$$= -2,250 + 100x \text{ lb.} \qquad (7)$$

and the bending moment

$$M(x) = (2750)(20-x) - (2000)(15-x)$$

$$- (1000)(5-x) - \int_x^{20} 100\,(s-x)\, ds$$

$$= 2,250x - 50x^2 \text{ lb-ft} \qquad (8)$$

Collecting the above results, the expressions for the shear force in the beam are

$$V(x) = -2,250 + 100x \text{ lb}, \quad 0 < x < 5 \text{ ft}$$

$$= -1,250 + 100x \text{ lb}, \quad 5 < x < 15 \text{ ft}$$

$$= 750 + 100x \text{ lb}, \quad 15 < x < 20 \text{ ft}$$

and for the bending moment

$$M(x) = 2,250x - 50x^2 \text{ ft-lb}, \quad 0 < x < 5 \text{ ft}$$

$$= 5,000 + 1,250x - 50x^2 \text{ ft-lb}, \quad 5 < x < 15 \text{ ft}$$

$$= 35,000 - 750x - 50x^2 \text{ ft-lb}, \quad 15 < x < 20 \text{ ft.}$$

CABLES

• PROBLEM 3-12

The cord LMN is kept in place by a 100kN-force located at point N. Calculate the elevation of point M.

Solution: The reactions at point L can be determined from the free-body diagram of fig. 2.

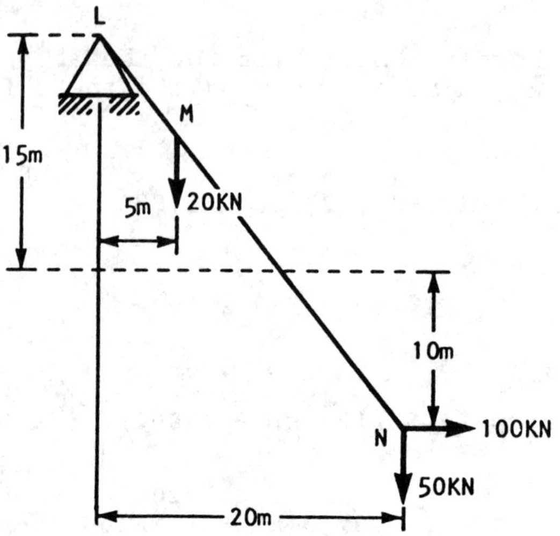

Fig. 1

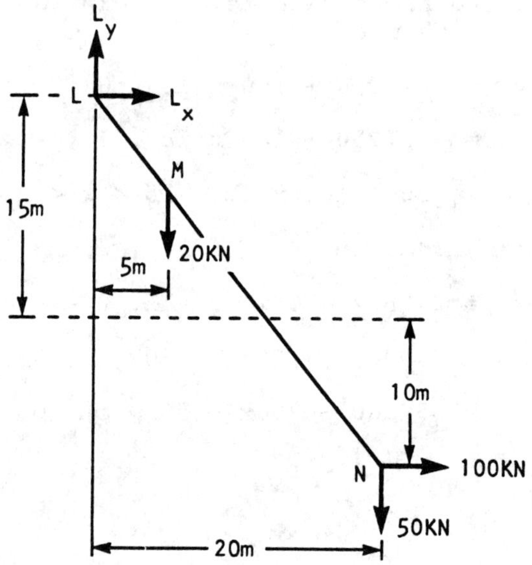

Fig. 2

$$+\uparrow\Sigma F_y = 0 : \quad -50\text{kN} - 20\text{kN} + L_y = 0$$

or

$$L_y = 70\text{kN}$$

$$+\circlearrowleft \Sigma M_N = 0 : \quad -(70\text{kN})(20\text{m}) - L_x(25\text{m}) + (20\text{kN})(15\text{m}) = 0$$

or

$$L_x = -44\text{kN}$$

In order to calculate the elevation at point M the section of the cord LM is taken as a free body and the the forces acting on it, are shown in fig. 3.

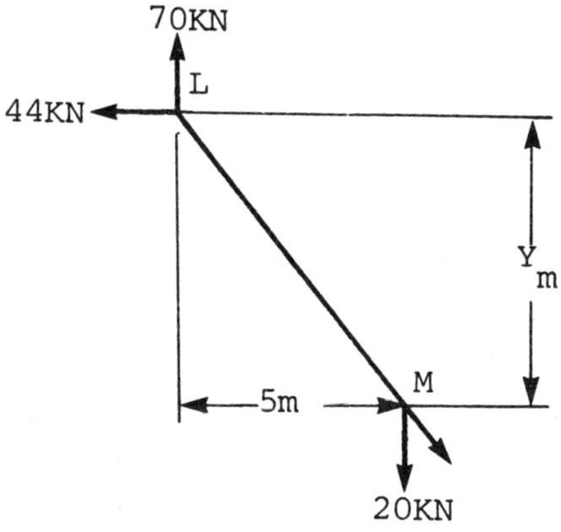

Fig. 3

$$+\circlearrowleft \Sigma M_M = 0 : \quad (44kN)Y_M - (70kN)(5m) = 0$$

or

Y_m = 7.95m below L

• PROBLEM 3-13

Determine a geometric relation between the changes in length of wire AC and BC if the common point C remains on the same vertical line, but is displaced by a small amount δ in the deformation. Assume that the deformation is compatible.

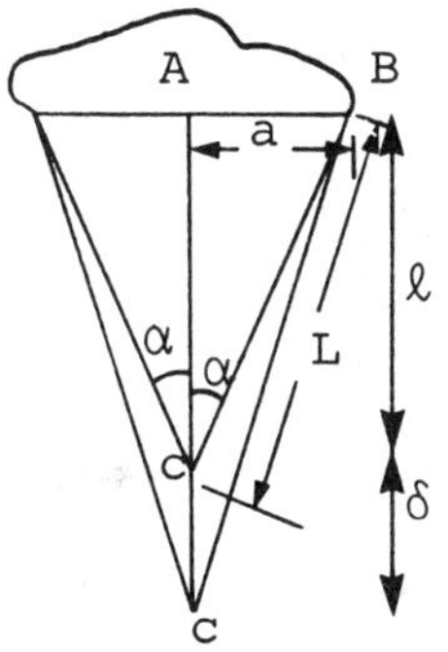

Solution: Denote AB by a, the length from A to C by ℓ, and the length from B to C by L. During the compatible deformation, point C moves down to a new position while the triangle ABC remains a right triangle. This means L and ℓ should satisfy the Pythagorean theorem:

$$L^2 = \ell^2 + a^2 \qquad (1)$$

where L and ℓ are variables during the deformation. Differentiating Equation (1) with respect to ℓ yields

$$2L\frac{dL}{d\ell} = 2\ell$$

$$\frac{dL}{d\ell} = \frac{\ell}{L}. \qquad (2)$$

However, since $\frac{\ell}{L} = \cos\alpha$, after multiplying both sides of equation (2) by $d\ell$, we obtain

$$dL = \cos\alpha\ d\ell. \qquad (3)$$

Now, let L_I and L_F be the lengths from B to C before and after the deformation, respectively. Also, denote the original length from A to C as ℓ_0. We can now integrate equation (3) to yield the desired relation.

$$\int_{L_I}^{L_F} dL = \int_{\ell_0}^{\ell_0\delta} \cos\alpha\ d\ell$$

$$L_F - L_I = \Delta L = \cos\alpha\ [\ell_0 + \delta - \ell_0] = \cos\alpha\delta.$$

Therefore, $\Delta L = \delta\cos\alpha$, which is the relation asked for.

• **PROBLEM 3-14**

A rope attached at point Q passes over a pulley at P making a sag of 2m, and the end of the cable at R carries a force F. If the weight per unit length of the rope is 5.0 N/m, calculate the length of the rope PQ and the magnitude of the force F. The weight of the rope section PR can be neglected.

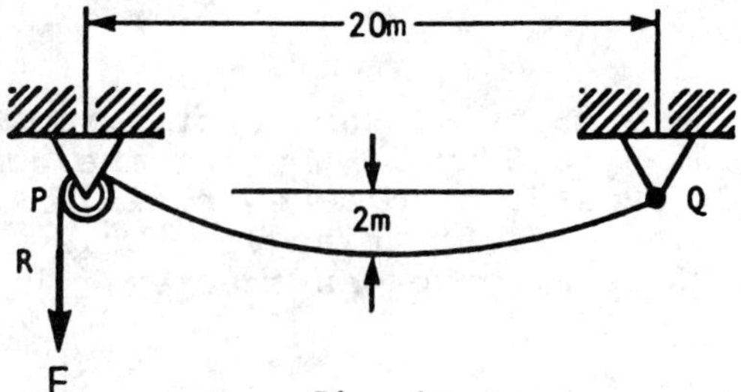

Fig. 1

Solution: The rope is assumed to sag, such that the center of the sag corresponds to the lowest point on the rope. At this point S, the rope is cut and the portion PS is made a free-body as shown in Fig. 2. The concentrated weight of the portion PS now acts at halfway between P and S and its magnitude is; W = (5N/m)(10m) = 50N.

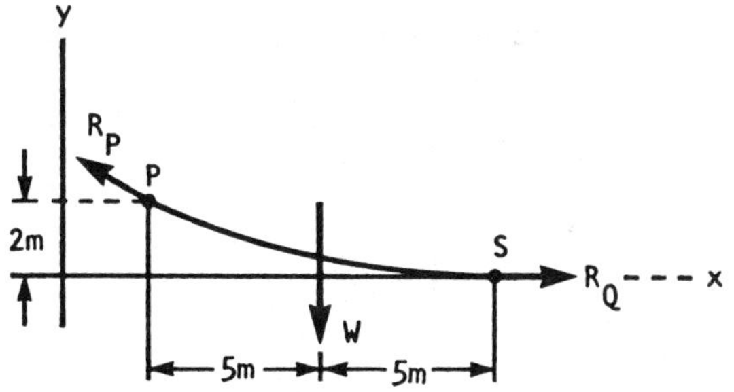

Fig. 2

$$+\circlearrowleft \; \Sigma M_P = 0 : \quad -(50N)(5m) + R_Q(2m)$$

or $\quad R_Q = 125N$

From Fig. 3, $R_P = \sqrt{W^2 + R_Q^{\,2}}$

$$= \sqrt{(50N)^2 + (125N)^2}$$

or

$$R_P = 134.6N$$

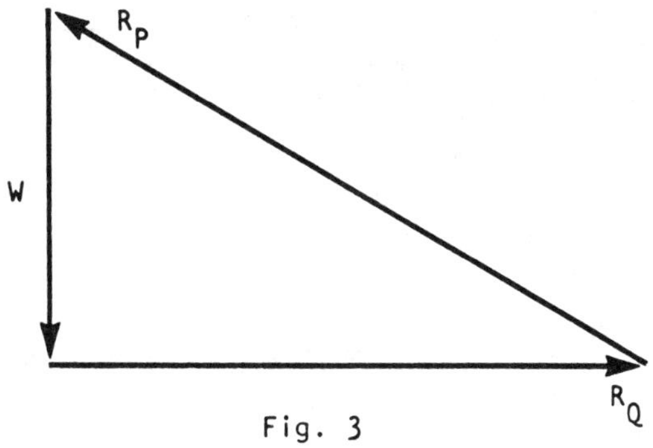

Fig. 3

Thus $F = R_P = 136.6N$

To obtain the length of the cable, a parabolic curve is assumed. Using the relation

$$L_{PQ} = 2\left\{X_{PQ}\left[1 + \frac{2}{3}\left(\frac{Y_{PQ}}{X_{PQ}}\right)^2\right]\right\}$$

where Y_{PQ} is the length of the sag and X_{PQ} is the horizontal distance between P and Q, from the length of a parabolic section,

$$L_{PQ} = 2\left\{(10m)\left[1 + \frac{2}{3}\left(\frac{2m}{10m}\right)^2\right]\right\}$$

$$= 20.53m$$

• PROBLEM 3-15

A chain is supported at two points on the same horizontal level and 12 ft. apart. Its slope at a point of support is $\frac{3}{4}$ and its density is given by the formula, for $s > 0$ where s is the length of the curve measured from the vertex, as $(1 + as)^{-1/2}$. Find the differential equation for the curve of the chain, and then, in the case $a = 0$, find the curve itself.

Solution: If the burden on a cable is distributed uniformly, the differential equation for the curve of the cable is:

$$\frac{dy}{dx} = \frac{L}{P}$$

where L is the burden and P is the pull on the chain in the horizontal direction at the point (x, y).

In the given problem, the weight of the cable is distributed along the cable; hence the nature of the burden is different. If the density of the cable is $\mu(s)$ where s is the length of the curve measured from the vertex, the load L is given by:

$$L = \int_0^s \mu(s)\, ds.$$

Thus the required differential equation is

$$\frac{dy}{dx} = \frac{1}{P}\int_0^s \mu(s)\, ds.$$

But we require a function of x on the right hand side, not a function of s. Remembering that the differential arc-length of a curve is given by:

$$ds = \sqrt{1 + (y')^2}\, dx,$$

we differentiate the differential equation to obtain

$$dy' = \frac{\mu(s)}{P} \sqrt{1 + (y')^2}\, dx$$

in accordance with the rules for differentiation under an integral sign. From the two equations

$$y' = \frac{1}{P} \int_0^s \mu(s)\, ds$$

$$dy' = \frac{\mu(s)}{P} \sqrt{1 + (y')^2}\, dx$$

we can sometimes eliminate s and obtain a differential equation for the curve of the chain.

Thus, in the given problem,

$$y' = \frac{1}{P} \int_0^s \frac{1}{\sqrt{1 + as}}\, ds = \frac{2}{ap} \{[1 + as]^{1/2}\}_0^s$$

$$= \frac{2}{ap} \{(1 + as)^{1/2} - 1\}$$

$$dy' = \frac{\sqrt{1 + (y')^2}}{p\sqrt{1 + as}}\, dx.$$

We eliminate $\sqrt{1 + as}$ by noting that $\sqrt{1 + as}$ is common to both equations. Then:

$$dy' = \frac{2\sqrt{1 + y'^2}}{P\,(aPy' + 2)}\, dx.$$ Separating variables,

$$\frac{aPy' + 2}{\sqrt{1 + y'^2}}\, dy' = \frac{2}{P}\, dx.$$

Solving, $aP\{\sqrt{1 + y'^2} - 1\} + 2\log\{y' + \sqrt{1 + y'^2}\}$

$$= \frac{2x}{P}.$$

This is the differential equation describing the curve of the chain.

If $a = 0$, the density $\frac{1}{\sqrt{1 + as}} = 1$ which means that the cable is of uniform density. The differential equation reduces to:

$$\log\left(y' + \sqrt{1 + y'^2}\right) = \frac{x}{P}.$$

Use the condition $y'(6) = \frac{3}{4}$ (the point of support is the mid-point of the distance separating the two points) to find $P = 6/\log 2$.

Then $\log\left\{y' + \sqrt{1 + y'}\right\} = \dfrac{6x}{\log 2}$

may be expressed as

$$\frac{dy}{dx} = \frac{1}{2}\left(e^{(x/6)\log 2} - e^{-(x/6)\log 2}\right).$$

Separating variables and applying the initial condition $y(0) = 0$ to the resulting integral:

$$y = \frac{3}{\log 2}\left(e^{(x/6)\log 2} + e^{-(x/6)\log 2} - 2\right)$$

or

$$y = \frac{3}{\log 2}\left(2^{x/12} - 2^{-x/12}\right)^2.$$

• PROBLEM 3-16

A bridge of weight w per unit length is to be hung from cables of negligible weight as illustrated in Figure 1. Determine the shape of the suspension cable such that the vertical cables, which are equally spaced, will support equal weights. Assume the vertical cables are closely spaced, the end points of the cable are $y = 0$, $x = \pm 1/2\, D$, and τ_0 is the maximum tension in the suspension cable.

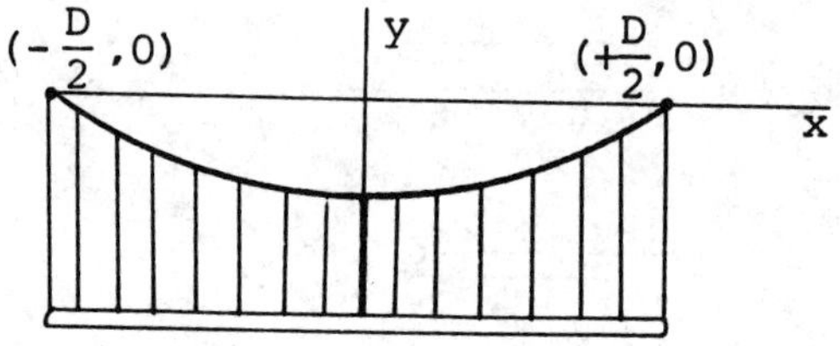

Fig. 1

Solution: To determine the shape of the suspension cable, we have to develop an equation for it in the x-y coordinate system (Fig. 2).

Consider a small segment of the suspension cable of length ds, with tensions T_1 and T_2 at either end. A downward force w' per unit length acts on the cable, resulting in a downward force on the small segment of w'ds, as shown in Figure 2.

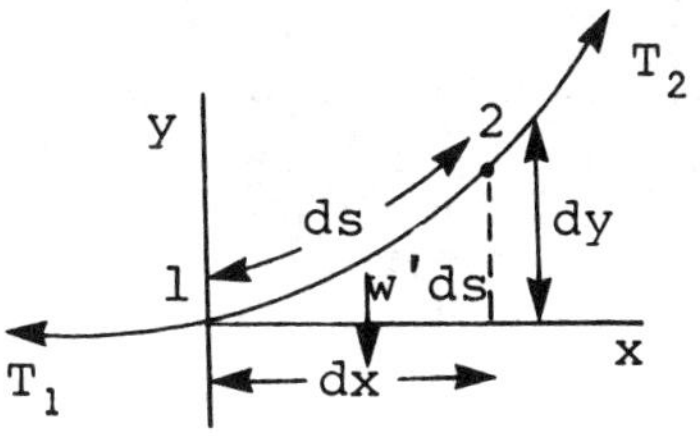

Fig. 2

Applying the conditions of equilibrium to the small segment, we have

(for $\Sigma F_x = 0$): $$T_2 \left(\frac{dx}{ds}\right)_2 - T_1 \left(\frac{dx}{ds}\right)_1 = 0 \qquad (1)$$

(for $\Sigma F_y = 0$): $$T_2 \left(\frac{dy}{ds}\right)_2 - T_1 \left(\frac{dy}{ds}\right)_1 - w'ds = 0 \qquad (2)$$

where the quantities in parentheses are to be evaluated at the corresponding ends of the segment.

From equation (1) we see that

$$T_2 \left(\frac{dx}{ds}\right)_2 = T_1 \left(\frac{dx}{ds}\right)_1$$

hence: $$T \left(\frac{dx}{ds}\right) = \text{constant} = B \qquad (3)$$

Equation (2) can be rewritten as

$$\frac{d}{ds} \left(T \frac{dy}{ds}\right) = w' \qquad (4)$$

and using the chain rule, equation (4) can be written

$$\frac{d}{ds} \left(T \frac{dy}{dx} \frac{dx}{ds}\right) = w' \qquad (5)$$

Substituting (3) into (5) gives $\frac{d}{ds} \left(B \frac{dy}{dx}\right) = w'$

or $$\frac{d}{ds} \left(\frac{dy}{dx}\right) = \frac{w'}{B} \qquad (6)$$

Now, the downward force w'ds is equal to the weight of a length dx of the bridge, so that

$$w'ds = w\ dx.$$

or $$w' = w \frac{dx}{ds} \qquad (7)$$

Substituting (7) into (6) we get

$$\frac{d}{ds}\left(\frac{dy}{dx}\right) = \frac{d}{dx}\ \frac{dx}{ds}\left(\frac{dy}{dx}\right) = \frac{w}{B}\ \frac{dx}{ds}$$

or $$\frac{d^2y}{dx^2} = \frac{w}{B} \qquad (8)$$

Integrating once with respect to x,

$$\frac{dy}{dx} = \frac{w}{B}\,x + C_1 \qquad (9)$$

where C_1 is a constant of integration which can be evaluated using the boundary conditions illustrated in Figure 1, namely, at $x = 0$, $\frac{dy}{dx} = 0$; since the cable is assumed to be symmetrical making the slope of the cable zero at its center. Therefore, from equation (9), $C_1 = 0$. Integrating a second time with respect to x,

$$y = \frac{1}{2}\,\frac{w}{B}\,x^2 + C_2 \qquad (10)$$

The second constant of integration, C_2, can be evaluated from the end point boundary conditions, at $x = \pm \frac{D}{2}$, $y = 0$.

Thus, $$y = \frac{1}{2}\,\frac{w}{B}\left(\frac{D}{2}\right)^2 + C_2 = 0$$

and $$C_2 = -\frac{1}{8}\,\frac{w\,D^2}{B}.$$

Giving $$y = \frac{1}{2}\,\frac{w}{B}\,x^2 - \frac{1}{8}\,\frac{w}{B}\,D^2 = \frac{1}{2}\,w\left(x^2 - \frac{D^2}{4}\right)\left(\frac{1}{B}\right) \qquad (11)$$

Now, we also have from Figure 2,

$$ds^2 = dx^2 + dy^2 = \left[1 + \left(\frac{dy}{dx}\right)^2\right] dx^2$$

or $$ds = \sqrt{1 + \left(\frac{dy}{dx}\right)^2}\ dx \qquad (12)$$

Substituting (9) into (12) yields (with $C_1 = 0$),

$$\frac{ds}{dx} = \sqrt{1 + \left(\frac{wx}{B}\right)^2} \qquad (13)$$

Combining (3) and (13)

$$T = B\sqrt{1 + \left(\frac{wx}{B}\right)^2} \qquad (14)$$

Letting τ_0 be the maximum tension when $x = \pm \frac{D}{2}$, equation (14) becomes

$$\tau_0 = B\sqrt{1 + \left(\frac{w\ D}{2B}\right)^2} \qquad (15)$$

Squaring, $$\tau_0^2 = B^2\left[1 + \left(\frac{w\ D}{2B}\right)^2\right]$$

$$\tau_0^2 = B^2 + \frac{w^2D^2}{4}$$

Solving for B we get

$$B = \sqrt{\tau_0^2 - \frac{w^2D^2}{4}} \qquad (16)$$

and finally, substituting this value for B in equation (11) we have the equation for the shape of the suspension cable:

$$y = \frac{1}{2}\, w \left(x^2 - \frac{D^2}{4}\right)\left(\tau_0^2 - \frac{w^2D^2}{4}\right)^{-\frac{1}{2}} \qquad (17)$$

which is the equation of a parabola.

• **PROBLEM** 3-17

Find the equation of the catenary, i.e., the equilibrium configuration of a cable loaded by its own weight.

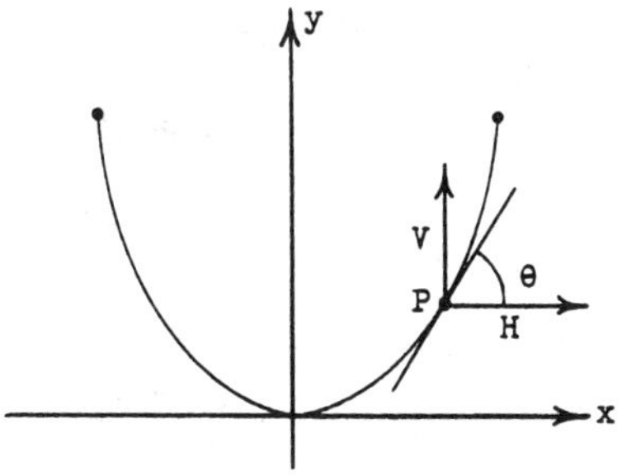

Solution: Let V and H be the vertical and horizontal components of the tension on the cable at point P, while θ is the angle of inclination.

Then $\tan\theta = \frac{V}{H}$ or

$V = H \tan \theta$. Let w be the weight of the cable per unit length. The weight of an element of cable ds will then be w ds. Now

$$(ds)^2 = (dx)^2 + (dy)^2, \text{ thus}$$

$$\cos \theta = \frac{dx}{ds} \quad \text{or} \quad ds = \frac{dx}{\cos \theta} .$$

We may, therefore, write the weight of ds as

$$w \frac{dx}{\cos \theta} . \tag{a}$$

The equilibrium condition means that H is constant, but that the vertical component V at x minus the vertical component at x + dx is equal to the weight of ds = $\sqrt{dx^2 + dy^2}$. Thus

$$d (H \tan \theta) = \frac{w\, dx}{\cos \theta},$$

$$H \frac{d}{dx} (\tan \theta) = \frac{w}{\cos \theta},$$

$$H \sec^2 \theta \frac{d\theta}{dx} = \frac{w}{\cos \theta},$$

$$H \frac{d\theta}{dx} = w \cos \theta .$$

Separating variables,

$$\frac{d\theta}{\cos \theta} = \frac{w}{H} dx.$$

Taking the definite integral

$$\int_0^\theta \frac{d\theta}{\cos \theta} = \frac{w}{H} \int_0^x dx \tag{b}$$

where we assume that $\theta = 0$ at $x = 0$, i.e., the lowest point is at the origin of our coordinate system.

Now, using a standard table of integrals, we integrate (b) to obtain

$$\log \left| \tan \left(\frac{\theta}{2} + \frac{\pi}{4} \right) \right| = \frac{w}{H} x$$

or,

$$\tan \left(\frac{\theta}{2} + \frac{\pi}{4} \right) = e^{\frac{w}{H} x} . \tag{c}$$

We note that if (c) is true then

$$\cot\left(\frac{\theta}{2} + \frac{\pi}{4}\right) = \frac{1}{e^{\frac{w}{H}x}} = e^{-\frac{w}{H}x}. \qquad \text{(d)}$$

But $\tan\left(\frac{\theta}{2} + \frac{\pi}{4}\right) - \cot\left(\frac{\theta}{2} + \frac{\pi}{4}\right) = 2\tan\theta.$

Hence $$\tan\theta = \frac{1}{2}\left(e^{\frac{w}{H}x} - e^{-\frac{w}{H}x}\right). \qquad \text{(e)}$$

Finally, $\tan\theta = \frac{dy}{dx}$

Substituting in (e)

$$\frac{dy}{dx} = \frac{1}{2}\left(e^{\frac{w}{H}x} - e^{-\frac{w}{H}x}\right).$$

Integrating $y = \frac{H}{2w}\left(e^{\frac{w}{H}x} - e^{-\frac{w}{H}x}\right) + C$

where C is the constant of integration. Since $y = 0$ when $x = 0$ we find that $C = 0$.

The equation of the catenary is

$$y = \frac{H}{2w}\left(e^{\frac{w}{H}x} - e^{-\frac{w}{H}x}\right).$$

CHAPTER 4

FRICTION

STATIC FRICTION

• PROBLEM 4-1

In Fig. 1 a beam LO weighing 50N rests against a wall, the lower part of the beam extending 15m from the base of the wall. Calculate the minimum value of the force F that can be applied to keep the beam in equilibrium. The coefficient of friction at L and O is 0.30.

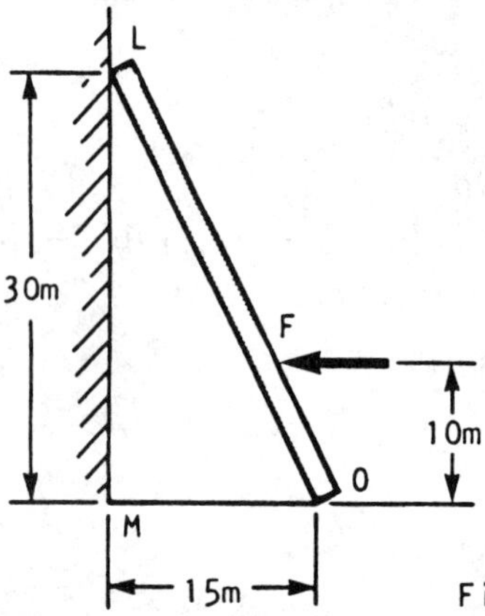

Fig. 1

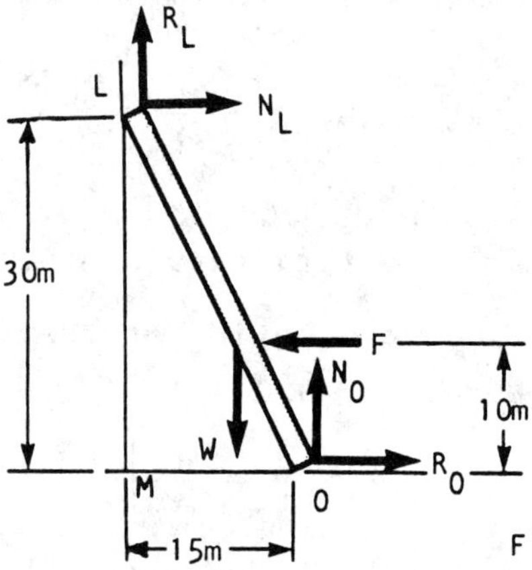

Fig. 2

Solution: Fig. 2 shows the free body diagram of the beam.

The conditions for translational equilibrium are:

$$\overset{+}{\rightarrow}\Sigma F_x = -F + R_o + N_L = 0 \qquad (1)$$

$$+\uparrow\Sigma F_y = N_o + R_L - W = 0 \qquad (2)$$

Rotational equilibrium demands that the sum of torques about any point be equal to zero. Torques about point M are:

$$+\ \Sigma\tau_M = 0 = F(10m) - N_L(30m) - R_L(15m) + W(7.5m) \quad (3)$$

Substitute $R_o = \mu N_o$, $R_L = \mu N_L$ in equations (1) and (2). These become:

$$\Sigma F_x = -F + \mu N_o + N_L = 0 \qquad (4)$$

$$\Sigma F_y = N_o + \mu N_L - W = 0 \qquad (5)$$

Equations (3), (4), and (5) are three linear equations in the three unknowns N_O, N_L, and F. These can be solved by the method of determinants or addition-subtraction and substitution of equations. Here we will use the latter. Solve equation (5) for N_O, substitute the result in equation (4), and solve resulting equation for N_L which is then substituted in equation (3) along with $R_L = \mu N_L$.

$$N_O = W - \mu N_L \tag{6}$$

$$N_L = \frac{\mu W - F}{(\mu^2 - 1)} \tag{7}$$

Equation (3), solved for F, becomes:

$$F = \frac{30\mu W + 15\mu^2 W - 7.5W(\mu^2 - 1)}{30 + 10\mu^2 - 10 + 15\mu}$$

This, for $\mu = 0.3$ and $W = 50N$, gives:

$$F = 33.8N$$

• **PROBLEM 4-2**

A ladder of length 10 m and mass 10 kg. leans against a frictionless vertical wall at an angle of 60° from the horizontal. The coefficient of static friction between the horizontal floor and the foot of the ladder is $\mu_s = 0.25$. A man of mass 70 kg starts up a ladder. How far along the ladder does he get before the ladder begins to slide down the wall?

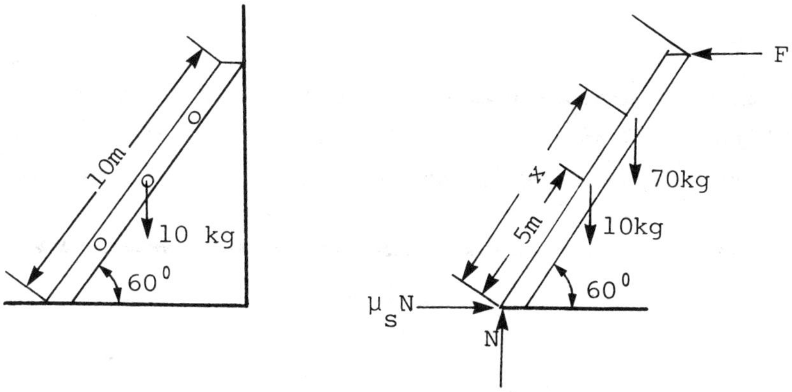

Solution: The forces on the ladder are shown. The horizontal force on the foot of the ladder is equal to $\mu_s N$ only at the instant before the ladder begins to slide. We wish to find the position of the man at this instant. Because the vertical wall is frictionless, it can exert a force on the ladder only perpendicular to itself as shown.

Since the net force on the ladder is zero, we find, taking vertical components,

$$N - 10g - 70g = 0$$

and, from horizontal components,

$$\mu_s N - F = 0$$

Let x be the distance of the man along the ladder from the foot at the instant the ladder begins to slide. Equate the torque about the foot of the ladder to zero.

$$\frac{L}{2} \cdot mg \cos\theta + xMg \cos\theta - FL \sin\theta = 0$$

From the first equation above,

$$N = 80\ g \text{ newtons}$$

From this and the second equation,

$$F = 0.25 \cdot 80\ g = 20\ g \text{ newtons}$$

From this and the third equation,

$$\frac{1}{2} m\, Lg \cos\theta + xMg \cos\theta - 20gL \sin\theta = 0$$

$$x = \frac{20\not{g}L \sin\theta - \frac{1}{2} mL\not{g} \cos\theta}{M\,\not{g} \cos\theta}$$

$$= \frac{200 \times 0.866 - \frac{1}{2} \times 10 \times 10 \times \frac{1}{2}}{70 \times \frac{1}{2}} = 4.2 \text{ meters.}$$

• PROBLEM 4-3

Find the largest angle θ at which a gradually increasing force $\vec{f}$ can be applied so that the 100 lb block will tip ($\mu_s = 0.5$).

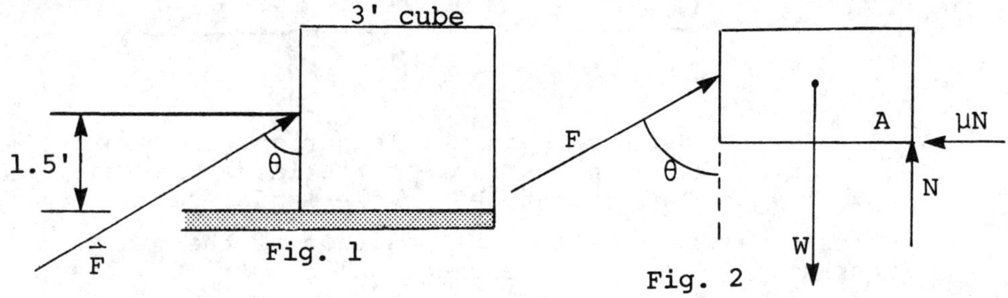

Fig. 1

Fig. 2

Solution: Newton's laws govern this motion and must be applied in the vertical and horizontal directions separately. In order to apply Newton's laws correctly, we will draw a force diagram. Figure two shows the force locations when the block just starts to tip.

In the horizontal direction,

$$\Sigma F_x = f \sin \theta - \mu N = 0 \qquad (1)$$

$$f \sin \theta = 0.5 \text{ N.}$$

Vertically we have

$$\Sigma F_y = f \cos \theta + N - 100 = 0 \qquad (2)$$

$$f \cos \theta = 100 - N.$$

The results of equations (1) and (2) leave us with two equations and three unknowns. Thus we must apply Newton's laws again using torques, clockwise torques being positive about point A.

$$\Sigma \tau_A = 0 = -100(1.5) + f \cos \theta \, (3) + f \sin \theta \, (1.5) \qquad (3)$$

$$0 = -150 + 3(100 - N) + 0.5N(1.5)$$

$$N = \frac{150}{2 \cdot 25} = 66.67 \text{ lbs.}$$

Knowing N, we can solve for θ as follows:

$$\tan \theta = \frac{f \sin \theta}{f \cos \theta} = \frac{.5 \text{ N}}{100 - N} \qquad (4)$$

$$\tan \theta = \frac{33.34}{33.34} = 1.$$

$$\theta = 45°.$$

• PROBLEM 4-4

The wheels of a small wagon are separated by a distance d, and the center of mass is a distance h above the ground. The wagon is at rest on a hill of slope angle θ, and between the wheels and the surface of the hill the coefficient of static friction is μ. How steep a hill can the wagon rest on without tipping over or sliding?

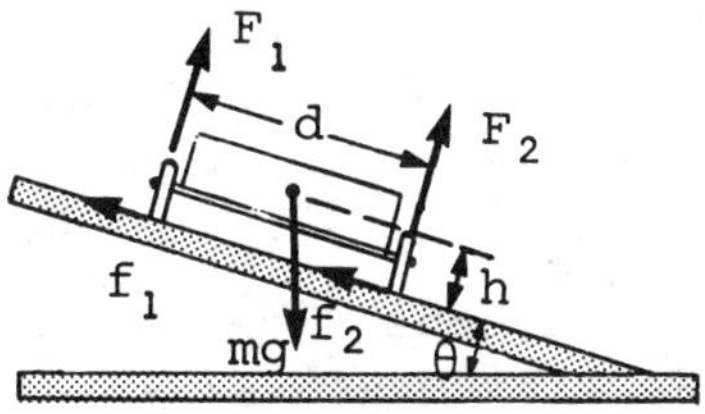

Fig. 1

Solution: The wagon is acted on by three forces, gravity and two contact forces at the wheels. Both the direction and the magnitudes of the contact forces are unknown. Consequently, we have four unknown force components, which are denoted by F_1, F_2, f_1, and f_2 in Fig. 1. Since we have only three equlibrium equations, this problem normally is statically indeterminate in the sense that f_1 and f_2 cannot be determined separately. However, we are now interested only in the condition when the wagon is on the verge of sliding or on the verge of tipping. In these cases, we have additional relations imposed between the variables. When the wagon is on the verge of sliding, the friction forces are fully developed at both wheels and, as we recall, are then simply related to the force components normal to the plane. If we let the symbols F_1, F_2, f_1, and f_2 denote the forces which obtain when the wagon is on the verge of sliding, we have

$$f_1 = \mu F_1 \qquad \text{and} \qquad f_2 = \mu F_2.$$

If we combine these relations with the general conditions for equilibrium,

$$F_x = - f_1 - f_2 + mg \sin \theta_1 = 0$$

$$F_y = F_1 + F_2 - mg \cos \theta_1 = 0,$$

we obtain μ mg cos θ_1 = mg sin θ_1 or

$$\tan \theta_1 = \mu.$$

In other words, in order to prevent sliding, we must have $\theta < \theta_1$.

When the wagon is on the verge of tipping, the forces F_1 and f_1 will be zero, since then the contact between the upper wheel and the plane will be broken. Then, if we consider the torque with respect to the contact point of the lower wheel, we see that for the total torque to be zero, the lever arm of the gravitational force with respect to this point must be zero. This will occur when the angle of inclination of the plane has the value given by

$$\tan \theta_2 = \frac{d}{2h} ,$$

and tipping will be prevented if $\theta < \theta_2$. If $\theta_1 = \theta_2$, it follows that tipping and sliding will occur simultaneously when $\mu = d/2h$. If μ is less than 2/dh, sliding will occur before tipping as the angle θ is increased. The opposite, of course, occurs if μ is larger than d/2h.

• PROBLEM 4-5

In Fig. 1, a door hinge is subjected to a load T, placed 5m from the 2m-diameter hinge pipe. If the weight of the hinge is negligible, calculate the least distance x where another load of equal magnitude F can be placed without the hinge moving on the pipe. The coefficient of static friction between the pipe and the hinge is 0.20.

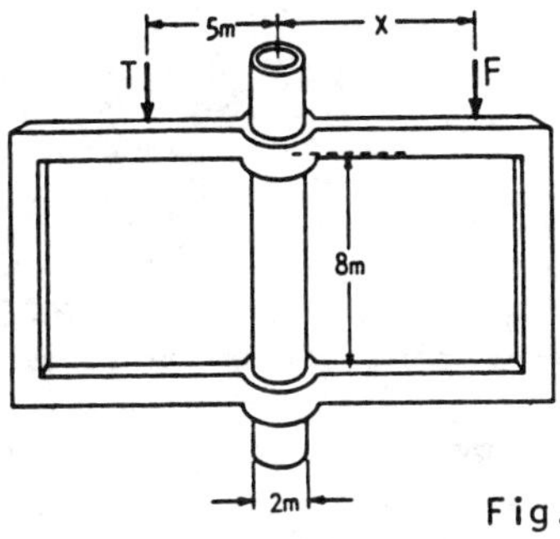

Fig. 1

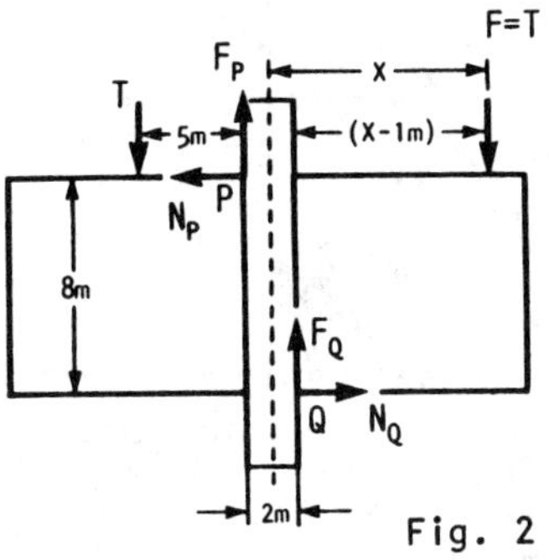

Fig. 2

Solution: It is clear that when the load is placed too close to the pipe, there will not be enough frictional force between the pipe and the hinge. Thus, the hinge will slip. However, when the load is well removed from the pipe, this will cause sufficient torque on the hinge so that the normal forces and the friction will be large enough to support the load.

The free body diagram of figure 2, shows how these forces act. Assume the load is placed at the minimum distance x from the axis of the pipe so that the minimum possible frictional force that can support the load occurs. That is

$$F_P + F_Q - 2T = 0 \qquad (1)$$

But

$$F = \mu N_P = 0.20N_P$$

$$F = \mu N_Q = 0.20N_Q \qquad (2)$$

A further equilibrium condition is that:

$$\overset{+}{\rightarrow} \Sigma F_x = 0 : \quad N_Q - N_P = 0$$

or

$$N_Q = N_P \qquad (3)$$

From equation 1,

$$0.20\ N_P + 0.20N_Q = 2T$$

$$0.20N_P + 0.20N_P = 2T$$

$$0.40N_P = 2T$$

$$N_P = 5T$$

Summing the torque about Q gives

$$+\circlearrowright \Sigma M_Q = 0 : \quad N_P(8m) - F_P(1m) + T(6m) - T(x - 1m) = 0$$

$$8N_P - 1F_P + 6T - Tx - 1T = 0$$

$$8(5T) - (1)(0.20)(5T) + 6T - 1T = Tx$$

Solving for x gives

$$x = 44\text{m}$$

• PROBLEM 4-6

In Fig. 1, a man applies a force F of 880N to a belt, wrapped four times around a pipe PQ, to draw water from a well. Given that the belt is just about to slip, calculate a) the coefficient of static friction between the belt and the pipe, b) F, if the belt is wrapped twice around the pipe.

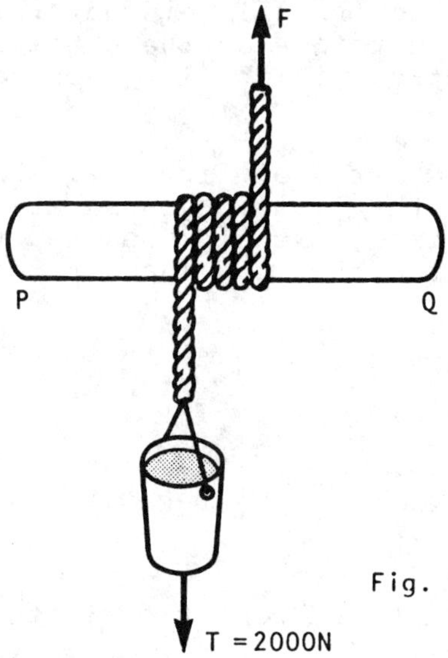

Fig. 1

Solution: The equation relating the tensions at the beginning and end of the belt wrapped around the pipe is given as

$$\ell n \frac{T}{F} = \mu_s \theta \qquad (1)$$

where μ_s is the coefficient of static friction and θ is the angle subtended by the belt when in contact with the pipe. θ is given in radians and is multiplied by the number of turns that the belt is wrapped on the pipe.

Thus

$$\theta = 4(2\pi \text{ rad}) = 25.1 \text{ rad}.$$

From equation (1)

$$\mu_s = \ell n\left(\frac{T}{F}\right)\left(\frac{1}{25.1 \text{ rad}}\right)$$

$$= \left[\ell n\left(\frac{2000\text{N}}{880\text{N}}\right)\right]\left(\frac{1}{25.1 \text{ rad}}\right)$$

$$\mu_s = 0.033$$

b) From equation (1)

$$\frac{T}{F} = e^{\mu_s \theta}$$

or

$$F = \frac{T}{e^{\mu_s \theta}} \qquad (2)$$

Now θ = 2 turns (2π rad) = 12.6 rad.

Calculating F from equation (2) gives

$$F = \frac{2000\text{N}}{e^{(0.033)(12.6)}}$$

$$= \frac{2000}{1.52} = 1315.8\text{N}$$

• PROBLEM 4-7

A capstan is used to lower a crate from an elevated loading platform to the ground. When held stationary on the incline, the crate applies a force F_2 of 5000N to the rope wound with two turns about the capstan, while it takes a force F_1 of 100N to hold the crate in place.

a) Calculate the coefficient of friction between the rope and the capstan.

b) Using this coefficient of friction, compute the force F_2 that will be held when the rope is wound with three turns.

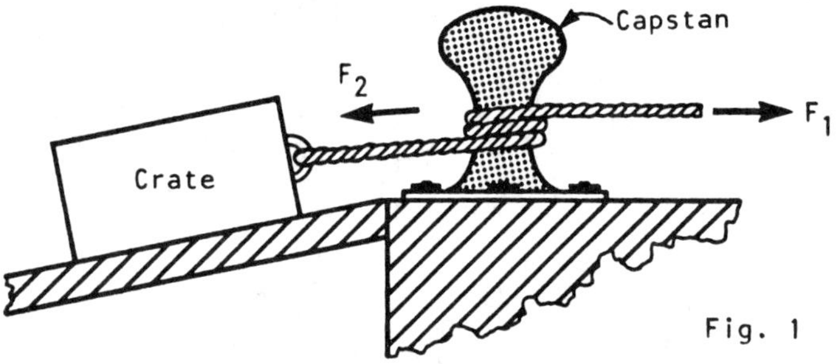

Fig. 1

Solution: The equation relating the tensions at the beginning and end of a rope wrapped around a fixed circular cylinder is

$$\ln \frac{T_2}{T_1} = \mu(\theta_2 - \theta_1). \qquad (1)$$

Since the rope is wrapped two full turns around the capstan, we have

$$(\theta_2 - \theta_1) = 2(2\pi) = 4\pi \text{ rad.}$$

Given are:

$$F_1 = 100\text{N}, \quad F_2 = 5000\text{N}$$

so

$$4\pi\mu = \ln \frac{5000}{100} = \ln 50 = 3.91.$$

Solving for μ, we get

$$\mu = \frac{3.91}{4\pi} = 0.31$$

Coefficient of friction in a rope wrapped around a cylinder. To analyze the problem when the rope is wrapped three times about the capstan, we first modify equation (1) by taking antilogs, yielding:

$$\frac{T_2}{T_1} = e^{\mu(\theta_2 - \theta_1)}$$

So, using $\theta_2 - \theta_1 = 3(2\pi) = 6\pi$ rad. we get

$$T_2 = (100N)e^{(0.31)(6\pi)}$$

$$= 34{,}493N.$$

• PROBLEM 4-8

A 200N.m torque is applied to the flywheel rotating on its axle P in Figure 1. Calculate the least force T, that must be applied by the wheel to a belt around it, in order to prevent slippage of the belt. The value of μ_s is 0.20.

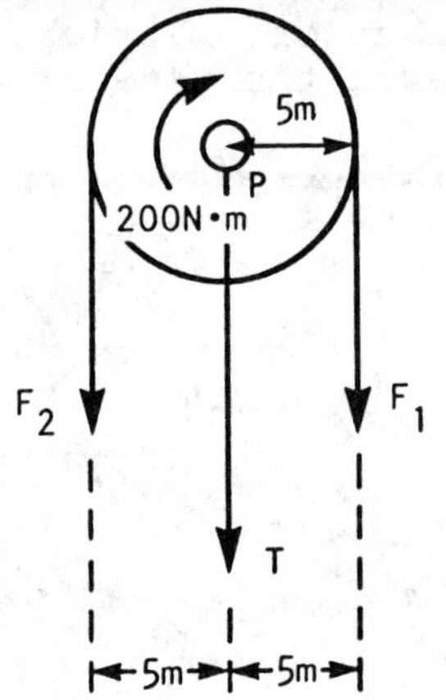

Fig. 1

Solution: The values of F_1 and F_2 corresponds to impending slippage, thus the relation

$$\frac{F_2}{F_1} = e^{\mu_s \theta} \quad (1)$$

will be used. The angle of contact θ, between the belt and the flywheel is given as

$$\theta = \pi \text{ radians} = 3.14 \text{ rad.}$$

Equation (1) becomes

$$\frac{F_2}{F_1} = e^{(0.20)(3.14)} = 1.87$$

or $F_2 = 1.87F_1$ (2)

Taking moments about point P gives

$$+\circlearrowright \Sigma M_P = 0 : \quad -F_2(5m) + F_1(5m) + 200N\cdot m = 0$$

$$(F_2 - F_1)(5m) = 200N\cdot m$$

or $F_2 - F_1 = 40N$

From equation (2) $F_2 = 1.87F_1$

Thus,

$$1.87F_1 - F_1 = 40N$$

$$0.87F_1 = 40N$$

or $F_1 = 46N$

and

$$F_2 = (46N)(1.87)$$

$$F_2 = 86.02N$$

To calculate T, sum all vertical forces equal to zero

$$+\uparrow\Sigma F_y = 0 : \quad -F_2 - T - F_1 = 0$$

$$T = -(86.02N) - (46N)$$

$$= -132.02N$$

The minus sign indicates that the force T would act in the opposite direction as that shown in Fig. 1. Thus

$$T = 132.02N$$

• PROBLEM 4-9

A small pump is mounted on a concrete slab which, in turn, rests on four concrete posts. Two of the posts have settled slightly, so that the slab must be leveled. The foreman decides to raise the low side with two large wedges located as drawn on the figure. Fortunately, there is a concrete floor on which the bottom wedge can be fastened. If μ_s between the two wedges is 0.4 and between the wedge and the concrete 0.6, and if the motor slab transmits a 1000 lb vertical force to the wedge, determine the force P required for lifting the slab.

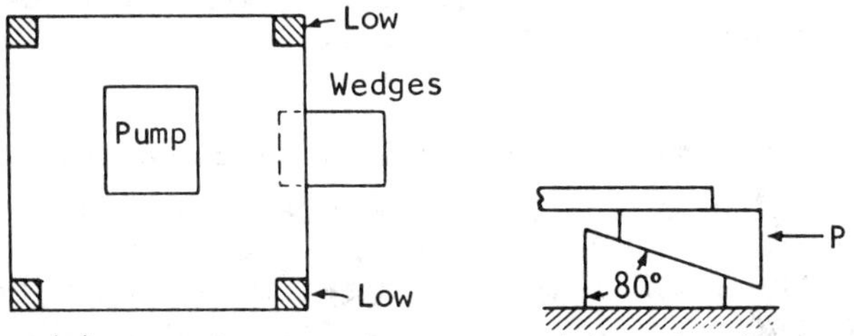

(a) Top view Fig. 1 (b) Side view

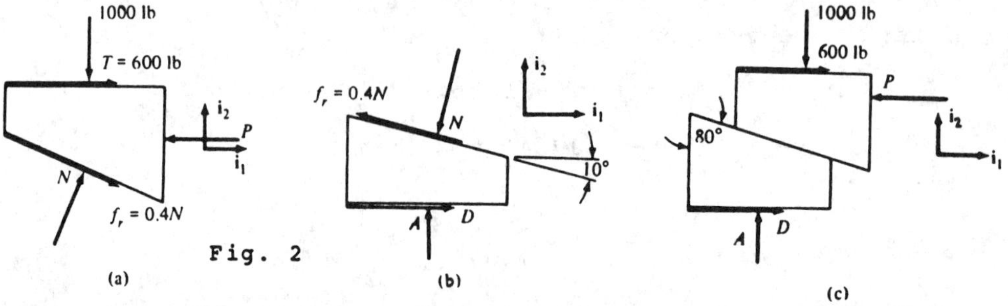

Fig. 2

(a) (b) (c)

Solution: The free body diagrams are shown in figure 2. Notice that part (c), showing the two wedge system, has only the external forces shown. The forces between the wedges are equal and opposite and cancel out by Newton's Third Law. If the sum of all the external forces are set equal to zero, the system is in equilibrium. That is, nothing moves. Any greater force will start lifting the slab.

$$\overset{+}{\rightarrow} \sum F_{i_1} = 0; \quad 600 + D - P = 0 \tag{1}$$

$$+\uparrow \sum F_{i_2} = 0; \quad -1000 + A = 0$$

so $\qquad A = 1000 \text{ lb.}$

The bottom wedge, as described, is fastened to the floor. The sum of the forces must be equal to zero throughout. This means that the force D is entirely reactive and does not depend on any coefficient of friction or the normal force A. The equilibrium equations are, from diagram (2b),

$$\overset{+}{\rightarrow} \sum F_{i_1} = 0; \quad D - 0.4\ N \cos 10° - N \sin 10° = 0, \tag{2}$$

$$+\uparrow \sum F_{i_2} = 0; \quad A - N \cos 10° + 0.4N \sin 10° = 0$$

or, $\quad 1000 - N \cos 10° + 0.4\ N \sin 10° = 0. \qquad (3)$

Solving equations (2) and (3) simultaneously for D yields

$$D = 1000 \frac{0.4 \cos 10° + \sin 10°}{\cos 10° - 0.4 \sin 10°}$$

$$= 621 \text{ lb.}$$

Putting this value into equation (1) yields

$$P = 1221 \text{ lb.}$$

The force P needed by this method is greater than the 1000 lbs. needed for a direct lift.

• PROBLEM 4-10

A flat leather belt is wrapped a quarter turn around a fixed wooden circular cylinder. The lower end holds a weight of 100 lb. If the coefficient of static friction between the leather and the wood is 0.2, determine the P sufficient to start moving the weight upward.

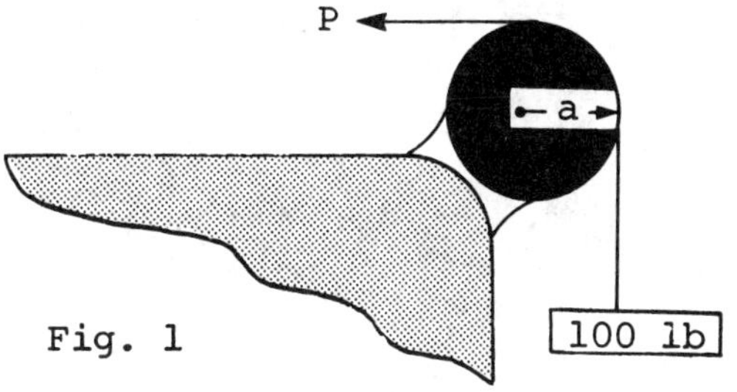

Fig. 1

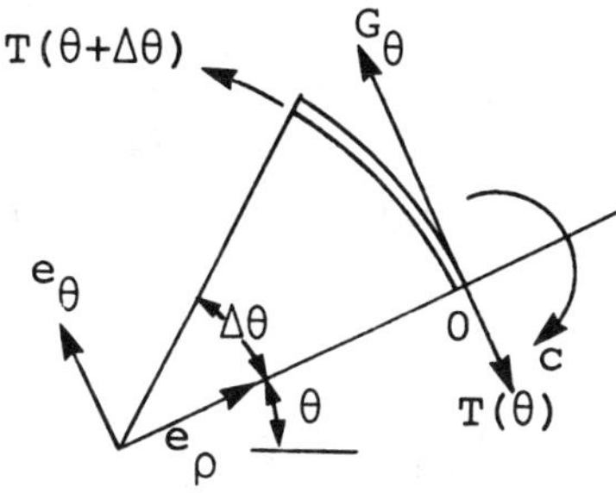

Fig. 2

<u>Solution</u>: Analyze this problem by considering a small portion of the belt as shown in figure 2. Let T(θ) denote the tension in the belt at an angle θ. Replace the forces of contact between the belt and the cylinder by a single force

$$\vec{G} = G_\rho \hat{e}_\rho + G_\theta \hat{e}_\theta$$

at the point O. From the condition of equilibrium, $\sum \vec{F}_0 = 0$, we have

$$T(\theta + \Delta\theta)[\cos \Delta\theta \hat{e}_\theta - \sin \Delta\theta \hat{e}_\rho] - [T(\theta) - G_\theta]\hat{e}_\theta + G_\rho \hat{e}_\rho = 0. \quad (1)$$

From this vector equation we obtain the scalar equations

$$T(\theta + \Delta\theta)\cos \Delta\theta - T(\theta) = -G_\theta \quad (2)$$

$$T(\theta + \Delta\theta)\sin \Delta\theta = G_\rho. \quad (3)$$

Relate G_θ and G_ρ through the friction equation,

$$G_\theta = -\mu_s G_\rho. \quad (4)$$

Combining equations (4), (3) and (2) yields

$$T(\theta + \Delta\theta)\cos \Delta\theta - T(\theta) = \mu T(\theta + \Delta\theta)\sin \Delta\theta. \quad (5)$$

Now, if equation (5) is divided by Δθ, and taking the limit as $\Delta\theta \to 0$, remembering that

$$\lim_{\Delta\theta \to 0} (\cos \Delta\theta) = 1;$$

$$\lim_{\Delta\theta \to 0} \left(\frac{\sin \Delta\theta}{\Delta\theta}\right) = 1,$$

the result is

$$\frac{dT}{d\theta} = \mu T. \quad (6)$$

Now that this expression has been found, one can use the calculus to find $T = f(\theta)$. Let θ_1 and θ_2 ($\theta_2 > \theta_1$) be the values of θ defining the beginning and the end of the contact area, then $T(\theta_2) > T(\theta_1)$.

Let $T(\theta_2) = T_2$ and $T(\theta_1) = T_1$ and integrate equation (6):

$$\int_{T_1}^{T_2} \frac{dT}{T} = \mu \int_{\theta_1}^{\theta_2} d\theta$$

$$\ln \frac{T_2}{T_1} = \mu(\theta_2 - \theta_1)$$

or, $$T_2 = T_1 e^{\mu(\theta_2 - \theta_1)}.$$

Identifying $T_1 = 100$ lbs. and $\theta_2 - \theta_1 = \pi/2$ yields

$$P = 100e^{(0.2)(\pi/2)} = 137.0 \text{ lb.}$$

• **PROBLEM 4-11**

> Two steel plates are held together by a bolt and nut with a double square thread which has a mean diameter of 20mm and a pitch of 3mm. With what force are the plates held together if a moment of 60N is applied to the nut and the coefficient of friction is 0.25? To separate the plates, what force is needed to loosen the nut?

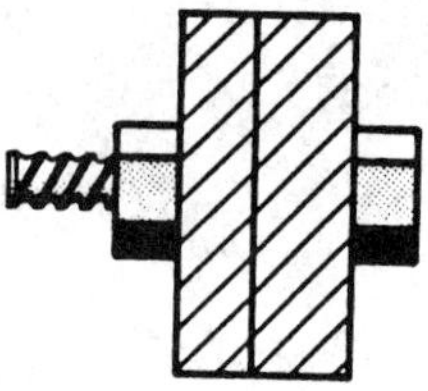

Fig. 1

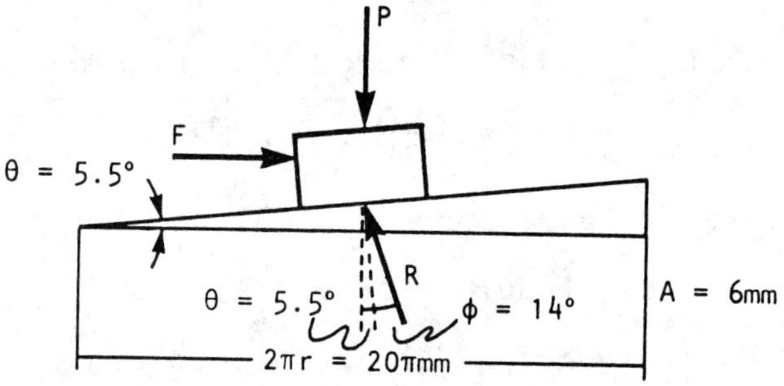

Fig. 2

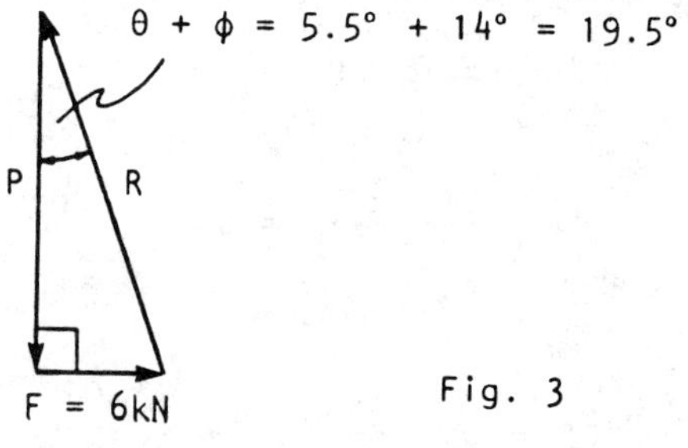

Fig. 3

Solution: (a) The mean diameter of the screw is 20mm, then the mean circumference of the screw is given by:

$$\pi D = 2\pi r = 20\pi \text{mm}.$$

The lead, A (advancement of the nut during one turn), is equal to twice the pitch since the screw is double-threaded.

$A = 2(3\text{mm}) = 6\text{mm}$

The lead angle, θ, of the screw, or the angle of an equivalent inclined plane is given by:

$$\tan\theta = \frac{A}{2\pi r} = \frac{6\text{mm}}{20\pi\text{mm}} = 0.095 \qquad \theta = 5.5°.$$

The friction angle, ϕ, is given by

$$\tan\phi = \mu = 0.25 \qquad \phi_s = 14.0°.$$

A free body diagram of the screw is shown in figure (1).

R in the diagram is the reaction force whose components are the normal and frictional forces of the pieces of wood on the screw. The frictional angle, ϕ, is the angle between R and the normal component of R. In this situation, the angle which R makes with the vertical will be given by $(\theta + \phi)$.

The force F acting on the bolt is applied tangentially to the thread. Its value can be obtained by equating its moment about the axis of the thread, Fr, to the applied torque.

$$F(10\text{mm}) = 60\text{N}\cdot\text{m}$$

$$F = \frac{40\text{N}\cdot\text{m}}{5\times10^{-3}\text{m}} = 6000\text{N} = 6\text{kN}.$$

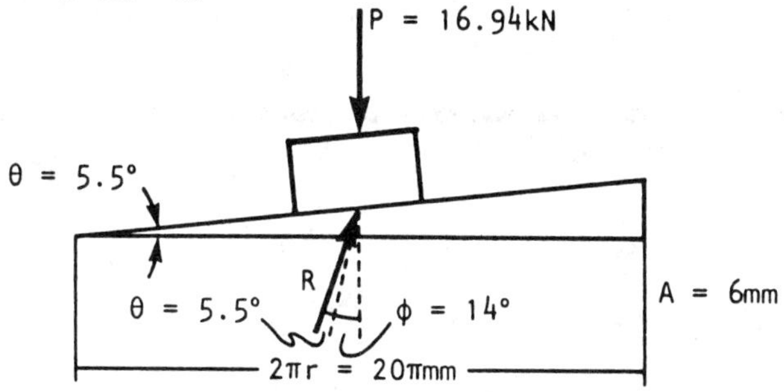

Fig. 4

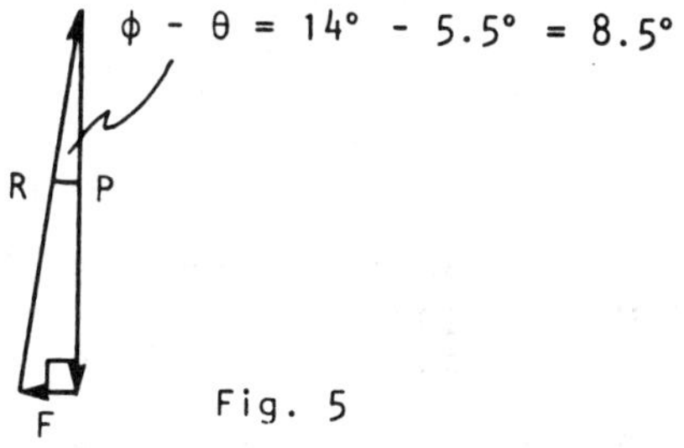

Fig. 5

The force, P, exerted on the plates can be found by using the force triangle in figure (2), which was obtained from the free-body diagram of figure (1).

$$\tan(\theta + \phi) = \frac{F}{P}$$

$$P = \frac{F}{\tan(\theta+\phi)} = \frac{6kN}{\tan(19.5°)} = 16.94kN$$

(b) The free-body diagram for the loosening of the nut is shown in figure (3). The sense of the force F is reversed, and the sense of the frictional force component of R has been reversed. In this situation, the angle made by R with the vertical is now given by $(\phi-\theta)$. A force triangle for this situation is shown in figure (4)

$$\phi - \theta = 14.0° - 5.5° = 8.5°$$

from figure (5).

$$F = P \tan(\phi - \theta)$$

$$F = (16.94kN)\tan(8.5°)$$

$$F = 2.532kN$$

The torque required to loosen the nut will then be given by Fr

$$\text{Torque} = Fr = (2.532kN)(10mm)$$

$$= (2.532 \times 10^3 N)(10 \times 10^{-3} m)$$

$$= 25.32 N\cdot m$$

Double-threaded screw.

• PROBLEM 4-12

Find the magnitude T of the couple which must be exerted on the drum for motion to impend. $(\mu_s = 0.1)$.

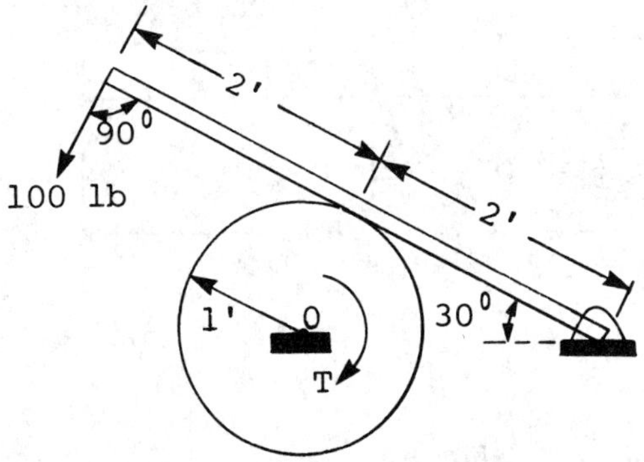

Solution: For motion to start, the moment couple on the drum must be equal to the friction force of the rod against the drum multiplied by the drum radius. $(\Sigma M_0 = 0)$.

Since the rod is in equilibrium we can take torques about the fulcrum to find the normal force of the drum on the rod.

$$\Sigma \tau_{fulcrum} = 0 = -100 \text{ lb}(4 \text{ ft}) + N(2 \text{ ft}) \qquad (1)$$

$N = 200$ lb.

From N, one can find the frictional force f by writing:

$f = \mu N = 0.1(200)$ (2)

$f = 20$ lb.

Therefore, the torque produced by the couple is

$T = 20 \text{ lb} \times 1 \text{ ft} = 20$ ft-lb.

KINETIC FRICTION

• PROBLEM 4-13

> Show that the tangent of the minimum angle of an inclined plane with the horizontal at which a block of wood will start sliding from rest is equal to the coefficient of static friction. Show that the tangent of the angle of an inclined plane with the horizontal at which a block of wood will move with uniform velocity down the plane is equal to the coefficient of kinetic friction. Would these results be true on the surface of the moon?

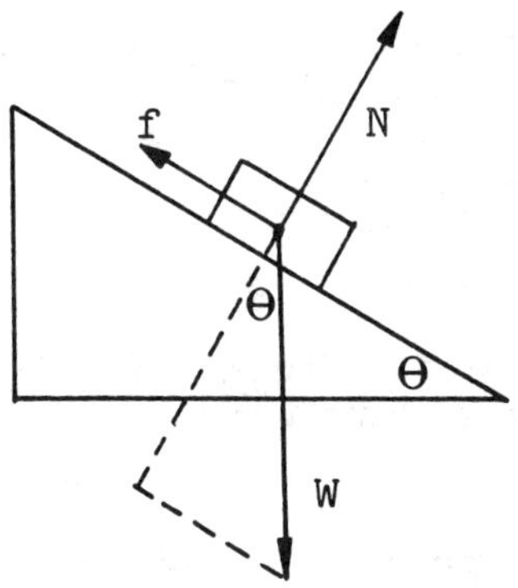

Fig. 1

Solution: All of the forces acting on the block are shown in figure (1).

The block is in equilibrium only if the vector sum of these forces equals zero. This occurs when the force components parallel to the plane and also those perpendicular to the plane sum to zero. Therefore, the component of weight down the plane must equal the force of friction up the plane. The normal force must equal the component of weight perpendicular to the plane. Angles θ in Fig. (1) are equal since their sides are mutually perpendicular.

$$\sum F_{\parallel} = f - W \sin\theta = 0 \tag{1}$$

$$\sum F_{\perp} = N - W \cos\theta = 0. \tag{2}$$

Therefore,

$$W \sin \theta = f \tag{3}$$

$$W \cos \theta = N. \tag{4}$$

Divide corresponding sides of Eq. (3) by Eq. (4) to get:

$$\frac{\sin \theta}{\cos \theta} = \frac{f}{N}.$$

Now $\quad f = \mu_s N, \quad$ and $\quad \dfrac{\sin \theta}{\cos \theta} = \tan \theta.$

Therefore,

$$\tan \theta = \mu_s. \tag{5}$$

As the angle of the inclined board is increased, the static friction force increases with the component of weight parallel to the plane. As soon as the latter exceeds the maximum value of the former, the block accelerates due to the unbalanced force. The angle just before slip occurs should be substituted in Eq. (5) to yield the static coefficient of friction.

The same argument applies in determining the kinetic coefficient of friction. But instead of the object not being touched, it is set into motion by an external force. The angle at which the block moves at constant velocity down the plane is the angle which produces equilibrium of forces parallel to the plane. Due to the state of motion, this happens at a smaller angle than the one for static friction. That is,

$$\mu_{kinetic} < \mu_{static}.$$

Since W = Mg and W cancel in above derivation, no change would occur if this were done on the moon.

• PROBLEM 4-14

Two blocks of equal mass are connected by a string which passes over a frictionless pulley. If the coefficient of friction is μ_k, what angle θ must the plane make with the horizontal so that each block will move with constant velocity once it is set in motion?

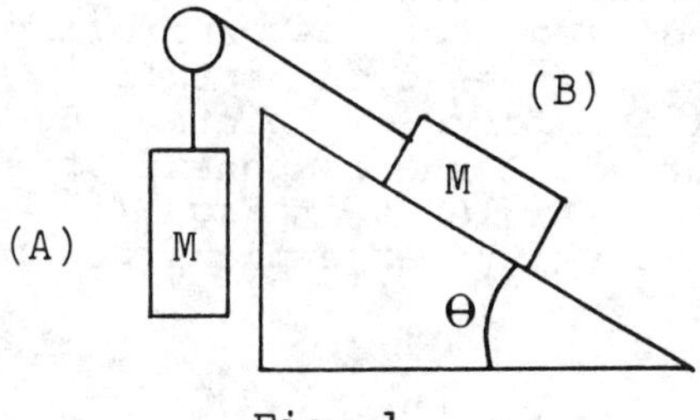

Fig. 1

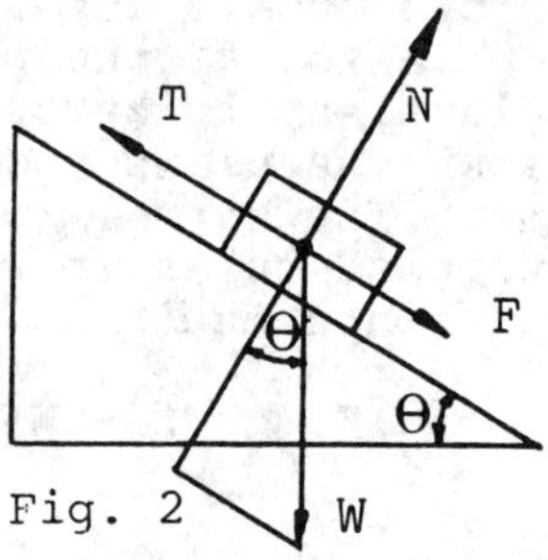

Fig. 2

Solution: The blocks will continue to move at constant velocity only if the forces on each block add up to zero.

There are two methods of approach in solving this problem. First, each block may be considered a system in itself. The sum of the forces on each block must then be set equal to zero. Second, the two blocks together may be considered a system. Only the forces external to the system (the string tension is considered an internal force) are then summed. Here, the first method will be used.

The two forces on block A are the tension in the string, T, and the weight, W = Mg. Equilibrium implies that the acceleration in the equation F = MA is zero, so the forces on the body must be:

$$\uparrow + \sum F_y = T - W = 0$$

or

$$T = W. \tag{1}$$

The string tension on block B must be equal and opposite to the component of its weight down the plane plus the friction force which also acts down the plane. That is,

$$+\nwarrow \sum F = T - W\sin\theta - f = 0. \tag{2}$$

A force diagram of block on inclined plane is shown in fig. 2.

The only forces which act on the block are: the tension T, the friction f, the normal force of the inclined plane on the block N, and the weight W. The two angles shown as θ are equal, since their sides are mutually perpendicular.

The sum of the forces perpendicular to the plane is:

$$\sum F_{\perp} = N - W\cos\theta = 0$$

or

$$N = W\cos\theta. \tag{3}$$

Friction is related to the normal force by

$$f = \mu_k N. \tag{4}$$

By substituting eq. (3) into eq. (4), the friction force f becomes

$$f = \mu_k W\cos\theta. \tag{a}$$

Substituting eq. (1) into eq. (2), we get

$$f = W - W\sin\theta. \tag{b}$$

Equating eq. (a) and (b),

$$\mu_k W\cos\theta = W - W\sin\theta, \quad \text{and}$$

$$W - W\sin\theta - {}_k W\cos\theta = 0 \tag{5}$$

$$1 - \sin\theta - \mu_k\cos\theta = 0.$$

Let $\cos\theta = -\sqrt{1 - \sin^2\theta}$ and rearrange

$$1 - \sin\theta = \mu_k\sqrt{1 - \sin^2\theta}.$$

Here square both sides, rearrange and factor into:

$$(1 + \mu_k^2)\sin^2\theta - 2\sin\theta + (1 - \mu_k^2) = 0. \qquad (6)$$

This is a quadratic equation with the sin θ as the unknown variable. The quadratic formula gives:

$$\sin\theta = \frac{2 \pm \sqrt{4 - 4(1 + \mu_k^2)(1 - \mu_k^2)}}{2(1 + \mu_k^2)}$$

$$\sin\theta = \frac{1 \pm \sqrt{\mu_k^4}}{(1 + \mu_k^2)} = \frac{1 \pm \mu_k^2}{1 + \mu_k^2}. \qquad (7)$$

The solution corresponding to the + sign is:

$$\sin\theta = 1; \quad \text{therefore,} \quad \theta = 90°.$$

This reduces to a simple pulley problem. When the - sign is used, the angle depends upon the coefficient of kinetic friction and is:

$$\theta = \arcsin\left(\frac{1 - \mu_k^2}{1 + \mu_k^2}\right) \qquad (8)$$

• PROBLEM 4-15

A skid weighing 500N and carrying 9500N of paper, is being pulled up onto a loading dock. The static and kinetic coefficients of friction are $\mu_s = 0.30$ and $\mu_k = 0.20$, respectively. What is the magnitude of the force F (1) when the skid begins to move upward; (2) while the skid is moving; (3) to prevent the skid from sliding downwards?

Fig. 1

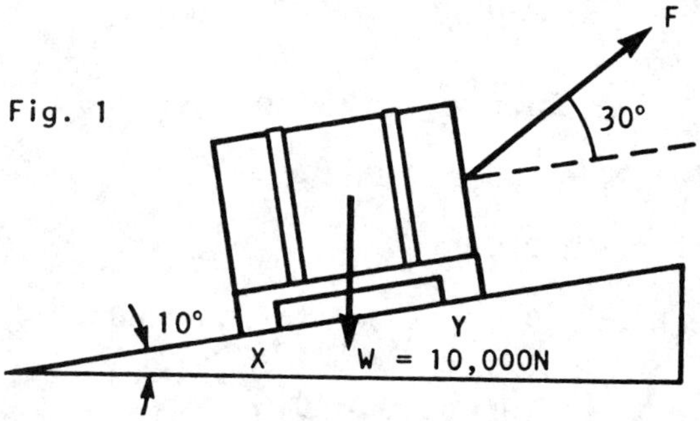

Solution: This problem may be solved by resolving forces into normal and tangential compounds with respect to the incline, to find values of the frictional force. However, this procedure actually does more than necessary. A more direct, geometrical procedure is possible. It is based on the fact that the combined normal and frictional forces, that is, the reaction of the incline on the skid, is directed at an angle ϕ defined by the friction coefficient μ and the equation $\tan\phi = \mu$.

The magnitude of the combined weight W of the skid and paper is

$$W = 500N + 9500N = 10kN$$

The free body diagram for part (1) is shown in figure 2.

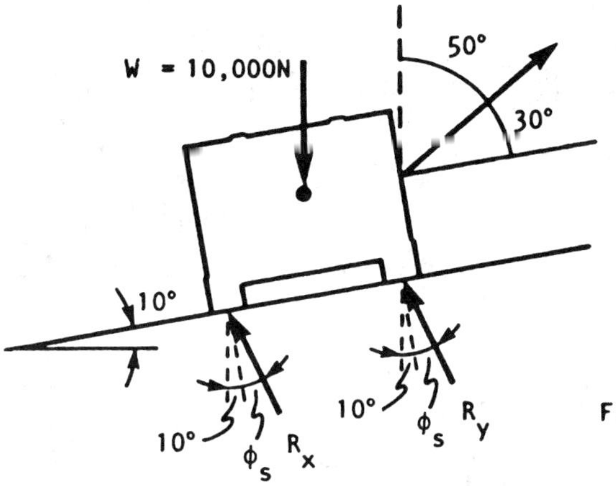

Fig. 2

Noting that R_x and R_y have the same direction (same angle of friction ϕ_s), draw a force triangle including the weight W, the force F and the sum $R = R_x + R_y$. The triangle is closed since there is no acceleration. The direction of R will be different for each part of this problem. For part (1),

$$\tan\phi_s = \mu_s = 0.30$$

$$\phi_s = 16.7°$$

This angle is measured from the normal, as shown, so the actual direction of R is given by 10° + 16.7° = 26.7°. The force triangle, then, is

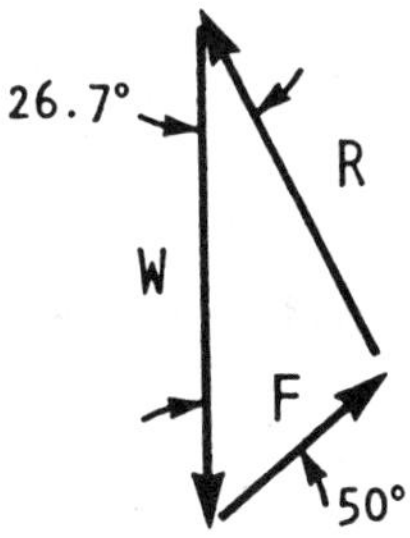

Fig. 3

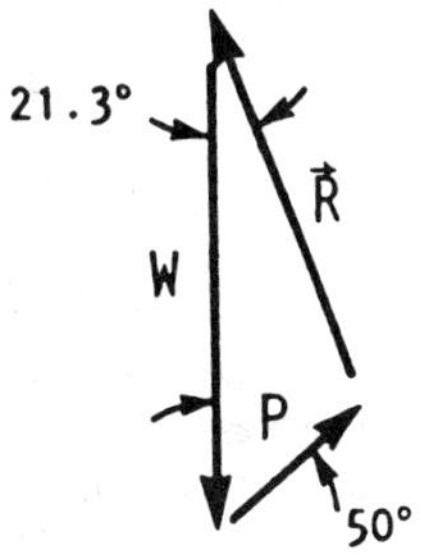

Fig. 4

Using the law of sines,

$$\frac{F}{\sin 26.7°} = \frac{W}{\sin[180° - (50° + 26.7°)]}$$

$$F = .478\text{kN, direction shown}$$

The situation is similar for part (2) except now the kinetic coefficient must be used. Therefore,

$$\tan\phi_k = \mu_k = 0.20$$

$$\phi_k = 11.3° .$$

And the direction of $\vec{R}$ is given by 10° + 11.3° = 21.3° Consequently,

$$\frac{F}{\sin 21.3°} = \frac{W}{\sin[180° - (50° + 21.3°)]}$$

$$F = .392\text{kN, direction shown.}$$

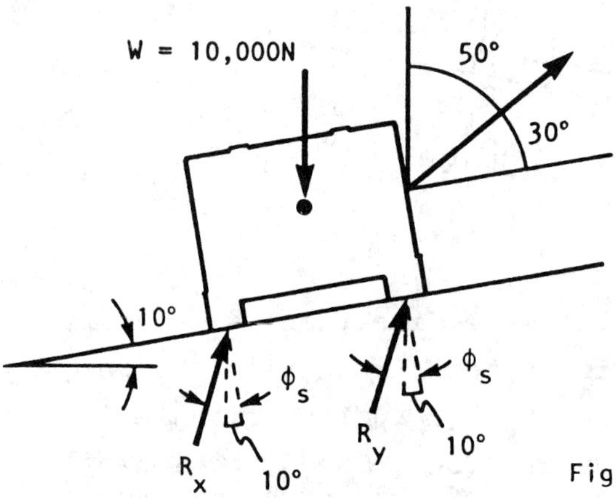

Fig. 5

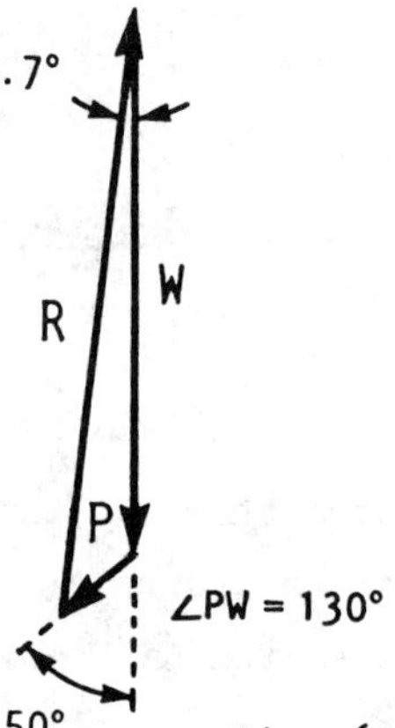

Fig. 6

For part (3), consider the frictional force to be opposite the downward motion of the skid, so that the sense of R_x and R_y has changed considerably (see figure 5). Again, $\phi_s = 16.7°$, but now the resulting direction of $\vec{R}$ is given by 16.7° - 10° = 6.7°. The force triangle for this situation is shown in figure 6.

Using the law of sines,

$$\frac{F}{\sin 6.7°} = \frac{W}{\sin[180° - (130° + 6.7°)]}$$

$$F = 1.701\text{kN.}$$

However, notice in figure 6 that F is directed from the head of the vector $\vec{W}$ to the tail of the vector $\vec{R}$, as was done in figures 3 and 4, and is now pointing downwards. In other words, it would take a push of more than 1.701kN along the direction of $\vec{F}$ to start the skid moving. Thus, the answer to (3) is that no force $\vec{F}$ is needed to keep the skid from sliding down; it will not slide down under its own weight.

• **PROBLEM 4-16**

A 300 lb load is to be raised by a rope passed over a pulley which is freely rotatable about a 4" diameter shaft. Assume equal static and kinetic coefficients of friction between pulley and shaft of 0.25. Find the minimum force F required to (a) raise the load, and (b) to hold the load in place. (c) If F is to be applied horizontally, what is its minimum magnitude to raise the load?

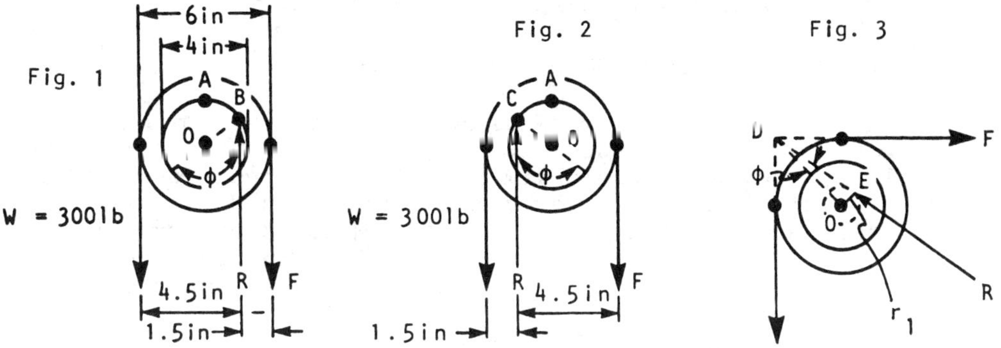

Solution: The point of contact between pulley and shaft shifts from A to B as F is increased beyond 300 lb, because the pulley rolls on the shaft. When motion is impending in figure 1, the angle of friction, ϕ, between reaction R and the radius line OB is given by:

$$\tan\phi = \mu = 0.25$$

But for small angles,

$$\tan\phi \simeq \sin\phi \simeq \mu$$

Therefore, the perpendicular distance, r_f, from the center O to the line of action of R is given by

$$r_f \simeq r\mu$$

where r is the radius of the shaft. Substituting in values for r and μ yields:

$$r_f \simeq (2 \text{ in.})(0.25) = 0.50 \text{ in.}$$

The only external forces on the system are W and F where only F is unknown. The perpendicular distances from point B to forces W and F are known. So by taking the sum of the moments about point B and setting it equal to zero, the unknown force F can be found. It is set equal to zero because the system is in equilibrium.

$$+\circlearrowright \Sigma M_B = (4.50 \text{ in})(300 \text{ lb}) - (1.50 \text{ in.})F = 0$$

and $F = 900$ lb downward.

b) Vertical Force F to Hold the Load.

As the force F is decreased, the pulley rolls around the shaft and contact takes place at C. The perpendicular distance (r_f) between O and the line of action of R will be the same because $r_f \approx r\mu$, and both r and μ remained constant. A free-body diagram of this situation is shown in figure 2. Summation of the moments about point C yields:

$$+\circlearrowright \Sigma M_C = (1.50 \text{ in})(300 \text{ lb}) - (4.50 \text{ in.})F = 0$$

and $$F = 100 \text{ lb downard}$$

c) Horizontal Force F to Raise the Load.

A free-body diagram of this situation is shown in figure 3. Since the three forces, W, F, and R are not parallel, they must be concurrent for the conditions of equilibrium to exist. The direction of R is determined from the fact that its line of action must pass through the point of intersection, D, of W and F, in order for the forces W, F, and R to be concurrent. The perpendicular distance from 0 to the force R is r_f and is equal to 0.25 in, as before. So

$$OE = r_f = 0.25 \text{ in.}$$

OD is the hypotenuse of a right isosceles triangle with legs equal to 3 in.

$$OD = \sqrt{3^2 + 3^2} = 4\text{-}1/4 \text{ in.}$$

The angle, θ, can be found through the relationship

$$\sin\theta = \frac{OE}{OD} = \frac{0.25 \text{ in.}}{4.25 \text{ in.}} = 0.0588$$

$$\theta = \arcsin(0.0588) = 3.4°$$

A force triangle can now be drawn. It is shown in figure 4.

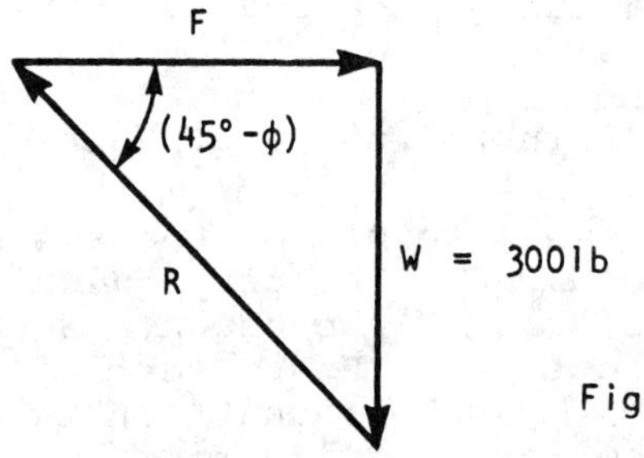

Fig. 4

Using the force triangle, we obtain

$$F = W \cot(45°-\theta)$$

$$F = (300 \text{ lb}) \cot(41.6°)$$

$$F = 338 \text{ lb to the right.}$$

CHAPTER 5

RIGID BODIES

CENTER OF MASS AND CENTER OF GRAVITY

• PROBLEM 5-1

A deep water acoustic transducer receives and transmits underwater sound vibrations. A typical transducer sketched in the figure weighs 25 lb and has its center of weight at $\vec{R} = (0.6\hat{i} + 3\hat{k})$ in. If the base section weighs 10 lb and has its center of weight at $\vec{R} = (0.1\hat{j} + 0.5\hat{k})$ in., where is the center of weight of the top section?

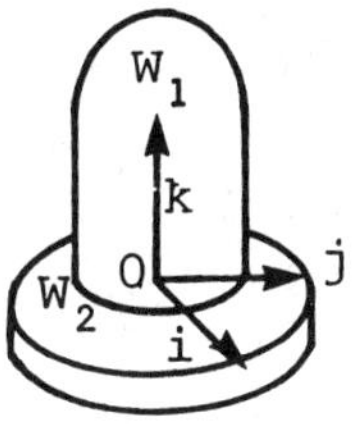

Solution: The equation for the center of mass of the whole body is

$$\vec{R} = \frac{W_1\vec{R}_1 + W_2\vec{R}_2}{W_1 + W_2}$$

All values are given in the question except $\vec{R}_1$. Substituting,

$$0.6\hat{i} + 3\hat{k} = \frac{15\vec{R}_1 + 10(0.1\hat{j} + 0.5\hat{k})}{25}$$

Solving this equation for $\vec{R}_1$, the center of weight of the top section, we find

$$\vec{R}_1 = (\hat{i} - 0.067\hat{j} + 4.67\hat{k}) \text{ in.}$$

• **PROBLEM 5-2**

Find the centroid of a right circular cone formed by rotating the line formed by $2x + y = 4$ between the x and y axes about the y axis.

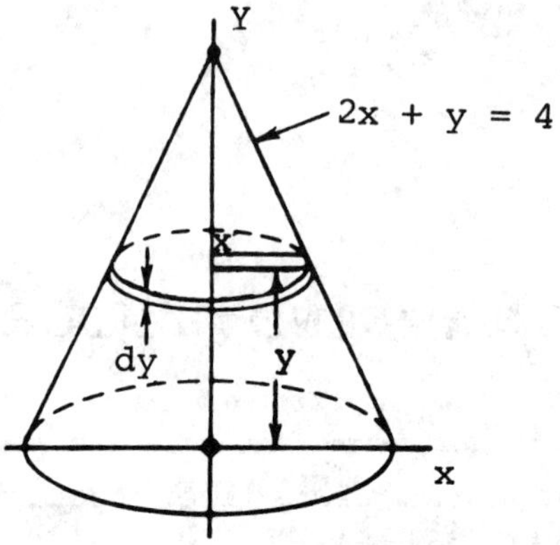

Solution: The centroid is on the y axis. Hence $\bar{x} = \bar{z} = 0$. The area of an element $= \pi r^2 = \pi x^2 = \pi\left(2 - \frac{y}{2}\right)^2$.

$$\bar{y} = \frac{\int_a^b \pi y \left(x_2 - x_1\right)^2 dy}{\int_a^b \pi\left(x_2 - x_1\right)^2 dy} .$$

$x_1 = 0$ (the axis). $x_2 = 2 - \frac{y}{2}$.

$$\bar{y} = \frac{\int_0^4 y\pi\left(2 - y/2\right)^2 dy}{\int_0^4 \pi\left(2 - y/2\right)^2 dy} = \frac{\int_0^4 \left(4y - 2y^2 + \frac{y^3}{4}\right)dy}{\int_0^4 \left(4 - 2y + \frac{y^2}{4}\right)dy}$$

Since π is a multiplying factor in both the numerator and denominator, it may be canceled.

$$\bar{y} = \frac{4y^2/2 - 2y^3/3 + y^4/16]_0^4}{4y - y^2 + y^3/12]_0^4} = 1 .$$

An analysis of this result indicates that the centroid of a right circular cone is 1/4 the distance from its base, since the height of the cone is 4.

• PROBLEM 5-3

Find the center of mass of a uniform plane sector of angle θ, of a circle of radius R.

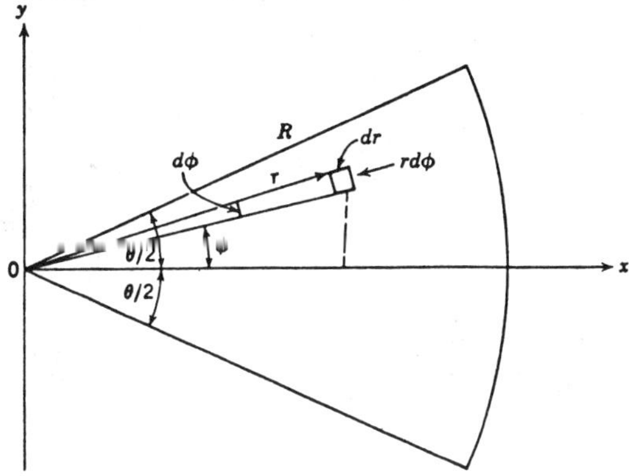

Solution: Let the x axis be chosen to pass through the axis of symmetry of the section as shown in the figure. Then the y coordinate of CM is $\bar{y} = 0$, and the problem is reduced to finding $\bar{x}$, the x coordinate of the CM.

Consider an elementary area of the sector at a distance r from the center O and having a width r dφ and length dr. If the pass per unit area is σ, then the mass of the element of area is σr dφ dr. The x distance of this element of area from the y-axis is r cos φ where φ is the angle which r makes with the x-axis. From the figure, it is seen that the limits of integration on r are O to R and on φ are − θ/2 to + θ/2. Hence,

$$\bar{x} = \frac{\int_{-\theta/2}^{\theta/2} \int_0^R r^2 \cos\phi \, dr \, d\phi \, \sigma}{\int_0^R \int_{-\theta/2}^{\theta/2} r \, dr \, d\phi \, \sigma}$$

$$= \frac{\frac{R^3}{3}\left[\sin\left(\frac{\theta}{2}\right) - \sin\left(\frac{-\theta}{2}\right)\right]}{\frac{R^2}{2}\left[\frac{\theta}{2} + \left(\frac{\theta}{2}\right)\right]}$$

$$= \frac{4R^3 \sin(\theta/2)}{3R^2\,\theta} = \frac{4R}{3\theta}\sin(\theta/2)$$

It can be readily appreciated that for bodies having some degree of symmetry, one must use judgment as to how to place the axes relative to an axis or plane of symmetry. With a little care, considerable time and effort may be saved.

(a) Locate the centroid of the T-section shown in Fig. 1.

(b) Also locate the centroid of the volume of the cone and hemisphere shown in Fig. 2, the values of r and h being 6 in. and 18 in., respectively.

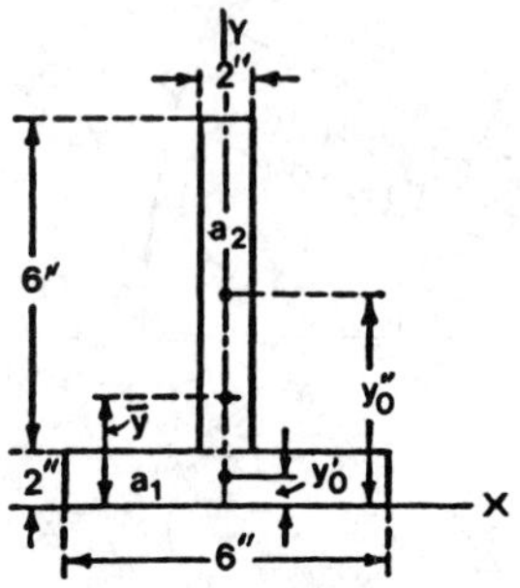

Fig. 1

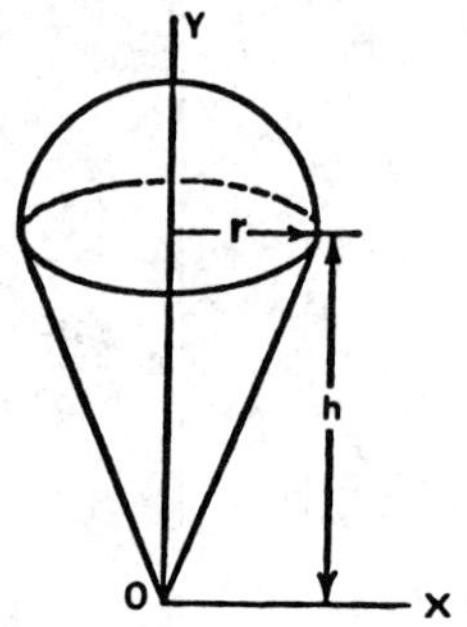

Fig. 2

Solution: (a) If axes be selected as indicated it is evident from symmetry that x = 0. By dividing the given area into areas a_1 and a_2 and by taking moments about the bottom edge of the area, $\overline{y}$ may be found as follows:

$$A\overline{y} = \Sigma(ay_0),$$

$$\overline{y} = \frac{12 \times 1 + 12 \times 5}{6 \times 2 + 6 \times 2} = 3 \text{ in.}$$

(b) The axis of symmetry will be taken as the y-axis. From symmetry then $\overline{x} = 0$. By taking the x-axis through the apex of the cone as shown, the equation $V\overline{y} = \Sigma(vy_0)$ becomes

$$\frac{1}{3}\pi r^2 h + \frac{2}{3}\pi r^3)\overline{y} = \frac{1}{3}\pi r^2 h \times \frac{3}{4} h + \frac{2}{3}\pi r^3 (h + \frac{3}{8} r) .$$

That is,

$$\frac{1}{3}\pi r^2 (h + 2r)\overline{y} = \frac{1}{3}\pi r^2 \left(\frac{3}{4} h^2 + 2rh + \frac{3}{4} r^2\right).$$

Therefore,

$$\overline{y} = \frac{\frac{3}{4} h^2 + 2rh + \frac{3}{4} r^2}{h + 2r}$$

$$= \frac{\frac{3}{4} \times (18)^2 + 2 \times 6 \times 18 + \frac{3}{4} \times (6)^2}{18 + 2 \times 6} = 16.2 \text{ in.}$$

• PROBLEM 5-5

Find the center of mass of a homogeneous right circular solid cone of vertical height h and base radius a.

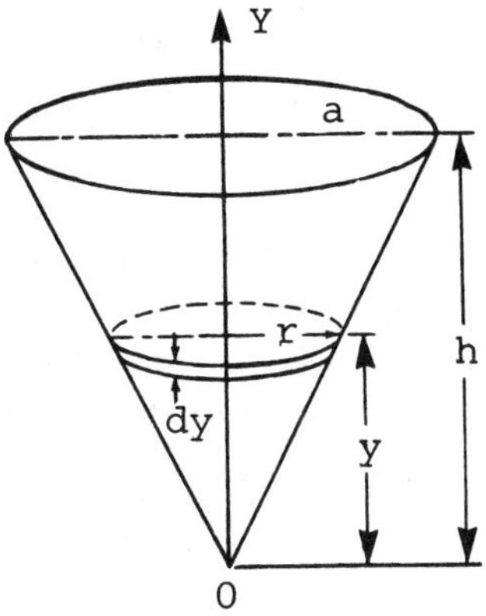

Solution: A well-made choice of the coordinate system exploiting the symmetry of the problem often reduces the complexity of the calculation of the center of mass. For the cone, exploiting this symmetry means that the y-axis can be placed immediately along the cone's center passing through the center of the base and the apex. Further simplification is obtained by placing the origin at the apex as shown in the figure.

Consider a horizontal thin slab of the cone whose thickness is dy, whose mean radius is r, and whose height above the origin is y. If the density or mass per unit volume of the cone is ρ, then the mass dm of the thin slab is

$$dm = \pi r^2 \rho dy$$

The radius r of the slab is proportional to its height y, and from the figure it can be seen that

$$\frac{r}{a} = \frac{y}{h} \qquad \text{or} \qquad r = \frac{ay}{h}$$

Thus,
$$dm = \frac{\pi \rho a^2 y^2}{h^2} dy$$

and the total mass M of the cone is

$$M = \frac{\pi \rho a^2}{h^2} \int_0^h y^2 dy = \frac{\pi \rho a^2 h}{3}$$

The vertical distance of the center of mass from O is

$$\bar{y} = \int_0^h \frac{y\, dm}{M} = \frac{\pi \rho a^2}{Mh^2} \int_0^h y^3 dy$$

Integrating and substituting the value for M, gives

$$\bar{y} = \frac{3\pi a^2h^4}{4h^2\pi\rho a^2h} = \frac{3h}{4}$$

or, the center of mass is three-fourths of the altitude measured from the apex.

AREAS AND VOLUMES

• PROBLEM 5-6

Determine by means of integration the area of the triangular region bounded by the lines y = 0, y = 3x, x = 2.

Solution: If we consider the element of area dA= dy dx shown in the figure, then

$$A = \int_{x=0}^{x=2}\int_{y=0}^{y=3x} dy\, dx$$

$$= \int_0^2 [y]_0^{3x} dx$$

$$= \int_0^2 3x\, dx$$

$$= \left[\frac{3x^2}{2}\right]_0^2$$

$$= 6.$$

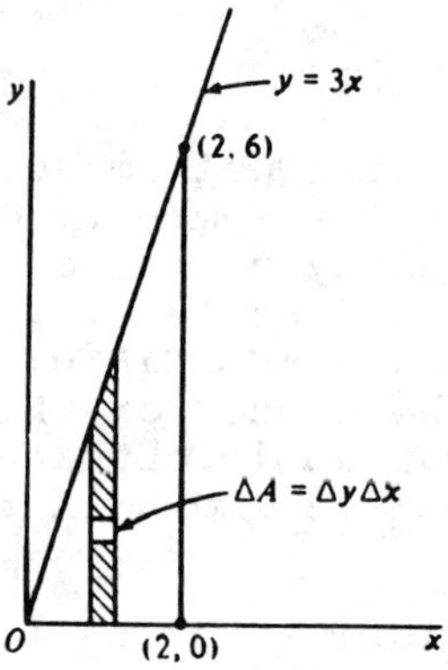

figure

• PROBLEM 5-7

Determine the area of the circular sector bounded by the lines x = 0, y = 0, and $x^2 + y^2 = a^2$.

Solution: Start with the area element dA = ρ dθ dρ. Let's integrate first with respect to ρ and then with respect to θ. We find

$$A = \int_{\theta=0}^{\pi/2}\int_{\rho=0}^{a} \rho\, d\rho\, d\theta$$

$$= \int_0^{\pi/2} \left[\frac{\rho^2}{2}\right]_0^a d\theta$$

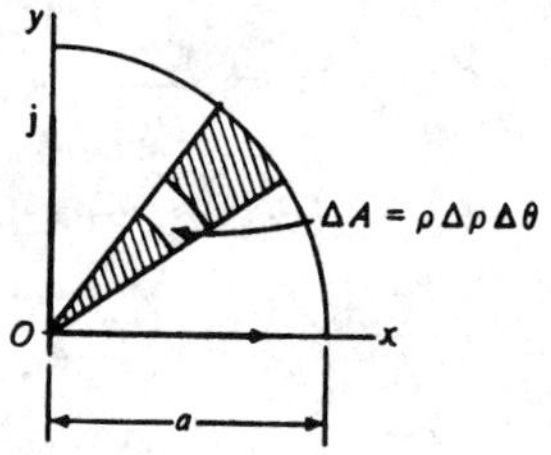

1

$$= \int_0^{\pi/2} \frac{1}{2} a^2 \, d\theta$$

$$= \frac{1}{4} \pi a^2$$

Note that the first integration yields a pie-shaped element (Figure 1). If we first integrate with respect to θ, we obtain a circular element as indicated in Figure 2. However, we can easily verify that the value obtained for the area is the same when the order of integration is reversed. Thus,

$$A = \int_{\rho=0}^{a} \int_{\theta=0}^{\pi/2} d\theta \, \rho \, d\rho$$

$$= \frac{\pi}{2} \int_0^a \rho \, d\rho$$

$$= \frac{1}{4} \pi a^2 .$$

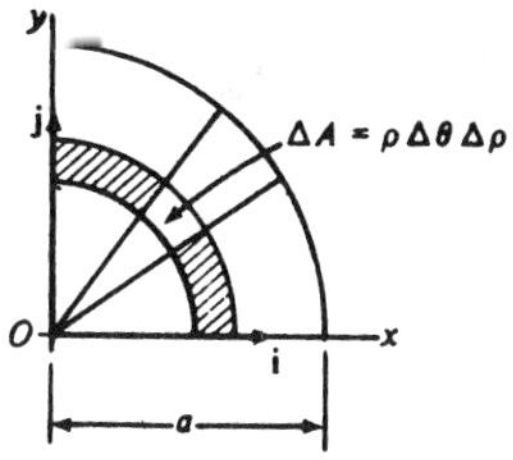

2

• **PROBLEM 5-8**

> Determine the volume of a cone $z^2 = x^2 + y^2$ between $z = 0$ and $z = h$.

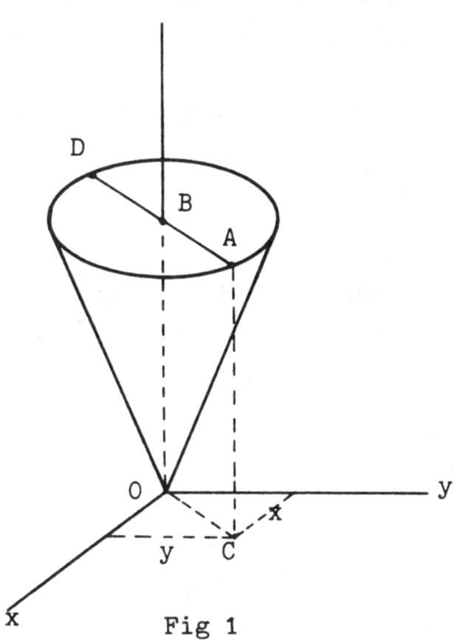

Fig 1

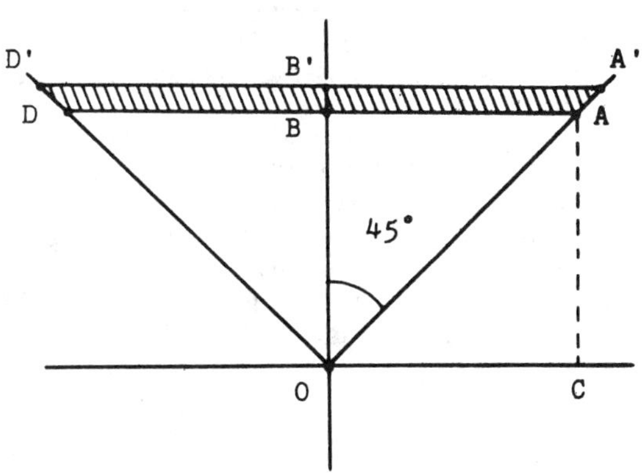

Fig 2

Solution: The cone is sketched in figure 1 for arbitrary height z.

A is an arbitrary point on the surface of the cone with rectangular coodinates x, y, z. Let

$R = \overline{AB} = \overline{OC} = \sqrt{x^2 + y^2}.$

It follows from the equation $z^2 = x^2 + y^2$, that

$$R = z \ (= \overline{OB}). \tag{1}$$

and $\angle BOA = 45°$. In order to find the volume of the cone, imagine that the cone is sliced into many circular parallel disks. A typical disk of this kind is indicated by A B D A' B' D" in figure 2, where dz = BB' is its thickness. The volume of this typical disk is

$$dV = \pi R^2 dz.$$

The volume of the cone from z = 0 to z = h is, therefore, the integral,

$$V = \int_0^h \pi R^2 dz. \tag{2}$$

To evaluate the integral, we need to know R as a function of z. In this case of a 45°-cone, this function is given by eq. (1). We have then,

$$V = \int_0^h \pi z^2 \, dz = \frac{\pi}{3} z^3 \Bigg|_{z=0}^{z=h} = \frac{\pi}{3} h^3 .$$

• PROBLEM 5-9

Determine the volume of the tetrahedon bounded by the planes x = 0, y = 0, z = 0, and x + y + z = 1.

Solution: Consider an element of volume dV = dx dy dz. If we first integrate with respect to z from z = 0 to z = 1 - x - y, we obtain a vertical element. By making the second integration with respect to x from x = 0 to x = 1 - y, we obtain a thin slice across the tetrahedron. Then by integrating with respect to y, the total volume can be obtained. Thus,

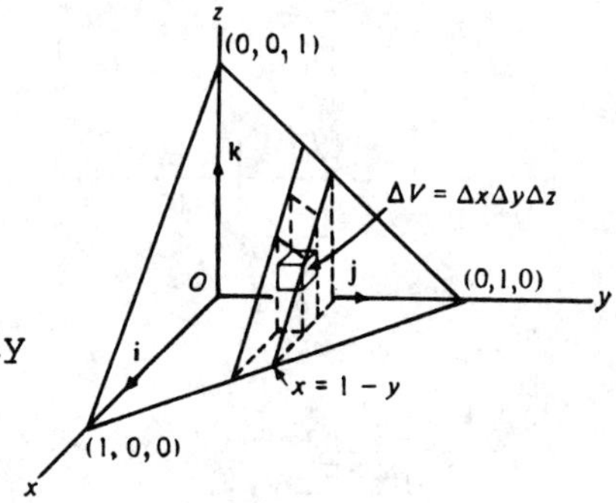

$$V = \int_{y=0}^{1} \int_{x=0}^{1-y} \int_{z=0}^{1-x-y} dz\, dx\, dy$$

$$= \int_{y=0}^{1} \int_{x=0}^{1-y} (1 - x - y)\, dx\, dy$$

$$= \int_0^1 [(1 - y^2) - \frac{1}{2} (1 - y)^2]\, dy$$

$$= \int_0^1 \frac{1}{2}(1 - y)^2\, dy$$

$$= \frac{1}{6} .$$

CENTROIDS OF AREAS AND VOLUMES

• PROBLEM 5-10

Determine the centroid of area of the plane area below if $A_1 = \frac{1}{3}$ sq. ft and $\vec{R}_1 = \left(\frac{3}{5}\hat{i} + \frac{3}{16}\hat{j}\right)$ ft.

Solution: The centroid location is given by the following equations:

$$R_x = \frac{R_{x_1} A_1 + R_{x_2} A_2}{A_1 + A_2}$$

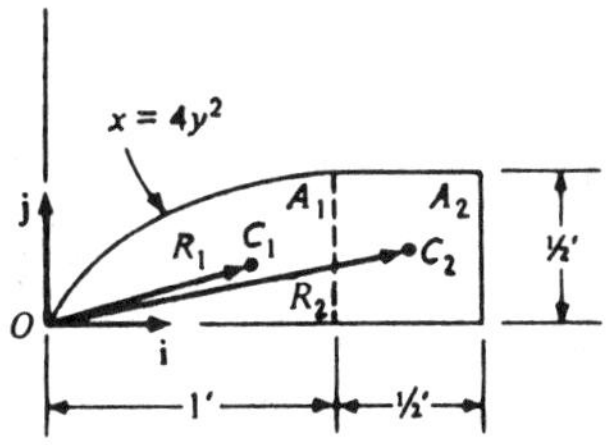

$$R_y = \frac{R_{y_1} A_1 + R_{y_2} A_2}{A_1 + A_2}$$

$$R_z = 0.$$

We know $A_1 = \frac{1}{3}$ sq. ft., $R_{x_1} = \frac{3}{5}$ ft., and $R_{y_1} = \frac{3}{16}$ ft.

The other values needed are easily calculated from the information in the figure. The results are tabulated here:

i	A_i	R_{xi}	A_iR_{xi}	R_{yi}	A_iR_{yi}
1	$\frac{1}{3}$	$\frac{3}{5}$	$\frac{1}{5}$	$\frac{3}{16}$	$\frac{1}{16}$
2	$\frac{1}{4}$	$\frac{5}{4}$	$\frac{5}{16}$	$\frac{1}{4}$	$\frac{1}{16}$
Sum	$\frac{7}{12}$		$\frac{41}{80}$		$\frac{1}{8}$

The tabulated values when substituted into the equations yield

$$R_x = \frac{\frac{4}{8}\ \frac{1}{6}}{\frac{7}{12}} = \frac{123}{140}$$

$$R_y = \frac{\frac{1}{8}}{\frac{7}{12}} = \frac{3}{14}$$

$$\vec{R} = \left(\frac{123}{140}\,\hat{i} + \frac{3}{14}\,\hat{j}\right)\ \text{ft.}$$

• PROBLEM 5-11

> Find the center of gravity of the area included by a semicircle and a diameter, and also of the semicircular arc. Use the theorems of Pappus-Guldinus.

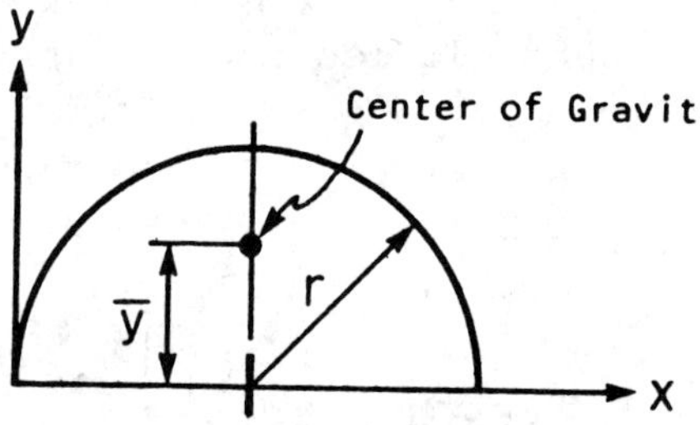

Solution: To find the C. G. of the semicircular area, the area is

$$\frac{\pi r^2}{2},$$

and the volume of the sphere (the revolved shape) is

$$\frac{4\pi r^3}{3}.$$

Applying the theorem,

$$\frac{4\pi r^3}{3} = \frac{\pi r^2}{2}(2\pi\bar{y})$$

from which

$$\bar{y} = \frac{4r}{3\pi}$$

To find the C. G. of the arc, the length of the arc is πr, and the surface area of the revolved shape is $4\pi r^2$. Using the theorem,

$$4\pi r^2 = \pi r(2\,\bar{y});\ \text{and}\ \bar{y} = \frac{2r}{\pi}$$

• PROBLEM 5-12

> Find the center of gravity of a cone of radius r and altitude h. (measure from the vertex.)

Solution: We are looking for a point, such that, if the mass of the cone were concentrated at that point, say $(\bar{x}, \bar{y}, \bar{z})$, the turning effect of this mass with respect to a given point is the same as that obtained in the following expression:

$$m\bar{x} = M_{yz}, \quad m\bar{y} = M_{zx}, \quad m\bar{z} = M_{xy},$$

where M_{yz} is the moment (turning effect) of the cone with respect to the plane formed by the y- and z-axes. The rest of the moments have the same definition with respect to their corresponding planes.

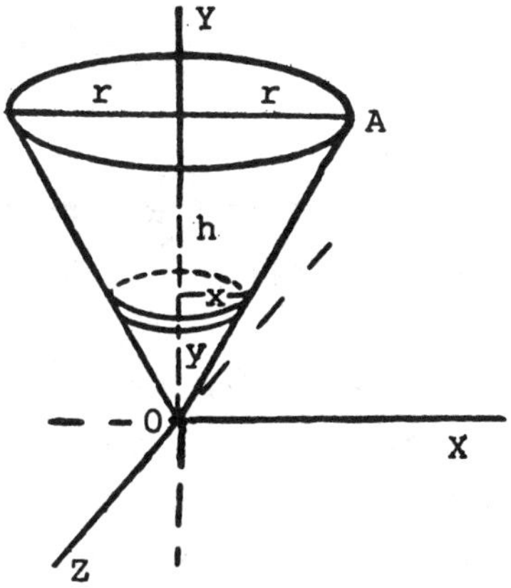

The whole cone can be thought of as made up of n small particles such that $m = \sum_{i=1}^{n} m_i$, and hence,

$$M_{yz} \cong \sum_{i=1}^{n} x_i m_i, \quad M_{zx} \cong \sum_{i=1}^{n} y_i m_i \quad \text{and} \quad M_{xy} \cong \sum_{i=1}^{n} z_i m_i .$$

We can furthermore assume that these particles are so small that the summation becomes integration. We express the masses in terms of their corresponding infinitesimal volumes and densities, ρ. Thus, we obtain:

$$M_{yz} = \int_v \rho x dv, \quad M_{zx} = \int_v \rho y dv \quad \text{and} \quad M_{xy} = \int_v \rho z dv$$

where $\int_v f(x,y,z)dv$ indicates that integration performed over the entire volume.

If the density of each element is the same, we say that the object, in this case, the cone, is homogeneous, and we can take ρ out of the integral sign. Assuming constant density,

$$m\bar{x} = \rho v\bar{x} = \rho\int x dv,$$

or,

$$v\bar{x} = \int x dv,$$

and

$$\bar{x} = \frac{\int x dv}{v} = \frac{\int x dv}{\int dv}.$$

Similarly,

$$\bar{y} = \frac{\int y dv}{\int dv}, \quad \bar{z} = \frac{\int z dv}{\int dv}.$$

As can be seen from the figure, the cone is symmetric about the y-axis. Therefore we can conclude that there are equal distributions of mass. The net moments of these particles on the y-z and x-y planes are zero -- suggesting $(0,\bar{y},0)$ is the point in question. Now we try to express the equation of the cone in mathematical form to find the value of $\bar{y}$.

We can find the equation of the line OA in the diagram in terms of the radius r and height h. The equation of a straight line is: $y = mx + b$, where m is the tangent of the angle made by the line with the positive x-axis and b is the intercept. In this case,

$$y = \frac{hx}{r}.$$

If we assume this line rotating about the y-axis, the generated volume is the desired cone. The small strip of rectangle with width dy drawn from the y-axis horizontally, one of its corners to touch a point (x,y) on the line, generates a disc with approximate volume

$$\Delta v = \pi x^2 \Delta y,$$

since x is its radius and it has a thickness of Δy. If the cone is composed of such discs, we can find the entire volume in integral form:

$$v = \int \pi x^2 dy.$$

Returning to the previous equations, we write:

$$\bar{y} = \frac{\int y dv}{\int dv}.$$

We obtain:

$$\bar{y} = \int yx^2 dy \Big/ \int x^2 dy,$$

and, using the expression: $y = \frac{hx}{r}$, we substitute $\frac{r^2}{h^2} y^2$ for x^2 in the above expression.

$$\bar{y} = \int_0^h y^3 dy \Big/ \int_0^h y^2 dy$$

$$= 3/4\ h.$$

INERTIA AND THE MOMENTS AND PRODUCTS OF INERTIA

• PROBLEM 5-13

Find the moment of inertia of the channel section shown in Fig. 1 with respect to the line XX. Find also the moment of inertia with respect to the parallel centroidal axis.

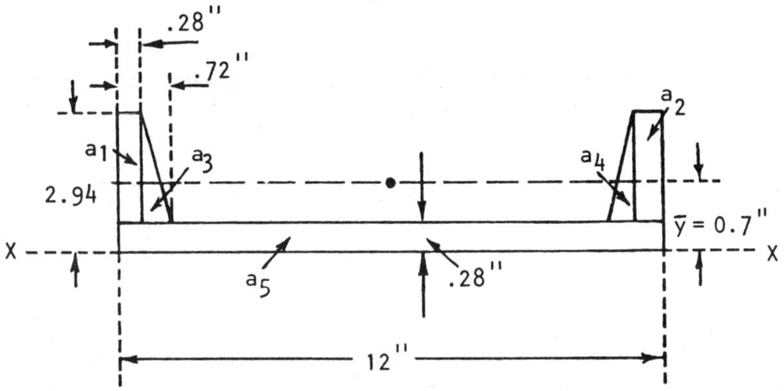

Fig. 1

Solution: The area may be divided into triangles and rectangles as shown in the figure. The values used in the solution may be put in tabular form as shown below, where a denotes the area of any part, y_0 the distance of the centroid of the part from the line XX, I_0 the moment of inertia of the part with respect to its own centroidal axis parallel to XX, and I'_x the moment of inertia of the part with respect to the axis XX.

Part	a	y_0	ay_0	I_0	ay_0^2	$I'_x = I_0 + ay_0^2$
a_1	0.745	1.61	1.20	0.44	1.93	2.37
a_2	0.745	1.61	1.20	0.44	1.93	2.37
a_3	0.585	1.17	0.68	0.23	0.80	1.03
a_4	0.585	1.17	0.68	0.23	0.80	1.03
a_5	3.360	0.14	0.47	0.02	0.07	0.09
	6.02 $in.^2$		4.23 $in.^3$			6.89 $in.^4$

Thus the moment of inertia I_x of the area with respect to the XX axis is

$$I_x = \Sigma I'_x = 6.89 \text{ in.}^4$$

Further, the total area is $A = \Sigma a = 6.02$ in.^{2}, and the moment of the area with respect to the XX axis is $\Sigma(ay_0) =$ 4.23 in.^{3} Hence the distance $\bar{y}$, of the centroid of the area from the XX axis is

$$\bar{y} = \frac{\Sigma(ay_0)}{} = \frac{4.23}{6.02} = 0.70 \text{ in.}$$

Therefore, the moment of inertia with respect to a line through the centroid and parallel to XX is given by the equation

$$I_x = I_x - Ad^2 = 6.89 - 6.02 \times (0.70)^2 = 3.94 \text{ in.}^4$$

• PROBLEM 5-14

Locate the horizontal centroidal axis, XX, of the T-section shown in Fig. 1, and find the moment of inertia of the area with respect to this centroidal axis.

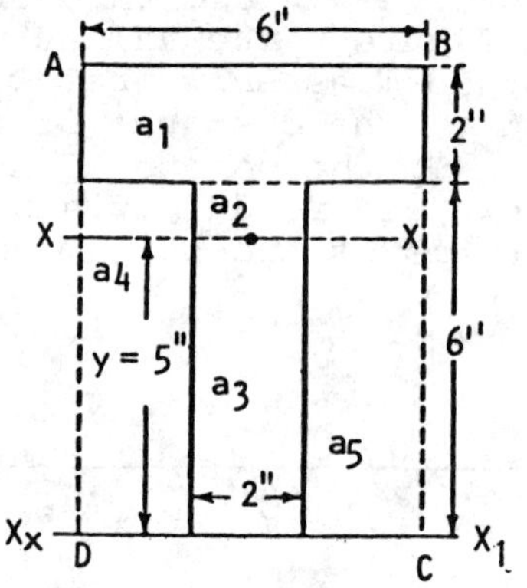

Fig. 1

Solution: First Method.--The distance, $\bar{y}$, of the centroid of the area from the axis X_1X_1 may be found from the equation

$$A\bar{y} = \Sigma(ay_0) .$$

Thus

$$\bar{y} = \frac{12 \times 7 + 12 \times 3}{12 + 12} = 5 \text{ in.}$$

The moment of inertia with respect to the XX axis is the sum of the moments of inertia of the three parts a_1, a_2 and a_3, with respect to that axis. Thus,

$$\bar{I}_x = \frac{1}{12} \times 6 \times (2)^3 + 12 \times (2)^2 + \frac{1}{3} \times 2 \times (1)^3$$

$$+ \frac{1}{3} \times 2 \times (5)^3$$

$$= 4 + 48 + 0.67 + 83.33 = 136 \text{ in.}^4$$

Second Method.--The moment of inertia of the T-section may also be determined as follows: First find the moment of inertia of th T-section with respect to the axis X_1X_1 by subtracting the moments of inertia of the parts a_4 and a_5 from the moment of inertia of the rectangular area ABCD and then find $\bar{I}_x$ for the T-section by use of the parallel-axis theorem. Thus, the moment of inertia, I_{x_1}, of the T-section with respect to the X_1X_1 axis is

$$\bar{I}_{x_1} = \frac{1}{3} \times 6 \times (8)^3 - 2 \times \frac{1}{3} \times 2 \times (6)^3 = 736 \text{ in.}^4,$$

and

$$\bar{I}_{x_1} = I_{x_1} - Ad^2 = 736 - 24 \quad (5)^2 = 136 \text{ in.}^4$$

• PROBLEM 5-15

Determine the moment of inertia about the z axis of the body shown in the figure. Both the rod and the block are homogeneous. The rod weighs 8 lb and the block weighs 16 lb.

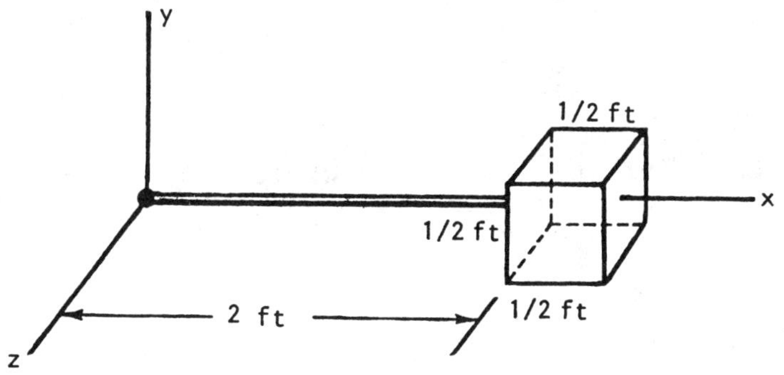

Solution: Both the rod and the block will be considered

separately. The moment of inertia of the rod about the z axis is

$$I_{zz} = \frac{M\ell^2}{3}$$

and in this case, $I_{zz} = \frac{(8/32.2)(2)^2}{3}$

$$= 0.33 \text{ lb-sec}^2\text{-ft}.$$

The moment of inertia of the block about an axis through its centroid is equal to

$$I_{zz_c} = \frac{Ma^2}{6}$$

$$= \frac{(16/32.2)(0.5)^2}{6} = 0.21.$$

Now the parallel axis theorem may be employed to determine the moment of inertia of this block about the z axis:

$$I_{zz} = I_{zz_c} + d^2M$$

or, in this case,

$$I_{zz} = 0.21 + (2.25)^2 \frac{16}{32.2}$$

$$= 2.43 \text{ lb-sec}^2\text{-ft}.$$

The moment of inertia of the whole body about the z axis will be simply the algebraic sum of the moment of inertia of each of the bodies. Thus,

$$I_{zz} = 0.33 + 2.43$$

$$= 2.76 \text{ lb-sec}^2\text{-ft}.$$

It is desired to determine the moment and product of inertia of an artificial earth satellite about each of the centroidal axes. Since the body is not at all homogeneous, a numerical integration must be performed, and this is done by breaking the body up into a number of discrete parts and tabulating the mass and coordinate of each of the parts. This is done in Table 1. For this problem, determine the moment of inertia about the z axis and the product of inertia about the yz axes.

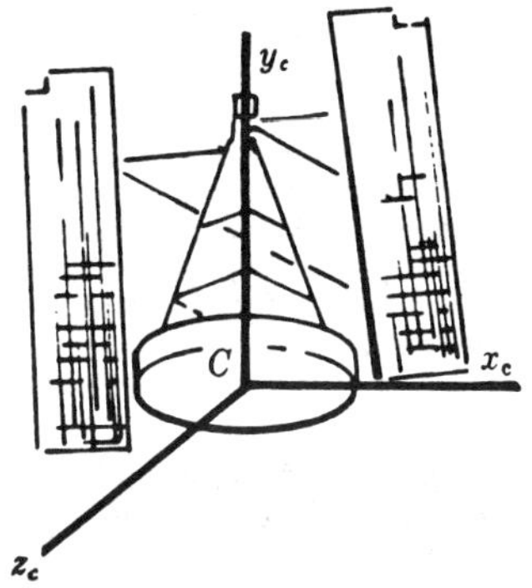

TABLE 1 NUMERICAL INTEGRATION

n	m_i	x_i	y_i	z_i
1	2	2	3	1
2	3	3	4	-2
3	1	2	-1	-2
4	2	1	-3	2
5	1	-3	-2	1
6	3	-2	3	4
7	3	-4	2	-3
8	2	-2	-3	-1

Solution: Often a complicated body must be considered in segments to enable the analysis to proceed. Given in Table I are the coordinates and mass of eight such segments.

The moment of inertia about the z axis is, in summation form,

$$I_{zz} = \sum_i (x_i^2 + y_i^2)m_i$$

and the products of inertia with respect to the yz axes are

$$I_{yz} = \sum_i y_i z_i m_i$$

In order to perform the summation, it might be best to establish Table 2.

TABLE 2.

For I_{zz}					For I_{yz}	
n	x_i^2	y_i^2	$(x_i^2 + y_i^2)$	$m_i(x_i^2 + y_i^2)$	z_iy_i	$m_iz_iy_i$
1	4	9	13	26	3	6
2	9	16	25	75	-8	-24
3	4	1	5	5	2	2
4	1	9	10	20	-6	-12
5	9	4	13	13	2	2
6	4	9	13	39	12	36
7	16	4	20	60	-6	-18
8	4	9	13	26	3	6
				264		-2

In this way, the moment and product of inertia are obtained by simply performing the desired summations. Thus,

$$I_{zz} = 264$$

and $I_{yz} = -2.$

• PROBLEM 5-17

Find the moment of inertia of the given circular area with respect to its axis 0, and also with respect to a diameter.

Solution: To find the moment of inertia with respect to the axis, a polar method is best suitable. The area of the annular element is $2\pi rdr$, and when integrated from the origin to $r = R$,

$$I_0 = \int_0^R (r^2)(2\pi rdr) = \frac{\pi r^4}{2} .$$

The moment of inertia about a diameter is equal to that about the x or y axis. Since

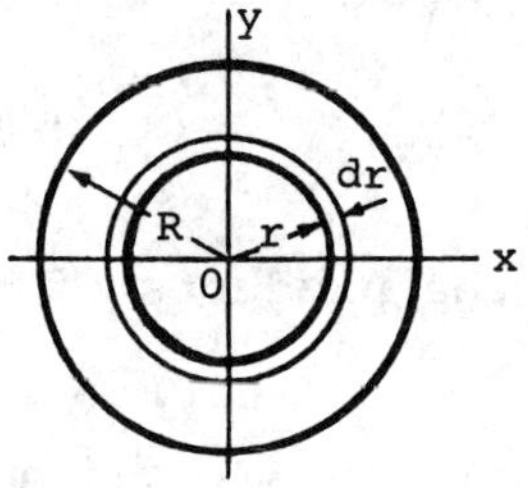

$$I_x + I_y = I_0 , \text{ and } I_x = I_y ,$$

$$2 I_x = I_0 = \frac{\pi r^4}{2} ,$$

from which

$$I_x = \frac{\pi r^4}{4} .$$

• **PROBLEM 5-18**

By using the parallel axis theorem, verify the following moment of inertia: a uniform sphere of mass M and radius r about a tangent to the sphere: $I = \frac{7}{5} Mr^2$; a uniform cylinder of mass M and radius R about a tangent to the cylinder parallel to the cylindrical axis: $I = \frac{3}{2} MR^2$;

a uniform sphere attached to a rigid rod about an axis perpendicular to the rod at point A as shown in the Figure:

$$I = \frac{1}{3} MD^2 + \frac{2}{5} Mr^2 + M(r + D)^2.$$

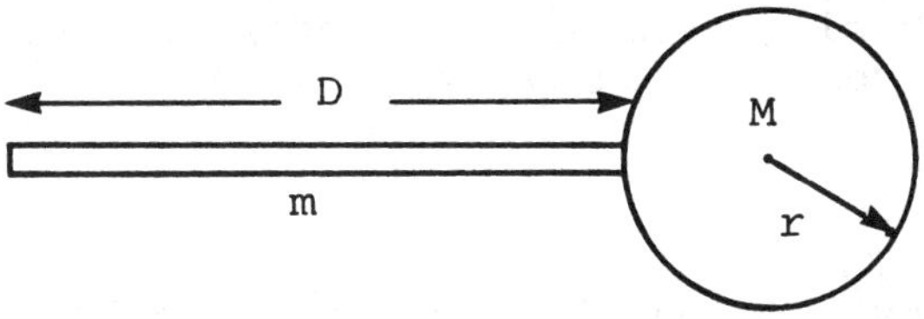

Solution: Parallel axis theorem states that

$$I = I_{cm} + Mh^2,$$

where I is the moment of inertia of a rigid body about an axis, I_{cm} is the moment of inertia about an axis through the center of mass and parallel to the axis of I, h is distance between these two parallel axes, and M is the mass of the body.

(a) A uniform sphere of mass M and radius r. From tables of moments of inertia, $I_{cm} = (2/5) Mr^2$. Because $h = r$, we have

$$I = \frac{2}{5} Mr^2 + Mr^2 = \frac{7}{5} Mr^2.$$

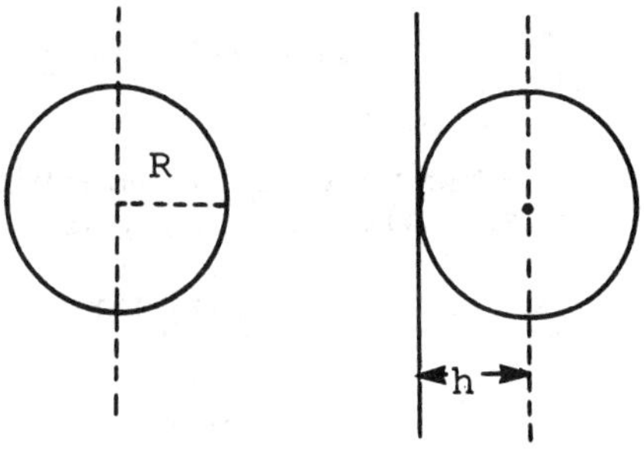

(b) A uniform cylinder of mass M, radius R. From tables of moments of inertia, $I_{cm} = (1/2) MR^2$. Because $h = R$, we have

$$I = \frac{1}{2} MR^2 + MR^2 = \frac{3}{2} MR^2.$$

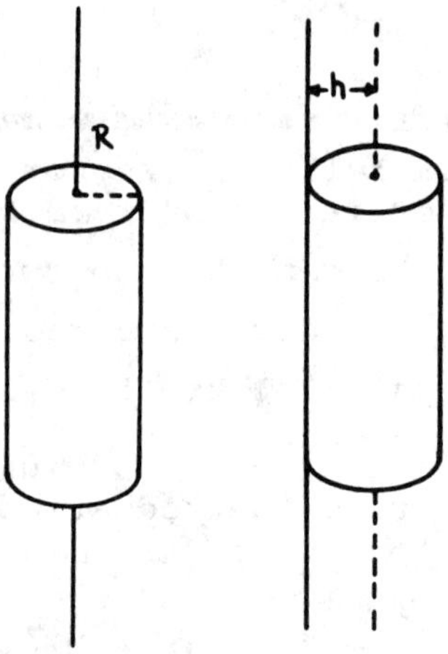

(c) A uniform sphere attached to a thin rod. For a thin rod of mass m and length D, the moment of inertia about an axis passing through the center and perpendicular to the rod is given by (see tables)

$$I_{cm}^{(rod)} = mD^2/12.$$

The moment of inertia about an axis passing through one end and perpendicular to the rod is, according to parallel axis theorem,

$$I^{(rod)} = \frac{mD^2}{12} + m\left(\frac{D}{2}\right)^2 - \frac{1}{3}\,mD^2. \qquad (1)$$

For the attached sphere, we have

$$I_{cm}^{(sphere)} = (2/5)\ Mr^2$$

$$h = r + D$$

and, therefore,

$$I^{(sphere)} = \frac{2}{5}\,Mr^2 + M\,(r + D)^2. \qquad (2)$$

The moment of inertia of the whole system is simply the sum of Eqs. (1) and (2),

$$I = I^{(rod)} + I^{(sphere)}$$

$$= \frac{1}{3}\,mD^2 + \frac{2}{5}\,Mr^2 + M\,(r + D)^2.$$

• PROBLEM 5-19

A solid cylinder of radius R rolls on a flat surface. Find the moment of inertia I_s of the cylinder about its line of contact with the surface.

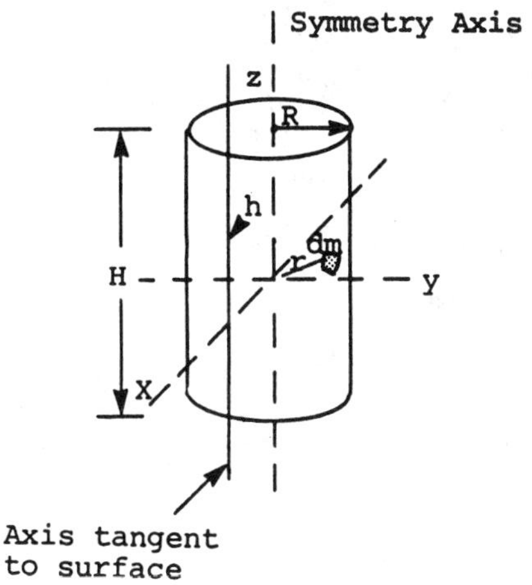

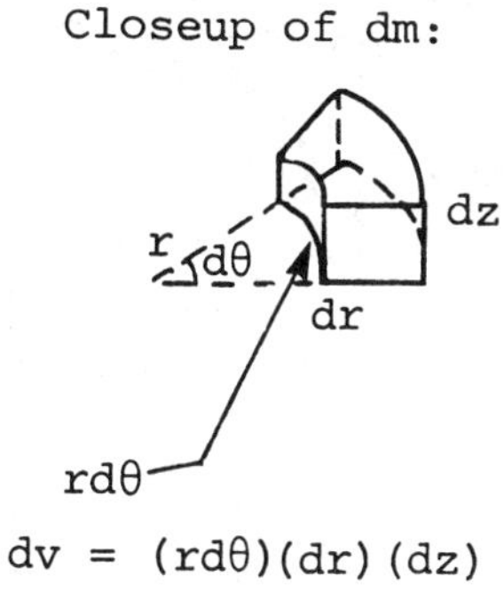

Solution: The definition of moment of inertia for a continuous mass distribution is dependent upon which axis we wish to calculate the moment of inertia about. In this case, we want to calculate I_s, the moment of inertia about an axis parallel to the symmetry axis of the cylinder, but tangent to the surface of the cylinder. To simplify the required integrations, we may equivalently calculate the moment of inertia, I_o, of the cylinder about its symmetry axis, and then employ the parallel axis theorem to find I_s.

The moment of inertia I_o is (with reference to the figure) defined as:

$$I_o = \int r^2\, dm \tag{1}$$

where r is the perpendicular distance between the symmetry axis of the cylinder and the mass element dm. Also note that dm, the mass contained in a differential volume dv, is the mass per volume contained in the cylinder times the volume dv, or

$$dm = \left(\frac{M}{\pi R^2 H}\right) r\, dr\, d\theta\, dz \tag{2}$$

where M is the total mass of the cylinder, and $\pi R^2 H$ is the volume of the cylinder. Combining (2) and (1), we obtain

$$I_o = \int \left(\frac{M}{\pi R^2 H}\right) r^3\, dr\, d\theta\, dz \tag{3}$$

where the integral is over the volume of the cylinder. Performing the integral:

$$I_o = \left(\frac{M}{\pi R^2 H}\right) \int_{-\frac{H}{2}}^{\frac{H}{2}} \int_0^{2\pi} \int_0^R r^3\, dr\, d\theta\, dz$$

$$I_o = \left(\frac{M}{\pi R^2 H}\right)\left(\frac{R^4}{4}\right)\int_{-\frac{H}{2}}^{\frac{H}{2}}\int_0^{2\pi} d\theta\, dz$$

$$I_o = \left(\frac{M}{\pi R^2 H}\right)\left(\frac{R^4}{4}\right)(2\pi)\int_{-\frac{H}{2}}^{\frac{H}{2}} dz$$

$$I_o = \left(\frac{M}{\pi R^2 H}\right)\left(\frac{R^4}{4}\right)(2\pi)(H)$$

$$I_o = \frac{1}{2} M R^2 \qquad (4)$$

This is the moment of inertia about the symmetry axis. To find the moment of inertia about the axis tangent to the surface of the cylinder, use the parallel axis theorem.

$$I_Q = I_G + M\Delta s^2$$

where I_Q is any axis parallel to the centroidal axis G, M is the total mass and Δs is the distance between the P and Q axes. In this problem, $I_Q = I_s$ $I_G = I_0$ and $\Delta s = R$.

Therefore, $I_s = I_0 + MR^2$

$$I_s = \tfrac{1}{2}MR^2 + MR^2$$

$$I_s = \frac{3}{2} MR^2 .$$

• PROBLEM 5-20

Find the product of inertia of a right triangle with respect to both of its sides, and also with respect to axes passing through the centroid of the triangle and being parallel to the sides of the triangle.

Solution: This problem lends itself to the use of the parallel-axis theorem. For this purpose, locate the sides of the triangle along x-y coordinate axes and apply an x'-y' set of axes through the center of gravity of the differential area. Assuming for the differential area ydx, the product of inertia

$$dP_{xy} = dP_{x'y'} + \bar{x}_0\ \bar{y}_0\ ydx\ .$$

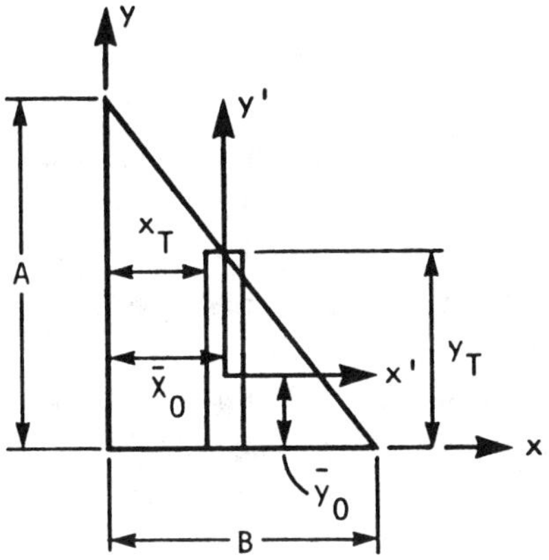

From the geometry of the triangle,

$$Y = A\left(1 - \frac{x}{B}\right) ,$$

and $dP_{x'y'} = 0$ because the differential area is symmetrical with respect to the x'y' axes.

Since $\bar{x}_0 = x$ and $\bar{y}_0 = \frac{1}{2}y$, integrating dP_{xy} from x = 0 to x = B

$$P_{xy} = \int_0^B \bar{x}_0\ \bar{y}_0\ (ydx)$$

$$P_{xy} = \int_0^B \frac{A^2x}{2}\left[1 - \frac{x}{B}\right]^2 dx = \frac{A^2B^2}{24}$$

To find the product of inertia P_c through the centroid of the triangle,

$$P_{xy} = P_c + \bar{x}\ \bar{y}\ A$$

from the parallel-axis theorem. P_{xy} has been found as

$$\frac{A^2B^2}{24} ,$$

and the coordinates of the centroid are known as

$$\bar{x} = \frac{B}{3}\ ;\ \bar{y} = \frac{A}{3}\ .$$

Consequently,

$$\frac{A^2B^2}{24} = P_c + \left[\frac{B}{3}\right]\left[\frac{A}{3}\right]\frac{AB}{2}$$

from which

$$P_c = \frac{-A^2B^2}{72}$$

• PROBLEM 5-21

For the rectangular parallelepiped shown in the figure, determine I_{xx} and I_{yz}. The xyz-system has its origin at the center of mass. Employing the parallel-axis theorem, determine $I_{x_1x_1}$ and $I_{y_1z_1}$. The $x_1y_1z_1$-coordinate system is located along the edges of the body as shown.

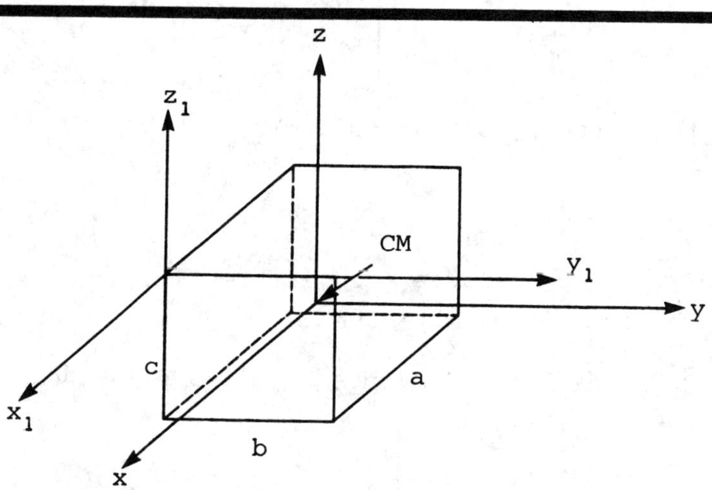

Solution: From the definition of the moment of inertia,

$$I_{xx} = \int_V \rho(y^2 + z^2)dV = \int_V \rho y^2 \, dV + \int_V \rho z^2 \, dV$$

Substituting dV = ac(dy) and dV = ab(dz), respectively, in the last two integrals of the above equation, we have

$$I_{xx} = \int_{-b/2}^{b/2} \rho acy^2 \, dy + \int_{-c/2}^{c/2} \rho abz^2 \, dz = \frac{m}{12}(b^2 + c^2)$$

since the total mass $m = \rho abc$.

The product of inertia I_{yz} is defined as

$$I_{yz} = -\int_V \rho yz \, dV = -\int_{-c/2}^{c/2}\int_{-b/2}^{b/2}\int_{-a/2}^{a/2} \rho yz \, dx \, dy \, dz$$

$$= - \int_{-c/2}^{c/2} \int_{-b/2}^{b/2} \rho xyz \; dy \; dz \Bigg]_{-a/2}^{a/2}$$

$$= - \int_{-c/2}^{c/2} \rho az \; dz \; \frac{y^2}{2} \Bigg]_{-b/2}^{b/2} = 0$$

This is the expected result since yz is a plane of symmetry. Employing the parallel-axis theorem,

$$I_{x_1x_1} = I_{xx} + md_x^2 \qquad \text{where } d_x = \left[\left(\frac{b}{2}\right)^2 + \left(\frac{c}{2}\right)^2\right]^{\frac{1}{2}}$$

Therefore, $I_{x_1x_1} = \frac{m}{12}(b^2 + c^2) + m\left(\frac{b^2}{4} + \frac{c^2}{4}\right) = \frac{m}{3}(b^2 + c^2)$

Similarly, $I_{y_1z_1} = I_{yz} + my_{CM}z_{CM} = 0 + m\left(\frac{b}{2}\right)\left(\frac{c}{2}\right) = \frac{mbc}{4}$.

• PROBLEM 5-22

Show that the moment of inertia of a body about any axis is equal to the moment of inertia about a parallel axis through the center of mass plus the product of the mass of the body and the square of the distance between the axes. This is called the parallel-axes theorem.

Prove also that the moment of inertia of a thin plate about an axis at right angles to its plane is equal to the sum of the moments of inertia about two mutually perpendicular axes concurrent with the first and lying in the plane of the thin plate. This is called the perpendicular-axes theorem.

FIGURE A

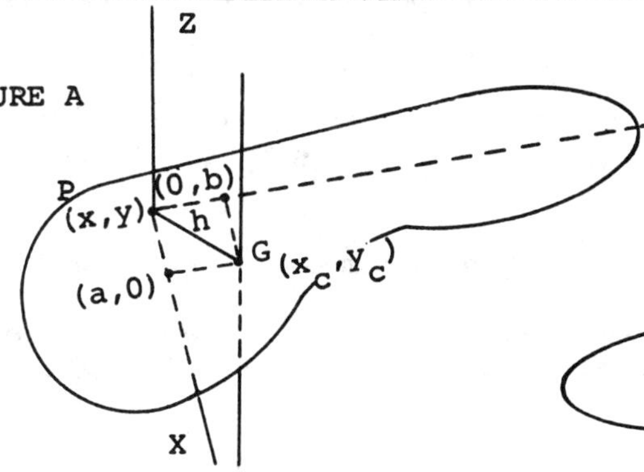

FIGURE B

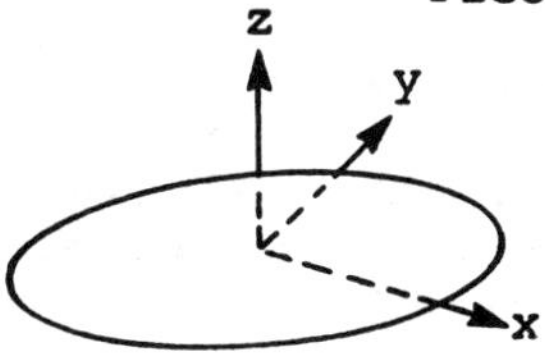

Solution: Let I be the moment of inertia of the body about an arbitrary axis and I_G the moment of inertia about the parallel axis through the center of mass G, the two axes being distance h apart. (See fig. (A)).

By definition of the center of mass of a body relative to an arbitrary axis through a point P, we obtain

$$I = \sum_i m_i r_i^2 = \sum_i m_i \left(x_i^{\,2} + y_i^{\,2} \right)$$

$$= \sum_i m_i x_i^{\,2} + \sum_i m_i y_i^{\,2}$$

where the sum is carried out over all mass particles m_i of the body, and $r_i^{\,2}$ is the distance from P to m_i. Now,

$$x_i = x_i' + a$$

$$y_i = y_i' + b$$

as shown in figure (A). Here, $\left(x_i', y_i' \right)$ locates m_i relative to G, the center of mass. Then

$$I = \sum_i m_i \left(x_i' + a \right)^2 + \sum_i m_i \left(y_i' + b \right)^2$$

$$I = \sum_i m_i \left(x_i'^{\,2} + y_i'^{\,2} \right) + \sum_i m_i \left(a^2 + b^2 \right) + 2a \sum_i m_i x_i' + 2b \sum_i m_i y_i'$$

But $x_i'^{\,2} + y_i'^{\,2} = r_i'^{\,2}$ and $a^2 + b^2 = h^2$, whence

$$I = \sum_i m_i r_i'^{\,2} + \sum_i m_i h^2 + 2a \sum_i m_i x_i' + 2b \sum_i m_i y_i'$$

By definition of the center of mass, however,

$$\sum_i m_i x_i' = \sum_i m_i y_i' = 0, \qquad \text{and}$$

$$I = \sum_i m_i r_i'^{\,2} + \sum_i m_i h^2 = I_G + Mh^2$$

where $M \left(= \sum_i m_i \right)$ is the net mass of the body. This is the parallel-axes theorem. Although we derived this theorem in 2 dimensions, it is equally applicable in three dimensions.

Take, in the case of the thin plate, the axes in the plane of the plate as the x- and y-axes, and the axis at right angles to the plane as the z-axis (see fig. (B)).

Then the moment of inertia of the plate about an axis perpendicular to the plate (the z-axis) is

$$I_z = \sum_i m_i r_i^2$$

where r_i locates m_i relative to O. But

$$r_i^2 = x_i^2 + y_i^2$$

where x_i and y_i are the x and y coordinates of m_i. Then

$$I_z = \sum_i m_i x_i^2 + \sum_i m_i y_i^2$$

But $\sum_i m_i x_i^2 = I_y$ and $\sum_i m_i y_i^2 = I_x$, whence

$$I_z = I_x + I_y$$

This is the perpendicular axes theorem.

• PROBLEM 5-23

Given a box of uniform density and mass m(a)Find the mass moments and products of inertia with respect to the three sides of the box.(b)Find the moment of inertia with respect to a diagonal.

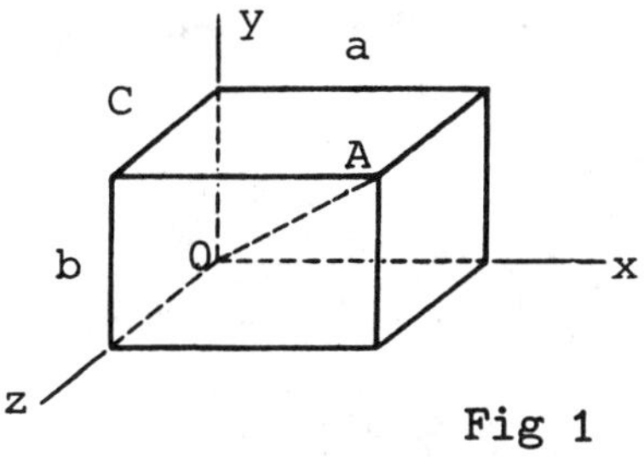

Fig 1

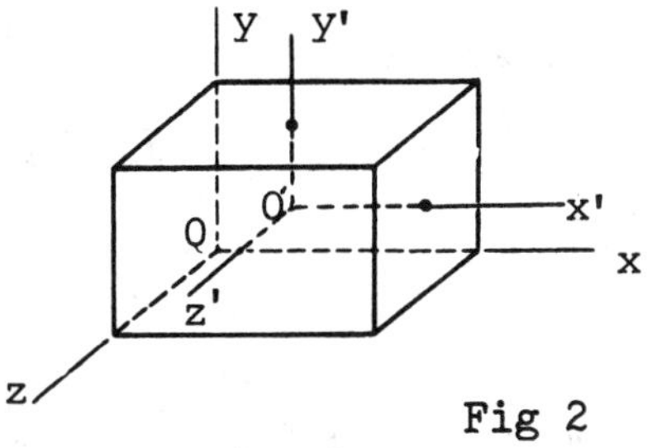

Fig 2

Solution: Locate the box in a coordinate system x-y-z as shown in fig.1, and start by considering the centroidal axes (x', y', z') parallel to the first set of axes and centered at O', the center of the cube. Now, apply the parallel axis theorem to obtain the moments:

$$I_x^0 = I_{x'}^{0'} + m(\bar{y}^2 + \bar{z}^2)$$

$$I_y^0 = I_{y'}^{0'} + m(\bar{x}^2 + \bar{z}^2)$$

$$I_z^0 = I_{z'}^{0'} + m(\bar{x}^2 + \bar{y}^2)$$

where $\bar{x}$, $\bar{y}$ and $\bar{z}$ are the coordinates of O' with respect to O' and the x, y, z axes. These equations yield

$$I_x^0 = \frac{1}{12} m(b^2 + c^2) + m\left(\frac{1}{4} b^2 + \frac{1}{4} c^2\right)$$

$$I_x^0 = \frac{1}{3} m(b^2 + c^2).$$

Similarly, $I_y^0 = \frac{1}{3} m(c^2 + a^2)$, $I_z^0 = \frac{1}{3} m(a^2 + b^2)$.

Due to symmetry, the products of inertia with respect to the centroidal axes x', y', and z' are zero, and these axes are principal axes of inertia. Using the P.A.T.,

$$P_{xy}^0 = P_{x'y'}^{0'} + mxy = 0 + m\left(\tfrac{1}{2}a\right)\left(\tfrac{1}{2}b\right)$$

$$P_{xy}^0 = \tfrac{1}{4}mab.$$

Similarly, $P_{xz}^0 = \frac{1}{4}mac$, $P_{yz}^0 = \frac{1}{4}mbc$.

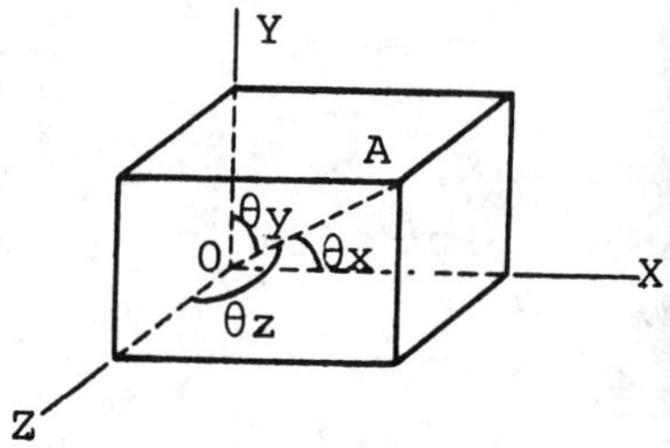

Fig 3

For the moment of inertia of the body through the diagonal OA, make use of the following expression:

$$I_{OA} = I_x \lambda_x^2 + I_y \lambda_y^2 + I_z \lambda_z^2 - 2P_{xy} \lambda_x \lambda_y$$

$$- 2P_{yz} \lambda_y \lambda_z - 2P_{zx} \lambda_z \lambda_x$$

where λ_x, λ_y, λ_z are the direction cosines of the axis OA. Figure 3 shows the angles involved. The cosines may be expressed as

$$\lambda_x = \cos\theta_x = \overline{OH}/\overline{OB} = a/(a^2 + b^2 + c^2)^{\frac{1}{2}}$$

$$\lambda_y = b/(a^2 + b^2 + c^2), \qquad \lambda_z = c/(a^2 + b^2 + c^2).$$

Substituting these values and the moments and products found into the expression for I_{OA} above yields

$$I_{OA} = \frac{1}{a^2 + b^2 + c^2}\left[\frac{1}{3} m(b^2 + c^2)a^2 + \frac{1}{3} m(c^2 + a^2)b^2\right.$$

$$\left. + \frac{1}{3} m(a^2 + b^2)c^2 - \frac{1}{2} m\, a^2 b^2 - \frac{1}{2} m\, b^2 c^2 - \frac{1}{2} m\, c^2 a^2\right],$$

$$I_{OA} = \frac{m}{6} \frac{a^2 b^2 + b^2 c^2 + c^2 a^2}{a^2 + b^2 + c^2}.$$

• PROBLEM 5-24

Evaluate the inertial coefficients for a thin uniform spherical shell of mass density σ per unit area and thickness d, for an axis through the center of the sphere.

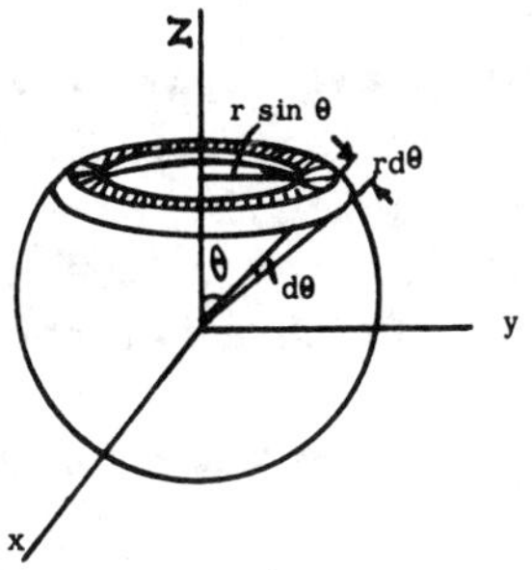

Solution: The angular momentum $\vec{L}$ of a rotating rigid body of mass M is related to its angular **velocity** $\vec{\omega}$ by:

$$\begin{pmatrix} L_x \\ L_y \\ L_z \end{pmatrix} \qquad \begin{pmatrix} I_{xx} & I_{xy} & I_{xz} \\ I_{yx} & I_{yy} & I_{yz} \\ I_{zx} & I_{zy} & I_{zz} \end{pmatrix} \qquad \begin{pmatrix} \omega_x \\ \omega_y \\ \omega_z \end{pmatrix}$$

where 'I's are the inertial coefficients given by

$$I_{xy} = \int dm\, xy$$

$$I_{xx} = \int dm\, (y^2 + z^2), \text{ etc.}$$

The off diagonal elements are zero for the spherical shell as a result of its symmetry. This follows, because for every contribution xy to the integral for I_{xy}, there is an equal but opposite contribution $(-x)y$ on the sphere. Hence, the integral of xy over a sphere is zero.

The diagonal elements are equal since the three axes are equivalent as far as the geometry of the shell is concerned. The ring shown in the figure has a mass

$$dm = \sigma(2\pi \ r \sin\theta)(r\, d\theta)$$

hence $$I_{zz} = \int dm\,(y^2+x^2) = 2\pi\ \sigma\ r^2 \int_0^{\pi} d\theta \sin\theta \ (r^2 \sin^2\theta)$$

$$= 2\pi\ \sigma\ r^4 \int_{-1}^{1} d(\cos\theta)[1 - \cos^2\theta]$$

$$= 2\pi\ \sigma\ r^4 \left(\frac{4}{3}\right) = \frac{8}{3}\pi\sigma r^4$$

$$= (4\pi\ r^2\ \sigma)\ \frac{2}{3}\ r^2 = \frac{2}{3}\ Mr^2$$

Therefore, $$I_{xx} = I_{yy} = I_{zz} = \frac{2}{3}\ Mr^2.$$

A solid right circular cone has a base of r = a, and altitude h as shown in fig.1. Find the mass moment of inertia with respect to (a) the x (central) axis, (b) the y axis, (c) an axis passing through the centroid and parallel to the y axis.

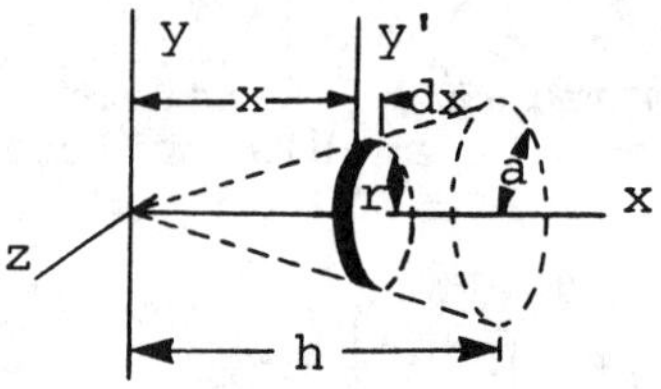

Fig. 1

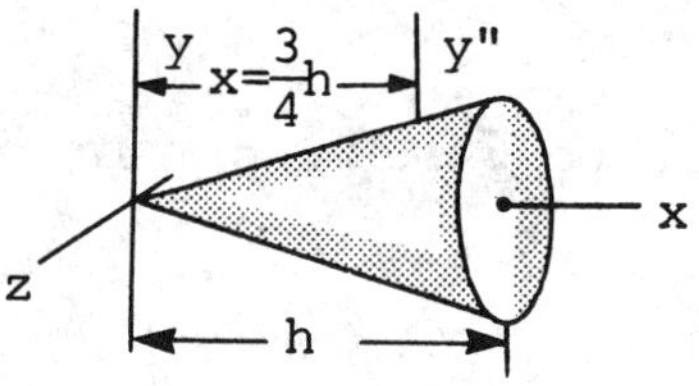

Fig. 2

Solution: Examine the differential element of mass shown in figure 1. Using the relationship, r = ax/h, yields

$$dm = \rho\pi r^2\,dx = \rho\pi\,\frac{a^2}{h^2}\,x^2\,dx, \text{ where } \rho \text{ is density.}$$

For the moment of inertia about the x-axis, consider the moment of the differential elements about the x-axis and integrate to obtain the total moment. The differential element is a circle with the axis through the center; thus,

$$dI_x = \frac{1}{2}\,r^2\,dm = \frac{1}{2}\left(a\,\frac{x}{h}\right)^2\left(\rho\pi\,\frac{a^2}{h^2}\,x^2\,dx\right)$$

$$= \frac{1}{2}\,\rho\pi\,\frac{a^4}{h^4}\,x^4\,dx.$$

Integrating from x = 0 to x = h results in

$$I_x = \int dI_x = \frac{1}{2}\,\rho\pi\,\frac{a^2}{h^2}\int x^4\,dx = \frac{1}{10}\,\rho\pi\,a^4h.$$

The total mass of the cone, m, is

$$m = \int dm = \rho\pi\,\frac{a^2}{h^2}\int_0^h x^2\,dx = \frac{1}{3}\,\rho\pi\,a^2h.$$

This gives $I_x = \frac{3}{10}\,ma^2.$

For the moment of inertia about the y-axis, I_y, use the same differential element. One can apply the parallel axis theorem which states that

$$dI_y = dI_{y'} + x^2\,dm.$$

$dI_{y'}$ is the moment of the differential disk about the axis y' shown in figure 1; it is

$$dI_{y'} = \frac{1}{4} r^2 dm.$$

This gives $dI_y = \left(\frac{1}{4} r^2 + x^2\right) dm$; substituting the expressions for r and dm,

$$dI_y = \left(\frac{1}{4} \frac{a^2}{h^2} x^2 + x^2\right) \left(\rho\pi \frac{a^2}{h^2} x^2 \, dx\right)$$

$$= \rho\pi \frac{a^2}{h^2} \left(\frac{a^2}{4h^2} + 1\right) x^4 \, dx.$$

Integrating,

$$I_y = \int dI_y = \rho\pi \frac{a^2}{h^2} \left(\frac{a^2}{4h^2} + 1\right) \int x^4 \, dx$$

$$= \rho\pi \frac{a^2}{h^2} \left(\frac{a^2}{4h^2} + 1\right) \frac{h^5}{5} .$$

This may be rewritten as

$$I_y = \frac{3}{5} \left(\frac{1}{4} a^2 + h^2\right) \frac{1}{3} \rho\pi a^2 h$$

or, $$I_y = \frac{3}{5} m \left(\frac{1}{4} a^2 + h^2\right) .$$

Apply the parallel axis theorem for the moment of inertia about y" (figure 2).

$$I_y = I_{y''} + m\bar{x}^2 , \qquad \bar{x} = 3/4 \, h$$

$$I_{y''} = I_y - m\bar{x}^2$$

$$= \frac{3}{5} m \left(\frac{1}{4} a^2 + h^2\right) - m \left(\frac{3}{4} h\right)^2$$

$$I_{y''} = \frac{3}{20} m \left(a^2 + \frac{1}{4} h^2\right) .$$

• PROBLEM 5-26

For the cylinder shown in figure 1, determine the moment of inertia about the y-axis passing through the center of mass, and the product of inertia about the xy axis. Also determine the moment of inertia, using the parallel axis theorem, about the y-axis that passes through the edge of the cylinder (tangent to the surface) as shown in figure 4.

Solution: First the moment of inertia about the y axis will be determined. The integral to be evaluated is

$$I_{yy} = \int_V \gamma (x^2 + z^2) \, dV$$

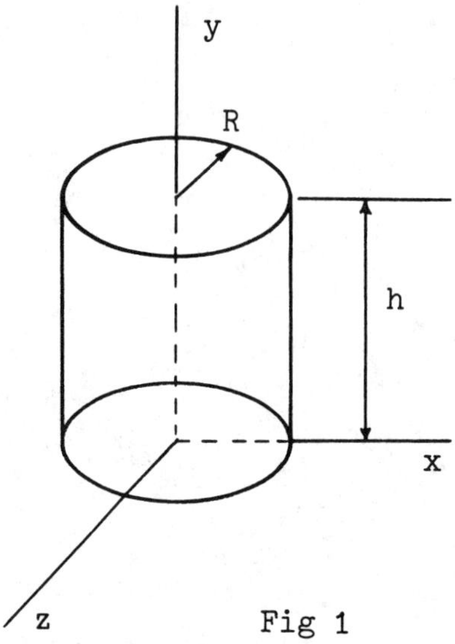

Fig 1

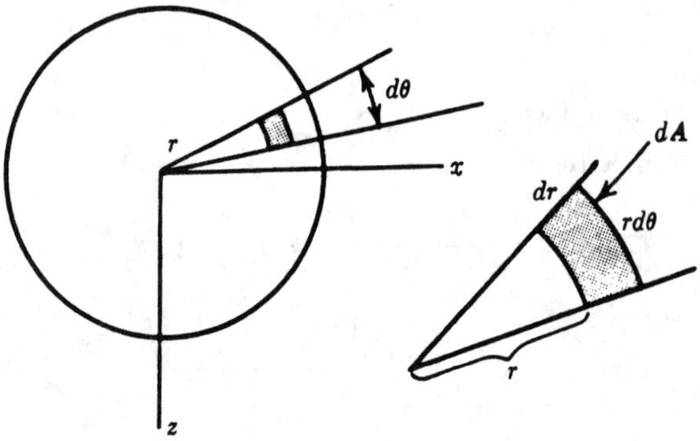

Figure 2

where γ is the mass density and $dm = \gamma\ dV$. Because of the circular symmetry in this problem, one might be tempted to try using polar coordinates. The cylinder, as viewed by looking down the y axis, is shown in Fig. 2.

If polar coordinates are to be used,

$$x^2 + z^2 = r^2$$

The increment of volume is

$$dV = hr\ dr\ d\theta$$

By substituting into the integral for the moment of inertia,

$$I_{yy} = \int_A \gamma r^3 h\ dr\ d\theta$$

or $$I_{yy} = \int_0^R \int_0^{2\pi} \gamma h r^3\ d\theta\ dr,$$

and $$I_{yy} = \int_0^R \gamma 2\pi h r^3\ dr$$

$$= \frac{\gamma \pi h R^4}{2}$$

The moment of inertia may be determined as a function of the mass of the cylinder. The mass of the cylinder may be determined by integrating the volume:

$$M = \int_V \gamma\ dV$$

Thus, $$M = \int_0^R \int_0^{2\pi} \gamma (hr\ d\theta\ dr)$$

and $M = \gamma\pi h R^2$

Therefore, the moment of inertia as a function of the mass is

$$I_{yy} = \frac{MR^2}{2} \tag{1}$$

This is an important relationship to remember because this problem appears very often in dealing with the motion of rigid bodies, specifically wheels and disks.

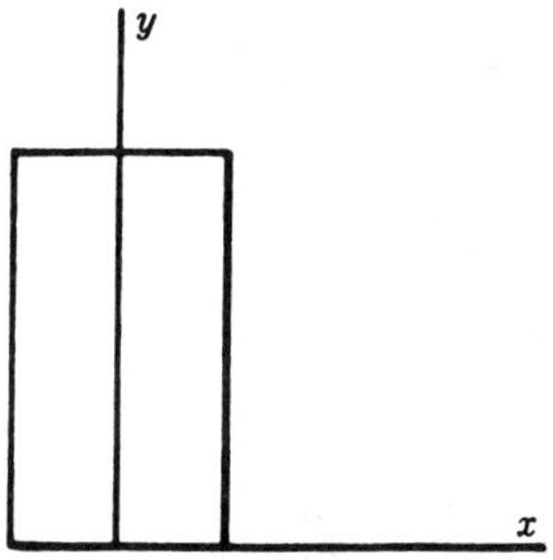

Figure 3

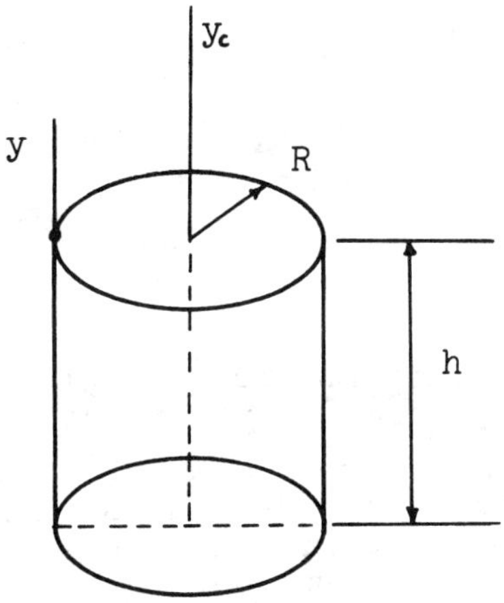

Fig 4

Now it is desired to determine the product of inertia with respect to the xy axes. Looking at the body in the xy plane (Fig. 3) one notes immediately that the y axis is an axis of symmetry, so that the product of inertia with respect to the xy axes will be zero.

Now the moment of inertia is known for the vertical axis y_c that passes through the center of mass shown in figure 4 from equation (1)

$$I_{yy_c} = \frac{MR^2}{2}$$

The parallel axis theorem states that

$$I_{yy} = I_{yy_c} + d^2M$$

where d is the perpendicular distance between the y_c and the y axis; in this case, d is equal to R. Therefore,

$$I_{yy} = \frac{MR^2}{2} + R^2M$$

or $$I_{yy} = \frac{3MR^2}{2} .$$

What is the rotational inertia about an axis through the center of a 25-kg solid sphere whose diameter is 0.30 m?

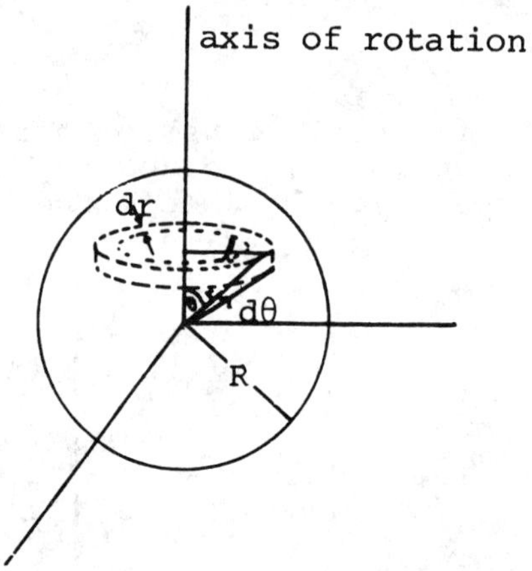

Solution: For a rigid body rotating with angular speed ω about a fixed axis, the kinetic energy is $K = \frac{1}{2} mv^2 = \frac{1}{2} m(\omega r)^2$. Each particle of this body can be considered as contributing to the total kinetic energy. The angular velocity, ω, of all the particles is the same but their distance r from the axis of rotation varies. Therefore, the total kinetic energy can be written as

$$K = \frac{1}{2} \left(m_1 r_1^2 + m_2 r_2^2 + \ldots\right)\omega^2 = \frac{1}{2}\Sigma\left(m_i r_i^2\right)\omega^2$$

where the summation is taken over all the particles in the rigid body. The rotational inertia, I, is defined as

$$I = \Sigma m_i r_i^2 \ .$$

As can be seen from the above equations, the rotational energy of a body, for a given angular speed ω, depends on the mass of the body and the way that mass is distributed around the axis of rotation. Since most rigid bodies are not composed of discrete point masses but are continuous distributions of matter, the summation for I in the above equation becomes an integration. Let the body be divided into infinitesimal elements of mass dm at a distance r from the axis of rotation. Then the rotational inertia is

$$I = \int r^2 \, dm$$

where the integral is taken over the whole body.

For a solid sphere of radius R,

$$dm = \rho\ dV$$

where ρ is the density of the sphere and dV is an infinitesimal volume. For dV, take a circle of radius ℓ and of thickness r $d\theta$, where ℓ is the distance from the axis of rotation. We have

$$dV = (2\pi\ell)(dr)(rd\theta) = 2\pi\ (r \sin\theta)\ r\, drd\theta$$

$$= 2\pi\ r^2 \sin\theta\ dr\ d\theta$$

$$\rho = \frac{m}{V} = \frac{m}{\frac{4}{3}\pi R^3}$$

Then

$$I = \int \ell^2\ dm = \int (r \sin\theta)^2\ \rho dV$$

$$= \int_{\theta=0}^{\pi} \int_{r=0}^{R} (r^2 \sin^2\theta) \left(\frac{m}{\frac{4}{3}\pi R^3}\right) (2\pi\ r^2 \sin\theta\ dr\ d\theta)$$

$$= \int_{\theta=0}^{\pi} \int_{r=0}^{R} \frac{3m}{2R^3}\ r^4 \sin^3\theta\ dr\ d\theta$$

$$= \int_{\theta=0}^{\pi} \left\{ \left. \frac{3m}{2R^3} \frac{r^5}{5} \sin^3\theta \right]_{r=0}^{R} \right\} d\theta$$

$$= \frac{3}{10}\ m\ R^2 \int_{\theta=0}^{\pi} \sin^3\theta\ d\theta$$

$$= \frac{3}{10}\ m\ R^2 \int_{\theta=0}^{\pi} \sin\theta\ (1 - \cos^2\theta) d\theta$$

Let $x = -\cos\theta$ and $dx = \sin\theta\ d\theta$. Then

$$I = \frac{3}{10}\ m\ R^2 \int_{x=-\cos 0°}^{x=-\cos\pi} (1 - x^2) dx$$

$$= \frac{3}{10}\ m\ R^2 \left. \left(x - \frac{x^3}{3}\right) \right]_{-1}^{1}$$

$$= \frac{3}{10}\ m\ R^2 \left(1 - \frac{1}{3} + 1 - \frac{1}{3}\right) = \left(\frac{3}{10}\ m\ R^2\right)\left(\frac{4}{3}\right) = \frac{2}{5}\ mR^2$$

For the given sphere, the mass is 25 kg and the radius is 0.30 m/2 = 0.15 m. Its rotational inertia is then

$$I = \frac{2}{5}\ mR^2 = \frac{2}{5}\ (25\ \text{kg})(0.15\ \text{m})^2 = 0.22\ \text{kg-m}^2.$$

(a) Find the centroid of the parabolic section shown in Fig. 1. (b) Also, detemine the moments of inertia of the shaded area shown with respect to each of the coordinate axes. (c) Using results of part (b), determine the radius of gyration of the shaded area with respect to each of the coordinate axes.

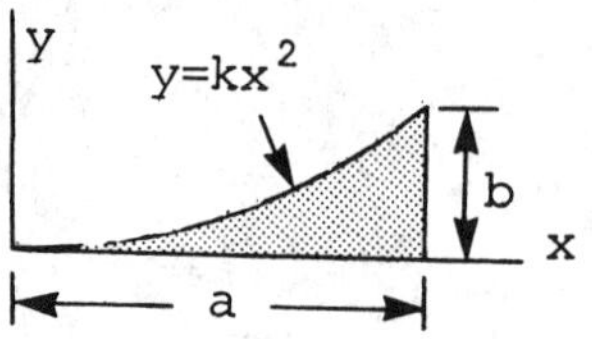

Fig. 1

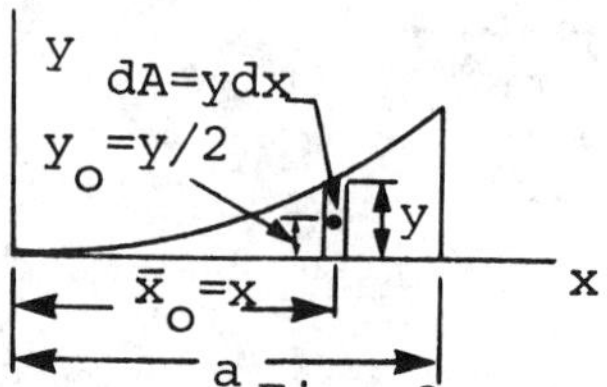

Fig. 2

Solution: Given $y = kx^2$, the value of k can be obtained from the point $x = a$, $y = b$. Hence, $k = b/a^2$.

Hence, $$y = \frac{b}{a^2} x^2 \tag{1}$$

or $$x = \frac{a}{b^{1/2}} y^{1/2} \tag{2}$$

Choose a vertical differential element as shown in figure 2. Then the total area A is

$$A = \int dA = \int ydx = \int_0^a \frac{b}{a^2} x^2 \, dx = \left[\frac{b}{a^2}\frac{x^3}{3}\right]_0^a$$

$$= \frac{ab}{3} \tag{3}$$

The moment of the differential element with respect to the y axis is $\bar{x}_o \, dA$. The moment of the total area with respect to this axis is, therefore,

$$\int \bar{x}_o \, dA = \int xy \, dx = \int_0^a x \left(\frac{b}{a^2} x^2\right) dx$$

$$= \left[\frac{b}{a^2}\frac{x^4}{4}\right]_0^a = \frac{a^2 b}{4} . \tag{4}$$

The centroid $\bar{x}$ is that point which, when multiplied by the area, gives the same result for the moment as the integration of the differential moments $x_o \, dA$.

Thus, $\bar{x}A = \int \bar{x}_o \, dA$

$$\bar{x}\,\frac{ab}{3} = \frac{a^2b}{4}$$

$$\bar{x} = \frac{3}{4}\,a. \tag{5}$$

Similarly, the moment of the differential element with respect to the x axis is $\bar{y}_o$ dA, and the moment of the total area is

$$\int \bar{y}_o \; dA = \int \frac{y}{2}\, y\, dx = \int_0^a \frac{1}{2}\left(\frac{b}{a^2}\, x^2\right) dx$$

$$= \frac{ab^2}{10} \quad .$$

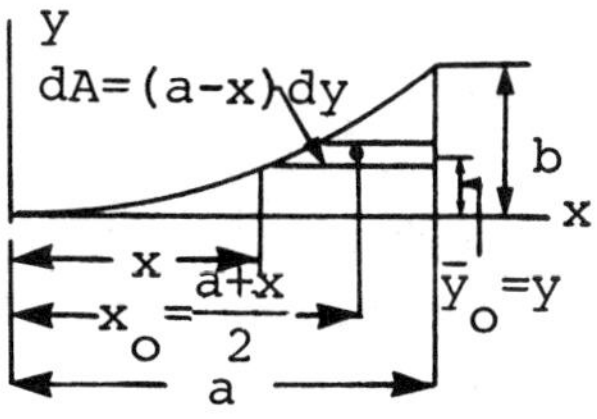

Fig. 3

Thus, $\quad \bar{y}_o = \int \bar{y}_{e\ell} \; dA$

$$\bar{y}\,\frac{ab}{3} = \frac{ab^2}{10}$$

$$\bar{y} = \frac{3}{10}\; b. \tag{6}$$

The same result will be obtained by using a horizontal element as shown in Fig. 3. For example, the moments with respect to the coordinate axes are:

$$\int \bar{x}_o \; dA = \int \frac{a+x}{2}\,(a - x)\; dy = \int_0^b \frac{a^2 - x^2}{2}\; dy$$

$$= \frac{1}{2}\int_0^b \left(a^2 - \frac{a^2}{b}\, y\right) dy = \frac{a^2b}{4} \; ,$$

$$\int \bar{y}_o \; dA = \int y\,(a - x)\; dy = \int y \left(a - \frac{a}{b^{\frac{1}{2}}}\, y^{\frac{1}{2}}\right) dy$$

$$= \int_0^b \left(ay - \frac{a}{b^{\frac{1}{2}}}\; y^{3/2}\right) dy = \frac{ab^2}{10} \; .$$

Hence, equations (5) and (6) follow.

(b) Using the vertical differential element, compute the moment of inertia. Since all portions of this element are not at the same distance from the x-axis, treat the elements as a thin rectangle.

The moment of inertia of the element with respect to x-axis is

$$dI_x = \frac{1}{3}\, y^3\; dx = \frac{1}{3}\left(\frac{b}{a^2}\, x^2\right)^3 = \frac{1}{3}\;\frac{b^3}{a^6}\; x^6\; dx$$

$$I_x = \int dI_x = \int_0^a \frac{1}{3}\,\frac{b^3}{a^6}\; x^6\; dx = \frac{ab^3}{21} \quad . \tag{7}$$

Similarly,

$$dI_y = x^2 \, dA = x^2 \, (y \, dx) = x^2 \left(\frac{b}{a^2} x^2\right) dx$$

$$= \frac{b}{a^2} x^4 \, dx$$

$$I_y = \int dI_y = \int_0^a \frac{b}{a^2} x^4 \, dx = \frac{a^3 b}{5} . \tag{8}$$

The radii of gyration are

$$kx^2 = \frac{I_x}{A} = \frac{ab^3/21}{ab/3} = \frac{b^2}{7}$$

$$kx = \sqrt{\frac{1}{7}} \, b,$$

$$ky^2 = \frac{I_y}{A} = \frac{a^3 b/5}{ab/3} = \frac{3}{5} a^2$$

$$ky = \sqrt{\frac{3}{5}} \, a. \tag{9}$$

• PROBLEM 5-29

Calculate the moment of inertia of a solid right-circular cone about a slant height.

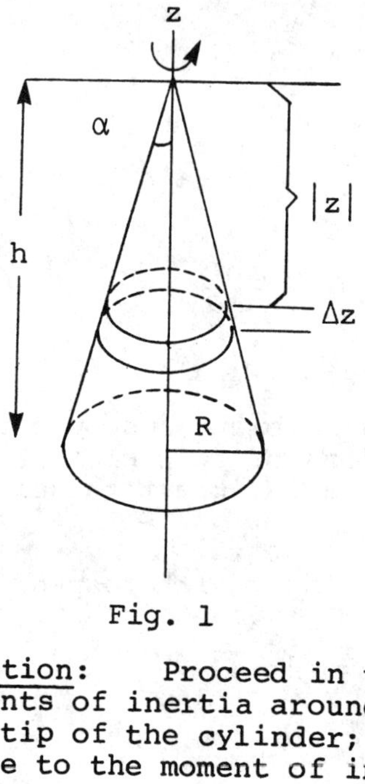

Fig. 1

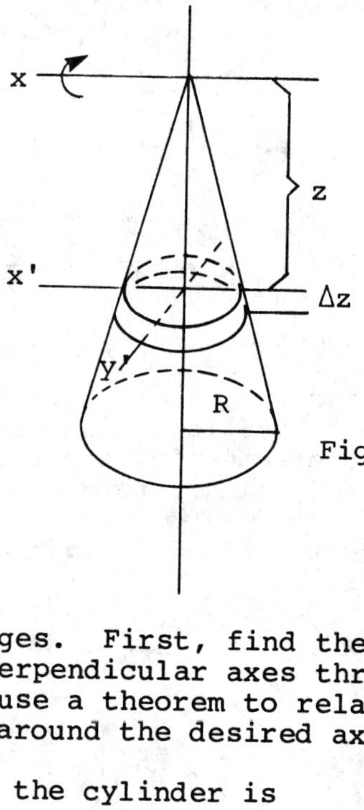

Fig. 2

Solution: Proceed in two stages. First, find the moments of inertia around two perpendicular axes through the tip of the cylinder; then, use a theorem to relate these to the moment of inertia around the desired axis.

See Fig. 1. The volume of the cylinder is

$(1/3)\pi R^2 h$; its density is $3m/(\pi R^2 h)$. Each disk of thickness Δz has radius Rz/h, area $\pi(Rz/h)^2$, and volume $\pi(Rz/h)^2\Delta z$. Its moment of inertia is $\frac{1}{2}\Delta m r^2$, where $r = Rz/h$, so we have $\Delta I = \frac{1}{2}(3mz^2\Delta z/h^3)(Rz/h)^2$. Add all the I's for all the disks which make up the cylinder and this results in the total moment of inertia for rotation around the z-axis:

$$I = \Sigma\Delta I = 3mR^2/(2h^5)\int_0^h z^4\, dz$$

$$= [3mR^2/(2h^5)][z^5/5]_0^h = 0.3\ mR^2.$$

Next, find I_x, obtained by spinning the cone around an axis through its tip perpendicular to the axis. See fig. 2.

Construct a slice as before. Find its moment of inertia for spinning around the axis x' by the perpendicular axis theorem:

$$\Delta I_{x'} + \Delta I_{y'} = \Delta I_z = \tfrac{1}{2}\Delta m r^2,$$

but $\Delta I_{x'} = \Delta I_{y'}$ due to symmetry; hence, $\Delta I_{x'} = \frac{1}{4}\Delta m r^2$. Further, the parallel axis theorem tells us that the moment of inertia around the x-axis is given by $\Delta I_x = \Delta I_{x'} + \Delta m z^2$, since z is the distance between the parallel x and x' axes. Here this becomes

$$\Delta I_x = \tfrac{1}{4}\Delta m r^2 + \Delta m z^2 = [\tfrac{1}{4}(Rz/h)^2 + z^2]\cdot 3mz^2\Delta z/h^3.$$

Now, add these contributions to I_x:

$$I_x = \Sigma\Delta I_x = 3m(\tfrac{1}{4}R^2 + h^2)/h^5\int_0^h z^4\, dz$$

$$= [3m\ (\tfrac{1}{4}R^2 + h^2)/h^5]\,[z^5/5]_0^h$$

$$= 0.6\ m\ (\tfrac{1}{4}R^2 + h^2).$$

The inertia around another axis, forming angles δ, β, γ with respect to a set of principal axes, may be found by

$$I = I_x\cos^2\delta + I_y\cos^2\beta + I_z\cos^2\gamma.$$

In this case, our desired axis forms the angle α with the z-axis and $(90°-\alpha)$ with the x-axis. Thus,

$$I = I_x\sin^2\alpha + I_z\cos^2\alpha$$

$$= 0.6\ (\tfrac{1}{4}R^2 + h^2)\sin^2\alpha + 0.3\ mR^2\cos^2\alpha\ .$$

Eliminate $R = h\tan\alpha$:

$$I = m\left[0.6\left(\tfrac{1}{4}\tan^2\alpha + 1\right)h^2 \sin^2\alpha + 0.3\, h^2 \tan^2\alpha \cos^2\alpha\right]$$

$$= 0.3\, mh^2 \left[\tfrac{1}{2}\tan^2\alpha + 2 + 1\right] \sin^2\alpha$$

$$= 0.15\, mh^2 \sin^2\alpha(\tan^2\alpha + 6).$$

GENERAL THEOREMS

• PROBLEM 5-30

Show that, if particle i were not part of a rigid body, but moved along an arbitrary path in space, $\dot{z}_i$ and $\ddot{z}_i$ would always be parallel to the z-axis and $\dot{\rho}_i$ and $\ddot{\rho}_i$ would always lie parallel to the xy plane.

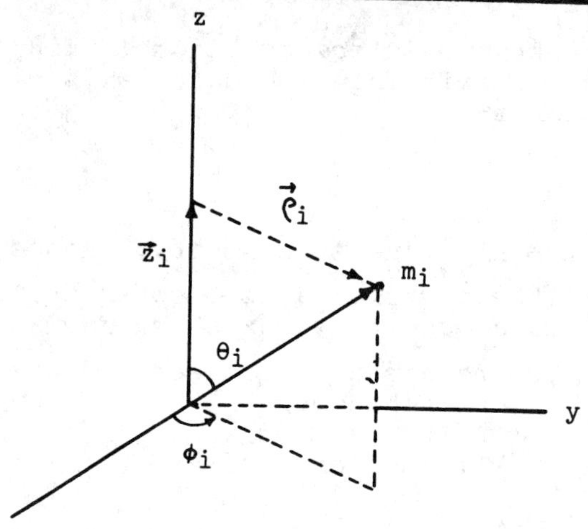

Solution: In terms of the coordinate unit vectors $\vec{i}$, $\vec{j}$, and $\vec{k}$, the vectors $\vec{z}_i$ and $\vec{\rho}_i$ can be written as

$$\vec{z}_i = z_i \vec{k}$$

$$\vec{\rho}_i = \rho_i \cos\phi_i \vec{i} + \rho_i \sin\phi_i \vec{j} = A\vec{i} + B\vec{j}$$

Here are introduced the abbreviations $A = \rho_i \cos\phi_i$ and $B = \rho_i \sin\phi_i$. Since $\vec{i}$, $\vec{j}$ and $\vec{k}$ are constants (both in magnitude and direction), the first and second time derivatives of $\vec{z}_i$ and $\vec{\rho}_i$ are

$$\dot{\vec{z}}_i = \dot{z}_i \vec{k}, \qquad \ddot{\vec{z}}_i = \ddot{z}_i \vec{k},$$

$$\dot{\vec{\rho}}_i = \dot{A}\vec{i} + \dot{B}\vec{j}, \qquad \ddot{\vec{\rho}}_i = \ddot{A}\vec{i} + \ddot{B}\vec{j}.$$

It is obvious that both $\dot{\vec{z}}_i$ and $\ddot{\vec{z}}_i$ are parallel to $\vec{k}$, and to the z-axis, no matter what $\dot{z}_i$ and $\ddot{z}_i$ are. It is also obvious that both $\dot{\vec{\rho}}_i$ and $\ddot{\vec{\rho}}_i$ are perpendicular to $\vec{k}$, and, therefore, parallel to the xy plane, regardless of what $\dot{A}$ and $\dot{B}$ or $\ddot{A}$ and $\ddot{B}$ may be. (Consider the scalar product,

$$\vec{k} \cdot \dot{\vec{\rho}}_i = \dot{A}\,(\vec{k} \cdot \vec{i}) + \dot{B}\,(\vec{k} \cdot \vec{j}) = 0$$

because $\vec{k} \cdot \vec{i} = 0$ and $\vec{k} \cdot \vec{j} = 0$. This means, of course, that $\dot{\vec{\rho}}_i$ is perpendicular to $\vec{k}$. Similar arguments apply to $\ddot{\vec{\rho}}_i$.)

• PROBLEM 5-31

For every particle i in the rotating rigid body as shown in the figure, show that $\dot{\vec{z}}_i = 0$, $\ddot{\vec{z}}_i = 0$, $|\vec{\rho}_i|$ = constant and that the tip of the vector $\vec{\rho}_i$ describes a circle around the axis of rotation.

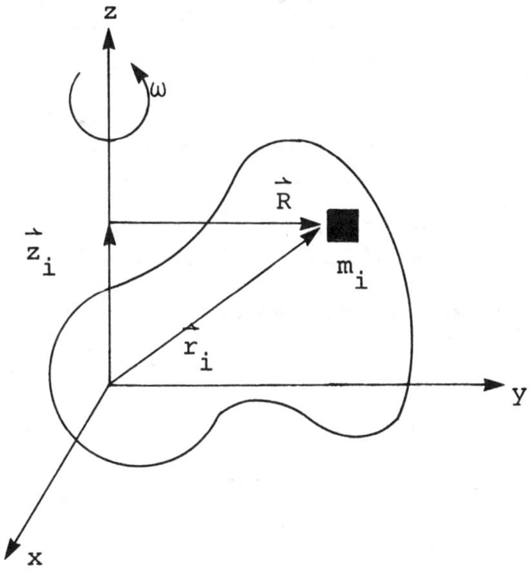

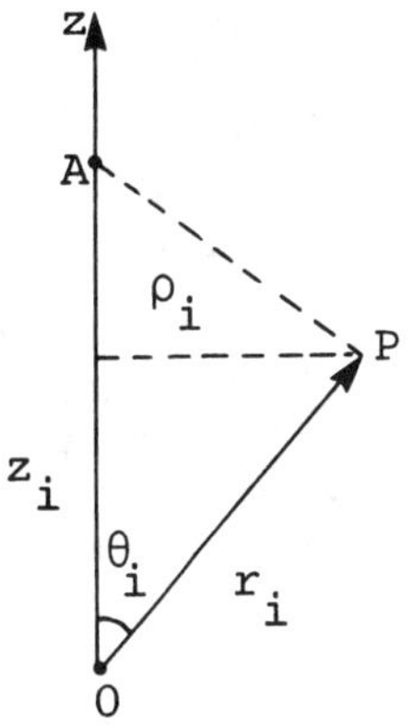

Solution: Let θ_i be the angle between $\vec{r}_i$ and the z-axis, and $\vec{k}$ be a unit vector along the z-axis. Then, the vector $\vec{z}_i$ can be written as

$$\vec{z}_i = z_i\vec{k} = r_i \cos \theta_i \, \vec{k}.$$

Since $\vec{k}$ is constant (both in magnitude and direction), it follows that z_i is constant, and, therefore, $\dot{\vec{z}}_i = 0$ and

$\ddot{\vec{z}}_i = 0$, if we can prove that

$$z_i = r_i \cos\theta_i = \text{constant}. \qquad (1)$$

Recall the definition of a rigid body; the particles in a rigid body always maintain the same positions with respect to one another. Consider three points O, P, A in the rigid body, where O and A are on the rotational axis (z-axis) and P is at the particle m_i. Since this is a rigid body, the triangle OPA maintains its size and shape. In particular, the angle θ_i and the side $r_i = \overline{OP}$ remain unchanged as the body rotates. The proof of statement (1) follows immediately.

Furthermore,

$$\rho_i = r_i \sin\theta_i = \text{constant},$$

because both r_i and θ_i are constants. Finally, the only possible motion of the point P (the tip of vector $\vec{\rho}_i$) is a circle (of radius ρ_i) around the axis, as it is obvious from the figure above in which r_i, θ_i, ρ_i and z_i all remain unchanged as the body rotates.

• PROBLEM 5-32

Show that the magnitude of ℓ_{iz} does not depend on the position chosen for the location of the coordinate system's origin, O, provided that this origin is on the axis of rotation.

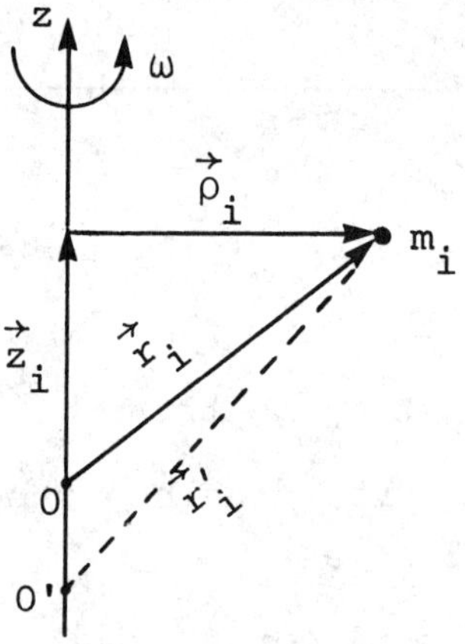

Solution: Let the rotation axis be the z-axis. As the rigid body rotates, the particle m_i moves in a circle around the axis. The circle is perpendicular to the axis and its radius is ρ_i, which is the perpendicular distance of m_i to the rotation axis. The speed of the circular motion is

$$v_i = \rho_i \omega \tag{1}$$

where ω is the angular velocity of rotation.

The angular momentum of m_i is, according to the definition, the vector product of the position vector $\vec{r}_i$ and its linear momentum $m_i \vec{v}_i$,

$$\vec{\ell}_i = \vec{r}_i \times m_i \vec{v}_i \quad .$$

Or, using $\quad \vec{r}_i = \vec{z}_i + \vec{\rho}_i$,

$$\vec{\ell}_i = m_i \vec{z}_i \times \vec{v}_i + m_i \vec{\rho}_i \times \vec{v}_i \quad . \tag{2}$$

The first term is a vector which is perpendicular to $\vec{z}_i$ or the z-axis. Since we are concerned only with the z-component of the angular momentum, this term does not contribute to ℓ_{iz}. The second term is a vector which is perpendicular to both $\vec{\rho}_i$ and $\vec{v}_i$ and, according to the Right-Hand-Rule, it is in the direction of positive z-axis. Therefore, we have

$$\ell_{iz} = m_i \rho_i v_i \sin 90° = m_i \rho_i v_i = m_i \rho_i^2 \omega \quad . \tag{3}$$

Here we have used Eq. (1) and the fact that the angle between $\vec{\rho}_i$ and $\vec{v}_i$ is always 90°. Now, since ρ_i is the perpendicular distance of m_i to the rotation axis, it is <u>independent</u> of the choice of the reference point O, as shown in the figure above. Therefore, according to Eq. (3), ℓ_{iz} is independent of the choice of the origin O so long as it is on the rotation axis.

CHAPTER 6

RECTILINEAR KINEMATICS

• PROBLEM 6-1

A stone is dropped into a well and the splash is heard two seconds later. If sound travels 1100 ft/sec, what is the depth of the well?

Solution: The total time to hear the splash is made up of t_1, the time required for the stone to hit the water and t_2, the time required for the sound to travel back. t_1 satisfies:

$$h = \frac{1}{2} g t_1^2 \tag{1}$$

where $g = 32.2\ \text{ft/sec}^2$ is the acceleration due to gravity and there is no initial velocity. t_2 satisfies

$$h = v_s t_2$$

where v_s is the velocity of sound. In both cases the distance travelled is h, thus we may equate the two expressions

$$\frac{1}{2} g t_1^2 = v_s t_2 .$$

Upon substitution of available numerical values we find that

$$t_2 = \frac{\frac{1}{2} g t_1^2}{v_s} = \frac{16.1}{1100} t_1^2 = 0.0146\ t_1^2 .$$

The total time required is 2 sec. Thus,

$$t_1 + t_2 = 2$$

and $t_1 + .0146\ t_1^2 = 2$

or $0.146\ t_1^2 + t_1 - 2 = 0.$

This equation is solved by means of the quadratic formula, which yields two values for t_1

$$t_{1,1}\ t_{1,2} = \frac{-1 \pm \sqrt{1 - 4(.0146)(-2)}}{2(.0146)} .$$

The two values are

$t_1 = 1.94$ sec, $- 70.4$ sec.

Negative time has no meaning here therefore we reject it and keep the first of the above values. The depth of the well can now be found from Eq. (1) to be

$$h = \tfrac{1}{2}(32.2)(1.944)^2$$

$$h = 60.6 \text{ ft.}$$

• PROBLEM 6-2

In the two pulley systems shown in Fig. 1a, determine the velocity and acceleration of block 3 when blocks 1 and 2 have the velocities and acceleration shown.

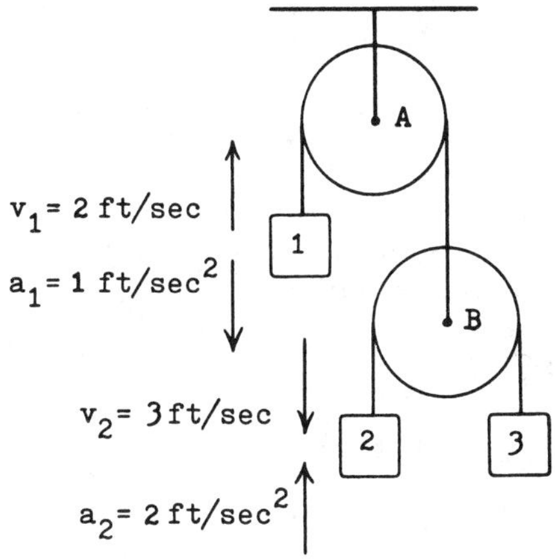

fig.1a

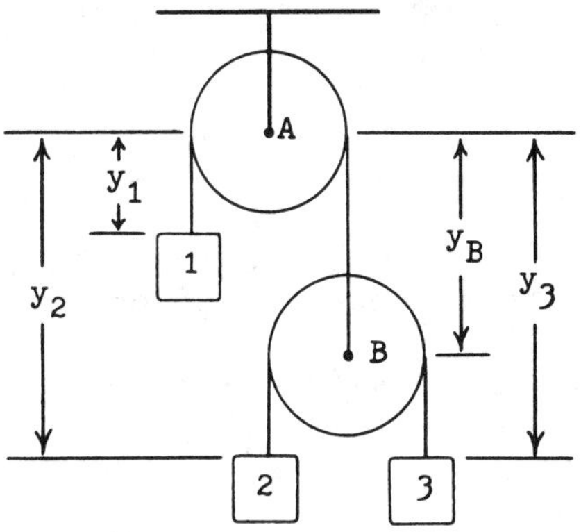

fig.1b

Solution: Figure 1b defines the positions relative to the fixed pulley A needed to solve solve this problem.

The lengths of the cord are assumed to be constant and the section of cord over the pulleys to remain constant, thus yielding the following constraint equations

$$y_1 + y_B = K_1, \qquad \text{(a)}$$

$$(y_2 - y_B) + (y_3 - y_B) = K_2,$$

or $$y_2 + y_3 - 2y_B = K_2. \qquad \text{(b)}$$

Differentiating each equation (a) and (b) twice yields:

$$\dot{y}_1 + \dot{y}_B = 0 \qquad \text{(c)}$$

$$\ddot{y}_1 + \ddot{y}_B = 0 \quad \text{(d)}$$

$$\dot{y}_2 + \dot{y}_3 - 2\dot{y}_B = 0 \quad \text{(e)}$$

$$\ddot{y}_2 + \ddot{y}_3 - 2\ddot{y}_B = 0 \quad \text{(f)}$$

These equations lead directly to the desired solutions upon substitution of the information from Figure 1a. If it is supposed that down is positive, then

$$v_1 = \dot{y}_1 = -2 \text{ ft/sec}, \qquad v_2 = \dot{y}_2 = 3 \text{ ft/sec},$$

$$\text{(g)}$$

$$a_1 = \ddot{y}_1 = 1 \text{ ft/sec}^2, \qquad a_2 = \ddot{y}_2 = -2 \text{ ft/sec}^2.$$

First solve (c) and (d) to find the motion of pulley B.

$$-2 \text{ ft/sec} + \dot{y}_B = 0$$

$$\dot{y}_B = 2 \text{ ft/sec} \quad \text{(h)}$$

$$1 \text{ ft/sec}^2 + \ddot{y}_B = 0$$

$$\ddot{y}_B = -1 \text{ ft/sec}^2. \quad \text{(i)}$$

Now using (g), (h) and (i) in equations (e) and (f), it is possible to obtain the desired results:

$$3 \text{ ft/sec} + \dot{y}_3 - 2\,(2 \text{ ft/sec}) = 0$$

$$\dot{y}_3 = 1 \text{ ft/sec}$$

$$-2 \text{ ft/sec}^2 + \ddot{y}_3 - 2\,(-1 \text{ ft/sec}^2) = 0$$

$$\ddot{y}_3 = 0$$

Thus, block 3 is moving down with a constant velocity

$$v_3 = 1 \text{ ft/sec}.$$

• PROBLEM 6-3

A particle moves from point A to B under the influence of a variable force. It has an acceleration of 5 ft/s^2 for 6 s, and then its acceleration increases to 7 ft/s^2 until its has reached the velocity of 100 ft/s. For sometime its velocity remains constant, then the particle starts to decelerate and within 10 s stops at B. The total travel time from A to B is 46 s. Draw the a-t, v-t, and x-t curves, and determine the distance between points A and B.

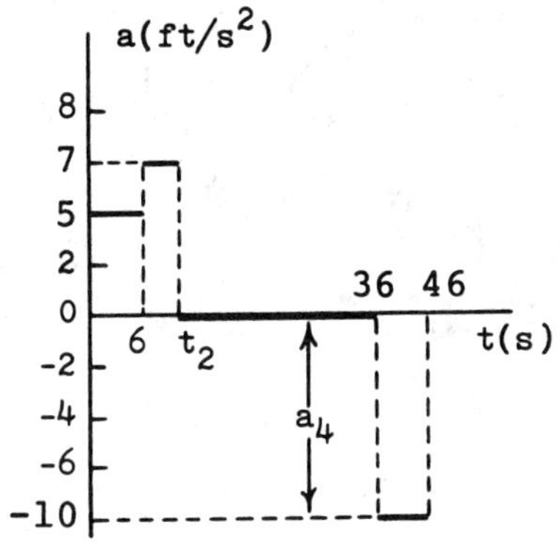

fig.1

<u>Solution:</u> <u>Acceleration-Time Curve</u>. Much of the information for this aspect of the problem is already given. It is known that $0 < t < 6$ $a = 5\ ft/s^2$, $6 < t < t_2$ $a = 7\ ft/s^2$, $t_2 < t < t_3$ $a = 0$, $t_3 < t < 46s$ a is not known. Since $v = 100\ ft/s$ at $t = t_2$ it is possible to solve for t_2 using

$$v_f = v_0 + a(t_f - t_0), \tag{a}$$

$$100\ ft/s = v(6) + 7\ ft/s^2\ (t_2 - 6s).$$

Now to solve for v(6), use (a) again

$$v_{(t=6)} = 0 + 5\ ft/s^2 \cdot (6s)$$

$$v(6) = 30\ ft/s. \tag{b}$$

Substituting (b) into (a) yields

$$100\ ft/s - 30\ ft/s = 7\ ft/s^2\ (t_2 - 6s)$$

$$t_2 - 6s = \frac{70}{7}\ s$$

$$t_2 = 6s + 10s = 16s.$$

t_3 is easily obtained. Since the total trip takes 46 seconds and deceleration starts 10 seconds from the end, $t_3 = 36$ s. Now the value for the deceleration again follows from equation (a),

$$0 = 100\ ft/s + a(46s - 36s)$$

$$a = -\frac{100}{10}\frac{ft}{s^2} = -10\ ft/s^2.$$

The curve containing this information is illustrated in Figure 1.

<u>Velocity-Time Curve</u>. Since the acceleration is either constant or zero, and velocity is the integral of acceleration, the velocity curve consists of straight lines connecting points whose abscissa are the times when the ac-

celeration changes (t = 0, 6, 16, 36, 46s) and whose ordinates are the total area under the a-t curve up to that value of t.

$t = 0 \quad v = 0$ $\qquad\qquad$ (0, 0)

$t = 6, v = (5)(6)$ $\qquad\qquad$ (6, 30)

$t = 16, v = (5)(6) + (7)(10)$ $\qquad\qquad$ (16, 100)

$t = 36, v = (5)(6) + (7)(10)$ $\qquad\qquad$ (36, 100)

$t = 46, v = (5)(6) + (7)(10) + (-10)(10)$ $\qquad\qquad$ (46, 0)

This curve is shown in Figure 2.

Position-Time Curve. Position is the integral of velocity, and the points to be connected on the x-t curve will be determined by the same method as those on the v-t curve. However, now the velocity is constant only between t = 16s and t = 36s. The only straight line segment will be present over this interval of the x-t curve. The other sections will have parabolic curves, turning upward for t < 16s, and downwards for t > 36s. The defining points are:

$t = 0, \quad x = 0$ $\qquad\qquad$ (0, 0)

$t = 6, \quad x = \frac{1}{2}(6)(30)$ $\qquad\qquad$ (6, 90)

$t = 16, \quad x = \frac{1}{2}(6)(30) + \frac{1}{2}(10)(30 + 100)$ $\qquad\qquad$ (16, 740)

$t = 36, \quad x = 740 + (20)(100)$ $\qquad\qquad$ (36, 2740)

$t = 46, \quad x = 3740 + \frac{1}{2}(10)(100)$ $\qquad\qquad$ (46, 3240)

This curve is shown in Fig. 3. Also, the result of x (t = 46s) = 3240 ft yields the distance between points A and B.

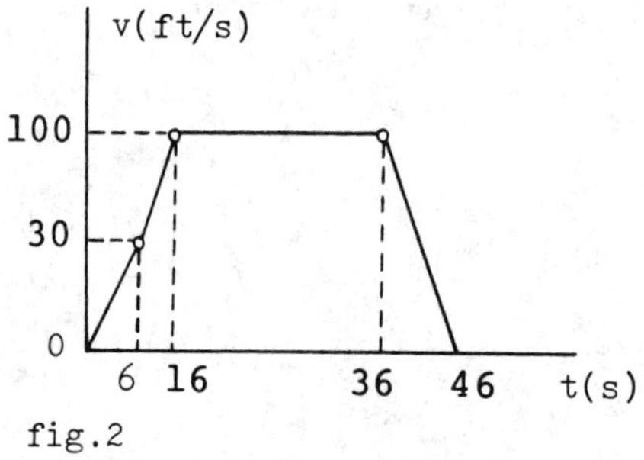

fig.2

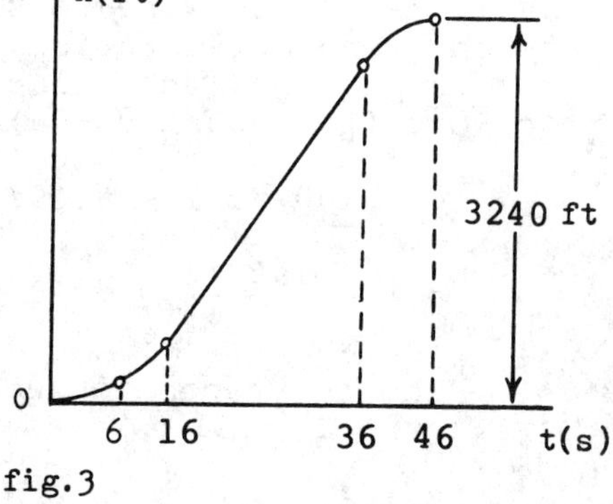

fig.3

• **PROBLEM 6-4**

The acceleration of the mass of a spring-mass system is given as a function of the displacement:

$$a(x) = -(k/m)x.$$

If the mass is started from rest at a displacement of $-\rho$ determine the velocity of the mass as a function of the displacement.

Solution: If we note that $a(x) = d^2x/dt^2$ we can write

$$\frac{d^2x}{dt^2} + \frac{k}{m}x = 0$$

from the given acceleration. This has the general solution

$$x = A \sin \omega t + B \cos \omega t \qquad (1)$$

where A, B are arbitrary constants, t is the time and $\omega = \sqrt{k/m}$ is the natural frequency. A and B are found by noting that at $t = 0$ the displacement x is $-\rho$ and the velocity dx/dt is zero (at rest). The first condition gives

$$-\rho = B. \qquad (2)$$

To apply the second condition we first find the velocity

$$v = \frac{dx}{dt} = A\omega \cos \omega t - B\omega \sin \omega t \qquad (3)$$

At $t = 0$ this gives

$$0 = A \qquad (4)$$

Now if we substitute Eqs. (4) and (2) into Eqs. (3) and (1) we find, after re-arranging them

$$-\left(\frac{x}{\rho}\right) = \cos \omega t$$

$$\left(\frac{v}{\rho\omega}\right) = \sin \omega t$$

Now, if we square each of these, add them together and observe that $\cos^2 \omega t + \sin^2 \omega t = 1$, we get

$$\frac{x^2}{\rho^2} + \frac{v^2}{\rho^2\omega^2} = \cos^2 \omega t + \sin^2 \omega t = 1$$

$$x^2 + \frac{v^2}{\omega^2} = \rho^2$$

$$\frac{v^2}{\omega^2} = (\rho^2 - x^2)$$

$$v^2 = (\rho^2 - x^2)\omega^2$$

$$v^2 = (\rho^2 - x^2)(k/m)$$

• PROBLEM 6-5

A braking device in a rifle is designed to reduce recoil. It consists of a piston attached to the barrel. The piston moves in a cylinder filled with oil. When the shot is fired the barrel and the piston recoil at an initial velocity V_0. The oil under pressure passes through tiny holes in the piston causing the piston and the barrel to decelerate at a rate proportional to their velocity $a = -kv$. Express (a) v in terms of t, (b) x in terms of t, (c) v in terms of x. Draw the corresponding motion curves.

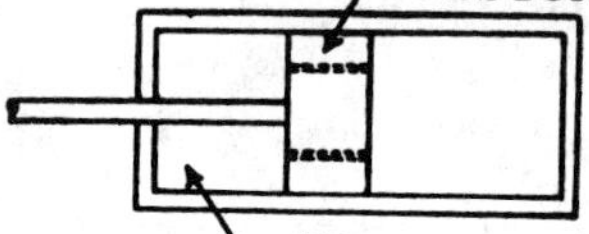

FIG. 1

Solution: (a) The basic definition of acceleration is

$$a = \frac{dv}{dt} \quad . \tag{a}$$

Substituting the expression given for the acceleration yields

$$- kv = \frac{dv}{dt} \; . \tag{b}$$

Rearranging to separate variables yields

$$- k\, dt = \frac{dv}{v} \; . \tag{c}$$

Now it is possible to integrate equation (c) to get

$$- k \int_0^t dt = \int_{v_0}^{v} \frac{dv}{v}$$

$$- kt = \ln \frac{v}{v_0} \quad . \tag{d}$$

If antilogs of equation (d) are taken, we arrive at the desired form

$$v = v_0\, e^{-kt} \tag{e}$$

This curve is shown in Figure 2.

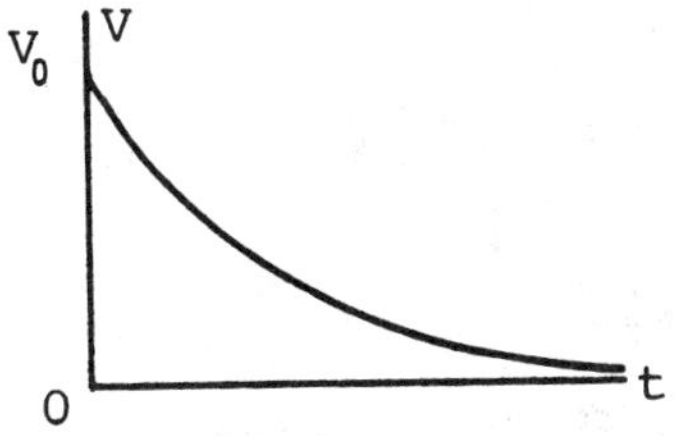

Fig. 2

(b) Again, starting with a basic definition, this time for v, we have

$$v = \frac{dx}{dt} \tag{f}$$

Substitution of the expression (e) into (f) yields

$$v_0 \, e^{-kt} = \frac{dx}{dt} \,,$$

$$v_0 \, e^{-kt} \, dt = dx. \tag{g}$$

Integrating (g)

$$v_0 \int_0^t e^{-kt} \, dt = \int_0^x dx \,,$$

$$\frac{-v_0}{k} \, e^{-kt} \Big|_0^t = x,$$

$$\frac{-v_0}{k} \, (e^{-kt} - e^0) = x,$$

$$x = \frac{v_0}{k} \, (1 - e^{-kt}). \tag{h}$$

This curve is shown in Figure 3.

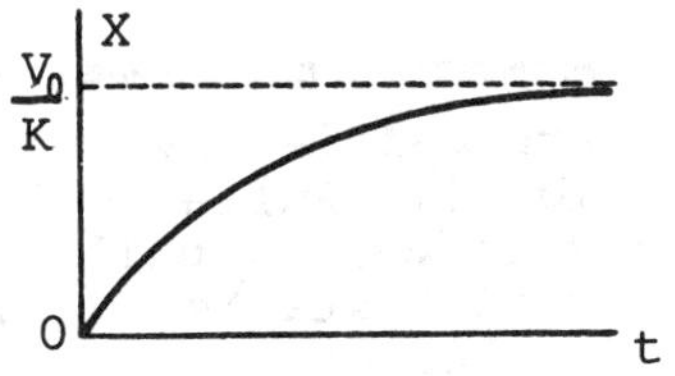

Fig. 3

(c) Part c may be done in two ways:

Using the chain rule for differentials, equation (a) may be rewritten

$$a = \frac{dv}{dt} = \frac{dx}{dt}\frac{dv}{dx} = v\,\frac{dv}{dx} \,. \tag{i}$$

Substituting a = - kv in equation (i), we get

$$- kv = v \frac{dv}{dx}$$

or $$\frac{dv}{dx} = - k,$$

$$dv = - k\ dx.$$

Integrating to solve yields

$$v - v_0 = - kx$$

or $$v = v_0 - kx.$$

As an alternate method, equations (e) and (h) can be combined. From (e) we know

$$e^{-kt} = \frac{v}{v_0} .$$

Substituting for the exponential in (h) yields

$$x = \frac{v_0}{k}\left(1 - \frac{v}{v_0}\right)$$

or $$kx = v_0 - v$$

or $$v = v_0 - kx.$$

This curve is illustrated in Figure 4.

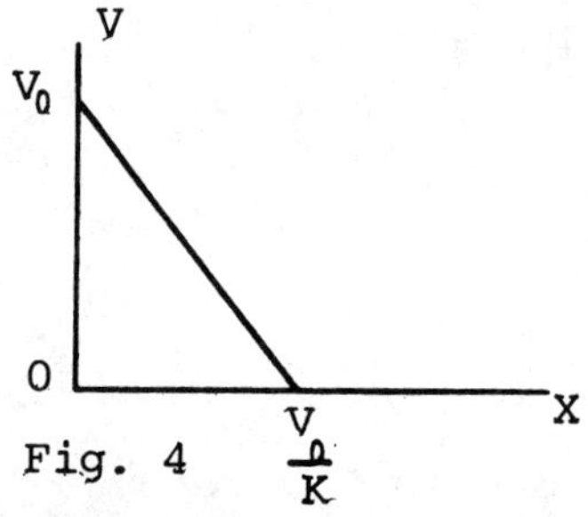

Fig. 4

• PROBLEM 6-6

Two loads A and B are carried by a system of three pulleys C, D, E as shown Fig. 1. Pulley D is moving downward with a constant velocity of 2 m/s, pulleys C and E are fixed. At a moment t = 0 load A starts to move (initial velocity v = 0) downward from the position M with the constant acceleration. Load A has velocity 8 m/s when it passes through N, MN = 4m. Calculate the change in elevation, the velocity and the acceleration of load B when A passes through N.

Solution: To solve this problem, analyze the motion of load A and pulley D. Then solve for the motion of B based on the fact that A, D and B are connected by a cord (which we assume does not stretch) of fixed length and will there-

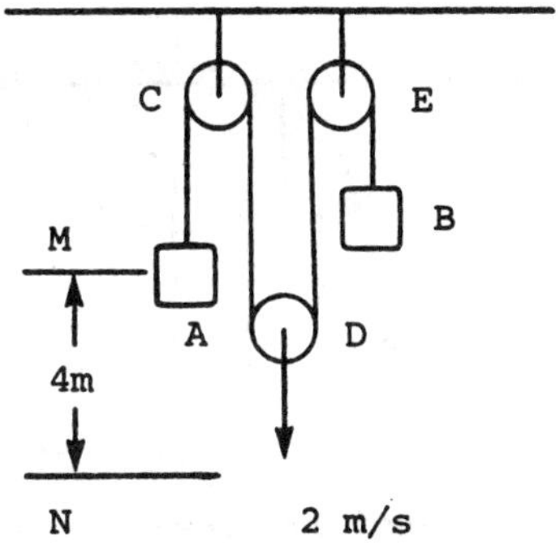

Fig. 1

fore move as a unit. A coordinate system is chosen such that the origin O is at the horizontal surface and the positive direction is chosen downward.

Load A: Given are the distance moved by A, the initial and the final velocities, and the fact that the acceleration is constant. The equation that combines this information is

$$2ad = v_f^2 - v_0^2. \qquad \text{(a)}$$

The given information is inserted into this equation to determine the unknown acceleration.

$$2a(4m) = (8m/s)^2 - (0)^2$$

$$a_A = \frac{64}{8}\ m/s^2 = 8.0\ m/s^2. \qquad \text{(b)}$$

It is necessary to determine the amount of time for the acceleration. The equation to be used is:

$$v_f = v_0 + at. \qquad \text{(c)}$$

$$8\ m/s = 0 + (8\ m/s^2)t$$

$$t = 1.0\ s. \qquad \text{(d)}$$

Pulley D: Since the velocity of this pulley is known to be constant, by using

$$d = vt \qquad \text{(e)}$$

it is possible to find how far it moves in 1 second during which load A is accelerating.

$$d_D = (2\ m/s)(1\ s)$$

$$d_D = 2\ m. \qquad \text{(f)}$$

Load B: The facts necessary to determine the acceleration of load B have been established. Observing Figure 2

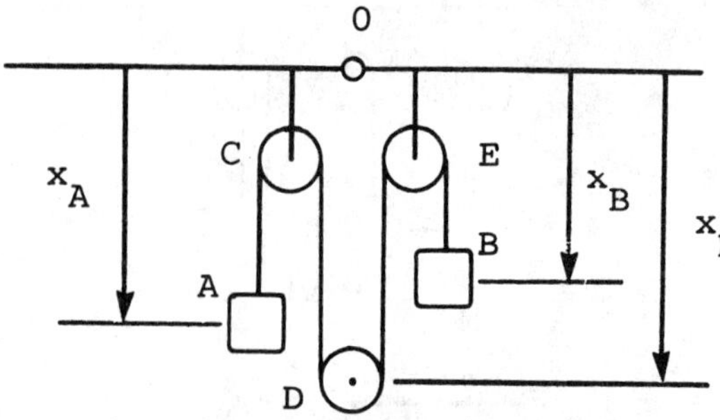

Fig. 2

and remembering the cord restraint, one should recognize that

$$X_A + 2X_D + X_B = \text{Constant} \quad \text{(g)}$$

Now consider the times t = 0 and t = 1.0 s, we have

$$\Delta X_A + 2\Delta X_D + \Delta X_B = 0. \quad \text{(h)}$$

From information given in the question

$\Delta X_A = 4\text{m}.$

From equation (f) we know

$\Delta X_D = 2\text{m}.$

Substituting these into (h) and solving yields

$$\Delta X_B = X_B - (X_B)_0 = -4\text{ m} - 2(2\text{ m})$$
$$= -8\text{ m},$$

that is, load B moves 8 m upward.

Differentiating equation (g) twice yields

$$v_A + 2v_D + v_B = 0 \quad \text{(i)}$$

$$a_A + 2a_D + a_B = 0. \quad \text{(j)}$$

Substituting the velocities at 1.0 second gives

$$8\text{ m/s} + 2(2\text{ m/s}) + v_B = 0$$
$$v_B = -12\text{ m/s (upwards)}.$$

And substituting the acceleration found in equation (b) into (j) yields

$$8\text{ m/s}^2 + 2(0) + a_B = 0$$
$$a_B = -8\text{ m/s}^2\text{ (upwards)}.$$

The motion of a particle along a straight line is described by the equation $x = t^3 - 3t^2 - 45t + 50$, where x is expressed in feet and t in seconds. Compute a) the time at which v(t) = 0, b) the position and distance traveled by the particle at that time, c) the acceleration of the particle at that time, d) the distance traveled by the particle from t = 4 sec. to t = 6 sec.

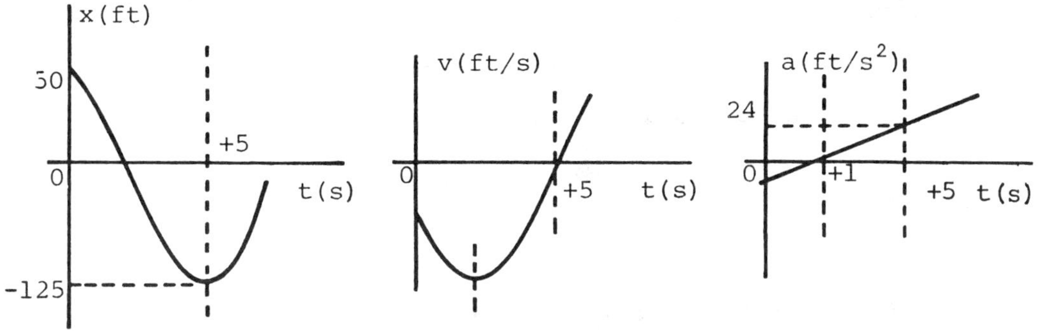

Solution: We know from basic concepts in mechanics that if the position of a particle as a function of time is x = f(t), then the velocity =

$$v = \frac{dx}{dt} = \frac{d\ f(t)}{dt}$$

and the acceleration =

$$a = \frac{dv}{dt} = \frac{d^2\ f(t)}{dt^2} .$$

Therefore $$x = t^3 - 3t^2 - 45t + 50 \qquad (a)$$

$$v = 3t^2 - 6t - 45 \qquad (b)$$

$$a = 6t - 6. \qquad (c)$$

These equations can be applied to solve the above problems.

(a) Setting v = 0 in equation (b) we obtain

$$3t^2 - 6t - 45 = 0.$$

Applying the quadratic formula we obtain the roots t = - 3 s and t = + 5 s. Since the negative value represents the time before the particle started moving, it is rejected. The velocity is zero at the end of 5 seconds.

(b) The position at t = 5 s can now be obtained by substituting t = + 5 sec into equation (a):

$$x(5) = (5)^3 - 3(5)^2 - 45(5) + 50 = - 125 \text{ ft}.$$

The initial position is $x(0) = +50$ ft, also from equation (a). Before asserting that the distance traveled is

$$x(5) - x(0) = -125 - 50 = -175 \text{ ft}, \qquad \text{(d)}$$

we must be certain that the particle did not retrace its path at some time between $t = 0$ s and $t = 5$ s. In order to do so, the particle would have to stop (at least momentarily) before reversing its motion. Since there was no solution for $v = 0$ between 0 and 5 seconds, the particle must have moved directly from $x(0)$ to $x(5)$. Thus, equation (d) is the correct answer to the distance traveled by the particle.

(c) Substituting $t = 5$ s into equation (c) yields

$$a = 6(5) - 6 = +24 \text{ ft/s}^2.$$

(d) To solve this, consider the fact that the velocity equals zero at one point during the interval $t = 4$ s to $t = 6$ s. Then separate the interval into two parts, $t = 4$ sec to $t = 5$ sec and $t = 5$ sec to $t = 6$ sec.

From $t = 4$ sec to $t = 5$ sec we have

$$x(4) = (4)^3 - 3(4)^2 - 45(4) + 50 = -114 \text{ ft}$$

from equation (a) and

$$d_1 = x(5) - x(4) = -125 \text{ ft} - (-114 \text{ ft})$$

$$= -11 \text{ ft}.$$

From $t = 5$ sec to $t = 6$ sec we have

$$x(6) = (6)^3 - 3(6)^2 - 45(6) + 50 = -112 \text{ ft}$$

hence $d_2 = x(6) - x(5) = -112 \text{ ft} - (-125) \text{ ft} = +13 \text{ ft}.$

Note that d_2 and d_1 are of opposite sign thus proving that the particle did retrace its path.

The total distance traveled is

$$D = |d_1| + |d_2| = 11 \text{ ft} + 13 \text{ ft} = 24 \text{ ft}.$$

CHAPTER 7

CURVILINEAR KINEMATICS

PROJECTILE MOTION

• PROBLEM 7-1

The total speed of a projectile at its greatest height, v_1, is $\sqrt{\frac{6}{7}}$ of its total speed when it is at half its greatest height, v_2. Show that the angle of projection is 30°.

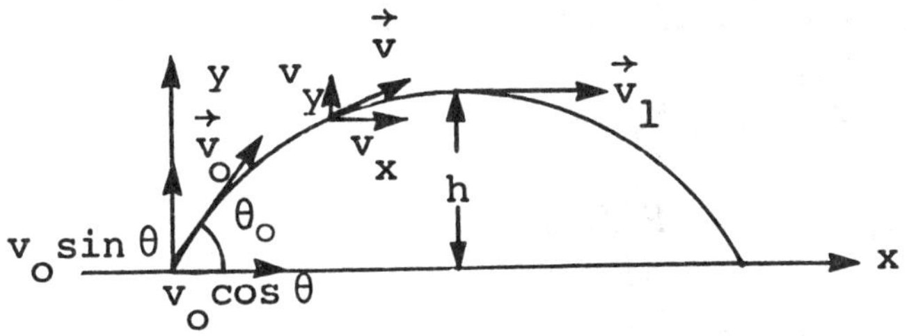

Solution: When a particle is projected as shown in the figure, the component of the velocity in the x-direction stays the same at all times, $v_x = v_0 \cos\theta_0$, since there is no acceleration in that direction, owing to the fact that there is no horizontal component of force acting on the projectile. It is assumed that the force due to air friction is small enough to be neglected.

In the y-direction, the upward velocity is initially $v_0 \sin\theta_0$ and gradually decreases, due to the acceleration g acting downward. At its greatest height, h, the upward velocity is reduced to zero. The kinematic relation for constant acceleration which does not involve time is used to find the greatest height of the trajectory. It is

$$v_f^2 = v_i^2 + 2as \qquad (1)$$

In this case $v_f = 0$, v_i, the initial velocity, is $v_0 \sin\theta_0$, $a = -g$ and $s = h$. Then

$$0 = (v_0 \sin\theta_0)^2 - 2gh \quad \text{or} \quad h = \frac{(v_0 \sin\theta_0)^2}{2g} .$$

The total velocity at the highest point is, therefore, the x-component only. That is, $v_1 = v_0 \cos\theta_0$. At half the greatest height, $h/2 = (v_0 \sin\theta_0)^2/4g$, the velocity in the y-direction, v_y, is obtained from the equation (1) with $v_f = v_y$, $v_i = v_0 \sin\theta_0$, $a = -g$, and $s = h/2$.

$$v_{y^2}^2 = (v_0 \sin\theta_0)^2 - 2g\,\frac{h}{2}$$

$$= (v_0 \sin\theta_0)^2 - \tfrac{1}{2}(v_0 \sin\theta_0)^2$$

$$= \tfrac{1}{2}(v_0 \sin\theta_0)^2 \,. \tag{2}$$

In addition, there is also the ever-present x-component of the velocity $v_0 \cos\theta_0$. Hence the total velocity at this point is obtained by the Pythagorean theorem,

$$v_2^2 = v_x^2 + v_{y^2}^2 = (v_0 \cos\theta_0)^2 + \tfrac{1}{2}(v_0 \sin\theta_0)^2$$

$$= (v_0 \cos\theta_0)^2 + \tfrac{1}{2}v_0^2(1 - \cos^2\theta_0)$$

$$= \tfrac{1}{2}v_0^2 + \tfrac{1}{2}(v_0 \cos\theta_0)^2 \,. \tag{3}$$

Here we used the trigonometric identity $\sin^2\theta + \cos^2\theta = 1$. However, we are given that

$$v_1 = \sqrt{\frac{6}{7}}\,v_2 \qquad \text{or} \qquad \frac{v_1^2}{v_2^2} = \frac{6}{7} \,.$$

Therefore,

$$\frac{(v_0 \cos\theta_0)^2}{\tfrac{1}{2}v_0^2 + \tfrac{1}{2}(v_0 \cos\theta_0)^2} = \frac{6}{7} \,;$$

or $\quad 7(v_0 \cos\theta_0)^2 = 3v_0^2 + 3(v_0 \cos\theta_0)^2$,

or $\quad 4\cos^2\theta_0 = 3$. One can therefore say that

$$\cos\theta_0 = \frac{\sqrt{3}}{2} \qquad \text{or} \qquad \theta_0 = 30° .$$

• PROBLEM 7-2

A projectile, fired with velocity v_0 and angle α_0 from the horizontal, is to impact at point P as shown below. What is the range measured along the straight line connecting the firing point and the target?

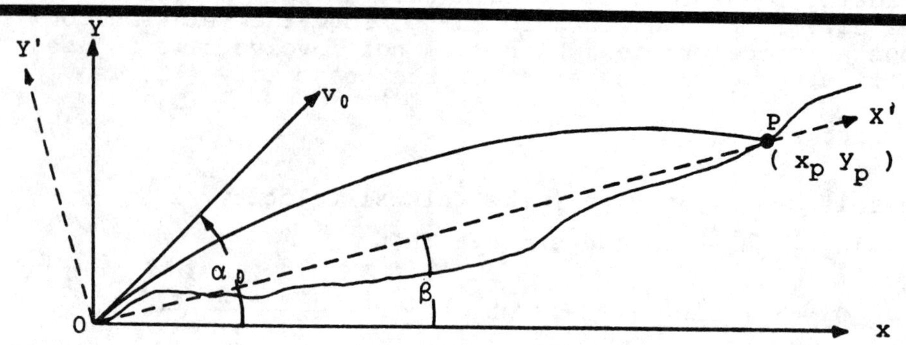

Solution:

$$R_\beta = \frac{x_p}{\cos\beta} = \frac{2}{g\cos\beta}(v_0\cos\alpha_0)^2 \ (\tan\alpha_0 - \tan\beta)$$

The horizontal and vertical components of the initial velocity v_0 are, respectively,

$$v_x = v_0 \cos\alpha_0, \qquad v_y = v_0 \sin\alpha_0$$

If we let t_p be the time it takes to reach point P, the horizontal distance covered is

$$x_p = (v_0 \cos\alpha_0)t_p. \tag{1}$$

Since there is no horizontal acceleration, the vertical distance covered in time t_p is

$$y_p = (v_0 \sin\alpha_0)t_p - \tfrac{1}{2}gt_p^2 \tag{2}$$

where $-g$ is the downward acceleration due to gravity. Eq. (1) gives the elapsed time as

$$t_p = x_p/(v_0 \cos\alpha_0).$$

Substitute this along with the relationship $y_p = x_p \tan\beta$ into Eq. (2) to give

$$x_p \tan\beta = (v_0 \sin\alpha_0)\frac{x_p}{v_0\cos\alpha_0} - \tfrac{1}{2}g\frac{x_p^2}{v_0^2\cos^2\alpha_0}$$

$$\tan\beta = \tan\alpha_0 - \tfrac{1}{2}\frac{g\,x_p}{v_0^2\cos^2\alpha_0}$$

$$x_p = 2(v_0\cos\alpha_0)^2\ (\tan\alpha_0 - \tan\beta)/g.$$

The range along a line from the firing point and the target is now obtained from

$$R_\beta = x_p/\cos\beta.$$

• PROBLEM 7-3

An airplane is traveling horizontally at 480 mph at a height of 6400 ft. The airplane drops a bomb aimed at a stationary target on the ground. To an observer on the aircraft, what angle must the target make with the vertical, when the bomb is dropped, for the bomb to hit the target? Neglect air resistance. (See the figure.)

Suppose that the target is a ship which is steaming at 20 mph away from the aircraft along its line of flight. What alterations would need to be made to the previous calculations?

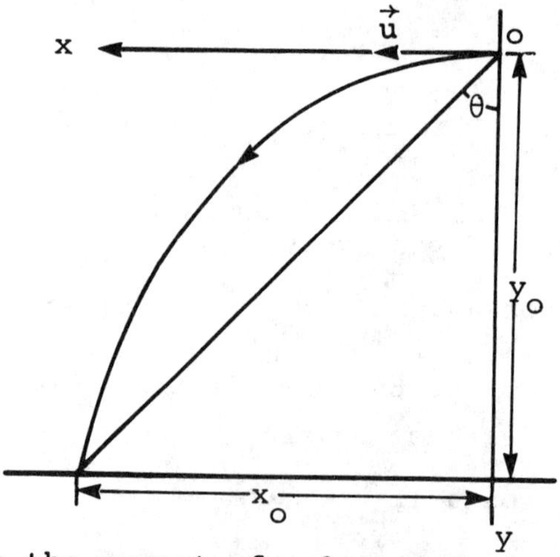

Solution: At the moment of release of the bomb, time t = 0, the airplane is at the point which is taken as the origin of the coordinate system, traveling in the positive x-direction with a speed u of 480 mph.

$$480 \text{ mph} = 480\ \frac{\text{mile}}{\text{hr}} \times \frac{5280 \text{ ft}}{\text{mile}} \times \frac{1 \text{ hr}}{60 \text{ min}} \times \frac{1 \text{ min}}{60 \text{ sec}}$$

$$= 704 \text{ ft/sec.}$$

The bomb has the same initial speed.

There is no acceleration in the x-direction, for no horizontal force acts on the system. Hence, after time t, when the bomb strikes the target, the distance traveled by the bomb in this direction is given by the kinematic equation for constant velocity, $x_0 = ut$.

The airplane and bomb have no initial speed in the y-direction, but the acceleration g acts in this direction. After time t, the downward distance traveled by the released bomb will be, using the kinematic equation for constant acceleration

$$y_o = v_{o_y} t + \tfrac{1}{2}gt^2.$$

Since $v_{o_y} = 0$, this becomes $y_0 = \tfrac{1}{2}gt^2$.

But $y_0 = 6400$ ft in this problem, and the time it takes the object to fall this distance is therefore

$$t = \sqrt{\frac{2y_0}{g}} = \sqrt{\frac{2 \times 6400 \text{ ft}}{32 \text{ ft} \cdot \text{s}^{-2}}} = 20 \text{ s.}$$

Thus, in the same time, the object moves a horizontal distance

$$x_0 = ut = 704 \text{ ft} \cdot \text{s}^{-1} \times 20 \text{ s} = 14{,}080 \text{ ft}$$

and $$\tan\theta = \frac{x_0}{y_0} = \frac{14{,}080 \text{ ft}}{6{,}400 \text{ ft}} = 2.2 \quad \text{or} \quad \theta = 65.5°.$$

The bomb should be released when the target is seen at an angle of 65.5° to the vertical.

If the target is moving, the relative velocity between plane and ship is the important velocity. For, relative to the ship, the bomb has an initial velocity $\vec{v}_{BS} = \vec{v}_{BW} + \vec{v}_{WS}$, where $\vec{v}_{BW}$ is the initial velocity of the bomb relative to the water, and $\vec{v}_{WS}$ the velocity of the water relative to the ship. Since the velocity of the ship relative to water $(\vec{v}_{SW})$ is given as 20 mph, the $\vec{v}_{WS}$ = - 20 mph. Thus

$$v_{BS} = (480 - 20)\text{mph} = 460 \text{ mph}$$

$$= 460 \frac{\text{mile}}{\text{hr}} \times \frac{5280 \text{ ft}}{\text{mile}} \times \frac{1 \text{ hr}}{60 \text{ min}} \times \frac{1 \text{ min}}{60 \text{ sec}}$$

$$= 674.6 \text{ ft/sec},$$

The foregoing analysis can thus be carried out once more, with v_{BS} in place of u. Thus

$$x_0 = v_{BS}t = 674.6 \text{ ft/sec} \times 20 \text{ sec} = 13,492 \text{ ft}$$

$$\tan \theta' = \frac{13,492 \text{ ft}}{6,400 \text{ ft}} = 2.1$$

and the bomb should now be released when the target is seen at an angle $\theta' = 64.5°$ to the vertical.

• PROBLEM 7-4

A gun is fired at a moving target. The bullet has a projectile speed of 1200 ft/sec. Both gun and target are on level ground. The target is moving 60 mph away from the gun and it is 30,000 ft from the gun at the moment of firing. Determine the angle of elevation θ needed for a direct hit.

Solution: First we note that 60 mph = 88 fps, and that the horizontal and vertical components of the initial velocity are, respectively,

$$v_{ox} = 1200 \cos \theta, \qquad v_{oy} = 1200 \sin \theta .$$

For the projectile to hit the target after flying a time t we note that the horizontal distance covered by the projectile must equal the distance covered by the target plus the original separation. Since there is no horizontal acceleration, the equation is:

$$30,000 + v_t t = v_{ox} t,$$

where v_t is the target velocity. Thus

$$30,000 + 88t = 1200 (\cos \theta)t. \qquad (1)$$

During the same time the net vertical distance covered by the projectile will be zero. Thus we use the formula

$$s_y = v_{oy}t - \tfrac{1}{2}gt^2 = 0$$

where g = 32.2 is the acceleration due to gravity acting downward. This gives

$$\tfrac{1}{2}gt^2 = 1200 \sin \theta t$$

$$16.1t = 1200 \sin \theta. \qquad (2)$$

Now we must solve the simultaneous system of Eqs. (1) and (2). First we re-arrange Eq. (2) to give

$$\sin \theta = \frac{16.1}{1200} t_0 \qquad (3)$$

Now from trigonometry we have

$$\cos \theta = \sqrt{1 - \sin^2 \theta} = \left[1 - \left(\frac{16.1}{1200}\right)^2 t^2\right]^{\frac{1}{2}}.$$

This is substituted into Eq. (1) to give

$$30{,}000 + 88t = 1200t \left[1 - \left(\frac{16.1}{1200}\right)^2 t^2\right]^{\frac{1}{2}}$$

This equation is non linear and not readily solvable. An iterative procedure will lead us to the solution. As a first approximation, we let t in the bracket equal zero and solve for t outside the radical

$$30{,}000 + 88t_1 = 1200t_1$$

$$t_1 = 26.98.$$

Now, substitute this new value for t_1 into the recursion formula,

$$30{,}000 + 88t_n = 1200\, t_n \left[1 - \left(\frac{16.1}{1200}\right)^2 t_{n-1}^2\right]^{\frac{1}{2}}$$

we get for our second trial value

$$30{,}000 + 88t_2 = 1200[.932]t_2$$

$$t_2 = 29.11.$$

Proceeding in the same way substitute t_2 into the recursion formula and solve for t_3. After 4 further trials one gets $t_6 = 29.61$, the next trial gives $t_7 = 29.61$ so we may stop the procedure here with confidence that t = 29.6 sec. Further accuracy, if desired, would require more iterations.

Then from Eq. (3) we find

$$\theta = \sin^{-1}\left[\frac{(16.1)(29.6)}{1200}\right] = 23.4°.$$

• **PROBLEM 7-5**

A stone is thrown from the top of a 200 m building with an initial velocity of 150 m/s at an angle of 30° with the horizontal line. Neglecting the air resistance, determine a) the horizontal distance from the building to the point where the stone lands, b) the maximum height above the ground reached by the stone.

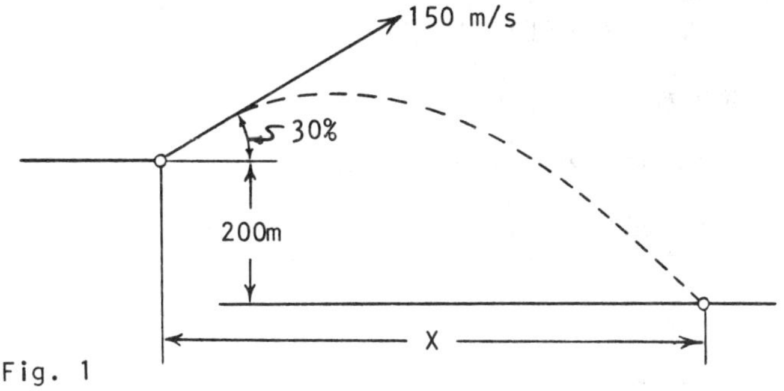

Fig. 1

Solution: The best way to solve projectile problems is to consider the horizontal and vertical motions separately. The initial velocity given may be decomposed as such:

$$v_{ox} = v_o \cos\theta$$

$$v_{ox} = 150 \text{ m/s } (\cos 30°) = + 129.9 \text{ m/s}$$

$$v_{oy} = v_o \sin\theta$$

$$v_{oy} = 150 \text{ m/sec } (\sin 30°) = + 75 \text{ m/s.}$$

The vertical motion is uniformly accelerated motion. The horizontal motion is uniform motion. The horizontal velocity remains constant. Thus, the equation for the horizontal distance is simply

$$x = (v_{ox})\, t \quad . \qquad \text{(a)}$$

The time is not known yet. However, it is clear that the same time taken for the horizontal motion is taken for the vertical motion. Thus, we will solve the vertical motion equation

$$y = v_{oy}\, t + \tfrac{1}{2} a_y\, t^2 \qquad \text{(b)}$$

for the time. Substituting values into (b) yields

$$- 200 \text{ m} = (75 \text{ m/s})\, t + \tfrac{1}{2}(- 9.8 \text{ m/s}^2)\, t^2 \qquad \text{(c)}$$

Rearranging (c) yields the quadratic equation

$$t^2 - 15.3\, t - 40.8 = 0.$$

The quadratic formula

$$t = \frac{15.3 \pm \sqrt{(15.3)^2 - 4(1)(-40.8)}}{2}$$

yields the only positive root of this equation

$t = 17.62$ seconds.

Now this value may be used in equation (a) to get

$x = (129.9 \text{ m/s})(17.62 \text{ s})$

$x = 2289 \text{ m}$

for the desired horizontal distance.

The projectile reaches the maximum height at the moment when $v_y = 0$. Applying the equation

$$v_y = v_{oy} + at$$

yields $0 = + 75 \text{ m/s} - 9.8 \text{ m/s}^2 \, t$

or $\quad t = \frac{-75}{-9.8} \text{ s} = 7.7 \text{ s}.$

Now, substituting this value into equation (b) to solve for y_{max} yields

$$y_{max} = + 75 \text{ m/s } (7.7 \text{ s}) - 4.9 \text{ m/s}^2 \, (7.7 \text{ s})^2 = 287 \text{ m}.$$

However, this describes the distance above the roof which the stone reaches rather than its elevation from the ground. The elevation above the ground is found by adding to y_{max} the height of the building. Hence, the greatest elevation

$= 200 \text{ m} + 287 \text{ m} = 487 \text{ m}.$

• PROBLEM 7-6

A rocket is launched from point B and its flight is tracked by an optical instrument from point A. Express the velocity in terms of s, θ and $\dot{\theta}$.

Solution: From trigonometry we note that

$x = s \tan \theta.$

Differentiate this with respect to time to obtain

$$\frac{dx}{dt} = s \sec^2 \theta \frac{d\theta}{dt} \quad .$$

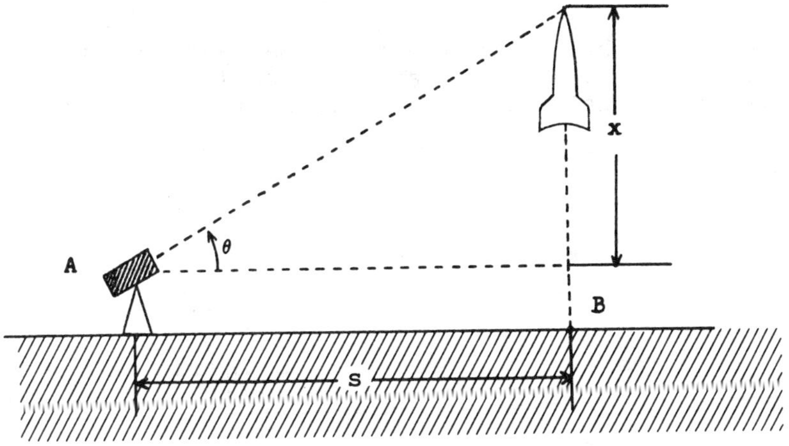

Since the velocity $v = dx/dt$, if we let $\dot{\theta} = d\theta/dt$. This becomes

$$v = s\,\dot{\theta}\,\sec^2\theta.$$

• PROBLEM 7-7

> In a gymnasium with a ceiling 30 ft high, a player throws a ball towards a wall 80 ft away. If he releases the ball 5 ft above the floor with initial velocity v_0 of 55 ft/sec, determine the highest point at which the ball could strike the wall.

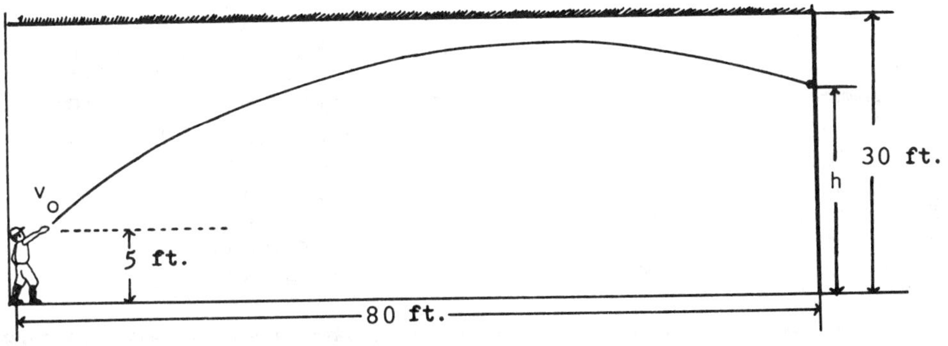

Solution: The highest point where the ball could strike the wall is limited by the ceiling height. This in turn limits the vertical component of the release velocity, v_{oy}. The latter is found by specifying that it will be just enough to cause the maximum height reached during flight to equal 25 ft.; that is, the ceiling height less the height of the release point. At the maximum point the vertical velocity is zero. Thus we use the formula

$$v_y^2 = v_{oy}^2 - 2gs_v = 0$$

where $g = 32.2$ is the acceleration of gravity acting downward. From this

$$v_{oy} = \sqrt{2gs} = \sqrt{64.4(25)} = 40.1 \text{ ft/sec.}$$

To find the height where the ball strikes we first find the time required to reach the wall. This requires, v_{ox} the horizontal component of velocity, which is found from the Pythagorean relation

$$v_o^2 = v_{ox}^2 + v_{oy}^2 \quad \text{or} \quad v_{ox}^2 = v_o^2 - v_{oy}^2.$$

Thus, upon substitution

$$v_{ox} = \sqrt{(55)^2 - (40.1)^2} = 37.6 \text{ ft/sec.}$$

Since there is no horizontal acceleration the time required to hit the wall at this velocity is found from

$$s_h = v_{ox}t = 80 \text{ ft.}$$

Therefore

$$t = \frac{80}{37.6} = 2.1 \text{ sec.}$$

Now, the vertical height at which the ball strikes after this time is found from

$$h = v_{oy}t - \tfrac{1}{2}gt^2 + s_o$$

where s_o is the height of the release point. Substitution then gives

$$h = 40.1(2.1) - \tfrac{1}{2}(32.2)(2.1)^2 + 5$$

$$h = 18.2 \text{ ft.}$$

• PROBLEM 7-8

A projectile of relatively small dimensions is fired with a velocity v_0 from a gun at an angle of elevation θ with the horizontal, as shown in the figure. Neglecting air resistance, determine the maximum height h and the distance r from the gun to the impact point P if we assume the gun is aimed down a hill whose surface has an angle β with the horizontal.

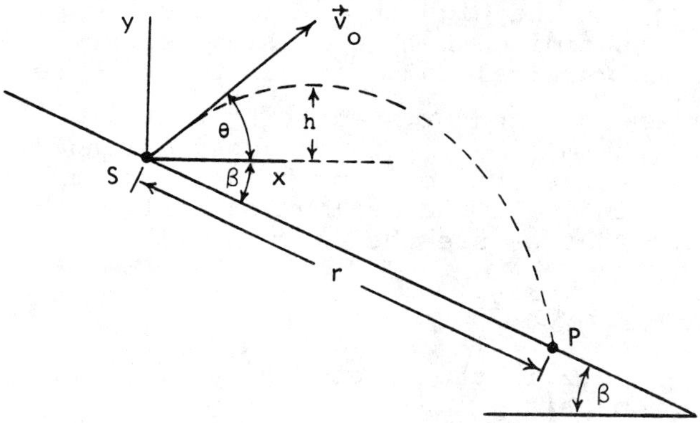

Solution: The x axis is chosen horizontally through the starting point S. The projectile has an initial horizontal speed $v_0 \cos\theta$ to the right and an initial vertical speed $v_0 \sin\theta$ upward. Since air resistance is neglected, there will be no horizontal acceleration. Hence, the horizontal speed is constant. In the vertical direction, however, there is an acceleration g due to gravity, which is directed downward and is negative if up is assumed positive. The

equations expressing the above facts are

$$\ddot{x} = 0 \qquad \text{and} \qquad \ddot{y} = -g. \qquad \text{(a)}$$

Integration yields,

$$\dot{x} = C_1 \qquad \text{and} \qquad \dot{y} = -gt + C_2 .$$

However, $\dot{x}$ has a constant value $v_0 \cos\theta$. Also, at $t = 0$, it is known that $\dot{y} = v_0 \sin\theta$. So,

$$\dot{x} = v_0 \cos\theta \qquad \text{and} \qquad \dot{y} = -gt + v_0 \sin\theta.$$

Another integration yields

$$x = (v_0 \cos\theta)t + C_3 \text{ and } y = \frac{-gt^2}{2} + (v_0 \sin\theta)t + C_4.$$

But, at $t = 0$, it is known that $x = 0$ and $y = 0$. Therefore, $C_3 = C_4 = 0$. The equations of motion are

$$x = (v_0 \cos\theta)t + C \text{ and } y = \frac{-gt^2}{2} + (v_0 \sin\theta)t. \qquad \text{(b)}$$

The coordinates of the impact point P are $x = r\cos\beta$ and $y = -r\sin\beta$. Placing these values into equation (b) we will get two equations in the unknowns, r and t, and we will be able to solve for r. Thus,

$$r\cos\beta = (v_0 \cos\theta)t \quad \text{and} \quad -r\sin\beta = \frac{-gt^2}{2} + (v_0 \sin\theta)t.$$

From the first equation,

$$t = \frac{r\cos\beta}{v_0 \cos\theta} .$$

By substituting this value in the second equation, we obtain

$$-r\sin\beta = \frac{-g}{2}\left(\frac{r\cos\beta}{v_0\cos\theta}\right)^2 + v_0 \sin\theta \frac{r\cos\beta}{v_0\cos\theta} .$$

Eliminating one power of r and rearranging the equation yields

$$r\left(\frac{g}{2}\ \frac{\cos^2\beta}{v_0^2\cos^2\theta}\right) = \sin\beta + \frac{\sin\theta\cos\beta}{\cos\theta}\ ,$$

$$r = \frac{2v_0^2}{g}\left(\frac{\cos^2\theta\sin\beta}{\cos^2\beta}\right) + \frac{v_0^2}{g}\ \frac{2\cos^2\theta\sin\theta\cos\beta}{\cos^2\beta\ \cos\theta}$$

$$= \frac{2v_0^2}{g}\cos^2\theta\sec\beta\tan\beta + \frac{v_0^2\sin 2\theta}{g\cos\beta}\ . \qquad (c)$$

To obtain the maximum height, set $\dot{y} = 0$ and solve as done elsewhere in this chapter. Here, however, a different technique will be used. This involves first finding the equation for $y = f(x)$ and then using $dy/dx = 0$ for the maximum height condition.

Solving for t in the left side of equation (b) and substituting for t on the right side of (b) yields:

$$y = \frac{-g}{2}\left(\frac{x}{v_0\cos\theta}\right)^2 + (v_0\sin\theta)\frac{x}{v_0\cos\theta}$$

$$= \frac{-g}{2v_0^2\cos^2\theta}x^2 + (\tan\theta)x \qquad (d)$$

Now, use $dy/dx = 0$ to find the value of x at which the curve reaches its maximum height or at which the slope is zero. Thus,

$$\frac{dy}{dx} = \frac{-g}{v_0^2\cos^2\theta}x + \tan\theta = 0$$

$$x = \frac{-\tan\theta\ v_0^2\cos^2\theta}{-g} = \frac{v_0^2}{g}\sin\theta\cos\theta$$

This is the x-coordinate of the maximum height location. Substituting this value into equation (d):

$$y = \frac{-g}{2v_0^2\cos^2\theta}\left(\frac{v_0^2}{g}\sin\theta\cos\theta\right)^2 + \tan\theta\left(\frac{v_0^2}{g}\sin\theta\cos\theta\right)$$

$$= -\frac{v_0^2}{2g}\sin^2\theta + \frac{v_0^2}{g}\sin^2\theta$$

$$= \frac{v_0^2\sin^2\theta}{2g}\ .$$

The reader may check this result by using the more common (and easier!) method.

• PROBLEM 7-9

A mortar at point A must lob shells over the cliff shown. (a) How far back must it be placed to shoot cleanly over the cliff? (b) How far beyond the edge will the shell strike?

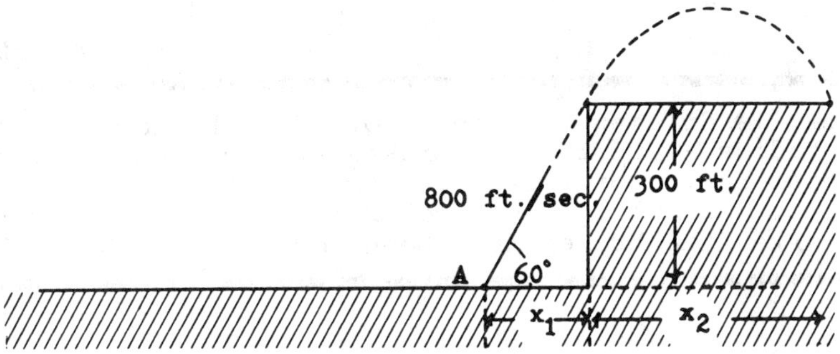

Solution: The horizontal and vertical components of the initial velocity are

$$v_{ox} = 800 \cos 60° = 400 \text{ ft/sec}$$

$$v_{oy} = 800 \sin 60° = 692 \text{ ft/sec.}$$

The time t required to reach the edge of the cliff, i.e., to reach a height of 300 ft., is found from

$$y = v_{oy}t - \tfrac{1}{2} gt^2 = 300$$

where g = 32.2 is the acceleration due to gravity acting downward. Upon substitution we obtain the equation

$$16.1t^2 - 692t + 300 = 0.$$

This is solved by the quadratic formula to give two results

$$t_1, t_2 = \frac{692 \pm \sqrt{(692)^2 - 4(16.1)(300)}}{2(16.1)}$$

$$t_1, t_2 = 0.438, 42.5 \text{ sec.}$$

The lower of these is the time required to rise to 300 ft. and the higher of these is the time required to rise and fall back to 300 ft. The distance x_1 is found by using the first value and the horizontal velocity component;

i.e.: $x_1 = v_{ox}t = 400(.438) = 175$ ft.

The distance x_2 is found by using the second value of time and the horizontal velocity component, i.e.:

$$x_2 = v_{ox}(t_2 - t_1)$$

$$= v_{ox}t_2 - v_{ox}t_1$$

$$= v_{ox}t_2 - x_1$$

$$= 400(42.5) - 175 = 16{,}825 \text{ ft.}$$

• PROBLEM 7-10

A race car is driving at a speed of 80 mi/h along a curved road of radius 2000 ft. The brakes are applied causing constant deceleration of the car, after 8s the speed decreases to 50 mi/h. Calculate the acceleration of the car at the instant after the brakes have been applied.

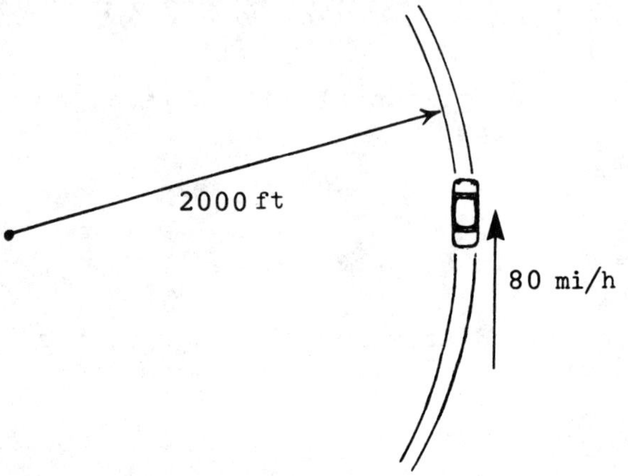

Solution: Because the car is moving on a curved road, it is essential to consider not only the acceleration that results in a decrease of speed (tangential acceleration), but also the centripetal acceleration which keeps the car moving on an arc.

Tangential Component. The standard equation for constant acceleration is

$$a = \frac{\Delta v}{\Delta t} \quad . \tag{a}$$

However, the units, as given in the problem, are mixed. Thus, it is necessary to re-express, the velocities involved:

$$80 \frac{mi}{h} = \left(80 \frac{mi}{h}\right)\left(\frac{5280 \text{ ft}}{1 \text{ mi}}\right)\left(\frac{1 \text{ h}}{3600 \text{ s}}\right) = 117 \text{ ft/s}$$

and $$50 \frac{mi}{h} = (50)(5280)\left(\frac{1}{3600}\right) \text{ ft/s} = 74 \text{ ft/s}.$$

At this point, apply equation (a) with Δt = the braking time,

$$a_T = \frac{\Delta v}{\Delta t} = \frac{74 \text{ ft/s} - 117 \text{ ft/s}}{8 \text{ s}}$$

$$a_T = -5.38 \text{ ft/s}^2.$$

Centripetal Acceleration. The standard equation for centripetal acceleration is

$$a_C = \frac{v^2}{r} \qquad \text{(b)}$$

The question asks for the acceleration immediately after the brakes have been applied, that is, when the velocity is still (approximately) 80 mi/h. Hence,

$$a_C = \frac{(117 \text{ ft/s})^2}{2000 \text{ ft}} = 6.84 \text{ ft/s}^2.$$

<u>Magnitude and Direction of Acceleration.</u> The magnitude and direction of the resultant $\underline{a}$ of the components a_C and a_T are

$$\tan a = \frac{a_C}{a_T} = \frac{6.84 \text{ ft/s}^2}{-5.38 \text{ ft/s}^2} \qquad a = -51.8°$$

$$a = \frac{a_C}{\sin a} = \frac{a_T}{\cos a} = -8.7 \text{ ft/s}^2$$

To calculate the magnitude only:

$$a = \sqrt{a_C^2 + a_T^2} = \sqrt{(+\ 6.84\text{ft/s})^2 + (-\ 5.38\text{ft/s})^2}$$

$$= 8.7 \text{ ft/s}^2.$$

• PROBLEM 7-11

Assume a particle moves in the x-y plane such that its x and y coordinates at any time t are governed by

$$\ddot{x} - 2b\dot{y} + \lambda x = 0 \qquad \text{(a)}$$

$$\ddot{y} + 2b\dot{x} + \lambda y = 0. \qquad \text{(b)}$$

Find the equation of the motion.

<u>Solution</u>: Let us introduce the complex variable

$$z = x + iy$$

where i is the imaginary $\sqrt{-1}$. Multiplying (b) by i and then adding to (a) yields

$$\ddot{z} + 2\,ib\dot{z} + \lambda z = 0 \qquad \text{(c)}$$

Since (c) is a homogeneous equation with constant coefficients we assume a solution of the form

$$z = e^{ikt}.$$

Substituting this into (c) we obtain the auxiliary equation

$$k^2 + 2bk - \lambda = 0.$$

Solving by the quadratic formula

$$k = \frac{-2b \pm \sqrt{4b^2 + 4\lambda}}{2}$$

$$k = -b \pm \sqrt{b^2 + \lambda} \quad .$$

Thus our general solution is

$$z = e^{-ibt} (Ae^{imt} + Be^{-imt}) \qquad \text{(d)}$$

where $m = \sqrt{b^2 + \lambda}$ and A and B are arbitrary constants. We now set the initial conditions as

$$x(0) = a, \qquad y(0) = 0$$

$$\dot{x}(0) = 0 \qquad \dot{y}(0) = V.$$

Thus $\quad z(0) = a, \qquad \dot{z}(0) = iV.$

From (d)

$$z(0) = e^0(Ae^0 + Be^0)$$

$$a = A + B. \qquad \text{(e)}$$

and $\dot{z}(t) = -ibe^{-ibt}(Ae^{imt} + Be^{-imt})$

$$+ e^{-ibt}(imAe^{imt} - imBe^{-imt}),$$

$$\dot{z}(0) = -ibe^0(Ae^0 + Be^0) + ime^0(Ae^0 - Be^0),$$

$$iV = -ib(A + B) + im(A - b). \qquad \text{(f)}$$

From (e) $A = a - B$ which we substitute in (f):

$$iV = -ib(a - B + B) + im(a - 2B),$$

$$V = -ba + m(a - 2B),$$

$$B = \frac{a}{2} - \frac{V + ba}{2m}$$

and $\quad A = \dfrac{a}{2} + \dfrac{V + ba}{2m} \quad .$

The final solution is

$$z = e^{-ibt}((R + S)e^{imt} + (R - S)e^{-imt})$$

where $\quad R = \dfrac{a}{2}, \; S = \dfrac{(V + ba)}{2m} \quad .$

Using Euler's identity

$$(e^{i\theta} = \cos\theta + i\sin\theta),$$

$$z = e^{-ibt}\,(2R \cos mt + 2i\,S \sin mt),$$

$$z = e^{-ibt}\left[a \cos mt + i\left(\frac{V + ba}{m}\right) \sin mt\right].$$

We now return to parametric equations:

Let $X(t) = a \cos mt$

$$Y(t) = \left(\frac{V + ba}{m}\right) \sin mt;$$

then $$\frac{X^2}{a^2} + \frac{Y^2}{\left(\frac{V + ba}{m}\right)} = 1$$

which is the equation of an ellipse.

Since $x(t) = e^{-ibt}\,X(t)$

$$y(t) = e^{-ibt}\,Y(t)$$

the motion of the particle is on an ellipse which is itself rotating about the origin with angular velocity b.

This describes, among other things, the foucault pendulum and the motion of an orbiting electron in a transverse magnetic field (Zeeman effect).

• PROBLEM 7-12

A particle moves along the path $r = 3\phi$ so that $\phi = 2t^3$. Time is in seconds, ϕ is in radians, and r is in feet. Determine the velocity of the particle when $\phi = 0.5$ rad.

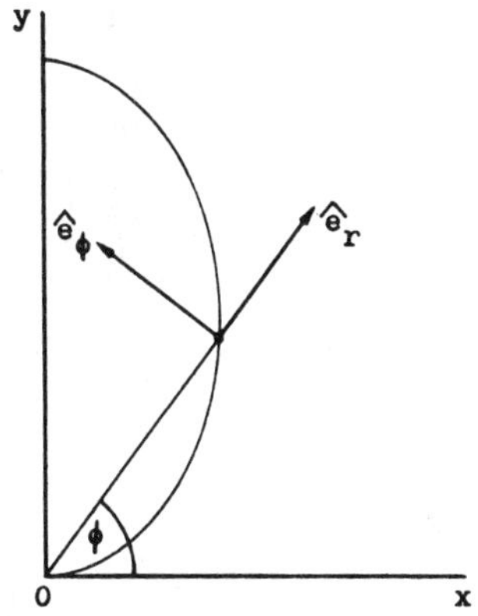

fig.1

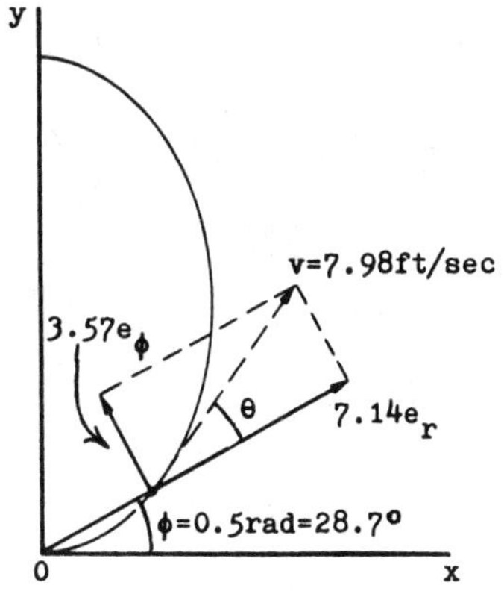

fig.2

Solution: The path $r = 3\phi$ is illustrated in figure 1 along with the unit vectors $\hat{e}_r$ and $\hat{e}_\phi$. The velocity vector

$$\vec{v} = \dot{r}\,\hat{e}_r + r\,\dot{\phi}\hat{e}_\phi$$

When ϕ is 0.5 rad, solving the equation $\phi = 2t^3$ for t yields the following result:

$$t = \sqrt[3]{0.5/2} = 0.63 \text{ sec}$$

Also $\dot{\phi} = \frac{d}{dt}(2t^3) = 6t^2$

and $\dot{r} = \frac{d}{dt}[3(2t^3)] = 18t^2$

When t = 0.63 sec, $\dot{\phi} = 6(0.63)^2 = 2.38$ rad/sec and $\dot{r}$ = 7.14 ft/sec. Hence

$$\vec{v} = 7.14\,\hat{e}_r + (3 \times 0.5)(2.38)\hat{e}_\phi = 7.14\hat{e}_r + 3.57\hat{e}_\phi$$

This vector is shown in Fig. 2. Note that

$$|\vec{v}| = \sqrt{7.14^2 + 3.57^2} = 7.98 \text{ ft/sec}$$

and $\theta = \tan^{-1}\frac{3.57}{7.14} = 26.6°$.

The angle between the x axis and the velocity vector is

$$\beta = \theta + \phi = 55.3°.$$

This problem can be checked by using rectangular coordinates. Note that $x = r\cos\phi$ and $y = r\sin\phi$. Hence, in terms of time t, the coordinates become

$$x = r\cos\phi = 3\phi\cos\phi = 3(2t^3)\cos 2t^3$$

$$y = r\sin\phi = 3\phi\sin\phi = 3(2t^3)\sin 2t^3$$

Also $\dot{x} = 18t^2\cos 2t^3 + 6t^3(-\sin 2t^3)(6t^2)$

$$\dot{y} = 18t^2\sin 2t^3 + 6t^3(\cos 2t^3)(6t^2)$$

When t = 0.63 sec,

$$\dot{x} = 18(0.63)^2\cos 0.5 - 36(0.63)^5\sin 0.5$$

$$= 6.27 - 1.71 = 4.56 \text{ ft/sec}$$

$$\dot{y} = 18(0.63)^2\sin 0.5 + 36(0.63)^5\cos 0.5$$

$$= 3.43 + 3.13 = 6.56 \text{ ft/sec.}$$

Hence, by rectangular components,

$$\vec{v} = 4.56\hat{i} + 6.56\hat{j}$$

Then $|\vec{v}| = \sqrt{4.56^2 + 6.56^2} = 7.98$ ft/sec and $\beta = 55.3°$.

• PROBLEM 7-13

Determine the acceleration of the particle from the previous problem when $\phi = 0.5$ rad.

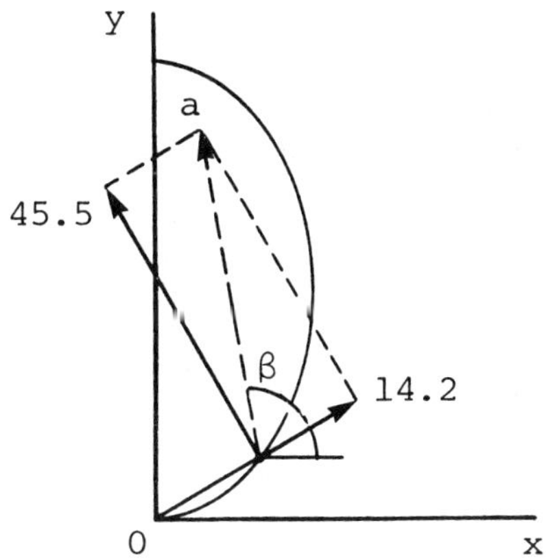

Solution: The equation for acceleration in polar coordinates is

$$\vec{a} = (\ddot{r} - r\dot{\phi}^2)\hat{e}_r + (r\ddot{\phi} + 2\dot{r}\dot{\phi})\hat{e}_\phi$$

where $\hat{e}_r$ and $\hat{e}_\phi$ are, respectively, the unit vectors outward along the line for which $\phi = 0.5$ rad and perpendicular to that line.

From the preceding example $\dot{\phi} = 6t^2$ and $\dot{r} = 18t^2$. The second time derivatives are $\ddot{\phi} = 12t$ and $\ddot{r} = 36t$. When these values are substituted, the result is

$$\vec{a} = [36t - (6t^3)(6t^2)^2]\hat{e}_r + [(6t^3)(12t) + 2(18t^2)(6t^2)]\hat{e}_\phi$$

$$= (36t - 216t^7)\hat{e}_r + (288t^4)\hat{e}_\phi$$

When t = 0.63 sec, this expression becomes

$$\vec{a} = 14.2\hat{e}_r + 45.5\hat{e}_\phi$$

This vector is shown in figure 1. Its magnitude is

$$|\vec{a}| = \sqrt{14.2^2 + 45.5^2} = 47.5 \text{ ft/sec} .$$

The angle β between the x axis and the resultant acceleration vector is

$$\beta = (0.5)(57.3°) + \tan^{-1}\frac{45.5}{14.2} = 101°.$$

As in the last problem, this result can be checked by using rectangular coordinates. From the previous results

$$\dot{x} = 18t^2 \cos 2t^3 - 36t^5 \sin 2t^3$$

$$\dot{y} = 18t^2 \sin 2t^3 + 36t^5 \cos 2t^3$$

The second time derivatives are

$$\ddot{x} = 36t \cos 2t^3 - 18t^2(\sin 2t^3)(6t^2) - 180t^4 \sin 2t^3 - 36t^5(\cos 2t^3)(6t^2)$$

$$= (36t - 216t^7)\cos 2t^3 - 288t^4 \sin 2t^3$$

$$\ddot{y} = 36t \sin 2t^3 + 18t^2(\cos 2t^3)(6t^2) + 180t^4 \cos 2t^3 - 36t^5(\sin 2t^3)(6t^2)$$

$$= (36t - 216t^7)\sin 2t^3 + 288t^4 \cos 2t^3.$$

Using the fact that the arguments of the trig functions $2t^3$ is equal to $\phi = 0.5$ rad at $t = 0.63$ sec, $\ddot{x}$ and $\ddot{y}$ are easily evaluated. The results are $\ddot{x} = -9.27$ and $\ddot{y} = 46.6$. The magnitude of the acceleration is

$$|a| = \sqrt{(-9.27)^2 + 46.6^2} = 47.5 \text{ ft/sec}^2.$$

The angle β between the x axis and the acceleration vector is

$$\beta = \tan^{-1} \frac{46.6}{-9.27} = 101°.$$

• PROBLEM 7-14

A particle moves along the curve $y = x^{3/2}$ such that its distance from the origin, measured along the curve, is given by $s = t^3$. Determine the acceleration when $t = 2$ sec. Units used are inches and seconds.

Solution: By differentiating the distance function we obtain

$$\dot{s} = 3t^2 \text{ and at } t = 2 \text{ sec, } \dot{s} = 12 \text{ in./sec}$$

$$\ddot{s} = 6t \text{ and at } t = 2 \text{ sec, } \ddot{s} = 12 \text{ in./sec}^2$$

This gives us the tangential acceleration. We still need to find the normal acceleration using the equation

$$a_n = \frac{\dot{s}^2}{\rho} .$$

Also by differentiating the equation of the path we obtain

$$\frac{dy}{dx} = \frac{3x^{\frac{1}{2}}}{2} \qquad \text{and} \qquad \frac{d^2y}{dx^2} = \frac{3x^{-\frac{1}{2}}}{4} .$$

The curvature is given by

$$\frac{1}{\rho} = \frac{\frac{d^2y}{dx^2}}{\left[1 + \left(\frac{dy}{dx}\right)^2\right]^{3/2}} = \frac{\frac{3x^{-\frac{1}{2}}}{4}}{\left[1 + \left(\frac{3x^{\frac{1}{2}}}{2}\right)^2\right]^{3/2}}$$

To determine the x coordinate at t = 2 sec we relate x and s, through differential geometry, by the expression

$$ds = \sqrt{dx^2 + dy^2} = \sqrt{1 + \left(\frac{dy}{dx}\right)^2}\, dx = \sqrt{1 + \frac{9}{4}x}\, dx$$

Integration yields

$$\int_0^s ds = \int_0^x \sqrt{1 + \frac{9}{4}x}\, dx \quad \text{or} \quad s = \frac{8}{27}\left[1 + \frac{9}{4}x\right]^{3/2}$$

When t = 2 sec, s = 8 in. and hence

$$8 = \frac{8}{27}\left[1 + \frac{9}{4}x\right]^{3/2} \qquad \text{or} \qquad x = \frac{32}{9} \text{ in.}$$

Substitution in the expression for curvature gives

$$\frac{1}{\rho} = \frac{\frac{3}{4}\left[\frac{32}{9}\right]^{-\frac{1}{2}}}{\left[1 + \frac{9}{4}\left(\frac{32}{9}\right)\right]^{3/2}} = \frac{1}{48\sqrt{2}} .$$

The normal component of the acceleration is

$$a_n = (12 \text{ in/sec})^2 \left(\frac{1}{48\sqrt{2}} \text{ in}^{-1}\right)$$

$$= 2.12 \text{ in/sec}^2$$

So the total acceleration vector at t = 2 sec is:

$$\vec{a} = 12\hat{e}_t + 2.12\hat{e}_n \text{ in./sec}^2 .$$

• PROBLEM 7-15

A particle moves along the path $y = x^2$ in such a way that $x = 2t$. Determine the velocity and the acceleration in terms of time t.

Solution: The coordinates x and y in terms of time are $x = 2t$ and $y = 4t^2$. Then

$$\dot{x} = 2 \qquad \text{and} \qquad \dot{y} = 8t$$

$$\ddot{x} = 0 \qquad \text{and} \qquad \ddot{y} = 8$$

From differential geometry, $ds = \sqrt{(dx)^2 + (dy)^2}$ where s and ds are measured along the path of the particle. Hence,

$$\frac{ds}{dt} = \sqrt{\left(\frac{dx}{dt}\right)^2 + \left(\frac{dy}{dt}\right)^2} = \sqrt{\dot{x}^2 + \dot{y}^2} = \sqrt{4 + 64t^2}$$

So the velocity has a magnitude of $\sqrt{4 + 64t^2}$ and its direction is along the tangent for which the slope is $\frac{dy}{dx} = \frac{\dot{y}}{\dot{x}} = 4t.$

The acceleration component along the tangent line is

$$\ddot{s} = \frac{d}{dt}\left(\frac{ds}{dt}\right) = \frac{d}{dt}\,(4 + 64t^2)^{\frac{1}{2}} = \frac{64t}{\sqrt{4 + 64t^2}} \,.$$

The centripetal acceleration is given by

$$a = \frac{\dot{s}^2}{r} \,. \tag{a}$$

However, r = r(t) has not been determined yet and developing an expression for the radius of curvature will require some digression.

The curvature at a point, which we call K, is equal to 1/r. K is defined as the rate of change of direction of a curve. Draw a curve with a tangent at a point which has slope θ with the horizontal.

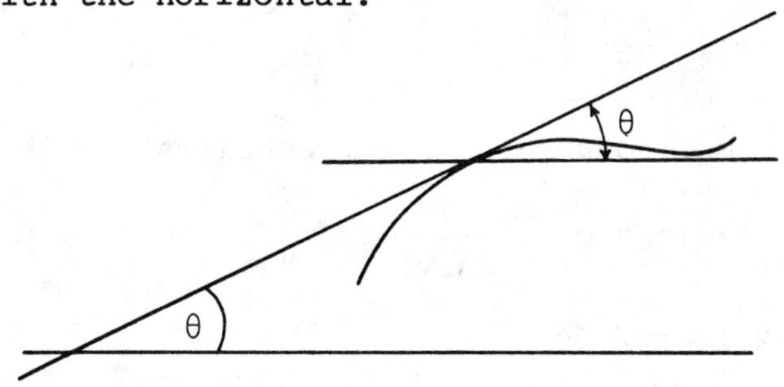

$$\theta = \tan^{-1} \frac{dy}{dx}$$

$$\frac{d\theta}{dx} = \frac{d}{dx}\,(\tan^{-1} y') = \frac{1}{1 + (y')^2}\,\frac{dy'}{dx}$$

$$= \frac{y''}{1 + (y')^2} \tag{b}$$

Again using $(ds)^2 = (dx)^2 + (dy)^2$

$$\frac{ds}{dx} = \left[1 + \left(\frac{dy}{dx}\right)^2\right]^{\frac{1}{2}} \tag{c}$$

The change of direction of this curve would be

$$K = \frac{d\theta}{ds} \quad .$$

To solve for K we divide equation (c) into equation (b) to give us

$$K = \frac{d\theta}{ds} = \frac{y''}{[1 + (y')^2]^{3/2}}$$

so $$\frac{1}{r} = \frac{y''}{[1 + (y')^2]^{3/2}} \quad . \tag{d}$$

We will now develop some equations relating space derivatives to time derivatives.

$$\frac{dy}{dx} = \frac{dy}{dt}\frac{dt}{dx} = \frac{\dot{y}}{\dot{x}} \tag{e}$$

$$\frac{d^2y}{dx^2} = \frac{d}{dx}\left(\frac{dy}{dx}\right) = \left[\frac{d}{dt}\left(\frac{dy}{dx}\right)\right]\frac{dt}{dx}$$

$$= \frac{d}{dt}\left(\frac{\dot{y}}{\dot{x}}\right)\frac{1}{\dot{x}} \quad ,$$

$$\frac{d^2y}{dx^2} = \frac{\dot{x}\,\ddot{y} - \dot{y}\,\ddot{x}}{\dot{x}^3} \tag{f}$$

Now substituting (e) and (f) into (d) we get:

$$\frac{1}{r} = \frac{\dot{x}\,\ddot{y} - \dot{y}\,\ddot{x}}{\dot{x}^3\left[1 + \left(\frac{\dot{y}}{\dot{x}}\right)^2\right]^{3/2}}$$

$$\frac{1}{r} = \frac{\dot{x}\,\ddot{y} - \dot{y}\,\ddot{x}}{(\dot{x}^2 + \dot{y}^2)^{3/2}} \tag{g}$$

Now go back to equation (a) for the centripetal acceleration a_c.

$$\frac{\dot{s}^2}{\rho} = (4 + 64t^2)\,\frac{\dot{x}\,\ddot{y} - \dot{y}\,\ddot{x}}{(\dot{x}^2 + \dot{y}^2)^{3/2}}$$

$$= (4 + 64t^2)\,\frac{2(8) - 8t(0)}{(4 + 64t^2)^{3/2}} = \frac{16}{\sqrt{4 + 64t^2}}$$

The acceleration vector then becomes

$$\vec{a} = \frac{64t}{\sqrt{4 + 64t^2}}\,\hat{e}_t + \frac{16}{\sqrt{4 + 64t^2}}\,\hat{e}_n$$

where $\hat{e}_t$ is the unit vector in the direction of the path and $\hat{e}_n$ is the unit vector normal to the path.

A body moves on the curve illustrated in the figure from point A to point B with a constant acceleration. The velocity of the particle at A is 10 ft/sec and 10 seconds later at B it is 50 ft/sec. What is the total acceleration of the particle at point B?

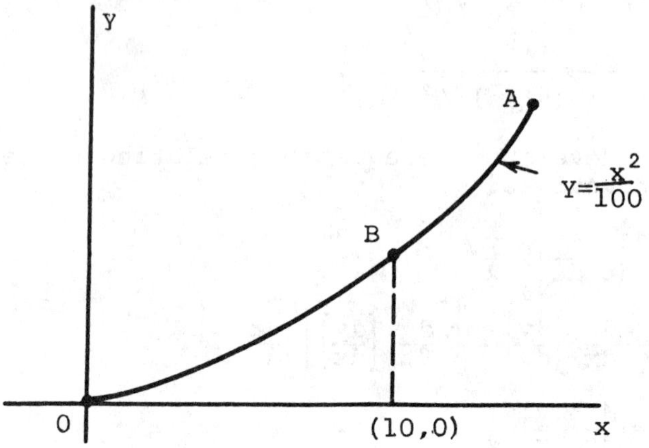

Solution: The acceleration of the particle is given in vector form as

$$\vec{a} = \frac{dv}{dt}\hat{e}_t - \frac{v^2}{\rho}\hat{e}_n$$

Since the velocity at A and B are known and the time for the particle to move from A to B is known, the acceleration along the path can be determined from an integration:

$$v = v_0 + \int_0^t a_t \, dt$$

Since the tangential acceleration is a constant,

$$50 = 40 + a_t(10)$$

from a direct integration and substitution, and the tangential acceleration becomes

$$a_t = 4 \text{ ft/sec}^2$$

The radius of curvature can be determined from the standard form:

$$\rho = \frac{[1 + (y')^2]^{3/2}}{y''}$$

The first and second time derivatives of y with respect to x are

$$y' = \frac{x}{50}$$

$$y'' = \frac{1}{50}$$

so that the radius of curvature becomes

$$\rho = \frac{[1 + (x/50)^2]^{3/2}}{(1/50)}$$

at $x = 10$, $\rho = \dfrac{[1 + (1/5)^2]^{3/2}}{(1/50)}$

Solving for ρ, $\qquad \rho = 53$ ft

Now, at point B, the velocity of the particle is 50 fps; therefore, acceleration of the particle becomes

$$\vec{a} = 4\hat{e}_t - \frac{(50)^2}{53}\hat{e}_n$$

or $\qquad \vec{a} = 4\hat{e}_t - 47.2\hat{e}_n$

and the magnitude of the acceleration is

$$a = \sqrt{4^2 + 47.2^2} = 47.3 \text{ ft/sec}^2$$

• PROBLEM 7-17

An airplane is attempting to fly with a constant speed, v, from point A ≡ (a, 0), on the x axis of the accompanying figure, to the origin, 0. A wind blowing with speed w in the positive y direction will greatly affect the flight of the plane. The pilot, who is not familiar with vectors, always points his plane toward 0, thinking this to be the shortest way. Find the path that he is actually taking due to his mistake.

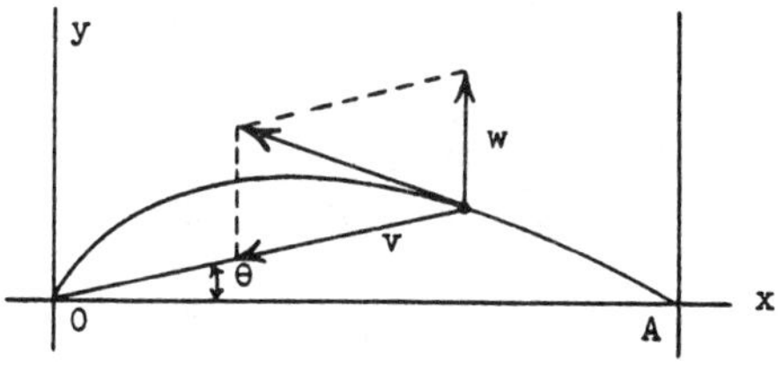

Solution: As velocitites add vectorially, we know that the pilot should have aimed his plane so that it would always have a velocity component in the negative y direction of magnitude w. In this manner he would have ended up traveling along the straight line from A to 0. Since the pilot actually aimed his plane directly at 0, the wind caused him to drift in the y direction and thus, he was forced to take a longer path. The path taken resembles the curve in the above figure and we are asked to determine the equation of that curve.

Let $x = x(t)$ and $y = y(t)$ describe the airplane's coordinates as functions of the time. The velocity with which the plane traveled, v, directed towards 0, is shown in the figure, as are the wind velocity and the resultant of the two. Using the angle θ as defined in the figure, we can write

$$x'(t) = - v \cos \theta \tag{1}$$

and $$y'(t) = w - v \sin \theta \ . \tag{2}$$

Substituting for $\cos \theta$ and $\sin \theta$ in terms of x and y, we have

$$x'(t) = - \frac{vx}{(x^2 + y^2)^{1/2}} \tag{3}$$

and $$y'(t) = w - \frac{vy}{(x^2 + y^2)^{1/2}} \ . \tag{4}$$

Knowing that $\frac{dy}{dx} = \frac{y'(t)}{x'(t)}$ we find

$$\frac{dy}{dx} = \frac{vy - w(x^2 + y^2)^{1/2}}{vx} \ . \tag{5}$$

This equation is homogeneous; that is, a substitution of mx for x, and my for y leads to the same equation multiplied by m to some power, one, in this case. For equations of this type we remember that the substitution $y = zy$ allows us to separate variables. We find

$$\frac{dx}{x} + \frac{v\,dz}{w(1 + z^2)^{1/2}} = 0 \ . \tag{6}$$

Writing the second term as

$$\frac{1 + \frac{z}{(1 + z^2)^{1/2}}}{z + (1 + z^2)^{1/2}} \quad \frac{v\,dz}{w}$$

we can integrate the expression. Then

$$\frac{v}{w} \ln [z + (1 + z^2)^{1/2}] + \ln x = C \ .$$

When $x = a$, initially, $z = y = 0$. This gives $C = \ln a$. Therefore,

$$z + (1 + z^2)^{1/2} = \left(\frac{x}{a}\right)^{-\frac{w}{v}} \tag{7}$$

Solving for z gives

$$z = \frac{1}{2}\left[\left(\frac{x}{a}\right)^{-\frac{w}{v}} - \left(\frac{x}{a}\right)^{\frac{w}{v}}\right] \qquad (8)$$

which leaves

$$y = \frac{a}{2}\left[\left(\frac{x}{a}\right)^{1-\frac{w}{v}} - \left(\frac{x}{a}\right)^{1+\frac{w}{v}}\right] \qquad (9)$$

This describes the much longer path the pilot had to take because he was not familiar with vectors.

CHAPTER 8

RIGID BODY KINEMATICS

ANGULAR MOTION

• PROBLEM 8-1

Write the angular velocity vector for a top, shown in the figure, as it spins about its fixed axis AO. The magnitude of the angular velocity is $\omega = 24$ radians/sec.

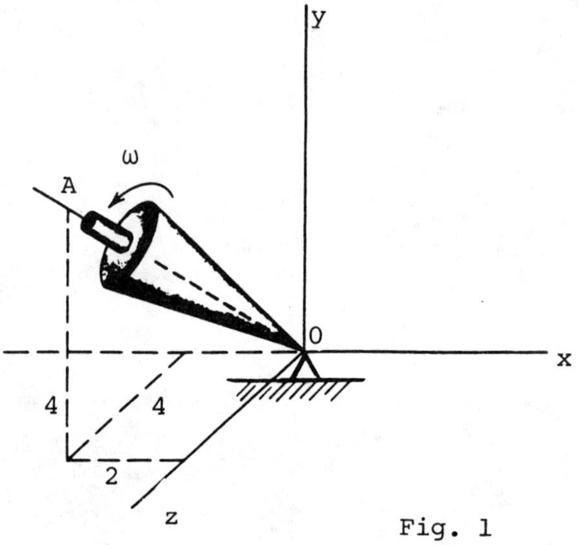

Fig. 1

Solution: By noting the definition of the angular velocity vector, it is observed from the diagram that the vector will lie along line OA in the direction from O to A. The distance OA from figure 2 is

$$OA = \sqrt{4^2 + 4^2 + 2^2} = 6$$

Therefore
$$\omega_x = -24\left(\frac{2}{6}\right) = -8$$
$$\omega_y = 24\left(\frac{4}{6}\right) = 16$$
$$\omega_z = 24\left(\frac{4}{6}\right) = 16$$

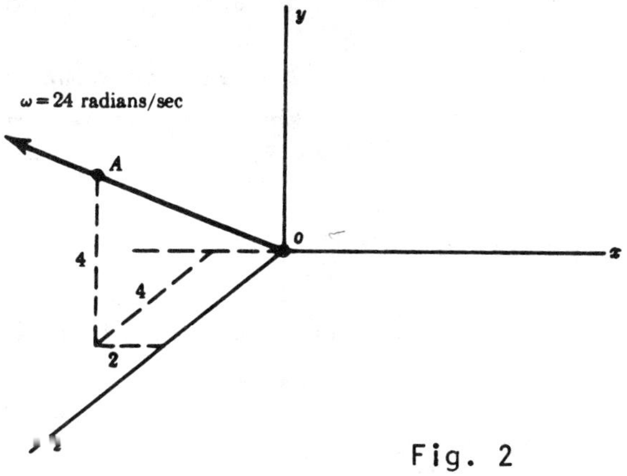

Fig. 2

so that $\vec{\omega} = 8\hat{i} + 16\hat{j} + 16\hat{k}.$

• PROBLEM 8-2

Two pulleys connected with a belt can function together in three positions as shown in the figure. Shaft A starts to move (initial angular velocity $\omega_A = 0$) with an angular acceleration $\alpha_A = 5$ rad/s . For each of the three positions of the belt calculate the time needed for shaft B to reach 600 rpm.

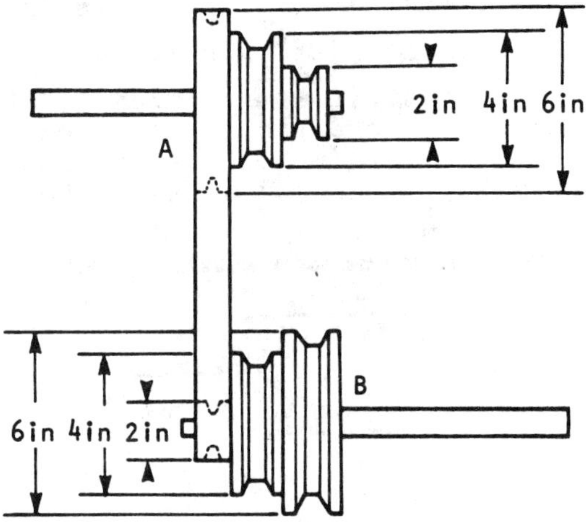

Solution: Use the basic relation for rotating bodies to solve this problem, namely,

$$\omega_A = \alpha_A t$$

where ω_A is the angular velocity of shaft A, α_A is the angular acceleration of shaft A and t is the elapsed time. There is no initial velocity in the problem. In each case, the angular velocity of shaft B is to be 600 rev/min or,

$$\omega_B = 600 \frac{\text{rev}}{\text{min}} \times 2\pi \frac{\text{rad}}{\text{rev}} \times \frac{1 \text{ min}}{60 \text{ sec}} = 62.83 \frac{\text{rad}}{\text{sec}}.$$

The acceleration of shaft A is given as $\alpha_A = 5$ rad/sec², while the angular velocity of shaft A can be expressed in terms of ω_B and the diameters as

$$\omega_A D_A = \omega_B D_B \text{ or } \omega_A = \omega_B D_B / D_A.$$

If we substitute this into the first equation we obtain,

$$t = \frac{\omega_B D_B}{\alpha_A D_A} = \frac{62.83}{5} \frac{D_B}{D_A} = 12.57 \frac{D_B}{D_A}.$$

Now, in the first arrangement $D_B/D_A = 1/3$ and so t = 4.19 sec. In the middle arrangement $D_B/D_A = 1$ and t = 12.57 sec. In the third arrangement $D_B/D_A = 3$ and t = 37.7 sec.

• PROBLEM 8-3

A cylinder which is 4 ft in diameter rotates with an angular velocity of 3 rad/sec counterclockwise when viewed from the right-hand end of the horizontal axis. As shown in the Figure, the $\hat{i}$, $\hat{j}$, $\hat{k}$ triad is chosen so that $\hat{i}$ is directed along the horizontal centerline. The $\hat{j}$ axis is shown vertical at the instant, but the $\hat{j}$ and $\hat{k}$ axes rotate with the body. Using $\hat{j}$ as a vector $\vec{\rho}$ fixed in the cylinder, determine $d\hat{j}/dt$ for each of the following conditions: a) when the axis of rotation is along the centerline of the cylinder, as in Fig. a and b) when the axis of rotation is along the bottom edge of the cylinder, as in Fig. b.

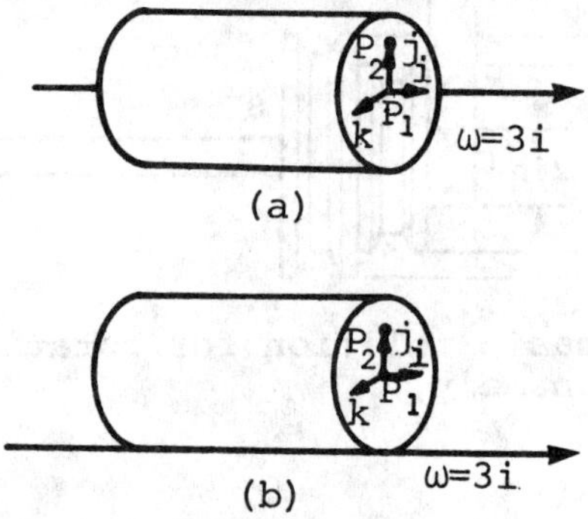

Fig. 1

Solution: From Figure a, we have $\underset{\sim}{\omega} = 3\underset{\sim}{i}$

and $\frac{d\hat{j}}{dt} = \vec{\omega} \times \hat{j} = 3\hat{i} \times \hat{j} = 3\hat{k}$ ft/sec

This velocity is a vector having a magnitude of 3 ft/sec in the positive $\hat{k}$ direction. It could also be thought of as $\vec{v}_{P_2} - \vec{v}_{P_1}$, where P_1 and P_2 are shown in the Figure a. Point P_1, being on the axis of rotation, has zero velocity. Since P_2 is 1 ft above the axis, the magnitude of its velocity is $r\omega = (1)(3) = 3$ ft/sec, out of the paper toward the reader, or in the positive $\hat{k}$ direction. So both methods yield the same result.

b) Although the axis of rotation is moved, the value of dj/dt is the same as for part a), because both ω and and j are the same. Hence,

$$\frac{d\hat{j}}{dt} = \hat{\omega} \times \hat{j} = 3\hat{i} \times \hat{j} = 3\hat{k} \text{ ft/sec}$$

This velocity can also be thought of as $\vec{v}_{P_2} - \vec{v}_{P_1}$, where points P_1 and P_2 are, respectively, 2 ft and 3 ft vertically above the axis of rotation. So, $\vec{v}_{P_2} = (3)(3)\hat{k} = 9\hat{k}$ ft/sec and $\vec{v}_{P_1} = (2)(3)\hat{k} = 6\hat{k}$ ft/sec. Their difference is $d\hat{j}/dt = 3\hat{k}$ ft/sec.

This example shows that the time rate of change of a vector fixed in a rotating body is independent of the location of the axis of rotation, provided that the axis is parallel to the vector $\vec{\omega}$.

• PROBLEM 8-4

A bar pivoting about one end has an angular velocity of 8 rad/sec clockwise when it is subjected to a constant angular deceleration, which is to act for a time interval until the bar has an angular displacement of 10 rad counterclockwise from its position at the instant at which the deceleration was first applied. The bar will have moved through a total angular distance of 26 rad in this same time interval. What will be the angular velocity at the end of the time interval?

Solution: The initial angular velocity is 8 rad/sec clockwise. If clockwise is called negative, then $\omega_0 = -8$ rad/sec.

Since the deceleration α must be counterclockwise, its magnitude is called + C. The bar will continue to move clockwise through some angle θ until it comes to rest, and will then start to move counterclockwise. The bar will then return through the angle θ to its original position, and will continue counterclockwise for an additional 10 rad. Hence, its total angular travel will be $(2\theta + 10)$ rad. Since this total angle is given as 26 rad, the magnitude of θ must be 8 rad.

We have $$d\theta = \omega dt \qquad (1)$$

$$d\omega = \alpha dt \qquad (2)$$

Hence dividing (1) by (2)

$$\alpha d\theta = \omega d\omega \tag{3}$$

integrating $$\int_{\theta_0}^{\theta} \alpha d\theta' = \int_{\omega_0}^{\omega} \omega' d\omega \tag{4}$$

$$\omega^2 = \omega_0^2 + 2C(\theta - \theta_0)$$

We are now in a position to find C or α. During the clockwise motion, the speed decreases to zero, and we have

$$(0)^2 = (-8)^2 + 2C(-8)$$

Hence, $\alpha = +4$ rad/sec^2.

The final angular velocity ω can be found by considering the motion either from the initial position or from the position of reversal of motion. In the first case, the equation is

$$\omega^2 = (-8)^2 + 2(+4)(+10) = 64 + 80 = 144$$

In the second case, the equation is

$$\omega^2 = (0)^2 + 2(+4)(18) = 144$$

Hence, ω is 12 rad/sec counterclockwise.

• PROBLEM 8-5

A pin on the end of a retractable arm slides in the slot as shown in the figure. Suppose the arm moves from A to B with a constant angular velocity $\dot{\theta} = c$, write the acceleration in polar vector form as a function the angle θ.

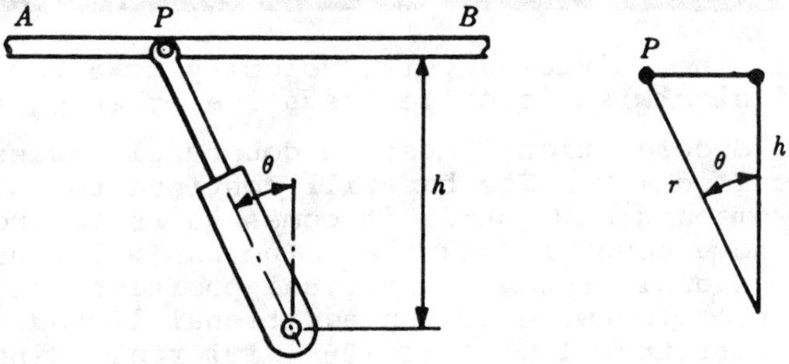

Figure

Solution: From the figure it is easily seen that

$$h = r\cos\theta$$

and therefore, $r = h \sec \theta$.

The first time derivative of r is

$$\dot{r} = h \sec \theta \tan \theta \frac{d\theta}{dt}.$$

Noting that $\frac{d\theta}{dt} = \dot{\theta} = c$

$\dot{r}$ becomes $\dot{r} = hc \sec \theta \tan \theta,$

Similarly, the second time derivative of r is

$$\ddot{r} = hc^2 \sec \theta \, (\tan^2 \theta + \sec^2 \theta).$$

The acceleration equation is

$$\vec{a} = (\ddot{r} - r\dot{\theta}^2)\hat{e}_r + (r\ddot{\theta} + 2\dot{r}\dot{\theta})\hat{e}_\theta$$

We have $\dot{\theta} = c$, $\ddot{\theta} = 0$ and r, $\dot{r}$, $\ddot{r}$ determined above yielding

$$\vec{a} = \{hc^2 \sec \theta \, (\tan^2 \theta + \sec^2 \theta) - hc^2 \sec \theta\} \hat{e}_r$$

$$+ \{2hc^2 \sec \theta \tan \theta\} \hat{e}_\theta$$

or $$\vec{a} = \{2hc^2 \sec \theta \tan^2 \theta\}\hat{e}_r + \{2hc^2 \sec \theta \tan \theta\}\hat{e}_\theta.$$

• PROBLEM 8-6

A race car travels on a circular track at 300 ft radius at a speed of 60 mph. An observer stands at a point 150 ft from the center of the circle and watches the car with binoculars. What will be the angular speed of his line of sight when the radial line to the auto makes an angle of 30° with the line connecting the observer and the center of the circle?

Fig. 1

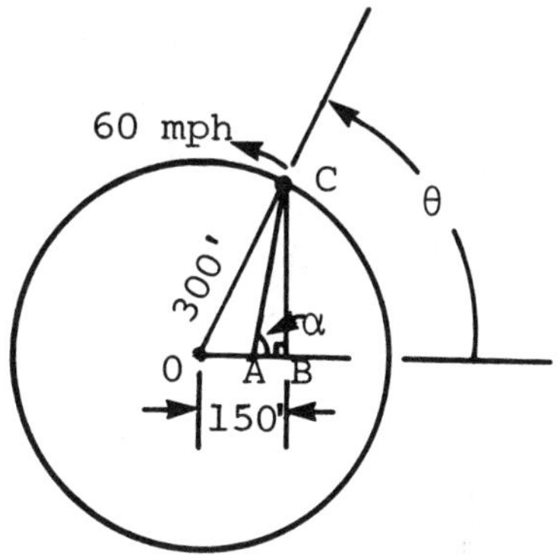

Solution: In Fig. 1 the car is at C and the observer is at A. His line of sight is AC and its rate of change or angular speed is $\dot{\alpha}$. The problem is to find $\dot{\alpha}$ when θ is 30°. This can be done by writing an equation for α in terms of θ, then by differentiating it with respect to time and finally substituting $\theta = 30°$. By definition of the trigonometric functions we have

$$\tan\alpha = \frac{CB}{AB} \tag{1}$$

$$\sin\theta = \frac{CB}{300} \tag{2}$$

$$\cos\theta = \frac{OB}{300} = \frac{AB + 150}{300} \tag{3}$$

solving (2) and (3) for CB and AB respectively and substituting into (1) we have,

$$\tan\alpha = \frac{300\sin\theta}{300\cos\theta - 150}. \tag{4}$$

Differentiating (4) yields

$$(\sec^2\alpha)\dot{\alpha} = \frac{(300\cos\theta - 150)(300\cos\theta)\dot{\theta}}{(300\cos\theta - 150)^2} + \frac{300\sin\theta\,(300\sin\theta)\dot{\theta}}{(300\cos\theta - 150)^2}. \tag{5}$$

Now $\dot{\theta}$ is related to the car's speed by

$$\dot{\theta} = \frac{v}{R}$$

$$= \frac{88\text{ fps}}{300\text{ ft}}$$

$$= 0.293\text{ rad/sec} \tag{6}$$

For $\theta = 30°$,

$$CB = 300\sin 30° = 150\text{ ft} \tag{7}$$

$$OB = 300\cos 30° = 259.8\text{ ft}$$

$$AB = OB - 150 = 109.8\text{ ft} \tag{8}$$

Inserting values for CB and AB from (7) and (8) into (1) yields,

$$\alpha = \tan^{-1}\left(\frac{150}{109.8}\right)$$

$$= 53.8°. \tag{9}$$

Finally, values for $\dot{\theta}$ and α from (6) and (9) can be substituted into (5) with $\theta = 30°$:

$$2.87\ \dot{\alpha} = \frac{[(259.8 - 150)(259.8) + (150)(150)][0.293]}{(259.8 - 150)^2}.$$

Solving for $\dot{\alpha}$ we get

$\dot{\alpha} = 0.433$ rad/sec.

• PROBLEM 8-7

A chalk mark is made on the lower edge of a tire of radius R. The tire is rolling with speed v_0 without slipping along a horizontal road. During a time t, a line from the axle to the chalk mark turns through an angle θ such that $\theta = \omega t = v_0 t/R$. Find the x and y coordinates of the chalk mark at any time t or angle θ and also the velocity and acceleration of the chalk mark. (The path of the chalk mark is a cyloid).

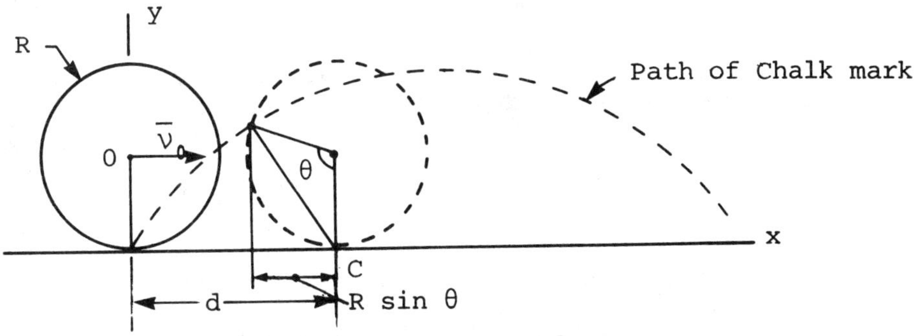

Solution: In the figure, the chalk mark is initially at the origin of the coordinate system and in contact with the ground. After the tire has rolled through an angle θ, the chalk mark is at A. The tire center has moved a distance d. But for no slippage,

$$d = \text{arc length (AC)}.$$

From the arc formula AC = Rθ and therefore,

$$d = R\theta. \tag{1}$$

Letting x and y be the coordinates of A we have,

$$x = d - R \sin \theta, \tag{2}$$

$$y = R - R \cos \theta.$$

Substituting from (1),

$$x = R(\theta - \sin \theta) \tag{3}$$

$$y = R(1 - \cos \theta). \tag{4}$$

If (3) and (4) are differentiated with respect to time, the results are the X and Y components of velocity. Differentiating (3), we have

$$\dot{x} = v_x = R\ [\dot{\theta} - (\cos\ \theta)\dot{\theta}]. \tag{5}$$

But $\dot{\theta}$ is the angular velocity of the wheel, ω. Making this substitution,

$$v_x = R\ [\omega - (\cos\ \theta)\omega]$$

and, factoring ω, yields

$$v_x = R\omega\ (1 - \cos\ \theta). \tag{6}$$

Now differentiating (4), we get

$$\dot{y} = v_y = R\ (\sin\ \theta)\dot{\theta}.$$

Substituting ω for $\dot{\theta}$, this becomes

$$v_y = R\omega\ \sin\ \theta. \tag{7}$$

The magnitude of the total velocity vector is found by the theorem of Pythagorus

$$v = \sqrt{v_x^2 + v_y^2}.$$

Substituting from (6) and (7), we get

$$v = \sqrt{[R\omega\ (1 - \cos\ \theta)]^2 + [R\omega\ \sin\ \theta]^2}$$

$$v = \sqrt{(R\omega)^2\ (1 - 2\ \cos\ \theta + \cos^2\ \theta) + (R\omega)^2\ \sin^2\ \theta}$$

$$v = R\omega\ \sqrt{1 - 2\ \cos\ \theta + \cos^2\ \theta + \sin^2\ \theta}.$$

But with the identity

$$\cos^2\ \theta + \sin^2\ \theta = 1, \tag{8}$$

the velocity magnitude becomes

$$v = R\omega\ \sqrt{2\ (1 - \cos\ \theta)}. \tag{9}$$

The direction of the velocity vector is tangent to the path being traveled (the cycloid) or, equivalently, perpendicular to line AC (since C is the instant center for the wheel).

Now, if (6) and (7) are differentiated with respect to time, the results are the X and Y components of acceleration. Differentiating (6), we obtain

$$\dot{v}_x = a_x = (R\omega \sin \theta)\dot{\theta} \tag{10}$$

where ω is constant because v_0 is constant. Since $\dot{\theta} = \omega$ (10) can be written as

$$a_x = R\omega^2 \sin \theta. \tag{11}$$

Differentiating (7), we get

$$\dot{v}_y = a_y = (R\omega \cos \theta)\dot{\theta} \tag{12}$$

Again with $\dot{\theta} = \omega$ this becomes

$$a_y = R\omega^2 \cos \theta. \tag{13}$$

The magnitude of the total acceleration vector is found by the theorem of Pythegorus.

$$a = \sqrt{a_x^2 + a_y^2}$$

Substituting from (11) and (13), we have

$$a = \sqrt{(R\omega^2 \sin \theta)^2 + (R\omega^2 \cos \theta)^2}$$

$$a = \sqrt{(R\omega^2)^2 \sin^2 \theta + (R\omega^2)^2 \cos^2 \theta}$$

$$a = R\omega^2 \sqrt{\sin^2 \theta + \cos^2 \theta}\,.$$

But, from (8), this simplifies to

$$a = R\omega^2. \tag{14}$$

It is interesting to note that the acceleration of the chalk mark is constant in magnitude as the tire rolls. The direction of the acceleration vector, however, changes continuously. The angle it makes with the vertical is

$$\alpha = \tan^{-1}\left(\frac{a_x}{a_y}\right)$$

Substituting from (11) and (13), we determine that

$$\alpha = \tan^{-1}\left(\frac{R\omega^2 \sin \theta}{R\omega^2 \cos \theta}\right)$$

Canceling $R\omega^2$,

$$\alpha = \tan^{-1} (\tan \theta)$$

$$\alpha = \theta$$

Thus, initially, the acceleration vector is vertical (pointing up) since $\theta = 0$. At $\theta = 180°$ the chalk mark is at the top of the cycloid and the acceleration vector is again vertical (this time pointing straight down). The acceleration vector at A, in fact, always points towards the center of the tire.

• PROBLEM 8-8

As shown, a rope is wrapped around the cylinder A and the pulley B, and another rope from B is attached to block C. Cylinder A has angular velocity 3 rad/sec counterclockwise and angular acceleration 6 rad/sec^2 counterclockwise. Also, the center of A has a linear velocity and acceleration 4 ft/sec and 6 ft/sec^2 downwards. Determine the linear velocity and acceleration of block C.

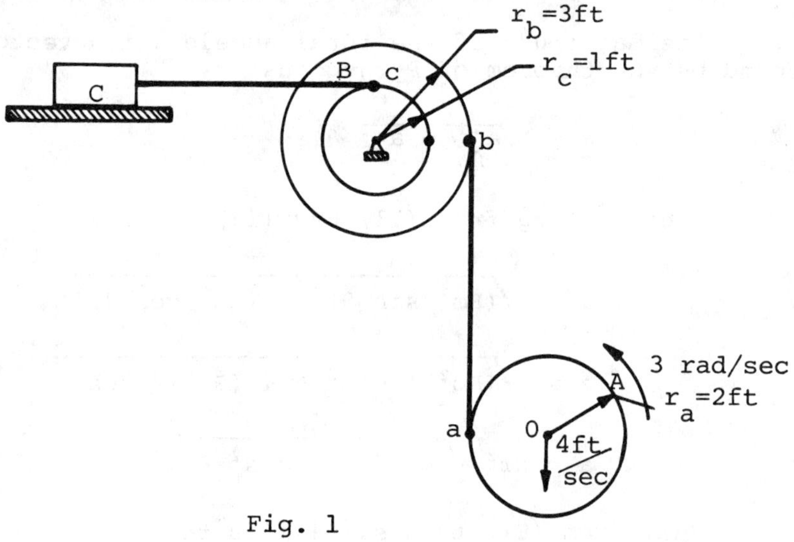

Fig. 1

Solution: To solve for velocity, let a and O be points on cylinder A, and let b and c be points on cylinder B.

Since the motion of cylinder A is completely specified, begin with that cylinder and we will work our way through cylinder B to block C. The velocity of point a is first found by relating it to point O.

$$\overline{v}_a = \overline{v}_O + \overline{v}_{aO}. \quad (1)$$

From the given information,

$$\overline{v}_O = 4 \text{ ft/sec} \downarrow. \quad (2)$$

The velocity of a relative to O is given by the formula

$$v_{aO} = (aO)\,\omega_A.$$

Distance aO is the radius of cylincer A. So,

$$v_{aO} = (2\text{ ft})\ 3\text{ rad/sec}$$

$$= 6\text{ ft/sec.}$$

The direction of $\bar{v}_{aO}$ must be perpendicular to line aO and agree in sense with $\bar{\omega}_A$. Therefore,

$$\bar{v}_{aO} = 6\text{ ft/sec} \downarrow. \qquad (3)$$

Substituting (2) and (3) into (1),

$$\bar{v}_a = 4 \downarrow + 6 \downarrow$$

$$= 10\text{ ft/sec} \downarrow.$$

The ropes are assumed to be unstretchable. It is also assumed that the ropes do not slip over the cylindrical surfaces at a, b or c. From these two assumptions, it follows that all points on the vertical rope have the same velocity as point a. Furthermore, it follows that point b has this same velocity.

Since the center of cylinder B is stationary, two simple velocity formulas can be written for points b and c:

$$v_b = r_b\ \omega_B \qquad (4)$$

$$v_c = r_c\ \omega_B \qquad (5)$$

Solving (4) and (5) for ω_B and equating, we have

$$\frac{v_b}{r_b} = \frac{v_c}{r_c}.$$

Rewriting,

$$v_c = \frac{r_c}{r_b}\ v_b.$$

Substituting known values,

$$v_c = \frac{1\text{ ft}}{3\text{ ft}}\ (10\text{ ft/sec})$$

$$= 3.33\text{ ft/sec.}$$

Because of the two assumptions noted above, all points on the horizontal rope will have the same velocity and the velocity of block C will be the same as point d:

$$\bar{v}_C = \bar{v}_d$$

$$= 3.33\text{ ft/sec} \rightarrow.$$

The direction of $\bar{v}_C$ is ascertained by noting that $\bar{\omega}_B$ is clockwise due to the downward motion of b. Thus, block C is pulled to the right.

To solve for acceleration, note that point a has two components of acceleration due to the rotation of A: one in the vertical or tangential direction and one in the normal direction along aO. The acceleration of a can also be related to that of O as was done for velocity.

$$\vec{a}_a = \vec{a}_a^{\,n} + \vec{a}_a^{\,t} = \vec{a}_O + \vec{a}_{aO}^{\,n} + \vec{a}_{aO}^{\,t}. \tag{6}$$

Equating the components of acceleration in the tangential (vertical) direction in Equation (6), we have

$$\bar{a}_a^{\,t} = \bar{a}_O + \bar{a}_{aO}^{\,t}. \tag{7}$$

From the given data,

$$\bar{a}_O = 6 \text{ ft/sec}^2 \downarrow. \tag{8}$$

The tangential component of the acceleration of a relative to O is given by the formula

$$a_{aO}^{\,t} = (aO)\ \dot{\omega}_A. \tag{9}$$

From the given data

$$a_O = 2 \text{ ft}$$

$$\bar{\dot{\omega}}_A = 6 \text{ rad/sec}^2 \ \text{ counterclockwise.}$$

Substituting these values into (9), we have

$$a_{aO}^{\,t} = (2 \text{ ft})\ (6 \text{ rad/sec}^2)$$

$$= 12 \text{ ft/sec}^2 \downarrow. \tag{10}$$

Now from (7), (8) and (10) we get

$$\bar{a}_a^{\,t} = 6\downarrow + 12\downarrow$$

$$= 18 \text{ ft/sec}^2 \downarrow. \tag{11}$$

This is the same acceleration as that of point b. Points b and c move on circular paths and, so, the simple formulas apply:

$$a_b^{\,t} = r_b\ \dot{\omega}_B \tag{12}$$

$$a_c^{\,t} = r_c\ \dot{\omega}_B \tag{13}$$

Solving (12) and (13) for $\dot{\omega}_B$ and equating, we get

$$\frac{a_b^t}{r_b} = \frac{a_c^t}{r_c}$$

Rewriting, $a_c^t = \frac{r_c}{r_b} a_b^t$

Substituting known values,

$$a_c^t = \frac{1 \text{ ft}}{3 \text{ ft}} \; 18 \text{ ft/sec}^2$$

$$= 6 \text{ ft/sec}^2$$

The tangential component of acceleration of point c is the same as that of any point on the horizontal rope from C to c and the same as that of block C.

The direction of block C's acceleration is determined from the direction of $\bar{\dot{\omega}}_B$. Referring to equation (12), we are reminded that the direction of $\bar{\dot{\omega}}_B$ must agree with the direction of $\bar{a}_b^t$. The latter is downward. So, $\bar{\dot{\omega}}_B$ is clockwise. From this, it follows that $\bar{a}_d^t$ is to the right and the acceleration of the block is also:

$$\bar{a}_C = 6 \text{ fps}^2 \rightarrow.$$

It should be noted that it is possible for the block acceleration to be to the left. This would be the case if C was moving to the right and slowing down.

• PROBLEM 8-9

The essential elements of a "planetary gearing" are shown in the figure. The wheels 1 and 2 with radii r_1 and r_2 and angular velocities ω_1 and ω_2, respectively, impart their motion to a third wheel 3, the so called "planet pinion," of radius

$$r_3 = \frac{1}{2} (r_2 - r_1).$$

When friction is not sufficient to ensure rolling, the three wheels are geared, and the bearing of the planet pinion is carried by an arm of length O_0' of length $\ell = \frac{1}{2} (r_1 + r_2)$ which rotates about O with an angular velocity ω. Determine ω in terms of ω_1 and ω_2.

Solution: The instantaneous M_1, M_2, and M of gears 1, 2, and of the arm ℓ all coincide with O. The points P, Q where the planet pinion contacts gears 1 and 2 therefore have velo-

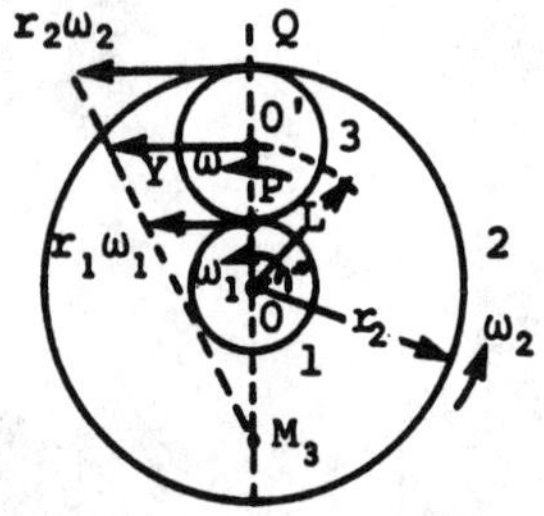

cities $\omega_1 r_1$ and $\omega_2 r_2$ normal to the arm. Since P and Q can also be regarded as points of the planet pinion 3, the instantaneous center M_3 of the planet pinion lies on the joining line PQ, and its location is determined by the fact that the distances M_3P and M_3Q must have the ratio $r_1\omega_1$ to $r_2\omega_2$. The center O' of the planet pinion 3 has the speed

$$v = \frac{1}{2}(r_1\omega_1 + r_2\omega_2)$$

and also belongs to the arm ℓ, whose angular velocity is determined by

$$\omega = \frac{v}{\ell} = \frac{r_1\omega_1 + r_2\omega_2}{r_1 + r_2}.$$

If one of the gears 1, 2 is held fixed, we have $\omega_1 = 0$ or $\omega_2 = 0$, and consequently

$$\omega = \frac{r_2}{r_1 + r_2}\omega_2 \qquad \text{or} \qquad \omega = \frac{r_1}{r_1 + r_2}\omega_1,$$

respectively, whereas when the arm is locked, $\omega = 0$ and thus $\omega_2/\omega_1 = -r_1/r_2$.

INSTANTANEOUS CENTER METHOD

• PROBLEM 8-10

The crank AB rotates clockwise with a constant angular velocity ω_{AB} = 3000 rpm. It is connected with the piston P through the rod BC as shown in Fig. 1. For the crank position as indicated compute a) the velocity of the piston P and the angular velocity ω_{BC} of the connecting rod;

b) using the method of instantaneous center of rotation compute a), c) determine the angular acceleration of the rod BC and the acceleration of point C.

Fig. 1

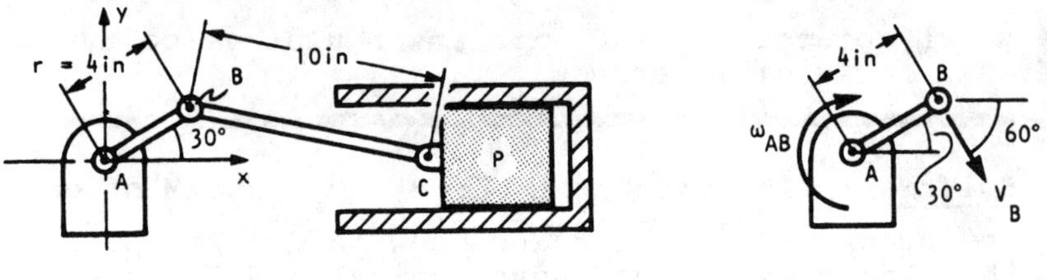

Solution: a) Motion of crank AB. The crank AB rotates with constant angular velocity $\vec{\omega}_{AB}$ around pt. A. Expressing $\vec{\omega}_{AB}$ in radians/sec and observing that $\vec{V}_B = \vec{\omega}_{AB} \times \vec{r}_{AB}$, we calculate $\vec{V}_B$.

$$\vec{\omega}_{AB} = 3000 \frac{\text{rev}}{\text{min}} \frac{2\pi \text{ rad}}{\text{rev}} \frac{1 \text{ min}}{60 \text{ sec}} \vec{k} = -314.2 \vec{k}$$

$$\vec{V}_B = \vec{\omega}_{AB} \times \vec{r}_{AB} = 314.2 \vec{k} \times (4 \cos 30 \vec{i} + 4 \sin 30 \vec{j})$$

$$\vec{V}_D = 628.4 \vec{i} - 1088.4 \vec{j} = 1256.8 \frac{\text{in}}{\text{s}} \angle 60°$$

The motion of the connecting rod BC can be considered as the vector sum of a rotation and a translation (see Fig. II).

$$\vec{V}_C = \vec{V}_B + \vec{V}_{C/B} \quad \text{where} \quad \vec{V}_{C/B} = \vec{\omega}_{BC} \times \vec{r}_{BC}$$

$$\text{and} \quad \vec{V}_C = V_C\vec{i}, \ \vec{\omega}_{BC} = \omega_{BC}\vec{k}$$

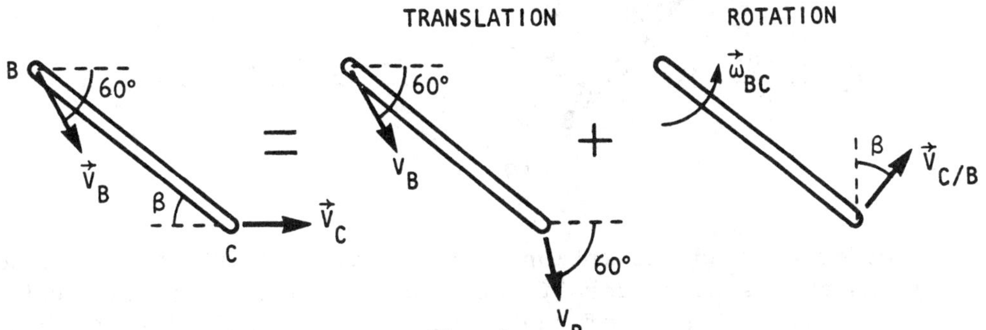

Fig. 2

In order to obtain the position vector $\vec{r}_{BC}$ and the angle β must be found. From the law of sines

$$\frac{\sin 30°}{10} = \frac{\sin \beta}{4} \qquad \beta = 11.5°$$

$$\vec{V}_{C/B} = \omega_{BC}\vec{k} \times (10 \cos 11.5°\vec{i} - 10 \sin 11.5°\vec{j})$$

$$V_{C/B} = 2\omega_{BC}\vec{i} + 9.8\ \omega_{BC}\ \vec{j}$$

$$V_C\vec{i} = 628.4\,\vec{i} - 1088.4\,\vec{j} + 2\,\omega_{BC}\vec{i} + 9.8\,\omega_{BC}\vec{j}$$

Equating $\vec{j}$ components

$$1088.4\,\vec{j} = 9.8\,\omega_{BC}\vec{j}$$

$$111.1 = \omega_{BC} \qquad \vec{\omega}_{BC} = 111.1\,\vec{k}\ \text{rad/s}$$

Equating $\vec{i}$ components

$$V_C\vec{i} = 628.4\,\vec{i} + 2(111.1)\,\vec{i}$$

$$V_C = 850.6 \qquad \vec{V}_C = 850.6\,\vec{i}\ \text{in/s} \ : \text{velocity of piston}$$

b)

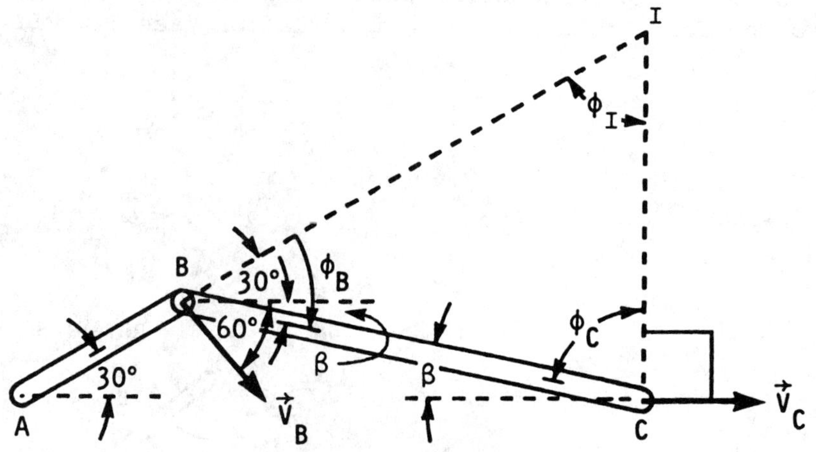

Fig. 3

In order to find the motion of connecting rod BC by the method of instantaneous center of rotation the interior angles and lengths of each side of the constructed triangle (see Fig. 3) must be found. The interior angles are

$$\phi_B = 30° + \beta \qquad \phi_C = 90° - \beta$$

$$\phi_B = 30° + 11.5° = 41.5° \qquad \phi_C = 90° - 11.5° = 78.5°$$

$$\phi_I = 180° - (41.5° + 78.5°) = 60°$$

The lengths are found by the law of sines

$$\frac{\sin\phi_B}{CI} = \frac{\sin\phi_I}{BC} = \frac{\sin\phi_C}{BI}$$

$$CI = 10\,\frac{\sin 41.5°}{\sin 60°} = 7.65 \text{ in} \qquad BI = 10\,\frac{\sin 78.5°}{\sin 60°} = 11.32 \text{ in}$$

At the instant shown in Fig. I points B and C on the connecting rod rotate about point I. The velocity of point B is known therefore the angular velocity can be calculated.

$$V_B = \ell_{BI}\omega_{BC} \qquad \omega_{BC} = \frac{V_B}{\ell_{BI}} = \frac{1256.8}{11.32} = 111.0\,\frac{\text{rad}}{\text{s}} \text{ CCW}$$

The velocity of point C is found similarly:

$$V_C = \ell_{CI}\omega_{BC} = 7.65(111.0) = 849.3\,\frac{\text{in}}{\text{s}} \text{ to the right.}$$

c) The acceleration of rod BC will be analyzed as the sum of translational rotational motions (see Fig. IV).

$$\vec{a}_C = \vec{a}_B + \vec{a}_{C/B} = \vec{a}_B + (a_{C/B})_n + (a_{C/B})_t$$

where $\vec{a}_C = a_C\vec{i}$ and n denotes the normal component and t denotes the tangential component of the relative acceleration.

Since crank AB has no angular acceleration there is only a normal acceleration component given by

$$\vec{a}_B = \vec{\omega}_{AB} \times (\vec{\omega}_{AB} \times \vec{r}_{AB}) = \vec{\omega}_{AB} \times \vec{V}_B$$

$$\vec{a}_B = -314.2\,\vec{k} \times (628.4\,\vec{i} - 1088.4\,\vec{j})\,\frac{1 \text{ ft}}{12 \text{ in}}$$

$$\vec{a}_B = -28498\,\vec{i} - 16454\,\vec{j} = -32907\,\frac{\text{ft}}{\text{s}^2}\angle 30°$$

The components of the relative acceleration are

$$(\vec{a}_{C/B})_n = \vec{\omega}_{BC} \times (\vec{\omega}_{BC} \times \vec{r}_{BC})$$

$$(\vec{a}_{C/B})_n = 111.1\,\vec{k} \times [111.1\,\vec{k} \times (9.8\,\vec{i} - 2\,\vec{j})]\,\frac{1 \text{ ft}}{12 \text{ in}}$$

$$(\vec{a}_{C/B})_n = -10080\,\vec{i} + 2057\,\vec{j} = 10288\,\frac{\text{ft}}{\text{s}^2}\angle -11.5$$

$$(\vec{a}_{C/B})_t = \vec{\alpha}_{BC} \times \vec{r}_{BC} \qquad \text{where} \qquad \vec{\alpha}_{BC} = \alpha_{BC}\,\vec{k}$$

$$(a_{C/B})_t = \alpha_{BC}\,\vec{k} \times (9.8\,\vec{i} - 2\,\vec{j})\,\frac{1 \text{ ft}}{12 \text{ in}}$$

$$(a_{C/B})_t = \alpha_{BC}.167\,\vec{i} + \alpha_{BC}.817\,\vec{j}$$

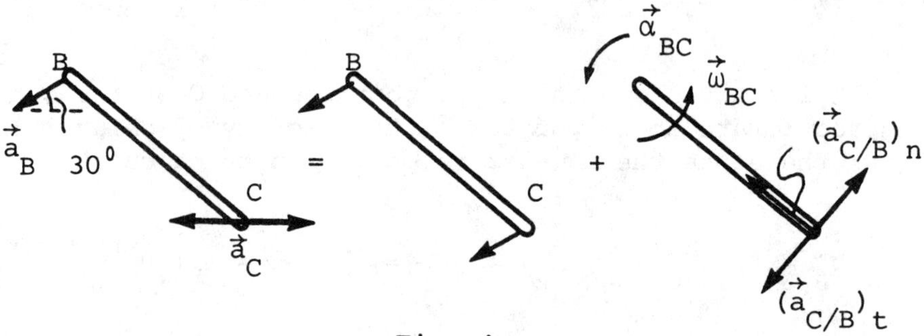

Fig. 4

$$a_C\vec{i} = -28498\,\vec{i} - 16454\,\vec{j} - 10080\,\vec{i} + 2057\,\vec{j} + \alpha_{BC}.167\,\vec{i} + \alpha_{BC}.817\,\vec{j}$$

Equating $\vec{j}$ components

$$\alpha_{BC}.817\,\vec{j} = 16454\,\vec{j} - 2057\,\vec{j}$$

$$\alpha_{BC} = 17622 \qquad \vec{\alpha}_{BC} = 17622\,\frac{\text{rad}}{\text{s}^2}\,\vec{k}$$

Equating $\vec{i}$ components

$$a_C\vec{i} = -28498\,\vec{i} - 10080\,\vec{i} + .167(17622)\,\vec{i}$$

$$a_C = -35635 \qquad \vec{a}_C = -35635\,\frac{\text{ft}}{\text{s}^2}\,\vec{i}$$

• PROBLEM 8-11

In Figure 1a is shown a slider crank mechanism in schematic form. While the crank OA of length R rotates counterclockwise about point O with a constant angular velocity $\vec{\omega}_A = C\hat{k}$ rad/sec, the connecting rod of length l drives the crosshead represented by a point B back and forth with a reciprocating motion. Using the instantaneous center method, determine the angular velocity of the connecting rod and the linear velocity of the point B on the crosshead for any angle θ.

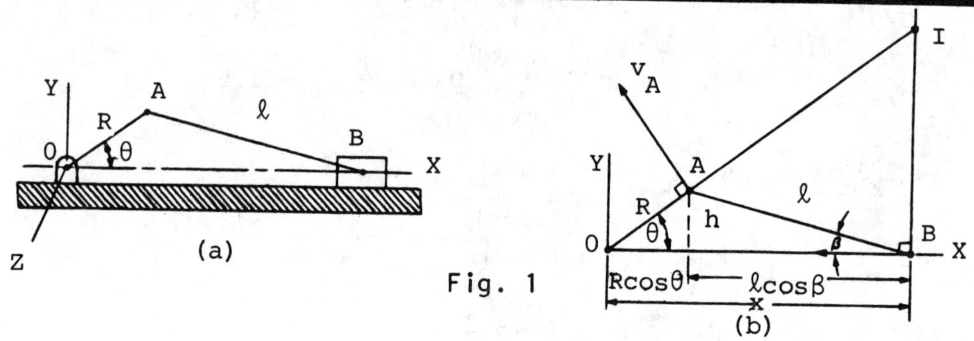

Fig. 1

Solution: As shown in Fig. 1b, the instantaneous center I for the motion of the connecting rod is at the intersection of lines drawn perpendicular to the velocities of the two ends of the connecting rod. The point B is on both the connecting rod and the crosshead, and it has the same velocity on either member. Being on the crosshead, its velocity must be horizontal. Hence, I is located somewhere on the vertical line through B, since this line is perpendicular to the velocity of B. Point A is on both the crank and the connecting rod, and has the same velocity on either member. This velocity must be perpendicular to the crank. The instantaneous center I is somewhere on the line through A that is perpendicular to the velocity of A. This perpendicular is, of course, the extension of the crank. The instantaneous center I is therefore located in the position shown.

In the right triangle OIB in Fig. 1b, the length of the base OB is $x = R \cos \theta + \ell \cos \beta$. This length can be expressed in terms of the one variable θ by considering the two small right triangles with the common altitude h. In the left-hand triangle, $h = R \sin \theta$; and in the other triangle $h = \ell \sin \beta$. Therefore, $\sin \beta = R/\ell \sin \theta$, and

$$\cos \beta = \sqrt{1 - \sin^2 \beta} = \sqrt{1 - \frac{R^2}{\ell^2} \sin^2 \theta}$$

Hence, $$x = R \cos \theta + \ell \sqrt{1 - \frac{R^2}{\ell^2} \sin^2 \theta}$$

To use the instantaneous center method, it is necessary to determine the position vector from I to A, $\vec{\rho}_{IA}$. Its magnitude is $\frac{x}{\cos \theta} - R$, and it makes an angle θ with the horizontal. Hence,

$$\vec{\rho}_{IA} = \left(\frac{R \cos \theta + \ell \sqrt{1 - \frac{R^2}{\ell^2} \sin^2 \theta}}{\cos \theta} - R \right)$$

$$[- (\cos \theta)\hat{i} - \sin \theta)\hat{j}]$$

$$= - \ell \sqrt{1 - \frac{R^2}{\ell^2} \sin^2 \theta}\, \hat{i} - \ell \tan \theta \sqrt{1 - \frac{R^2}{\ell^2} \sin^2 \theta}\, \hat{j}.$$

The linear velocity of point A, considered as a point on the crank is

$$\vec{v}_A = \vec{\omega}_{OA} \times \vec{\rho}_{OA} = C\hat{k} \times [(R \cos \theta)\hat{i} + (R \sin \theta)\hat{j}]$$

$$= -(RC\sin\theta)\hat{i} + (RC\cos\theta)\hat{j}.$$

Point A, considered as a point on the connecting rod, is rotating about the instantaneous center I with an angular velocity $\omega_{AB}\hat{k}$, which is assumed to be counterclockwise. If the computed value of ω_{AB} is found to be negative, it will mean that the rotation of the connecting rod about I is clockwise instead of counterclockwise as assumed. The velocity of point A derived from the rotation of the connecting rod is

$$\vec{v}_A = \omega_{AB}\hat{k} \times \vec{\rho}_{IA}$$

$$= \omega_{AB}\hat{k} \times \left[-\ell\sqrt{1 - \frac{R^2}{\ell^2}\sin^2\theta}\,\hat{i} - \ell\tan\theta\sqrt{1 - \frac{R^2}{\ell^2}\sin^2\theta}\,\hat{j} \right]$$

$$= \omega_{AB}\ell\tan\theta\sqrt{1 - \frac{R^2}{\ell^2}\sin^2\theta}\,\hat{i} - \omega_{AB}\ell\sqrt{1 - \frac{R^2}{\ell^2}\sin^2\theta}\,\hat{j}.$$

The value of ω_{AB} can now be determined by equating the two values of $\vec{v}_A$, as follows:

$$-(RC\sin\theta)\hat{i} + (RC\cos\theta)\hat{j}$$

$$= \omega_{AB}\ell\tan\theta\sqrt{1 - \frac{R^2}{\ell^2}\sin^2\theta}\,\hat{i} - \omega_{AB}\ell\sqrt{1 - \frac{R^2}{\ell^2}\sin^2\theta}\,\hat{j}.$$

Equating the coefficient of $\hat{i}$ on the left-hand side to the coefficient of $\hat{i}$ on the right-hand side of the preceding equation, we get

$$-RC\sin\theta = \omega_{AB}\ell\tan\theta\sqrt{1 - \frac{R^2}{\ell^2}\sin^2\theta}$$

Hence,
$$\omega_{AB} = -\frac{RC\cos\theta}{\ell\sqrt{1 - \frac{R^2}{\ell^2}\sin^2\theta}}$$

The reader should make sure that the same result is obtained if the coefficients of $\hat{j}$ are equated. The significance of the minus sign has already been explained.

To find the velocity of the point B on the crosshead, use point B as a point on both the crosshead and the connecting rod. As a point on the latter, its velocity is

$\vec{v}_B = \vec{\omega}_{AB} \times \vec{\rho}_{IB}$. The vector $\vec{\rho}_{IB}$ is vertically downward, and its magnitude is

$$x \tan\theta = \left[R\cos\theta + \ell\sqrt{1 - \frac{R^2}{\ell^2}\sin^2\theta}\right]\tan\theta$$

Hence,

$$\vec{v}_B = -\frac{RC\cos\theta}{\ell\sqrt{1 - \frac{R^2}{\ell^2}\sin^2\theta}}\hat{\kappa} \times \left[-\left(R\sin\theta + \ell\tan\theta\sqrt{1 - \frac{R^2}{\ell^2}\sin^2\theta}\right)\hat{j}\right]$$

$$= \left(-\frac{R^2C\sin\theta\cos\theta}{\ell\sqrt{1 - \frac{R^2}{\ell^2}\sin^2\theta}} - RC\sin\theta\right)\hat{i}$$

The velocity of B is horizontally to the left and has the magnitude shown.

The magnitude of $\vec{v}_B$ could have been found by taking a time derivative of the value of x derived earlier in this example. Thus,

$$v_B = \dot{x} = -R\sin\theta\,\dot{\theta} + \frac{1}{2}\left(1 - \frac{R^2}{\ell^2}\sin^2\theta\right)^{-1/2}\left(\frac{-2R^2\dot{\theta}}{\ell^2}\sin\theta\cos\theta\right)$$

$$= -RC\sin\theta - \frac{R^2C\sin\theta\cos\theta}{\ell\sqrt{1 - \frac{R^2}{\ell^2}\sin^2\theta}}$$

• PROBLEM 8-12

The quadric crank mechanism is shown in Figure 1. The crank AB is rotating with an angular velocity of 5 rad/sec counter-clockwise. Determine a) the linear velocities of the points B and C and the angular velocities of the bar BC and the crank DC. Various dimensions of the crank mechanism are shown in the Figure. b) Determine the linear accelerations of points B and C, as well as the angular accelerations of the bar BC and crank CD using the same data and assuming that there is no angular acceleration for crank AB. c) Determine the linear velocities of the points B and C, as well as the angular velocities of the bar BC and the crank CD, using the instantaneous center method.

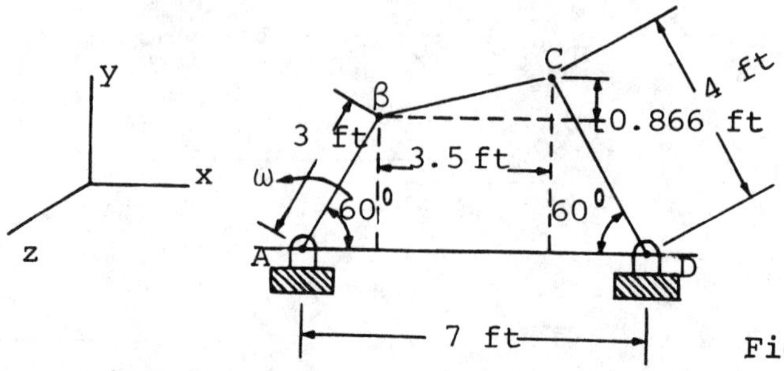

Fig. 1

Solution: a) Point B is on a rotating body. Hence, the linear velocity of point B is $\vec{v}_B = \vec{\omega}_{AB} \times \vec{\rho}_{AB}$. Here, $\vec{\omega}_{AB} = +5\hat{k}$ rad/sec, because according to the right-hand rule counterclockwise rotation about the Z axis is in the positive $\hat{k}$ direction. Also, $\vec{\rho}_{AB} = 3\cos 60°\hat{i} + 3\sin 60°\hat{j} = (1.5\hat{i} + 2.6\hat{j})$ ft. Performing the operation indicated by the cross product, we get

$$\vec{v}_B = (+5\hat{k}) \times (1.5\hat{i} + 2.6\hat{j}) = (7.5\hat{j} - 13.0\hat{i}) \text{ ft/sec.}$$

The same result is obtained by using the relation $v_B = r\omega = (3(5) = 15$ ft/sec and realizing that point B moves to the left and upward along a perpendicular to the crank AB. The vector form is, therefore,

$$-15\cos 30°\hat{i} + 15\sin 30°\hat{j} = (-13.0\hat{i} + 7.5\hat{j}) \text{ ft/sec.}$$

To determine the motion of the bar BC, use the relative-velocity equation

$$\vec{v}_C = \vec{v}_B + \vec{v}_{C/B} = \vec{v}_B + \vec{\omega}_{BC} \times \vec{\rho}_{BC}$$

Since point C must move perpendicular to bar CD, due to the pivot at D, v_C is along a line which makes an angle of 30° with the horizontal. If we assume that bar BC rotates counterclockwise and the point C moves towards the right and upward, the known information is

$$\vec{\omega}_{BC} = +\omega_{BC}\hat{k}$$

$$\vec{\rho}_{BC} = (3.5\hat{i} + 0.866\hat{j}) \text{ ft}$$

$$\vec{v}_C = v_C\cos 30°\hat{i} + v_C\sin 30°\hat{j} = 0.866v_C\hat{i} + 0.5v_C\hat{j}$$

When these values are substituted, together with the value of v_B previously determined, the equation becomes

$$0.866\,v_C\hat{i} + 0.5\,v_C\hat{j} = -13.0\hat{i} + 7.5\hat{j} + (+\omega_{BC}\hat{k}) \times (3.5\hat{i} + 0.866\hat{j})$$

$$= -\ 13.0\hat{i} + 7.5\hat{j} + 3.5\ \omega_{BC}\hat{j} - 0.866\ \omega_{BC}\hat{i}$$

Equating the coefficients of the $\hat{i}$ terms on both sides of this equation and then equating the $\hat{j}$ coefficients, we obtain the simultaneous equations

$$0.866\ v_C = -\ 13.0 - 0.866\ \omega_{BC}$$

$$0.5\ v_C = +\ 7.5 + 3.5\ \omega_{BC}$$

The values found by solving these equations are $\omega_{BC} = -\ 3.75$ rad/sec and $v_C = -\ 11.2$ ft/sec. Hence, the bar BC rotates clockwise (instead of counterclockwise as assumed), and the point C moves downward and to the left along a line that makes an angle of 30° with the horizontal.

Finally, since there is pure rotation of the crank DC, it is known that

$$\vec{v}_C = \vec{\omega}_{BC} \times \vec{\rho}_{DC}$$

Here, $\vec{v}_C = 0.866\ (-\ 11.2)\hat{i} + 0.5\ (-\ 11.2)\hat{j} = 9.68\hat{i} - 5.6\hat{j}$

$$\vec{\omega}_{CD} = +\ \omega_{CD}\hat{k}$$

$$\vec{\rho}_{DC} = -\ 2.0\hat{i} + 3.46\hat{j}.$$

Hence, $-9.68\hat{i} - 5.6\hat{j} = (+\ \omega_{CD}\hat{k}) \times (-\ 2.0\hat{i} + 3.46\hat{j})$

$$= -\ 2.0\ \omega_{CD}\hat{j} - 3.46\ \omega_{CD}\hat{i}$$

By comparing the coefficients of $\hat{i}$ on both sides, we get

$$\omega_{CD} = \frac{-9.68}{-3.46} = +\ 2.8 \text{ rad/sec}$$

The crank DC rotates counterclockwise, as assumed, with an angular velocity of 2.8 rad/sec. This result can also be obtained by comparing the coefficients of $\hat{j}$. Thus,

$$\omega_{CD} = \frac{-\ 5.6}{-\ 2.0} = +\ 2.8 \text{ rad/sec.}$$

Now that the formal vector solution has been given, let us repeat the solution by using a vector triangle and solve the relative-velocity equation by applying the trigonometric law of sines. The equation given before was

$$\vec{v}_C = \vec{v}_B + \vec{v}_{C/B}.$$

We know that the magnitude of the velocity of B is 15 ft/sec and that the point moves upward and to the left along a line making an angle of 30° with the horizontal. We also know that the velocity of C is perpendicular to the crank CD and its magnitude is 4 ω_{CD}, but ω_{CD} is still unknown. Thus, we know that the velocity of C is along a line at 30° with the horizontal, although the sense is unknown. It is known, too, that the velocity of C relative to B is perpendicular to the bar BC, and that its magnitude is equal to the product of the length of the bar and ω_{BC}. The length of bar BC is $\sqrt{0.866^2 + 3.5^2} = 3.61$ ft. Also, $\vec{v}_{C/B}$ is directed along a line whose slope is - 3.5/0.866, since it is perpendicular tc bar BC, whose slope is 0.866/3.5.

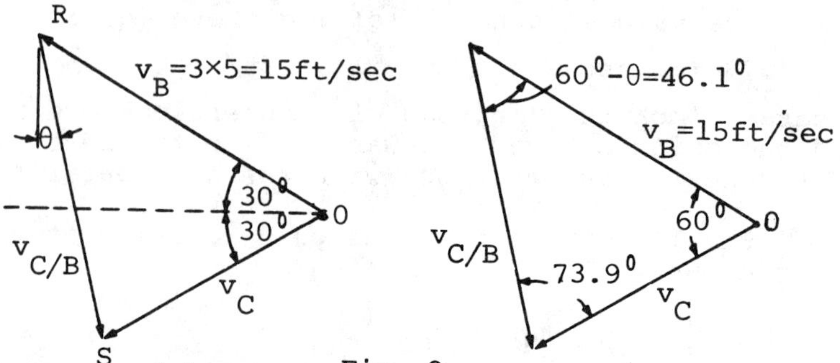

(a) Fig. 2 (b)

The vector equation is indicated graphically in Fig. 2a, where v_B is drawn first from any point O because it is known completely. Then through the arrow end R of $\vec{v}_B$ is drawn a line with the slope - 3.5/0.866. Somewhere on this line will be located the arrow end of $\vec{v}_{C/B}$. From point O is drawn a line making an angle of 30° with the horizontal, as shown. This line meets the line along which $v_{C/B}$ is drawn at point S. The direction of $v_{C/B}$ is determined by the angle θ. Thus, $\theta = \tan^{-1} 0.866/3.5 = 13.9°$. In Fig. 2b the arrows are drawn so that the velocity of B added to the velocity of C relative to B is equal to the velocity of C. The angles are indicated, as well as the magnitudes of the velocities. By the sine law,

$$\frac{15}{\sin 73.9°} = \frac{v_{C/B}}{\sin 60°} = \frac{v_C}{\sin 46.1°}$$

The solutions are $v_C = 11.2$ ft/sec and $v_{C/B} = 13.5$ ft/sec.

Finally, $\omega_{CB} = v_{C/B}/\rho_{BC} = 13.5/3.61 = 3.75$ rad/sec clockwise. The sense of the angular velocity is determined by noting that the sense of $\vec{v}_{C/B}$ is downward and to the right and reasoning that point C must be rotating clockwise about B. Similarly, $\omega_{CD} = v_C/\rho_{DC} = 11.2/4 = 2.8$ rad/sec counterclockwise. The sense of $\vec{\omega}$ is determined by noting that the sense of $\vec{v}_C$ is downward and to the left and reasoning that C must be turning counterclockwise about D.

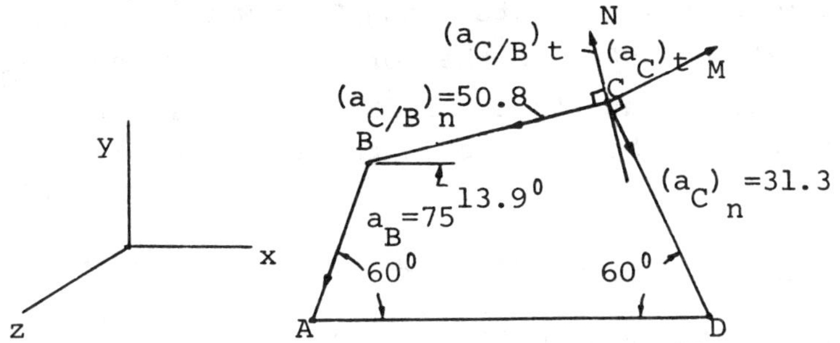

b) The mechanism is redrawn in Fig. 2. The acceleration of point B has only a normal component, because there is no angular acceleration of the crank AB. Its magnitude is $\rho_{AB}\ (\omega_{AB})^2$, which is the same as the magnitude of $\vec{\omega}$ x ($\vec{\omega}$ x $\vec{\rho}$). Thus, the magnitude of $\vec{a}_B$ is 3 $(5)^2$ = 75 ft/sec^2 and this acceleration is from B toward A.

The acceleration of point C is expressed by the equation

$$\vec{a}_C = \vec{a}_B + (\vec{a}_{C/B})_t + (\vec{a}_{C/B})_n$$

The point C, which is a point on the bar BC and also a point on the crank CD, has the same acceleration on either member. When C is used as a point on CD, the magnitude of the normal component $(\vec{a}_C)_n$ is $\rho_{DC}\ (\omega_{CD})^2 = 4\ (2.8)^2 = 31.3$ ft/sec^2, and this component is from C to D. Also, the magnitude of the tangential component $(\vec{a}_C)_t$ is $\rho_{DC}\ \alpha_{CD} = 4\ \alpha_{CD}$, and this component is directed along line CM, which is perpendicular to the crank CD. Its sense is as yet unknown.

The magnitude of $(\vec{a}_{C/B})_n$ is $\rho_{BC}\ (\omega_{BC})^2 = 3.61\ (3.75)^2 = 50.8$ ft/sec^2, and it is from C toward B. The magnitude of $(\vec{a}_{C/B})_t$ is $\rho_{BC}\alpha_{BC} = 3.61\ \alpha_{BC}$, and this component is directed along the line CN, which is perpendicular to the bar BC. Its sense is as yet unknown.

If $\vec{a}_C$ is replaced by its normal and tangential components, the equation for the acceleration of C becomes

$$(\vec{a}_C)_n + (\vec{a}_C)_t = \vec{a}_B + (\vec{a}_{C/B})_t + (\vec{a}_{C/B})_n$$

Three of the five vectors in this equation are known completely, as shown in Fig. 2. The other two, $(\vec{a}_C)_t$ and $(\vec{a}_{C/B})_t$, are known in direction but not in magnitude. Since only two parts are unknown, a solution is possible. The two unknown quantities are actually α_{CD} and α_{BC}, which occur in the expressions for the magnitudes of $(\vec{a}_C)_t$ and $(\vec{a}_{C/B})_t$. If we assume that $(\vec{a}_C)_t$ acts along CM upward and to the right at an angle of 30° with the horizontal and that $(\vec{a}_{C/B})_t$ acts along CN upward and to the left at an angle of 13.9° with the vertical, then the vectors in the above equation expressed in $\hat{i}$, $\hat{j}$ notation are

$$\vec{a}_B = 75\ (-\cos 60°\hat{i} - \cos 30°\hat{j}) = -37.5\hat{i} - 65.0\hat{j}$$

$$(\vec{a}_C)_n = 31.3\ (\cos 60°\hat{i} - \cos 30°\hat{j}) = 15.65\hat{i} - 27.1\hat{j}$$

$$(\vec{a}_{C/B})_n = 50.8\ (-\cos 13.9°\hat{i} - \sin 13.9°\hat{j}) = 49.2\hat{i} - 12.2\hat{j}$$

$$(\vec{a}_{C/B})_t = 3.61\alpha_{BC}(-\sin 13.9°\hat{i} + \cos 13.9°\hat{j}) = -0.866\alpha_{BC}\hat{i} + 3.5\alpha_{BC}\hat{j}$$

$$(\vec{a}_C)_t = 4\alpha_{CD}\ (\cos 30°\hat{i} + \sin 30°\hat{j}) = 3.464\alpha_{CD}\hat{i} + 2.0\alpha_{CD}\hat{j}$$

The vector equation for the acceleration a_C becomes

$$15.65\hat{i} - 27.1\hat{j} + 3.464\alpha_{CD}\hat{i} + 2.0\alpha_{CD}\hat{j}$$
$$= -37.5\hat{i} - 65.0\hat{j} - 0.866\alpha_{BC}\hat{i} + 3.5\alpha_{BC}\hat{j} - 49.2\hat{i} - 12.2\hat{j}$$

Collecting terms and equating the coefficients of the $\hat{i}$ terms and $\hat{j}$ terms, respectively, we obtain

$$3.464\alpha_{CD} + 0.866\alpha_{BC} = -102.4$$

$$2.0\alpha_{CD} - 3.5\alpha_{BC} = -50.1$$

The results obtained by solving these last two equations are

$$\alpha_{CD} = -29.0 \text{ rad/sec}^2 \quad \text{and} \quad \alpha_{BC} = -2.26 \text{ rad/sec}^2$$

The minus signs indicate that the directions assumed for the two tangential components were incorrect. Hence, the component $(\vec{a}_C)_t$ actually acts downward and to the left, and the crank has a counterclockwise angular acceleration $\vec{\alpha}_{CD}$ whose magnitude is 29.0 rad/sec². Also, the component $(\vec{a}_{C/B})_t$ actually acts downward and to the right, and the bar BC has a clockwise angular acceleration $\vec{\alpha}_{BC}$ whose magnitude is 2.26 rad/sec².

The acceleration $\vec{a}_C$ can be found by using either side of the vector equation. Naturally, both values should be the same. From the left-hand side,

$$\vec{a}_C = 15.65\hat{i} - 27.1\hat{j} + 3.464\ (-\ 29.0)\hat{i} + 2.0\ (-\ 29.0)\hat{j}$$

$$= (-\ 84.7\hat{i} - 85.1\hat{j})\ \text{ft/sec}^2$$

From the right-hand side of the vector equation,

$$\vec{a}_C = -\ 37.5\hat{i} - 65.0\hat{j} - 0.866\ (-\ 2.26)\hat{i} + 3.5\ (-\ 2.26)\hat{j}$$

$$-\ 49.2\hat{i} - 12.2\hat{j}$$

$$= (-\ 84.7\hat{i} - 85.1\hat{j})\ \text{ft/sec}^2$$

In summary, the acceleration $\vec{a}_C$ has an x component of magnitude 84.7 ft/sec^2 to the left and a y component of magnitude 85.1 ft/sec^2 downward.

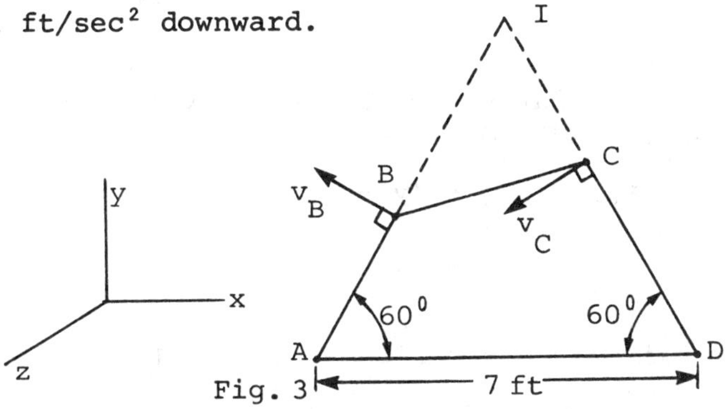

Fig. 3

c) The mechanism is redrawn in Fig. 3. The velocity $\vec{v}_B$ is perpendicular to the crank AB, and the velocity $\vec{v}_C$ is perpendicular to the crank CD. Hence, the instantaneous center I is at the intersection of extensions of AB and CD. Since the triangle AID in this particular case is equilateral, sides AI and ID are each 7 ft long. Also, BI = 4 ft and IC = 3 ft.

The magnitude of the velocity $\vec{v}_B$ of B as a point on AB is 3 x 5 = 15 ft/sec. The velocity of B as a point on BC, has the same magnitude but is also equal to (IB)ω_{BC}. Hence, the angular velocity of bar BC about I is 15/4 = 3.75 rad/sec, and it must be clockwise because point B is moving clockwise about I.

The magnitude of the velocity $\vec{v}_C$ of C as a point on BC is 3 x 3.75 = 11.2 ft/sec, and this velocity is directed as shown in Figure 3. The magnitude of the velocity of C as a point on CD is 4ω_{CD}. Hence, the magnitude of the angular velocity of crank CD is 11.2/4 = 2.8 rad/sec, and this velocity is counterclockwise to conform to the sense of $\vec{v}_C$.

A disk of radius R rolls on a horizontal rail. The arm of length r of the rectangular bracket shown is fitted to a bearing on its axis, and the other arm is free to slide over the end of the guiding rail. Determine the two pole curves for the bracket.

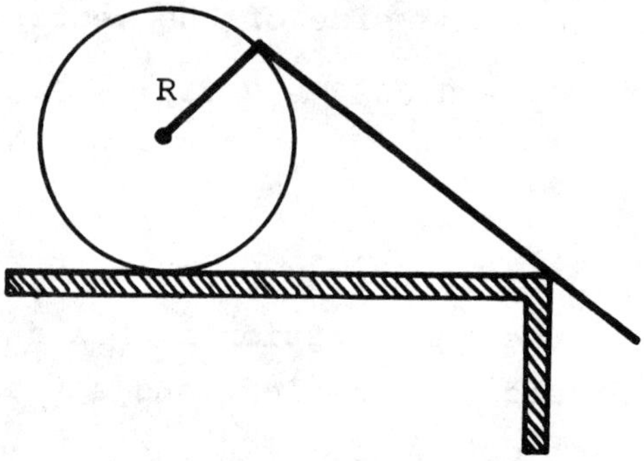

Solution: The velocity of a point P at (x,y) in a body moving in a plane relative to the velocity of a reference point A in the body is given by

$$v_{Px} = v_{Ax} - (y - y_A)\omega$$
$$v_{Py} = v_{Ay} + (x - x_A)\omega \qquad (1)$$

where $\omega = \dot{\theta}$ is the angular velocity of the body about point A.

In this problem, point A is taken to be the end of the bracket attached to the hub of the wheel.

The space pole curve or space centrode is the curved traced out in a space fixed coordinate system by the instantaneous center. The body pole curve or body centrode is the same curve with respect to a coordinate system fixed in the body. By definition the instantaneous center is the point in space which is instantaneously at rest about which the body turns. Let the point P at (x,y) be the instantaneous center. Then by definition $v_x = v_y = 0$ for P. So, from equation (1)

$$0 = v_A - (y - y_A)\omega \qquad 0 = (x - x_A)\omega .$$

Since $v_{Ay} = 0$ and $v_{Ax} = v_A$.

Thus $x = x_A = -b \qquad y = y_A + v_A/\omega . \qquad (2)$

In order to relate v_A and ω consider the velocity at the origin, O.

$$v_{Ox} = v_O \cos 2\phi \qquad v_{Oy} = -\, v_O \sin 2\phi \qquad x_O = 0 \qquad y_O = 0$$

$$y_A = r \qquad x_A = -\, b.$$

Using these relations in equation (1) yields,

$$v_{Ox} = v_O \cos 2\phi = v_{Ax} - (y_O - y_A)\omega \;\; = v_A + r\omega$$

$$v_{Oy} = -\, v_O \sin 2\phi \;\; = (x_O - x_A)\omega = b\omega.$$

Solving for v_A and ω yields

$$v_A = v_O \cos 2\phi - r\omega \qquad \text{and} \qquad \omega = -\frac{v_O \sin 2\phi}{b}. \qquad (3)$$

Substituting in the first equation for ω gives,

$$v_A = v_O \left(\cos 2\phi + \frac{r}{b} \sin 2\phi\right).$$

Next, substitute for v_O in the equation for ω.

$$\omega = -\frac{\sin 2}{b} \, \frac{v_A}{\cos 2\phi + \frac{r}{b} \sin 2\phi} = -\frac{v_A \sin 2\phi}{b \cos 2\phi + r \sin 2\phi}$$

$$= \frac{-v_A}{r + b \cot 2\phi}.$$

This gives the value of v_A/ω which can be used in equation (2) to yield

$$x = -\, b \qquad y = r - (r + b \cot 2\phi) = -\, b \cot 2\phi. \qquad (4)$$

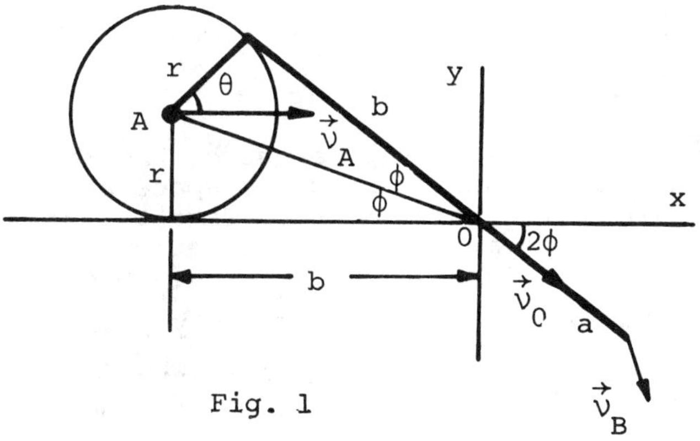

Fig. 1

From Figure 1 $\tan \phi = r/b$. Thus,

$$x = -\, r \cot \phi \qquad\qquad y = -\, r \cot \phi \cot 2\phi. \qquad (5)$$

These equations give the position of the instantaneous

center as a function of ϕ. To find the space pole curve, use y as a function of x. To find this, note $\cot \phi = -\frac{x}{r}$ and $\cot 2\phi = \frac{1 - \tan^2\phi}{2 \tan \phi} = \frac{1}{2} \cot \phi \left(1 - \frac{1}{\cot^2\phi}\right)$. Substituting these expressions in the equation for y yields

$$y = x \frac{1}{2} \left(-\frac{x}{r}\right) \left(1 - \frac{r^2}{x^2}\right) = -\frac{x^2}{2r} + \frac{r}{2}.$$

This can be rewritten as

$$y - \frac{r}{2} = -\frac{x^2}{2r} .$$

When $x = 0$, $y = \frac{r}{2}$. Thus, this is the equation of a parabola which has its vertex at $\left(0, \frac{r}{2}\right)$. The minus sign means the parabola opens downward. The situation is shown in Figure 2. The parabola is the space pole curve. By the definition of a parabola, the directrix is a line parallel to the x-axis at a distance from the vertex equal to the distance of the vertex from the focus. In this case the focus is at the origin and the directrix is at $y = r$. Therefore the directrix is the line traced out by the point A, i.e., the hub of the wheel.

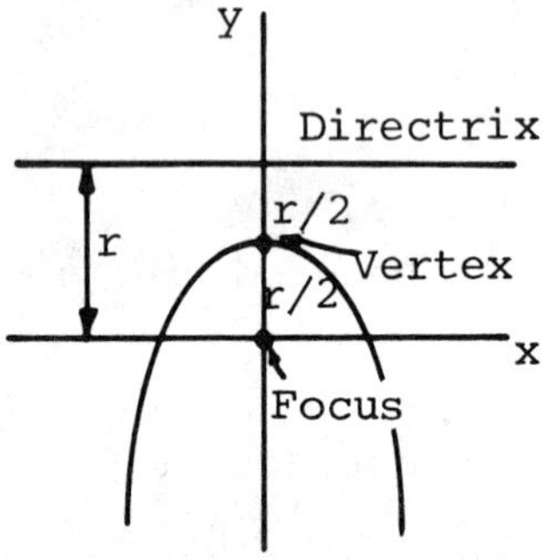

Fig. 2

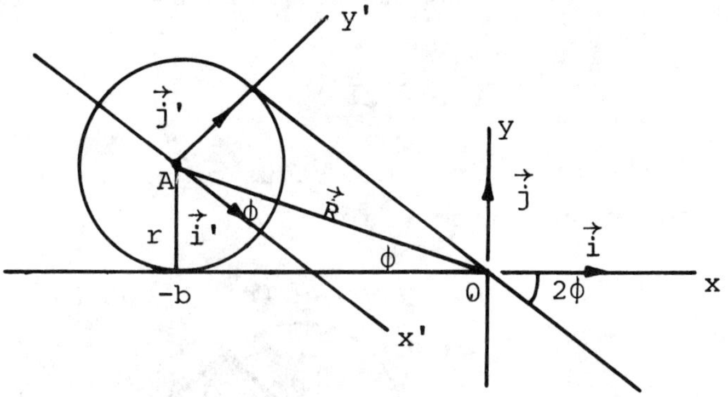

Fig. 3

Next, find the body pole curve. Do this by making a coordinate transformation from the space system, x,y to a system fixed in the body x', y'. Take the origin of the prime system to be at A with the x'-axis parallel to the long arm as shown in Figure 3. In the space system, the instantaneous center is at $\vec{r} = x\,\vec{i} + y\,\vec{j}$ while in the body system it is at $\vec{r}\,' = x'\,\vec{i}\,' + y'\,\vec{j}\,'$. These are related by by the relation $\vec{r}\,' = \vec{R} + \vec{r}$, where in the body system, $\vec{R} = R \cos \phi\,\vec{i}\,' + R \sin \phi\,\vec{j}\,'$. The problem now is to write $\vec{r}$ in the prime system, which means we have to relate $\vec{i}$ and $\vec{j}$ to $\vec{i}\,'$ and $\vec{j}\,'$. From Figure 3,

$$\vec{i} = \cos 2\phi\,\vec{i}\,' + \sin 2\phi\,\vec{j}\,'$$

$$\vec{j} = -\sin 2\phi\ \vec{i}' + \cos 2\phi\ \vec{j}'.$$

Using these relations in the equation for $\vec{R} + \vec{r}$ yields

$$\vec{r}' = (+ x \cos 2\phi + R \cos \phi - y \sin 2\phi)\vec{i}'$$

$$+ (x \sin 2\phi + y \cos 2\phi + R \sin \phi)\vec{j}'$$

But $\vec{r}' = x'\ \vec{i}' + y'\ \vec{j}'$ so a comparison of the two gives us x' and y'.

$$x' = x \cos 2\phi + R \cos \phi - y \sin 2\phi$$

From Figure 3, $b = R \cos \phi$, $r = R \sin \phi$. Using these relations and equation (4), x' becomes

$$x' = - b \cos 2\phi + b + b \cot 2\phi \sin 2\phi$$

$$= b\ (1 - \cos 2\phi + \cos 2\phi) = b = r \cot \phi.$$

Similarly for y'

$$y' = x \sin 2\phi + y \cos 2\phi + R \sin \phi = - b \sin 2\phi$$

$$- b \cot 2\phi \cos 2\phi + r$$

$$= r - b \left(\sin 2\phi + \frac{\cos^2 2\phi}{\sin 2\phi}\right) = r - \frac{b}{\sin 2\phi}$$

$$= r - \frac{r \cos \phi}{\sin \phi\ 2 \sin \phi \cos \phi}$$

$$= r \left(1 - \frac{1}{2 \sin^2 \phi}\right) = \frac{r}{2} \left(\frac{2 \sin^2 \phi - 1}{\sin^2 \phi}\right)$$

$$= \frac{r}{2} \left(\frac{\sin^2 \phi - \cos^2 \phi}{\sin^2 \phi}\right) = \frac{r}{2}\ (1 - \cot^2 \theta).$$

To summarize

$$x' = r \cot \phi \qquad\qquad y' = \frac{r}{2}\ (1 - \cot^2 \phi). \qquad (6)$$

These equations give the location of the instantaneous center with respect to the body system and are analogus to equation (5). Again, we want to find y' as a function of x'.

$$y' = \frac{r}{2} \left(1 - \frac{x'^2}{r^2}\right) \quad \text{or} \quad y' - \frac{r}{2} = - \frac{x'^2}{2r}.$$

So the body pole curve is also a parabola with the focus at A, the vertex at $y' = r/2$ and the directrix along the long arm of the body.

Notice that the two curves always have the instantaneous center in common and therefore the body pole curve rolls on the space pole curve.

• PROBLEM 8-14

A mechanical system consists of the crank rotating with angular velocity ω about O, the piston with a translatory motion of velocity v in the direction AO and the connecting rod which joins the crank and the piston and executes a plane motion. Determine the instantaneous centers of each link. Also, express the speed of the piston in terms of the angular velocity ω of the crank shaft and the distance z between O and the point where the axis of connecting rod cuts OB.

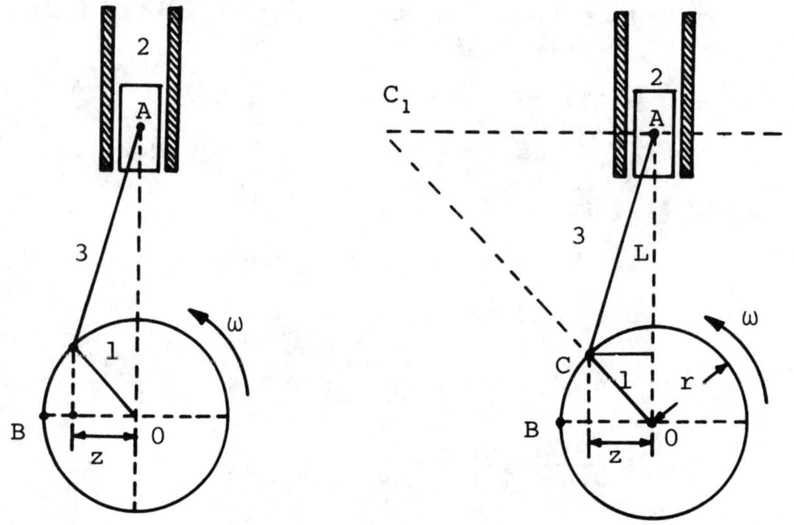

Fig. 1

Solution: Instantaneous centers can be found by drawing perpendicular lines to the velocity vector at any two points on a linkage. Drawing perpendiculars of velocity vector of points A and C, we get C_1 as the instantaneous center of link AC. Note that C_1 is the point where the ray $\vec{OC}$ meets the horizontal line through A as shown in Figure 1. Since the crank shaft is pinned at O, its instantaneous center is a permanent center of rotation. Thus center of rotation of OC is O.

The piston can only move in a straight line, upward or downward. Therefore, the instantaneous center of rotation of the piston lies at infinity.

In order to find $\vec{v}_A$ at an instant, we will use the instantaneous center for connecting rod AC. The instantaneous center is defined as the point from which the velocity of a point of a body is given by $r_P \omega_{AC}$ where r_P is the distance from the point to the center and ω_{AC} is the angular velocity of the crank shaft.

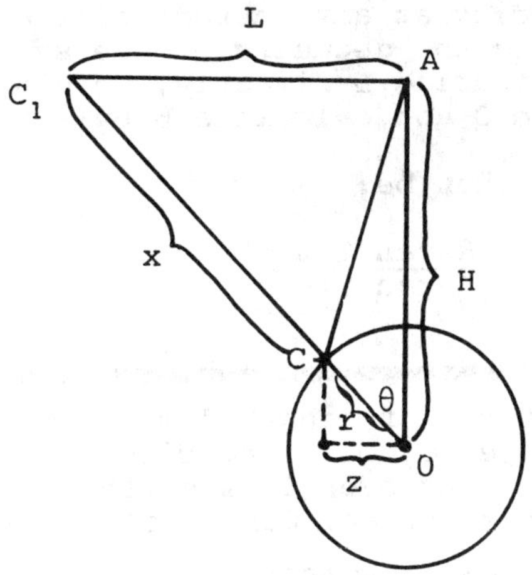

Fig. 2

The situation is shown in Figure 2. The distance from point C to C_1, the instantaneous center, is x. This distance x can be found by geometry to be $\frac{L}{\sin\theta} - r$.

The distance from A to C_1 is L. Therefore, ω_{AC} will be given by:

$$\omega_{AC} = \frac{v_A}{L} \qquad (1)$$

and also by $\omega_{AC} \quad \frac{v_C}{x}$. (2)

Equating these equations and substituting for x we get

$$\frac{v_{AB}}{L} = \frac{v_C}{\frac{L}{\sin\theta} - r}$$

but v_C is given by $r\omega$ where ω is the angular velocity of the crank shaft. Also, r can be shown to equal $\frac{z}{\sin\theta}$. Substituting these values yields:

$$\frac{\omega\left(\frac{z}{\sin\theta}\right)}{\frac{L}{\sin\theta} - \frac{z}{\sin\theta}} = \frac{v_A}{L}$$

Upon solving for v_A we find:

$$v_A = \frac{L\ \omega\ z}{L - z}$$

The velocity at any instant of the piston can be found by knowing the geometry of the system at that instant. Rather than measuring L directly, we can measure H, the distance from A to O and calculate L by: $L = H \tan\theta$.

v_A would then be:

$$v_A = \frac{H \tan\theta\ \omega\ z}{H \tan\theta - z}$$

• PROBLEM 8-15

A thin straight rod of length ℓ moves so that it always rests on the edge, C, of a semicircular guidance of radius R with end A sliding over this surface. Assuming that $\phi = 30°$ and $v_C = 2$ ft/sec. Construct the instantaneous center and the velocities of points A and B. With $\ell = 2R$ and an arbitrary angle ϕ, calculate the position of the instantaneous center, the angular velocity of the rod, the speed of A and B, and both pole curves.

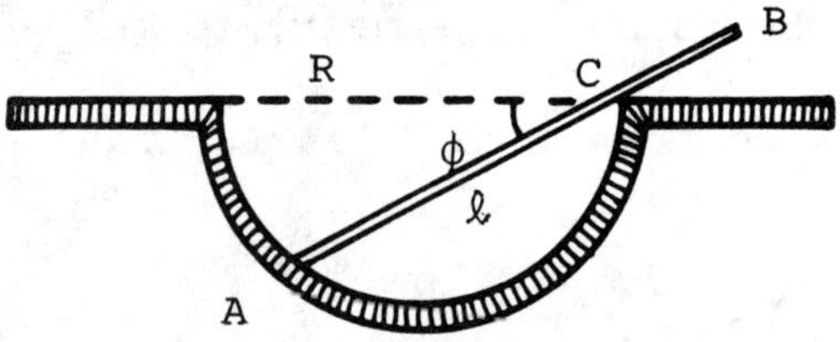

Solution: We are given $v_C = 2$ ft/sec, $\phi = 30°$ and from the diagram $\theta + 2\phi = 90°$. The general motion of a lamina in the plane is shown in Figure 2.

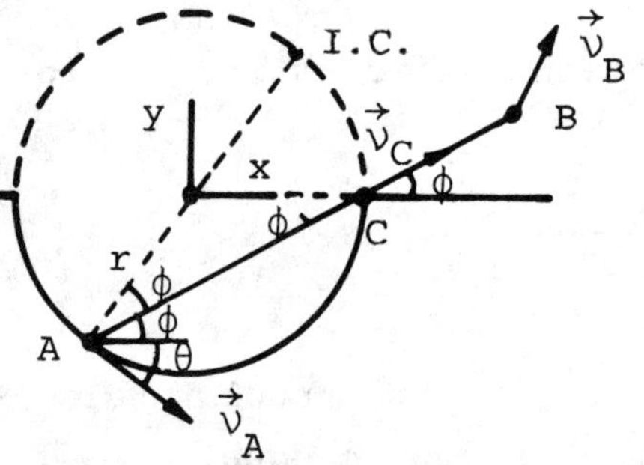

Fig. 1

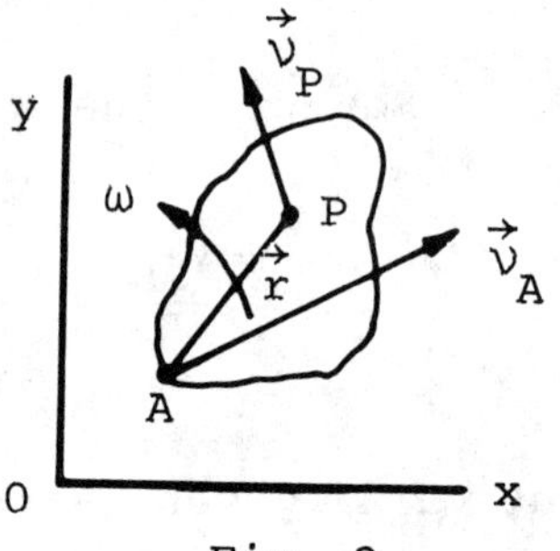

Fig. 2

Point A is taken as the reference point. $\vec{\omega}$ is the angular velocity of the lamina about point A. The point P located at (x,y) is some other point in the body. The velocity of P relative to A is $\vec{\omega} \times \vec{r}$; where $\vec{r}$ is the position of P relative to A. The velocity of P in the fixed system is $\vec{v}_P = \vec{v}_A + \vec{\omega} \times \vec{r}$ or in components:

$$v_{Px} = v_{Ax} - (y - y_A)\omega$$
$$v_{Py} = v_{Ay} + (x - x_A)\omega. \qquad (1)$$

By definition, the instantaneous center is the point in space which is instantaneously at rest and about which the body is turning. Thus, if P is the instantaneous center,

$$v_{Px} = 0,\ v_{Py} = 0.$$

In this problem, take point A to be the end of the rod in contact with the semi-circle. Its velocity is tangent to the semi-circle at the point of contact. Since the rod remains in contact with the fixed point C, its velocity at C must be in the direction of the rod.

From Figure 1 $x_A = -r \cos 2\phi \qquad y_A = -r \sin 2\phi$

$$v_{Ax} = v_A \cos \theta \qquad v_{Ay} = -v_A \sin \theta \ (2)$$

$$v_{Cx} = v_C \cos \phi \qquad v_{Cy} = v_C \sin \phi.$$

Since $\theta = 90 - 2\phi$,

$$\cos \theta = \sin 2\phi \quad \text{and} \quad \sin \theta = \cos 2\phi.$$

From equations (2) the equation for $\vec{v}_A$ becomes

$$v_{Ax} = v_A \sin 2\phi \qquad v_{Ay} = -v_A \cos 2\phi \qquad (3)$$

Now take point C as the general point. C is located at $x_C = r$, $y_C = 0$, therefore, using equations (2) and (3), equation (1) becomes,

$$v_{Cx} = v_C \cos \phi = v_{Ax} - (y_C - y_A)\omega = v_A \sin 2\phi - r\,\omega \sin 2\phi$$

$$= (v_A - r\,\omega) \sin 2\phi \qquad (4)$$

$$v_{Cy} = v_C \sin \phi = v_{Ay} + (x_C - x_A)\omega = -v_A \cos 2\phi$$

$$+ (r + r \cos 2\phi)\omega.$$

These are two simultaneous equations for the unknowns ω and v_A. Solving gives

$$\omega = \frac{v_C \sin \phi + v_A \cos 2\phi}{r\ (1 + \cos 2\phi)} \qquad v_A = v_C \frac{\cos \phi}{\sin 2\phi} + r\omega.$$

Substitution of the first equation into the second gives

$$v_A = v_C \frac{1}{2 \sin \phi} + \frac{v_C \sin \phi + v_A \cos 2\phi}{1 + \cos 2\phi} \quad \text{or}$$

$$v_A = v_C \frac{1 + \cos 2\phi + 2 \sin^2 \phi}{2 \sin \phi}.$$

But from the trignometric identities $1 + \cos 2\phi = \sin^2\phi + \cos^2\phi + \cos^2\phi - \sin^2\phi = 2 \cos^2\phi$. So,

$$v_A = v_C \frac{2 \cos^2\phi + 2 \sin^2\phi}{2 \sin \phi} = \frac{v_C}{\sin \phi} \tag{5}$$

Equation (5) gives the relation between the velocity of points A and C. Substitution of equation (5) into the equation for ω gives

$$\omega = v_C \frac{\sin \phi + \cos 2\phi/\sin \phi}{2 r \cos^2\phi}.$$

Since $\sin^2\phi + \cos 2\phi = \cos^2\phi$, the equation for ω becomes

$$\omega = \frac{v_C}{2 r \sin \phi}. \tag{6}$$

Equations (5) and (6) give the values of v_A and ω in terms of v_C and ϕ. Next we want to find v_B. Let $CB = d = \ell - 2 r \cos \phi$

$$x_B = r + d \cos \phi \qquad y_B = d \sin \phi, \tag{7}$$

Then using equations (2), (3) and (7), equation (1) becomes

$$v_{Bx} = v_A \sin 2\phi - (d \sin \phi + r \sin 2\phi)\omega$$

$$v_{By} = -v_A \cos 2\phi + (r + d \cos \phi + r \cos 2\phi)\omega.$$

Substitution for v_A and ω from equations (5) and (6) and using $d = \ell (1 - \cos\phi)$ gives

$$v_{Bx} = v_C \{2 \cos \phi - [(1 - \cos\phi) + \cos \phi]\}$$

$$= v_C (2 \cos \phi - 1)$$

$$v_{By} = \frac{v_C}{\sin \phi} \{- \cos 2\phi + \left[\frac{1}{2} (1 + \cos 2\phi\right] + (1 - \cos \phi) \cos \phi\}. \tag{8}$$

The last factor for v_{Bx} can be written in terms of functions of $\phi/2$.

$$2 \cos \phi - 1 = 2 (\cos^2\phi/2 - \sin^2\phi/2) - 1 = 2 \cos^2\phi/2$$

$$- 2 (1 - \cos^2\phi/2) - 1 = 4 \cos^2\phi/2 - 3.$$

The trigonometric factor in equation (8) can also be simplified.

$$\frac{1}{\sin\phi}\{\ \} = \frac{1}{2\sin\phi}\{-2\cos\phi + 1 + \cos 2\phi$$

$$+ 2\cos\phi\,(1 - \cos\phi)\}$$

$$= \frac{1}{2\sin\phi}\{1 - \cos 2\phi + 2\cos\phi\,(1 - \cos\phi)\}$$

$$= \frac{1}{\sin\phi}\{\sin^2\phi + \cos\phi\,(1 - \cos\phi)\}$$

$$= \sin\phi + \frac{\cos\phi\,\sin\phi/2}{\cos\phi/2}$$

$$= \frac{2\sin\phi/2\,\cos^2\phi/2 + \cos^2\phi/2\,\sin\phi/2 - \sin^3\phi/2}{\cos\phi/2}$$

$$= \tan\frac{\phi}{2}\,(4\cos^2\frac{\phi}{2} - 1).$$

With these results, equation (8) becomes

$$v_{Bx} = v_C\,(4\cos^2\frac{\phi}{2} - 3)$$

$$v_{By} = v_C\tan\frac{\phi}{2}\,(4\cos^2\frac{\phi}{2} - 1). \qquad (9)$$

Next, find v_B by using $v_B{}^2 = v_{Bx}{}^2 + v_{By}{}^2$. This gives

$$v_B{}^2 = \frac{v_C{}^2}{\cos^2\frac{\phi}{2}}\{16\cos^6\phi/2 - 24\cos^4\phi/2 + 9\cos^2\phi/2$$

$$+ 16\cos^4\phi/2\,\sin^2\phi/2 - 8\cos^2\phi/2\,\sin^2\phi/2$$

$$+ \sin^2\phi/2\}$$

$$= \frac{v_C{}^2}{\cos^2\phi/2}\{16\cos^4\phi/2\,(1 - \sin^2\phi/2)$$

$$- 24\cos^2\phi/2\,(1 - \sin^2\phi/2) + 8\cos^2\phi/2$$

$$+ 16\cos^4\phi/2\,\sin^2\phi/2 - 8\cos^2\phi/2\,\sin^2\phi/2 + 1\}$$

$$= \frac{v_C{}^2}{\cos^2\phi/2}\{16\,(\cos^4\phi/2 + \cos^2\phi/2\,\sin^2\phi/2$$

$$- \cos^2\phi/2) + 1\}.$$

Since $\sin^2\phi/2 = 1 - \cos^2\phi/2$, the term in the parentheses is zero. So, we obtain

$$v_B = \frac{v_C}{\cos \phi/2}. \tag{10}$$

From equation (1) and the discussion following it, the instantaneous center is at the point I.C. with coordinates x, y such that,

$$0 = v_{Ax} - (y - y_A)\omega$$

$$0 = v_{Ay} + (x - x_A)\omega$$

or $\quad y = y_A + v_{Ax}/\omega \qquad$ and $x = x_A - v_{Ay}/\omega$.

Using equations (2), (5), and (6), these become

$$x_{IC} = - r \cos 2\phi + \frac{v_C}{\sin \phi} \frac{\cos 2\phi}{v_C} 2 r \sin \phi = r \cos 2\phi$$

$$(11)$$

$$y_{IC} = - r \sin 2\phi + \frac{v_C}{\sin \phi} \frac{\sin 2\phi}{v_C} 2 r \sin \phi = r \sin 2\phi.$$

For the special case $\phi = 30°$ and $v_C = 2$ ft/sec, we can use equations (11), (5) and (10) to find the I.C. and v_A and v_B. These give,

$x_{IC} = r/2 = .5\, r \qquad\qquad v_A = 4$ ft/sec

$y_{IC} = \frac{\sqrt{3}}{2} r = .866\, r \qquad\qquad v_B = 2.07$ ft/sec.

The space pole curve or the space centrode is the curve traced out in space, i.e., with respect to the coordinate system fixed in space. From equation (11), note:

$$x_{IC}^2 + y_{IC}^2 = r^2 \cos^2 2\phi + r^2 \sin^2 2\phi = r^2.$$

Thus the space pole curve is a circle of radius r centered at the origin. When y = 0, $x_{IC} = r$, and $y_{IC} = 0$. So, the I.C. starts at B and moves along the circle towards the y axis as ϕ increases. Note that the I.C. can also be constructed geometrically if the velocity of two points of the body are known. Since at any instant, the body is rotating about the I.C., the line from the I.C. to a point in the body must be perpendicular to the velocity at that point. Thus, if the velocities of two points of the body are known, the intersection of the perpendiculars to the velocities at those points is the I.C.

The body pole curve or the body centrode is the curve traced out with respect to a coordinate system attached to the body. In this case we take a system with the x' axis along the rod and the y' axis perpendicular to the rod and centered at A. From Figure 1, the position of the I.C. in this coordinate system is

$$x'_{IC} = 2\ r \cos\phi \qquad y'_{IC} = 2\ r \sin\phi .$$

Squaring each coordinate and adding gives

$$\left(x'_{IC}\right)^2 + \left(y'_{IC}\right)^2 = (2r)^2 .$$

So the body pole curve is also a circle, but of radius 2r centered at A. Note that at any instant the curves are in contact at the I.C. As ϕ increases, the body curve rolls on the space curve.

The position of the I.C. in the prime coordinate system can also be obtained directly from its position in the unprimed system by a coordinate transformation.

RELATIVE SPATIAL MOTION

• PROBLEM 8-16

Shown in figure 1 is a slider-crank mechanism. Wheel A rotates with a constant angular velocity of $\dot{\alpha}$, and the block B moves in a linear path. Determine the velocity of the block B as a function of the angle α.

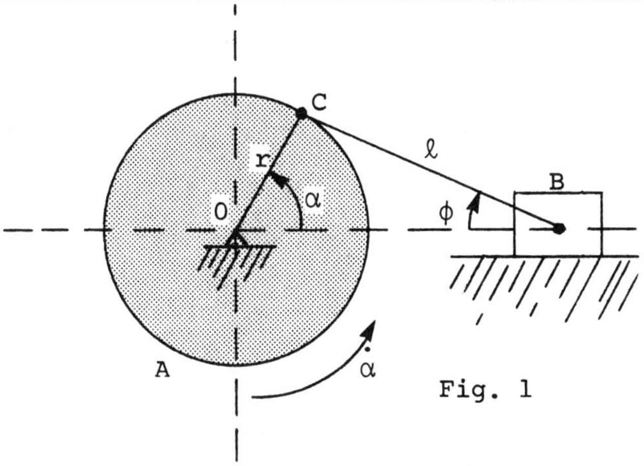

Fig. 1

Solution: Consider the vector equation

$$\vec{v}_B = \vec{v}_C + \vec{v}_{B/C} .$$

By observation, it is noted that both the magnitude and direction of the velocity at point C are known. Also it is noted that the directions of both $\vec{v}_B$ and $\vec{v}_{B/C}$ are known. That is, $\vec{v}_B$ is horizontal and $\vec{v}_{B/C}$ is always perpendicular to the rigid rod CB. Thus, even though the magnitude of $\vec{v}_B$ is not known, it may be determined by constructing a vector diagram such as figure 2.

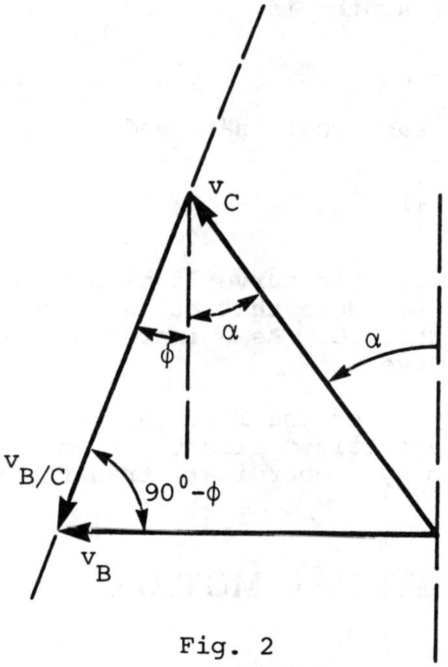

Fig. 2

By the law of sines,

$$\frac{v_B}{\sin(\phi + \alpha)} = \frac{v_C}{\sin[(\pi/2) - \phi]}$$

and $$v_B = v_C \frac{\sin(\phi + \alpha)}{\cos \phi}.$$

After some trigonometric reduction, using the identity:

$$\sin(A + B) = \sin A \cos B + \sin B \cos A$$

$$v_B = v_C (\sin \alpha + \cos \alpha \tan \phi).$$

It would be nice to find ϕ in terms of α; this can be done by noting that

$$r \sin \alpha = \ell \sin \phi$$

so that $$\sin \phi = \left(\frac{r}{\ell}\right) \sin \alpha$$

and $$\cos \phi = (1 - \sin^2 \phi)^{1/2}$$

$$\tan = \frac{\sin\phi}{\cos\phi}.$$

Therefore, with the aid of the binomial expansion,

$$\tan\phi = \left(\frac{r}{\ell}\right)\sin\alpha\left[1 - \frac{1}{2}\left(\frac{r}{\ell}\right)^2\sin^2\alpha + \text{(higher-order terms)}\right]$$

or $$\tan\phi = \left(\frac{r}{\ell}\right)\sin\alpha - \frac{1}{2}\left(\frac{r}{\ell}\right)^2\sin^3\alpha + \ldots$$

Now, for many situations, $r < \ell$, and since $\sin\alpha \leq 1$, it is seen that the second term can be disregarded. Finally,

$$v_B = v_C\left[\sin\alpha + \cos\alpha\left(\frac{r}{\ell}\right)\sin\alpha\right]$$

and since $v_C = r\dot{\alpha}$, then

$$v_B = r\dot{\alpha}\left[\sin\alpha + \frac{1}{2}\left(\frac{r}{\ell}\right)\sin 2\alpha\right].$$

• PROBLEM 8-17

A spool of wire is located on a horizontal surface. The outside diameter of the spool is 8 in, the inside 5 in. The end of the wire is pulled to the right with the constant velocity 3 in/s causing the spool to roll without sliding. Determine: a) the angular velocity of the spool b) the linear velocity of the spool c) the length of wire which is unwound or wound per second.

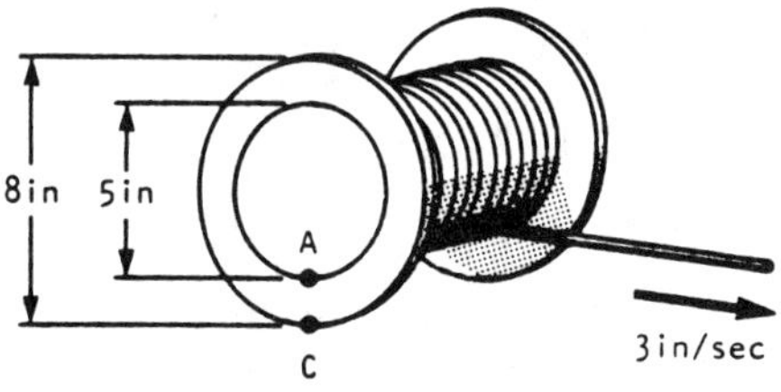

Fig. 1

Solution: "Rolling without sliding" at point C means that there is no relative motion between the two coincident particles at C, one on the ground and one on the spool. The wire, which is assumed to be unstretchable, has rolling contact with the spool at A, (No slipping of the wire across the spool occurs at A.) Therefore, the velocity of a particle on the spool at A is 3 in/sec to the right.

a) To find the angular velocity of the spool we can write the relative velocity equation between particles A and C on the spool:

$$\bar{v}_A = \bar{v}_C + \bar{v}_{A/C}. \quad (1)$$

But $\bar{v}_C = 0$ and

$$v_{A/C} = (AC)\omega.$$

Substituting into (1) we have

$$v_A = (AC)\omega \quad (2)$$

and, solving for ω, we have the desired quantity:

$$\omega = v_A/AC \quad (3)$$

$$= 3/1.5$$

$$= 2 \text{ rad/sec}.$$

Since angular velocity is a vector quantity, the direction of $\vec{\omega}$ must also be determined. $\vec{\omega}$ must agree in sense with $\vec{v}_{A/C}$. Because $\vec{v}_{A/C} = \vec{v}_A = 3$ in/sec to the right, $\vec{\omega}$ must be clockwise. So,

$$\vec{\omega} = 2 \text{ rad/sec.} \circlearrowright$$

Equation (2) can also be written directly from the fact that C is the instantaneous center of the cylinder.

b) Relating the velocity of the spool center to particle C, we have

$$\vec{v}_0 = \vec{v}_c + \vec{v}_{0/c} \quad (4)$$

$$\vec{v}_0 = 0 + (OC)\omega$$

$$= (4 \text{ in}) \ 2 \text{ rad/sec}$$

$$= 8 \text{ in/sec}.$$

Since C is the instant center and $\vec{\omega}$ is clockwise, the cylinder must be moving to the right.

$$\vec{v}_0 = 8 \text{ in/sec.} \rightarrow$$

c) Wire is being wound on the drum since

$$\vec{v}_0 > \vec{v}_D.$$

In one second O moves 8 in. while D moves 3 in., both to the right. Therefore, 5 in. of wire is wound per second.

In Fig. 1a the bar AB is horizontal in the phase shown. It is rotating about A with an angular velocity $\vec{\Omega} = 3\hat{k}_0$ rad/sec and an angular acceleration $\vec{\alpha} = 6\hat{k}_0$ rad/sec^2. Determine the angular velocity and angular acceleration of bar CD, which is rotating about C and is attached at D to a collar which is free to slide on AB.

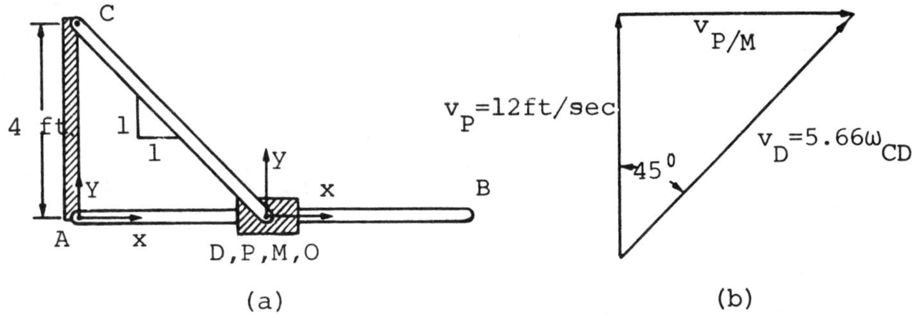

Fig. 1

Solution: Choose fixed axes XYZ at point A and a rotating set xyz at point P on the horizontal bar. Let point P coincide with D on the sloping bar. The mating point M will be fixed on AB at D. The equation for velocities with the known items below the corresponding vectors is

$$v_D = v_P + v_{P/M} + v_M$$

$$\vec{v}_D = \rho_{CD}\,\omega_{CD}$$

$$\vec{v}_P = \rho_{AP}\,\omega_{AB} \uparrow$$

$$= (4)\ (3)\uparrow = 12\uparrow \text{ ft/sec.}$$

$$\vec{v}_{P/M} = v_{P/M} \rightarrow$$

$$\vec{v}_M = 0.$$

Since D is on the sloping member, the magnitude of its absolute velocity is $\rho_{CD}\omega_{CD} = 4\sqrt{2}\ \omega_{CD} = 5.66\omega_{CD}$.

The vector triangle for velocities is shown in Fig. 1b. From it can be deduced the relation

$$\frac{v_\rho}{v_D} = \cos 45° \qquad \text{or} \qquad \frac{12}{5.66\omega_{CD}} = 0.707$$

Hence, ω_{CD} = 3 rad/sec counterclockwise.

The acceleration equation with the known information shown under the vectors is

$$(\vec{a}_D)_t + (\vec{a}_D)_n = (a_P)_t + (a_P)_n + \vec{a}_{P/M} + \vec{a}_M + 2\Omega \times v_{P/M}.$$

$(\vec{a}_D)_t = 5.66\ \alpha_{CD}$

$(\vec{a}_D)_n = 5.66\ \omega^2_{CD}$

$= (5.66)\ (3)^2 = 50.9\ \text{ft/sec}^2$

$(\vec{a}_P)_t = (4)\ (6) = 24\ \uparrow$

$(\vec{a}_P)_n = (4)\ (3)^2 = 36\ \leftarrow$

$\vec{a}_{P/M} = a_{P/M}\ \rightarrow$ (assumed to the right)

$\vec{a}_M = 0$

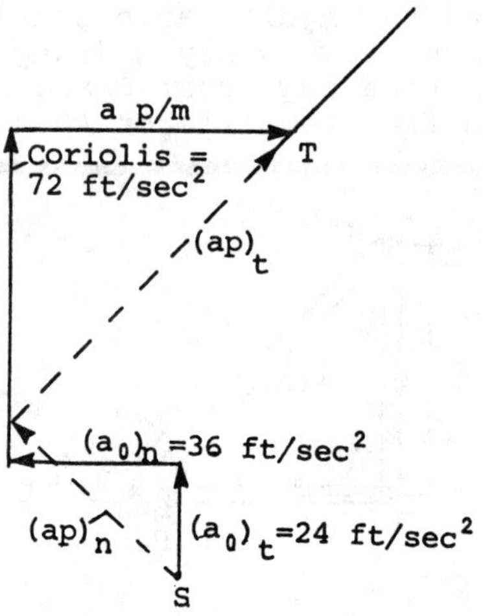

(C)

$2\vec{\Omega} \times \vec{v}_{P/M} = (2)\ (3)\ (12) = 72\ \uparrow$

where the fact that $v_{P/M} = \sqrt{v_D^2 - v_P^2} = 12$

and $\vec{v}_{P/M} \perp \vec{\Omega}$ has been used.

We can now set up the two component equations.

x-component:

$$(5.66\ \alpha_{CD})\ .707 - 50.9\ (.707) = -\ 36 + a_{P/M}$$

y-component:

$$(5.66\ \alpha_{CD})\ .707 + (50.9)\ (.707) = 24 + 72.$$

So there are two equations in two unknowns and α_{CD} may be solved for, yielding

$\alpha = 15.0$ rad/sec

$a_{P/M} = +\ 60$ (correct sense was assumed for each; the angular acceleration is counter-clockwise).

• PROBLEM 8-19

Point 1 on the rod shown in the figure possesses a velocity $\vec{v}_1$ and acceleration $\vec{a}_1$ to the right at the instant shown. Determine the instantaneous velocity of the point 2 as well as the angular velocity and the angular acceleration of the rod.

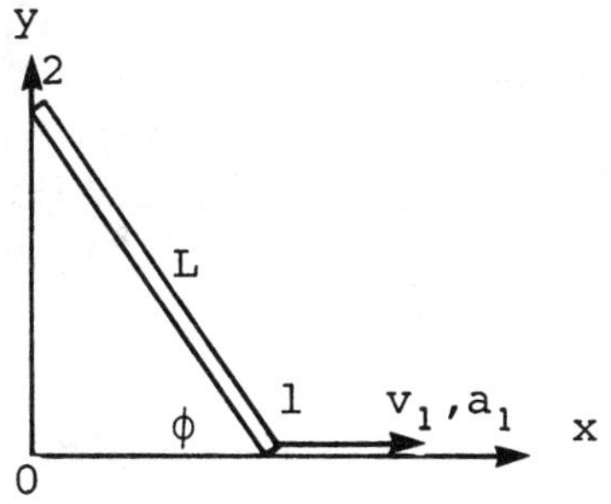

Solution: In the coordinate system OXY, the coordinates of particles 1 and 2 are $\vec{r}_1 = (L \cos \phi, 0)$ and $\vec{r}_2 = (0, L \sin \phi)$.

The velocity vectors are

$$\vec{v}_1 = (v_1, 0) \text{ and } \vec{v}_2 = (0, v_2).$$

If $\vec{\omega} = \omega_x \hat{i} + \omega_y \hat{j} + \omega_z \hat{h}$ to the angular velocity, then we have

$$\vec{v}_1 - \vec{v}_2 = \vec{\omega} \times (\vec{r}_1 - \vec{r}_2).$$

Thus,

$$v_1 \hat{i} - v_2 \hat{j} = (\omega_x \hat{i} + \omega_y \hat{j} + \omega_z \hat{h}) \times L (\cos \phi \hat{i} - \sin \phi \hat{j}).$$

Evaluating the cross product and equating the coefficients of the unit vectors, we obtain three equations

$$v_1 = \omega_z L \sin \phi \qquad (a)$$

$$v_2 = -\omega_2 L \cos \phi \qquad (b)$$

$$0 = -\omega_x L \sin \phi - \omega_y L \cos \phi \qquad (c)$$

This set of equations shows that $\omega_x = \omega_y = 0$ and $\omega_z = \frac{v_1}{L \sin \phi} = \frac{-v_2}{L \cos \phi}$

Since $\vec{v}_1$ and $\vec{v}_2$ vectors lie in the same plane as $\vec{v}_1 - \vec{v}_2$, the symmetry dictates that $\vec{\omega}$ has only a z component.

Consequently, the angular velocity of the rod is

$$\vec{\omega} = \omega_z \hat{k} = \frac{v_1}{L \sin \phi} \hat{k}$$

$$\vec{v}_2 = -\frac{v_1}{\sin \phi} \cos \phi \hat{j} = -\frac{v_1}{\tan \phi} \hat{j}.$$

It is interesting to determine the components of v_1 and v_2 along the rod; these must clearly be equal if the rod is rigid. Forming the scalar product of velocity with a unit vector colinear with the rod, we have

$\vec{v}_1 \cdot (\cos\phi\hat{i} - \sin\phi\hat{j}) = (\omega_z L \sin\phi\hat{i}) \cdot (\cos\phi\hat{i} - \sin\phi\hat{j})$

$= \omega_z L \sin\phi \cos\phi$

$\vec{v}_2 \cdot (\cos\phi\hat{i} - \sin\phi\hat{j}) = (-\omega_z L \cos\phi\hat{j}) \cdot (\cos\phi\hat{i} - \sin\phi\hat{j})$

$= \omega_z L \cos\phi \sin\phi.$

The acceleration of point 1 relative to point 2 is given by

$$\vec{a}_1 - \vec{a}_2 = \vec{\omega} \times (\vec{\omega} \times (\vec{r}_1 - \vec{r}_2) + \dot{\vec{\omega}} \times (\vec{r}_1 - \vec{r}_2)$$

$$a_1\hat{i} - a_2\hat{j} = \omega_z\hat{k} \times [\omega_z\hat{k} \times L(\cos\phi\hat{i} - \sin\phi\hat{j})] + (\omega_x\hat{i} + \omega_y\hat{j} + \omega_z\hat{k}) \times L(\cos\phi\hat{i} - \sin\phi\hat{j}).$$

Expanding and equating coefficients of the unit vectors:

$$a_1 = -\omega_z^2 L \cos\phi + \omega_z L \sin\phi \qquad \text{(d)}$$

$$a_2 = -\omega_z^2 L \sin\phi - \dot{\omega}_z L \cos\phi \qquad \text{(e)}$$

$$0 = -\dot{\omega}_x L \sin\phi = \dot{\omega}_y L \cos\phi \qquad \text{(f)}$$

From Eq. (d),

$$\dot{\omega}_z = \frac{a_1}{\sin L\,\phi} + \frac{\omega_z^2}{\tan\phi}$$

Equation (f) is satisfied for all ϕ if $\dot{\omega}_x = \dot{\omega}_y = 0$.

• PROBLEM 8-20

Let us assume that the joystick of a computer-game rotates with a constant angular velocity $\vec{\omega}_1 = 0.25 \frac{\text{rad}}{\text{s}} \vec{j}$. Simultaneously the stick is being raised with a constant angular velocity $\vec{\omega}_2 = 0.40 \frac{\text{rad}}{\text{s}} \vec{k}$ relative to the center 0, see Fig. 1. The length of the joystick is $\ell = 0.15$ m, determine a) the angular velocity $\vec{\omega}$ of the stick, b) the angular acceleration $\vec{a}$ of the stick, c) the velocity $\vec{V}$ of the tip of the stick, and d) the acceleration $\vec{a}$ of the tip of the stick.

Solution: a) The angular velocity of the joystick is merely the sum of the two components $\vec{\omega}_1$ and $\vec{\omega}_2$.

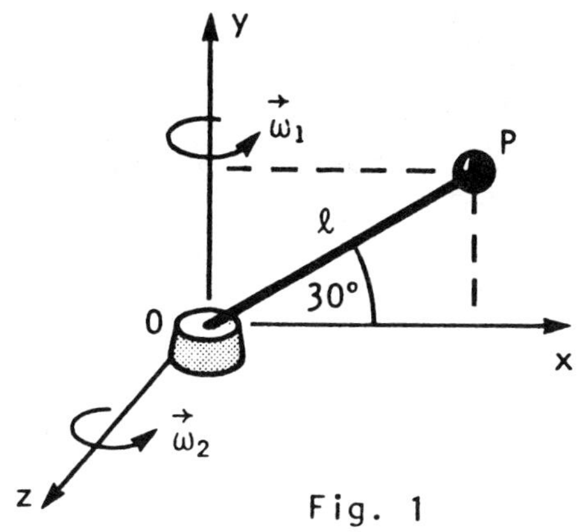

Fig. 1

$$\vec{\omega} = \vec{\omega}_1 + \vec{\omega}_2 = (.25\,\vec{j} + .40\,\vec{k})\,\frac{\text{rad}}{\text{s}}$$

b) The angular acceleration is found by differentiating the equation for angular velocity:

$$\vec{\omega} = \omega_1\vec{j} + \omega_2\vec{k}$$

$$\vec{\alpha} = \dot{\vec{\omega}} = \dot{\omega}_1\vec{j} + \omega_1\frac{d\vec{j}}{dt} + \dot{\omega}_2\vec{k} + \omega_2\frac{d\vec{k}}{dt}$$

$$\dot{\omega}_1\vec{j} = \vec{\alpha}_1 = 0 \qquad \dot{\omega}_2\vec{k} = \vec{\alpha}_2 = 0 \qquad \omega_1\frac{d\vec{j}}{dt} = 0$$

$$\vec{\alpha} = \omega_2\frac{d\vec{k}}{dt}$$

The angular accelerations $\vec{\alpha}_1$ and $\vec{\alpha}_2$ are zero because the problem stated that the joystick moved with constant angular velocity. The fact that $\frac{d\vec{j}}{dt}$ is zero essentially means that the direction of ω_1 is constant (ω_1 is always pointed in the $\vec{j}$ direction). On the other hand the direction of $\vec{\omega}_2$ is changing (i.e. $\frac{d\vec{k}}{dt} \neq 0$) due to the influence of $\vec{\omega}_1$. This is equivalent to saying that the angular acceleration $\vec{\alpha}$ with respect to a fixed reference frame is equal to the angular acceleration of the joystick in a reference frame rotating with constant angular velocity $\vec{\omega}_1$ plus the relative acceleration of that frame with respect to the fixed frame.

$$\vec{\alpha}_f = \dot{\vec{\omega}}_f = \dot{\vec{\omega}}_{2_r} + \vec{\omega}_{1_f} \times \vec{\omega}_{2_r} = \omega_2\frac{d\vec{k}}{dt}$$

$$\dot{\omega}_{2_r} = \vec{\alpha}_2 = 0$$

where the subscripts f and r denote fixed and rotating reference frame

$$\vec{\alpha}_f = .25\,\vec{j} \times .40\,\vec{k} = .1\,\vec{i}\,\frac{\text{rad}}{s^2}$$

c) The velocity of the tip of the joystick is equal to the angular velocity crossed with the position vector for the tip.

$$\vec{V} = \vec{\omega} \times \vec{r} = \vec{\omega} \times (\ell \cos 30°\vec{i} + \ell \sin 30°\vec{j})$$

$$\vec{V} = (.25\,\vec{j} \times +.40\,\vec{k}) \times (.13\,\vec{i} + .075\,\vec{j})$$

$$\vec{V} = (-.030\,\vec{i} + .052\,\vec{j} - .0325\,\vec{k})\,\frac{m}{s}$$

d) The acceleration of the tip is equal to the sum of the normal and tangential components

$$\vec{a} = \vec{a}_n + \vec{a}_T \qquad \text{where} \qquad \begin{aligned} \vec{a}_n &= \vec{\omega} \times (\vec{\omega} \times \vec{r}) = \vec{\omega} \times \vec{V} \\ \vec{a}_t &= \vec{\alpha} \times \vec{r} \end{aligned}$$

$$\vec{a}_n = (.25\,\vec{j} + .40\,\vec{k}) \times (-.030\,\vec{i} + .052\,\vec{j} - .0325\,\vec{k})$$

$$\vec{a}_n = -.0289\,\vec{i} - .012\,\vec{j} + .0075\,\vec{k}$$

$$\vec{a}_t = .1\,\vec{i} \quad (.13\,\vec{i} + .075\,\vec{j}) = .0075\,\vec{k}$$

$$\vec{a} = (-.0289\,\vec{i} - .012\,\vec{j} + .015\,\vec{k})\,\frac{m}{s^2}$$

• PROBLEM 8-21

The vertical yoke in Fig. 1 is turning with a constant angular speed of 6 rad/sec about its vertical centerline. Within the yoke is a disk, which is rotating about a horizontal axis that is supported by the yoke. For convenience, it is assumed that the x and X axes are coincident and lie in the axis of rotation of the disk, and also that the y and Y axes are coincident and lie in the vertical centerline of the yoke. The disk is then in the yz, or YZ, plane.
a) Determine the velocity of point P which is located on the disk as shown. b) Assuming the disk has constant speed, determine the acceleration of point P.

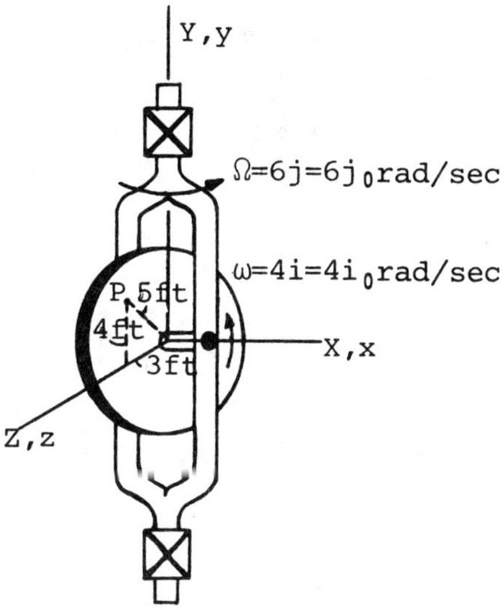

Fig. 1

Solution: For any rotating vector $\vec{P} = x\hat{i} + y\hat{j} + z\hat{k}$. The velocity $\vec{v}_P = \dot{P} = \vec{v}_{P/M} + \vec{v}_M$ where $\vec{v}_P$ is the absolute velocity of P(i.e. velocity relative to XYZ axes). $v_{P/M}$ is velocity of P relative to the point M on the moving body which is coincident with P at the instant and v_M is the velocity of M relative to the fixed XYZ axes.

The yoke is the rotating body whose angular velocity relative to the fixed X, Y, Z axes is $\vec{\Omega} = 6\hat{j}_0$ rad/sec, where $\hat{j}_0$ is the unit vector along the Y axis. The point M is coincident with P at the instant, but is considered fixed in the yoke. It may be imagined that the yoke is extended to include point M. The absolute velocity of M is $\vec{v}_M = \vec{\Omega} \times \vec{\rho}$, where $\Omega = 6\hat{j}_0$ and ρ relative to the fixed set of axes is equal to $4\hat{j}_0 + 3\hat{k}_0$. Hence,

$$\vec{v}_M = 6\hat{j}_0 \times (4\hat{j}_0 + 3\hat{k}_0) = 18\hat{i}_0.$$

The velocity $v_{P/M}$ is the velocity of P relative to the embedded x, y, z axes. In this case, it is due to the rotation of P about the x axis with an angular velocity $\vec{\omega} = 4\hat{i}$. Hence, $\vec{v}_{P/M} = \vec{\omega} \times \vec{\rho}$, where $\vec{\rho}$ is the position vector of P relative to the x, y, z set. Since $\vec{\rho} = +3\hat{k} + 4\hat{j}$,

$$\vec{v}_{P/M} = (4\hat{i}) \times (3\hat{k} + 4\hat{j}) = -12\hat{j} + 16\hat{k}.$$

To an observer at M, which is fixed in the yoke, the point P, would appear to be moving downward and forward.

The y and Y axes are collinear, as are the z and Z axes. Hence, instantaneous values are $\hat{j} = \hat{j}_0$ and $\hat{k} = \hat{k}_0$. So, $\vec{v}_{P/M} = -12\hat{j}_0 + 16\hat{k}_0$. The absolute velocity of P is found by adding the vectors for v_M and $v_{P/M}$. Thus,

$$\vec{v}_P = (18\hat{i}_0 - 12\hat{j}_0 + 16\hat{k}_0) \text{ ft/sec.}$$

b) The disk has no angular acceleration. Hence, $\vec{a}_{P/M}$, which is the acceleration of P moving about the x axis, has only a normal component. Its magnitude is $5(4)^2 = 80$ ft/sec^2. Since it is along $\vec{\rho}$ toward the origin, it can be written as

$$-\frac{3}{5}(80)\hat{k} - \frac{4}{5}(80)\hat{j} = -48\hat{k} - 64\hat{j}.$$

It could also be calculated from the relation

$$(\vec{a}_{P/M})_n = \vec{\omega} \times (\vec{\omega} \times \vec{\rho}) = (4\hat{i}) \times [4\hat{i} \times (3\hat{k} + 4\hat{j})]$$

$$= -48\hat{k} - 64\hat{j}.$$

In determining $\vec{a}_M$, there is no angular acceleration $\vec{\alpha}$ of the rotating body, which is the yoke. Hence, the tangential component of $\vec{a}_M$ is zero.

Its normal component, with $\hat{j} = \hat{j}_0$ and $\hat{k} = \hat{k}_0$, is

$$\vec{\Omega} \times (\vec{\Omega} \times \vec{\rho}) = 6\hat{j}_0 \times [6\hat{j}_0 \times (3\hat{k}_0 + 4\hat{j}_0)] = 6\hat{j}_0$$

$$\times (+18\hat{i}_0) = -108\hat{k}_0.$$

This component could also be determined by dropping a perpendicular from M, or P, to the vertical centerline about which the yoke is rotating with an angular speed of 6 rad/sec. Since the circle on which P is turning at the instant has a radius of 3 ft, the component of the acceleration is $3(6)^2 = 108$ ft/sec^2 in the negative Z direction.

Since there is a moving point in a rotating frame being considered, a Coriolis acceleration also exists. The Coriolis term is

$$2\vec{\Omega} \times \vec{v}_{P/M} = 2(6\hat{j}_0) \times (16\hat{k}_0 - 12\hat{j}_0) = 192\hat{i}_0 \text{ ft/sec}^2.$$

As indicated earlier, we have drawn the axes so that at the instant $\hat{i} = \hat{i}_0$, $\hat{j} = \hat{j}_0$, and $\hat{k} = \hat{k}_0$. Hence, the sum of all the vectors is

$$\vec{a}_P = -48\hat{k}_0 - 64\hat{j}_0 - 108\hat{k}_0 + 192\hat{i}_0$$

$$= (+192\hat{i}_0 - 64\hat{j}_0 - 156\hat{k}_0) \text{ ft/sec}^2.$$

A disk of radius b is fixed to a rotating turntable. The disk is rotating in the radial plane of the turntable with a constant angular velocity of $\dot{\phi}$, as shown in Figure 1. The turntable itself is rotating with a constant angular velocity of $\dot{\Omega}$. a) Determine both the velocity and acceleration of point A. b) Solve the the absolute velocity and acceleration using a coordinate system that is fixed to the rod OB Figure 2. The disk is then rotating with respect to the reference system.

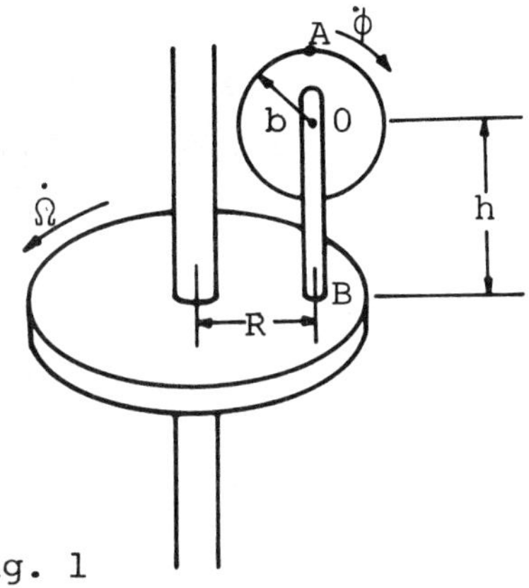

Fig. 1

Solution: a) First we must define a moving coordinate system. We shall define the moving system in the following manner: The origin of the system will be placed at point O, and the axes themselves will be considered to be fixed to the disk and rotating with the disk. The following list gives a term-by-term evaluation of each quantity of the relative velocity and acceleration equation, $\vec{v} = \vec{v}_0 + \vec{\omega} \times \vec{\rho} + \vec{v}_r$

$\vec{a} = \vec{a}_0 + \vec{\omega} \times (\vec{\omega} \times \vec{\rho}) + \dot{\vec{\omega}} \times \vec{\rho} + 2\vec{\omega} \times \vec{v}_r + \vec{a}_r$, with a short explanation of each.

$$\vec{v}_0 = - R\dot{\Omega}\hat{k}.$$

This is the absolute velocity of the origin of the moving system. (Note that all quantities are written in terms of unit vectors in the moving system.)

$$\vec{a}_0 = - R\dot{\Omega}^2\hat{i}.$$

This is the absolute acceleration of the origin of the moving system.

$$\vec{\omega} = \dot{\vec{\phi}} + \dot{\vec{\Omega}}$$

$$= - \dot{\phi}\hat{k} + \dot{\Omega}\hat{j}.$$

The angular velocity of the moving system in the position shown is a combination of two rotations: a rotation about the y axis (for this position only) due to the rotation of the turntable, and a rotation about the z axis due to the spin of the disk.

$$\dot{\vec{\omega}} = \ddot{\vec{\phi}} + \ddot{\vec{\Omega}}$$

But $$\ddot{\vec{\Omega}} = 0$$

$$\ddot{\vec{\phi}} = \dot{\vec{\Omega}} \times \dot{\vec{\phi}}$$

$$= (\dot{\Omega}\hat{j}) \times (-\dot{\phi}\hat{k}) = -\dot{\Omega}\dot{\phi}\hat{i}$$

so $$\dot{\omega} = -\dot{\Omega}\dot{\phi}\hat{i}.$$

The time derivative of the angular velocity vector is obtained by direct differentiation. The vector $\ddot{\vec{\Omega}}$ is equal to zero because $\dot{\vec{\Omega}}$ is neither changing in magnitude nor direction. But, since $\dot{\vec{\phi}}$ is changing direction by virtue of its angular velocity of rotation $\dot{\vec{\Omega}}$, it does have a time derivative, given by $\dot{\vec{\Omega}} \times \dot{\vec{\phi}}$.

$$\vec{\rho} = b\hat{j}.$$

This is the radius vector to point A.

$$\vec{v}_r = 0$$

$$\vec{a}_r = 0.$$

Since the moving system is fixed to the wheel, an observer in the moving system will see point A as a stationary point. Thus, the relative velocity and acceleration will be equal to zero.

Upon substituting these quantities into the relative velocity and acceleration equations, we get

$$\vec{v} = \dot{\phi}b\hat{i} - R\dot{\Omega}\hat{k}$$

$$\vec{a} = (-R\dot{\Omega}^2)\hat{i} + (-b\dot{\phi}^2)\hat{j} + (-2b\dot{\phi}\dot{\Omega})\hat{k}.$$

b) A term-by-term evaluation of all quantities of the general acceleration equation follows.

$$\vec{\omega} = \Omega\hat{j}$$

$$\dot{\vec{\omega}} = 0.$$

In this case, since the frame is not connected to the disk, the reference frame will rotate with the flywheel only.

$$\vec{v}_r = b\dot{\phi}\hat{i}$$

$$\vec{a}_r = -b\dot{\phi}^2\hat{j}$$

$$\vec{\rho} = b\hat{j}$$

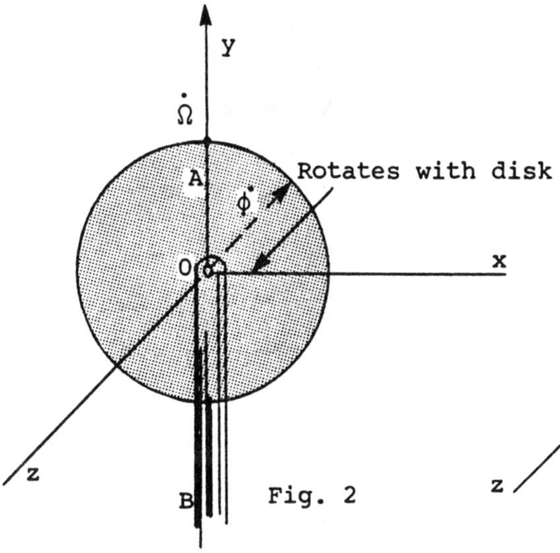

Fig. 2

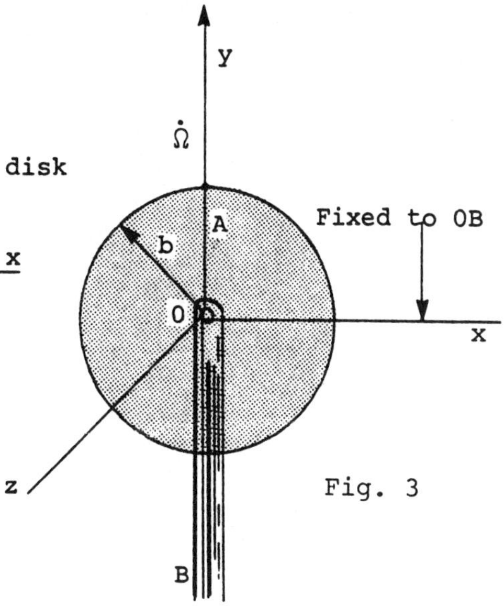

Fig. 3

$$\vec{a}_0 = - R\dot{\Omega}^2\hat{i}.$$

In this case, the relative velocity and acceleration are not equal to zero, since the disk is rotating with respect to the reference system. Recall that the relative velocity and acceleration of point A is the velocity and acceleration of A as seen by an observer sitting on the moving system.

Upon substitution into the general acceleration equation,

$$\vec{a} = (- R\dot{\Omega}^2)\hat{i} - b\dot{\phi}^2\hat{j} - 2b\dot{\phi}\dot{\Omega}\hat{k}$$

and thus it is seen that the acceleration of point A agrees with the acceleration as determined in part (a).

• PROBLEM 8-23

The link AB shown in the xy plane in Fig. 1 is rotating about the Z axis of the fixed set of X, Y, Z axes with an angular velocity $\vec{\omega}_{link} = + 2\hat{k}_0$ rad/sec and an angular acceleration $\vec{\alpha}_{link} = + 4\hat{k}_0$ rad/sec^2. The disk is pinned to the link at B, and a set of axes designated as x, y, z is embedded in the disk. The disk and these axes are turning with a velocity $\vec{\Omega}_{disk} = + 3\hat{k}$ rad/sec and an acceleration $\vec{\alpha}_{disk} = -12\hat{k}$ rad/sec^2. A straight slot in the disk is located so that a line from the center of the disk and perpendicular to the slot is 2 in. long and makes an angle of 60° with the x axis. In the slot, and coincident with the end of the 2-in. perpendicular line, is a particle P which is moving relative to the disk with a speed of 4 in./sec upward and to the left and with an acceleration of 5 in./sec^2 upward and to the left. Determine the absolute acceleration of the point P relative to the fixed X, Y, Z axes.

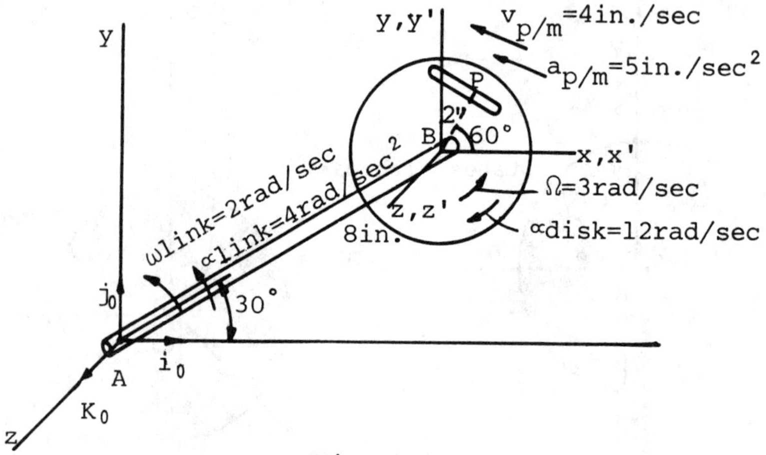

Fig. 1

Solution: The acceleration equation is

$$\vec{a}_P = \vec{a}_B + \vec{a}_{P/M} + \vec{a}_M + 2\vec{\Omega} \times \vec{v}_{P/M}.$$

In this example, $\vec{a}_B$ has a normal component. Its magnitude is $8(2)^2 = 32.0$ in./sec^2, and it is downward and to the left at an angle of 30° with the horizontal. It can be found in either of the following ways:

$$(\vec{a}_B)_n = -\ 32.0 \cos 30°\hat{i}_0 - 32.0 \sin 30°\hat{j}_0 = -\ 27.7\hat{i}_0 - 16.0\hat{j}_0$$

or

$$(\vec{a}_B)_n = \vec{\omega}_{link} \times (\vec{\omega}_{link} \times \vec{R})$$
$$= 2\hat{k}_0 \times \{2\hat{k}_0 \times [(8)(0.866)\hat{i}_0 + (8)(0.5)\hat{j}_0]\}$$
$$= -\ 27.7\hat{i}_0 - 16.0\hat{j}_0.$$

Also, $\vec{a}_B$ has a tangential component. Its magnitude is $8(4) = 32$ in./sec^2, and it is upward and to the left at an angle of 60° with the horizontal. Hence,

$$(\vec{a}_B)_t = -\ 32 \cos 60°\hat{i}_0 + 32 \sin 60°\hat{j}_0 = -16.0\hat{i}_0 + 27.7\hat{j}_0$$

or

$$(\vec{a}_b)_t = \vec{\alpha}_{link} \times \vec{R} = 4\vec{k}_0 \times [(8)(0.866)\hat{i}_0 + (8)(0.5)\hat{j}_0]$$
$$= -\ 16.0\hat{i}_0 + 27.7\hat{j}_0.$$

The acceleration $\vec{a}_{P/M}$, which is the acceleration of P relative to the coincident point M that is fixed on the disk, is given as 5 in./sec^2 upward to the left at an angle of 30° with the horizontal. Hence,

$$\vec{a}_{P/M} = -\ 5 \cos 30°\hat{i} + 5 \sin 30°\hat{j} = -\ 4.33\hat{i} + 2.5\hat{j}$$

The acceleration $\vec{a}_M$ is the acceleration of M relative to the x', y', z' axes which pass through B and always are parallel to the fixed X, Y, Z axes. The magnitude of the normal component of $\vec{a}_M$ is $2(3)^2 = 18.0$ in./sec^2, and this component is from M to B. So,

$$(\vec{a}_M)_n = -18.0 \cos 60° \hat{i}_0 - 18.0 \sin 60° \hat{j}_0 = -9.0\hat{i}_0 - 15.6\hat{j}_0$$

The magnitude of the tangential component of $\vec{a}_M$ is $2(12) = 24.0$ in./sec^2, and this component is downward and to the right at angle of 30° with the x' axis. Hence,

$$(\vec{a}_M)_t = +24.0 \cos 30° \hat{i}_0 - 24.0 \sin 30° \hat{j}_0 = 20.8\hat{i}_0 - 12.0\hat{j}_0$$

In the Coriolis component, which is $2\vec{\Omega} \times \vec{v}_{P/M}$, it is known that $\vec{\Omega} = 3\hat{k}$. Also, since $v_{P/M}$ is given as 4 in./sec upward and to the left at an angle of 30° with the x axis.

$$\vec{v}_{P/M} = -4 \cos 30° \hat{i} + 4 \sin 30° \hat{j} = -3.46\hat{i} + 2.0\hat{j}$$

Hence, $2\vec{\Omega} \times \vec{v}_{P/M} = 2(3\hat{k}) \times (-3.46\hat{i} + 2.0\hat{j}) = -12.0\hat{i} - 20.8\hat{j}$

This result can also be obtained by noting that the magnitude of the Coriolis component is $2(3)(4) = 24$ in./sec^2 and, according to the sense of $\vec{\Omega}$, the component is from P to B.

In this example, since the x, y, z set coincides with the x', y', z' set, it follows that $\hat{i} = \hat{i}_0$ and $\hat{j} = \hat{j}_0$. The expression for $\vec{a}_P$ can be found by adding the tangential and normal components just calculated. Thus,

$$\begin{aligned}\vec{a}_P &= (\vec{a}_B)_t + (\vec{a}_B)_n + \vec{a}_{P/M} + (\vec{a}_M)_t + (\vec{a}_M)_n + 2\vec{\Omega} \times \vec{v}_{P/M} \\ &= -16.0\hat{i}_0 + 27.7\hat{j}_0 - 27.7\hat{i}_0 - 16.0\hat{j}_0 - 4.33\hat{i}_0 \\ &\quad + 2.5\hat{j}_0 + 20.8\hat{i}_0 - 12.0\hat{j}_0 - 9.0\hat{i}_0 - 15.6\hat{j}_0 \\ &\quad - 12.0\hat{i}_0 - 20.8\hat{j}_0 \\ &= (-48.2\hat{i}_0 - 34.2\hat{j}_0) \text{ in./sec}^2\end{aligned}$$

For the gear arrangement shown in Fig. I spur gears 1 and 2 rotate with the same angular velocity and their center moves at a velocity of 1.5 m/s. If rack 3 is stationary find a) the angular velocity of the spur gears, b) the velocity of rack 4, c) the velocity of point A on gear 2, d) the velocities in parts (b) and (c) using the method of instantaneous center of rotation.

Fig. 1

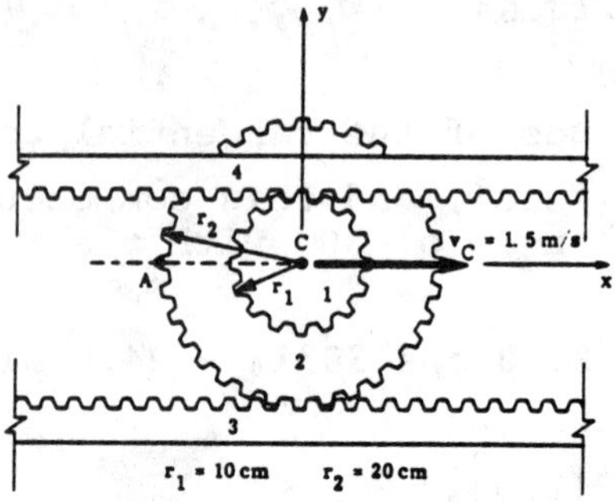

Solution: a) There is pure rolling between gear 2 and rack 4 consequently the distance traveled by the gear center C is equivalent to the distance the gear has rolled along the rack.

$$X_c = -r_2\theta \tag{1}$$

where c denotes the gear center and $r_1\theta$ is the arc length of that portion of the gear which has been in contact with the rack. A relationship between the angular and linear velocities can be obtained by differentiating equation (1):

$$V_c = \frac{dX_c}{dt} = -r_2\frac{d\theta}{dt} = -r_2\omega \tag{2}$$

$$\omega = \frac{-V_c}{r_2} = \frac{1.5 \text{ m/s}}{.2 \text{ m}} = -7.5 \text{ rad/s}$$

$$\omega = 7.5 \text{ rad/s CW}$$

b) Since rack 4 must be free to move its velocity is equi-

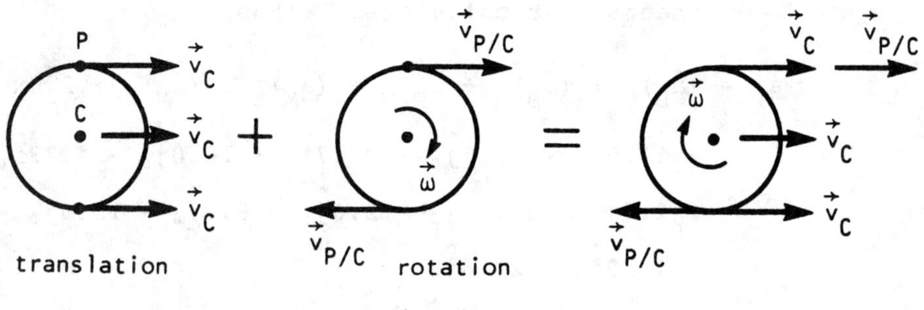

Fig. 2

valent to that of the contact point on gear 1. The velocity of the contact point V_p can be expressed as the vector sum of a translation of point C and a rotation about point C (see Fig. II).

$$\vec{V}_p = \vec{V}_c + \vec{V}_{p/c} \qquad \vec{V}_{p/c} = \vec{\omega} \times \vec{r}_p$$

$$\vec{V}_p = 1.5\,\vec{i} - 7.5\,\vec{k} \times .1\,\vec{j}$$

$$\vec{V}_p = 1.5\,\vec{i} + .75\,\vec{i} = 2.25\,\frac{m}{s}\,\vec{i} = 2.25\,\frac{m}{s}\angle 0°$$

c) The velocity of point A is found by using the same approach as in part (b) (see Fig. III).

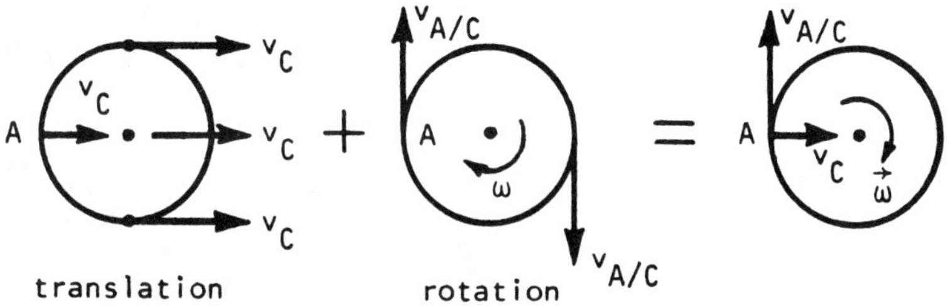

Fig. 3

$$\vec{V}_A = \vec{V}_c + \vec{V}_{A/c} \qquad \vec{V}_{A/c} = \vec{\omega} \times \vec{r}_A$$

$$\vec{V}_A = 1.5\,\vec{i} - 7.5\,\vec{k} \times (-.2\,\vec{i})$$

$$\vec{V}_A = 1.5\,\vec{i} + 1.5\,\vec{j} = 2.12\angle 45°$$

d) The direction of velocities $\vec{V}_A$ and $\vec{V}_p$ are known therefore their magnitudes can be found by the method of instantaneous center of rotation. The perpendiculars to the directions of velocities $\vec{V}_{c}$, $\vec{V}_A$ and $\vec{V}_p$ are constructed as in Fig. IV. At the instant shown in Fig. IV points A, P and C are rotating

Fig. 4

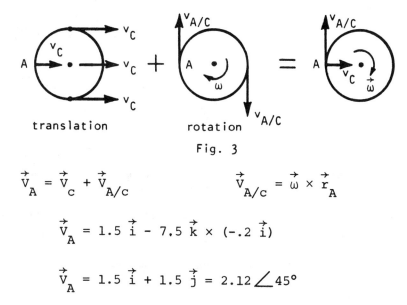

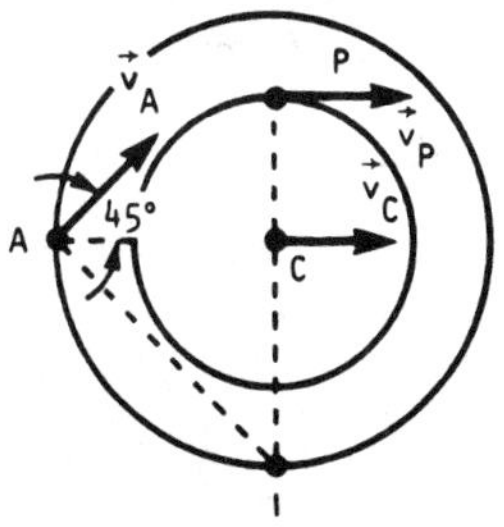

about point I. Since V_c is known this angular velocity, ω_I, can be determined.

$$V_c = r_2\omega_I \qquad\qquad \omega_I = \frac{V_c}{r_2} = \frac{1.5}{.2} = 7.5 \text{ rad/sec}$$

The magnitudes of the other velocities can be found once the lengths AI and PI are known.

$$\cos 45° = \frac{r_2}{AI} \qquad\qquad PI = r_1 + r_2 = .2 + .1 = .3 \text{ m}$$

$$AI = \frac{.2}{\cos 45°} = .283 \text{ m} \qquad\qquad V_p = PI\ \omega_I$$

$$V_A = AI\ \omega_I \qquad\qquad V_p = .3(7.5) = 2.25 \text{ m/s}$$

$$V_A = .283(7.5) = 2.12 \text{ m/s} \qquad\qquad \vec{V}_p = 2.25 \text{ m/s} \angle 0°$$

$$\vec{V}_A = 2.12 \angle 45°$$

• PROBLEM 8-25

For the gear arrangement in problem 8-24 $\vec{v}_c = 1.5$ m/s $\vec{i}$ and $\vec{a}_c = 4$ m/s^2 $\vec{i}$. Find a) the angular acceleration of the spur gears and b) the acceleration of rack 4 and point A (see Fig. 1).

Fig. 1

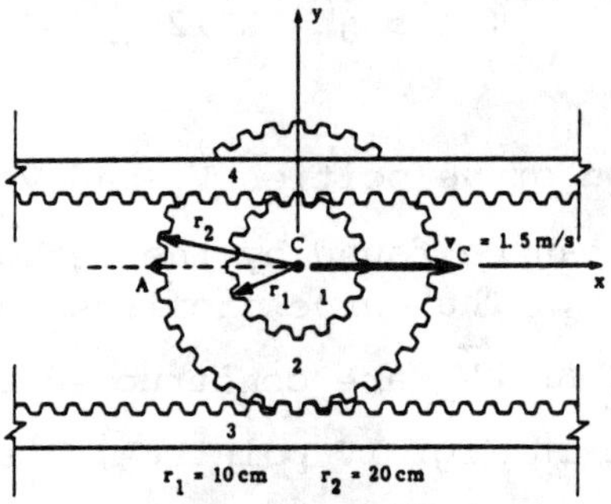

Solution: a) From the previous problem we saw that velocity V_c could be expressed as the first time derivative of θ; $V_c = -r\frac{d\theta}{dt} = -r\omega$. Differentiating a second time will yield a relationship between the angular and linear accelerations of the spur gears.

$$a_c = -r\frac{d\omega}{dt} = -r\alpha$$

$$\alpha = \frac{-a_c}{r_2} = \frac{4\ m/s^2}{.2\ m} = 20\ \frac{rad}{s^2}\ CW$$

b) The acceleration of rack 4 is equivalent to that of the contact point on gear 1. Once again the motion will be resolved into the sum of a translation and a rotation (see Fig. 2).

Fig. 2

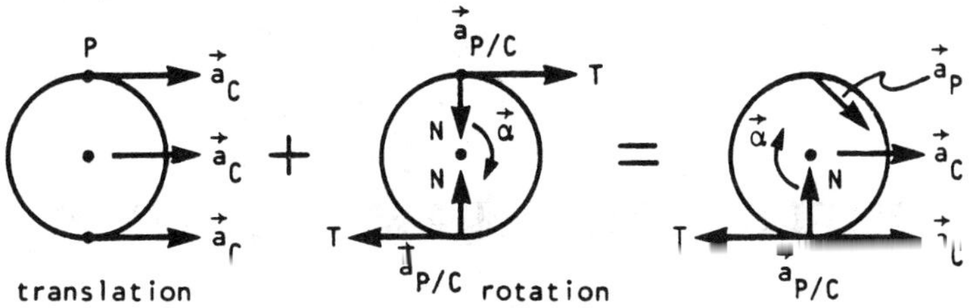

$$\vec{a}_p = \vec{a}_c + (\vec{a}_{p/c})_T + (\vec{a}_{p/c})_N$$

$$(\vec{a}_{p/c})_T = \vec{\alpha} \times \vec{r}_p = -20\ \vec{k} \times .1\ \vec{j} = 2\ \vec{i}$$

$$(\vec{a}_{p/c})_N = \vec{\omega} \times (\vec{\omega} \times \vec{r}_p) \qquad \omega = -7.5\ \vec{k} \text{ from problem 8-24}$$

$$(\vec{a}_{p/c})_N = -7.5\ \vec{k} \times (-7.5\ \vec{k} \times .1\ \vec{j}) = -5.62\ \vec{j}$$

$$\vec{a}_p = 4\ \vec{i} + 2\ \vec{i} - 5.62\ \vec{j} = 6\ \vec{i} - 5.62\ \vec{j} = 8.22\ \frac{m}{s^2} \angle 43.2°$$

Again resolving into translational and rotational components (see Fig. 3), we solve for acceleration at point a:

Fig. 3

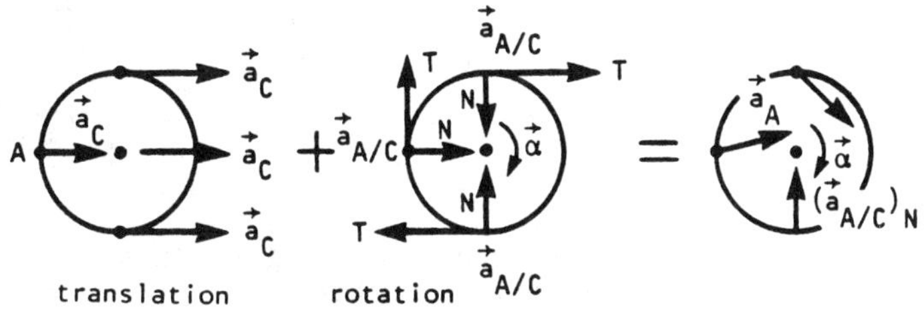

$$\vec{a}_A = \vec{a}_c + (\vec{a}_{A/c})_T + (\vec{a}_{A/c})_N$$

$$(\vec{a}_{A/c})_T = \vec{\alpha} \times \vec{r}_A = -20\ \vec{k} \times (-.2\ \vec{i}) = 4\ \vec{j}$$

$$(\vec{a}_{A/c})_N = \vec{\omega} \times (\vec{\omega} \times \vec{r}_A) = -7.5\ \vec{k} \times [-7.5\ \vec{k} \times (-.2\ \vec{i})]$$

$$= 11.25\ \vec{i}$$

$$\vec{a}_A = 4\ \vec{i} + 11.25\ \vec{i} + 4\ \vec{j} = 15.25\ \vec{i} + 4\ \vec{j} = 15.77\ \frac{m}{s^2} \angle 14.2°$$

CHAPTER 9

PARTICLE KINETICS: FORCE, ACCELERATION

NEWTON'S SECOND LAW OF MOTION

• PROBLEM 9-1

If the pulleys have negligible mass and there is no friction, show that: (a) the acceleration of the blocks A and B are g/7 and 2g/7; (b) the tension in the string is 5.71 lb_f .

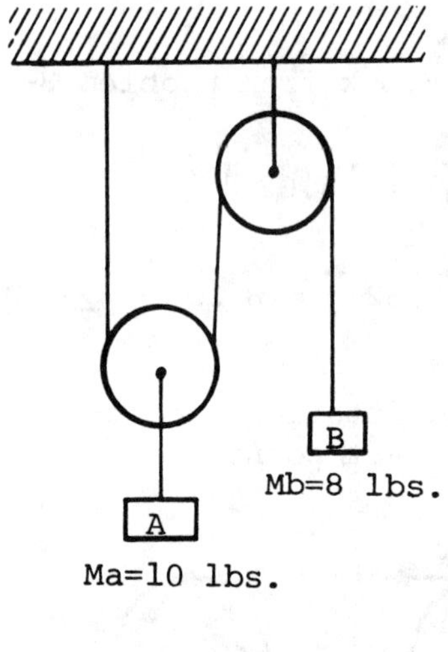

Fig. 1

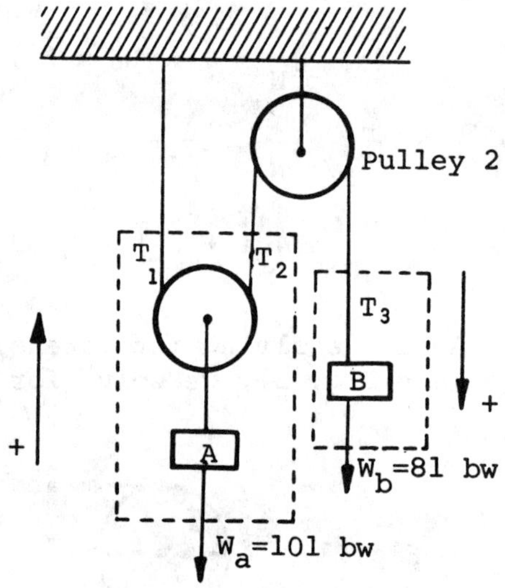

Fig. 2

Solution: We will apply Newton's Second Law (net $\vec{F} = m\vec{a}$) to the various masses to determine the accelerations. Draw regions of isolation around the weights A and B as shown and take the positive direction downward on weight B. Clearly A moves upward for positive motion of B.

First write the force equations for A and B

$$T_1 + T_2 - W_A = m_A a_A \tag{1}$$

$$W_B - T_3 = m_B a_B . \tag{2}$$

Since the rope with T_1 is fixed and does not stretch, each unit mass on that side of pulley 1 has zero acceleration. If each unit mass on the T_2 side undergoes an acceleration = a_B , then a_A must be the average of the accelerations on both sides or

$$a_A = \frac{1}{2} a_B \text{ .} \qquad (3)$$

Since we have massless, frictionless pulleys, all of the tensions are equal

$$T_1 = T_2 = T_3 = T \text{ .} \qquad (4)$$

Substituting (3) and (4) in (1) and (2)

$$2T - W_A = \frac{1}{2} m_A a_B \qquad (5)$$

$$W_B - T = m_B a_B \quad . \qquad (6)$$

Solving (6) for T and substituting into (5) yields after rearrangement

$$a_B = \frac{2W_B - W_A}{2m_B + \frac{1}{2}m_A}$$

But W = mg , m = w/g , therefore

$$a_B = \frac{2W_B - W_A}{2W_B/g + \frac{1}{2} W_A/g} = \left(\frac{2W_B - W_A}{2W_B + \frac{1}{2} W_A}\right) g$$

$$a_B = \frac{2\cdot 8 - 10}{2\cdot 8 + 5} g = \frac{2}{7} g. \qquad (7)$$

From (3) $\qquad a_A = \frac{1}{2} a_B = \frac{1}{7} g$.

From (6)

$$\begin{aligned} T &= W_B - m_B a_B \\ &= 8 - \frac{8}{g}(\frac{2}{7} g) \\ &= 8 - \frac{16}{7} = 5.71 \text{ lbW.} \end{aligned}$$

• PROBLEM 9-2

In the modified Atwood machine shown, both pulleys have negligible rotational inertia but the movable pulley does have mass m. Suppose the mass m_3 is large enough to give a downward acceleration a' . Find the acceleration a, $m_1 > m_2$, in terms of m_1, m_2, a' and g.

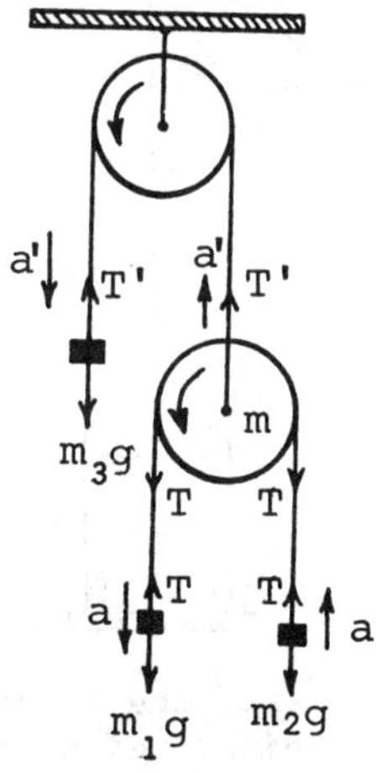

Solution: Concentrate on the pulley on the right and the two masses suspended from it. The mass m_1 is subjected to an upward force T and a downward force $m_1 g$. Since it is moving down, the net force on it is $m_1 g - T$. But it also has a net acceleration $a - a'$ if we include the upward acceleration of the pulley a'. Using $F = ma$, we get

$$m_1 g - T = m_1(a - a') \quad . \tag{1}$$

Similarly, for mass m_2, the net upward force is $T - m_2 g$, whereas its net upward acceleration is $a + a'$. Hence one can say,

$$T - m_2 g = m_2(a + a') \; . \tag{2}$$

Adding (1) and (2), yields

$$m_1 g - T + T - m_2 g = m_1 a - m_1 a' + m_2 a + m_2 a'$$

or $$m_1 g - m_2 g + m_1 a' - m_2 a' = m_1 a + m_2 a$$

or $$g(m_1 - m_2) + a'(m_1 - m_2) = a(m_1 + m_2)$$

or $$(g + a')(m_1 - m_2) = a(m_1 + m_2)$$

or $$a = \frac{m_1 - m_2}{m_1 + m_2}(g + a')$$

which gives us the acceleration of mass m_1.

• PROBLEM 9-3

A body A on a horizontal frictionless table is connected to a string which runs over a pulley and carries a platform on its other end, as shown in Fig. 1. A body C is placed on the platform. The weights of A, B, and C are 10 lb, 2 lb, and 3 lb, respectively. What is the acceleration of A when the system is released from rest, and what is the tension S in the string? Determine the contact force between B and C.

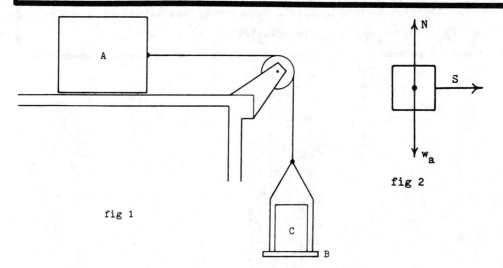

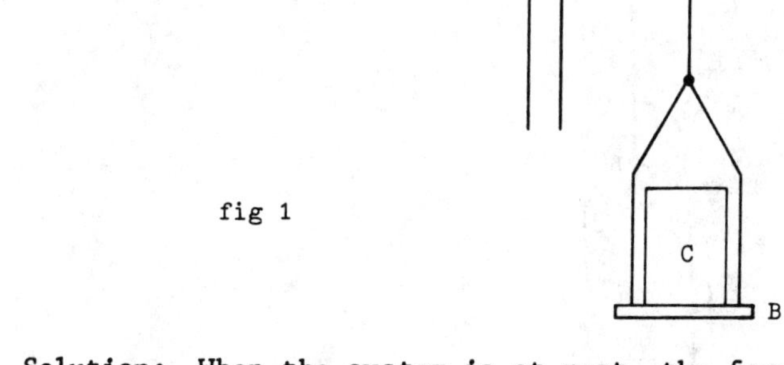

fig 1

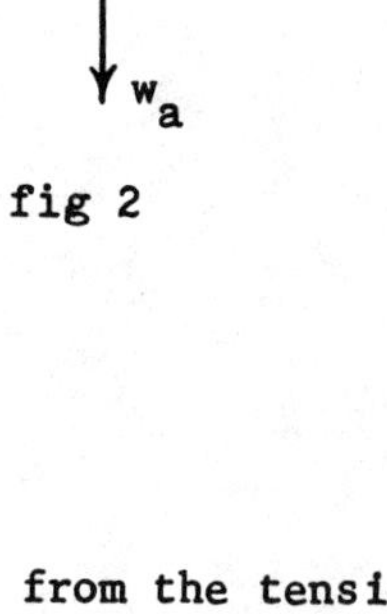

fig 2

Solution: When the system is at rest, the force on A, from the tension in the string, is equal to the combined weight of C and B. This problem will show that when C and B are allowed to move under the influence of gravity the tension in the string is greatly reduced. We start by isolating the various bodies and indicating all the forces acting upon them. First we consider A. The forces on it are shown in Fig. 2. (The weight of A is balanced by the contact force N from the table.) Assume that the acceleration a of A is toward the right. The equation of motion is then

$$S = m_a a \; . \tag{1}$$

Next, we consider B + C as a unit that moves downward, also with an acceleration a. The forces on B + C are shown in Fig. 3. The equation of motion is

$$(w_b + w_c) - S = (m_b + m_c)a \; , \tag{2}$$

where we have here taken forces and accelerations positive when directed downward. The two unknown quantities a and S are obtained from these two equations. Thus, from Eqs. (1) and (2), we get

$$w_b + w_c = (m_a + m_b + m_c)\, a$$

$$a = \frac{w_b + w_c}{m_a + m_b + m_c} = \frac{m_b + m_c}{m_a + m_b + m_c}\, g \; , \tag{3}$$

since w is mg. Inserting numerical values, we obtain

$$A = \frac{2 + 3}{15}\, g = \frac{g}{3} \; .$$

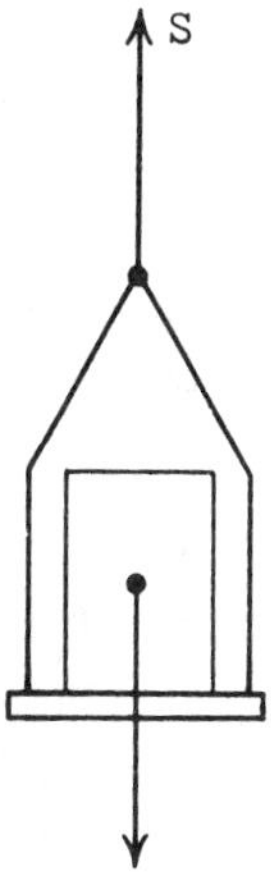

We can solve for the tension, S, from either equation (1) or (2). Putting the acceleration for (3) into (2) yields

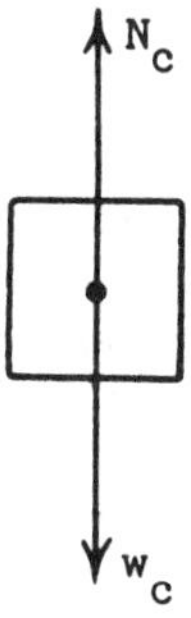

fig 4

$$\begin{aligned} S &= (w_b + w_c) - (m_b + m_c)\left[\frac{w_b + w_c}{m_a + m_b + m_c}\right] \\ &= (w_b + w_c)\left[1 - \frac{m_b + m_c}{m_a + m_b + m_c}\right] \\ &= (w_b + w_c)\left[\frac{m_a}{m_a + m_b + m_c}\right] \\ &= m_a\left[\frac{w_b + w_c}{m_a + m_b + m_c}\right] \; . \end{aligned}$$

So,

fig 3

$$S = \tfrac{2}{3}\,(w_b + w_c) = \frac{10}{3}\ \text{lb.}$$

In other words, the tension in the string is in this case only two-thirds of the weight of B + C.

Finally, to determine the contact force between B and C, we isolate body C, as shown in Fig. 4. We then find that

$$w_c - N_c = m_c a \; ,$$

$$N_c = w_c - m_c a = m_c(g - a) = \frac{2m_c g}{3} = \tfrac{2}{3}\, w_c \; .$$

Note carefully that the tensions and contact forces are different in the equilibrium and nonequilibrium cases, as demonstrated in this example.

The system of two blocks A and B and two pulleys C and D is assembled as shown (Fig. 1). Neglecting the friction and the mass of the pulleys and assuming that the whole system is initially at rest, determine the acceleration of each block and the tension in each cord.

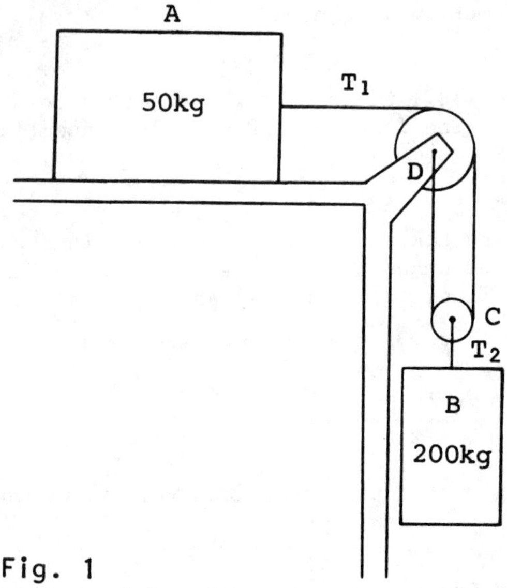

Fig. 1

Solution: Since the pulleys are massless, the cord ACD has the same tension throughout, say T_1. The tension in cord BC can be called T_2. The effect of having the cord double back around pulley C produces

$$d_B = \tfrac{1}{2} d_A , \tag{1}$$

where d_B, d_A are the distances moved, respectively, by blocks B and A. Then differentiating twice

$$a_B = \tfrac{1}{2} a_A \tag{2}$$

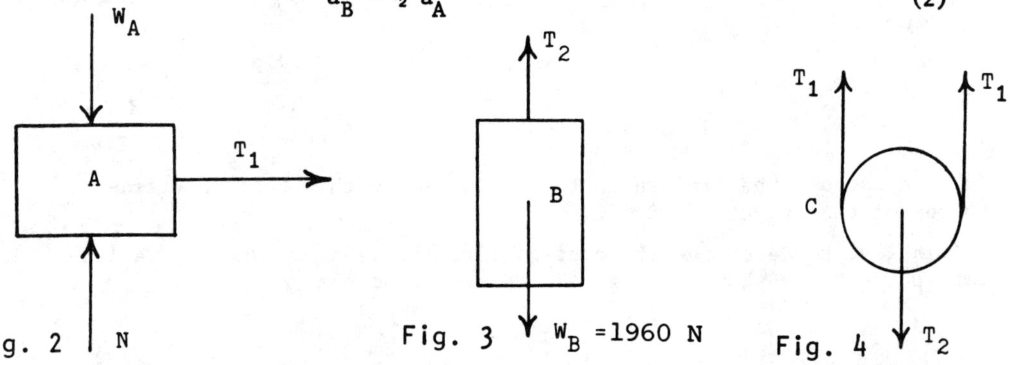

Fig. 2 Fig. 3 Fig. 4

Now we shall analyze block A, B, and pulley C as free bodies. The free body diagram of block A is given in Figure 2. W_A and N, the weight and normal force respectively, cancel, so from Newton's Second Law we have

$$T_1 = 50\, a_A . \tag{3}$$

The free body diagram of block B is in Figure 3.

$$W_B = mg = (200\ \text{kg})(9.8\ \text{m/s}^2) = 1960\text{N}.$$

Again using Newton's Law

$$W_B - T_2 = 1960\ \text{N} - T_2 = 200\text{kg}\ a_B \qquad (4)$$

Pulley C (Figure 3) is massless, therefore the sum of the forces on it must be zero.

$$2T_1 - T_2 = 0 \qquad (5)$$

Equations (2), (3), (4) and (5) now form a system of 4 equations. Substituting for a_A from (2) into (3) yields $T_1 = 100\text{kg}\ a_B$.

Using this and equation (4) to substitute into (5) yields

$$200\text{kg}\ a_B - (1960\ \text{N} - 200\text{kg}\ a_B) = 0$$

$$200\text{kg}\ a_B + 200\text{kg}\ a_B = 1960\ \text{N}$$

$$a_B = \frac{1960\ \text{N}}{500\ \text{kg}} = 3.92\ \text{m/s}^2 . \qquad (6)$$

Putting (6) back into (2) gives us

$$a_A = 7.84\ \text{m/s}^2 \qquad (7)$$

Now the tensions follow easily; from (3) and (6)

$$T_1 = 392\ \text{N} \qquad (8)$$

and from (5) and (8)

$$T_2 = 784\ \text{N}. \qquad (9)$$

Equations (6), (7), (8) and (9) are the answers to the problem stated.

• PROBLEM 9-5

Two springs S_1 and S_2 of equal lengths $L = 0.5$ m, but with different spring constants $K_1 = 50$n/m and $K_2 = 100$ n/m, are joined and fastened between two supports A and B which are a distance 2L apart, as shown in Fig.1. A body C of mass m = 2.5 kgm is fastened to the springs at their junction and is pulled downward vertically until the length of each spring has doubled. The body is then released. What is the initial acceleration of the body?

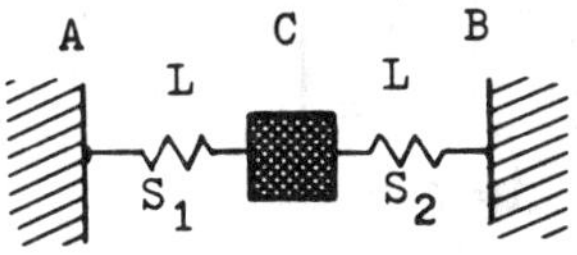

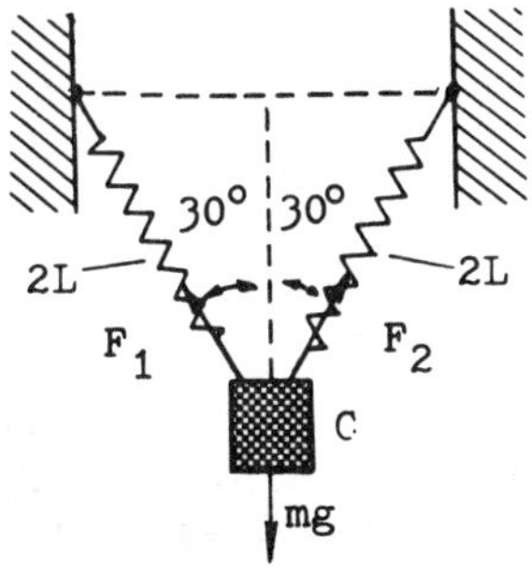

Solution: We describe the acceleration in terms of its horizontal x- and vertical y-components. The body C is acted upon by three forces, the weight mg and the two spring forces, which have the magnitudes $F_1 = K_1(2L - L) = K_1L$ and $F_2 = K_2L$. The angle between the springs and the vertical is 30° since the right triangles outlined in Figure 1 have hypotenuse 2L and lie opposite to leg L. We then project the forces in the x- and y-directions, respectively, and obtain

$$ma_x = K_2 L \sin 30^\circ - K_1 L \sin 30^\circ = \frac{K_2 - K_1}{2} L$$

$$ma_y = (K_2 + K_1)L \cos 30^\circ - mg = \frac{(K_2 + K_1)L \cdot \sqrt{3}}{2} - mg .$$

That is,

$$a_x = \frac{K_2 - K_1}{2m} L \quad \frac{(50)\,(0.5)}{(2)\,(2.5)} = 5\text{m/sec}^2 ,$$

$$a_y = \frac{(K_1 + K_2)L\sqrt{3}}{2m} - g \simeq 16.3\text{m/sec}^2 .$$

• PROBLEM 9-6

Two springs, S_1 and S_2, of negligible mass, with spring constants K_1 and K_2, respectively, are arranged to support a body A. In Fig. 1 the springs are coupled in "series" and in Fig. 2 they are in "parallel". What are the extensions of the individual springs in these two cases as a result of the force of gravity on A? Determine also the equivalent spring constant in the two cases.

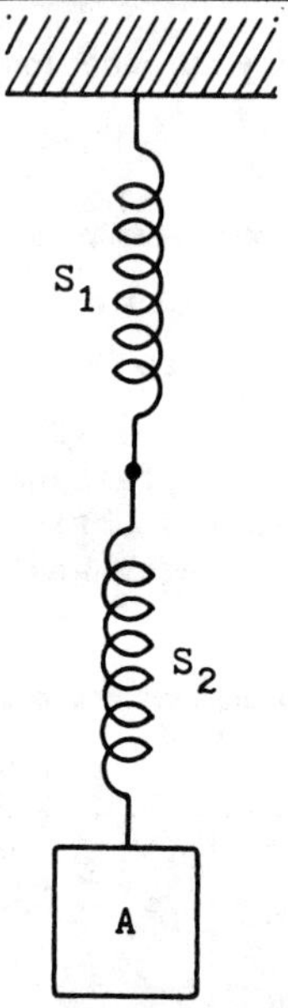

fig 1

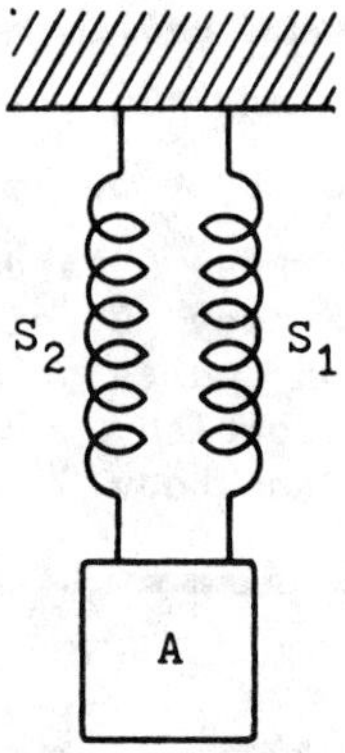

fig 2

Solution: In the first case the body is acted upon by two forces, the weight mg, and the spring force F_2. The spring S_2, in turn, is acted upon by F_2 at the lower end and by a contact force at the junction with S_1 at the upper end. Since the weight of the spring is negligible, the forces at the ends of the spring must cancel each other, that is, $F_1 = F_2$. Similarly, we find that spring S_1 is acted upon by opposite forces of magnitude F_2 at the two ends of the spring, as shown in Fig. 3. Finally, the support from which the spring S_1 is hanging is acted upon by a force F_2 downward. We have indicated all the forces acting upon the individual parts of the system and can now impose the equilibrium conditions that the net force on A must be zero, i.e., $F_2 = Mg$. The extensions of the two springs can now be expressed in terms of mg and the spring constants:

$$x_1 = \frac{F_2}{K_1} = \frac{Mg}{K_1} \ , \quad x_2 = \frac{F_1}{K_2} = \frac{Mg}{K_2} \ .$$

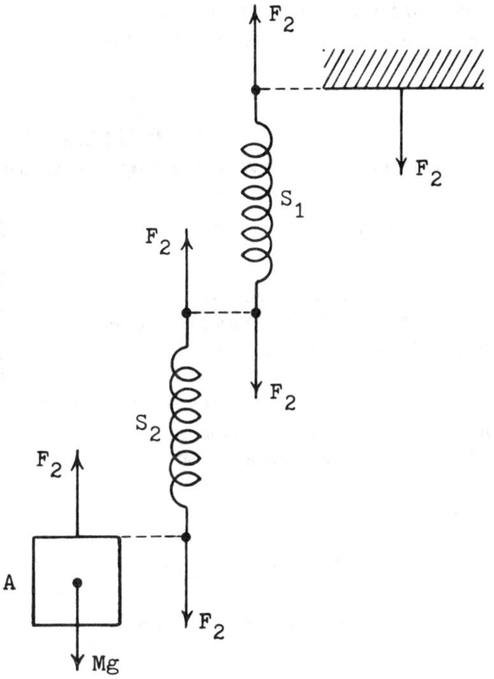

fig 3

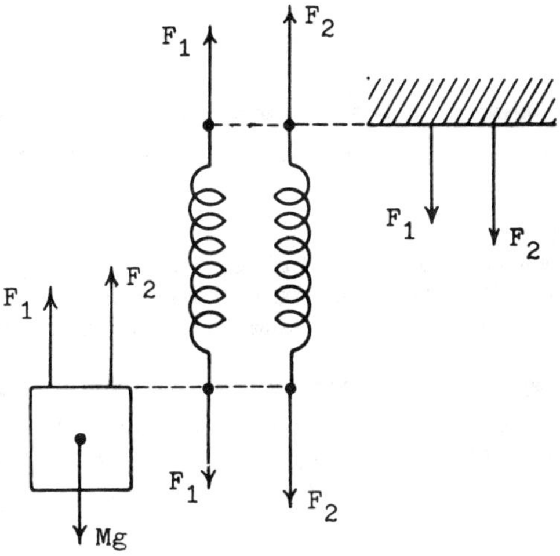

fig 4

The total extension of the two springs is $x_1 + x_2 = Mg(1/K_1 + 1/K_2)$. In other words, the equivalent spring constant, $K_a = Mg/(x_1 + x_2)$, then becomes such that

$$\frac{1}{K_a} = \frac{1}{K_1} + \frac{1}{K_2} \ .$$

In case (b), the extensions of the two springs are the same:

$$x_1 = x_2 = x_b \ .$$

Body A is now acted upon by three forces, the weight and the two spring forces F_1 and F_2, as illustrated in Fig. 4, where the forces on the other parts of the system are also shown. We now have the relations

$$F_1 + F_2 = Mg \ ,$$

and

$$F_1 = K_1 x_b \ , \quad F_2 = K_2 x_b \ .$$

The condition for equilibrium of A then becomes

$$x_b(K_1 + K_2) = Mg ,$$

$$x_b = \frac{Mg}{K_1 + K_2} .$$

Consequently the equivalent spring constant in this case becomes

$$K_b = K_1 + K_2 .$$

In the special case when $K_1 = K_2 = K$, the equivalent spring constants in cases (a) and (b) become $K_a = K/2$ and $K_b = 2K$, respectively. In other words, the parallel coupling of the springs corresponds to a stiffness which is four times as large as that in the series coupling of the spring.

• PROBLEM 9-7

A tug of war is held between two teams of five men each. Each man weighs 160 lb. and each man's pull on the rope can be described as:

$$F = (200 \text{ lb.})\ e^{-t/\tau} ,$$

where the mean tiring time τ is 10 sec for team A, and 20 sec for team B. If the mass of the rope is 50 lb., find the motion, that is, the final velocity of the teams. What assumption leads to this absurd result?

Solution: Since we have the forces that act in this system and we have to find the motion, we can apply Newton's Second Law of motion as:

$$\Sigma F = F_A - F_B = M_{rope} \frac{dv}{dt}$$

$$50\ dv = (F_A - F_B)dt = [5(200)e^{-t/10} - 5(200)e^{-t/20}]dt \qquad (1)$$

Since we can assume that $v = 0$ when $t = 0$, v can be determined by integrating equation (1):

$$50 \int_0^v dv = 1000 \int_0^t (e^{-t/10} - e^{-t/20})dt$$

$$50v - 0 = 1000[-10e^{-t/10} - (-20e^{-t/20}) + 10 - 20]$$

$$v = 20[-10e^{-t/10} + 20e^{-t/20} - 10] \qquad (2)$$

Since $v = dx/dt$, the substitution can be made, and integrated again the equation becomes:

$$\int_0^x dx = 200 \int_0^t (-e^{-t/10} + 2e^{-t/20} - 1)dt$$

$$x - 0 = 200[-(-10e^{-t/10}) + 2(-20e^{-t/20}) - t - 10 + 40 - 0]$$

$$x = 200[10e^{-t/10} - 40e^{-t/20} - t + 30]$$

$$x = -200t + 2000[e^{-t/10} - 4e^{-t/20} + 3] .$$

Factoring the last term gives:

$$x = -200t + 2000(1 - e^{-t/20})(3 - e^{-t/20}) \qquad (3)$$

As $t \to \infty$, the final velocity can be found from equation (2) as follows:

$$v = 2-[-10e^{-\infty} + 20e^{-\infty} - 10]$$

$$v = -200 \text{ ft/sec.}$$

Here the minus sign simply means that the motion is in the direction of team B, which one would expect since team B tires less easily than team A. Obviously, this large a final velocity is absurd. It comes about because we assumed a force which was independent of the velocity. Another reason for this result is that we neglected the mass of the teams. Obviously if one team didn't let go of the rope, (their feet slipped along the ground) the system acceleration would be much different. This type of problem shown the importance of a good system model in obtaining meaningful results.

• PROBLEM 9-8

A particle of mass $m = 2\text{kg}$ starts from rest at the origin of an inertial coordinate system at time $t = 0$. A force $\vec{F} = 2\hat{i} + 4t\hat{j} + bt^2\hat{k}$ is applied to it. Find the acceleration, velocity, and position of the particle for any later time.

Solution: Newton's Second Law for this case will read as follows:

$$\vec{F} = m\ddot{\vec{r}}$$

$$(2\hat{i} + 4t\hat{j} + 6t^2\hat{k})\text{N} = (2\text{kg})\ddot{\vec{r}}$$

Hence the acceleration will be given by the expression

$$\ddot{r} = (\hat{i} + 2t\hat{j} + 3t^2\hat{k})\text{m/sec}^2$$

The three components of velocity may be found by integrating the three components of acceleration separately and using the initial velocity $\vec{v}_1$. The x-component is found as follows.

$$v_x - v_{1x} = \int_0^t dt = t \Big|_0^t = t \text{ m/sec}$$

but the initial x-component of the velocity v_{1x} is equal to zero from the statement of the problem, so

$$v_x = t \text{ m/sec} .$$

Similarly, the y- and z-components of velocity are, since $v_{y1} = v_{z1} = 0$,

$$v_y = 2\int_0^t t\,dt = \frac{2t^2}{2}\Big|_0^t = t^2 \text{ m/sec}$$

$$v_z = 3\int_0^t t^2 dt = \frac{3t^3}{3}\Big|_0^t = t^3 \text{ m/sec} .$$

Hence the vector velocity is given by the expression

$$v = t\hat{i} + t^2\hat{j} + t^3\hat{k} \text{ m/sec} .$$

Similarly the three components of displacement may be found by integrating the three components of velocity separately and by using the given information that the initial displacement r_1 is equal to zero. The results are

$$x = \frac{t^2}{2} \text{ m}$$

$$y = \frac{t^3}{3} \text{ m}$$

$$z = \frac{t^4}{4} \text{ m}$$

so that the displacement as a function of time is given by the expression

$$\vec{r} = \frac{t^2}{2}\hat{i} + \frac{t^3}{3}\hat{j} + \frac{t^4}{4}\hat{k} \text{ m}$$

• PROBLEM 9-9

At $t = 0$, a particle of mass 1 kg has a velocity $v_0 = 3\hat{j}$ m/sec and is at a position $r_0 = 2\hat{k}$ m. Find the acceleration, velocity, and position of the particle as a function of time if it is acted on by the two forces

$$F_1 = 2\hat{i} + 3\hat{j} + 4\hat{k} \text{ and } F_2 = 1\hat{i} - 2\hat{j} - 3\hat{k} \text{ simultaneously.}$$

Solution: In this problem we meet the basic problem of classical mechanics: to find the functions which specify the positions of a particle $\vec{r}$, the speed of the particle $\vec{v} = d\vec{r}/dt$ and the acceleration $\vec{a} = d\vec{v}/dt$. Newton's 2nd law describes the motion of particles being acted upon by one or more forces. Thus, the total force that acts upon the particle is the sum of the two forces F_1 and F_2, and the particle will move with an acceleration given by the relation:

$$\vec{a} = \frac{\vec{F}_1 + \vec{F}_2}{m}.$$

Given are values for $\vec{F}_1$, $\vec{F}_2$ and m in the statement of the problem. This leaves $\vec{a}$ as the only unknown, allowing us to write an expression for the acceleration directly as

$$\vec{a} = \frac{\vec{F}_1 + \vec{F}_2}{m} \qquad (1)$$

$$\vec{a} = \frac{(2+1)\vec{i} + (3-2)\vec{j} + (4-3)\vec{k}}{1}$$

$$\vec{a} = 3\vec{i} + \vec{j} + \vec{k} \quad \text{m/sec}^2 \qquad (2)$$

if the forces are measured in Newtons. The acceleration is related to the velocity since it is the time derivative of the velocity

$$\vec{a} = \frac{d\vec{v}}{dt}. \qquad (3)$$

Decomposing,

$$a_x = \frac{dv_x}{dt}$$

$$a_y = \frac{dv_y}{dt}$$

$$a_z = \frac{dv_z}{dt}$$

where a_x, a_y, a_z are the components of the acceleration on the three direction x,y,z and the same for v_x, v_y, v_z.

$$\vec{a} = a_x\hat{i} + a_y\hat{j} + a_z\hat{k}$$

and

$$\vec{v} = v_x\hat{i} + v_y\hat{j} + v_z\hat{k}.$$

So we can write:

$$\vec{a} = \frac{d\vec{v}}{dt}.$$

Rearranging terms in equation (3) and substituting in the value for the acceleration yields

$$d\vec{v} = \vec{a}\,dt = (3\vec{i} + \vec{j} + \vec{k})dt. \qquad (4)$$

If both sides of equation (4) are integrated between the limits of the beginning of the problem $(t = 0)$ and again at any later time t, we have

$$\int_{\vec{v}_0}^{\vec{v}} d\vec{v} = \int_0^t (3\vec{i} + \vec{j} + \vec{k})dt$$

$$\vec{v} - \vec{v}_0 = 3t\vec{i} + t\vec{j} + t\vec{k} - 0 .$$

Substituting in our value for $\vec{v}_0$ yields

$$\vec{v} = 3t\vec{i} + (t+3)\vec{j} + t\vec{k}. \tag{5}$$

Since velocity is the time derivative of the displacement, we can follow an argument similar to that used in finding the velocity when finding displacement. This yields

$$\vec{v} = \frac{d\vec{r}}{dt}$$

or

$$d\vec{r} = \vec{v}\, dt .$$

Again, integrating from $t = 0$ to $t = t$ we have

$$\int_{\vec{r}_0}^{\vec{r}} d\vec{r} = \int_0^t 3t\vec{i} + (t+3)\vec{j} + t\vec{k})dt$$

$$\vec{r} - \vec{r}_0 = \frac{3}{2}t^2\,\vec{i} + (\frac{1}{2}t^2 + 3t)\vec{j} + \frac{1}{2}t^2\,\vec{k} - 0$$

Substituting in our given value for $\vec{r}_0$ produces

$$\vec{r} = \frac{3}{2}t^2\,\vec{i} + (\frac{1}{2}t^2 + 3t)\vec{j} + (\frac{1}{2}t^2 + 2)\vec{k} \quad \text{meters.}$$

• PROBLEM 9-10

A particle is projected up an inclined plane with an initial velocity v_0 of 100 cm/sec and an initial angle θ_0 of 135 degrees between the velocity and the line of maximum slope, as shown in Fig. 1. Neglecting friction, what is the particle velocity when θ has the values 90, 45, and 0 degrees? Use path coordinates.

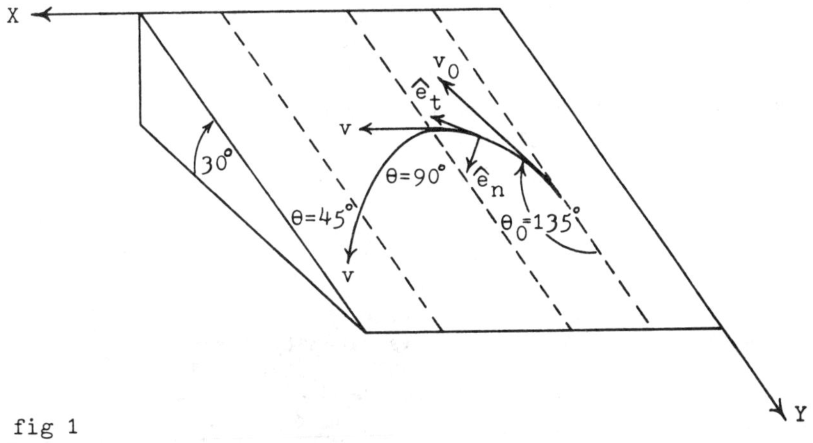

fig 1

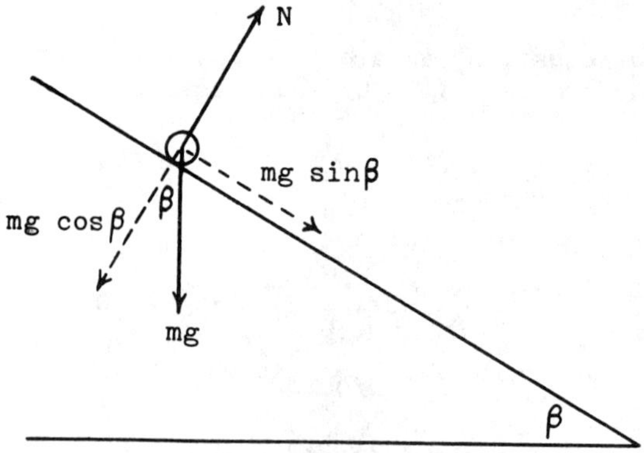

fig 2

<u>Solution</u>: Motion occurs entirely in a single plane. The forces acting on the particle (gravity and the normal reaction) must be resolved into components F_t and F_n, tangent and normal to the path, in the plane of motion. No acceleration occurs normal to the plane of the incline because the particle is in equilibrium in this direction. Examine Figure 2, which shows the forces acting on the particle. N and mg cos β add up to zero so the only unbalanced force is mg sin β . Now examine Figure 3 to see just what F_t and F_n are

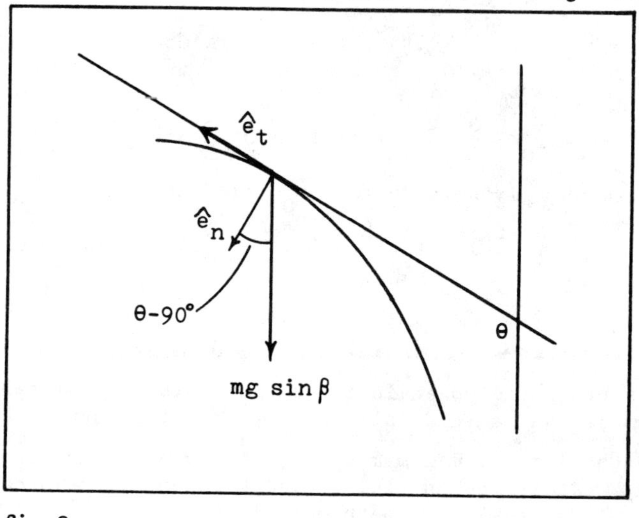

fig 3

$$F_t = mg \sin \beta \cos \theta$$

$$F_n = mg \sin \beta \cos(90^\circ - \theta)$$

$$= mg \sin \beta \sin \theta \quad .$$

The equations of motion may be written at this point. The motion tangent to the path is given by

$$F_t = mg \sin \beta \cos \theta = m\ddot{s} = m \frac{dv}{dt}$$

or (1)

$$\frac{dv}{dt} = g \sin \beta \cos \theta$$

where β is the angle of the incline.

The motion normal to the path is given by

$$F_n = mg \sin \beta \sin \theta = \frac{m(\dot{s})^2}{\rho} = \frac{mv^2}{\rho}$$

or

$$\frac{v^2}{\rho} = g \sin \beta \sin \theta \ . \qquad (2)$$

In order to effect the change to path coordinates, the following substitutions into equations (1) and (2) may be made:

$$\frac{dv}{ds}\frac{ds}{dt} = v\frac{dv}{ds}$$

for dv/dt and $-d\theta/ds$ for $1/\rho$ (note the relative direction of change of s and θ to account for the minus sign). The equations of motion become

$$v\frac{dv}{ds} = g \sin \beta \cos \theta$$

$$v^2 \frac{d\theta}{ds} = -g \sin \beta \sin \theta \ .$$

Dividing the first expression above by the second and separating variables,

$$\frac{dv}{v} = -\cot \theta \, d\theta \ .$$

Integrating,

$$\ln v = -\ln \sin \theta + \ln C$$

or

$$v = \frac{C}{\sin \theta} \ .$$

The constant of integration is determined by substituting the initial condition $\theta = \theta_0 = 135$ degrees, $v = v_0 = 100$ cm/sec. Thus $C = 70.7$ cm/sec, and the velocity at any θ is given by

$$v = \frac{70.7}{\sin \theta} \text{ cm/sec.}$$

• PROBLEM 9-11

A force $\vec{F} = t\hat{i} + t^2\hat{j} + t^3\hat{k}$ measured with respect to the inertial coordinate system is applied to a particle of mass 1 kg which is initially at rest at the origin of the coordinate system. Find the acceleration, velocity, and displacement of the particle as functions of time, expressing your results in vector form.

Solution: What does it mean when a force is measured with respect to the inertial coordinate system? It means that the laws of Newton are valid with respect to these systems, which are those whose axes are either fixed or experience constant-velocity motion only. First, write down all of the information given in the statement of the problem.

$$mg = 1 \text{ kg}, \ v_0 = 0 \ , \ r_0 = 0$$

Newton's 2nd law describes the motion of any body which is being acted upon by a force. It states that the acceleration gained by a mass m by a force $\vec{F}$ is given by:

$$\vec{a} = \frac{\vec{F}}{m} \ .$$

Substituting in the given value for m, we can write an expression for the acceleration directly as:

$$\vec{a} = t\vec{i} + t^2\vec{j} + t^3\vec{k} \ \text{m/sec}^2. \qquad (1)$$

Acceleration is defined as the time derivative of the velocity. Thus,

$$\vec{a} = \frac{d\vec{v}}{dt}$$

or

$$d\vec{v} = \vec{a}\,dt = (t\vec{i} + t^2\vec{j} + t^3\vec{k})\,dt \; . \qquad (2)$$

Integrating both sides of equation (2) between limits evaluated at the beginning of the problem $(t = 0)$ and again at some later time $(t = t)$ gives

$$\int_{v_0}^{v} d\vec{v} = \int_{t_0}^{t} (t\vec{i} + t^2\vec{j} + t^3\vec{k})\,dt$$

$$\vec{v} - \vec{v}_0 = \tfrac{1}{2}\,t^2\vec{i} + \tfrac{1}{3}\,t^3\vec{j} + \tfrac{1}{4}\,t^4\vec{k}$$

or, since $v_0 = 0$

$$\vec{v} = \tfrac{1}{2}\,t^2\vec{i} + \tfrac{1}{3}\,t^3\vec{j} + \tfrac{1}{4}\,t^4\vec{k} \quad \text{m/sec.} \qquad (3)$$

Remembering that velocity is the time derivative of the displacement, we can proceed to find the displacement of the particle as a function of time in a similar manner as follows:

$$d\vec{v} = \vec{v}\,dt \; .$$

Integrating both sides, we have:

$$\int_{r(0)}^{r(t)} d\vec{v} = \int_{0}^{t} \vec{v}\,dt$$

$r(0) = 0$ (the particle starts at the origin)

$$\vec{r}(t) = \int_0^t (\tfrac{1}{2}\,t^2\vec{i} + \tfrac{1}{3}\,t^3\vec{j} + \tfrac{1}{4}\,t^4\vec{k})\,dt$$

$$\vec{r} = \frac{1}{6}\,t^3\vec{i} + \frac{1}{12}\,t^4\vec{j} + \frac{1}{20}\,t^5\vec{k} \; , \text{ meters.}$$

• PROBLEM 9-12

To a particle of mass m, initial velocity v_0, apply the force:

$$F(t) = \begin{cases} 0 & \text{for } t < t_0 \\ p_0/\delta t & t_0 \le t \le t_0 + \delta t \\ 0 & t > t_0 + \delta t \end{cases}$$

a) Find $v(t)$ and $x(t)$.

b) Show that as $\delta(t) \to 0$ the motion approaches constant velocity with an abrupt change of velocity at $t = t_0$ with amount p_0/m.

<u>Solution</u>: a) We solve the problem for the three periods of time because of different behavior of the force.

I $\qquad F(t) = 0$ for $t < t_0$.

From the Second Law of Newton, one can find that the acceleration is

$$a = F/m \; , \quad \text{but} \quad a = dv/dt \; .$$

or

$$dv = a t$$

$$dv = F/m\,dt. \qquad (1)$$

Integrating yields:

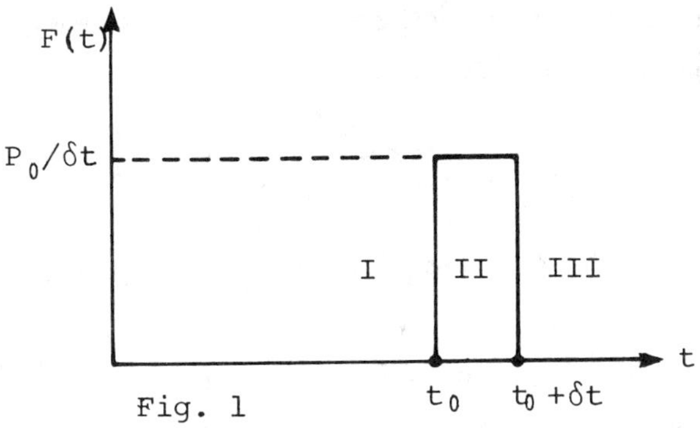

Fig. 1

$$\int_{v_0}^{v} dv = 1/m \int_{0}^{t} F\,dt = 0 \quad \text{for} \quad t < t_0$$

$$v(t) - v_0 = 0$$

$$v(t) = v_0$$

and $$v(t) = dx(t)/dt.$$

Then $$dx(t) = v_0\,dt$$

$$x(t) = v_0 t\ , \text{ letting } \ x(0) = 0\ .$$

II $$F(t) = p_0/\delta t \quad \text{for} \quad t_0 \le t \le t_0 + \delta t\ .$$

Integrating the relation (1), yields:

$$\int_{v_0}^{v} dv = 1/m \int_{t_0}^{t} p_0/\delta t\ dt$$

$$v - v_0 = \frac{p_0}{m} \cdot \frac{t - t_0}{\delta t}$$

$$v(t) = v_0 + \frac{p_0(t - t_0)}{m\,\delta t} \tag{2}$$

and

$$\int_{x_0}^{x} dx = \int_{t_0}^{t} [v_0 + \frac{p_0(t' - t_0)}{m\,\delta t}]\ dt'$$

but: $x_0 = v_0 t_0$. Hence:

$$x - v_0 t_0 = v_0(t-t_0) + \frac{p_0(t'-t_0)^2}{2m\,\delta t}\Bigg|_{t_0}^{t}\ ,\ x(t) = v_0 t + \frac{p_0}{2m\,\delta t}(t-t_0)^2. \tag{3}$$

III $$F(t) = 0 \quad \text{for} \quad t > t_0 + \delta t\ .$$

From relation (2), with $t = t_0 + \delta t$ yields:

$$v(t) = v_0 + p_0\frac{(t_0 + \delta t - t_0)}{m\ \delta t}$$

$$v(t) = v_0 + \frac{p_0}{m} \tag{4}$$

that is a constant in time.

To find x, again integrate, using the value of x when $t = t_0 + \delta t$ from equation (4) as the lower limit.

$$\int_{x(t_0+\delta t)}^{x} dx = \int_{t_0+\delta t}^{t} (v_0 + \frac{p_0}{m})\, dt$$

$$x - v_0 t_0 - \frac{p_0}{2m\,\delta t}(t_0 + \delta t - t_0)^2 = (v_0 + \frac{p_0}{m})[t - (t_0 + \delta t)]$$

$$x - v_0 t_0 - \frac{p_0\,\delta t}{2m} = (v_0 + \frac{p_0}{m})t - (v_0 + \frac{p_0}{m})t_0 - (v_0 + \frac{p_0}{m})\,\delta t$$

$$x = \frac{p_0\,\delta t}{2m} + v_0 t + \frac{p_0 t}{m} - \frac{p_0 t_0}{m} - \frac{p_0\,\delta t}{m}\ .$$

Rearranging and collecting terms again gives:

$$x = v_0 t - \frac{p_0\,\delta t}{2m} + \frac{p_0}{m}(t - t_0)\ . \qquad (5)$$

b) Recalling that there is a constant velocity of v_0 when $t < t_0$ and another constant velocity of $v_0 + p_0/m$ when $t > t_0 + \delta t$, it can be understood that as $\delta t \to 0$, $v \to v_0 + p_0/m$. Thus the velocity changes abruptly from v_0 to $v_0 + p_0/m$ at t_0 as $\delta t \to 0$.

• PROBLEM 9-13

An electron is released from rest at the cathode of a vacuum tube containing two parallel plates. The potential between cathode and anode is V volts. Determine the minimum uniform magnetic induction field required to prevent the electron from reaching the anode. Also determine the trajectory of the electron. Consider the electric field between cathode and anode to be uniform.

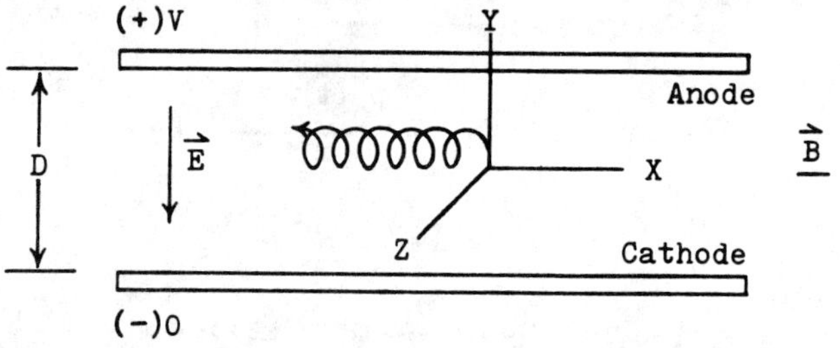

fig 1

Solution: Assume $\vec{B} = B\hat{k}$ and $\vec{E} = -(V/D)\hat{j}$. The electric field pulls the elctron upward, but as the electron gains velocity the magnetic field will deflect it sideways. The motion of the electron will be similar to that shown in Fig. 1.

The Lorentz force is $F = -e(\vec{E} + \vec{v} \times \vec{B})$. Applying Newton's second law

$$-e\vec{E} - e(\vec{v} \times \vec{B}) = m\vec{a}\ .$$

The equations of motion are therefore,

$$-e\dot{y}B = m\ddot{x}$$

$$\frac{eV}{D} + e\dot{x}B = m\ddot{y}$$

$$0 = m\ddot{z}$$

Because the electron starts from rest, $\dot{x}(0) = \dot{y}(0) = \dot{z}(0) = 0$. Selecting the initial position as the origin of the coordinate system: $x(0) = y(0) = z(0) = 0$. The solution of the Z-motion is $z = 0$. Integrating the X-equation of motion once,

$$-eyB = m\dot{x} + C_1 .$$

On the basis of the initial conditions, $C_1 = 0$, and

$$\dot{x} = -\frac{eB}{m} y = -\omega y \quad , \quad \omega = \frac{eB}{m} .$$

Substituting this value into the equation of motion for the Y-direction,

$$\frac{eV}{D} - \omega eBy = m\ddot{y} .$$

Dividing by m and rearranging,

$$\ddot{y} + \omega^2 y = \frac{eV}{Dm} .$$

The solution of the homogeneous part of this equation is

$$y = C_2 \sin \omega t + C_3 \cos \omega t$$

and the solution to the inhomogeneous part is

$$y = \frac{eV}{Dm\omega^2} .$$

The complete solution is therefore

$$y = \frac{eV}{Dm\omega^2} + C_2 \sin \omega t + C_3 \cos \omega t .$$

From the initial conditions,

$$y(0) = \frac{eV}{Dm\omega^2} + C_3 = 0$$

$$C_3 = -\frac{eV}{Dm\omega^2} .$$

At $t = 0$,

$$\dot{y}(0) = C_2\omega = 0$$

and therefore $C_2 = 0$. The solution for y is therefore

$$y = \frac{eV}{Dm\omega^2} (1 - \cos \omega t) .$$

Recall that

$$\dot{x} = -\omega y = -\omega\left[\frac{eV}{Dm\omega^2}(1 - \cos \omega t)\right] .$$

Integrating,

$$x = -\frac{eV}{Dm\omega}\left(t - \frac{\sin \omega t}{\omega}\right) + C_4 .$$

Applying the initial condition, at $t = 0$, $x = 0$: $C_4 = 0$. Thus

$$x = -\frac{eV}{Dm\omega^2}(\omega t - \sin \omega t)$$

We can see that the equation for x and y show that while y increases from 0 to $y_{max} = 2eV/Dm\omega^2$ sinusoidally, x decreases monotonically (although not steadily), thus producing the two dimensional curve in Fig. 1.

To determine the minimum magnetic induction field to cause the electron to just miss the anode, the condition $\dot{y} = 0$ is employed,

$$\dot{y} = \frac{eV}{Dm\omega} \sin \omega t = 0$$

which is satisfied when $\omega t = 0, \pi, 2\pi, \ldots, n\pi$. The values of $0, 2\pi, \ldots, 2n\pi$ correspond to the return of the electron to the cathode, while $\omega t = \pi, 3\pi, \ldots, (2n-1)\pi$ correspond to the electron at the anode. When $\omega t = \pi$, $y = D$, and

$$D = \frac{2eV}{Dm\omega^2}$$

or

$$\omega = \left(\frac{2eV}{mD^2}\right)^{\frac{1}{2}}$$

Since $\omega = eB/m$,

$$B = \left(\frac{2mV}{eD^2}\right)^{\frac{1}{2}}$$

the mimimum field so that the electron just misses the anode.

• PROBLEM 9-14

A particle falls toward the earth and is acted upon solely by the force of gravitation. The height from which the particle began falling, h, is so great that the approximation $F = mg$ cannot be used. Describe the motion.

Solution: Since the distance the particle falls is so large, Newton's law of gravitation must be used in the form

$$F = -\frac{kmM}{y^2} \tag{1}$$

where y is measured from the center of the earth, M is the earth's mass, and m is the mass of the particle. Newton's second law becomes

$$m\frac{dv}{dt} = -\frac{kmM}{y^2} . \tag{2}$$

To write this in terms of constants we are more familiar with, when the particle reaches the earth's surface, $y = R$, we know that $dv/dt = -g$. Hence $k = gR^2/M$ and (2) changes into

$$\frac{dv}{dt} = -\frac{gR^2}{y^2} . \tag{3}$$

As $v = dy/dt$, we can write

$$\frac{dv}{dt} = \frac{dv}{dy}\frac{dy}{dt} = v\frac{dv}{dy} \tag{4}$$

which, upon substitution in (3), becomes

$$v\,dv = -\frac{gR^2}{y^2}\,dy . \tag{5}$$

Integration of this separated differential equation gives

$$v^2 = \frac{2gR^2}{y} + C . \tag{6}$$

The initial condition $y = h$ when $v = 0$ leads to

$$v^2 = 2gR^2\left(\frac{1}{y} - \frac{1}{h}\right) . \tag{7}$$

As h approaches an infinite height, the velocity with which the particle arrives at earth, $y = R$, becomes

$$v = \sqrt{2g\,R} . \tag{8}$$

We can now solve the velocity equation for v, taking the negative root because y is decreasing. After separating variables we find

$$\frac{dy}{\sqrt{\frac{1}{y} - \frac{1}{h}}} = -\sqrt{2gR^2}\,dt . \tag{9}$$

If we write the equation as

$$\frac{\sqrt{h}\ y\ dy}{\sqrt{hy - y^2}} = -\sqrt{2gR^2}\ dt\ , \tag{10}$$

the integral on the left can be found in tables. It can be shown that

$$t = \frac{1}{R}\sqrt{\frac{h}{2g}}\left[\sqrt{hy - y^2} - \frac{h}{2}\sin^{-1}\left(\frac{2y - h}{h}\right)\right] + C\ . \tag{11}$$

The condition $y = h$ when $t = 0$ gives

$$t = \frac{1}{R}\sqrt{\frac{h}{2g}}\left[\sqrt{hy - y^2} + \frac{h}{2}\cos^{-1}\left(\frac{2y - h}{h}\right)\right] \tag{12}$$

since $C = \frac{1}{R}\sqrt{\frac{h}{2g}}\frac{h\pi}{4}$. It is not possible to solve (12) explicitly for y so (7) and (12) must suffice to describe the motion.

MOTION OF THE CENTER OF MASS

• PROBLEM 9-15

Two particles of mass, m_1 (= 3 Kg) and m_2 (= 1 Kg) are attached to the ends of a rigid massless bar 40 cm. long. The system is placed, with the bar vertical and m_1 on top, on a frictionless plane. It is then released. How far from the initial position of m_2 will the mass m_1 be when it hits the plane?

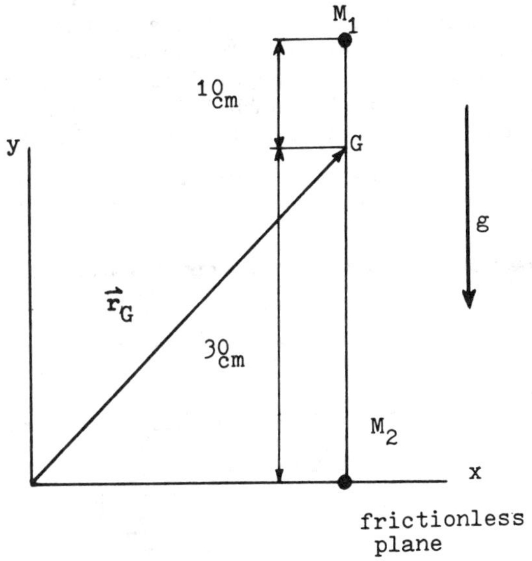

Solution: When a system of particles is acted upon by external forces, its center of mass moves in a manner as if the total mass were concentrated at that point. The motion of the center of mass, G , is governed by the equation

$$\left(\Sigma\ m_i\right)\ \ddot{\vec{r}}_G = \Sigma\,\vec{F}_i \tag{1}$$

In this problem we are concerned about the motion of G in the x (horizontal) direction. The x component of equation (1) is

$$(m_1 + m_2)\ddot{x}_G = \Sigma F_{xi} . \qquad (2)$$

Since the plane is frictionless, the external forces acting on the system are: (1) The gravity forces $m_1 g$ and $m_2 g$ on the two particles, and (2) a vertical reaction force at m_2. Thus there is no horizontal external force acting on the system. Equation (2) therefore becomes

$$(m_1 + m_2)\ddot{x}_G = 0 \qquad (3)$$

i.e., $\ddot{x}_G = 0$. Upon integrating this we find that the horizontal component of velocity of G is

$$\dot{x}_G = \text{constant} .$$

Since G was initially at rest

$$\dot{x}_G = 0 \qquad (4)$$

i.e., G does not move in the horizontal direction at all. Upon integrating equation (4) we get

$$x_G = \text{constant} \qquad (5)$$

Therefore, after being released the two particles move in such a manner that the center of mass G moves in a vertical straight line. When the bar reaches a horizontal position, G will be at the point where m_2 was initially. Since G is 10 cm from m_1, m_1 will strike the surface 10 cm from the initial contact point of m_2.

• PROBLEM 9-16

An unsymmetrical dumbbell consists of two balls, $w_1 = 1$ lb and $w_2 = 3$ lb, which are connected by a massless rod such that there is a separation of 2 ft between their centers. The dumbbell is at rest on a frictionless table; two horizontal forces of 3 lb and 4 lb are applied at t = 0, as shown. The axis of the dumbbell is initially along the x-axis, with the small weight at the origin. The forces $\vec{F}_1$ and $\vec{F}_2$ remain constant in magnitude and direction regardless of the dumbbell's motion. What are the coordinates of the center of mass of the dumbbell after 3 sec?

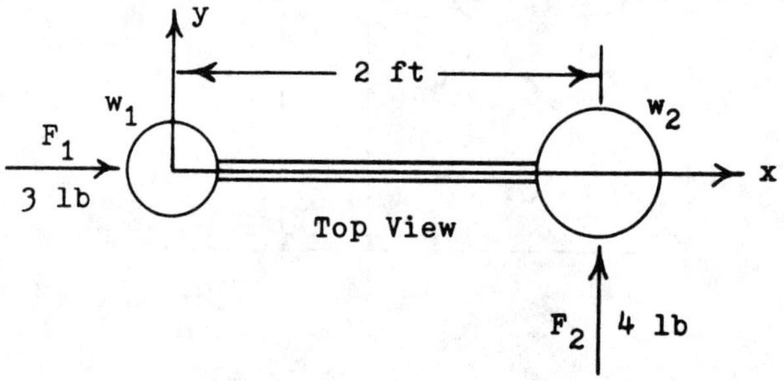

Solution: The center of mass of the body is located a distance [1/(1+3)]· 2 = 0.5 ft from the 3-lb body. Originally, then, the coordinates of the center of mass are $X_0 = 1.5$, $Y_0 = 0$.

The motion of the center of mass is governed by

$$M \frac{dV_x}{dt} = F_x , \quad M \frac{dV_y}{dt} = F_y ,$$

where F_x and F_y are the components of the external force, and V_x and V_y are the components of the velocity of the center of mass. Since $M = (1+3)/32 = 1/8$ slug, $F_x = 3$ lb, and $F_y = 4$ lb, we obtain the relations

$$\frac{dV_x}{dt} = 3 \cdot 8 = 24 \, , \quad \frac{dV_y}{dt} = 4 \cdot 8 = 32 \, .$$

These in turn yield $dX/dt = V_x = 24t$ and $dY/dt = V_y = 32t$, since the velocity is zero at $t = 0$. Similarly, the coordinates of the center of mass can be obtained:

$$X - X_0 = 24 \frac{t^2}{2} = 12t^2 \quad , \quad Y - Y_0 = 32 \frac{t^2}{2} = 16t^2 \, .$$

With $t = 3$, $Y_0 = 0$, and $X_0 = 1.5$, the center-of-mass coordinates are then $X = 108 + 1.5 = 109.5$ ft and $Y = 144$ ft.

FRICTIONAL FORCES

• PROBLEM 9-17

A weight, W, is found not to slide off a rotating horizontal disk if the weight is closer than 2 ft. to the center. The rotation rate is 21 rpm. What is the coefficient of friction between the weight and the disk?

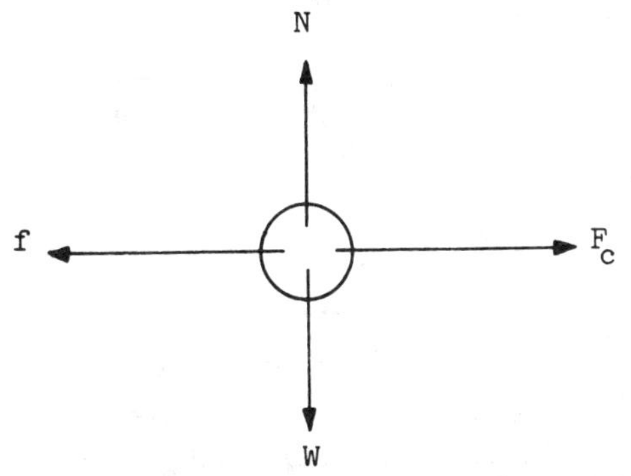

Solution: Draw the free-body diagram for the weight W. The condition that the weight does not slide off on the rotating disk is that the sum of the forces that act upon it are zero (equilibrium condition).

$$\Sigma F_{horizontal} = 0 = f - F_c$$

Therefore $f = F_c$. Since $f = \mu N$ and $F_c = mr\omega^2$, (we can write)

$$\mu N = mr\omega^2 \tag{1}$$

where N is normal force. Substituting in given values for r and ω, and remembering that ω must be written in units of radians/sec.

$$\mu N = m \times (2 \text{ ft.}) \times \left[21 \text{ rev/min} \times 2\pi \text{ rad/rev} \times \frac{1}{60 \text{ sec/min}}\right]^2$$

$$\mu N = m(9.67 \text{ ft./sec.}^2) \tag{2}$$

Using Newton's law to sum forces in the vertical direction, where acceleration is zero,

$$\Sigma F_{vertical} = 0 = N - W$$

Therefore $\quad N = W = mg$.

Now we can substitute the above into equation (2) as follows:

$$\mu mg = m(9.67 \text{ ft./sec}^2)$$

$$\mu = \frac{9.67 \text{ ft./sec}^2)}{g} = \frac{9.67}{32.2}$$

$$\mu = 0.3 \quad .$$

• PROBLEM 9-18

As shown in the figure, $M_A = 2Kg$, $M_B = 8Kg$, $M_C = 4Kg$. If the coefficient of friction between A and B is 0.6 and zero between B and the table, show that block A will slide along the block B. Find the accelerations of blocks A and B relative to the table.

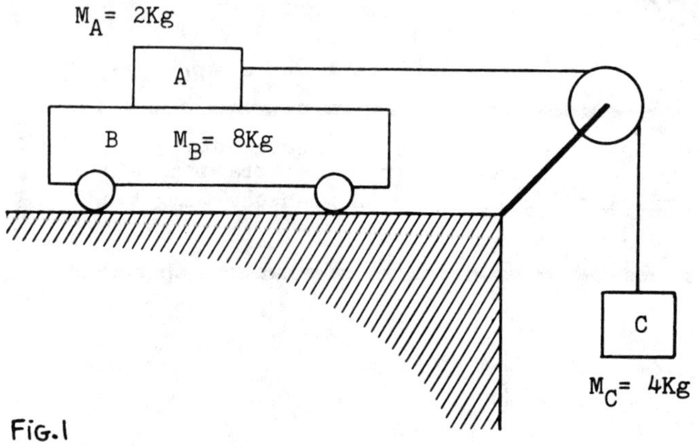

FiG.1

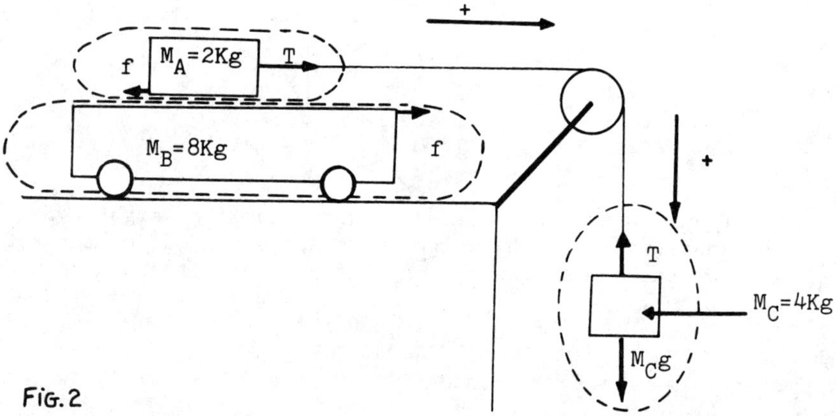

FiG.2

Solution: This problem calls for an analysis of the motions of the three masses using Newton's Second Law. First, draw spheres of isolation around the three masses as shown.

The force of friction f acts backwards on M_A and forwards on M_B by Newton's Third Law. T, the tension in the string, acts in the directions shown, again by the Third Law. Here the assumption is made that the pulley is massless and frictionless. The key to this solution lies in analyzing the force of friction between the two blocks. The maximum value of f will be $f_{max} = M_A g \mu$ where μ = coefficient of friction. Thus $f_{max} = (2kg)(9.8\ m/sec^2)(.6) = 11.76N$.

First determine if this friction force is kinetic, thus constant, or static, and thus a maximum value. In order for the friction to be static, block M_A would not move relative to M_B, i.e. the accelerations would be equal. Then the friction force pulling B forward would have to be 4 times the total force on A.

$$f = 4(T - f)$$

$$5f = 4T$$

$$T = \frac{5}{4} \cdot f = 14.7N\ .$$

Of course the tension would also be smaller since this is the limiting case for static friction . No move can be done with this now, but use it as a check after solving the problem, assuming that the blocks do move in relation to each other.

Now write one equation for each of the three masses.

$$T - f = M_A a_A\ ;\ T - 11.76 = 2a_A \tag{1}$$

$$f = M_B a_B\ ;\ 11.76 = 8a_B \tag{2}$$

$$M_C g - T = M_C a_C\ ;\ 39.2 - T = 4a_C\ . \tag{3}$$

Solving (2) for a_B gives $a_B = 1.47\ m/sec^2$. We establish that $a_C = a_A$ because the string will not stretch. Thus we must solve (1) and (3) for a_A:

$$T - 11.76 = 2a_A \tag{4}$$

$$39.2 - T = 4a_C. \tag{5}$$

Adding these two equations yields

$$6a_A = 27.44$$

$$a_A = 4.57\ m/sec^2\ .$$

To check, put this value into equation (4) $T = 20.9N$ is obtained. Thus, the solution is consistent with the assumption that the friction if kinetic.

• PROBLEM 9-19

A uniform straight rigid bar of mass m and length L is placed in a horizontal position across the top of two identical cylindrical rollers, rotating as shown in Figure 1. Axes of the rollers are a distance $2d$ apart. If μ is the coefficient of friction between each cylinder surface and the bar, show that if the bar is displaced a distance x from its central position, the net horizontal force on the bar is $F = -mg\mu x/d$. Show that the bar will execute simple harmonic motion with a period of $P = 2\pi\sqrt{d/\mu g}$.

Solution: The bar moves in the plane of the paper under the action of the frictional forces acting on it from the rollers. Figure 2 shows the free-body diagram of the bar. Displacement of the center of gravity G of the

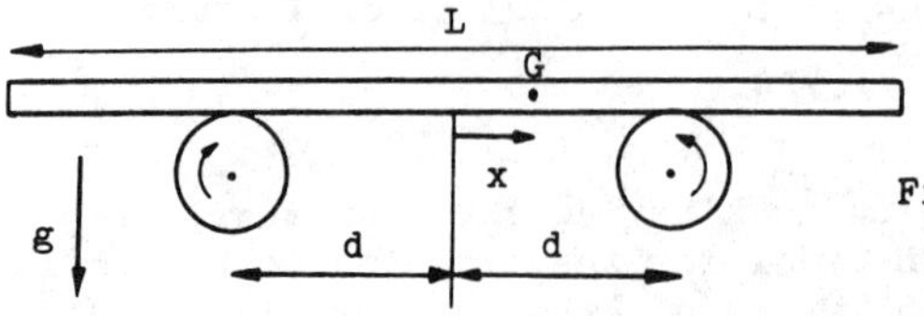

Figure 1

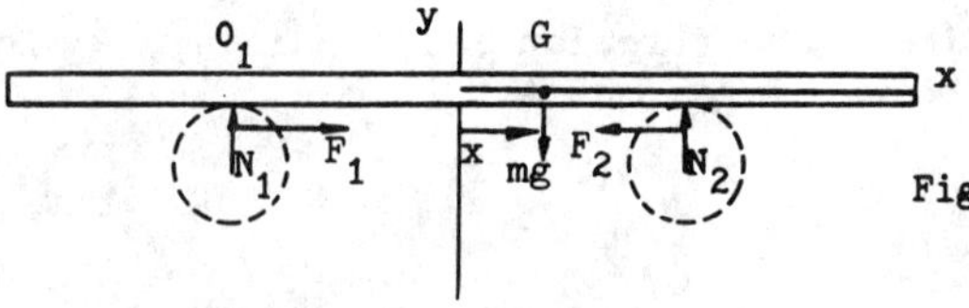

Figure 2

bar is x , positive to the right, measured from the midway point between the rollers, 0 . Equations of motion of the bar are

$$m\ddot{x} = \Sigma F_x = F_1 - F_2 \tag{1}$$

$$m\ddot{y} = 0 = \Sigma F_y = N_1 + N_2 - mg \tag{2}$$

where N_1 and N_2 are the normal forces and F_1 and F_2 are the friction forces on the bar. For a given displacement x of the bar, N_1 and N_2 are found by a static analysis. Taking moments about 0_1

$$N_2(2d) = mg(x+d) ,$$

or

$$N_2 = mg(1+x/d)/2 \quad . \tag{3}$$

From equation (2)

$$N_1 = mg - N_2 = mg(1-x/d)/2 \quad . \tag{4}$$

Therefore the net horizontal force on the bar is

$$\Sigma F_x = F_1 - F_2 = \mu(N_1 - N_2)$$

or

$$F = - mg\mu x/d \qquad \text{Ans.} \tag{5}$$

From equation (5) we note that when G is to the right of the midpoint 0 , the net force on the bar is to the left, and vice versa. Thus we should expect that the bar, once disturbed from its equilibrium position (G at 0; x = 0), would oscillate about this position. Substituting equation (5) into (1) we obtain the equation of motion for the bar

$$m\ddot{x} = - mg\mu x/d$$

or

$$\ddot{x} + g\mu x/d = 0 , \tag{6}$$

This equation is of the form

$$\ddot{x} + \omega^2 x = 0 \tag{7}$$

which describes simple harmonic motion of a particle at angular frequency ω , in radians per unit of time. Comparing equations (6) and (7) we see that

$$\omega^2 = g\mu/d$$

or

$$\omega = \sqrt{g\mu/d} \quad .$$

The period of oscillation is

$$P = 2\pi/\omega = 2\pi\sqrt{d/\mu g} \quad . \qquad \text{Ans.}$$

An Eskimo is about to push along a horizontal snowfield a sled weighing 57.6 lbs carrying a baby seal weighing 70 lbs which he has killed while hunting. The coefficient of static friction between sled and seal is 0.8 and the coefficient of kinetic friction between sled and snow is 0.1. Show that the maximum horizontal force that the Eskimo can apply to the sled without losing the seal is 114.8 lbs. Calculate the acceleration of the sled when this maximum horizontal force is applied.
Note: $\vec{f}_2$ points to the right. The sled is moving to the right, and the seal's inertia tends to keep it at rest. Thus any relative motion of the seal with respect to the sled would have the seal moving to the left and, therefore, the friction would oppose this motion and point to the right.

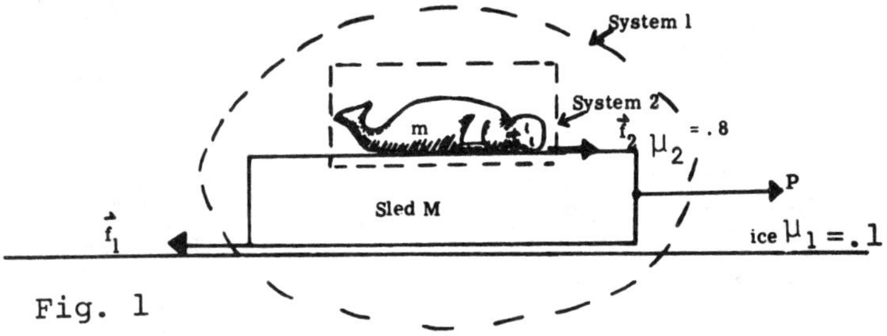

Fig. 1

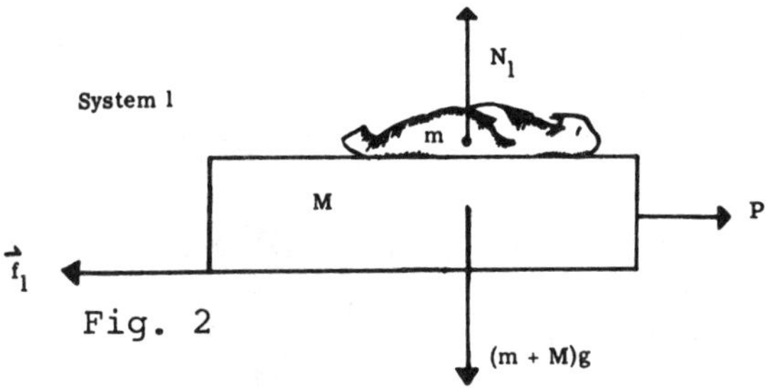

Fig. 2

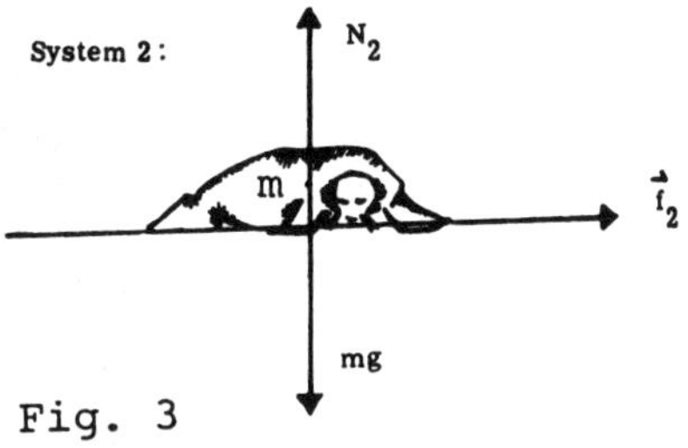

Fig. 3

Solution: The seal will slide off the sled when the acceleration provided by $\vec{P}$ is greater than the maximum acceleration which $\vec{f}_2$ can provide to the seal. (See Fig. 1). The acceleration of the seal is the same as the acceleration of the seal-sled system, as long as the seal does not fall off the sled. Using Newton's Second Law to calculate the latter, we obtain

$$P - f_1 = (m + M)a \qquad (1)$$

where a is the acceleration of the system, taken as positive in the direction of P. The frictional force law is

$$f_1 = \mu_1 N_1 \qquad (2)$$

where μ_1 is the coefficient of kinetic friction between sled and ice, and N_1 is the normal force of the ice on the sled-seal system. Substituting (2) in (1)

$$P - \mu_1 N_1 = (m + M)a$$

Since the system is in vertical equilibrium

$$N_1 = (m + M)g$$

and

$$P - \mu_1(M + m)g = (M + m)a \quad .$$

Finally,

$$a = \frac{P - \mu_1(m + M)g}{(m + M)} \tag{3}$$

is the acceleration of the sled-seal system.

Now, applying the Second Law to the system consisting of seal alone (see Fig. 3) we obtain

$$f_2 = ma \tag{4}$$

where a is the acceleration in (3). But since we require the seal to remain at rest on the sled,

$$f_2 \leq \mu_2 N_2 \tag{5}$$

where μ_2 is the coefficient of static friction between sled and seal, and N_2 is the normal force of the sled on the seal. Inserting (5) in (4)

$$ma \leq \mu_2 N_2 \tag{6}$$

Substituting (3) in (6)

$$m\left(\frac{P - \mu_1(m + M)g}{(m + M)}\right) \leq \mu_2 N_2$$

Since the seal is in equilibrium vertically,

$$N_2 = mg$$

and

$$m\left(\frac{P - \mu_1(m + M)g}{(m + M)}\right) \leq \mu_2 mg$$

Solving for P

$$P - \mu_1(m + M)g \leq \mu_2(m + M)g$$

$$P \leq (\mu_1 + \mu_2)(m + M)g$$

$$P \leq (.9)(57.6 + 70)\text{lbs.}$$

$$P \leq 114.8 \text{ lbs.}$$

The maximum value of P is 114.8 lbs. Also the maximum acceleration is

$$a = \frac{P - \mu_1(m + M)g}{m + M}$$

$$a = \frac{114.8 \text{ lbs} - .1(57.6 \text{ lbs} + 70 \text{ lbs})}{(57.6 \text{ lbs} + 70 \text{ lbs})/32.2 \text{ f/s}^2}$$

$$a = 25.75 \text{ f/s}^2 \quad .$$

• PROBLEM 9-21

A horizontal force, $|F| = bt$ (where t is time in seconds), is applied to a block of wood of mass m at rest on a horizontal surface. The coefficient of static friction is μ_s and coefficient of kinetic friction is μ_k. Find the acceleration of the block of wood as a function of time.

Solution: The free-body diagram is shown in fig. 1. Until the applied force overcomes the static friction, the acceleration is zero. The time t_1 at which static friction is overcome is obtained by equating the applied force to the maximum static frictional force; i.e.,

$$bt_1 = \mu_s mg$$

from which

$$t_1 = \frac{\mu_s mg}{b} .$$

In the moment that the force $F_1 = bt$ is higher than the static friction, that is at time t_1 the motion begins, kinetic friction takes over and the block accelerates. To determine the acceleration use Newton's Law,

$$F = ma . \qquad (1)$$

Now the force that acts upon the block is:

$$F = bt - \mu_k mg .$$

Substitution into eq. (1) gives

$$a = (bt - \mu_k mg)/m$$

for

$$t > t_1 = \frac{\mu_s mg}{b}$$

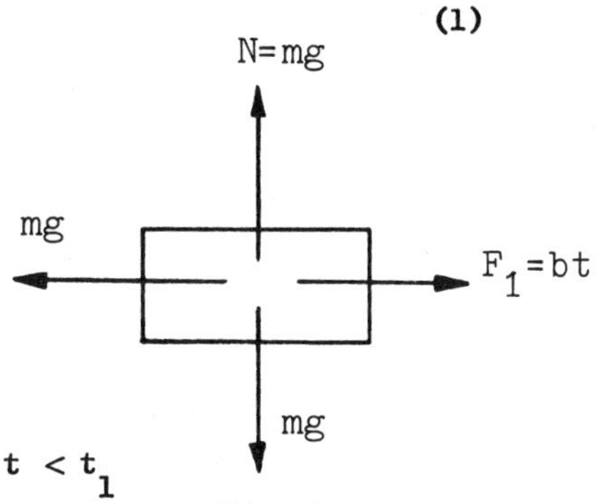

Fig.1

The answer is

$$a = \begin{cases} 0 & t < t_1 \\ \frac{b}{m} t - \mu mg & t \geq t_1 \end{cases}$$

where $t_1 = \frac{\mu_s mg}{b}$.

• PROBLEM 9-22

> The coefficient of kinetic friction is determined to be a function of velocity given in Figure 1. Determine, as a function of time, the velocity of a 10-g mass acted upon by a horizontal force of 2.94×10^3 dynes.

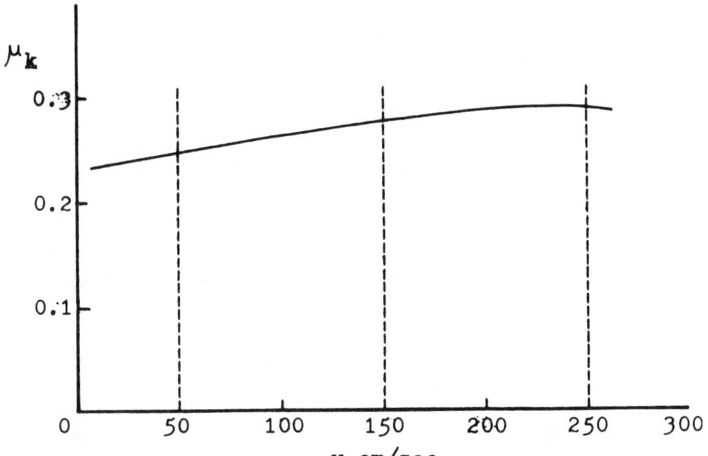

fig 1

Solution: There are a number of different approaches to this problem. The difficulty lies in expressing $\mu_k = f(v)$. Analytical techniques exist preferably with the help of a digital computer to transform empirical data into a polygonal approximation. These could provide a relatively accurate solution but are not absolutely necessary for good results. Another method is to break up the analysis into segments and use Simpson's approximation, a trapezoidal approximation, or even a central value approximation. Such methods are easy and desired accuracy can be achieved simply by reducing the increments taken. This can be done more readily with the aid of a digital computer.

The last of these methods is simplest and for brevity of calculation wide intervals should be used. Obviously, when applying this method to critical applications, or if the frictional coefficient varies much more rapidly, the reader should work harder and achieve more accurate results.

From Newton's Second Law, the equation of motion is,

$$F - \mu_k N = ma_x \tag{1}$$

Substituting $N = mg$ and solving for a_x,

$$a_x = \frac{F}{m} - \mu_k g = 294 - \mu_k(980). \tag{2}$$

For the central value approximation we will assume that μ_k remains constant over a limited range of velocity and that its value is that of μ_k in the center of each range. Then our values and the results for a_x are

$$\begin{aligned} 0 < v_x < 100 \text{ cm/sec}; \ \mu_k &= 0.25 \text{ and } a_x = 49 \text{ cm/sec}^2 \\ 100 < v_x < 200 \text{ cm/sec}; \ \mu_k &= 0.28 \text{ and } a_x = 20 \text{ cm/sec}^2 \\ 200 < v_x < 300 \text{ cm/sec}; \ \mu_k &= 0.29 \text{ and } a_x = 10 \text{ cm/sec}^2 \end{aligned} \tag{3}$$

Now to solve for v as a function of time one must find the inverse $t = g(v)$. The reason for this procedure is that μ_k and a_x are defined for definite velocity limits. Using the equation

$$v_{x_f} = v_{0x} + a_x t \tag{4}$$

we have

$$t = \frac{v_{x_f} - v_{0x}}{a_x} .$$

This may be applied to each velocity interval

a) $0 < v_x < 100$ cm/secl $v_{0x} = 0$, $v_{x_f} = 100$ cm/sec

$$t_1 = 100/49 = 2.04 \text{ sec}$$

b) $100 < v_x < 200$ cm/sec; $v_{0x} = 100$ cm/sec, $v_{x_f} = 200$ cm/sec

$$t_2 = \frac{200 - 100}{20} = 5.0 \text{ sec}$$

c) $200 < v_x < 300$ cm/sec; $v_{0x} = 200$ cm/sec, $v_{x_f} = 300$ cm/sec

$$t_3 = \frac{300 - 200}{10} = 10.0 \text{ sec.}$$

Using equation (4) and the data from (3), an answer to the question can be formulated.

$$v = \begin{cases} 49t \text{ cm/sec} & 0 < t < 2.04 \text{ sec} \\ 100 \text{ cm/sec} + 20t \text{ cm/sec} & 2.04 < t < 7.04 \text{ sec} \\ 200 \text{ cm/sec} + 10t \text{ cm/sec} & 7.04 < t < 17.04 \text{ sec.} \end{cases}$$

A plot of this function is easily made since we have formed an approximation that the velocity increase linearly between the (t,v) coordinates; (0,0), (2.04,100), (7.04,200) and (17.04,300), (Figure 2).

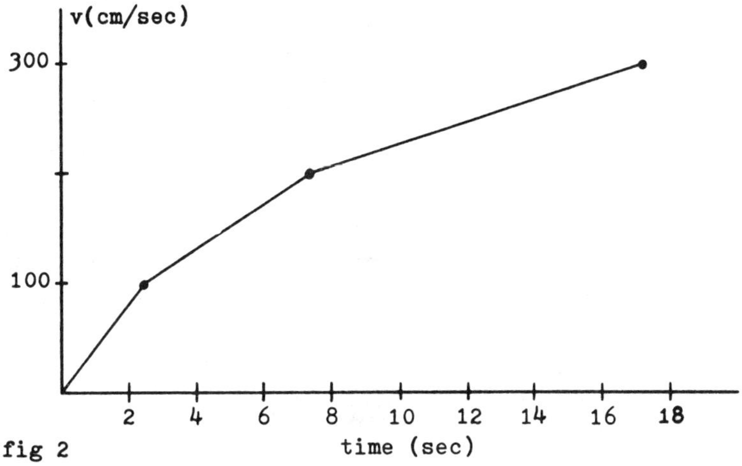

fig 2

UNIFORM CIRCULAR MOTION

• PROBLEM 9-23

An automobile weighing 3400 lb is driven by a man weighing 150 lb. It is moving on a circular curve in a highway; the curve has a radius of 2000 ft. If the automobile is moving with a velocity of 60 mph, how much centrifugal force does the man experience? Find the frictional force between the wheels and the road.

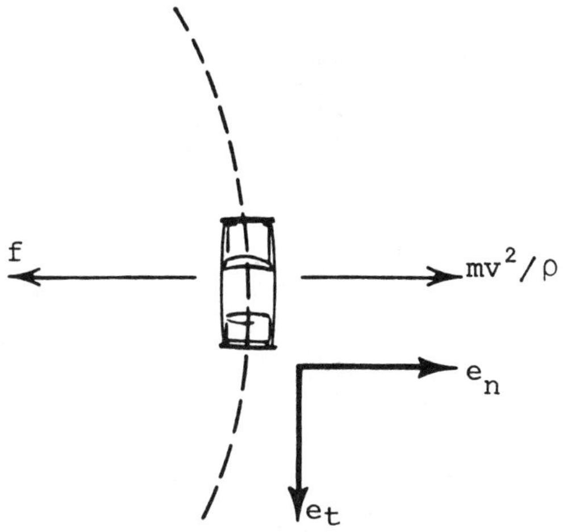

<u>Solution</u>: The centrifugal force acting on the driver is nothing more than the inertia force on the man in the normal direction. The centri-

fugal force on the driver will be equal to mv^2/ρ. In this case,

$$\text{Centrifugal force} = \frac{(150/32.2)(88)^2}{2000}$$

$$= 18.1 \text{ lb}$$

where 60 mph = 88 ft/sec.

The force of the wheels on the road must be equal to the centripetal force for the entire man-car system, assuming that the road is flat. Proceeding with that assumption

$$f = mv^2/\rho$$

$$= \frac{\left(\frac{3400 + 150}{32}\right)(88)^2}{2000}$$

$$f = 426 \text{ lb.}$$

• PROBLEM 9-24

The string of a conical pendulum is 10 ft long and the bob has a mass of ½ slug. The pendulum is rotating at ½ rev·s^{-1}. Find the angle the string makes with the vertical, and also the tension in the string.

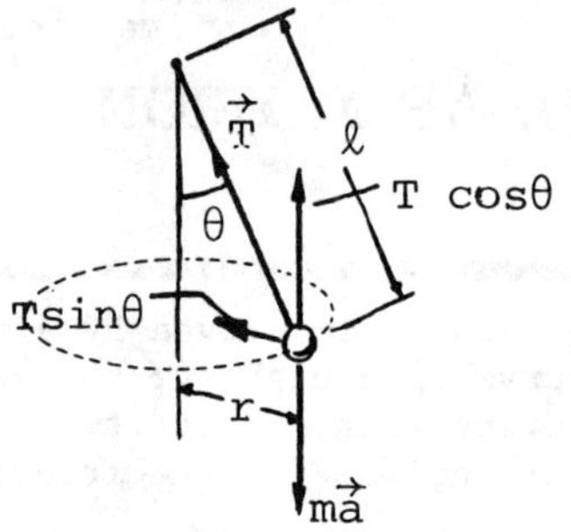

Solution: Let r be the radius of the horizontal circle traversed by the bob of mass m, ℓ be the length of the string, and $\vec{T}$ be the tension which the string exerts on the mass. The forces acting on the bob are the weight $\vec{mg}$ downward and the tension $\vec{T}$ at an angle θ to the vertical. Resolve $\vec{T}$ into horizontal and vertical components. Applying Newton's Second Law to the horizontal direction of motion

$$F_{net} = ma$$

where a is the horizontal acceleration of m, and F_{net} is the net horizontal force on m. Since m is in uniform circular motion, T sin θ provides the centripetal force necessary to keep the bob in the circle. Thus T sin θ $= mv^2/r$, where v is the velocity of the bob. But v is the distance traveled in 1 s. That is, $v = n \times 2\pi r = 2\pi rn$, where n is the angular speed in rev·s^{-1}. Also, from the figure, $\sin\theta = r/\ell$.

$$\therefore T = \frac{4\pi^2 rmn^2}{r/\ell} = 4\pi^2 m\ell n^2$$

$$= 4\pi^2 \times \tfrac{1}{2} \text{ slug} \times 10 \text{ ft} \times \left(\tfrac{1}{2} \text{ s}^{-1}\right)^2$$

$$= 49 \text{ lb.}$$

The bob stays in the same horizontal plane, so that the vertical forces must balance. Thus, from Newton's Second Law, $T \cos\theta = mg$.

$$\therefore \quad \cos\theta = \frac{mg}{4\pi^2 m\ell n^2}$$

$$= \frac{32}{4\pi^2 \times 10 \text{ ft} \times \left(\tfrac{1}{2} \text{ s}^{-1}\right)^2} = 0.327;$$

$$\therefore \quad \theta = 71^\circ.$$

• PROBLEM 9-25

The outside curve on a highway forms an arc whose radius is 150 ft. If the roadbed is 30 ft. wide and its outer edge is 4 ft. higher than the inner edge, for what speed is it ideally banked?

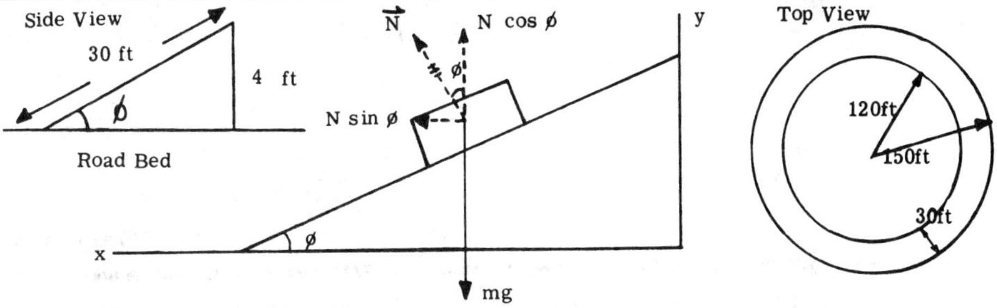

Solution: We wish to relate the velocity of the car to ϕ, the banking angle. Note that the car is undergoing circular motion, hence its acceleration in the x-direction is $a = \frac{v^2}{R}$, where R is its distance from the center of the circle (see figure). Applying Newton's Second Law, F=ma, to the x component of motion,

$$ma = N \sin\phi$$

But $a = v^2/R$ and

$$\frac{mv^2}{R} = N \sin\phi \tag{1}$$

The acceleration of the car in the y-direction is zero, since it remains on the road. Applying the Second Law to this component of motion,

$$N \cos\phi = mg \tag{2}$$

Dividing (1) by (2),

$$\frac{\frac{mv^2}{R}}{mg} = \frac{N \sin\phi}{N \cos\phi} = \tan\phi$$

$$\tan\phi = \frac{v^2}{Rg}$$

Hence $\qquad v = \sqrt{Rg \tan\phi}$

Now, note that the width of the road bed is much smaller than the inner radius of the road. Hence, we may approximate R as the inner radius.

$$R \simeq 150 \text{ ft}$$

$$v = \sqrt{(150 \text{ ft})(32 \text{ ft/s}^2) \tan \phi}$$

From the figure,

$$\sin \phi = 4/30$$

$$\cos^2 \phi = 1 - \sin^2 \phi = \frac{900}{900} - \frac{16}{900} = \frac{884}{900}$$

Hence $$\cos \phi = \frac{\sqrt{884}}{30}$$

and $$\tan \phi = \frac{\frac{4}{30}}{\frac{\sqrt{884}}{30}} = \frac{4}{\sqrt{884}} = .1345$$

Therefore $$v = \sqrt{(150 \text{ ft})(32 \text{ ft/s}^2)(.1345)}$$

$$v = \sqrt{645.6 \text{ ft}^2/\text{s}^2}$$

$$v = 25.41 \text{ ft/s}$$

• PROBLEM 9-26

a. A particle of mass m and charge q is injected into a uniform magnetic field, $\vec{B}$. If the particle velocity is initially perpendicular to the field, determine its trajectory.
b. Determine the cyclotron frequency of an electron in the atmosphere.

Solution: The motion of a charged particle in a pure magnetic field is easily separated into two parts: motion in the direction of the field, which is unchanged, and circular motion perpendicular to the field. This is described by the force expression for such a particle:

$$\vec{F} = q(\vec{v} \times \vec{B}) .$$

The cross product states that $\vec{F}$ is perpendicular to both $\vec{B}$ and $\vec{v}$, so that there is no acceleration in the direction of $\vec{B}$. But a force which is perpendicular to the motion produces circular motion. Of course, whether the trajectory is circular or helical depends on the direction of the initial velocity with respect to the magnetic field.

In this problem, $\vec{v}$ is initially perpendicular to $\vec{B}$ so $v_{||}$, the parallel component, remains zero throughout. The trajectory then, is a circle of radius still to be determined.

For $\vec{v}$ perpendicular to $\vec{B}$ the force is

$$F = qvB .$$

The particle, moving about a circle, arranges its velocity and radius so as to balance its centrifugal force, mv^2/r, with the centripetal force. Thus,

$$qvB = mv^2/r$$

and

$$r = mv/qB \ .$$

The time to travel a single revolution at the constant speed v is

$$T = \frac{2\pi r}{v} = \frac{2\pi mv}{vqB} = \frac{2\pi m}{qB} \text{ sec.}$$

The circular frequency is

$$f = \frac{2\pi}{T} = \frac{qB}{m} \ .$$

It is of interest to note that the period and frequency are independent of the speed of the particle as well as the radius of the orbit. The above frequency is called the cyclotron frequency. The cyclotron, a device for accelerating charged particles, employs the principles here discussed to contain the particles.

The geomagnetic field at the earth's surface has an average strength of 4×10^{-5} Wb/m^2, or 0.4 G. The electronic charge and mass are, respectively, 1.6×10^{-19} C and 9.1×10^{-31} kg. The cyclotron frequency is, therefore,

$$\omega = \frac{(1.6 \times 10^{-19})(4 \times 10^{-5})}{9.1 \times 10^{-31}} = 7.03 \times 10^{6} \text{ rad/sec.}$$

• PROBLEM 9-27

A pendulum consists of a 3-ft string, fixed at one end and carrying a 4-lb body at the other. As the pendulum swings back and forth in a vertical plane, the velocity is found to be 10 ft/sec when the angular deflection is 45°. At this instant, what is the tension in the string?

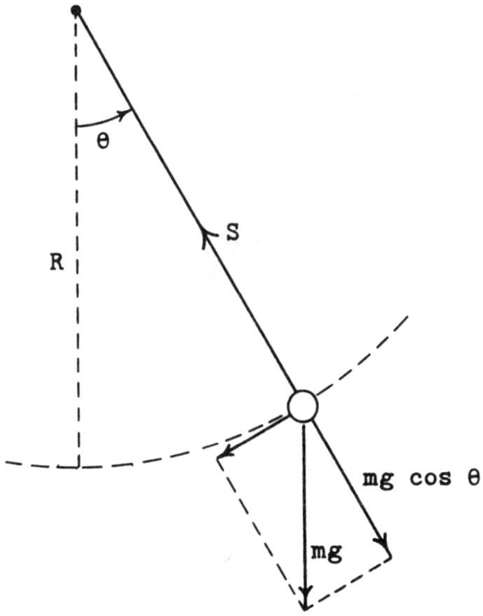

Solution: There are two forces acting upon the body at the end of the string, its weight mg and the tension S in the string. The acceleration toward the center of the circular path of the body is v^2/R. The resulting force component in this direction is $S - mg\cos\theta$, as indicated in the figure. Consequently, we have

$$S - mg\cos\theta = \frac{mv^2}{R}$$

or

$$S = \frac{mv^2}{R} + mg\cos\theta .$$

Using the numerical values $v = 10$ ft/sec and $m = 4/32$ slug, $mg = 4$ lb, $R = 3$ ft, and $\theta = 45°$, we obtain

$$S = \frac{4}{32} \cdot \frac{10^2}{3} + 4\frac{\sqrt{2}}{2} \simeq 7.01 \text{ lb.}$$

• PROBLEM 9-28

A pendulum consists of a 3 ft string fixed at one end and carrying a 4 lb body at the other end. Let the pendulum move conically, so that the body at the end of the string moves in a horizontal rather than a vertical circle, and the string generates the surface of a cone, as shown. (Such motion is employed in centrifugal regulators.) What speed is required to make the angle between the string and the vertical 45°, and what is the corresponding tension in the string?

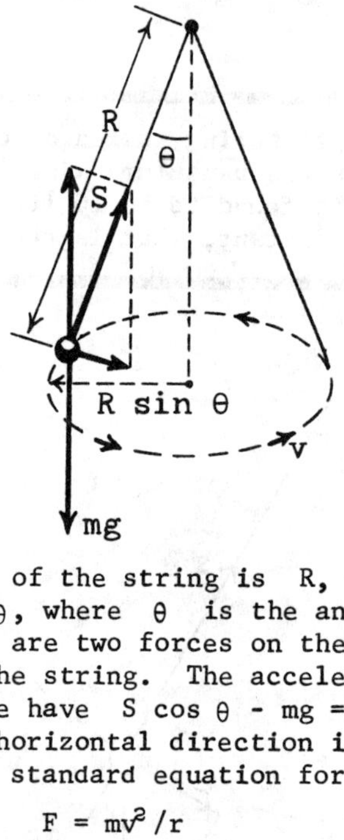

Solution: If the length of the string is R, the radius of the horizontal circle is $R\sin\theta$, where θ is the angle between the string and the vertical. There are two forces on the body, its weight mg and the tension $\vec{S}$ in the string. The acceleration in the vertical direction is zero, and we have $S\cos\theta - mg = 0$, or $S = mg/\cos\theta$. The acceleration in the horizontal direction is toward the center of the circle and using the standard equation for circular motion,

$$F = mv^2/r$$

the following is produced:

$$S\sin\theta = mv^2/R\sin\theta$$

or

$$v^2 = \frac{RS\sin^2\theta}{m} = Rg\tan\theta\sin\theta .$$

With the numerical values $R = 3$ ft, $g \simeq 32$, and $\theta = 45°$, we obtain

$$v \simeq \sqrt{3\cdot 32\cdot 1\cdot 0.7} \simeq 8.2 \text{ ft/sec.}$$

The corresponding tension in the string is

$$S = \frac{mg}{\cos\theta} \simeq 5.7 \text{ lb.}$$

• PROBLEM 9-29

A string connecting a 5-lb ball and a 10-lb block is passed over an ideal pulley of negligible radius, as shown in Figure 1. The pulley is then rotated about the axis a - a, which is assumed to pass through both the center of the pulley and the center of gravity of the 10-lb weight. If the amount of string on the ball side of the pulley is 3 ft. what must be the constant speed of the ball to keep the 10-lb block from falling?

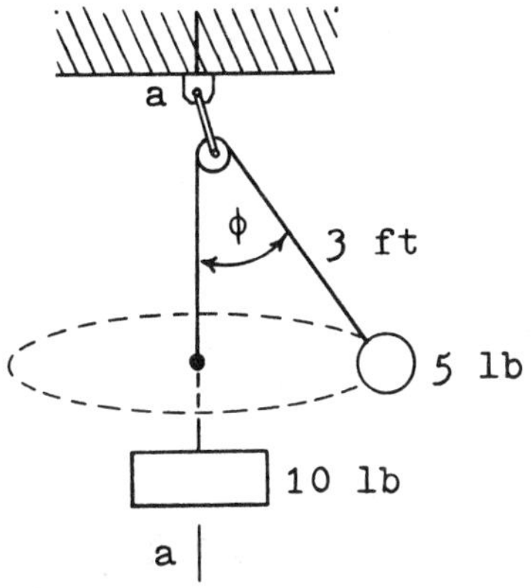

fig 1

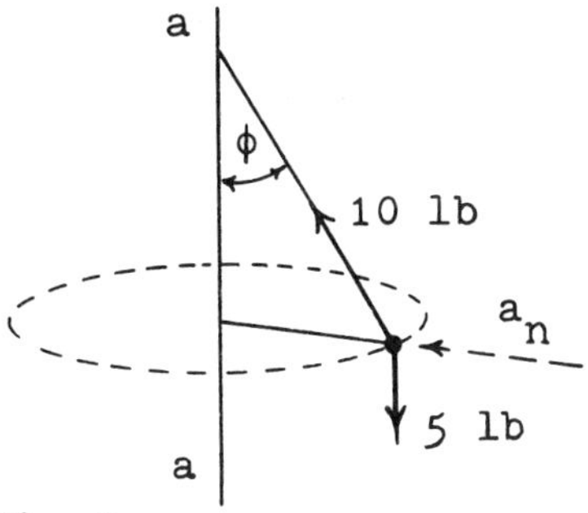

fig 2

Solution: It is clear that if the block must not accelerate, the tension in the string must be equal to the weight of the ball, that is, T = 10 lb.

The free-body diagram of the ball is shown in Figure 2. Since the speed of the ball is to be constant, the angle φ must be constant and the acceleration of the ball must be directed radially inward.

The angle φ is fixed by the fact that the ball experiences no net vertical force so

$$(10 \text{ lb}) \cos \varphi = 5(\text{lb})$$

$$\varphi = \cos^{-1}(\tfrac{1}{2}) = 60^\circ .$$

The value of a_n is found using Newton's Second Law,

$$(10 \text{ lb}) \sin \theta = ma_n ,$$

where

$$m = W/g = (5/32.2) \text{ slugs},$$

so

$$a_n = \left(\frac{32.2}{5}\right)(10) \sin 60^\circ$$

$$a_n = 55.8 \text{ ft/sec}^2 .$$

With the value of a_n known, the velocity of the ball may be determined. Since a_n is the centripetal acceleration,

$$a_n = v^2/r .$$

The radius of the ball's path is

$$r = (3 \text{ ft})(\sin 60^\circ) = 2.60 \text{ ft}.$$

The velocity, then, is

$$v = \sqrt{(55.8 \text{ ft/sec}^2)(2.60 \text{ ft})}$$

$$= 12.0 \text{ ft/sec}.$$

A sphere of weight w = 4 lbs attached to two wires BC and AC as shown (Fig 1) revolves in a horizontal circle at a constant speed. The distance AB is 5 ft.
Determine: a) the speed v for which the tension in each wire is equal b) the value of that tension.

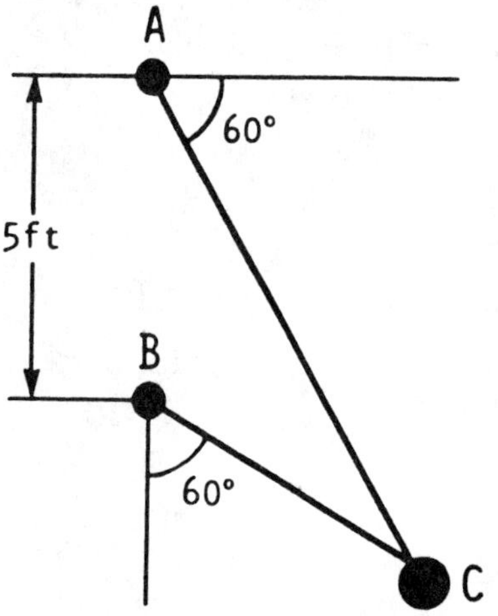

Fig. 1

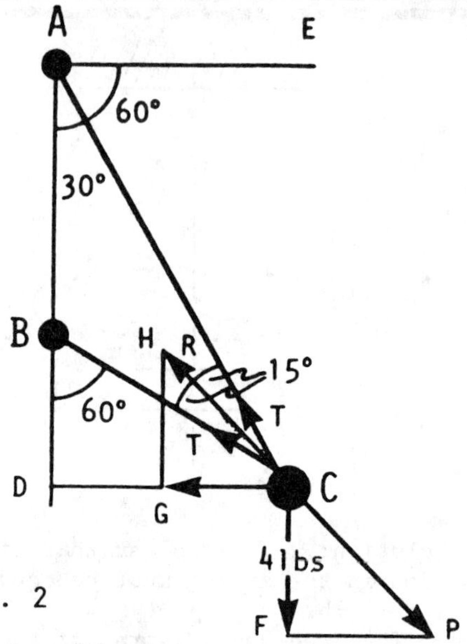

Fig. 2

Solution: Let us first consider the geometry of the problem. $< EAC$ is given to be 60°, so $< BAC$ must be 30°. If $< DBC$ is 60°, then $< BCD = 30^\circ$. Also, if $< DAC = 30^\circ$, then $< ACD = 60^\circ$, or $< ACB = 30^\circ$. Since the tension in both the wires is equal, say T, then the resultant of these forces, R, must bisect the angle between them. This is because the resultant is located symmetrically with respect to the two equal forces. That is, the $< HCB$ and $< HCA$ are both equal to 15°. Thus R is 45° above horizontal. The force R can be resolved into two components, CG and GH. In order to maintain equilibrium, the upward force GH must be equal to the downward force (weight of the sphere). Thus GH is 4 lbs. Since CGH is a 45°-45°-90° triangle, CG is also 4 lbs. This force CG is the centripetal force, F_C, which keeps the sphere moving in a horizontal circle. Using the formula,

$$F_C = \frac{mv^2}{r} \tag{1}$$

$F_C = 4$ lb, m = 4/32 slugs, v and r are unknown. The radius of the circle in which the sphere is moving is CD, which can also be evaluated geometrically. In $\triangle ABC$, $< BAC = < ACB = 30^\circ$, the side AB = side BC = 5 ft. In $\triangle BCD$, if BC = 5 ft, then $CD = 5 \cos 30^\circ = 5 \times .866 = 4.33$ ft. After rewriting equation (1) as

$$v^2 = \frac{F_C \cdot r}{m},$$

we have,

$$v^2 = \frac{4 \times 4.33}{32} = \frac{4 \times 4.33 \times 32}{4} = 138.56$$

Therefore $v = \sqrt{138.56} = 11.77$ ft/sec. **Answer**

Upon re-examining figure (2), notice that

$$R^2 = \overline{CH}^2 = \overline{CG}^2 + \overline{GH}^2 = 4^2 + 4^2 = 16 + 16 = 32.$$

$$R = \sqrt{32} = 5.66 \text{ lb.}$$

However, this resultant R is the sum of the components of T, and T along the direction CH. These two components are T cos 15° each. Thus

$$R = T\cos 15^\circ + T\cos 15^\circ = 2T\cos 15^\circ$$

or

$$5.66 = 2T \times .9659$$

or, rearranging,

$$T = \frac{5.66}{2 \times .9659} = 2.93 \text{ lbs.}$$ Answer.

• **PROBLEM 9-31**

Prove that the radial acceleration experienced by a body moving in circular motion is equal to v^2/r, using calculus.

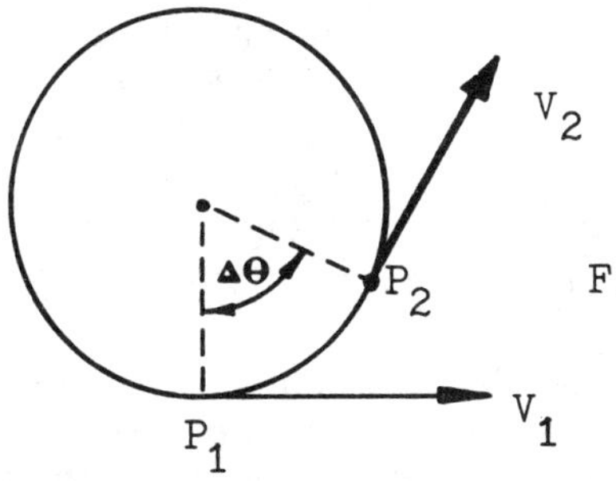

Figure 1

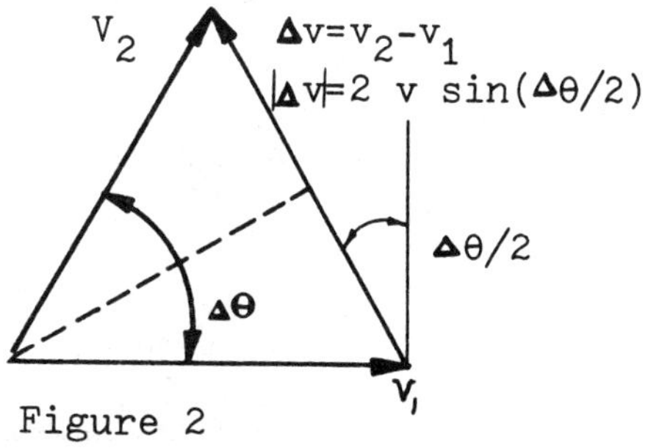

Figure 2

Solution: It is known from the study of calculus that the average acceleration $\vec{a} = \Delta v/\Delta t$ becomes the instantaneous acceleration in the limit that $\Delta t \to 0$. That is

$$\vec{a} = \lim_{\Delta t \to 0} (\Delta\vec{v}/\Delta t),$$

where

$$\frac{\Delta\vec{v}}{\Delta t} = (\vec{v}_2 - \vec{v}_1)/(t_2 - t_1)$$

Consider a body moving with constant speed in a circle of radius R. The average velocity between points P_1 and P_2 in figure 1 can be determined. The points are separated by an angle $\Delta\theta$ and the instantaneous velocities $\vec{v}_2$ and $\vec{v}_1$ are shown.

The change in velocity between these points $\Delta\vec{v} = \vec{v}_2 - \vec{v}_1$ is shown by the triangle method in figure 2. Since $\vec{v}_1$ and $\vec{v}_2$ are perpendicular to the radii which meet with angle $\Delta\theta$, $\vec{v}_1$ and $\vec{v}_2$ meet with same angle $\Delta\theta$ as shown. It is clear from the figure that $\Delta\vec{v}$ has magnitude

$$|\Delta v| = 2v\sin(\Delta\theta/2)$$

and points in a direction which makes an angle of $\Delta\theta/2$ with the vertical. Where $|\vec{v}_1| = |\vec{v}_2| = v$ the angular velocity of the body is

$$\omega = v/R$$

and the time Δt that it takes the body to move from P_1 to P_2 is

$$\Delta t = \Delta\theta/\omega = R\Delta\theta/v. \tag{1}$$

So the magnitude of the average acceleration is

$$\bar{a} = \frac{2v \sin(\Delta\theta/2)}{R\Delta\theta/v} = \frac{v^2}{R} \frac{\sin(\Delta\theta/2)}{\Delta\theta/2} .$$

The value of the limit of $\sin x/x$ as $x \to 0$ is exactly one, so

$$\bar{a} = \lim_{\Delta\theta \to 0} = v^2/R .$$

It is clear from equation (1) as $\Delta t \to 0$, that $\Delta\theta \to 0$ so it can be determined that the magnitude of the instantaneous acceleration is v^2/R. Since the average acceleration yields an angle of $\Delta\theta/2$ with the perpendicular to $\vec{v}_1$, as Δt, $\Delta\theta \to 0$ the direction of the instantaneous acceleration is perpendicular to $\vec{v}_1$, hence, directed towards the center of the circle.

CENTRAL FORCES AND THE CONSERVATION OF ANGULAR MOMENTUM

• PROBLEM 9-32

A sphere of mass m = 3kg is attached to an elastic cord, the spring constant of the cord k = 120 n/m. At the position P (see Fig 2) the velocity of the sphere $\vec{V}_a$ is perpendicular to OP, V_a = 5 m/s and its distance from the original position O (when the cord is unstretched) is a = 0.8m.
Determine: a) the maximum distance from the origin O attained by the sphere b) the corresponding speed of the sphere.

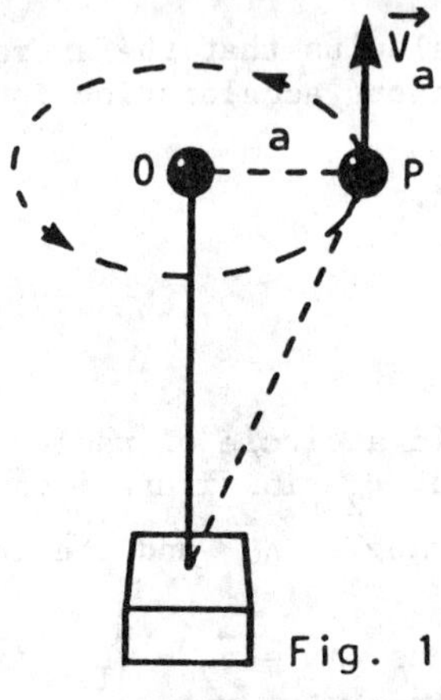

Fig. 1

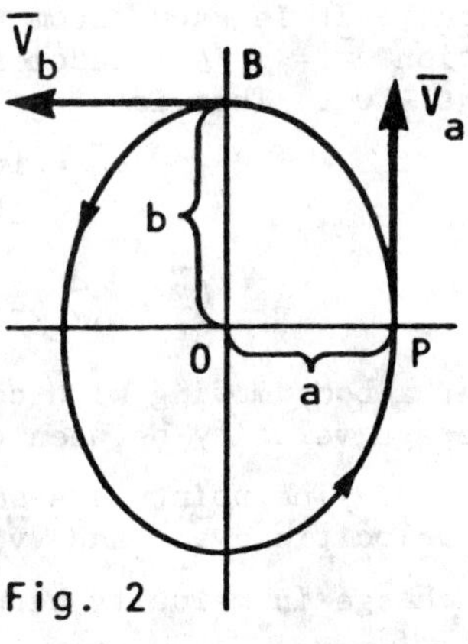

Fig. 2

Solution: Here is a system composed of a ball that is at the origin O when the elastic cord is unstretched and in position P when the cord is stretched. Since the system is conservative, one can apply to this system the conservation laws of energy and angular momentum. In this way, one may have two relations for the two quantities that must be determined: b and v_b (fig. 2).

When seen from the top, the sphere is initially at P with a = 0.8m and moving with a speed v_a = 5m/sec. It then travels to the point B where its speed becomes v_b and its distance OB = b is an extremum which will turn out to be a maximum too. Since the energy is conserved in this process, the sum of potential energy and kinetic energy at P and B can be written as

$$PE_a + KE_a = PE_b + KE_b$$

or $$\tfrac{1}{2} ka^2 + \tfrac{1}{2} mv_a^2 = \tfrac{1}{2} kb^2 + \tfrac{1}{2} mv_b^2 \; . \qquad (1)$$

Here we use the fact that the potential energy of a stretched string is $\frac{1}{2} kx^2$. Dropping the factor $\frac{1}{2}$ throughout and substituting the numerical values $k = 120$ N/m, $a = .8$m, $v_a = 5$m/sec, $m = 3$kg, yields

$$120(.8)^2 + 3(5)^2 = 120^2 + 3v_b^2$$

$$120(.64) + 3(25) = 120b^2 + 3v_b^2$$

$$76.8 + 75 = 120b^2 + 3v_b^2$$

$$120b^2 + 3v_b^2 = 151.8$$

or dividing by 3, $40^2 + v_b^2 = 50.6$ (2)

We now have an equation for b and v_b. Next the conservation law of angular momentum may be used. The angular momentum of a particle of mass m, tangential velocity v, and radial distance r is given by $L = mvr$. Since the force is always radial, the torque on the particle is zero, so the angular momentum remains constant. The angular momentum at P equals that at B. That is,

$$mv_a a = mv_b b \; .$$

Dropping m, and letting $v_a = 5$m/sec, $a = .8$m, yields $5(.8) = v_b b$

or $$v_b = \frac{4}{b} \quad \text{or} \quad v_b^2 = \frac{16}{b^2} \; .$$

Substituting this value of v_b^2 in equation (2) we find,

$$40b^2 + \frac{16}{b^2} = 50.6$$

or multiplying by b^2,

$$40b^4 + 16 = 50.6b^2$$

or $$40b^4 - 50.6b^2 + 16 = 0 \; .$$

This is a quadratic equation in b^2. Solving it by using the standard formula, yields

$$b^2 = \frac{50.6 \pm \sqrt{(50.6)^2 - 4(40)16}}{2 \times 40}$$

$$= \frac{50.6 \pm \sqrt{2560.36-2560}}{80} = \frac{50.6 \pm \sqrt{0.36}}{80} = \frac{50.6 \pm 0.6}{80}$$

$$= \frac{51.2}{80} \text{ or } \frac{50}{80} = 0.64 \text{ or } 0.625 \; .$$

Thus $b = \pm\sqrt{0.64}$ or $\pm\sqrt{0.625} = \pm 0.8$ or ± 0.79.

We found four values for b: ± 0.79 and ± 0.8 because we had a biquadratic equation in b. These four values correspond to the four extremum positions that the ball can have: P, B, R, A (fig. 2). Obviously the answer, 0.8m, represents the maximum distance of the sphere from the origin.

To find the velocity v_b, recall that

$$v_b = \frac{4}{b} = \frac{4}{0.8} = 5\text{m/sec}$$

A small ball rolls along a horizontal circle inside a bowl at a speed V_o (Fig 1). The inside surface of the bowl is obtained by rotating the curve OA about the y axis. Assuming that the speed of the ball V_o is proportional to the distance x from the y axis to the ball determine the curve OA (Fig 2) of the inside surface of the bowl.

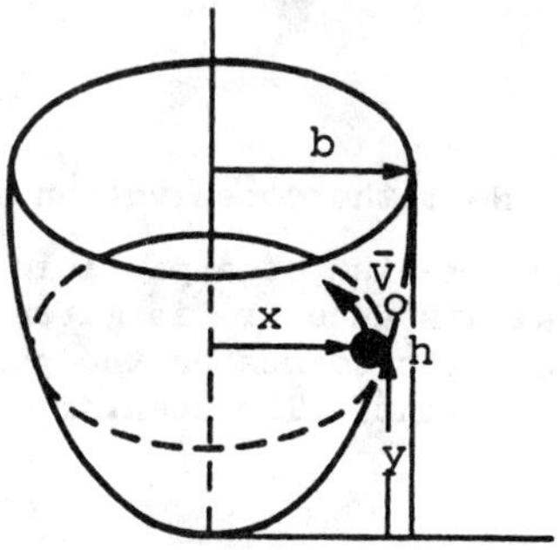

Fig. 1

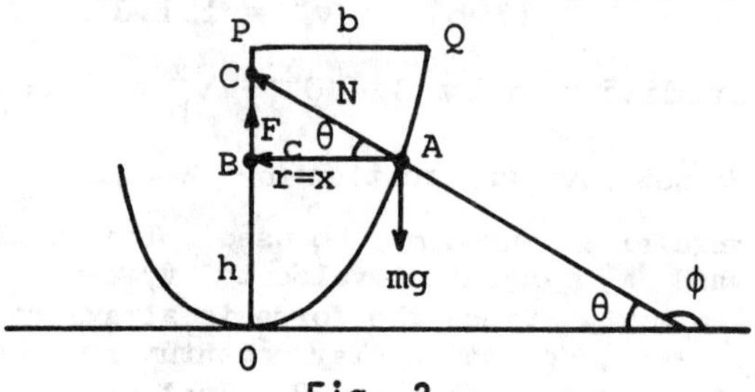

Fig. 2

Solution: Since the ball rolls along a horizontal circle of radius $r = x$ the force applied on it by the bowl must be perpendicular to the surface. This force should have no component along the surface of the bowl, otherwise the ball will ride up and down. The normal reaction of the surface must counterbalance the force of gravity, mg, on the ball and also provide the necessary centripetal force, $F_C = mv_0^2/r$, to keep it rolling along the circle of radius r.

Resolving the force into two components along AB and BC, note that F_{BC} must be equal and opposite to mg. Also, F_{AB} must be the centripetal force F_C. From ΔABC,

$$\tan\theta = \frac{BC}{AB} = \frac{mg}{F_C} = \frac{mg}{mv_0^2/r} = \frac{gr}{v_0^2} .$$

Since $r = x$, $\tan\theta = \frac{gx}{v_0^2}$. (1)

It is also required that speed v_0 must be proportional to the distance x . Or, $v_0 \propto x$. Or, $v_0^2 = kx^2$ where k is a constant of proportionality. Substituting this in equation (1),

$$\tan\theta = \frac{gx}{kx^2} = \frac{g}{kx} .$$

From differential calculus, H is established that the tangent of the angle between the normal to a curve and the positive x-axis is $-dx/dy$. That is $\tan\varphi = -dx/dy$. Since θ and φ are supplementary angles, $\tan\theta = -\tan\varphi = dx/dy$. Thus $\tan\theta = dx/dy = g/kx$, or $kx\,dx = g\,dy$. Integrating both sides,

$$\int kx\,dx = \int g\,dy$$

or

$$\frac{kx^2}{2} = gy + c \qquad (2)$$

where c is a constant of integration. Now the curve passes through the origin, x = 0 when y = 0 . Therefore, 0 = 0 + c or c = 0 . Furthermore, the curve also goes through the point x = b, y = h. With these values in equation (2)

$$\frac{kb^2}{2} = gh \;, \quad \text{or} \quad k = \frac{2gh}{b^2} \;.$$

If we replace k by this value in equation (2), we get

$$\frac{2gh}{b^2} \cdot \frac{x^2}{2} = gy \;.$$

Cancelling 2 and g , $y = hx^2/b^2$ is produced which is the desired equation of the curve. We also note that this is the equation of a parabola with its vertex at the origin and its central axis along the y-axis.

A liquid rotating uniformly has its velocity proportional to its distance from the axis of rotation. It thus assumes a paraboloidal surface.

• PROBLEM 9-34

A 2800-lb automobile moves along a highway down into a valley. The highway's path is a parabola, $y = x^2/1000$. What is the normal force on the car as it passes through the nadir of the curve, coordinate (0,0), at 60 mph?

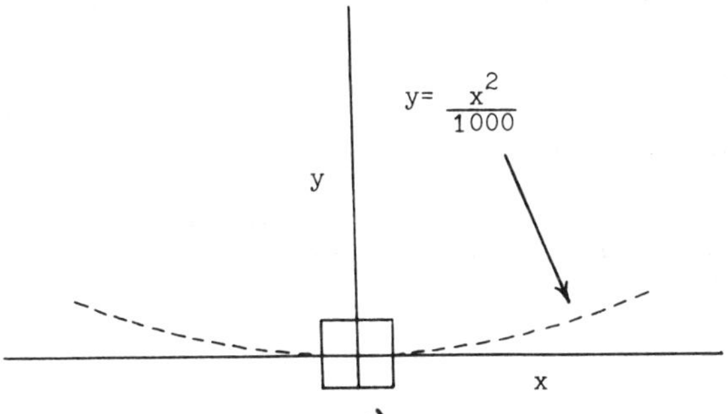

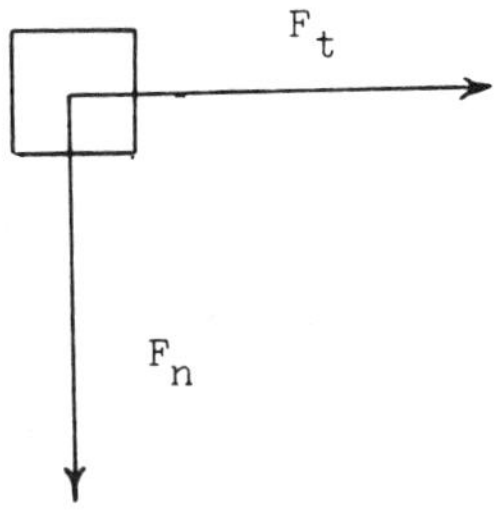

Solution: In a non-inertial situation it is not neccessarily true that the normal force is equal to the weight, as this problem illustrates.

It should be clear that the car is being guided along a curved path by the road, and that this means the road causes a normal acceleration of the car which is directly caused by variations in the normal force because of the curved slope of the road. At the moment in question, that acceleration will be vertical and that vertical acceleration is provided by the normal force upwards exceeding the weight downwards.

By Newton's law, the normal acceleration will be

$$a_n = (N - W)/m \;. \qquad (1)$$

We approximate the motion to be circular at the moment considered so that we can set the normal acceleration equal to v^2/r , so

$$v^2/r = (N - W)/m \;. \qquad (2)$$

Here the mathematical statement of the opening comments indicates that if the road has a curved slope $(r \neq \infty)$, the normal force and the weight are not equal.

The radius of curvature is given by

$$r = \frac{(1 + y'^2)^{3/2}}{y''} . \qquad (3)$$

For $y = x^2/1000$, $y' = x/500$, $y'' = 1/500$ so

$$r = \frac{[1 + (x/500)^2]^{3/2}}{1/500} ,$$

and at (0,0)

$$r = 500 \text{ ft}. \qquad (4)$$

Placing all the known information into equation (2), yields the normal force, $v = 60$ mph $= 88$ ft/sec and $m = w/g = 2800/32.2$ slugs ,

$$(88)^2/500 = \frac{(N - 2800)}{2800/32.2}$$

$$N = 1347 + 2800$$
$$= 4147 \text{ lb}.$$

• PROBLEM 9-35

A particle of mass m moves according to

$$x = x_0 + at^2$$
$$y = bt^3$$
$$z = ct .$$

Find the angular momentum L at any time t. Find the force $\vec{F}$, and from it the torque $\vec{N}$, acting on the particle. Verify that these quantities satisfy

$$\frac{d\vec{L}}{dt} = \vec{r} \times \vec{F} = \vec{N}$$

Solution: Required are the definitions, in vector component terms, of angular momentum, force, and torque. The presumption is that the center of rotation is the origin. By definition

$$\vec{L} = \vec{r} \times m \frac{d\vec{r}}{dt} \qquad (1)$$

where $\vec{r} = x\hat{i} + y\hat{j} + z\hat{k}$ is the radius vector. (2)

Substituting the given values of x, y and z:

$$m\frac{d\vec{r}}{dt} = m\frac{d}{dt}(x_0 + at^2)\hat{i} + m\frac{d}{dt}(bt^3)\hat{j} + m\frac{d}{dt}(ct)\hat{k}$$
$$= 2mat\hat{i} + 3mbt^2\hat{j} + mc\hat{k} \qquad (3)$$

Substituting (2) and (3) and the given values into (1)

$$\vec{L} = \begin{vmatrix} \hat{i} & \hat{j} & \hat{k} \\ x_0+at^2 & bt^3 & ct \\ 2mat & 3mbt^2 & mc \end{vmatrix}$$

Expanding yields:

$$\vec{L} = [mbct^3 - 3mbct^3]\hat{i}$$
$$-[(x_0+at^2)mc - 2mact^2]\hat{j}$$
$$+[(x_0+at^2)3mbt^2 - 2mabt^4]\hat{k}$$
$$= (-2mbct^3)\hat{i} + (mact^2 - mx_0c)\hat{j} \tag{4}$$
$$+ (mabt^4 + 3mx_0bt^2)\hat{k} .$$

From Newton's Second Law

$$\vec{F} = m\vec{a} = m\frac{d^2}{dt^2}(x\hat{i} + y\hat{j} + z\hat{k}) .$$

Substituting the given values of x, y and z we get for each of the components

$$F_x = m\frac{d^2}{dt^2}(x_0+at^2) = 2ma$$
$$F_y = m\frac{d^2}{dt^2}(bt^3) = 6mbt \tag{5}$$
$$F_z = m\frac{d^2}{dt^2}(ct) = 0 .$$

Now, to compute the torque which by definition is $\vec{r} \times \vec{F}$:

$$\vec{r} \times \vec{F} = \begin{vmatrix} \hat{i} & \hat{j} & \hat{k} \\ x_0 + at^2 & bt^3 & ct \\ 2ma & 6mbt & 0 \end{vmatrix} \tag{6}$$

where the components of $\vec{F}$ in the last row of the determinant come from (5).

Again, expanding the determinant (6) yields:

$$\vec{r} \times \vec{F} = (-6mbt)(ct)\hat{i} - (-2ma)(ct)\hat{j} + [(x_0+at^2)(6mbt) - (2ma)(bt^3)]\hat{k}$$
$$= -6mbct^2\hat{i} + (2mact)\hat{j} + (4mabt^3 + 6mx_0bt)\hat{k} . \tag{7}$$

Now we wish to show (7) is the same as $\frac{d}{dt}\vec{L}$. From (4)

$$\frac{d}{dt}\vec{L} = \frac{d}{dt}(-2mbct^3)\hat{i} + \frac{d}{dt}(mact^2 - mx_0c)\hat{j}$$
$$+ \frac{d}{dt}(mabt^4 + 3mx_0bt^2)\hat{j} .$$

Computing the derivatives:

$$\frac{d\vec{L}}{dt} = (-6mbct^2)\hat{i} + (2mact)\hat{j} + (4mabt^3 + 6mx_0bt)\hat{k}$$

which agrees with (7).

• PROBLEM 9-36

A particle of mass m is repelled from the origin by a force $f = k/x^3$ where x is the distance from the origin. Solve the equation of motion if the particle is initially at rest at a distance x_0 from the origin.

Solution: We must use $\vec{F} = m\vec{a}$ and solve the resulting differential equation, using the initial conditions:

$$x(t = 0) = x_0 \ ; \ \dot{x}(t = 0) = 0 \ . \tag{1}$$

First $m\ddot{x} = \frac{k}{x^3}$ from Newton's Second Law, or

$$\ddot{x} = \frac{k}{mx^3} \ .$$

This is a second order, nonlinear, homogeneous differential equation. Note that there are only terms in x and its time derivatives; there are no terms in t alone.

A substitution which will reduce this equation to a first order equation is

$$\dot{x} = p \ . \tag{2}$$

Then

$$\ddot{x} = \frac{d^2x}{dt^2} = \frac{dp}{dt} = p\,\frac{dp}{dx} \ . \tag{3}$$

The last expression comes from the realization that

$$p\,\frac{dp}{dx} = \frac{dx}{dt}\,\frac{d}{dx}\left(\frac{dx}{dt}\right) = \frac{d^2x}{dt^2} \ .$$

Now, the differential equation in reduced form is:

$$p\,\frac{dp}{dx} = \frac{k}{mx^3} \ . \tag{4}$$

This equation is separable. Rearranging terms yields:

$$p\,dp = \frac{kdx}{mx^3} \ .$$

Integrating both sides yields:

$$\frac{p^2}{2} = -\frac{k}{2mx^2} + c_1 \ , \quad c_1 = \text{constant}$$

or

$$p^2 = -\frac{k}{mx^2} + c_2 \ , \quad c_2 = 2c_1 \ .$$

Substituting from (2)

$$\dot{x}^2 = -\frac{k}{mx^2} + c_2 \ . \tag{5}$$

Evaluating c_2 by using the boundary conditions, eq.(1)

$$0 = -\frac{k}{mx_0^2} + c_2 \ ; \ c_2 = \frac{k}{mx_0^2} \ .$$

Substituting into (5) and solving for $\dot{x}$:

$$\dot{x} = \frac{dx}{dt} = \pm\sqrt{\frac{k}{m}\left(\frac{1}{x_0^2} - \frac{1}{x^2}\right)}$$

$$= \pm K\left(\frac{1}{x_0^2} - \frac{1}{x^2}\right)^{1/2} \quad \text{where} \quad K = \sqrt{\frac{k}{m}} \tag{6}$$

$$\frac{dx}{dt} = \pm K\,\frac{(x^2 - x_0^2)^{1/2}}{x_0 x} \ . \tag{7}$$

Separating variables in (7)

$$dt = \pm\frac{dx}{\frac{K}{x_0 x}(x^2 - x_0^2)^{1/2}}$$

$$= \pm \frac{x_0}{K} \frac{x\,dx}{(x^2 - x_0^2)^{1/2}} \; . \qquad (8)$$

Integrating both sides yields

$$t = \pm \frac{x_0}{K} (x^2 - x_0^2)^{1/2} + c_3 \; . \qquad (9)$$

Evaluating the integration constant c_3 through substitution of (1) yields $c_3 = 0$. Solving (9) for x and substituting (6) yields

$$x = \pm \left(x_0^2 + \frac{kt^2}{mx_0^2}\right)^{1/2} .$$

Since $x = x_0$ when $t = 0$ we choose the positive square root for our solution.

• PROBLEM 9-37

Ring A of mass 8lbs is attached to a spring of constant 40 lb/ft (Fig 1). When the system is at rest the length of the spring is 18 in and the distance between the y axis and the ring r = 12 in. The system is set into motion with V_θ = 15 ft/sec and V_r = 0.

Determine: a) the maximum distance between the origin and the ring, b) the corresponding velocity.

Fig. 1

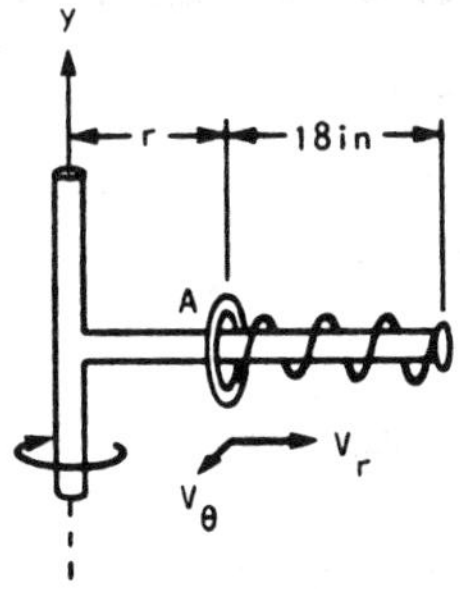

Solution: When the ring is set in motion it is given a certain kinetic energy. This energy is conserved, even though part of it becomes the potential energy of the compressed spring when the ring reaches its maximum distance from the origin.

Considering the total energy it can be established that:

$$PE_1 + KE_1 = PE_2 + KE_2$$

or $0 + \frac{1}{2} mv_1^2 = \frac{1}{2}kx^2 + \frac{1}{2}mv_2^2$. (1)

Here x is the compression in the spring. If r is the maximum distance of the ring from the origin, then the length of the spring is (2.5-r)ft and hence, its change in length, $x = 1.5-(2.5-r) = r - 1$ ft. Here all lengths and distances have been expressed in feet. The velocity v_1 = 15 ft/sec, m = 8/32, and k = 40 lb/ft, so equation (1) can be expressed (dropping the factor ½ throughout) as,

$$\frac{8}{32}(15)^2 = 40(r-1) + \frac{8}{32} v_2^2 \; .$$

Multiplying by 32/8,

$$225 = 160(r-1)^2 + v_2^2 \; . \qquad (2)$$

Since we cannot find r and v_2 from (2), we develop another equation by using the law of conservation of angular momentum. The radial velocity $v_r = 0$ in both the initial and final positions, the conservation of angular momentum can be written as

$$mr_1v_1 = mr_2v_2 \ .$$

Here $r_1 = 1$ ft., $v_1 = 15$ ft./sec., and neglecting m's,

$$1(15) = r_2v_2 \ , \quad \text{or} \quad v_2 = \frac{15}{r_2} \tag{3}$$

Here $r_2 = r$, the maximum distance of the collar. The relations (2) and (3) together yield an equation for the extremal values of r.

$$225 = 160(r^2 - 2r + 1) + (15)^2/r^2$$

$$225\ r^2 = 160\ r^2(r^2 - 2r + 1) + (15)^2$$

$$225\ r^2 = 160(r^4 - 2r^3 + r^2) + 225$$

$$225\ r^2 - 160\ r^4 + 320\ r^3 - 160\ r^2 - 225 = 0$$

or

$$-160\ r^4 + 320\ r^3 + 65\ r^2 - 225 = 0$$

Dividing by -32, $5r^4 - 10r^3 - 2r^2 + 7 = 0$

To solve this fourth order equation, recall that $r = 1$ ft. is one extremal position, hence, it is a root of the equation. Thus $(r-1)$ is a factor of the left hand terms. Dividing, by $(r-1)$ yields,

$$
\begin{array}{rl}
r-1)\ 5r^4 - 10r^3 - 2r^2 + 7 & (5r^3 - 5r^2 - 7r - 7 \\
\underline{5r^4 - 5r^3} & \\
- 5r^3 - 2r^2 & \\
\underline{- 5r^3 + 5r^2} & \\
- 7r^2 \quad + 7 & \\
\underline{- 7r^2 + 7r} & \\
-7r + 7 & \\
\underline{-7r + 7} & \\
0 &
\end{array}
$$

That is, $(r-1)(5r^3 - 5r^2 - 7r - 7) = 0$. Dropping the factor $(r-1)$ there remains the cubic equation

$$5r^3 - 5r^2 - 7r - 7 = 0.$$

Unfortunately it cannot be factored, and one should try to solve it by successive approximations. By physical constraints, it is already established that r must be between 1 and 2.5. Also the positive term $5r^3$ must cancel all the remaining negative terms. Assuming $r = 2$ to a reasonable answer, the numerical value of the expression is

$$5(2)^3 - 5(2)^2 - 7(2) - 7 = 5(8) - 5(4) - 7(2) - 7 = -1$$

The negative result suggests that r should be increased slightly, perhaps to $r = 2.2$. Again,

$$5(2.2)^3 - 5(2.2)^2 - 7(2.2) - 7 = 53.24 - 24.2 - 15.4 - 7 = 6.64$$

The positive result suggests that a lower value of r should be used. 2.1 should be tried as the next approximation. Again,

$$5(2.1)^3 - 5(2.1)^2 - 7(2.1) - 7 = 46.305 - 22.050 - 14.7 - 7 = 2.55$$

So we should try a smaller value, say 2.05. Continuing in this fashion leads to r = 2.029. Thus within a reasonable degree of accuracy, r = 2.029 = 24.35 inches is the maximum distance.

To find the corresponding velocity, go back to equation (3), $v_2 = \frac{15}{r_2}$.

With $r_2 = 2.029$, we find $v_2 = \frac{15}{2.029} = 7.39$ ft/sec

FALLING BODIES AND DAMPING PROBLEMS DEPENDING ON THE VELOCITY

• PROBLEM 9-38

A ball bearing is released from rest and drops through a viscous medium. The retarding force acting on the ball bearing has magnitude kv, where k is a constant depending on the radius of the ball and the viscosity of the medium, and v is the bearing's velocity. Find the terminal velocity acquired by the ball bearing and the time it takes to reach a speed of half the terminal velocity.

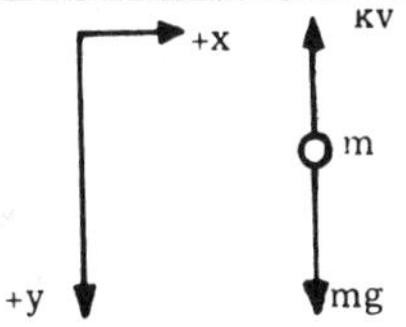

Solution: As the ball falls through the medium, it is accelerated by gravity and the viscous force. To find the acceleration of the bearing, we use Newton's Second Law to relate the net force on the ball to its acceleration. Taking the positive direction downward (see figure).

$$mg - kv = ma,$$

where a is the acceleration produced at any time. The initial value of a is g, since at the moment of release v = 0. As the value of v increases, the acceleration decreases until, when $v = v_0$, the terminal velocity, a = 0. Thus $mg - kv_0 = 0$. Therefore

$$v_0 = (m/k)g.$$

In order to find out at what time $v = v_0$, we must calculate v as a function of t (or vice versa).

At any time t, it will be found that $mg - kv = ma = m(dv/dt)$. or

$$(mg - kv)dt = mdv$$

$$dt = \frac{mdv}{mg-kv} = \frac{mdv}{m(g-(k/m)v)}$$

$$dt = \frac{dv}{g-(k/m)v} \; ; \quad \text{and} \quad \int_0^t dt = \int_0^v \frac{dv}{g-(k/m)v} \; .$$

where, in the integration limits, v = 0 at t = 0 and v = v at t = t. Hence

$$t = \int_0^v \frac{dv}{g-(k/m)v}$$

Letting $u = g - (k/m)v$

$$du = -k/m\ dv$$

To find the new integration limits, we realize that when $v = 0$, $u = g$ and when $v = v$, $u = g-(k/m)v$, whence

$$t = \int_g^{g-(k/m)v} \frac{-m/k\ du}{u} = \frac{-m}{k}\ \ell n\left(|u|\right)\ \Big|_g^{g-(k/m)v}$$

$$t = \frac{-m}{k}\{\ell n\left(|g-(k/m)v|\right) - \ell n(g)\}$$

$$t = -\frac{m}{k}\ \ell n\left(\left|\frac{g-(k/m)v}{g}\right|\right)$$

$$t = -m/k\ \ell n|1-(k/mg)v| \qquad (1)$$

The time to acquire half the terminal velocity, T, is thus found by inserting $v = v_0/2$ in (1)

$$T = -\frac{m}{k}\ \ell n\left|1 - \frac{k}{mg}\cdot\frac{mg}{2k}\right| = -\frac{m}{k}\ \ell n\left|\frac{1}{2}\right| = +\frac{m}{k}\ \ell n|2| = 0.69\ \frac{m}{k}\ .$$

• PROBLEM 9-39

A body of mass m is projected upwards in the air with velocity v_0. Air resistance is proportional to the square of the velocity. Find the motion of the body as it rises.

Solution: Let the y axis be positive in the upward direction. The forces acting on the body, gravity and air resistance, are both in the negative y direction. Gravitational attraction is of magnitude mg, and air resistance of magnitude c^2v^2, where c is a constant. Newton's second law of motion tells us

$$m\frac{dv}{dt} = -mg - c^2v^2\ . \qquad (1)$$

Separating the variables leads to

$$\frac{m\ dv}{mg + c^2v^2} = -dt\ . \qquad (2)$$

Before the constants begin getting in the way, let us write (2) as

$$\frac{dv}{(\mu^2 + v^2)} = -k^2\ dt \qquad (3)$$

where $k^2 = c^2/m$, and $\mu^2 = g/k^2$. This integrates to

$$\frac{1}{\mu}\tan^{-1}\frac{v}{\mu} = -k^2t + C_1\ . \qquad (4)$$

The initial condition $v(0) = v_0$, substituted in (4) changes the expression to

$$\frac{1}{\mu}\left(\tan^{-1}\frac{v}{\mu} - \tan^{-1}\frac{v_0}{\mu}\right) = -k^2t\ . \qquad (5)$$

The further substitutions, $\gamma = \tan^{-1}\frac{v_0}{\mu}$, and $\lambda = \frac{g}{\mu}$, allow us to express v as

$$v = \mu\tan(\gamma - \lambda t)\ . \qquad (6)$$

A second integration, and the assumption that $y(0) = 0$, gives

$$y = \frac{\mu}{\lambda}\ln\left[\frac{\cos(\gamma - \lambda t)}{\cos\gamma}\right]\ . \qquad (7)$$

Equations (6) and (7) describe the ascending motion of the body. From (6) we learn that when the time reaches $t = \gamma/\lambda$, the velocity is zero and the body has reached its maximum height. Substituting $t = \gamma/\lambda$ in (7) we find that height to be $\mu/\lambda \ln \sec \gamma$. Note that after this height is attained, the equations we found no longer hold. As the body falls, the air resistance is in the positive y direction and the differential equation, (1), must be appropriately changed. We will not solve the problem of this body falling with air resistance, but an interesting result is that the time it would take to fall is longer than the time it took to reach the maximum height. Also, the final velocity with which it would strike the earth would be less than v_0.

• PROBLEM 9-40

A body of mass m, with initial velocity v_0, falls vertically. If the initial position is denoted s_0, determine the position and velocity of the body at the end of t seconds. First, assume the body is acted upon by gravity alone. Then, do the problem again assuming air resistance proportional to the square of the velocity.

Solution: Taking the positive s direction to be vertically downward, the force of gravity acting on the particle is mg. From Newton's second law we have

$$m \frac{dv}{dt} = mg \tag{1}$$

which separates and integrates to be

$$v = gt + C_1 \tag{2}$$

We can evaluate C_1, using the initial condition $v(0) = v_0$. Therefore,

$$v = gt + v_0 \tag{3}$$

Since $v = ds/dt$, a second integration gives

$$s = \tfrac{1}{2} gt^2 + v_0 t + C_2 \tag{4}$$

The other condition, $s(0) = s_0$ leaves

$$s = \tfrac{1}{2} gt^2 + v_0 t + s_0 \tag{5}$$

Equations (3) and (5) describe the motion for the first part of the problem.

We are then asked to assume that a retarding force, proportional to the square of the velocity, say $-mk^2v^2$, also acts upon the body. In this case Newton's second law, after dividing by m, leaves

$$\frac{dv}{dt} = g - k^2v^2 \tag{6}$$

Again, the variables separate. Hence,

$$\frac{dv}{g - k^2v^2} = dt \tag{7}$$

The integral of the left-hand side can be found in tables. We then have

$$\frac{1}{k\sqrt{g}} \tanh^{-1} \frac{kv}{\sqrt{g}} = t + C_3 \tag{8}$$

or

$$v = \frac{\sqrt{g}}{k} \tanh k\sqrt{g}\,(t + C_3) \tag{9}$$

The condition $v(0) = (v_0)$ gives us

$$C_3 = \frac{1}{k\sqrt{g}} \tanh^{-1} \frac{kv_0}{\sqrt{g}} \tag{10}$$

The solution comprised of (9) and (10) reaches a limit as t increases without bound. The hyperbolic tangent of a very large argument approaches one; therefore,

$$\lim_{t\to\infty} v = \frac{\sqrt{g}}{k} \tag{11}$$

This terminal velocity could also have been arrived at by examining equation (6). When the terminal velocity is reached, $dv/dt = 0$ and (6) gives us

$$0 = g - k^2 v_T^2 \tag{12}$$

or

$$v_T = \frac{\sqrt{g}}{k}$$

It remains for us now to find $s(t)$. Equation (9), substituting ds/dt for v, becomes

$$\frac{ds}{dt} = \frac{\sqrt{g}}{k} \tanh k\sqrt{g}\,(t + C_1) \tag{13}$$

Again using tables, we find

$$s = \frac{1}{k^2} \ln \cosh k\sqrt{g}\,(t + C_1) + C_2 \tag{14}$$

From $s(0) = s_0$ it can be shown that

$$C_2 = s_0 - \frac{1}{k^2} \ln \cosh k\sqrt{g}\, C_1 \tag{15}$$

The motion of a body in this resistive medium, air, is given by equations (9), (10), (14), and (15)

CHAPTER 10

PARTICLE KINETICS: WORK-ENERGY

WORK

• PROBLEM 10-1

Derive an expression for the work done by the force in a spring whose free end moves from A to B, as shown.

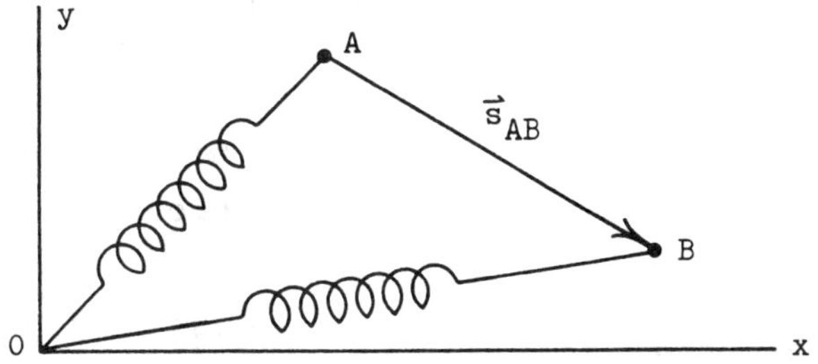

Solution: We consider the spring in two different deformed positions, as shown. We let the deformation of the spring at A be called s_1 and that at B (the amount of stretch from the unstretched position) be called s_2. We assume a flexible spring so that the internal force acts "along" the spring at all times.

It can be seen that, during the deformation, the force varies in magnitude and also in direction with respect to $\vec{S}_{AB}$. However, the displacement $\vec{S}_{AB}$ can be accomplished by a summation at each intermediate position of the spring of infinitesimal displacements along the spring and perpendicular to the spring, respectively. The work done by the spring force during the infinitesimal displacement perpendicular to the spring is zero. We let ds be the magnitude of an infinitesimal displacement along the spring. The work done by the spring force F during these displacements is

$$U = \int_{s_1}^{s_2} F\, ds.$$

But it is known that the force in a spring varies as the

negative of the deformation; that is, if the spring is elongated, the force acts toward the center of the spring, but if the spring is compressed, the force acts away from the center of the spring. Substitution yields

$$U = \int_{s_1}^{s_2} - ks\, ds$$

where k is the spring constant and s is the deformation. Upon performing the integration, we get

$$U = -\tfrac{1}{2}k(s_2^2 - s_1^2).$$

• PROBLEM 10-2

> What average force is necessary to stop a bullet of mass 20 gm and speed 250 m/sec as it penetrates wood to a distance of 12 cm?

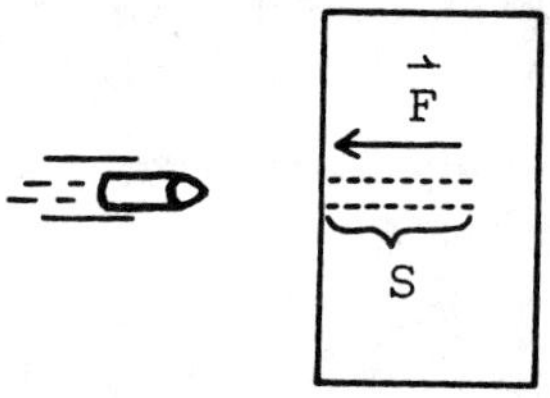

Solution: As it travels through the block, the bullet experiences an average force, $\vec{F}_{avg}$, which retards its motion. By the work-energy theorem, the work done by the net force on an object equals the change in kinetic energy of the object. Hence

$$\int \vec{F} \cdot d\vec{s} = \tfrac{1}{2}mv_f^2 - \tfrac{1}{2}mv_0^2$$

But we only know $\vec{F}$ as an average value. Hence

$$\int \vec{F} \cdot d\vec{s} \simeq \vec{F}_{avg} \cdot \Delta\vec{s} = \tfrac{1}{2}mv_f^2 - \tfrac{1}{2}mv_0^2$$

By definition

$$\vec{F}_{avg} \cdot \Delta\vec{s} = |\vec{F}_{avg}|\,|\Delta\vec{s}| \cos\theta = \tfrac{1}{2}mv_f^2 - \tfrac{1}{2}mv_0^2$$

where θ is the angle between $\vec{F}_{avg}$ and Δs, 180° in this problem. Whence

$$-|\vec{F}_{avg}|\,|\Delta\vec{s}| = \tfrac{1}{2}mv_f^2 - \tfrac{1}{2}mv_0^2$$

$$|\vec{F}_{avg}| = \frac{\tfrac{1}{2}mv_0^2 - \tfrac{1}{2}mv_f^2}{|\Delta\vec{s}|}$$

Hence, $$|\vec{F}_{avg}| = \frac{\tfrac{1}{2}(.02\text{kg})(250\text{ m/s})^2 - 0}{.12\text{ m}} = 5.2 \times 10^3 \text{ nt}$$

This force is nearly 30,000 times the weight of the bullet.

The initial kinetic energy, $\frac{1}{2}mv^2 = 620$ joules, is largely wasted in heat and in work done in deforming the bullet.

• PROBLEM 10-3

A body of mass m has an initial velocity v_0 directed up a plane that is at an inclination angle θ to the horizontal. The coefficient of sliding friction between the mass and the plane is μ. What distance d will the body slide up the plane before coming to rest?

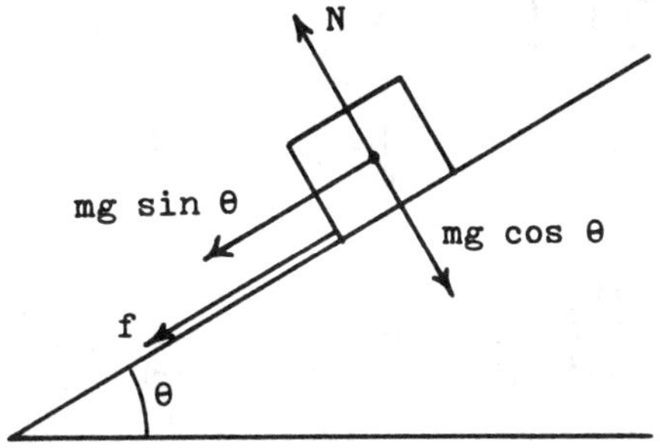

Fig. 1

Solution: The forces on the body, resolved in the plane and perpendicular to the plane are shown in Fig. 1.

The motion is perpendicular to the normal force N and the mg cos θ component of gravity: they do not work on the block. The other two forces mg sin θ and $f = \mu N = \mu mg\cos\theta$, are along the path of motion and do work. The amount of which is equal to their magnitudes, which are constant, times the distance d the body travels;

$$W = -mg\sin\theta\, d - \mu mg\cos\theta\, d.$$

This quantity of work is equal to the energy loss, from the body's initial kinetic energy,

$$\Delta KE = -\frac{1}{2}mv_0^2$$

$$-\frac{1}{2}mv_0^2 = -mg\sin\theta\, d - \mu mg\cos\theta\, d \qquad \text{(a)}$$

$$d = \frac{v_0^2}{2g(\mu\cos\theta + \sin\theta)}.$$

The purist may say this analysis misleadingly puts the non-conservative force, friction, on equal footing with the conservative force, gravity. The results, however, are identical. Using the most general energy conservation law,

$$W_{nc} = \Delta E + \Delta V$$

yields equation (a) again. $W_{nc} = -\mu mg\, d\cos\theta$, $\Delta E = -\frac{1}{2}mv_0^2$ and $\Delta V = mg\,\Delta h = mg\, d\sin\theta$ so

$$-\mu mg\, d\cos\theta = -\frac{1}{2}mv_0^2 + mg\, d\sin\theta.$$

• **PROBLEM 10-4**

A body is attached to one end of a string of length $R = 1_m$ which has its other end fixed on an incline plane as shown in the figure. The mass of the body is 2 kg. If the coefficient of sliding friction between the mass and the plane is $\mu = 0.25$, how much kinetic energy does the body lose in traversing the semicircular path from A to B?

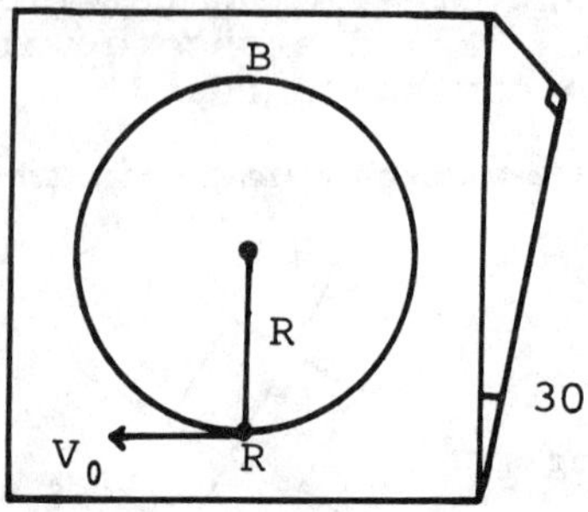

Solution: The kinetic engery lost by the body in this problem will consist of two parts. The first part is lost in heat due to friction and is equivalent to the work done on the body by the frictional force.

This work W_f can be represented mathematically as:

$$\oint_A^B F_f \cdot ds$$

F_f is the friction force and d_s is a differential path length.

As the body moves from A to B, the normal force on the body from the inclined plane is constant, therefore, F_f is constant. (F_f is proportional to the normal force F_n. $\rightarrow F_f = \mu Fn$.)

The integral is then easily evaluated

$$\oint_A^B F_f \cdot ds = F_f \, \pi R$$

πR is the distance traveled from A to B.

Therefore, the kinetic energy lost is $F_f \, \pi R = \mu F_n \, \pi R$. To find F_n, first draw a free body diagram.

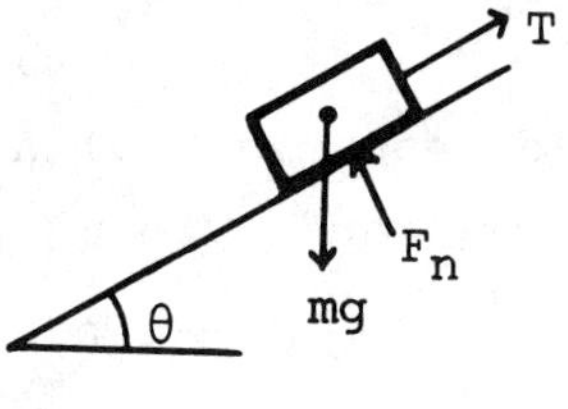

It is clear that $F_n = mg \cos \theta$.

Therefore, $W_f = \mu \cos \theta \; mg \; \pi R$

The second part of the kinetic energy is lost to gravitional potential energy.

This piece is equal to the body's weight times the change in its height above ground level.

The change in height is 2R sin θ and the energy change becomes 2 mg R sin θ.

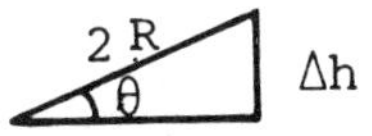

The total kinetic energy change is the sum of the parts.

$$\Delta KE = \mu \cos\theta \, mg \, \pi R + 2 \, mg \, R \sin\theta$$

$$\Delta KE = mg \, R \, [\mu \, \pi \cos\theta + 2 \sin\theta]$$

Substituting the given quantities into the expression for ΔKE yields

$$\Delta KE = 2kg \; (9.8 \; m/s^2) \; (1m) \; [.25 \; (\pi) \cos(30°) + 2 \sin 30°]$$

$$\Delta KE = 19.6 \text{ Joules } [1.68]$$

$$\Delta KE = 32.9 \text{ Joules}$$

• PROBLEM 10-5

A force system is composed of a conservative spring force and a nonconservative friction force. The mass is 10 g and it rests on a horizontal surface, μ = 0.095. If the mass is initially displaced 20 cm,elongating the spring (k = 1000 dynes/cm, relaxed length 100 cm) and is released, determine the manner in which the total mechanical energy of the system varies with distance moved until motion ceases.

Fig. 1

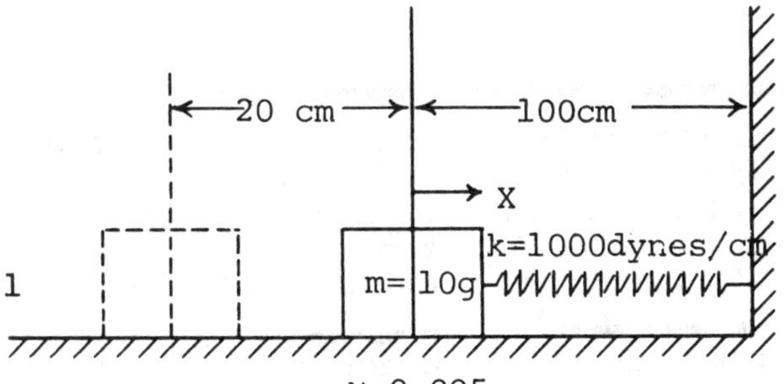

Solution: The mass, initially displaced to x_0 = 20 cm has potential energy

$$PE = \tfrac{1}{2} k \, x_0^2 = \tfrac{1}{2} \; (10^3 \text{ dynes/cm}) (20 \text{ cm})^2 \qquad (1)$$

$$PE = 2 \times 10^5 \text{ ergs.}$$

As the mass is released, this energy is converted into kinetic energy and into work done against friction. The work done against friction is

$$W = f \cdot s$$

where $f = \mu mg$, the force of friction and s is the total distance traveled. Thus

$$W = \mu mg\ s = (0.095)(10\ g)(980\ cm/sec^2)s \qquad (2)$$

$$W = (931)s\ (ergs).$$

The mechanical energy remaining is thus,

$$E = PE - W = 2 \times 10^5 - 931s. \quad (ergs) \qquad (3)$$

where s is the distance traveled in cm from the initial position.

$E = 0$ when $2 \times 10^5 = 931\ s$, or

$$s = \frac{2 \times 10^5}{931} = 214.8 \text{ cm before stopping.}$$

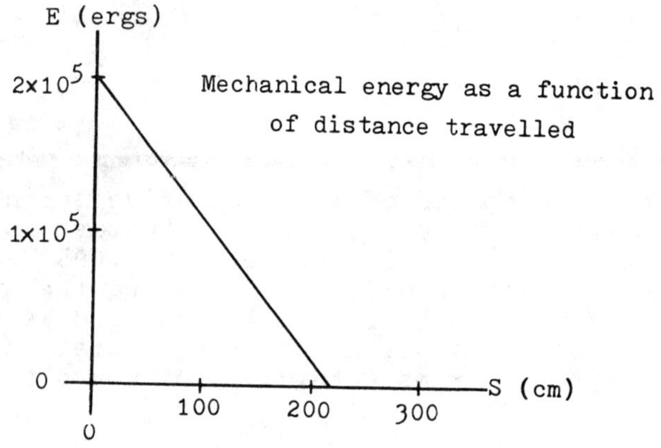

fig 2

• PROBLEM 10-6

In order to protect it during shipment a delicate item weighing 4 oz. is packed in excelsior. From a static test on excelsior, the force-deflection curve shown was obtained. Determine the maximum height from which the package may be dropped if the force on the item is not to exceed 12 lbs.

Solution: By examining the diagram we find that for a force of 12 lb, the deflection is 2.5 inches. The packing material may undergo a deflection of 2.5 inches at maximum impact. Since

$$W = \int F \cdot ds,$$

The area under such a force-deflection curve is proportional to the work done in deflecting the material.

The proportionality constant relating the work done to the area under the F-S curve is equal to the

$$\frac{x_{scale}}{\text{square side}} \times \frac{y_{scale}}{\text{square side}} \quad \text{or:}$$

$$\frac{\frac{1}{2}\text{ in}}{1\text{ square side}} \times \frac{2\text{ lb}}{1\text{ square side}} = \frac{1\text{ in - lb}}{\text{square}}$$

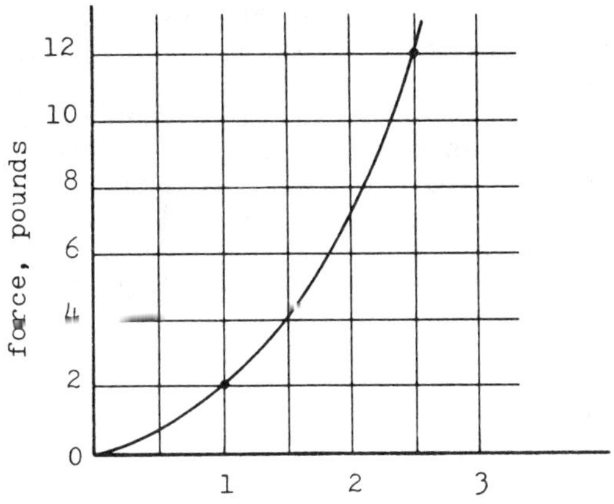

Estimating the area under the curve of the 12 lb-2.5 in point, we count about 8 ½ squares. These 8.5 squares are worth

$$8\tfrac{1}{2}\text{ squares} \times \frac{1\text{ in - lb}}{\text{square}} = 8.5\text{ in-lb}$$

or $$\frac{8.5}{12}\text{ ft-lbs.}$$

This work is actually done by the 4-oz object when it drops from the unknown height. The potential energy lost during the fall is mgh where mg is the weight of the object, ¼ lb, and h is the desired height. Equating work and potential energy, we have

$$\frac{8.5}{12} = \frac{1}{4}h$$

or $$h = \frac{8.5 \times 4}{12}\text{ ft} = 8.5 \times 4\text{ in} = 34\text{ inches.}$$

So the delicate object can survive a 34 inch drop.

• PROBLEM 10-7

A wooden block of mass M is resting on a horizontal surface. The coefficient of friction is μ. One end of a spring with spring constant k is attached to the block, and the other end to a solid wall. The spring is unstretched. A bullet of mass m hits the block and becomes embedded in it. Find the velocity of the bullet before impact in terms of the maximum compression x of the spring and M, m, k, μ, and g.

Solution: The collision time during which the bullet and block collide is generally very short, so that any non-impulsive forces such as the frictional force and the spring force would not affect the outcome of the collision to any appreciable extent. Therefore, we may assume that the total linear momentum of the system of bullet and block will be conserved to a good approximation, during the short collision time. Furthermore, it should be noted that the block will not have moved significantly during the impact. We will apply conservation of momentum to the system of bullet and block to obtain an expression relating the initial velocity of the bullet to the common velocity of the block and bullet just after impact. Next, we apply the work-energy theorem to obtain another expression relating the common velocity of the block and bullet to the maximum compression of the spring. The common velocity is then eliminated between the two expressions to obtain an equation relating the initial velocity of the bullet to the maximum compression of the spring.

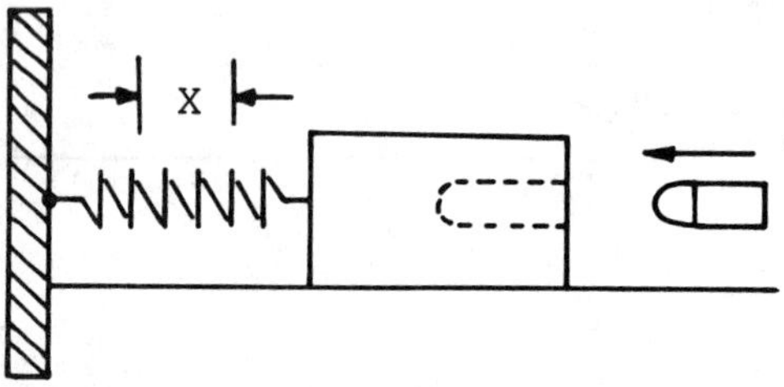

The conservation of momentum of this system may be expressed as,

$$P(\text{Before}) = P(\text{After})$$

$$mv + M \cdot 0 = (m + M)V_c.$$

V_c is the common velocity of the bullet and block just after impact.

The equation simplifies to

$$mv = (m + M)V_c. \qquad (1)$$

Now, we apply the work-energy theorem to the composite system of bullet, block and spring.

We state the theorem in the following form:

The work done by the non-conservative forces is equal to the sum of the change in the kinetic energy of the system plus the change in the potential energy of the system. Applying the theorem to our situation we note that the frictional force is the only non-conservative force present.

Now, the work done by a force equals,

$$W = \int_a^b \vec{F} \cdot d\vec{r} = \int_a^b F \cos\theta \, ds. \qquad (2)$$

Since the force friction is pointed in the direction opposite to that of the block's motion, the angle θ must equal 180°.

The force of friction's magnitude is given by

$$F = \mu N$$

where N is the normal force. But N = weight of composite system, or

$$N = (m + M) g$$

where g is the free fall acceleration constant. Substituting these results into equation 2:

$$W = \int_a^b F \cos \theta \, ds = - \int_a^b \mu (m + M) g \, ds$$

and noting that all the terms are constants, we factor them out of the integral to obtain:

$$W = - \mu (m + M) g \int_a^b ds \qquad (3)$$

The remaining integral is the total distance traveled by the block. Before the impact, the spring was unstretched. Therefore, the distance traveled by the block equals the amount by which the spring is compressed from its equilibrium length.

Let x be a positive quantity representing the amount by which the spring has been compressed and also, the distance traveled by the block. Equation 3 becomes,

$$W = - \mu (m + M) gx \qquad (4)$$

representing the work done by the friction force.

The change in the kinetic energy of the system is equal to,

$$\Delta E_k = E_k \text{ (after)} - E_k \text{ (before)}.$$

Substituting for the mass and velocities before and after, we obtain

$$\Delta E_k = \tfrac{1}{2} (m + M) 0^2 - \tfrac{1}{2} (m + M) V_c^2$$

$$\Delta E_k = - \tfrac{1}{2} (m + M) V_c^2. \qquad (5)$$

The change in the potential energy of the system is equal to the change in the (elastic) potential energy of the spring.

$$\Delta E_p = E_p \text{ (after)} - E_p \text{ (before)} \qquad (6)$$

$$\Delta E_p = \tfrac{1}{2} kx^2 - \tfrac{1}{2} k0^2$$

$$\Delta E_p = \frac{1}{2} kx^2. \qquad (7)$$

The work-energy theorem may be expressed as

$$W = \Delta E_k + \Delta E_p .$$

Substituting equations 4, 5 and 7 into this equation yields,

$$- \mu(m + M)gx = - \frac{1}{2} (m + M)V_c^2 + \frac{1}{2} kx^2. \qquad (8)$$

Solving equation 1 for V_c

$$V_c = \frac{mv}{(m + M)}$$

and substituting this expression into equation 8:

$$- \mu(m + M)gx = - \frac{1}{2} (m + M)\left(\frac{mv}{m + M}\right)^2 + \frac{1}{2} kx^2.$$

Rearranging terms,

$$\frac{1}{2} (m + M) \left(\frac{mv}{m + M}\right)^2 = \mu(m + M)gx + \frac{1}{2} kx^2,$$

$$\frac{1}{2} \frac{m^2}{m + M} v^2 = \mu(m + M)gx + \frac{1}{2} kx^2,$$

$$v^2 = \frac{(m + M)}{m^2} [2\mu(m + M)gx + 2 \cdot \frac{1}{2} kx^2].$$

Factoring (m + M) from within the brackets,

$$v^2 = \frac{(m + M)^2}{m^2} \left[2\mu gx + \frac{kx^2}{(m + M)}\right] .$$

Taking the square root of both sides of the equation yields the desired result:

$$v = \frac{(m + M)}{m} \left[2\mu gx + \frac{kx^2}{(m + M)}\right]^{\frac{1}{2}} .$$

• PROBLEM 10-8

A particle of mass m is attached to the end of a string and moves in a circle of radius r on a frictionless horizontal table. The string passes through a frictionless hole in the table and, initially, the other end is fixed.

a) If the string is pulled so that the radius of the circular orbit decreases, how does the angular velocity change if it is ω_0 when $r = r_0$?

b) What work is done when the particle is pulled slowly in from a radius r_0 to a radius $r_0/2$?

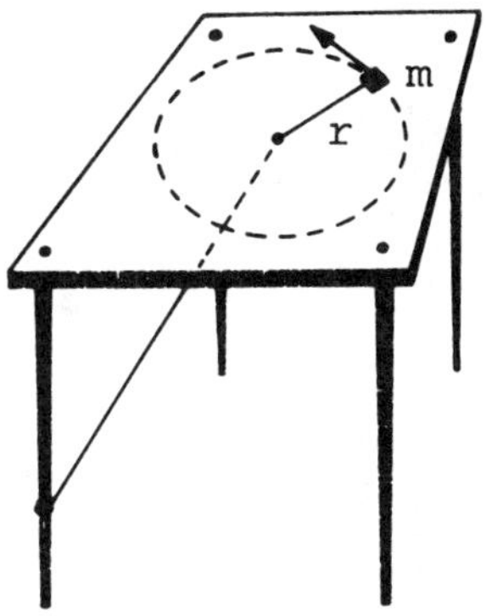

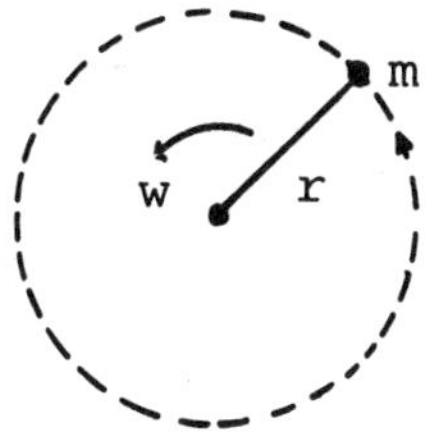

Fig. 1

Solution: a) In this problem, no external torques act on the particle, therefore, angular momentum is conserved. Angular momentum, L, equals $\vec{r} \times m\vec{v}$ and, for this problem, $L = mr^2\omega$ since the motion is circular and $r\vec{\omega} = \vec{v}$.

If the strings length is altered from r_0 to r, the angular rotation changes from ω_0 to ω.

However, $L = L_0$ (angular momentum is conserved) and it follows that $mr^2\omega = mr_0^{\ 2}\ \omega_0$.

Rearranging the above expression and dividing by m yields

$$\omega = \frac{r_0^{\ 2}\ \omega_0}{r^2} = \left(\frac{r_0}{r}\right)^2\ \omega_0 .$$

b) The work done shortening the string is equal to the integral of the tension in the string times an infinitesimal distance integrated over the total length change. Mathematically, this is written as

$$W = \int_{r_0}^{r_0/2} T \cdot dr$$

where W is work and T is tension.

The tension in the string is merely the centripetal force in this problem.

$$T = \frac{mv^2}{r} = mr\omega^2$$

Integrating, $\quad W = \int_{r_0}^{r_0/2} mr\omega^2 dr.$

Using the expression for ω from part a),

$$W = \int_{r_0}^{r_0/2} mr\left[\left(\frac{r_0}{r}\right)^2 \omega_0\right]^2 dr$$

$$W = mr_0^4 \omega_0^2 \int_{r_0}^{r_0/2} \frac{dr}{r^3} dr$$

$$W = mr_0^4 \omega_0^2 \left[-\frac{1}{2} \left(\frac{1}{r^2} \right) \right]_{r_0}^{r_0/2}$$

$$W = mr_0^4 \omega_0^2 \left[\left(-\frac{1}{2} \right) \left\{ \left(\frac{2}{r_0} \right)^2 - \left(\frac{1}{r_0} \right)^2 \right\} \right]$$

$$W = mr_0^4 \omega_0^2 \left(-\frac{3}{2} \right) r_0^{-2}$$

$$W = -\frac{3}{2} mr_0^2 \omega_0^2$$

The negative sign indicates that the work is done on the particle, which is as it should be.

• PROBLEM 10-9

A particle moves 180° around a semicircular path of radius R. It is attracted towards its starting point A by a force proportional to its distance from A. At the final point, B, the force towards A is F_0. Calculate the work done against this force when the particle moves around the semicircle from A to B.

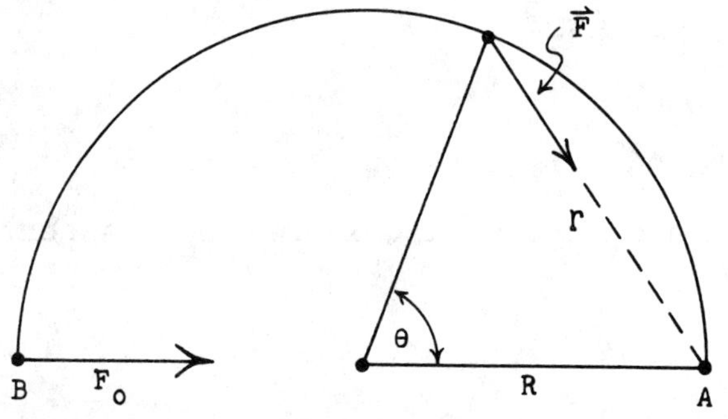

fig 1

Solution: To solve this problem integrate

$$W = -\int_A^B \vec{F} \cdot d\vec{s}$$

The geometry of this problem is shown in Figures (1) and (2).

The force may be written as

$\vec{F} = k\vec{r}$, $\vec{F}$ and $\vec{r}$ are shown in Figure 1.

$F_B = F_0 = k(2R)$

therefore, $k = \frac{F_0}{2R}$. r may also be expressed using the law of cosines as

$$r^2 = R^2 + R^2 - 2R^2 \cos \theta$$

$$r^2 = 2R^2 (1 - \cos \theta)$$

$$r = R \sqrt{2(1 - \cos \theta)}.$$

Thus, $F = kr = \frac{F_0 R}{2R} \sqrt{2(1 - \cos \theta)} = F_0 \sqrt{\frac{(1 - \cos \theta)}{2}}$.

Using the trig identtity

$$\frac{1 - \cos 2\theta}{2} = \sin^2 \theta$$

$$F = F_0 \sin \theta/2.$$

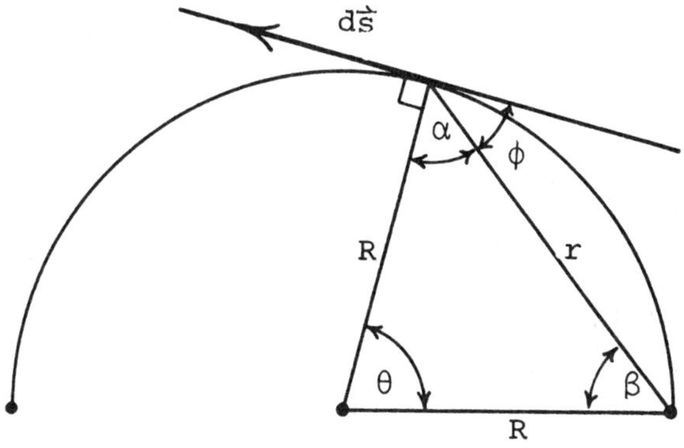

Fig. 2

The dot product $\vec{F} \cdot d\vec{s}$ equals $- Fds \cos \phi$. From Figure 2, noting that the triangle is isosceles it can be seen that $\alpha = \beta$, $\alpha + \beta + \theta = 180°$ and $\alpha + \phi = 90°$. From these three equations we get $\alpha = 90° - \phi$, $2\alpha + \theta = 180°$,

$$- 2\phi + \theta = 0, \qquad \phi = \theta/2.$$

Finally, expressing ds as $ds = R\, d\theta$, the work integral is

$$W = F_0 R \int_0^{\pi} \sin \theta/2 \cos \theta/2 \, d\theta.$$

Since $d(\sin \theta/2) = \frac{1}{2} \cos \theta/2 \, d\theta$

$$W = 2F_0 R \int_0^{\pi} (\sin \theta/2)\, d(\sin \theta/2)$$

$$= 2F_0R \frac{\sin^2 \theta/2}{2} \Bigg|_0^{\pi}$$

$$= F_0 R (1 - 0)$$

$$W = F_0R.$$

An alternate (and easier method) is to recognize that a spring exerts force proportional to its change in length, and use the spring potential energy equation.

If $|\vec{r}|$ is allowed to be Δx then F at any point is equal to $\frac{1}{2} k|\vec{r}|^2$. From the spring potential energy equation:

$$\Delta E = \frac{1}{2} k (\Delta x_f^2 - \Delta x_i^2)$$

$$\Delta x_f = |\vec{r}_f| = 2R$$

$$\Delta x_i = |\vec{r}_i| = 0.$$

Substituting in these particular values for $\Delta x_f + \Delta x_i$ yields:

$$\Delta E = \frac{1}{2} k ((2R^2) - 0)$$

$$\Delta E = \frac{1}{2} k(4R^2).$$

ΔE is equal to the work done on the spring. Since:

$$F_0 = k (2R)$$

$$W = \frac{1}{2} k (2r)(2R)$$

$W = F_0 R \rightarrow$ the same result as before!

• PROBLEM 10-10

A particle in the xy-plane is attracted toward the origin by a force $F = k/y$. Calculate the amount of work done when the particle moves from $(x, y) = (0, a)$ to $(2a, 0)$:

(a) Along the path $(0, a)$ to $(2a, a)$, then $(2a, a)$ to $(2a, 0)$.

(b) Along the ellipse of semiaxes a, 2a (i.e., $x = 2a \sin \theta$, $y = a \cos \theta$).

Solution: (a) $F = k/y$

$$\text{Work} = \int \vec{F}_0 \, d\vec{r} = \int F \, dr \cos \theta.$$

From Figure 1,

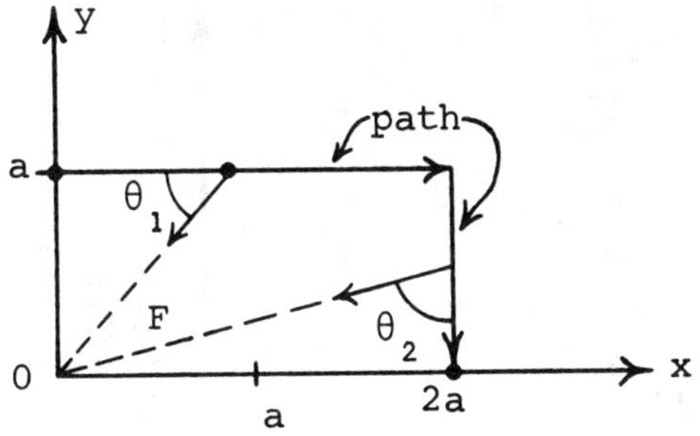

Fig. 1

$$\cos \theta_1 = \frac{x}{\sqrt{x^2 + y^2}} .$$

1) From (0, a) to (2a, a). y = a, dr = dx

$$\text{Work}_1 = -\int_0^{2a} \frac{k}{a} \frac{x}{\sqrt{x^2 + a^2}}\, dx = -\frac{k}{a} \int_0^{2a} \frac{x\, dx}{\sqrt{x^2 + a^2}}$$

(minus sign, because force is opposing displacement).

Changing variables to

$$u = \sqrt{x^2 + a^2}$$

$$du = \frac{2x\, dx}{2\sqrt{x^2 + a^2}} = \frac{x\, dx}{u} .$$

$$\text{Work}_1 = -\frac{k}{a} \int \frac{u\, du}{u} = -\frac{k}{a} u = -\frac{k}{a} \sqrt{x^2 + a^2} \Big]_0^{2a} = k\,(1 - \sqrt{5}).$$

2) From (2a, a) to (2a, 0). x = 2a, dr = dy

$$\cos \theta_2 = \frac{y}{\sqrt{(2a)^2 + y^2}} .$$

$$\text{Work}_2 = \int_a^0 \frac{k}{y} \frac{y}{\sqrt{(2a)^2 + y^2}}\, dy = k \int_a^0 \frac{1}{\sqrt{(2a)^2 + y^2}}\, dy$$

$$\text{Work}_2 = k\, \ell n(y + \sqrt{(2a)^2 + y^2}) \Big]_0^a = k\, [\ell n a(1 + \sqrt{5}) - \ell n\, 2a]$$

$$\text{Work}_2 = k\, \ell n \left(\frac{1 + \sqrt{5}}{2}\right) .$$

$$\text{Total Work} = W_1 + W_2 = k \left[(1 - \sqrt{5}) + \ell n \left(\frac{1 + \sqrt{5}}{2}\right) \right].$$

(b) Ellipse $\frac{x^2}{2a^2} + \frac{y^2}{a^2} = 1$

Fig. 2 elliptic path.

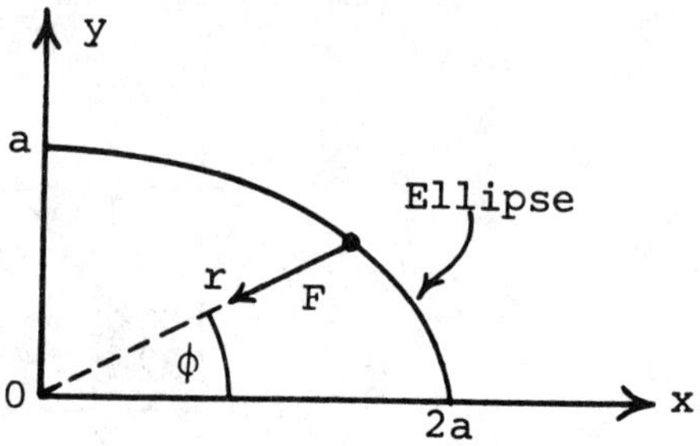

Fig. 2

$$r^2 = x^2 + y^2 \tag{1}$$

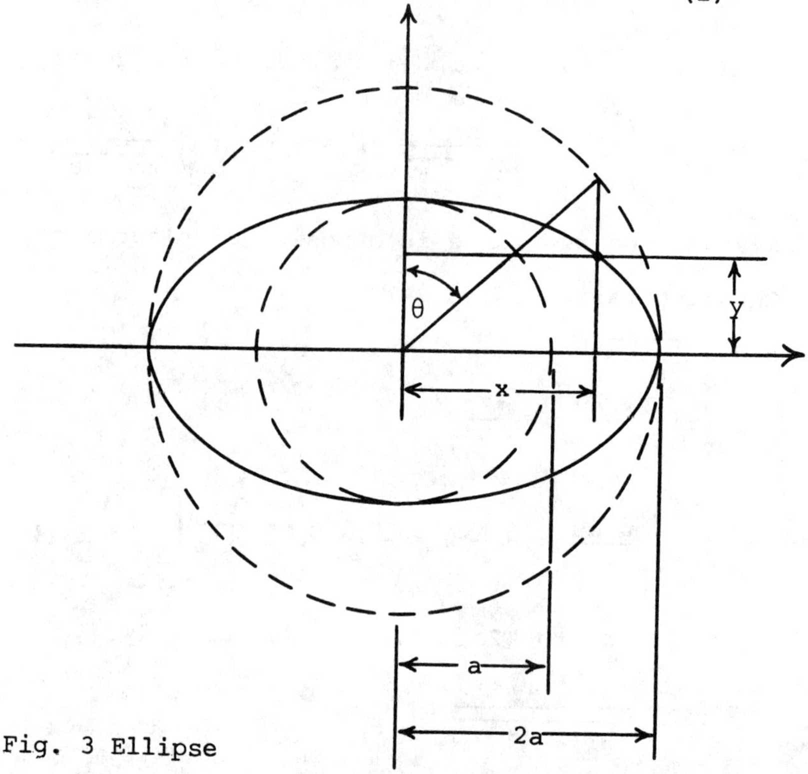

Fig. 3 Ellipse

For an ellipse, the equation can be given in terms of inscribed and circumscribed circles and a central angle θ (Fig. 3). Then

$$\frac{x}{2a} = \sin\theta, \qquad \frac{y}{a} = \cos\theta \tag{2}$$

and $\sin^2\theta + \cos^2\theta = 1$.

$$\text{Work} = \int \vec{F} \cdot d\vec{r} = \int F_x\, dx + \int F_y\, dy \,, \quad F = \frac{k}{y}$$

$$F_x = F\cos\phi, \qquad F_y = F\sin\phi$$

$$F_x = F\,\frac{x}{r} \qquad F_y = F\,\frac{y}{r}\,. \tag{3}$$

Combining equations (1), (2) and (3);

$$F_x = -\frac{k}{a\cos\theta} \times \frac{2\,a\sin\theta}{\sqrt{(2a\sin\theta)^2 + a^2\cos^2\theta}}$$

$$= -\frac{2\,k\sin\theta}{a\,(\sqrt{4\sin^2\theta + \cos^2\theta})\cos\theta}$$

$$F_y = -\frac{k}{a\cos\theta} \times \frac{a\cos\theta}{\sqrt{(2a\sin^2\theta) + a^2\cos^2\theta}}$$

$$= \frac{-k}{a\sqrt{4\sin^2\theta + \cos^2\theta}} .$$

With, $x = 2a\sin\theta,$ $\qquad dx = 2a\cos\theta\, d\theta$

$y = a\cos\theta$ $\qquad dy = -a\sin\theta\, d\theta$

$$\text{Work} = \int_0^{\pi/2} \frac{-2k\sin\theta}{a\cos\theta\sqrt{4\sin^2\theta + \cos^2\theta}} \times 2a\cos\theta\, d\theta +$$

$$\int_0^{\pi/2} \frac{-k}{a\sqrt{4\sin^2\theta + \cos^2\theta}} (-a\sin\theta\, d\theta)$$

$$\text{Work} = k\int_0^{\pi/2} - \left(\frac{-4\sin\theta}{\sqrt{4\sin^2\theta + \cos^2\theta}} + \frac{\sin\theta}{\sqrt{4\sin^2\theta + \cos^2\theta}}\right) d\theta$$

$$\text{Work} = k\int_0^{\pi/2} - \frac{3\sin\theta\, d\theta}{\sqrt{1 + 3\sin^2\theta}}$$. From table of integrals

$$\int \frac{\sin\theta\, d\theta}{\sqrt{1 + p^2\sin^2\theta}} = -\frac{1}{p}\arcsin\left(\frac{p\cos\theta}{\sqrt{1+p^2}}\right) .$$

$$W = -3k\left[-\frac{1}{\sqrt{3}}\arcsin\left(\frac{\sqrt{3}\cos\theta}{\sqrt{1+3}}\right)\right]_0^{\pi/2} = \frac{3k}{\sqrt{3}}\left(0 - \frac{\pi}{3}\right)$$

$$W = -\frac{k\pi}{\sqrt{3}} .$$

• PROBLEM 10-11

Solve the ballistic pendulum problem illustrated below. Given are a bullet of known mass m_1, a block of mass m_2, and the distance the block rises after impact h. Find the velocity of the bullet, v_1.

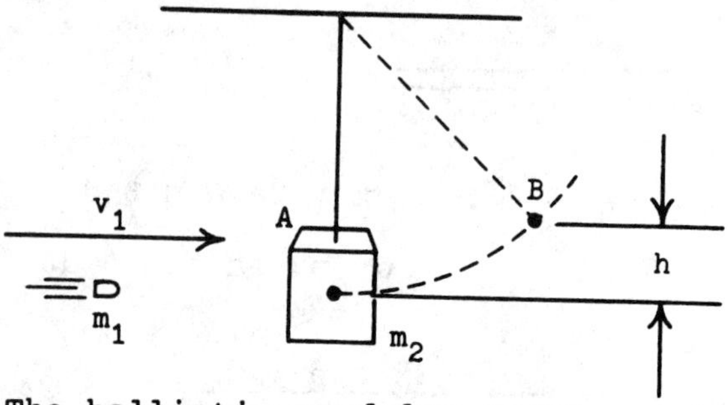

Solution: The ballistic pendulum problem naturally divides into two parts of analysis: The totally inelastic collision when the bullet imbeds itself into the block, and the rise of the bullet and the block together due to the velocity imparted by the collision.

The collision is inelastic so we are restricted to the always applicable conservation of momentum equation.

$$\underset{\text{before collision}}{\underline{m_1 v_1 + m_2 v_2}} = \underset{\text{after collision}}{\underline{m_1 V_1 + m_2 V_2}}$$

For the ballistic pendulum, $v_2 = 0$ and $V_1 = V_2$ since the bullet imbeds itself in the block. So

$$m_1 v_1 = (m_1 + m_2)V. \qquad (a)$$

Now we must determine V by consideration of the rise of the pendulum. The equation used now is conservation of energy

$$KE_i + PE_i = KE_f + PE_f.$$

At the top of the rise, the system is not moving, so $KE_f = 0$. We use the available arbitrariness to set $PE_i = 0$. Therefore, we have

$$\tfrac{1}{2}(m_1 + m_2)V^2 = (m_1 + m_2)\, g\, h. \qquad (b)$$

Equation (a) and (b) contain all the analyses needed for this problem. Solving for v_1 yields

$$v_1 = \left(\frac{m_1 + m_2}{m_1}\right)\sqrt{2\,gh}. \qquad (c)$$

Just to put some perspective on this highly practical equation, we will provide some pertinent data: m_1 weighs 0.10 lb, m_2 weighs 25 lb and h = 4 in.

$$v_1 = \left(\frac{0.10/32.2 + 25/32.2}{0.10/32.2}\right) \sqrt{2 \cdot 32.2 \cdot 4/12}$$

$$= \left(\frac{25.1}{0.10}\right) \sqrt{21.4} = 1163 \text{ ft/sec.}$$

• PROBLEM 10-12

A skier with a weight of 150 lbs. stands at point A at the top of a 60 ft. ski slope. The ski slope goes down into a round valley with a radius of curvature of 30 ft. Neglecting friction find (1) the force exerted on the skier by the ski slope at point B. (2) Find the minimum radius of curvature at point C for which the skier will not fly up into the air.

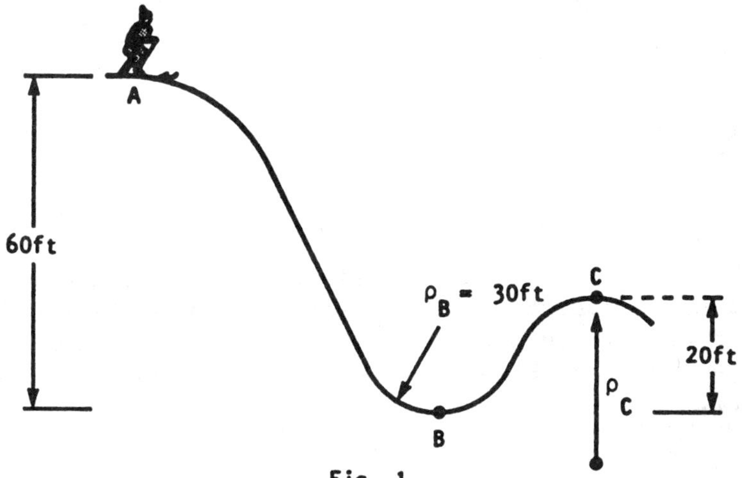

Fig. 1

Solution: At point (B) the two forces acting on the skier are the weight force and the normal force (see Fig. 2). We are interested in solving for the normal force. At point (B) there is only a centripetal acceleration $a_n = \frac{v^2}{\rho_B}$. Using the second law of motion stated by Newton, we obtain $a_n = \frac{(N - W)}{m}$. N can be determined if we first obtain a value for (V) at point B

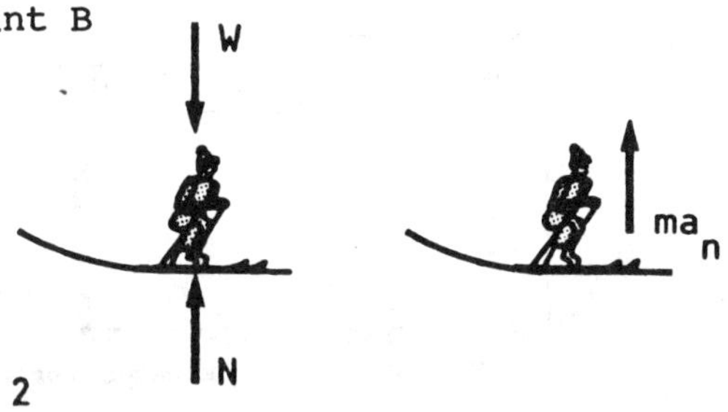

Fig. 2

Since we are neglecting friction, the principle of Conservation Energy could give a value for V.

$$KE_A + PE_A = KE_B + PE_B$$

Using point (B) as a reference point, we calculate the potential energies of the system.

$$h_B = 0$$

$$PE_B = mgh_B = 0$$

$$PE_A = mgh_A$$

Assuming there are no special forces of contraint, the minimum radius of curvature at (C) for which the skier will not fly into the air is when N = 0. Using conservation of energy between points (B) and (C) will give us value for V at point (C).

• PROBLEM 10-13

A block of mass m, initially at rest, is dropped from a height h onto a spring whose force constant is k. Find the maximum distance y that the spring will be compressed. See figure.

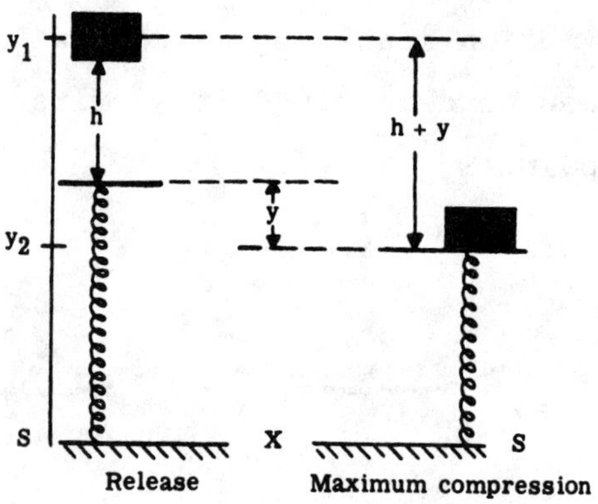

The total fall of the block is h + y.

Solution: The general procedure used in solving any problem in mechanics is to calculate all the forces acting on the system and then derive the equation of motion of the system.

An easier way to do mechanics problems involves the use of conservaion principles. These laws are not applicable to all problems, but when they are, they simplify the calculation of the solution tremendously.

In this problem, we may use the principle of conservation of energy. We relate the energy of the block before it was released to the block's energy at the point of maximum compression (see figure). At the moment of release, the kinetic energy is zero. At the moment when maximum compression occurs, there is also no kinetic energy.

As shown in the figure, the reference level for gravitational potential energy is the surface S. The initial gravitational potential energy of m is mgy_1. At the point of maximum compression, the potential energy of m is mgy_2. However, at this point, the spring is compressed a distance y and also has elastic potential energy $\frac{1}{2} ky^2$. Hence, equating the energy at the point of release to the energy at the point of maximum compression,

$$mgy_1 = mgy_2 + \frac{1}{2} ky^2$$

$$mg(y_1 - y_2) = \frac{1}{2} ky^2$$

But $y_1 - y_2 = h + y$ and

$$mg(h + y) = \frac{1}{2} ky^2$$

$$y^2 = \frac{2\ mg}{k}(h + y)$$

$$y^2 - \left(\frac{2\ mg}{k}\right) y - \frac{2\ mgh}{k} = 0$$

Therefore, using the quadratic formula to solve for y,

$$y = \frac{1}{2}\left[\frac{2mg}{k} \pm \sqrt{(2mg/k)^2 + (8mgh/k)}\right] .$$

• PROBLEM 10-14

The spring shown in the Figure has a stiffness of 100 lb/ft and an unstretched length of 15 in. If the system is released from rest in the position shown, how far will the weight deflect? The weight is equal to 20 lb.

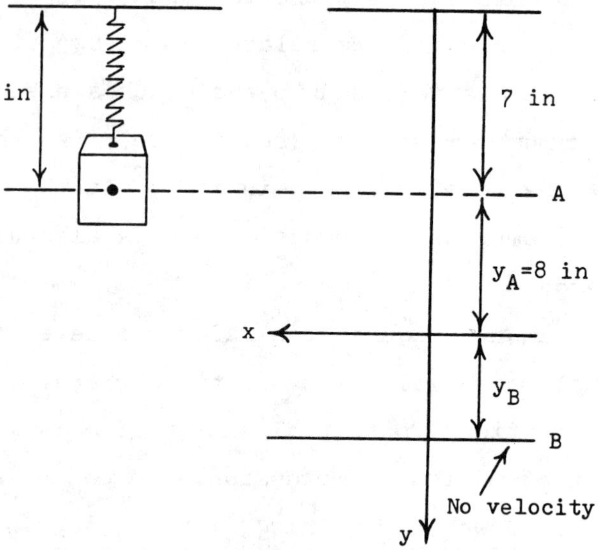

Solution: Using the coordinate system shown in the Figure, the potential-energy function for the gravitational force will be - Wy, and the potential-energy function for the spring will be $ky^2/2$. Thus the potential-energy function for the net force on the body is

$$V = \tfrac{1}{2}\ ky^2 - Wy.$$

Since the kinetic energy of the body may be written as $m\dot{y}^2/2$, the total energy of the system becomes

$$E = T + V$$

$$= \tfrac{1}{2}m\dot{y}^2 + \tfrac{1}{2}ky^2 - Wy.$$

If the system is released from rest at point A, the velocity of the body at point A will be zero. The position of the mass at A is - 8.0 in., or - 0.67 ft. Thus the total energy at A becomes

$$E_A = \tfrac{1}{2}(100)(-\ 0.67)^2 - (20)(-\ 0.67)$$

$$= 35.9\ \text{ft-lb}.$$

At point B the velocity of the body will be zero, and the position of the body at that point is the desired unknown y_B. Thus the total energy at B is

$$E_B = \tfrac{1}{2}(100)y_B^2 - 20y_B$$

and since energy must be conserved in the system, $E_A = E_B$ and

$$35.9 = 50y_B^2 - 20y_B$$

By solving the quadratic for y_B,

$$y_B = \frac{+\ 20 \pm \sqrt{400 + 4(50)(35.9)}}{100}$$

$$= 1.07 \text{ ft}, \quad -.67 \text{ ft}.$$

The negative root refers to the initial position and is discarded here. The other, 1.07 ft = 12.8 in, gives a total deflection of 20.8 in.

• PROBLEM 10-15

A 30-lb coupling from an automatic orange picker slides on a vertical shaft. A spring with an original length of 5 inches and spring constant of 4 lbs/in is connected to the coupling. The coupling starts its motion at position (1). It drops under its own weight 10" to position (2) making the spring extend. Neglect friction. What is the velocity of the coupling when it reaches position (2)?

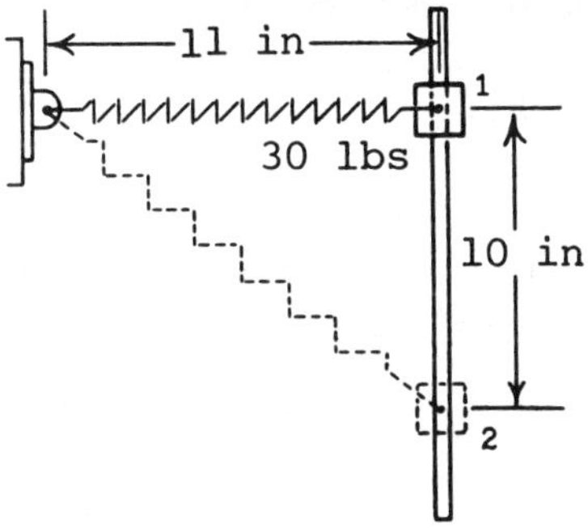

Solution: Since there is no friction in this problem it can be assumed that energy is conserved as the mass moves from point 1 to point 2. At point 1 the mass is at rest and its kinetic energy is 0. If point 2 is used as a potential energy reference point, then the potential energy at 1 is mg(10 in). The spring energy at 1,

$$SE_1 = \tfrac{1}{2}\, k\, (\text{Length}_{\text{stretched}} - \text{Length}_{\text{unstretched}})^2 .$$

SE_1 depends on how much the spring is stretched or compressed in order to attach it to the mass. Mathematically the relations look like this:

$$KE_1 = 0$$

$$PE_1 = mg(10 \text{ in})$$

$$SE_1 = \tfrac{1}{2} k \ (11 \text{ in} - 5 \text{ in})^2 = \tfrac{1}{2} k(6 \text{ in})^2.$$

At point 2 potential energy is 0, and the spring energy is equal to $\frac{1}{2} k \ (L - 4 \text{ in})^2$ where L is the spring's length when the mass is at point 2. Kinetic energy is unknown, but equal to $\frac{1}{2} mv^2$.

$$PE_2 = 0$$

$$KE_2 = \tfrac{1}{2} mv^2$$

$$SE_2 = \tfrac{1}{2}k \ (L - 5 \text{ in})^2.$$

Putting all of this together yields the following equation:

$$KE_1 + PE_1 + SE_1 = KE_2 + PE_2 + SE_2$$

$$(mg)(10 \text{ in}) + \tfrac{1}{2} k(6 \text{ in})^2 = \tfrac{1}{2} mv^2 + \tfrac{1}{2} k(L - 5 \text{ in})^2$$

To find v we solve the equation.

$$v^2 = \frac{2[(10 \text{ in})(mg) + \frac{1}{2}k[(6 \text{ in})^2 - (L - 5 \text{ in})^2]]}{M} \times \frac{1 \text{ ft}}{12 \text{ in}}$$

$$v = \sqrt{\frac{2[(10 \text{ in})(mg) + \frac{1}{2}k[(6 \text{ in})^2 - (L - 5 \text{ in})^2]]}{M} \times \frac{1 \text{ ft}}{12 \text{ in}}}$$

There is only one thing left to do, find L.

From the Pythagorean theorem

$$L = \sqrt{(10 \text{ in})^2 + (11 \text{ in})^2}$$

$$L = 14.87 \text{ in}$$

Now substituting this value for L into the equation yields v.

$$v = \sqrt{\frac{2[(10 \text{ in})(30 \text{ lbs}) + \frac{1}{2}(4)[(6 \text{ in})^2 - (14.87 \text{ in} - 5 \text{ in})^2]]}{30 \text{ lbs}/32.2 \text{ ft/s}^2} \times \frac{1 \text{ ft}}{12 \text{ in}}}$$

$$v = \sqrt{31.69 \text{ ft}^2/\text{s}^2}$$

$v = \pm 5.63$ ft/s (+ going down
− going up)

• PROBLEM 10-16

A small mass m moves around a circular track with no friction. The weight of the mass is 15 lb. A spring is connected to the mass and to a fixed point, 0. It has a spring constant $k = 10.0$ lb/in. The unstretched length of the spring is 12 in. (Other dimensions on accompanying figure). If the mass has a velocity of 20 in/sec to the right at point A, what will the velocity of the mass at point B be?

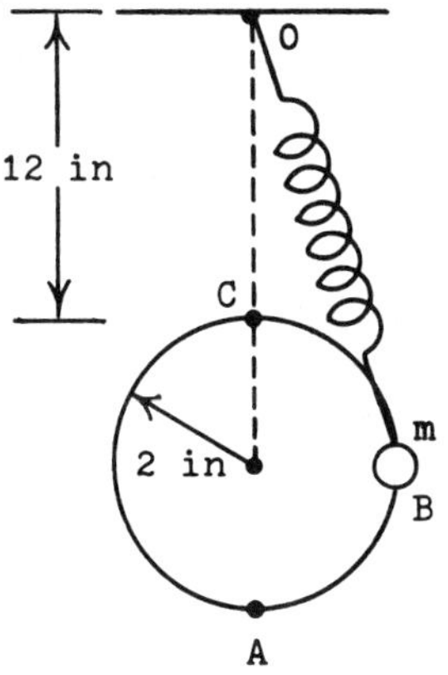

fig 1

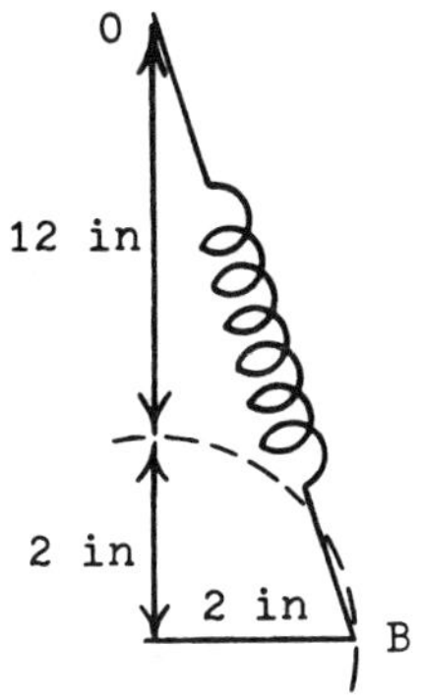

fig 2

Solution: Since there is no friction, this problem can be solved by the conservation of energy. The basic equation for such an approach is

$$KE_i + PE_i = KE_f + PE_f \quad . \qquad \text{(a)}$$

The sources of potential energy in this problem are gravity (PE = mgh) and the spring ($PE = \frac{1}{2}ks^2$). If we set the gravity potential energy equal to zero at point A (we are always allowed this measure of arbitrariness) we have

$$PE_i = \tfrac{1}{2} \cdot (10\ \text{lb/in})(4\ \text{in})^2 = 80\ \text{lb-in.} \qquad \text{(b)}$$

The value $S_A = 4$ in. was determined from $\overline{OA} = 16$ in. minus the relaxed spring length, 12 in.

The potential energy at position B is,

$$PE_f = \left(\frac{15}{32.2}\ \text{slugs}\right)(32.2\ \text{ft/sec}^2)(2\ \text{in}) + \tfrac{1}{2}\ (10\ \text{lb/in})(S_B)^2.$$

For the distance, S_B, examine Figure 2. The hypotenuse of the right triangle is the full spring length. That value is

$$S = \sqrt{14^2 + 2^2}\ \text{in}$$

and $S_B = S - 12\ \text{in} = 2.14$ in.

We have, then

$$PE_f = (15\ \text{lb})(2\ \text{in}) + \tfrac{1}{2}\ (10\ \text{lb/in})(2.14\ \text{in})^2 \qquad \text{(c)}$$

$$= 52.9\ \text{lb-in.}$$

Equations (c), (b) and (a) can now be put together.

$$\frac{1}{2}\left(\frac{15\text{ lb}}{386\text{ in/sec}^2}\right)(20\text{ in/sec})^2 + 80\text{ lb-in}$$

$$= \frac{1}{2}\left(\frac{15}{386}\right)v_B^{\,2} + 52.9\text{ lb-in.}$$

Note that we now have to convert $g = 32.2\text{ ft/sec}^2$ to 386 in/sec^2 to keep the units consistent. Solving for v_B,

$$\frac{1}{2}\left(\frac{15}{386}\right)v_B^{\,2} = (80 - 52.9) + \frac{1}{2}\left(\frac{15}{386}\right)(400)$$

$$v_B^{\,2} = 2\left(\frac{386}{15}\right)(27.1) + 400$$

$$v_B^{\,2} = 1794.7\text{ in}^2\text{/sec}^2$$

$$v_B = 42.4\text{ in/sec.}$$

• PROBLEM 10-17

A small block of mass m slides along the frictionless loop-the-loop track shown in the figure. (a) If it starts from rest at P, what is the resultant force acting on it at Q? (b) At what height above the bottom of the loop should the block be released so that the force it exerts against the track at the top of the loop equals its weight?

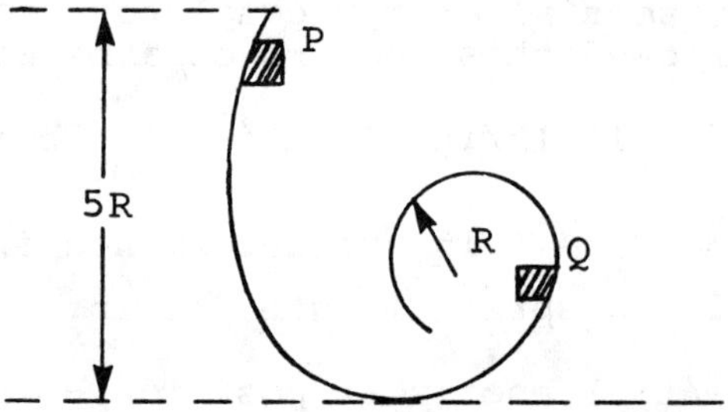

<u>Solution:</u> (a) Point Q is at a height R above the ground. Thus, the difference in height between points P and Q is 4R, and the difference in gravitational potential energy of the block between these two points is:

$$mgh_2 - mgh_1 = mg(h_2 - h_1) = mg(4R)$$

$$= 4\ mgR$$

Since the block starts from rest at P, its kinetic energy at Q is equal to its change in potential energy, 4 mgR; by the principle of conservation of energy

$$\tfrac{1}{2}\,mv^2 = 4\ mgR$$

$$v^2 = 8\ gR$$

At Q, the only forces acting on the block are its weight, mg, acting downward, and the force N of the track on the block, acting in the radial direction. Since the block is moving in a circular path

$$N = \frac{mv^2}{R} = \frac{8\ mgR}{R} = 8\ mg$$

The loop must exert a force on the block equal to eight times the block's weight.

(b) For the block to exert a force equal to its weight against the track at the top of the loop:

$$\frac{mv'^2}{R} = 2\ mg, \qquad v'^2 = 2gR$$

This is the case because gravity exerts a downward force mg on the block. Thus, in order to keep the block moving in a circular path, the rest of the force (= mg) must be exerted by the loop-the-loop. Therefore:

$$mgh = \tfrac{1}{2}\ mv'^2$$

$$h = \frac{v'^2}{2g} = \frac{2gR}{2g} = R$$

This is the height above the point where $v'^2=2gR$, that is, above the top of the loop. The height above the bottom of the loop is, then, $h=R+2R=3R$.

• PROBLEM 10-18

A small bead of mass m is free to move without friction on a vertical hoop of radius R. The bead moves under the influence of the hoop constraint, gravity, and a spring which has one end attached to a pivot a distance R/2 above the center of the hoop. The spring, of force constant k, is at its relaxed length when the bead is at the top of the circle.

a) Find the potential energy of the bead as a function of the bead's angular position as measured from the center of the circle; draw a potential-energy diagram of V vs θ.

b) What minimum kinetic energy must the bead have at the top position in order to go all the way around the hoop?

c) If the bead starts from the top with this kinetic energy, what force does the hoop exert on the bead at the top and at the bottom points on the hoop?

Solution: The potential energy of the bead consists of two parts: a) gravitational potential energy and b) spring potential energy.

If the top position is defined as a reference point (gravitational potential energy equals zero), then the gravitational potential energy is equal to -mgh where h is the bead's depth below the top of the rim.

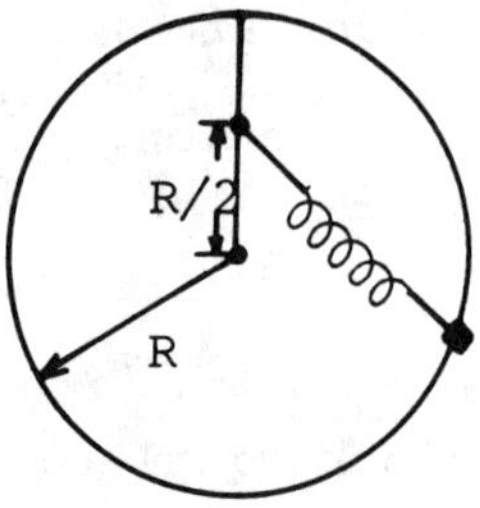

Fig. 1

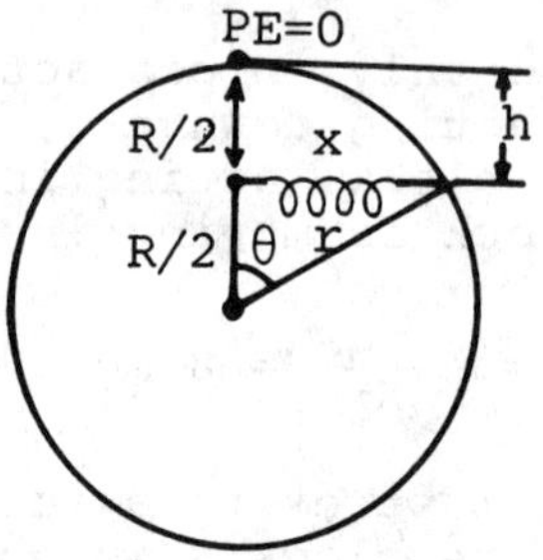

Fig. 2

h is equal to r - r cos θ = r (1 - cos θ), so the potentail energy due to gravity is

$$- mg\ r\ (1-\cos\theta).$$

The spring potential energy is equal to $1/2\ k\ (x - r/2)^2$ where x is the stretched length and r/2 is the unstretched length. From the law of cosines,

$$x^2 = \left(\frac{r}{2}\right)^2 + (r)^2 - 2\frac{r}{2}\ r\cos\theta$$

$$x^2 = \frac{5}{4}(r^2) - r^2\cos\theta$$

$$x = \frac{1}{2} r\sqrt{5 - 4\cos\theta}$$

$$x = \frac{1}{2} r\sqrt{1 + 4(1 - \cos\theta)}$$

Now, the spring potential energy is

$$\frac{1}{2} k\left(\frac{1}{2} r\sqrt{1 + 4(1 - \cos\theta)} - \frac{r}{2}\right)^2.$$

Bringing the r/2 out of parentheses yields

$$\frac{1}{2} k\frac{r^2}{4}\left(\sqrt{1 + 4(1 - \cos\theta)} - 1\right)^2$$

or, $\frac{1}{8} k r^2 (\sqrt{1 + 4(1 - \cos\theta)} - 1)^2$.

The total potential energy is the sum of the parts.

$$V = P.E. = \frac{1}{8} k r^2 (\sqrt{1 + 4(1 - \cos\theta)} - 1)^2 - mgr(1 - \cos\theta)$$

Graphically, the spring P.E., gravity P.E. and cheir sum V look like this:

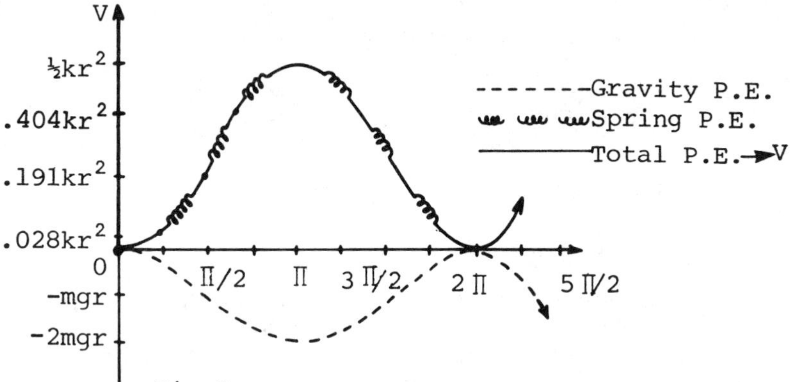

Fig.3

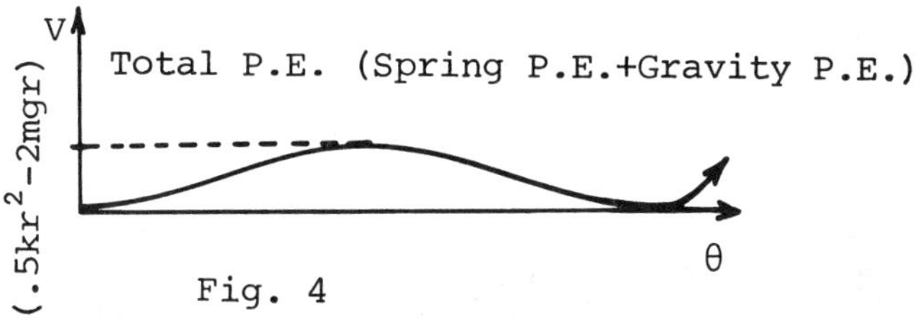

Fig. 4

In order for the bead to pass all the way around the circle, its initial kinetic energy must be slightly greater than the maximum potential energy. That is,

$$KE > \frac{1}{2} Kr^2 - 2\ mgr$$

for the bead to travel all the way around.

If the bead has this kinetic energy at the top of the hoop, the net force on the bead must equal the mass times the centripetal acceleration. A free body diagram is shown in figure -(5). Note that at the top of the hoop the spring does not exert any force on the bead. Since the weight of the bead is pulling down, the force applied by the hoop must be directed upward to keep it on the hoop -

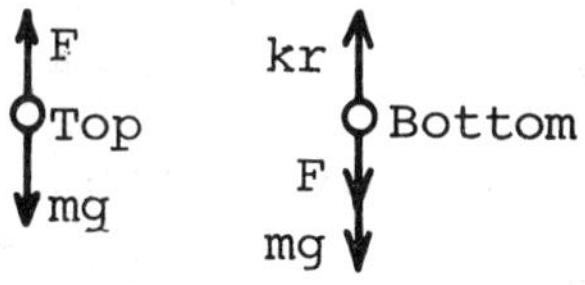

Fig. 5

Applying Σ F - mα,

i) $mg - F = \frac{mv^2}{r}$ (F = normal force from hoop)

ii) $KE = \frac{1}{2} mv^2 = \frac{1}{2} Kr^2 - 2\ mgr$

Solving ii) for mv^2 and inserting i) yields:

$$mv^2 = Kr^2 - 4\ mgr$$

$$mg - F = \frac{Kr^2 - 4\ mgr}{r}$$

$$F = 5\ mg - Kr$$

(with down as the positive direction).

At the bottom of the hoop, since the original K.E. just equals the potential energy, $V \approx 0$ and the force on the bead is the spring force minus the weight. The spring force is $K\ (3r/2 - r/2) = Kr$ upwards. For equilibrium, the hoop must push down on the bead if down is defined as the positive direction,

$$F + mg - Kr = 0$$

and $F = Kr - mg.$

CONSERVATIVE FORCES

• PROBLEM 10-19

Determine the force in the plane of motion acting on a particle resting on an inclined plane using the potential energy for formulation.

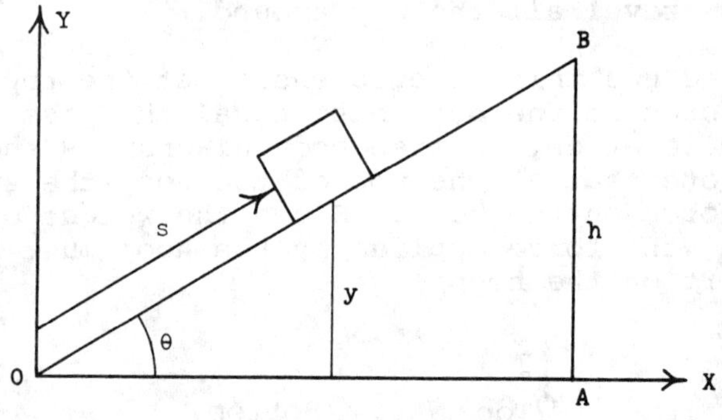

Solution: Consider the system shown in the Figure. The gravitational potential energy on the mass shown is $V = mgy$. To find the force in the plane, we need to change the independent variable from y to s. From the diagram, it can be seen that $y = s \sin \theta$ so

$$V = mgs \sin \theta\ .$$

The conservative force F_s acting on the mass in direction of increasing s (up the inclined plane) is

$$F_s = - \frac{\partial V}{\partial s} = - mg \sin \theta\ .$$

Thus the force on the particle is $mg \sin \theta$, directed down the plane. The same result can be obtained by resolving the gravitational force mg into components along, and at right angles to the plane.

• PROBLEM 10-20

Determine which of the following forces are conservative and find the potential energy for those which are.

(a) $F_x = 6abyz^3 - 20bx^3y^2$,

$F_y = 6abxz^3 - 10bx^4y$, $F_z = 18abxz^3y$

(b) $F_x = 18abyz^3 - 20bx^3y^2$,

$F_y = 18abxz^3 - 10bx^4y$, $F_z = 6abxyz^2$

(c) $\vec{F} = \hat{x}\, F_x(x) + \hat{y}\, F_y(y) + \hat{z}\, F_z(z)$.

Solution: If a force is conservative, there exists a potential function V such that

$$F_x = -\frac{\partial V}{\partial x}, \quad F_y = -\frac{\partial V}{\partial y}, \quad F_z = -\frac{\partial V}{\partial z}.$$

From the theory of partial differential equation,

$$V_x = -\int F_x\, dx + f_x(y, z)$$

$$V_y = -\int F_y\, dy + f_y(x, z)$$

$$V_z = -\int F_z\, dz + f_z(x, y).$$

$f_{x,y,z}$ must be chosen so that $V_x = V_y = V_z$. If this cannot be done the force is not conservative.

Proceeding then, for part (a)

(a) $F_x = 6abyz^3 - 20bx^3y^2 \quad \therefore \; V_x = -6abyxz^3 + 5bx^4y^2 + f(y,z)$

$F_y = 6abxz^3 - 10bx^4y \quad \therefore \; V_y = -6abxyz^3 + 5bx^4y^2 + f(x,z)$

$F_z = 18\, abxz^2y \quad \therefore \; V_z = -6abxz^3y + f(x,y).$

If $V_x = V_y = V_z$ then $f(x, y) = 5bx^4y^2$

$f(x, z) = 0$

$f(y, z) = 0.$

Thus $V = 5bx^4y^2 - 6abyxz^3$ and is conservative.

Part (b)

$$F_x = 18abyz^3 - 20abx^3y^2, \qquad V_x = -18abxyz^3 + 5bx^4y^2 + f(y,z)$$

$$F_y = 18abxz^3 - 10bx^4y, \qquad V_y = -18abxyz^3 + 5bx^4y^2 + f(x,z)$$

$$F_z = 6abxyz^2 \qquad V_z = 2abxyz^3 + f(x,y).$$

Any function possible for f(x,y) still leaves $V_z \neq V_x = V_y$ so this force is not conservative.

Part (c)

Under special conditions, the potential energy is equal to the work integral. That is

$$V = -\int_{r_0}^{r} \vec{F} \cdot d\vec{s}.$$

The necessary condition is that the integral be independent of the path from r_0 to r. Vector analysis shows that this is true if

$\vec{\nabla} \times \vec{F} = 0.$ For this given force,

$$\vec{\nabla} \times \vec{F} = \left(\frac{\partial}{\partial y} - \frac{\partial}{\partial x}\right) F_x(x)\hat{x} + \left(\frac{\partial}{\partial z} - \frac{\partial}{\partial x}\right) F_y(y)\hat{y}$$

$$+ \left(\frac{\partial}{\partial x} - \frac{\partial}{\partial y}\right) F_z(z)\hat{z} = 0,$$

given the functional dependences of the $\vec{F}$ components. Thus,

$$V = -\left[\int_{x_0}^{x} F_x(x)dx + \int_{y_0}^{y} F_y(y)dy + \int_{z_0}^{z} F_z(z)dz\right].$$

POWER AND EFFICIENCY

• **PROBLEM 10-21**

An electric motor is used to lift a 1600-lb block at a rate of 10 fpm. How much electric power (watts) must be supplied to the motor, if the motor lifting the block is 60 percent efficient? The weight W is 1600 lb.

Solution: The power requirement of the motor is the power exerted in lifting the block. The power exerted by the motor is the same as the power exerted by the force F in lifting the block. This is

$$P = Fv.$$

Since the block is moving at a constant velocity,

$$F = 1600 \text{ lb}$$

so that the power exerted by the motor is

$$P = Fv = (1600)\left(\frac{10}{60}\right) = 267 \text{ ft-lb/sec}$$

or, in terms of horsepower,

$$P = \frac{267}{550} = 0.485 \text{ hp.}$$

This is the power required to lift the block at the velocity required, and this is the power that the motor has to produce. However, since the motor is only 60 percent efficient, the power input to the motor must be greater. Thus,

$$P_{in} = \frac{P_{out}}{\varepsilon}$$

or $$P_{in} = \frac{0.485}{0.60} = 0.81 \text{ hp}$$

and the power input in terms of the electric power supplied to the motor is

$$P_{in} = 0.81 \times 746 = 605 \text{ watts.}$$

• PROBLEM 10-22

A chairlift is designed to transport 90 skiers of average mass 75 kg from base A to summit B at an inclination of 30° with an average speed of 80 m/min. Determine: a) the average power required; b) the required capacity of the motor if mechanical efficiency is 85% and a 300% overload is to be built in.

Solution: Power is defined as rate of energy change. In this case, the motor is giving up mechanical and electrical energy to increase the gravitational potential energy of the skiers. Neglecting friction losses, the power from the motor equals the power gained by the skiers. Since gravitational potential energy is given by E = mgh, the power gained is $P = \Delta\ (mgh)/\Delta t = mg\ \frac{\Delta h}{\Delta t}$.

Here, m = (75 kg) (90) = 6750 kg. With a foward velocity of 80 m/min. = 8/6 m/sec., the rate of height increase is:

$$\frac{\Delta h}{\Delta t} = \frac{8}{6} \text{ m/sec.)} \sin 30° = \frac{4}{6} \text{ m/sec.}$$

Thus, the power required is

$$\text{Power} = (6750 \text{ kg})\ (9.8 \text{ m/sec}^2)\left(\frac{4}{6} \text{ m/sec}\right) = 44.1 \text{ kW}$$

A 300% overload means that motor must be able to provide 4 times this figure or 176.4 kW. However, with an 85% efficiency, the actual power rating need is

$$\text{Power} = \frac{176.4\text{ kW}}{.85} = 207.5\text{ kW}.$$

• PROBLEM 10-23

> The crate shown weighs 1500 lbs. It is raised by means of a motor and counterweight. The counterweight exerts a constant force of 1000 lbs. The lifting ropes are inextensible. Calculate (a) the power needed to lift the crate at a constant rate of 25 ft/s, (b) the power used at the instant the crate is moving 25 ft/s and accelerating 5 ft/s² upward.

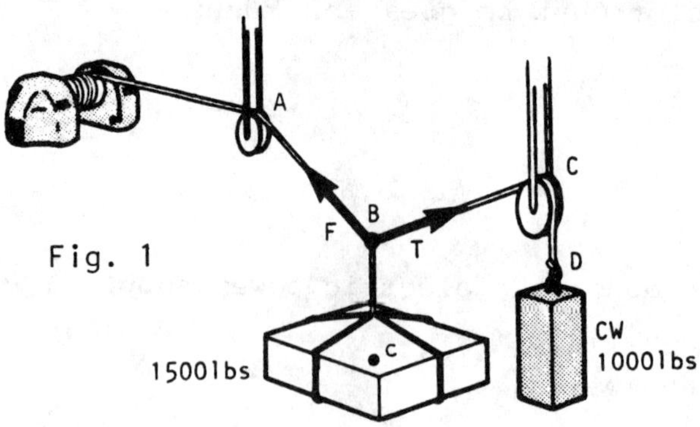

Fig. 1

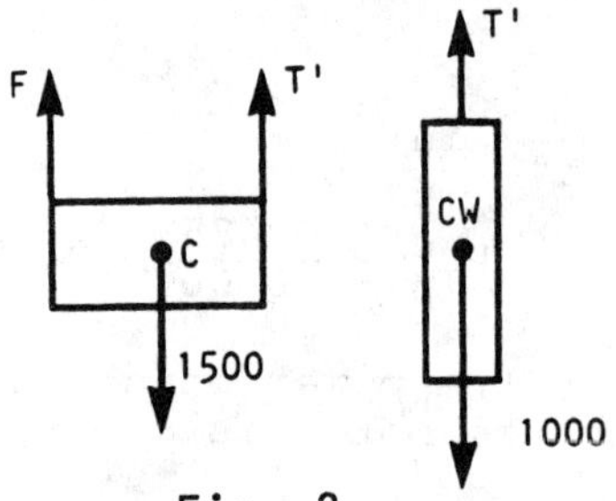

Fig. 2

Solution: The quantities $\vec{F}$ and $\vec{V}$ have the same direction. The equation expressing power is $P = \frac{d}{dt}F_d = FV$ for a constant force. The force $\vec{F}$ in rope AB must be determined for the two situations given.

a) Free body diagrams for of the crate and counterweight is shown in figure 2. By Newton's first law, the net force on the crate is zero when it is moving with constant velocity V. Assigning up as the positive direction we write:

$$\Sigma F_{vertical} = 0 \tag{1}$$

$$T - 1000 = 0 \tag{2}$$

$$F + T - 1500 = 0 \tag{3}$$

T can be eliminated from the force equilibrium equation (3). Solving for F we find F = 500 lbs. Power is

$$P_a = F_{vertical}V$$

$$P_a = (500\text{ lb})(25\text{ ft/sec})$$

$$P_a = 12{,}500 \text{ ft-lb/sec}$$

b) When the crate is accelerating upward, there is an accelerating force that gives rise to a dynamic equilibrium. Using Newton's Second Law

$$\Sigma F_{vertical} = ma_{vertical}$$

The dynamic equilibrium for the counterweight

$$T - W_{cw} = m_{cw} a_{cw}$$

The dynamic equilibrium for the crate

$$F + T - W_c = m_c a_c$$

Solving the equation for F,

$$T = m_{cw} a_{cw} + W_{cw}$$

$$F + (m_{cw} a_{cw} + W_{cw}) - W_c = m_c a_c$$

$$F = (m_c a_c + w_c) - (m_{cw} a_{cw} + W_{cw})$$

The counterweight and crate are connected by the inextensible rope BCD, so $-a_{cw} = a_c = a$

$$F = (m_c a + W_c) - (m_{cw}(-a) + W_{cw})$$

$$F = W_c[\frac{a}{g} + 1] - W_{cw}[\frac{-a}{g} + 1]$$

$$F = (W_c - W_{cw}) + (W_c + W_{cw})\frac{a}{g}$$

Substituting the values for a, g, W_c and W_{cw} yields

$$F = (1500 \text{ lb} - 1000 \text{ lb})$$

$$+ (1500 \text{ lb} + 1000 \text{ lb}) 5\text{ft/sec}^2 / 32.2 \text{ ft/sec}^2$$

$$F = 465.84 \text{ lbs.}$$

The power in this case = P_b

$$P_b = FV$$

$$P_b = (465.84 \text{ lb})(25 \text{ ft/sec})$$

$$P_b = 11{,}646 \text{ ft-lb/sec}$$

Using a conversion factor, 550 ft-lbs/sec = 1 hp

$$P_a = (12,500 \frac{\text{ft-lb}}{\text{sec}})(1 \text{ hp}/550 \text{ ft-lb/sec}) = 22.7 \text{ hp}$$

$$P_b = (11,205 \frac{\text{ft-lb}}{\text{sec}})(1 \text{ hp}/550 \text{ ft-lb/sec}) = 20.4 \text{ hp.}$$

FORCES AND EQUILIBRIUM

• PROBLEM 10-24

A frictionless one-dimensional spring-mass system consists of a mass of 8 g attached by means of a linear spring (k = 20 dyne/cm) to a rigid wall. Determine and plot the velocity displacement characteristics for the system at mechanical energies: 1000, 490 and 160 dyne-cm.

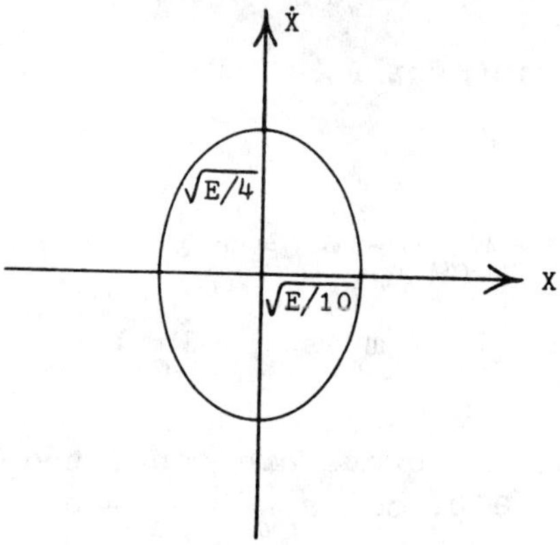

Solution: The governing equation of an oscillating spring-mass system is

$$\ddot{x} + \frac{k}{m} x = 0$$

where $\ddot{x}$ is the acceleration, x is the displacement, k is the spring rate and m is the mass. The general solution of this equation is

$$x = A \cos \sqrt{\frac{k}{m}}\, t + B \sin \sqrt{\frac{k}{m}}\, t \qquad (1)$$

where A and B are arbitrary constants that are to be found from the initial conditions. For the present case we stipulate that at t = 0 the displacement is zero and the kinetic energy is $E = \frac{1}{2} m\dot{x}^2$. To apply these we first differentiate (1) to obtain the velocity

$$\dot{x} = - A \sqrt{\frac{k}{m}}\ 8m \sqrt{\frac{k}{m}}\, t + B \sqrt{\frac{k}{m}} \cos \sqrt{\frac{k}{m}}\, t \, . \qquad (2)$$

If we apply the zero displacement condition we find that A = 0. Then the energy condition is found from

$$KE = \frac{1}{2} m\dot{x}^2$$

$$KE = \frac{1}{2} m \; B^2 \left(\frac{k}{m}\right) \cos^2 \left(\sqrt{\frac{k}{m}}\, t\right).$$

at t = 0 $\cos^2 \left(\sqrt{\frac{k}{m}}\, t\right) = 1$, $KE = KE_{max}$ and PE = 0

therefore at t = 0 KE_{max} = total mechanical energy = E

$$E = \frac{1}{2} mB^2 \frac{k}{m}$$

$$B^2 = \frac{2E}{k}$$

Thus $B = \sqrt{\frac{2E}{k}}$

and the displacement and velocity solutions become

$$x = \sqrt{\frac{2E}{k}} \sin \sqrt{\frac{k}{m}}\, t$$

$$\dot{x} = \sqrt{\frac{2E}{m}} \cos \sqrt{\frac{k}{m}}\, t.$$

If we square these, add them, and observe that $\sin^2 \alpha + \cos^2 \alpha = 1$, we find that

$$\dot{x}^2 = \frac{2E - x^2 k}{m} \quad .$$

Now for m = 8 g and k = 20 dyne/cm this becomes

$$\dot{x}^2 = \tfrac{1}{4} (E - 10x^2).$$

This is further rearranged to give

$$\frac{(\dot{x})^2}{(E/4)} + \frac{x^2}{(E/10)} = 1$$

and it is seen that this is the equation of an ellipse when the co-ordinate axes are chosen to be x and $\dot{x}$. The semi-axes of the ellipse would be $\sqrt{E/4}$ along the $\dot{x}$ direction and $\sqrt{E/10}$ along the x direction, as shown.

• PROBLEM 10-25

A mass m is supported by two linear springs (k_1, k_2) located between rigid supports. Determine the equilibrium position of the mass in terms of the spring constants, the undeformed spring lengths ℓ_1 and ℓ_2, and the distance L between supports.

Solution: The deformation δ of each spring may be described in terms of the variable y, the vertical position of the mass above the lower support:

δ = (final length) - (undeformed length)

$\delta_1 = (L - y) - \ell_1$

$\delta_2 = y - \ell_2.$

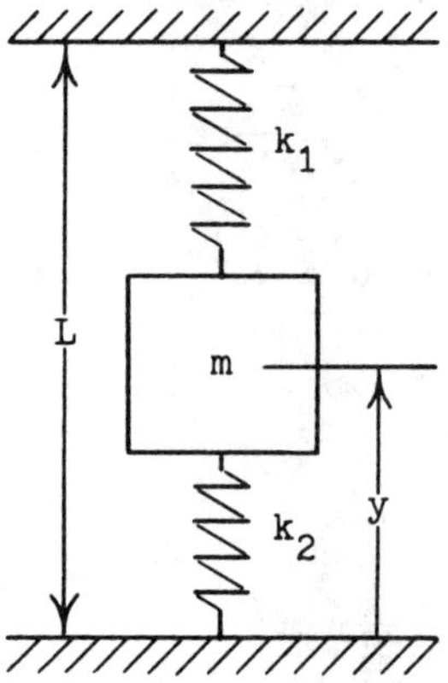

The system potential energy is determined by adding the gravitational and elastic components:

$$V(y) = mgy + \tfrac{1}{2}k_1\delta_1^2 + \tfrac{1}{2}k_2\delta_2^2$$

$$= mgy + \tfrac{1}{2}k_1(L - y - \ell_1)^2 + \tfrac{1}{2}k_2(y - \ell_2)^2.$$

Next we set dV/dy = 0 to obtain the equilibrium position.

$$\frac{dV}{dy} = 0 = mg - k_1(L - y - \ell_1) + k_2(y - \ell_2)$$

Solving for y (which is now y_{eq}) the equilibrium position is

$$y_{eq} = \frac{k_1(L - \ell_1) + k_2\ell_2 - mg}{k_1 + k_2}.$$

Since $d^2V/dy^2 = k_1 + k_2$, a positive number for all y, the equilibrium is stable at y_{eq}.

• PROBLEM 10-26

The potential energy for the force between two atoms in a diatomic molecule is approximately

$$V(x) = -\frac{a}{x^6} + \frac{b}{x^{12}} .$$

a and b are positive constants and x is the distance between the atoms.

(a) Find the force.

(b) Assuming one atom is heavy and remains at rest, describe the possible motions for the other atom.

(c) Find the equilibrium distance and the period of small oscillations about the equilibrium distance for the light atom of mass m.

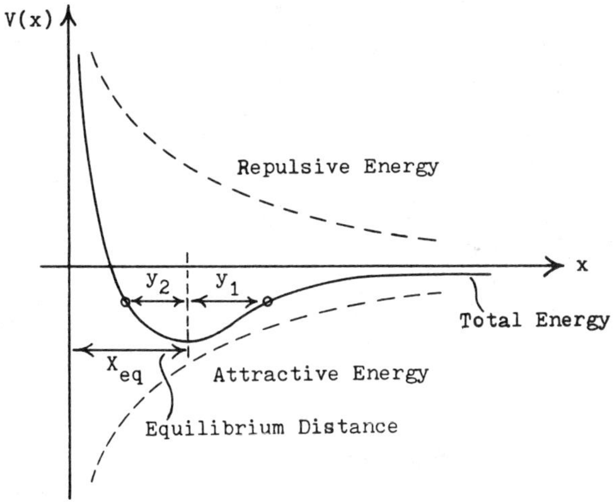

fig 1 Potential Trough

Solution: The potential energy is given as

$$V(x) = -ax^{-6} + bx^{-12} \qquad \text{(see Fig. 1)}.$$

The force is the first derivative of V(x).

a) $$F(x) = -\frac{d\,V(x)}{dx} = -6ax^{-7} + 12bx^{-13} \qquad \text{(see Fig. 2)}$$

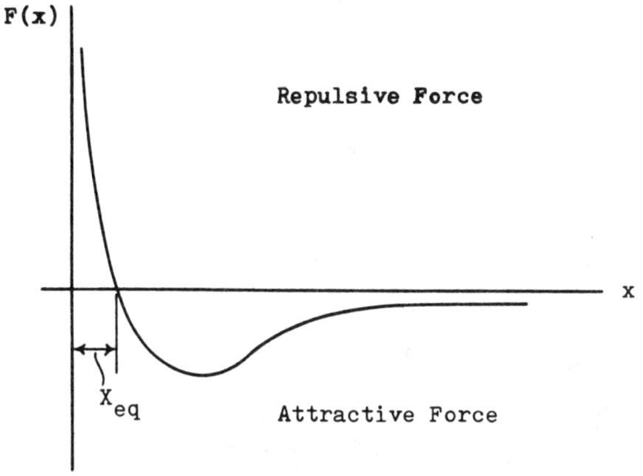

fig 2 Force

b) The particle is in equilibrium if the potential energy is a minimum, i.e., the force is zero. The particle oscillates about its equilibrium position within limits given by the potential trough; (see y_1 and y_2, Fig. 1), depending on the amount of energy (heat) supplied. Because the trough is not symmetrical, oscillation is larger to one side. The particle can escape if the energy supplied is sufficient to "move it out of the trough". This energy is less than the integral under the force curve from x_{eq} to infinity.

Since $W = -\int \vec{F} \cdot d\vec{s}$.

c) To find x_{eq}, set the force equal to zero

$$F(x) = -6ax^{-7} + 12bx^{-13} = 0 \therefore$$

Solving for x

$$x_{eq}^6 = \frac{12\ b}{6\ a} \quad \therefore \quad x_{eq} = (2\ b/a)^{1/6}.$$

The force near the equlibrium position can be expanded into a Taylor series

$$F(x) = F(x_{eq}) + F^1(x_{eq})(x - x_{eq}) + \frac{F^{11}(x_{eq})(x - x_{eq})^2}{2!} + \ldots$$

$F(x_{eq}) = 0$ from above.

Calling $y = x - x_{eq}$, the amplitude of small oscillations, y^2 is small compared to y, hence

$$F(x) \overset{\circ}{=} F^1(x_{eq})\ y$$

i.e. for small oscillations the force is a linear function of the displacement y.

For small oscillations, the force acts as a linear spring with spring constant

$$k = F^1(x_{eq})$$

$$F^1(x) = \frac{d}{dx}(-6ax^{-7} + 12bx^{-13}) = +42ax^{-8} - 156bx^{-14}$$

Substituting $x_{eq} = (2b/a)^{1/6}$ for x

$$k = F^1(x_{eq}) = +42a(2b/a)^{-8/6} - 156b(2b/a)^{-14/6}$$

$$= 42a\left(\frac{2^8b^8}{a^8}\right)^{-1/6} - 156b\left(\frac{2^{14}\ b^{14}}{a^{14}}\right)^{-1/6}$$

$$= 42\left(\frac{2^8\ b^8}{a^{8+6}}\right)^{-1/6} - 156\left(\frac{2^{14}\ b^{14-6}}{a^{14}}\right)^{-1/6}$$

$$= 42\left(\frac{2^8\ b^8}{a^{14}}\right)^{-1/6} - 156\left(\frac{2^8\ b^8}{a^{14}}\right) 2^{-1}$$

$$= (42 - 78)\left(\frac{a^{14}}{2^8\ b^8}\right)^{1/6}$$

$$= -36\left(\frac{1}{2^{8/6}}\right)\left(\frac{a^{14}}{b^8}\right)^{1/6} = -36\frac{1}{2^{12/6}\ 2^{-4/6}}\left(\frac{a^{14}}{b^8}\right)^{1/6}$$

$$= -\frac{36}{4}\ 2^{2/3}\left(\frac{a^{14}}{b^8}\right)^{1/6}$$

$$k = -9\,(2^{2/3})\left(\frac{a^{14}}{b^8}\right)^{1/6}.$$

So, dropping the smaller terms in the Taylor series expansion, we have

$$F(x) = -ky \qquad \text{where}$$

$$k = 9(2^{2/3})(a^{14}/b^8)^{1/6}.$$

The motion is governed by the differential equation

$\ddot{x} + \frac{k}{m}x = 0$ where the frequency $\omega = \sqrt{\frac{k}{m}}$ and the period is

$$T = \frac{2\pi}{\omega} = \frac{2\pi}{\sqrt{\frac{k}{m}}}$$

$$T = \frac{2\pi}{\sqrt{\dfrac{9 \times 2^{2/3}\left(\frac{a^{14}}{b^8}\right)^{1/6}}{m}}}$$

$$T = \frac{2\pi}{\left[\dfrac{9(2^{2/3})\left(\frac{a^{14}}{b^8}\right)^{1/6}}{m}\right]^{\frac{1}{2}}}$$

$$= \frac{2\pi}{3}\left(\frac{m^{\frac{1}{2}}}{2^{1/3}\left(\frac{a^7}{b^4}\right)^{1/6}}\right)$$

$$= \frac{2\pi}{3}\left(\frac{m^3\ b^4}{4\ a^7}\right)^{1/6}$$

• PROBLEM 10-27

According to Yakawa's theory of nuclear forces, the attractive force between a neutron and a proton has a potential

$$V(r) = \frac{Ke^{-\alpha r}}{r} \qquad K < 0.$$

(a) Find the force.
(b) Discuss types of motion possible for mass m under such a force.
(c) Find L and E for motion on a circle of radius a.
(d) Find the period of circular motion and the period of small radial oscillations.

Solution: (a) The force is equal to the negative gradient of the potential function,

$$\vec{F} = -\vec{\nabla}V.$$

Although the gradient may be expressed in terms of any coordinate system, the mathematics involved is considerably simplified if we express the gradient in terms of polar coordinates. The gradient expressed in terms of polar coordinates is

$$\vec{\nabla} = \hat{r}\frac{\partial}{\partial r} + \hat{\theta}\frac{1}{r}\frac{\partial}{\partial \theta} + \hat{\phi}\frac{1}{r \sin\theta}\frac{\partial}{\partial \phi} .$$

The force equals,

$$\vec{F} = -\vec{\nabla}V(r) = -\left(\hat{r}\frac{\partial}{\partial r}\right) V(r)$$

since $\dfrac{\partial V(r)}{\partial \theta} = 0$ and $\dfrac{\partial V(r)}{\partial \phi} = 0.$

Differentiating, we obtain

$$\vec{F} = -\hat{r}\frac{\partial}{\partial r}\left(\frac{Ke^{-\alpha r}}{r}\right) = -\hat{r}\left[\frac{K}{r}\frac{\partial}{\partial r}e^{-\alpha r} + e^{-\alpha r}\frac{\partial\left(\frac{K}{r}\right)}{\partial r}\right]$$

$$\vec{F} = -\hat{r}\left[-\alpha e^{-\alpha r}\frac{K}{r} - \frac{K}{r^2}e^{-\alpha r}\right]$$

$$= \hat{r}\left[\alpha e^{-\alpha r}\frac{K}{r} + \frac{K}{r^2}e^{-\alpha r}\right]$$

factoring $\dfrac{Ke^{-\alpha r}}{r^2}$ from each term, we have

$$\vec{F} = \hat{r}\frac{Ke^{-\alpha r}}{r^2}(\alpha r + 1).$$

With the exception of K, all the other scalars occurring in the expression $\vec{F}$ are positive. Therefore, the product of all the scalars is negative and, since the unit vector $\vec{r}$ is directed from the origin toward the particle, the force, $\vec{F}$, is directed from the particle toward the origin. Thus, the force, $\vec{F}$, is an attractive force with a magnitude,

$$F = |K|\frac{e^{-\alpha r}}{r^2}(\alpha r + 1)$$

(b) Since the force is radial only, the entire motion of the particle is confined to a plane containing the origin and the velocity vectors.

In other words, $\vec{L}$, the angular momentum, is constant in magnitude and direction. The angular momentum of the particle about the origin is

$$\vec{L} = \vec{r} \times m\vec{v} \tag{1}$$

where $\vec{L}$ is perpendicular to both $\vec{r}$ and $\vec{v}$. Since $\vec{L}$ has a fixed orientation and since both $\vec{r}$ and $\vec{v}$ are perpendicular to $\vec{L}$, the motion of the particle is confined to a plane which contains both $\vec{r}$ and $\vec{v}$ and which is perpendicular to $\vec{L}$.

Having established that the motion of the particle is confined to single plane, we will establish, in what is to follow, a general but qualitative discussion of the types of motion possible for the particle. This will be done by analyzing the effective potential energy diagrams for the particle.

Because a potential energy function exists, the mechanical energy of the particle is conserved. Thus, we write,

Mechanical Energy = Kinetic Energy + Potential Energy

$$E = \frac{1}{2} mv^2 + K \frac{e^{-\alpha r}}{r} \quad . \tag{2}$$

The location of the particle is completed as specified by the polar coordinates r and θ, both of which are measured within the plane, which is perpendicular to $\vec{L}$ and which contains the origin and the particle.

The velocity of the particle expressed in polar coordinates is

$$\vec{v} = \dot{r}\,\hat{r} + r\,\dot{\theta}\,\hat{\theta}$$

The speed squared is

$$v^2 = \dot{r}^2 + r^2\,\dot{\theta}^2$$

Substituting the expression for v^2 into equation 1:

$$E = \frac{1}{2} m\,(\dot{r}^2 + r^2\dot{\theta}^2) + K \frac{e^{-\alpha r}}{r} \quad \text{or}$$

$$E = \frac{1}{2} m\dot{r}^2 + \frac{1}{2} mr^2\dot{\theta}^2 + K \frac{e^{-\alpha r}}{r} \quad . \tag{4}$$

The second term on the right hand side of the equation may be expressed in terms of the angular momentum, the mass and the radial distance.

$$\vec{L} = \vec{r} \times m\vec{v} = (r\hat{r}) \times m(\dot{r}\hat{r} + r\dot{\theta}\hat{\theta})$$

Noting that $\hat{r} \times \hat{r} = 0$, and $\hat{r} \times \hat{\theta} = \hat{K}$ where $\hat{K}$ is a unit vector perpendicular to both $\hat{r}$ and $\hat{\theta}$, after distributing the vector terms, we obtain

$$\vec{L} = rm\dot{r}\hat{r} \times \hat{r} + mr^2\dot{\theta}\hat{r} \times \hat{\theta}$$

$$\vec{L} = mr^2\dot{\theta}\hat{K}.$$

The magnitude of $\vec{L}$ is constant and is

$$L = mr^2\dot{\theta}.$$

Dividing both sides of the equation by mr^2 we obtain

$$\dot{\theta} = \frac{L}{mr^2} \quad . \tag{5}$$

Substituting this expression into the second term of equation 4 yields,

$$E = \frac{1}{2}\, m\dot{r}^2 + \frac{1}{2}\, mr^2 \left(\frac{L}{mr^2}\right)^2 + K\,\frac{e^{-\alpha r}}{r}$$

Simplifying,

$$E = \frac{1}{2}\, m\dot{r}^2 + \frac{L^2}{2mr^2} + K\,\frac{e^{-\alpha r}}{r} \quad . \tag{6}$$

Let us define as the effective potential energy

$$V'(r) = \frac{L^2}{2mr^2} + \frac{Ke^{-\alpha r}}{r} \quad . \tag{7}$$

It will be shown how a plot of $V'(r)$ against r yields information on the possible types of motion of the particle.

Factoring out $\frac{(-K)}{\alpha r^2}$ from both terms of equation 7 we obtain,

$$V'(r) = \frac{(-K)}{\alpha r^2}\left(\frac{\alpha L^2}{2m\ (-K)} - \alpha r\ e^{-\alpha r}\right)$$

Define a constant

$$\sigma \equiv \frac{\alpha L^2}{2m\ (-K)} \tag{8}$$

and substitute it into the expression above for $V'(r)$, then

$$V'(r) = \frac{(-K)}{\alpha r^2}\ (\sigma - \alpha r\ e^{-\alpha r}) \tag{9}$$

Note that, since $(-K) > 0$, $\alpha > 0$ and $m > 0$, that $\sigma > 0$, if $L \neq 0$. Furthermore, $\frac{(-K)}{\alpha r^2} > 0$. Thus the sign of $V'(r)$ will be the same as the factor

$$(\sigma - \alpha re^{-\alpha r}).$$

Both the sign and the roots of $V'(r)$ are identical to those of the factor. In other words, the range of values of r corresponding to positive or negative $V'(r)$ are the same as those of the factor. We will study the behavior of the factor, which is more tractable than the original function $V'(r)$, in order to determine the behavior of $V'(r)$. We will show that if $\sigma > \frac{1}{e}$, then $V'(r)$ has no roots and is positive for all values of r; and if $\sigma < \frac{1}{e}$, then $V'(r)$ has two roots and therefore is negative for a finite range of values of r. The situation corresponding to $\sigma = \frac{1}{e}$ is one in which $V'(r)$ is positive for all values of r, except for $r = \frac{1}{\alpha}$ where $V'\left(\frac{1}{\alpha}\right) = 0$.

The factor $(\sigma - \alpha r e^{-\alpha r})$ can be written in a more convenient form if $x = \alpha r$; then

$$\sigma - \alpha r e^{-\alpha r} = \sigma - x e^{-x}.$$

We now sketch the factor $\sigma - xe^{-x}$ versus x for values of σ greater than, less than and equal to $\frac{1}{e}$. Figure 1 is a sketch of the factor for $\sigma = \frac{1}{e}$.

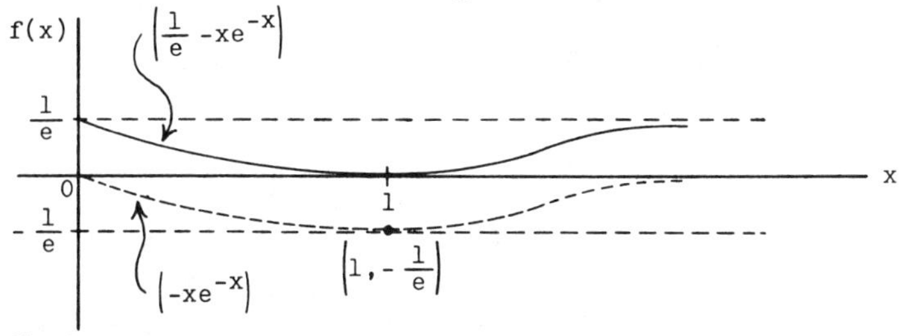

fig 1

Note that when $\sigma = \frac{1}{e}$, $\left(\frac{1}{e} - xe^{-x}\right)$ has only one root and is positve elsewhere. The root occurs when $x^* = 1$. But since $\alpha r^* = x^* = 1$, $r^* = \frac{1}{\alpha}$.

Figure 2 is a sketch of $V'(r)$ for $\sigma = \frac{1}{e}$, where equation 9 has been sketched with $\frac{1}{e} = \sigma = \frac{\alpha L^2}{2m\ (-\ K)}$.

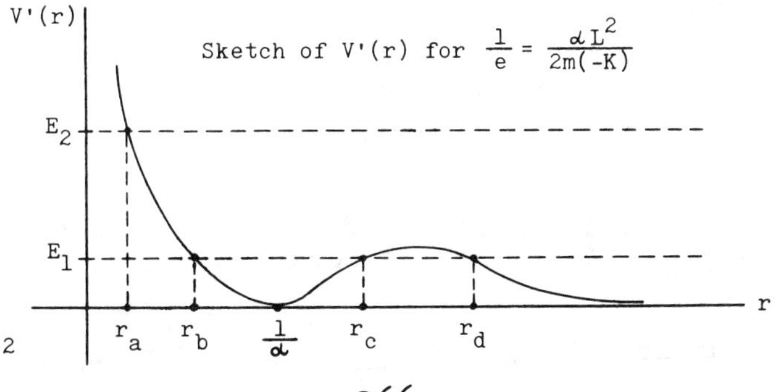

fig 2

Equation 9 with $\sigma = \frac{1}{e}$ is

$$V'(r) = \frac{(-K)}{\alpha r^2}\left(\frac{1}{e} - \alpha r e^{-\alpha r}\right).$$

Digressing for the moment, let us re-write equation 6 by substituting equation 7 for the last two terms in equation 6.

$$E = \frac{1}{2} m\dot{r}^2 + \frac{L^2}{2mr^2} + \frac{Ke^{-\alpha r}}{r} = \frac{1}{2} m\dot{r}^2 + V'(r)$$

$$E = \frac{1}{2} m\dot{r}^2 + V'(r). \tag{10}$$

We see in this equation that when $V'(r) = E$, then

$$E = \tfrac{1}{2} m\dot{r}^2 + E$$

and thus $\frac{1}{2} m\dot{r}^2 = 0$; but then $\dot{r} = 0$ and so $\dot{r} = 0$. This means that the particle's radial speed is zero and that the particle at that moment is moving with a tangential velocity only providing $L \neq 0$. (It is moving in a closed loop, with velocity tangent to its path).

Again looking at equation 10 we see that if $V'(r) > E$, then $\dot{r}^2 < 0$; a result that is not allowed since it must be a real number.

In Figure 2, the quantities E_1 and E_2 represent the mechanical energy of the particle for two independent situations. The quantities r_a, r_b, r_c and r_d represent the values of r for which $V'(r) = E$ and thus where $\dot{r} = 0$. These points are called the turning points. When the particle reaches a turning point, it immediately reverses its radial direction of motion (after momentarily coming to rest in its radial motion), and then proceeds to move radially in the opposite direction. For example, if the mechanical energy of the particle is E_1 and also has a value of r less than r_c, it will oscillate radially between the extremes $r = r_c$ and $r = r_b$. The region r less than r_b is forbidden because there $V'(r)$ would be greater than E_1.

On the other hand, if the mechanical energy of the particle is E_2 or if the particle has a mechanical energy of E_1 and also a value of r greater than r_d, then the particle's radial motion is not confined to finite range of values of r. Upon encountering the turning point (r_a or r_d, respectively) it reverses its motion and continues to recede from the origin.

A special but interesting situation arises whenever the radial mechanical energy is zero. Looking at Figure 2 we see that this occurs for $r = \frac{1}{\alpha}$ only and that both the mechanical and effective potential energy are zero. This implies that $r = 0$, since $E = V\left(\frac{1}{\alpha}\right) = 0$. Thus, the particle orbits the origin in a circular path of radius $r = \frac{1}{\alpha}$. From equation 5 we write,

$$\dot{\theta} = \frac{L}{mr^2}$$

and since all the quantities on the right hand side of the equation are constant, it follows that the angular velocity is constant. The particle travels about the origin in uniform circular motion. This completes the basic description of possible motions since changing the value of σ moves the factor up or down but does not change the shape from which the variations of motion are derived.

Having discussed qualitatively the general types of motion possible with $L \neq 0$, consider the case when $L = 0$. From equation 5

$$\dot{\theta} = \frac{L^2}{2mr^2} = \frac{0^2}{2mr^2} = 0$$

at all times. But if $\dot{\theta} = 0$ at all times then θ = constant and the motion of the particle is confined to a straight line which passes through the origin.

When $L = 0$, $V' = \frac{Ke^{-\alpha r}}{r}$ (see equation 7). The sketch for $V'(r)$ when $L = 0$ is shown in Figure 3.

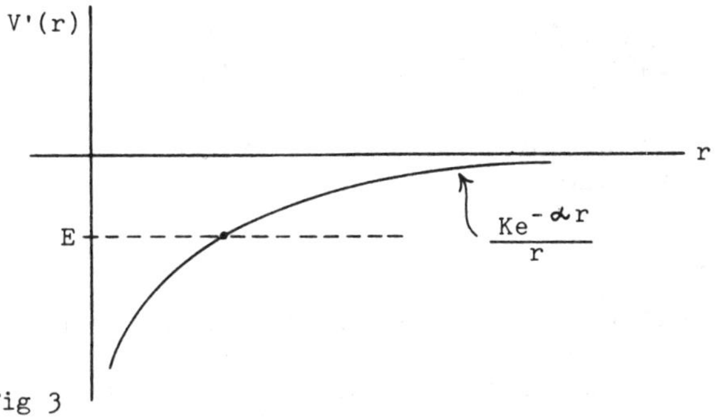

fig 3

Note that if the mechanical energy of the particle is zero or positive, the particle will continue to recede indefinitely if it is moving away from the origin. If its mechanical energy is negative, the particle will reflect back toward the origin.

(c) The angular momentum may be obtained from equation 5. Therefore $\dot{\theta} = \frac{L}{mr^2}$, and since the particle is moving in a circular path of radius a, $\dot{\theta} = \frac{L}{ma^2}$. Thus, the particle moves with constant angular velocity. However we should express both L and E in terms of the given constants K, α, a, and m. To accomplish this it is necessary to obtain an independent relation involving $\dot{\theta}$ and those constants. According to Newton's Second Law of Motion, the resultant force on the particle moving with uniform circular motion equals the mass of the particle times it centripetal acceleration:

$$\vec{F} = m\vec{A}, \tag{11}$$

where $\vec{A} = -a\dot{\theta}^2\hat{r}$, and (from part a) substituting r = a we obtain

$$\vec{F} = (1 + \alpha a)\frac{Ke^{-\alpha a}}{a^2}\hat{r}.$$

Substituting these expressions for $\vec{A}$ and $\vec{F}$ into equation 11, yields

$$(1 + \alpha a)\frac{Ke^{-\alpha a}}{a^2}\hat{r} = m(-a\dot{\theta}^2\hat{r})$$

$$\frac{(1 + \alpha a)Ke^{-\alpha a}}{a^2} = -ma\dot{\theta}^2.$$

Dividing both sides by - ma yields,

$$-\frac{(1 + \alpha a)Ke^{-\alpha a}}{ma^3} = \dot{\theta}^2. \tag{12}$$

Substituting the expression for $\dot{\theta}$ from equation 5, $\dot{\theta} = \frac{L}{ma^2}$ into equation (12):

$$-\frac{(1 + \alpha a)\ Ke^{-\alpha a}}{ma^3} = \left(\frac{L}{ma^2}\right)^2 = \frac{L^2}{m^2a^4}$$

Multiplying both sides of the equation by m^2a^4, we obtain the result we sought,

$$-ma(1 + \alpha a)Ke^{-\alpha a} = L^2. \tag{12a}$$

L^2 is expressed in terms of the constants.

Equation 6 allows for the mechanical energy of the particle

$$E = \frac{1}{2}m\dot{r}^2 + \frac{L^2}{2mr^2} + \frac{Ke^{-\alpha r}}{r}.$$

Since the particle moves in a circular path of radius a, r must be constant, thus $\dot{r} = 0$. Substituting this result and the expression for L^2 (equation 12a) into equation 6, we obtain

$$E = \frac{-ma(1 + \alpha a)Ke^{-\alpha a}}{2ma^2} + \frac{Ke^{-\alpha a}}{a}$$

$$E = -(1 + \alpha a)\frac{Ke^{-\alpha a}}{2a} + \frac{Ke^{-\alpha a}}{a}$$

factoring $\frac{Ke^{-\alpha a}}{2a}$

$$E = (-\ (1 + \alpha a) + 2)\ \frac{Ke^{-\alpha a}}{2a}$$

$$E = (1 - \alpha a)\ \frac{Ke^{-\alpha a}}{2a}\ .$$

(d) The product of the angular velocity of the particle, undergoing uniform circular motion, and the period of time it takes to complete one revolution equals 2π radians. Therefore,

$$2\pi = \dot{\theta}\ T \tag{13}$$

From equation 12:

$$\frac{-\ (1 + \alpha a)Ke^{-\alpha a}}{ma^3} = \dot{\theta}^2 .$$

Taking the square root of both sides of the equation,

$$\left[\frac{-\ (1 + \alpha a)Ke^{-\alpha a}}{ma}\right]^{\frac{1}{2}} = \dot{\theta}\ .$$

Substituting this expression for $\dot{\theta}$ in equation 13:

$$2\pi = \left[\frac{-\ (1 + \alpha a)Ke^{-\alpha a}}{ma^3}\right]^{\frac{1}{2}} T$$

$$T = 2\pi \left[\frac{-\ (1 + \alpha a)Ke^{-\alpha a}}{ma^3}\right]^{-\frac{1}{2}} \tag{14}$$

The period, T corresponds to the particle having a minimum mechanical energy. However, if its mechanical energy is slightly higher than the minimum, then the particle will execute radial oscillations about the point $r = a$, while moving in a quasi-circular path of radius a with a period approximated by the expression given in equation 14.

Assuming that the radial displacements are small, we may then expand $V'(r)$ in a Taylor series about the point $r = a$, retaining terms up to second order in $(r - a)$.

Taylor series expansion of $V'(r)$ about $r = a$:

$$V'(r) = V'(a) + \left(\frac{dV'}{dr}\right)_a (r - a) + \frac{1}{2}\left(\frac{d^2V'}{dr}\right)_a (r - a)^2 + \ldots$$

retain terms up to second order in $(r - a)$,

$$V'(r) \overset{\circ}{=} V'(a) + \left(\frac{dV'}{dr}\right)_a (r - a) + \frac{1}{2}\left(\frac{d^2V'}{dr}\right)_a (r - a)^2$$

Noting that V'(r) is a minimum at r - a, and therefore that $\left(\frac{dV'}{dr}\right)_a = 0$ we write,

$$V'(r) = V'(a) + 0\ (r - a) + \frac{1}{2}\left(\frac{d^2V'}{dr^2}\right)_a (r - a)^2$$

$$V'(r) = V'(a) + \frac{1}{2}\left(\frac{d^2V'}{dr^2}\right)_a (r - a)^2. \qquad (15)$$

The mechanical energy of the particle is given by equation 10:

$$E = \tfrac{1}{2}\ m\dot{r}^2 + V'(r)$$

Substituting the expression for V'(r) from equation 16 we find,

$$E = \frac{1}{2}\ m\dot{r}^2 + V'(a) + \frac{1}{2}\left(\frac{d^2V'}{dr^2}\right)_a (r - a)^2$$

Subtracting V'(a) from both sides of this equation and defining $R = \left(\frac{d^2V'}{dr}\right)_a$:

$$E - V'(a) = \tfrac{1}{2}\ m\dot{r}^2 + \tfrac{1}{2}\ R\ (r - a)^2. \qquad (16)$$

Both E and V'(a) are constants, therefore E - V'(a) is a constant, which will be called E'.

Substituting E' for the expression on the right hand side of equation 17:

$$E' = \tfrac{1}{2}\ m\dot{r}^2 + \tfrac{1}{2}\ R(r - a)^2$$

The general solution of this first order differential equation is

$$r(t) = \left[\frac{2E'}{R}\right]^{\frac{1}{2}} \sin\left[\left(\frac{R}{m}\right)^{\frac{1}{2}} t + \phi_0\right] + a \qquad (17)$$

where t is a particular instant of time and ϕ_0 is a constant chosen to satisfy the initial conditions.

As long as the radial displacements are small, then the approximations made leading up to equation 17 are justified, and the particle undergoes simple harmonic radial motion about the point r = a. The period for this harmonic motion, implied by equation 13, is

$$T_r = \frac{2\pi}{\left(\frac{R}{m}\right)^{\frac{1}{2}}} = 2\pi\left(\frac{m}{R}\right)^{\frac{1}{2}} \qquad (18)$$

where the constant R was deined earlier to be,

$$R = \left(\frac{d^2V'}{dr^2}\right)_a .$$

To obtain the period of small radial oscillations, first differentiate the expression given by equation 7 for V'(r) twice and evaluate it for r = a. Finally, the quantity will be substituted into equation 18 to obtain the period, T_r.

Equation 7 yields

$$V'(r) = \frac{L^2}{2mr^2} + \frac{Ke^{-\alpha r}}{r} .$$

Differentiating both sides with respect to r:

$$\frac{d\ V'(r)}{dr} = -\ 2\ \frac{L^2}{2mr^3} - \frac{Ke^{-\alpha r}}{r^2} - \frac{\alpha Ke^{-\alpha r}}{r}$$

$$\frac{dV'(r)}{dr} = -\ \frac{L^2}{mr^3} - \frac{Ke^{-\alpha r}}{r^2} - \frac{\alpha Ke^{-\alpha r}}{r}$$

Differentiating again with respect to r:

$$\frac{d^2V'(r)}{dr^2} = 3\ \frac{L^2}{mr^4} + 2\ \frac{Ke^{-\alpha r}}{r^3} + \frac{\alpha Ke^{-\alpha r}}{r^2} + \frac{\alpha Ke^{-\alpha r}}{r^2} + \frac{\alpha^2 Ke^{-\alpha r}}{r}$$

Now evaluating the expression at r = a the following is obtained:

$$\left(\frac{d^2V'}{dr^2}\right)_a = 3\ \frac{L^2}{ma^4} + 2\ \frac{Ke^{-\alpha a}}{a^3} + \frac{\alpha Ke^{-\alpha a}}{a^2} + \frac{\alpha Ke^{-\alpha a}}{a^2} + \frac{\alpha^2 Ke^{-\alpha a}}{a} .$$

Combining the four terms which have the common factor $\frac{Ke^{-\alpha a}}{a}$ into a single term, we obtain

$$\left(\frac{d^2V'}{dr^2}\right)_a = 3\ \frac{L^2}{ma^4} + \left(\frac{2}{a^2} + \frac{\alpha}{a} + \frac{\alpha}{a} + \alpha^2\right) \frac{Ke^{-\alpha a}}{a} . \qquad (19)$$

L^2 was evaluatdd in part c where

$$L^2 = -\ ma\ (1 + \alpha a)Ke^{-\alpha a} .$$

Substituting this expression into equation 19:

$$\left(\frac{d^2V'}{dr^4}\right)_a = 3\ \frac{[-\ ma(1 + \alpha a)Ke^{-\alpha a}]}{ma^4} + \left(\frac{2}{a^2} + \frac{2\alpha}{a} + \alpha^2\right) \frac{Ke^{-\alpha a}}{a}$$

$$= -\ 3\ \frac{(1 + \alpha a)Ke^{-\alpha a}}{a^3} + \left(\frac{2}{a^2} + \frac{2\alpha}{a} + \alpha^2\right) \frac{Ke^{-\alpha a}}{a}$$

Factoring the common product term $\frac{Ke^{-\alpha a}}{a}$ yields:

$$\left(\frac{d^2V'}{dr^2}\right)_a = \left(-3\frac{(1+\alpha a)}{a^2} + \frac{2}{a^2} + \frac{2\alpha}{a} + \alpha^2\right)\frac{Ke^{-\alpha a}}{a} .$$

Rewriting the terms in parentheses with the common denominator a,

$$\left(\frac{d^2V'}{dr^2}\right)_a = \left(\frac{-3(1+\alpha a) + 2 + 2\alpha a + \alpha^2 a^2}{a^2}\right)\frac{Ke^{-\alpha a}}{a}$$

Combining terms, we obtain

$$\left(\frac{d^2V'}{dr^2}\right)_a = \left(\frac{-3 - 3\alpha a + 2 + 2\alpha a + \alpha^2 a^2}{a^2}\right)\frac{Ke^{-\alpha a}}{a}$$

$$\left(\frac{d^2V'}{dr^2}\right)_a = \left(\frac{-1 - \alpha a + \alpha^2 a^2}{a^2}\right)\frac{Ke^{-\alpha a}}{a} .$$

Rearranging terms and combining common factors,

$$\left(\frac{d^2V'}{dr^2}\right)_a = -\frac{(1 + \alpha a - \alpha^2 a^2)\,Ke^{-\alpha a}}{a^3}$$

Substituting this expression into equation 18, we obtain for the period of small radial oscillations,

$$T_r = 2\pi\left[\frac{m}{\frac{-(1+\alpha a - \alpha^2 a^2)Ke^{-\alpha a}}{a^3}}\right]^{\frac{1}{2}}$$

$$= 2\pi\left[\frac{ma^3}{-(1+\alpha a - \alpha^2 a^2)Ke^{-\alpha a}}\right]^{\frac{1}{2}} .$$

Inverting the quantity in brackets and changing the sign of the exponent yields,

$$T_r = 2\pi\left[\frac{-(1+\alpha a - \alpha^2 a^2)Ke^{-\alpha a}}{ma^3}\right]^{\frac{1}{2}} .$$

CHAPTER 11

IMPULSE AND MOMENTUM

IMPULSIVE FORCES IN ELASTIC IMPACTS

• PROBLEM 11-1

Two ice skaters holding the ends of a 20-ft inextensible rope, as shown in Fig. 1, pull on the rope so that they are drawn toward each other. One skater weighs 150 lb and the other skater weighs 200 lb. Friction between the skates and the ice can be considered negligible. The surface of the ice is horizontal. a) Determine the relationship between the speeds of the two skaters as they approach each other if they were at rest when they started to pull. b) If the distance between the two skaters is decreasing at the rate of 14 ft/sec after they start to move, what will be the speeds of the two skaters when they meet?

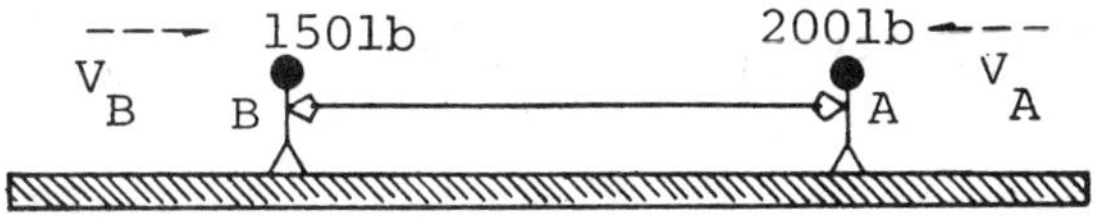

Fig. 1

Solution: a) Consider the two skaters and the rope as a system. Then the only forces external to the system are vertical. Hence, linear momentum is conserved horizontally. Since initial momentum is zero, the final momentum must be zero. If we choose vectors to the right as positive, then the final momentum in scalar form becomes

$$m_B\ v_B - m_A\ v_A = 0.$$

Thus, $150v_B - 200v_A = 0$

and $v_B = \frac{4}{3}v_A.$

b) To determine the absolute speed of each skater, write the relative-speed equation along the horizontal. If senses to the right are taken as positive, this equation is

$$-\ v_A = -\ v_{A/B} + v_B$$

where $v_{A/B}$ is the speed of skater A as viewed by skater B and is the rate at which the distance between the skaters is decreasing. Hence,

$$- v_A = - 14 + v_B$$

$$- v_A = - 14 + \frac{4}{3} v_A$$

Solving this equation, we find that $v_A = + 6$. The sense of v_A was already taken into account in the relative motion equation, so $v_A = 6$ ft/sec to the left. Then, $v_B = 8$ ft/sec to the right

• PROBLEM 11-2

The force-time curve for a chemical rocket is given in the figure, with the rectangular areas used to approximate the actual area. Assuming the total mass to remain constant, determine the velocity of the rocket at burnout (t_b) if its initial velocity is zero and the rocket is directed vertically upward.

Solution: There are two principal contributors to the total impulse, neglecting air drag: the propulsive force and the gravitational attraction of the earth. The equation of impulse and momentum including both influences is

$$\int_0^{t_b} F_t \, dt + \int_0^{t_b} (- mg) \, dt = mv_b - 0 \qquad (1)$$

where g has been assumed constant and the first integral on the left side above is the area under the curve in the figure. Solving for v_b,

$$v_b = \frac{\int_0^{t_b} F_t \, dt - mgt_b}{m}.$$

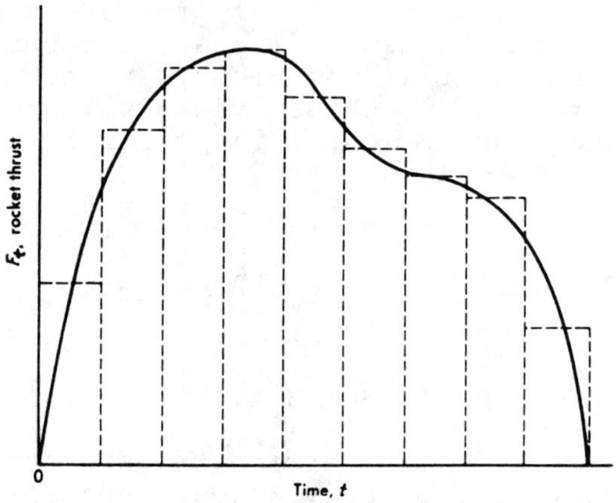

In the case of variable mass, Eq. (1) can be employed for each time interval in the figure by assuming F_t and m constant for a given interval. The velocity thus determined at the end of one interval represents the initial velocity for the next. If the thrust and mass are expressible as reasonably simple mathematical functions of time, the impulse may often be analytically determined.

On the fourth of July, an old 2,500 kg. cannon fires a 20 kg. shell with an initial velocity of 400 m/s at an angle of 20°. The cannon is chained to a post by a tauntly stretched chain. The cannon does not have a recoil mechanism, and the barrel is rigidly attached to the wheels. It takes the the shell 10m secs to leave the barrel after ignition. Find

(a) the horizontal resultant force R_h exerted by the chain and the vertical resultant force R_v exerted by the ground.

(b) If the chain breaks when the horizontal resultant force $\frac{R_h}{2}$, what is the recoil velocity of the cannon (neglect friction).

Fig. 1

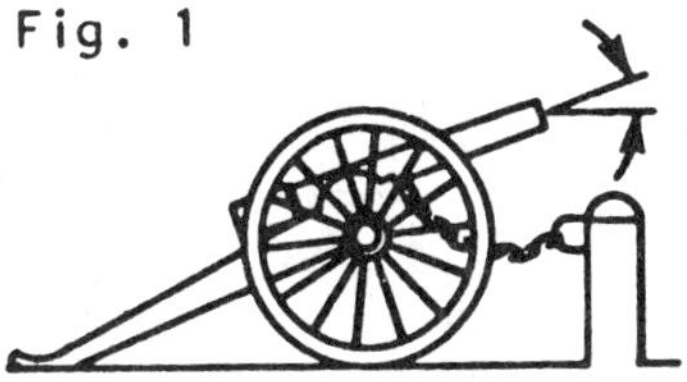

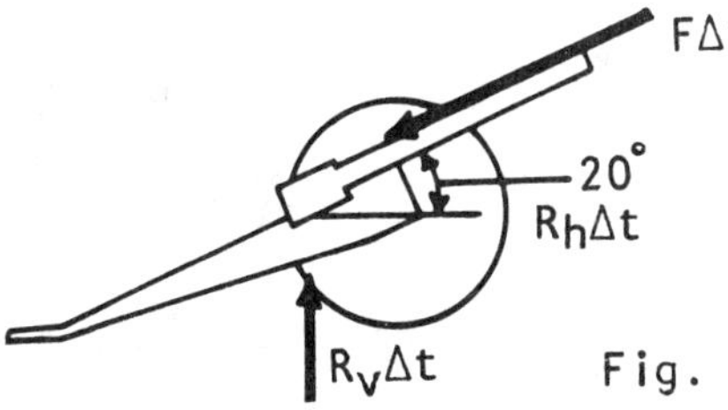

Fig. 2

Solution: Using the principal of impulse and momentum, we are able to find the impulse FΔt exerted by the gun on the shell. Applying the impulse term to the gun and dividing by the time it takes the shell to leave the gun, we can find R_h exerted by the chain, and R_v exerted by the ground. The impulse approximation can be applied here because of the predominant forces caused by firing of the shell and the speed with which the shell leaves the gun.

The conservation of impulse momentum equation for the shell is:

$$(M_s \vec{V}_s)_1 + (\text{Imp})_{1-2} = (M_s \vec{V}_s)_2$$

$$0 + \vec{F}_{22}\Delta t = (20 \text{ kg})(400 \text{ m/sec})$$

$$\vec{F}_{12}\Delta t = 8000 \text{ kg-m/sec} = 8000 \text{ N-sec}$$

$$\vec{F}_{12} = \frac{\vec{F}_{12}\Delta t}{\Delta t} =$$

$$\vec{F}_{12} = \frac{8000}{.01} \frac{\text{N - sec}}{\text{sec}} = 800{,}000 \text{ N}$$

Resolving $\vec{F}_{12}$ into force components and using Newton's Third Law, we find the equal and opposite resultant forces R_h and R_v caused by the impulsive forces

$$R_{h_{1-2}} = \vec{F}_{12} \cos 20° = -751{,}754 \text{ N}$$

$$R_{V_{1-2}} = \vec{F}_{12} \sin 20° = -273,616 \text{ N}$$

To find the final recoil velocity, we must again use the principal of impulse and momentum. Since impulse obeys Newton's Third Law and is a time averaged quantity, we are able to say that at the instant the chain breaks the acting impulse is $\frac{R_{1-2h}\Delta t}{2}$ (figure 3). The final recoil velocity is

$$(m_c v_c)_1 + \frac{R_{h_{1-2}} \Delta t}{2} = (m_c v_c)_2$$

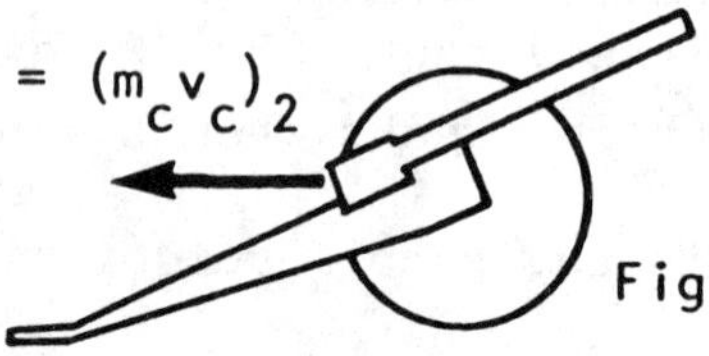

Fig. 3

thus calculated:

$$(M_c \vec{V}_c)_1 + \left(\frac{Imp}{2}\right)_{1-2} = (M_c \vec{V}_c)$$

$$0 + \frac{R_{h_{1-2}} \Delta t}{2} = (M_c \vec{V}_c)_2$$

$$\frac{(-751754 \text{ N}) \times (.01 \text{ sec})}{2} = 2,500 \text{ kg } V_c$$

$$V_c = -1.5 \text{ m/sec}$$

• PROBLEM 11-4

A 25 kg. box leaves the end of a slide with a speed of 6m/sec, and lands on a 50 kg. sled. If the sled starts from rest, and slides without friction, find the final velocity of the sled.

Solution: This is a system of two bodies in two states: initial and final.

It is known that $m_b = 25$ kg $v_{bi} = 6$ m/sec.

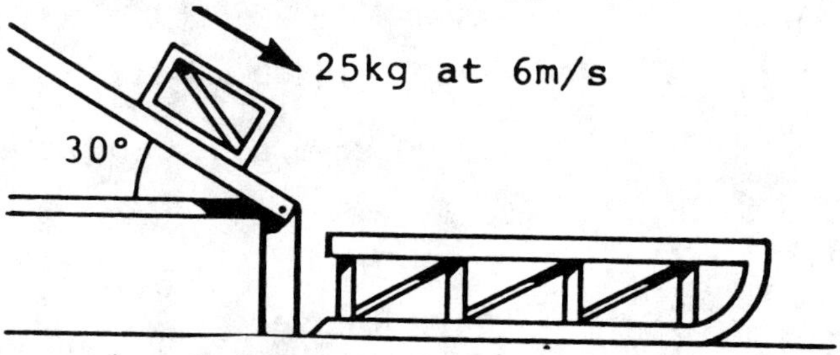

Fig. 1

where b is assigned for box and i for initial state.

Also, $m_s = 25$ kg

$v_{si} = 0$ where s is assigned for sled.

For final state $v_{bf} = v_{sf} = v_f$ because the package

will be on the sled. To determine the final speed v_f, use the conservation of linear momentum. Because the momentum is a vector, there are two components: horizontal and vertical. It is known that when the box lands on the sled, an upward force on the box-sled system will be exerted by the surface on which the sled rests. However, this will not change the horizontal component of momentum in which we are interested. (The vertical component of the momentum of the system will be changed by this external force in such a way that the vertical component of the final velocity is zero.)

Since the horizontal component of momentum of the system is conserved, write:

$$p_{horiz,\, b,i} + p_{horiz,\, s,i} = p_{horiz,\, b,f} + p_{horiz,\, s,f} \quad (1)$$

In terms of velocities, noting that the initial horizontal component of the package velocity is $v_{p,i}$ cos 30° and also that the final velocities of package and cart will be identical, write this momentum equation:

$$m_b\ (v_{b,i} \cos 30°) + m_s\ (0) = m_b\ v_{bf} + m_s\ v_{sf}$$

or $\quad m_b\ v_{bi} \cos 30° = (m_b + m_s)\ v_f$

Solving for the final velocity,

$$v_f = \frac{m_b}{m_b + m_s}\ v_{bi} \cos 30°.$$

Substituting the given numerical values yields

$$v_f = \left(\frac{25\ \text{kg}}{25\ \text{kg} + 50\ \text{kg}}\right) \left(6\ \frac{m}{s}\right) \cos 30° = 1.7319 \frac{m}{s};$$

i.e., the final velocity of the cart is 0.742 $\frac{m}{s}$.

• PROBLEM 11-5

A railway gun, initially at rest, whose mass is 70,000 kg fires a 500-kg artillery shell at an angle of 45° and with a muzzle velocity of 200 m/sec. Calculate the recoil velocity of the gun.

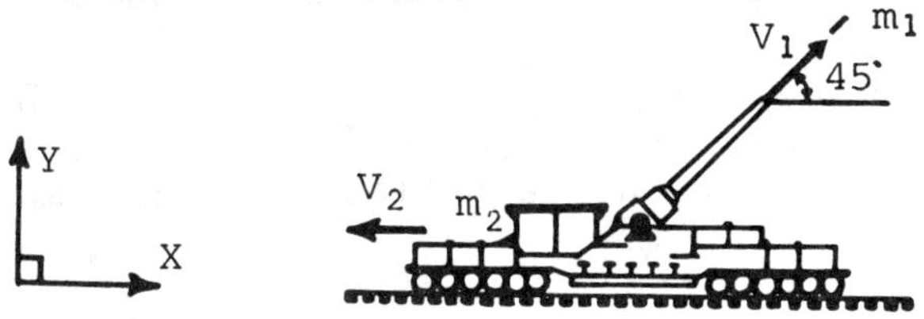

Solution: Designate the shell with subscript 1 and the gun with subscript 2.

Since there are no external forces acting on the system (gun + shell) in the x-direction, p_x (before) = p_x (after) = p_{1x} (after) + p_{2x} (after). 0 = p_{1x} (after) - p_{2x} (after) because the gun will move to the left.

$$0 = m_1 v'_{1x} - m_2 (v'_{2x}).$$

So, $$m_2 v'_{2x} = m_1 v'_{1x}.$$

But, $$v'_{1x} = v'_1 \cos 45°.$$

Therefore, express p_{1x} (after) as

$$p_{1x} \text{ (after)} = m_1 v'_1 \cos 45°$$

$$p_{1x} \text{ (after)} = (500 \text{ kg}) (200 \text{ m/s}) (0.707)$$

$$= 7.07 \times 10^4 \text{ kg - m/sec.}$$

This must be numerically equal, but oppositely directed to p_{2x} (after) from eq. (1). So, p_{2x} (after) $= m_2 v'_{2x}$ $= - 7.07 \times 10^4$ kg - m/sec.

And $$v'_{2x} = \frac{- 7.07 \times 10^4 \text{ kg - m/sec}}{7 \times 10^4 \text{ kg}}$$

$$v'_{2x} \approx 1 \frac{m}{s}$$

Ignore the vertical component of the recoil momentum due to the extremely large value of the earth's mass compared to the railway gun.

• PROBLEM 11-6

A ball weighing 8 lbs. slides along a frictionless surface. The ball strikes a wall with a velocity of 30 fps and rebounds with the same velocity. If the time of contact between the ball and the wall is a quarter-second, determine the maximum force of the wall acting on the ball, assuming that the force-time curve is triangular.

Solution: The impulse-momentum equation may be employed since the change in velocity of the particle is known and since it is desired to determine the force-time characteristic of the force of the wall acting on the ball during the impact:

$$\int_A^B F\, dt = mv_B - mv_A$$

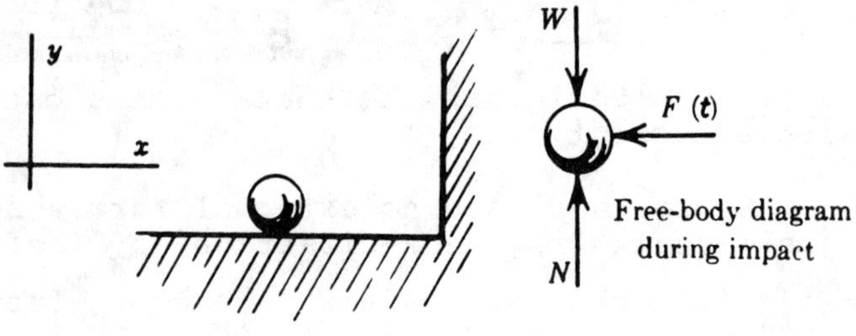

(a) Fig 1 (b)

where A is the time immediately preceding impact and B is the time immediately after impact.

Since the velocity before and after impact is known, the impulse may be obtained by evaluating the change in momentum. First, the velocity before and after impact is

$$v_A = 30\bar{i} \text{ fps}$$

$$v_B = -30\bar{i} \text{ fps}$$

Dropping the vector notation since motion occurs in one dimension only, the impulse-momentum equation becomes

$$\int_A^B F\,dt = \frac{8}{32.2}[(-30) - (+30)]$$

$= -14.9$ lb-sec. The negative sign indicates F(t) is to the left.

The free-body diagram of the particle during impact is shown in Fig. 1(b). Since the net force in the y direction is equal to zero, momentum will be conserved in the y direction. In this case, the velocity is zero and will remain zero during the motion. The force F(t) is the force that the wall exerts on the ball; thus, it acts during contact only. Experiments have shown that the time distribution of that force during impact is roughly of the form shown in Fig. 2(a). In this case it is assumed for analytical purposes that the force-time distribution is triangular, as shown in Fig. 2(b).

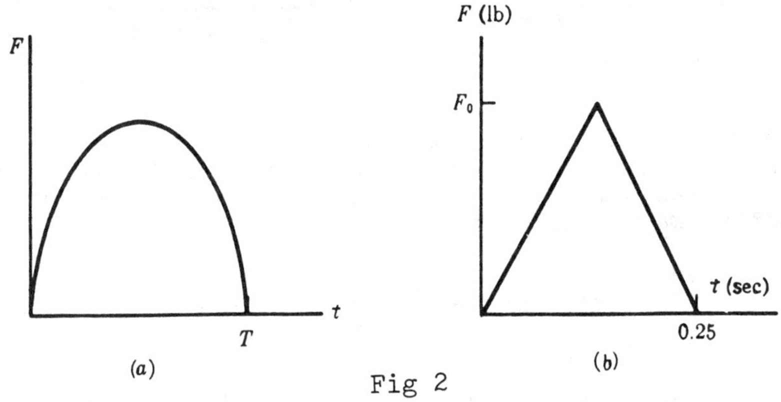

Fig 2

But the impulse is just the area under the force-time curve, so that

$$\text{Impulse} = -\frac{1}{2}(0.25)(F_0)$$

$$= -0.125\,F_0$$

The minus sign must be chosen because F(t) acts in the negative x direction. However, we already know what the impulse must be for the ball to rebound as it does. Thus,

$$- 0.125 F_0 = - 14.9$$

and, solving for F_0,

$$F_0 = 119 \text{ lb.}$$

• PROBLEM 11-7

A space vehicle weighing 1000 lb is traveling with a velocity of 1200 fps (in the positive x direction relative to the reference system shown in Fig. 1). One of the control rockets on the vehicle ignites and creates a force $F_1(t)$, as shown in the graph of Fig. 2. The force F_1 acts in the positive y direction and ignites at t = 0. At t = 4 sec, a second rocket ignites and creates a force $F_2(t)$, as shown; F_2 acts in a positive x direction. Determine the velocity of the vehicle after 8 sec have elapsed. Assume that the vehicle is far enough out into space that any gravitational or attractive force from any of the planets may be neglected.

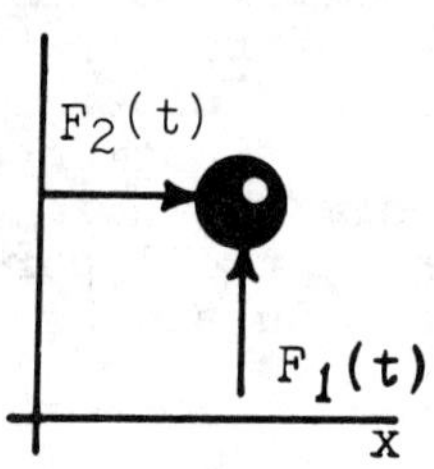

Fig 1

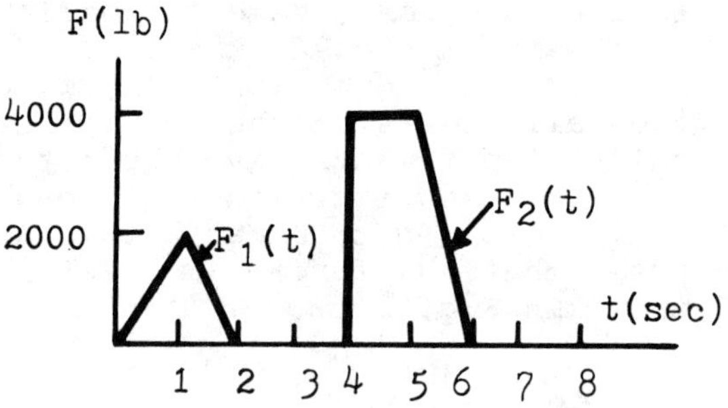

Fig 2

Solution: Since the forces on the vehicle are given as functions of time, and since it is desired to determine the velocity of the vehicle, it appears that the impulse-momentum equation may be employed:

$$\int_0^8 \vec{F}\, dt = m\vec{v}_B - m\vec{v}_A.$$

It is seen that the only unknown quantity in the preceding equation is the velocity of the vehicle v_B after the impulse. The impulse due to the force may be easily evaluated; it is equal in each direction to the area under the force-time curve. In this case, there are forces in two directions, and, since the impulse-momentum equation is a vector equation, care must be taken to preserve the directions when writing the equation:

$$\int_0^8 \vec{F}\, dt = \int_0^8 F_1 \hat{j}\, dt + \int_0^8 F_2 \hat{i}\, dt$$

$$= \left[\frac{1}{2}(2000)(2)\right]\hat{j} + \left[(4000)(1) + \frac{1}{2}(4000)(1)\right]\hat{i}$$

$$= 6000\,\hat{i} + 2000\,\hat{j} \text{ lb.}$$

Note that F_1 is acting in the positive y direction and F_2 is acting in the positive x direction and that the appropriate unit vectors are associated with each impulse in the preceding equations. The other terms in the impulse-momentum equation may be evaluated:

$$m = \frac{1000}{32.2} = 31 \text{ slugs}$$

$$\vec{v}_A = 1200\,\hat{i} \text{ fps.}$$

Making the appropriate substitutions into the impulse-momentum equation,

$$6000\,\hat{i} + 2000\,\hat{j} = 31\,[\vec{v}_B - 1200\,\hat{i}].$$

Solving for the velocity of the vehicle after the impulse,

$$31\,\vec{v}_B = 6000\,\hat{i} + 2000\,\hat{j} + 37{,}200\,\hat{i}.$$

Thus, $\vec{v}_B = 1394\,\hat{i} + 65\,\hat{j}$ fps.

• PROBLEM 11-8

In order to test the ability of a chain to resist impact, the chain is hung from a 250 kg. block. The chain also has a metal plate hanging from the final link. A 50 kg. weight is released from a height of 2m above the plate. It drops to hit the plate. Find the impulse exerted by the weight if the impact is perfectly elastic and the block is supported by

(a) by two perfectly rigid columns
(b) perfectly elastic springs
(c) find the energy absorbed by the chain in each case.

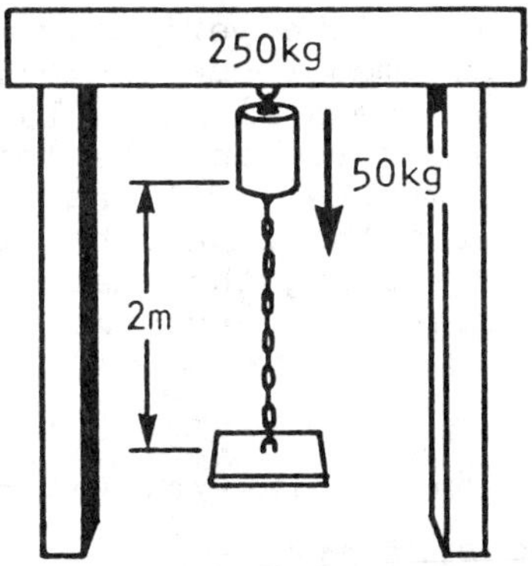

Fig. 1

Solution: The impulse momentum relationship says that the impulse the chain will experience is equal and opposite to the change in momentum of the weight. The chain is assumed to be massless, and, consequently, this same impulse will be transmitted to the block (although the impulse will differ in cases (a) and (b). Plastic impact implies the weight will not rebound.

Regard downward velocity as positive. Subscripts w and b will be used to designate weight and block, respectively. Second subscripts i and f will represent initial and final conditions (i.e., immediately before and after impact), respectively.

Part (a). With the columns perfectly rigid (and the block also assumed to be rigid), the entire system will have zero velocity after the impact. Thus, the change in momentum of the weight is

$$P_{wf} - P_{wi} = m_w v_{wf} - m_w v_{wi} = m_w(0) - m_w v_{wi} = - m_w v_{wi}$$

since the final velocity of the weight is zero. Thus the impulse on the chain is $-(P_{cf} - P_{ci}) = m_c v_{ci}$ (1) since it is equal and opposite to the impulse on the weight.

By the time of impact the weight will have fallen a distance h (= 2.0 meters) and its kinetic energy will be given by

$$K_{wi} = \frac{1}{2} m_w v_{wi}^2 = m_w gh. \tag{2}$$

Solving this equation for v_{wi} gives

$$v_{wi} = \sqrt{2\,gh}. \tag{3}$$

From Eqs. (1) and (3) the impulse on the chain is $m_w v_{wi} = m_w\sqrt{2\,gh}$ and, substituting numerical values, its magnitude is $50 \text{ kg} \sqrt{2\,(9.81 \text{ m/s}^2)(2.0 \text{ m})} = 313.21 \text{ N·s}$.

(b) With the columns like perfectly elastic springs (but the block still assumed to be rigid), the block-chain-weight system will now have a finite velocity after impact. (Since the gravitational force and the force supplied by the columns are non-impulsive, they will have no effect on the change of momentum during the impact because the time involved is negligible.) Neglecting the mass of the chain, from the conservation of linear momentum write $m_w v_{wi} + m_b v_{bi} = m_w v_{wf} + m_b v_{bf}$.

Since the collision is plastic, $v_{wf} = v_{bf}$. Call this final velocity of the system v_f.

Knowing that $v_{wi} = \sqrt{2\ gh}$ from above, and that $v_{bi} = 0$,

we have $\quad m_w \sqrt{2\ gh} + m_b(0) = m_w\ v_f + m_b\ v_f$

Solving for v_f obtains

$$v_f = \frac{m_w \sqrt{2\ gh}}{m_w + m_b}. \qquad (4)$$

The impulse exerted on the chain is

$$-(P_{wf} - P_{wi}) = -(m_w\ v_f - m_w\ v_i) = m_w\ (v_i - v_f).$$

Using Eqs. (3) and (4) for v_i and v_f gives for the impulse on the chain,

$$-(P_{wf} - P_{wi}) = m_w \left(\sqrt{2\ gh} - \frac{m_w \sqrt{2\ gh}}{m_w + m_b} \right) = m_w \left(\sqrt{2\ gh}\ \frac{m_b}{m_b + m_w} \right) \qquad (5)$$

Substituting numerical values gives for the impulse,

$$50\ \text{kg} \sqrt{2 \left(9.81\ \frac{\text{m}}{\text{s}^2}\right) 2.0\ \text{m}} \left(\frac{250\ \text{kg}}{250\ \text{kg} + 50\ \text{kg}} \right) = 261\ \text{N}\cdot\text{s}$$

Because the mass of the chain is assumed to be zero, conservation of momentum requires that the change of momentum of the block be the negative of that experienced by the weight. Thus, it could have been said that the impulse exerted on the chain is equal to the change in the block's momentum.

$$P_{bf} - P_{bi} = m_b\ v_{bf} - m_b\ v_{bi}$$

$$= m_b\ v_f - m_b(0)$$

or using Eq. (4),

$$= \frac{m_b\ m_w \sqrt{2\ gh}}{m_w + m_b}$$

which is the same as the result obtained previously.

(c) Since no change in the potential energy of the block or the cylinder occurs during the impact, we can write from conservation of energy:

$$\Delta E\ (= \text{Energy absorbed by the chain}) = (K_{wi} - K_{wf}) + (K_{bi} - K_{bf}). \qquad (6)$$

For case (a) $K_{wf} = 0$, $K_{bi} = 0$, $K_{bf} = 0$ and, by Eq.(2), $K_{ci} = m_w gh$. Hence,

$$\Delta E = m_w gh = 50 \text{ kg} \left(9.81 \frac{\text{m}}{\text{s}^2}\right)(1.5 \text{ m}) = 368 \text{ J}.$$

For case (b) $K_{bi} = 0$, $K_{wi} = m_w gh$ (as before), $K_{wf} = \frac{1}{2} m_c v_f^2$

and $K_{bf} = \frac{1}{2} m_b v_f^2$ where, by Eq.(4), $v_f = \dfrac{m_w \sqrt{2gh}}{m_w + m_b}$.

Using these values in (6) gives:

$$\Delta E = m_w gh - \frac{1}{2} m_w \left(\frac{m_w \sqrt{2gh}}{m_w + m_b}\right)^2 - \frac{1}{2} m_b \left(\frac{m_w \sqrt{2gh}}{m_w + m_b}\right)^2$$

$$= m_w gh - \frac{m_w^2 gh}{(m_w + m_b)^2} (m_w + m_b)$$

$$= m_w gh \left[1 - \frac{m_w}{(m_w + m_b)}\right] = m_w gh \left(\frac{m_b}{m_w + m_b}\right)$$

$$= \frac{m_w m_b}{m_w + m_b} gh.$$

Substituting numerical values,

$$\Delta E = \frac{(50)(250)}{50 + 250)} \text{ kg} \left(9.81 \frac{\text{m}}{\text{s}^2}\right) (2.0 \text{ m}) = 817.5$$

• PROBLEM 11-9

Ball (1) is suspended from a thin regid rod. An identical ball (2) is dropped a height (h) along the length of the rod. Ball (2) attains a velocity v_0 before it makes contact with ball (1). Assume $e = 1$ and there is no friction between the ball surfaces, determine the velocity of the balls (1) and (2) at the instant after impact.

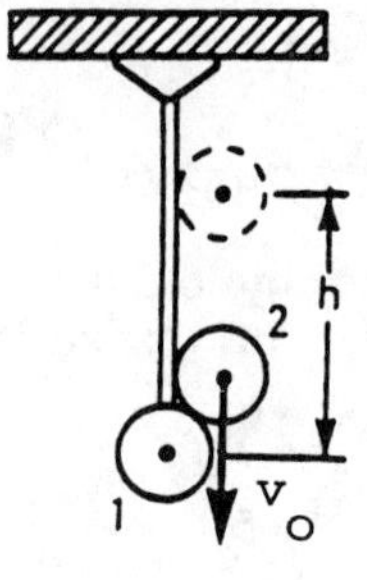

Fig. 1

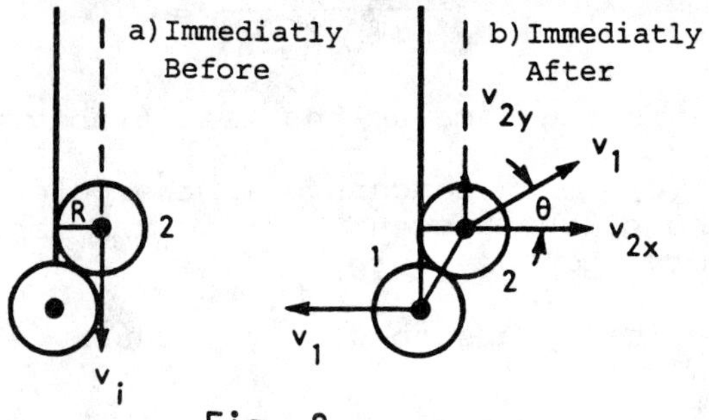

Fig. 2

Solution: We wish to find v_1, v_2 and θ, given v_2. The three equations to be used are derived from the laws of I. Conservation of momentum, II. conservation of energy, III. conservation of angular momentum.

I. The y-component of momentum is not conserved as external forces are acting in the y-direction. In the x-direction, however, momentum is conserved so we may write: P_x (before) $= P_x$ (after). From Figure 2 note that P_x (before) $= 0$ and P_x (after) $= mv_{2x} + m(-v_1)$. The string constrains ball B to move in the -x direction. The masses are the same, thus, $v_1 = v_{2x}$, but $v_{2x} = v_2 \cos\theta$, so

$$v_1 = v_2 \cos\theta. \tag{1}$$

II. The collision is totally elastic (e = 1), therefore, the kinetic energy is conserved. Write: KE (before) = KE (after)

$$\frac{1}{2} mv_0^2 = \frac{1}{2} mv_2^2 + \frac{1}{2} mv_1^2 \tag{2}$$

Dividing by m/2 yields

$$v_0^2 = v_2^2 + v_1^2. \tag{3}$$

Substituting v_1 from equation (1) into equation (3) obtains

$$v_0^2 = v_2^2 + v_2^2 \cos^2\theta. \tag{4}$$

Factoring out the v_2^2,

$$v_0^2 = v_2^2 (1 + \cos^2\theta), \text{ hence,}$$

$$v_2^2 = v_0^2/(1 + \cos^2\theta).$$ Taking the square root yields

$$v_2 = v_0/\sqrt{1 + \cos^2\theta}. \tag{5}$$

III. The angular momentum about an origin at the center of ball 1 is conserved as no external torque acts about this point. From Figure 3 note that L (before) $= mv_0R$. The

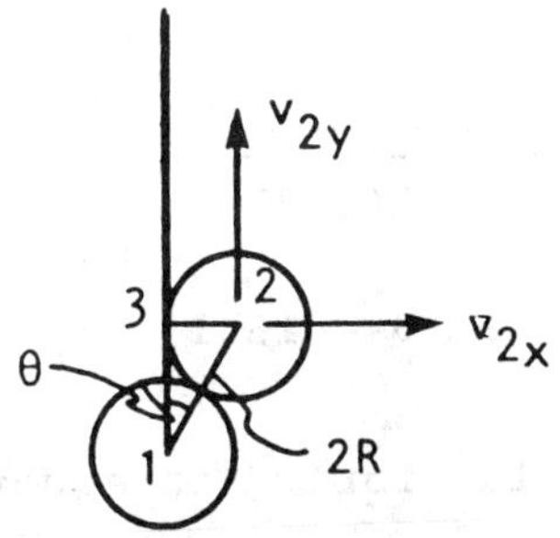

Fig. 3

triangle 1,2,3 has side $\overline{3,2}$ equal to R and hypotenuse 2R, thus, $\alpha = \sin^{-1} \frac{R}{2R} = \sin^{-1} \frac{1}{2} = 30°$; $\overline{3,1} = 2R \cos 30° = \sqrt{3}R$.

Hence, $$L \text{ (after)} = mv_{2x} (\sqrt{3}R) - mv_{2y}R. \tag{6}$$

Setting L (before) = L (after) and substituting $v_{Ax} = v_2 \cos \theta$ and $v_{2y} = v_2 \sin \theta$ obtains

$$mv_i R = mv_2 (\sqrt{3}R) \cos \theta - mv_2 (R) \sin \theta$$

dividing by mR and substituting for v_0 from equation (5) yields

$$v_2 \sqrt{1 + \cos^2 \theta} = \sqrt{3}\, v_2 \cos \theta - v_2 \sin \theta \tag{7}$$

$$\sqrt{1 + \cos^2 \theta} = \sqrt{3} \cos \theta - \sin \theta. \tag{8}$$

Squaring,

$$1 + \cos^2 \theta = 3 \cos^2 \theta - 2\sqrt{3} \sin \theta \cos \theta + \sin^2 \theta. \tag{9}$$

Now, $3 \cos^2 \theta + \sin^2 \theta = 2 \cos^2 \theta + 1$, so

$$1 + \cos^2 \theta = 2 \cos^2 \theta + 1 - 2\sqrt{3} \sin \theta \cos \theta. \tag{10}$$

Subtracting of $(1 + \cos^2 \theta)$ from both sides yields

$$0 = \cos^2 \theta - 2\sqrt{3} \sin \theta \cos \theta \tag{11}$$

$$\sin \theta = \frac{1}{2\sqrt{3}} \cos \theta \tag{12}$$

$$\tan \theta = \frac{1}{2\sqrt{3}} \tag{13}$$

$$\theta = \tan^{-1} \frac{1}{2\sqrt{3}} = 16.1°. \tag{14}$$

Substituting $\theta = 16.1°$ into equation (5) yields:

$$v_2 = \frac{v_i}{\sqrt{1 + \cos^2 (16.1°)}}, \text{ or} \tag{15}$$

$$v_2 = 0.721\, v_0 \text{ at } \theta = 16.1° \tag{16}$$

Substituting this value for v_2 in equation (1) yields:

$$v_1 = v_2 \cos(16.1°), \text{ or} \quad (17)$$

$$v_1 = 0.693\ v_i \quad (18)$$

in the -x direction.

IMPULSIVE FORCES IN INELASTIC IMPACTS

• PROBLEM 11-10

A 6000 kg plane lands on the deck of a carrier at a speed of 200 km/hr relative to the carrier and is stopped in 3.0 sec. Determine the average horizontal force exerted by the carrier on the plane if: a) the carrier is at rest, b) the carrier is moving at a speed of 15 knots in the same direction as the airplane (1 knot = 0.514 m/sec).

Solution: The force exerted on the plane equals the change in momentum relative to the carrier divided by the stopping time.

$$F = \frac{\Delta P}{\Delta t} \quad (1)$$

$$P_0 = mv_0 = 6 \times 10^3 \text{ kg} \left(200 \frac{\text{km}}{\text{hr}}\right)\left(0.2778 \frac{\text{m/sec}}{\text{km/hr}}\right)$$

$$P_0 = 3.334 \times 10^5 \text{ kg m/sec} \quad (2)$$

The final momentum of the plane is zero since its velocity is zero relative to the carrier. Thus,

$$F = \frac{P_f - P_0}{\Delta t} = \frac{0 - 3.334 \times 10^5 \text{ kg m/sec}}{3.0 \text{ sec}} \quad (3)$$

$$F = -1.111 \times 10^5 \text{ N} = -111.1 \text{ kN. (a)} \quad (4)$$

The answer to part (b) will be the same since the change in momentum relative to the carrier is the same in both cases. The negative sign indicates it is opposite to the direction of motion.

• PROBLEM 11-11

A 10-lb block in Fig. 1 rests on a horizontal plane for which the coefficient of sliding friction is 0.4 and the coefficient of static friction is 0.5. The force given by $P = (3t^2 + 2)$ lb, where t is in seconds, is applied at time t = 0. Determine the speed of the block 4 sec later.

Solution: The free-body diagram is shown in Fig. 2. Since force P is applied when t = 0, it must be determined first if the block slides immediately or if there is a certain

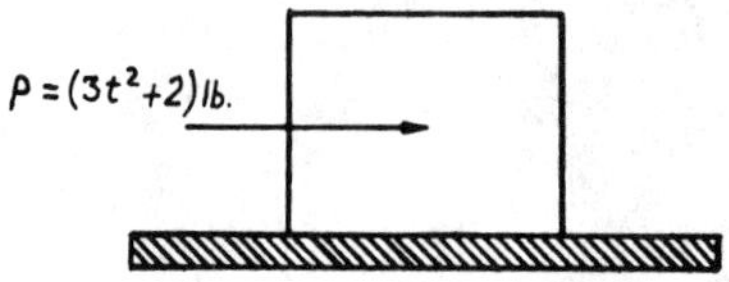

Fig 1

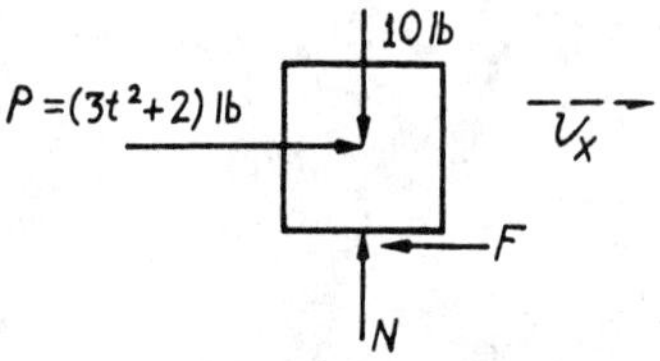

Fig 2

lapse of time before sliding takes place. It can be seen that at t = 0 the magnitude of the force P is 2 lb, and this force cannot overcome the maximum static frictional force of 5 lb.

$$(F = \mu_s N = (.5)(10 \text{ lb.}) = 5 \text{ lbs}).$$

Thus, at impending motion, $3t^2 + 2$ must equal the static frictional force of 5 lb. So, $3t^2 + 2 = 5$, or t = 1 sec, before motion impends. After 1 sec the frictional force becomes 4 lb. The impulse-momentum relation is

$$\int F_x \, dt = \Delta mv.$$

Putting in values,

$$\int_1^4 [(3t^2 + 2) - 4] \, dt = (mv_x)_f - (mv_x)_i$$

and $$\int_1^4 (3t^2 - 2) \, dt = mv_{xf}$$

since $v_{xi} = 0$ at t = 1 sec.

Integrating, $$t^3 - 2t \Big|_1^4 = \frac{10 \text{ lbs.}}{32.2 \text{ m/s}^2} v_{xf}$$

$$v_{xf} = 183.5 \text{ ft/s}.$$

• PROBLEM 11-12

(A) and (B) are identical frictionless poolballs. The direction and magnitude of their respective velocities is shown. If e = 0.85 find the direction and magnitude of their velocities immediately after impact.

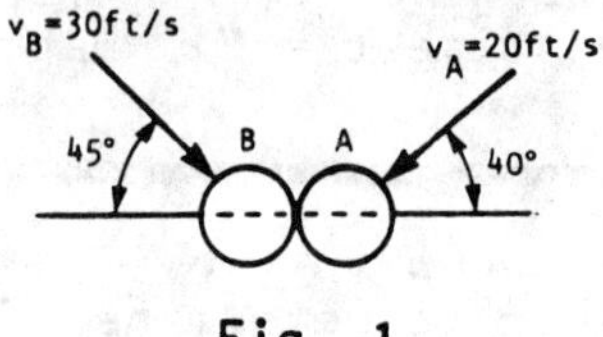

Fig. 1

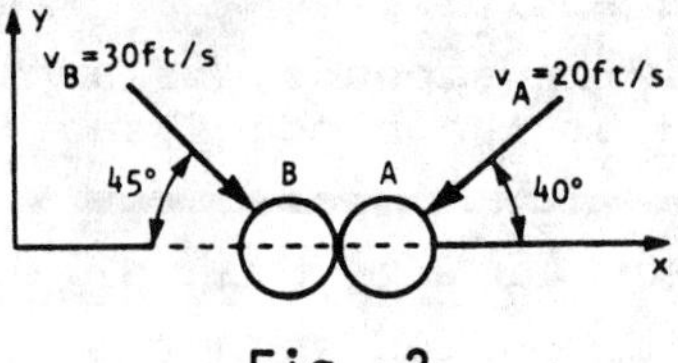

Fig. 2

Solution: The first step in solving this problem is to make a suitable choice of a coordinate system. This is partially facilitated by remembering that, during the collision, the impulsive forces acting between the balls are directed along a line joining their centers. This is called the line of impact. The coordinate axes are then chosen parallel and perpendicular to the line of impact, as shown in Fig. 2.

The velocity vectors before the collision can thus be decomposed into their x and y components:

$$(v_A)_x = -(v_A \cos 40°) = -15.32 \text{ ft/s}$$
$$(v_A)_y \quad (v_A \sin 40°) = -12.86 \text{ ft/s}$$

$$(v_B)x = +(v_B \cos 45°) = +21.21 \text{ ft/s}$$
$$(v_B)_y = +(v_B \cos 45°) = -21.21 \text{ ft/s}$$

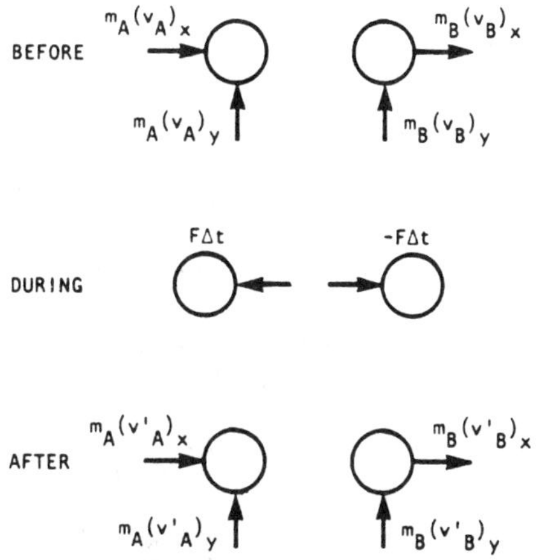

Fig. 3

The effects of the collision are pictured in Fig. 3. To analyze the motion perpendicular to the line of impact, only the y components are considered, and the principle of impulse and momentum is applied to each ball separately. Since no vertical impulsive force acts during the impact, the vertical component of the momentum and, hence, the vertical component of the velocity of each ball is unchanged.

$$(\vec{v}'_A)_y = -12.85 \text{ ft/sec downward}$$

$$(\vec{v}'_B)_y = -21.21 \text{ ft/sec downward}$$

The motion parallel to the line of impact, however, is affected by the impulsive forces.

The internal impulses of the system are FΔt and - FΔt and, according to Newton's Third Law, they cancel in this problem. Therefore, the total momentum of the balls in the x-direction is conserved:

$$m_A (v_A)_x + m_B (v_B)_x = m_A (v'_A)_x = m_B (v'_B)_x$$

$$m(-15.32) + m(+21.21) = m (v'_A)_x + m (v'_B)_x$$

$$(v'_A)_x + (v'_B)_x = 5.89 \qquad (1)$$

Given that the coefficient of restitution e = 0.85, another relation between the **relative** velocities is

$$(v'_B)_x - (v'_A)_x = e [(v_A)_x - (v_B)_x]$$

$$(v'_B)_x - (v'_A)_x = (0.85) \cdot [(-15.32) - (21.21)]$$

$$(v'_B)_x - (v'_A)_x = -31.05 \qquad (2)$$

The velocity components are found by solving equations (1) and (2) simultaneously; thus,

$(v'_A)_x = 12.58$ $\qquad$ $(v'_B)_x = -18.47$

$(\vec{v}'_A) = 12.58$ ft/s to the right $\qquad$ $(\vec{v}'_B)_x = 18.47$ ft/s to the left

The resultant velocity vectors are found by adding the velocity components of each ball as in Fig. 4.

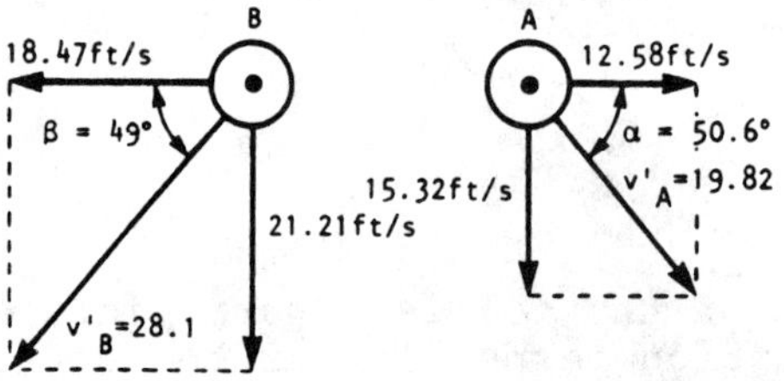

Fig. 4

$\vec{v}'_A = 19.82$ ft/s at an angle of 50.6° with the horizontal towards the lower right

$\vec{v}'_B = 28.1$ ft/s at an angle of 49.0° with the horizontal towards the lower left

• PROBLEM 11-13

A 600 lb. telephone pole is being driven into the ground until resistance to penetration equals 20,000 lbs. This is achieved by dropping a 1,000 lb. weight 6 ft. onto the top of the pole. How much deeper will the pole be driven into the ground by the next blow if it meets with a resisting force of 20,000 lb. Assume a completely inelastic collision.

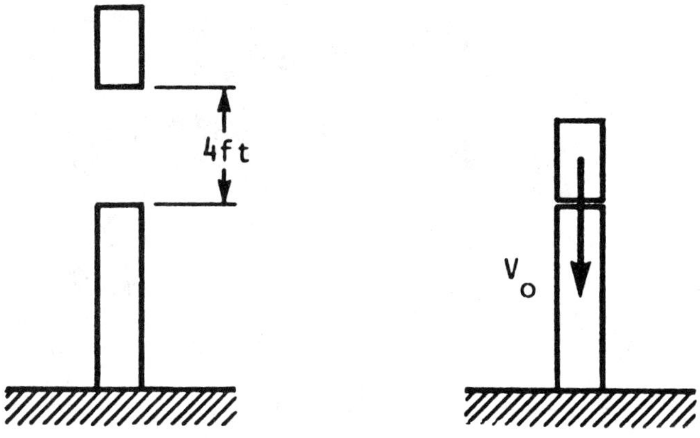

Fig. 1

Solution: In the first part of the problem, use the law of conservation of momentum to find the initial velocity of the telephone pole plus the weight after the inelastic collision. Then, use the work-energy theorem to find the penetration depth.

In falling from a height h, the initial potential energy of the weight, mgh, is converted into kinetic energy, $\frac{1}{2} mV_0{}^2$, so $\frac{1}{2} mV_0{}^2 = mgh$.

Solving for V_0 using elementary algebra yields:

$$V_0{}^2 = 2\ gh$$

$$V_0 = \sqrt{2\ gh}$$

$$V_0 = \sqrt{2\ (32\ ft/sec^2)\ (6\ ft)}$$

$$V_0 = 19.60\ ft/sec$$

The momentum of the weight just before the collision is $p_0 = m_w V_0$.

Just after the collision, the weight and the telephone pole have a commom final velocity, V, so that

$$p = (m_w + m_{tp})\ V = p_0, \text{ or}$$

$$(m_w + m_{tp})V = mwV_0. \quad \text{Thus,}$$

$V = \frac{m_w}{(m_w + m_{tp})} V_0$. Multiplying top and bottom by g, remembering $W = mg$, then $V = \frac{Ww}{(Ww + Wtp)} V_0$

and $V = \frac{1000\ lb}{1600\ lb}(19.60\ ft/sec)$. Hence

$$V = 12.25\ ft/sec.$$

The initial kinetic energy of the telephone pole plus

the weight is:

$$KE_0 = \frac{1}{2} (m_w + m_{tp}) V^2, \text{ so}$$

$$KE_0 = \frac{1}{2}\left(\frac{1600 \text{ lb}}{32.2 \text{ ft/sec}^2}\right)(12.25 \text{ ft/sec})^2$$

$$KE_0 = 3728.3 \text{ ft. lb}$$

This kinetic energy is equal to the work done by the ground stopping the pile. Assuming that the resistance force is constant and directed against the motion,

$$F \cdot y = KE_0. \quad \text{Thus, } (20{,}000 \text{ lb}) \cdot y = 3728.3 \text{ ft.lb},$$

or $$y = \frac{3728.3}{20000} \text{ ft. and}$$

$$y = 0.186 \text{ ft., or}$$

$$y = 2.24 \text{ in.}$$

An alternate way of solving is to make the following assumptions: a) the resisting force is constant while the telephone pole is moving; b) the weight and pole move as a unit after the inelastic collision. Now the problem is broken up into 2 parts: the collision phase and the movement phase.

Since the collision phase takes place over a very short interval, assume that momentum is perfectly conserved. In this case, momemtum before the collision, $m_w V_w$ is equal to momemtum after the collision $(m_w + m_{tp}) V_{tpw}$ where m is mass, V is velocity and the subscripts tp & w refer to the telephone pole and weight respectively. (remember the weight and the telephone pole move as a unit after the collision.)

To find V_w, use the energy equation. The weight's kinetic energy just before the collision is equal to its potential energy at its maximum height.

$$m_w gh = \frac{1}{2} m_w V_w^2; \qquad h = \text{height}$$

$$g = \text{acceleration constant of gravity}$$

$$V_w = \sqrt{2gh}$$

Now, using this expression for V_w in the momentum equation,

$$m_w (\sqrt{2gh}) = (m_w + m_{tp}) V_{tpw}$$

$$V_{tph} = \frac{m_w (\sqrt{2gh})}{(m_w + m_{tp})} \qquad (1)$$

Next is the movement phase. S, the distance traveled before the pole stops, is the information desired. Use the acceleration-displacement equation,

$$S = V_0 \Delta t + \frac{1}{2} a (\Delta t)^2, \qquad (2)$$

to find it.

Here, if movement down is chosen to be positive, $V_0 = V_{tpw}$, and $a = -\frac{F}{(m_{tp} + m_w)}$, from Newton's Second Law, with $F = 20{,}000$ lb. The minus sign indicates it is decelerating.

There is still one more unknown, Δt. To find it, use the Impulse Equation.

$$F\Delta t = \Delta\ (mV).$$

Because the mass remains constant,

$$F\Delta t = (m_{tp} + m_w)\ \Delta V.$$

Since the force acts until the telephone pole and weight are at rest, $\Delta V = V_{init} - V_{final} = V_{tpw} - 0\ V_{tpw}$

$$\Delta t = \frac{(m_{tp} + m_w)\ V_{tpw}}{F}.$$

Substituting given values in equation (1),

$$V_{tpw} = \frac{1000 \text{ lb}/g\ \sqrt{2(32.2 \text{ ft/s}^2)\ 6 \text{ ft}}}{(600 \text{ lb} + 1000 \text{ lb})/g}$$

$$V_{ph} = 12.29 \text{ ft/sec}$$

$$\Delta t = \frac{[(600 \text{ lb} + 1000 \text{ lb})/32.2 \text{ ft/s}^2]\ (12.29 \text{ ft/s})}{20{,}000 \text{ lb.}}$$

$$= .03052 \text{ sec}$$

It takes .03052 sec. for the weight and pole to stop.

From Eq. (2),

$$S = 12.29 \text{ ft/sec } (.03052 \text{ sec})$$

$$- \frac{1}{2}\ \frac{24{,}000 \text{ lb}}{(600 \text{ lb} + 1000 \text{ lb})/32.2 \text{ ft/s}^2}\ (.03052 \text{ sec})^2$$

$$S = .3751 \text{ ft.} - .1875 \text{ ft.}$$

$$S = .1876 \text{ ft}$$

$$S = 2.25 \text{ in.}$$

The telephone pole travels 2.25 in. when the resisting force is 20,000 lb.

It is found that it takes approximately 0.2 sec for a 4 kg mass, given an initial velocity up an incline of 2 m/sec, to stop. If the angle of the incline is 30 degrees with the horizontal, determine the average coefficient of sliding friction. Refer to Fig. 1.

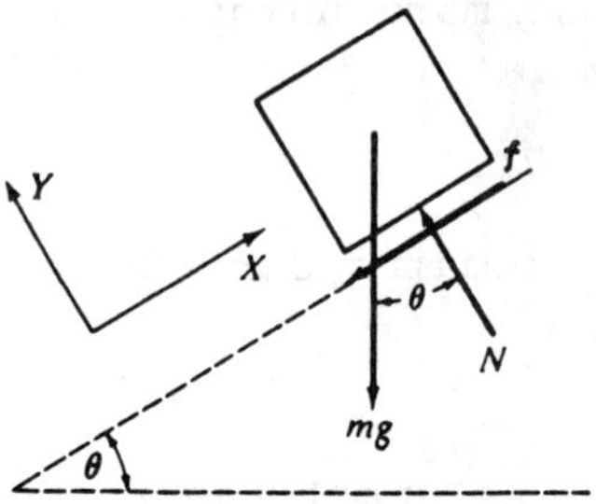

Fig. 1

Solution: For convenience, choose a coordinate system parallel to the incline. The forces on the mass are drawn in Fig. 1. All motion occurs along the x-axis, therefore, the equation of impulse and momentum is:

$$\int F_x \, dt' = \Delta p_x \qquad (1)$$

$$F_x = -f - mg \sin\theta = -\mu_k N - mg \sin\theta, \qquad (2)$$

where μ_k is the kinetic coefficient of friction.

To find the normal force N, use Newton's Third Law. It states that action and reaction pairs must be opposite and equal.

Therefore, $N = mg \cos\theta$. (3)

Combining equations (2) and (3) yields

$$F_x = -\mu_k N - mg \sin\theta$$

and
$$F_x = -\mu_k\, mg \cos\theta - mg \sin\theta. \qquad (4)$$

Substitute equation (4) into equation (1) to obtain

$$\int_{t_i}^{t_f} (-\mu_k\, mg \cos\theta - mg \sin\theta)\, dt' = \Delta p_x,$$

where dt' is the differential and t is the time.

The change in momentum in the x-direction, Δp_x, is

given by

$$\Delta p_x = \Delta(mv_x) = m(\Delta v_x) = (mv_f)_x - (mv_i)_x$$

Thus, $\int_{t_i}^{t_f} (-\mu_k \text{ mg} \cos\theta - \text{mg} \sin\theta)\, dt' = (mv_f)_x - (mv_i)_x$

Integrating,

$$(-\mu_k \text{ mg} \cos\theta - \text{mg} \sin\theta) \Big[t' \Big]_{t_i}^{t_f} = (mv_f)_x - (mv_i)_x$$

$$(-\mu_k \text{ mg} \cos\theta - \text{mg} \sin\theta)\Delta t = (mv_f)_x - (mv_i)_x.$$

m is in all the terms, so it cancels out.

Dividing by Δt, $-\mu_k g \cos\theta - g \sin\theta = \dfrac{v_{fx} - v_{ix}}{\Delta t}$

$$-\mu_k g \cos\theta = \frac{v_{fx} - v_{ix}}{\Delta t} + g \sin\theta$$

$$-\mu_k = \frac{v_{fx} - v_{ix}}{g \cos\theta\ \Delta t} + \frac{g \sin\theta}{g \cos\theta}$$

$$\mu_k = -\left(\frac{v_{fx} - v_{ix}}{g \cos\theta\ \Delta t} + \tan\theta\right)$$

$$\mu_k = \frac{v_{ix} - v_{fx}}{g \cos\theta\ \Delta t} - \tan\theta$$

Substituting given values yields

$$\mu_k = \frac{2\ \frac{m}{s} - 0}{(9.8\ m/s^2)\ (\cos 30°)\ (0.2s)} - \tan 30°,$$

so $\mu_k = 0.6$.

Note that the answer didn't depend on the mass of the body.

Since an average μ_k is sought, we shall regard it as a constant. The integral thus becomes

$$m\ (-\mu_k g \cos\theta - g \sin\theta)\Delta t = m\ (v_x - v_{0x})$$

Solving for μ_k,

$$\mu_k = \frac{v_{0x} - v_x}{g(\cos\theta)\Delta t} - \tan\theta$$

For the data given $\mu_k \cong 0.6$.

IMPULSIVE TORQUE AND ANGULAR MOMENTUM

• PROBLEM 11-15

A slender rod having a weight W and a length L is rotating in a horizontal plane about a vertical axis through one end with an angular velocity of ω_0. A small weight W_1 is attached to the outer end of the rod. What will the angular velocity of the rod be immediately after the small weight breaks off?

Solution: Since no external torque acts on the rod, conservation of angular momentum applies to the entire system. The initial angular momentum is

$$I_0\omega_0.$$

$$I_0 = I_{rod} + I_{small\ weight},$$

so
$$I_0 = \frac{1}{3}ML^2 + mL^2,$$

where $M = \frac{W}{g}$ is the mass of the rod and $m = \frac{W_1}{g}$ is the mass of the small weight. After the weight breaks off, the moment of inertia, $I = \frac{1}{3}ML^2$, is changed, so the final angular momentum of the rod is

$$I\omega = \left(\frac{1}{3}ML^2\right)\omega.$$

However, the final angular momentum of the system is the angular momentum of the rod plus the angular momentum of the ball (see the diagram below).

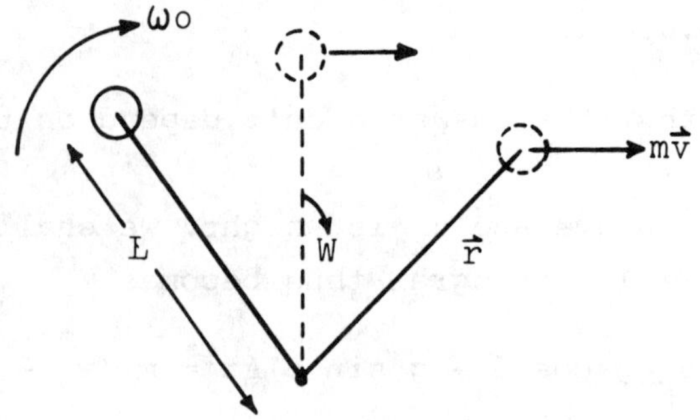

Upon breaking off, the ball moves with linear velocity,

$v = \omega_0 L$. The angular momentum of the ball is $\vec{r} \times m\vec{v} = Lm\vec{v}$ $= L^2 m\omega_0$. Thus, the result is angular momentum before = angular momentum after.

$$\left(\frac{1}{3} ML^2 + mL^2\right)\omega_0 = \left(\frac{1}{3} ML^2\right)\omega + mL^2\omega_0$$

which yields $\omega = \omega_0$; therefore, the angular velocity of the rod is unchanged.

• PROBLEM 11-16

A wheel rolls without slipping along a horizontal plane under the action of a horizontal force P of 50 lbs, as shown in Fig. 1. The wheel weighs 300 lb, and its center of mass is at its geometric center O. Its radius is 2 ft, and its radius of gyration with respect to an axis through the center O is 1.5 ft. Determine the time required to increase the velocity of its center from 4 ft/sec to the right to 8 ft/sec.

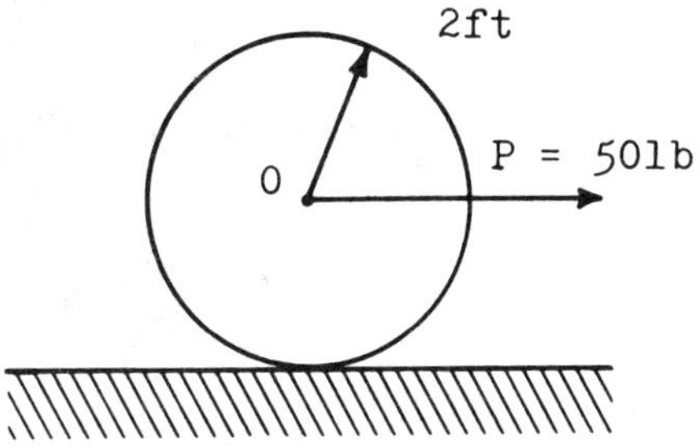

Fig 1

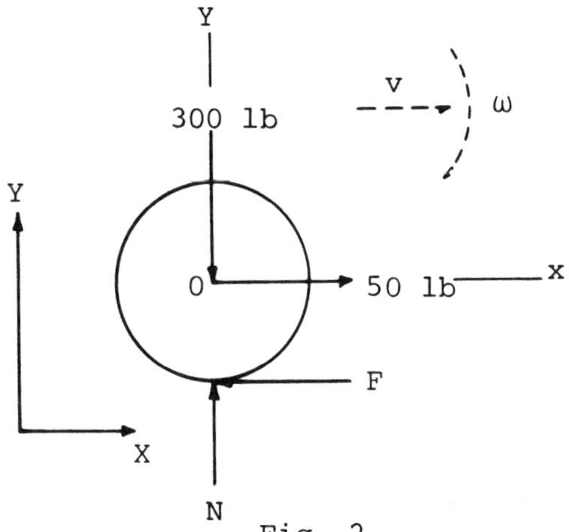

Fig. 2

Solution: A free-body diagram of the wheel is shown in Fig. 2. We choose the positive senses of $\vec{\omega}$ and $\vec{v}$ as shown in Fig. 2. The equations of motion will be written in the impulse-momentum form, and the center of the wheel will be chosen as the origin of the moving, nonrotating axes.

Linear Impulse-momentum equation:

$$\int F_x \, dt = m\Delta v$$

$$\int_0^1 (50 - F)\, dt = \frac{300}{g}(8 - 4) \qquad (1)$$

Rotational Impulse-momentum equation:

$$\int M_0 \, dt = I\Delta\omega$$

$$\int_0^t 2F \, dt = \frac{300}{g} (1.5)^2 \left(\frac{8}{2} - \frac{4}{2}\right) \qquad (2)$$

Doing the integrations indicated yields

$$50t - Ft = 37.27$$

and $$2Ft = 41.93.$$

Solving these two equations simultaneously, the result is that t = 1.16 sec.

An alternate form of the second equation of motion would result if we choose the origin of moving axes at the point where the wheel touches the ground. For a nonslipping wheel this point is an instantaneous center. Hence, its velocity is zero, and equation (2) becomes

$$\int_0^t (2 \times 50) \, dt = \left[\frac{300}{g} (1.5)^2 + \frac{300}{g} (2)^2\right] \left(\frac{8}{2} - \frac{4}{2}\right)$$

We can then find the time directly from this single equation. Thus,

$$100t = \frac{300}{g} (6.25)(2)$$

and t = 1.16 sec.

• PROBLEM 11-17

A circular platform weighs 500 lb and has a diameter of 12 ft. It initially rotates at 30 rpm on a vertical shaft. A 150 lb man originally sitting on the edge at rest, relative to the disk, starts running along the edge with a constant speed of 12 ft/sec relative to the disk in the same direction that the disk is moving. What will be the final angular speed of the disk?

Solution: Since no external torque acts on the system, the angular momentum of man plus disk remains constant. Write L (before) = L (after). L (before) = $I_{disk}\omega_0 + I_{man}\omega_0$. Here ω_0 is the original disk rotation speed and

$$I_{man} = m R^2 \qquad I_{disk} = \frac{1}{2} M R^2$$

$$m = \frac{\text{weight of man}}{g} \qquad M = \frac{\text{weight of disk}}{g}.$$

Now, L (before) $= \frac{1}{2} M R^2 \omega_0 + m R^2 \omega_0$ (1)

L (after) = L (disk) + L (man). Thus,

$$L \text{ (after)} = \frac{1}{2} M R^2 \omega + \frac{1}{2} m R^2 \omega_m, \quad (2)$$

where ω is the new disk rotation speed and ω_m is the man's rotation speed.

But $\omega_m = \omega + \frac{v}{R}$, relative to the ground and v is the speed at which the man runs, relative to the disk.

So, $\omega_m = \omega + \frac{12 \text{ ft/sec}}{6 \text{ ft}} = \omega + 2 \frac{\text{rad}}{\text{sec}}$, and

$$L \text{ (after)} = \frac{1}{2} M R^2 \omega + m R^2 \left(\omega + 2 \frac{\text{rad}}{\text{sec}}\right). \quad (3)$$

Now, $\omega_0 = 30 \frac{\text{rev}}{\text{min}} \left(2\pi \frac{\text{rad}}{\text{rev}}\right)\left(\frac{1}{60} \frac{\text{sec}}{\text{min}}\right) = 3.14 \frac{\text{rad}}{\text{sec}}$, so substituting in equation (1) yields

$$\left[\frac{1}{2} M R^2 + m R^2\right]\left(3.14 \frac{\text{rad}}{\text{sec}}\right) = \frac{1}{2} M R^2 \omega + m R^2 \left(\omega + 2 \frac{\text{rad}}{\text{sec}}\right) \quad (4)$$

$$(M + 2m)(3.14) = M\omega + 2 m\omega + m (4) \quad (5)$$

$$3.14 (M + 2m) - 4 m = (M + 2m)\omega \quad (6)$$

$$\omega = 3.14 \frac{\text{rad}}{\text{sec}} - \frac{4 m}{(M + 2m)} \frac{\text{rad}}{\text{sec}}. \quad (7)$$

Thus, the angular velocity has been reduced by an amount $\frac{4 m}{(M + 2m)}$. Substitute the weights given, as the factor of g will divide out; thus,

$$\omega = 3.14 - \frac{4 (150)}{(500 + 300)}$$

$$\omega = 3.14 - 0.75 = 2.39 \text{ rad/sec}. \quad (8)$$

GREEN'S FUNCTION TECHNIQUE

• PROBLEM 11-18

Solve the equation of the damped, forced harmonic oscillator by Green's function technique.

Solution: The equation to solve is

$$\frac{d^2x}{dt^2} + 2\lambda \frac{dx}{dt} + \omega_0^2 x = \frac{f(t)}{m}. \tag{1}$$

Assume f(t) = 0 except for an instantaneous impulse at t = s while the system is at rest. For t > s the system will be governed by the homogeneous equation.

$$\frac{d^2x}{dt^2} + 2\lambda \frac{dx}{dt} + \omega_0^2 x = 0. \tag{2}$$

The solution of (2) is well known from previous chapters to be

$$x(t) = c_1 e^{-\lambda t} \cos \omega t + c_2 e^{-\lambda t} \sin \omega t$$

where $\omega^2 = \omega_0^2 - \lambda^2$. Assume that at the blow

$$x(s) \approx 0$$

$$v(s) = \frac{I}{m}.$$

These two conditions determine the constants c_1 and c_2

$$0 = e^{-\lambda s} (c_1 \cos \omega s + c_2 \sin \omega s)$$

$$\frac{I}{m} = e^{-\lambda s} \omega(- c_1 \sin \omega s + c_2 \cos \omega s)$$

$$+ (- \lambda e^{-\lambda s}) (c_1 \cos \omega s + c_2 \sin \omega s).$$

Solving this yields

$$x(t) = \frac{I}{m\omega} e^{-\lambda(t-s)} \sin \omega(t-s); \quad (t > s). \tag{3}$$

Now, suppose there is a continuous distribution of impulses dI = f(s) ds. The cumulative effect of these impulses will be obtained by integrating (3) after letting

$$x \to dx$$

$$I \to dI.$$

Thus, $$x(t) = \int_{t_0}^{t} \frac{f(s)\, ds}{m\omega} e^{-\lambda(t-s)} \sin \omega(t-s). \tag{4}$$

Define $$G(t, s) = \frac{1}{\omega} e^{-\lambda(t-s)} \sin \omega(t-s)$$

for $t > s$ as the Green's function. (4) may be written

$$x(t) = \int_{t_0}^{t} G(t, s) \frac{f(s)}{m} ds.$$

A formal way of viewing the instantaneous impulse is to write

$$f(t) = \delta (t - s).$$

Then the Green's function may be obtained by solving

$$\frac{d^2G}{dt^2}(t, s) + 2\lambda \frac{dG}{dt}(t, s) + \omega_0^2 G(t, s) = \frac{\delta (t - s)}{m}.$$

• PROBLEM 11-19

Solve the differential equation of a particle of mass m moving in a resistive medium (coefficient R) under the influence of an external force f(t) by use of Green's function technique.

Solution: From Newton's Second Law:

$$m \frac{dv}{dt} = - Rv + f(t) \qquad (1)$$

where $v = v(t)$ is the velocity of the particle, and $- Rv$ is the resistive force.

Suppose the force f(t) exists only for a short period of time, say from s to $s + \Delta s$. After $s + \Delta s$ the motion is governed by:

$$m \frac{dv}{dt} = - Rv$$

or $$\frac{dv}{v} = - \frac{R}{m} dt$$

by separating variables. Integrating and then exponentiating,

$$v = Ae^{-(R/m)t} \qquad (2)$$

where A is the arbitrary constant of integration.

Now, integrate (1) between s and $s + \Delta s$:

$$m \int_{s}^{s+\Delta s} \frac{dv}{dt} dt = - R \int_{s}^{s+\Delta s} vdt + \int_{s}^{s+\Delta s} f(t)\, dt. \qquad (3)$$

The integral $\int_s^{s+\Delta s} f(t)dt = I$ is the impulse of the force. If Δs is small, v(t) will not change much in its integral, so assume v(t) is constant.

$$\int_s^{s+\Delta s} v(t)dt \cong v(s) \int_s^{s+\Delta s} dt \cong v(s)\ \Delta s \cong 0.$$

Assume the particle was at rest at t = s. Thus, (3) becomes

$$m\ [v(s + \Delta s) - v\ (s)] = 0 + I$$

$$v\ (s + \Delta s) = \frac{I}{m}.$$

By continuity with solution (2) which takes over at $t = s + \Delta s$,

$$v\ (s + \Delta s) = Ae^{-\frac{R}{m}(s + \Delta s)}$$

$$\cong Ae^{-\frac{Rs}{m}}.$$

Thus,
$$v(t) = \begin{cases} 0 & t \leq s \\ \frac{I}{m}\ e^{-(R/m)(t-s)}, & t > s \end{cases} \qquad (4)$$

Now, suppose the particle suffers two "blows", one at $t = s_1$ and one at $t = s_2$. Then a similar argument yields

$$v(t) = \begin{cases} 0 & t < s_1 \\ \frac{I_1}{m}\ e^{-\frac{R}{m}(t-s_1)} & s_1 < t < s_2 \\ \frac{I_1}{m}\ e^{-\frac{R}{m}(t-s_1)} + \frac{I_2}{m}\ e^{-\frac{R}{m}(t-s_2)} & t > s_2 \end{cases}$$

where $I_1 = \int_{s_1}^{s_1+\Delta s} f(t)dt$ and

$$I_2 = \int_{s_2}^{s_2+\Delta s} f(t)dt.$$

Generalizing to the case where f(t) is in effect for duration Δs at times $s_1, s_2, \ldots, s_n$, the superposition solution is

$$v(t) = \sum_{k=1}^{n} \frac{I_k}{m} e^{-\frac{R}{m}(t-s_k)} \qquad (5)$$

where $$I_k = \int_{s_k}^{s_k+\Delta s} f(t)\,dt.$$

The next step is to imagine that the continuous action of f(t) may be represented as the sum of impulses, $dI = f(s)\,ds$. But the sum of a continuous variable requires the integral. Thus, (5) becomes

$$v(t) = \int_{t_0}^{t} \frac{f(s)\,ds}{m} e^{-\frac{R}{m}(t-s)} \qquad (6)$$

for $t > t_0$.

Now, define

$$G(t, s) = \begin{cases} 0 & (t < t_0) \\ \frac{1}{m} e^{-\frac{R}{m}(t-s)} & (t > t_0) \end{cases}$$

Then the response may be written

$$v(t) = \int_{t_0}^{t} G(t, s)\, f(s)\, ds. \qquad (7)$$

G (t, s) is called the Green's function of the problem. Equation (7) is the characteristic form of the solution when Green's function technique is used.

CHAPTER 12

SYSTEMS OF PARTICLES

CENTER OF MASS

• PROBLEM 12-1

Consider a system made up of two particles of equal mass, both initially at rest at the origin. Particle one is subjected to no forces and remains at the origin. Particle two is subjected to a constant force $\vec{F}$ in the x-direction. Find an equation for the motion of the center of mass of the system. Show that it satisfies $M\ddot{R} = F^{(e)}$, where $F^{(e)}$ is the total external force on the system.

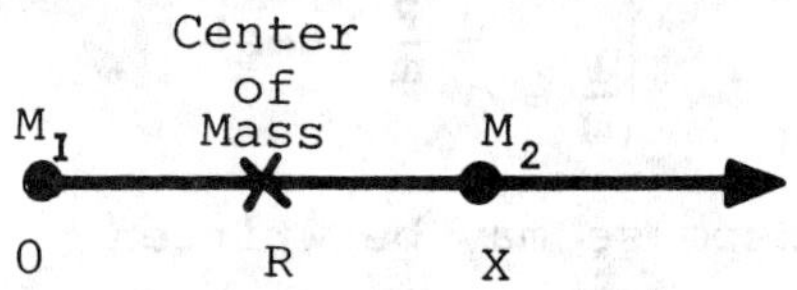

Solution: Let x be the coordinate of the second particle at time t. For the first particle x = 0 for all time. The center of mass of the system is then, according to the definition,

$$R = (m_1 x_1 + m_2 x_2)/(m_1 + m_2)$$

$$= (m.0 + m.x)/2m = x/2.$$

The acceleration of the second particle is given by Newton's Second Law,

$$\ddot{x} = F/m$$

The acceleration of the center of mass is then,

$$\ddot{R} = \ddot{x}/2 = F/2m, \text{ or } 2m\ddot{R} = F.$$

The last equation is consistent with $M\ddot{R} = F^{(e)}$, because the total mass of the system is $M = m_1 + m_2 = 2m$, and the total external force on the system is $F^{(e)} = F$.

• **PROBLEM 12-2**

Two particles of equal mass move according to the equations $\vec{r}_1 = t\hat{i} + 2t^2\hat{j} + 4\hat{k}$ and $\vec{r}_2 = 3t^3\hat{i} - t\hat{j} + t^4\hat{k}$. Find the position $\vec{R}$, the velocity $\dot{\vec{R}}$, and the acceleration $\ddot{\vec{R}}$ of the center of mass as a function of time.

Solution: From the definition of center of mass we have

$$\vec{R} = \frac{m_1 \vec{r}_1 + m_2 \vec{r}_2}{m_1 + m_2} = \frac{m \vec{r}_1 + m \vec{r}_2}{2\ m} = \frac{1}{2}(\vec{r}_1 + \vec{r}_2)$$

$$= \frac{1}{2}(t\hat{i} + 2t^2\hat{j} + 4\hat{k}) + \frac{1}{2}(3t^3\hat{i} - t\hat{j} + t^4\hat{k})$$

$$= \left(\frac{1}{2}t + \frac{3}{2}t^3\right)\hat{i} + \left(t^2 - \frac{1}{2}t\right)\hat{j} + \left(2 + \frac{1}{2}t^4\right)\hat{k}$$

To obtain the velocity and the acceleration of the center of mass, we simply take the derivatives of R with respect to t once and twice,

$$\dot{\vec{R}} = \left(\frac{1}{2} + \frac{9}{2}t^2\right)\hat{i} + \left(2t - \frac{1}{2}\right)\hat{j} + 2t^3\hat{k},$$

$$\ddot{\vec{R}} = 9t\hat{i} + 2\hat{j} + 6t^2\hat{k}.$$

Here, we have used the "power rule" in calculus,

$$\frac{d}{dt}(a\ t^n) = n\ a\ t^{n-1}.$$

• **PROBLEM 12-3**

At t = 0, three particles are in a position as shown in Figure 1. At this instant, particle 1 has a mass of 1 and a velocity of 3 upward. Particle 2 has a mass of 5 and a velocity of 2 in the positive x direction, and particle 3 has a mass of 4 and a velocity of 6 in the negative x direction. If there are no external forces on this system, determine (a) the subsequent motion of the center of mass, and (b) the total kinetic energy of the system of particles. The particles move in a horizontal plane.

Solution: Since there are no external forces acting on the system, momentum is conserved. Thus,

$$mv_c = \text{constant}$$

where m is the total mass of the system and v_c is the velocity of the center of mass.

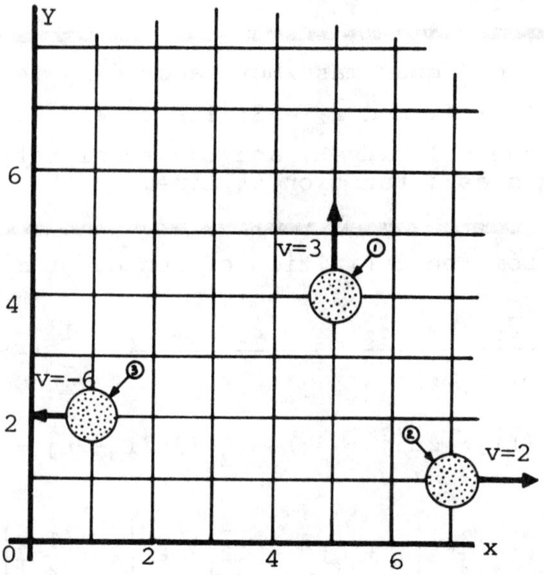

Fig. 1

The first problem is to determine the constant. This is done by using the definition of the center of mass of the system and the fact that the velocity of each of the particles is known at t = 0.

$$mv_c = \sum_i m_i v_i$$

$$= (1)\,(3\hat{j}) + (5)\,(2\hat{i}) + (4)\,(-6\hat{i})$$

$$= -\,14\hat{i} + 3\hat{j}$$

The total mass m is

$$m = m_1 + m_2 + m_3 .$$

$$= 1 + 5 + 4$$

$$= 10$$

Finally, the velocity of the center of mass is

$$v_c = \frac{-14\hat{i} + 3\hat{j}}{10} = -\,1.4\hat{i} + 0.3\hat{j}$$

Now by definition $v_c = \dfrac{dr_c}{d\,t}$.

To find the position , r_c, of the mass center as a function of time, we can separate variables and integrate

$$\int_0^t dr_c = \int_0^t v_c\ dt = \int_0^t [-1.4\hat{i} + 0.3\hat{j}]\ \ dt$$

$$r_c(t) - r_0 = -1.4t\hat{i} + 0.3t\hat{j}$$

$$r_c(t) = r_0 + [-1.4t\hat{i} + 0.3t\hat{j}].$$

r_0 is the position of the mass center at t = 0. This can be evaluated directly.

$$r_0 = x_0\hat{i} + y_0\hat{j}$$

$$x_0 = \frac{1(5) + 5(7) + 4(1)}{10}$$

$$y_0 = \frac{1(4) + 5(1) + 4(2)}{10}$$

so that $x_0 = 4.4$

$y_0 = 1.7$

The position of the center of mass is then

$$r_c(t) = 4.4\hat{i} + 1.7\hat{j} + [-1.4t\hat{i} + 0.3t\hat{j}]$$

$$r_c(t) = [-1.4t + 4.4]\ \hat{i} + [0.3t + 1.7]\ \hat{j}.$$

The kinetic energy of this system may be obtained directly, as the velocity of each of the particles is known:

$$T = \sum_i \frac{1}{2} m_i v_i^2$$

and, for this system,

$$T = \frac{1}{2}(1)(3)^2 + \frac{1}{2}(5)(2)^2 + \frac{1}{2}(4)(6)^2$$

and $T = 86.5$

• **PROBLEM** 12-4

Consider the system of particles described below:

$m_1 = 5$ g $\quad v_1 = 3\hat{j} + 4\hat{k}$ cm/sec

$m_2 = 6$ g $\quad v_2 = 2\hat{j} + 5\hat{k}$ cm/sec

$m_3 = 4$ g $\quad v_3 = \hat{j} + 3\hat{k}$ cm/sec

$m_4 = 4$ g $\quad v_4 = -4\hat{j} - 5\hat{k}$ cm/sec

$m_5 = 1$ g $\quad v_5 = -15\hat{j} - 2\hat{k}$ cm/sec

Determine the kinetic energy of the system by a) finding the sum of the individual kinetic energies of each particle and by b) finding the velocity of the mass center and using relative velocities of the particles with respect to the mass center.

Solution: a) The kinetic energy of each particle is given by $\frac{1}{2}mv^2$ where m is the mass of the particle and v is the magnitude of the velocity of the particle. The magnitude of a vector is given by the square root of the sum of the squares of each component. Thus

$$(v_1)^2 = 3^2 + 4^2 = 25$$

$$(v_2)^2 = 2^2 + 5^2 = 29$$

$$(v_3)^2 = 1^2 + 3^2 = 10$$

$$(v_4)^2 = (-4)^2 + (-5)^2 = 41$$

$$(v_5)^2 = (-15)^2 + (-2)^2 = 229$$

The total kinetic energy of the system, T, is just the sum of the individual kinetic energies of each particle.

$$T = \sum_{i=1}^{5} m_i v_i^2 = \frac{1}{2} m_1 (v_1)^2 + \frac{1}{2} m_2 (v_2)^2 + \frac{1}{2} m_3 (v_3)^2 + \frac{1}{2} m_4 (v_4)^2$$

$$+ \frac{1}{2} m_5 (v_5)^2$$

Substituting in the given values yields

$$T = \frac{1}{2}(5)\ 25 + \frac{1}{2}(6)\ 29 + \frac{1}{2}(4)\ 10 + \frac{1}{2}(4)\ 41$$

$$+ \frac{1}{2}(1)\ 229$$

$$T = 366 \text{ ergs}$$

b) Another way to find the total kinetic energy is to find the velocity of the mass center, and the velocity of the particles relative to the mass center. Then the total kinetic energy will be given by finding the kinetic energy value of the total mass traveling at this velocity, and adding this to the kinetic energy of each particle traveling at its relative velocity towards the mass center. If M equals the total mass of the system, v_{cm} equals the velocity of the mass center and u_1, u_2, u_3, u_4, u_5 represent the represent the relative velocities of each particle with respect to the mass center, then the total kinetic energy will be given by

$$T = \frac{1}{2} M (v_{cm})^2 + \frac{1}{2} m_1 (u_1)^2 + \frac{1}{2} m_2 (u_2)^2 + \frac{1}{2} m_3 (u_3)^2$$

$$+ \frac{1}{2} m_4 (u_4)^2 + \frac{1}{2} m_5 (u_5)^2 \quad (1)$$

The position, r_{cm}, of the mass center is given by

$$\vec{r}_{cm} = \frac{m_1\vec{r}_1 + m_2\vec{r}_2 + m_3\vec{r}_3 + m_4\vec{r}_4 + m_5\vec{r}_5}{m_1 + m_2 + m_3 + m_4 + m_5} \quad (2)$$

where $\vec{r}_1$, $\vec{r}_2$, $\vec{r}_3$, $\vec{r}_4$, $\vec{r}_5$ are the position vectors of the particles and the denominator is equal to M.

Differentiating with respect to time yields the velocity of the mass center.

$$\frac{d\vec{r}_{cm}}{dt} = \vec{v}_{cm} = \frac{m_1\vec{v}_1 + m_2\vec{v}_2 + m_3\vec{v}_3 + m_4\vec{v}_4 + m_5\vec{v}_5}{m_1 + m_2 + m_3 + m_4 + m_5} \quad (3)$$

Equation (3) represents a vector equation and therefore it means there are three scalar equations, one for each component of velocity. Since in this system there are only y and z components of velocity, only 2 equations can be used. In the y direction

$$v_{cm_y}\hat{j} = \frac{(5)(3)\hat{j} + 6(2)\hat{j} + 4(1)\hat{j} + 4(-4)\hat{j} + (1)(-15)\hat{j}}{5 + 6 + 4 + 4 + 1}$$

$$v_{cm_y}\hat{j} = 0\hat{j}$$

In the z direction

$$v_{cm_z}\hat{k} = \frac{(5)(4)\hat{k} + (6)(5)\hat{k} + (4)(3)\hat{k} + (4)(-5)\hat{k} + 1\ (-2)\hat{k}}{20}$$

$$v_{cm}\hat{k} = 2\hat{k}$$

The total velocity of the mass center is then

$$\vec{v}_{cm} = v_{cm_y}\hat{j} + v_{cm_z}\hat{k} = 2\hat{k}$$

The relative velocity of each component with respect to the mass center is

$$\vec{u} = \vec{v} - \vec{v}_{cm}$$

Therefore

$$\vec{u}_1 = 3\hat{j} + 4\hat{k} - 2\hat{k} = 3\hat{j} + 2\hat{k}$$
$$\vec{u}_2 = 2\hat{j} + 5\hat{k} - 2\hat{k} = 2\hat{j} + 3\hat{k}$$
$$\vec{u}_3 = 1\hat{j} + 3\hat{k} - 2\hat{k} = 1\hat{j} + 1\hat{k}$$
$$\vec{u}_4 = -4\hat{j} - 5\hat{k} - 2\hat{k} = -4\hat{j} - 7\hat{k}$$
$$\vec{u}_5 = -15\hat{j} - 2\hat{k} - 2\hat{k} = -15\hat{j} - 4\hat{k}$$

Also,

$$(u_1)^2 = 3^2 + 2^2 = 13$$
$$(u_2)^2 = 2^2 + 3^2 = 13$$
$$(u_3)^2 = 1^2 + 1^2 = 2$$
$$(u_4)^2 = (-4)^2 + (-7)^2 = 65$$
$$(u_5)^2 = (-15)^2 + (-4)^2 = 241$$

$$(v_{cm})^2 = 0^2 + 2^2 = 4$$

Since M = 20, we can substitute in equation (1).

$$T = \frac{1}{2}(20)(4) + \frac{1}{2}(5)(13) + \frac{1}{2}(6)(13) + \frac{1}{2}(4)(2) + \frac{1}{2}(4)(65) + \frac{1}{2}(1)(241)$$

T = 366 ergs, which is the same value we obtained before.

• PROBLEM 12-5

Consider a system consisting of N number of particles. The motion of the system can be described by the equation

$$M\ddot{R} = \sum_{\substack{i=1 \\ (i \neq j)}}^{n} \sum_{j=1}^{n} Fij + \sum_{i=1}^{n} Fi^e$$

where M, R are the total mass and the center of mass of the system, Fij is the force exerted on particle i by particle j and Fi^e is the external force acting on the ith particle. Show that the double sum in the equation is equal to zero.

Solution: The double sum,

$$\sum_{i=1}^{n} \sum_{j=1}^{n} \vec{F}ij\ (i \neq j) = \vec{F}_{12} + \vec{F}_{13} + \vec{F}_{14} + \ldots + \vec{F}_{1n}$$
$$+ \vec{F}_{21} + \vec{F}_{23} + \vec{F}_{24} + \ldots + \vec{F}_{2n}$$

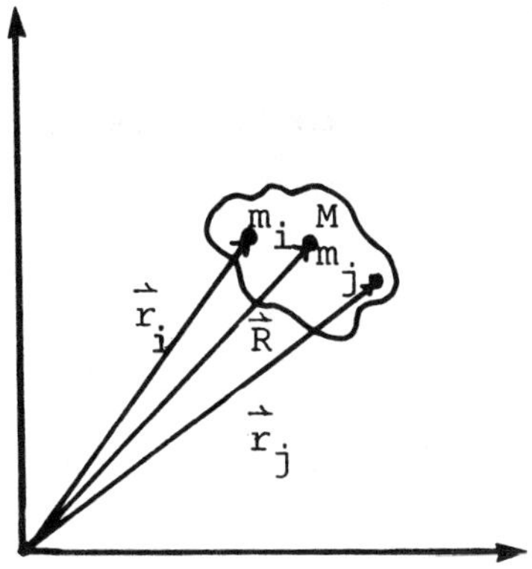

$$+ \vec{F}_{31} + \vec{F}_{32} + \vec{F}_{34} + \dots + \vec{F}_{3n}$$

$$+ \dots\dots\dots\dots\dots\dots\dots\dots\dots\dots$$

$$\dots$$

$$+ \vec{F}_{n1} + \vec{F}_{n2} + \vec{F}_{n3} + \dots + \vec{F}_{n,\, n-1}$$

is equal to zero because of Newton's Third Law, which says that the interaction forces between two particles are equal in magnitude but opposite in direction. For examples:

$$\vec{F}_{12} = -\vec{F}_{21},\ \vec{F}_{13} = -\vec{F}_{31},\ \vec{F}_{2n} = -\vec{F}_{n2},\ \text{etc.}$$

All the terms in the double sum are paired off and cancel out, like

$$\vec{F}_{12} + \vec{F}_{21} = 0,\ \vec{F}_{13} + \vec{F}_{31} = 0,\ \vec{F}_{2n} + \vec{F}_{n2} = 0,\ \text{etc.}$$

• PROBLEM 12-6

Relate the kinetic energy of a system of particles in an inertial reference frame O to that in a reference frame which is moving with a velocity $\dot{\vec{s}}$ with respect to the inertial frame.

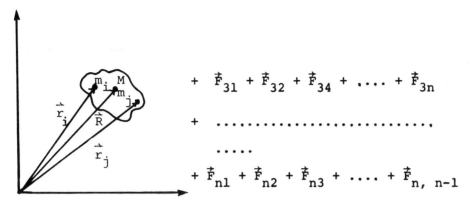

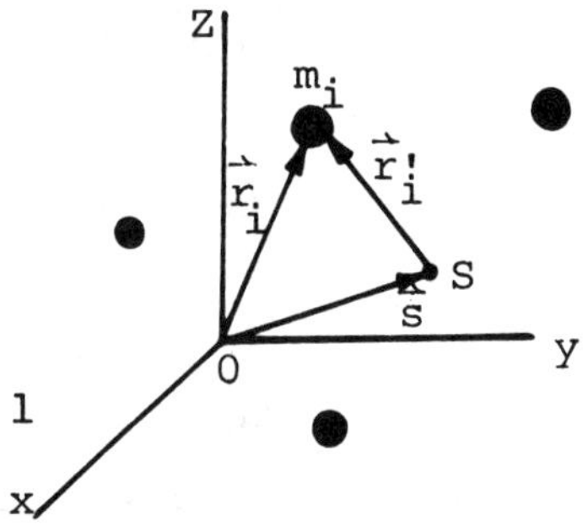

Fig. 1

Solution: We shall start with the equation for the total kinetic energy as measured by an observer in the inertial frame O:

$$T = \sum_{i=1}^{n} \frac{1}{2} m_i \vec{v}_i \cdot \vec{v}_i = \frac{1}{2} \sum_{i=1}^{n} m_i \dot{\vec{r}}_i \cdot \dot{\vec{r}}_i . \qquad (1)$$

Next let us find a relationship between the position vectors $\vec{r}_i$ and $\vec{r}_i{}'$. From Fig. 1 we get:

$$\vec{r}_i = \vec{r}_i{}' + \vec{s}. \tag{2}$$

Differentiating eq. (2) with respect to time we obtain:

$$\dot{\vec{r}}_i = \dot{\vec{r}}_i{}' + \dot{\vec{s}}, \tag{3}$$

and substituting eq. (3) into eq. (1) we get:

$$T = \frac{1}{2} \sum_{i=1}^{n} m_i \left[\left(\dot{\vec{r}}_i + \dot{\vec{s}}\right) \cdot \left(\dot{\vec{r}}_i + \dot{\vec{s}}\right)\right]. \tag{4}$$

Evaluating the dot product, we have:

$$T = \frac{1}{2} \sum_{i=1}^{n} m_i \dot{r}_i{}'^2 + \frac{1}{2} \sum_{i=1}^{n} m_i \dot{s}^2 + \sum_{i=1}^{n} m_i \dot{\vec{r}}_i{}' \cdot \dot{\vec{s}}. \tag{5}$$

In eq. (5), $\dot{\vec{s}}$ has the same value for every term in the summation. We can therefore remove this term from the summation, this yields:

$$T = \frac{1}{2} \sum_{i=1}^{n} m_i \dot{r}_i{}'^2 + \frac{1}{2} \left(\sum_{i=1}^{n} m_i\right) \dot{s}^2 + \dot{\vec{s}} \cdot \sum_{i=1}^{n} m_i \dot{\vec{r}}_i{}' \tag{6}$$

We can simplify eq. (6) by noting that

$$\sum_{i=1}^{n} m_i = M \quad \text{and} \quad \sum_{i=1}^{n} m_i \dot{\vec{r}}_i = \frac{d}{dt} \sum_{i=1}^{n} m_i \vec{r}_i{}' = M\dot{\vec{R}},$$

where M is the total mass of the system and R' is the position vector of the center of mass relative to s. Substituting these expressions into eq. (6) we obtain the desired equation,

$$T = \frac{1}{2} \sum_{i=1}^{n} m_i \dot{r}_i{}'^2 + \frac{1}{2} M\dot{s}^2 + M\dot{\vec{R}}' \cdot \dot{\vec{s}}. \tag{7}$$

The first term in eq. (7) is the total kinetic energy relative to the point s and the second term is the kinetic energy of a particle of mass M moving with a speed $\dot{s}$. The last term in eq. (7) does not lend itself to a simple physical interpretation, but this term is equal to zero if $\dot{\vec{R}}' = 0$, i.e., point s is the center of mass of the system of particles.

• PROBLEM 12-7

Summarize the laws of motion of a system of particles relative to an arbitrary point S. Simplify the equation of motion if the arbitrary point s is located at the center of mass of the system of particles.

Solution: The equations of motion for a system of particles relative to an arbitrary point s are:

$$M\ddot{\vec{s}} + M\ddot{\vec{R}} = \vec{F} + \sum_{i=1}^{n} \sum_{\substack{j=1 \\ j\neq i}}^{n} \vec{f}_{ij} \qquad (1)$$

and $$\dot{\vec{L}} = \sum_{i=1}^{n} \vec{r}_i \times \vec{F}_i = M\vec{R} \times \ddot{\vec{s}} + \sum_{i=1}^{n} \sum_{\substack{j=1 \\ j\neq i}}^{n} \vec{r}_i \times \vec{f}_{ij}, \qquad (2)$$

where $M = \sum_{i=1}^{n} m_i$ = the total mass,

$\vec{F} = \sum_{i=1}^{n} \vec{F}_i$ = the resultant external force,

$\vec{R} = \sum_{i=1}^{n} m_i \vec{r}_i/M$ = the position of the center of mass relative to s,

$\vec{f}_{ij}$ = the internal force on particle i arising from its interaction with the jth particle,

$\vec{r}_i$ = the position vector of particle i relative to s,

and $\vec{L}$ = the total angular momentum about the point s.

If the internal forces are equal and opposite and act along the line joining the interacting particles then the last term in eqs (1) and (2) is equal to zero and we have:

$$M\ddot{\vec{s}} + M\ddot{\vec{R}} = \vec{F} \qquad (3)$$

and $$\dot{\vec{L}} = \sum_{i=1}^{n} \vec{r}_i \times \vec{F}_i - M\vec{R} \times \ddot{\vec{s}}. \qquad (4)$$

We can simplify these general equations if the arbitrary point s is the center of mass of the system of particles. In this case $\vec{R} = 0$ and eqs. (3) and (4) reduce to:

$$M\ddot{\vec{s}} = \vec{F} \qquad (5)$$

and $$\dot{\vec{L}} = \sum_{i=1}^{n} \vec{r}_i \times \vec{F}_i. \qquad (6)$$

Eq. (5) indicates that the center of mass moves as if the resultant external force acted directly on the center of mass. Eq. (6) shows that the net external torque calculated with respect to the center of mass is equal to the time rate of change of the angular momentum calculated with respect to the center of mass.

• PROBLEM 12-8

A 40 lb. missile moves with a velocity of 150 ft./sec. It is intercepted by a laser beam which causes it to explode into two fragments (A) and (B) which weigh 25 lbs. and 15 lbs. respectively. If the fragments travel as shown immediately after explosion, find the velocity of each fragment.

Solution: There are no external forces in this system since the explosion is caused by an internal force. Therefore linear momentum of the system is conserved.

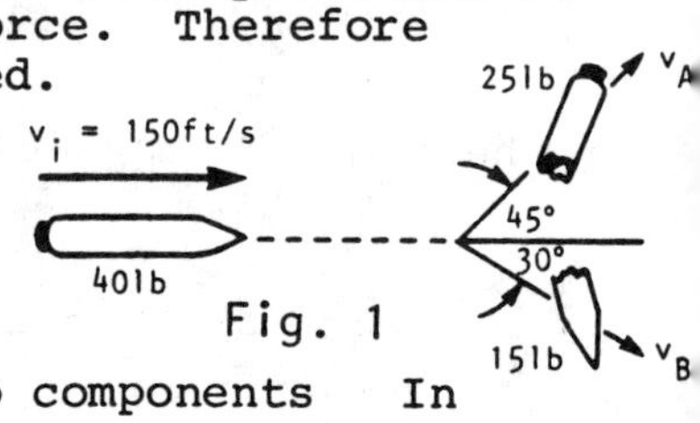

Fig. 1

Momentum before = Momentum after

$$(m_A + m_B)\, v_0 = m_A\, v_A + m_B\, v_B$$

This equation can be broken up into components In the x - direction

$$(m_A + m_B)\, v_0 = m_A\, v_A \cos 45° + m_B\, v_B \cos 30°$$

and in the y - direction

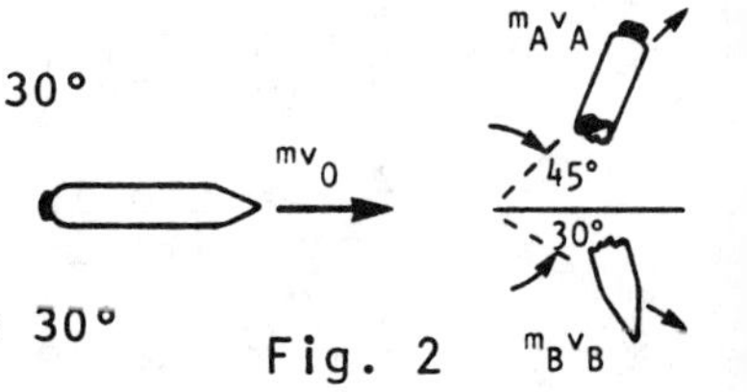

Fig. 2

$$(m_A + m_B)\,(0) = m_A\, v_A \sin 45° - m_B\, v_B \sin 30°$$

Substituting in numerical values yields a set of simultaneous equations.

$$[\frac{40}{g}](150) = (\frac{25}{g})v_A \cos 45° + (\frac{15}{g})v_B \cos 30°$$

$$[\frac{40}{g}](0) = (\frac{25}{g})v_A \sin 45° - [\frac{15}{g}]v_B \sin 30°$$

Cancelling out g, and then solving for v_A and v_B yields:

v_A = 43.92 ft/sec 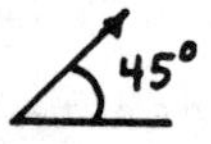

v_B = 146.41 ft/sec 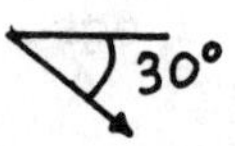

• PROBLEM 12-9

During a hockey practice session, pucks (B) and (C) lie side by side without motion on the ice. Puck (A) is shot, and collides with pucks (B) and (C). The pucks (A), (B), and (C) scatter in directions shown in the diagram. If puck (A) is moving with a velocity of 5m/sec at the moment of impact, find the velocities of puck (A), (B), (C) after collision. The collision is perfectly elastic and the ice acts as a frictionless surface.

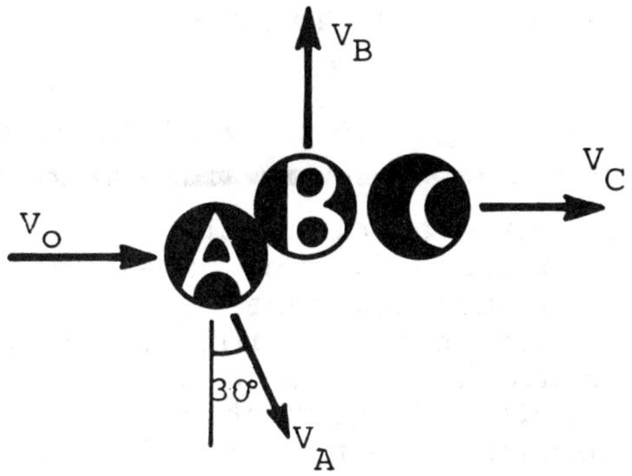

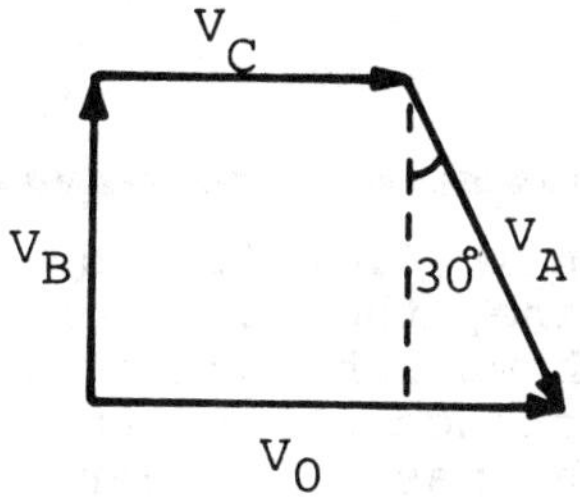

Solution: From the principle of conservation of momentum we have

$$m_A \vec{v}_0 = m_A \vec{v}_A + m_B \vec{v}_B + m_C \vec{v}_C,$$

or, because of equal mass of pucks,

$$\vec{v}_o = \vec{v}_A + \vec{v}_B + \vec{v}_C.$$

In component form,

$$\begin{cases} v_B - v_A \cos 30° = 0 & (1). \\ v_C + v_A \sin 30° = v_0 = 5. & (2) \end{cases}$$

Because of the assumption of elastic collision there also exists the conservation of kinetic energy,

$$KE_{After} = KE_{Before}$$

$$\frac{1}{2} m \left(v_A{}^2 + v_B{}^2 + v_C{}^2\right) = \frac{1}{2} m\, v_0{}^2,$$

or $\quad v_A{}^2 + v_B{}^2 + v_C{}^2 = v_0{}^2 = 25. \qquad (3)$

First square equation (1) and (2) and add them. This yields

$$v_B{}^2 + v_C{}^2 + v_A{}^2 + 2v_A\,(v_C \sin 30° - v_B \cos 30°) = 25.$$

Subtracting equation (3) yields

$$2v_A\,(v_C \sin 30° - v_B \cos 30°) = 0$$

or after dividing by $2v_A$ and solving for v_C

$$v_C = v_B \cot 30° = 1.732\, v_B \qquad (4)$$

Equations (1), (2) and (4) can be solved simultaneously. The results are

$$v_A = 2.50 \text{ m/s}, \; v_B = 2.16 \text{ m/s}, \; v_C = 3.749 \text{ m/s}.$$

• PROBLEM 12-10

Shown in Fig. (1) is a schematic diagram of an electrical machine, the rotor of which is unbalanced. The unbalance is shown schematically by the concentrated mass m placed at a distance e from the axis of rotation of the rotor. Both m, the mass of the rotor, and e are known from a previous analysis. The mass of the remainder of the machine is M. The machine is placed on four mounts, shown as two in the figure. Determine the motion of the machine in the lateral direction, assuming that the mounts offer no resistance to the motion of the machine in the lateral direction for small oscillations. To do this, use the mathematical model of the machine established in Figure (2). The mass M moves on the rod shown with no friction between the mass and the rod. Determine the motion of the large mass, that is, the motion of the machine when the angular velocity of the rotor is ω.

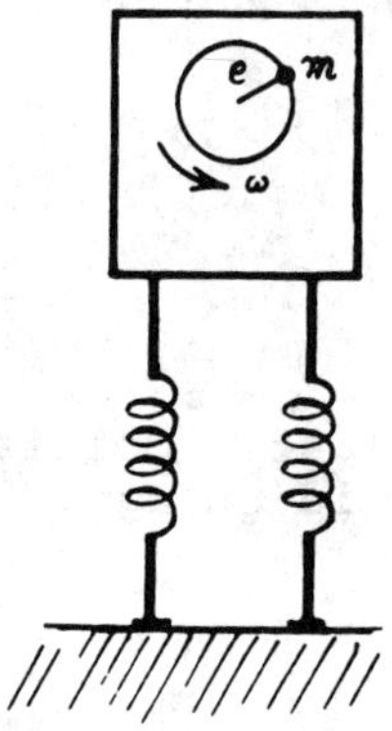

Fig. 1

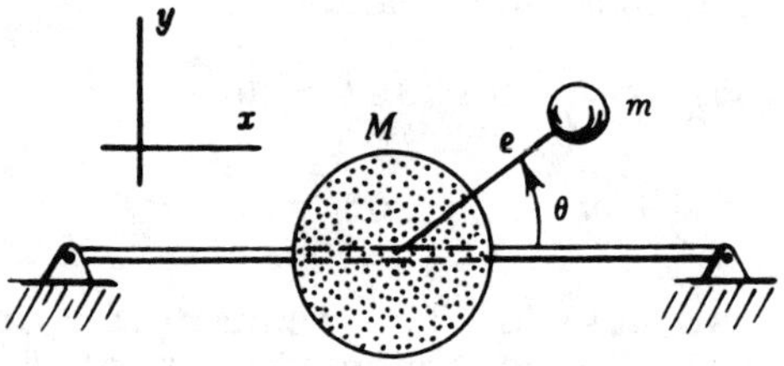

Fig. 2

Solution: In the system in Figure (2), there are no external forces in the x - direction. Consequently, linear momentum in the x - direction must be conserved. This implies that if the system started from rest, the center of mass of the system would be at rest until an outside force acted on the system. Since $\dot{x}_{cm}$, the velocity of the mass center in the x - direction is 0, the total momentum of the system must equal zero, and by the law of conservation of momentum it must remain constant.

The total momentum of the system is given by

$$M\dot{x}_1 + m\dot{x}_2 = 0 \qquad (1)$$

where $\dot{x}_1$ is the horizontal velocity of M, and $\dot{x}_2$ is the horizontal velocity of m. Now, $\dot{x}_2$ can be expressed in terms of $\dot{x}_1$ by the relative velocity equation

$$\dot{x}_2 = \dot{x}_1 + \dot{x}_{2/1}. \qquad (2)$$

The velocity of m with respect to M, $\dot{x}_{2/1}$, in the x - direction, is found in Figure (2) to be

$$\dot{x}_{2/1} = - e \omega \sin \theta .$$

The negative sign is included because of the counter-clockwise rotation of m. Equation (2) becomes

$$\dot{x}_2 = \dot{x}_1 - e \omega \sin \theta$$

and upon substitution in Equation (1)

$$M\dot{x}_1 + m (\dot{x}_1 - e \omega \sin \theta) = 0$$

Solving for the velocity of M,

$$\dot{x}_1 = \frac{m e \omega \sin \theta}{M + m} \tag{3}$$

To find the displacement of the large mass M, as a function of time, we can use the relationship $\theta = \omega t$ and $\dot{x}_1 = \frac{dx_1}{dt}$ and integrate from t = 0 to t = t.

Equation (3) can be rewritten as

$$\dot{x}_1 = \frac{dx_1}{dt} = \frac{m e \omega \sin \omega t}{M + m}$$

Separating variables and adding limits

$$\int_0^{x_1} dx_1 = \int_0^t \frac{m e \omega \sin \omega t}{M + m} dt$$

$$x_1 = \frac{- m e \cos \omega t}{M + m} + \frac{m e}{M+m} \tag{4}$$

This gives the lateral displacement of the machine due to the inbalance. The second term of equation (4) is a constant. Its magnitude notes the distance away from the median position of the mass M, at time t = 0. The situation is shown in Figure (3). The dashed line is the median position of mass M.

To make equation (4) in terms of the distance away from the median position, set the median position at a displacement of x_1. As seen in Figure (3) - b, at the median position $x_1 = \frac{m e}{M+m}$. Therefore subtract this from equation (4)

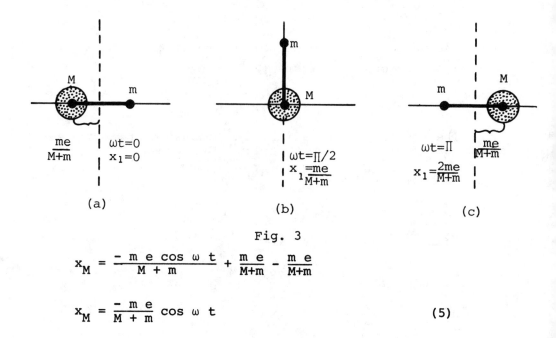

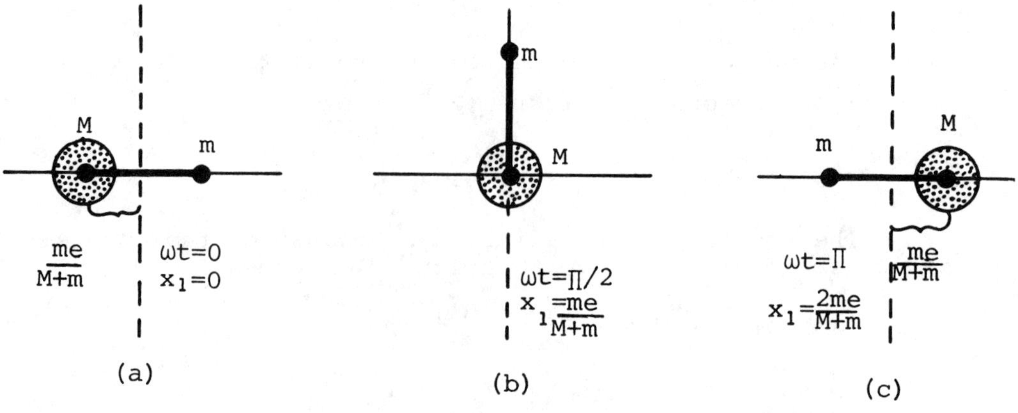

Fig. 3

$$x_M = \frac{-\ m\ e\ \cos\ \omega\ t}{M\ +\ m} + \frac{m\ e}{M+m} - \frac{m\ e}{M+m}$$

$$x_M = \frac{-\ m\ e}{M\ +\ m} \cos\ \omega\ t \tag{5}$$

Equation (5) now describes the motion of the machine, where x_M is the displacement of the machine from its median position.

Of course, in the situation of a real machine, the mounts would offer some lateral resistance, and also the magnitude of the displacement of M would be relatively small, since both m and e are likely to be much smaller than M.

• PROBLEM 12-11

Consider the earth as a system of particles. With respect to an inertial reference frame, is the linear momentum of the earth conserved? Consider the solar system as a system of particles. Is the linear momentum of the solar system conserved?

Solution: Momentum conservation in a system of particles is a very useful concept. However, one must be very careful when defining the system. The system definition should depend on the problem to be solved, and in particular the accuracy desired in the solution. For example, if one is trying to find the difference casued by the moon between the earth's average orbital path around the sun and its actual orbital path, then the approximation of the earth and moon as a system of particles in an inertial space gives very good results. However, for greater accuracy one should include the effects of the sun and the planets.

The chief point to remember is that linear momentum is never exactly conserved in a system unless every body which exerts a force on any body in the system is included in the system. (This means, that because of gravity forces the only system of particles where momentum conservation holds exactly is the system containing every particle in the universe.) Approximations can be made by dropping bodies from the system if the forces they exert are small in comparison with those exerted by other particles.

NEWTON'S SECOND LAW OF MOTION

• PROBLEM 12-12

Block A in the figure is pushed horizontally with a force of 5 lb. Block A weighs 10 lb and block B weighs 20 lb. The coefficient of friction between A and B is 0.2 and between B and the floor it is zero (no friction). a) Determine the acceleration of each block. b) Determine the time when A lines up with the right edge of B.

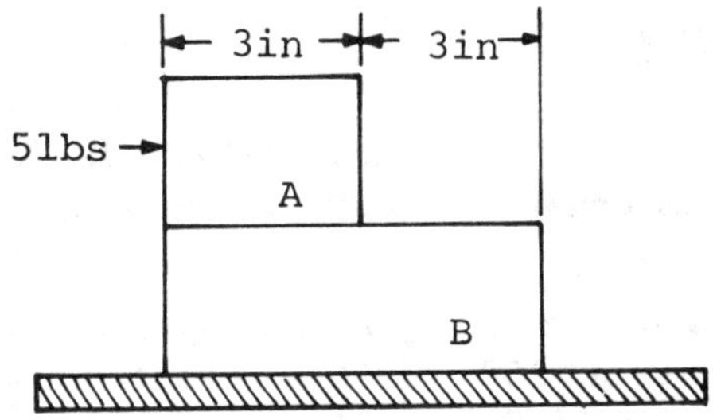

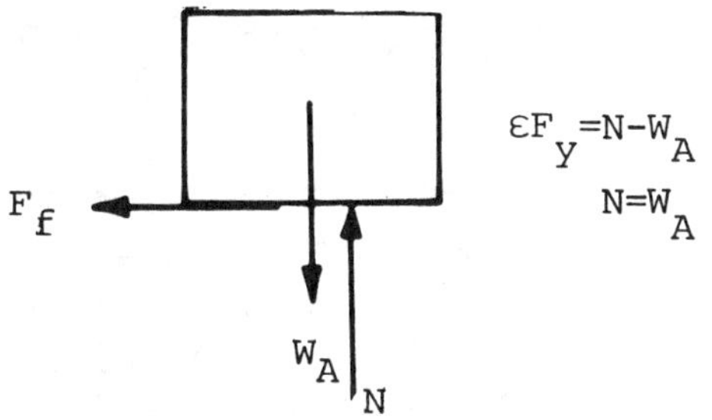

Solution: (a) The net force on block A is 5 lb minus the force of friction, which gives it an acceleration a_A.

or $\quad$ net $F = 5 - F_f = m\, a_A$

where F_f is given by μN, where the normal force N is equal to w_A. This can be seen from a free body diagram of A.

$$\varepsilon F_y = 0 = N - w_A$$

$$N = w_A$$

or $\quad 5 - .2 \times 10 = \frac{10}{32}\, a_A$

or $\quad 5 - 2 = \frac{10}{32}\, a_A \quad$ or $\quad \frac{3 \times 32}{10} = a_A$

or $\quad a_A = 9.6\ \text{ft/sec}^2.$

Similarly, the only force acting on block B is the force of friction between the two blocks, which is .2 x 10 = 2 lb. Thus,

$$2 = \frac{20}{32} a_B$$

or $$a_B = \frac{2 \times 32}{20} = 3.2 \text{ ft/sec}^2.$$

(b) To calculate the time of travel of block A through a distance of 3" $= \frac{1}{4}$ ft., note that its acceleration with respect to the lower block is $9.6 - 3.2 = 6.4$ ft/sec². Thus, using

$$s = v_0 t + \frac{1}{2} at^2,$$

$$\frac{1}{4} = 0 + \frac{1}{2} \times 6.4 \ t^2$$

or $$\frac{2}{4 \times 6.4} = t^2$$

or $$t = \sqrt{\frac{1}{2 \times 3.2}} = .395 \text{ sec.}$$

• PROBLEM 12-13

Two bodies A and B of masses $m_a = 2$kg and $m_b = 1$kg are connected by a string of negligible mass. They are placed on an inclined plane, angle of inclination 30°, with A located down the plane from B. The coefficients of friction are $\mu_a = 0.1$, $\mu_b = 0.2$. What is the tension in the string as the bodies slide down the plane.

Fig. 1

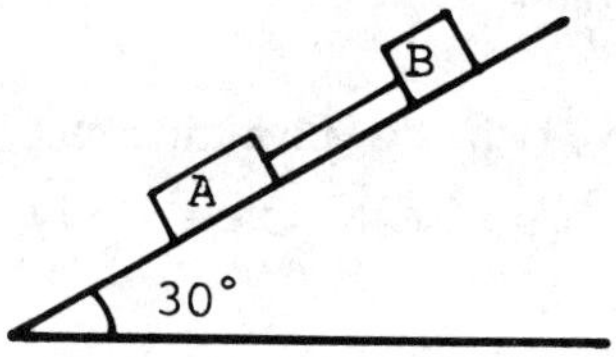

Solution: We note that the lower body, A, has the smaller coefficient of friction, hence it will slide more easily than B and thus keep the string under tension. Consequently, the distance between A and B (equal to the length of the string) will remain constant. This, in turn, requires that both bodies experience the same acceleration.

The free-body diagrams for A and B are shown in Figs. 2 and 3.

The two bodies connected by a string act as a system that will slide down with an acceleration given by the Second Law of Newton:

$$a = \frac{F}{m}$$

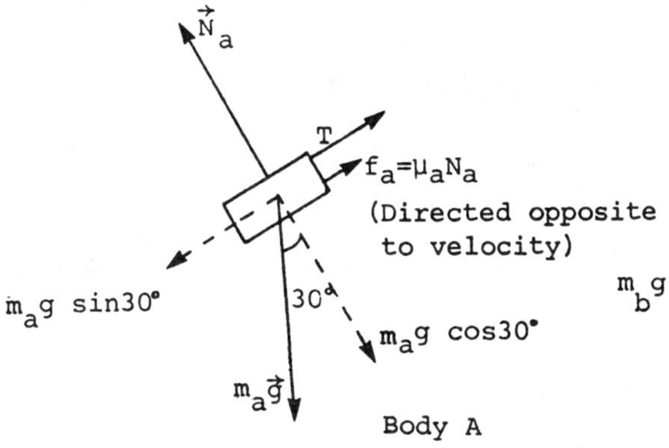

Fig. 2

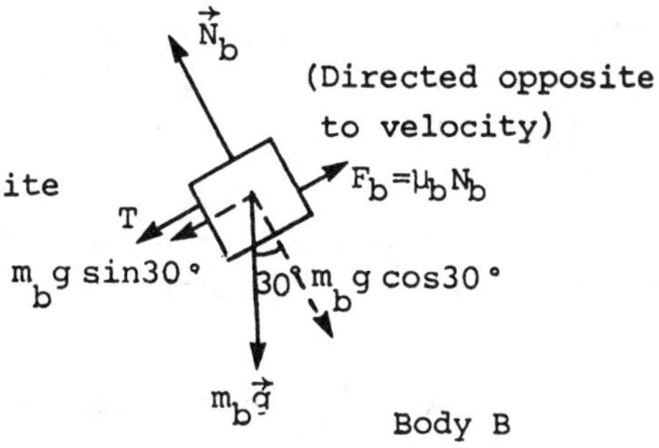

Fig. 3

As shown in Fig. 2 and Fig. 3, T is the tension in the string, and it is the same for both bodies. This tension in the string is a reaction of the force that acts upon a body (Newton's Third Law of Motion).

We have decomposed the weight mg of the body into two components: one along the inclined plane mg sin 30° and one perpendicular to the plane mg cos 30°. The perpendicular component is canceled by the normal reaction of the inclined plane, N.

Thus $N_a = m_a g \cos 30°$

and $N_b = m_b g \cos 30°$

From the frictional force law the frictional force, f, (which is directed up the slope, parallel to the plane but opposite to the direction of the velocity), has a magnitude $f = \mu N$. Where μ is the coefficient of sliding friction. Using the preceding equations for N_a and N_b, for the two bodies:

$$\left.\begin{aligned} f_a &= \mu_a\, m_a g \cos 30° \\ \text{and} \quad f_b &= \mu_b\, m_b g \cos 30° \end{aligned}\right\} \qquad (1)$$

By Newton's Second Law $\vec{a} = \vec{F}/m$ where $\vec{F}$ is the resultant of all forces acting on the body. Here, the total force that acts upon the body is the force that is along (parallel) the plane, and it is the algebric sum of the parallel component of weight, the friction force and the tension in the string. In each case $\vec{a}$, and hence $\vec{F}$, is directed down the slope (parallel to the plane).

From Figs. 2 and 3 we see

$$\left.\begin{aligned} F_a &= m_a g \sin 30° - f_a - T \\ \text{and} \quad F_b &= m_b g \sin 30° - f_b + T. \end{aligned}\right\} \qquad (2)$$

Substituting the f values from Eqs. (1) into Eqs. (2) gives:

$$F_a = m_a g\,(\sin 30° - \mu_a \cos 30°) - T$$

and

$$F_b = m_b g\,(\sin 30° - \mu_b \cos 30°) + T \qquad (3)$$

Using a = F/m and Eqs. 3 we can then express the accerleration of each body

$$A_a = \frac{F_a}{m_a} = g\,(\sin 30° - \mu_a \cos 30°) - \frac{T}{m_a}$$

$$A_b = \frac{F_b}{m_b} = g\,(\sin 30° - \mu_b \cos 30°) + \frac{T}{m_b}. \qquad (4)$$

Since these accelerations must be identical we can equate them to get:

$$g\,(\sin 30° - \mu_a \cos 30°) - \frac{T}{m_a}$$

$$= g\,(\sin 30° - \mu_b \cos 30°) + \frac{T}{m_b} \qquad (5)$$

which we now use to solve for T.

Cancelling the g sin 30° that is on both sides of Eq. 5 and rearranging terms gives

$$T\left(\frac{1}{m_b} + \frac{1}{m_a}\right) = g\,(\mu_b - \mu_a) \cos 30° \qquad (6)$$

and since $\frac{1}{m_b} + \frac{1}{m_a} = \frac{m_a + m_b}{m_a\, m_b}$ we can divide each side of Eq. 6 by this term to finally get

$$T = \frac{m_a\, m_b}{m_a + m_b}\, g\,(\mu_b - \mu_a) \cos 30°. \qquad (7)$$

Substituting numerical values

$$T = \frac{(2\text{kg})(1\text{kg})}{(2\text{kg} + 1\text{kg})}\left(9.81\ \frac{\text{m}}{\text{s}^2}\right)(0.2 - 0.1) \cos 30° = 0.566\ \text{N}.$$

An Alternate Approach:

The entire assembly (2 bodies + massless string) must, by Newton's Second Law, experience an acceleration

$$a = F\ \text{system}/(m_a + m_b) \qquad (7)$$

where F system = $(m_a + m_b)\ g \sin 30° - f_a - f_b$

or, using Eqs. (1):

$$F \text{ system} = (m_a + m_b)\ g \sin 30° - (\mu_a m_a + \mu_b m_b)\ g \cos 30°. \quad (8)$$

Thus Eq's 7 and 8 result in:

$$a = g \sin 30° - \frac{(\mu_a m_a + \mu_b m_b)}{m_a + m_b}\ g \cos 30°. \quad (9)$$

Now this same acceleration must be experienced by each part of the system. Looking in particular at body A, we can equate its acceleration, given by the first of Eqs. 4 to that of the system given by Eq. 9:

$$g \sin 30° - g\mu_a \cos 30° - \frac{T}{m_a} = g \sin 30°$$

$$- \frac{(\mu_a m_a + \mu_b m_b)}{m_a + m_b}\ g \cos 30° \quad (10)$$

cancelling g sin 30° and solving for T

$$T = g \cos 30° \left[- \mu_a + \frac{\mu_a m_a + \mu_b m_b}{m_a + m_b} \right] m_a. \quad (11)$$

We can show that Eq. 11 is identical to Eq. 7 by simplifying the expression inside the brackets:

$$\left[- \mu_a + \frac{\mu_a m_a + \mu_b m_b}{m_a + m_b} \right] = \frac{-\mu_a (m_a + m_b) + \mu_a m_a + \mu_b m_b}{m_a + m_b}$$

$$= \frac{(\mu_b - \mu_a)\ m_b}{m_a + m_b}.$$

Thus Eq. 11 becomes

$$T = g \cos 30° \ \frac{(\mu_b - \mu_a)}{m_a + m_b}\ m_a m_b,$$

which is Eq. 7.

• PROBLEM 12-14

Consider the system of particles m_1 and m_2 shown in Fig. (1). If the masses slide under the influence of gravity, derive an expression for the acceleration of the mass center along the incline (x-direction) in terms of m_1, m_2, θ, and the coefficients of sliding friction μ_1 and μ_2. Also do it for the frictionless case ($\mu_1 = \mu_2 = 0$). Then calculate the acceleration of the mass center, the acceleration of each mass, and the cord tension if $m_1 = m_2 = 0.1$ kg, and $\theta = 45°$ in the following situations:

a) $\mu_1 = 0.1$; $\mu_2 = 0.2$

b) $\mu_1 = \mu_2 = 0.2$

c) $\mu_1 = 0.2$; $\mu_2 = 0.1$

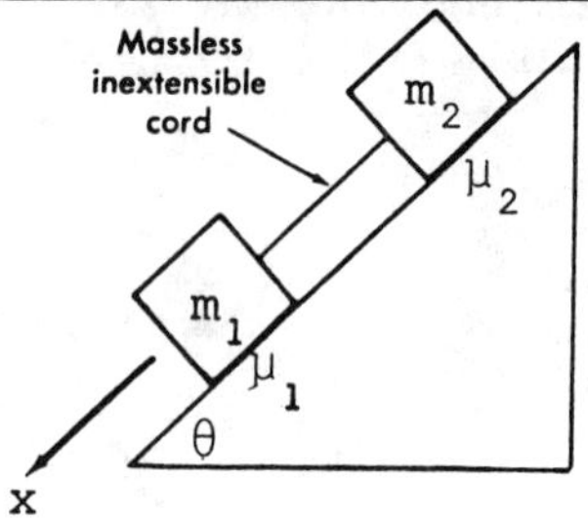

Fig. 1

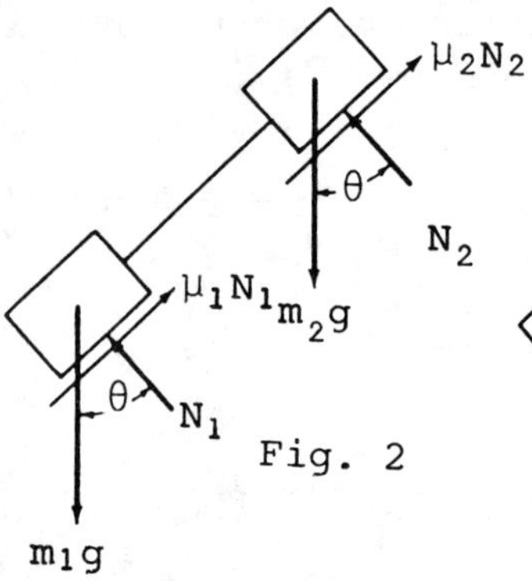

Fig. 2

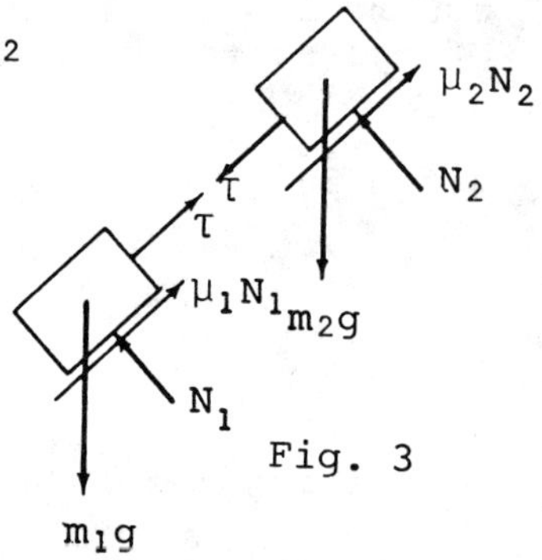

Fig. 3

Solution: A free body diagram of the system containing both masses is shown in Fig. (2). Any motion, and therefore any acceleration, must occur along the inclined plane. The sum of the external forces along the plane must equal the total mass of the system multiplied by the acceleration of the mass center along the plane according to Newton's Second Law of Motion.

$$m_1 g \sin\theta - \mu_1 m_1 g \cos\theta + m_2 g \sin\theta - \mu_2 m_2 g \cos\theta$$

$$= (m_1 + m_2)\, a_{cm}$$

Solving for a_{cm}, the acceleration of the center of mass yields:

$$a_{cm} = \frac{(m_1 + m_2)\, g \sin\theta - (\mu_1 m_1 + \mu_2 m_2)\, g \cos\theta}{m_1 + m_2} \qquad (1)$$

For the frictionless case, $\mu_1 = \mu_2 = 0$ and then

$$a_{cm} = \frac{(m_1 + m_2)\, g \sin\theta - 0}{(m_1 + m_2)} = g \sin\theta$$

For the rest of the problem, we must examine the motion of each block separately. A separate free body diagram for each block is shown in Figure (3). Applying

$\Sigma F = m a$ for each block yields:

For mass 1

$$- \tau - \mu_1 m_1 g \cos\theta + m_1 g \sin\theta = m_1 a_1 \qquad (2)$$

For mass 2

$$\tau - \mu_2 m_2 g \cos\theta + m_2 g \sin\theta = m_2 a_2 \qquad (3)$$

In situation (a) mass 2 will be slowed down by a greater frictional force than mass 1, since $\mu_2 > \mu_1$. Therefor it will retard the motion of mass 1 through the tension in the cord. Since the cord is in tension, $a_1 = a_2$. Now in equations (2) and (3) we have 2 unknowns, $a_1 = a_2$, and τ. Substituting in numerical values and solving the set of equations yields:

$$a_1 = a_2 = 5.9 \text{ m/sec}^2$$

$$\tau = 0.036 \text{ newton.}$$

The acceleration of the mass center in this case should be the same as that for each mass, since there is no motion relative to the mass center because both masses move at the same acceleration. Substitution of values in equation (1) yields

$$a_{cm} = 5.9 \text{ m/sec}^2 \quad \text{as expected.}$$

In situation (b), $\mu_1 = \mu_2 = 0.2$, therefore the forces on each mass are identical and the acceleration of each mass is identical, $a_1 = a_2$. In this situation, we can assume that the tension, τ, is tensile or zero. Plugging numerical values into equation (2) or (3) yields:

$$a_1 = a_2 = 5.5 \text{ m/sec}^2.$$

Since again, both masses have identical accelerations, the acceleration of the mass center should also be 5.5 m/sec^2. Use of numerical values in equation (1) yields the same value.

In situation (c) the force of friction is greater on mass 1 than on mass 2, since $\mu_1 > \mu_2$. Thus, mass 2 will be accelerated faster, and therefore the cord cannot be in tension. Since the cord cannot sustain compression, we conclude that $\tau = 0$. Use of numerical values in equations (2) and (3) yields:

$$a_1 = 5.5 \text{ m/sec}^2$$

$$a_2 = 6.2 \text{ m/sec}^2.$$

As stated, mass 2 will move closer towards mass 1.

Equation (1) yields the acceleration of the mass center.

$$a_{cm} = 5.9 \text{ m/sec}^2$$

It is observed that both masses experience motion relative to the mass center in this situation because a_1 and a_2 are positive.

• PROBLEM 12-15

Two blocks, A and B, are attached by a cord wrapped around a frictionless, massless pulley, C. The coefficient of friction between each block and the horizontal platform is 0.3. Block A weighs 10 lb, block B weighs 20 lb. If the platform rotates about the vertical axis shown, determine the angular speed at which the blocks start to slide radially.

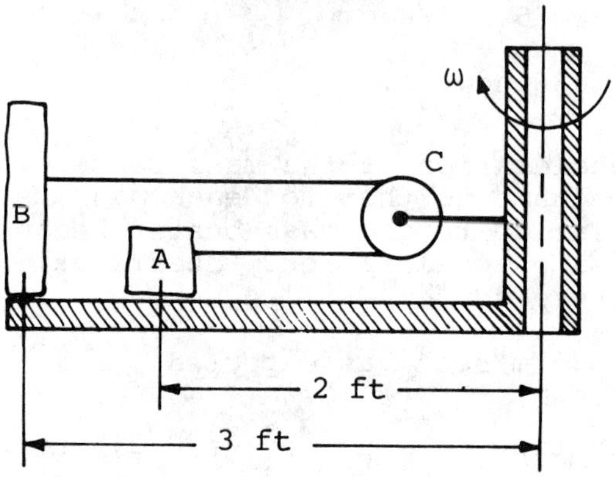

Solution: The two blocks will start sliding radially when the centrifugal force on B is sufficient to overcome (a) the force of friction between block B and the table and (b) the tension in the cord which is equal to the centrifugal force on block A plus the force of friction between block A and the table. Using the formula for centrifugal force:

$$F_C = \frac{m\,v^2}{r} = \frac{W\,v^2}{gr} = \frac{W\,\omega^2\,r^2}{gr} = \frac{W\,\omega^2\,r}{g}$$

and for friction: $F_f = \mu\,W$,

where W is the weight of a block, ω is the angular velocity, r the radius, g = 32 ft/sec^2, the acceleration due to gravity, and μ the coefficient of static friction, we can write the desired equation as:

$$\frac{W_B\,\omega^2\,r_B}{g} = \mu\,W_B + \mu\,W_A + \frac{W_A\,\omega^2\,r_A}{g}$$

where the subscripts A and B refer to the two blocks.

$$\frac{W_B \omega^2 r_B}{g} - \frac{W_A \omega^2 r_A}{g} = \mu \left(W_A + W_B\right)$$

$$\frac{\omega^2 \left(W_B r_B - W_A r_A\right)}{g} = \mu \left(W_A + W_B\right)$$

$$\omega^2 = \frac{\mu g \left(W_A + W_B\right)}{W_B r_B - W_A r_A}$$

$$\omega = \sqrt{\frac{\mu g \left(W_A + W_B\right)}{W_B r_B - W_A r_A}} .$$

Substituting the numerical values, yields

$$\omega = \sqrt{\frac{.3 \times 32 (10 + 20)}{20 \times 3 - 10 \times 2}} = \sqrt{7.2} \doteq 2.68 \text{ radians/sec.}$$

• PROBLEM 12-16

Shown in Fig. 1 is a system of particles. Relative to the inertial coordinate system O, the particles are acted on by external torques. What is the torque of the system of particles about the point s which moves with an acceleration $\ddot{\vec{s}}$ relative to point O?

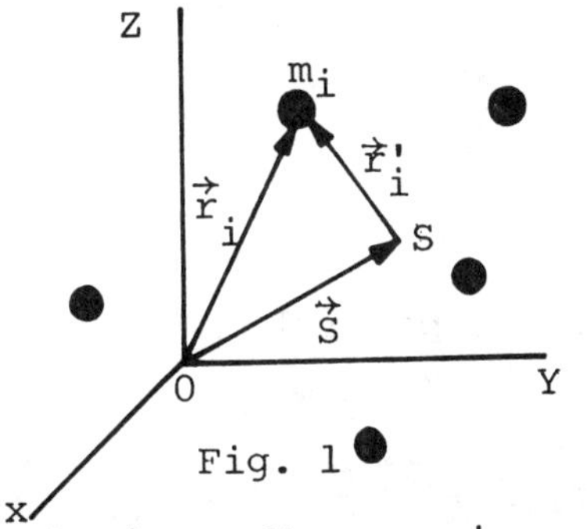

Fig. 1

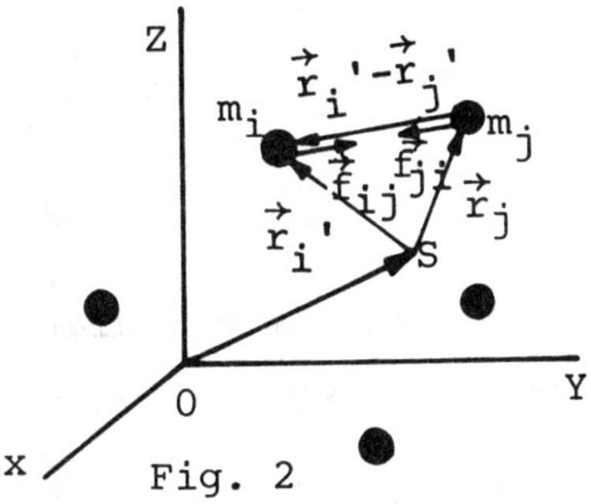

Fig. 2

Solution: We are given that reference frame O is an inertial reference frame and reference frame s is a non-inertial reference frame. Thus first we have to determine the total force, (both real and inertial force), acting on a particle in reference frame s. Once we have accomplished this we can determine the torque about the point s.

We start with Newton's Second Law applied to particle m_i in the inertial frame O:

$$m_i \ddot{\vec{r}}_i = \vec{F}_i + \sum_{\substack{j=1 \\ j \neq i}}^{n} \vec{f}_{ij} \qquad (1)$$

where $\vec{F}_i$ is the external force acting on particle m_i and $\vec{f}_{ij}$ is the force on particle m_i caused by particle m_j.

Next let us determine the equation of motion in the accelerated frame s. From Fig. 1 we obtain:

$$\vec{r}_i = \vec{r}_i{}' + \vec{s} \tag{2}$$

where $\vec{r}_i{}'$ is the position vector of particle m_i relative to s and $\vec{s}$ is the position vector of point s relative to O. Let us differentiate eq. (2) twice with respect to time, this yields:

$$\ddot{\vec{r}}_i = \ddot{\vec{r}}_i{}' + \ddot{\vec{s}}. \tag{3}$$

Next let us substitute eq. (3) into eq. (1) and rearrange the resulting equation:

$$m_i \ddot{\vec{r}}_i{}' = \vec{F}_i - m_i \ddot{\vec{s}} + \sum_{\substack{j=1 \\ j \neq i}}^{n} \vec{f}_{ij} \tag{4}$$

Notice in the accelerated frame we have the additional term $- m_i \ddot{\vec{s}}$. This is the inertial force or fictitious force term. Thus with respect to the accelerated frame the total force acting on particle m_i is given by:

$$\vec{F}_i{}' = \vec{F}_i - m_i \ddot{\vec{s}} + \sum_{\substack{j=1 \\ j \neq i}}^{n} \vec{f}_{ij} \tag{5}$$

We can now calculate the torque on particle m_i about the point s, obtaining:

$$\vec{N}_i{}' = \vec{r}_i{}' \times \vec{F}_i{}' = \vec{r}_i{}' \times \vec{F}_i - \vec{r}_i{}' \times m_i \ddot{\vec{s}} + \sum_{\substack{j=1 \\ j \neq i}}^{n} \vec{r}_i{}' \times \vec{f}_{ij}, \tag{6}$$

and to obtain the total torque acting on the system of particles we have to sum over all n particles in the system, this yields:

$$\vec{N}' = \sum_{i=1}^{n} \vec{N}_i{}' = \sum_{i=1}^{n} \vec{r}_i{}' \times \vec{F}_i - \sum_{i=1}^{n} \vec{r}_i{}' \times m_i \ddot{\vec{s}}$$

$$+ \sum_{i=1}^{n} \sum_{\substack{j=1 \\ j \neq i}}^{n} \vec{r}_i{}' \times \vec{f}_{ij}. \tag{7}$$

Consider the last term in eq. (7). This term consists of pairs of term of the form

$$\vec{r}_i' \times \vec{f}_{ij} + \vec{r}_j' \times \vec{f}_{ji}. \tag{8}$$

If the forces obey Newton's Third Law then $\vec{f}_{ij} = - \vec{f}_{ji}$ and expression (8) becomes:

$$(\vec{r}_i' - \vec{r}_j') \times \vec{f}_{ij}$$

This term is equal to zero if f_{ij} is parallel to the vector $r_i' - r_j'$, that is if the forces obey the strong form of Newton's Third Law. Thus the last term in eq. (7) is equal to zero and eq. (7) reduces to:

$$\vec{N}' = \sum_{i=1}^{n} \vec{r}_i' \times \vec{F}_i - \sum_{i=1}^{n} \vec{r}_i' \times m_i \ddot{\vec{s}} \quad \text{or}$$

$$\vec{N}' = \sum_{i=1}^{n} \vec{r}_i' \times \vec{F}_i - \left(\sum_{i=1}^{n} m_i \vec{r}_i' \right) \times \ddot{s}. \tag{9}$$

We can simplify eq. (9) by using the definition of the center of mass, namely

$$M \vec{R}' = \sum_{i=1}^{n} m_i \vec{r}_i' \tag{10}$$

where M is the total mass of the system and $\vec{R}'$ is the position vector of the center of mass relative to point s. Substituting eq. (10) into eq. (9) we obtain:

$$\vec{N}' = \sum_{i=1}^{n} \vec{r}_i' \times \vec{F}_i - M \vec{R}' \times \ddot{\vec{s}}. \tag{11}$$

Eq. (11) is the desired result. The last term in eq. (11) is sometimes called the fictitious torque, this term will be zero if the center of mass is located at the point S or if $\ddot{\vec{s}}$ is parallel to the vector $\vec{R}'$. For these cases the equation $\dot{\vec{L}} = \vec{N}^e$ is also valid in a non-inertial reference frame.

• PROBLEM 12-17

Four blocks weighing 4, 3, 2, & 1 lbs. are hanging from weightless, frictionless pulleys as shown. When all are released from rest find the acceleration of each block.

Solution: To solve problems of this type one proceeds as follows: a) Draw free body diagrams of each moving particle. b) Set up acceleration constraint equations. c) Sum the

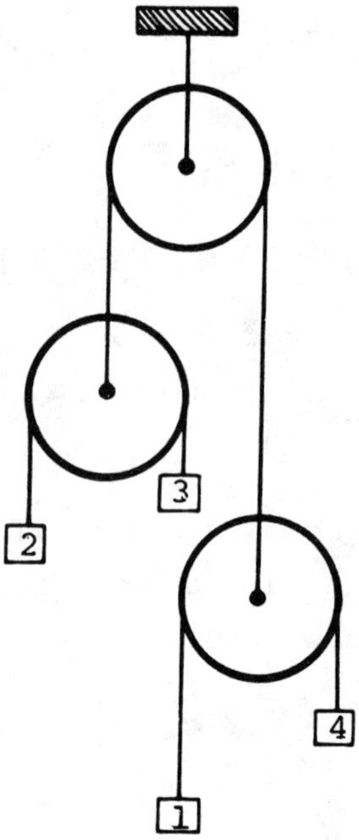

forces on each particle and set them equal to the particle's mass times its acceleration. (Newton's Second Law) d) Solve for the accelerations.

a) Free Body Diagrams

i) The free body diagram for blocks 1, 2, 3, 4.

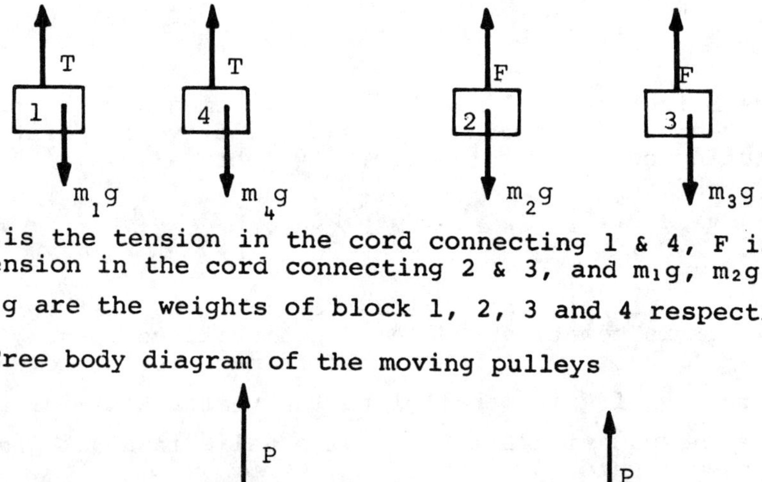

T is the tension in the cord connecting 1 & 4, F is the tension in the cord connecting 2 & 3, and m_1g, m_2g, m_3g and m_4g are the weights of block 1, 2, 3 and 4 respectively.

ii) Free body diagram of the moving pulleys

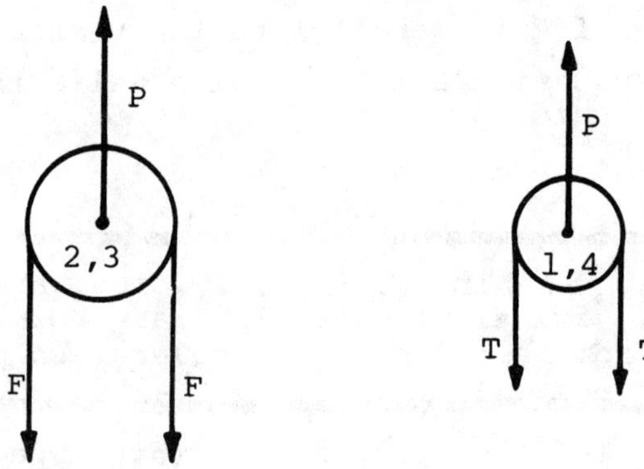

2, 3 is the pulley masses 2 and 3 are connected to, 1, 4 is the pulley masses 1 and 4 are connected to, and P is the tension in the cord connecting the two pulleys.

b) Constraint equations

The important constraints in this problem are:

i) The acceleration of pulley 1, 4 is equal in magnitude to the acceleration of pulley 2, 3; but in the opposite direction.

ii) The average acceleration of blocks 1 and 4 is the acceleration of pulley 1, 4.

iii) The average acceleration of blocks 2 and 3 is the acceleration of pulley 2, 3.

Mathematically the relations look like this:

i) $a_{2,3} = - a_{1,4}$

ii) $\dfrac{a_1 + a_4}{2} = a_{1,4}$

iii) $\dfrac{a_2 + a_3}{2} = a_{2,3}$

Combining the three relations yields,

$$\frac{a_2 + a_3}{2} = - \left(\frac{a_1 + a_4}{2}\right)$$

$$a_2 + a_3 = - (a_1 + a_4).$$

c) Summing the forces

Adding the forces on each block and setting them equal to the block's acceleration multiplied by its mass yields: (with down as the positive direction)

For block 1: $m_1 g - T = m_1 a_1$

For block 2: $m_2 g - F = m_2 a_2$

For block 3: $m_3 g - F = m_3 a_3$

For block 4: $m_4 g - T = m_4 a_4$

All accelerations are assumed to be downward (positive). If the sign of the acceleration comes out to be positive, the assumption is correct. If any accelerations come out negative, it means the assumption was wrong, and it is accelerating upward.

Doing the same thing for the pulleys, (where the mass is assumed to be 0):

$2T - P = 0$ For pulley 1, 4

$2F - P = 0$ For pulley 2, 3

Equating the two gives:

$$2T - P = 2F - P$$

or $T = F$.

d) Solving.

From part b one obtains 1 relation containing 4 unknowns, from part c, 6 relations and 3 additional unknowns. The seven equations together form a solvable system in seven unknowns. The seven equations are:

$$a_2 + a_3 = -(a_1 + a_4)$$

$$m_1 g - T = m_1 a_1$$

$$m_2 g - F = m_2 a_2$$

$$m_3 g - F = m_3 a_3$$

$$m_4 g - T = m_4 a_4$$

$$T = F$$

$$P = 2T$$

To solve, proceed as follows:

i) Solve each of the block force equations for T & F.

For block 1 $T = m_1 g - m_1 a_1$ (1)

For block 2 $F = m_2 g - m_2 a_2$ (2)

For block 3 $F = m_3 g - m_3 a_3$ (3)

For block 4 $T = m_4 g - m_4 a_4$ (4)

ii) Setting equation (1) equal to eq. (4);

$$m_1 g - m_1 a_1 = m_4 g - m_4 a_4$$

$$m_1 g + m_4 a_4 - m_4 g = m_1 a_1$$

$$a_1 = g + \frac{m_4}{m_1} a_4 - \frac{m_4}{m_1} g$$

and setting eq. (2) equal to (3), gives

$$m_2 g - m_2 a_2 = m_3 g - m_3 a_3$$

$$m_2 g + m_3 a_3 - m_3 g = m_2 a_2$$

$$a_2 = g + \frac{m_3}{m_2} a_3 - \frac{m_3}{m_2} g$$

iii) Substituting those values for a_1 and a_2 in $(a_2 + a_3) = -(a_1 + a_4)$ gives:

$$(a_2 + a_3) = -(a_1 + a_4)$$

$$g + \frac{m_3}{m_2} a_3 - \frac{m_3}{m_2} g + a_3 = -g - \frac{m_4}{m_1} a_4 + \frac{m_4}{m_1} g - a_4 \quad (5)$$

iv) Now, if an expression for a_3 in terms of a_4 can be found, the problem is solved.

This equation can be found by setting eq. (3) equal to eq. (4), remembering T = F.

So, $m_3 g - m_3 a_3 = m_4 g - m_4 a_4$

$$m_3 g - m_4 g + m_4 a_4 = m_3 a_3$$

$$a_3 = g - \frac{m_4}{m_3} g + \frac{m_4}{m_3} a^4$$

Putting this expression for a_3 into eq. (5) gives;

$$g + \frac{m_3}{m_2}\left(g - \frac{m_4}{m_3} g + \frac{m_4}{m_3} a^4\right) - \frac{m_3}{m_2} g + g \frac{m_4}{m_3} g + \frac{m_4}{m_3} a_4$$

$$= -g - \frac{m_4}{m_1} a_4 + \frac{m_4}{m_1} g - a_4$$

$$g + \frac{m_3}{m_2} g - \frac{m_4}{m_2} g + \frac{m_4}{m_2} a_4 - \frac{m_3}{m_2} g + g - \frac{m_4}{m_3} g + \frac{m_4}{m_3} a_4$$

$$= -g - \frac{m_4}{m_1} a_4 + \frac{m_4}{m_1} g - a_4 \quad (6)$$

The only unknown in eq. (6) is a_4. Since $\frac{m_1}{m_2} = \frac{m_1 g}{m_2 g} = \frac{w_1}{w_2}$, we can substitute the given weights for the masses in eq. (6). This gives:

$$g + \frac{3}{2g} - \frac{4}{2g} + \frac{4}{2a_4} - \frac{3}{2g} + g - \frac{4}{3g} + \frac{4}{3a_4} = -g - \frac{4}{1a_4} + \frac{4}{1g} - a_4 \quad (7)$$

Bringing all terms containing a_4 to the right side of eq. (7) gives:

$$g + \frac{3}{2g} - \frac{4}{2g} - \frac{3}{2} g + g - \frac{4}{3g} + g - \frac{4}{1g} = -\frac{4}{1a_4} - a_4 - \frac{4}{2a_4} - \frac{4}{3a_4}$$

$$g\left(1+\frac{3}{2}-2-\frac{3}{2}+1-\frac{4}{3}+1-4\right) = a_4\left(-4-1-2-\frac{4}{3}\right)$$

$$-4.33\,g = -8.33\,a_4$$

$$a_4 = \frac{4.33}{8.33}\,(32.2\ \text{ft/s}^2)$$

$$a_4 = 16.74\ \text{ft/s}^2$$

Then $a_3 = g - \frac{4}{3g} + \frac{4}{3a_4}$

$$a_3 = -\frac{1}{3}g + \frac{4}{3}\,(16.74\ \text{ft/s}^2)$$

$$a_3 = 11.6\ \text{ft/s}^2$$

Also $a_2 = g + \frac{3}{2}a_3 - \frac{3}{2}g$

$$a_2 = -\frac{1}{2}g + \frac{3}{2}\,(11.6)$$

$$a_2 = 1.29\ \text{ft/s}^2$$

Finally, $a_1 = g + \frac{4}{1}a_4 - \frac{4}{1g}$

$$a_1 = -3g + 4\,(16.74)$$

$$a_1 = -29.6\ \text{ft/s}^2$$

Results:

Block 1 goes up with $a_1 = 29.6\ \text{f/s}^2$

Block 2 goes down with $a_2 = 1.29\ \text{f/s}^2$

Block 3 goes down with $a_3 = 11.6\ \text{f/s}^2$

Block 4 goes down with $a_4 = 16.74\ \text{f/s}^2$

• **PROBLEM 12-18**

Two uniform bars A B and B C are pinned together and to the walls as shown. A B weighs $m_{AB}g$ and B C weighs $m_{BC}g$. Determine the angular acceleration of A B immediately after pin C is removed.

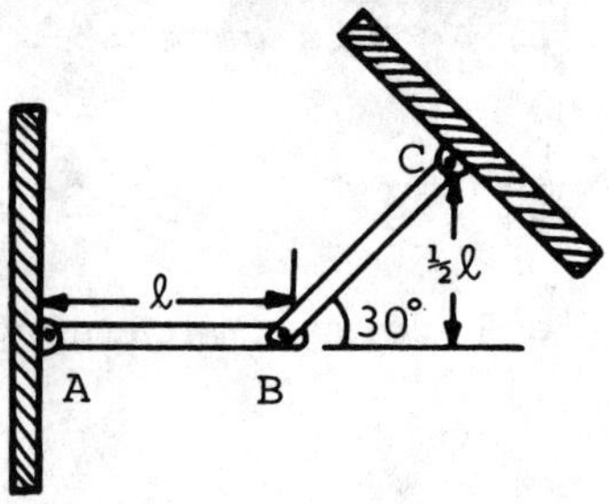

Solution: When the pin is released at C the two bars will start to fall. Bar A B will be pulled by its weight and pushed by a component of B C's weight at point B.

The free body diagrams for each bar just as the pin is released look like this:

Fig. a)

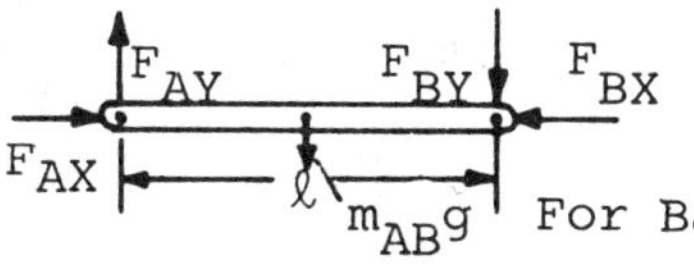

For Bar AB

Sign Convention

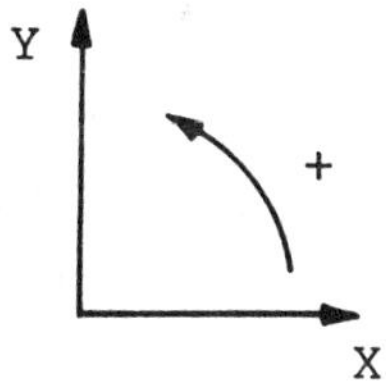

Fig. b)

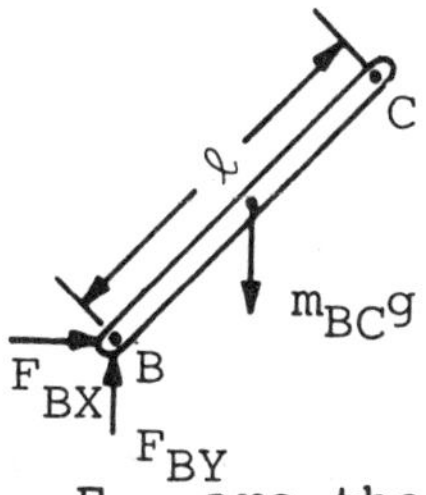

For Bar BC

F_{Ax}, F_{Ay} are the horizontal and vertical components the reaction force at pin A, $m_{AB}g$ and $m_{BC}g$ are the bar weights ($m \rightarrow$ mass, $g \rightarrow$ acceleration due to gravity) and F_B is the internal force between the bars at pin B.

From (Fig. a) summing forces and moments according to Newton's Second Law, (sign convention as shown)

i) $\Sigma F_x = F_{Ax} - F_{Bx} = m_{AB}\ a_x$

ii) $\Sigma F_y = F_{Ay} - m_{AB}g - F_{By} = m_{AB}\ a_y$

iii) $\Sigma M_B = -\ F_{Ay}\ (\ell) + m_{AB}g\ (\ell/2) = I_B\ \alpha_A$

I_B is the moment of inertia of bar A B about point B. The bar is exactly horizontal at the instant the pin is released. The component of the angular acceleration in the x direction is $(\ell/2)\ \alpha \sin\theta$. The angle is zero. Therefore

$$a_x = (\ell/2)\ \alpha \sin (0) = 0.$$

The component of the angular acceleration in the y direction is $(\ell/2)\ \alpha \cos\theta$. The angular acceleration about y is $-\ \alpha_A$. The angle is zero. Therefore

$$a_y = (\ell/2)\ (-\ \alpha) \cos (90) = (\ell/2)\ (-\ \alpha),$$

since the motion is pure rotation about pt. A. It follows that i) and ii) become

i) $F_{Ax} = F_{Bx}$

ii) $F_{Ay} = m_{AB}g + F_{By} - m_{AB}\frac{\ell}{2}\ \alpha_A$

Combining the previous equation with iii) from above yields,

iii) $I_B\ \alpha_A = -F_{By}\ell - m_{ABg}\ell + m_{ABg}\frac{\ell}{2} + m_{AB}\frac{\ell^2}{2}\ \alpha_A$

$$\alpha_A\left(I_B - m_{AB}\frac{\ell^2}{2}\right) = -\left(F_{By}\ \ell + m_{AB}\ g\ \frac{\ell}{2}\right)$$

iv) $\alpha_A = \dfrac{-\left(F_{By}\ \ell + m_{AB}g\ \frac{\ell}{2}\right)}{\left(I_B - m_{AB}\frac{\ell^2}{2}\right)}$

Now for bar B C, (from Fig. b), summing forces and moments yields:

v) $\Sigma F_x = F_{Bx} = m_{BC}\ a_x$

vi) $\Sigma F_y = F_{By} - m_{BC}g = ma_y$

vii) $\Sigma M_B = m_{BC}\ g\ \cos 30°\ \frac{\ell}{2} = I_B'\ \alpha_B$

I_B' = moment of inertia of bar B C around pt B.

From rigid body Kinetics, $a_x = a_{Bx} + \frac{\ell}{2}\ \alpha_B \sin 30°$

$$a_y = a_{By} + \frac{\ell}{2}\ \alpha_B \cos 30°.$$

The acceleration of bar A B in the x direction, a_x, is zero. Therefore:

$a_{Bx} = 0$. Also, $a_{By} = \ell\ \alpha_A$ $\therefore$

$a_x = \frac{\ell}{2}\alpha_B \sin 30°$ $\qquad a_y = \ell\ \alpha_A + \frac{\ell}{2}\ \alpha_B \cos 30°$

v) and vi) now become,

v) $F_{Bx} = m_{BC}\left(\frac{\ell}{2}\ \alpha_B \sin 30°\right)$

vi) $F_{By} - m_{BC}g = m_{BC}\left(\ell\ \alpha_A + \frac{\ell}{2}\ \alpha_B \cos 30°\right)$

iv), vi) and vii) make up a solvable system in four

unknowns, F_{Ay}, F_{By}, α_A and α_B:

$$\text{iv)}\quad \alpha_A = \frac{-\left(F_{By}\,\ell + m_{AB}g\,\frac{\ell}{2}\right)}{\left(I_B - m_{AB}\,\frac{\ell^2}{2}\right)}$$

$$\text{vi)}\quad F_{By} = m_{BC}g + m_{BC}\left(\ell\,\alpha_A + \frac{\ell}{2}\,\alpha_B\cos 30°\right)$$

$$\text{vii)}\quad \alpha_B = \frac{m_{BC}g\cos 30°\left(\frac{\ell}{2}\right)}{I_B'}\,.$$

Using iv), vi) and vii) together,

$$F_{By} = m_{BC}g + m_{BC}\left(\frac{\ell}{2}\,\frac{\left(m_{BC}g\cos 30°\left(\frac{\ell}{2}\right)\right)}{I_B'} - \frac{\ell\left(F_{By}\,\ell + m_{AB}g\,\frac{\ell}{2}\right)}{\left(I_B - m_{AB}\,\frac{\ell^2}{2}\right)}\right)$$

$$F_{By}\left(1 + \frac{m_{BC}\,\ell^2}{\left(I_B - m_{AB}\,\frac{\ell^2}{2}\right)}\right) = m_{BC}g + m_{BC}\left(\frac{\left(\frac{\ell}{2}\right)^2 m_{BC}g\cos 30°}{I_B'} - \frac{\frac{\ell^2}{2}\,m_{AB}g}{\left(I_B - m_{AB}\,\frac{\ell^2}{2}\right)}\right)$$

$$F_{By} = \frac{m_{BC}\,g + m_{BC}\left(\frac{(\ell/2)^2\,m_{BC}g\cos 30°}{I_B'} - \frac{\ell^2/2\;m_{AB}\,g}{(I_B - m_{AB}\,\ell^2/2)}\right)}{\left(1 + \frac{m_{BC}\,\ell^2}{(I_B - m_{AB}\,\ell^2/2)}\right)}$$

Substituting this result into iv) yields

$$\alpha_A = -\frac{\left[\frac{\left(m_{BC}g + m_{BC}\left(\frac{((\ell/2)^2\,m_{BC}g\cos 30°}{I_B'} - \frac{\ell^2/2\;m_{AB}g}{(I_B - m_{AB}\,\ell^2/2)}\right)\right)\ell}{\left(1 + \frac{m_{BC}\,\ell^2}{(I_B - m_{AB}\,\ell^2/2)}\right)} + m_{AB}g\,\ell/2\right]}{(I_B - m_{AB}\,\ell^2/2)}$$

Note: $I_B = 5/12\ m_{AB}\ \ell^2$, $I_B' = 5/12\ m_{BC}\ \ell^2$ } from integral table

From this problem one can easily see why it is common practice to substitute numbers for symbols in the earliest equations before proceeding to a solution. By holding on to general symbols as long as possible the solution obtained is more valuable since it can be used for many different problems but the difficulty in computation is greater. One must weigh the difficulty against the value of the answer so that time is not wasted.

ANGULAR MOMENTUM AND TORQUE

• PROBLEM 12-19

Show that the rate of change of the total angular momentum of a system of particles is equal to the resultant torque exerted by all external forces which act on the system.

Solution: For a system of particles, the angular momentum is the vector sum of the individual angular momenta, or

$$\vec{L} = \sum_{i=1}^{n} (\vec{r}_i \times m_i\ \vec{v}_i) \qquad (1)$$

Taking the time derivative yields

$$\frac{d\vec{L}}{dt} = \sum_{i=1}^{n} \left(\frac{d\vec{r}_i}{dt} \times m_i\ \vec{v}_i \right) + \sum_{i=1}^{n} \left(\vec{r}_i \times m_i\ \frac{d\vec{v}_i}{dt} \right)$$

The first term vanishes since $d\vec{r}_i/dt = \vec{v}_i$ and $\vec{v}_i \times \vec{v}_i = |v_i|\ |v_i| \sin 0° = 0$. The second term $m_i\ d\vec{v}_i/dt = m_i\ \vec{a}_i$ represents the total force acting on particle i. This total force is made up of all the external forces F_i acting on particle i plus all of the internal forces exerted by any other particle in the system. However, all of these internal forces act in pairs and cancel each other out according to Newton's Third Law.

This leaves $\dfrac{d\vec{L}}{dt} = \sum_{i=1}^{n} \vec{r}_i \times \vec{F}_i$

where $\sum_{i=1}^{n} \vec{r}_i \times \vec{F}_i$ is the resultant torque exerted by all external forces which act on the system.

• PROBLEM 12-20

Is it possible for the resultant torque $\vec{N}^e$ by all external forces $\vec{F}^e$ on the system to be equal to zero if $\vec{F}^e$ is not equal to zero? If the particles in the system interact with the magnetic force, will the equation $\dot{\vec{L}} = \vec{N}^e$ necessarily be valid? Will $M\ddot{\vec{R}} = \vec{F}^e$ necessarily be valid?

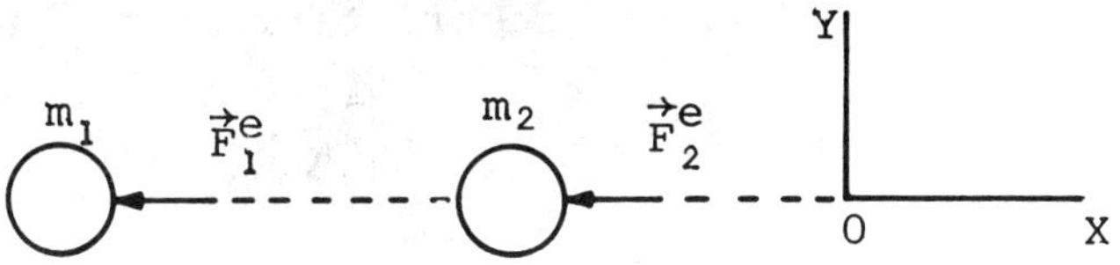

Fig. 1

Solution: We shall answer the first question by considering the specific two particle system shown in Fig. 1.

First, evaluate the resultant external torque N^e about the origin O. From the definition for torque it is known that

$$\vec{N}^e = \vec{r}_1 \times \vec{F}_1{}^e + \vec{r}_2 \times \vec{F}_2{}^e \tag{1}$$

where $\vec{r}_1$ is the vector from the origin O to the particle m_1 and $\vec{r}_2$ is the vector from O to the particle m_2. But in eq. (1) $\vec{r}_1 \times \vec{F}_1{}^e = 0$, since the line of action of $\vec{F}_1{}^e$ passes through the point O, similarly $\vec{r}_2 \times \vec{F}_2{}^e = 0$. Thus for the case shown in Fig. 1 the net external torque about the point O is equal to zero.

Next, let us consider the resultant external force acting on the system,

$$\vec{F}^e = \vec{F}_1{}^e + \vec{F}_2{}^e. \tag{2}$$

Since the two forces are parallel to each other, the net external force is not equal to zero. It is therefore possible for $\vec{N}^e = 0$ even if $\vec{F}^e \neq 0$.

Let us now answer the second question. Is the equation

$$\dot{\vec{L}} = \vec{N}^e \tag{3}$$

valid if the particles interact with the magnetic force?

From our study of mechanics we know that the time rate of change of the total angular momentum $\dot{\vec{L}}$ is given by the equation

$$\dot{\vec{L}} = \sum_{i=1}^{n} \vec{r}_i \times \vec{F}_i^{\,e} + \sum_{i=1}^{n} \sum_{\substack{j=1 \\ j \neq i}}^{n} \vec{r}_i \times \vec{F}_{ij} \qquad (4)$$

In eq. (4) the first term is just the resultant external torque $\vec{N}^e$ and the second term is the total internal torque. Thus eq. (3) is valid whenever the second term is equal to zero.

Let us consider the case where we have only two particles, i.e. n = 2. Then the second term in eq. (4) becomes

$$\sum_{i=1}^{2} \sum_{\substack{j=1 \\ j \neq i}}^{2} \vec{r}_i \times \vec{F}_{ij} = \vec{r}_1 \times \vec{F}_{12} + \vec{r}_2 \times \vec{F}_{21}, \qquad (5)$$

where $\vec{F}_{12}$ is the internal force exerted on particle 1 by by particle 2 and $\vec{F}_{21}$ is the internal force exerted on particle 2 by particle 1.

Now according to Newton's Third Law of Motion we have that $\vec{F}_{12} = - \vec{F}_{21}$. Making this substitution in eq. (5) we obtain:

$$\vec{r}_1 \times \vec{F}_{12} + \vec{r}_2 \times \vec{F}_{21} = (\vec{r}_1 - \vec{r}_2) \times \vec{F}_{12}. \qquad (6)$$

We notice that eq. (6) is equal to zero if $\vec{F}_{12}$ is parallel to $\vec{r}_1 - \vec{r}_2$, see Fig. 2. To summarize, eq. (3) is valid if the forces obey Newton's Third Law and if the forces are directed along a line joining the two particles.

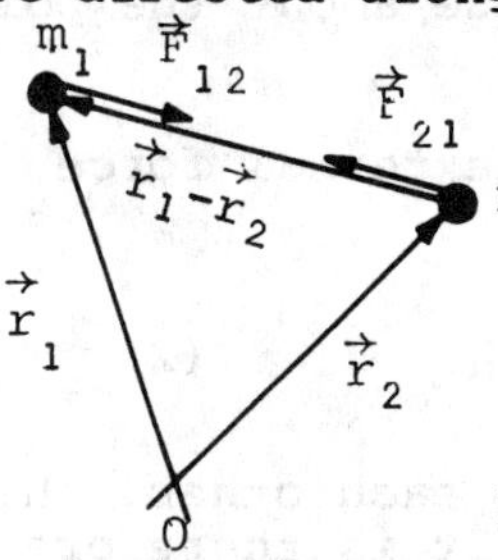

Fig. 2

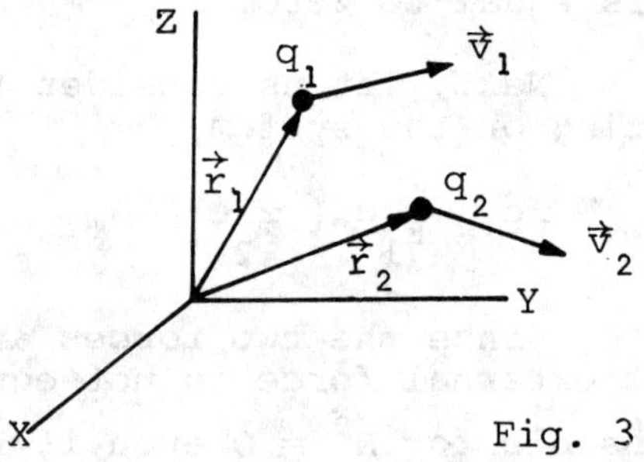

Fig. 3

Let us show that the magnetic force does not satisfy this condition. In Fig. 3 we have two particles moving at some arbitrary velocities. The magnetic force exerted on particle 1 by particle 2 is given by:

$$\vec{F}_{12} = q_1 \vec{v}_1 \times \vec{B}_2, \qquad (7)$$

where $\vec{B}_2 = \frac{\mu_o}{4\pi} \frac{q_2 \vec{v}_2 \times (\vec{r}_1 - \vec{r}_2)}{|\vec{r}_1 - \vec{r}_2|^3}$ is the magnetic

field at the position of particle 1 due to the motion of particle 2 (non-relativistic case). Also the force exerted on particle 2 by particle 1 is given by:

$$\vec{F}_{21} = q_2 \vec{v}_2 \times \vec{B}_1, \tag{8}$$

where $\vec{B}_1 = \frac{\mu_o}{4\pi} \frac{q_1 \vec{v}_1 \times (\vec{r}_2 - \vec{r}_1)}{|\vec{r}_2 - \vec{r}_1|^3}$.

Next let us substitute the expression for B_2 into eq. (7), this yields:

$$\vec{F}_{12} = \frac{\mu_o}{4\pi} \frac{q_1 \quad q_2}{|\vec{r}_1 - \vec{r}_2|^3} \{\vec{v}_1 \times [\vec{v}_2 \times (\vec{r}_1 - \vec{r}_2)]\}. \tag{9}$$

Eq. (9) can be simplified by using the vector identity

$$\vec{A} \times (\vec{B} \times \vec{C}) = \vec{B} \, (\vec{A} \cdot \vec{C}) - \vec{C} \, (\vec{A} \cdot \vec{B}).$$

Letting $\vec{A} = \vec{v}_1$, $\vec{B} = \vec{v}_2$ and $\vec{C} = \vec{r}_1 - \vec{r}_2$ we obtain

$$\vec{F}_{12} = \frac{\mu_o}{4\pi} \frac{q_1 \quad q_2}{|\vec{r}_1 - \vec{r}_2|^3} \{\vec{v}_2 \, [\vec{v}_1 \cdot (\vec{r}_1 - \vec{r}_2)] - (\vec{r}_1 - \vec{r}_2) \, (\vec{v}_1 \cdot \vec{v}_2)\}. \tag{10}$$

The force on particle 2 can be obtained in a similar way, yielding

$$\vec{F}_{21} = \frac{\mu_o}{4\pi} \frac{q_1 \quad q_2}{|\vec{r}_2 - \vec{r}_1|^3} \{\vec{v}_1 \, [\vec{v}_2 \cdot (\vec{r}_2 - \vec{r}_1)] - (\vec{r}_2 - \vec{r}_1) \, (\vec{v}_2 \cdot \vec{v}_1)\}. \tag{11}$$

Studying equs. (10) and (11) we can make the following observations:

Consider the second term in the bracket. This is a component of force directed along a line joining the two particles. This term produces no difficulty.

Next consider the first term in each bracket. $\vec{F}_{12}$ has a component of force in the direction of $\vec{v}_2$ while $\vec{F}_{21}$ has a component in the direction of $\vec{v}_1$. These two components do not obey Newton's Third Law nor are they directed along a line joining the two particles.

We can therefore conclude that the equation $\dot{\vec{L}} = \vec{N}^e$ is not valid if the particles interact with the mangetic force.

Next let us consider whether the equation

$$M \ddot{\vec{R}} = \vec{F}^e \tag{12}$$

is valid when the particles interact with the magnetic force. In general we have the relationship

$$M \ddot{\vec{R}} = \vec{F}^e + \sum_{i=1}^{n} \sum_{\substack{j=1 \\ j \neq i}}^{n} \vec{F}_{ij}, \tag{13}$$

where F_e is the resultant external force on the system of particles and F_{ij} is the force exerted on particle i by particle j. Thus eq. (12) will be valid whenever the second term in eq. (13) is equal to zero.

For the case of two particles, the second term becomes:

$$\sum_{i=1}^{2} \sum_{\substack{j=1 \\ j \neq i}}^{2} \vec{F}_{ij} = \vec{F}_{12} + \vec{F}_{21}. \tag{14}$$

We see that eq. (14) will equal zero when the internal forces obey Newton's Third Law of motion, that is $\vec{F}_{12} = - \vec{F}_{21}$.

We already showed in the discussion following equation (11) that when the particles interact via the magnetic force, Newton's Third Law is not valid. Thus eq. (12) is not necessarily valid, if the particles in the system interact with the magnetic force.

• PROBLEM 12-21

Consider the system of particles which is made up of the turbine rotor of the single jet engine of an airplane which is banking in a horizontal circular turn. Assume that the center of mass of the rotor lies on its axis of rotation and that the center of the circle around which the airplane moves is at rest in an inertial reference system. Does the net external torque on the rotor, about any arbitrary point on its axis of symmetry, equal the rate of change of angular momentum about that point? Is there any single point on the axis of symmetry about which the net external torque equals the rate of change of angular momentum?

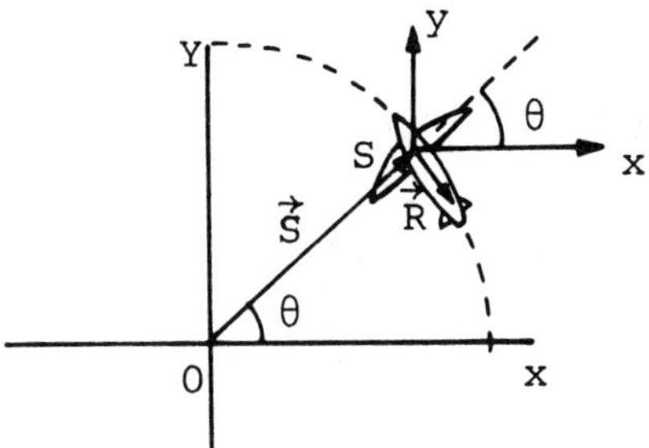

Fig. 1

Solution: This problem can be solved by using the general equation of motion for rotational motion about an arbitrary point s fixed on the axis inside the airplane, namely

$$\dot{\vec{L}} = \vec{N} - M\vec{R} \times \ddot{\vec{s}} \tag{1}$$

In eq. (1), $\vec{L}$ is the total angular momentum about the point s, $\vec{N}$ is the total external torque about the point s, and $\vec{R}$ is the position vector of the center of mass of the rotor relative to s. We want to determine the condition when the last term in eq. (1) is equal to zero.

First let us find an expression for the vector $\ddot{\vec{s}}$. From Fig. 1 we observe that $\vec{s}$ is equal to:

$$\vec{s} = r\,\hat{a}_r. \tag{2}$$

Differentiating eq. (2) with respect to time we obtain:

$$\dot{\vec{s}} = \dot{r}\,\hat{a}_r + r\,\dot{\hat{a}}_r = \dot{r}\,\hat{a}_r + r\,\dot{\theta}\,\hat{a}_\theta. \tag{3}$$

But for circular motion $\dot{r} = 0$ and eq. (3) reduces to:

$$\dot{\vec{s}} = r\,\ddot{\theta}\,\hat{a}_\theta. \tag{4}$$

Differentiating eq. (4) with respect to time once more yields

$$\ddot{\vec{s}} = \dot{r}\dot{\theta}\hat{a}_\theta + r\ddot{\theta}\hat{a}_\theta + r\dot{\theta}\dot{\hat{a}}_\theta = (\dot{r}\dot{\theta} + r\ddot{\theta})\,\hat{a}_\theta - r\dot{\theta}^2\hat{a}_r. \tag{5}$$

We again notice that $\dot{r} = 0$ and eq. (5) reduces to:

$$\ddot{\vec{s}} = r\ddot{\theta}\hat{a}_\theta - r\dot{\theta}^2\hat{a}_r. \tag{6}$$

Next we have to find an expression for $\vec{R}$. From Fig. 1 notice that $\vec{R}$ is a vector in the $-\,\hat{a}_\theta$ direction and therefore

$$\vec{R} = - R\,\hat{a}_\theta . \tag{7}$$

Now evaluate the last term in eq. (1).

$$M\,\vec{R} \times \ddot{\vec{s}} = M\,(- R\,\hat{a}_\theta) \times (r\ddot{\theta}\hat{a}_\theta - r\dot{\theta}^2\hat{a}_r) \tag{8}$$

Evaluating the vector cross product gives

$$M\,\vec{R} \times \ddot{\vec{s}} = - M\,R\,r\dot{\theta}^2\,\hat{k} . \tag{9}$$

Eq. (9) is only equal to zero if $R = 0$. The only point on the axis of symmetry about which the total external torque is equal to the rate of change of the total angular momentum is if the point s is the center of mass of the system.

• PROBLEM 12-22

Two masses are rotating on a weightless rod in a horizontal plane, as shown in Fig. 1. Both are at a distance d from the axis of rotation. The smaller mass is given as m, and the larger mass is given as 2m. The angular velocity of the system is a constant, and is equal to ω_1. If the two masses are released and allowed to move out to the end of the rod, what will happen to the angular velocity of the rod?

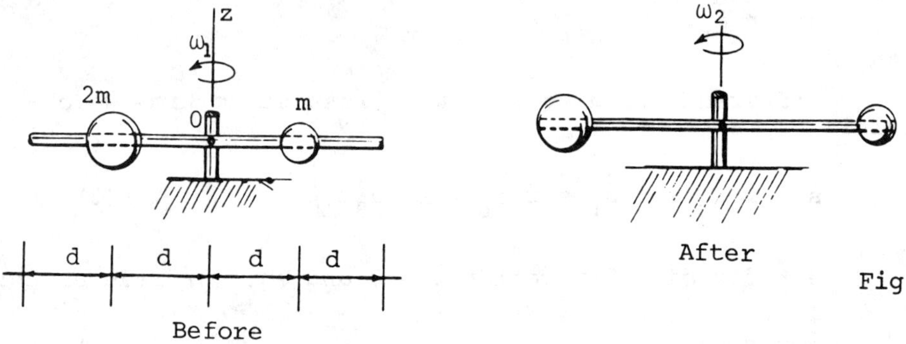

Fig. 1

Solution: In this case, we choose the fixed point, O, as our reference point. Since in this situation, there are no external torques about point O, we can say that the angular momentum about point O in our system (the 2 masses and the weightless bar) is conserved. The angular momentum of a particle is $L = m\,v\,r$, where m is the mass of the particle, v is the velocity, and r its distance from a reference point. The velocity of a particle about that reference point is equal to ωr where ω is the angular velocity of the particle. Therefore, $L = m\,(\omega r)r = m\,\omega\,r^2$. Now let us apply conservation of angular momentum.

Angular Momentum Before = Angular Momentum After

$$2m\omega_1\,I^2 + m\omega_1\,I^2 = 2m\omega_2\,(2I)^2 + m\omega_2(2I)^2$$

where ω_1, and ω_2 are the angular velocities of the system before and after, respectively. Solving for ω_2 yields

$$\omega_2 = \frac{\omega_1}{4}$$

so that when the two masses move to the end of the rod, the angular velocity will decrease by a factor of 4. One might suspect that this problem could be solved by using the fact that the kinetic energy of the system before release is equal to the kinetic energy of the system after release. But if we use this, we find to our surprise that the angular velocity of the rod after release is different from the value as calculated above. It turns out that kinetic energy is not conserved for this system because the internal forces in the system must do work in decelerating the masses when they reach the end of the rod.

One might also be tempted to use the conservation of momentum theorem for a system of particles, but momentum is not conserved because there is an external force on the system, namely, the force that the reaction exerts on the bar. The center of mass of the system is not at the point O, and therefore has an acceleration, since it is moving in a circular path about O. Therefore, since the center of mass has an acceleration, there must be a force causing that acceleration.

• PROBLEM 12-23

The gravitional field of the moon attracts the oceans of the earth, causing tides. The friction of the tides against the land exerts a torque on the earth which gradually slows down the rotation of the earth about its axis. At the same time, the radius of the path of the moon about the earth is observed to increase slowly with time. Using conservation of angular momentum, show qualitatively how these two observations can be related. Is mechanical energy conserved in this process? Predict the ultimate result of this process. Assume that the center of mass of the moon follows a circular path about the center of mass of the earth and that the center of mass of the earth is fixed in an inertial reference frame.

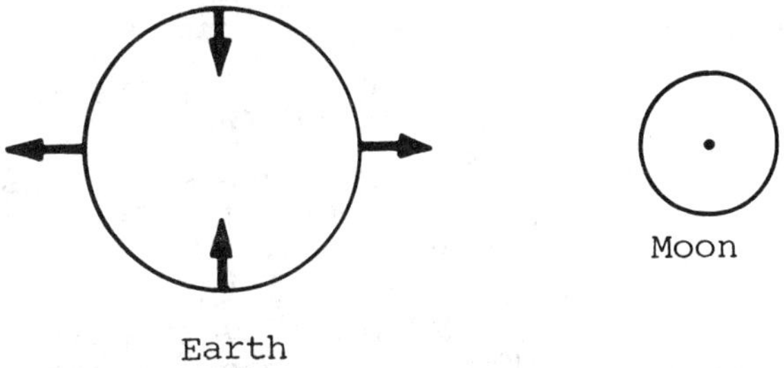

Solution: When tidal forces are acting on a body, such as the moon's on the earth, they cause an elongation along the axis facing the other body and a simultaneous compression around a belt perpendicular to this axis.

This produces two high tides and two low tides in a given 24 hour period due to the rotation of the earth. The work done in repeated stretching and compressing of the earth is done at the expense of the rotational kinetic energy of the earth. This is gradually slowing down the earth in its rotational motion. The moon, on the other hand has already developed a permanent elongation and manages to keep this longest axis always facing towards the earth. By doing so, it is already in its lowest energy position, and is thus tidally coupled to the earth. From the earth we always see only one face of the moon, and its period of rotation and revolution are equal.

The earth is undergoing a similar process. It is losing its rotational kinetic energy due to tidal friction. However, the angular momentum of the earth - moon system is conserved, because no outside torques are acting on the system. The total angular momentum L is the sum of the angular momentum of earth around its own axis, L_E, the angular momentum of moon around its own axis L_M, and the angular momentum of the moon around the earth, L_{ME} or,

$$L = L_E + L_M + L_{ME}.$$

But $L_E = I_E \omega_E$, $L_M = I_M \omega_M$

and $L_{ME} = m_M r v$

where I_E and I_M are moments of inertia for the earth and the moon around their own axes, and ω_E and ω_M are their respective angular velocities around their axes. Also m_M is the mass of the moon, r is the earth - moon distance and v is the velocity of the moon around the earth. Thus,

$$L = I_E \omega_E + I_M \omega_M + m_M r v.$$

Since L, I_E, I_M are constants and ω_E is decreasing, the sum of last two terms must increase. Since ω_M is not expected to increase, that is, the moon is not going to speed up its rotation, only $m_M r v$ is going to increase. Since m_M is a constant, only r v is going to increase. This is dynamically possible if r increases and v actually decreases in order to maintain the necessary gravitational - force centripetal - force balance. Ultimately, the earth will also settle into a comfortable position with its longest axis always directed towards the moon. At that time, the distance between the earth and the moon will be much more than what it is today.

This, of course, does not take into account the effect of the sun's tidal forces on the earth and the moon.

• **PROBLEM 12-24**

It is known that the tides raised on a planet by its sun decreases the planets angular velocity of rotation. Find a formula relating ω, the angular velocity of rotation of the planet, to the orbit radius r for later or earlier times. Assume planet radius a, with orbit radius r_0 and ω_0 at present time. Apply the formula to the earth to find the change in the orbit distance when the length of day becomes equal to the present year. Ignore the moon's affect in this problem.

Solution: The total angular momentum of the earth consists of the spin angular momentum (daily rotation) and the orbital angular momentum (revolution about the sun). The total angular momentum remains constant because the sun's gravitational force acting on the earth is a central force and therefore produces no torque. This conservation of total angular momentum can be written as

$\Delta L_s = -\Delta L_0$, L_s = spin angular momentum and L_0 = orbital angular momentum. The change in spin angular momentum is

$$\Delta L_s = \frac{2}{5} m a^2 (\omega - \omega_0).$$

We are making the approximation here that the moment of inertia of the earth is equal to $\frac{2}{5} m a^2$, the moment of inertia of a sphere of radius a. In reality, the earth's moment of inertia is less than $\frac{2}{5} m a^2$ because the earth's mass is concentrated near its center.

The change in orbital angular momentum is

$$\Delta L_0 = m v r = m v_0 r_0$$

$$= 2 \pi m \left(\frac{r^2}{Y} - \frac{r_0^{\,2}}{Y_0} \right) \qquad (1)$$

where Y is the length of the year (period). Thus,

$$2 \pi m \left(\frac{r^2}{Y} - \frac{r_0^{\,2}}{Y_0} \right) = \frac{2}{5} m a^2 (\omega_0 - \omega). \qquad (2)$$

Now the variation of the length of the year Y as a function of r must be found. From Kepler's Laws

$$Y^2 \propto r^3,$$

setting up ratios

$$\frac{Y^2}{r^3} = \frac{Y_0^{\ 2}}{r_0^{\ 3}}$$

$$Y^2 = \frac{Y_0^{\ 2}}{r_0^{\ 3}} r^3$$

$$Y = \left(\frac{Y_0^{\ 2}}{r_0^{\ 3}}\right)^{\frac{1}{2}} r^{\frac{3}{2}} .$$

Substituting this into equation (2) gives

$$2 \pi m \left[\frac{r^{1/2}}{\left(Y_0^{\ 2}/r_0^{\ 3}\right)^{1/2}} - \frac{r_0^{\ 2}}{Y_0}\right] = \frac{2}{5} m a^2 (\omega_0 - \omega) .$$

Solving for r gives

$$r = \frac{Y_0^{\ 2}}{r_0^{\ 3}} \left(\frac{a^2}{5\pi} (\omega_0 - \omega) + \frac{r_0^{\ 2}}{Y_0}\right)^2$$

$$= r_0 \left[1 + \frac{1}{5\pi} \left(\frac{a}{r_0}\right)^2 (\omega_0 - \omega) Y_0\right]^2 .$$

Using the following data,

$a = 3.96 \times 10^3$ miles

$r_0 = 9.26 \times 10^7$ miles $\qquad \left(\frac{a}{r_0}\right)^2 = 1.83 \times 10^{-9}$

$Y_0 = 365$ days $\qquad \omega_0 = \frac{2\pi}{\Delta \text{ day}} = 365 \left(\frac{2\pi}{Y_0}\right)$

when $\omega = \frac{2\pi}{Y_0} = \frac{2\pi}{365 \text{ day}} = 0.0172$ radians/day

$$r = r_0 \left[1 + \frac{2}{5} (1.83 \times 10^{-9}) (365 - 1)\right]^2$$

$$= r_0 [1 + 2.66 \times 10^{-7}]^2$$

Since the second term in parenthesis is much smaller then one, we can use the binomial theorem approximation to evaluate the terms in parenthesis. The binomial approximation is

$$(1 + x)^n \approx 1 + nx \text{ for } x << 1.$$

Using this theorem in the preceding equation yields

$$r \approx r_0[(1 + 2\,(2.66 \times 10^{-7})].$$

So that

$$r - r_0 = 2\,r_0\,(2.66 \times 10^{-7}) = 2\,(9.26 \times 10^7)(2.66 \times 10^{-7})$$

$$= 49.3 \text{ miles.}$$

• PROBLEM 12-25

Thin disk B can roll without slipping on the horizontal plane. From its top hangs disk A whose radius is one-half that of B. Disk A weighs 8.05 ℓb and disk B weighs 32.2 ℓb. Determine the initial angular acceleration of each disk when a horizontal force of 8 ℓb. is applied to the center of disk B.

Fig. 1

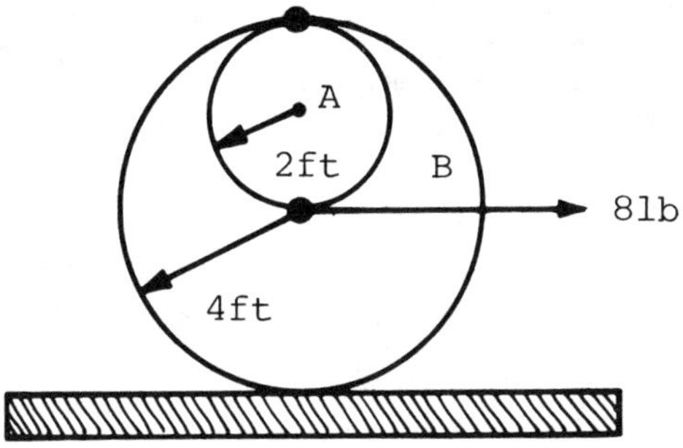

Solution: The center of mass of a disk is its center 0. The motion of the center of mass is determined by the linear momentum theorem.

The equation for the rotational motion about the center of mass is given by the angular momentum theorem.

In this particular case the initial net forces are horizontal.

These theorems, applied to our problem, can be interpreted as follows, for each disk, where the horizontal distance to the right is considered positive.

Theorem of Linear Momentum $\left(\frac{d}{dt} M \dot{\vec{r}}\right) = \vec{F}$

$Ma_c = F$ where F is the net external force on the disk a_c is the acceleration of the center of the disk (its center of mass) and M is the total mass of the disk

Theorem of Angular Momentum $(I_c \dot{\vec{\omega}} = \vec{\tau}_c)$

$I\alpha = \tau$

where τ is the net clockwise torque about the center (center of mass) of the disk

α is the (clockwise) angular acceleration of the disk

and $I \left(= \frac{1}{2} M r^2 \right)$ is the moment of inertia of the disk (of radius r) about its center of mass.

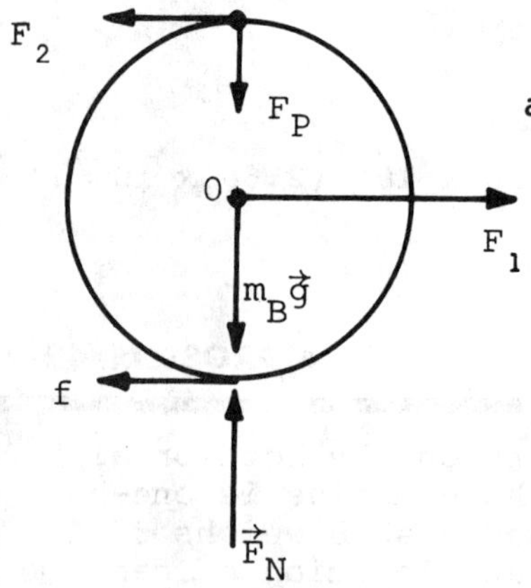

Forces on disk B

Fig. 2

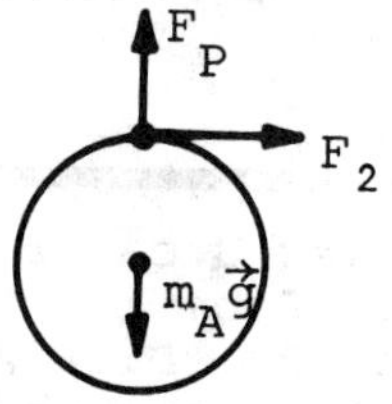

Forces on disk A

Fig. 3

To further particularize this problem, draw free-body diagrams of the two disks and use subscripts A and B to differentiate quantities pertaining to the individual disks. The vertical forces (mg, and the vertical components of interaction with the plane [$\vec{F}_n$] and with each other [F_p] sum to zero for each disk. They cause no torque about its center of mass and will henceforth be disregarded.

Applying the momentum theorem to disk B:

$$M_B \, a_{CB} = F_1 - F_2 - f \qquad (1)$$

and $\frac{1}{2} M_B \, r_B^{\,2} \, \alpha_B = f \, r_B - F_2 \, r_B$, by the moment eq. $\tau = I\alpha$, which, after dividing by $r_B/2$ becomes

$$M_B \, r_B \, \alpha_B = 2f - 2F_2 \qquad (2)$$

where F_1 is the applied force (8 lb)

F_2 is the interaction force with disk A

and f is the frictional interaction with the plane.

And similarly to the disk A:

$$M_A \, a_{CA} = F_2 \qquad (3)$$

and $\frac{1}{2} M_A \, r_A^{\,2} \, \alpha_A = F_2 \, r_A$ or, dividing by $r_A/2$

$$M_A \, r_A \, \alpha_A = 2 \, F_2 \qquad (4)$$

Our desired quantities are α_A and α_B but in our four equations we have the additional unknowns a_{CB}, F_2, f, and

a_{CA}. We thus need two more independent equations relating these quantities.

The instantaneous horizontal acceleration of a point on the top of the disk (= a_{TB}) can be expressed in terms of linear acceleration (of the center of the disk) and angular acceleration about the center as follows:

$$a_{TB} = a_{CB} + r_B \alpha_B \qquad (5)$$

(We could have taken as our starting point an expression for the velocity of a point at the top:

$v_{TB} = v_{CB} + \omega r_B$, and obtain Eq. 5 by taking its time derivative, or we could even have started by looking at the result of infinitesimal linear and angular displacements.)

Since disk B rolls without slipping we can regard the bottom as an instantaneous pivot point. Knowing that the center is halfway to the top yields

$$a_{TB} = 2a_{CB} \qquad (6)$$

We now have two more equations but we have introduced another unknown, a_{TB}. We can eliminate a_{TB}, using Eqs. 5 and 6 but since the point at the top is common to both disks we will first write an equation for disk A similar to Eq. 5 for disk B:

$$a_{TB}\ (= a_{TA}) = a_{CA} + r_A \alpha_A \qquad (7)$$

We now have 7 equations and 7 unknowns and we can proceed to solve for α_A and α_B. (We might note Eq. 6 has no equivalent for disk A since the bottom of A is not instantaneously fixed.)

From Eqs. 5 and 6 $\quad 2a_{CB} = a_{CB} + r_B \alpha_B$

$$\text{or} \quad a_{CB} = r_B \alpha_B \qquad (8)$$

From Eqs. 6 and 7 $\quad 2a_{CB} = a_{CA} + r_A \alpha_A$

and combining this with Eq. 8 gives

$$2r_B \alpha_B = a_{CA} + r_A \alpha_A$$

$$\text{or} \quad a_{CA} = 2r_B \alpha_B - r_A \alpha_A \qquad (9)$$

We now use Eqs. 8 and 9 to eliminate a_{CB} and a_{CA} from Eqs. 1 and 3 thus obtaining:

$$M_B r_B \alpha_B = F_1 - F_2 - f \qquad (10)$$

and $\quad M_A (2r_B \alpha_B - r_A \alpha_A) = F_2 \qquad (11)$

(Our working equations, with four unknowns, are now Eqs. 2, 4, 10 and 11.)

We now solve for f in Eq. 10: $f = F_1 - F_2 - M_B\, r_B\, \alpha_B$ and replace f in Eq. 2 by this value to obtain

$$M_B\, r_B\, \alpha_B = 2\,(F_1 - F_2 - M_B\, r_B\, \alpha_B) - 2\,F_2$$

or, collecting terms,

$$3\, M_B\, r_B\, \alpha_B = 2\,F_1 - 4\,F_2 \qquad (12)$$

Now we will use Eq. 4 to eliminate F_2 from Eq. 11 to give

$$M_A\,(2r_B\, \alpha_B - r_A\, \alpha_A) = \frac{1}{2}\, M_A\, r_A\, \alpha_A$$

Dividing by $M_A\, r_A$, noting $r_B/r_A = 2$, and collecting terms:

$$8\alpha_B = 3\alpha_A \qquad (13)$$

Using Eq. 4 to eliminate F_2 from Eq. 12:

$$3\, M_B\, r_B\, \alpha_B = 2\,F_1 - 2\,M_A\, r_A\, \alpha_A \qquad (14)$$

Using 13 to eliminate α_B from Eq. 14

$$3\, M_B\, r_B \left(\frac{3}{8}\,\alpha_A\right) = 2\,F_1 - 2\,M_A\, r_A\, \alpha_A$$

$$\alpha_A \left(\frac{9}{8}\, M_B\, r_B + 2\,M_A\, r_A\right) = 2\,F_1$$

or $$\alpha_A = F_1 \left(\frac{9}{16}\, M_B\, r_B + M_A\, r_A\right)^{-1}$$

Substituting numerical values, and noting that

M_B = 32.2 lb (mass) = 1 slug and M_A = 8.05 lb (mass) $= \frac{1}{4}$ slug

$$\alpha_A = 8\text{ lb}\left[\frac{9}{16}\,(1\text{ slug})(4\text{ ft}) + \left(\frac{1}{4}\text{ slug}\right)(2\text{ ft})\right]^{-1}$$

$= 2.91\ \frac{\text{rad}}{\text{s}^2}$ C w (A negative value would represent C C w acceleration.)

Using this value for α_A in Eq. 13 gives for α_B:

$$\alpha_B = \frac{3}{8}\,\alpha_A = \frac{3}{8}\left(2.91\ \frac{\text{rad}}{\text{s}^2}\right) = 1.09\ \frac{\text{rad}}{\text{s}^2}\text{ C w}$$

Two homogeneous cylinders are constrained as shown in the figure. Each weighs 805 lb. The coefficient of friction at all surfaces is 0.50. Determine the angular acceleration of each cylinder when a clockwise moment of 860 lb. ft. is applied to the upper cylinder.

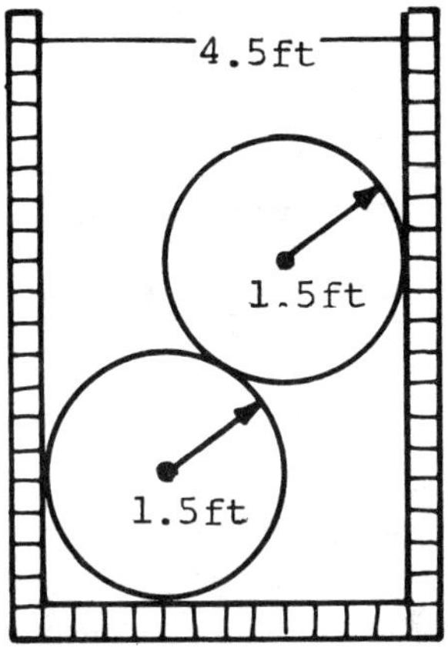

Solution: This problem is solved using Newton's Second Law in its rotational and linear forms. The procedure is as follows: 1) Draw free body diagrams to clearly indicate the important forces. 2) Take the sum of forces in the x and y directions to get expressions for the reactions at the walls. 3) Take the sum of the torques on each body to find the body's angular acceleration.

1) Free body diagrams.

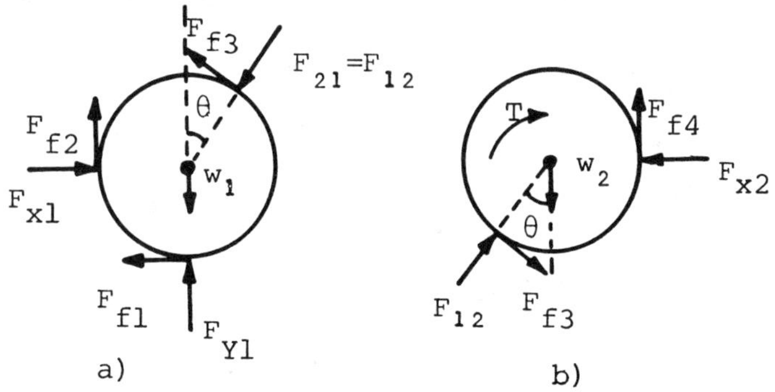

Fig. a) is the free body diagram of the lower cylinder w_1 is its weight, F_{x1} and F_{y1} are the horizontal and verticle wall reactions respectively, $F_{12} = F_{21}$ (by Newton's 3rd Law) is the contact force between the upper and lower cylinders. F_{fn} is the friction force at point n. θ is the angle between the vertical and the cylinders' line of tangency.

Fig. b) is the free body diagram of the upper cylinder, with all symbols similarly defined. T is the applied torque.

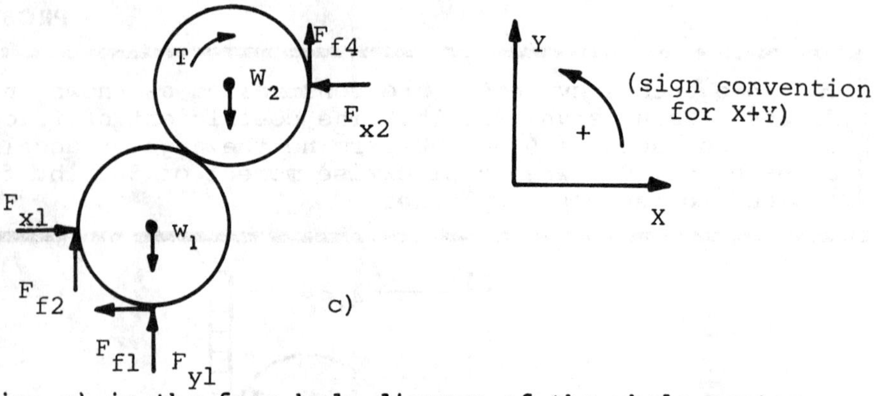

Fig. c) is the free body diagram of the whole system.

2) Summing forces in the x & y directions

ΣF_x (forces to the right are positive) $= m\,a_x$

ΣF_y (forces up are positive) $= m\,a_y$

Since the system is constrained horizontally by the walls $a_x = 0$ and $\Sigma F_x = 0$.

Summing forces for the whole system,

$$\Sigma F_x = F_{x1} - F_{f1} - F_{x2} = 0$$

$$F_{x1} = F_x2 + F_{f1}$$

and $\Sigma F_y = F_{f2} + F_{y1} + F_{f4} - (w_1 + w_2) = (m_1 + m_2)\,a_y.$

Summing forces on the lower cylinder,

$$\Sigma F_x = F_{x1} - F_{f1} - F_{12} \sin\theta - F_{f3} \cos\theta = 0$$

$$\Sigma F_y = F_{f2} + F_{y1} - w_1 - F_{12} \cos\theta + F_{f3} \sin\theta$$

$$= m_1\,a_y = m\,a_y$$

$m_1\,a_{1y} = m\,a_y$ because the cylinder's masses, $m_1 + m_2$ are the same (i.e., m), and also because the cylinder's acceleration in the y direction must be the same or the cylinders will come apart and the driving force will disappear on both cylinders.

On the upper cylinder,

$$\Sigma F_x = F_{12} \sin\theta - F_{x2} + F_{f3} \cos\theta = 0$$

$$F_{12} \sin\theta + F_{f3} \cos\theta = F_{x2}$$

$$\Sigma F_y = m a_y = F_{f4} = F_{f3} \sin\theta + F_{12} \cos\theta - w_2$$

$$f_{f4} - F_{f3} \sin\theta + F_{12} \cos\theta - w_2 = m a_y$$

But $a_y = 0$ because the center of mass of the system has no acceleration in the y-direction.

Restating the force equations:
in the x direction

i) $F_{x1} = F_{x2} + F_{f1}$

ii) $F_{x1} = F_{f1} + F_{12} \sin\theta + F_{f3} \cos\theta$

iii) $F_{x2} = F_{12} \sin\theta + F_{f3} \cos\theta$

Note that ii is the sum of i + iii, the three equations can be simplified to only two equations. This means that only 2 of the 3 equations are independent. So only 2 free body diagrams are needed. The same is true for the y directions, where the two independent relations are: (with $a_y = 0$)

iv) $F_{f2} + F_{y1} + F_{f4} - (w_1 + w_2) = 0$

v) $F_{f2} + F_{y1} - w_1 - F_{12} \cos\theta + F_{f3} \sin\theta = 0$

since $w_1 = w_2$ (iv) becomes $F_{f2} + F_{y1} + F_{f4} - (2\,w) = 0$
Remembering the general formula, $F_f \le \mu N$, where F_f is the friction force, μ is the coefficient of friction, and N is the normal force, we can write;

$$F_{f1} \le \mu F_{y1}$$

$$F_{f2} \le \mu F_{x1}$$

$$F_{f3} \le \mu F_{12}$$

$$F_{f4} \le \mu F_{x2}$$

3) Summing moments

For this problem it is easiest to sum moments about the center of the cylinder. For each cylinder,

$\Sigma M_{cp} = I_{0n} \alpha_n$ where I_{0n} = moment of inertia about central axis of cylinder n, α_n = angular acceleration of cylinder n. Where n = 1 for cylinder 1, and n = 2 for cylinder 2.

For the lower cylinder, counterclockwise moments positive, $I_{01} = I_{02} = I_0 = \frac{1}{2} m r^2$ because the mass and radii of each cylinder is the same, yields;

$$\Sigma M_{cp} = I_0 \alpha_1$$

Recalling that the moment is zero of any force that acts through the point one is taking moments about, because the moment a r m is zero.

vi) $-(F_{f2})r + (F_{f3})r - (F_{f1})r = I_0 \alpha_1.$

For the upper cylinder,

vii) $(F_{f4})r + (F_{f3})r - T = I_0 \alpha_2$

Now we can solve for α_1 and α_2, if both cylinders rotate then the friction forces take on maximum values and the friction equations become:

$$F_{f1} = \mu F_{y1}$$

$$F_{f2} = \mu F_{x1}$$

$$F_{f3} = \mu F_{12}$$

$$F_{f4} = \mu F_{x2}$$

Substituting into equation vi yields,

$$-\mu F_{x1}r + \mu F_{12}r - \mu F_{y1}r = I_0 \alpha_1$$

$$\mu r(-F_{x1} + F_{12} - F_{y1}) = I_0 \alpha_1$$

Putting these friction equations into eq. (vii) gives,

$$\mu F_{x2}r + \mu F_{12}r - T = I_0 \alpha_2$$

$$\mu r(F_{x2} + F_{12}) - T = I_0 \alpha 2$$

Rewriting i), iii), iv) & v)

i) $F_{x1} = F_{x2} + \mu F_{y1}$

iii) $F_{x2} = F_{12} \sin\theta + \mu F_{12} \cos\theta$

iv) $\mu F_{x1} + F_{y1} + \mu F_{x2} - 2w = 0$

v) $\mu F_{x1} + F_{y1} - w - F_{12}\cos\theta + \mu F_{12}\sin\theta = 0$

This is a system of 6 equations in 6 unknowns.

First, find θ.

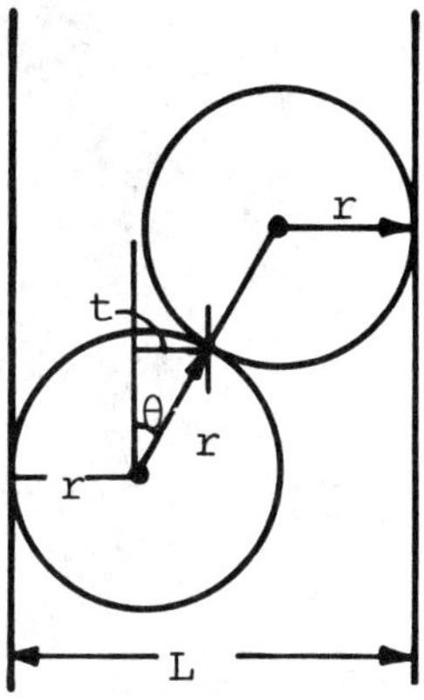

Since the contact point is located at L/2 when the circles are of equal size,

$$\theta = \sin^{-1}\frac{t}{r}$$

$$\theta = \sin^{-1}\left[\frac{(L/2 - r)}{r}\right]$$

In this problem $\theta = \sin^{-1}\left[\frac{\frac{4.5'}{2} - 1.5'}{1.5'}\right]$

$$\theta = 30°$$

$$\sin\theta = \frac{1}{2} \quad \cos\theta = \frac{\sqrt{3}}{2} = .8660$$

Equation v) can be rewritten, putting in $\sin\theta = \frac{1}{2}$, $\cos\theta = .866$, $\mu = 0.5$, and $w = 805$ lbs.

$$.5\,F_{x1} + F_{y1} - 805 - .866\,F_{12} + (.5)\left(\frac{1}{2}\right)F_{12} = 0$$

$$.5\,F_{x1} + F_{y1} - .616\,F_{12} = 805$$

(viii) $.5\,F_{x1} + F_{y1} = 805 + .616\,F_{12}$

Equation (iv) can be written;

$$.5\,F_{x1} + F_{y1} + .5\,F_{x2} - 2\,(805) = 0$$

(ix) $.5\,F_{x1} + F_{y1} = 1610 - .5\,F_{x2}$

Equating (viii) and (ix), gives

$$805 + .616\,F_{12} = 1610 - .5\,F_{x2}$$

$$.616\,F_{12} + .5\,F_{x2} = 805$$

Multiplying this equation by 2 yields

$$1.23\,F_{12} + F_{x2} = 1610$$

$$F_{x2} = 1610 - 1.23\,F_{12}$$

Setting this equation equal to equ. (iii) yields

$$1610 - 1.23\,F_{12} = F_{12}\sin\theta + \mu\,F_{12}\cos\theta$$

Putting in known values for μ, $\sin\theta$ + $\cos\theta$ gives

$$1610 - 1.23\ F_{12} = .5\ F_{12} + (.5)\ F_{12}\ (.866)$$

$$1610 - 1.23\ F_{12} = .933\ F_{12}$$

$$1610 = 2.16\ F_{12}$$

$$F_{12} = 745.4 \text{ lbs.}$$

So since $F_{x2} = 1610 - 1.23\ F_{12}$

$$F_{x2} = 639.2 \text{ lbs.}$$

Inserting the value of F_{x2} into eq. i) gives

$$\text{(x)}\quad F_{x1} = 639.2 + .5\ F_{y1}$$

And putting the value of F_{x2} into eq. (ix) yields

$$.5\ F_{x1} + F_{y1} = 1263.4$$

$$F_{x1} + 2\ F_{y1} = 2526.8$$

(xi) $$F_{x1} = 2526.8 - 2\ F_{y1}$$

Equating (x) and (xi) gives,

$$639.2 + .5\ F_{y1} = 2526.8 - 2\ F_{y1}$$

$$2.5\ F_{y1} = 1887.6$$

$$F_{y1} = 755.04 \text{ lbs.}$$

and $F_{x1} = F_{x2} + \mu\ F_{y1}$

$$F_{x1} = 639.2 + (.5)\ (755.04)$$

$$F_{x1} = 1016.7 \text{ lbs.}$$

Now to compute α_1, use moment equation (vi);

$$\frac{\mu r}{I_0}\ (-\ F_{x1} + F_{12} - F_{y1}) = \alpha_1$$

$$\alpha_1 = \frac{(.5)r}{I_0}\ (-\ 1016.7 + 745.4 - 755.04)$$

$$\alpha_1 = \frac{(.5)r}{(.5)m\ r^2}\ (-\ 1026.34)$$

$$\alpha_1 = \frac{-\ 1026.34}{m\ r} = \frac{-\ 1026.34}{\left(\frac{W}{g}\right)\ r} = \frac{-\ 1026.34}{W\ r}\ g$$

$$= \frac{-1026.34 \ (32.2)}{(805) \qquad (1.5)}$$

$$\alpha_1 = -27.37 \frac{rad}{s^2}$$

This acceleration is the maximum possible acceleration of the bottom cylinder, the acceleration when it is slipping, or rotating with respect to the walls. Note that this is a negative acceleration. The negative sign means that if the cylinder ever starts to rotate in the direction it is being driven by the top cylinder (the positive direction) it will immediately start to decelerate. This leads to the conclusion that the bottom cylinder stays at rest and its angular acceleration is zero.

If the bottom cylinder is at rest the top cylinder may still rotate.

From the previous force equations for the upper cylinder:

$$F_{12} \sin\theta + F_{f3} \cos\theta = F_{x2}$$

$$F_{f4} + F_{12} \cos\theta = F_{f3} \sin\theta + w$$

Summing moments on the upper cylinder yields:

$$F_{f4}\, r + F_{f3}\, r - 860 \text{ f - lbs} = I_0\, \alpha$$

(remember counterclockwise moments are positive)

Allowing F_{f3} and F_{f4} to take on maximum values, $F_{f3} = \mu F_{12}$ $F_{f4} = \mu F_{x2}$.

The force moment equations then become:

a) $F_{12} \sin\theta + \mu F_{12} \cos\theta = F_{x2}$

b) $\mu F_{x2} + F_{12} \cos\theta = \mu F_{12} \sin\theta + w$

c) $\mu r F_{x2} + \mu r F_{12} - 860 \text{ f - lbs} = I_0\, \alpha$

from a) $F_{x2} = (\sin\theta + \mu\cos\theta)\, F_{12}$

$$= (.5 + .5\ (.866))\ F_{12}$$

$$F_{x2} = .93\ F_{12}$$

From b) using the above result

$$\mu\ (.93\ F_{12}) + F_{12}\ (\cos\theta - \mu\sin\theta) = w$$

$$F_{12} = \frac{w}{.93\ \mu + \cos\theta - \mu \sin\theta} = .92\ w$$

Then $F_{12} = .92\ w$ and $F_{x2} = .86\ w$. Using this result in the moment equation, c), yields:

$$.5\ (1.5\ \text{ft})\ .86\ w + .5\ (1.5\ \text{ft})\ .92\ w - 860\ \text{f - lbs}$$

$$= I_0\ \alpha$$

$$\alpha = \frac{.5\ (1.5\ \text{ft})\ .86\ w + .5\ (1.5\ \text{ft})\ .92\ w - 860\ \text{f- lbs}}{I_0}$$

$$I_0 = \frac{1}{2}\ m\ r^2 = \frac{1}{2}\ \frac{805\ \text{lbs}}{32.2\ \text{f/s}^2}\ (1.5\ \text{ft})^2 = 28.13\ \text{slug - ft}^2$$

$$\alpha = \frac{1.34\ \text{ft}\ (805\ \text{lbs}) - 860\ \text{ft - lbs}}{28.13\ \text{slug - ft}^2}$$

$$\alpha = 7.63\ \text{r/s}^2$$

This acceleration is in the counter-clockwise direction, opposite the direction of the applied torque. Once again the friction forces are so great that the cylinder is not allowed to turn.

By assuming the cylinders were moving, (and friction forces were maximum) we found the cylinders were decelerated. This means that the cylinders remain at rest.

$$\alpha_{upper} = 0,\ \alpha_{lower} = 0.$$

CHAPTER 13

COLLISIONS

ELASTIC COLLISIONS

• PROBLEM 13-1

A neutron with mass m_1 strikes an atomic nucleus with mass m_2 centrally. If the collision is elastic, determine the ratio of the loss in the kinetic energy of the neutron to its original energy. Assume the laws of classical mechanics apply. Calculate the energy loss if the collision occurs with a hydrogen atom. What would be the loss if lead were used as a target?

Solution: a) The ratio of energy lost by a neutron to its original energy is:

$$E = \frac{\frac{1}{2} m_1 v_1^2 - \frac{1}{2} m_1 v_1'^2}{\frac{1}{2} m_1 v_1^2}$$

$$E = \frac{v_1^2 - v_1'^2}{v_1^2} = 1 - \left(\frac{v_1'}{v_1}\right)^2 . \qquad (1)$$

Let primes indicate velocity after collision, and subscripts 1, 2 indicate neutron and atomic nucleus respectively. Assume target nucleus is initially at rest, i.e. v = 0. Since the collision is elastic, kinetic energy and linear momentum are both conserved during the collision:

$$\frac{1}{2} m_1 v_1^2 + 0 = \frac{1}{2} m_1 v_1'^2 + \frac{1}{2} m_2 v_2'^2 \qquad (2)$$

$$m_1 v_1 + 0 = m_1 v_1' + m_2 v_2' \qquad (3)$$

Solve Equation (3) for v_2' and substitute the result in Equation (2).

First, $\quad v_2' = \dfrac{m_1 (v_1 - v_1')}{m_2}$,

substituting,

$$\frac{1}{2} m_1 v_1^2 = \frac{1}{2} m_1 v_1'^2 + \frac{1}{2} m_2 \frac{m_1^2}{m_2^2} (v_1 - v_1')^2$$

$$v_1^2 - v_1'^2 = \frac{m_1}{m_2} (v_1 - v_1')^2$$

$$(v_1 - v_1')(v_1 + v_1') = \frac{m_1}{m_2} (v_1 - v_1')^2$$

$$v_1 + v_1' = \frac{m_1}{m_2} (v_1 - v_1')$$

$$1 + \frac{v_1'}{v_1} = \frac{m_1}{m_2} \left(1 - \frac{v_1'}{v_1}\right).$$

From which,

$$\frac{v_1'}{v_1} = \frac{m_1 - m_2}{m_1 + m_2} . \qquad (4)$$

Substitute Equation (4) in Equation (1) to get:

$$E = 1 - \left(\frac{m_1 - m_2}{m_1 + m_2}\right)^2 . \qquad (5)$$

b) When the neutron strikes a hydrogen nucleus, a proton, $m_1 = m_2$ in Equation (5) and

E = 100% energy transfer.

c) For a lead target, $m_2 = 207$, $m_1 = 1$ and Equation (5) gives:

$$E = 1 - \left(\frac{1 - 207}{1 + 207}\right) = 0.019 = 1.9\%.$$

• PROBLEM 13-2

Suppose a particle of mass m, initially travelling with velocity $\vec{v}$, collides elastically with a particle of mass m initially at rest. Prove that the angle between the velocity vectors of the two particles after the collision is 90°.

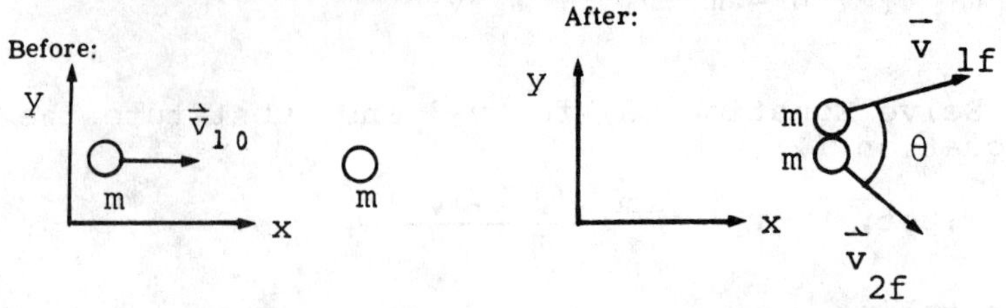

Solution: Whenever we are confronted with a collision problem, we may apply the law of conservation of momentum if no external forces act on the system. External forces are forces which act upon the system being considered (in our case, the system consists of the 2 masses) that are due to the environment external to the system. (i.e. friction). Because no external forces are acting in this problem, we may use this conservation law, ,which is written mathematically as

$$\vec{p}_{10} + \vec{p}_{20} = \vec{p}_{1_f} + \vec{p}_{2_f} \tag{1}$$

In this equation, $\vec{p}_{10}$ and $\vec{p}_{20}$ are the initial momenta of particles 1 and 2, respectively, and $\vec{p}_{1_f}$ and $\vec{p}_{2_f}$ are the final monenta of particles 1 and 2, respectively. Note two things: first, momentum is defined as the product of the mass of a particle and its velocity (i.e. $\vec{p} = m\vec{v}$) and, secondly, because $\vec{v}$ is a vector (i.e. it has direction and magnitude, as in 50 mph EAST), $\vec{p}$ is also a vector.

Applying equation (1), and noting that $\vec{v}_{20}$ is 0 because particle 2 is initially at rest, we may write

$$\vec{p}_{10} = \vec{p}_{1_f} + \vec{p}_{2_f} \tag{2}$$

Now, we have still not used one last bit of information. The collision is elastic, and, whenever this is the case, kinetic energy is conserved. The law of kinetic energy conservation may be written mathematically as

$$\frac{1}{2} m_1 v_{10}^2 + \frac{1}{2} m_2 v_{20}^2 = \frac{1}{2} m_1 v_{1_f}^2 + \frac{1}{2} m_2 v_{2_f}^2 \tag{3}$$

where v_{10}, v_{1_f}, v_{20}, v_{2_f} have the same meaning as previously. Using this equation, we obtain

$$\frac{1}{2} m_1 v_{10}^2 = \frac{1}{2} m_1 v_{1_f}^2 + \frac{1}{2} m_2 v_{2_f}^2 \tag{4}$$

But $m_1 = m_2 = m$, hence

$$\frac{1}{2} m v_{10}^2 = \frac{1}{2} m v_{1_f}^2 + \frac{1}{2} m v_{2_f}^2 \tag{5}$$

$$v_{10}^2 = v_{1_f}^2 + v_{2_f}^2 \tag{6}$$

Rewriting equation (2), we find

$$m \vec{v}_{10} = m \vec{v}_{1_f} + m \vec{v}_{2_f} \tag{7}$$

Dividing both sides of (7) by m,

$$\vec{v}_{10} = \vec{v}_{1f} + \vec{v}_{2f} \tag{8}$$

We now want to express equation (8) in terms of magnitudes. To find the magnitude of a vector, we multiply it by itself, using the dot product, and take the square root. The dot product of 2 vectors is defined as

$$\vec{A} \cdot \vec{B} = AB \cos \theta$$

where A is the magnitude of $\vec{A}$, B is the magnitude of $\vec{B}$, and θ is the angle between $\vec{A}$ and $\vec{B}$. Note that

$$\vec{A} \cdot \vec{A} = (A)(A) = A^2$$

because, in this case, $\theta = 0°$ and $\cos 0° = 1$.

Now, dotting each side of equation (8) into itself, we find

$$\vec{v}_{10} \cdot \vec{v}_{10} = \left(\vec{v}_{1f} + \vec{v}_{2f}\right) \cdot \left(\vec{v}_{1f} + \vec{v}_{2f}\right) \tag{9}$$

$$\vec{v}_{10} \cdot \vec{v}_{10} = \vec{v}_{1f} \cdot \vec{v}_{1f} + 2\vec{v}_{1f} \cdot \vec{v}_{2f} + \vec{v}_{2f} \cdot \vec{v}_{2f}$$

or $$v_{10}^2 = v_{1f}^2 + v_{2f}^2 + 2\vec{v}_{1f} \cdot \vec{v}_{2f} \tag{10}$$

Subtracting equation (6) from (10)

$$\begin{aligned} v_{10}^2 &= v_{1f}^2 + v_{2f}^2 + 2\vec{v}_{1f} \cdot \vec{v}_{2f} \\ - \quad v_{10}^2 &= v_{1f}^2 + v_{2f}^2 \\ \hline 0 &= 2\vec{v}_{1f} \cdot \vec{v}_{2f} \end{aligned}$$

or $$\vec{v}_{1f} \cdot \vec{v}_{2f} = 0 \tag{11}$$

$$v_{1f}\, v_{2f} \cos \theta = 0 \tag{12}$$

But because $v_{1f}, v_{2f} \neq 0$, this means that $\cos\theta = 0$ or $\theta = 90°$, as was to be shown.

• PROBLEM 13-3

A particle of mass m_1, momentum p_{1i} collides elastically with a particle of mass m_2, momentum p_{2i} going in the opposite direction. If m_1 leaves the collision at an angle θ_1 with its original direction, find p_{1f}.

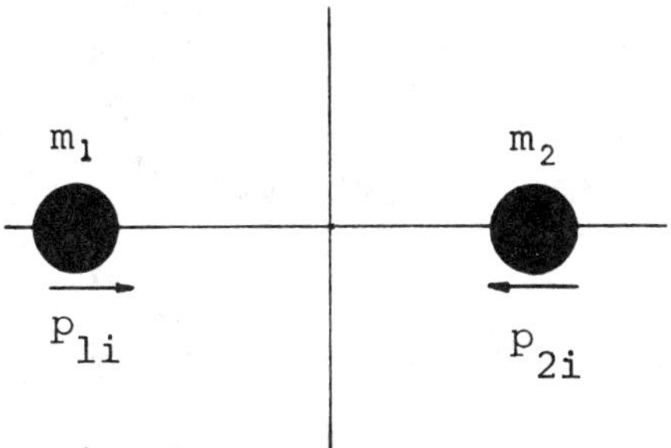

Fig. 1

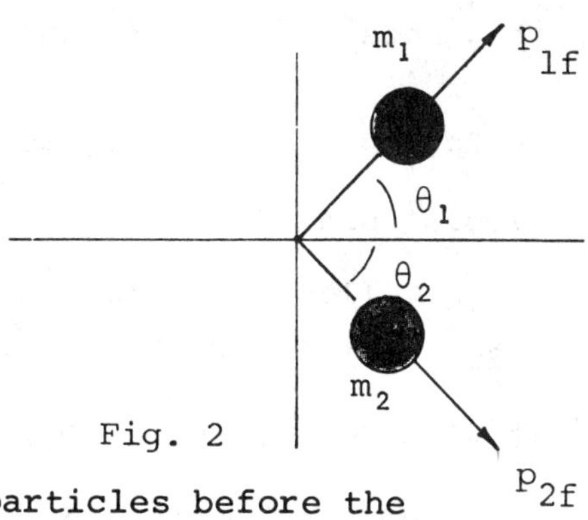

Fig. 2

Solution: Figure 1 shows the particles before the collision travelling in opposite directions along the x-axis, and Figure 2 shows the particles after the collision.

The Principle of Conservation of Linear Momentum requires that the total momentum in the x-direction before the collision must equal the total momentum in the x-direction after the collision. Thus, we may write:

$$p_{1i} - p_{2i} = p_{1f} \cos\theta_1 + p_{2f} \cos\theta_2 \tag{1}$$

where p_{2i} is negative because it is directed toward the left. Similarly, one may write an equation for conservation of linear momentum in the y-direction:

$$0 = p_{1f} \sin\theta_1 - p_{2f} \sin\theta_2. \tag{2}$$

Since the collision is elastic, kinetic energy is also conserved. Thus,

$$\frac{1}{2}\frac{p_{1i}^2}{m_1} + \frac{1}{2}\frac{p_{2i}^2}{m_2} = \frac{1}{2}\frac{p_{1f}^2}{m_1} + \frac{1}{2}\frac{p_{2f}^2}{m_2} \tag{3}$$

where we have used $v = p/m$ to express the velocity in terms of the momentum.

By letting $\gamma = m_2/m_1$ and multiplying through equation (3) by $2m_2$, Equation (3) reduces to:

$$\gamma\, p_{1i}^2 + p_{2i}^2 = \gamma\, p_{1f}^2 + p_{2f}^2. \tag{4}$$

From Equation (1) isolate the term containing p_{2f}:

$$p_{2f} \cos\theta_2 = p_{1i} - p_{2i} - p_{1f} \cos\theta_1. \tag{5}$$

Also, rewriting Equation (2):

$$p_{2f} \sin\theta_2 = p_{if} \sin\theta_1. \tag{6}$$

Squaring both sides of Equations (5) and (6) and adding:

$$p_{2f}^2 \cos^2 \theta_2 + p_{2f}^2 \sin^2 \theta_2 = p_{1i}^2 + p_{2i}^2 + p_{1f}^2 \cos^2 \theta_1$$

$$- 2p_{1i}\, p_{2i} - 2p_{1i}\, p_{1f} \cos \theta_1 + 2p_{2i}\, p_{1f} \cos \theta_1$$

$$+ p_{1f}^2 \sin^2 \theta_1 . \qquad (7)$$

Collecting terms, and using $\sin^2 \theta + \cos^2 \theta = 1$:

$$p_{2f}^2 = p_{1i}^2 + p_{2i}^2 - 2p_{1i}\, p_{2i} + p_{1f}^2$$

$$+ 2p_{1f} \cos \theta_1 \,(p_{21} - p_{1i}) . \qquad (8)$$

Now, substituting (8) into (4) and collecting terms in p_{1f}:

$$(1 + \gamma)\, p_{1f}^2 + 2 \cos \theta_1 \,(p_{2i} - p_{1i})\, p_{1f}$$

$$+ \left[(1 - \gamma)\, p_{1i}^2 - 2p_{1i}\, p_{2i}\right] = 0 . \qquad (9)$$

Since this is quadratic in p_{1f}, we may solve for p_{1f} by using the quadratic formula:

$$p_{1f} =$$

$$\frac{-2 \cos \theta_1 (p_{2i} - p_{1i})}{2(1 + \gamma)} \pm$$

$$\frac{\sqrt{4 \cos^2 \theta_1 (p_{2i} - p_{1i})^2 - 4(1 + \gamma)\left[(1 - \gamma)p_{1i}^2 - 2p_{1i}p_{2i}\right]}}{2(1 + \gamma)} \qquad (10)$$

This may be simplified to:

$$(1 + \gamma)p_{1f} = (p_{1i} - p_{2i}) \cos \theta_1 \pm$$

$$\sqrt{(p_{2i} - p_{1i})^2 \cos^2 \theta_1 - (1 + \gamma)\left[(1 - \gamma)p_{1i}^2 - 2p_1 p_{2i}\right]} \qquad (11)$$

• PROBLEM 13-4

A uniform cube having a weight W and a side s is sliding with a speed u on a smooth horizontal floor when its front edge strikes an upraised tile. If the impact is perfectly elastic, determine the components of the velocity of the mass center immediately after impact. What is the angular velocity after impact?

Solution: Since the force acting on the cube passes through the edge that strikes the tile (0 as shown in Figure 1), this point may be considered a rotational

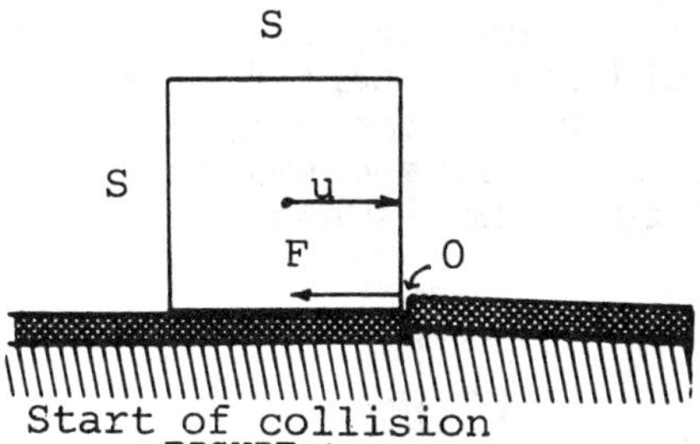

Start of collision
FIGURE 1

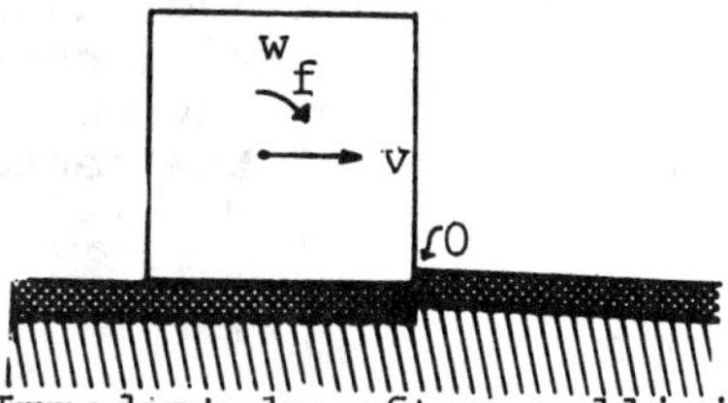

Immediately after collision
FIGURE 2

center and conservation of angular momentum can be applied. The torque, $\vec{F} \times \vec{r}$, is equal to zero about point 0.

Of course, we are assuming that the point 0 does not move in this problem. But in the sense that the collision is impulsive, this is true. An alternate method of solution will treat the impulsive nature of the collision directly.

At any rate, the initial angular momentum relative to the point 0 is simply,

$$(L_0)_i = m\ u\ r = \frac{W}{g}\ u\ \frac{s}{2}.$$

Immediately after the impact, the velocity of the cube has changed impulsively to v, which we take to be in the same direction as u; and rotational velocity ω_f as shown in Figure 2. The moment of inertia about 0, I_0, equals the sum of the centroidal moment, which is $\frac{1}{6}$ of the mass multiplied by the length of the cube squared, the product of the mass, and the square of half the length of the diagonal. Thus,

$$I_0 = \frac{1}{6}\frac{W}{g}\ s^2 + \frac{W}{g}\left(\frac{s}{\sqrt{2}}\right)^2 = \frac{2}{3}\frac{W}{g}\ s^2.$$

The final angular moment after impact, is

$$(L_0)_f = I\ \omega_f + m\ v\ r$$

$$(L_0)_f = \frac{2}{3}\frac{W}{g}\ s^2\ (\omega_f) + \frac{W}{g}\frac{s}{2}\ v.$$

Equating $(L_0)_i$ and $(L_0)_f$ we get

$$\frac{W}{g}\frac{s}{2}\ u = \frac{2}{3}\frac{W}{g}\ \omega_f\ s^2 + \frac{W}{g}\frac{s}{2}\ v. \qquad (1)$$

Since the impact is elastic, e = 1. The resulting velocity, for a collision with an immovable object, (the tile), is

$$v_0 = -\ e\ u = -\ u.$$

However, in the impulsive approximation, we have been using (with linear and angular velocities changing suddenly but the cube not having rotated or translated yet), the linear velocity of any point of the cube is equal to the linear velocity of the center of the cube. Thus,

$$v_0 = v$$

and $v = -u$.

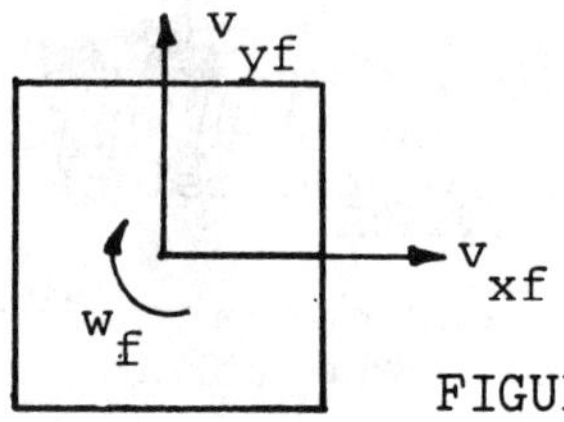

FIGURE 3

This gives for Equation (1),

$$\frac{s}{2} u = \frac{2}{3} s^2 \omega_f - \frac{s}{2} u$$

$$\omega_f = \frac{3}{2} \frac{u}{s}.$$

The positive sign indicates the guess was correct in drawing ω_f clockwise in Figure 2.

The actual motion of the object is shown in Figure 3.

Find v_{xf} and v_{yf} using kinematic relations.

$$v_{0x} = v_{xf} - \frac{s}{2} \omega_f$$

or $$v_{xf} = - u + \frac{s}{2} \left(\frac{3}{2} \frac{u}{s}\right) = - \frac{1}{4} u,$$

and $$v_{oy} = v_{yf} - \frac{s}{2} \omega_f$$

or $$v_{yf} = 0 + \frac{s}{2} \left(\frac{3}{2} \frac{u}{s}\right) = \frac{3}{4} u.$$

Note the minus sign for v_{xf} shows an incorrect guess for its direction in Figure 3. Also, note how only in this last step, was the impulse condition weakened to combine the linear and rotational velocities as they do in real life.

Alternate Solution: If you found all these impulse arguments confusing, this method which uses the impulse approach more rigorously may be clearer, but be forewarned that the arguments are still subtle.

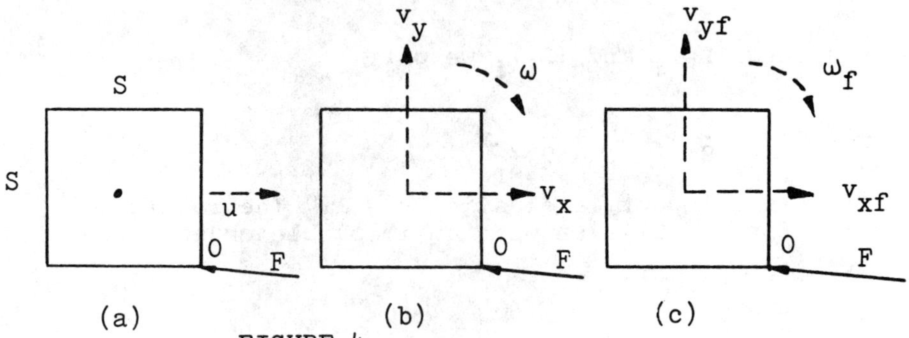

FIGURE 4

Now, divide up the already brief collision time into two segments. The first is the deformation period, during which the gross translation moment of the block is arrested, and only rotation about the point 0 occurs. Also, deformations (shear, twist, compression etc) of the block are at a maximum. The second is the restoration period, during which the block attains its final state of rotational, plus translational, motion. Figure 4(a) shows the cube just at the start of the deformation period. Figure 4(b) shows the cube at the end of the deformation, the velocities shown are those which result due to rotation

$$\left(v_x = \left(\frac{s}{2}\right)\omega,\ v_y = \left(\frac{s}{2}\right)\omega\right)$$

there is no translation velocity shown. Figure 4(c) shows the cube at the end of the restoration period with v_{xf}, v_{yf} and ω_f to be determined independently.

The impact equations for the deformation period, if it is assumed that the impulse is caused by an impulsive force F with components F_x and F_y, are

$$-\int_0^{t_D} F_x dt = \frac{W}{g}(v_x - u) = \frac{W}{g}\left(\frac{s}{2}\omega - u\right)$$

$$+\int_0^{t_D} F_y dt = \frac{W}{g}(v_y - 0) = \frac{W}{g}\frac{s}{2}\omega$$

$$\frac{s}{2}\int_0^{t_D} F_x dt - \frac{s}{2}\int_0^{1D} F_y dt = \frac{1}{6}\frac{W}{g} s^2 (\omega - 0).$$

Note that the moment of inertia of a cube about a centroidal axis parallel to a side is 1/6 ms^2. Substitute the impulse values from the first two equations into the third equation to obtain

$$\frac{s}{2}\left(-\frac{W}{g}\right)\left(\frac{s}{2}\omega - u\right) - \frac{s}{2}\frac{W}{g}\frac{s}{2}\omega = \frac{1}{6}\frac{W}{g}s^2\omega.$$

From this equation,

$$s\omega = 3/4\ u.$$

Hence,

$$v_x = v_y = \frac{s}{2}\omega = \frac{3}{8}u.$$

Also, the impulse in the x direction is

$$\frac{W}{g}(v_x - u) = \frac{W}{g}\left(\frac{3}{8}u - u\right) = -\frac{5W}{8g}u$$

and the impulse in the y direction is

$$\frac{W}{g}\frac{s}{2}\omega = \frac{3}{8}\frac{W}{g}u.$$

The conditions at the end of the restitution period are shown in Figure 4, where the subscripts indicate final velocities. Since the impact is perfectly elastic, e = 1 and the impulses have the same values as during deformation. Here, the coefficient of restitution is being defined as the ratio of the linear impulse during restitution to the linear impulse during deformation. The two definitions of e used in this problem may be shown to be equivalent. The equations for the restitution period are

$$-\frac{5}{8}\frac{W}{g}u = \frac{W}{g}\left(v_{xf} - \frac{3}{8}u\right)$$

$$+\frac{3}{8}\frac{W}{g}u = \frac{W}{g}\left(v_{yf} - \frac{3}{8}u\right)$$

$$\left(\frac{5}{8}\frac{W}{g}u\right)\frac{s}{2} - \left(\frac{3}{8}\frac{W}{g}u\right)\frac{s}{2} = \frac{1}{6}\frac{W}{g}s^2(\omega_f - \omega).$$

The first of these equations yields

$$v_{xf} = -1/4\ u$$

or 1/4 u to the left. The second equation yields

$$v_{yf} = +3/4\ u$$

or 3/4 u upward. Divide the third equation by (W/g)s to obtain

$$\frac{5}{16}u - \frac{3}{16}u = \frac{1}{6}s\omega_f - \frac{1}{6}s\omega.$$

Since $s\omega$ has been found to equal 3/4 u,

$$\frac{5}{16}u - \frac{3}{16}u = \frac{1}{6}s\omega_f - \frac{1}{8}u.$$

Hence, $\omega_f = \frac{3}{2}\frac{u}{s}.$

INELASTIC COLLISIONS

• PROBLEM 13-5

Two similar cars, A and B are connected rigidly together and have a combined mass of 4 kgm. Car C has a mass of 1 kgm. Initially, A and B have a speed of 5 m/sec and C is at rest, as shown in Figure 1.

a) Suppose the collision between A and C is perfectly inelastic. What is the final speed of the system?

b) Suppose the collision between A and C is perfectly elastic, and between B and C is perfectly inelastic.

What is the final speed of the system? Compare with a) and explain.

c) Suppose both collisions A and C, and B and C are perfectly elastic. What are the speeds of C after the first and second collisions?

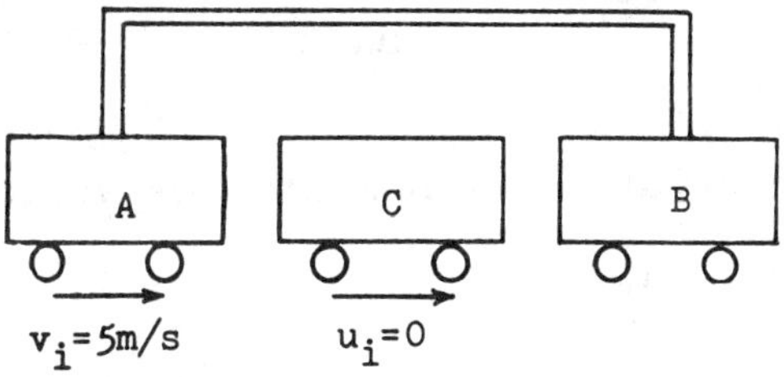

FIGURE 1

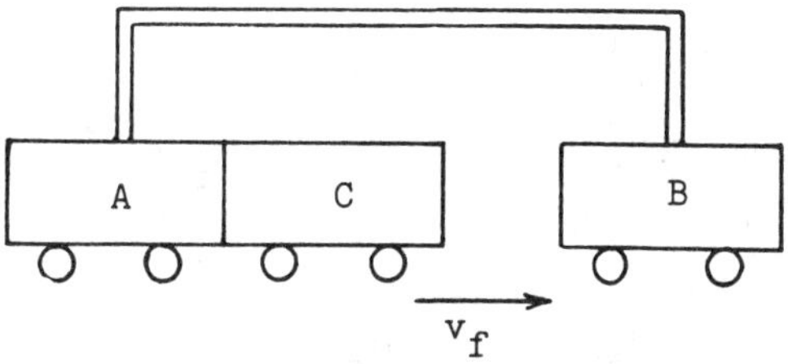

FIGURE 2

Solution: a) A perfectly inelastic collision means that the two colliding bodies stick together and move with the same velocity after the collision, as shown in Figure 2. From the Principle of Conservation of Linear Momentum, we may write

Total Momentum Before Collision = Total Momentum After Collision.

Thus,

$$4\ (v_i) + 0 = (4 + 1)\ (v_f). \tag{1}$$

Solving for the final velocity, v_f:

$$v_f = \frac{4}{5} v_i = \frac{4}{5} (5 \text{ m/sec}) = 4 \text{ m/sec}.$$

b) (1) If the collision between A and C is perfectly elastic, kinetic energy is conserved as well as linear momentum. See Figure 3.

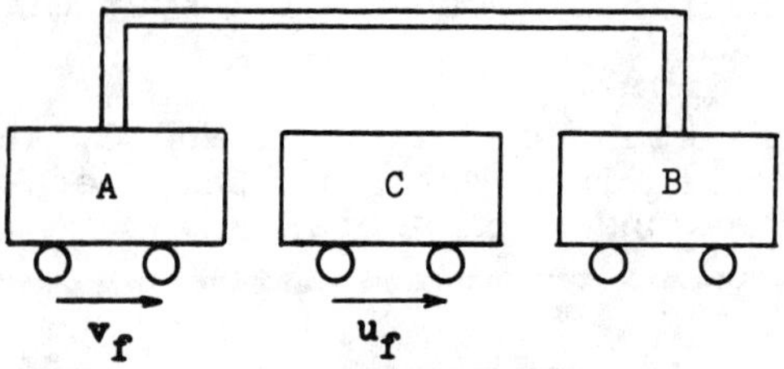

FIGURE 3

From conservation of momentum we write:

$$4v_i = 4v_f + u_f, \tag{2}$$

and from conservation of kinetic energy, we obtain:

$$\frac{1}{2}(4)\ v_i^{\,2} = \frac{1}{2}(4)\ v_f^{\,2} + \frac{1}{2}(1)\ u_f^{\,2}. \tag{3}$$

Thus, from (2), $u_f = 4\ (v_i - v_f)$

and substituting into (3):

$$4v_i^{\,2} = 4v_f^{\,2} + 16\ (v_i - v_f)^2, \tag{4}$$

rearranging,

$$5v_f^{\,2} - (8v_i)\ v_f + 3v_i^{\,2} = 0. \tag{5}$$

The last equation is quadratic in v_f and may be solved by using the quadratic formula, or by factoring. Since it is somewhat easier, let us solve (5) for v_f by factoring. Equ. (5) may be written:

$$(5f_v - 3v_i)\ (v_f - v_i) = 0. \tag{6}$$

Thus, the roots of (6) are:

$$v_{f1} = v_i = 5\ \frac{m}{sec},\ \text{and}\ v_{f2} = \frac{3}{5}\ v_i = 3\ \frac{m}{sec}.$$

Then, again using (2) we find the corresponding values of u_f:

$$u_{f1} = 4\ (v_i - v_{f1}) = 4\ (5 - 5) = 0\ \text{m/sec.}$$

and $u_{f2} = 4\ (5 - 3) = 8$ m/sec.

Since $u_f = 0$ is physically impossible (it would require car A to pass through car C), we must discard the root v_{f1} as extraneous. The final answer is, therefore:

$v_f = 3$ m/sec for cars A and B,

and $u_f = 8$ m/sec for car C.

(2) When car C catches up with car B, there will be a second collision which, in this case, is inelastic. See Figure 4.

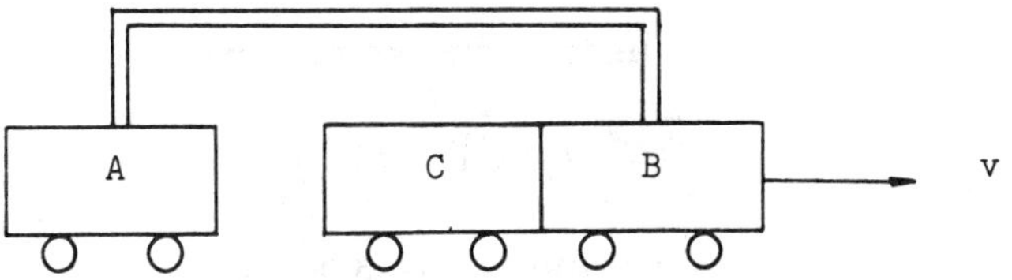

FIGURE 4

From conservation of momentum write:

$$4v_f = u_f = (4 + 1)\ v. \tag{7}$$

Thus, $v = \frac{1}{5}(12 + 8) = 4$ m/sec. is the final velocity of the system. It is seen that the final velocity of the system is the same in part b) as it is in part a), even though there is an additional elastic collision in part b). This result is to be expected because there is no kinetic energy lost from the system during the elastic collision of b), and the kinetic energy lost during the inelastic collisions of parts a) and b) is the same.

c) Using the results of part (1) of b), and the notation shown in Figure 5,

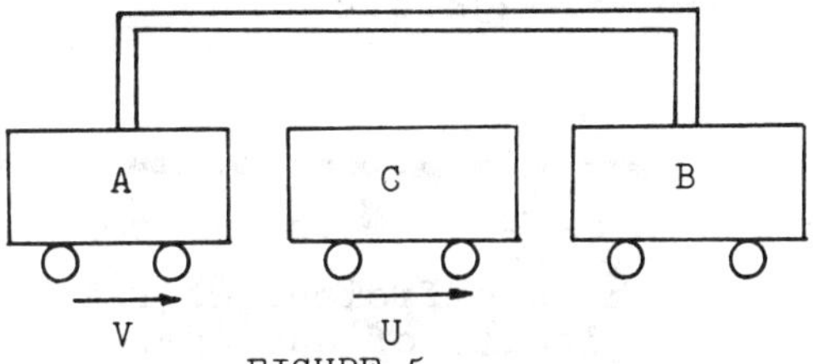

FIGURE 5

We can write for conservation of momentum:

$$4v_f + u_f = 4V + U \tag{8}$$

and for conservation of kinetic energy:

$$\frac{1}{2}(4)v_f^2 + \frac{1}{2}u_f^2 = \frac{1}{2}(4)V^2 + \frac{1}{2}U^2 . \tag{9}$$

Now, rearranging (8) and (9) and inserting numerical values ($v_f = 3$ m/sec, $u_f = 8$ m/sec) yields a pair of simultaneous equations.

$$20 = 4V + U \tag{10}$$

and $$100 = 4V^2 + U^2 . \tag{11}$$

Solving (10) for $U = 20 - 4V$, and substituting into

(11) yields

$$20\,V^2 - 160V + 300 = 0 \tag{12}$$

This may be reduced to

$$V^2 - 8V + 15 = 0 \tag{13}$$

and solved for using the quadratic formula.

$$V = \frac{8 \pm \sqrt{64-60}}{2} = 4 \pm 1$$

or $V_1 = 5$ m/sec. and $V_2 = 3$ m/sec

from which, using (10) again,

$U_1 = 0$ m/sec and $U_2 = 8$ m/sec.

The second set of velocities ($V_2 = 3$ m/sec, $U_2 = 8$ m/sec) is physically impossible and is discarded. The final velocities are therefore

$V = 5$ m/sec for cars A and B,

and $U = 0$ m/sec for car C,

so that after two elastic collisions, the system is back to its initial conditions, except that the cars have been displaced in the direction of motion of A and B. If allowed to continue, the whole process will be repeated again and again.

• PROBLEM 13-6

Two particles with masses m_1 and m_2 collide as shown in Figure 1. If the coefficient of restitution is 0.5, what is the velocity of each particle after the collision?

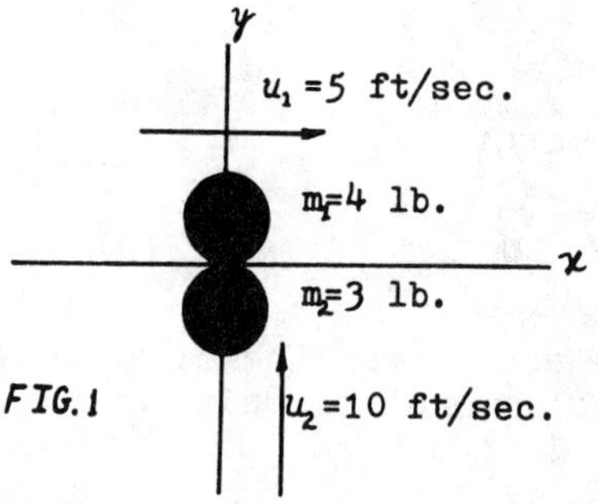

FIG. 1

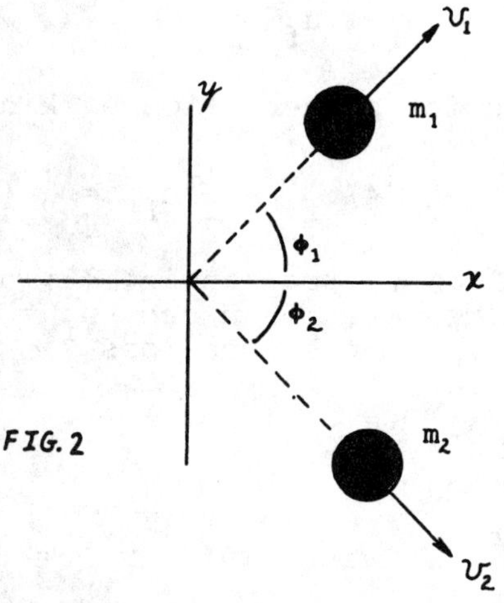

FIG. 2

Solution: Assuming the particles are smooth (frictionless), the only forces acting during the impact, as shown in Figure 1, act in the y-direction, since the line of impact is along the y-axis, Because there are no forces in the x-direction, the x-components of momentum, and therefore the x-components of velocity remain unchanged after the impact. Using the notation shown in Figure 2:

$$v_{1x} = u_1 = 5 \text{ ft/sec.}$$

and $v_{2x} = 0$

In the y-direction, applying the principle of conservation of linear momentum we have:

$$m_2 u_2 = m_1 v_{1y} - m_2 v_{2y} \tag{1}$$

Also, using the definition of the coefficient of restitution:

$$e = \frac{\text{relative velocity of separation}}{\text{relative velocity of approach}},$$

yields, for the y-component of motion:

$$e = 0.5 = \frac{v_{1y} + v_{2y}}{u_2} \tag{2}$$

Thus, from (1) and (2) we have, upon inserting numerical values:

$$\left(\frac{3 \text{ lb.}}{32 \text{ ft/sec}^2}\right) (10 \text{ ft/sec})$$

$$= \left(\frac{4 \text{ lb.}}{32 \text{ ft/sec}^2}\right) v_{1y} - \left(\frac{3 \text{ lb.}}{32 \text{ ft/sec}^2}\right) v_{2y} \tag{3}$$

(Note that the acceleration due to gravity, which converts the weight of the particle into mass, appears in every term and can be cancelled out. In general, one can use either mass or weight in the momentum equation.)

and $(.5)\ (10 \text{ ft/sec}) = v_{1y} + v_{2y}$, **from (2).** (4)

Solving (3) and (4) simultaneously, the y-components of velocity are found to be:

$v_{1y} = 6.43$ ft/sec (traveling upward)

and $v_{2y} = -1.43$ ft/sec (traveling upward)

Since v_{2y} was assumed downward, as shown in Figure 2, the negative value for v_{2y} implies that m_2 is actually traveling upward after the impact.

We can now find the magnitude of the final velocity of m_1:

$$v_1 = \sqrt{(v_{1x})^2 + (v_{1y})^2}$$

$$= \sqrt{(5)^2 + (6.43)^2} = 8.15 \text{ ft./sec.}$$

The direction, ϕ_1, of m_1 is:

$$\phi_1 = \tan^{-1}\left(\frac{v_{1y}}{v_{1x}}\right) = \tan^{-1}\left(\frac{6.43}{5}\right) = 52° \ .$$

Of course, the final velocity of m_2 was seen before to be in the upward y-direction at 1.43 ft/sec.

• PROBLEM 13-7

Two balls hang down from massless cords so that their centers are at equal heights and they just touch at the point on a line between their centers. Ball A has mass 2m, ball B has mass m. Ball A is raised so that its string makes an angle of 35° with the vertical. It is released from rest, hits ball B making B swing up to a maximum position in which its angle is 35° from the vertical. Determine the coefficient of restitution e.

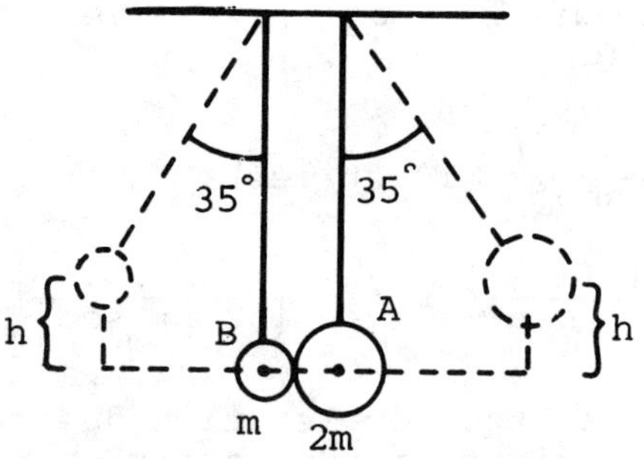

Fig. 1

Solution: Let h be the vertical distance through which each ball moves. The kinetic energy of A at its lowest position, just before collision, must equal its potential energy at the starting position:

$$\frac{1}{2} M_A V_A^2 = M_A gh, \qquad (1)$$

where M_A = 2m, the mass of ball A.

The maximum potential energy of B must equal its kinetic energy just after collision,

$$mgh = \frac{1}{2} m V_B'^2 \qquad (2)$$

A prime on V indicates velocity after collision, subscripts indicate the body.

Since ball B rises to the same angle as ball A was dropped from h in equation (1) equals h is equation (2). Further, the masses cancel in each equation, thus, the right side of Equation (1) equals the left side of Equation (2);

$$V_A = V_B'. \qquad (3)$$

During collision, energy is lost in the form of heat, sound, and possibly sparks; so kinetic energy is not conserved in the collision. A measure of this lost energy is given by comparing the relative velocities of the spheres before and after collision. This is done by defining the coefficient of restitution as:

$$e = \frac{V_B' - V_A'}{V_A - V_B}. \qquad (4)$$

Since there is no horizontal component of tension in the strings during collision, it follows that linear momentum is conserved in the horizontal direction.

Linear momentum before collision equals the linear momentum after collision. That is,

$$M_A V_A + 0 = M_A V_A' + M_B V_B'. \qquad (5)$$

Substitute $M_A = 2m$, $M_B = m$, and use Equation (3) to get:

$$V_A' = V_B'/2. \qquad (6)$$

When Equations (3) and (6) are substituted into Equation (4), the result is:

$$e = \frac{V_B' - V_B'/2}{V_B' - 0} = \frac{1}{2}.$$

• PROBLEM 13-8

As shown in Figure 1, a 2-lb sphere moving to the right with a velocity of 6 ft/sec strikes a 6-lb slender bar which is 5 ft long and is moving with a mass-center velocity of 2 ft/sec to the left. The bar has no initial angular velocity. The coefficient of restitution e is 0.4, and the objects are on a smooth horizontal plane. Determine the velocity of the sphere, the mass-center velocity of the bar, and the angular velocity of the bar immediately after impact.

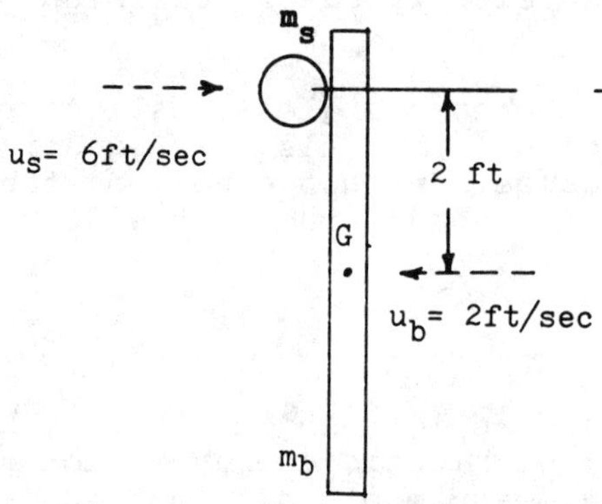

FIG. 1

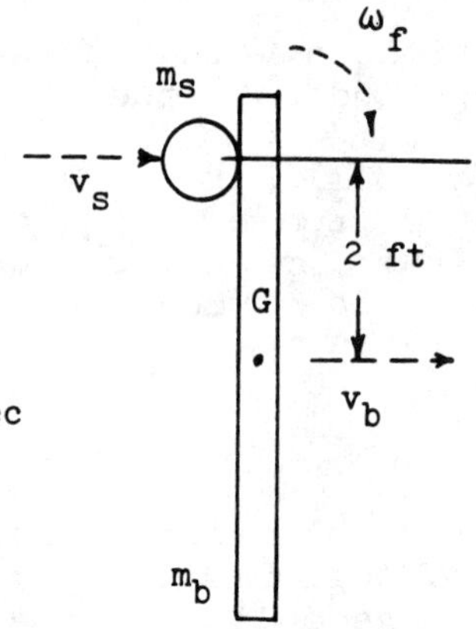

FIG. 2

Solution: The final conditions are shown in Figure 2, where the three unknowns are the linear velocity v_s of the sphere, the linear velocity v_b of the mass center of the bar, and the angular velocity ω_f of the bar.

The conservation of linear momentum is represented by the equation

$$m_s u_s + m_b u_b = m_s v_s + m_b v_b .$$

If directions to the right are called positive, this equation becomes

$$\frac{2}{g} (+ 6) + \frac{6}{g} (- 2) = \frac{2}{g} v_s + \frac{6}{g} v_b$$

or $\quad v_s + 3 v_b = 0.$ (a)

Considering the system of bar and sphere it is clear that there are no external impulsive forces acting on the system. Hence, the angular momentum about any point is conserved. The equation for conservation of angular momentum about the mass center G is, if clockwise rotations are called positive,

$$m_s u_s r + 0 = m_s v_s r + \frac{1}{12} m_b \ell^2 \omega_f$$

where r is the distance from the sphere to point G, and ℓ is the length of the bar. Note that $1/12\ m_b \ell^2$ is the moment of inertia of the bar about point G.

Substituting in numerical values yields:

$$\frac{2}{g} (+ 6) (2) + 0 = \frac{2}{g} v_s (2) + \frac{1}{12} \frac{6}{g} (5)^2 \omega_f$$

or $4v_s + 12.5\,\omega_f = 24$ (b)

A third equation, which is based on the definition of the coefficient of restitution, will now be found. The instantaneous final velocity of the point of the bar where the ball strikes is $v_b + 2\omega_f$. Its initial velocity is -2, because the bar is initially translating to the left. Since e = 0.4, the equation becomes

$$v_b + 2\omega_f - v_s = -\,0.4[-2 - (+\,6)]$$

or $v_b + 2\omega_f - v_s = 3.2.$ (c)

Substitute $v_s = {}^{-}3v_b$ from Eq. (a) in Eqs. (b) and (c) to obtain

$$4(-\,3v_b) + 12.5\omega_f = 24$$

and $v_b + 2\ _f - (-\,3v_b) = 3.2.$

Solving these two equations simultaneously, we find that $v_s = +0.324$, or 0.324 ft/sec to the right, and $\omega_f = +\,1.815$, or 1.815 rad/sec clockwise. Of course, $v_b = -v_s/3 = -\,0.108$, or 0.108 ft/sec to the left.

• PROBLEM 13-9

A sphere with mass m, moving with velocity u downward to the right at an angle θ with the horizontal strikes an immovable horizontal plane. What will be its velocity v after impact, if the plane is smooth and the coefficient of restitution is e?

Solution: Linear momentum is conserved in the horizontal direction because there is no component of force acting on the mass horizontally, given that the surface is smooth (frictionless). Thus,

$$mu \cos\theta = mv \cos\theta'. \qquad (1)$$

In the vertical direction, the coefficient of restitution may be expressed as

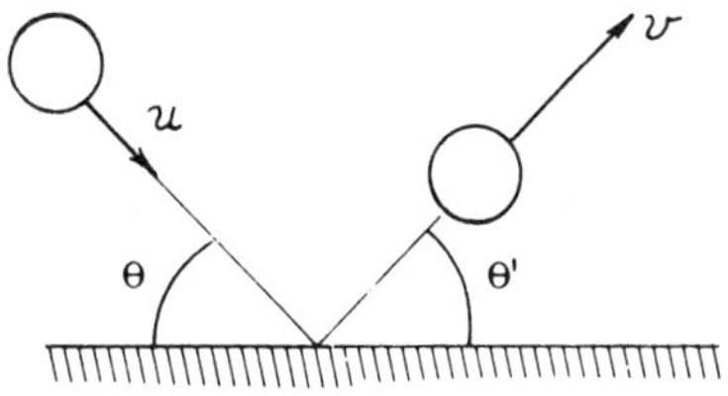

$$e = \frac{\text{velocity of separation}}{\text{velocity of approach}}.$$

Thus, in the vertical direction:

$$e = \frac{v \sin\theta'}{u \sin\theta}. \qquad (2)$$

From (1) we can solve for cos θ':

$$\cos\theta' = \frac{u}{v}\cos\theta. \qquad (3)$$

Using,

$$\sin\theta' = \sqrt{1 - \cos^2\theta'} \tag{4}$$

then, $$\sin\theta' = \sqrt{1 - \left(\frac{u}{v}\right)^2 \cos^2\theta}. \tag{5}$$

Substituting (5) into (2),

$$eu\sin\theta = v\sin\theta' = v\sqrt{1 - \left(\frac{u}{v}\right)^2 \cos^2\theta}, \tag{6}$$

rearranging,

$$e\left(\frac{u}{v}\right)\sin\theta = \sqrt{1 - \left(\frac{u}{v}\right)^2 \cos^2\theta}.$$

Squaring,

$$e^2\left(\frac{u}{v}\right)^2 \sin^2\theta = 1 - \left(\frac{u}{v}\right)^2 \cos^2\theta.$$

Collecting terms,

$$\left(\frac{u}{v}\right)^2 (e^2\sin^2\theta + \cos^2\theta) = 1.$$

Solving for v,

$$v = u\sqrt{e^2\sin^2\theta + \cos^2\theta}. \tag{7}$$

• PROBLEM 13-10

A mass m strikes a smooth flat surface. If the coefficient of restitution is e, determine the angle of rebound, ϕ, and the change in kinetic energy, ΔT, in terms of the initial velocity u and the angle of incidence, θ.

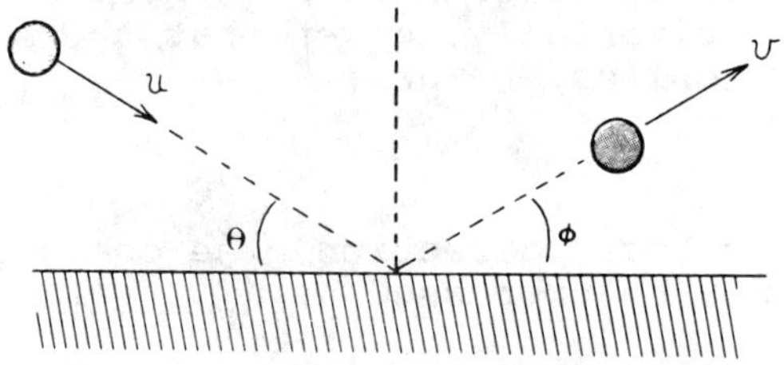

Solution: Since the surface is smooth, or frictionless, there will be no component of force acting on the mass in the horizontal direction during the impact. Therefore, the horizontal component of linear momentum must remain constant. Thus, applying the principle of conservation of momentum to the horizontal components we may write:

$$mu\cos\theta = mv\cos\phi. \tag{1}$$

In the vertical direction, the definition of the coefficient of restitution becomes:

$$e = \frac{\text{velocity of separation}}{\text{velocity of approach}} = \frac{v\sin\phi}{u\sin\theta}. \tag{2}$$

Rewriting (2) and (1),

$$e u \sin \theta = v \sin \phi \quad (3)$$
and
$$u \cos \theta = v \cos \phi \quad (4)$$

and dividing (3) by (4), yields:

$$e \tan \theta = \tan \phi. \quad (5)$$

from (5) it may be seen that, for $0 < e < 1$, the angle of rebound ϕ must be less than the angle of incidence. Only if the collision is perfectly elastic ($e = 1$) will the angle of rebound equal the angle of incidence.

To find the loss of kinetic energy, first find the velocity after the collision. From (1) we have:

$$v = u \frac{\cos \theta}{\cos \phi}. \quad (6)$$

Note also the trigonometric identity:

$$\sec^2 \phi = \frac{1}{\cos^2 \phi} = 1 + \tan^2 \phi$$

or
$$\frac{1}{\cos \phi} = \sqrt{1 + \tan^2 \phi}. \quad (7)$$

Substituting (7) into (6):

$$v = u \cos \theta \sqrt{1 + \tan^2 \phi}. \quad (8)$$

Now, the initial kinetic energy, T_i, is:

$$T_i = \frac{1}{2} mu^2 \quad (9)$$

and the kinetic energy after the collision, T_f, is:

$$T_f = \frac{1}{2} mv^2 = \frac{1}{2} mu^2 \cos^2 \theta \left(1 + \tan^2 \phi\right). \quad (10)$$

The change in kinetic energy is, therefore:

$$\Delta T = T_f - T_i = \frac{1}{2} mu^2 [\cos^2 \theta (1 + \tan^2 \phi) - 1]. \quad (11)$$

From (5) we have:

$$\tan^2 \phi = e^2 \tan^2 \theta \quad (12)$$

so that:

$$\Delta T = \frac{1}{2} mu^2 [\cos^2 \theta + e^2 \cos^2 \theta \tan^2 \theta - 1]. \quad (13)$$

Again, using $\tan \theta = \sin \theta / \cos \theta$ and $\cos^2 \theta = 1 - \sin^2 \theta$, (13) becomes:

$$\Delta T = \frac{1}{2} mu^2 [1 - \sin^2 \theta + e^2 \sin^2 \theta - 1] \quad (14)$$

which reduces to:

$$\Delta T = -\frac{1}{2} mu^2 \sin^2 \theta (1 - e^2). \quad (15)$$

Since $e \leq 1$, ΔT will be zero or negative. For a perfectly elastic collision ($e = 1$) $\Delta T = 0$ and there will be no loss of kinetic energy. For any inelastic collision ($e < 1$) $\Delta T < 0$ and the kinetic energy will decrease.

• PROBLEM 13-11

A billiard shot is made whereby the cue ball, C, traveling parallel to the right-hand cushion strikes one object ball, B, and then rebounds off the right-hand cushion to strike a second ball, A, in direct central impact. The impact between C and B is fully elastic. The coefficient of restitution for the rebound off the cushion is 0.80. Ignoring spin and friction, find the distance, a, as a function of the radius R of each which will permit the shot to be made. Both balls have equal mass.

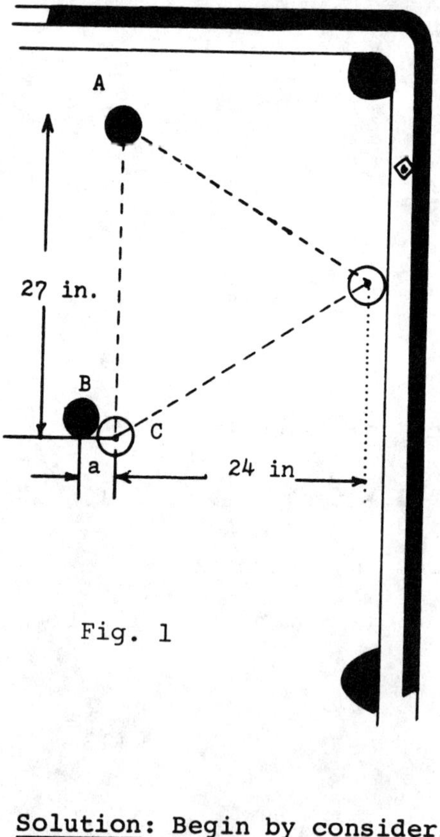

Fig. 1

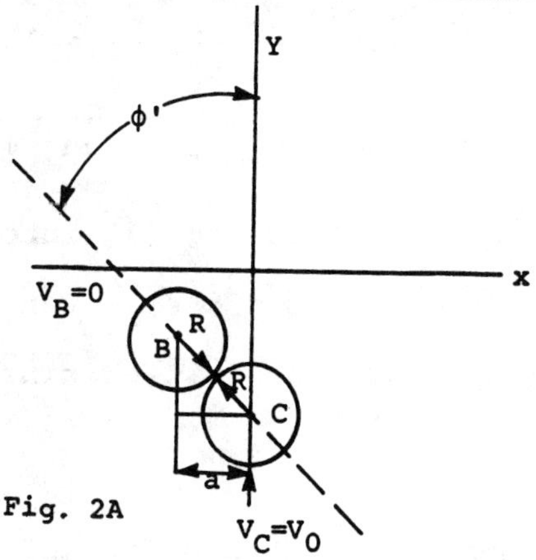

Fig. 2A

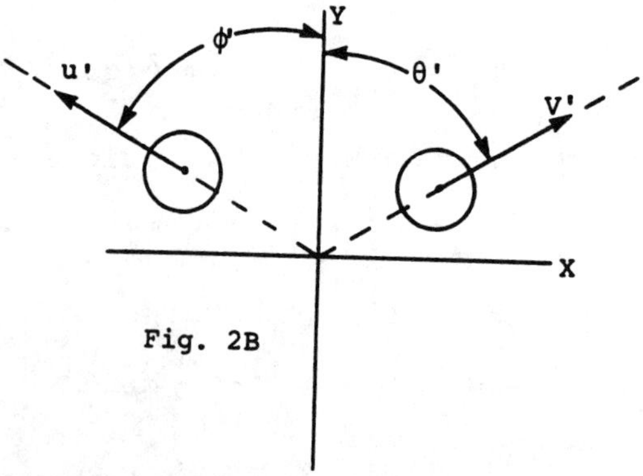

Fig. 2B

Solution: Begin by considering the collision between the cue ball C and the object ball B. Velocities and directions are defined by figure 2a, which shows the balls C and B just before the collision, and figure 2b after the collision.

Applying the principle of conservation of linear momentum to the collision process, we get for the x-component of motion:

$$0 = - u' \sin \phi' + v' \sin \theta' \qquad (1)$$

and for the y-component:

$$v_0 = u' \cos \phi' + v' \cos \theta'. \qquad (2)$$

Also, since the impact between C and B is fully elastic, kinetic energy is conserved and we can write:

$$v_0^2 = u'^2 + v'^2 \qquad (3)$$

In equations (1), (2), and (3), the mass has already been cancelled out, since it is assumed that all of the billiard balls have identical masses.

Since the impact between C and B is frictionless, the only force acting on B is along the line of impact, from center of C to center of B. Therefore the change in the momentum of B occurs along the line of impact, and the velocity u after the impact must be in the same direction as the line of impact. Thus, from the triangle in Figure 1a,

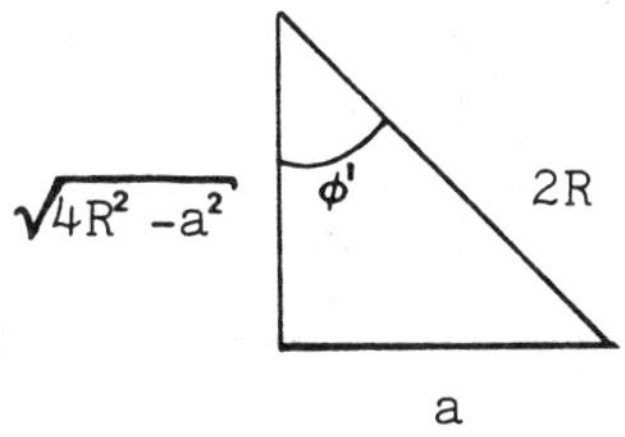

Fig. 1A

we have, $\sin \phi' = a/2R$ and $\cos \phi' = \dfrac{\sqrt{4R^2 - a^2}}{2R}$ (4)

Now, solve Eqs. (1), (2), (3) and (4) for θ' in terms of a and R.

Squaring (2),

$$v_0^2 = v'^2 \cos^2 \theta' + u'^2 \cos^2 \phi + 2v' u' \cos \theta' \cos \phi'. \qquad (5)$$

Since $\cos^2 \theta' = 1 - \sin^2 \theta'$, (5) becomes (6)

$$v_0^2 = v'^2 - v'^2 \sin^2 \theta' + u'^2 \cos^2 \phi'$$

$$+ 2v' u' \cos \theta' \cos \phi'. \qquad (7)$$

Next, squaring (1), $u'^2 \sin^2 \phi' = v'^2 \sin^2 \theta'$, (8)

and from (3) we have $v_0^2 - v'^2 = u'^2$. (9)

Combining (7), (8) and (9) yields

$$u'^2 = - u'^2 \sin^2 \phi' + u'^2 \cos^2 \phi'$$

$$+ 2u'^2 \frac{\cos \theta'}{\sin \theta'} \sin \phi' \cos \phi' \qquad (10)$$

and, cancelling the u'^2,

$$1 = -\sin^2 \phi' + \cos^2 \phi' + 2\,\frac{\sin \phi' \cos \phi'}{\tan \theta'}. \qquad (10)$$

Using (6) again,

$$1 = -\sin^2 \phi' + 1 - \sin^2 \phi' + 2\,\frac{\sin \phi' \cos \phi'}{\tan \phi'}, \qquad (11)$$

or $$\tan \theta' = \frac{1}{\tan \phi'} = \frac{a}{\sqrt{4R^2 - a^2}}. \qquad (12)$$

Next, consider the collision of the cue ball C with the cushion. Refer to Figure 3 for notation.

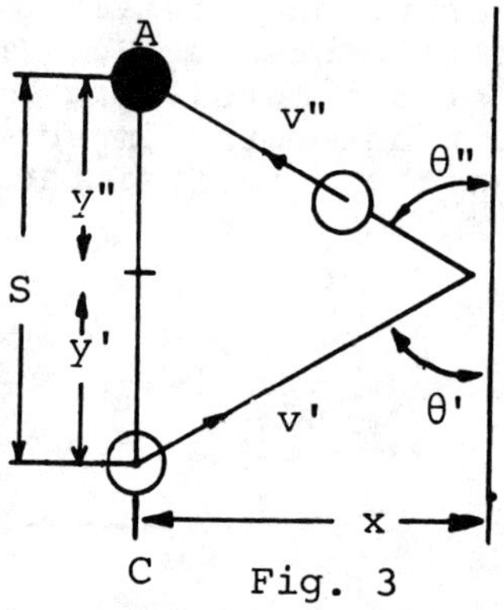

Fig. 3

Since the collision with the cushion is assumed frictionless, the linear momentum in the y-direction is constant. Thus,

$$v' \cos \theta' = v'' \cos \theta''. \qquad (13)$$

In the x-direction, apply the definition of the coefficient of restitution:

$$e = 0.80 = \frac{v'' \sin \theta''}{v' \sin \theta'} \qquad (14)$$

or $$0.8\, v' \sin \theta' = v'' \cos \theta''. \qquad (15)$$

Dividing (15) by (13),

$$0.8 \tan \theta' = \tan \theta''. \qquad (16)$$

From Figure 3,

$$\tan \theta' = \frac{x}{y'} \text{ and } \tan \theta'' = \frac{x}{y''} \qquad (17)$$

also, $$y' + y'' = S \qquad (18)$$

Combining (16) and (17),

$$(0.8)\ \frac{x}{y'} = \frac{x}{y''} \tag{19}$$

or $$0.8y'' = y'. \tag{20}$$

Substituting (20) into (18),

$$0.8y'' + y'' = S \tag{21}$$

or $$y'' = \frac{S}{1.8} \tag{22}$$

and $$y' = \frac{0.8}{1.8}\,S = \frac{4}{9}\,S. \tag{23}$$

Finally, from (17) and (12):

$$\tan\theta' = \frac{x}{y'} = \frac{9x}{4S} = \frac{a}{\sqrt{4R^2 - a^2}} \tag{24}$$

or $$\frac{9x}{4S}\sqrt{4R^2 - a^2} = a.$$

Squaring,

$$a^2 = (4R^2 - a^2)\ \frac{81\ x^2}{16\ S^2} \tag{25}$$

or $$a = \sqrt{\frac{4R^2\left[\dfrac{81\ x^2}{16\ S^2}\right]}{1 + \dfrac{81\ x^2}{16\ S^2}}} \tag{26}$$

Letting x = 24 inches and S = 27 inches:

$$a = \frac{4R}{\sqrt{5}} = 1.79\ R \tag{27}$$

• **PROBLEM 13-12**

The conditions prior to an oblique central impact are illustrated in Figure 1. If the line of impact is parallel with the x-axis, determine the final velocity of each mass and the loss of kinetic energy for a) e = 0, b) e = 0.5, and c) e = 1.

Solution: Figure 2 shows the positions of the masses at impact, with the line of impact (the line joining the centers of the two masses) parallel to the x-axis.

Figure 3 shows the conditions just after the impact.

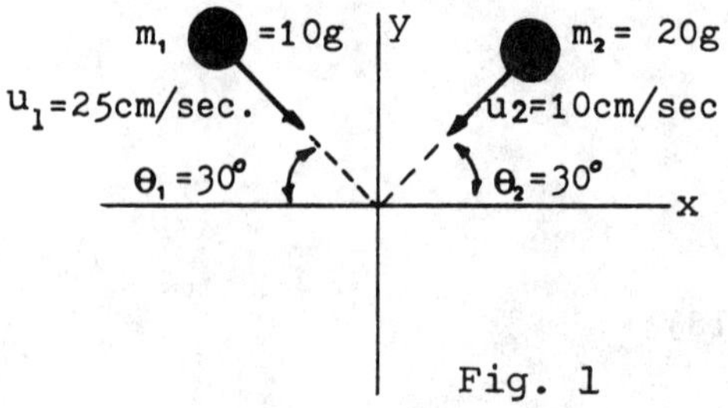

Fig. 1

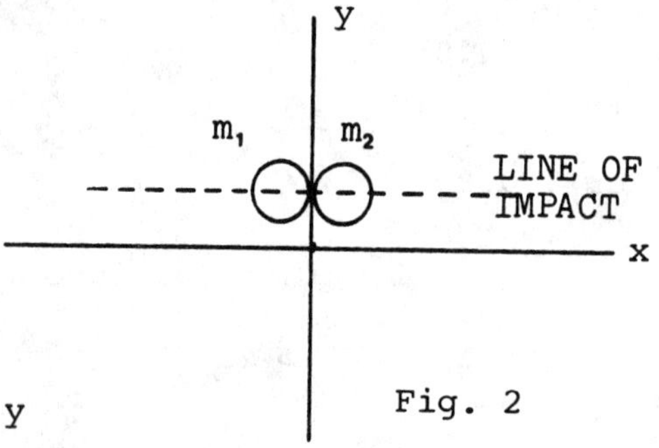

Fig. 2

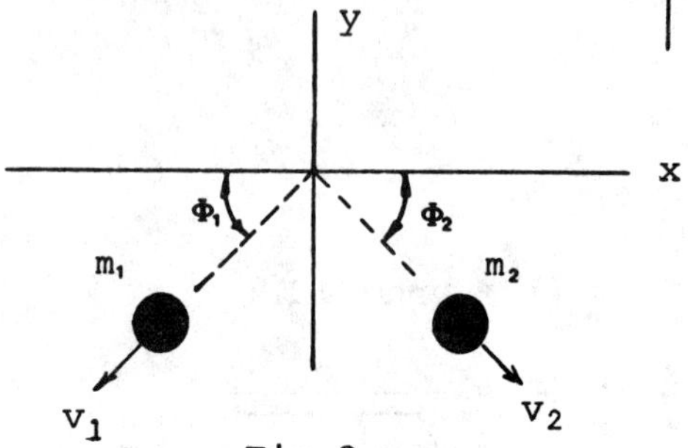

Fig.3

If it is assumed that the masses are smooth (frictionless), then during impact, the only force acting on each mass will be parallel with the x-axis. There will be no forces in the y-direction. This means that the momentum of each mass in the y-direction does not change, hence the y-components of velocity do not change during the impact. Thus, for all cases:

$$u_{1y} = -u_1 \sin \theta_1 = -(25)(\sin 30°) = -12.5 \text{ cm/sec}$$

and $$u_{2y} = -u_2 \sin \theta_2 = -(10)(\sin 30°) = -5 \text{ cm/sec.}$$

Case I. $e = 0$

Since the coefficient of restitution is zero, this requires that the velocity of separation in the x-direction must also be zero. The velocity of separation can be written:

$$v_{2x} - v_{1x} = 0$$

or $$v_{2x} = v_{1x}$$

and the two masses therefore have the same x-component of velocity.

Applying the principle of conservation of momentum to the motion in the x-direction:

$$m_1 (u_1 \cos \theta_1) - m_2 (u_2 \cos \theta_2) = (m_1 + m_2) v_x.$$

Substituting the numerical data given:

$$V_x = \frac{(10g)(25\ cm/sec)(\cos 30°) - (20g)(10\ cm/sec)(\cos 30°)}{(10 + 20g)}$$

or $\quad V_x = 1.44$ cm/sec.

The result is positive, therefore both masses are moving toward the right after the collision. Using the vertical components of velocity, $u_{1y} = -12.5$ cm/sec and $u_{2y} = -5$ cm/sec, we can now find the magnitudes and directions of the velocities of the two masses after impact. For the magnitudes of the velocities:

$$v_1 = \sqrt{V_x^2 + u_{1y}^2} = \sqrt{(1.44)^2 + (-12.5)^2} = 12.58 \text{ cm/sec}$$

and $\quad v_2 = \sqrt{(1.44)^2 + (-5)^2} = 5.20$ cm/sec.

To find the direction of travel of m_1 after the impact, note Figure 4:

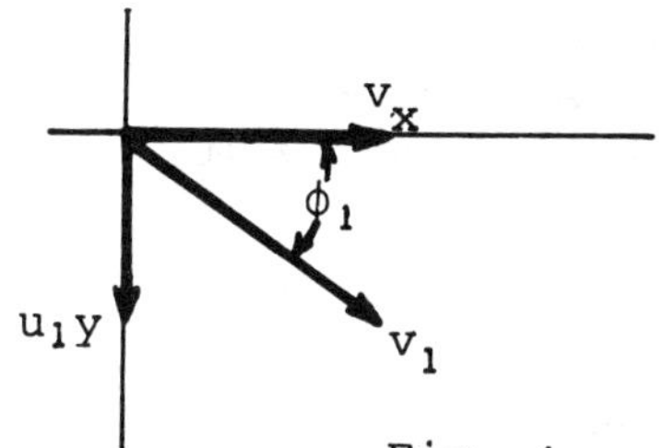

Fig. 4

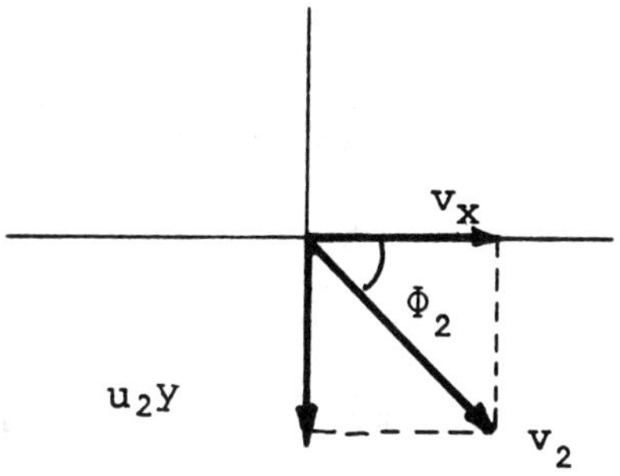

Fig. 5

Thus,

$$\phi_1 = \tan^{-1}\left(\frac{u_{1y}}{v_{1x}}\right)$$

or $$\phi_1 = \tan^{-1}\left(\frac{12.5}{1.44}\right) = 83.4°.$$

Also, from Figure 5 for m_2:

$$\phi_2 = \tan^{-1}\left(\frac{u_{2y}}{v_{2x}}\right)$$

or $$\phi_2 = \tan^{-1}\left(\frac{5.0}{1.44}\right) = 73.9°.$$

To find the loss of kinetic energy, first find the initial kinetic energy:

$$T_i = \frac{1}{2} m_1 u_1^2 + \frac{1}{2} m_2 u_2^2$$

or $$T_i = \frac{1}{2}(10g)(25\ cm/sec)^2 + \frac{1}{2}(20g)(10\ cm/sec)^2$$

$$T_i = 4125 \text{ ergs}.$$

Next, find the final kinetic energy:

$$T_f = \frac{1}{2} m_1 v_1^2 + \frac{1}{2} m_2 v_2^2$$

or $$T_f = \frac{1}{2}(10g)(12.58 \text{ cm/sec})^2 + \frac{1}{2}(20g)(5.20 \text{ cm/sec})^2$$

$$T_f = 1062 \text{ ergs.}$$

The loss of kinetic energy is, therefore,

$$\Delta T = T_i - T_f = 4125 - 1062 = 3063 \text{ ergs.}$$

The percent loss of energy is:

$$\frac{\Delta T}{T_i} = \frac{3063}{4125} \times 100\% = 74\%.$$

Case II. $e = 0.5$

The y-components of velocity again remain unchanged after the impact, as in Case I. In the x-direction, we have from the principle of conservation of momentum (see Figures 1 and 3 for notation):

$$m_1 u_1 \cos\theta_1 - m_2 u_2 \cos\theta_2 = -m_1 v_{1x} + m_2 v_{2x}.$$

Inserting numerical data, this becomes:

$$(10g)(25 \text{ cm/sec})(\cos 30°) - (20g)(10 \text{ cm/sec})(\cos 30°)$$

$$= -10\, v_{1x} + 20\, v_{2x}$$

or $$-10\, v_{1x} - 20\, v_{2x} = 43.3. \qquad (1)$$

Also, from the definition of the coefficient of restitution,

$$e = \frac{\text{relative velocity of separation}}{\text{relative velocity of approach}} = \frac{v_{2x} + v_{1x}}{u_{1x} + u_{2x}}$$

and inserting numerical data for the x-component of motion:

$$0.5 = \frac{v_{2x} + v_{1x}}{(25 \cos 30°) + (10 \cos 30°)}$$

or $$v_{1x} + v_{2x} = 15.2. \qquad (2)$$

Solving equations (1) and (2) simultaneously for v_{1x} and v_{2x} yields:

$$v_{1x} = 8.7 \text{ cm/sec (toward the left)}$$

$$v_{2x} = 6.5 \text{ cm/sec (toward the right).}$$

The magnitudes of the velocities after the impact are:

$$v_1 = \sqrt{v_{1x}^2 + u_{1y}^2} = \sqrt{(8.7)^2 + (12.5)^2} = 15.2 \text{ cm/sec}$$

and: $v_2 = \sqrt{(6.5)^2 + (5.0)^2} = 8.20$ cm/sec.

The directions are found as before:

$$\phi_1 = \tan^{-1}\left(\frac{u_{1y}}{v_{1x}}\right) = \tan^{-1}\left(\frac{12.5}{8.7}\right) = 55.2°$$

and $$\phi_2 = \tan^{-1}\left(\frac{u_{2y}}{v_{2x}}\right) = \tan^{-1}\left(\frac{5.0}{6.5}\right) = 37.6°.$$

The final kinetic energy is:

$$T_f = \frac{1}{2} m_1 v_1^2 + \frac{1}{2} m_2 v_2^2$$

$$T_f = \frac{1}{2} (10g)(15.2 \text{ cm/sec})^2 + \frac{1}{2} (20g)(8.20 \text{ cm/sec})^2$$

$$T_f = 1828 \text{ ergs}$$

The loss of kinetic energy is:

$$\Delta T = T_i - T_f = 4125 - 1828 = 2297 \text{ ergs.}$$

or $$\frac{\Delta T}{T_i} = \frac{2297}{4125} \times 100\% = 56\%.$$

Case III. $e = 1$

The y-components of velocity are again unchanged after the collision, as in Case I. In the x-direction for this case there will be conservation of both momentum and kinetic energy. From conservation of momentum we can write (see Figure 3):

$$m_1 u_1 \cos\theta_1 - m_2 u_2 \cos\theta_2 = - m_1 v_{1x} + m_2 v_{2x}.$$

Inserting numerical values yields:

$$-10v_{1x} + 20v_{2x} = 43.3. \qquad (3)$$

From the coefficient of restitution:

$$1.0 = \frac{v_{2x} + v_{1x}}{(25 \cos 30°) + (10 \cos 30°)}$$

or $$v_{1x} + v_{2x} = 30.3. \qquad (4)$$

Solving equations (3) and (4) simultaneously for v_{1x} and v_{2x} gives:

$$v_{1x} = 18.8 \text{ cm/sec} \quad \text{(toward the left)}$$

and $$v_{2x} = 11.5 \text{ cm/sec} \quad \text{(toward the right)}.$$

The velocities of the masses after the collision become:

$$v_1 = \sqrt{(v_{1x})^2 + (u_{1y})^2} = \sqrt{(18.8)^2 + (12.5)^2} = 22.6 \text{ cm/sec}$$

and $$v_2 = \sqrt{(v_{2x})^2 + (u_{2y})^2} = \sqrt{(11.5)^2 + (5.0)^2}$$

$$= 12.54 \text{ cm/sec}$$

The directions are found as before:

$$\phi_1 = \tan^{-1}\left(\frac{u_{1y}}{v_{1x}}\right) = \tan^{-1}\left(\frac{12.5}{18.8}\right) = 34°$$

and $$\phi_2 = \tan^{-1}\left(\frac{u_{2y}}{v_{2x}}\right) = \tan^{-1}\left(\frac{5.0}{11.5}\right) = 23.5°$$

The final kinetic energy is:

$$T_f = \frac{1}{2} m_1 v_1^2 + \frac{1}{2} m_2 v_2^2$$

$$T_f = \frac{1}{2}(10g)(22.6 \text{ cm/sec})^2 + \frac{1}{2}(20g)(12.54 \text{ cm/sec})^2$$

$$T_f = 4126 \text{ ergs}.$$

Thus, it can be seen that the final energy is the same as the initial energy, except for a small error due to rounding off numerical values.

CHAPTER 14

VARIABLE MASS SYSTEMS

• **PROBLEM** 14-1

Suppose that a two-stage rocket starts from rest with a mass m_1. At burn-out of the first stage engine, the mass is m_2. If the exhaust velocity is v_0, the rocket velocity after the first stage engines quit is

$$v = v_0 \ln \left(\frac{m_1}{m_2}\right) .$$

Before the second stage engines are ignited, part of the mass m_2 is discarded - the mass of the first stage engines. The mass of the second stage is m_A when the engines start and m_B when the engines shut down. What is the terminal velocity of the second stage?

Solution: The velocity gained by firing the second stage follows a similar equation to that of the equation for the first stage, with suitable entries for initial and final mass,

$$\text{Velocity gained is} \quad v_0 \ln \left(\frac{m_A}{m_B}\right) .$$

This velocity is added onto the terminal velocity of the first stage, giving a total velocity

$$v = v_0 \left[\ln \left(\frac{m_1}{m_2}\right) + \ln \left(\frac{m_A}{m_B}\right) \right] .$$

Adding logarithms is equivalent to taking the logarithms of the product, so

$$v = v_0 \ln \left(\frac{m_1}{m_2} \frac{m_A}{m_B}\right)$$

• **PROBLEM** 14-2

A train is traveling at a constant speed of 60 mph and is picking up water from a trough between the rails. If it picks up 15,000 lb of water in 400 yd, determine the exact force and horsepower required.

Solution: The speed of the engine is 60 mph or 88 ft/sec. It goes 400 yd in (400 × 3)/88 sec so that the time rate of picking up water is (15,000 × 88)/(400 × 3) = 1100 lbw/sec. The force F which the engine must exert to pick up this water is given by Newton's second law as

$$F = \frac{d(mv)}{dt} = v\frac{dm}{dt} = \frac{v}{g}\frac{dm}{dt} = \frac{88}{32} \times 1100 = 3025 \text{ lbw}$$

since the velocity v is constant. The additional horsepower which the engine must develop to maintain its constant speed and pick up the water is equal to the force exerted times the velocity. The conversion factor from ft-lb/sec to horsepower is 550 ft-lb/sec = 1 hp.

The additional power required is then

$$\frac{Fv}{550} = \frac{3025 \times 88}{550} = 484 \text{ hp}.$$

Now, it might be thought that the additional force necessary could be found from the relationship between the work done by the force F and the increase in kinetic energy of the water. An amount of kinetic energy

$$K.E. = \frac{1}{2}\frac{w}{g}v^2 = \frac{1}{2} \times \frac{15000}{32} \times 88 \times 88 = 1.815 \times 10^6 \text{ ft. lbw}$$

is picked up in a distance of 1200 ft, so the force required for this is

$$F = \frac{1.815 \times 10^6}{1200} = 1512.5 \text{ lbw}.$$

This is exactly half the value obtained by using the rate of change of momentum. The discrepancy lies in the fact that the engine has done more work than that required to give the water kinetic energy. Some energy goes into heat and is exactly equal to the kinetic energy received. Thus, the total amount of energy received by the water in the 400 yd is double that given by the kinetic energy, or 3.63×10^6 ft lbw. With this value for the energy the force comes out correctly. For the situation in which a mass in motion captures another object, or whenever one mass is added to or subtracted from another, the force exerted cannot be obtained directly from the relationship between the work done and the increase in kinetic energy.

• PROBLEM 14-3

A rocket has an initial mass of 2×10^4 kg, a mass ratio of 3, a burning rate of 100 kg/sec, and an exhaust velocity of 980 m/sec. The rocket is fired vertically from the surface of the earth. How long after ignition of the engines will the rocket leave the ground?

Solution: The impulsive force exerted on the rocket by the burning fuel is

$$f = \frac{d}{dt}(mv) = v\frac{dm}{dt}$$

When this exceeds the weight of the remaining mass of the rocket

$$W = mg = \left(m_0 - \frac{dm}{dt}t\right)g,$$

and the rocket will lift off. Setting (1) equal to (2) gives

$$v\frac{dm}{dt} = \left(m_0 - \frac{dm}{dt}t\right)g.$$

Putting in the given quantities,

$$980\ \frac{m}{sec}\left(\frac{100\ kg}{sec}\right) = \left(2 \times 10^4\ kg - \frac{100\ kg}{sec}t\right)\frac{9.8\ m}{sec^2}.$$

Cancelling approriate units and numbers,

$$100\ (100)\ sec = 2 \times 10^4\ sec - 100\ t$$

$$100\ t = (2 \times 10^4 - 10^4)\ sec$$

$$t = 100\ sec.$$

• PROBLEM 14-4

Fluid leaving a nozzle impinges horizontally on a dividing vane in such a way that 1/3 of the stream is deflected upward through 90° and 2/3 of the stream is deflected downward through 90°. The nozzle delivers fluid to the vane at 12 slugs/sec with a velocity of 20 ft/sec. Assuming no losses as the fluid traverses the vane, determine the force necessary to keep the vane stationary.

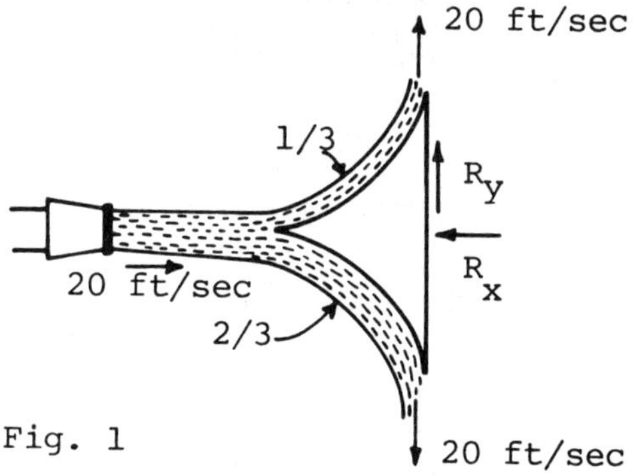

Fig. 1

Solution: The momentum per second of the water striking the vane is (12 slugs/sec)(20 ft/sec) = 240 lb. Since the water leaves the vane in the vertical direction only, this must be counteracted by the force R_x, so

$$R_x = -240\ lb.$$

There is no vertical momentum impingent on the vane, so R_y is equal and opposite to the difference in the momentum rates upwards and downwards; that is,

$$R_y = -[(20 \text{ ft/sec})(2/3)(12 \text{ slugs/sec}) + (20 \text{ ft/sec})(1/3)(12 \text{ slugs/sec})]$$

$$= -80 \text{ lb.}$$

So, $$R = \sqrt{R_x^2 + R_y^2} = \sqrt{(-240)^2 + (-80)^2} = 253 \text{ lb.}$$

$$\theta = \tan^{-1}\left(\frac{-80}{-240}\right) = 18.40.$$

• PROBLEM 14-5

A spaceship of mass m has velocity $\vec{v}$ in the positive x direction of an inertial reference frame. A mass dm is fired out the rear of the ship with constant exhaust velocity $(-v_0)$ with respect to the spaceship.

(a) Using conservation of momentum, show that

$$\frac{dv}{v_0} = \frac{dm}{m}$$

(b) By integration, find the dependence of v on m if v_1 and m_1 are initial values.

(c) Can the acceleration be constant if dm/dt, the burning rate, is constant?

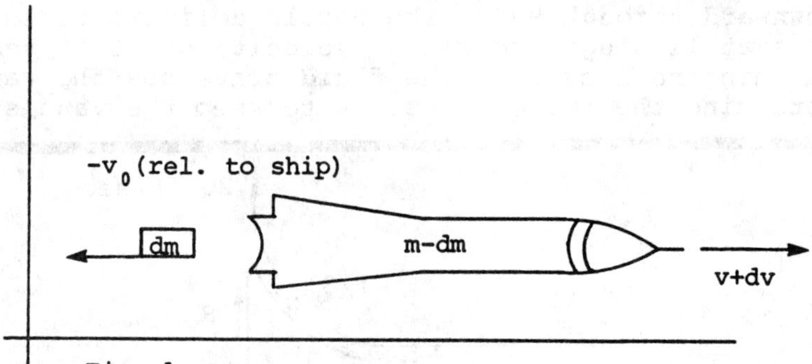

Fig. 1

Solution: (a) The momentum before the fuel is expelled is

$$p_{before} = mv. \quad (1)$$

After the fuel is expelled, fuel and rocket have a combined momentum

$$p_{after} = dm\,(v + dv - v_0) + (m - dm)(v + dv). \quad (2)$$

Setting the momenta before and after equal to each

other and expanding and cancelling terms gives

$$mv = dm\,v + dm\,dv - dm v_0 + mv + m\,dv - dm\,v - dm\,dv$$

Collecting the remaining terms into dm and dv terms:

$$0 = -v_0\,dm + m\,dv$$

which can be reorganized to:

$$\frac{dv}{v_0} = \frac{dm}{m} \qquad (3)$$

(b) The formula relating m and v at all times is obtained by integrating equation 3.

$$\int_{v_1}^{v} \frac{dv}{v_0} = \int_{m_1}^{m} - \frac{dm}{m}\ .$$

The negative sign before dm signifies that, as dm is expelled, there is a decrease in m.

The indefinite integrals and limits are:

$$\frac{1}{v_0}\,[v]_{v_1}^{v} = [-\ln m]_{m_1}^{m}$$

$$v - v_1 = v_0 \ln\,[m_1/m] \qquad (4)$$

$$v = v_1 + v_0 \ln\,[m_1/m]\ .$$

Operate with $\frac{d}{dt}$ on both sides:

$$\frac{dv}{dt} = 0 + v_0 \left(\frac{m}{m_1}\right)\left(-\frac{m_1}{m^2}\right)\frac{dm}{dt}$$

$$\frac{dv}{dt} = -\frac{v_0}{m}\left(\frac{dm}{dt}\right)\ . \qquad (5)$$

But, for $\frac{dm}{dt}$ constant, m(t) is proportional to $\left(\frac{dm}{dt}\right)$ t, so, even if $\frac{dm}{dt}$ is a constant, the acceleration will vary with time.

• PROBLEM 14-6

A fireboat draws water from the bay through a vertical inlet and sprays it out over the bow horizontally at 30 ft/sec. The diameter of the nozzle of the fire hose is 7 in., and it can be assumed that there is no accumulation of water in the fireboat. Assume the density of water to be 62.4 lb/cu ft. Determine the horizontal force from the propellers necessary to keep the boat stationary.

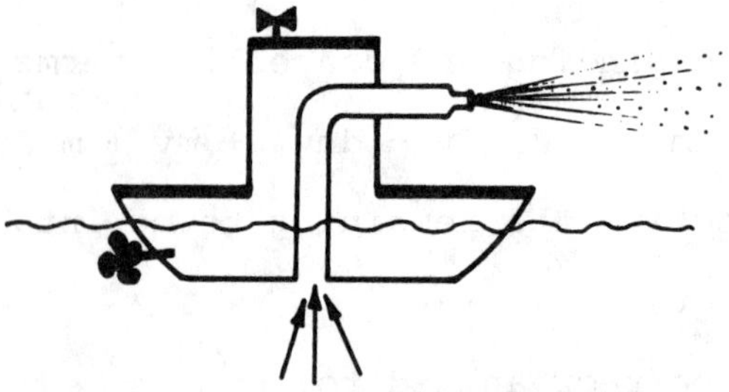

Solution: Assuming no accumulation of water in the fireboat, the rate of mass in must equal the rate of mass out. If the nozzle is 7 in. in diameter and water is emitted at 30 ft/sec, then the rate at which mass flows through the system is

$$\frac{dm}{dt} = Av\gamma \text{ slugs/sec}$$

where A is the area of the nozzle, v is the speed of the fluid flow through the nozzle, and γ is the mass density of the fluid. Thus,

$$\frac{dm}{dt} = \frac{\pi}{4}\left(\frac{7}{12}\right)^2 (30) \left(\frac{62.4}{g}\right) = 15.5 \text{ slug/sec.}$$

where $g = 32.2$ ft/sec^2, the acceleration due to gravity.

To determine the necessary thrust of the propellers, we will consider the change in momentum of the fluid in the horizontal direction and assume forward to be positive.

The momentum of the fluid coming into the boat is vertical, so the change in the horizontal momentum is just the momentum of the water coming out of the nozzle.

The force is equal to the rate of change of momentum.

$$F = \frac{d\,(mv)}{dt} = v\,\frac{dm}{dt}$$

since the horizontal velocity, v, of the water coming out is constant.

This must be equal to the force from the propellers so

$$R_x = (30 \text{ ft/sec})(15.5 \text{ slugs/sec}) = 465 \text{ lb.}$$

Hence, the propellers must develop 465 lb of thrust to maintain the boat in a stationary position.

• PROBLEM 14-7

As shown in the figure, a chute discharges gravel at the rate of 1000 lb/min onto a conveyor belt moving with a velocity of 8 ft/sec. The gravel falls onto the belt with a velocity of 12 ft/sec. What force is necessary to keep the conveyor belt moving at constant speed when the weight of gravel on the belt is 500 lb?

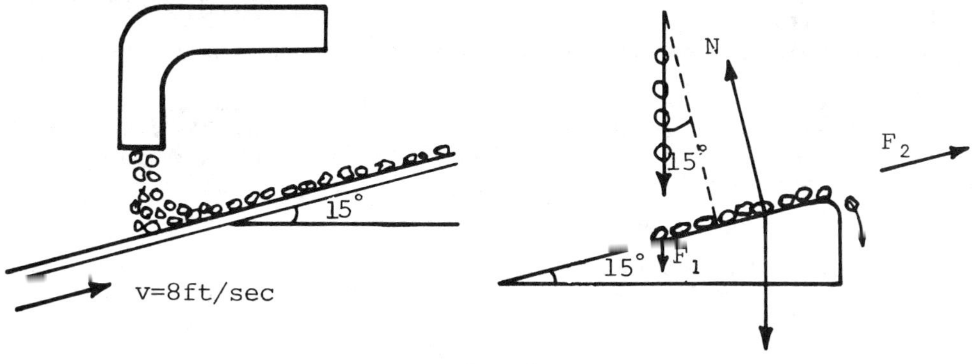

Fig. 1

Solution: Figure 1 shows the four forces acting on the gravel on the conveyor belt. F_1 is the impulsive force delivered by the falling gravel, and F_2 is the sought-after answer.

To obtain constant velocity of the belt, the sum of the components of the four forces along the incline must equal the change in momentum, dp/dt, of the belt and gravel as more gravel is added.

$$\frac{dp}{dt} = \frac{d\ mv}{dt} = v\ \frac{dm}{dt}$$

since the velocity is constant, therefore,

$$F_2 - F_1 \sin 15° - mg \sin 15° = v\ \frac{dm}{dt}$$

F_1 is equal to the impulse delivered by the falling gravel. The average F_1 per second is

$$F_1 = \left(\frac{1000\ \text{lbs}/60\ \text{sec}}{32.2\ \text{lbs/slug}}\right) 12\ \text{ft/sec}^2$$

$$= 6.21\ \text{lb.}$$

Thus, the average F_2 is

$$F_2 - \sin 15° \ (6.21 + 500)\ \text{lbs}$$

$$= \frac{8\ \text{ft}}{\text{sec}} \left(\frac{1000\ \text{lbs}}{60\ \text{sec}}\right)\left(\frac{1}{32.2\ \text{ft/sec}^2}\right)$$

$$F_2 = [0.259\ (506.21) + 4.14]\ \text{lbs.}$$

$$F_2 = [131.11 + 4.14]\ \text{lbs.}$$

$$F_2 = 135.25\ \text{lbs.}$$

• PROBLEM 14-8

A chain of mass m per unit length and total length ℓ lies in a pile on the floor. At time t = 0 a force P is applied and the chain is raised with constant velocity v. Find P(t).

Solution: We consider the portion of the chain which is moving as a constant velocity system which is accumulating mass at a constant rate. That rate is equal to its mass per unit length times its velocity (hence, $\dot{m}_1 = mv$) with a velocity (relative to the moving system) of $\vec{u}_1 = +v$. Of course, $\vec{u}_2 = 0$ and $\dot{m}_2 = 0$ (the velocity at rate of mass ejected from the system). Thus, the total external force acting on the system equals the applied force P minus the gravity force acting on the chain above the floor:

$$\Sigma F = P - mgvt.$$

Hence, using the variable mass system equation,

$$m\vec{a} = \Sigma\vec{F} + \dot{m}_2\vec{u}_2 - \dot{m}_1\dot{u}_1$$

and we have

$$0 = P - mgvt - (mv)(+v)$$

or $$P = mgvt + mv^2 = mv(v + gt).$$

• PROBLEM 14-9

A jet engine in a test stand takes in 200 lb/sec. of air and exhausts it at a speed of 1800 ft/sec. How much thrust is developed by the engine?

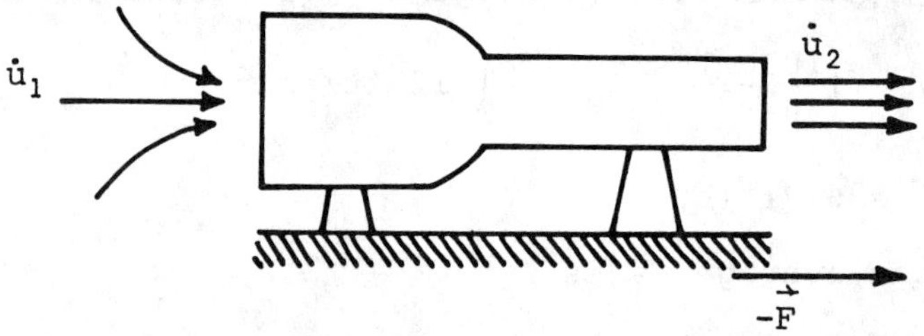

Solution: The intake and exhaust mass flows may be assumed equal (neglecting the small quantity of fuel consumed relative to the air flow) and equal to 200/32.2 slugs/sec. Intake velocity is assumed to be negligible also (i.e., $u_1 = 0$). Since the engine is stationary, its mass is not relevant and its acceleration is zero. (i.e., a = 0).

Thus, in the equation for systems of variable mass with inflow of $\dot{m}_1$ (and relative velocity u_1) and exhaust of $\dot{m}_2$ (and relative velocity of u_2), we have:

$$m\vec{a} = \Sigma\vec{F} + \dot{m}_2\vec{u}_2 - \dot{m}_1\vec{u}_1.$$

Substituting the values for $\dot{m}_1 = \dot{m}_2 = \frac{200}{32.2}$, $\vec{u}_1 = 0$, $\vec{u}_2 = 1{,}800$ fps and, solving for $-\vec{F}$,

$$-\vec{F} = \frac{200}{32.2}(1800) = 11{,}180 \text{ lbs.}$$

force where $\vec{F}$ is a force to the right applied by the test stand to the engine and $-\vec{F}$ is the engine's thrust.

• PROBLEM 14-10

As shown in Figure 1, a chain AB which is 14 ft. long hangs over the top of two slopes. Which end will reach the top, and what will the velocity of the chain be at that instant?

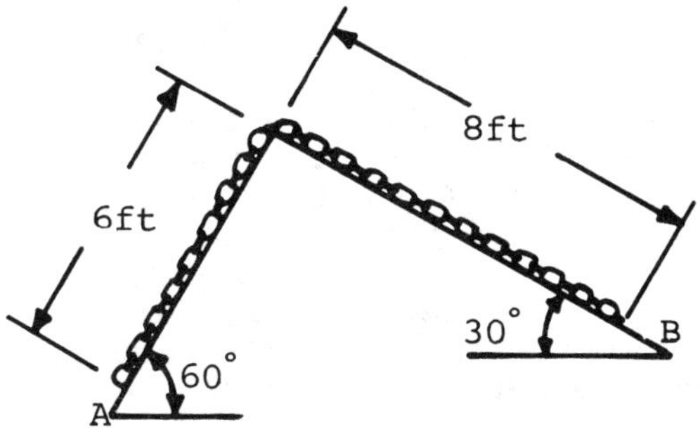

Fig. 1

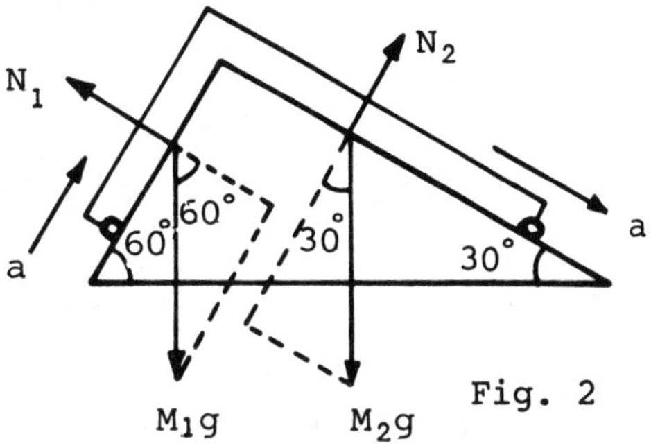

Fig. 2

Solution: We draw the external forces on the two parts of the chain. See fig.2

Consider the chain as a single unit with positive accelerations and forces along the slope in the direction of the $\vec{a}$ arrows.

The sum of accelerating forces

$$f = m_2 g \sin 30° - m_1 g \sin 60° = (m_1 + m_2)a. \qquad (1)$$

If $f > 0$, end A moves to the top; if $f < 0$, end B moves to the top. By substituting the initial conditions (where λ is the linear density of the chain),

$$f = \lambda g(8(0.5) - 6(0.866)) = \lambda g(4 - 5.19) = -1.19\,\lambda g.$$

So $f < 0$, and end B will move up.

We can use this value in equation (1) to find the original acceleration, but since the force on the left side increases and the force on the right side decreases, the acceleration increases and does not remain constant. Therefore, this analysis will not help us to find the velocity of the chain when end B reaches the top. To use Newton's Second Law would require use of variable mass equations for the chain on each side of the peak. An easier method will be to use the law of conservation of energy. The chain is being acted on only by a conservative force (gravity) since we have assumed that there is no friction on the chain. Thus, the change in potential energy equals the change in kinetic energy.

$$mg\Delta h = \frac{1}{2} mv^2 \qquad (2)$$

where Δh is the change in height of the mass center of the chain. The change in kinetic energy is simply $(1/2)mv^2$ since it is initially at rest. The v in equation two will be the desired result, the velocity when end B is at the peal. Rearranging equation (2) yields:

$$v = \sqrt{2g\Delta h}. \qquad (3)$$

The mass center of the piece of the chain on the left side is a distance of 3 sin 60° ft. away from the peak. For the other side of the chain, the mass center is 4 sin 30° ft. away from the peak. To find the mass center of the whole chain, multiply the weight of each side by its mass center, add these 2 terms together and divide this result by the total weight of the chain. Let d equal the distance from the peak of the total chain's mass center. Then,

$$d = \frac{3 \sin 60°[6\lambda] + 4 \sin 30° \; [8\lambda]}{14\lambda}$$

$$d = 2.256 \text{ ft.}$$

Now, when B reaches the top, the whole chain will be on the left side. Therefore, the distance from the chain's mass center to the peak will be equal to 7 sin 60°.

The Δh in equation (2) will be the difference in these two values.

$$\Delta h = 7 \sin 60° - d = 6.062 - 2.256$$

$$\Delta h = 3.806 \text{ ft.}$$

By substituting Δh into equation (3), we can find the velocity

$$v = \sqrt{2(32.2 \text{ ft/sec}^2)(3.806 \text{ ft})}$$

$$v = 15.66 \text{ ft/sec.}$$

• PROBLEM 14-11

In the first second of its flight, a rocket ejects 1/60 of its mass with velocity of 6800 ft/sec. What is the acceleration of the rocket?

Solution: The acceleration of the rocket at any time is

$$\frac{d\vec{v}}{dt} = \frac{\vec{F}_{ext}}{M} + \frac{\vec{u}}{M}\frac{dM}{dt} \qquad (1)$$

where M is the instantaneous mass of the rocket, $\vec{F}_{ext}$ is the net external force on the rocket-fuel system, $\vec{u}$ is the exhaust velocity of the fuel, and dM/dt is the time rate of change of the mass of the rocket.

In our case, $\vec{F}_{ext} = - Mg\hat{\jmath}$ where $\hat{\jmath}$ is a unit vector pointing up from the surface of the earth. Also $\vec{u} = - u\hat{\jmath}$. Then

$$\frac{d\vec{v}}{dt} = - \frac{Mg}{M}\hat{\jmath} - \frac{u}{M}\hat{\jmath}\frac{dM}{dt} = - \hat{\jmath}\left(g + \frac{u}{M}\frac{dM}{dt}\right)$$

Now, after the first second of flight,

$$M = M_0 - \frac{M_0}{60} = \frac{59}{60} M_0$$

where M_0 is the mass of the rocket at t = 0. Also,

$$\frac{dM}{dt} = - \frac{1/60\ M_0}{1\ sec} = - \frac{M_0}{60\ sec}$$

where the negative sign accounts for the fact that M is decreasing as time increases. Hence,

$$\frac{d\vec{v}}{dt} = - \hat{\jmath}\left[\frac{32f}{s^2} + \left(\frac{6800\ f/s}{\frac{59}{60} M_0}\right)\left(- \frac{M_0}{60\ sec}\right)\right]$$

$$= - \hat{\jmath}\left(32\ f/s^2 - 115.2\ f/s^2\right)$$

$$= 83.2\ f/s^2\ \hat{\jmath}$$

• PROBLEM 14-12

Coal drops at the rate of 25 slugs per second from a hopper onto a horizontal moving belt which transports it to the screening and washing plant. If the belt travels at the rate of 10 ft per second, what is the horsepower of the motor driving the belt? Assume that 5% of the energy available is used in overcoming friction in the pulleys. (See figure.)

Solution: We take the belt and hopper as our system. By using the principle of conservation of momentum, we can relate the net external force on the system (provided by the motor) to the time rate of change of the system momentum $\vec{P}$, or

$$\frac{d\vec{P}}{dt} = \vec{F}_{ext} \tag{1}$$

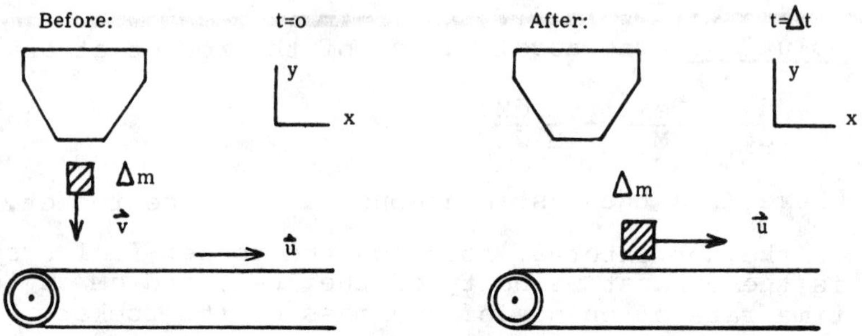

We examine a small element of mass, Δm, leaving the hopper and falling on the belt. (See figure.) At $t = 0$, the momentum of Δm in the x direction is

$$\vec{P}_0 = (\Delta m)(0)$$

and, at $t = \Delta t$, it is

$$\vec{P}_f = \Delta m\, \vec{u}$$

Hence $\Delta\vec{P} = \vec{P}_f - \vec{P}_0 = \Delta m\, \vec{u}$

or $$\frac{\Delta\vec{P}}{\Delta t} = \vec{u}\,\frac{\Delta m}{\Delta t}$$

Taking the limit as $\Delta t \to 0$,

$$\frac{d\vec{P}}{dt} = \vec{u}\,\frac{dm}{dt} \tag{2}$$

Comparing (2) with (1)

$$\vec{F}_{ext} = \vec{u}\,\frac{dm}{dt}$$

The power provided by the motor must be the time rate of change of the work, W, done on the belt by the motor. Hence,

$$P = \frac{dW}{dt} = \frac{d}{dt}\left[\int \vec{F}_{ext} \cdot d\vec{s}\right]$$

where $d\vec{s}$ is an element of path traversed by the belt. Assuming $\vec{F}_{ext}$ = constant, and noting that $d\vec{s}/dt = \vec{u}$, we obtain

$$P = \frac{d}{dt}\left[\vec{F}_{ext} \cdot \int d\vec{s}\right] = \frac{d}{dt}\left(\vec{F}_{ext} \cdot \vec{s}\right)$$

$$P = \vec{F}_{ext} \cdot \frac{d\vec{s}}{dt} = \vec{F}_{ext} \cdot \vec{u}$$

Since $\vec{F}_{ext}$ and $\vec{u}$ are parallel, we find, using (2)

$$P = \left(F_{ext}\right)(u) = u^2 \frac{dm}{dt}$$

$$P = (10 \text{ ft/s})^2 (25 \text{ s}^1/\text{s}) = 2500 \text{ lb} \cdot \text{ft/s} \qquad (3)$$

This power must be supplied by the motor to keep the belt moving at a uniform rate, assuming that none of it is dissipated in friction. Now, if 5% of the power supplied by the motor is used in overcoming friction, only 95% remains to power the belt. Since the power needed to move the belt is given by (3), we obtain,

$$P = 95\% \; P'$$

or $$P' = \frac{100}{95} P = \left(\frac{100}{95}\right)(2500 \text{ lb} \cdot \text{ft/s})$$

$$P' = 2631.6 \text{ ft} \cdot \text{lb/s}$$

Since $\quad 1 \text{ hp} = 550 \text{ ft} \cdot \text{lb/s}$

$$P' = 4.8 \text{ hp}$$

Taking friction into account, the rate of working of the motor is 4.8hp.

• PROBLEM 14-13

A gravel shipment is being transferred from one train to another. It leaves the hopper car at the first train at a rate of 300 lb/sec and hits the transfer chute at point G with a velocity of v_g = 30 ft./sec. The gravel leaves the chute at point W with a velocity V_W = 20 ft/sec at an angle of 15° with the horizontal. The total weight of the gravel and chute is represented as force F, acting at point D with a magnitude of 700 lbs. Find reaction of the roller support at point W and the components of the reaction at hinge point V.

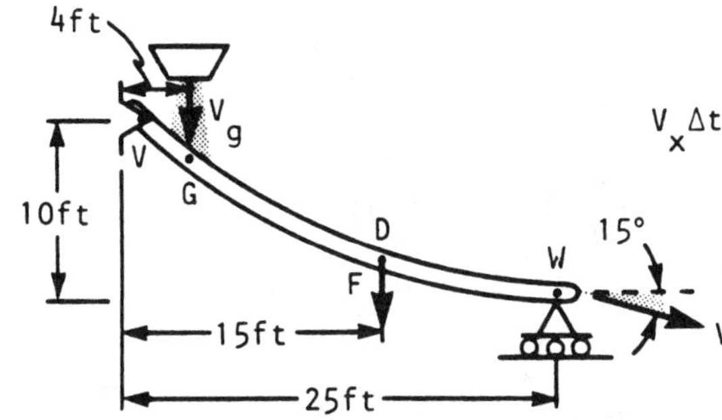

Fig. 1

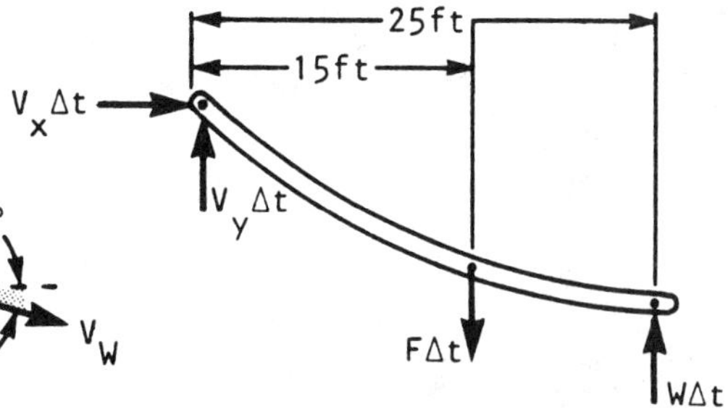

Fig. 2

Solution: The reaction forces at points V and W are in static equilibrium with the weight of the chute, the gravel sliding on it, and impulsive force caused by the loss in velocity of the gravel as it slides down the chute. The momentum of the gravel on the chute does not contribute to the conservation of momentum equation because it is constant.

Using the principle of impulse and momentum for a small time interval Δt, we may set the change in the momentum of the gravel, the weight of the chute and the gravel equal to the reactions forces at V and W.

$$\Delta m\,(V_g - V_w) = \Delta m V_{wx}\hat{\imath} + \Delta m (V_{wy} - V_{gy})\hat{\jmath}$$

Substituting the necessary quantities, we obtain for the x-component:

$$V_x \Delta t = \Delta M V_w \text{ COS } 15° \tag{1}$$

for the y-component:

$$V_y \Delta t - F\Delta t + W\Delta t = \Delta M[-V_w \sin 15° - (V_g)] \tag{2}$$

We may apply the principle of impulse and momentum to rotational systems. Here we consider an equation for impulse torque angular momentum. We first take the moments about V:

$$-15(F\Delta t) + 25(W\Delta t) = \Delta M[(r_w \times V_w) - (r_g \times V_g)]$$

$$= (\Delta M)10V_w \text{ COS } 15° - (\Delta M)25V_w \sin 15° - (-4\Delta M)V_g$$

Given: $F = 700$ lbs., $V_g = 30$ ft/sec, $V_w = 20$ ft/sec

$$\frac{\Delta M}{\Delta t} = 300\ \frac{\text{lb}}{\text{sec}} \cdot \frac{1}{32.2}\,\frac{\text{slugs}}{\text{lb}} = 9.32\ \frac{\text{slugs}}{\text{sec}}$$

Rewriting equations (1), (2), and (3)

$$V_x = (9.32 \text{ slugs/sec})(20 \text{ ft/sec}) \text{ COS } 15°$$

$$-15(700 \text{ lb}) + 25W = (9.32\ \frac{\text{slugs}}{\text{sec}})[(10)(20) \text{ ft/sec}) \text{ COS } 15°$$

$$- 25(20\frac{\text{ft}}{\text{sec}}) \sin 15° + 4(30)\frac{\text{ft}}{\text{sec}}$$

$$V_y - 700 \text{ lb} + W = (9.32\ \frac{\text{slugs}}{\text{sec}})\ (-20 \text{ ft/sec}) \sin 15°$$

$$+ 30 \text{ ft/sec}]$$

From this we calculate

$V_x = 180.0$

$W = 491.6$

$V_y = 208.4$

• PROBLEM 14-14

A chain of length ℓ ft. and total weight W lb. is partially hanging over the corner of a frictionless table. If it is released from rest with the segment of length, a, hanging, what will its speed be when the last link clears the top?

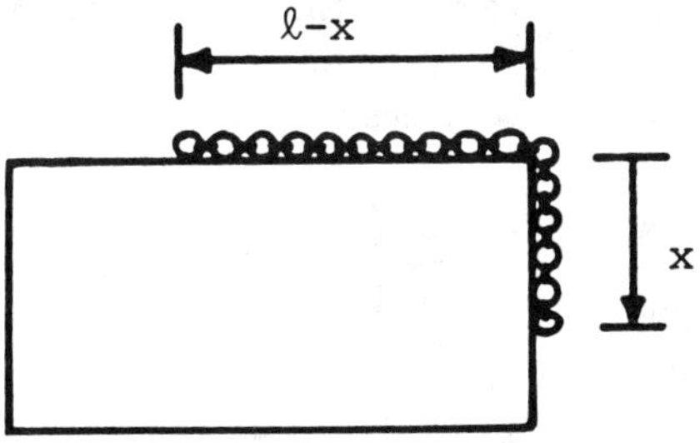

Fig. 1

Solution: Notice Figure 1 in which the length of chain hanging over the corner is x (initial value, $x_0 = a$).

Consider the chain as a single body. The accelerating force (neglecting friction) equals the weight of the chain hanging over the edge.

$$F = W\left(\frac{x}{\ell}\right) = mg\left(\frac{x}{\ell}\right)$$

Applying F = ma, we have

$$\frac{mg\,x}{\ell} = ma = m\frac{dv}{dt} = m\left(\frac{dv}{dx}\right)\left(\frac{dx}{dt}\right)$$

Using the chain rule for derivatives gives

$$\frac{mg\,x}{\ell} = m\frac{dv}{dx}\frac{dx}{dt} = mv\frac{dv}{dx} .$$

The differential equation is solved by definite integration as the variables are separated. Thus,

$$v\frac{dv}{dx} = \frac{xg}{\ell}$$

$$\int_0^v v\,dv = \frac{g}{\ell}\int_a^\ell x\,dx.$$

One must check that the limits of both integrations correspond. Initially, when x = a, the velocity is 0.

Finally, when $x = \ell$, the velocity is the sought after v. On integration of both sides,

$$\left[\frac{v^2}{2}\right]_0^v = \frac{g}{\ell}\left[\frac{x^2}{2}\right]_a^\ell$$

or, $$\frac{v^2}{2} - 0 = \frac{g}{2\ell}[\ell^2 - a^2]$$

Then, $$v = \sqrt{\frac{g}{\ell}(\ell^2 - a^2)}\ .$$

• PROBLEM 14-15

A vertically launched missile has an initial mass m_i, including fuel, at time $t = 0$. The missile consumes its fuel supply at a rate $c = dm/dt$. It is expelled at a constant velocity v_m relative to the missile. Derive an expression for the velocity of the missile at time $t = t_1$. Neglect air resistance.

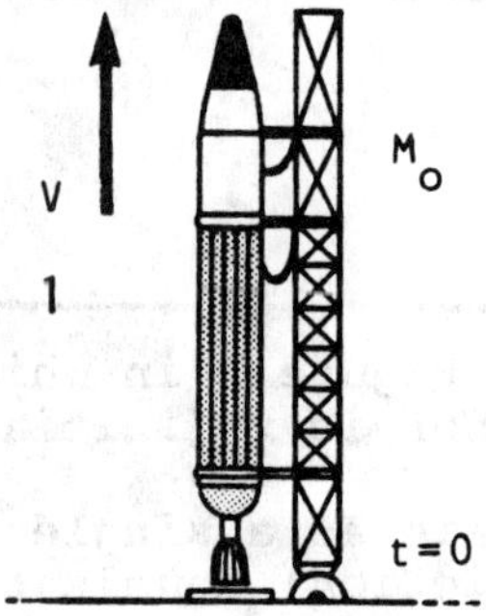

Fig. 1

Solution: Though the missile and fuel form a closed system, they are acted upon constantly by an external gravitation field. Thus, the principle of impulse and momentum may be applied to the system of missile and fuel

$$p_m + F_e \Delta t = p_f,$$

where F_e is only the external gravitation force (mg) and not the force of the ejected gas upon the rocket.

At time t, the mass of the missile and remaining fuel is $m = m_i - qt$, and the velocity is u. During the time interval Δt, a mass of fuel $\Delta m = q\ \Delta t$ is expelled with a speed v_m relative to the rocket. Denoting by v_e the absolute velocity of the expelled fuel we apply the principle of impulse and momentum between time t and time $t + \Delta t$.

We write

$$(m_i - ct)u - g(m_i - ct)\Delta t$$
$$= (m_i - ct - c\Delta t)(u + \Delta u) - c\Delta t(v_m - u)$$

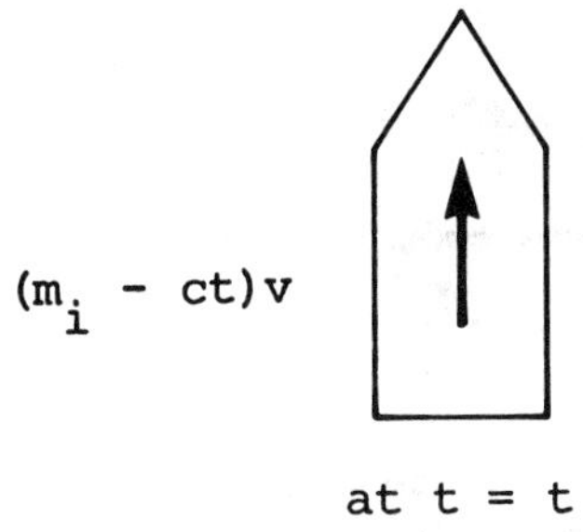

Fig. 2

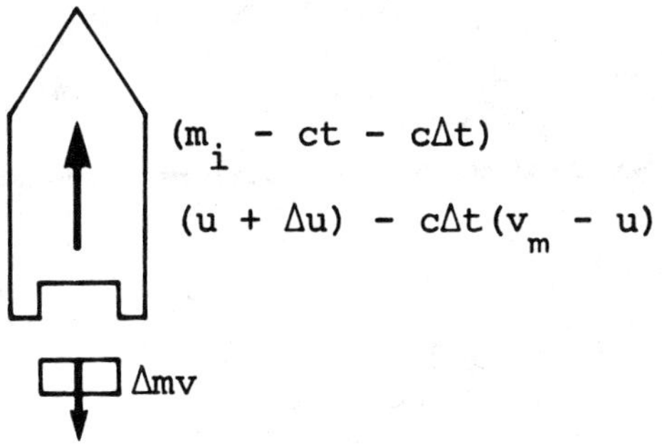

$\Delta mv_e = c\Delta t(v_m - u)$ Fig. 3

at $t = t + \Delta t$

Dividing through by Δt, and letting Δt approach zero, we obtain

$$- g(m_i - ct) = (m_i - ct) = du/dt - cv_m$$

Separating variables and integrating from $t = 0$, $u = 0$ to $t = t$, $u = u$.

$$du = \left(\frac{cv_m}{m_i - ct} - g\right) dt$$

$$\int_0^u du = \int_0^t \left(\frac{cv_m}{m_i - ct} - g\right) dt$$

$$u = [- v_m \ln(m_i - ct) - gt]_0^t$$

$$u = v_m \ln \left(\frac{m_i}{m_i - ct}\right) - gt .$$

Remark. The mass remaining at time t_f, after all the fuel has been expended, is equal to the mass of the missile $m_m = m_i - ct$ and the maximum velocity attained by the rocket is $v_{max} = v_m \ln(m_i/m_m) - gt_f$. Assuming that the fuel is expelled in a relative short period of time, the term gt_f is small and we have

$$v_{max} \cong v_m \ln(m_i/m_m)$$

If a missile is to be placed into orbit outside of the gravitational effect of the earth it must achieve a velocity of 11,180 m/s. If we assume $v_m = 3000$ m/s we get a value for m_1/m_m of 41.5 kg. This means that in order to place 1 kg of material into orbit, 41.5 kg of a fuel that gives a velocity of 3000 m/s must be burned. Therefore 207,500 kg of fuel must be burned to place the 5000 kg payload into orbit.

• PROBLEM 14-16

A 10 foot long chain is placed on a 4 foot high, frictionless table so that one end just reaches the floor.

With what velocity will the other end slide off the table?

Solution: Let M be the mass of that part of the chain which has not yet reached the ground. Let m be the mass of the part of the chain already lying on the floor. Let δ be the mass per foot of the chain. As mass must be conserved, we have

$$\frac{dM}{dt} = -\frac{dm}{dt}. \tag{1}$$

Let x be the length of the chain still on the table. Newton's second law, force equals rate of change of momentum, expressed in terms of these variables, becomes

$$F = \frac{d(Mv)}{dt} + v\frac{dm}{dt}, \tag{2}$$

since at time t the part of the chain that reaches the floor has a velocity v. Expanding the derivative gives

$$F = M\frac{dv}{dt} + v\frac{dM}{dt} + v\frac{dm}{dt}; \tag{3}$$

by using (1), this is simplified to

$$F = M\frac{dv}{dt}. \tag{4}$$

The chain not yet on the floor has a length (x + 4) feet and, therefore, a mass $(x + 4)\delta$.

The part of the chain lying on the floor is supported by the floor; the part on the table is similarly supported. The only unbalanced force, then, is the gravitational attraction of the four feet of chain between the table top and the floor. This force can be expressed as $4g\delta$. Equation (4) becomes

$$(x + 4)\,\delta\frac{dv}{dt} = 4g\delta \tag{5}$$

or $$(x + 4)\frac{dv}{dt} = 4g. \tag{6}$$

Since $v = -\frac{dx}{dt}$,

and $$\frac{dv}{dt} = \frac{dv}{dx}\frac{dx}{dt} = -v\frac{dv}{dx} \tag{7}$$

equation (6) becomes

$$v\,dv = -\frac{4g}{x + 4}\,dx. \tag{8}$$

This separated equation integrates to

$$v^2 = -8g\ln(x + 4) + C. \tag{9}$$

Here, the initial condition is that when v = 0, x = 6. This gives C, so

$$v^2 = 8g \ln \left(\frac{10}{x + 4}\right) \tag{10}$$

We are asked for the velocity with which the other end slides off the table. Since this occurs when x = 0, we find

$$v^2 = 8g \ln 2.5 \tag{11}$$

and $$v = \sqrt{8g \ln 2.5} \text{ ft/sec.} \tag{12}$$

$$v = 15.4 \text{ ft./sec.}$$

• PROBLEM 14-17

A vertical chain weighs w lb/ft of length and is ℓ ft long. It is held in such a way that its lower end just touches a horizontal table top. If the upper end of the chain is released, determine the force on the table at any instant during the fall.

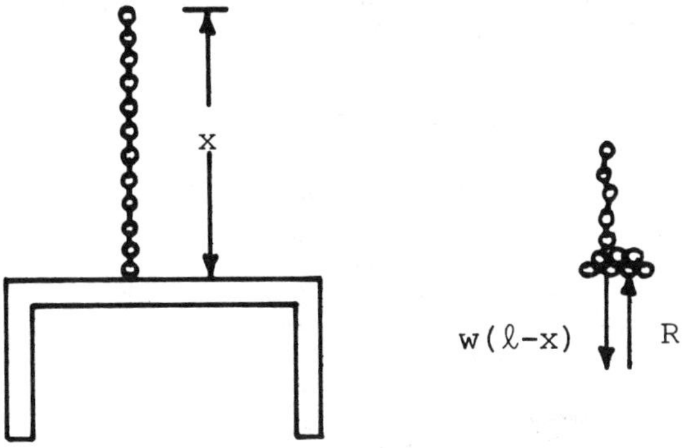

Solution: Let x be the length in feet of the part of the chain above the table at time t in seconds, and let F be the reaction in pounds of the table on the chain at time t. We shall use the portion of the chain at rest on the table as the free body. The forces acting on this free body are its weight, which is w(ℓ - x) down, and the reaction F up. The mass of this free body is increasing at the rate of (w/g)v, where v is the speed of the chain in feet per second at any time t. The velocity of the particles added to the parent mass, relative to the parent mass, is u = - v - 0 = - v. We are here assuming that the upward sense is positive. The most general form of Newton's 2nd law is

$$R = \frac{d}{dt}(mv) \text{ or } R = m\frac{dv}{dt} - u\frac{dm}{dt}$$

Hence, $$F - w(\ell - x) = 0 + v\left(\frac{w}{g}v\right)$$

Since the part of the chain above the table is undergoing

free fall, its speed at any time t, which is the speed of the top of the chain, is

$$v = \sqrt{2g(\ell - x)}$$

Hence, $F = w(\ell - x) + w/g[2g(\ell - x)]$

or, $F = 3w(\ell - x)$, upward

The force exerted by the chain on the table is equal and opposite to F. The force on the table is $3w(\ell - x)$ downward.

We can see that the maximum value of the downward force on the table is $3w\ell$ and it is developed just before the end of the chain reaches the table.

• PROBLEM 14-18

A scoop of mass m_1 is attached to an arm of length ℓ and negligable weight and can swing freely in the vertical arc of radius ℓ.

The scoop is released from an angle of 45°, swings down and grabs a load of sand of mass m_2 with no frictional losses. To what angle with the vertical does the scoop rise?

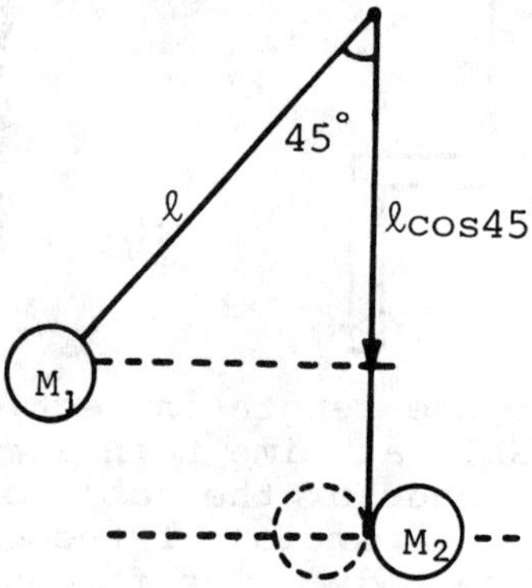

Fig. 1

Solution: The way to solve this problem is by making use of the laws of conservation of energy and momentum, as applicable.

Using conservation of energy for figure 1, up to the moment of collision,

$$m_1 g\ (\ell - \ell \cos 45°) = \frac{1}{2} m v_1{}^2 .$$

We find that the velocity of m_1 just before impact is:

$$v_1 = \sqrt{2g\ell\ (1 - \cos 45°)}$$

During the impact, which is of very short duration, momentum is conserved (but energy is not).

$$m_1 v_1 = (m_1 + m_2)v_2.$$

The velocity of the combined masses just after impact is

$$v_2 = \frac{m_1}{m_1 + m_2}\left(\sqrt{2g\ell(1 - \cos 45°)}\right).$$

After impact, the scoop and sand rise together to angle θ, conserving energy as they do. Therefore,

$$\frac{1}{2}(m_1 + m_2)\, v_2{}^2 = (m_1 + m_2)\, g(\ell - \ell \cos \theta).$$

Putting in v_2 from the last step,

$$\frac{1}{2}(m_1 + m_2)\left[\left(\frac{m_1}{m_1 + m_2}\right)^2 (2g\ell(1 - \cos 45°)\right]$$

$$= (m_1 + m_2)\, g\ell(1 = \cos \theta).$$

Simplifying, we have

$$\left(\frac{m_1}{m_1 + m_2}\right)^2 (1 - \cos 45°) = (1 - \cos \theta).$$

Solving for cos θ,

$$\left(\frac{m_1}{m_1 + m_2}\right)^2 \left(1 - \frac{\sqrt{2}}{2}\right) - 1 = -\cos \theta$$

$$\cos \theta = 1 - \left(\frac{m_1}{m_1 + m_2}\right)^2 (.293)$$

$$\theta = \cos^{-1}\left[1 - \left(\frac{m_1}{m_1 + m_2}\right)^2 (0.293)\right]$$

• PROBLEM 14-19

A jet is flying horizontally with a velocity v of 700 Km/h. The jet has a mass of 10,000 kg, an air intake rate of 80 kg/s and an air discharge velocity of 700 m/s relative to the plane. (a) Find the total air drag due to friction. (b) If drag is proportional to v^2, find the speed of the plane if the intake rate is increased 10% to 88 kg/s.

Solution: (a) The intake and exhaust mass flows are equal at 80 kg/s (i.e., $\dot{m}_1 = \dot{m}_2 = 80$ kg/s). The aircraft velocity is 700 Km/h = 194.4 m/s, thus, the intake flow enters with that velocity relative to the aircraft (i.e., $\vec{u}_1 = 194.4$ m/s).

Since at steady flight the engine thrust equals air friction drag, we have, from the variable mass system equation,

$$m\vec{a} = \Sigma\vec{F} + \dot{m}_2\vec{u}_2 - \dot{m}_1\vec{u}_1,$$

and, noting that $\vec{a} = 0$ due to steady flight,

$$-\vec{F} = \dot{m}\,(\vec{u}_2 - \vec{u}_1) = 80(700 - 194.4) = 40,448\ v.$$

The minus sign indicates that the force of the aircraft onto the engine (i.e., the drag) opposes the direction of aircraft velocity.

(b) The new cruising speed will be at that point where the engine thrust again equals air friction drag. Now, since drag is assumed to be proportional to the square of speed, we have that the drag is:

$$-F = 40,448\left(\frac{v}{194.4}\right)^2$$

Note that the exhaust (relative) velocity is still equal to 700 m/sec, but the intake (relative) velocity (u_1) is now equal to the new cruising speed. Thus, we have

$$-F = \dot{m}\,(\vec{u}_2 - \vec{u}_1)$$

and, noting that the new $\dot{m}$ = 88 kg/s,

$$40,448\left(\frac{v}{194.4}\right)^2 = 88\ (700 - v),$$

which is a quadratic equation for v having a positive root: v = 268.12 m/s = 965.22 Km/h.

• PROBLEM 14-20

A lunar landing craft is hovering over the moon's surface. Assume that one-third of its weight is fuel, that the exhaust velocity from the rocket engine is 1500 m/sec, and that the acceleration of gravity on the moon is one-sixth of that on earth. Find out how long it will be before the craft runs out of fuel and crashes.

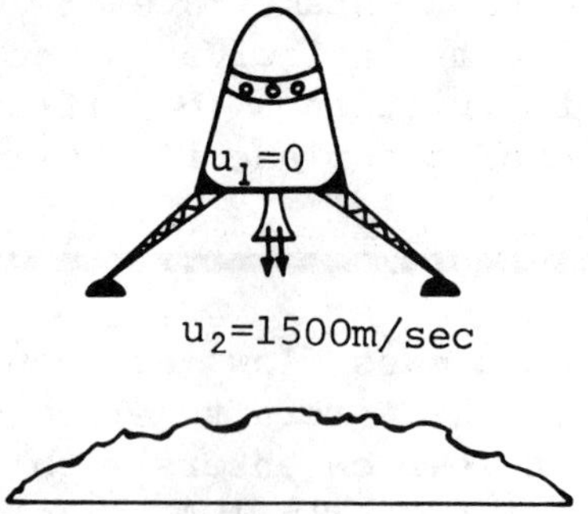

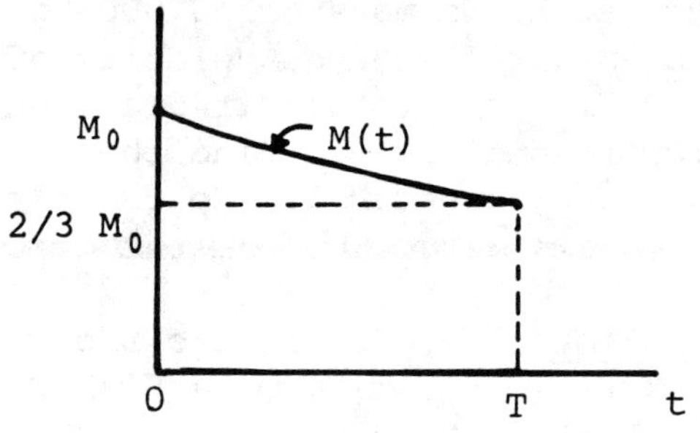

Solution: Since the craft is hovering, we have a = 0 in the variable mass system equation; hence the thrust (and, thus, the burn rate $\dot{m}$ must vary as the fuel is consumed, as shown on the graph of craft mass vs time.

$$m\vec{a} = \Sigma\vec{F} + \dot{m}_2\vec{u}_2 - \dot{m}_1\vec{u}_1$$

where ΣF = the force of gravity on the craft mass.

$$\Sigma F = -g_m\, m(t)$$

and $a = 0$, $u_1 = 0$ (the intake velocity), $\dot{m}_2 = -\frac{dm}{dt}$.
Thus, substituting,

$$0 = -g_m\, m - u_2 \frac{dm}{dt}$$

or $$g_m\, m = -u_2 \frac{dm}{dt},$$

and $$\frac{g_m}{u_2}\, dt = -\frac{dm}{m}, \text{ or } \frac{g_m}{u_2}\int_0^t dt = -\int_{m_0}^{m} \frac{dm}{m}.$$

Integrating both sides yields

$$\frac{g_m}{u_2}\, t = -\ln\left(\frac{m}{m_0}\right) = \ln\left(\frac{m_0}{m}\right)$$

or $$t = \frac{u_2}{g_m}\ln\left(\frac{m_0}{m}\right) = \frac{1500}{9.8/6}\ln\left(\frac{m_0}{\frac{2}{3}m_0}\right) = 372.4 \text{ sec.}$$

• PROBLEM 14-21

A rocket sled on a horizontal track weighs W-lb when empty and carries W_0-lb of fuel. It burns fuel at the rate of w lb/sec and emits exhaust horizontally with speed v_0 ft/sec relative to the sled. If the sled starts from rest at t = 0, derive an expression for its velocity at any time t later. What will the velocity of the sled be when all the fuel is spent? Neglect all resistances.

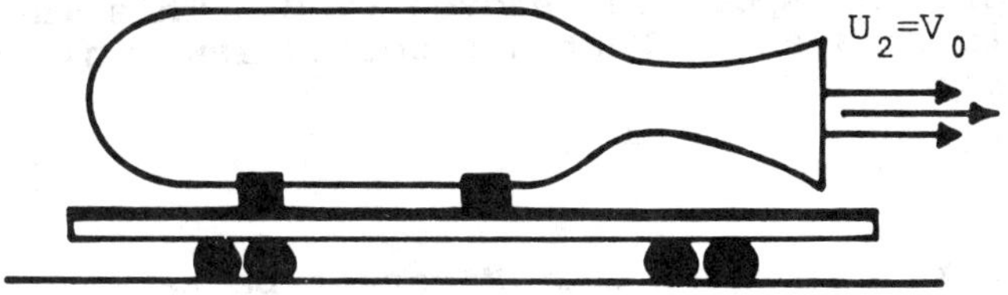

Solution: (a) Since the rate of fuel burn is constant, the expression for the variable weight of the rocket sled is

$$W(t) = W + W_0 - wt,$$

where $\frac{dW}{dt} = w = \frac{W_0}{t}$, and T is the total time to burn out.

Due to the absence of friction aerodynamic drag, etc., we have that $\Sigma F = 0$ in the variable mass system equation.

$$m\vec{a} = \Sigma\vec{F} + \dot{m}_2\vec{u}_2 - \dot{m}_1\vec{u}_1$$

Also, $\vec{u}_1 = 0$, since there are no intake ports and $\dot{m}_2 = \frac{w}{g}$. Thus, substituting,

$$\frac{W(t)}{g}\, a = \frac{w}{g}\, v_0 .$$

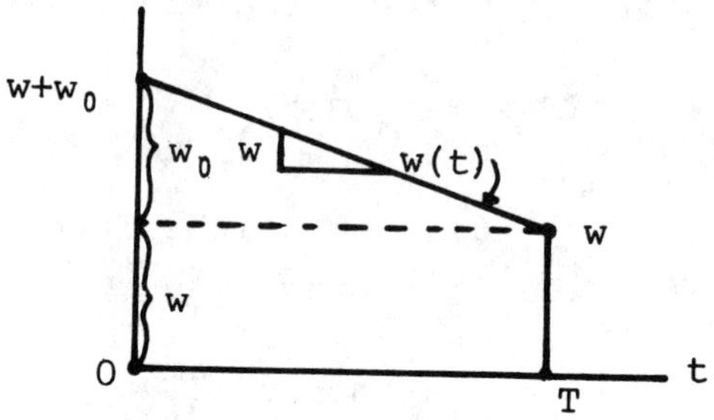

Hence, $$a \equiv \frac{dv}{dt} = \frac{w\, v_0}{W + W_0 - wt}$$

or $$dv = v_0 \frac{w\, dt}{W + W_0 - wt}$$

Integrating both sides, we have

$$\int_0^v dv = v \int_0^t \frac{w\, dt}{W + W_0 - wt} = -\, v_0 \ln \frac{W + W_0 - wt}{W + W_0}$$

or, $$v = v_0 \ln \left(\frac{W + W_0}{W + W_0 - wt}\right) .$$

(b) At $t = T$, $W(t) = W + W_0 - wt = W$. Hence, the final velocity is

$$v_f = v_0 \ln \left(\frac{W + W_0}{W}\right) = v_0 \ln \left(1 + \frac{W_0}{W}\right) .$$

• PROBLEM 14-22

A chain is wound on a spool fixed at the top of a frictionless inclined plane. The plane has a length ℓ and is at an angle θ with the horizontal. One end of the chain is initially at the lower edge of the plane, but is free to fall straight down. Find the velocity of this end after it has fallen y feet, assuming a link of the chain has no velocity until it leaves the spool.

Solution: Let the mass of the section of chain that has already left the spool be denoted by M. This mass will not remain constant. Newton's second law, in its most general form, is

$$F = \frac{d\,(Mv)}{dt} \tag{1}$$

and it must be kept in mind that M cannot be taken outside of the derivative. We must have

$$F = M\frac{dv}{dt} + v\frac{dM}{dt} . \tag{2}$$

If δ is the mass per unit length of the chain, (2) can be expressed

$$F = \delta(\ell + y)\frac{dv}{dt} + v\delta\frac{d(\ell + y)}{dt} \qquad (3)$$

or $$F = \delta(\ell + y)\frac{dv}{dt} + v\delta\frac{dy}{dt}\,. \qquad (4)$$

The forces acting along the line of motion of the chain are the gravitational attraction for the portion that left the inclined plane, $y\delta g$, and the attraction for that part still on the plane, $\ell\delta\, g \sin\theta$. From (4) we have

$$\delta g(\ell \sin\theta + y) = \delta(\ell + y)\frac{dv}{dt} + \delta v^2 \qquad (5)$$

where $\frac{dy}{dt}$ has been replaced by v. Since $\frac{dv}{dt}$ can be written as $v\frac{dv}{dy}$, equation (5) is equivalent to

$$g(\ell \sin\theta + y) = (\ell + y)\, v\frac{dv}{dy} + v^2 \qquad (6)$$

or $$v\frac{dv}{dy} + \frac{v^2}{\ell + y} = \frac{g(\ell \sin\theta + y)}{\ell + y} \qquad (7)$$

which we recognize as a Bernoulli type. The substitution $z = v^2$ allows us to obtain the first-order linear equation

$$\frac{dz}{dy} + \frac{2z}{\ell + y} = \frac{2g(\ell \sin\theta + y)}{\ell + y} \qquad (8)$$

which has an integrating factor $(\ell + y)^2$. The solution to (8) is

$$z = \frac{c}{(\ell + y)^2} + \frac{2g}{(\ell + y)^2}\left[y\ell^2 \sin\theta + \frac{y^2}{2}\ell(\sin\theta + 1) + \frac{y^3}{3}\right] \qquad (9)$$

but the initial condition states that when y = 0, v and, therefore, z are also zero. This gives us c = 0 and (9) becomes

$$z = \frac{2g}{(\ell + y)^2}\left[y\ell^2 \sin\theta + \frac{y^2\ell}{2}(\sin\theta + 1) + \frac{y^3}{3}\right] \qquad (10)$$

or $$v^2 = \frac{g[2y^3 + 3\ell y^2(\sin\theta + 1) + 6\ell^2 y \sin\theta]}{3(\ell + y)^2}\,. \qquad (11)$$

Equation (11) expresses the velocity of the lower end as a function of the distance fallen.

• PROBLEM 14-23

A 3.60×10^4-kg rocket rises vertically from rest. It ejects gas at an exhaust velocity of 1800 m/sec at a mass rate of 580 kg/sec for 40 sec before the fuel is expended. Determine the upward acceleration of the rocket at times t = 0,20, and 40 sec.

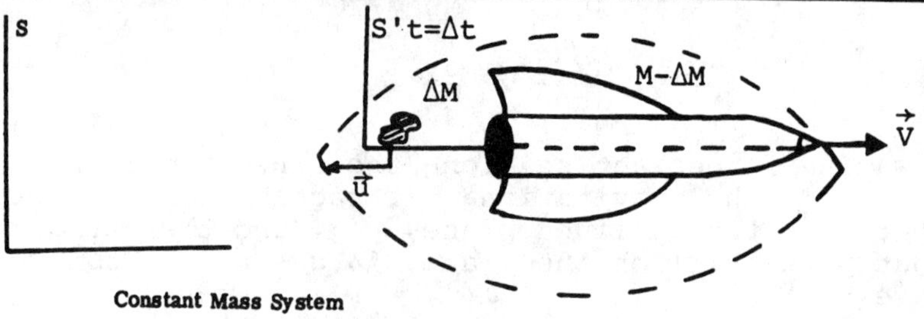

Constant Mass System

Solution: Suppose a rocket is travelling with a velocity $\vec{v}$ relative to a stationary cooordinate system (S), and emits fuel at velocity $\vec{u}$ with respect to the rocket (S') (see figure). The exhaust velocity with respect to S, $\vec{w}$, is the exhaust velocity with respect to S' plus the velocity of S' with respect to S or

$$\vec{w} = \vec{u} + \vec{v}.$$

Now that we know all the velocities with respect to a stationary frame, we may use the law of conservation of momentum in this frame to find the velocity of the rocket after a mass Δm has been emitted.

At t = 0, no fuel has been emitted and the initial momentum of the fuel-rocket system is

$$\vec{P}_0 = M\vec{v}.$$

At t = Δt, a mass ΔM (where ΔM > 0) of fuel has been emitted and travels with velocity $\vec{w}$. The rocket now has mass M - ΔM and travels with a velocity $\vec{v} + \Delta\vec{v}$. The momentum is then

$$\vec{P}_f = (M - \Delta M)(\vec{v} + \Delta\vec{v}) + \Delta M(\vec{w})$$

or $$\vec{P}_f = (M - \Delta M)(\vec{v} + \Delta\vec{v}) + \Delta M(\vec{u} + \vec{v})$$

Hence, the change in momentum is

$$\Delta\vec{P} = \vec{P}_f - \vec{P}_0 = (M - \Delta M)(\vec{v} + \Delta\vec{v}) + \Delta M(\vec{u} + \vec{v}) - M\vec{v}$$

$$= M\vec{v} - \Delta M\vec{v} + M\Delta\vec{v} - \Delta M\Delta\vec{v} + \vec{u}\Delta M + \vec{v}\Delta M - M\vec{v}$$

$$= M\Delta\vec{v} + \vec{u}\,\Delta M - \Delta M\Delta\vec{v}$$

Then $$\frac{\Delta\vec{P}}{\Delta t} = M\frac{\Delta\vec{v}}{\Delta t} + \vec{u}\frac{\Delta M}{\Delta t} - \frac{\Delta M\Delta\vec{v}}{\Delta t}$$

Taking the limit as $\Delta t \to 0$

$$\frac{d\vec{P}}{dt} = \lim_{\Delta t \to 0} \frac{\Delta\vec{P}}{\Delta t} = M\frac{d\vec{v}}{dt} + \vec{u}\frac{dM}{dt}$$

because $\frac{\Delta M\Delta\vec{v}}{\Delta t} \to 0$ as $\Delta t \to 0$, whence

$$\frac{d\vec{P}}{dt} = M\frac{d\vec{v}}{dt} + \vec{u}\frac{dM}{dt}$$

But, for a constant mass system, which our's is, (see figure) the time rate of change of momentum of the system equals the net external force on the system and

$$M \frac{d\vec{v}}{dt} + \vec{u} \frac{dM}{dt} = \vec{F}_{ext} \qquad (1)$$

In this equation, $d\vec{v}/dt$ is the rocket's acceleration, M is the instantaneous mass of the rocket, $\vec{F}_{ext}$ is the net force on the fuel-rocket system, and dM/dt is the rate of change of the rocket's mass. Note that in (1), dM/dt > 0 due to our derivation. Hence, we may replace dM/dt by - dM/dt if we redefine dM/dt to be less than zero. Then

$$M \frac{d\vec{v}}{dt} - \vec{u} \frac{dM}{dt} = \vec{F}_{ext}$$

$$\text{or} \quad M \frac{d\vec{v}}{dt} = \vec{F}_{ext} + \vec{u} \frac{dM}{dt} \qquad (2)$$

We define $\vec{u} \frac{dM}{dt}$ as the rocket's thrust. Solving for $\frac{d\vec{v}}{dt}$,

$$\frac{d\vec{v}}{dt} = \frac{\vec{F}_{ext}}{M} + \frac{\vec{u}}{M} \frac{dM}{dt} \qquad (3)$$

For our problem, $\vec{F}_{ext} = - Mg\hat{j}$, where $\hat{j}$ is a unit vector in the positive y direction. Furthermore, if the rocket is propelled straight up, $\vec{u} = - u\hat{j}$. Hence,

$$\frac{d\vec{v}}{dt} = - g\hat{j} - \frac{u}{M} \hat{j} \frac{dM}{dt} = - \hat{j} \left(g + \frac{u}{M} \frac{dM}{dt}\right) \qquad (4)$$

Note that since M is a function of time, $d\vec{v}/dt$ will also be time dependent.

At t = 0, $M = 3.60 \times 10^4$ kg

$$\text{and} \quad \frac{d\vec{v}}{dt} = - \hat{j} \left[\frac{9.8m}{s^2} + \frac{(1800\ m/s)}{(3.6 \times 10^4\ kg)} \left(- \frac{580\ kg}{s}\right)\right]$$

$$\frac{d\vec{v}}{dt} = - \hat{j}\ (9.8\ m/s^2 - 29\ m/s^2)$$

$$\frac{d\vec{v}}{dt} = 19.2\ m/s^2\ \hat{j}$$

At t = 20 secs, $M = 3.6 \times 10^4\ kg - (580\ kg/s)(20s)$

$= 3.6 \times 10^3\ kg - 11.6 \times 10^3\ kg$

$= 24.4 \times 10^3\ kg$

$$\text{and} \quad \frac{d\vec{v}}{dt} = - \hat{j} \left[\frac{9.8\ m}{s^2} + \frac{(1800\ m/s)}{(2.44 \times 10^4\ kg)} \left(- \frac{580\ kg}{s}\right)\right]$$

$$\frac{d\vec{v}}{dt} = - \hat{j}\ (9.8\ m/s^2 - 42.79\ m/s^2) = 32.99\ m/s^2\ \hat{j}$$

At t = 40 secs, $M = 3.6 \times 10^4\ kg - (580\ kg/s)(40s)$

$$= 3.6 \times 10^4\ kg - 2.32 \times 10^4\ kg$$

$$= 1.28 \times 10^4\ kg$$

and $$\frac{d\vec{v}}{dt} = -\hat{j}\left[\frac{9.8\ m}{S^2} + \frac{(1800\ m/s)}{(1.28 \times 10^4\ kg)}\left(-\frac{580\ kg}{S}\right)\right]$$

$$\frac{d\vec{v}}{dt} = -\hat{j}\ (9.8\ m/s^2 - 81.56\ m/s^2)$$

$$\frac{d\vec{v}}{dt} = 71.76\ m/s^2\ \hat{j}$$

In this example we have neglected air friction and the variation of $\vec{g}$ with altitude.

• PROBLEM 14-24

A water jet directs a stream of water at a turbine. The stream has a cross-sectional area A and a velocity V_1. The turbine deflects the stream through an angle α. This causes the turbine to move to the right with a constant velocity U. If the water maintains a constant speed along the turbine find (a) the components of the force acting on the stream by the turbine, (b) the velocity U which produces a maximum power.

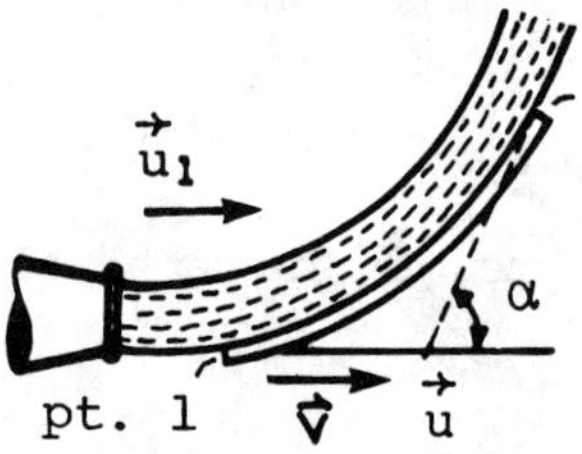

Fig. 1

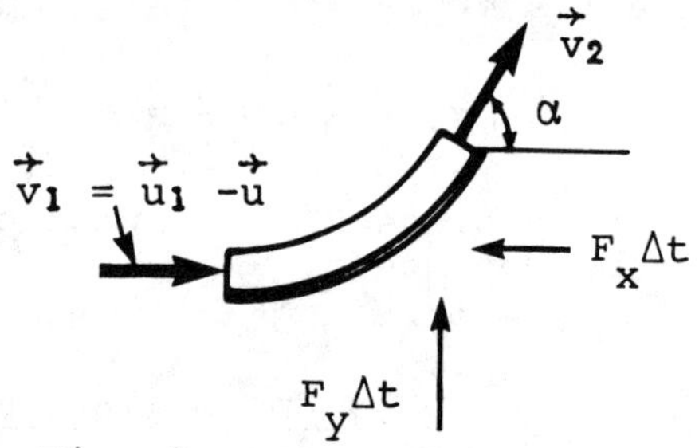

Fig. 2

Solution:

If we use a reference frame moving with the blade at a constant velocity U, the particles striking the block will have a relative velocity $v_1 = (u_1 - u)$ and will leave the blade with relative velocity v_2. Given that the particles move along the blade with uniform velocity, the relative velocities v_1 and v_2 are equal; $v_1 = v_2 = v$. Representing the density of water with ρ, the mass of the particles striking the during time interval Δt, is $\Delta M = A\rho(u_1 - u)\Delta t$; a mass of particles equal to this leaves the blade during interval Δt. The principle of impulse momentum may be applied to the particles striking the blade, and those particles in contact with the blade.

Since v_1 and v_2 are equal to v, and the constant momentum of the water in contact with the blade does not impart an impulse, the components of the impulse momentum equation can be written:

$$\text{x component:} \quad \Delta mV - F_x \Delta t = \Delta MV \cos \alpha$$

$$\text{y component:} \quad F_y \Delta t = (\Delta M) V \sin \alpha$$

F_x, F_y are the components of the blade reaction force deflecting the stream. Setting $\frac{\Delta M}{\Delta t} = A_\rho (u - U)$ and $v = (u_1 - U)$ we write:

$$F_x = A_\rho (U_1 - U)^2 (1 - \cos\alpha)$$

$$F_y = A_\rho (U_1 - U)^2 \sin \alpha$$

Figure (2) shows the respective directions of F_x, F_y.

The power of the system is equal to the velocity of the blade increased by F_x force component exerted on the blade. Power $= F_x V = A\rho (u_1 - u)^2 (1 - \cos\alpha) U$.

Differentiating the power with respect to U and setting the derivative equal to zero, we obtain

$$\frac{d\ (\text{power})}{dU} = A\rho (u_1^2 - 4u_1 U + 3U^2)(1 - \cos\alpha) = 0$$

Solving the quadratic for U results in

$$U = u_1, \qquad U = 1/3\ u_1$$

The first root, u_1 gives a power of zero; it is the minimum for the power. The second root $U = 1/3\ U_1$ gives the velocity for maximum power.

• PROBLEM 14-25

Make the following assumptions:

(1) The one stage rocket thrust is vertical, and horizontal force will not be taken into account. (2) The acceleration of gravity does not vary with height. (3) The rate of change of mass of the rocket is constant until burnout and v_0 is the exhaust velocity of the rocket.

Find the following:

(a) The equation of the vertical ascent of a rocket. (b) If the initial velocity and mass are zero and m_1, and the final mass of rocket is m_2, what is the velocity of the rocket at burnout? Express the result in terms of time at burnout. (c) If the initial height is zero, what is the altitude of the rocket at burnout?

Solution: Using Assumption 3, set

$$\frac{dm}{dt} = -k \qquad (1)$$

so that k is a positive constant. Taking the positive upward, the equation of motion of the rocket is

$$ma = v_0 k - mg$$

or
$$\frac{dv}{dt} = \frac{v_0 k}{m} - g \qquad (2)$$

where v_0 is the exhaust velocity of the rocket. Multiply both sides of Eq. (2) by dt, and substitute for dt from Eq. (1) to obtain

$$dv = -v_0 \frac{dm}{m} + \frac{g}{k}\, dm.$$

If the initial velocity is zero, the velocity at a later time will be found by summing or integrating these velocity increments. Integrate this equation from an initial mass m_1 to a mass m at a later time. The velocity at this later time is

$$v = -\int_{m_1}^{m} \left(\frac{v_0}{m}\right) dm + \int_{m_1}^{m} \frac{g}{k}\, dm$$

$$v = -v_0 \int_{m_1}^{m} \frac{dm}{m} + \frac{g}{k} \int_{m_1}^{m} dm$$

$$v = -v_0 \ln m \Big|_{m_1}^{m} + \frac{g}{k}\, m \Big|_{m_1}^{m}$$

$$v = -v_0 (\ln m - \ln m_1) + \frac{g}{k}(m - m_1)$$

$$v = -(v_0 \ln m) + (v_0 \ln m_1) + \frac{g}{k} m - \frac{g}{k} m_1 . \qquad (3)$$

If the final mass of the rocket at burnout is m_2, the vertical velocity v_b at burnout is given by

$$v_b = \left(v_0 \ln \frac{m_1}{m_2}\right) - g \frac{(m_1 - m_2)}{k}$$

where the upper limit of integration m in Eq. (3) is replaced by m_2.

If t_b is the time to burnout and $R = m_1/m_2$ is the mass ratio of the rocket, then

$$k = \frac{\Delta m}{\Delta t}$$

$$t_b = \frac{(m_1 - m_2)}{k}$$

and the velocity at burnout is:

$$v_b = (v_0 \ln R) - gt_b. \quad (4)$$

Now let us find the height of the rocket above the ground at burnout. Let z be the vertical coordinate. Set v = dz/dt in Eq. (3), multiply through by dt, and substitute for dt from Eq. (1) to obtain the equation

$$dz = + \left(\frac{v_0}{k} \ln m\right) dm - \frac{g}{k^2} m\, dm + \left(\frac{g}{k^2} m_1 - \frac{v_0}{k} \ln m_1\right) dm.$$

If the initial height is zero, the height z_b at burnout will be found by summing or integrating these increments of height. Integrate this equation from an initial mass m_1 to a final mass m_2 at burnout using this indefinite integral:

$$\int (\ln m) dm = (m \ln m) - m.$$

$$z = \int_{m_1}^{m_2} \frac{v_0}{k} \ln m\, dm - \int_{m_1}^{m_2} \frac{g}{k^2} m\, dm$$

$$+ \int_{m_1}^{m_2} \left(\frac{g}{k^2} m_1 - \frac{v_0}{k} \ln m_1\right) dm$$

$$z = \frac{v_0}{k} \Big[(m \ln m) - m \Big]_{m_1}^{m_2} - \frac{g}{k^2} \frac{m^2}{2} \Big]_{m_1}^{m_2}$$

$$+ \left(\frac{g}{k^2} m_1 - \frac{v_0}{k} \ln m_1\right) m \Big]_{m_1}^{m_2}$$

$$z = \frac{v_0}{k} [m_2 \ln m_2 - m_2 - m_1 \ln m_1 + m_1]$$

$$- \frac{g}{2k^2} (m_2{}^2 - m_1{}^2) + \left(\frac{gm_1}{k^2} - \frac{v_0 \ln m_1}{k}\right) (m_2 - m_1)$$

Once again, if the time of burnout is

$$t_b = \frac{m_1 - m_2}{k} \quad \text{and } R = \frac{m_1}{m_2}$$

then after rearrangement, the altitude at burnout can be expressed as:

$$z_b = v_0 t_b - \frac{1}{2} g t_b^2 - \frac{m_2 v_0}{k} \ln R.$$

• PROBLEM 14-26

Show that the maximum height achieved by the rocket in the preceding problem is

$$z_{max.} = \frac{v_0^2 \left(\ln \frac{m_1}{m_2}\right)^2}{2g} - v_0 t_b \left[\frac{\ln \frac{m_1}{m_2}}{1 - \left(\frac{m_1}{m_2}\right)} - 1\right]$$

Solution: The rocket will usually be designed to have vertical velocity at burnout, so that after burnout it will continue to climb, but now without power. From the law of conservation of energy, the additional height h which the rocket will climb after burnout is given by equation

$$mgh = \frac{1}{2} m v_b^2,$$

where m is mass, v_b is the burnout velocity. So

$$h = \frac{v_b^2}{2g}$$

The total height is

$$z_{max} = z_b + h.$$

z_b from the preceding problem is

$$z_b = v_0 t_b - \frac{1}{2} g t_b^2 - \frac{m_2 v_0}{k} \ln R.$$

v_b from the preceding problem is

$$v_b = \left(v_0 \ln \frac{m_1}{m_2}\right) - g t_b$$

Now, $$z_{total} = v_0 t_b - \frac{1}{2} g t_b^2 - \frac{m_2 v_0}{k} \ln \frac{m_1}{m_2} + \frac{\left(v_0 \ln \frac{m_1}{m_2} - g t_b\right)^2}{2g}$$

Expanding and simplifying as follows,

$$z_{total} = v_0 t_b - \frac{1}{2} g t_b^2 - \frac{m_2 v_0}{k} \ln \left(\frac{m_1}{m_2}\right) + \frac{v_0^2}{2g} \ln^2 \left(\frac{m_1}{m_2}\right) - t_b v_0 \ln \left(\frac{m_1}{m_2}\right) + \frac{1}{2} g t_b^2$$

$$z_{total} = v_0 t_b - v_0 t_b \ln \left(\frac{m_1}{m_2}\right) - \frac{m_2 v_0}{k} \ln \left(\frac{m_1}{m_2}\right) + \frac{v_0^2}{2g} \ln^2 \left(\frac{m_1}{m_2}\right)$$

Since $t_b = \frac{m_1 - m_2}{k}$, multiply the third term by $\frac{t_b}{t_b}$,

$$z_{total} = v_0 t_b - v_0 t_b \ln \frac{m_1}{m_2} - v_0 t_b \frac{m_2 \ln \frac{m_1}{m_2}}{k \left(\frac{m_1 - m_2}{k}\right)} + \frac{v_{02}}{2g} \ln^2 \left(\frac{m_1}{m_2}\right)$$

Now multpy the third term by $\left(\frac{\frac{1}{m_1}}{\frac{1}{m_1}}\right)$

$$z_{total} = v_0 t_b - v_0 t_b \ln \frac{m_1}{m_2} - v_0 t_b \left(\frac{\left(\frac{m_2}{m_1}\right) \ln \frac{m_1}{m_2}}{1 - \frac{m_2}{m_1}}\right) + \frac{v_0^2}{2g} \ln^2 \left(\frac{m_1}{m_2}\right)$$

Factoring out $v_0 t_b$ yields:

$$z_{total} = v_0 t_b \left(1 - \ln \frac{m_1}{m_2} - \frac{\left(\frac{m_2}{m_1}\right) \ln \frac{m_1}{m_2}}{1 - \frac{m_2}{m_1}}\right) + \frac{v_0^2}{2g} \ln^2 \left(\frac{m_1}{m_2}\right)$$

By multiplying $1 - \ln \frac{m_1}{m_2}$ by $\frac{1 - \frac{m_2}{m_1}}{1 - \frac{m_2}{m_1}}$, the factor of $v_0 t_b$ can be simplified.

$$z_{total} = v_0 t_b \left(1 - \frac{\ln \frac{m_1}{m_2}}{1 - \frac{m_2}{m_1}}\right) + \frac{v_0^2}{2g} \ln^2 \left(\frac{m_1}{m_2}\right)$$

Factoring out - 1 in the first term yields:

$$z_{total} = \frac{v_0^2}{2g} \ln^2 \left(\frac{m_1}{m_2}\right) - v_0 t_b \left(\frac{\ln \frac{m_1}{m_2}}{1 - \frac{m_2}{m_1}} - 1\right).$$

Since $R = \frac{m_1}{m_2}$, this is equivalent to

$$z_{total} = \frac{v_0^2}{2g} \ln^2 R - v_0 t_b \left(\frac{\ln R}{1 - \left(\frac{1}{R}\right)} - 1\right).$$

This is the maximum altitude reached by a single-stage sounding rocket in a uniform gravitational field.

CHAPTER 15

NEWTONIAN GRAVITATION

GRAVITATIONAL FIELD AND GRAVITATIONAL POTENTIAL

• **PROBLEM 15-1**

Show that the acceleration of a particle toward a very massive uniform sphere is independent of the mass of the particle.

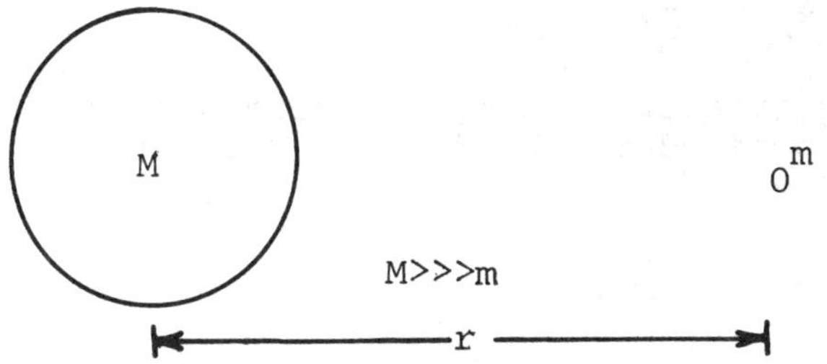

Fig. 1

Solution: This problem can be solved by applying Newton's Laws of Motion. Consider a very massive uniform sphere of mass, M, separated from a small particle of mass, m, by a distance r, M > > > m. The situation is shown in the accompanying figure. There will be a gravitational force on the particle due to the sphere given by Newton's Law of Universal Gravitation.

$$F_g = \frac{G\,M\,m}{r^2} \tag{1}$$

This force will cause an acceleration by the particle, whose magnitude can be found by applying Newton's Second Law.

$$F_g = m\,a_m$$

$$a_m = \frac{F_g}{m} \tag{2}$$

Substituting equation (1) into (2) yields

$$a_m = \frac{G\,M}{r^2}. \tag{3}$$

From Newton's Third Law, we also know that there is a force equal and opposite to the force in equation (1) which acts on the massive sphere. However, due to the sphere's enormous mass, M, this acceleration will be negligible in comparison to the acceleration given by equation (3). Therefor, the resultant acceleration will be just equation (3), and it can be seen that the acceleration of the particle does not depend on its mass, m.

While we have shown qualitatively what the particle's acceleration is, we can also show it quantitatively. The acceleration of the large sphere is given by

$$a_M = \frac{F_g}{M} = \frac{\frac{G\,M\,m}{r^2}}{M} = \frac{G\,m}{r^2}. \tag{4}$$

The resultant acceleration is then $a_m + a_M$ since they are both being accelerated toward each other.

$$a = a_m + a_M = \frac{G\,M}{r^2} + \frac{G\,m}{r^2}. \tag{5}$$

Since M is much greater than m, the second term in equation (5) is negligible compared to the first. Equation (5) then reduces to equation (3)

$$a = a_m + a_M = \frac{G\,M}{r^2}. \tag{6}$$

We can see by inspection of equation (6) that the resultant acceleration of the particle toward the sphere is independent of the particle's mass, m.

This is why all objects near the earth are accelerated at the same rate, g, by the earth's gravitational field. The acceleration g is independent of the object's mass.

• PROBLEM 15-2

A newly discovered planet has twice the density of the earth, but the acceleration due to gravity on its surface is exactly the same as on the surface of the earth. What is its radius?

Solution: This problem must be approached carefully. We must express the acceleration due to gravity in terms of the density and the radius of the planet. If the radius is R and the mass of the planet M, then the acceleration due to gravity on its surface is found from Newton's Second Law, F = ma. Consider an object of mass m on the surface of the planet. Then the only force on m is the gravitational force F, and

$$F = \frac{GMm}{R^2}$$

But a is the acceleration of m due to the planet's gravitational field, or g_p. Then

$$g_v = \frac{GM}{R^2}$$

Assuming the planet is spherical, its volume is the volume of a sphere of radius R:

$$V = \frac{4}{3}\pi R^3$$

Since Mass = Volume X Density.

$$M = \frac{4\pi R^3 \rho}{3}$$

where ρ (the Greek letter rpho) is the density of the planet. Therefore

$$g_v = \frac{G\frac{4}{3}\pi R^3\rho}{R^2}$$

$$= \frac{4\pi}{3} GR\rho$$

Similarly, the acceleration due to gravity on the surface of the earth is

$$g = \frac{4}{3}\pi G R_e \rho_e$$

where ρ_e is the density of the earth, and R_e is its radius. If

$$g_v = g$$

Then

$$\frac{4}{3}\pi G R\rho = \frac{4}{3}\pi G R_e \rho_e$$

Canceling $\frac{4}{3}\pi G$ on both sides

$$R\rho = R_e \rho_e$$

If the density of the planet is twice that of the earth,

$$\rho = 2\rho_e$$

So $$R2\rho_e = R_e \rho_e$$

Whence $$R = \tfrac{1}{2} R_e$$

$$= \tfrac{1}{2} \times 6.38 \times 10^6 \text{ m}$$

$$= 3.19 \times 10^6 \text{ m}$$

The radius of the planet is one half of the radius of the earth, or 3.19×10^6 meters.

• PROBLEM 15-3

A hole is bored through a diameter of the earth (assume constant density). Show that the force on a particle in the hole at a distance r from the center is proportional to r and in the direction toward the center. Show that $\ddot{r} = -gr/R$, where R is the radius of the earth and g is the acceleration of gravity at the surface of the earth. Assuming $r = R \cos 2\pi k/T$ is a solution, where T is the period of the motion, show that: (a) the time required for a particle dropped from the surface of the earth into the hole to the center is $\sqrt{\pi^2 R/4g}$; (b) its speed at the center of the earth is $\sqrt{gR}$; (c) the speed with which the particle emerges from the other side of the earth is zero.

Solution: As the particle passes through the hole, the only matter exerting an unbalanced force is the matter within a sphere of radius r whose center is coincident with the earth's. r is the particle's distance from the earth's center. This is in accordance with Gauss' Law. The physical reason is that the pull of all the matter in the shell from r to the earth's surface cancels to zero when integrated over the shell.

The force on the particle, according to Newton's Law of Gravitation, is $F = \frac{-G\,m\,M}{r^2}$. G is the universal gravity constant, m is the particle mass and M is the mass of the body affecting the particle. Inside the surface of the body, M is $\rho\left(\frac{4}{3}\pi r^3\right)$. ($\rho$ is density and $\frac{4}{3}\pi r^3$ is the volume of M.)

Substituting this into Newton's Law of Gravitation yields:

$$F = \frac{-G\,m\,\rho\left(\frac{4}{3}\pi r^3\right)}{r^2}$$

$$F = -\left[G\,m\,\rho\left(\frac{4}{3}\pi\right)\right]r$$

or, $F = -k\,r$ where $k = G\,m\,\rho\left(\frac{4}{3}\pi\right)$.

The force is proportional to r and, since the density is uniform, the center of the earth is the mass center and the force points in its direction.

At the surface of the earth, r = R and F = - m g. Therefor, it follows that

$$m\,g = k\,R$$

and $$k = \frac{m\,g}{R}$$

At any point, then, $F = -m\,g\left(\frac{r}{R}\right)$.

By Newton's Second Law, $F = m\ddot{r}$.

$$-m\,g\left(\frac{r}{R}\right) = m\ddot{r}$$

$$\ddot{r} = -g\left(\frac{r}{R}\right)$$

$\ddot{r} + g/R\,(v) = 0$ is the familiar harmonic oscillator equation. If $r = R\cos\left(\frac{2\pi t}{T}\right)$ is a solution of this equation then:

$$T = \frac{2\pi}{\omega} = \frac{2\pi}{\sqrt{g/R}}$$

r, then, is equal to R cos $\left(\frac{2\quad\pi\qquad t}{2\pi/\sqrt{g/R}}\right)$

$$r = R\cos\left(\sqrt{g/R}\ t\right)$$

a) At the center of the earth, r = 0; thus, cos $\sqrt{g/R}$ t must be zero.

$\sqrt{g/R}$ t then equals $\pi/2$

$$\sqrt{g/R}\ t = \pi/2$$

$$t = \sqrt{\frac{\pi^2 R}{4g}}$$ (This is the time required to the earth's center.)

b) $\dot{r} = \frac{dr}{dt}$ is the particle's velocity

$$\dot{r} = \sqrt{g/R}\ \ R\sin\sqrt{g/R}\ t.$$

At the center of the earth, cos $\sqrt{g/R}$ t = 0 and sin $\sqrt{g/R}$ t = 1.

The particle's speed is then $v = |\dot{r}|$

$$v = \sqrt{g/R}\ \ R = \sqrt{gR}.$$

c) When r = R, cos $\sqrt{g/R}$ t = 1 and sin $\sqrt{g/R}$ t = 0;

therefore, $v = \sqrt{gR}\sin\sqrt{g/R}\ t = 0$.

• PROBLEM 15-4

Show by using Gauss' Law that the force exerted on a mass placed in a deep hole in the earth is due to the matter in the spherical portion below the hole. Find the force on 1 kg at a depth of 1/4 and 1/2 of the radius of the earth, assuming that the density of the earth is constant.

Solution: For this problem, take a sphere of radius r, where r is the distance from the earth's center to the particle, as the Gaussian surface. According to Gauss' Law for Gravitation,

$$\frac{1}{4\pi G}\oint \vec{g}\cdot\vec{ds} = -\ m$$

where m is the mass inside of the Gaussian surface, $\vec{g}$ is gravity potential, $\vec{ds}$ is the area normal, and G is the universal gravity constant.

Since $\vec{g}$ points towards the earth's center and $\vec{ds}$ is anti-parallel to it,

$$\vec{g} \cdot \vec{ds} = g \, ds \cos 180° = - g \, ds.$$

$\vec{g} \cdot \vec{ds}$ over the Gaussian surface is then

$$\oint \vec{g} \cdot \vec{ds} = - g \oint ds$$

$$= - g \, 4 \pi r^2$$

From Gauss' Law,

$$\frac{- g \, 4 \pi r^2}{4 \pi G} = -m$$

$$\frac{g \, r^2}{G} = m.$$

m is the mass under the particle (enclosed by the Gaussian surface), so the mass above has no effect.

$$m = \rho \, (V) = \rho \left(\frac{4}{3} \pi r^3\right)$$

where ρ is density and V is the volume of the Gaussian sphere.

Combining the last two relations yields

$$\frac{g \, r}{G} = \rho \left(\frac{4}{3} \pi r^3\right)$$

$$g = \rho G \left(\frac{4}{3} \pi\right) r$$

$$g = k \, r$$

At the earth's surface, $r = r_o$ and $g = 9.8 \text{ m/s}^2$

$$k = \frac{g}{r_o} = \frac{9.8 \text{ m/s}^2}{r_o}$$

$$g = \left(\frac{9.8 \text{ m/s}^2}{r_o}\right) r$$

When $r = \frac{3}{4} r_o$,

$$g = \frac{(9.8 \text{ m/s}^2) \, 3 \, r_o}{4 \, r_o}$$

$$g = \frac{3(9.8 \text{ m/s}^2)}{4} = 7.35 \text{ m/s}^2$$

The force on 1 kg is its weight, W.

W = mg = 7.35 Newton's.

When $r = \frac{1}{2} r_o$,

$$g = \frac{(9.8\ m/s^2)\ \frac{1}{2} r_o}{r_o}$$

$$g = 4.9\ m/s^2.$$

The force on 1 kg is W = mg = 4.9 Newton's.

• PROBLEM 15-5

The earth acts on any body with a gravitational force inversely proportional to the square of the distance of the body from the center of the earth. Calculate the escape velocity from the earth, i.e., the speed with which a vertically moving body must leave the earth's surface in order to coast along without being pulled back to the earth. Also find the time it takes for a rocket projected upward with this escape velocity to attain a height above the earth's surface equal to its radius. In both cases ignore the effect of any other heavenly bodies and take the earth's radius as 6.38×10^6 m.

Solution: Assuming that the body moves along a radial trajectory after leaving the earth, it continually experiences a gravitational force

$$F = -\frac{Gmm_e}{r^2}$$

where m and m_e are the mass of the body and the earth, respectively. By Newton's Second Law, F = ma, and the acceleration of m is then

$$a = -\frac{Gm_e}{r^2}$$

But, being that the trajectory is radial,

$$a = \frac{dv}{dr}\frac{dr}{dt} = v\frac{dv}{dr} = -\frac{Gm_e}{r^2}$$

where v is the object's velocity.

$$v\ dv = -G\ m_e \frac{dr}{r^2} \qquad (1)$$

Now, we want a projectile to leave the earth ($r = R_e$) with a velocity v_e and to reach a destination at which it no longer feels the effect of the earth's gravitational force. If it reaches this point , the body can have no velocity and yet still not be accelerated towards the earth.

Since the gravitational force is zero at ∞, we require that v = 0 at r = ∞. Hence

$$\int_{v_e}^{0} v\,dv = -G\,m_e \int_{R_e}^{\infty} \frac{dr}{r^2}$$

$$-\frac{v_e^2}{2} = Gm_e \left(\frac{1}{r}\right)_{R_e}^{\infty}$$

$$v_e^2 = \frac{2Gm_e}{R_e} \qquad (2)$$

At the surface of the earth

$$\frac{Gm_e}{R_e^2} = g$$

and $\quad v_e^2 = 2gR_e$

$$v_e = \sqrt{2gR_e} = \sqrt{2\times 9.81\ \text{m}\cdot\text{s}^2 \times 6.38\times 10^6\ \text{m}}$$

$$v_e = 11.2\times 10^3\ \text{m/s}^2$$

The second part of the problem asks us to find out how long it takes for an object to reach a distance r = $2R_e$ from the center of the earth if its initial velocity is v_e. In order to do this, we must find r as a function of t. Going back to (1)

$$v\,dv = -Gm_e \frac{dr}{r^2}$$

But, now, v = v at r = r, and v = v_e at r = R_e and

$$\int_{v_e}^{v} v\,dv = -G\,m_e \int_{R_e}^{r} \frac{dr}{r^2}$$

$$\frac{v^2 - v_e^2}{2} = Gm_e \left(\frac{1}{r} - \frac{1}{R_e}\right)$$

$$v^2 = v_e^2 + \frac{2Gm_e}{r} - \frac{2Gm_e}{R_e} \qquad (3)$$

However, from (2)

$$v_e^2 = \frac{2Gm_e}{R_e}$$

Therefore, from (3)

$$v^2 = \frac{2Gm_e}{r}$$

$$v = \frac{\sqrt{2Gm_e}}{r^{\frac{1}{2}}}$$

To find v as a function of t, note that

$$v = \frac{dr}{dt}$$

Therefore, $$\frac{\sqrt{2Gm_e}}{r^{\frac{1}{2}}} = \frac{dr}{dt}$$

$$r^{\frac{1}{2}}\, dr = \sqrt{2Gm_e}\, dt$$

Since $r = R_e$ when $t = 0$, and $r = 2R_e$ when $t = t$,

$$\int_{Re}^{2R_e} r^{\frac{1}{2}}\, dr = \sqrt{2Gm_e} \int_0^t dt$$

$$\frac{2}{3}\, r^{3/2} \Big|_{R_e}^{2R_e} = \sqrt{2Gm_e}\;\; t$$

$$(2R_e)^{3/2} - (R_e)^{3/2} = \frac{3}{2}\sqrt{2Gm_e}\;\; t$$

or $$t = \frac{2[(2R_e)^{3/2} - (R_e)^{3/2}]}{3\sqrt{2Gm_e}} = \frac{2R_e^{\;3/2}[(2)^{3/2} - 1]}{3\sqrt{2Gm_e}}$$

$$t = \frac{2(6.38 \times 10^6\ m)^{3/2}[(2)^{3/2} - 1]}{3\sqrt{(2)(6.67{\times}10^{-11}\ N{\cdot}m^2/kg^2)(5.98{\times}10^{24}\ kg)}}$$

$$t = \frac{2(1.61 \times 10^{10}\ m^{3/2})(1.83)}{3(2.82 \times 10^7\ N^{\frac{1}{2}} \cdot m/kg^{\frac{1}{2}})}$$

$t = 696.53$ s $= 11.61$ minutes

This is the time required for the object to reach $r = 2R_e$.

i) Find the gravitational field at a distance x from an infinite plane sheet of surface density σ.

ii) Compare this result with the field just outside a spherical shell of the same surface density.

Solution: Gauss' Law for Gravitation can be written as

$$\frac{1}{4\pi G} \oint \vec{g} \cdot \vec{ds} = -m.$$

G is the universal gravitation constant, $\vec{g}$ is the gravity potential, $\vec{ds}$ is a vector with magnitude equal to a differential surface area and direction perpendicular to the differential, and - m is the mass enclosed by the surface of integration.

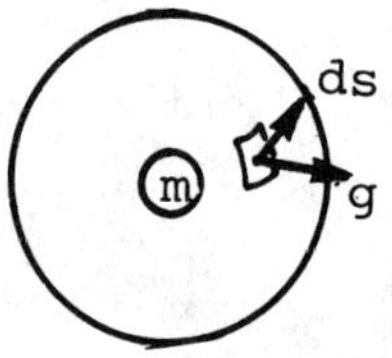

In words, the law states that the mass inside an arbitrary surface is proportional to the integral of $\vec{g} \cdot \vec{ds}$ over the entire surface.

The key step in solving a Guassian integration problem is to pick a surface that makes $\oint \vec{g} \cdot \vec{ds}$ "simple." This means don't make the surface too complex and orient it conveniently with respect to $\vec{g}$. A little ingenuity goes a long way here.

A convenient surface to pick for the plane sheet is a cylinder of radius r_0 perpendicular to the sheet.

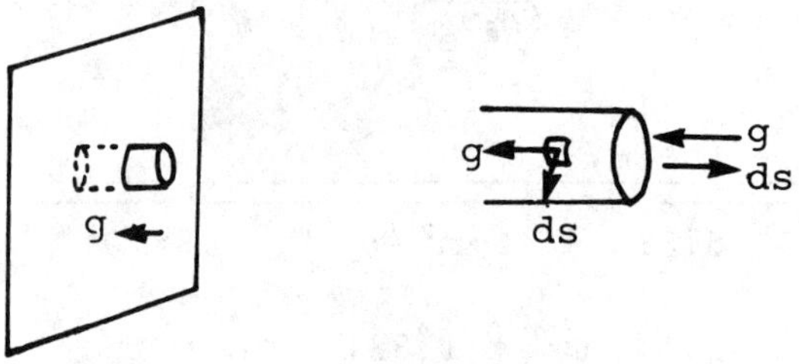

$\vec{g}$ will be perpendicular to the sheet, therefore, on the sides of the cylinders, $\vec{ds}$ will be perpendicular to $\vec{g}$ and $\vec{g} \cdot \vec{ds} = g\, ds \cos\theta$ (θ is the angle between g and ds)

$$\vec{g} \cdot \vec{ds} = g\, ds \cos 90°$$

$\vec{g} \cdot \vec{ds} = 0$ on the sides of the cylinder.

At the end caps, $\vec{g}$ is anti-parallel to $\vec{ds}$; therefore, $\vec{g} \cdot \vec{ds} = g\,(ds) \cos 180°$

$$\vec{g} \cdot \vec{ds} = -g\,ds$$

Integrating over both end caps (the integral over the sides is zero),

$$\oint \vec{g} \cdot \vec{ds} = -2\pi r_0^2\, g$$

and, from Gauss' Law,

$$-m = \frac{1}{4\pi G}(-2\pi\, r_0^2\, g)$$

$$g = \frac{2Gm}{r_0^2}.$$

m is simply the mass density σ multiplied by the cap area πr_0^2.

g then is equal to :

$$g = \frac{2\,G\,\pi\,r_0^2\,\sigma}{r_0^2}$$

$$g = 2\,\pi\,G\,\sigma.$$

For a spherical shell, pick a sphere incrementally larger than the Gaussian surface.

Note that $\vec{g}$ is anti-parallel to $\vec{ds}$ over the whole surface. Therefore,

$$\vec{g} \cdot \vec{ds} = g\,ds \cos 180°$$

$$= -\,g\,ds$$

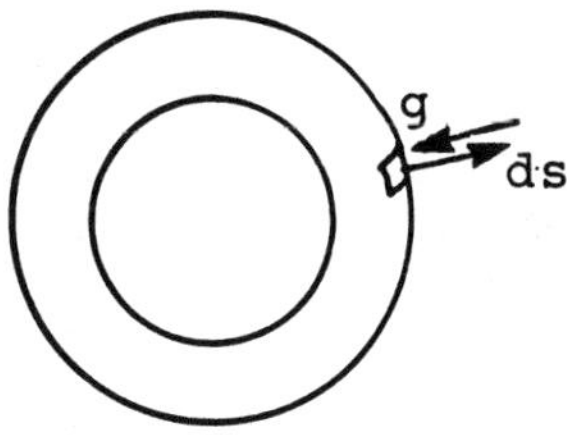

and $\oint \vec{g} \cdot \vec{ds} = \oint -\,g\,ds$

$$\oint \vec{g} \cdot \vec{ds} = -\,g \quad ds$$

$$\oint \vec{g} \cdot \vec{ds} = (4\,\pi\,r_0^2)\,(-g)$$

r_0 is the sphere's radius.

Now, $\frac{1}{4\pi G}(-4\,\pi\,r_0^2\,g) = -\,m$ from Gauss' Law,

and $m = 4\,\pi\,r_0^2\,\sigma$ as before.

Finally, combining the two,

$$\frac{-g r_0^2}{G} = -4 \pi r_0^2 \sigma$$

$$g = 4 \pi \sigma G$$

Therefore, the gravitational field just outside a thin spherical shell is twice as great as the field outside an infinite plane sheet of the same surface density.

• PROBLEM 15-7

Derive Newton's universal law for gravitation from Kepler's laws as well as the supposition that the force depends on the product of the masses of two bodies interacting.

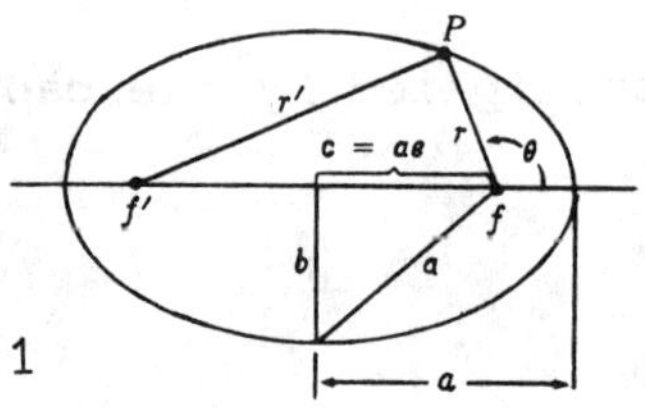

Fig 1

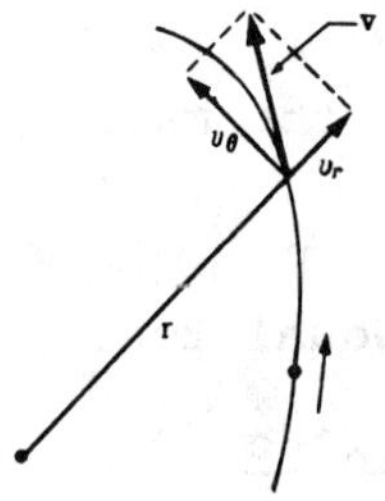

Fig 2

Solution: Kepler's first law, that the planets move in ellipses with the sun at one focus, establishes the shape of the orbit we will analyze.

Let us review the mathematical description of an ellipse. In Fig. 1 is shown an ellipse with one of its foci at the origin of a coordinate system. The distance from the center of the ellipse to the outer extremity is called the semimajor axis a, and the curve is such that r + r' = 2a, where in the figure the foci are at f and f'. The polar equation for an ellipse is

$$r = \frac{a(1 - e^2)}{1 + e \cos \theta} = \frac{b^2/a}{1 + e \cos \theta} , \qquad (1)$$

where e is the eccentricity and, from the figure, we have

$$\frac{b}{a} = \sqrt{1 - e^2} .$$

The assumption, not totally accurate, we are making here is that the sun does not move under the gravitational influence of the planet orbiting around it. The assumption is not necessary, but it does simplify things without leading to any inaccuracies.

Kepler's second law indicates to us that the gravity is a central force, therefore (r, θ), the polar coordinates used above centered on f, are particularly appropriate for this problem. The second law states: Areas swept out by the radius vector from the sun to the planet in equal times are equal. Analysis shows that this follows from the angular momentum staying constant and that fact is a result of the force being central (i.e. acting directly along a line connecting the masses).

The force then has a component only in the r-direction, and the problem is to compute this force from the known trajectory. The polar components of the velocity, $v_\theta = \omega r$ and $v_r = dr/dt$, are indicated in Fig. 2. In the case of a circle, when r remains constant, the velocity has only a v_θ-component, and we know that the acceleration resulting from this component in the radial direction is $\omega^2 r$ directed toward the center, where $\omega = d\theta/dt$. Now, when the radius r does vary, we have a velocity component $v_r = dr/dt$, and a contribution of d^2r/dt^2 to the radial acceleration. Consequently, the radial force is

$$F_r = m\frac{d^2r}{dt^2} - m\omega^2 r. \qquad (2)$$

We must now merely combine r from Eq. (1) with (2) and convince ourselves that, infeed, the force varies as $1/r^2$. Although the problem is straightforward, there is some algebraic work involved.

The second term in equation (2) may be evaluated using Kepler's second law. As noted this law indicates that the angular momentum $\ell = mr^2\omega$ is constant. Thus, $\omega = \ell/mr^2$ and $m\omega^2 r = \ell^2/mr^3$.

We would like to make a change of coordinates from (r, θ) to (u, θ), where $u = 1/r$, because r appears so often in the denominator. Equation (1) is then $r = 1/u = f(\theta)$. Next, we need to transform d^2r/dt^2.

First,

$$\frac{dr}{dt} = \frac{dr}{d\theta}\frac{d\theta}{dt} = \omega\frac{dr}{d\theta} = \frac{\ell}{m}u^2\frac{dr}{d\theta}$$

also $$\frac{dr}{d\theta} = \frac{d(1/u)}{d\theta} = -\frac{1}{u^2}\frac{du}{d\theta}.$$

Combining we get

$$\frac{dr}{dt} = -\frac{\ell}{m}\frac{du}{d\theta}$$

Differentiating again

$$\frac{d^2r}{dt^2} = -\frac{\ell}{m}\frac{d}{d\theta}\left(\frac{du}{d\theta}\right)\frac{d\theta}{dt} = -\frac{\ell}{m}\omega\frac{d^2u}{d\theta^2}$$

or $$\frac{d^2r}{dt^2} = -\left(\frac{\ell}{m}\right)^2 u^2\frac{d^2u}{d\theta^2}.$$

Now we can rewrite equations (1) and (2):

$$u = \frac{a}{b^2}(1 + e\cos\theta) \qquad (3)$$

$$F = -\frac{\ell^2}{m}u^2\left(\frac{d^2u}{d\theta^2} + u\right) \quad . \qquad (4)$$

From equation (3) we can evaluate $\frac{d^2u}{d\theta^2}$:

$$\frac{d^2u}{d\theta^2} = -\frac{a}{b^2}e\cos\theta$$

substituting

$$F = -\frac{\ell^2}{m}u^2\left(-\frac{a}{b^2}e\cos\theta + \frac{a}{b^2}(1 + e\cos\theta)\right)$$

$$= -\frac{\ell^2}{m}u^2\left(\frac{a}{b^2}\right).$$

This gives us the desired result,

$$F = -\frac{\ell^2}{m}\frac{a}{b^2}\frac{1}{r^2} \qquad (5)$$

that is, F varies as a function of $1/r^2$. However, ℓ, m, a and b all vary for different planets, so we have not yet found a universal law.

Proceeding further: the area of an elliptical orbit is πab.

From the general theory of central force orbits, the area is equal to $\frac{\ell T}{2m}$. Equating, we get

$$T = \frac{2m\,\pi ab}{\ell} \quad . \qquad (6)$$

If we set the coefficient in equation (5) $\frac{\ell^2 a}{mb^2}$ = K a constant, we can use this K to rewrite equation (6)

$$T^2 = 4m^2\pi^2a^2\frac{b^2}{\ell^2} = 4m^2\pi^2a^2\left(\frac{a}{Km}\right)$$

$$T^2 = \frac{4m\,\pi^2}{K}a^3 .$$

Kepler's third law says that T^2 = const a^3 for all the planets, thus, we have shown that K is a constant for all planets and

$$F = -\frac{K}{r^2} .$$

Using the supposition indicated, we will write

$$F = -G\frac{M_s m_p}{r^2}$$

and Kepler's third law indicates G is a universal constant. This is now Newton's universal law of gravitation.

• PROBLEM 15-8

Prove that the gravitational force at all points inside a thin, uniform, spherical shell is zero.

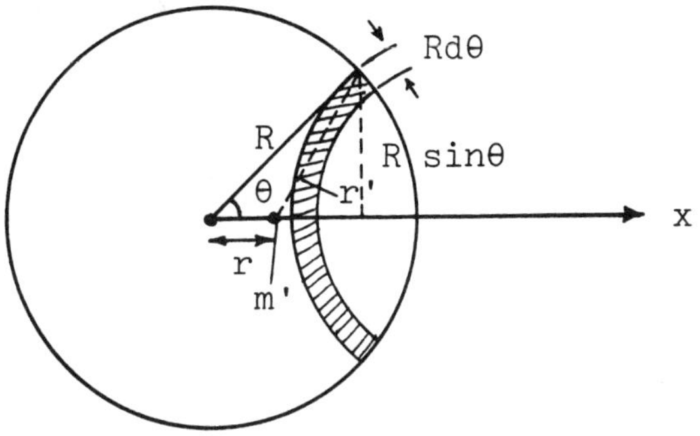

Fig 1

Solution: To find the gravitational force of attraction between a point mass and an extended body, divide the body into infinitesimally small parts, apply Newton's law of gravitational attraction, $F = G\frac{mm'}{r^2}$, then sum of the forces on the point mass due to all parts of the extended body. This process generally involves an integration over the volume of the extended body. We begin by dividing the shell into narrow rings as shown in Fig. 1.

Let us assume that the radius of the thin shell is R, its total mass is M, and the point mass m' is at a distance r from the center of the shell. Then, the narrow ring is of the width Rdθ, where dθ is the angular width of the ring, and the mass of the narrow ring is a fraction of the total mass as may be found from:

$$\text{mass of ring, } dm = \left(\frac{\text{area of ring}}{\text{area of shell}}\right)(M). \qquad (1)$$

Thus,
$$dm = \frac{(2\pi R \sin\theta)(Rd\theta)}{4\pi R^2} M \qquad (2)$$

or
$$dm = \tfrac{1}{2} M \sin\theta \, d\theta. \qquad (3)$$

The force exerted on m' due to a small segment of the narrow ring will be directed toward that segment, as shown in Fig. 2.

Since the angle α (the angle between the x-axis and the direction of the force) will be the same for all segments of the ring, it may be seen that the force components perpendicular to the x-axis will add to zero because of symmetry when we consider all segments of the ring.

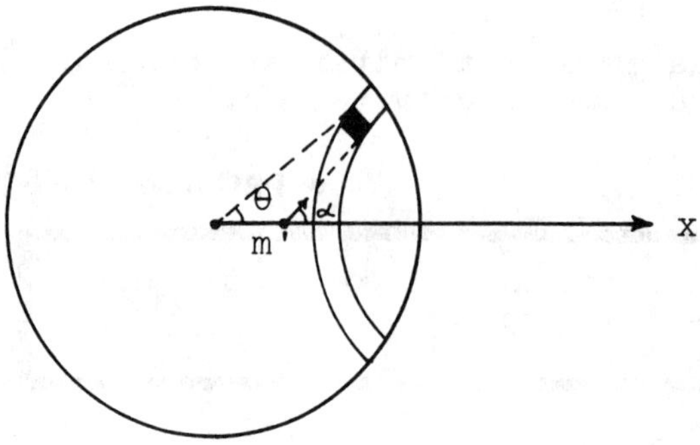

Fig 2

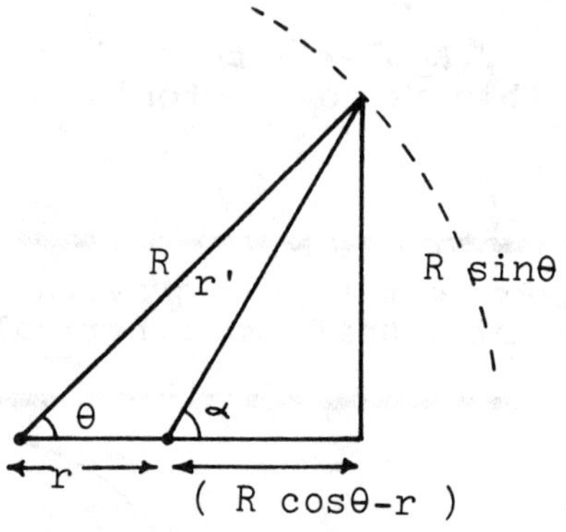

Fig 3

The force components parallel to the x-axis will all be in the same direction and will add up to yield

$$dF_x = G \frac{m' \, dm}{r'^2} \cos \alpha \qquad (4)$$

for one narrow ring.

To find the net force in the x-direction due to all narrow rings comprising the entire spherical shell, we integrate equation (4) to get:

$$F_x = G \, m' \int \frac{\cos \alpha}{r'^2} \, dm \qquad (5)$$

or $$F_x = G \, m' \int \frac{M}{2} \frac{\cos \alpha \sin \theta}{r'^2} \, d\theta. \qquad (6)$$

Equation (6) may be rewritten in terms of a single variable, then integrated. The easiest choice for a variable of integration is the angle θ. Thus, we may write, from the triangle in Fig. 3.

$$r' = \sqrt{R^2 + r^2 - 2rR \cos \theta} \qquad (7)$$

or $$\cos \alpha = (R \cos \theta - r)/r'. \qquad (8)$$

Substituting (7) and (8) into (6) we have:

$$F_x = Gm' \frac{M}{2} \int_0^{\pi} \frac{\sin \theta \, (R \cos \theta - r)}{(R^2 + r^2 - 2rR \cos \theta)^{3/2}} \, d\theta \qquad (9)$$

where the limits of integration are from zero to π in order to cover all points on the shell. The usual substitution for this type of integral is

$$u = R \cos \theta - r \qquad (10)$$

and $du = - R \sin \theta \, d\theta$.

$$F_x = -\frac{Gm'M}{2R}\int_{R-r}^{-R-r} \frac{u\,du}{(R^2 - r^2 - 2ru)^{3/2}}\,. \tag{11}$$

This integral may be found in standard integral tables. The result of integration is:

$$F_x = -\frac{Gm'M}{2R}\,\frac{1}{2r^2}\left[\sqrt{R^2 - r^2 - 2ru} + \frac{(R^2 - r^2)}{\sqrt{R^2 - r^2 - 2ru}}\right]\Bigg|_{R-r}^{-R-r} \tag{12}$$

$$F_x = -\frac{Gm'M}{4Rr^2}\left[\sqrt{R^2 + 2rR + r^2} - \sqrt{R^2 - 2rR + r^2} + (R^2 - r^2)\left(\frac{1}{\sqrt{R^2 + 2rR + r^2}} - \frac{1}{\sqrt{R^2 - 2rR + r^2}}\right)\right]. \tag{13}$$

Collecting terms:

$$F_x = -\frac{Gm'M}{4Rr^2}\left[(R + r) - (R - r) + (R^2 - r^2)\left(\frac{1}{R + r} - \frac{1}{R - r}\right)\right] \tag{14}$$

$$F_x = -\frac{Gm'M}{4Rr^2}\,[2r - 2r]$$

or $F_x = 0$.

Thus, we see that the gravitation force at any point inside of a thin uniform spherical shell is zero.

• PROBLEM 15-9

Calculate, using the calculus, the gravitational potential and force from: (1) a homogeneous spherical shell of mass M, (2) a homogeneously stratified sphere, (3) make some general conclusions concerning the composition of the earth.

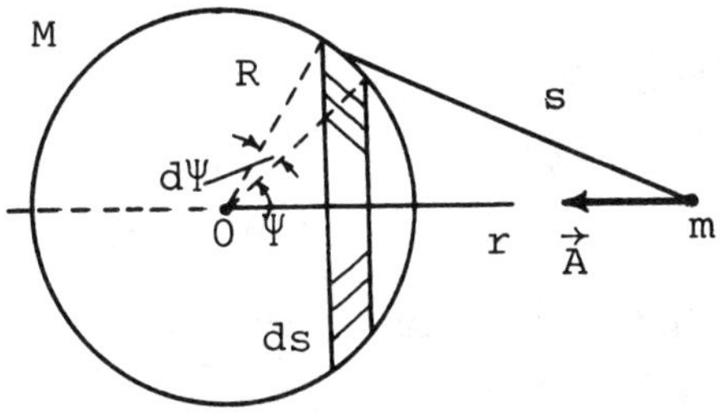

Fig 1

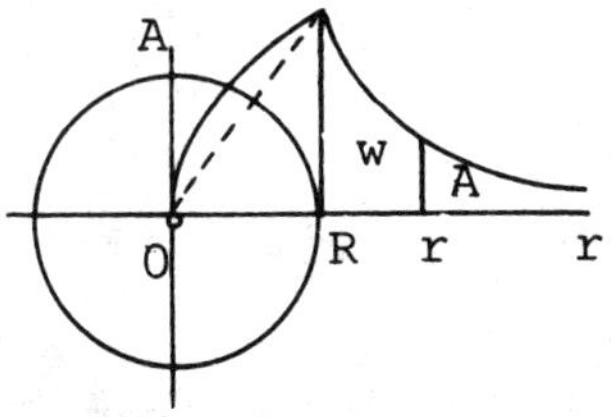

Fig 2

Solution: Figure 1 shows a homogeneous distribution of mass over the surface of a sphere with center O and radius R. A particle m is at distance r from O. If M is the total mass, then the fraction on an element of surface dS will be

$$dM = \frac{M}{4\pi R^2}\, dS. \tag{1}$$

If the narrow band in Fig. 1 is used as a surface element, all parts of this element are the same distance s from the particle. Since its area is:

$$dS = 2\pi R^2 \sin\psi \, d\psi,$$

instead of (1) we have

$$dM = \frac{M}{2} \sin\psi \, d\psi \,. \tag{2}$$

By virtue of symmetry, the corresponding attractive force $d\vec{A}$ is directed along the line joining m and O.

Using $V = -G\frac{Mm}{s}$ and equation (2) to write dV, the force has the differential potential

$$dV = -G\frac{Mm}{2}\; \frac{\sin\psi \, d\psi}{s} \quad . \tag{3}$$

The cosine theorem applied to the triangle with sides s, R, r in Fig. 1 gives

$$s^2 = R^2 + r^2 - 2Rr\cos\psi \; ,$$

and hence by differentiating with respect to ψ and multiplying through by $d\psi$ it follows that

$$s\, ds = Rr \sin\psi \, d\psi,$$

so that (3) can also be written in the form

$$dV = -G\frac{Mm}{2Rr}\, ds.$$

The potential of the resultant attractive force A is therefore

$$V = -G\frac{Mm}{2Rr}\int_{s_1}^{s_2} ds = -G\frac{Mm}{2Rr}(s_2 - s_1), \tag{4}$$

where s_2 is the maximum and s_1 the **minimum value of s.**

If the particle lies outside the sphere, we have $s_2 = r + R$ and $s_1 = r - R$, so that $s_2 - s_1 = 2R$ and by (4)

$$V = -G\frac{Mm}{r} \qquad (r \geq R).$$

The magnitude of the attractive force, being equal to $\frac{dV}{dr}$, is then equal to

$$A = G \frac{Mm}{r^2} \qquad (r \geq R) \qquad (5)$$

and corresponds with that which m would experience if the mass M were concentrated at the center of the sphere. If, on the other hand, the particle is located within the sphere, we have $s_2 = R + r$ and $s_1 = R - r$ so that $s_2 - s_1 = 2r$. Now, since the potential

$$V = - G \frac{Mm}{R} \qquad (r \leq R)$$

is constant, the resultant attraction vanishes.

A homogeneously stratified sphere, i.e., a sphere whose density depends only on the distance from the center, can be subdivided into homogeneous elements in the form of thin spherical shells. Therefore, the attractive force outside the sphere is still given by (5), provided that M is now interpreted as the total mass. In the interior, on the other hand, only those shells not containing the particle contribute to the attraction. If their total mass is denoted by M', we have in analogy to (5)

$$A = G \frac{M'm}{r^2} \qquad (r \leq R) \qquad (6)$$

where M' is to be calculated.

In the case of a homogeneous sphere we have

$$\frac{M'}{M} = \frac{r^3}{R^3} \quad ,$$

and therefore, by (6),

$$A = G \frac{Mm}{R^3} r \qquad (r \leq R). \qquad (7)$$

The attraction in the interior is thus proportional to the distance from the center. As shown in Figure 2, the dotted line for A(r), $r < R$.

In order to calculate the gravitational field of the Earth, it can as a first approximation be treated as a homogeneously stratified sphere. Expression (5) is therefore valid above the surface of the Earth. Now, denoting the magnitude of the attraction on the surface by the usual W, it follows from

$$G \frac{Mm}{R^2} = W = mg \qquad (8)$$

that (5) can also be written in the form

$$A = W \frac{R^2}{r^2} \qquad (r \geq R). \qquad (9)$$

For a satellite moving around the Earth at a distance $r = 2R$ from the center, the magnitude of the attraction is accordingly $A = W/4$.

From (8), the mass of the Earth is given by

$$M = \frac{R^2 g}{G} .$$

This can be used for the calculation of the mean density ρ_m. If, in addition to the data for g and G we use the numerical value R = 6367 km = 3956 miles for the mean radius of the Earth, we obtain ρ_m = 5.5 gm/cm³ = 0.20 lb/in³. However, geological investigations have revealed that the mean density of the Earth's crust is only ρ_a = 2.7 gm/cm³ = 0.10 lb/in³. We thus come to the conclusion that the density becomes very large near the center of the Earth and the variation of the attractive force in the interior is given rather by the solid curve than by the dotted line in Fig. 2.

• PROBLEM 15-10

What is the relation between the total energy and the angular momentum for a 2-body system, each body executing a circular orbit about the system center of mass?

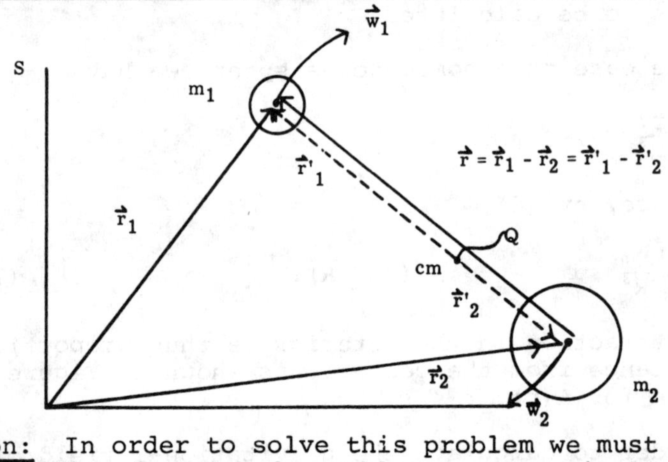

Solution: In order to solve this problem we must transform the given 2-body problem to an equivalent one-body problem. To do this, we find the equation of motion of each mass shown in the figure. Using Newton's Second Law and his Law of Universal Gravitation, we obtain:

$$\vec{F}_{12} = m_2\ddot{\vec{r}}_2 = \frac{Gm_1m_2(\vec{r}_1 - \vec{r}_2)}{|\vec{r}_1 - \vec{r}_2|^3} \tag{1}$$

$$\vec{F}_{21} = m_1\ddot{\vec{r}}_1 = -\frac{Gm_1m_2(\vec{r}_1 - \vec{r}_2)}{|\vec{r}_1 - \vec{r}_2|^3} \tag{2}$$

where $\vec{F}_{12}$ is the force exerted on 2 by 1, and similarly for $\vec{F}_{21}$. Rewriting (1) and (2)

$$\ddot{\vec{r}}_2 = \frac{1}{m_2} \frac{Gm_1 m_2(\vec{r}_1 - \vec{r}_2)}{|\vec{r}_1 - \vec{r}_2|^3} \tag{3}$$

$$\ddot{\vec{r}}_1 = -\frac{1}{m_1} \frac{Gm_1 m_2(\vec{r}_1 - \vec{r}_2)}{|\vec{r}_1 - \vec{r}_2|^3} \tag{4}$$

Subtracting (3) from (4), and using the figure to realize that $\vec{r}_1 - \vec{r}_2 = \vec{r}$, we obtain

$$\ddot{\vec{r}}_1 - \ddot{\vec{r}}_2 = \ddot{\vec{r}} = -\frac{1}{m_1}\frac{Gm_1 m_2 \vec{r}}{r^3} - \frac{1}{m_2}\frac{Gm_1 m_2 \vec{r}}{r^3}$$

or $$\ddot{\vec{r}} = -\frac{Gm_1 m_2 \vec{r}}{r^3}\left(\frac{1}{m_1} + \frac{1}{m_2}\right)$$

Defining the reduced mass μ as

$$\frac{1}{\mu} = \frac{1}{m_1} + \frac{1}{m_2} \tag{5}$$

we find $$\mu\ddot{\vec{r}} = -\frac{Gm_1 m_2 \vec{r}}{r^3} \tag{6}$$

This equation is a one body equation describing the motion of a particle of mass μ under the influence of a gravitational force.

Now, to further reduce the problem, assume that each mass shown in the figure rotates in a circular orbit with the given angular velocitites about Q, the center of mass. Using Newton's Second Law for each mass,

$$\frac{m_1 v_1'^2}{r_2'} = \frac{Gm_1 m_2}{r^2} \tag{7}$$

$$\frac{m_2 v_2'^2}{r_2'} = \frac{Gm_1 m_2}{r^2} \tag{8}$$

where the primed variables are measured with respect to the point Q. By definition of the center of mass

$$m_1\vec{r}_1' + m_2\vec{r}_2' = 0$$

or $$\vec{r}_1' = -\frac{m_2\vec{r}_2'}{m_1} \tag{9}$$

Furthermore, $\vec{r} = \vec{r}_1' - \vec{r}_2'$

or $$\vec{r}_2' = \vec{r}_1' - \vec{r} \tag{10}$$

Inserting (10) in (9), and solving for $\vec{r}_1'$

$$\vec{r}_1' = -\frac{m_2}{m_1}\vec{r}_2' = -\frac{m_2}{m_1}(\vec{r}_1' - \vec{r})$$

$$\vec{r}_1'\left(1 + \frac{m_2}{m_1}\right) = \frac{m_2}{m_1}\vec{r}$$

$$\vec{r}_1' = \frac{m_2/m_1\ \vec{r}}{1 + m_2/m_1} = \frac{m_2\vec{r}}{m_1 + m_2} \qquad (11)$$

But using (5)

$$\mu = \frac{m_1 m_2}{m_1 + m_2}$$

Hence, (9) becomes $\vec{r}_1' = \frac{\mu}{m_1}\vec{r}$ (12)

Similarly, $\vec{r}_2' = -\frac{\mu}{m_2}\vec{r}$ (13)

Hence

$$r_1' = \frac{\mu}{m_1} r$$

(14)

$$r_2' = \frac{\mu}{m_2} r$$

Using (14) in (7) and (8)

$$\frac{m_1^2 v_1^{\prime 2}}{\mu r} = \frac{Gm_1m_2}{r^2}$$

$$\frac{m_2^2 v_2^{\prime 2}}{\mu r} = \frac{Gm_1m_2}{r^2}$$

or

$$\frac{m_1 v_1^{\prime 2}}{2} = \frac{\mu\ Gm_2}{2r}$$

$$\frac{m_2 v_2^{\prime 2}}{2} = \frac{\mu\ Gm_1}{2r}$$

Therefore, the net kinetic energy of the system relative to the center of the mass is

$$T = \tfrac{1}{2} m_1 v_1^{\prime 2} + \tfrac{1}{2} m_2 v_2^{\prime 2} = \frac{\mu G}{2r}(m_1 + m_2) \qquad (15)$$

$$T = \frac{Gm_1m_2}{2r} \qquad (16)$$

by definition of μ. The total energy is

$$E = T + V = \frac{Gm_1m_2}{2r} - \frac{Gm_1m_2}{r}$$

$$E = -\frac{Gm_1m_2}{2r} \qquad (17)$$

To remove the variable r, we replace it with the angular momentum $\vec{J}$ as follows. The total system angular momentum is (relative to Q)

$$\vec{J} = \vec{r}_1' \times m_1\vec{v}_1' + \vec{r}_2' \times m_1\vec{v}_2'$$

Since $\vec{r}_1'$ and $\vec{v}_1'$ are perpendicular, and similarly for $\vec{r}_2'$ and $\vec{v}_2'$, we obtain

$$J = m_1r_1'v_1' + m_2r_2'v_2' \qquad (18)$$

From (12) and (13)

$$\vec{v}_1' = \frac{\mu}{m_1}\vec{v} \qquad \vec{r}_1' = \frac{\mu}{m_1}\vec{r}$$

$$\vec{v}_2' = -\frac{\mu}{m_2}\vec{v} \qquad \vec{r}_2' = -\frac{\mu}{m_2}\vec{r}$$

or

$$v_1' = \frac{\mu}{m_1}v \qquad r_1' = \frac{\mu}{m_1}r$$

$$(19)$$

$$v_2' = \frac{\mu}{m_2}v \qquad r_2' = \frac{\mu}{m_2}r$$

Using (19) in (18)

$$J = (m_1)\left(\frac{\mu}{m_1}r\right)\left(\frac{\mu}{m_1}v\right) + (m_2)\left(\frac{\mu}{m_2}r\right)\left(\frac{\mu}{m_2}v\right)$$

$$J = \frac{\mu^2}{m_1}vr + \frac{\mu^2}{m_2}vr = \mu^2 vr\left(\frac{1}{m_1} + \frac{1}{m_2}\right)$$

By definition of μ

$$J = \mu vr \qquad (20)$$

We now eliminate v in (20) so that we may substitute (20) in place of r in (17). We know, from (15) that

$$T = \tfrac{1}{2}m_1v_1'^2 + \tfrac{1}{2}m_2v_2'^2 = \frac{\mu G}{2r}(m_1 + m_2) \qquad (15)$$

Substituting for v_1', v_2' from (19) in (15),

$$T = \tfrac{1}{2}m_1\left(\frac{\mu^2}{m_1^2}v^2\right) + \tfrac{1}{2}m_2\left(\frac{\mu^2}{m_2^2}v^2\right) = \frac{\mu G}{2r}(m_1 + m_2)$$

$$T = \frac{\mu^2v^2}{2m_1} + \frac{\mu^2v^2}{2m_2} = \frac{\mu G}{2r}(m_1 + m_2)$$

$$T = \frac{\mu^2 v^2}{2}\left(\frac{1}{m_1} + \frac{1}{m_2}\right) = \frac{\mu G}{2r}(m_1 + m_2)$$

By definition of μ

$$T = \frac{\mu v^2}{2} = \frac{\mu G}{2r}(m_1 + m_2) \qquad (21)$$

Hence $$v^2 = \frac{G}{r}(m_1 + m_2) \qquad (22)$$

Using (22) in (20)

$$J = \mu r \sqrt{\frac{G\left(m_1 + m_2\right)}{r}}$$

or $J^2 = \mu^2 r G(m_1 + m_2)$

Therefore $$r = \frac{J^2}{\mu^2 G(m_1 + m_2)} \qquad (23)$$

Inserting (23) in (17)

$$E = -\frac{Gm_1m_2}{2}\left(\frac{\mu^2 G(m_1 + m_2)}{J^2}\right)$$

$$E = -\frac{G^2 m_1m_2(m_1 + m_2)\mu^2}{2J^2}$$

Finally,

$$m_1m_2(m_1 + m_2)\mu^2 = m_1m_2(m_1 + m_2)\left(\frac{m_1m_2}{m_1 + m_2}\right)\mu$$

or $m_1m_2(m_1 + m_2)\mu^2 = m_1^2m_2^2\,\mu$

Then $$E = -\frac{G^2\,\mu\,m_1^2m_2^2}{2J^2}$$

• PROBLEM 15-11

Find the equation of motion of a particle moving with respect to an inertial polar coordinate system under a central gravitational force.

Solution: Consider a particle moving with respect to an inertial polar coordinate system. The position of the particle in the plane is given by polar coordinates r and θ (Fig. 1). In this general case the velocity vector $\vec{v}$ of particle one need not be tangent to the circle of radius r. The velocity vector may be resolved into radial and tangential components. Using

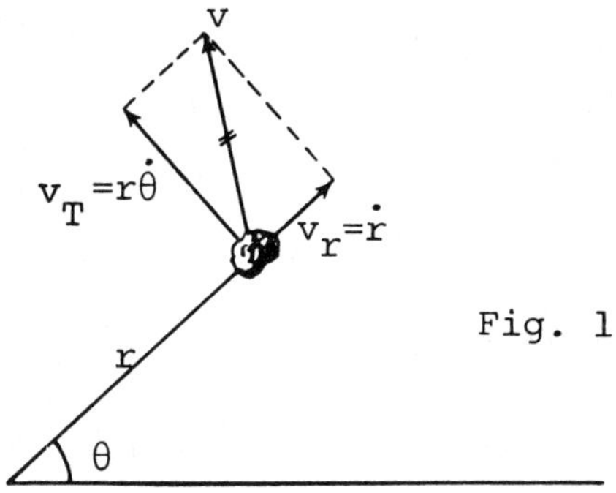

Fig. 1

$$\vec{v} = \dot{r}\hat{r} + r\dot{\theta}\hat{\theta}$$

and the Pythagorean theorem, we may write

$$v^2 = \dot{r}^2 + r^2\dot{\theta}^2. \tag{1}$$

Suppose that the particle is attracted to the origin of the coordinate system by a central gravitational force. We shall assume that the body at the origin which causes this attraction has very much more mass than the particle so that the body at the origin can be considered to be at rest.

If no external forces act on the two particles of this system, then the angular momentum of the system is conserved. Particle one is the only part of the system which can have angular momentum because it is the only part which is moving. The magnitude of the angular momentum of particle one about the origin is given by the expression

$$\ell = mr^2\dot{\theta} \tag{2}$$

where ℓ is a constant.

If there are no frictional or dissipative forces acting on the system the total energy is conserved. The kinetic energy of the particle is given by

$$T = \tfrac{1}{2}mv^2 = \tfrac{1}{2}m(\dot{r}^2 + r^2\dot{\theta}^2).$$

The potential energy is given by:

$$V = -\,GMm/r,$$

where m is the mass of the particle, M is the mass of the body at the origin, and G is the gravitational constant . The total energy of the particle is given by

$$E = T + V = \tfrac{1}{2}m(\dot{r}^2 + r^2\dot{\theta}^2) - \frac{GMm}{r} \tag{3}$$

where E is a constant.

Using Eqs. (2) and (3), we shall now show that particle one moves along a path which is a conic section.

Solve Eq. (2) for $\dot{\theta}$.

$$\dot{\theta} = \frac{\ell}{mr^2} = \frac{d\theta}{dt} \qquad (4)$$

Substitute this into Eq. (3).

$$E = \tfrac{1}{2}m\left[\dot{r}^2 + r^2\left(\frac{\ell}{mr^2}\right)^2\right] - \frac{GMm}{r}$$

Solve this equation for $\dot{r} = dr/dt$.

$$\frac{dr}{dt} = \sqrt{\frac{2E}{m} + \frac{2GM}{r} - \frac{\ell^2}{m^2r^2}}\,. \qquad (5)$$

Equations (4) and (5) tell us how r and θ change in a small increment of time dt. In a time dt, r will change by an amount dr and θ will change by an amount dθ. In any given time interval dt the relation between the changes dr and dθ can be calculated from Eqs. (4) and (5)

$$dt = \frac{dr}{\sqrt{2E/m + 2GM/r - \ell^2/m^2r^2}} = \frac{mr^2\ d\theta}{\ell}$$

or, after some manipulation,

$$d\theta = \frac{\ell\ dr}{r\sqrt{2mEr^2 + 2GMm^2\ r - \ell^2}}\,. \qquad (6)$$

We can find a relation between θ and r by integrating this equation.

$$\int_{\theta_1}^{\theta} d\theta = \int_{r_1}^{r} \frac{\ell\ dr}{r\sqrt{2mEr^2 + 2GMm^2r - \ell^2}}$$

where r_1 and θ_1 are the values of r and θ at some initial time. The left side of this equation is easily integrated to give $\theta - \theta_1$. The right side is an integral whose form we have not encountered before. It is necessary to consult a table of integrals. The solution gives

$$\theta - \theta_1 = \arcsin \frac{GMm^2r - \ell^2}{r(G^2M^2m^4 + 2mE\ell^2)^{\frac{1}{2}}}\Bigg|_{r_1}^{r}$$

The symbol arcsin A means "the angle whose sine is A."

Evaluating the right side of the last equation at the limits,

$$\theta - \theta_1 = \arcsin \frac{GM\ m^2r - \ell^2}{r(G^2M^2m^4 + 2mE\ell^2)^{\frac{1}{2}}}$$

$$- \arcsin \frac{GMm^2r_1 - \ell^2}{r_1(G^2M^2m^4 + 2mE\ell^2)^{\frac{1}{2}}}\,. \qquad (7)$$

Equation (7) gives a formal solution to the problem, that is, the functional relationship $\theta = f(r)$. It is useful but not at all transparent in giving information about the trajectory of a particle. We can go further by the simplifcation of choosing θ_1 to be equal to the second arcsine term on the right of (7) minus $\pi/2$. This is easily done by assuming a rotation of the axes so that θ_1 assumes the proper value. With this substitution we get

$$\theta - \left(-\frac{\pi}{2}\right) = \arcsin \frac{GMm^2r - \ell^2}{r(G^2M^2m^4 + 2mE\ell^2)^{\frac{1}{2}}} \quad .$$

Take the sine of both sides of this equation.

$$\sin\left(\theta + \frac{\pi}{2}\right) = \cos\theta = \frac{GMm^2r - \ell^2}{r(G^2M^2m^4 + 2mE\ell^2)^{\frac{1}{2}}} \quad .$$

We can now multiply both sides of this equation by the denominator of the right side and solve for r.

$$\ell^2 = r[GMm^2 - (G^2M^2m^4 + 2mE\ell^2)^{\frac{1}{2}} \cos\theta]$$

or
$$r = \frac{\ell^2/GMm^2}{1 - (1 + 2\ell^2E/G^2M^2m^3)^{\frac{1}{2}} \cos\theta} \quad . \tag{8}$$

Comparing this to the equation for an ellipse

$$r = \frac{b^2/a}{1 + e\cos\phi}$$

we can see that this is the equation of a conic section in which

$$e = \left(1 + \frac{2\ell^2E}{G^2M^2m^3}\right)^{\frac{1}{2}}$$

and $\phi = (\theta - \pi)$.

It is possible to tell in any individual case whether the path of particle one is a hyperbola or an ellipse, etc., by calculating e and by using the criteria

	ORBIT	
eccentricity	shape	nature
$e < 1 < 0$	ellipse	(closed)
$e = 1$	parabola	(open)
$e > 1$	hyperbola	(open)
$e = 0$	circle	(closed)

ORBITS

• PROBLEM 15-12

The eccentricity of the orbit of a, particle of mass m moving around a fixed mass M is given by

$$e = \left(1 + \frac{2\ell^2 E}{G^2 M^2 m^3}\right)^{\frac{1}{2}}$$

where ℓ is the angular momentum, E is the total energy of the system. Show that if E is positive the orbit is a hyperbola; if E is zero, the orbit is a parabola; and if E is negative, the orbit is an ellipse. Under what condition is the orbit a circle?

Solution: The general equation in polar coordinates for the conic sections is

$$r = \frac{ep}{1 + e \cos \theta} \qquad (1)$$

where e is the eccentricity and p is the distance from the focus of the curve to the directrix.

The following curves are identified by the value of the eccentricity:

Circle: $e = 0$

Ellipse: $0 < e < 1$

Parabola: $e = 1$

Hyperbola: $e > 1$

In the given expression for the eccentricity in terms of the total energy, if E is positive the e must be greater than 1, hence the curve is a hyperbola.

If $E = 0$, then $e = 1$ and the curve is a parabola.

If E is less than zero, but greater than $-\frac{G^2M^2m^3}{2\ell^2}$, then $1 > e > 0$, and the orbit will be an ellipse.

Finally, if $E = -\frac{G^2M^2m^3}{2\ell^2}$, $e = 0$, and the orbit will be a circle.

• PROBLEM 15-13

Show that the energy of an elliptic orbit can be written as

$$E = -\frac{GMm}{r_{max} + r_{min}} .$$

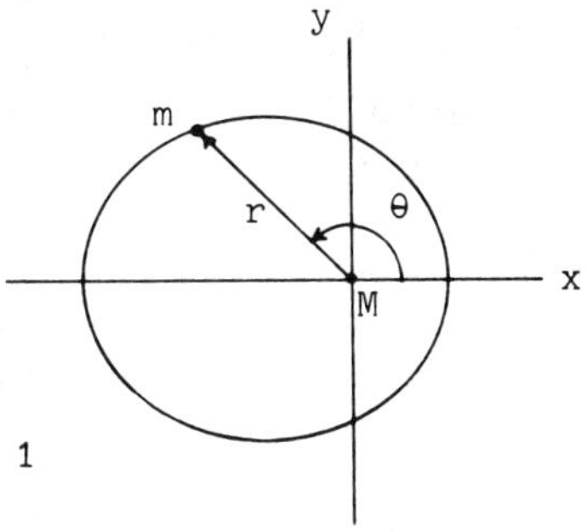

Fig 1

Solution: We shall use the known solution of an elliptical orbit to answer this question. The equation for the orbit is

$$\frac{1}{r} = \frac{1}{\alpha} + \frac{\varepsilon}{\alpha}\cos\theta \quad , \tag{1}$$

where $\alpha = \dfrac{\ell^2}{GMm\mu}$ and $\varepsilon = \sqrt{1 + \dfrac{2E\ell^2}{G^2M^2m^2\mu}}$.

In the above equations, the angular momentum $\ell = \mu r^2\dot{\theta}$ is constant and $\mu = \dfrac{Mm}{M + m}$ is the reduced mass. Remember that ε is the eccentricity and for an elliptical orbit $0 < \varepsilon < 1$.

Fig. 1 illustrates such an orbit. The axis of symmetry is along the x-axis and θ is measured with respect to the positive x-axis. From Fig. 1 we see that the minimum value of r occurs when $\theta = 0$ and the maximum value of r occurs when $\theta = \pi$ radians. Substituting these values into Eq. (1) and simplifying the resultant algebraic equations we obtain:

$$r_{min} = \frac{\alpha}{1 + \varepsilon} \quad \text{and} \tag{2}$$

$$r_{max} = \frac{\alpha}{1 - \varepsilon} \quad . \tag{3}$$

Adding Eqs. (2) and (3) yields:

$$r_{max} + r_{min} = \frac{\alpha}{1 - \varepsilon} + \frac{\alpha}{1 + \varepsilon} = \frac{2\alpha}{1 - \varepsilon^2} \tag{4}$$

Next, solve Eq. (4) for $1 - \varepsilon^2$. This yields,

$$1 - \varepsilon^2 = \frac{2\alpha}{r_{max} + r_{min}} \quad . \tag{5}$$

Substituting the values for α and ε into Eq. (5) we obtain the desired result:

$$1 - \left(1 + \frac{2E\ell^2}{G^2M^2m^2\mu}\right) = \frac{2\ell^2}{GMm\mu}\left(\frac{1}{r_{max} + r_{min}}\right) \quad \text{or}$$

$$E = -\frac{GMm}{r_{max} + r_{min}} \quad . \tag{6}$$

Note that $r_{max} + r_{min}$ is equal to twice the length of the major axis of the elliptical orbit.

• PROBLEM 15-14

The eccentricity of a particle's orbit in a polar coordinate system can be written as

$$e = \left(1 + \frac{2\ell^2 E}{G^2M^2m^3}\right)^{\frac{1}{2}} .$$

Show that the energy of an orbit is given by the expression

$$E = \frac{G^2M^2m^3}{2\ell^2}(e^2 - 1) = \frac{GMm}{2\mu e}(e^2 - 1),$$

where $\mu = \dfrac{\ell^2}{GMm^2e}$. For circular orbits, show that this reduces to $E = -\frac{1}{2}mv^2$.

Solution: Starting with the given expression for eccentricity, we can solve for the total energy, E. Squaring,

$$e^2 = 1 + \frac{2\ell^2E}{G^2M^2m^3} . \qquad (1)$$

Transposing terms,

$$\frac{2\ell^2}{G^2M^2m^3} E = e^2 - 1 .$$

Finally, dividing both sides by the coefficient of E,

$$E = \frac{G^2M^2m^3(e^2 - 1)}{2\ell^2} . \qquad (2)$$

Multiplying both numerator and denominator of (2) by e, and regrouping terms:

$$E = \frac{GMm^2e}{\ell^2} \cdot \frac{GMm(e^2 - 1)}{2e} \qquad (3)$$

This can be written as:

$$E = \frac{GMm(e^2 - 1)}{2\mu e} \qquad (4)$$

where $\dfrac{GMm^2e}{\ell^2} = \dfrac{1}{\mu}$.

For circular orbits, the eccentricity, $e = 0$. Thus,

$$E = -\frac{G^2M^2m^3}{2\ell^2} \qquad (5)$$

and the angular momentum $\ell = mvr$, where v is the velocity of the particle and r is the radius of the orbit.

Equation (5) then reduces to:

$$E = -\frac{1}{2}\frac{G^2M^2m^3}{m^2v^2r^2} = -\frac{1}{2}\frac{G^2M^2m}{v^2r^2} . \quad (6)$$

But, for circular motion of a mass m about a fixed mass M, the force of gravitational attraction equals the centripetal force (equilibrium condition):

$$\frac{GMm}{r^2} = m\frac{v^2}{r} \quad (7)$$

or $$GM = rv^2. \quad (8)$$

Finally, substituting (8) into (6), we get:

$$E = -\frac{1}{2}\frac{r^2v^4 m}{v^2 r^2} = -\frac{1}{2}mv^2 .$$

• PROBLEM 15-15

Show that the period of an elliptic orbit can be written in the form

$$T = \frac{2\pi GM}{(-2E/m)^{3/2}} .$$

Solution: This problem can be solved by obtaining an expression for the semi-major axis, a, and substituting this expression into the equation for the period of an elliptical orbit, namely

$$T = \frac{2\pi}{\sqrt{G(M+m)}} a^{3/2} . \quad (1)$$

To obtain the expression for the semi-major axis we shall use the equation for the trajectory of an elliptical orbit,

$$\frac{1}{r} = \frac{1}{\alpha} + \frac{\varepsilon}{\alpha}\cos\theta , \quad (2)$$

where $\alpha = \frac{\ell^2}{GMm\mu}$ and $\varepsilon = \sqrt{1 + \frac{2E\ell^2}{G^2M^2m^2\mu}}$.

From Eq. (2) we can calculate the minimum value of r and the maximum value of r. The minimum value of r occurs when $\theta = 0$, and substituting this value into Eq. (2) yields

$$r_{min} = \frac{\alpha}{1+\varepsilon} . \quad (3)$$

Also note that the maximum value of r occurs when $\theta = \pi$. Substituting this value into Eq. (2), yields

$$r_{max} = \frac{\alpha}{1-\varepsilon} . \quad (4)$$

But the semi-major axis is equal to

$$a = \frac{1}{2}(r_{min} + r_{max}). \tag{5}$$

Substituting Eqs. (3) and (4) into Eq. (5) and simplifying, we get:

$$a = \frac{\alpha}{1 - \varepsilon^2} . \tag{6}$$

Next, substitute the equations for α and ε into Eq. (6), this yields

$$a = \frac{\ell^2}{GMm\mu}\left[\frac{1}{1 - \left(1 + \frac{2E\ell^2}{G^2M^2m^2\mu}\right)}\right] = -\frac{GMm}{2E} . \tag{7}$$

Now substitute Eq. (7) into Eq. (1) and obtain the alternative expression for the period,

$$T = \frac{2\pi}{\sqrt{G(M + m)}}\left(-\frac{GMm}{2E}\right)^{3/2} . \tag{8}$$

Eq. (8) can be further simplified if we assume that $m << M$. Making this assumption yields,

$$T = \frac{2\pi}{\sqrt{GM}}\left(-\frac{GMm}{2E}\right)^{3/2} \quad \text{or}$$

$$T = \frac{2\pi GM}{(-2E/m)^{3/2}} . \tag{9}$$

SATELLITE PROBLEMS

• PROBLEM 15-16

Scientists from NASA are planning to put a satellite into orbit with the following conditions: When all fuel has been consumed, the satellite will be 400 miles from the surface of the earth and will have a velocity of 30,000 ft/sec perpendicular to the radial line through it. Determine the eccentricity e of the proposed orbit. Also, find the apogee of the orbit.

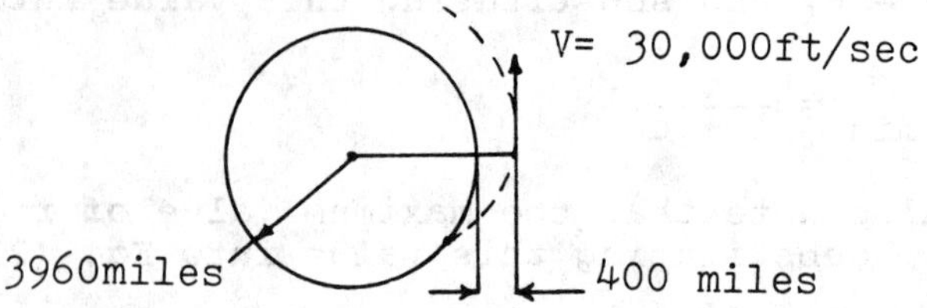

Fig 1

Solution: At first, we do not know the type of path the satellite will follow. However, the general equation is

$$r = \frac{C^2/GM}{1 + e \cos \theta}$$

where $C = r^2\dot{\theta}$

$$G = 3.44 \times 10^{-8} \frac{ft^3}{slug\text{-}sec^2}$$

M = mass of the earth = 4.09×10^{23} slugs

e = eccentricity.

In Fig. 1 are shown the initial conditions, with the velocity perpendicular to the reference line. Since $\theta = 0$ at the start of the free fall orbit, $v = r\dot{\theta}$ and $C = r^2\dot{\theta} = rv$. Hence,

$$C = (3960 + 400)(5280)(30{,}000) = 6.92 \times 10^{11} ft^2/sec.$$

Substituting this value in the general equation with $r = 4360 \times 5280$ and $\cos \theta = 1$, we obtain

$$4360 \times 5280 = \frac{(6.92 \times 10^{11})^2/(3.44 \times 10^{-8} \times 4.09 \times 10^{23})}{1 + e} .$$

Hence, $e = 0.476$ and the path is elliptical.

The apogee occurs just opposite the point of closest approach, $\theta = 0$, that is when $\theta = \pi$ or $\cos \theta = -1$ so

$$r_A = \frac{C^2/GM}{1 - e} = \frac{(6.92 \times 10^{11})^2}{(1 - 0.476)(3.44 \times 10^{-8})(4.09 \times 10^{23})}$$

$$= 6.48 \times 10^7 \text{ ft or } 12{,}270 \text{ miles.}$$

• PROBLEM 15-17

A comet moves into orbit parallel to the surface of the earth at a distance of 800 km. It has a speed of 33,160 km/hr. Find (a) the maximum distance achieved by the comet. The comet completes one orbit when it is perpendicularly struck at point 1 by a meteorite. After impact, the comet stays in orbit, but has a new point of closest approach which is not less than 300 km from the earth's surface. (b) What is the maximum possible angle of deflection caused at the moment of impact?

Solution: Maximum Distance: In Figure (1), point 2 is the farthest distance from the earth. Line r_1 shows the distance from point (2) to the earth's center. For free flight between points (1) + (2), the principle of conservation of energy may be used to calculate r_1.

$$T_1 + V_1 =$$

$$T_2 + V_2 \qquad (1)$$

$$\tfrac{1}{2}mv_0^2 - \frac{GMm}{r_0} =$$

$$\tfrac{1}{2}mv_1^2 - \frac{GMm}{r_1} \qquad (2)$$

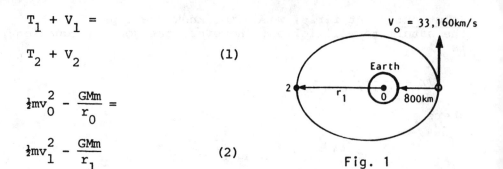

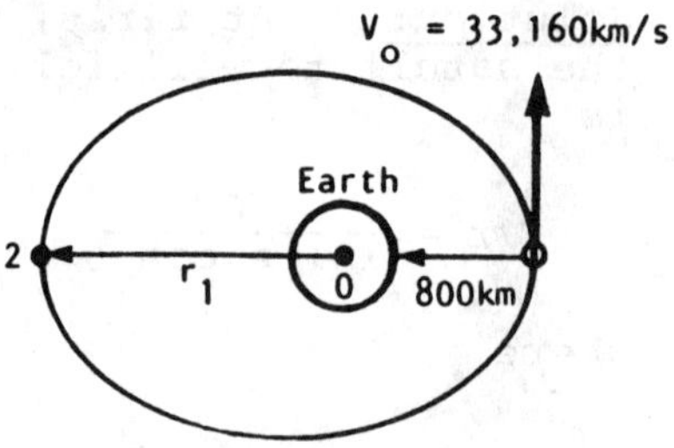

Fig. 1

The only acting force is gravity. It is a central force. The angular momentum of the comet is conserved about point 0, the center of the earth.

Writing the equations for the angular momentum at points (1) and (2), we find an expression for v_1:

$$r_0 m v_0 = r_1 m v_1; \; v_1 = v_0 \frac{r_0}{r_1} \qquad (3)$$

Substituting the expression for v_1 into equation (2), dividing out the term for the comet mass, and using algebra to relate distance r_1 to the known variable, we have:

$$\tfrac{1}{2}v_0^2\left[1 - \frac{r_0^2}{r_1^2}\right] = \frac{GM}{r_0}\left(1 - \frac{r_0}{r_1}\right); \; 1 + \frac{r_0}{r_1} = \frac{2GM}{r_0 v_0^2} \qquad (4)$$

$$\frac{1}{r_1} = \frac{2GM}{r_0^2 v_0^2} - \frac{1}{r_0} \qquad (5)$$

The value of the radius of the earth 6370 km, must be used in our calculations

$r_0 = 6370$ km + 800 km = 7170 km = 7.17×10^6m

$v_0 = 33{,}160$ km/h = $(3.316 \times 10^7$m$)/3.6 \times 10^3$s = 9.21×10^3m/s

$GM = gR^2 = (9.81 \text{ m/s}^2)(6.37 \times 10^6)^2 = 3.98 \times 10^{14} \text{m}^3\text{s}^2$

Substituting these values into equation (5) yields

$$r_1 = 23.24 \times 10^6 \text{m}$$

the maximum distance from the earth is

$$23.24 \times 10^6\text{m} - 6.37 \times 10^6\text{m} = 16.87 \times 10^6\text{m} = 16{,}870 \text{ km}$$

Maximum Angle of Deflection from Impact.

When the comet is struck perpendicularly at point 1, the resulting angle of deflection will determine a new point of closest approach. Let us call it point 3. The closest point 3 can be is r_{min} = 6370 km + 300 km = 6670 km.

The motion imparted by the meteorite is perpendicular to the velocity of the comet. It thus does not directly add to or take away from the momentum of the comet, and the principles of conservation of energy and conservation of momentum hold for this system. Using these principles between points 2 and 3 (see fig. 2) will yield a value for θ.

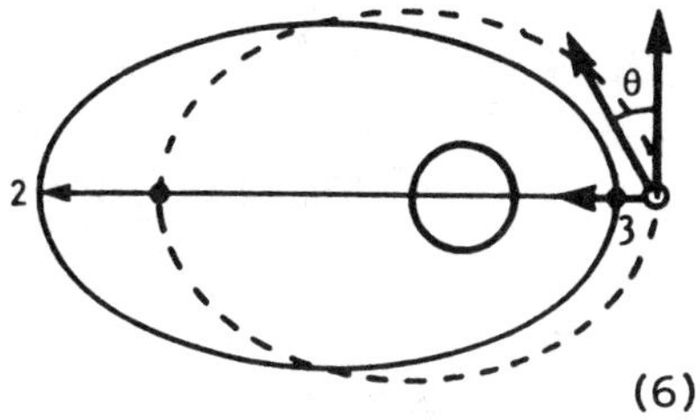

Fig. 2

$$\tfrac{1}{2}mv_0^2 - \frac{GMm}{r} = \tfrac{1}{2}mv_{max}^2 - \frac{GMm}{r_{min}} \qquad (6)$$

$$r_0 m v_0 \sin\theta_0 = r_{min} m v_{max} \qquad (7)$$

Solving equation (6) for v_{max} and substituting that value into equation (7), we may solve of θ_0. Using the given quantities v_0 and GM from part (a) and calculating

$$\frac{r_0}{r_{min}} = 1.075, \text{ we find}$$

$$\sin\theta_0 = .97486\ , \quad \theta_0 = 90° - 12.9° = 77.1°$$

maximum possible deflection = - 12.9°

• PROBLEM 15-18

The satellite shown in Fig. 1 is fired from point E on the surface of the earth. The burnout point, or the point at which all fuel has been expended, is at S. Assume that the initial velocity v_0 at this point is parallel to the earth's surface, or perpendicular to a radius extended, and that the distance to S from the earth's center C is 4500 miles. Determine the magnitude of v_0 in order that the orbit on which the satellite then travels will be (a) a circle and (b) a parabola.

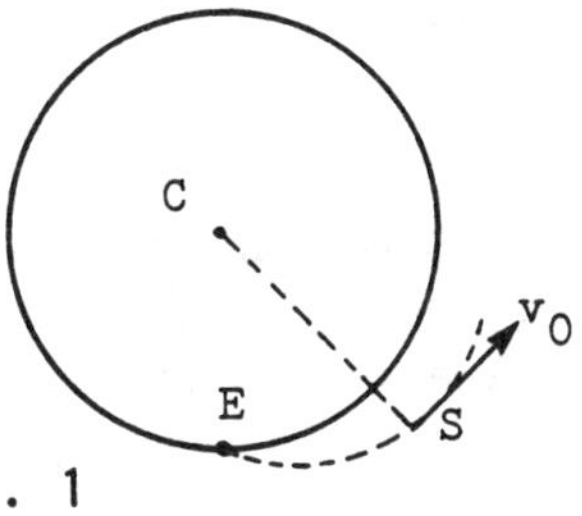

Fig. 1

Solution: Since the velocity is perpendicular to the radius at burnout, the angular momentum then (and afterwards) is:

$$L = mrv_0 = (4500 \text{ miles})(5280 \text{ ft/mile})\, mv_0$$

where m is the mass of the satellite and r is the radius at burnout - 4500 miles.

One form for the general equation for a particle in a gravitational orbit is:

$$r = \frac{L^2/GMm^2}{1 + e\cos\theta} \ . \qquad (1)$$

For a circular orbit, e = 0, therefore $r = \dfrac{r^2 v_c^2}{GM}$

$$\text{or} \quad v_c = \left(\frac{GM}{r}\right)^{\frac{1}{2}}$$

$$= \left[\frac{(3.44 \times 10^{-8}\ ft^3/slug\text{-}sec)(4.09 \times 10^{23}\ slugs)}{(4500)(5280)ft}\right]^{\frac{1}{2}}$$

v_0 = 24,300 ft/sec.

In the orbit equation (1), θ = 0 is the point of nearest approach by the satellite. The parabolic orbit asked for will take the satellite steadily farther from the earth where the eccentricity e = 1. So taking θ = 0 at S we have

$$r = \frac{r^2 v_p^{\ 2}/GM}{2}\ ; \qquad v_p = \left(\frac{2GM}{r}\right)^{\frac{1}{2}} = 1.4 \quad v_c = 34,400 \text{ ft/sec.}$$

• **PROBLEM 15-19**

A comet enters into orbit at a distance of 800 km from the earth's surface. It was traveling with a velocity of 33,160 km/hr parallel to the earth's surface, when it entered the orbit. Determine a) the furthest distance the comet will travel away from earth's surface (b) the periodic time of the comet.

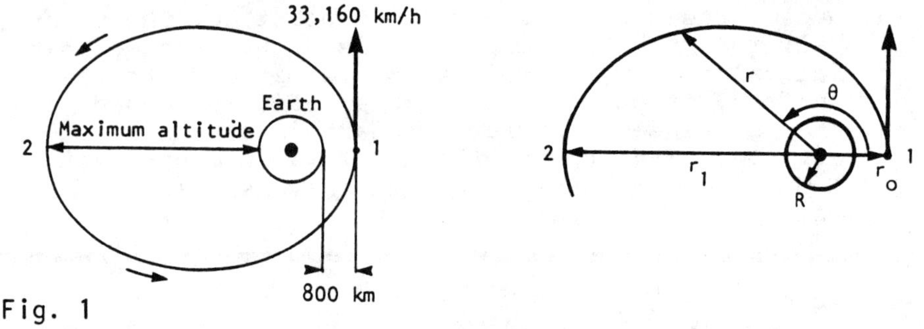

Fig. 1

Solution:

a) Maximum Distance: Upon entering orbit, the earth's gravitational field is the only acting force. It is a centrally directed attractive force giving rise to a motion governed by the following equation;

$$r_1 = \frac{h^2/GM}{1 + e\cos\theta} \tag{1}$$

Traveling parallel to the surface of the earth, the radial velocity component of the comet at point (1) of entry is zero. The angular momentum quantity is $h = r_0 v_0$. Using the radius of the earth R = 6370 km, we can calculate:

$$r_0 = 6370 \text{ km} + 800 \text{ km} = 7170 \text{ km} = 7.170 \times 10^6 \text{m}$$

$$v_0 = 33,160 \text{ km/hr.} = \frac{3.316 \times 10^7 \text{m}}{3.6 \times 10^3 \text{s}} = 9.21 \times 10^3 \text{m/s}$$

$$h = r_0 v_0 = (7.17 \times 10^6 m)(9.21 \times 10^3 m/s) = 6.60 \times 10^{10} m^2/s$$

$$h^2 = 4.36 \times 10^{21}\ m^4/s^2$$

Given $GM = gR^2$; R is the radius of the earth we find:

$$GM = gR^2 = (9.81\ m/s^2)(6.37 \times 10^6 m)^2 = 3.98 \times 10^{14}\ \frac{m^3}{s^2}$$

$$\frac{h^2}{GM} = \frac{4.36 \times 10^{21}\ m^4/s^2}{3.98 \times 10^{14}\ m^3/s^2} = 1.095 \times 10^7\ m$$

Substituting this value into (1), we have:

$$r = \frac{1.095 \times 10^7\ m}{1 + e\cos\theta} \qquad (2)$$

At point (1) $\theta = 0$, $r = r_0 = 7.17 \times 10^6 m$ e can be calculated

$$7.17 \times 10^6 m = \frac{1.09 \times 10^7\ m}{1 + e\cos 0°} \qquad e = .5276$$

At point (2), the maximum distance from the earth, $\theta = 180°$. Using equation (2) we compute the maximum

$$\text{distance } r_1 = \frac{1.09 \times 10^7}{1 + (.5276)(-1)} = 2.31 \times 10^7\ m$$

The maximum distance from the surface of the earth is $23.1 \times 10^6\ m = 6.37 \times 10^6 = 16.73 \times 10^6\ m = 16{,}730$ km

b) Periodic Time

Points (1) and (2) are the perigee and apogee respectively of the elliptical orbit of the comet. Using the equations for the semi major and semiminor axes of the orbit (see Figure 2)

$$a = \tfrac{1}{2}(r_0 + r_1) = \tfrac{1}{2}(7.17 + 16.73)(10^6)m = 11.95 \times 10^6 m$$

$$b = \sqrt{r_0 r_1} = \sqrt{(7.17)(16.73)}(10^6)m = 10.95 \times 10^6\ m$$

$$\tau = \frac{2\pi ab}{h} = \frac{2\pi(11.95 \times 10^6)(10.95 \times 10^6)}{6.60 \times 10^{10}}$$

$$= 1.25 \times 10^4\ \text{sec}$$

$$\tau = 1.25 \times 10^4\ \text{sec}$$

$$= 208.3\ \text{min}$$

$$= 3\ \text{hr}\ 28.3\ \text{min}$$

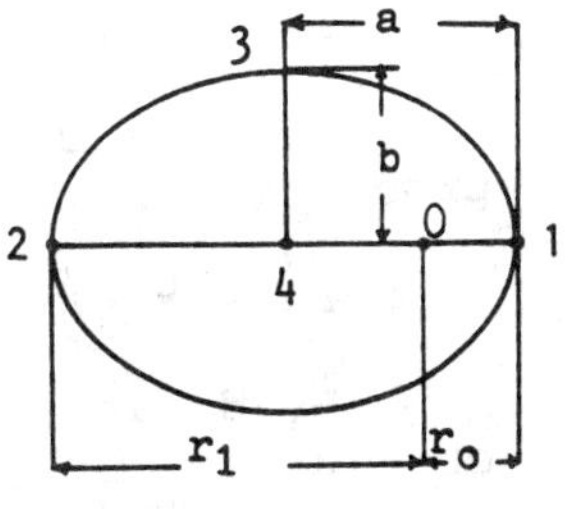

Fig. 2

• **PROBLEM 15-20**

Show that the incremental area ΔA swept out in an incremental time Δt by the position vector to a satellite from the center of attraction is a constant. Show that this result is valid for any central force field, not only the inverse square field.

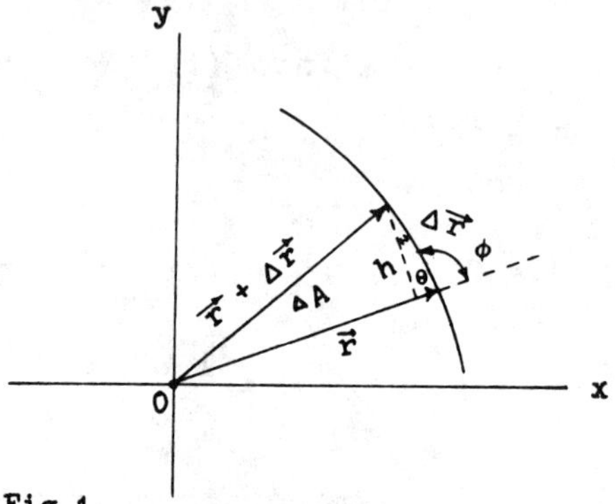

Fig.1

Solution: In order to do this problem, we must obtain an expression for $\dot{A}$, the incremental area per unit time swept out by the radius vector $\vec{r}$. We shall find that $\dot{A}$ is a constant if the angular momentum $\vec{L}$ is a constant.

Let us assume that the mass of the satellite is much smaller than the mass of the planet. Then, the center of mass of the system is located at the position of the planet, point O in FIG. 1. To obtain an expression for ΔA, consider two successive positions of the satellite as shown in Fig. 1. This area is the triangular area bounded by $\vec{r}$, $\Delta\vec{r}$, and $\vec{r} + \Delta\vec{r}$, and therefore

$$\Delta A = \tfrac{1}{2} rh. \tag{1}$$

But the height h of the triangle is equal to

$$h = \Delta r \sin\theta = \Delta r \sin\phi \tag{2}$$

since $\phi = \theta + 90°$, or $\sin\phi = \sin\theta$.

Substituting Eq. (2) into Eq. (1) yields:

$$\Delta A = \tfrac{1}{2}\, r(\Delta r \sin\phi).$$

Eq. (3) can be rewritten in terms of the vector cross product

$$\Delta A = \tfrac{1}{2}\, |\vec{r} \times \Delta\vec{r}| \quad . \tag{4}$$

By dividing Eq. (4) by Δt and taking the limit, we obtain

$$\dot{A} = \tfrac{1}{2} \, |\vec{r} \times \vec{v}| \quad . \tag{5}$$

Multiplying and dividing Eq. (5) by m, the mass of the satellite, yields

$$\dot{A} = \frac{1}{2m} \, |\vec{r} \times m\vec{v}| = \frac{L}{2m} \, , \tag{6}$$

where L is the angular momentum. From Eq. (6) note that $\dot{A}$ is a constant if L is a constant.

By studying angular momentum and torque we know that the rate of change of the angular momentum about the center of mass is equal to the torque about the center of mass or

$$\dot{L} = \vec{r} \times \vec{F} \quad . \tag{7}$$

We can therefore deduce that L is a constant if $\vec{r} \times \vec{F}$ is equal to zero. But $\vec{r} \times \vec{F}$ is equal to zero for any central force since in this case the vectors $\vec{r}$ and $\vec{F}$ are antiparallel. Therefore $\dot{A}$ is a constant for any central force.

• PROBLEM 15-21

A 3000-lb space vehicle is to be launched into a circular parking orbit at an altitude of 100 miles. After making one revolution, a rocket engine on the vehicle fires and gives enough impulse to the vehicle so that it is put into a parabolic escape trajectory. Determine the velocity required to launch the vehicle into the circular orbit, the period of the orbit, the velocity required for escape, and the impulse that the rocket engines must impart to the vehicle in order to achieve the escape velocity.

Solution: The general equation

$$r = \frac{\ell^2/GM}{1 + e \cos \theta} \tag{1}$$

reduces by $\ell = r_0 v_0$, $e = 0$ for a circular orbit to

$$r_0 = r_0^2 \, v_0^2/GM.$$

So $v_0 = \sqrt{GM/r_0}$, where v_0 is the velocity of the circular orbit.

Using 4000 miles for the radius of the earth and $GM = 14.1 \times 10^{15}$ ft^3/sec^2,

$$v_0 = \sqrt{\frac{14.1 \times 10^{15}}{(4000 + 100)5280}}$$

$$= 25{,}500 \text{ ft/sec.}$$

The period for one rotation is

$$T = \frac{2\pi r_0}{v_0}$$

$$= \frac{(6.28)(4000 + 100)(5280)}{25{,}500}$$

$$= 5330 \text{ sec (or 89 min, or 1 hr 29 min).}$$

To find the escape velocity take the general equation (1) and substitute in:

$\ell = r_0 v_e$ where v_e equals the escape velocity,

eccentricity $e = 1$ for parabolic trajectory,

$\theta = 0$ for the start of the parabolic trajectory.

$$r_0 = \frac{r_0^2 v_e^2/GM}{1 + 1}$$

From this the velocity required for escape is

$$v_e = \sqrt{\frac{2GM}{r_0}}$$

or $\sqrt{2}$ times the circular orbit speed. Thus,

$$v_e = 1.41(25{,}500)$$

$$= 35{,}700 \text{ fps}$$

The impulse required to put the vehicle into an escape trajectory is just the change in momentum of the vehicle:

$$\text{Impulse} = \int F\,dt = mv_e - mv_0$$

$$= \left(\frac{3000}{32.2}\right)(35{,}700 - 25{,}500)$$

$$= 950{,}000 \text{ lb-sec.}$$

• PROBLEM 15-22

A comet approaches the sun on an elliptical orbit. Its nearest point of approach is 1.6×10^{11} m from the center of the sun. The corresponding velocity of this point is 4.0×10^4 m/sec. Predict the approximate number of years which will elapse before the comet will return to the same point.

Solution: From studying central force motion $f = f(1/r^2)$ we know that the period of the motion is given by the equation:

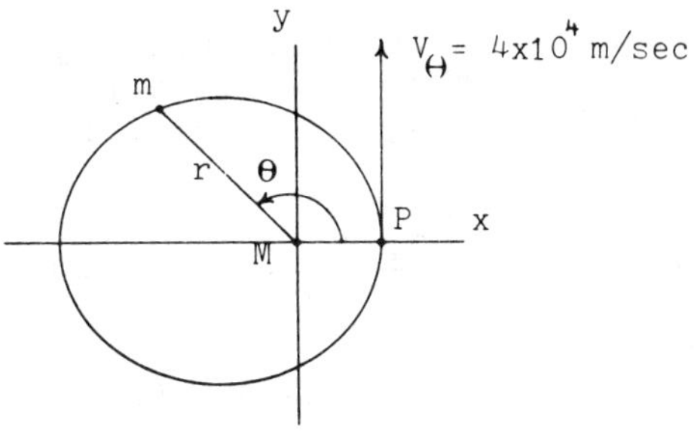

Fig 1

$$T = \frac{2\pi}{\sqrt{G(M + m)}} a^{3/2} \qquad (1)$$

where a is the semi-major axis of the elliptical orbit. We also know that the equation for the orbit is

$$\frac{1}{r} = \frac{1}{\alpha} + \frac{\varepsilon}{\alpha} \cos \theta , \qquad (2)$$

where $\alpha = \frac{\ell^2}{GMm\mu}$ and $\varepsilon = \sqrt{1 + \frac{2E\ell^2}{G^2M^2m^2\mu}}$.

In Eq. (2) $\ell = \mu r^2 \dot{\theta}$ is the angular momentum and $\mu = \frac{Mm}{M + m}$ is the reduced mass. To evaluate Eq. (1) we need to know the value of the semi-major axis. Do this by obtaining the equation for the trajectory of the comet.

First, evaluate α:

$$\alpha = \frac{\ell^2}{GMm\mu} = \frac{\mu^2 r^4 \dot{\theta}^2}{GMm\mu} = \frac{\mu r^4 \dot{\theta}^2}{GMm} = \frac{Mm\ r^4 \dot{\theta}^2}{(M + m)GMm}$$

$$= \frac{r^4 \dot{\theta}^2}{G\ (M + m)} . \qquad (3)$$

But m << M, i.e. the mass of the comet is much smaller than the mass of the sun. Thus Eq. (3) reduces to:

$$\alpha = \frac{r^4 \dot{\theta}^2}{GM} . \qquad (4)$$

Next, note that at the point of closest approach, point P in Fig. 1, the total velocity of the comet is in the θ direction. Therefore we have that at point P, $V_\theta = r_{min} \dot{\theta}$ and $r = r_{min}$. Using these values in Eq. (4) yields,

$$\alpha = \frac{r^2_{min} V^2_\theta}{GM} . \qquad (5)$$

Substituting all known values into Eq. (5) allows us to find the value of α.

$$\alpha = \frac{(1.6 \times 10^{11})^2 (4 \times 10^4)^2}{(6.72 \times 10^{-11})(1.97 \times 10^{30})} = 3.094 \times 10^{+11} \text{ m.} \qquad (6)$$

Next, evaluate ε, the eccentricity of the orbit. From Eq. (2) observe that the distance of closest approach occurs when $\theta = 0$. Using this value in Eq. (2) and solving for ε yields

$$\varepsilon = \frac{\alpha}{r_{min}} - 1 = \frac{3.094 \times 10^{11}}{1.6 \times 10^{11}} - 1 = 0.9338. \qquad (7)$$

Thus the equation for the orbit is

$$r = \frac{\alpha}{1 + \varepsilon \cos \theta} = \frac{3.094 \times 10^{11}}{1 + 0.9338 \cos \theta} \text{ m}. \qquad (8)$$

From Fig. 1 we observe that the semi-major axis is given by:

$$a = \frac{r_{min} + r_{max}}{2}. \qquad (9)$$

The value of $r_{min} = 1.6 \times 10^{11}$ m and the value of r_{max} can be obtained from Eq. (8). For $\theta = \pi$ radians Eq. (8) becomes

$$r_{max} = \frac{3.094 \times 10^{11}}{1 - 0.9338} = 46.72 \times 10^{11} \text{ m}.$$

Substituting the values for r_{min} and r_{max} into Eq. (9) we obtain

$$a = 24.16 \times 10^{11} \text{ m}. \qquad (10)$$

We can now determine the period of the comet. Substituting all values into Eq. (1) and assuming that the mass of the comet is much smaller than the mass of the sun yields:

$$T = \frac{2\pi}{\sqrt{G(M + m)}} a^{3/2} = \frac{2\pi}{\sqrt{GM}} a^{3/2}$$

$$= \frac{2\pi (2.416 \times 10^{12})^{3/2}}{\sqrt{(6.72 \times 10^{-11})(1.97 \times 10^{30})}}$$

or $\quad T = 2.05 \times 10^{9}$ sec $= 65$ years.

• PROBLEM 15-23

A satellite describes a circular orbit about the earth at an altitude of 500 miles. It ejects a pod tangent to its orbit. What must the speed of the pod be at that time so that it will strike the earth at an angle of 60° with respect to the surface?

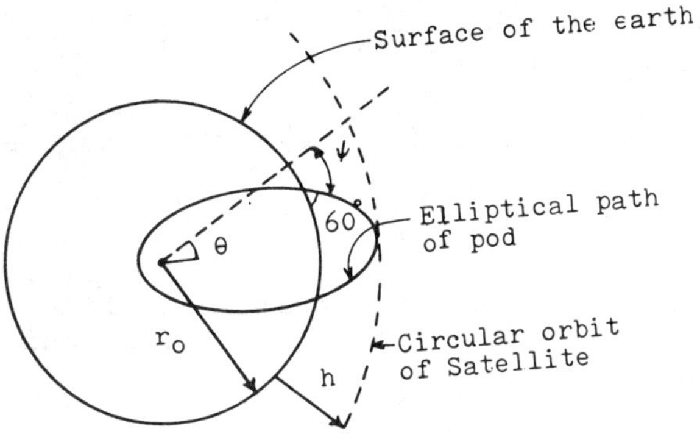

Fig 1

Solution: The path of the pod after ejection from the satellite must be an ellipse having one focus at the center of the earth, and intersecting the surface of the earth at the required angle. The geometry of the problem is shown in Figure 1.

The procedure for solving this problem will first be to find the eccentricity of the elliptical path. Then, using the eccentricity, find the velocity of the pod at apogee.

For an ellipse with one focus at the origin of a polar coordinate system,

$$r = \frac{a\ (1 - e^2)}{1 - e \cos \theta} \qquad (1)$$

where r is the distance from the origin to any point on the ellipse, a is the semimajor axis of the ellipse, e is the eccentricity, and θ is the polar angle with respect to the major axis of the ellipse.

The angle ψ between the radius vector and a line tangent to a curve is given by a calculus theorem as:

$$\tan \psi = \frac{r}{dr/d\theta} \ . \qquad (2)$$

Differentiating equation (1) with respect to θ:

$$\frac{dr}{d\theta} = \frac{a\ e \sin \theta\ (1 - e^2)}{(1 - e \cos \theta)^2} \qquad (3)$$

and substituting (3) into (2):

$$\tan \psi = \frac{r}{dr/d\theta} = \frac{1 - e \cos \theta}{e \sin \theta} \ . \qquad (4)$$

The coordinates of the point of intersection of the ellipse with the surface of the earth, assuming the surface is to be represented by a circle of radius r_0, can be found from equation (1) thus:

$$r_0 = \left(\frac{r_0 + h}{1 + e}\right)\left(\frac{1 - e^2}{1 - e \cos \theta}\right) = \frac{(r_0 + h)(1 - e)}{(1 - e \cos \theta)} \qquad (5)$$

since the semimajor axis $a = \frac{r_0 + h}{1 + e}$, and $1 - e^2 = (1 + e)(1 - e)$.

From equation (5) the angle θ can be found as

$$\cos\theta = \frac{r_0 e - h(1 - e)}{r_0 e}. \qquad (6)$$

Sin θ can be found from identities or by constructing a right triangle as shown in Fig. 2 using (6).

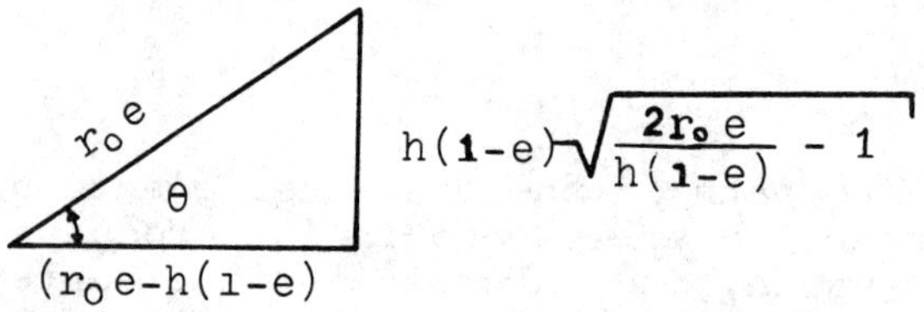

Figure 2

Thus, $$\sin\theta = \frac{h(1 - e)}{r_0 e}\sqrt{\frac{2r_0 e}{h(1 - e)} - 1}. \qquad (7)$$

And, from equation (4), substituting (6) and (7):

$$\tan\psi = \frac{1 - e\cos\theta}{e\sin\theta} = \frac{r_0 + h}{h} \frac{1}{\sqrt{\frac{2r_0 e}{h(1 - e)} - 1}}. \qquad (8)$$

Since $\psi = 90° - 60° = 30°$, $\tan\psi = \tan 30° = 1/\sqrt{3}$, equation (8) can be solved for the eccentricity.

$$e = \frac{3h + h\left(\frac{h}{r_0 + h}\right)^2}{(2r_0 + h)\left(\frac{h}{r_0 + h}\right)^2 + 3h}$$

Assuming r_0 = 4000 miles and h = 500 miles, we get e = 0.9385.

Now, the equation of motion for a body moving in an elliptical orbit around the earth is:

$$\frac{1}{r} = \frac{GM}{C^2} - D\cos\theta \qquad (4)$$

where G is the universal gravitational constant and M is the mass of the earth, $C = r^2\, d\theta/dt$, and D is a constant which can be evaluated from initial conditions. Let θ = 0, then:

$$r = r_0 + h;$$

$$C = r^2 \frac{d\theta}{dt} = rv_0 = (r_0 + h)v_0;$$

and $$D = \left(\frac{GM}{C^2} - \frac{1}{r}\right)_{r = r_0 + h}. \qquad (10)$$

Thus, equation (9) becomes:

$$\frac{1}{r} = \frac{GM}{(r_0 + h)^2 v_0} - \left(\frac{GM}{(r_0 + h)^2 v_0^2} - \frac{1}{(r_0 + h)} \right) \cos\theta. \quad (11)$$

Rearranging equation (1), we get:

$$\frac{1}{r} = \frac{1}{a(1 - e^2)} - \frac{e}{a(1 - e^2)} \cos\theta \quad (12)$$

and comparing the first terms on the right-hand sides of (11) and (12):

$$\frac{GM}{(r_0 + h)^2 v_0^2} = \frac{1}{a(1 - e^2)} = \frac{(1 + e)}{(r_0 + h)(1 - e^2)}$$

$$= \frac{1}{(r_0 + h)(1 - e)} \quad (13)$$

from which $$v_0 = \sqrt{\frac{GM(1 - e)}{(r_0 + h)}} \quad (14)$$

Finally, letting $GM = 1.466 \times 10^{16}$ ft^3/sec^2,

$e = 0.9385$, and $(r_0 + h) = 4500 \times 5280$ ft:

$$v_0 = 6160 \text{ ft/sec.}$$

• PROBLEM 15-24

A ballistic rocket is launched at the equator in an easterly direction, at an angle of 60° from the horizontal. Its initial speed is V_1 = 6 km/sec as measured with respect to the earth. What will be the great circle distance, around the earth covered by the rocket?

Solution: The rotating earth does not constitute an inertial frame. Our first job will be to find the velocity of the rocket in an inertial frame. With respect to the earth, the initial velocity is V_1 = 6 km/sec, the horizontal component of this velocity is $V_h = V_1 \cos 60° = 6$ km/sec $\times \frac{1}{2}$ = 3 km/sec. The vertical component is $V_v = V_1 \sin 60°$ = 6 km/sec × 0.866 = 5.196 km/sec. Now the earth rotates once in approximately 24 hours or 86,400 sec. Hence its angular velicity ω is $2\pi/86{,}400$ radians/sec. Since the radius of the earth is approximately r_e = 6371 km, the tangential velocity of a point on the equator is $V_t = r_e \omega$ = 6371×3671 × $2\pi/86{,}400$ km/sec = 0.4634 km/sec. Since the earth rotates toward the east, the total horizontal velocity component v_T of the ballistic rocket with respect to an inertial frame will be the sum of V_h and V_t.

$$v_T = V_h + V_t = 3.00 + 0.4634 = 3.4634 \text{ km/sec.}$$

The vertical component of the velocity in the inertial frame is the same as in the frame rotating with respect to the earth: $v_V = V_V = 5.196$ km/sec. The total velocity v in the inertial frame is

$$v = \sqrt{v_V^2 + v_T^2} = \sqrt{38.995}\ \text{km/sec} = 6.245\ \text{km/sec}.$$

Now we calculate the angular momentum ℓ and the total energy E of the rocket. The magnitude of the angular momentum is equal to the mass m of the rocket times the radius of the earth times the tangential component of the velocity of the rocket. Let us use meters instead of kilometers.

$$\ell = mr_e v_T = m \times 6.371 \times 10^6\ \text{m} \times 3.4634 \times 10^3\ \text{m/sec}$$

$$= 2.2065 \times 10^{10}\ \text{m}^2/\text{sec} \times \text{m}.$$

Now the total energy of the rocket is given by

$$E = m \left[\tfrac{1}{2}v^2 - \frac{GM}{r_e} \right]$$

where v is the total velocity of the rocket in the inertial frame computed earlier, M is the mass of the earth and is equal to 5.983×10^{24} kg, G is the gravitational constant and is equal to 6.67×10^{-11} N-m^2/kg^2. Using these values yields energy equal to

$$E = -\ 4.3185 \times 10^7\ \text{m}^2/\text{sec}^2 \times \text{m}.$$

Using these numerical values for angular momentum and total energy, we can calculate the eccentricity of the orbit from

$$e = \left[1 + \frac{2\ell^2 E}{G^2 M^2 m^3} \right]^{\frac{1}{2}}.$$

The mass m of the rocket will cancel in this expression. The result is

$$e = (0.7343)^{\frac{1}{2}} = 0.8571\ .$$

Now the equation of the orbit of the ballistic rocket above the earth is given by

$$r = \frac{\ell^2/GMm^2}{1 - e\cos\theta}\ ,$$

with $\dfrac{\ell^2}{GMm^2} = 1.220 \times 10^6$ m.

By setting the radius equal to the radius of the earth, we can find the angle θ between the apogee of the orbit and the launch and impact points. Thus θ is the half angle subtended at the origin of the inertial reference frame between the point of launch and the point of impact. Solving this equation for $\cos\theta$ and substituting values, we have

$$\cos\,\theta = \left[\frac{r_e - \ell^2/GMm^2}{er_e}\right] = 0.94335$$

or $\qquad \theta = 19°\ 23'$

and $\qquad 2\theta = 38°\ 46' = 0.6777$ radian.

The great circle distance s covered by the rocket is found from the equation

$$2\theta = \frac{s}{r_e}$$

from which $\qquad s = 2\theta r_e = 4.317 \times 10^6$ m .

This is equal to the distance around the earth covered by the rocket with respect to the inertial frame.

• PROBLEM 15-25

Show that a particle initially at a very great distance from the earth, with a small initial velocity, that is not directed radially toward the earth, will describe a parabolic path as it falls toward the earth.

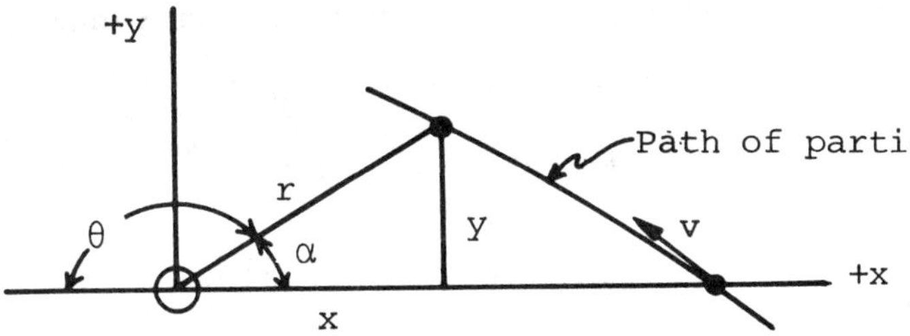

Solution: For central force motion, the governing equation is given by

$$\frac{1}{r} = K + C\cos\theta \tag{1}$$

where $\qquad K = \frac{GM}{h^2}.$

Assume that initially the particle is approximately on the positive x-axis a great distance from earth so that

$$\theta_0 \cong 180° \quad \text{and}$$

$$\frac{1}{r_0} \cong 0.$$

When these conditions are applied to Eq.(1), we get

$$0 \cong K - C$$

or, approximately,

$$C = K.$$

Therefore, one can write Eq.(1) as

$$\frac{1}{r} = K\,(1 + \cos\,\theta). \qquad (2)$$

Now, from the figure,

$$\theta = 180° - \alpha$$

and

$$\cos\,(\theta) = \cos\,(180° - \alpha)$$
$$= -\cos\,(\alpha)$$
$$= -\frac{x}{r}.$$

So, Eq.(2) may be written as

$$\frac{1}{r} = K\left(1 - \frac{x}{r}\right).$$

Multiplying through by r and then rearranging yields

$$1 + Kx = Kr.$$

Squaring both sides and substituting $r^2 = x^2 + y^2$ leads to

$$1 + 2Kx + K^2x^2 = K^2(x^2 + y^2).$$

Finally, solving for x yields

$$x = ay^2 + b \qquad (3)$$

where $a = \frac{K}{2}$,

and $b = -\frac{1}{2K}$.

Eq.(3) shows that the path is a parabola with symmetry about the x-axis.

CHAPTER 16

CENTRAL FORCES

• PROBLEM 16-1

Show that the velocity which a particle must have in order to escape to infinity from a distance r is equal to $\sqrt{2}$ times the velocity for a stable circular orbit at a distance r.

Solution: The conservation of mechanical energy states that the sum of the potential and kinetic energies of a particle at any point in a conservative field, like gravity, is the same as at any other point in the field. This fact enables us to set up an equation relating a point in the stable circular orbit to a point at infinity:

$$(PE)_r + (KE)_r = (PE)_\infty + (KE)_\infty \tag{1}$$

At infinity, ∞, we define the potential energy, (PE), to be zero.

The minimum escape velocity is one which will result in a zero velocity of the particle at infinity. That is, the pull of gravity on the satellite due to the earth brings it to rest at infinity. Therefore the right side of equation (1) is zero, and when we substitute for the left side we get:

$$\frac{-\,GmM}{r} + \frac{mv^2}{2} = 0 \tag{2}$$

or

$$\frac{GmM}{r} = \frac{mv^2}{2}\ . \tag{3}$$

If we divide the both sides of equation (3) by m and solve for v, we have:

$$v = \sqrt{2}\ \sqrt{GM/r}\ .$$

The velocity for a stable circular orbit at a distance r is:

$$v = \sqrt{GM/r}\ . \tag{5}$$

Hence, we see that if the speed of a satellite in

circular orbit is multiplied by the $\sqrt{2}$, we get the speed required to send the satellite to infinity.

• PROBLEM 16-2

Show that the angular momentum of a particle about a point 0 will be equal to zero if any one of the following conditions applies

(a) $\vec{P} = 0$

(b) the particle is at point 0

(c) $\vec{P}$ is parallel (or anti-parallel) to $\vec{r}$

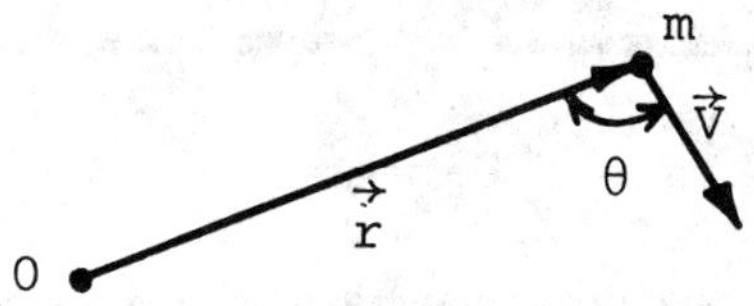

Solution: In general, for a particle of mass m and velocity $\vec{v} = (v_1\ \hat{i} + v_2\ \hat{j} + v_3\ \hat{k})$ located a distance r from point 0, $\vec{r} = (r_1\ \hat{i} + r_2\ \hat{j} + r_3\ \hat{k})$; angular momentum $\vec{L} = \vec{r} \times m\ \vec{v}$.

In scalar form, this equation becomes $L = r\ m\ v \sin\theta$ where θ is the angle between $\vec{r}$ and $\vec{v}$.

a) $\vec{P} = m\ \vec{v}$. If $\vec{P} = 0$ then m v must be 0.

$L = r\ (0) \sin\theta$

$L = 0$

b) If the particle is located at point 0 then $\vec{r} = 0$ and $r = 0$.

$(0)\ (m\ v) \sin\theta = 0$

$L = 0$

c) If $\vec{P}$ is parallel or anti-parallel to $\vec{r}$ then $\vec{v}$ is parallel or anti-parallel to $\vec{r}$ since $\vec{P}$ has the same direction as $\vec{v}$ because m is a scalar. If $\vec{v}$ is parallel or anti-parallel to $\vec{r}$ then $\theta = 0°$ or $180°$. In either case, $\sin\theta = 0$.

$r\ m\ v\ (0) = 0$

$L = 0$

• PROBLEM 16-3

Show that: (a) If the torque on a particle is zero, its angular momentum must be conserved, i.e., if the only force on a particle is due to a central force field, the angular momentum of the particle about the source of that field is constant. (b) This result holds also for a particle under the influence of a force which repels it from the origin of an inertial reference system.

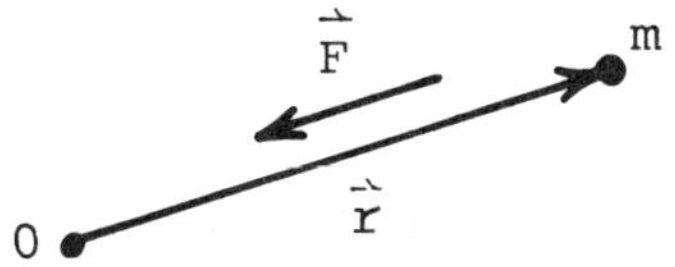

Solution: Given the central force system above, $\vec{r}$ is the radius vector from the origin to a particle of mass m, and $\vec{F}$ is the central force on the particle. The torque on the particle about point 0 is by definition $\vec{r} \times \vec{F}$. This torque is also proportional to the particle's rate of change of angular momentum.

The magnitude of $\vec{r} \times \vec{F}$ is equal to $r\,F \sin\theta$ where θ is the angle between the two vectors. For a central force problem, $\vec{F}$ is parallel or anti-parallel to $\vec{r}$ (parallel when $\vec{F}$ is repulsive, anti-parallel when $\vec{F}$ is attractive). In either case, $\sin\theta = \sin(0) = \sin(\pi) = 0$.

This leads to a zero torque about 0. If the torque is zero, the rate of change of angular momentum is zero and the angular momentum must be constant.

• PROBLEM 16-4

In order to keep a particle in a circular orbit, it is necessary to apply a constant centripetal force on the particle toward the center of the circle. Based on this statement, find the orbit of the satellite if it moves around the earth with a given tangential velocity v_T.

From this result, show that the square of the time required for one revolution about the circle is proportional to the cube of the radius of the circle. This is Kepler's Third Law of Planetary Motion.

Solution: To find the orbit of the satellite use Newton's Second Law, which will give us the acceleration and hence the radius r of the orbit. Newton's Second Law,

$$\vec{F} = M\vec{A}, \qquad (1)$$

shows that the acceleration vector is in the same direction as the unbalanced force vector. The acceleration vector must be stretched or contracted by the factor M to give the unbalanced force vector. The only unbalanced force on a satellite in orbit is the gravitational force of attraction:

$$\vec{F} = -\frac{G\,mM}{r^2}\,\hat{r} \tag{2}$$

where m is the satellite's mass, M the earth's mass, G the universal gravitational constant, r is the distance between the satellite and the earth, and $\hat{r}$ is a unit vector pointing from the earth to the satellite.

The centripetal acceleration is:

$$\vec{A} = \frac{-v^2}{r}\,\hat{r} \tag{3}$$

where v is the transverse velocity which, in this case is the same as the tangential velocity of the satellite.

Substituting (2) and (3) in (1) yields:

$$\frac{-GmM}{r^2} = m\left(\frac{-v^2}{r}\right). \tag{4}$$

Solving this equation for r shows that if v is a constant, then r is a constant which is the equation of a circle in polar coordinates. Therefore, the orbit is circular and

$$r - \frac{GM}{v^2}. \tag{5}$$

Now linear velocity, v, is related to angular velocity, ω, by the equation:

$$v = r\omega.$$

Angular velocity equals the angle turned through by the radius vector divided by the time required. For one revolution the angle is 2π radians and the time for one revolution is called the period, T. Substituting for ω in equation (5) gives us:

$$v = r\omega = r(2\pi/T). \tag{6}$$

Substituting the square of the above in equation (4), cancelling and rearranging gives:

$$T^2 = \frac{4\pi^2}{GM}\,r^3 \tag{7}$$

which is Kepler's Third Law of Planetary Motion.

• PROBLEM 16-5

Shown in Fig. 1 is the orbit of an artificial earth satellite. Because the earth is rotating on its polar axis, it appears to an observer on earth that the orbit plane is regressing. What is actually happening is that the orbit plane has a fixed orientation in space, and the earth is spinning relative to the orbit plane. Use the moment equation to show that this is what is actually happening physically. The line of action of the force of attraction that the earth exerts on the satellite passes through the center of the earth.

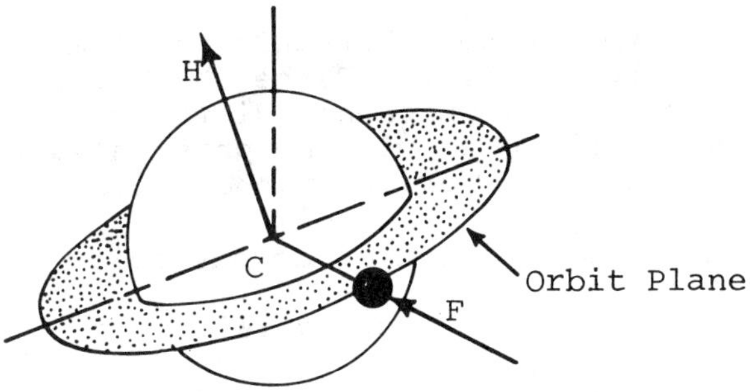

Solution: Consider the center of the earth, C, as being the reference point. Since the force of the earth acting on the satellite has a line of action that passes through the point C, the center of the earth, then the moment about the point C of the force on the particle is equal to zero:

$$\vec{M}_C = 0$$

This implies that the time derivative of the moment of momentum of the body about the point C is also equal to zero. Thus,

$$\vec{H}_C = \text{constant.}$$

If the moment of momentum of the satellite about the point C is equal to a constant, then it must be that the plane of motion (the orbit plane) of the vehicle has a fixed orientation in space. For, if the vehicle should move out of the orbit plane, then there would be a component of the moment of momentum perpendicular to the original moment-of-momentum vector. This means the net result will be that the moment-of-momentum vector will not retain its fixed orientation in space, but will change direction. Thus it is seen that the moment-of-momentum vector will be constant only if the motion of the satellite is in a plane that has a fixed orientation in space.

• PROBLEM 16-6

For an ellipse show that $e = \frac{r_{max} - r_{min}}{r_{max} + r_{min}}$ where r_{max} is the distance from a focus to the farthest point on the ellipse and r_{min} is the distance from the focus to the nearest point on the ellipse.

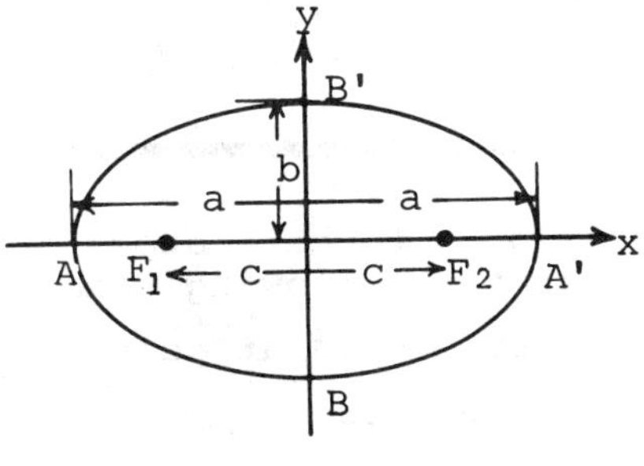

Solution: We presume that this problem refers to a two body orbit, therefore, let F_1 be the focus which is the center of force from which all distances are measured. F_2 is a mathematical point with no physical significance except that it is necessary for the mathematical definition of an ellipse.

The usual definition of eccentricity e defines

$$c = ea \qquad \text{(see Figure).} \qquad (1)$$

From inspection

r_{max} = the distance between points F_1 and A',

r_{min} = the distance between points F_1 and A.

Also $\quad F_1A' = c + a; \qquad F_1A = a - c$

$$\therefore \quad r_{max} = c + a; \qquad r_{min} = a - c.$$

Substituting (1) we have

$$r_{max} = ea + a = (1 + e)a \qquad (2)$$

$$r_{min} = ea - a = (1 - e)a \qquad (3).$$

We have two equations in two unknowns. Dividing (2) by (3) yields

$$\frac{r_{max}}{r_{min}} = \frac{(1 + e)}{(1 - e)} .$$

Cross multiplying

$$r_{max} (1 - e) = r_{min} (1 + e).$$

Expanding each side

$$r_{max} - r_{max} e = r_{min} + r_{min} e$$

Solving for e yields

$$e = \frac{r_{max} - r_{min}}{r_{max} + r_{min}} .$$

• PROBLEM 16-7

Show that the torque of a force $\vec{F}$ about 0 will be equal to zero if any one of the following conditions exist, (a) $\vec{F} = 0$, (b) the particle is at point 0, (c) $\vec{F}$ is parallel (or antiparallel) to $\vec{r}$.

Solution: Vector torque about the origin is defined as:

$$\text{torque} = \vec{r} \times \vec{F} = |\vec{r}||\vec{F}| \sin(\phi)\ \hat{U} \qquad (1)$$

An equivalent definition of torque is the following determinant:

$$\text{torque} = \begin{vmatrix} i & j & k \\ x & y & z \\ F_x & F_y & F_z \end{vmatrix}$$

where $\vec{r} = x\hat{i} + y\hat{j} + z\hat{k}$, the position vector of $\vec{F}$, and the latter is defined by:

$$\vec{F} = F_x\hat{i} + F_y\hat{j} + F_z\hat{k} .$$

The position vector is drawn from the origin to the initial end of $\vec{F}$.

(a) If $\vec{F} = 0$, then the magnitude of $\vec{F}$ in equation (1) is zero which makes the torque equal to zero. Another viewpoint is that $\vec{F}$ can be zero only if each of its components (F_x, F_y, F_z) is zero. In this case all the elements in the last row of the determinant become zero which makes the value of the determinant zero.

(b) If the particle on which the force acts is at the origin, then $\vec{r} = 0$ and hence the torque is zero. Also, looking at the determinant, all the elements in the second row are zero which makes the value of the determinant zero.

(c) If F is parallel to r, then the angle ϕ between them is either 0° or 180°. In either case, the sine of the angle ϕ is zero which makes the torque equal to zero. (See equation (1)). The elements in the second and third row of the determinant are proportional to each other which makes the determinant equal to zero.

• PROBLEM 16-8

A particle of mass m moves with a vector velocity $\vec{v}$ which is constant in magnitude and direction. A point 0 lies a perpendicular distance b from the path of the particle. Show that the angular momentum of the particle about 0 is constant.

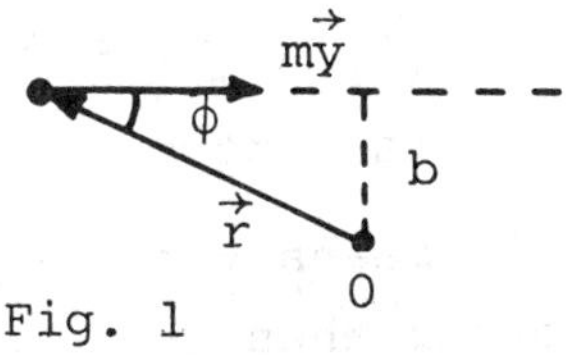

Fig. 1

Solution: By definition, the moment of a vector is: $\bar{r} \times$ (that vector). And the angular momentum (also called moment of linear momentum) is:

$$\vec{r} \times m\vec{v} = m\ |\vec{r}|\,|\vec{v}|\ \sin(\phi)\ \hat{U} \qquad (1)$$

$\vec{r}$ and $\vec{v}$ are the position and velocity vectors respectively and $\hat{U}$ is a unit vector in the sense of the right-hand rule. From Figure (1) observe that $|\vec{r}|\ \sin(\phi)$ is the perpendicular distance, b, from point 0 to the line containing the linear momentum vector, $m\vec{v}$. This is a constant independent of the particle's position along the line. Since m and $\vec{v}$ are also constants, the angular moment is a constant.

• PROBLEM 16-9

Show that, if the body at the origin does not have much more mass than the particle, the coordinate system determined by the central body is not inertial. What difficulties will this cause in the analysis of the motion of the particle?

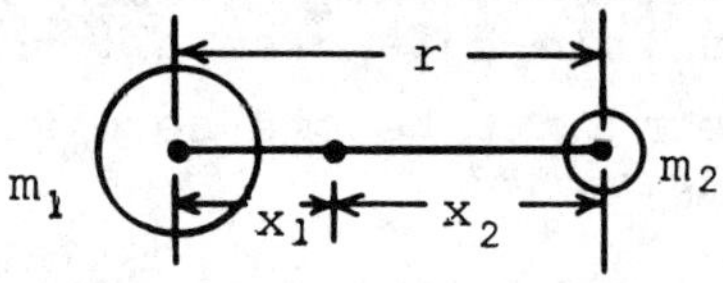

Solution: The system above consists of two masses m_1 and m_2 with m_2 in orbit about m_1. As the two masses orbit, they are also translating in inertial space, and, in accordance with Newton's Second Law, the mass center of the two bodies moves in pure translation. In other words, since there are no forces exerted on the two masses from outside of the system, their mass center is moving in pure translation and could be defined as the origin of an inertial coordinate system.

Each of the two bodies then has two components of motion, one is a constant component which is the same as the velocity of the mass center. The other is a constantly changing component (due to rotation which is dependent on the body's distance from the mass center). Since the origin of an inertial coordinate system is defined as the origin of a coordinate system with constant translational velocity, neither body can be used as an inertial origin unless its center coincides with the system's center of mass. That is, x_1 or $x_2 = 0$ and $r = x_2$ or x_1, respectively.

Remember that $x_2 = \dfrac{r\,m_1}{(m_1 + m_2)}$ and $x_1 = \dfrac{r\,m_2}{(m_1 + m_2)}$ from the definition of mass center. If $m_1 >>> m_2$ then $x_2 \approx r$ and $x_1 \approx 0$. This means that if one body is a great deal larger than the other then the mass center of the larger body is almost coincident with the mass center of the system and can be approximated as the origin of an inertial coordinate system.

If one were trying to solve for the motions of the bodies, this approximation would greatly simplify the procedure since the larger particle's motion with respect to the inertial coordinate system would be zero. Only the smaller body would have a rotational motion. Rotational motions are often hard to find mathematically, and having only one to solve for simplifies the problem.

When the approximation is not valid then one must solve for two rotational motions, one for each body, about the system's mass center. This could be a much more complex problem.

• PROBLEM 16-10

Show that angular momentum is not necessarily conserved about an origin other than the source by the central field.

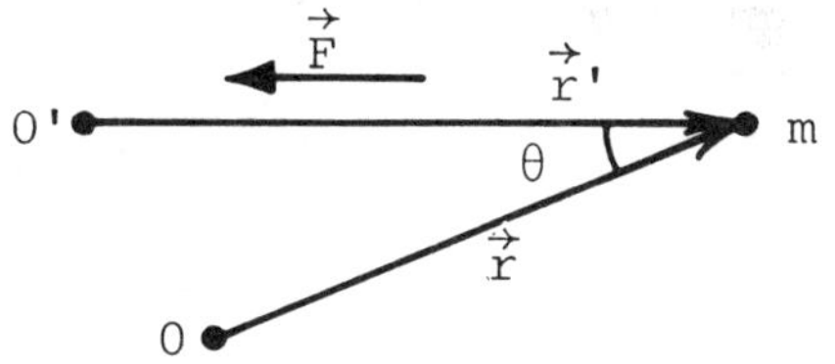

Solution: For the system above, $\vec{r}$ is radius vector from the origin 0 to a particle of mass m , $\vec{r}\,'$ is the radius vector from another origin 0' (which is the source of a central force field to the same particle) and $\vec{F}$ is the force acting on the particle from 0'.

By definition, torque is equal to $\vec{r} \times \vec{F}$ and has magnitude r F sin θ where θ is the angle between $\vec{r}$ and $\vec{F}$. Also by definition, the rate of change of angular momentum about an origin is equal to the torque about that origin.

$$T_0 = \frac{dL_0}{dt}$$

Where L_0 is angular momentum about the origin and T_0 is torque about the particular origin. It follows, then, that the magnitude of the rate of change of angular momentum about an origin is r F sin θ.

With respect to the origin 0, θ, $\vec{r}$ and $\vec{F}$ all have definite values, therefore, $\frac{dL_0}{dt} \neq 0$. However, if the origin is taken as the source of the central force field then $\vec{r} = \vec{r}\,'$ and r' is parallel to $\vec{F}$. Therefore, sin θ = 0 and

$$\frac{dL_{o'}}{dt} = 0.$$

This leads to the conclusion that angular momentum in a central force system is conserved only if the source of the central force field is taken as the origin of coordinates.

• PROBLEM 16-11

A small ball swings in a horizontal circle at the end of a cord of length ℓ_1 which forms an angle θ_1 with the vertical. The cord is slowly shortened by pulling it through a hole in its support until the free length is ℓ_2 and the ball is moving at an angle θ_2 from the vertical. a) Derive a relation between ℓ_1, ℓ_2, θ_1, and θ_2. b) If ℓ_1 = 600 mm, θ_1 = 30° and, after shortening, θ_2 = 60°, determine ℓ_2.

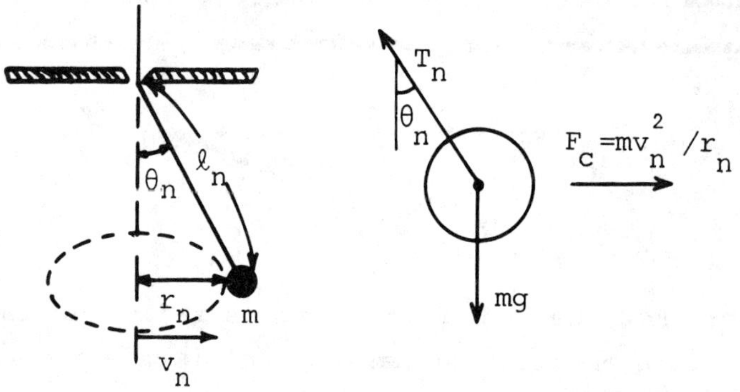

Solution: In the system shown, ℓ_n is the cord length, θ_n is the cone angle cut out by the cord as it revolves, r_n is the distance from the mass to the axis of rotation, and v_n is the velocity of the mass.

To find the relationship between ℓ_1, ℓ_2, θ_1 and θ_2 when the cord is shortened from ℓ_1 to ℓ_2, one should first analyze a free body diagram of the particle.

Since the particle is in a circular motion, the inward component of T must supply the centripetal force, $m\,v_n^2/r_n$.

$$T_n \sin\theta_n = \frac{m\,v_n^2}{r_n} \qquad (1)$$

Also, the vertical component of T must balance the weight.

$$T_n \cos\theta_n = m\,g \qquad (2)$$

Dividing (1) by (2) yields:

$$\text{Tan}\,\theta_n = \frac{m\,v_n^2}{m\,g\,r_n} = \frac{v_n^2}{g\,r_n} \qquad (3)$$

One more relationship is needed to solve the problem. Since no external torques act on the system, angular momentum L is conserved.

$$L_1 = L_2$$

$$m\, r_1\, v_1 = m\, r_2\, v_2$$

$$r_1\, v_1 = r_2\, v_2. \qquad (4)$$

If one rearranges (3),

$$g\, r_n^{\,3} \tan\theta_n = r_n^{\,2}\, v_n^{\,2}. \qquad (5)$$

From (4), it follows that $(r_1\, v_1)^2 = (r_2\, v_2)^2$. (6)

Combining (5) and (6), $g\, r_1^{\,3} \tan\theta_1 = g\, r_2^{\,3} \tan\theta_2$. (7)

Finally, from geometry, $r_n = \ell_n \sin\theta_n$ (8)

Using this relation in (7) and dividing both sides by g produces the desired result.

$$(\ell_1 \sin\theta_1)^3 \tan\theta_1 = (\ell_2 \sin\theta_2)^3 \tan\theta_2.$$

In this problem, $\ell_1 = .6\text{m}$, $\theta_1 = 30°$, $\theta_2 = 60°$

Solving for ℓ_2 yields

$$\ell_2 = \sqrt[3]{\frac{(\ell_1 \sin\theta_1)^3 \tan\theta_1}{\sin\theta_2^{\,3} \tan\theta_2}}$$

$$\ell_2 = \frac{\ell_1 \sin\theta_1}{\sin\theta_2} \sqrt[3]{\frac{\tan\theta_1}{\tan\theta_2}},$$

and substituting in the numbers yields

$$\ell_2 = \frac{.6\text{ m }(\sin 30°)}{\sin 60°} \sqrt[3]{\frac{\tan 30°}{\tan 60°}}$$

$$= \frac{.6\text{ m }(.5)}{(.87)} \sqrt[3]{\frac{.58}{1.73}}$$

$$= .24\text{ m}.$$

• **PROBLEM 16-12**

An asteroid is observed at a distance of 10^6 km from the earth. It is approaching the earth with a velocity of 10 km/sec in a direction which, if extended, would pass a perpendicular distance, b, of 20,000 km from the center of the earth. What will be the minimum distance between the asteroid and the earth? Will the asteroid return toward the earth again? Assume that the center of earth constitutes the origin of an inertial coordinate system.

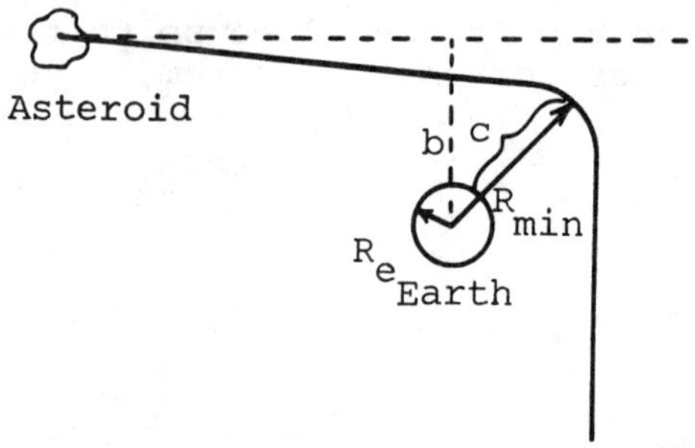

Solution: For a two-body gravitational system in which the central force is an inverse square force given by Newton's Law of gravitational attraction, the potential energy of the system may be expressed as

$$U(r) = -G\frac{Mm}{r} = -\frac{C}{r} \tag{1}$$

where r is the distance between the two bodes, M and m are the masses of the two bodies, and $G = 6.67 \times 10^{-11}$ N-m^2/kg^2 is the universal constant of gravitational attraction. In this problem, we shall let M be the mass of the earth, and m the mass of the satellite. The path of the satellite is a conic section which may be expressed in polar coordinates as:

$$r = \frac{r_0}{1 - \varepsilon \cos\theta} \tag{2}$$

where r_0 is a constant, and ε is a dimensionless parameter, called the eccentricity, which is a characteristic of the shape of the path. r_0 and ε are related to the angular momentum ℓ, and the total energy, E, both of which are constants of the motion. Thus,

$$\ell = mv_0b \tag{3}$$

and $$E = \frac{1}{2}mv_0^2 \tag{4}$$

the values when the asteroid is at a large distance from the earth.

r_0 represents the radius of the circular orbit for which the angular momentum is $\ell = mv_0b$. Thus, for circular motion, the gravitational force must equal the centripetal force:

$$G\frac{Mm}{r_0^2} = \frac{C}{r_0^2} = \frac{mv^2}{r_0}$$

$$v^2 = \frac{C}{mr_0}\ . \tag{5}$$

Then, the square of the angular momentum for the circular motion is:

$$\ell^2 = m^2v^2r_0^2 = mCr_0 \tag{6}$$

or $$r_0 = \frac{\ell^2}{mC} \quad . \tag{7}$$

Now, the eccentricity is given by

$$\varepsilon = \sqrt{1 + \frac{2E\ell^2}{mC^2}} \quad . \tag{8}$$

Calculating values we get:

$$E = \frac{1}{2} mv_0^2 = \frac{1}{2} (10^4 m/s)^2 \ (m)$$

$$C = GMm = (6.67 \times 10^{-11} \ N\text{-}m^2/kg^2)(5.98 \times 10^{24} \ kg)(m)$$

$$= 3.99 \times 10^{14} \ (m) \tag{10}$$

$$\ell = mv_0b = (10^4 \ m/s)(2 \times 10^7 \ m)(m)$$

$$= 2. \times 10^{11} (m) \tag{11}$$

$$r_0 = \frac{\ell^2}{mC} = \frac{(2 \times 10^{11})^2 (m)^2}{(4 \times 10^{14})(m)^2} = 10^8 \text{ meters} \tag{12}$$

and $$\varepsilon = \sqrt{1 + \frac{2E\ell^2}{mC^2}} =$$

$$= \sqrt{1 + \frac{(2)(.5 \times 10^8)(4 \times 10^{22})(m)^3}{(m)^3 \ (16 \times 10^{28})}} = 5.1. \tag{13}$$

Since the eccentricity is greater than unity, the path of the asteroid will be hyperbolic, so that the asteroid will pass by the earth and will not return again.

Also, from Eqn. (2), the minimum value of r occurs when $\cos\theta = -1$, thus

$$r_{min} = \frac{r_0}{1 + \varepsilon} = \frac{10^8}{1 + 5.1} = 1.64 \times 10^7 \text{ meters.}$$

The radius of the earth is 6.4×10^6 meter, so the asteroid passes within 10^7 meters or 10,000 kilometers from the surface of the earth.

• PROBLEM 16-13

Find the potential energy and the force, given the equation of the trajectory of a particle moving in a central force field:

$$r = r_0 e^{\theta} \ .$$

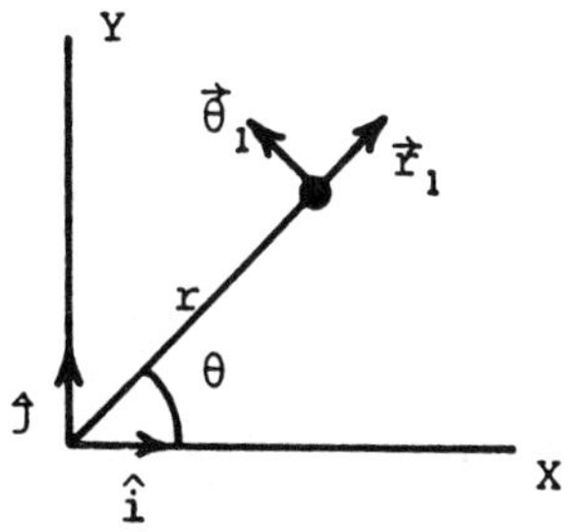

Solution: The motion of a particle under the influence of a central force requires that the trajectory of the particle lie in a plane. Assuming the plane to be the plane of a polar coordinate system, we first find the equations of motion of the particle. The position of the particle is given in polar coordinates by:

$$\vec{r} = r\,\vec{r}_1$$

where $\vec{r}_1$ is a unit vector in the direction of a line drawn from the origin to the position of the particle. Note that the angular position is θ, and that $\vec{\theta}_1$ is a unit vector normal to $\vec{r}_1$.

Then, the velocity is:

$$\vec{v} = \frac{d\vec{r}}{dt} = \frac{dr}{dt}\,\vec{r}_1 + r\,\frac{d\vec{r}_1}{dt}\,. \qquad (1)$$

By the chain rule for derivatives,

$$\frac{d\vec{r}_1}{dt} = \frac{\partial \vec{r}_1}{\partial r}\,\frac{dr}{dt} + \frac{\partial \vec{r}_1}{\partial \theta}\,\frac{d\theta}{dt} \qquad (2)$$

But, $\dfrac{\partial \vec{r}_1}{\partial r} = 0$ since $\vec{r}_1$ has a constant magnitude.

$$\text{Also,}\quad \frac{\partial \vec{r}_1}{\partial \theta} = \frac{\partial}{\partial \theta}\,(\vec{i}\cos\theta + \vec{j}\sin\theta)$$

$$= -\,\vec{i}\sin\theta + \vec{j}\cos\theta = \vec{\theta}_1 \qquad (3)$$

$$\text{Thus,}\quad \frac{d\vec{r}_1}{dt} = \dot{\vec{r}}_1 = \dot{\theta}\,\vec{\theta}_1 \qquad (4)$$

where we have written the unit vector $\vec{r}_1$ in terms of its rectangular components using the rectangular unit vectors $\vec{i}$ and $\vec{j}$. The last result in (3) is obtained from the requirement that $\vec{\theta}_1$ is normal to $\vec{r}_1$.

Thus, (1) becomes

$$\vec{v} = \dot{r}\,\vec{r}_1 + r\,\dot{\theta}\,\vec{\theta}_1\,. \qquad (5)$$

The acceleration is found from (5) by again differentiating with respect to time:

$$\vec{a} = \frac{d\vec{v}}{dt} = \frac{d}{dt}\,(\dot{r}\,\vec{r}_1 + r\,\dot{\theta}\vec{\theta}_1)$$

$$= \ddot{r}\vec{r}_1 + \dot{r}\dot{\vec{r}}_1 + \dot{r}\dot{\theta}\vec{\theta}_1 + r\ddot{\theta}\vec{\theta}_1 + r\dot{\theta}\dot{\vec{\theta}}_1 \qquad (6)$$

$$= \ddot{r}\vec{r}_1 + \dot{r}\dot{\theta}\vec{\theta}_1 + \dot{r}\dot{\theta}\vec{\theta}_1 + r\ddot{\theta}\vec{\theta}_1 + (r\dot{\theta})(-\,\dot{\theta}\vec{r}_1)$$

or $\vec{a} = (\ddot{r} - r\dot{\theta}^2)\vec{r}_1 + (2\dot{r}\dot{\theta} + r\ddot{\theta})\vec{\theta}_1$ (7)

where we have used $\dot{\vec{r}}_1 = \dot{\theta}\vec{\theta}_1$, and

$$\dot{\vec{\theta}}_1 = \frac{d\vec{\theta}_1}{dt} = \frac{\partial\vec{\theta}_1}{\partial r}\frac{dr}{dt} + \frac{\partial\vec{\theta}_1}{\partial\theta}\frac{d\theta}{dt}$$

$$= (0)\dot{r} + (-\vec{i}\cos\theta - \vec{j}\sin\theta)\dot{\theta} \quad (8)$$

$$= -\dot{\theta}\vec{r}_1.$$

From Newton's Second Law,

$$\vec{F} = m\vec{a}$$

we get:

$$\vec{F} = m\{(\ddot{r} - r\dot{\theta}^2)\vec{r}_1 + (2\dot{r}\dot{\theta} + r\ddot{\theta})\vec{\theta}_1\}. \quad (9)$$

Since $\vec{F}$ is a central force, it has only 2 radial components. Therefore, the radial component of $\vec{F}$ has magnitude f(r), where

$$f(r) = m(\ddot{r} - r\dot{\theta}^2) \quad (10)$$

and $2\dot{r}\dot{\theta} + r\ddot{\theta} = 0.$ (11)

Equations (10) and (11) are the equations of motion for a particle of mass m moving in a central force field.

Now we will find an expression for the kinetic energy of the particle as a function of its coordinates. Multiplying (11) by (mr)

$$m(2r\dot{r}\dot{\theta} + r^2\ddot{\theta}) = \frac{d}{dt}(mr^2\dot{\theta}) = \frac{dH}{dt} = 0, \quad (12)$$

where we have defined $H = mr^2\dot{\theta}$, as the angular momentum of the particle. Equ. (12) shows that since $\frac{dH}{dt} = 0$, the H = constant and (12) is a statement of the principle of conservation of angular momentum in a central force field.

Now, from (12) we have

$$\dot{\theta} = \frac{H}{mr^2} \quad (13)$$

and substituting (13) into (10),

$$f(r) = m\ddot{r} - \frac{H^2}{mr^3}. \quad (14)$$

Multiplying this by $\dot{r}$,

$$\dot{r}f(r) = m\dot{r}\ddot{r} - \frac{H^2}{mr^3}\dot{r} \quad (15)$$

which can also be written as:

$$\dot{r}f(r) = \frac{d}{dt}\left(\frac{1}{2}m\dot{r}^2\right) + \frac{d}{dt}\left(\frac{H^2}{2mr^2}\right). \tag{16}$$

Again, substituting for H from (13) into (16)

$$\dot{r}f(r) = \frac{d}{dt}\left(\frac{1}{2}m\dot{r}^2\right) + \frac{d}{dt}\left(\frac{1}{2}mr^2\dot{\theta}^2\right). \tag{17}$$

The kinetic energy, T, is given by:

$$T = \frac{1}{2}mv^2 = \frac{1}{2}m\dot{r}^2 + \frac{1}{2}mr^2\dot{\theta}^2 \tag{18}$$

where we have used (5) to find $v^2 = \vec{v} \cdot \vec{v}$.

Thus, combining (17) and (18),

$$\dot{r}f(r) = \frac{dT}{dt}\,. \tag{19}$$

This last equation shows that the rate of change of the kinetic energy is equal to the rate at which the central force does work on the particle.

From the principle of conservation of energy we have,

$$E = T + V = \text{constant} \tag{20}$$

where E is the total energy of the particle, and V is the potential energy. Substituting (18) into (20),

$$E = \frac{1}{2}m\dot{r}^2 + \frac{1}{2}mr^2\dot{\theta}^2 + V. \tag{21}$$

The trajectory of the particle was given as

$$r = r_0 e^{\theta} \tag{22}$$

and differentiating with respect to time:

$$\dot{r} = r_0 e^{\theta}\dot{\theta} = r\dot{\theta}. \tag{23}$$

Then, the total energy becomes:

$$E = \frac{1}{2}mr^2\dot{\theta}^2 + \frac{1}{2}mr^2\dot{\theta}^2 + V$$

or $$E = mr^2\dot{\theta}^2 + V. \tag{24}$$

The potential energy is therefore:

$$V = E - mr^2\dot{\theta}^2 \tag{25}$$

or $$V = E - \frac{H^2}{mr^2}$$

where we have again used $H = mr^2\dot{\theta}$.

The central force function, f(r) is found from the potential by

$$f(r) = -\frac{dV}{dr}. \tag{27}$$

Thus, $$f(r) = -\frac{d}{dr}\left(E - \frac{H^2}{mr^2}\right)$$

or $$f(r) = -\frac{2H^2}{mr^3} \tag{28}$$

and the required force function is seen to be a force of attraction with an inverse cubed dependence on the radial distance. In factor form the force law may be written:

$$\vec{F} = -\frac{2H^2}{mr^3}\vec{r}_1. \tag{29}$$

• PROBLEM 16-14

A particle is observed to move in a spiral orbit given by the equation $r = k\theta$ where k is a constant. Determine the central force law which leads to this orbit.

Solution: To solve this problem we shall use Newton's Second Law. In polar coordinates, F(r), the radial component of force is:

$$\mu(\ddot{r} - r\dot{\theta}^2) = F(r) \tag{1}$$

where $\mu = \frac{Mm}{m + M}$ is the reduced mass. Also note that for any central force problem the angular momentum $\vec{\ell}$ is a constant and that the magnitude of the angular momentum is given by the equation,

$$\ell = \mu r^2\dot{\theta}. \tag{2}$$

Substituting Eq. (2) into Eq. (1) yields

$$\mu\ddot{r} - \frac{\ell^2}{\mu}r^{-3} = F(r). \tag{3}$$

Eq. (3) is not suitable for determining F(r) because we do not know r as a function of time. We therefore have to perform a change of variables to obtain a differential equation for the orbit. Let

$$r = \frac{1}{u} \tag{4}$$

and we differentiate this equation with respect to t, this yields:

$$\dot{r} = -\frac{1}{u^2}\dot{u} = -\frac{1}{u^2}\dot{\theta}\frac{du}{d\theta}. \tag{5}$$

Substituting Eq. (2) into Eq. (5) and replacing r by

$\frac{1}{u}$, we get

$$\dot{r} = -\frac{\ell}{\mu}\frac{du}{d\theta} \quad . \tag{6}$$

Next, differentiate Eq. (6) once more with respect to t, we have

$$\ddot{r} = -\frac{\ell}{\mu}\frac{d}{dt}\left(\frac{du}{d\theta}\right) = -\frac{\ell}{\mu}\dot{\theta}\frac{d^2u}{d\theta^2} \quad . \tag{7}$$

Substituting Eq. (2) into Eq. (7) allows us to eliminate $\dot{\theta}$ from the above equation and we obtain,

$$\ddot{r} = -\frac{\ell^2}{\mu^2}u^2\frac{d^2u}{d\theta^2} \; . \tag{8}$$

We can now substitute Eqs. (4) and (8) into Eq. (3), this yields,

$$-\frac{\ell^2}{\mu}u^2\frac{d^2u}{d\theta^2} - \frac{\ell^2}{\mu}u^3 = F(u), \tag{9}$$

or $$\frac{d^2u}{d\theta^2} + u = -\frac{\mu}{\ell^2u^2}F(u) \; . \tag{10}$$

Next let us express Eq. (10) in terms of the variable r, thus we have:

$$\frac{d^2}{d\theta^2}\left(\frac{1}{r}\right) + \frac{1}{r} = -\frac{\mu}{\ell^2}r^2 F(r) \quad . \tag{11}$$

We will now use Eq. (11) to determine the force law for the spiral orbit given by the equation

$$r = k\theta \tag{12}$$

or $$\frac{1}{r} = k^{-1}\theta^{-1}. \tag{13}$$

Differentiating Eq. (13) twice with respect to θ we obtain

$$\frac{d}{d\theta}\left(\frac{1}{r}\right) = -k^{-1}\theta^{-2},$$

and $$\frac{d^2}{d\theta^2}\left(\frac{1}{r}\right) = +2k^{-1}\theta^{-3}. \tag{14}$$

But $\theta^{-1} = \frac{k}{r}$ and substituting this expression into Eq. (14) we obtain,

$$\frac{d^2}{d\theta^2}\left(\frac{1}{r}\right) = \frac{2k^2}{r^3} \; . \tag{15}$$

Finally to determine F(r) we will substitute Eq. (15) into Eq. (11) and solve for F(r):

$$\frac{2k}{r^3} + \frac{1}{r} = -\frac{\mu r^2}{\ell^2} F(r)$$

or $$F(r) = -\frac{\ell^2}{\mu}\left(\frac{2k}{r^5} + \frac{1}{r^3}\right). \tag{16}$$

Thus the force law is a combination of an inverse cube and inverse fifth power law.

• PROBLEM 16-15

Derive an expression for the differential scatter cross section of a mass m subject to the force

$$\vec{F} = \frac{K}{R^3}\,\hat{e}_R.$$

Solution: The force given is conservative and the associated potential energy is

$$V = \frac{K}{2R^2} = \frac{Ku^2}{2} \tag{a}$$

where the zero of potential energy is "at $R = \infty$."

The total energy E in the case of conservative central forces is given by

$$E = \frac{1}{2}\frac{H^2}{\mu}\left(\frac{du}{d\phi}\right)^2 + \frac{1}{2}\frac{H^2u^2}{\mu} + \frac{Ku^2}{2} \tag{b}$$

where H is the (constant) angular momentum, μ is the reduced mass, and ($u = 1/r$) and ϕ are the orbit parameters. In this problem, the force is defined with reference to a fixed force center in space. That is equivalent to saying that the force center has infinite mass which leads to replace μ by m, i.e., the mass and reduced mass are identical in this problem.

The energy equation may be written

$$E = \frac{1}{2}\frac{H^2}{m}\left(\frac{du}{d\phi}\right)^2 + \frac{1}{2}\left(\frac{H^2}{m} + K\right)u^2 \tag{c}$$

which corresponds to the form

$$E = \frac{1}{2}m\left(\frac{dx}{dt}\right)^2 + \frac{1}{2}kx^2$$

for a one-dimensional linear oscillator, if we make the following associations:

$$m \to H^2/m$$

$$k \to H^2/m + K$$

$$x \to u$$

$$t \to \phi.$$

By analogy, the solution of the path equation (c) is

$$u = A \cos \left[\left(\frac{H^2 + mK}{H^2} \right)^{\frac{1}{2}} \phi + \lambda \right]$$

where substitution for u into the energy equation gives $A = [2mE/(H^2 + mK)]^{\frac{1}{2}}$ and λ describes the orientation of the trajectory in the plane of the motion. The particle position is thus

$$\frac{1}{R} = \left(\frac{2mE}{H^2 + mK} \right)^{\frac{1}{2}} \cos \left[\left(\frac{H + mK}{H^2} \right)^{\frac{1}{2}} \phi + \lambda \right] .$$

Before and after scattering, the particle is a great distance from the scatterer ($1/R = 0$),

$$\cos \left[\left(\frac{H^2 + mK}{H^2} \right)^{\frac{1}{2}} \phi + \lambda \right] = 0$$

or
$$\left(\frac{H^2 + mK}{H^2} \right)^{\frac{1}{2}} \phi_0 + \lambda = \pm \frac{\pi}{2} .$$

The plus and minus correspond to the incoming ϕ_{in} and outgoing ϕ_{out} positions of the mass. Thus the change in ϕ, $\phi_{out} - \phi_{in}$ is

$$\Delta\phi = \left(\frac{\pi}{2} - \lambda \right) \left(\frac{H^2}{H^2 + mK} \right)^{\frac{1}{2}} - \left(- \frac{\pi}{2} - \lambda \right) \left(\frac{H^2}{H^2 + mK} \right)^{\frac{1}{2}}$$

or
$$\Delta\phi = \pi \left(\frac{H^2}{H^2 + mK} \right)^{\frac{1}{2}} .$$

The angle of scatter is defined as

$$\psi = \pi - \Delta\phi = \pi \left[1 - \left(\frac{H^2}{H^2 + mK} \right)^{\frac{1}{2}} \right].$$

Since the differential cross section is expressed in terms of the impact parameter s, we express ψ in terms of the same variable. The energy and angular momentum when the particle is very far removed from the force center are $E = \frac{1}{2} mv_0^2$ and $H = mv_0 s$. Thus

$$\psi = \pi \left[1 - \left(\frac{2mEs^2}{2mEs^2 + mK} \right)^{\frac{1}{2}} \right]$$

$$= \pi \left[1 - \left(\frac{s^2}{s^2 + K/2E} \right)^{\frac{1}{2}} \right] .$$

Solving for s^2,

$$\left(\frac{s^2}{s^2 + K/2E} \right)^{\frac{1}{2}} = \frac{1}{\pi} (\pi - \psi) ; \quad \frac{s^2}{s^2 + K/2E} = \frac{1}{\pi^2} (\pi - \psi)^2$$

$$s^2 \left[1 - \frac{1}{\pi^2}(\pi - \psi)^2\right] = \frac{1}{\pi^2}(\pi - \psi)\frac{K}{2E}$$

$$s^2 = \frac{1/\pi^2 \, (\pi - \psi)^2 \; K/2E}{1 - 1/\pi^2 \, (\pi - \psi)^2} = \frac{(\pi - \psi)^2 \; K/2E}{\pi^2 - (\pi - \psi)^2}$$

$$s^2 = \frac{K}{2E\left[\dfrac{\pi^2}{(\pi - \psi)^2} - 1\right]} \quad . \qquad \text{(d)}$$

The differential scatter cross section is defined as

$d\sigma = - 2\pi s ds = - \pi d(s^2)$.

Taking the differential of equation (d) we use

$$d\left[\frac{1}{f(u)}\right] = - \frac{1}{[f(u)]^2} \, d[f(u)]$$

to give us

$$d(s^2) = - \frac{K}{2E} \frac{1}{\left[\dfrac{\pi^2}{(\pi - \psi)^2} - 1\right]^2} - \frac{\frac{1}{2} \, \pi^2}{(\pi - \psi)^2} (- d\psi)$$

$$= - \frac{K}{4E}\left\{\frac{\pi^2 \, (\pi - \psi)}{[\pi^2 - (\pi - \psi)^2]^2}\right\} d\psi$$

and substitution gives us

$$d\sigma = \frac{\pi^3 \, K \, (\pi - \psi)}{2E \, [\pi^2 - (\pi - \psi)^2]^2} \, d\psi \quad .$$

CHAPTER 17

RIGID BODY KINETICS: FORCE, TORQUE

MASS MOMENTS OF INERTIA

• PROBLEM 17-1

What is the moment of inertia of a uniform rod of mass M and length D about an axis perpendicular to the rod through one end of the rod? Also show that the moment of inertia of the axis through the center of the rod is given by $I = \frac{1}{12} MD^2$.

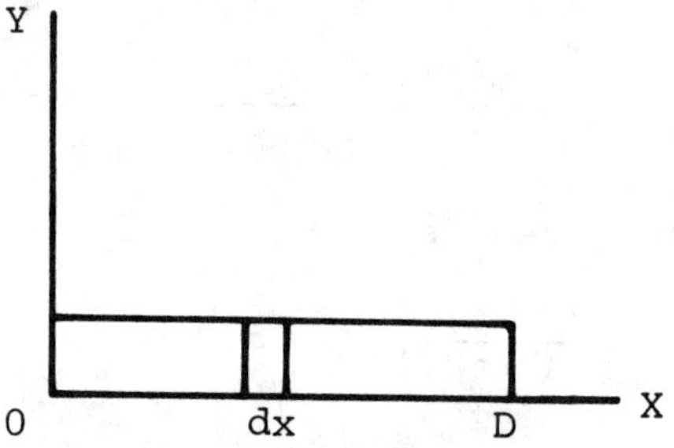

Solution: Consider the element of the rod of width dx and cross-section area A. Its mass is $dm = \rho\, Adx$. The mass of the rod is

$$M = \int dm = \int_0^D \rho\, Adx = \rho\, Ax \Big|_0^D = \rho\, AD. \qquad (1)$$

By definition the moment of inertia about the y-axis is

$$I_0 = \int x^2 dm = \rho\, A \int_0^D x^2 dx = \frac{1}{3}\, \rho\, AD^3. \qquad (2)$$

Substituting from equation (1) gives

$$I_0 = \frac{1}{3} MD^2.$$

The parallel axis theorem states $I_L = I_C + M_r^2$ where I_L is the moment of inertia about the line parallel to L through the center of mass and r is the perpendicular distance between the two lines.

Since the rod is uniform the center of mass is at D/2. So in this case r = D/2 and

$$I_C = I_L - Mr^2 = \frac{1}{3} MD^2 - M\frac{D^2}{4} = \frac{1}{12} MD^2.$$

• PROBLEM 17-2

Find the centroid of weight of the paraboloid $2z = x^2 + y^2$ between z = 0 and z = 2 ft when the weight density is given by r = 100 (1 + 0.2z) lb/ft³.

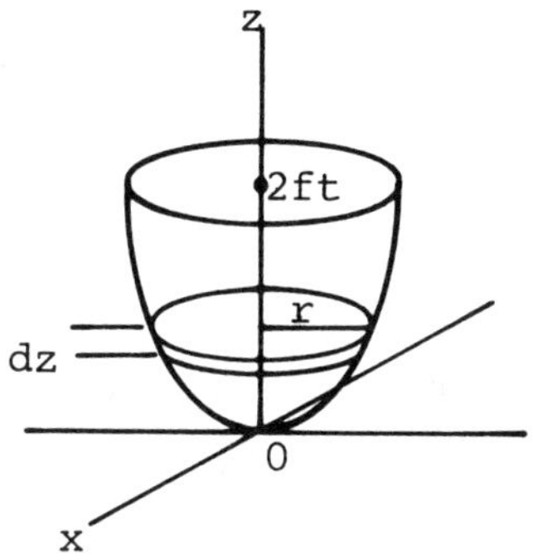

Fig. 1

Solution: The equation of the paraboloid is $2z = x^2 + y^2$ and the weight density is

$$\rho = 100\ (1 + .2z)\ \text{lb/ft}^3.$$

The centroid of weight is defined as $\bar{z}$, where

$$\bar{z} = \frac{1}{W}\int z dW. \tag{1}$$

It is first necessary to calculate the weight. Since $\rho = \frac{dW}{dV}$, the weight is

$$W = \int dW = \int \rho dV.$$

Consider the disk of radius r and thickness dz located at z which is shown in Figure 1. Its volume, dV, is

$$dV = \pi r^2 dz = \pi(x^2 + y^2)\ dz = 2\pi\ zdz \tag{2}$$

where we have used the equation of the paraboloid to eliminate x and y. Thus,

$$W = 2\pi \int_0^2 z\rho dz = 200\pi \int_0^2 (z + .2z^2)dz$$

$$= 200\pi \left(\frac{z^2}{2} + .2\frac{z^3}{3} \right) \Big|_0^2 = 200\pi \frac{7.6}{3} \text{ lb.} \qquad (3)$$

Next, find the centroid of weight. From equations (1) and (2)

$$\bar{z} = \frac{2\pi}{W} \int \rho z^2 dz = \frac{200\pi}{W} \int_0^2 (z^2 + .2z^3)\, dz$$

$$= \frac{200\pi}{W} \left(\frac{z^3}{3} + .2\frac{z^4}{4} \right) \Big|_0^2 = \frac{200\pi}{W} \frac{10.4}{3} \text{ ft.}$$

Substituting for the weight from equation (3) gives

$$\bar{z} = \frac{10.4}{3} \frac{3}{7.6} = 1.37 \text{ ft.}$$

• PROBLEM 17-3

A solid object with the shape of an isosceles triangle is shown in the figure. Given its mass m, base b, and height h, find its moment of inertia with respect to the following centroid axes: (a) ii′, (b) ℓℓ′ (c) mm′.

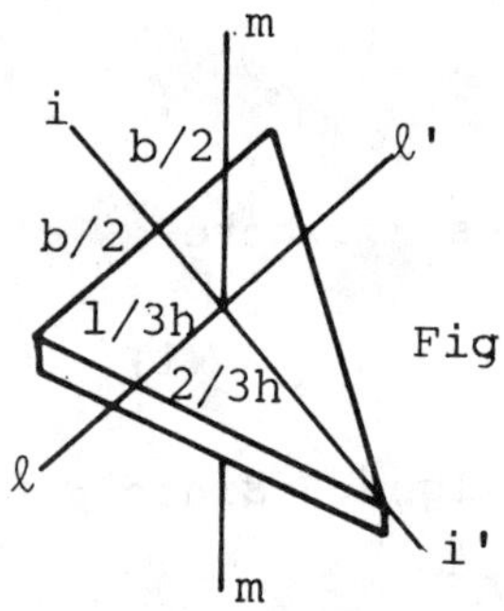

Fig. 1

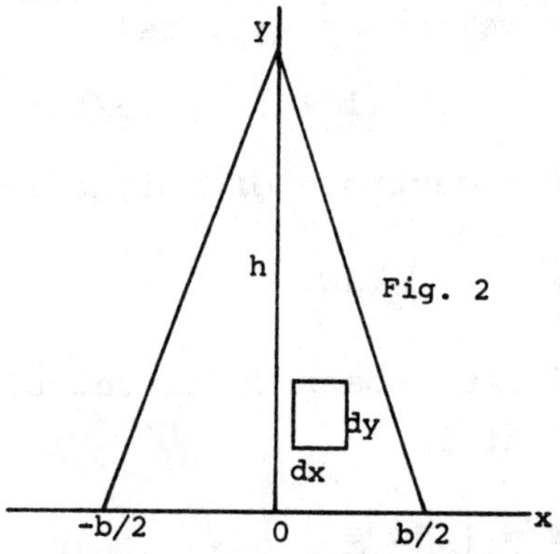

Fig. 2

Solution: Consider the element of area dxdy located at x,y shown in Figure 2. Its mass is $dm = \rho\, dA = \rho\, dxdy$ where ρ is the surface density. The equation of the line forming the right side of the triangle is $y = \frac{-2hx}{b} + h$ and for the left side $y = \frac{2hx}{b} + h$. For a given value of x, y ranges from zero to the line.

The mass of the plate is

$$m = \int dm = \rho \int dxdy = \rho \left\{ \int_0^{b/2} dx \int_0^{\frac{-2hx}{b} + h} dy \right.$$

$$\left. + \int_{\frac{-b}{2}}^{0} dx \int_0^{\frac{2hx}{b} + h} dy \right\} = \rho \left\{ \int_0^{b/2} \left(- \frac{2hx}{b} + h\right) dx \right.$$

$$\left. + \int_{\frac{-b}{2}}^{0} \left(\frac{2hx}{h} + h\right) dx \right\}$$

$$= \rho \left\{ \left(- \frac{hx^2}{b} + hx\right) \Bigg|_0^{\frac{b}{2}} + \left(\frac{hx^2}{b} + hx\right) \Bigg|_{\frac{-b}{2}}^{0} \right\}$$

$$= \frac{1}{2} \rho hb. \qquad (1)$$

Because of the symmetry of the triangle, it is easier to find the moments of inertia through O and then use the parallel axis theorem to find them through the center of mass.

$$I_{Ox} = \int y^2 \, dm = \rho \int y^2 \, dxdy$$

$$= \rho \left\{ \int_0^{\frac{b}{2}} dx \int_0^{\frac{-2hx}{b} + h} y^2 \, dy + \int_{\frac{-b}{2}}^{0} dx \int_0^{\frac{2hx}{b} + h} y^2 \, dy \right\} \qquad (2)$$

Note that if we let $x = -u$ the second integral becomes

$$\int_{\frac{b}{2}}^{0} (- du) \int_0^{\frac{-2hu}{b} + h} y^2 \, dy = \int_0^{\frac{b}{2}} du \int_0^{\frac{-2hu}{b} + h} y^2 \, dy.$$

This is the same as the first integral in equation (2). So,

$$I_{Ox} = 2 \rho \int_0^{\frac{b}{2}} dx \int_0^{\frac{-2hx}{b} + h} y^2 \, dy = \frac{2}{3} \rho \int_0^{\frac{b}{2}} \left(- \frac{2hx}{b} + h\right)^3 dx$$

$$= \frac{2}{3} \rho \int_0^{\frac{b}{2}} \left(- \frac{8h^3x^3}{b^3} + 3 \frac{4h^3x^2}{b^2} - 3 \frac{2h^3x}{b} + h^3\right) dx$$

$$= \frac{2h^3 \rho}{3} \left[- \frac{2x^4}{b^3} + 4 \frac{x^3}{b^2} - \frac{3x^2}{b} + x \right] \Bigg|_0^{\frac{b}{2}}$$

$$= \frac{1}{12} \rho h^3 b$$

$$= \frac{1}{b} m h^2, \qquad (3)$$

using equation (1).

For moments about the y-axis

$$I_{Oy} = \int x^2 dm = 2 \rho \int_0^{\frac{b}{2}} x^2 \int_0^{\frac{-2hx}{b} + h} dy$$

$$= 2 \rho \int_0^{\frac{b}{2}} \left(- \frac{2hx^3}{b} + h x^2 \right) dx$$

$$= 2 \rho \left(- \frac{hx^4}{2b} + \frac{hx^3}{3} \right) \Bigg|_0^{\frac{b}{2}}$$

$$= \frac{1}{48} \rho hb^3 = \frac{1}{24} mb^2. \qquad (4)$$

For the moment about the z-axis

$$I_{Oz} = \int r^2 dm = \int x^2 dm + \int y^2 dm = I_{Oy} + I_{Ox}$$

$$= \frac{1}{24} m (b^2 + 3h^2). \qquad (5)$$

The parallel axis theorem relates the moment of inertia about a line through O with the moment of inertia about a parallel line through the center of mass where R is the perpendicular distance between the lines.

$$I_O = I_C + m R^2$$

In this case

$$I_{ii'} = I_{Cy} = I_{Oy} = \frac{1}{24} m b^2 \qquad (6)$$

since ii' goes through both O and C, R = O.

$$I_{\ell\ell'} = I_{Cx} = I_{Ox} - m\left(\frac{h}{3}\right)^2 = \frac{1}{6} m h^2 - \frac{1}{9} m h^2 = \frac{1}{18} m h^2 \quad (7)$$

$$I_{mm'} = I_{Cz} = I_{Oz} - m\left(\frac{h}{3}\right)^2 = \frac{1}{24} m b^2 + \frac{1}{6} m h^2 - \frac{1}{9} m h^2$$

$$= \frac{1}{72} m (3 b^2 + 4 h^2). \quad (8)$$

Equations (6), (7), and (8) give the required moments about the axes through the center of mass.

• PROBLEM 17-4

A circular disk lies in the xy-plane with its center at the origin. The half above the x-axis has surface density σ, the half below the x-axis has density 2σ. Find the center of mass G and the moments parallel to the x, y, z axes through G and through the origin.

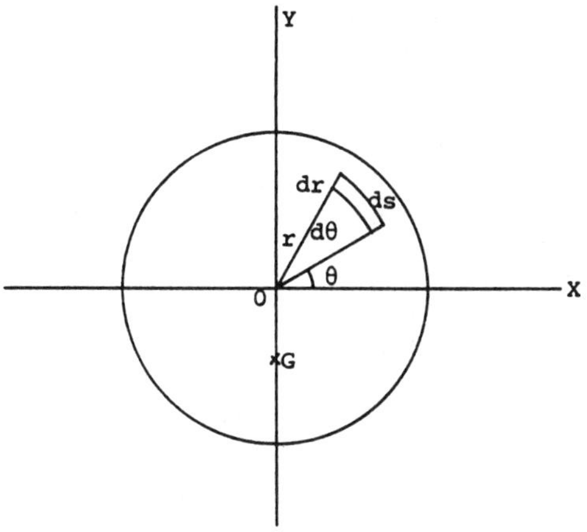

Solution: Since the disk has circular symmetry, the integrations will be simplified if polar coordinates are used.

The mass of the element at r, θ, is $dm = \rho\ da = \rho\ drds = \rho\ rdrd\theta$ where the surface density, $\rho = \sigma$ above the x-axis and 2σ below.

The mass of the disk is

$$m = \int dm = \iint \rho\ rdrd\theta = \int_0^a \int_0^\pi \sigma rdrd\theta + \int_0^a \int_\pi^{2\pi} 2\sigma rdrd\theta$$

(The integral must be separated into two parts since the density is not constant.)

$$m = \pi\sigma \int_0^a rdr + 2\pi\sigma \int_0^a rdr = \frac{\pi\sigma a^2}{2} + \frac{2\pi\sigma a^2}{2} = \frac{3\pi\sigma a^2}{2}. \quad (1)$$

By symmetry the center of mass must lie on the y-axis therefore, $\bar{x} = 0$.

$$\bar{y} = \frac{1}{m} \int y dm = \frac{1}{m} \int r \sin\theta \, dm = \frac{1}{m} \left\{ \sigma \int_0^a \int_0^\pi r^2 \sin\theta \, dr d\theta \right.$$

$$\left. + 2\sigma \int_0^a \int_\pi^{2\pi} r^2 \sin\theta \, dr d\theta \right\}.$$

Doing the r integration gives $\bar{y} = \frac{\sigma a^3}{3m} \left\{ \int_0^\pi \sin\theta \, d\theta \right.$

$$\left. + 2 \int_\pi^{2\pi} \sin\theta \, d\theta \right\},$$

$$= \frac{\sigma a^3}{3m} \left\{ -\cos\theta \Big|_0^\pi - 2\cos\theta \Big|_\pi^{2\pi} \right\}$$

$$= -\frac{2}{3} \frac{\sigma a^3}{m}.$$

From equation (1), this becomes

$$\bar{y} = -\frac{2}{3} \sigma a^3 \frac{2}{3\pi\sigma a^2} = -\frac{4}{9} \frac{a}{\pi}. \tag{2}$$

The moment of inertia about the z-axis through 0 is by definition

$$I_{0z} = \int r^2 dm = \int_0^a r^3 dr \left\{ \sigma \int_0^\pi d\theta + 2\sigma \int_\pi^{2\pi} d\theta \right\} \tag{3}$$

$$= \frac{3\pi\sigma a^4}{4} = \frac{1}{2} ma^2$$

where in the last step equation (1) is used.

The moment of inertia about the x-axis is

$$I_{0x} = \int y^2 dm = \int r^2 \sin^2\theta \, dm$$

$$= \int_0^a r^3 dr \left\{ \sigma \int_0^\pi \sin^2\theta \, d\theta + 2\sigma \int_\pi^{2\pi} \sin^2\theta \, d\theta \right\}.$$

Now $$\int \begin{Bmatrix} \sin^2\theta \\ \cos^2\theta \end{Bmatrix} d\theta = \frac{\theta}{2} \mp \frac{\sin 2\theta}{4}. \tag{4}$$

Since the 2nd term is zero at 0, π and 2π, I_{0x} becomes

$$I_{0x} = \frac{3\pi\sigma a^4}{8} = \frac{1}{4} ma^2. \qquad (5)$$

The moment of inertia about the y-axis is

$$I_{0y} = \int x^2 dm = \int r^2 \cos^2\theta\, dm.$$

Since the second term in equation (4) also vanishes in this case, the integration is the same as for I_{0x}, therefore

$$I_{0y} = I_{0x} = \frac{1}{4} ma^2. \qquad (6)$$

To find the moments of inertia about axes parallel to the x, y, z axes through the center of mass, G, the parallel axis theorem can be used. It states $I_0 = I_G + m R^2$ where R is the perpendicular distance between the axes O and G. In this case G is at (0, - 4a/9π, 0). So,

$$I_{Gz} = I_{0z} - m\left(\frac{4a}{9\pi}\right)^2 = \frac{1}{2} ma^2 - \frac{16}{81}\frac{ma^2}{\pi^2} = \left(\frac{81\pi^2 - 32}{162\pi^2}\right) ma^2$$

$$I_{Gx} = I_{0x} - m\left(\frac{4a}{9\pi}\right)^2 = \frac{1}{4} ma^2 - \frac{16}{81}\frac{ma^2}{\pi^2} = \left(\frac{81\pi^2 - 64}{324\pi^2}\right) ma^2$$

$$I_{Gy} = I_{0y} - m(0)^2 = \frac{1}{4} ma^2.$$

In the latter case the y-axis goes through both O and G and therefore, R = O.

• **PROBLEM 17-5**

Locate the center of gravity for the following regular pyramid. Dimensions: square base with 250 mm sides, 300 mm high. Composition: solid wood with steel sheets, 1 mm thick, on the triangular faces only. (Densities: steel 7850 Kg/m^3, wood 550 Kg/m.

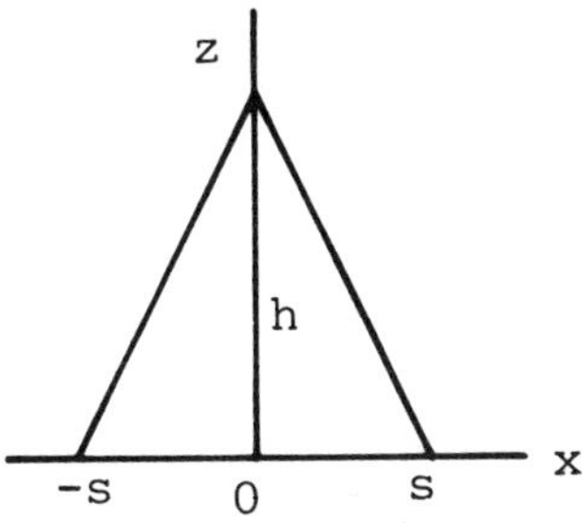

Fig. 1

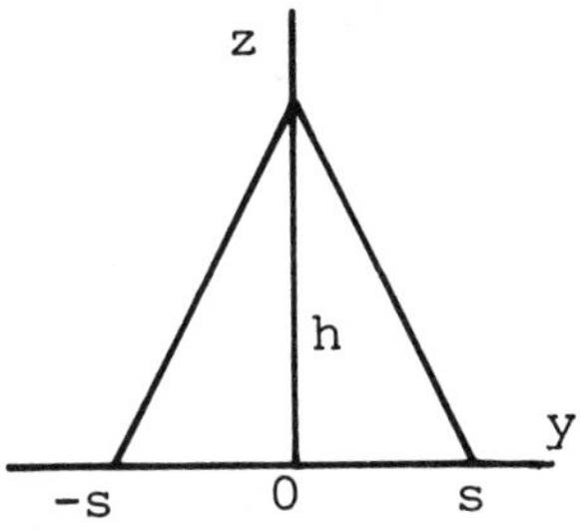

Fig. 2

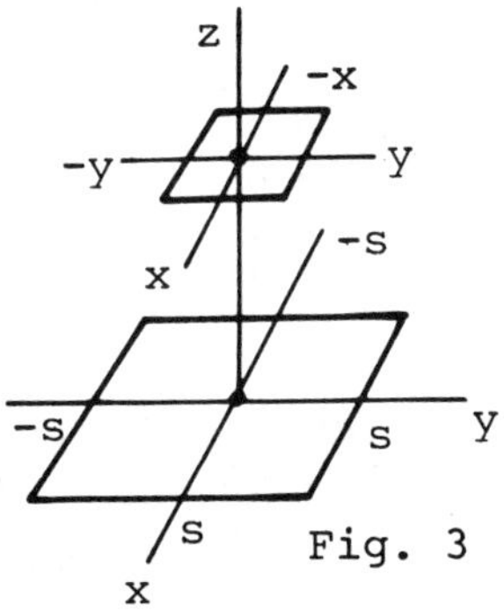

Fig. 3

Solution: From Figure 3 it is seen that at any height, z, the section of the pyramid parallel to the base is a square of area 4xy. But x and y are functions of z which are determined by the equations of the sides of the pyramid. From Figures 1 and 2 these are:

$$z = -\frac{h}{s}x + h \qquad \text{and} \qquad z = -\frac{h}{s}y + h. \qquad (1)$$

The mass of a pyramid with uniform density ρ is

$$m = \int \rho \, dv = \rho \int_0^h dz \int_{-y}^{y} dy \int_{-x}^{x} dx = 4\rho \int_0^h xy \, dz$$

using equations (1) for x and y,

$$= \frac{4\rho s^2}{h^2} \int_0^h (h^2 - 2hz + z^2)dz = \frac{4\rho s^2}{h^2} \left[h^2 z - hz^2 + \frac{z^3}{3} \right]_0^h$$

$$= \frac{4\rho s^2 h}{3} = \frac{1}{3} \rho \ell^2 h$$

where $\ell = 2s$, the length of a side of the base. By symmetry, the center of mass lies on the z axis and is given by

$$\bar{z} = \frac{1}{m} \int z dm = \frac{4\rho}{m} \frac{s^2}{h^2} \int_0^h z(h - z)^2 dz$$

$$= \frac{4\rho}{m} \frac{s^2}{h^2} \int_0^h (h^2 z - 2hz^2 + z^3)dz = \frac{4\rho}{m} \frac{s^2}{h^2} \left[\frac{h^2 z^2}{2} - \frac{2}{3} hz^3 + \frac{z^4}{4} \right]_0^h$$

$$\bar{z} = \frac{1}{12} \cdot \frac{4\rho s^2 h^2}{m} = \frac{1}{4} h.$$

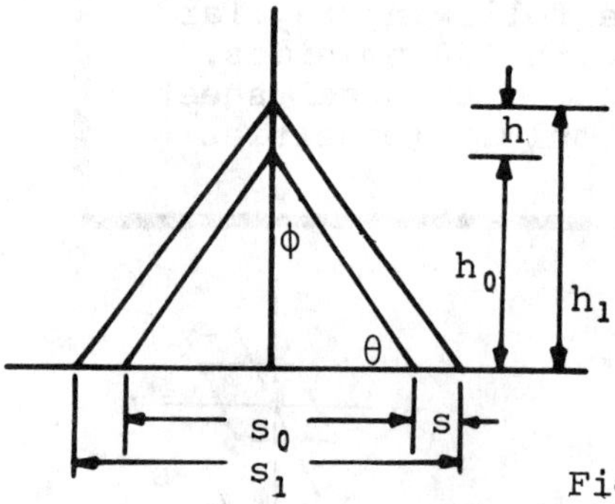

Fig. 4

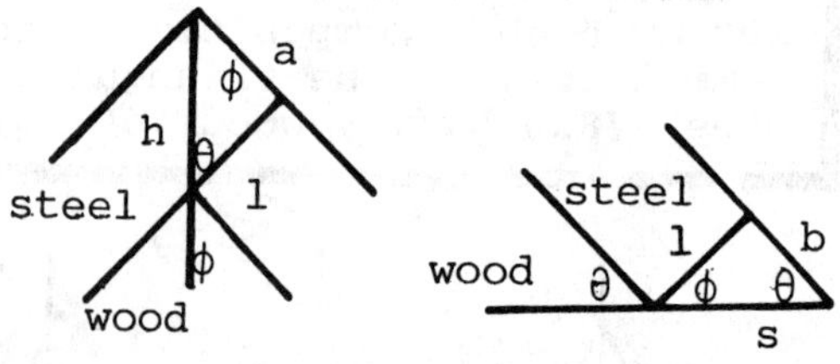

Fig. 5

For this problem, a section of the pyramid is shown in Figure 4. The outside sheets are 1 mm thick of steel with density $\rho_s = 7850\ Kg/_{m^3}$ and the inside of wood of density $\rho_w = 550\ Kg/_{m^3}$,

$h_1 = 300$ mm, $s_1 = 250$ mm.

From Figures 4 and 5

$$\tan\theta = \frac{h_1}{s_{1/2}}, \quad h_1 = h_0 + h$$

$$\cos\theta = \frac{1}{h}, \ \sin\theta = \frac{1}{s}, \ s_1 = s_0 + 2s$$

$$\tan\theta = \frac{600}{250} = 2.4$$

$$h = \frac{1}{\cos\theta} = 2.6 \text{ mm}, \ s = \frac{1}{\sin\theta} = 1{,}0833$$

$$h_0 = h_1 - h = 297.4 \text{ mm}, \ s_0 = s_1 - 25 = 247.8334 \text{ mm}.$$

We can now find the center of mass of the pyramid by considering it to be made up of three parts: a solid steel pyramid of height h_1, and base $s_1^{\,2}$, a steel "hole" of height h_0 and base $s_0^{\,2}$ and a wooden pyramid of height h_0 and base $s_0^{\,2}$.

$$\bar{z} = \frac{1}{m}(m_1 \bar{z}_1 + m_2 \bar{z}_2 + m_3 \bar{z}_3)$$

$$= \frac{h_{1/4}\, \rho_s v_1 - h_{0/4}\, \rho_s v_2 + h_{0/4}\, \rho_w v_2}{\rho_s v_1 - \rho_s v_2 + \rho_w v_2}$$

$$= \frac{1}{4}\,\frac{h_1 v_1 - h_0 v_2 (1 - \rho_w/\rho_s)}{v_1 - v_2 (1 - \rho_w/\rho_s)}$$

Since $h_1 = h_0 + h$ this becomes

$$\bar{z} = \frac{1}{4}\,\frac{h\, v_1 + h_0 [v_1 - v_2 (1 - \rho_w/\rho_s)]}{v_1 - v_2 (1 - \rho_w/\rho_s)}$$

$$= \frac{1}{4}\left[h_0 + \frac{h}{1 - \frac{v_2}{v_1}(1 - \rho_w/\rho_s)}\right].$$

Now $$\frac{v_2}{v_1} = \frac{1/3\, s_0^{\,2} h_0}{1/3\, s_1^{\,2} h_1} = \frac{(247.83)^2\ 297.4}{(250)^2\ 300} = .9742$$

$$1 - \rho_w / \rho_s = 1 - 550/7850 = .9299$$

$$1 - \frac{v_2}{v_1}(1 - \rho_w / \rho_s) = 1 - .9059 = .0941.$$

Therefore, $\bar{z} = \frac{1}{4}\left(297.4 + \frac{2.6}{.0941}\right)$ mm = 81.26 mm.

Note that in the limit when $\rho_w \to \rho_s$, i.e., the pyramid becomes uniform that $\bar{z} \to \frac{1}{4}(h_0 + h) = \frac{1}{4}h_1$ as it should.

If the pyramid had been all wood the center of mass would have been at $\bar{z} = \frac{h_1}{4} = 75$ mm. The presence of the steel plates raises the center of mass by 6.26 mm.

TRANSLATION AND ROTATIONAL MOTION

• PROBLEM 17-6

Show that the torque on a rigid body of mass M, due to a uniform gravitational field $\vec{g}$, is the same as that due to a single force $M\vec{g}$ applied to the center of mass of the rigid body.

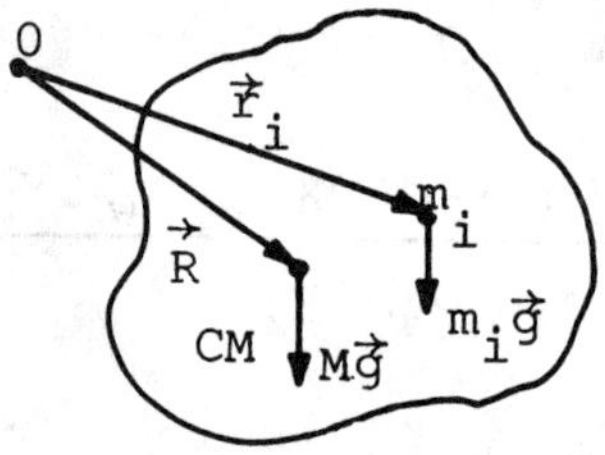

Solution: A rigid body may be regarded as a system of point masses whose relative positions are fixed. Thus the total mass of the body can be written as $M = \sum_i m_i$ (1)

where m_i is the mass of the ith particle as shown in the figure.

The torque $\vec{\tau}_i$ about the point O, of the point mass m_i due to gravitational field is

$$\vec{\tau}_i = \vec{r}_i \times m_i\vec{g}. \quad (2)$$

Thus the total torque τ of the rigid body about O is

$$\vec{\tau} = \sum_i \vec{r}_i \times m_i\vec{g}. \quad (3)$$

Since the gravitational field is uniform, we can write Eq. (3) as

$$\vec{\tau} = (\sum_i m_i \vec{r}_i) \times \vec{g}. \qquad (4)$$

However, the position vector $\vec{R}$, corresponding to the center of mass of the rigid body is defined by

$$M\vec{R} = \sum_i m_i \vec{r}_i . \qquad (5)$$

Substituting Eq. (5) in Eq. (4) we obtain

$$\vec{\tau} = \vec{R} \times M\vec{g}. \qquad (6)$$

This is exactly the torque due to a single force $M\vec{g}$ acting at the center of mass of the rigid body.

• PROBLEM 17-7

Show that in a uniform vertical gravitational field a rigid body can be supported at rest without rotation by a force applied to a single point of the body if this point lies on a vertical line which passes through the center of mass.

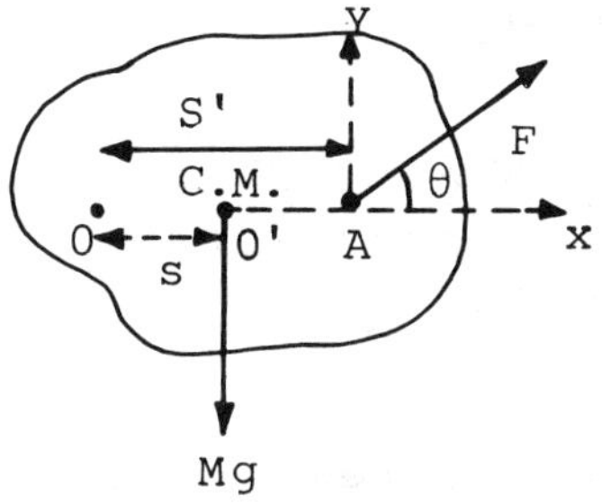

Solution: For a rigid body to be supported at rest, the total force and the total torque acting on the body must be equal to zero.

Suppose to maintain the body in equilibrium an arbitrary force F is applied to the body at the point A, as shown in the figure. For simplicity, assume that F is in the vertical xy plane, y being the axis along the vertical direction.

Since for translational equilibrium

$$\Sigma\vec{F} = 0, \text{ we have}$$

$$\vec{F} + M\vec{g} = 0. \qquad (1)$$

Taking components of Eq. (1) in the x and y directions

$$F \cos \theta = 0 \tag{2}$$

$$F \sin \theta - Mg = 0. \tag{3}$$

Eq. (2) give $\theta = 90°$ which implies that F must act vertically upward. Then from Eq. (3) we obtain

$$F = mg \tag{4}$$

which means that the applied force must be equal to the weight of the body.

For rotational equilibrium of the body, the total torque about any point must be zero, i.e.,

$$\Sigma \vec{\tau} = 0$$

Taking the torque of the two forces about the point O, (remembering F acts vertically upwards) yields

$$F s' - mg\, s = 0 \tag{5}$$

where s and s' are the normal distances of F and Mg from O. Using Eq. (4) in Eq. (5) yields

$$F (s' - s) = 0 \tag{6}$$

which means that s' = s or F must act along the same vertical line as the weight of the body. Thus, the body can be supported at rest without rotation by a force equal to its weight, applied to a point on a vertical line passing through the center of mass of the body.

• PROBLEM 17-8

A hunter enters a lion's den and stands on the end of a concealed uniform trapdoor of weight 50 lb freely pivoted at a distance x from the other end. Given that the hunter's weight is 150 lb, what fraction of the total length must x be in order that he and the end of the trapdoor shall start dropping into the depths with acceleration g when the trapdoor is released?

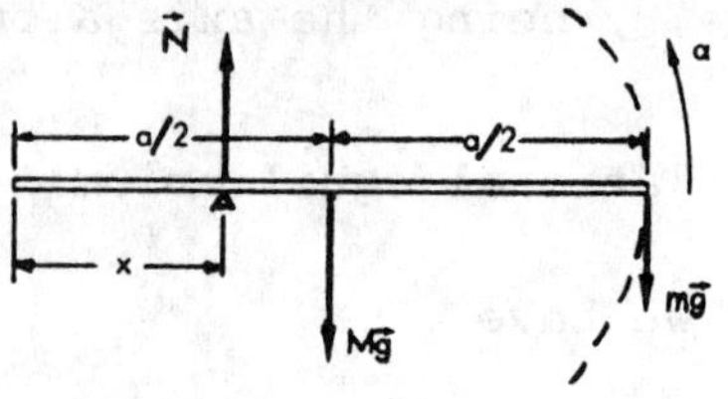

Solution: The forces acting on the trapdoor of length a are its weight Mg acting downward at the center, since it is uniform, the hunter's weight mg acting downward at the end, and the normal force N exerted by the pivot upward (see the figure). When we take moments about the pivot, counterclockwise moments being taken as positive, the moments causing rotational acceleration are

$-Mg\left(\frac{a}{2} - x\right) - mg(a - x)$, and thus using the rigid body

analog of Newton's second law, if T is the resultant torque acting on the rigid bar, I_T, the moment of inertia of the man-bar system about the pivot, and α its angular acceleration, then $T = I_T\alpha$ or

$$-Mg\left(\frac{a}{2} - x\right) - mg(a - x) = I_T\alpha$$

The moment of inertia of the trapdoor about a horizontal line parallel to the pivot and passing through the center of gravity is $\frac{1}{12} Ma^2$. By the parallel-axis theorem, if I_{cm} is the moment of inertia of the bar (or any rigid body) about its center of gravity and I_b is its moment of inertia about any axis, where h is the distance separating the two axes, then $I_b = I_{cm} + Mh^2$. Thus, I_b about the pivot is $\frac{1}{12} Ma^2 + M\left[(a/2) - x\right]^2$. The moment of inertia of the hunter about the pivot is $I_m = m(a - x)^2$. Hence

$$T = \left(I_b + I_m\right)\alpha \quad \text{for } I_T = I_b + I_m$$

$$-Mg\left(\frac{a}{2} - x\right) - mg(a - x) = \alpha\left[\frac{1}{12}Ma^2 + M\left(\frac{a}{2} - x\right)^2 + m(a - x)^2\right]. \qquad (1)$$

If the hunter and the end of the trapdoor are to have a linear acceleration g downward, then $-g = (a - x)\alpha$. Where $(a - x)$ is the radius of the circular arc path the man follows about the pivot (see the figure). Therefore, upon division of both sides of equation (1) by α and using the above relation between g and α,

$$M\left(\frac{a}{2} - x\right)(a - x) + m(a - x)^2 = \frac{1}{12}Ma^2 + M\left(\frac{a}{2} - x\right)^2 + m(a - x)^2.$$

$$M\left(\frac{a}{2} - x\right)(a - x) = \frac{1}{12}Ma^2 + M\left(\frac{a}{2} - x\right)^2$$

$$\frac{a^2}{2} - ax - \frac{ax}{2} + x^2 = \frac{1}{12}a^2 + \frac{a^2}{4} - ax + x^2$$

$$\left(\frac{a^2}{2} - \frac{a^2}{4}\right) - \frac{3}{2}ax + ax = \frac{1}{12}a^2$$

$$\frac{a^2}{4} - \frac{ax}{2} = \frac{1}{12}a^2$$

$$\therefore\ \frac{a}{2}\left(\frac{a}{2} - x\right) = \frac{1}{12}a^2 \qquad \therefore\ 3a^2 - 6ax = a^2 \qquad \therefore\ x = \frac{1}{3}a$$

The pivot must be located one-third of the length of the trapdoor from the end.

Figure 1 shows two pulleys holding a rigid bar KN and two weights in equilibrium. The string at point N is suddenly cut. Given the length L and mass m of the bar, find the initial acceleration of a) end K and b) end N of the bar.

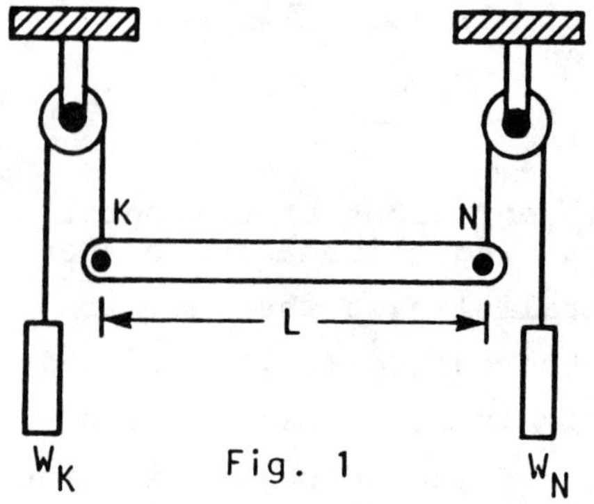

Fig. 1

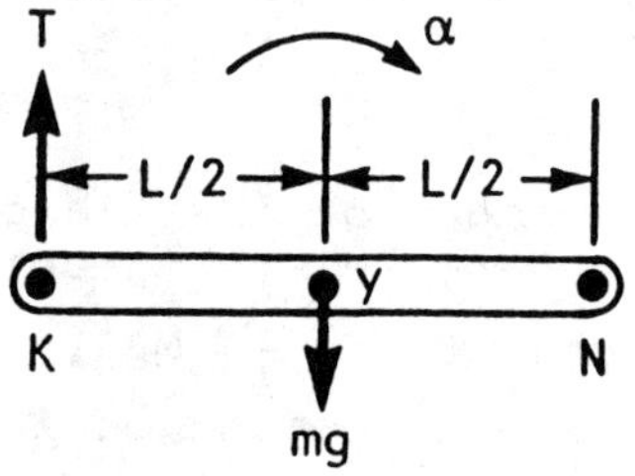

Fig. 2

Solution: Before the wire breaks, equilibrium conditions hold. For equilibrium:

$$\Sigma F_y = w_K + w_N - mg = 0$$

or $$w_K + w_N = mg$$

Taking moments about A, gives

$$\Sigma m_K, \; w_N L - mg\, L/2 = 0$$

or $$w_N = \frac{1}{2}\, mg$$

since $$w_K = mg - w_N$$

$$w_K = \frac{1}{2}\, mg$$

Thus $$w_K = w_N = \frac{mg}{2}$$

The weight of the bar acts at the center of gravity, at L/2 from K. Therefore, a moment after breaking, the following free-body diagram applies.

The forces and torques are not in balance, so the following equations of motion apply:

$$ma_y = mg - T$$

$$I\alpha = \text{Torque} = T\left(\frac{L}{2}\right)$$

where we have taken down to be positive for a_y, the acceleration of the center of mass of the bar. For the moment equation we have taken clockwise rotation to be positive.

T is the tension in the wire at A. With the acceleration motions, it is no longer equal to mg/2. The weight w_K has acceleration a_K, the same as that of point A. Therefore,

$$w_K - T = ma_K$$

or

$$\frac{mg}{2} - T = ma_K$$

is the equation of motion of the weight, yielding

$$T = \frac{mg}{2} - ma_A.$$

Thus, the equations of motion for the bar are

$$ma_y = mg - m_{y/2} + ma_K$$

or,

$$ma_y = \frac{mg}{2} + ma_K \tag{1}$$

and

$$I\alpha = \left(\frac{mg}{2} - ma_K\right)\frac{L}{2}.$$

Substituting the value of ma_A from equation (1) gives,

$$I\alpha = \left[\frac{mg}{2} - \left(ma_y - \frac{mg}{2}\right)\right]\frac{L}{2} \tag{2}$$

In addition to these equations, the kinematic relationships for the relative motion of the ends of the bar apply. That is,

$$a_K = a_y - \alpha\frac{L}{2} \tag{3}$$

$$a_N = a_y + \alpha\frac{L}{2}. \tag{4}$$

Substituting $I = \frac{1}{12} mL^2$, into equation (2) gives,

$$\frac{1}{12} mL^2\alpha = \left[\frac{mg}{2} - \left(ma_y - \frac{mg}{2}\right)\right]\frac{L}{2}$$

Solving for α, yields,

$$\alpha = \frac{12}{mL^2}\left[\frac{mg}{2} - \left(ma_y - \frac{mg}{2}\right)\right]\frac{L}{2}$$

or, $\alpha = \frac{6}{L}(g - a_y)$. (5)

Substituting equation (5) into (3) and (4) yields

$$a_K = 4a_y - 3g \quad (6)$$

$$a_N = -2a_y + 3g. \quad (7)$$

We may now put equation (6) back into (1) to solve for a_y:

$$a_y = \frac{g}{2} + 4a_y - 3g$$

$$a_y = \frac{5}{6} g.$$

Substitution back into equation (6) and (7) yields the results

$$a_K = \frac{1}{3} g$$

$$a_N = \frac{4}{3} g.$$

• PROBLEM 17-10

Figure 1 shows a thin, rectangular block, suspended in equilibrium by three ideal strings KE, FI, and HJ. The mass of the block is 60 kg, and the length of each string is 2 m. If string KE breaks, find a) the initial acceleration of the block and b) the initial tension in strings FI and HJ when the string is broken.

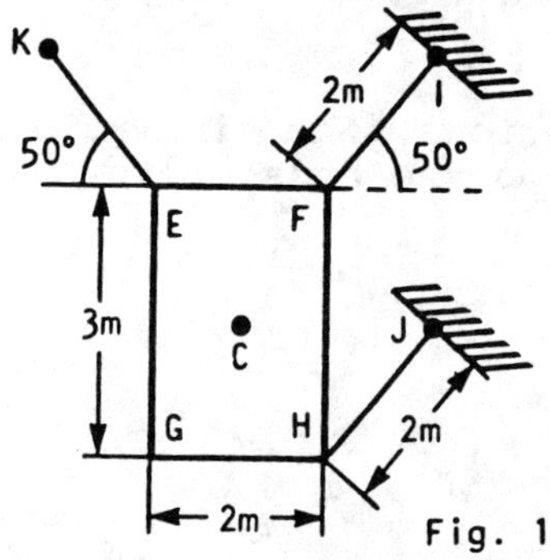

Fig. 1

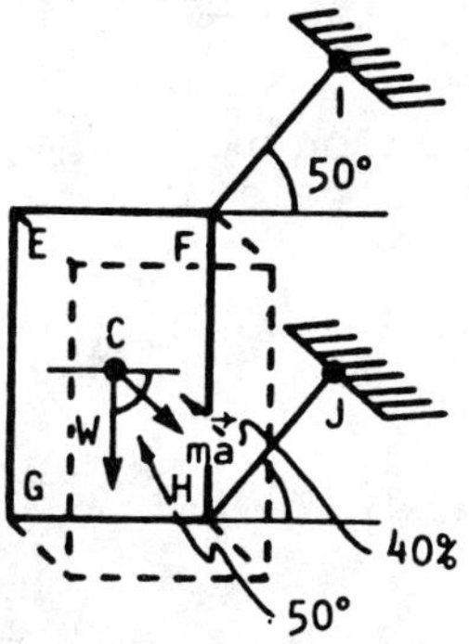

Fig. 2

Solution: When wire KE is cut, the plate center moves only under the force of gravity. Since the plate's motion is constrained by ropes FI and HJ, each of length 2 m, the corners F and H describe parallel circles (Fig. 2). The motion of the plate is a uniform translation along a circular path with a radius of 2 m. At the moment string KE is cut, the plate has zero velocity, and an acceleration $\vec{a}$ tangent to the circular path.

The acceleration is also perpendicular to the wires at F and H. That direction is 50° - 90° = -40°, downwards as shown in Figure 2.

This plate is considered while it is in translation. It is a dynamic system where the observed motion is caused by the resultant of the eternal forces of weight W and the tensions T_F and T_H exerted by the strings. By Newton's Second Law, the resultant force may be expressed $R = m\vec{a}$ of mass center C. Writing the force and motion in equations component form we have:

$$W \cos 50° = m\vec{a} \tag{1}$$

$$T_F + T_H - W \sin 50° = 0 \tag{2}$$

From (1), since W = mg, we find that

$$a = \frac{mg \cos 50°}{m} = g \cos 50° \tag{3}$$

Given that $g = 9.81$ m/sec^2 and cos 50° = .6428 equation (3) becomes

$$a = 9.81 \times .6428 = 6.31 \text{ m/sec}^2$$

$$\vec{a} = 6.31 \text{ m/sec}^2, \text{ 40° downwards}$$

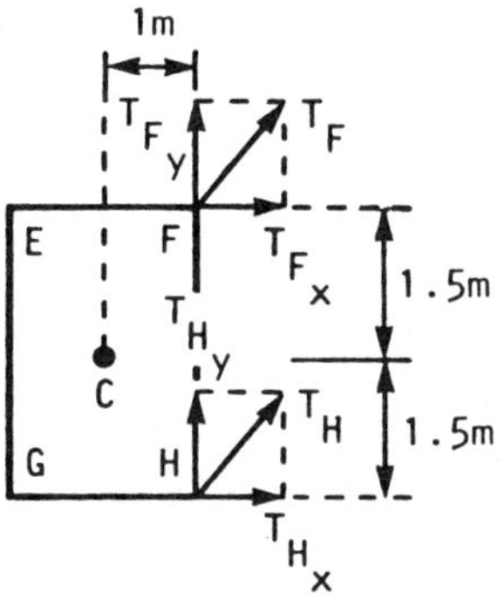

Fig. 3

To find the tensions in strings FI and HJ we considered the fact that the motion of the plate is pure translation. The acting moments are therefore in equilibrium. Writing the equilibrium equations for the moments around point G we may get an expression T_H and T_F:

$$T_{F_y} (1 \text{ m}) - T_{F_x} (1.5) + T_{H_y} (1 \text{ m}) + T_{H_x} (1.5) = 0$$

$$(T_F \sin 50°)(1 \text{ m}) - T_F \cos 50°)(1.5 \text{ m})$$

$$+ (T_H \sin 50°)(1 \text{ m}) + (T_H \cos 50°)(1.5 \text{ m}) = 0$$

Solving for T_H gives

$$T_H = +.1145\ T_F \tag{4}$$

Substituting this into equation (2), yields

$$T_F + 0.1145\ T_F - W \sin 50° = 0$$

or $\quad 1.1145\ T_F - W \sin 50° = 0$

$$T_F = (.687)(W) = (.687)\ mg$$

$$T_F = (.687)(60\ kg)(9.8\ m/sec^2) = 404.37\ N$$

And from (4)

$$T_H = (.1145)(404.37) = 46.30\ N$$

• PROBLEM 17-11

Suppose that a uniform pulley of mass M and radius R is free to turn about a frictionless axle through its center. Two masses M_1 and M_2 are suspended at the end of a weightless inextensible cord passing over the pulley, Fig. 1. Find the acceleration of this system.

Solution: If mass M_1 is larger than mass M_2, then M_1 will descend with some acceleration and M_2 will ascend with an equal acceleration. The pulley will rotate only if there is friction between the cord and the periphery of the pulley. If this is assumed to be so, the tensions in the cord on the two sides of the pulley must be different. They are denoted by T_1 and T_2, and from the direction of the accelerations it follows that T_1 must be greater than T_2. The upper arrow for T_1, Fig. 1, represents the direction of the force exerted by the cord on the pulley, and the lower arrow the direction of the force exerted by the cord on mass M_1.

If we apply Newton's second law to each of the masses M_1 and M_2 respectively, we have for the linear acceleration a

$$M_1g - T_1 = M_1a \tag{1}$$

$$T_2 - M_2g = M_2a \tag{2}$$

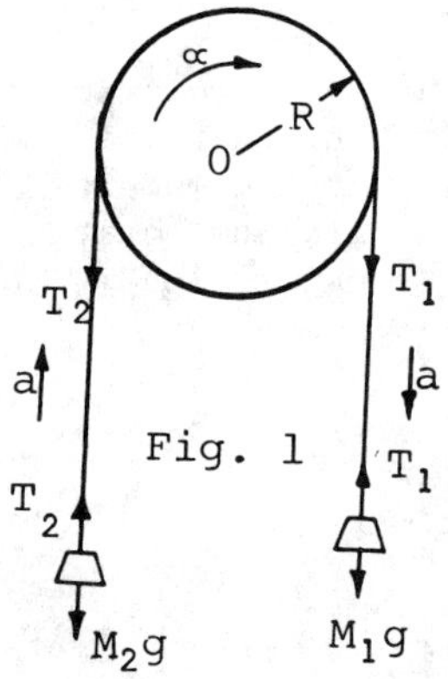

Fig. 1

Taking moments about the axis of rotation O

$$T_1R - T_2R = I_0\alpha = \frac{MR^2\alpha}{2} \tag{3}$$

where α is the angular acceleration of the pulley and I_0 the moment of inertia of the pulley about the axis through O. But α = a/R. Hence

$$T_1 - T_2 = \frac{Ma}{2} \tag{4}$$

Combining Equation (1) and (2), we have

$$(M_1 - M_2)g - (T_1 - T_2) = a(M_1 + M_2).$$

Using equation (4),

$$(M_1 - M_2)g = a\left(M_1 + M_2 + \frac{M}{2}\right)$$

or $$a = \frac{(M_1 - M_2)g}{M_1 + M_2 + \frac{M}{2}} = \frac{(M_1 - M_2)g}{[M_1 + M_2 + (I_0/R^2)]}$$

Thus the effect of the mass of the pulley is to increase the inertia or effective mass of the system by I_0/R^2 or $M/2$.

• PROBLEM 17-12

As shown in Fig. 1, a pulley is supported by a frictionless bearing at O, and a flexible, but inextensible, string is passed over the pulley and connected to two weights. The pulley has a weight of 100 lb and a radius of gyration of 1.5 ft. If the weights are released from rest at the same elevation, determine the vertical distance of separation after 1 sec.

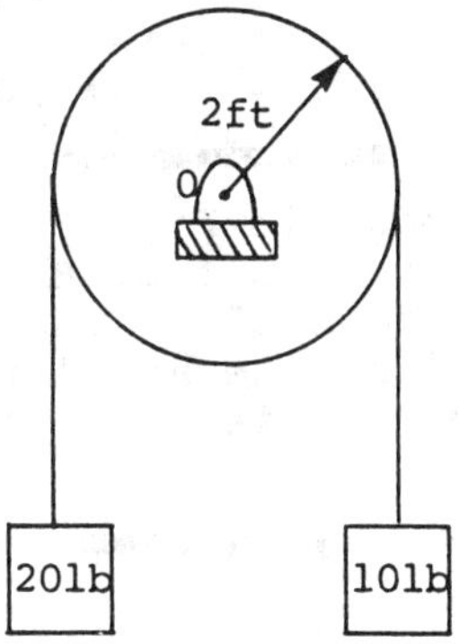

Fig. 1

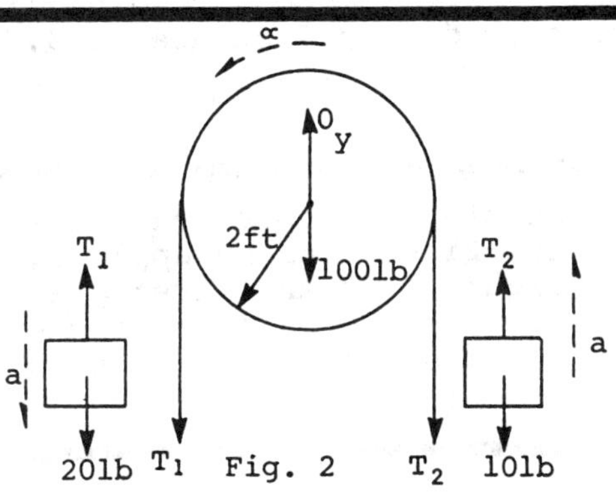

Fig. 2

Solution: Let us first find the constant acceleration of the two weights by drawing the free-body diagrams of the weights and the pulley, as shown in Fig. 2. Note that the senses of the kinematic quantities a and α are consistent. The equations of motion in the vertical direction for the two weights and the moment equation of motion for the pulley are

$$+\downarrow \ \Sigma F_v = -T_1 + 20 = \frac{20}{g} a$$

$$+\uparrow \ \Sigma F_v = T_2 - 10 = \frac{10}{g} a$$

$$\circlearrowleft + \ \Sigma M_0 = T_1(2) - T_2(2) = \frac{100}{g}(1.5)^2\alpha.$$

From the kinematics of the system, $a = 2\alpha$. So,

$$-T_1 + 20 = \frac{20}{g} a$$

$$T_2 - 10 = \frac{10}{g} a$$

$$T_1 - T_2 = \frac{56.3}{g} a$$

Solving these equations simultaneously for the linear acceleration of each block, we find that $a = 3.73$ ft/sec².

Since the weights started from rest, and then were constantly accelerated, the distance each block moved can be expressed as

$$d = \frac{1}{2} a t^2$$

$$= \frac{1}{2} (3.73)(1)^2 = 1.87 \text{ ft.}$$

The separation is twice this distance since the displacements are in opposite directions. Therefore the blocks are separated by a distance of 3.74 ft after one second.

• PROBLEM 17-13

Two large circular spur gears are shown in Fig. 1. A motor drives the first gear by applying the moment of M as shown in the figure. If the first gear weighs 28 lb and the second gear weighs 48 lb, what will be the angular acceleration of the second gear if the moment M is equal to 0.2t ft-lb?

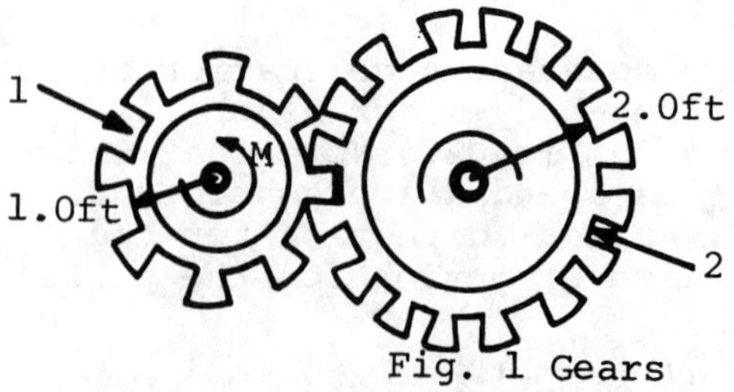

Fig. 1 Gears

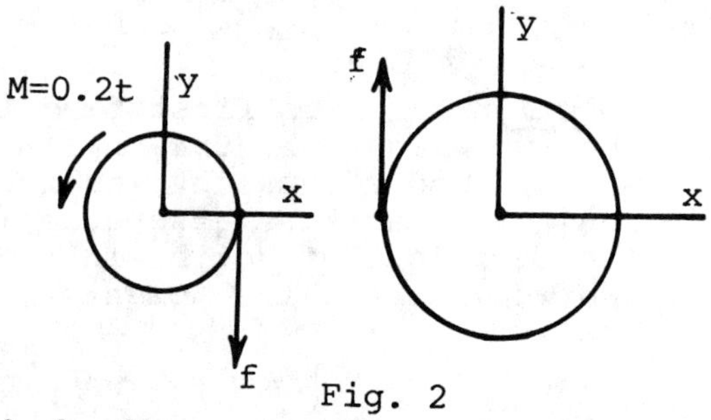

Fig. 2

Solution: First see the free body diagrams, Figure 2, showing the relevant (torque producing) forces of both gears. The force transmitted through the gears is given as f. In order to write the moment equations about the axis of rotation for each of the gears, the moment of inertia of each of the gears about the axis of rotation must be calculated using $I = \frac{1}{2} \frac{W}{g} mr^2$:

$$I_1 = \frac{1}{2}\left(\frac{28}{32.2}\right)(1)^2 = 0.435 \text{ lb-sec}^2\text{-ft}$$

$$I_2 = \frac{1}{2}\left(\frac{48}{32.2}\right)(2)^2 = 2.98 \text{ lb-sec}^2\text{-ft.}$$

We shall first write the equation of motion of the second gear, since this equation contains our desired unknown, the angular acceleration of the gear:

$$M_2 = I_2\dot{\omega}_2$$

$$-2f = 2.98\dot{\omega}_2.$$

Note that counterclockwise moments are positive. In this one equation there are two unknown quantities, f and $\dot{\omega}_2$, so that we must seek another equation in which both terms appear. The logical choice would be the moment equation for the first gear. The force f will appear in this equation:

$$M_1 = I_1\dot{\omega}_1$$

$$0.2t - f = 0.435\dot{\omega}_1.$$

By writing this equation, we have introduced another unknown, $\dot{\omega}_1$. However, the motions of the two gears are not independent; they are related by the following equation, $\omega_1 = -\frac{N_2}{N_1}\omega_2$, where N is the number of teeth. Therefore:

$$\dot{\omega}_1 = -2\dot{\omega}_2.$$

This equation comes from the fact that at the point of contact the accelerations of the gears are equal. These accelerations are equal to the angular accelerations of the gears multiplied by their respective radii.

The minus sign indicates that when the first gear is moving counterclockwise (in a positive direction, as we have defined it by the right-hand rule), the second gear is moving clockwise, or in a negative direction. The two equations of motion along with this condition form a set of three simultaneous equations, which may be solved for $\dot{\omega}_2$:

$$\dot{\omega}_2 = -0.0847t \text{ rad/sec}^2.$$

The minus sign indicates that the second gear has an angular acceleration in the clockwise direction.

• PROBLEM 17-14

A yo-yo, shown in Figure 1, is made of two uniform disks of equal mass and radius R, which are joined along a common axis through their centers by a smaller disk of negligible mass and radius r. A string is wrapped around the smaller disk and its other end is fixed. If the yo-yo is released, what will be the acceleration of its center as the string unwinds without slipping? Neglect the mass and thickness of the string. Also assume that the string is in a vertical direction. (Although in a real yo-yo there will be a slight horizontal component to the string's direction.)

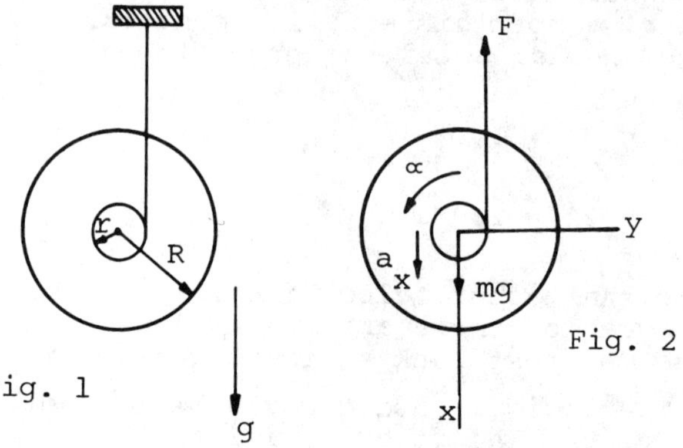

Fig. 1

Fig. 2

Solution: The problem deals with the motion of a rigid body, the yo-yo, in a plane. The equations governing its motion are

$$I\,\alpha = \Sigma T_i$$

$$m\,a_x = \Sigma F_{xi}$$

$$m\,a_y = \Sigma F_{yi}.$$

The free-body diagram of the system is shown in Figure 2. Let the mass of each large disk be m/2. Thus the total mass is m and its mass moment of inertia about the centroidal axis is

$$I = 2\left(\frac{1}{2}\,\frac{m}{2}\,R^2\right) = \frac{1}{2}\,mR^2. \quad (1)$$

There is no force in the horizontal direction therefore $a_y = 0$.

In the vertical direction we have the gravity force mg acting downwards and the string tension F acting upwards. The only torque about the centroidal axis is Fr. The remaining two equations therefore, are

$$I\,\alpha = Fr \quad (2)$$

$$m\,a_x = mg - F. \quad (3)$$

Further, since the string does not slip on the small disk

$$a_x = \alpha r \quad (4)$$

Equations (2), (3) and (4) contain three unknowns: F, α and a_x. We substitute for α from equation (4) into (2)

$$I\, a_x/r = Fr \qquad (5)$$

Next we eliminate F by adding r times equation (3) to (5), which yields

$$(m\,r + I/r)\; a_x = mgr$$

or $$a_x = \frac{mgr}{mr + I/r} \qquad (6)$$

Now, substituting the expression for I from equation (1) into (6) gives

$$a_x = \frac{mgr}{mr + \frac{mR^2}{2r}}$$

$$= \frac{g}{1 + \frac{R^2}{2r^2}}$$

Note: The string tension F may now be determined from equation (5), which gives

$$F = I\, a_x/r^2$$

Upon substituting the expression found for a_x, and after some simplification we get

$$F = \frac{mg}{1 + 2r^2/R^2}$$

which is less than mg, the weight of the yo-yo.

• PROBLEM 17-15

The drum and shaft of a winch have a mass m = 25kg, a radius of r = 0.2m and a radius of gyration k = 0.1m. A rope with mass m = 15 kg and length ℓ = 35m is wrapped around the drum; 5m of the rope is left to hang freely (fig.1). The system is at rest when the safety latch is released and it begins rotational motion. Neglecting friction determine the initial angular acceleration of the drum.

Solution: Gravity acts on the hanging portion of the rope, and the latter exerts a force on the drum through the wound portion of the rope. Thus the hanging rope is accelerated downwards and the drum, with the remainder of the rope, is given an angular acceleration. To solve the problem, use the two fundamental equations:

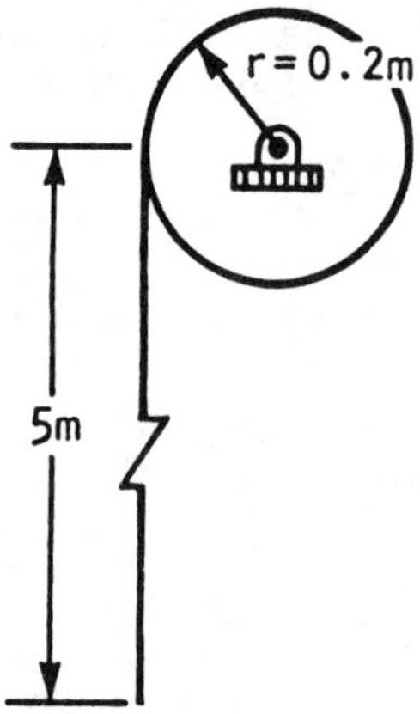

Fig. 1

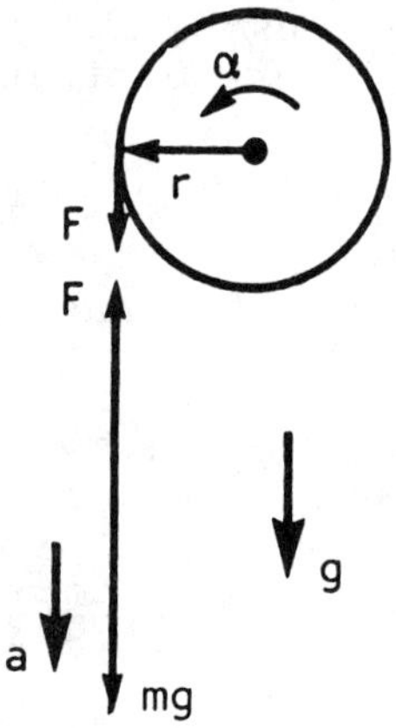

Fig. 2

$$m\ a = \Sigma F_i \tag{1}$$

applied to the hanging rope, and

$$I\ \alpha = \Sigma T_i \tag{2}$$

applied to the drum and the wound rope. In these equations m is the mass of the hanging portion of the rope and a is its acceleration. I is the total mass moment of inertia of the drum and the wound portion of the rope about their centroidal axis and α is their angular acceleration.

The free body diagrams of the hanging rope and the drum are shown in Figure 2. Positive directions are taken as the downward and counterclockwise directions. The force F shown in this figure is the tension in the rope where it comes off the drum. Note that it is not equal to the weight of the hanging rope, mg, since the rope is accelerating. There are two additional forces acting which are not shown in the free body diagram. These are the weight of the drum and the wound rope, and the bearing reaction force. These act through the drum's centroidal axis and therefore cause no torque about it. Equation (1) may now be written as

$$m\ a = m\ g - F. \tag{3}$$

The only torque on the drum is due to F and equation (2) becomes

$$I\ \alpha = F\ r. \tag{4}$$

Further, as the rope does not slip on the drum, the accelerations a and α are related by the equation

$$a = \alpha\ r. \tag{5}$$

Equations (3), (4) and (5) contain three unknowns: F, a and α. Substituting for a from (5) into (3), yields

$$m\alpha r = m\ g - F. \tag{6}$$

Now, to eliminate F, multiply both sides of equation (6) by r and add to (4), which yields

$$(I + mr^2)\ \alpha = mgr$$

or $$\alpha = \frac{mgr}{I + mr^2} \qquad (7)$$

The quantity $(I + mr^2)$ in equation (7) is the total effective mass moment of inertia of the drum, wound rope, and the hanging rope, referred to the drum's centroidal axis.

Numerical data for the problem is: $m = 2.5$ Kg, $r = 0.2$ m, $g = 9.81$ m/s^2. The mass moment of inertia of the drum and shaft is

$$(25 \text{ Kg})\ (0.1 \text{ m})^2 = 0.25 \text{ Kg - m}^2.$$

The mass of the wound rope is 12 Kg. This mass is distributed at a radius of 0.2 m. Thus its mass moment of inertia is

$$(12.5 \text{ Kg})\ (0.2 \text{ m})^2 = 0.5 \text{ Kg - m}^2.$$

Therefore $I = 0.25 + 0.5 = 0.75$ Kg - m^2.

Substituting the numerical values into equation (7) gives

$$\alpha = 5.77 \text{ radians/s}^2.$$

Note that the force F in the rope may now be calculated from equation (6), i.e.,

$$F = m\ (g - \alpha\ r)$$

which is less than mg, the weight of the hanging rope.

• PROBLEM 17-16

A cord is wound around a cylinder of mass M and radius R and the free end of the cord is passed over a fixed smooth peg with a mass M' attached to this end (see Figure 1). Suppose that mass M' moves upwards with an acceleration a'. Find a, the linear acceleration of the cylinder, a', and T, the tension in the cord in terms of M, M' and R.

Solution: The cylinder has an instantaneous axis of rotation about O, the point of contact of the cord and cylinder. Let the constant angular acceleration about O be α. Then the axis of the cylinder has a linear acceleration $a'' = \alpha R$ relative to O. Since the point O moves downwards with acceleration a', a (the absolute acceleration) is $a'' + a'$.

Consider first the angular acceleration and the torque about the center of mass of the cylinder. From Fig. 1 we see that

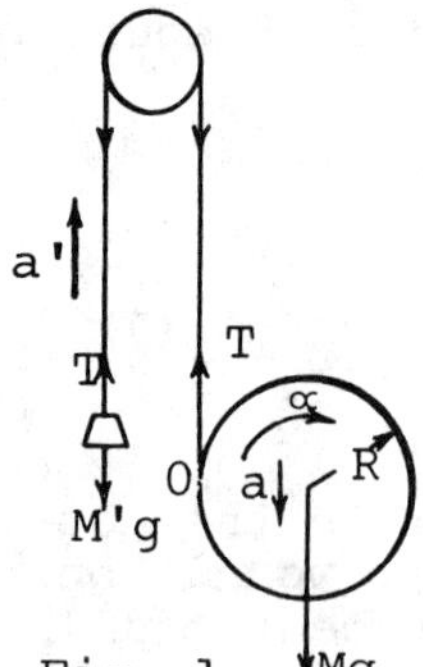

Fig. 1

$$\tau_{C.M.} = TR = I_{C.M.}\alpha = I_{C.M.}\frac{a''}{R}.$$

For a uniform cylinder $I_{C.M.} = MR^2/2$ so that

$$T = \frac{Ma''}{2}. \qquad (1)$$

From Newton's second law, $M\ddot{\vec{r}} = \Sigma\vec{F}$, for the linear acceleration of the center of mass of the cylinder

$$Mg - T = M(a'' + a'). \qquad (2)$$

Similarly, for mass M'

$$T - M'g = M'a'. \qquad (3)$$

These three equations are the dynamical equations of motion of the system. From them we obtain our required quantities. Substituting for T from equation (1) into equation (2) and (3) gives

$$g = \frac{3a''}{2} + a' \qquad (4)$$

and,
$$\frac{Ma''}{2} = M'a' + M'g$$

or
$$a' + g = \frac{M}{2M'}a''. \qquad (5)$$

Solving equations (1), (4) and (5)

$$a'' = \frac{4gM'}{M + 3M'} \qquad a' = \frac{g(M - 3M')}{M + 3M'} \qquad T = \frac{2gMM'}{M + 3M'}. \qquad (6)$$

and
$$a = \frac{9(M + M')}{M + 3M'}.$$

Note from the expression for a' in equation (6), that if $M' > \frac{1}{3}M$, the hanging mass will move downwards.

• PROBLEM 17-17

A cylinder rests on a horizontal rotating disc, as shown in the figures. Find at what angular velocity, ω, the cylinder falls off the disc, if the distance between the axes of the disc and cylinder is R, and the coefficient of friction $\mu > \frac{D}{h}$, where D is the diameter of the cylinder and h is its height.

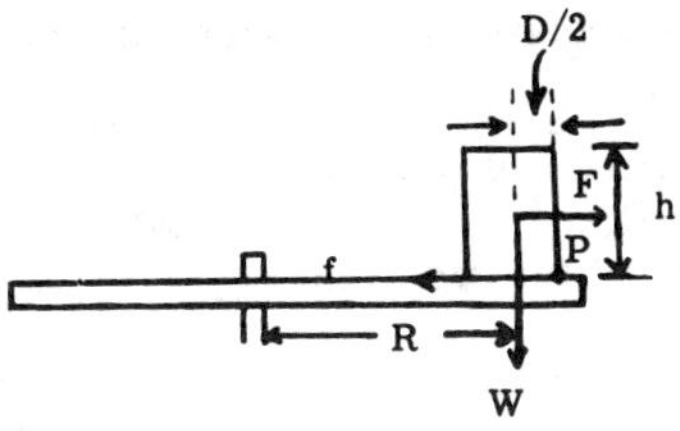

Figure A

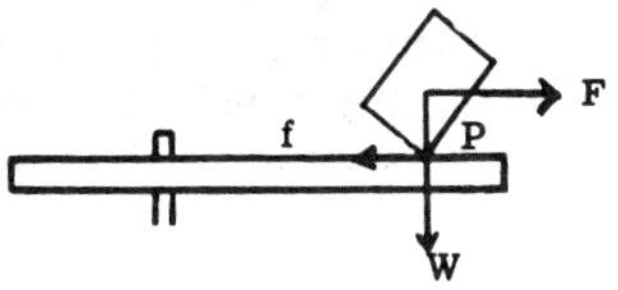

Figure B

Solution: The centripetal force that keeps the cylinder at rest on the disc is the frictional force f. According to Newton's third law of motion, the cylinder reacts with an equal and opposite force, F, which sometimes is referred to as the **centrifugal force,**

$$F = M\,\omega^2 R$$

where M is the mass of the cylinder. The cylinder can fall off either by slipping away (Fig. A) or by tilting about point P (Fig. B), depending on whichever takes place first. The critical angular speed, ω_1, for slipping occurs when F equals f;

$$F = f$$

$$M\,\omega_1^2\,R = \mu g\,M$$

where g is the gravitational acceleration. Hence

$$\omega_1 = \sqrt{\frac{\mu g}{R}}\,.$$

F tries to rotate the cylinder about P, but the weight W opposes it. The rotation becomes possible, when the torque created by F is large enough to take over the opposing torque caused by W;

$$F\,\frac{h}{2} = W\,\frac{D}{2}$$

$$F = W\,\frac{D}{h}$$

$$M\,\omega_2^2\,R = Mg\,\frac{D}{h}$$

giving $\omega_2 = \sqrt{\frac{D}{hR}}$

Since we are given that $\mu > \frac{D}{h}$, we see that

$$\omega_1 > \omega_2$$

and the cylinder falls off by rolling over at $\omega = \omega_2$.

A uniform rod of mass m and length 2a stands vertically on a rough horizontal floor and is allowed to fall. Assuming that slipping has not occurred, show that, when the rod makes an angle θ with the vertical,

$$\omega^2 = (3g/2a)(1 - \cos\theta)$$

where ω is the rod's angular velocity.

Also find the normal force exerted by the floor on the rod in this position, and the coefficient of static friction involved if slipping occurs when θ = 30°.

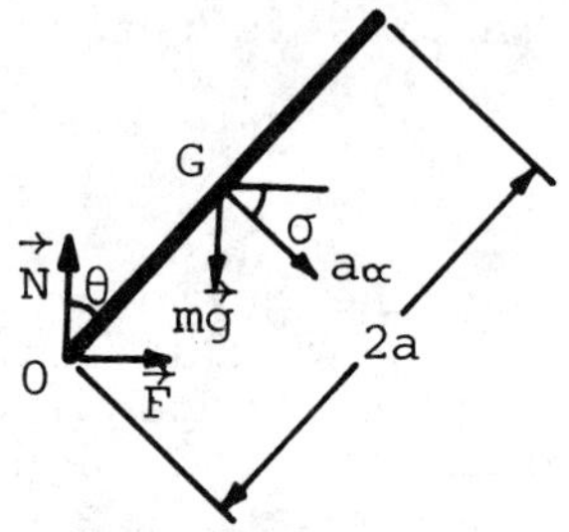

Solution: The forces acting on the rod are the weight $\vec{mg}$ acting downward and the normal force $\vec{N}$ and the frictional force $\vec{F}$ of magnitude μN exerted by the floor at the end O in contact with the floor. In order to find ω, we relate the net torque τ on the rod to the rod's angular acceleration α by using

$$\tau = I\alpha$$

Here, I is the rod's moment of inertia. We will then be able to solve for ω.

When one takes moments about O, the only force producing rotation about O is the weight of the rod. Hence

$$\tau = mga \sin\theta = I_0\,\alpha = \frac{4}{3}ma^2\,\alpha$$

Here, I_O is the rod's moment of inertia about O. Now,

$$\alpha = \frac{d\omega}{dt} = \frac{d\omega}{d\theta} \times \frac{d\theta}{dt} = \omega\,\frac{d\omega}{d\theta} = \frac{3}{4}\,\frac{g}{a}\sin\theta.$$

$$\int_0^\omega \omega\,d\omega = \int_0^\theta \frac{3}{4}\,\frac{g}{a}\sin\theta\,d\theta.$$

$$\left[\tfrac{1}{2}\omega^2\right]_0^\omega = \left[-\frac{3}{4}\,\frac{g}{a}\cos\theta\right]_0^\theta \text{ or } \omega^2 = \frac{3g}{2a}(1 - \cos\theta).$$

The center of gravity G has an angular acceleration α about O, and thus a linear acceleration aα at right angles to the direction of the rod. This linear acceleration can be split into two components, $a\alpha \cos\theta$ horizontally and $a\alpha \sin\theta$ vertically downward. The horizontal acceleration of the center of gravity is due to the force μN and the vertical acceleration is due to the net effect of the forces mg and N. Thus, using Newton's Second Law, and taking the positive direction downward,

$$mg - N = ma\alpha \sin\theta = \frac{3}{4} mg \sin^2\theta \qquad (1)$$

and $$F_{max} = \mu N = ma\alpha \cos\theta = \frac{3}{4} mg \sin\theta \cos\theta. \qquad (2)$$

From (1)

$$N = mg - \frac{3}{4} mg \sin^2\theta = \frac{mg}{4}\left(4 - 3\sin^2\theta\right).$$

But when $\theta = 30°$, slipping just commences. At this angle F has its limiting, maximum value of F_{max}.

We have

$$\mu_s = \frac{F_{max}}{N} = \frac{\frac{3}{4} mg \sin\theta \cos\theta}{\frac{mg}{4}\left(4 - 3\sin^2\theta\right)} = \frac{3 \sin\theta \cos\theta}{\left(4 - 3\sin^2\theta\right)}$$

$$= \frac{3 \times \frac{1}{2} \times (\sqrt{3}/2)}{4 - \frac{3}{4}} = \frac{3\sqrt{3}}{13} = 0.400.$$

• **PROBLEM 17-19**

A uniform meter stick whose mass is 150 gm is free to turn on a horizontal axis 10 cm from one of its ends. It is started from rest when the rod is at an angle of 45° above the horizontal, find the resultant reaction of the axis of rotation.

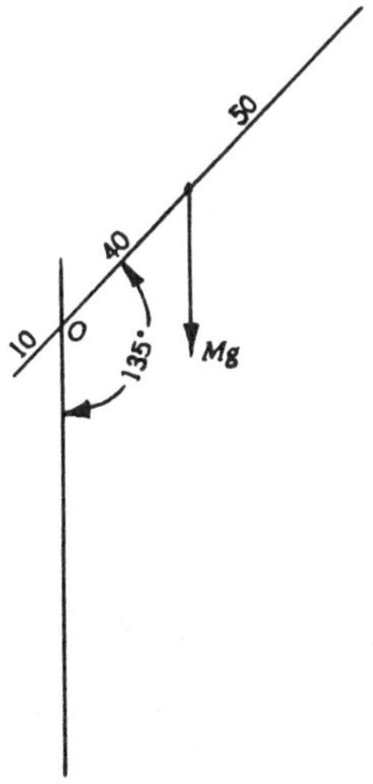

Solution: In this problem the reaction at O will be different from mg because the bar is allowed to rotate. The key relationships to use are: 1) When the bar is at rest the acceleration of the center of mass is equal to ($\vec{\alpha} \times \vec{r}$) where α is the angular acceleration of the bar and $\vec{r}$ is the vector from O to the mass center; and 2) the moment of inertia about any parallel axis is equal to the moment of inertia about the centroidal axis plus the bar's mass multiplied by the square of the distance between the two axes.

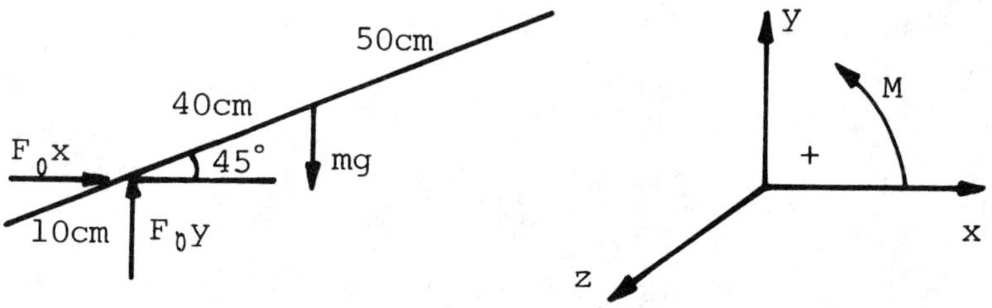

(Positive sign convention)

First summing forces in the x direction, $F_{0x} = ma_x$, (1) and summing forces in the y direction $F_{0y} - mg = ma_y$. (2) Next summing moments about point O mg (40 cm) cos 45° $= I_0\alpha$

As a vector $\vec{\alpha} = \dfrac{-\,mg\ (40\text{ cm})\cos 45°}{I_0}(\hat{k})$

and $\vec{r} = 40\text{ cm}\,(\sin 45°\,\hat{j} + \cos 45°\,\hat{i})$

The acceleration $\vec{a}$ is then:

$$\vec{a} = \vec{\alpha} \times \vec{r}$$

$$= \frac{-\,mg\ (40\text{ cm})\cos 45°}{I_0}(\hat{k})$$

$$\times\ 40\text{ cm}\,(\sin 45°\,\hat{j} + \cos 45°\,\hat{i})$$

But $\hat{k} \times \hat{j} = -\,\hat{i}$ and $\hat{k} \times \hat{i} = \hat{j}$. Therefore,

$$\vec{a} = \frac{mg\ (40\text{ cm})\cos 45°}{I_0}\left(40\text{ cm}\,(\sin 45°\,\hat{i} - \cos 45°\,\hat{j})\right)$$

$$\vec{a} = \frac{mg\ (40\text{ cm})^2}{I_0}\cos 45°\,(\sin 45°\,\hat{i} - \cos 45°\,\hat{j})$$

Hence, the components of $\vec{a}$ are:

$$a_x = \frac{mg\ (40\text{ cm})^2\cos 45°\sin 45°}{I_0}$$

and $a_y = -\dfrac{mg\ (40\ \text{cm})^2 \cos^2 45°}{I_0}$

Substituting these values of a_x and a_y into equations (1) and (2) gives

$$F_{0x} = m\left(\frac{mg\ (40\ \text{cm})^2 \cos 45° \sin 45°}{I_0}\right)$$

and $$F_{0y} = mg + m\left(\frac{-\ mg\ (40\ \text{cm})^2 \cos^2 45°}{I_0}\right)$$

or, $$F_{0y} = mg\left(1 - \frac{m\ (40\ \text{cm})^2 \cos^2 45°}{I_0}\right)$$

$I_0 = m\ (40\ \text{cm})^2 + I_{C.M.}$ by the parallel axis theorem, and $I_{C.M.} = \frac{1}{12}\ m\ \ell^2$ according to an inertia table. Hence,

$$I_0 = m\left((40\ \text{cm})^2 + \frac{(100\ \text{cm})^2}{12}\right)$$

$$I_0 = m\ (49.33\ \text{cm})^2$$

Substituting this value for I_0 simplifies the force expressions and they become:

$$F_{0x} = m\left(\frac{g\ (40\ \text{cm})^2 \cos 45° \sin 45°}{(49.33\ \text{cm})^2}\right)$$

$$F_{0y} = mg\left(1 - \frac{(40\ \text{cm})^2}{(49.33\ \text{cm})^2} \cos^2 45°\right)$$

Evaluating the expressions,

$$F_{0x} = .15\ \text{kg}\left(9.8\ \text{m/s}^2\ \frac{(40\ \text{cm})^2}{(49.33\ \text{cm}^2)}\ (.707)(.707)\right)$$

$$F_{0x} = .48\ \text{Newtons}$$

$$F_{0y} = (.15\ \text{kg})(9.8\ \text{m/s}^2)\left(1 - \frac{(40\ \text{cm})^2}{(49.33\ \text{cm})^2}\ (.707)^2\right)$$

$$F_{0y} = .99\ \text{Newtons}$$

The magnitude of the total resultant force is

$$F_t = \sqrt{F_{0x}^2 + F_{0y}^2}$$

$$F_t = \sqrt{(.48 \text{ Newtons})^2 + (.99 \text{ Newtons})^2}$$

$$= 1.10 \text{ Newtons}$$

The angle the resultant makes with the horizontal is given by:

$$\tan\theta = \frac{F_{0y}}{F_{0x}} = \frac{.99}{.48} = 2.0625$$

$$\theta = \arctan[2.0625] = 64.1°$$

The resultant reaction is given by a force of 1.10 Newtons directed at an angle 64.1° above the horizontal.

Remember that these results depended a great deal on the relationship between $\vec{\alpha}$ and $\vec{a}$. The relationship used in this problem is only good when the bar is at rest. If the bar is moving, the effects of centripetal acceleration must also be considered.

• PROBLEM 17-20

The bar AB in the figure is 10 ft long, is uniform, and weighs 48 lb. There is a pin at A, and the surface at B is smooth. Determine the maximum acceleration of the sled before the bar AB begins to rotate about the point A counterclockwise.

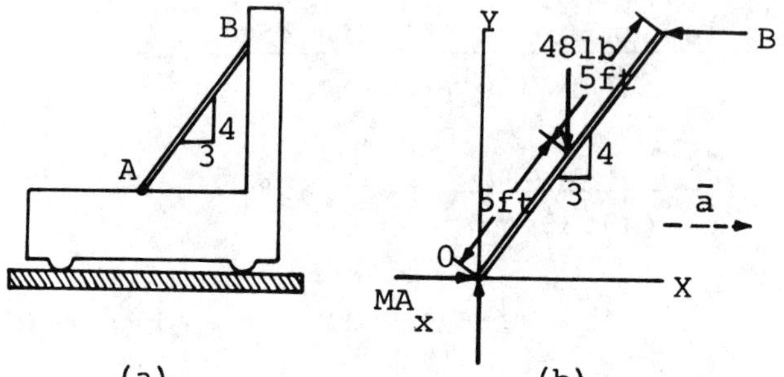

Solution: The free-body diagram in the figure shows the real forces acting on the bar as the cart accelerates to the right as indicated. The inertia of the bar also causes a fictitious force ma acting on the center of mass in a direction opposite a, to the left. Since we are analyzing the bar's motion in the cart's frame of reference we may use this expression and take moments about O:

$$(48 \text{ lb})(3 \text{ ft}) - (ma)\, 4 \text{ ft} - B\,(6 \text{ ft}) = 0.$$

The sum of the moments about O equals zero in the equation because there is no rotation yet. No rotation implies that $\omega = 0$, and hence $I\omega = 0$.

Rotation will occur when a is great enough and when B shrinks to zero representing the bar losing contact with the cart at B when the bar begins to rotate.

We may use that fact in the equation, leaving the sum equal to 0 at the moment B = 0 to give us the limiting value on a.

$$(48 \text{ lb})(3 \text{ ft}) - \left(\frac{48 \text{ lb}}{32.2 \text{ ft/sec}^2}\right) a \ (4 \text{ ft}) = 0$$

Solving for a, yields,

$$a = 24.1 \text{ ft/sec}^2.$$

• PROBLEM 17-21

A uniform wheel (Fig. 1) weighing 240 lb is rotating about a fixed axis with an angular velocity of 3.5 radians/sec. Determine the reactions at the bearing points A and B. The moment of inertia of the body about the axis of rotation is 10 lb-sec^2-ft. The rod is assumed to be weightless.

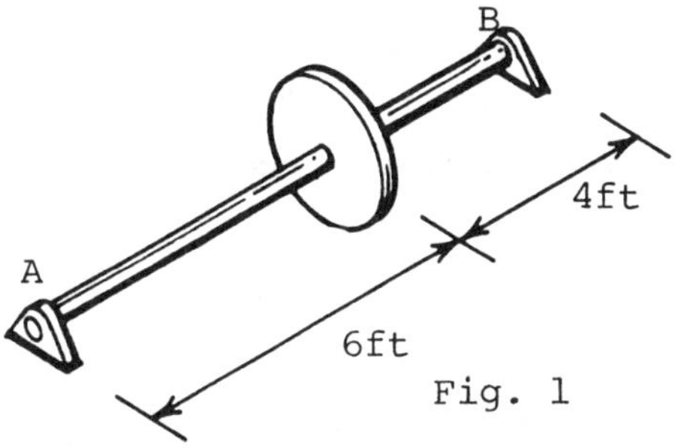

Fig. 1

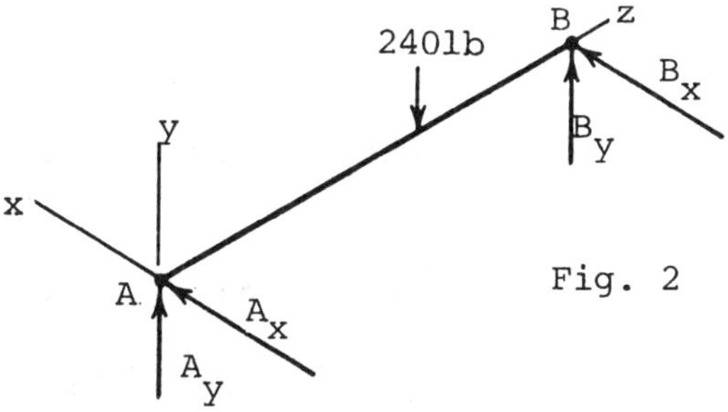

Fig. 2

Solution: First, a free-body diagram of the system along with the reference axis is drawn in Fig. 2. Note that the reference axes are defined so that the z axis coincides with the axis of rotation. The origin of the axes is at the left bearing reaction. The axes are assumed to be fixed to the body so that they are rotating with the body.

The equations of motion for a body rotating are $\vec{M} = \dot{\vec{H}}$, where $\vec{M}$ is the torque and $\vec{H}$ is the angular momentum. The general forms of the components of the angular momentum are

$$H_x = I_{xx}\omega_x - I_{xy}\omega_y - I_{xz}\omega_z$$

$$H_y = - I_{yx}\omega_x + I_{yy}\omega_y - I_{yz}\omega_z$$

$$H_z = - I_{zx}\omega_x - I_{zy}\omega_y + I_{zz}\omega_z.$$

When the body rotates about a fixed axis (the z-axis); we have $\omega_x = 0$, and $\omega_y = 0$. Then, above equations can be written as,

$$\vec{H} = - I_{xz}\omega_z \hat{i} - I_{yz}\omega_z \hat{j} + I_{zz}\omega_z \hat{k}.$$

then,
$$\dot{\vec{H}} = - I_{xz}\dot{\omega}_z \hat{i} - I_{yz}\dot{\omega}_z \hat{j} + I_{zz}\dot{\omega}_z \hat{k} - I_{xz}\omega_z \dot{\hat{i}} - I_{yz}\omega_z \dot{\hat{j}} + I_{zz}\omega_z \dot{\hat{k}}. \quad (1)$$

The time derivatives of the unit vectors are determined by noting that the axes are also rotating about the z-axis with an angular velocity $\vec{\omega} = \omega_z \hat{k}$. The time derivatives of the unit vectors are

$$\dot{\hat{i}} = \vec{\omega} \times \hat{i} = \omega_z \hat{j}$$

$$\dot{\hat{j}} = \vec{\omega} \times \hat{j} = - \omega_z \hat{i}$$

$$\dot{\hat{k}} = \omega \times \hat{k} = 0.$$

Substituting these into the expression for $\dot{\vec{H}}$, equation (1), gives us

$$\dot{\vec{H}} = [- I_{xz}\dot{\omega}_z + I_{yz}\omega_z^2] \hat{i} + [- I_{yz}\dot{\omega}_z - I_{xz}\omega_z^2] \hat{j} + I_{zz}\dot{\omega}_z \hat{k}.$$

Applying $\vec{M} = \dot{\vec{H}}$ gives us the following equations of motion

$$M_x = - I_{xz}\dot{\omega}_z + I_{yz}\omega_z^2 \quad (2)$$

$$M_y = - I_{yz}\dot{\omega}_z - I_{xz}\omega_z^2 \quad (3)$$

$$M_z = I_{zz}\dot{\omega}_z. \quad (4)$$

Since the z-axis is an axis of symmetry, both I_{xz} and I_{zy} are equal to zero. It should also be noted that since the center of mass of the body lies on the axis of rotation, there is no motion of the center of mass, and therefore the acceleration of the center of mass is zero at all times. Therefore the equations of motion are

$$\Sigma F_x = 0 \quad (5)$$

$$\Sigma F_y = 0 \quad (6)$$

and from (2) and (3)

$$\Sigma M_x = 0 \quad (7)$$

$$\Sigma M_y = 0. \quad (8)$$

Writing out explicitly the terms in equations (5)-(8):

$$A_x + B_x = 0$$

$$A_y + B_y - 240 = 0$$

$$-10\,B_y + 6\,(240) = 0$$

$$10\,B_x = 0$$

where the torques were taken about the origin and the right hand rule determined the signs.

The set of four equations can then be solved simultaneously to show that

$$A_x = 0$$

$$A_y = 96 \text{ lb}$$

$$B_x = 0$$

$$B_y = 144 \text{ lb}$$

But these reactions are nothing more than the static bearing reactions. That is to say, the reactions at the bearings when the body is in motion have the same value as the reactions at the bearings when the body is at rest. This is a very desirable condition because no dynamic or time-varying forces are transmitted to the supporting structures by this rotating body.

• PROBLEM 17-22

The system shown in Fig. 1 has a constant angular velocity of 10 radians/sec about the fixed axis. The first mass has a mass of 3 slugs and the second mass has a mass of 2 slugs. The bars are assumed to be weightless. Determine the dynamic bearing reactions as a function of time.

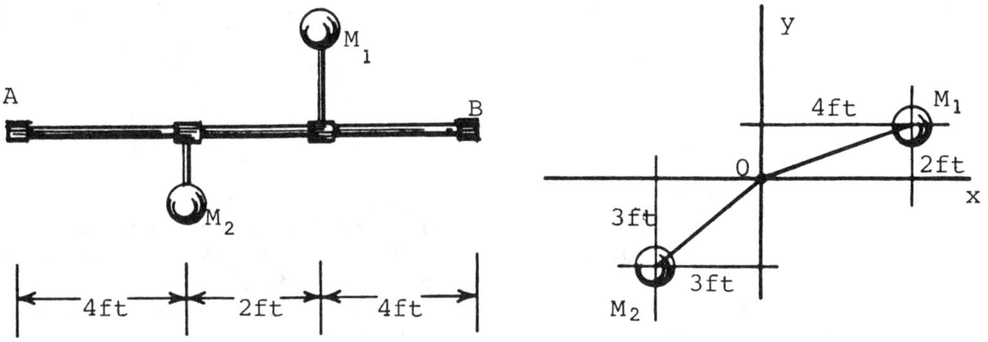

Fig. 1

Solution: The equations of motion for an object constrained to rotate about an axis are

$$M_x = - I_{xz}\dot{\omega}_z + I_{yz}\omega_z^{\ 2} \tag{1}$$

$$M_y = - I_{yz}\dot{\omega}_z - I_{xz}\omega_z^{\ 2} \tag{2}$$

$$M_z = I_{zz}\dot{\omega}_z \tag{3}$$

and

$$F_x = ma_{cm} \tag{4}$$

$$F_y = ma_{cm}. \tag{5}$$

The axes are assumed to be fixed to the body so that they are rotating with the body.

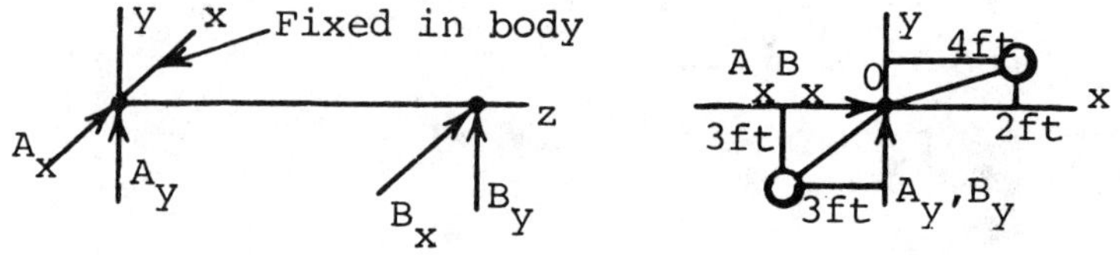

Fig. 2

First a free-body diagram of the system (Fig. 2) will be drawn to define the unknown bearing reactions and the reference axes. Since we are interested only in the dynamic bearing reactions, the weight of the body is not shown in the free-body diagram. In order to use the force equations, the acceleration of the center of mass of the system must be determined, and in order to determine the acceleration of the center of mass, the position of the center of mass must be found. The position of the center of mass relative to the reference axes shown in Fig. 1 may be calculated as follows:

$$x_c = \frac{2\ (-\ 3)\ +\ 3\ (4)}{5}$$

$$= 1.2 \text{ ft}$$

$$y_c = \frac{3\ (2)\ -\ 2\ (3)}{5}$$

$$= 0.$$

So, the center of mass lies right on the x axis. To use the force equations, we must know the acceleration of the center of mass. Since the system is moving with a constant angular velocity about the axis of rotation, the center of mass will be moving in a circular path. And since there is no angular acceleration, the acceleration of the center of mass will have a normal component only. Thus,

$$\ddot{\theta} = 0$$

$$\ddot{r} = - x_c\omega^2$$

$$= - 1.2\ (10)^2$$

$$= - 120\ \text{ft/sec}^2.$$

Now, in order to use the moment equations, we must know the product of inertia of the system with respect to the xz and the yz axes. Since the system is composed of point masses, we may perform a numerical integration to determine I_{xz} and I_{yz}:

$$I_{xz} = \sum_i x_i z_i m_i$$

$$= (4)\ (6)\ (3) + (-\ 3)\ (4)\ (2)$$

$$= 48.$$

and

$$I_{yz} = \sum_i y_i z_i m_i$$

$$= (2)\ (6)\ (3) + (-\ 3)\ (4)\ (2)$$

$$= 12.$$

Substituting these values into equation (1) and (2), remembering that $\omega_z = 0$.

$$M_x = 12\ (10)^2$$

$$M_y = - 48\ (10)^2.$$

Using the reaction forces indicated in Figure 2 and the right hand rule with the origin as indicated:

$$- 10\ B_y = 1200, \quad B_y = - 120\ \text{lb} \qquad (6)$$

$$10\ B_x = - 4800, \ B_x = - 480\ \text{lb}. \qquad (7)$$

From the force equations, (4) and (6),

$$A_x + B_x = (5)\ (-\ 120) \qquad (8)$$

$$A_y + B_y = 0. \qquad (9)$$

Combining (6), (7), (8) and (9) yields

$$A_y = 120\ \text{lb}$$

$$A_x = - 120\ \text{lb}.$$

What we have calculated here is the dynamic reaction components for the body at one particular position. But we are interested in writing the reactions as a function of time. We note that the reactions that we have determined here are rotating with the moving xy axes (Fig. 3). The magnitude of A is easily determined:

$$A = \sqrt{120^2 + 120^2}$$

$$= 170 \text{ lb.}$$

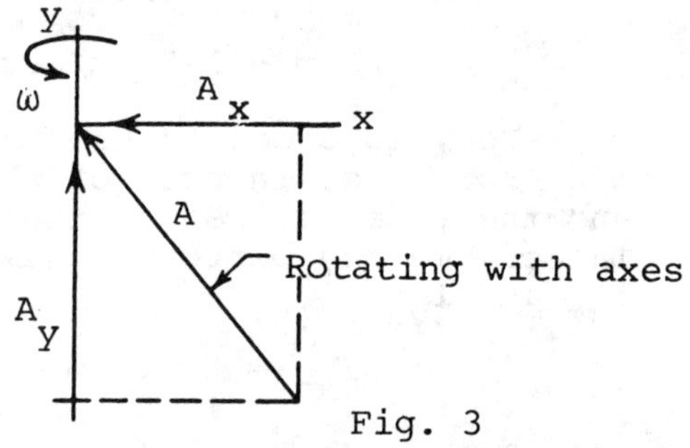

Fig. 3

But since A is rotating with the moving coordinate system, as shown in Fig. 3, we must write the horizontal and vertical components of the bearing reactions at A (referred to inertial space) as

$$A_V = 170 \cos(\omega t + \phi)$$

$$A_H = 170 \sin(\omega t + \phi).$$

The phase angle ϕ simply indicates that the starting position of the system may vary. By a similar argument, the magnitude of the reaction at B is

$$B = \sqrt{480^2 + 120^2}$$

$$= 494 \text{ lb}$$

and the horizontal and vertical components of the reaction at B are

$$B_V = 494 \cos(\omega t + \alpha)$$

$$B_H = 494 \sin(\omega t + \alpha).$$

The phase angle α indicates not only that the starting position of the system may vary, but also that the reaction at B is out of phase with the reaction at A; that is, they do not attain their maximum values at the same time.

• PROBLEM 17-23

The system of the previous example is to be dynamically balanced by adding two 32.2 lb weights in correction planes, each 2 ft from each bearing as shown in Figure 1. Determine the proper location of these weights in the x-y plane.

Solution: The data and reference axes are the same as in the previous problem. Hence, the center of mass before considering the corrective weights is at $x_c = 1.2$ ft, $y_c = 0$. In order for the dynamic bearing reactions to be zero, the center of mass must be on the z-axis and $I_{xz} = I_{yz} = 0$.

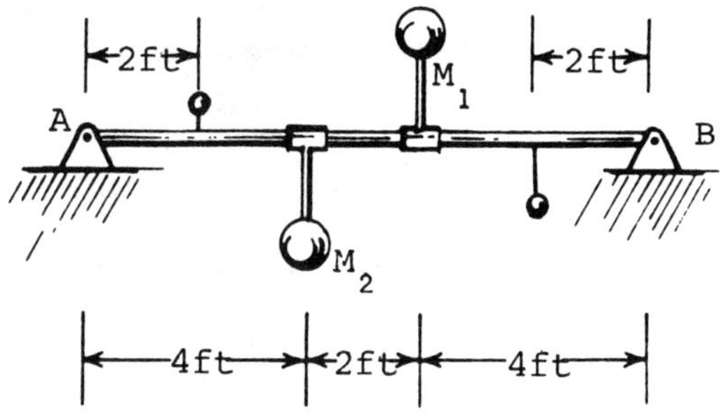

Fig. 1

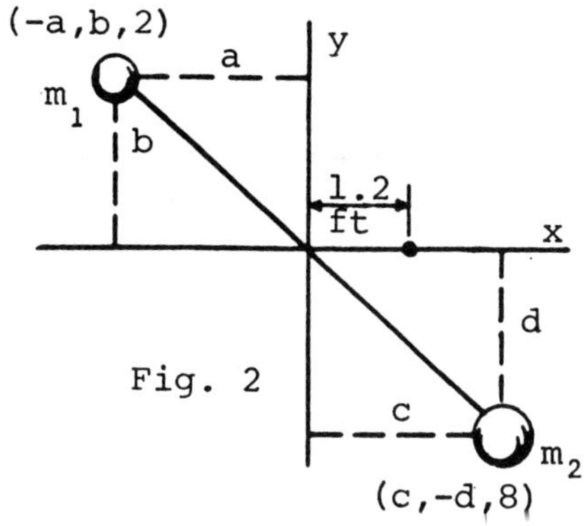

Fig. 2

If the positions shown in Figure 2, for the added weights is assumed, the necessary equations may be written.

For the center of mass:

$$y_c = 0$$

$$m_1\ b + m_2\ (-\ d) = 0$$

$$b - d = 0 \tag{1}$$

$$x_c = 0$$

$$5\ (1.2) + m_1\ (-\ a) + m_2\ (c) = 0$$

$$-\ a + c = 6 \tag{2}$$

since $m_1 = m_2 = 1.$

Also, $I_{xz} = 0$

$$4(3)\ (6) + (-\ 3)\ (4)\ (2) + (-\ a)\ (2)\ m_1 + (c)\ (8)\ m_2 = 0$$

$$-\ 2a + 8c = -48 \tag{3}$$

$$I_{yz} = 0$$

$$2(6)\ (3) + (-\ 3)\ (4)\ (2) + b\ (2)\ m_1 + (-\ d)\ (8)\ m_2 = 0$$

$$2b - 8d = -\ 12 \tag{4}$$

Here the data from the individual weights of the original system have been included.

Solving equations (1), (2), (3) and (4) yields

$$a = -\ 16$$

$$b = 2$$

$$c = -\ 10$$

$$d = 2.$$

The minus signs indicate the chosen directions of a and c in Figure 2 were wrong.

• PROBLEM 17-24

Determine the center of percussion for the baseball bat shown in Figure 1. The bat weighs 2 lb and it is 36 in. long. It is assumed to be swung at a point in the middle of the grip, 4 in. from the end, as shown.

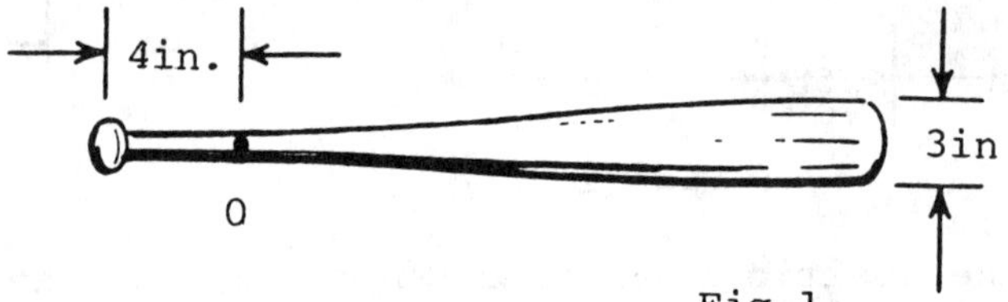

Fig.1

Solution: The distance from point O, where the bat is swung, to the center of percussion is given by

$$L = \frac{I_0}{md}$$

where I_0 is the moment about the pivot O, and d is the distance from O to the center of mass. Locate C, the center of mass, and find I_0.

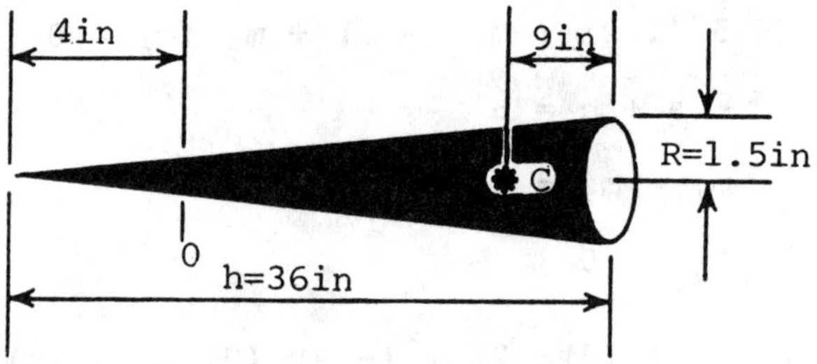

Fig. 2

Approximate the bat to be a cone with dimensions shown in Figure 2. It is known that, for a cone, C is a distance $\frac{1}{4}$ the height of the cone from the base. This too is shown in Figure 2. The moment of inertia of the cone about the point C is given by

$$I_C = \frac{3m}{80}\,(4\,R^2 + h^2)$$

and, using the parallel axis theorem

$$I_0 = I_C + md^2$$

$$= \frac{3\ (2/32.2)}{80}\left[4\left(\frac{1.5}{12}\right)^2 + \left(\frac{36}{12}\right)^2\right] + \left(\frac{2}{32.2}\right)\left(\frac{23}{12}\right)^2$$

$$= 0.238 \text{ lb-sec}^2\text{-ft}$$

Therefore the distance from O to the center of percussion is

$$L = \frac{I_0}{md}$$

$$= \frac{0.238}{(2/32.2)\ (23/12)}$$

$$= 2.00 \text{ ft}$$

$$= 24 \text{ in.}$$

Thus, the center of percussion for the bat is at a distance of 1 in. to the right of the center of mass, so that by hitting the ball with the bat at that particular point on the bat, theoretically the ballplayer should experience no force from the bat. One should realize, however, that the assumption that the bat was pinned at the point O was not quite correct, since obviously a bat cannot be gripped at a point.

• PROBLEM 17-25

The identical uniform cylindrical disks are mounted on bearings at the ends of a uniform beam AC. The beam is free to rotate about a fixed bearing perpendicular to AC through its center at B. The bearing at B is frictionless but the one at A is not. The bearing at C is frozen so that the disk at C cannot rotate with respect to the beam. Initially the beam is not rotating, but the disk on the left is rotating. Describe qualitatively what happens as time passes if the system is left undisturbed. When the disk at A has stopped rotating with respect to AC, what is the value of the angular velocity of the beam?

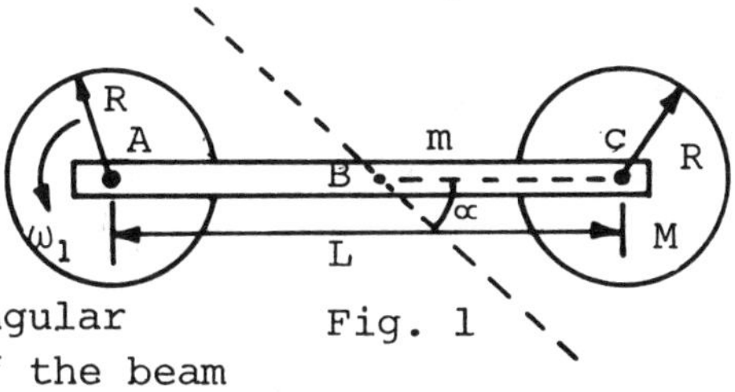

ωf=final angular velocity of the beam

Fig. 1

Solution: The initial angular momentum of the rotating disk is going to be transmitted to the bar by the friction forces present in the bearings. Since those forces are internal, the angular momentum of the entire system will remain constant (the external forces, friction at B and gravity are zero and have zero lever arm, respectively). Hence, as the clockwise rotation of disk A decreases, the entire object, the beam and 2 disks, will rotate counterclockwise.

We first apply the D'Alembert's principle to find a relationship between the angular acceleration α and the frictional torque τ.

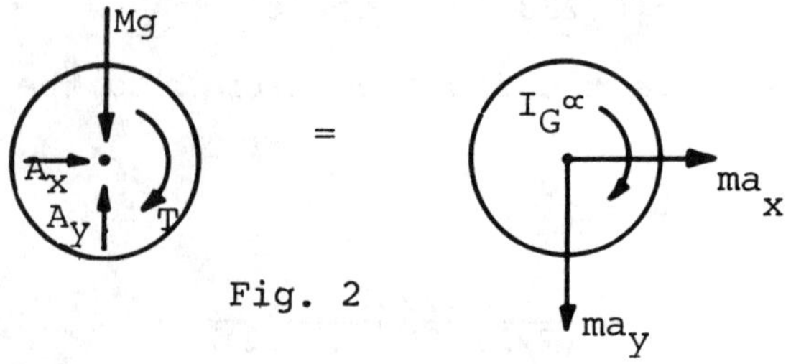

Fig. 2

From Figure 2,

$$\circlearrowright+ \quad \Sigma m = m_{eff}$$

$$\tau = I_G\alpha$$

For a disk $I_G = \frac{1}{2} MR^2$

Then, $$\tau = \frac{1}{2} MR^2\alpha \qquad (1)$$

Using the equation of angular motion for disk A, we have

$$\omega_2 = \omega_1 - \alpha t,$$

since ω_2 = final angular velocity = 0,

$$\therefore t = \frac{\omega_1}{\alpha}. \qquad (2)$$

Now applying the principle of angular momentum and impulse, we have

$$(I_1\omega_1)_{disk} + \tau\Delta t = (I_2\omega_f)_{system} \qquad (3)$$

substituting appropriate terms, we get

$$\frac{1}{2} MR^2\omega_1 + \tau\Delta t = \left\{ \frac{1}{12} mL^2 + \left[\frac{1}{2} MR^2 + M\left(\frac{L}{2}\right)^2\right] \times 2\right\} \omega_f$$

Note that I_2 is the moment of inertia of the whole system (disks plus the beam) about point B, as the beam rotates about thin point. Substituting values of τ and Δt yields

$$\frac{1}{2} MR^2\omega_1 + \frac{1}{2} MR^2\alpha \left(\frac{\omega_1}{\alpha}\right) = \left(\frac{1}{12} mL^2 + MR^2 + \frac{mL^2}{2}\right) \omega_f$$

simplifying further,

$$MR^2\omega_1 + MR^2\omega_1 = \left(\frac{mL^2}{6} + 2\ MR^2 + ML^2\right) \omega_f$$

which yields,

$$\omega_f = \frac{2\ MR^2 \omega_1}{\frac{mL^2}{6} + 2\ MR^2 + ML^2}$$

or $$\omega_f = \left[\frac{mL^2}{12}\left(\frac{1}{MR^2}\right) + \frac{1}{2}\left(\frac{L}{R}\right)^2 + 1\right]^{-1} \omega_1 .$$

• PROBLEM 17-26

An airplane propeller, of moment of inertia I, is subject to the driving torque $N = N_0\ (1 + \alpha \cos \omega_0 t)$, and a frictional torque due to air resistance $N_f = -\ b\dot{\theta}$.

Find its steady state motion.

Solution: The equation of rotational motion of a rigid body is given by

$$\dot{\vec{L}} = N_t \tag{1}$$

where L is the angular momentum and N_t is the total external torque about the axis of rotation.

If I is the moment of inertia of the propeller about the axis of rotation, then

$$L = I\dot{\theta}, \tag{2}$$

where $\dot{\theta}$ is the angular velocity of the propeller about the axis of rotation.

The total external torque acting on the propeller is

$$N_t = N + N_f = N_0\ (1 + \alpha \cos \omega_0 t) - b\dot{\theta}. \tag{3}$$

Using Eqs. (2) and (3) in Eq. (1) we obtain

$$I\ddot{\theta} = N_0\ (1 + \alpha \cos \omega_0 t) - b\dot{\theta}. \tag{4}$$

Dividing by I and rearranging terms we have

$$\ddot{\theta} + \frac{b}{I}\dot{\theta} = \frac{N_0}{I} + \frac{N_0 \alpha}{I} \cos \omega_0 t. \tag{5}$$

This is a non-homogeneous second order differential equation which can be solved by standard methods. Here we solve Eq. (5) by direct integration. First we integrate Eq. (5) once with respect to time t to obtain

$$\dot{\theta} + \frac{b}{I}\,\theta = \frac{N_0}{I}\,t + \frac{\alpha N_0}{I\omega_0}\sin\omega_0 t + A \qquad (6)$$

where A is a constant of integration.

This first order DE can be solved by multiplying by the integrating factor $e^{bt/I}$. Then the above equation takes the form

$$\frac{d}{dt}\left(\theta\, e^{\frac{b}{I}t}\right) = \frac{N_0}{I}\,t\, e^{\frac{b}{I}t} + \frac{\alpha N_0}{I\omega_0}\, e^{\frac{b}{I}t}\sin\omega_0 t + A\, e^{\frac{b}{I}t}. \qquad (7)$$

Integrating both sides with respect to time, (performing the integrations of the first two terms on the right hand side by parts) we obtain

$$\theta\, e^{\frac{b}{I}t} = \frac{N_0}{b}\, e^{\frac{b}{I}t}\left(t - \frac{I}{b}\right) + \frac{N_0\alpha}{I\omega_0}\,\frac{e^{\frac{b}{I}t}}{I^2\omega_0^{\ 2} + b^2}\,[b\, I \sin\omega_0 t - I^2\omega_0\cos\omega_0 t] + A\,\frac{I}{b}\, e^{\frac{b}{I}t} + C \qquad (8)$$

where the first, second and third terms on the right hand side of Eq. (8) are the integrals of the first, second and third terms of Eq. (7). C is a constant of integration.

Dividing Eq. (8) by $e^{\frac{b}{I}t}$ and rearranging terms we obtain

$$\theta = \frac{N_0}{b}\,t + \frac{N_0\alpha}{I^2\omega_0^{\ 3} + b^2\omega_0}\,[b\sin\omega_0 t - I\omega_0\cos\omega_0 t] + \frac{I}{b}\left[A - \frac{N_0}{b}\right] + C\, e^{-\frac{b}{I}t} \qquad (9)$$

In the steady state i.e., for $t \to \infty$, the last term in in Eq. (9) will drop out. Defining $\theta_0 = \frac{I}{b}\left(A - \frac{N_0}{b}\right)$ we have

$$\theta = \theta_0 + \frac{N_0}{b}\,t + \left[\alpha N_0 / \left(I^2\omega_0^{\ 3} + b^2\omega_0\right)\right]\left[b\sin\omega_0 t - I\omega_0\cos\omega_0 t\right]. \qquad (10)$$

• **PROBLEM 17-27**

Shown is a rigid ring to which are attached three particles whose relative weights are given. The ring is placed on a horizontal plane in this position with no angular velocity. What is the angular acceleration of the ring immediately after it is placed on the plane? Assume that the ring itself has negligible weight and does not slip on the plane.

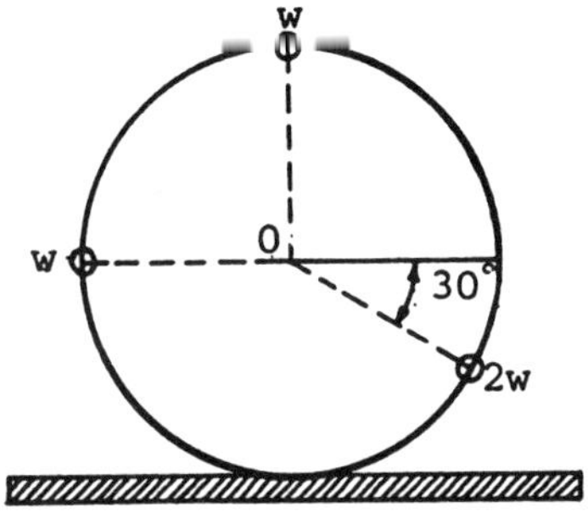

Fig. 1

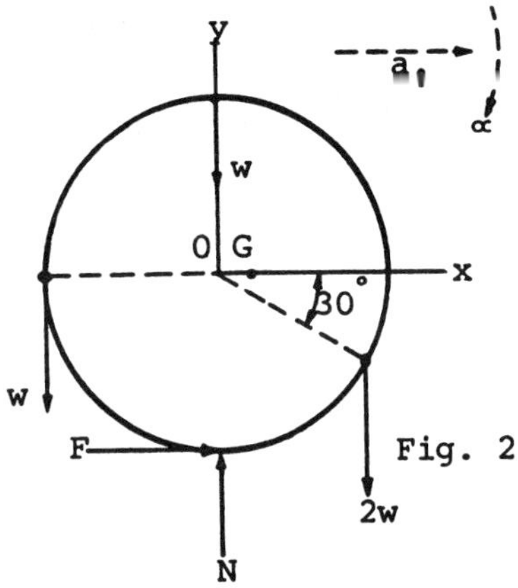

Fig. 2

Solution: The free-body diagram is shown in Fig. 2 with the assumed directions of the angular acceleration and the linear acceleration of the center of the ring indicated by dashed arrows. A nonrotating set of xy axes is chosen with the origin at O. For the indicated orientation of the particles, the coordinates of the center of mass G are x = 0.183r and y = 0.

With no slipping the components of the acceleration of the geometric center O of the ring are $a_{0_x} = r\alpha$ and $a_{0_y} = 0$. Also, having chosen a_0 to the right in Fig. 2, the angular acceleration α must be assumed clockwise to be physically consistent with the sense of a_0. Writing the moment equation and taking counterclockwise rotations to be positive yields

$$w\ r - 2\ w \cos 30°\ r + F\ r = I\ (-\alpha) = -\left(\frac{4\ w}{g}\ r^2\right)\alpha$$

or $$F - 0.73\ w = -\frac{4\ w}{g}\ r\alpha. \qquad \text{(a)}$$

The equation of motion in the x direction is

$$\vec{F} = \left(\frac{4\ w}{g}\right)\vec{a}_{G_x}. \qquad \text{(b)}$$

We know the relationship between a_0 and α. To find how a_g fits into this we note that

$$\vec{a}_G = \vec{a}_{G/0} + \vec{a}_0$$

where $\vec{a}_{G/0}$ is the acceleration of G relative to O. At the moment being considered, the ring has no angular velocity so $\vec{a}_{G/0}$ is perpendicular to the radius OG, i.e., downwards.

Therefore, $a_{G_x} = 0 + \vec{a}_0 = r\alpha$

so equation (b) becomes

$$F = \frac{4\ w}{g} r\alpha. \qquad \text{(c)}$$

Combining this equation and equation (a) yields

$$\frac{8\ w}{g} r\alpha = 0.73\ w$$

$$\alpha = \frac{2.94}{r}\ \text{rad/sec}^2,$$

where r is measured in feet.

• PROBLEM 17-28

The flatbed cart shown in Fig. 1 weighs 128.8 lb, and each of its two wheels weighs 16.1 lb. The linear acceleration of the car under the action of the 10-lb force may be estimated by assuming that the wheels are rigidly connected to the cart and that there is no friction between the wheels and the road. Calculate the acceleration of the cart (using this approximation) and compare it with the exact acceleration of the cart, taking into account the rotary motion of the wheels. The center of mass of the system lies on the axle.

Solution: By assuming that the wheels are connected to the cart, the acceleration of the cart is easily calculated by using the force equation for the x direction:

$$F_x = m\ddot{x}_c$$

$$10 = \left(\frac{128 + 2(16.1)}{32.2}\right)\ddot{x}_c$$

$$\ddot{x}_c = 2.0\ \text{ft/sec}^2.$$

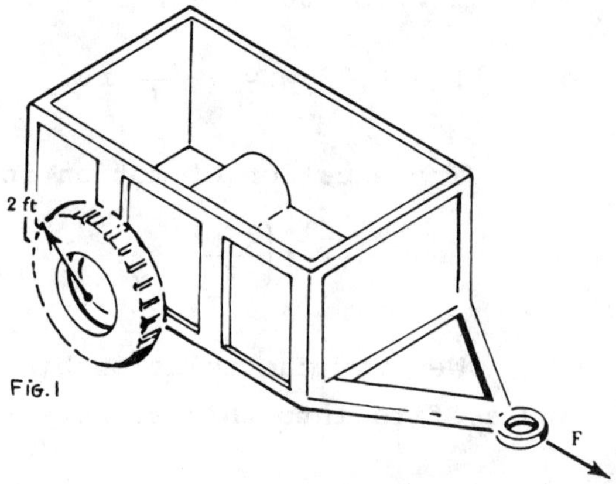

Fig. 1

Taking into account the angular motion of the wheels, first draw a free-body diagram of both the cart and the wheels (Fig. 2). The equation of motion for the cart in the x direction becomes

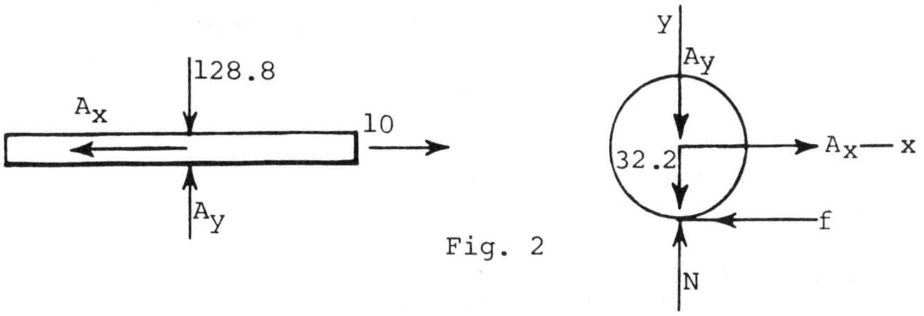

Fig. 2

$$\Sigma F = m_c \ddot{x}_c$$

$$10 - A_x = \frac{128}{32.2} \ddot{x}_c = 4 \ddot{x}_c .$$

and to eliminate A_x, we write the force equation for the wheels:

$$\Sigma F = m_\omega \ddot{x}_\omega = m_\omega \ddot{x}_c$$

$$A_x - f = \frac{2(16.1)}{72.2} \ddot{x}_c = \ddot{x}_c$$

where f is the total friction force on the wheels.

Now we must eliminate f by seeking another equation. The moment equation may be introduced:

$$M = I\dot{\omega}$$

$$- 2f = \frac{1}{2} \left(\frac{32.2}{32.2}\right) (2)^2 \dot{\omega}$$

$$- 2f = 2\dot{\omega} .$$

Solve the preceding three equations simultaneously for the linear acceleration of the center of mass of the cart by noting that $\ddot{x}_c = - 2\dot{\omega}$:

$$\ddot{x}_c = 1.82 \text{ ft/sec}^2 .$$

Thus, it is seen that by ignoring the rotary motion of the wheels, the linear acceleration of the cart can be calculated to within roughly 10 percent of the true value. This approximation becomes better when the weight of the cart becomes much larger than the weight of the wheels and vice versa.

• **PROBLEM 17-29**

A yo-yo rests on a level surface. A gentle horizontal pull (see the figure) is exerted on the cord so that the yo-yo rolls without slipping. Which way does it move and why?

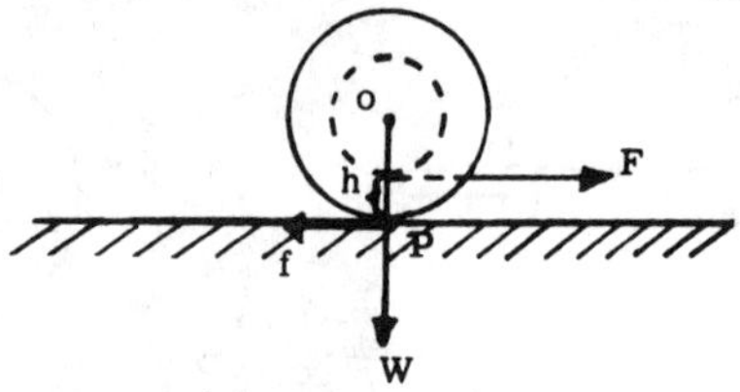

Solution: The forces acting on the yo-yo are the horizontal pull F and the frictional force f

$$f = \mu W$$

where μ is the coefficient of friction.

Instantaneous rotation takes place about an axis through the point of contact P (not about the center of the yo-yo, although it might appear so) since the instantaneous velocity of the contact point is zero. Therefore, the yo-yo rolls in the direction of the pull and its rotation is determined by the torque about P,

$$\tau = Fh.$$

It should be observed that the frictional force does not contribute to this torque, since f is acting at P.

• **PROBLEM 17-30**

A cable drum of inner and outer radii r and R is lying on rough ground, the cable being wound round the inner cylinder and being pulled off from the bottom at an angle θ to the horizontal. An inquiring student strolling by notes that when the cable is pulled by a workman, with θ a small angle, the drum rolls without slipping toward the workman. Whereas if θ is large, the drum rolls without slipping in the opposite direction. He works out a value for the critical angle θ_0 which separates the two types of motion. What is the value of θ_0?

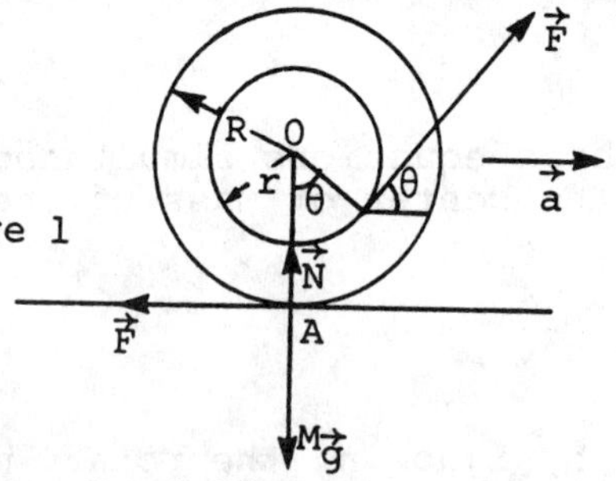

Figure 1

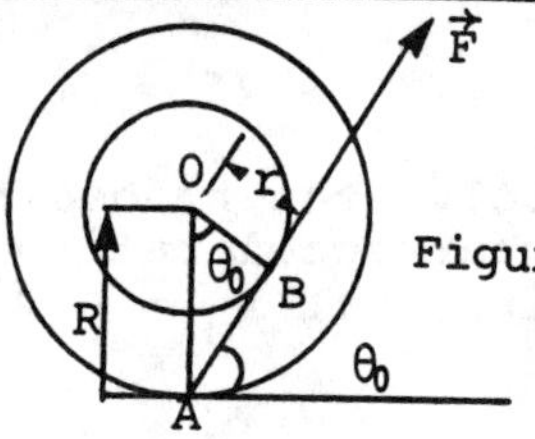

Figure 2

Solution: We define the drum's acceleration to be positive when the drum moves toward the workman. Hence the mathematical condition that the drum roll toward or away from the worker is that a (the acceleration) be positive or negative, respectively. The critical condition distinguishing the 2 types of motion is that $a = 0$. Therefore, to find the critical angle θ_0, we find a as a function of θ, set it equal to zero, and solve for θ.

We can use Newton's Second Law to relate the net force on the drum to its acceleration. (We do this for the vertical and horizontal directions separately.)

Figure (1) shows the drum with the forces acting on it. The force applied by the workman acts tangentially to the inner cylinder at such a position that the angle between this tangent and the horizontal is θ. It follows that the angle between the corresponding radius and the vertical is θ also, and that this radius is at right angles to the tangential force $\vec{F}$.

The other forces acting are the weight $\vec{Mg}$ of the cable drum, the normal forces exerted by the ground on the drum at the two points of contact, which combine into a resultant $\vec{N}$ passing through the center of gravity, and the frictional forces at the same points of contact which combine to form a single resultant force $\vec{f}$.

There is no movement in the vertical direction. Hence

$$N = Mg - F \sin \theta. \tag{1}$$

The forces in the horizontal direction produce an acceleration a. Thus

$$F \cos \theta - f = Ma. \tag{2}$$

Further, the moments of the forces about the center of mass O produce a rotational acceleration about that point. The only forces whose lines of action do not pass through the center of gravity are F and f. Hence

$$Fr - fR = I\alpha \tag{3}$$

where I is the moment of inertia of the drum about its center of mass.

At the points at which the drum touches the ground no slipping occurs. Therefore instantaneously these points are at rest. But all points of the drum have an acceleration a forward and in addition the points of contact, due to the rotation about the center of mass, have a further linear acceleration $R\alpha$ forward. Thus

$$a + \alpha R = 0$$

or

$$\alpha = - a/R \tag{4}$$

We wish to eliminate f from (2). Solving (3) for f

$$\frac{Fr - I\alpha}{R} = f \tag{5}$$

Substituting (4) in (5)

$$f = \frac{Fr + Ia/R}{R}$$

$$f = \frac{Fr}{R} + \frac{Ia}{R^2} \tag{6}$$

Inserting (6) in (2)

$$F \cos \theta - \frac{Fr}{R} - \frac{Ia}{R^2} = Ma$$

Solving for a

$$\frac{F \cos \theta - \frac{Fr}{R}}{M + \frac{I}{R^2}} = a$$

or $$a = \frac{F(\cos \theta - r/R)}{M + \frac{I}{R^2}} \tag{7}$$

Since we do not know F, we solve (1) for F and insert this in (7)

$$F = \frac{Mg - N}{\sin \theta}$$

and $$a = \frac{\left(\frac{Mg - N}{\sin \theta}\right)\left(\cos \theta - \frac{r}{R}\right)}{M + \frac{I}{R^2}} \tag{8}$$

The critical value of θ, (θ_0), is found by setting (8) equal to 0, whence

$$a = 0$$

$$\cos \theta_0 - \frac{r}{R} = 0$$

$$\cos \theta_0 = \frac{r}{R} \tag{9}$$

If $\cos \theta > r/R$, the drum rolls towards the workman, and vice versa.

This result could be obtained more easily by considering rotation about A, the line of the drum instantaneously at rest. The only force that does not pass through A is $\vec{F}$, the applied force. If the line of action of $\vec{F}$ cuts the ground to the left of A, the moment of $\vec{F}$ about A causes the drum to roll to the right. If the line of action of $\vec{F}$ cuts the ground to the right of A, the moment of $\vec{F}$ about A causes the drum to move to the left. If the line of action of $\vec{F}$ passes through A, the drum is stationary and θ has the critical value θ_0.

Figure (2) shows this situation. Since the line of action of $\vec{F}$ is tangential to the inner cylinder, OB and AB are at right angles and $\angle$ AOB is θ_0.

$$\therefore \cos \theta_0 = \frac{r}{R} .$$

• PROBLEM 17-31

A wooden log and a steel pipe were being rolled up a ramp. The person who was rolling them together slipped and let the log and the pipe roll down the ramp. The log and the pipe have the same radius and started from rest and roll with out slipping. The ramp has an incline of 20°. Find the distance between them after 3 seconds, assuming they still are on the ramp. Treat log a cylinder.

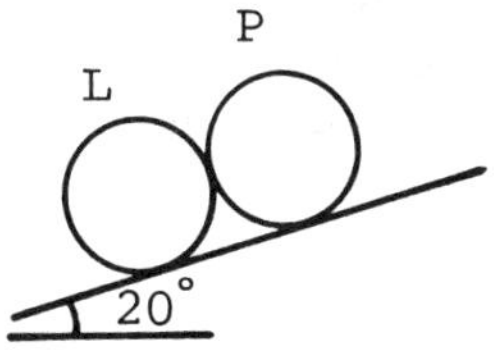

Solution: This problem can be solved by D'Alembert's principle. For cylinder C we draw the force and the effective force diagrams.

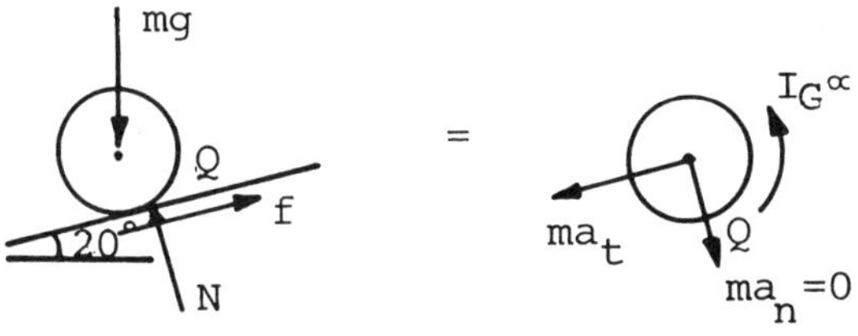

Force diagram Effective force diagram

Summing the moment at point Q, the point in contact with the plane, we have,

$$\circlearrowleft+ \quad \Sigma M_Q = \Sigma M_{eff} .$$

In the force diagram the normal force and the frictional force have moment arms of zero with respect to point Q. Therefore they don't contribute any moment. The perpendicular component of the gravitational force has a moment arm of r. This moment is producing a linear acceleration, a_t, of the mass center, which is a distance r from Q, and an angular acceleration, α, of the body about its mass center. Therefore:

$$mg \sin 20° \, (r) = ma_t \, (r) + I_G \alpha .$$

For a cylinder, $a_t = r\alpha$ and $I_G = \frac{1}{2} mr^2$. Therefore,

$$mg \sin 20° \ (r) = mr\alpha \ (r) + \frac{1}{2} mr^2\alpha.$$

Dividing through by mr and solving for α yields

$$\alpha = \frac{2g \sin 20°}{3r}. \qquad (1)$$

Similarly, for the pipe, P, we go through the same procedure,

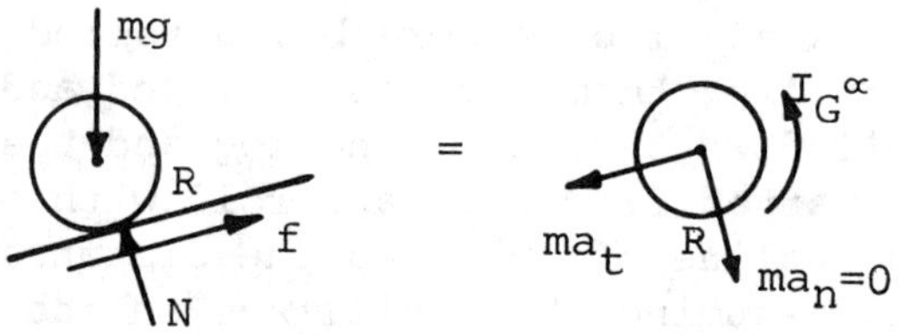

$$\circlearrowleft + \ \Sigma M_R = \Sigma M_{eff}.$$

Everything for the pipe is the same as with the solid cylinder except the value of the moment of inertia I_G.

$$mg \sin 20° \ (r) = ma_t \ (r) + I_G\alpha.$$

$a_t = r\alpha$ and for a thin wall pipe $I_G = mr^2$. Therefore, the above moment equation becomes

$$mg \sin 20° \ (r) = mr\alpha \ (r) + mr^2\alpha$$

simplifying and solving for α yields

$$\alpha = \frac{g \sin 20°}{2r}. \qquad (2)$$

The equation for angular motion is

$$\theta = \theta_0 + \omega_0 t + \frac{1}{2} \alpha t^2.$$

For the given problem $\theta_0 = \omega_0 = 0$. Therefore,

$$\theta = \frac{1}{2} \alpha t^2$$

For For the log, $\theta = \left(\frac{1}{2}\right) \frac{(2g \sin 20°)}{3r} (3)^2$ substituting $g = 9.81 \text{ m/sec}^2$, we get

$$\theta = \frac{10.07}{r} \text{ radians.}$$

For the pipe, $\theta = \frac{1}{2}\left[\frac{9.81 \times \sin 20°}{2r}\right](3)^2 = \frac{7.55}{r}$ rad.

The relationship between the linear distance s along the circumference of a circle and the angle θ is

$$s = r\theta.$$

Using this equation, the distance traveled by a cylinder is

$$s_L = r\theta_L$$

$$= r\,\frac{10.07}{r} = 10.07 \text{ m}.$$

Similarly for the pipe,

$$s_p = r\theta_p$$

$$= r\,\frac{7.55}{r} = 7.55 \text{ m}.$$

The distance between cylinder and pipe is therefore given by

$$s_C - s_p = 10.07 - 7.55 = 2.52 \text{ m}.$$

• PROBLEM 17-32

Study the dynamics of a uniform cylinder shown in Fig. 1 undergoing plane motion down an inclined plane subject to the following conditions: (1) the cylinder does not slip, and (2) the cylinder experiences slip.

Fig. 1

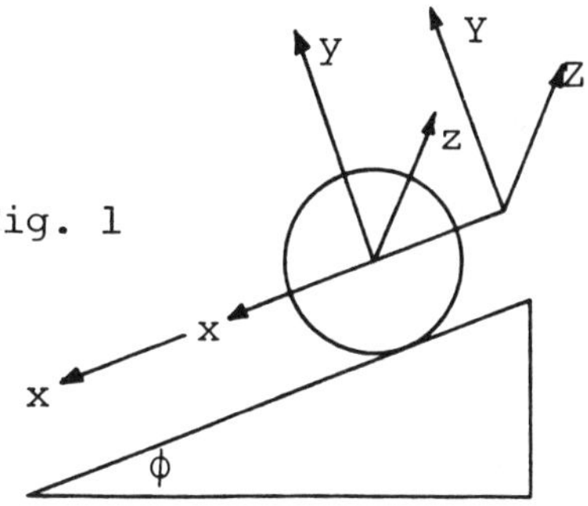

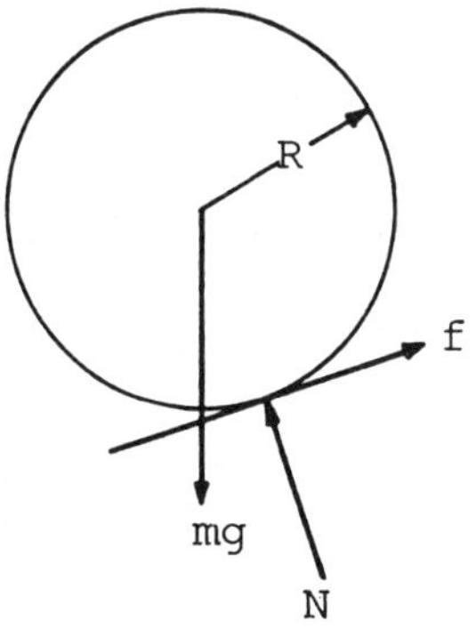

Fig. 2

Solution: The motion is general in as much as no point is fixed; simplification is attributable to the constraints associated with maintaining plane motion and the condition that the mass center of the cylinder remains a fixed distance from the inclined plane. Thus the mass center undergoes rectilinear motion while the elements of mass comprising the cylinder rotate about the mass center.

The expression relating torque $\vec{M}$ to angular momentum $\vec{H}$, $\vec{M} = \frac{d}{dt}\vec{H}$, is generally quite complicated since $\vec{H} = \overleftrightarrow{I}\vec{\omega}$, where $\overleftrightarrow{I}$ is a 3 x 3 tensor and $\vec{\omega}$ is a 3 x 1 vector. However, when plane motion occurs $\vec{\omega} = \omega_z \hat{k}$ only, so

$$
\begin{aligned}
M_{CM,x} &= I_{xz}\dot{\omega}_z - I_{yz}\omega_z^{\,2} \\
M_{CM,y} &= I_{yz}\dot{\omega}_z + I_{xz}\omega_z^{\,2} \qquad (1) \\
M_{CM,z} &= I_{zz}\dot{\omega}_z .
\end{aligned}
$$

But, for a uniform cylinder these equations are employed in considerably simplified form since $I_{xz} = I_{yz} = 0$. The governing equations are $\Sigma F_x = m\ddot{x}_{CM}$, $\Sigma F_y = m\ddot{y}_{CM}$, and $\Sigma M_{CM,z} = I_{zz}\dot{\omega}_z$.

The free-body diagram, Fig. 2, applies whether or not slip occurs. Summing forces,

$$\Sigma F_x = mg \sin\phi - f = m\ddot{x}_{CM} \qquad (2)$$

$$\Sigma F_y = N - mg\cos\phi = m\ddot{y}_{CM} = 0 \text{ (since } y_{CM} = \text{constant}). \quad (3)$$

The frictional force causes the only torque about the mass center:

$$M_{CM,z} = -fR = I_{zz}\dot{\omega}_z \qquad f = \frac{-I_{zz}\dot{\omega}_z}{R}. \qquad (4)$$

The treatment to this point has been general. Further development depends upon the conditions relating to slip.

(1) If no slip occurs, that is, if there is no relative velocity at the point of contact between the cylinder and the inclined surface, the X-displacement of the mass center is the developed circumference of the cylinder, $x_{CM} = -R\theta$, where θ represents the angular displacement of an arbitrary line on the cylinder. Consequently, the velocity and acceleration of the mass center are, respectively,

$$\dot{x}_{CM} = -R\dot{\theta} = -R\omega_z$$

$$\ddot{x}_{CM} = -R\ddot{\theta} = -R\dot{\omega}_z .$$

Equating the force of friction in the force and moment equations (2) and (4),

$$f = mg\sin\phi - m\ddot{x}_{CM} = -\frac{I_{zz}\dot{\omega}_z}{R} = \frac{I_{zz}\ddot{x}_{CM}}{R^2}$$

where $-\ddot{x}_{CM}/R$ has been substituted for ω_z. Solving for $\ddot{x}_{CM}$, we have

$$\ddot{x}_{CM} = \frac{mg \sin \phi}{m + I_{zz}/R^2}.$$

Since every term in the above expression is constant, it follows that $\ddot{x}_{CM}$ and therefore $\dot{\omega}_z$ are constant. Substituting $I_{zz} = \frac{1}{2}mR^2$ for a uniform cylinder, $\ddot{x}_{CM} = \frac{2}{3}g \sin \phi$.

Observe that the acceleration of the mass center is less than that of a non-rotating mass sliding without friction down the incline. This is because of the I_{zz}/R^2 term in the denominator which increases the effective mass being accelerated. The kinetic energy associated with the mass center must be less for the case of rotation because a portion of the initial system energy is transformed into rotational kinetic energy.

For the "no-slip" case, relative sliding does not occur and therefore no work is done in overcoming friction. Since we have a conservative system, the sum of the gravitational potential energy relative to some datum (for example, the base of the incline) and the kinetic energy remains constant.

$$\frac{1}{2}m\dot{x}_{CM}^2 + \frac{1}{2}I_{zz}\omega_z^2 - mgx_{CM} \sin \phi = E.$$

Substituting $\omega_{z,CM} = - x_{CM}/R$ and differentiating with respect to time,

$$m\dot{x}_{CM}\ddot{x}_{CM} + I_{zz}\dot{x}_{CM}\frac{\ddot{x}_{CM}}{R^2} - mg\dot{x}_{CM} \sin \phi = 0.$$

Dividing by $\dot{x}_{CM}$ and solving for $\ddot{x}_{CM}$, we derive the same result as before.

The frictional force required for zero slip can also be determined by making the appropriate substitutions for $\ddot{x}_{CM}$.

$$f = \frac{-I_{zz}\omega_z}{R} = \frac{-\frac{1}{2}mR^2}{R}\left(\frac{-\ddot{x}_{CM}}{R}\right) = \frac{1}{3} mg \sin \phi.$$

Thus the minimum coefficient of static friction required for zero slip (pure rolling) is

$$\mu = \frac{f}{N} = \frac{\frac{1}{3}mg \sin \phi}{mg \cos \phi} = \frac{1}{3} \tan \phi.$$

(2) Assuming now that relative motion occurs between the line of contact of the cylinder and inclined plane, we may apply $f = \mu_k N$. The X-equation thus becomes

$$mg \sin \phi - \mu_k N = m\ddot{x}_{CM}.$$

Substituting $N = mg \cos \phi$ and solving for the acceleration of the mass center, we have

$$\ddot{x}_{CM} = g \sin \phi - \mu_k g \cos \phi$$

and it is observed that for the case of slip as well as for no slip, the acceleration of the mass center remains constant. The angular acceleration is given by

$$\dot{\omega}_z = - \frac{fR}{I_{zz}} = - \frac{\mu_k mgR \cos \phi}{I_{zz}}.$$

If the cylinder begins its travel with zero translational and angular velocities, integration of $\ddot{x}_{CM}$ and $\dot{\omega}_z$ with respect to time yields

$$\dot{x}_{CM} = (g \sin \phi - \mu_k g \cos \phi)t$$

$$\omega_z = - \frac{\mu_k mgR \cos \phi}{I_{zz}} t.$$

While the ratio $- \dot{x}_{CM}/\omega_z R$ is equal to unity for the zero-slip case, it exceeds unity if slip occurs. This ratio can therefore serve as a slip parameter for this specific example:

$$\frac{- \dot{x}_{CM}}{\omega_z R} = \frac{g \sin \phi - \mu_k g \cos \phi}{\mu_k mgR^2 \cos \phi / I_{zz}} = \frac{(\tan \phi/\mu_k) - 1}{mR^2/I_{zz}}$$

Substituting $I_{zz} = \frac{1}{2}mR^2$,

$$\frac{- \dot{x}_{CM}}{\omega_z R} = \frac{\tan \phi}{2\mu_k} - \frac{1}{2}.$$

Clearly, for $\tan \phi > 3\mu_k$, $- \dot{x}_{CM}/\omega_z R > 1$, and slip occurs.

• PROBLEM 17-33

A thin disk of radius r and mass m is placed on a flat surface with a clockwise angular velocity ω_0, but no linear velocity, see Figure 1. With the coefficient of friction μ between the disk and the floor, determine (a) the time t_1 at which the disk will start rolling without sliding and (b) the linear and angular velocities of the disk at time t_1.

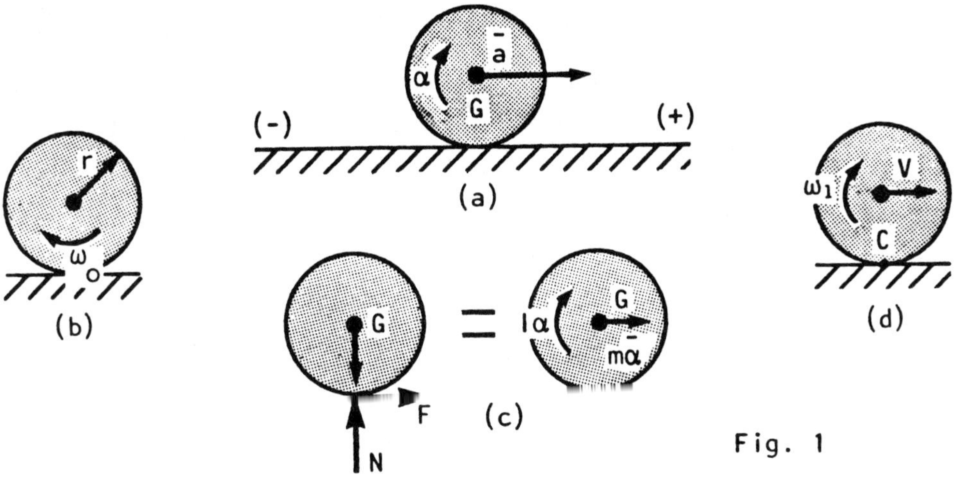

Fig. 1

Solution: We choose $\vec{a}$, the linear acceleration, and $\vec{\alpha}$, the angular acceleration, fig.1 (a), to be positive when $\vec{a}$ is in the right direction and $\vec{\alpha}$ is clockwise. The external forces acting on the hoop consist of the weight $\vec{W}$, the normal reaction $\vec{N}$, and the frictional force F. While the hoop is sliding, fig.1 (b) the magnitude of the friction is $\vec{F} = \mu N$. There is no vertical motion after the hoop touches down so N = W and F = $\mu W = \mu mg$.

The force F is in the opposite direction of the motion of the hoop at its point of contact with the plane, fig.1(c). Thus, F points horizontally to the right and is the only horizontal force, so.

$$F = \mu mg = ma$$

$$\vec{a} = +\ \mu g.$$

$\vec{F}$ also exerts a moment about the center of the hoop, opposite the positive rotation

$$-\ Fr = I\alpha$$

$$-\mu\, mgr = \frac{1}{2}\, mr^2\alpha \qquad \left(\text{since } I = \frac{1}{2}\, mr^2\right)$$

Solving for α yields

$$\alpha = \frac{-\ \mu g}{2r}.$$

Pure rolling occurs at the moment when the spinning has been reduced and the translational velocity has been increased so that

$$v\ (t) = r\omega\ (t).$$

From the accelerations,

$$V(t) = \overline{V}_0 + \overline{a}t = \mu gt$$

$$\omega(t) = \omega_0 + \alpha t = \omega_0 - \frac{\mu g}{r}\, t$$

When $t = t_1$, the hoop will start rolling. Then

$$v(t_1) = r\omega(t_1)$$

or

$$\mu g t_1 = r\left(\omega_0 - \frac{\mu g}{r} t_1\right)$$

Solving for t_1, gives,

$$t_1 = \frac{r\omega_0}{2\mu g}.$$

Substituting t_1 for $v(t)$ and $\omega(t)$,

$$v(t_1) = \mu g\left(\frac{r\omega_0}{2\mu g}\right) = \frac{1}{2} r\omega_0 \rightarrow$$

$$\omega(t_1) = \omega_0 - \frac{\mu g}{r}\left(\frac{r\omega_0}{2\mu g}\right) = \frac{1}{2}\omega_0. \circlearrowright$$

• PROBLEM 17-34

A billiard ball is struck by a cue as in figure (a). The line of action of the applied impulse is horizontal and passes through the center of the ball. The initial velocity $\vec{v}_0$ of the ball after impact, its radius R, its mass M, and the coefficient of friction μ between the ball and the table are all known.

(a) How far will the ball move before it ceases to slip on the table and starts to roll?

(b) What will its angular velocity be at this point?

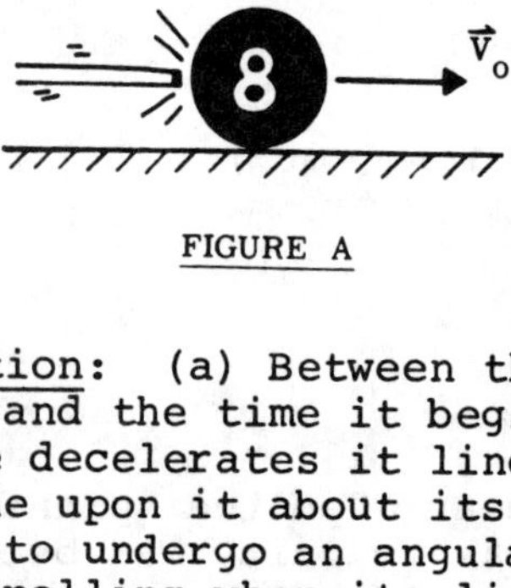

FIGURE A

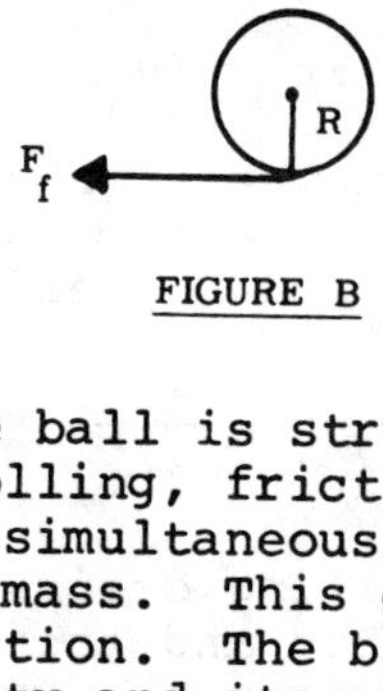

FIGURE B

Solution: (a) Between the time the ball is struck by the cue, and the time it begins pure rolling, friction with the table decelerates it linearly, but simultaneously exerts a torque upon it about its center of mass. This causes the ball to undergo an angular acceleration. The ball begins pure rolling when its linear velocity and its angular velocity have been decreased and increased respectively to the point at which the relation

$$V = R\omega$$

holds. We recognize this as the definition of linear velocity with respect to angular velocity for pure rolling. The force of friction on the ball is by definition:

$$F_f = \mu N = -\mu Mg$$

where N = Mg is the normal force between the ball and the table. The negative sign indicates that F_f is directed opposite to v_0.

The ball's linear acceleration is:

$$a = \frac{F_f}{M} = -\mu g.$$

Thus, its linear velocity at time t is given by:

$$v(t) = v_0 + at = v_0 - \mu gt$$

The torque on the ball is (see figure (b)):

$$\tau = F_f R = \mu MgR,$$

where the positive value signifies a torque pointing into the page.

Since we know that the moment of inertia of a solid sphere about an axis passing through the center is $I = \frac{2}{5} MR^2$, we can calculate its angular acceleration:

$$\tau = I\alpha, \qquad \alpha = \frac{\tau}{I} = \frac{\mu\ MgR}{\frac{2}{5}\ MR^2} = \frac{5}{2}\ \frac{\mu g}{R}$$

The ball's angular velocity at time t is given by:

$$\omega(t) = \omega_0 + \alpha t, \qquad \omega_0 = 0$$

$$\omega(t) = \alpha t = \frac{5}{2}\ \frac{\mu g}{R}\ t$$

To calculate the distance the ball will move before it begins pure rolling, we must first calculate how long it is after the ball has been struck that this occurs. Rolling begins when

$$v(t) = R\omega(t)$$

$$v_0 - \mu gt = R\left(\frac{5}{2}\ \frac{\mu g}{R}\right) t = \frac{5}{2}\mu gt$$

$$\frac{7}{2}\mu gt = v_0, \qquad t = \frac{2}{7}\ \frac{v_0}{\mu g}$$

The distance the ball travels is therefore, since the acceleration is constant,

$$s = v_0 t + \frac{1}{2} at^2 = v_0 t - \frac{1}{2} \mu g t$$

$$s = v_0 \left(\frac{2}{7} \frac{v_0}{\mu g}\right) - \frac{1}{2} \mu g \left(\frac{2}{7} \frac{v_0}{\mu g}\right)^2 = \frac{2}{7} \frac{v_0^2}{\mu g} - \frac{2}{49} \frac{v_0^2}{\mu g}$$

$$= \frac{12}{49} \frac{v_0^2}{\mu g}$$

Here v_0 is the ball's initial velocity.

(b) Its angular velocity at this point is:

$$\omega(t) = \omega_0 + \alpha t$$

$$\omega(t) = \alpha t = \frac{5}{2} \frac{\mu g}{R} \left(\frac{2}{7} \frac{v_0}{\mu g}\right) = \frac{5}{7} \frac{v_0}{R}$$

since the ball's initial angular velocity $\omega_0 = 0$.

• PROBLEM 17-35

A uniform, spherical bowling ball is projected without initial rotation along a horizontal bowling alley. How far will the ball skid along the alley before it begins to roll without slipping? Assume that the ball does not bounce. (See figure).

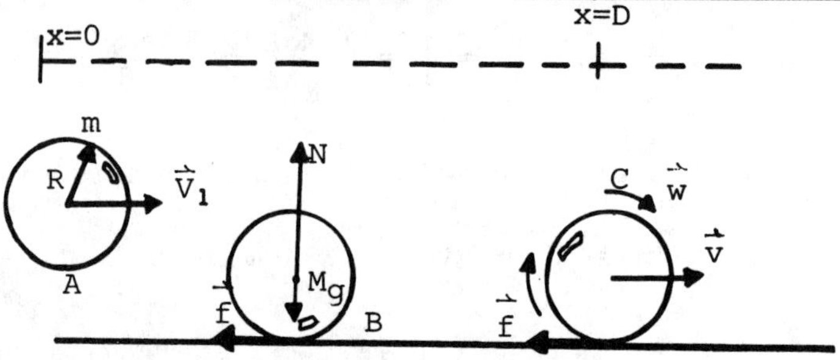

Solution: The ball is projected along the alley at point A. When it hits the alley at point B, the ball will roll and slip. At point C, the ball has begun to roll without slipping. In order to find the distance at which skidding (or slipping) stops, we must find the acceleration of the ball, and then solve for its position.

Applying Newton's Second Law to the horizontal direction of motion of the ball, we obtain

$$ma_x = -f \tag{1}$$

where a_x is the horizontal acceleration of the ball of mass m. Note that we take a_x to be positive in the positive x-direction. Applying the Second Law to the vertical direction

$$ma_y = N - mg$$

But a_y, the y acceleration of the ball, is zero since the ball doesn't bounce. Therefore

$$N = mg \tag{2}$$

In order to calculate torques, $\vec{\tau}$, about 0, we use

$$\vec{\tau} = I\vec{\alpha} \qquad (3)$$

where $\vec{\alpha}$ is the angular acceleration of the ball. But, the only torque acting on the ball is due to f. Hence

$$\vec{\tau} = \vec{r} \times \vec{f}$$

where $\vec{r}$ is the location of the point of application of $\vec{f}$. Since $\vec{r}$ and $\vec{f}$ are perpendicular (see figure),

$$\tau = Rf$$

From (3)

$$Rf = I\alpha \qquad (4)$$

where α is positive in a direction pointing into the plane of the figure. Now, if f is constant, (1) tells us that a_x is constant, since

$$a_x = \frac{-f}{m}$$

Hence, we may use the kinematics equations for constant acceleration to find

$$v_x = v_{x_0} \frac{-f}{m} t \qquad (5)$$

$$x = x_0 + v_{x_0} t \frac{-f}{2m} t^2 \qquad (6)$$

where x_0 and v_{x_0} are the initial position and velocity of the ball.

Similarly, we use (4) to solve for the angular velocity of the ball.

$$\alpha = \frac{Rf}{I}$$

But $\alpha = \frac{d\omega}{dt}$ where ω is the angular velocity of the ball. Therefore

$$\frac{d\omega}{dt} = \frac{Rf}{I}$$

$$\int_{\omega_0}^{\omega} d\omega = \frac{Rf}{I} \int_0^t dt$$

where ω_0 is the angular velocity **at** $t = 0$.

$$\omega = \omega_0 + \frac{Rf}{I} t \qquad (7)$$

While the ball is skidding, (7) and (5) are completely independent relations. However, when the ball starts rolling without slipping, they are related by

$$v_x = \omega R \qquad (8)$$

Substituting (7) and (5) in (8)

$$v_{x_0} - \frac{f}{m} t' = \omega_0 R + \frac{R^2 f}{I} t'$$

But $\omega_0 = 0$. Solving for t'

$$v_{x_0} = t'\left(\frac{R^2 f}{I} + \frac{f}{m}\right)$$

$$t' = \frac{v_{x_0}}{\frac{R^2 f}{I} + \frac{f}{m}} \qquad (9)$$

At $t = t'$, slipping stops. To find the position of the ball when slipping stops, we substitute (9) in (6)

$$x = x_0 - v_{x_0}\left[\frac{v_{x_0}}{\frac{R^2 f}{I} + \frac{f}{m}}\right] - \frac{f}{2m}\left[\frac{v_{x_0}^{\ 2}}{\left(\frac{R^2 f}{I} + \frac{f}{m}\right)}\right]$$

Furthermore, $x_0 = 0$ (see figure), and

$$x = \frac{v_{x_0}^{\ 2}}{f\left(\frac{R^2}{I} + \frac{1}{m}\right)} - \frac{-f\, v_{x_0}^{\ 2}}{2mf^2\left(\frac{R^2}{I} + \frac{1}{m}\right)^2} \qquad (10)$$

The frictional force law is

$$f = u_k N \qquad (11)$$

where N is the normal force of the alley on the ball, and u_k is the coefficient of kinetic friction. (We use the coefficient of kinetic friction because the ball skids from $t = 0$ to $t = t'$.) Substituting (2) in (11)

$$f = u_k mg$$

Substituting this in (10),

$$x = \frac{v_{x_0}^{\ 2}}{u_k mg\left(\frac{R^2}{I} + \frac{1}{m}\right)} - \frac{v_{x_0}^{\ 2}}{2u_k m^2 g\left(\frac{R^2}{I} + \frac{1}{m}\right)^2}$$

$$x = \frac{v_{x_0}^{\ 2}}{u_k mg\left(\frac{R^2}{I} + \frac{1}{m}\right)}\left\{1 - \frac{1}{2m\left(\frac{R^2}{I} + \frac{1}{m}\right)}\right\}$$

For a sphere $I = 2/5\ mR^2$ and $R^2/I = R^2/(2/5\ mR^2) = 5/2m$ then

$$x = \frac{v_{x_0}^{\ 2}}{u_k mg\left(\frac{5}{2m} + \frac{1}{m}\right)}\left\{1 - \frac{1}{2m\left(\frac{5}{2m} + \frac{1}{m}\right)}\right\}$$

$$x = \frac{2v_{x_0}^{\ 2}}{7u_k g}\left\{1 - \frac{1}{7}\right\} = \frac{12v_{x_0}^{\ 2}}{49\ u_k g}$$

This is the position at which slipping stops.

• PROBLEM 17-36

A homogeneous cylinder is given a horizontal velocity V_1 and a counterclockwise angular velocity $\omega_1 = V_1/R$ on the frictionless part of a horizontal surface. Beyond point A, the surface changes so that the friction coefficient to the right of A is μ (see Fig. 1).

After the cylinder passes A, it first will slip on the rough plane, but will eventually roll with a constant velocity without slipping. What is the condition satisfied when the body starts to roll without slipping, and what is the corresponding velocity of the center of mass?

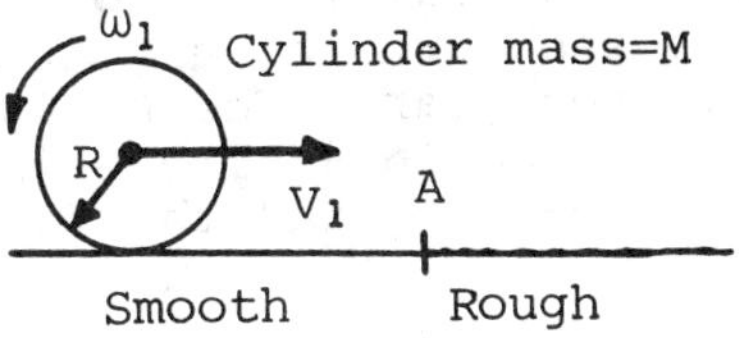

Fig. 1

Solution: The only force on the body in the direction of motion is the contact force, and consequently its line of action is in the plane. Therefore, the angular momentum of the cylinder with respect to a reference point in the plane, throughout the entire motion, remains constant.

The angular momentum of the cylinder taken about any point in the plane, while on the smooth surface, is

$$L = MV_1R - I_0\omega_1 = MV_1R\left(1 - \frac{k_0^2}{R^2}\right),$$

which is the sum of the angular momentum of the center of mass and the angular momentum with respect to the center of mass. Here we have introduced $\omega_1 = V_1/R$, as specified by the initial conditions of the problem. We have also considered the moment of inertia of the cylinder I_0, to be equal to Mk_0^2 where k_0 is the radius of gyration of the cylinder.

When the body rolls without slipping, the relation between the angular and translational velocity is $V_2 = \omega R$, and the angular momentum is

$$L' = MV_2R + I_0\omega = MV_2R\left(1 + \frac{k_0^2}{R^2}\right).$$

$I_0\omega$ is positive in this case since when the cylinder is rolling without slipping, it is rotating in a clockwise direction.

Then, equating L and L', we have

$$V_2 = \frac{1 - k_0^2/R^2}{1 + k_0^2/R^2} V_1 = \frac{1/2}{3/2} V_1 = \frac{1}{3} V_1,$$

since $I = \frac{1}{2} mR^2 = mk_0^2$ for a cylinder so that $k_0^2/R^2 = \frac{1}{2}$.

The final velocity is one-third of the original velocity.

• PROBLEM 17-37

A railroad car rounds a circular turn of radius R on unbanked rails. The radius of the turn is very much greater than the dimensions of the railroad car. The horizontal thrust necessary to keep the car moving in a circle is supplied to the flanges of the inner wheels. Find the forces exerted on the car by the rails.

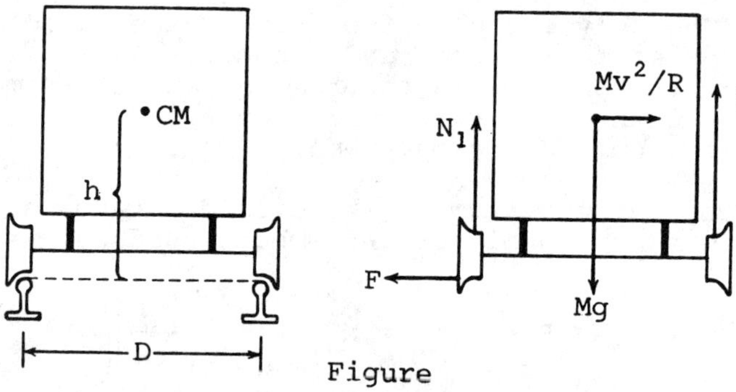

Figure

Solution: Since the radius of the curve is very much greater than the dimensions of the railroad car, the problem is equivalent to one in which the car is accelerated to the left in the figure with an acceleration v^2/R. Consider the car in a frame in which the car is at rest. In this accelerating frame a pseudo-force equal to Mv^2/R acts on the center of mass of the car to the right in the figure, where v is the speed of the car around the curve. Isolate the railroad car analytically as on the right. Using the symbols in the figure, equate the horizontal and vertical components of the forces to zero, using D'Alembert's principle for Mv^2/R.

$$N_1 + N_2 - Mg = 0$$

$$F - \frac{Mv^2}{R} = 0.$$

Take the torque about the outer rail.

$$DN_1 + h\frac{Mv^2}{R} - \frac{D}{2}Mg = 0.$$

From the last equation,

$$N_1 = M\left(\frac{g}{2} - \frac{hv^2}{DR}\right)$$

From this and the first equation,

$$N_2 = M\left(\frac{g}{2} + \frac{hv^2}{DR}\right)$$

From the second equation,

$$F = \frac{Mv^2}{R}$$

N_1, N_2, and F represent the sum of the force components applied to the railroad car by all the wheels on their respective sides.

• PROBLEM 17-38

A car, with its door open and free to swing on its hinges is accelerating with a constant acceleration a. Determine the angular acceleration of the door, relative to the car, as a function of the acceleration of the automobile.

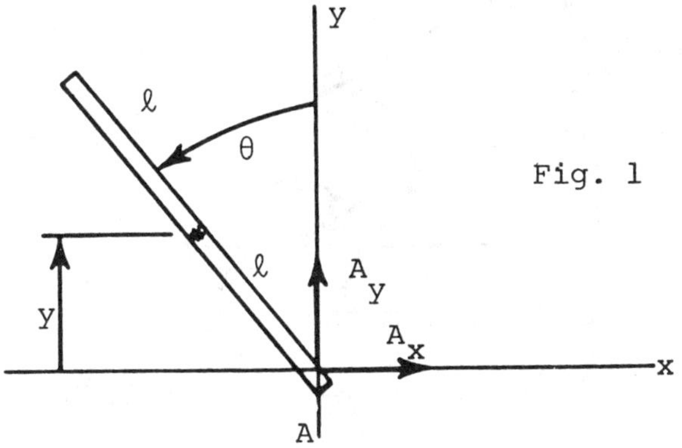

Fig. 1

Solution: Shown is a free-body diagram of the door, of mass m and length 2ℓ. Not shown is the fictitious force ma acting at the center of mass of the door. In the car's frame of reference we consider this force to be real and use it in developing a solution.

The forces A_x and A_y are unknown. When one takes moments about A, these two unknown forces passing through point A would not appear in the moment equation. The moment equation then, is

$$ma_y = I_A\alpha$$

$$(ma)\ \ell \cos\theta = I_A\alpha \text{ (since } y = \ell \cos\theta)$$

Solving for the angular acceleration of the door,

$$\alpha = \frac{ma\ \ell \cos\theta}{I_A}.$$

A box that weighs 50lb, has a height h = 1ft, length L = 3ft and width w = 1.5 ft. The box is oriented as shown in the fig. The corners of the box slice freely without friction along the surfaces. The box is release without an initial velocity. Determine (a) The angular acceleration of the box, (b) The reactions at A and B

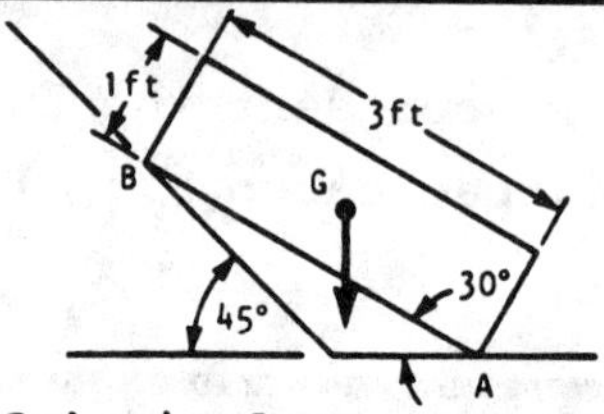

Solution: Due to the frictionless contact between the corners of the box and the surfaces, the reaction forces $\vec{R}_A$ and $\vec{R}_B$ must be perpendicular to the respective surfaces. The free-body diagram, therefore, is as shown in figure 1.

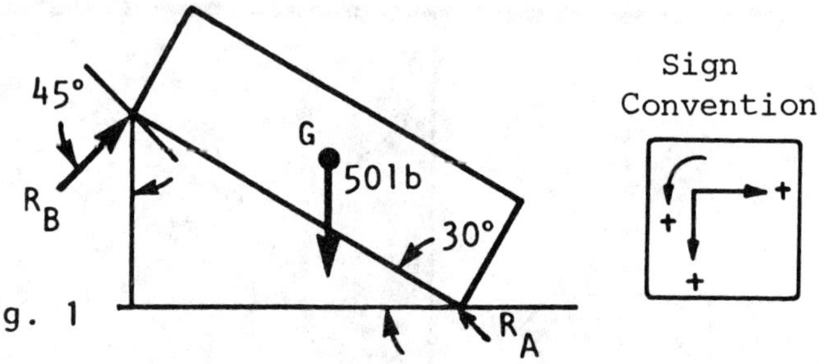

Fig. 1

The equations of motion are

$$\Sigma F_x: \quad R_B \cos 45° = \left(\frac{50}{32.2}\right) a_x = 1.55\, a_x \qquad (1)$$

$$\Sigma F_y: \quad - R_B \sin 45° - R_A + 50 = \left(\frac{50}{32.2}\right) a_y = 1.55\, a_y \qquad (2)$$

$$\Sigma M: \quad - R_B\,(2) + R_A\,(2 \cos 30°) = 1.29\, \alpha. \qquad (3)$$

$$\text{Here } I = \frac{1}{12} m(\ell^2 + h^2) = \frac{1}{12}\left[\frac{50}{32.2}\right](3^2 + 1^2) = 1.29 \text{ lb-ft-sec}^2.$$

There are 3 equations but 5 unknowns.

Kinematic relationships will offer two more equations.

Since the motion is constrained, the acceleration of G must be related to the angular acceleration α. To obtain this relation, consider the acceleration of ends A and B. See Figures 2 and 3.

$$\vec{a}_B = \vec{a}_A + \vec{a}_{B/A} = \vec{a}_A + 3\vec{\alpha}.$$

The law of sines applied to the vector addition triangle, Figure 3, yields:

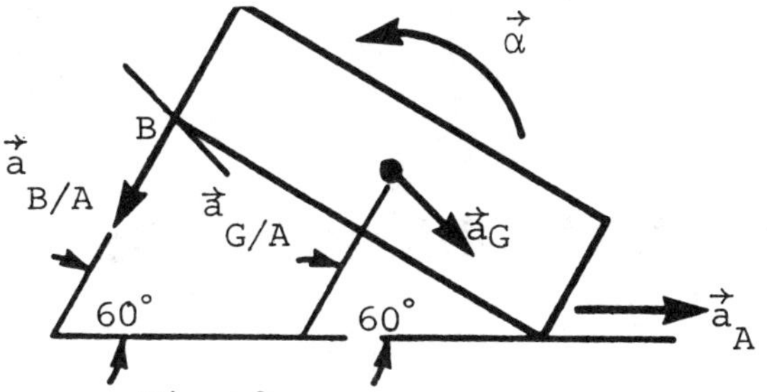

Fig. 2

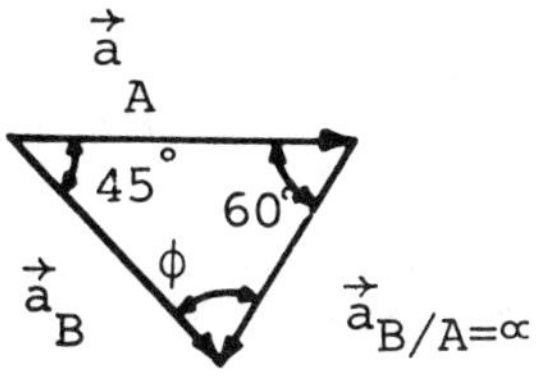

Fig. 3

$$a_A = 4.10\alpha \qquad (4)$$

$$a_B = 3.67\alpha.$$

Relating the acceleration of G to A:

$$\vec{a}_G = \vec{a}_A + \vec{a}_{G/A}$$

as shown in Figure 4, with $a_{G/A}=1.5\alpha$.

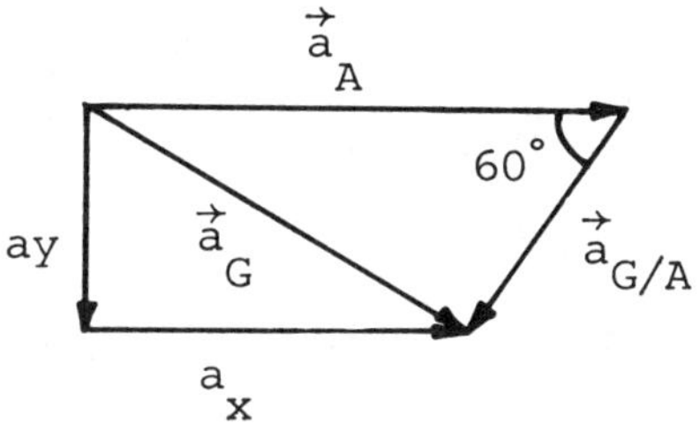

Fig. 4

The acceleration $\vec{a}_A$ is at an angle of 30° to the box. The accelerations $\vec{a}_{B/A}$ and $\vec{a}_{G/A}$ are both perpendicular, that is 90°, to the box. The angle between $\vec{a}_A$ and the other accelerations is then 180° - (90° + 30°) = 60°

Taking a_y and a_x from Figure 4

$$a_x = 4.10\alpha - 1.5\alpha \cos 60° = 3.35\alpha \qquad (5)$$

$$a_y = -1.5\alpha \sin 60° = -1.3\alpha. \qquad (6)$$

Equations (5) and (6) along with (1), (2) and (3) now provide the means to the solution desired. The algebraic solution results in:

$$\alpha = 3.95 \text{ rad/sec}^2$$

$$R_B = 29.0 \text{ lb}$$

$$R_A = 37.4 \text{ lb}.$$

CHAPTER 18

RIGID BODY KINETICS: WORK, IMPULSE

WORK DONE BY A FORCE AND A COUPLE

• **PROBLEM** 18-1

Two homogeneous cylinders made of the same material rotate about their own axes at the same angular velocity ω. What is the ratio between their kinetic energies if the radius of one cylinder is twice the radius of the other?

Solution: The kinetic energy of a rotating cylinder is

$$KE = \frac{1}{2} I\omega^2. \tag{1}$$

where I is the mass moment of inertia. For a cylinder

$$I = \frac{1}{2} mr^2 \tag{2}$$

where r is the radius. The mass, in turn, is equal to

$$m = \pi r^2 L\rho \tag{3}$$

where L is the length, πr^2 is the area of the cross-section and ρ is the density. Assume that both cylinders are of the same length and density.

Substituting eq. (3) into (2), and adding subscripts to distinguish them apart,

$$I_1 = \frac{1}{2}(\pi r_1{}^2 L\rho)r_1{}^2, \text{ and } I_2 = \frac{1}{2}(\pi r_2{}^2 L\rho)r_2{}^2$$

Therefore, the kinetic energies are

$$(KE)_1 = \frac{1}{2} I_1\omega^2 = \frac{1}{2}\left(\frac{1}{2}\pi r_1{}^4 L\rho\right)\omega^2$$

$$(KE)_2 = \frac{1}{2} I_2\omega^2 = \frac{1}{2}\left(\frac{1}{2}\pi r_2{}^4 L\rho\right)\omega^2$$

So $$\frac{(KE)_2}{(KE)_1} = \frac{\frac{1}{4}\pi r_2{}^4 L\rho\omega^2}{\frac{1}{4}\pi r_1{}^4 L\rho\omega^2} = \frac{r_2{}^4}{r_1{}^4}$$

If we let $r_2 = 2r_1$, then

$$\frac{(KE)_2}{(KE)_1} = \frac{\left(2r_1\right)^4}{r_1^4} = 16.$$

Therefore, the ratio is 16:1.

• PROBLEM 18-2

If the center of mass of the rigid body lies on its axis of rotation, show that the kinetic energy of the rigid body is given by $K = \frac{1}{2} m\dot{r}_0{}^2 + \frac{1}{2} I\omega^2$.

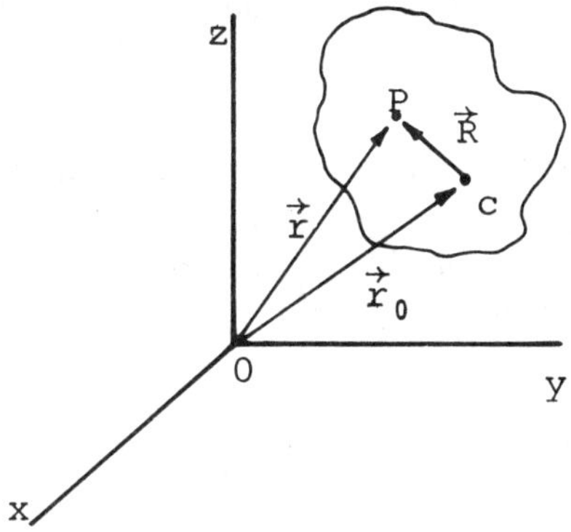

Solution: Consider the diagram shown in the Figure. The point P is a general point in the rigid body and the point C is the location of its center of mass. $\vec{r}$ and $\vec{r}_0$ are the locations of P and C respectively with respect to the fixed coordinate system x, y, z. $\vec{R}$ is the position of P with respect to C. By vector addition

$$\vec{r} = \vec{r}_0 + \vec{R}. \tag{1}$$

Differentiation of equation (1) with respect to time yields

$$\dot{\vec{r}} = \dot{\vec{r}}_0 + \dot{\vec{R}}. \tag{2}$$

Now consider an element of mass, dm, at point P. By definition its kinetic energy, dk, is

$$dk = \frac{1}{2} dm\, \dot{r}^2. \tag{3}$$

Substitution of equation (2) into equation (3) gives

$$dk = \frac{1}{2} dm\, (\dot{\vec{r}}_0 + \dot{\vec{R}})\cdot(\dot{\vec{r}}_0 + \dot{\vec{R}}) = \frac{1}{2} dm\, (\dot{r}_0{}^2 + 2\, \dot{\vec{r}}_0\cdot \dot{\vec{R}} + \dot{R}^2). \tag{4}$$

Integration of equation (4) gives the total kinetic energy of the body:

$$K = \int dk = \frac{1}{2} \int_V (\dot{r}_0{}^2 + 2\dot{\vec{r}}_0 \cdot \dot{\vec{R}} + \dot{R}^2)\ dm \qquad (5)$$

The integration is over the volume which contains the mass of the body. The most convenient coordinate system for the integration is the $\vec{R}$ system whose origin is at C with axis X, Y, Z, i.e., $dm = \rho dXdYdZ$, where ρ is the density. Then the first term in equation (5) becomes

$$\int \dot{r}_0{}^2 dm = \dot{r}_0{}^2 \int dm = m\dot{r}_0{}^2 \qquad (6)$$

$\dot{r}_0{}^2$ can be taken outside the integral since it doesn't depend on X, Y, Z. This is only the kinetic energy of the center of mass, assuming all the mass of the body is concentrated at C. The second term in equation (5) is

$$\int \vec{r}_0 \cdot \vec{R}\ dm = \vec{r}_0 \cdot \int \vec{R}\ dm.$$

In order to simplify this it is necessary to use the definition of the center of mass,

$$\vec{R}_{Cm} = \frac{1}{m} \int \vec{R}\ dm.$$

However, by definition in the $\vec{R}$ coordinate system $\vec{R}_{CM} = 0$ since the center of mass is at C. Therefore,

$\int \vec{R}\ dm = 0$, and after differentiating with respect to time

this becomes $\int \dot{\vec{R}}\ dm = 0$. Thus the middle term in equation (5) is zero.

The third term in equation (5) is $\int \dot{R}^2\ dm$, where $\dot{R}$ is the velocity of the point P with respect to the center of mass. However, the velocity of a point rotating with angular velocity $\vec{\omega}$ with respect to an axis through the center of mass can be written

$$\dot{\vec{R}} = \vec{\omega} \times \vec{R}.$$

Therefore, $\dot{R}^2 = \dot{\vec{R}} \cdot \dot{\vec{R}} = (\vec{\omega} \times \vec{R}) \cdot (\vec{\omega} \times \vec{R}).$ (7)

An identity from vector analysis is

$$(\vec{A} \times \vec{B}) \cdot (\vec{C} \times \vec{D}) = (\vec{A} \cdot \vec{C})(\vec{B} \cdot \vec{D}) - (\vec{A} \cdot \vec{D})(\vec{B} \cdot \vec{C}).$$

Using this, equation (7) becomes

$$\dot{\vec{R}}^2 = \omega^2 R^2 - 2\vec{\omega} \cdot \vec{R}.$$

So the third term is

$$\int \dot{\vec{R}}^2 \, dm = \int \vec{\omega}^2 \vec{R}^2 \, dm - 2 \int \vec{\omega} \cdot \vec{R} \, dm \qquad (8)$$

$$= \vec{\omega}^2 \int \vec{R}^2 \, dm - 2\vec{\omega} \cdot \int \vec{R} \, dm.$$

By definition of the movement of inertia of the body, $I = \int R^2 \, dm$, the first term in equation (8) becomes $I\omega^2$. The second term is zero for the same reason as the middle term of equation (5). So, there remains

$$\int \dot{R}^2 \, dm = I\omega^2. \qquad (9)$$

Substituting equations (6) and (9) in equation (5) gives the final expression for the total kinetic energy of the body:

$$K = \frac{1}{2} m \dot{r}_0^{\,2} + \frac{1}{2} I\omega^2.$$

This shows that the kinetic energy of the rigid body is made up of two parts: 1) the kinetic energy of the center of mass, 2) the rotational kinetic energy of the body rotating about an axis through the center of mass. This result holds in general and can be stated as follows: the total kinetic energy of a body is equal to the kinetic energy of the center of mass plus the kinetic energy of the body relative to the center of mass.

• PROBLEM 18-3

Suppose that the uniform disk shown in Figure 1 has a mass M = 500 gm, a radius R = 10 cm, and a thickness, t. A constant force of 20,000 dynes is applied at the end of the cord. If the disk has an initial angular velocity of 4 radians/sec, when the force is applied, find the angular velocity and rotational kinetic energy after 2 seconds.

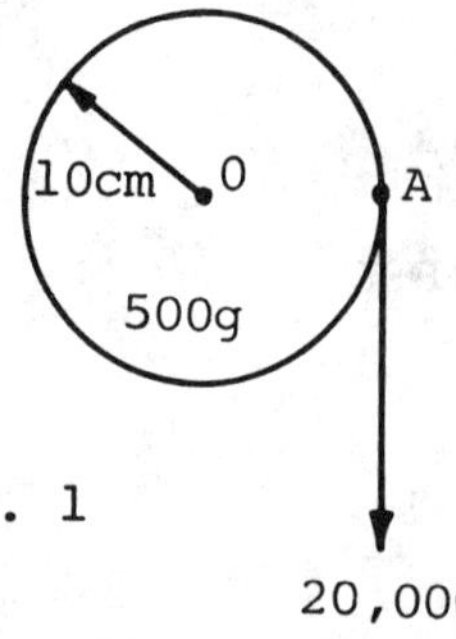

Fig. 1

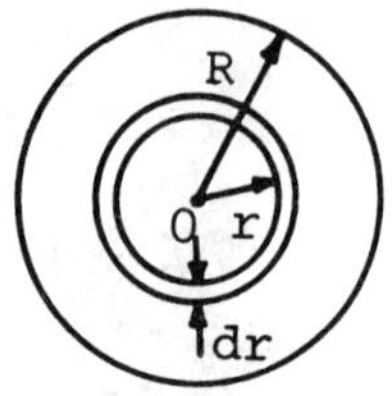

Fig. 2

Solution: First we must calculate the value of the moment of inertia I of the circular disk about its axis of rotation. The definition given for the moment of inertia is

$$I = \Sigma\, mr^2$$

where the summation is taken over all the particles composing the body. For a continuously distributed mass such as a uniform solid body, for example, the disk, one replaces the summation $\Sigma\, mr^2$ used for a discrete series of masses by an integral. Thus for a uniform solid the moment of inertia is given by

$$I = \int r^2 dm. \tag{1}$$

A useful relationship to remember when solving this integral, is that for a body with a uniform density ρ, the differential piece of mass, dm, is given by

$$dm = \rho dV. \tag{2}$$

Since the disk in this problem is uniform, its density is found by dividing its total mass by its total volume.

$$\rho = \frac{M}{V} = \frac{M}{\pi R^2 t} \tag{3}$$

where the volume of the disk is just the circular area multiplied by t, the thickness of the disk. Substitution of equations (2) and (3) in equation (1) yields

$$I = \int r^2\, dV = \int \frac{r^2 M}{\pi R^2 t}\, dV \tag{4}$$

The calculation of I for the disk about an axis through its center perpendicular to its plane can be made by considering the disk as made up of a large number of concentric circular rings of volume dV. One of these has a radius r, and a width of dr as shown in Figure 2. The volume element dV will be given by its circumference multiplied by its width and thickness, since to a first approximation, all points on this ring lie at the same distance from the center.

$$dV = 2\pi rt\, dr$$

Substituting this into equation (4) yields

$$I = \int \frac{r^2 M\, 2\pi rt}{\pi R^2 t}\, dr$$

which becomes

$$I = \frac{2M}{R^2} \int r^3\, dr \qquad (5)$$

The integration, will be a summing up all of the values for each circular ring. Integration from the limits of r = 0, to r = R yields

$$I = \frac{2M}{R^2} \int_0^R r^3\, dr = \frac{2M}{4R^2}\, r^4 \Big|_0^R$$

$$I = \frac{MR^2}{2}, \qquad (6)$$

which is the value of the moment of inertia about an axis through the center of and perpendicular to a disk of mass M and radius R. Notice that the value of I does not depend on the thickness, t, of the disk. Thus, this is also the moment of inertia of a cylinder through its central axis.

In our problem M = 500 gm and R = 10 cm so that

$$I = \frac{MR^2}{2} = 2.5 \times 10^4 \text{ gm cm}^2.$$

The torque τ exerted on the disk is

$$\tau = FR = 20{,}000 \times 10 = 2 \times 10^5 \text{ dyne cm}$$

From $\tau = I\alpha$, the angular acceleration of the disk α is

$$\alpha = \frac{2 \times 10^5 \text{ dyne cm}}{2.5 \times 10^4 \text{ gm cm}^2} = 8 \text{ radians/sec}^2.$$

If the angular velocity at zero time is ω_0 and at time t is ω, then, since $\alpha = d\omega/dt$, it follows that

$$\int_{\omega_0}^{\omega} d\omega = \int_0^t \alpha\, dt$$

or $$\omega - \omega_0 = \alpha t.$$

Since the initial angular velocity ω_0 in this problem is 4 radians/sec, then the angular velocity ω after 2 sec is

$$\omega = \omega_0 + \alpha t$$

$$= 4 + (8 \times 2) = 20 \text{ radians/sec.}$$

The rotational kinetic energy after 2 sec is

$$\frac{1}{2} I\omega^2 = \frac{1}{2} \times 2.5 \times 10^4 \times 400 = 5 \times 10^6 \text{ ergs.}$$

Let us show that the work done $W = \tau\theta$ is equal to the change in kinetic energy.

We must first find the angle θ through which the disk turns in the 2 sec. Since $\omega = d\theta/dt$, it follows that, if $\theta = 0$ at $t = 0$, then

$$\int_0^\theta d\theta = \int_0^t \omega\, dt = \int_0^t (\omega_0 + \alpha t)\, dt$$

or $$\theta = \omega_0 t + \frac{1}{2} \alpha t^2.$$

In our example

$$\theta = (4 \times 2) + \frac{1}{2} (8 \times 4) = 24 \text{ radians}$$

so that the work done by the force in the 2 sec is:

$$\tau\theta = (2 \times 10^5 \times 24) = 4.8 \times 10^6 \text{ ergs.}$$

This quantity should be equal to the change in kinetic energy during the 2 sec or

$$\tau\theta = \frac{1}{2} I\omega^2 - \frac{1}{2} I\omega_0{}^2$$

$$= \left(\frac{1}{2} \times 2.5 \times 10^4 \times 400\right) - \left(\frac{1}{2} \times 2.5 \times 10^4 \times 16\right)$$

$$= (5 \times 10^6) - (2 \times 10^5)$$

$$= 4.8 \times 10^6 \text{ ergs.}$$

which, of course, agrees with the value of $\tau\theta$ given above.

• PROBLEM 18-4

The moment of a couple is $M = \theta^2 + 2\theta$ ft-lb., where θ is measured in radians. If the couple rotates a shaft 60°, what work does it do?

Solution: The work done by a torque is given by

$V = \int_{\theta_1}^{\theta_2} \tau d\theta$, while the torque τ causes a rotation from θ_1 to θ_2. In the given situation,

$$V = \int_0^{\pi/3} (\theta^2 + 2\theta)\, d\theta = \left| \frac{\theta^3}{3} + \theta^2 \right|_0^{\pi/3}$$

$$= \frac{\pi^3}{81} + \frac{\pi^2}{9} = 1.48 \text{ ft-lb.}$$

• PROBLEM 18-5

A 4 ft. diameter wheel is free to rotate about a fixed axis through its center. A 10-lb. force acts to the right at the top of the wheel, and another 10 lb. force to the left at the bottom of the wheel. Determine the total work done by the pair of forces if they act while the wheel rotates through an angle of 90°.

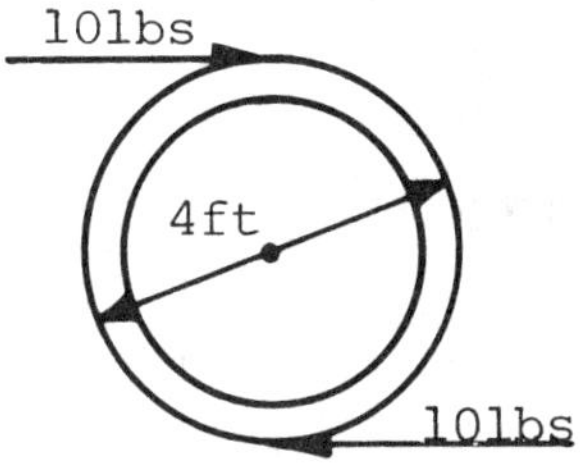

Solution: This problem can be solved if we recognize that the pair of forces constitute a couple which rotates the wheel through an angle of 90° or $\pi/2$ radians. Now, the work done by a couple is given by

$$W = M\,\Delta\theta,$$

where M is the moment of the couple and $\Delta\theta$ is the angle that the wheel rotates through. The moment of a couple is given by the magnitude of one of the forces multiplied by the perpendicular distance between the force pair. Hence,

$$M = 10 \text{ lb } (4\text{ft}) = 40 \text{ ft.-lb.}$$

We were given that $\Delta\theta = \pi/2$ radians. Substituting values into the equation yields

$$W = M\,\Delta\theta = 40\ (\pi/2)$$

$$W = 62.8 \text{ ft-lb.}$$

• PROBLEM 18-6

A particle of mass M is attached to a string (see the figure) and constrained to move in a horizontal plane (the plane of the dashed line). The particle rotates with velocity v_0 when the length of the string is r_0. How much work is done in shortening the string to r?

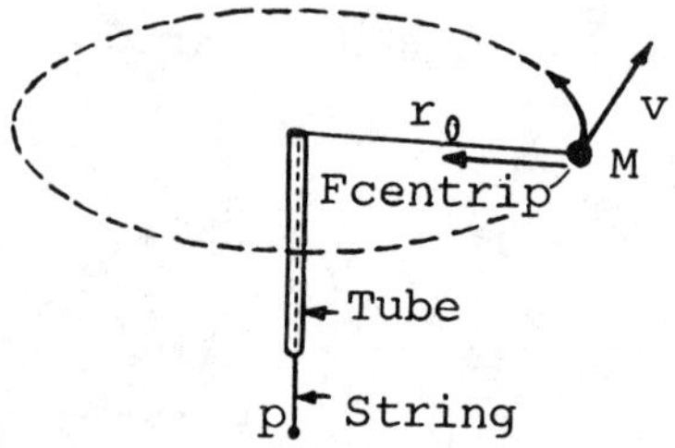

Solution: The string is stretched under the action of the radial centripetal force which keeps the mass M on its circular path. When we pull in the string we shorten r_0 by increasing the radial force $\vec{F}_{centrip}$ on M. As we know, a force can only produce a torque about the axis of rotation if it has a component perpendicular to the radius which locates the mass M. A purely radial force like $\vec{F}_{centrip}$ has no such component, therefore the angular momentum must remain constant as the string is shortened.

$$Mv_0r_0 = Mvr \tag{1}$$

The kinetic energy at r_0 is $\frac{1}{2}Mv_0^2$; at r it has been increased to

$$\frac{1}{2}Mv^2 = \frac{1}{2}Mv_0^2\left(\frac{r_0}{r}\right)^2$$

because $v = v_0r_0/r$ from above. It follows that the work W done from outside in shortening the string from r_0 to r is

$$W = \frac{1}{2}Mv^2 - \frac{1}{2}Mv_0^2 = \frac{1}{2}Mv_0^2\left[\left(\frac{r_0}{r}\right)^2 - 1\right] . \tag{2}$$

This can also be calculated directly as the work done by $\vec{F}_{centrip}$ along the distance $r_0 - r$;

$$W = \int_{r_0}^{r} \vec{F}_{centrip} \cdot d\vec{r} = -\int_{r_0}^{r} F_{centrip}\, dr$$

$$W = -\int_{r_0}^{r} dr\,\frac{Mv^2}{r} = -\int_{r_0}^{r} dr\,\frac{M}{r}\,\frac{v_0^2r_0^2}{r^2}$$

where we have used (1). Hence

$$W = -Mv_0^2r_0^2\int_{r_0}^{r}\frac{dr}{r^3}$$

$$W = \frac{Mv_0^2 r_0^2}{2r^2}\Bigg|_{r_0}^{r} = \frac{Mv_0^2 r_0^2}{2}\left(\frac{1}{r^2} - \frac{1}{r_0^2}\right)$$

$$W = \frac{1}{2} Mv_0^2 \left[\left(\frac{r_0}{r}\right)^2 - 1\right]$$

which is (2).

We see that the angular momentum acts on the radial motion as an effective repulsive force. We have to do extra work on the particle on bringing it from large distances to small distances if we require that the angular momentum be conserved in the process.

• PROBLEM 18-7

The Prony brake shown tests the horsepower of an engine. Friction between the wheel and the band causes the band and the attached arm to try to rotate clockwise. This rotation is resisted by the scale at the right. When the scale reading is 15.8 lb. the wheel is observed to have a speed of 210 rpm. What horsepower is being dissipated by the brake?

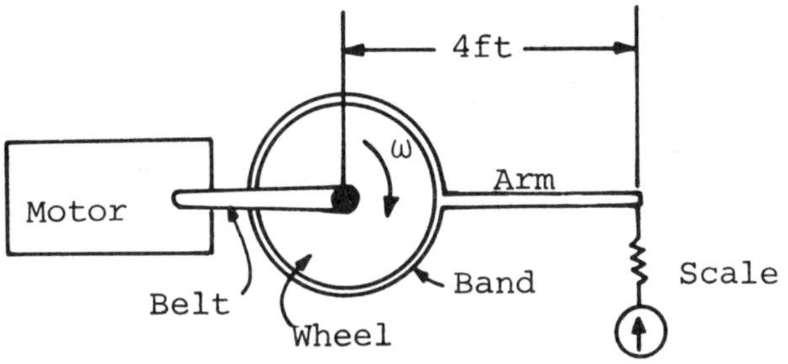

Fig. 1

Solution: In this problem the scale and motor are applying equal and opposite torques on the wheel. The power produced by the motor is lost in the friction of the band and wheel bearings.

From the power equation, the power lost in friction is:

$$P = T\omega,$$

where P is power, T is the torque applied by the band and ω is rotation speed in radians per second.

Since the arm and band do not move the equations of

statics hold. To find the torque applied to the wheel by the band, sum the torques on the arm and band about the center of the wheel. (Counterclockwise torques are defined as positive.)

$$F_{scale}\ \ell - T = 0,$$

where F_{scale} is the scale force, and ℓ is the distance from the scale to the center of the wheel.

$$T = F_{scale}\ \ell$$

combining the power equation and torque equation yields:

$$P = F_{scale}\ \ell\ \omega.$$

Substituting the given values for the problem yields the final result:

$$P = (15.8\ \text{lb})\ (4\ \text{ft}) \left(\frac{210\ \text{rotations}}{\text{minute}} \times \frac{1\ \text{minute}}{60\ \text{sec}} \times \frac{2\pi\ \text{radians}}{\text{rotation}} \right),$$

$$P = (63.2\ \text{ft-lb}) \left(21.99\ \frac{\text{rad}}{\text{sec}} \right)$$

$$P = 1389.4\ \frac{\text{ft-lb}}{\text{sec}}.$$

Remembering that $1\ \text{hp} = 550\ \frac{\text{ft-lb}}{\text{sec}}$,

$$P = 1389.\ \frac{\text{ft-lb}}{\text{sec}}\ \frac{1\ \text{hp}}{550\ \frac{\text{ft-lb}}{\text{sec}}}$$

$$P = 2.53\ \text{h p.}$$

This is the power lost in friction.

• PROBLEM 18-8

Show that the work done by a constant torque N_z about a fixed axis z is given by $W = N_z\theta$, where θ is the angular displacement of the rigid body in time t. If the rotating body has an initial angular velocity ω_1 and an initial angular displacement θ_1, show that $N_z\ (\theta - \theta_1) = \frac{1}{2} I\omega^2 - \frac{1}{2} I\omega_1{}^2$.

Solution: Consider the displacement of a rigid body about a fixed Z-axis due to a constant torque N_z about that axis.

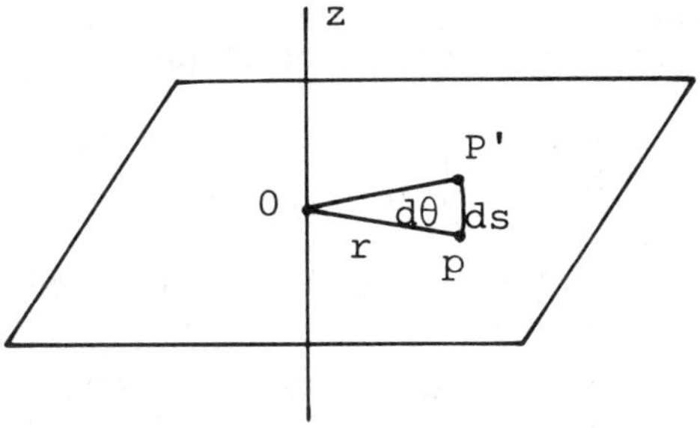

Fig. 1

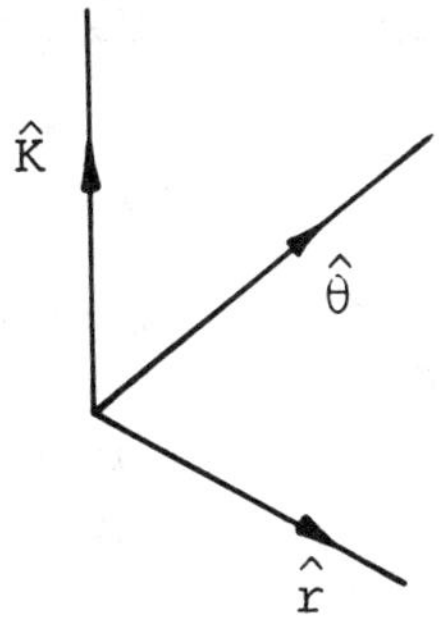

Fig. 2

Choose a plane perpendicular to the axis, within the body and a point P in the plane at a distance r from the axis, as shown in Figure 1. Under an infinitesimal displacement ds about the Z-axis, the point P moves to P'. Since by definition a rigid body is one in which any two points remain equidistant under a displacement, OP = OP' = r i.e., under the conditions of the problem, the displacement is circular so $ds = r\ d\theta$. (1)

The coordinate system is shown in Figure 2, where $\hat{r}$, $\hat{\theta}$, $\hat{k}$ are unit vectors in the r, θ, z directions. Since the displacement is in the $\hat{\theta}$ direction, equation (1) can be rewritten. First $d\vec{s} = ds\vec{\theta}$. Next, define the vector angular displacement as $d\vec{\theta} = d\theta\hat{k}$, i.e., a rotation of dθ about the Z-axis. The vector displacement is then $d\vec{s} = d\vec{\theta} \times \vec{r}$ since

$$d\vec{s} = d\theta\ (\hat{k} \times \vec{r}) = r\ d\theta\ (\hat{k} \times \hat{r}) = ds\hat{\theta}.$$

By definition the work done in an infinitesimal displacement $d\vec{s}$ by a force $\vec{F}$ is $dW = \vec{F}\cdot d\vec{s}$.

In this case $dW = \vec{F}\cdot d\vec{s} = \vec{F}\cdot(d\vec{\theta} \times \vec{r})$. (2)

The following is an identity from vector analysis:

$$\vec{A}\cdot\vec{B} \times \vec{C} = \vec{B}\cdot\vec{C} \times \vec{A}. \quad (3)$$

Letting $\vec{A} = \vec{F}$, $\vec{B} = d\vec{\theta}$ and $\vec{C} = \vec{r}$, equation (3) becomes

$$\vec{F}\cdot(d\vec{\theta} \times \vec{r}) = d\vec{\theta}\cdot\vec{r} \times \vec{F} = d\vec{\theta}\cdot\vec{N}, \quad (4)$$

where $\vec{N} = \vec{r} \times \vec{F}$ is the torque produced by $\vec{F}$. Substituting equation (4) into equation (2) gives

$$dW = \vec{N}\cdot d\vec{\theta} \quad (5)$$

Since in this case $\vec{N}\cdot d\vec{\theta} = N_z d\theta$, equation (5) becomes

$$dW = N_z d\theta. \quad (6)$$

For a constant torque producing a displacement of θ in time t, equation (6) yields for the total work done:

$$W = \int dW = \int_0^{\theta} N_z d\theta = N_z \int_0^{\theta} d\theta = N_z\theta. \qquad (7)$$

The next problem is to determine the angular position of the body acted on by the constant torque N_z at time t given that it was initially at θ_1, with angular velocity ω_1. For ciruclar motion, Newton's Second Law has the form $\vec{N} = I\vec{\alpha}$ where $\vec{N}$ is the torque, $\vec{\alpha}$ is the angular acceleration and I is the moment of inertia of the body. In this case since the torque is constant about the Z-axis the equation of motion becomes

$$N_z = I \frac{d^2\theta}{dt^2} = I \frac{d\omega}{dt}. \qquad (8)$$

However, by the chain rule of the calculus $\frac{d\omega}{dt} = \frac{d\omega}{d\theta} \cdot \frac{d\theta}{dt} = \omega \frac{d\omega}{d\theta}$. Substitution in equation (8) and integrating from θ_1 to $\theta(t)$ and ω_1 to $\omega(t)$ gives

$$\int_{\theta_1}^{\theta(t)} N_z \, d\theta = \int_{\omega_1}^{\omega(t)} I \, \omega d\omega.$$

For a constant torque and moment of inertia, this becomes

$$N_z \left(\theta(t) - \theta_1\right) = \frac{1}{2} I \left(\omega^2(t) - \omega_1^{\,2} \right). \qquad (9)$$

Note that this is the rotational analogue of the work-energy theorem. From equation (7), the left side of equation (9) is the work done on the system by the torque, $\vec{N}$. Since $\frac{1}{2} I\omega^2$ is the rotational kinetic energy of the body, the right side is the increase in the rotational kinetic energy of the body, the right side is the increase in the rotational kinetic energy of the body. Thus, the work done on (by) the body is equal to the increase (decrease) in its rotational kinetic energy

• PROBLEM 18-9

Gear (1) with a mass of 15 kg and a radius of gyration of 250 mm is turned by gear (2). Gear (2) has a mass of 5 kg. and radius of gyration of 65 mm. While the system is at rest a couple M of magnitude 8 N · M is applied to gear (2). There is no friction between the gear teeth. Determine (a) the number of revolutions turned by gear (2) before its angular velocity reaches 550 rpm (b) the force that is tangentically exerted by gear (2) on gear (1).

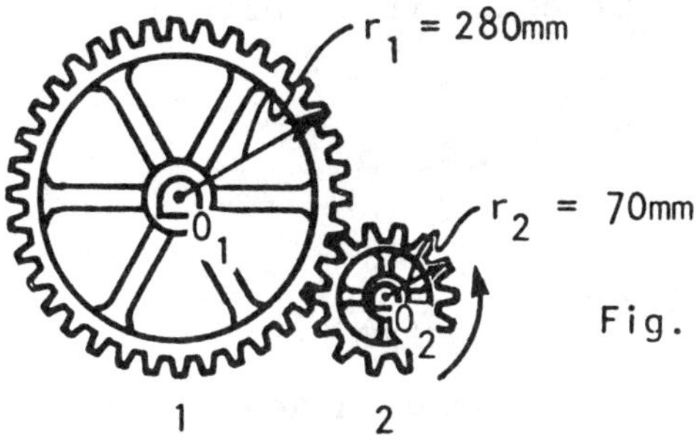

Fig. 1

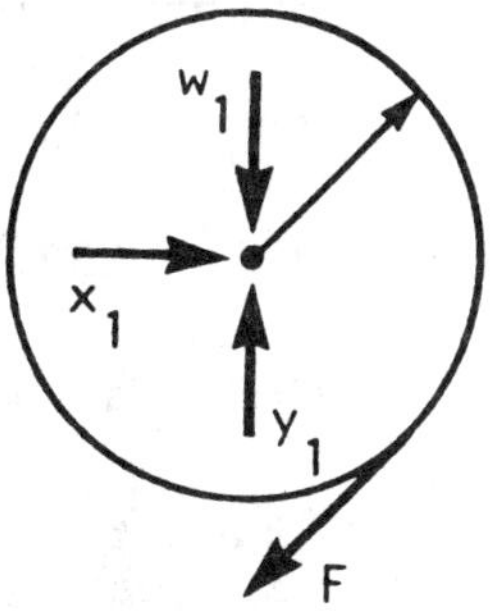

Fig. 2

Solution: The data given for the mass and radius of gyration for each of the wheels allows us to compute the moments of inertia, I_A and I_B.

$$I_A = m_A k_A{}^2 = (10\text{ kg})\ (0.250\text{ m})^2 = 0.9375\text{ kg - m}^2$$

$$I_B = m_B k_B{}^2 = (3\text{ kg})\ (0.065\text{ m})^2 - 0.0211\text{ kg - m}^2.$$

We can also obtain an important kinematic relationship by observing that the tangential speed of the two gears must be equal. Thus

$$r_A \omega_A = r_B \omega_B$$

$$\omega_A = \frac{r_B}{r_A}\omega_B = \frac{70\text{ mm}}{280\text{ mm}}\omega_B = 0.25\ \omega_B. \qquad (1)$$

To answer part (a) we use the work-energy condition which states that the amount of work done to a system is equal to the change in kinetic energy of the system. This is most useful when there are no dissipative forces (neglecting friction) and the work done on the system is easily found. These conditions are satsified here.

The work-energy equation is $u = T_f - T_i$. We have $T = \frac{1}{2} I_A \omega_A{}^2 + \frac{1}{2} I_B \omega_B{}^2$. Since the gears start at rest, $T_i = 0$, and the final state is $\omega_{Bf} = 550$ rpm. That is,

$$\omega_{Bf} = 600 \cdot \frac{1}{60} \cdot 2\pi = 57.59\text{ rad/s},$$

and $\omega_{Af} = 0.25\ \omega_B = 14.39$ rad/s.

This gives us for T_f:

$$T_f = \frac{1}{2}\ (0.9375\text{kg - m}^2)\ (14.39\text{ rad/s})^2 + \frac{1}{2}\ (0.0211\text{ kg - m}^2)$$

$$(57.59\text{ rad/s})^2$$

$$= 132.0\text{ J}$$

The work done which must equal this value is

$$U = M(\theta_{Bf} - \theta_{Bi}) = M\theta_B.$$

$$M\theta_B = 8\theta_B = 132.0 \text{ J}.$$

and $\theta_B = 16.5 \text{ rad} = 2.62 \text{ rev}.$

We can find the tangential force on gear A from gear B if we analyze gear A separately as in figure 2. Again $U = T_f$ ($T_i = 0$), but now $U = (F r_A)\,\theta_A$.

Since $\omega_A = \dfrac{d\theta_A}{dt}$, and $\omega_B = \dfrac{d\theta_B}{dt}$, equation (1) can be re-written as

$$r_A\theta_A = r_B\theta_B.$$

Therefore $U = F(\theta_B r_B) = F(16.5 \text{ rad})(0.07 \text{ m})$

$$= F(1.155\text{m}).$$

Thus $F(1.155\text{ m}) = \frac{1}{2} I_A\omega_{Af}{}^2 = \frac{1}{2}(0.9375 \text{ kg - m}^2)(14.39 \text{ rad/s})^2$

$$F = \frac{97 \text{ J}}{1.155 \text{ m}} = 84.04 \text{ N}.$$

• PROBLEM 18-10

Consider a wheel pulled forward by a force F_1 on its center of mass axis. F_1 is horizontal and the wheel rolls without slipping on a horizontal table.

1) Find the linear acceleration of its center of mass.
2) Show that the work done by the force is equal to the gain in kinetic energy. What does this indicate about the friction causing the rolling?
3) If the static friction coefficient is μ, what is the maximum value possible for the acceleration found in part 1)?

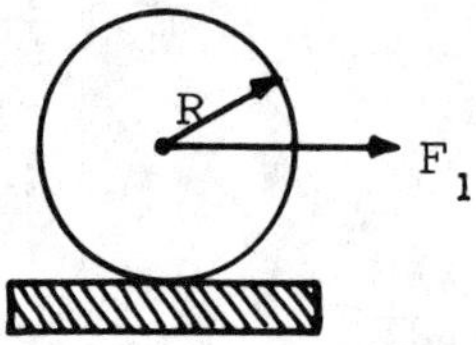

Fig. 1

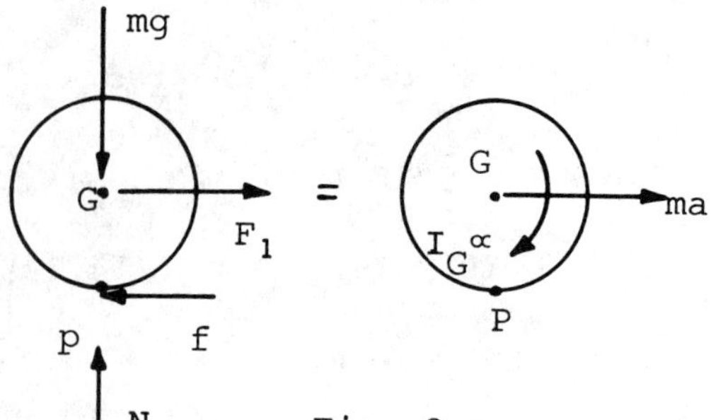

Fig. 2

Solution: First apply D'Alembert's principle to evaluate the acceleration.

By D'Alembert's principle, the external forces are equal to the effective forces (fig. 2). Taking moments about point P

$$\overset{+}{\curvearrowright} \quad \Sigma M_P = \Sigma M_{eff}$$

$$F_1 \ (R) = ma \ (R) + I_G \alpha. \tag{1}$$

For a wheel $I_G = mk_0{}^2$, (2)

where k_0 is the radius of gyration. Also,

$$a = \alpha R \tag{3}$$

since there is only rolling, no slipping.

Putting equations (2) and (3) into (1) yields

$$F_1 \ (R) = ma \ (R) + mk_0{}^2 \ (a/R).$$

$$F_1 \ (R) = \left(mR + \frac{mk_0{}^2}{R} \right) a$$

$$F_1 \ (R) = ma \ R \left(1 + \frac{k_0{}^2}{R^2} \right)$$

And, $$a = \frac{F_1}{m(1 + k_0{}^2/R^2)} \tag{4}$$

There are 2 components of kinetic energy. They are translational and rotational.

$$\text{KE of the wheel} = \frac{1}{2} mv^2 + \frac{1}{2} I_G \omega^2.$$

Now, $$v = at = \left[\frac{F_1}{m(1 + k_0{}^2/R^2)} \right] t$$

and, $$\omega = \frac{v}{R} = \frac{F_1 t}{mR \ (1 + k_0{}^2/R^2)}$$

$$\therefore \quad KE = \frac{1}{2} m \left[\frac{F_1 t}{m \ (1 + k_0{}^2/R^2)} \right]^2 + \frac{1}{2} mk_0{}^2 \left[\frac{F_1 t}{mR \ (1 + k_0{}^2/R^2)} \right]^2.$$

$$KE = \frac{F_1{}^2 \ t^2}{2m \ (1 + k_0{}^2/R^2)^2} + \frac{F_1{}^2 \ t^2 \ k_0{}^2}{2mR^2 \ (1 + k_0{}^2/R^2)^2},$$

$$KE = \frac{F_1^2 t^2}{2m(1 + k_0^2/R^2)^2} + \frac{F_1^2 t^2 k_0^2/R^2}{2m(1 + k_0^2/R^2)^2},$$

$$KE = \frac{F_1^2 t^2 (1 + k_0^2/R^2)}{2m(1 + k_0^2/R^2)^2}$$

$$KE = \frac{F_1^2 t^2}{2m(1 + k_0^2/R^2)} . \tag{5}$$

Equation (5) is the kinetic energy gained by the wheel. Now we must find the work done by the force.

$$W = \vec{F} \cdot \vec{S} = (F_1)(s)$$

s, the displacement, is $\frac{1}{2}at^2$.

$$\therefore \quad \text{Work} = F_1 \left(\frac{1}{2}at^2\right)$$

$$\text{Work} = F_1 \frac{1}{2}\left[\frac{F_1}{m(1 + k_0^2/R^2)}\right] t^2$$

$$\text{Work} = \frac{F_1^2 t^2}{2m(1 + k_0^2/R^2)} \tag{6}$$

This is the same as the kinetic energy (eq. 5).

Conclusion: The friction force causing the rolling does not do any work (there is no displacement involved with this force).

(3) From Fig. (2), set the external forces equal to the effective forces in the horizontal direction

$$\overset{+}{\rightarrow} \quad \Sigma F = \Sigma F_{eff}$$

$$F_1 - f = ma.$$

$$F_1 - \mu N = ma \tag{7}$$

Doing the same for the vertical direction gives

$$+\uparrow \; \Sigma F = \Sigma F_{eff}$$

$$N - mg = 0$$

$$N = mg.$$

So equation (7) becomes

$$F_1 - \mu mg = ma$$

or $$F_1 = ma + \mu mg \tag{8}$$

Taking moments about point P, and setting it equal to the effective moments (see fig. 2) yields

$$+\circlearrowright \ \Sigma M_P = \Sigma M_{eff}$$

$$F_1 (R) = ma (R) + I_G \alpha \tag{9}$$

Substituting equations (2) and (3) for I_G and α into equation (9) we get

$$F_1 (R) = ma (R) + \frac{mk_0^2 a}{R} .$$

Solving for F_1, yields:

$$F_1 = ma + \frac{mk_0^2 a}{R^2} . \tag{10}$$

Equating equations (10) and (8) gives

$$ma + \frac{mk_0^2 a}{R^2} = ma + \mu mg$$

$$a = (\mu mg) \left(\frac{R^2}{mk_0^2} \right)$$

$$a = \frac{\mu g R^2}{k_0^2}$$

• PROBLEM 18-11

A homogeneous bar of weight 8 lb and length 1 ft rotates on a horizontal plane with a constant angular velocity of 5 rad/sec, about the pivoted end O. A constant force K is applied normal to the bar at its free end. Determine (1) the expression for angular position of the rod as a function of t, (2) reactions at end O, and (3) the constant K necessary to stop the bar in one-half a revolution.

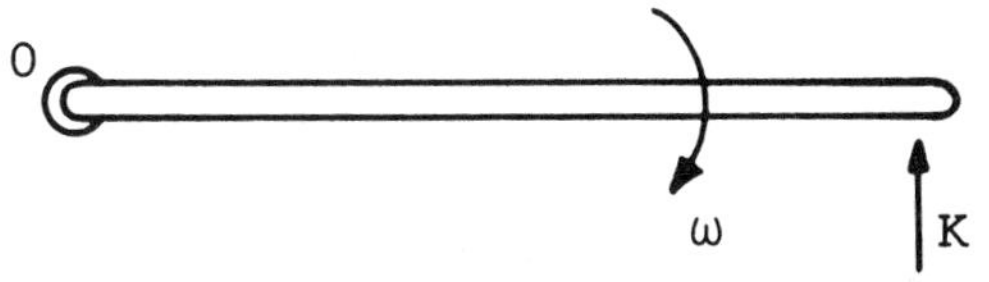

Solution: In order to obtain the expression for the angle ϕ, we must find the angular deceleration of the rod due to force K. Using D'Alembert's principle draw the force and effective force diagrams.

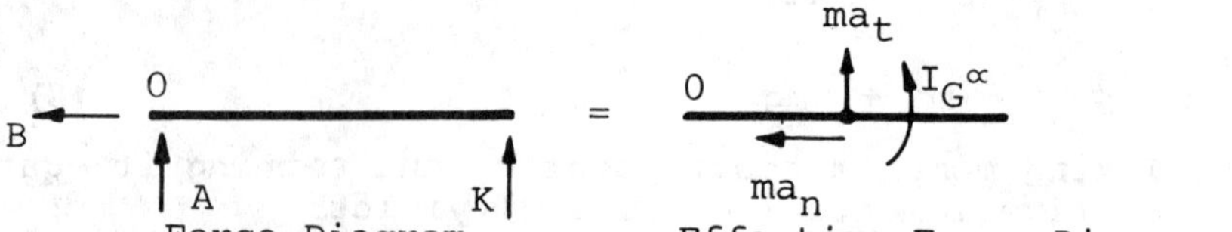

We equate the forces and moment on the bar to obtain, for the radial direction:

$$\Sigma F_r = \Sigma F_{r\ eff},$$

$$B = ma_n.$$

Now, $a_n = r\omega^2 = \frac{\ell}{2}\omega^2$

$$\therefore \quad B = m\frac{\ell}{2}\omega^2. \qquad (1)$$

Equate the forces in tangential direction:

$$\Sigma F_t = \Sigma F_{t\ eff}$$

$$A + K = ma_t.$$

Now, $a_t = r\alpha = \frac{\ell}{2}\alpha$

$$\therefore \quad A + K = \frac{m\ell\alpha}{2}. \qquad (2)$$

Sum the moment of forces at point O:

$$\Sigma M_0 = \Sigma M_{0eff}$$

$$K(\ell) = ma_t(\ell/2) + I_G\alpha.$$

$I_G = \frac{1}{12}m\ell^2$, moment of inertia of the rod about its centroidal axis. Substitute this value of I_G and $a_t = \frac{\ell}{2}\alpha$ into the above equation to obtain:

$$K(\ell) = m\frac{\ell}{2}\alpha(\ell/2) + \frac{1}{12}m\ell^2\alpha.$$

Simplifying and solving for K yields:

$$K = \frac{m\ell\alpha}{3} \qquad (3)$$

or $\alpha = \frac{3K}{m\ell}$. (4)

The equation of angular displacement in a rotational motion is given as

$$\phi = \phi_0 + \omega_0 t + \frac{1}{2}\alpha t^2 \tag{5}$$

For the given problem $\phi_0 = 0$, $\alpha = \frac{3K}{m\ell}$. Note that the force K opposes the motion of the rod hence the acceleration of the rod is negative. Equation (5) can be written as

$$\phi = \omega_0 t - \frac{1}{2}\frac{3K}{m\ell}t^2.$$

The reaction force K has already been found, and is expressed in equation (3). The other reaction force, A, can be found from eq. (2).

$$A + K = \frac{m\ell\alpha}{2}$$

Substituting the known value of K gives

$$A + \frac{m\ell\alpha}{3} = \frac{m\ell\alpha}{2}$$

or $$A = \frac{m\ell\alpha}{6}.$$

Note $A = \frac{m\ell\alpha}{6} = \frac{1}{2}\left(\frac{m\ell\alpha}{3}\right) = K/2$, independent of α.

For the bar to stop in one-half revolution we have $\phi = \pi$, $\omega_f = 0$. Now the expression for angular velocities can be written as

$$\omega_f^2 = \omega_0^2 + 2\alpha(\phi_f - \phi_0).$$

Since $\phi_0 = 0$, $\phi_f = \pi$, $\omega_f = 0$ $\alpha = -\frac{3K}{m\ell}$, we can substitute these values into the above equation, obtaining:

$$0 = \omega_0^2 - 2\frac{3K}{m\ell}(\pi).$$

Solving for K yields:

$$K = \frac{m\omega_0^2\ell}{6\pi}$$

Substituting $m = \frac{8}{32.2}$ slugs, $\omega_1 = 5\frac{\text{rad}}{\text{sec}}$, $\ell = 1$ ft yields

$$K = \frac{8}{32.2}\text{ slugs}\left[5\frac{\text{rad}}{\text{sec}}\right]^2(1\text{ ft}) \times \frac{1}{6\pi}$$

$$K = 0.330\text{ lb.}$$

CONSERVATION OF ENERGY

• PROBLEM 18-12

An 18 lb. homogeneous rod is pivoted as shown. If the bar falls from rest at the given position, what will be its angular speed at the lowest position?

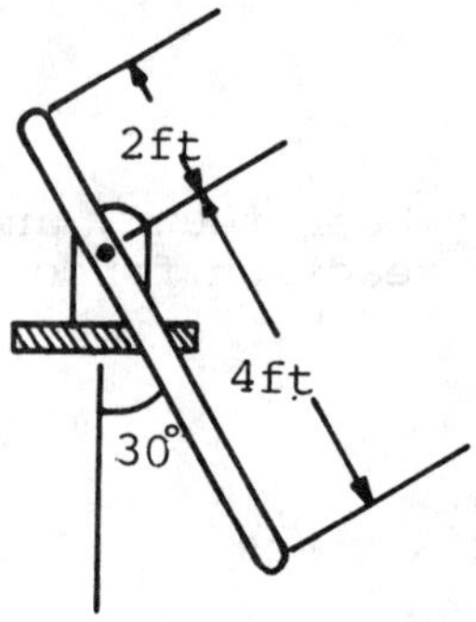

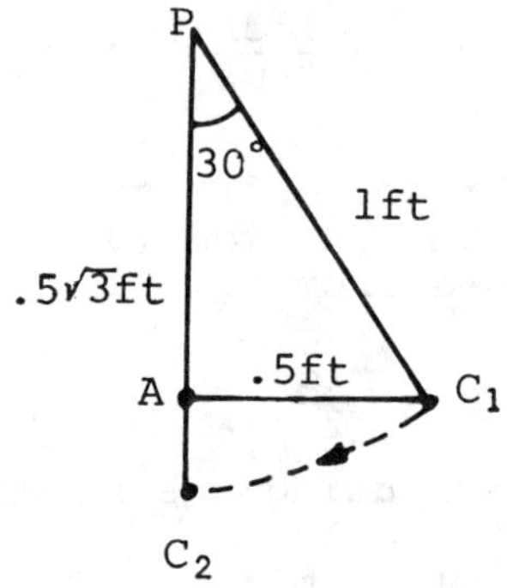

Solution: The problem can be solved by equating the loss of potential energy, mgh, to the gain in the kinetic energy of the system, $\frac{1}{2}I\omega^2$.

The potential energy decreases when the center of gravity of the rod swings from C_1 to C_2. The actual drop in the center of gravity is AC_2. Here, P is the pivot, and $PC_1 = PC_2 = 1$ ft. Since $PA = .5\sqrt{3}$ ft, $AC_2 = 1 - .5\sqrt{3}$ ft. Using the relation PE = mgh, yields

$$PE = mg \cdot (1 - .5\sqrt{3}),$$

where m is the mass of the rod.

On the other hand, kinetic energy is given by $\frac{1}{2}I\omega^2$, where I is the moment of inertia of the rod about the pivot, and ω is its final angular velocity. To find the moment of inertia of the rod, think of it as consisting of two segments 2 ft. and 4 ft. long. The moment of inertia of each segment is given by $\frac{1}{3}M\ell^2$, where M is the mass of each segment and ℓ is the length. The two-foot segment has mass $= \frac{m}{3}$ and the four-foot segment has $\frac{2m}{3}$. Thus,

$$I = \frac{1}{3} \cdot \frac{m}{3} \cdot (2)^2 + \frac{1}{3}\ \frac{2m}{3}(4)^2 = 4m.$$

Therefore, equating $\frac{1}{2}I\omega^2$ to mgh,

$$\frac{1}{2} \cdot 4m\ \omega^2 = mg\ (1 - .5\sqrt{3})$$

or, cancelling m on both sides, and setting g = 32,

$$\omega^2 = \frac{32\ (1 - .5\sqrt{3})}{2} \quad \text{or} \quad \omega = \sqrt{16\ (1 - .5\sqrt{3})}$$

or $\omega = 1.46$ rad/sec.

• PROBLEM 18-13

As shown in the figure, a rod is pivoted freely in vertical plane without friction at end O. Initially the rod is at position 1 at t = 0. Determine all the initial conditions at t = 0 for the rod to attain position 2 with zero velocity.

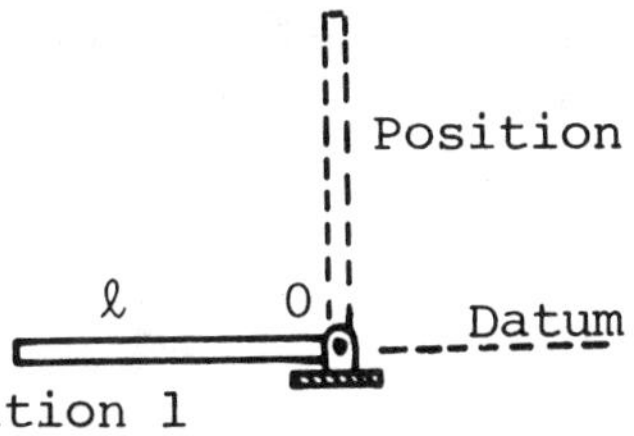

Solution: This problem can be solved by using the principle of work and energy choosing the datum as shown. The initial kinetic energy of the rod changes into potential energy when the rod rises from position (1) to position (2).

Now, $\Delta PE = (mg)\left(\frac{\ell}{2}\right)$

and $\Delta KE = -\left(\frac{1}{2}\,mv^2 + \frac{1}{2}\,I_G\omega_0{}^2\right)$.

Since $\Delta PE = -\Delta KE$, as the energy is conserved, we get

$$\frac{mg\ell}{2} = \frac{1}{2}\,mv^2 + \frac{1}{2}\left(\frac{1}{12}\,m\ell^2\right)\omega_0{}^2.$$

Now $v = \frac{\ell}{2}\,\omega_0$ so we get

$$\frac{mg\ell}{2} = \frac{1}{2}\,m\left(\frac{\ell}{2}\,\omega_0\right)^2 + \frac{1}{24}\,m\ell^2\,\omega_0{}^2$$

$$= \frac{4}{24}\,m\ell^2\,\omega_0{}^2\ .$$

Solving for ω_0 yields:

$$\omega_0 = \sqrt{\frac{3g}{\ell}}\ .$$

At position (1) there is only the vertical component of velocity; the horizontal component is zero. Therefore

$$v_{y_0} = \frac{\ell}{2}\,\omega_0 = \frac{\ell}{2}\sqrt{\frac{3g}{\ell}} = \sqrt{\frac{3g\ell}{4}}$$

$$v_{x_0} = 0.$$

• PROBLEM 18-14

The semi-circle of radius r shown in fig.1 is released from rest. Determine the angular velocity when it falls such that its plane section is parallel to the ground, figure 2. Assume there is no slipping.

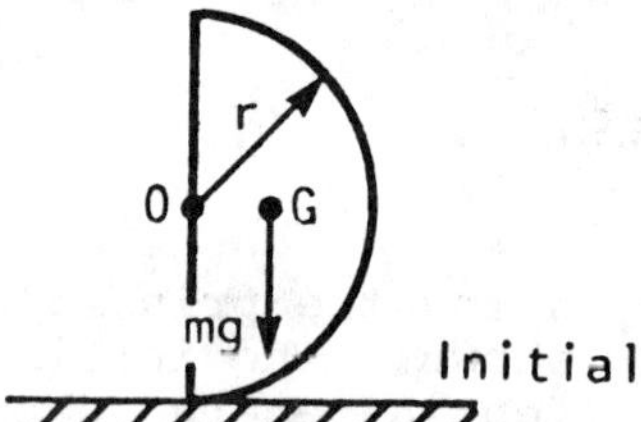

Fig. 1

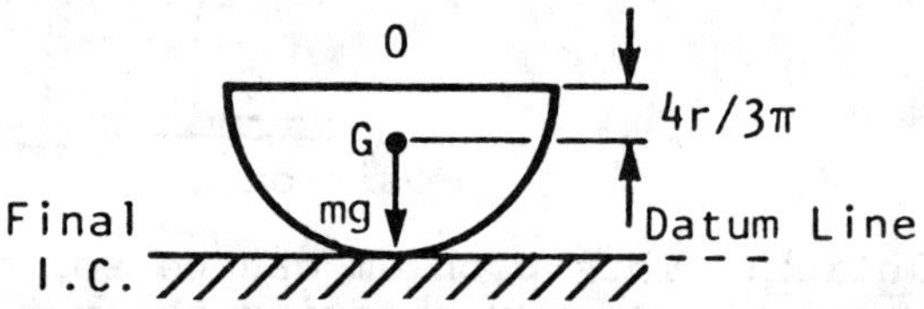

Fig. 2

Solution: This problem is solved by applying the conservation of energy equation. Part of the potential energy of the semi-circle is changed into kinetic energy. Thus, $KE_1 + PE_1 = KE_2 + PE_2$ choosing datum (zero potential energy) as shown,

$PE_1 = mg(r)$, $PE_2 = mg\left(\frac{4}{7}r\right)$, $KE_1 = 0$ since it is originally at rest, and $KE_2 = \frac{1}{2} I_{IC}\omega^2$ where I_{IC} is the moment of inertia of the semi-circle about an axis passing through the instantaneous center of rotation.

From the parallel-axis theorem;

$$I_{IC} = I_G + m\left(\frac{4}{7}r\right)^2,$$

where I_G is the moment of inertia about an axis through the center of gravity. For a semi-circle

$$I_G = 2/5\ mr^2 - m\left(\frac{4r}{3\pi}\right)^2$$

$$I_G = .220\ mr^2$$

Using this value, we can calculate I_{IC}.

$$I_{IC} = .220\ mr^2 + m\,(4/7r)^2 = .795\ mr^2$$

Substituting these values in the energy equation we get:

$$mg(r) = \frac{1}{2}(.795\ mr^2)\omega^2 + mg\left(\frac{4}{7}r\right).$$

Simplifying and solving for ω

$$\omega = \sqrt{\frac{4244\ mgr}{3975\ mr^2}} = \sqrt{\frac{1.06g}{r}}\ \text{rad/sec}$$

• **PROBLEM 18-15**

The 64.4 lb. homogeneous cylinder shown is rolling at 4 rad/sec clockwise when the spring shown is unstretched and attached at the top of the cylinder. What will be the angular speed when the cylinder has rolled without slipping through one-half turn?

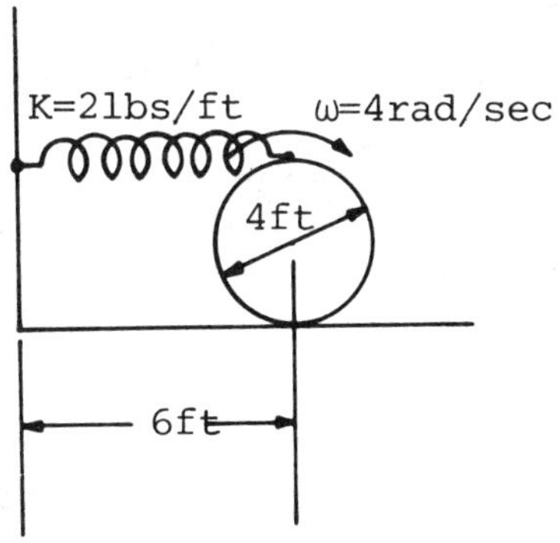

Solution: As the spring stretches, potential energy is stored there. Since there is no slippage, energy is conserved and any reduction in kinetic energy becomes potential energy in the spring. This can be written as follows:

$$\Delta KE = \Delta PE \qquad (1)$$

$$-\frac{1}{2}m\left(v_f^2 - v_0^2\right) = \frac{1}{2}k\left(x^2 - 0^2\right) \qquad (2)$$

The negative sign in equation (2) signifies that if kinetic energy is gained, potential energy is lost, and vice versa.

Since $v = r\omega$, $m = \frac{W}{g}$, and $x = \pi(\text{diam})/2$, we obtain:

$$-\frac{1}{2}\frac{W}{g}\left(r^2\omega_f^2 - r^2\omega_0^2\right) = \frac{1}{2}k\ (2\pi)^2$$

$$-\frac{1}{2}\left(\frac{64.4}{32.2}\right)\left[(2)^2\omega_f^2 - (2)^2\ (4)^2\right] = \frac{1}{2}(2)\ (2\pi)^2$$

$$-4\omega_f^2 + 64 = 4\pi^2$$

$$\vec{\omega}_f = 2.48 \text{ rad/sec clockwise.}$$

• PROBLEM 18-16

Starting from rest at the top, a small sphere rolls without slipping off a large fixed sphere. At what point will the small sphere leave the surface of the big sphere? (See figure).

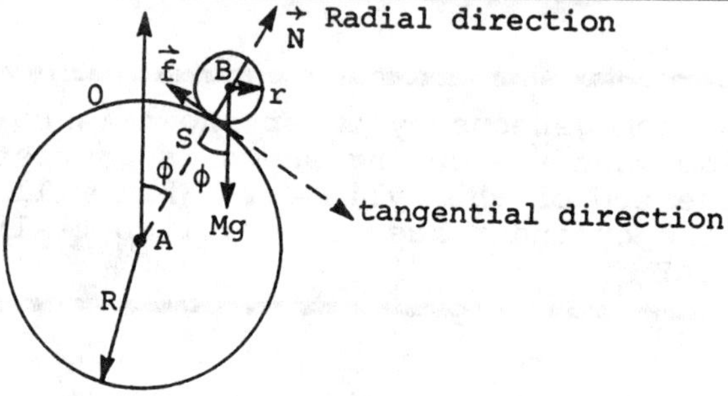

Solution: The small sphere will leave the surface of the large sphere when the normal force ($\vec{N}$) of the latter on the former is zero, for this means that contact has ceased. Applying Newton's Second Law to the tangential and radial components of motion of the small sphere (see figure), we obtain

$$F_{radial} = Ma_{radial} = Mg\cos\phi - N$$

$$F_{tangential} = Ma_{tangential} = mg\sin\phi - f$$

where a_{radial} is positive in the direction of $\vec{BA}$, and $a_{tangential}$ is positive in the direction of motion of the shpere. Furthermore, $a_{radial} = v^2/R+r$, since the small sphere is traveling along a circular arc. (Here, v is the latter's speed). Then

$$\frac{Mv^2}{R+r} = Mg\cos\phi - N$$

or

$$N = Mg\cos\phi - \frac{Mv^2}{R+r}.$$

We require N = 0, or

$$Mg\cos\phi = \frac{Mv^2}{R+r}.$$

Therefore,
$$\cos\phi = \frac{v^2}{(R+r)g}. \qquad (1)$$

We are not yet finished, since we don't know v in (1).

In order to find it, we may use the principle of conservation of energy to relate the energy of the small sphere at points O and S. Since the sphere starts from rest at O, it has only potential energy. Measuring potential energy from A, the energy at O is

$$E = Mg\ (R+r) \qquad (2)$$

At S, the sphere has potential and kinetic energy equal in amount to the energy at O. Hence

$$E = \frac{1}{2}\ Mv^2 + \frac{1}{2}\ I\omega^2 + Mg\ (R+r)\ \cos\ \phi \qquad (3)$$

where the first and second terms are the translational and rotational kinetic energies of the sphere respectively. Equating (2) and (3)

$$\frac{1}{2}\ Mv^2 + \frac{1}{2}\ I\omega^2 + Mg\ (R+r)\ \cos\ \phi = Mg\ (R+r). \quad (4)$$

Since the small sphere rolls without slipping, $v = \omega r$. Using (4), we obtain:

$$\frac{1}{2}\ Mv^2 + \frac{1}{2}\ I\ \frac{v^2}{r^2} = Mg\ (R+r)\ (1 - \cos\ \phi).$$

The moment of inertia of a shpere about its center is

$$I = \frac{2}{5}\ Mr^2.$$

Then $\frac{1}{2}\ Mv^2 + \frac{1}{2}\frac{2}{5}(Mr^2)\ \frac{v^2}{r^2} = Mg\ (R+r)\ (1 - \cos\ \phi)$

$$\frac{1}{2}\ Mv^2 + \frac{1}{5}\ Mv^2 = Mg\ (R+r)\ (1 - \cos\ \phi)$$

$$\frac{7}{10}\ v^2 = g\ (R+r)\ (1 - \cos\ \phi)$$

$$v^2 = \frac{10}{7}\ g\ (R+r)\ (1 - \cos\ \phi). \qquad (5)$$

Utilizing (5) in (1),

$$\cos\ \phi = \frac{10g\ (R+r)\ (1 - \cos\ \phi)}{7\ (R+r)\ g}$$

$\cos\ \phi = \frac{10}{7}\ (1 - \cos\ \phi).$ Solving for ϕ

$\frac{10}{7}\ \cos\ \phi + \cos\ \phi = \frac{10}{7}$ $\qquad \cos\ \phi\ \left(\frac{17}{7}\right) = \frac{10}{7}$

$\cos\ \phi = \frac{10}{17}$ and $\phi = \cos^{-1}\left(\frac{10}{17}\right) \approx 54°.$

A small sphere of radius r and mass m stands on an ideally rough cylinder of radius R. As a result of a disturbance, the sphere loses its equilibrium. Formulate the differential equations of motion. Determine: (1) the normal force N and friction force F as functions of position, (2) the position where the sphere leaves the cylinder.

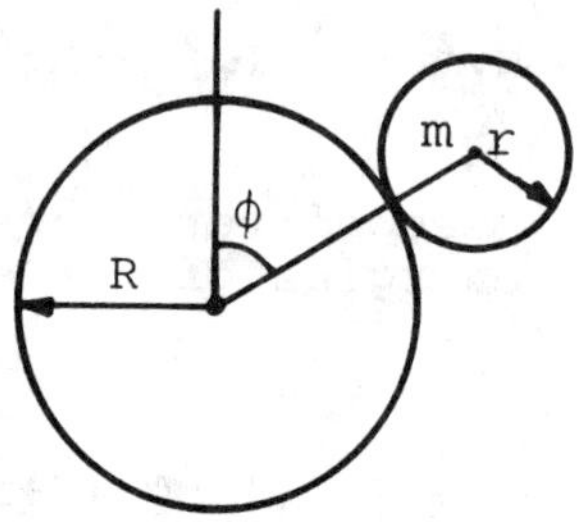

Fig. 1

Solution: In this problem we apply D'Alembert's principle along with the principle of Conservation of Energy. First, we draw the free body diagram and the effective force diagram for the dropping sphere.

By D'Alembert's principle,

$$\Sigma\vec{F}_{real} = \Sigma\vec{F}_{eff}.$$

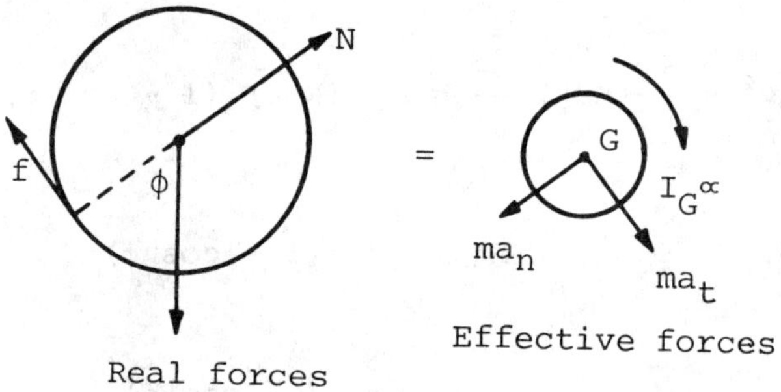

Fig. 2

Summing the real forces in the tangential direction and setting it equal to the effective force yields

$$-f + mg\sin\phi = ma_t$$

$$f = mg\sin\phi - ma_t \qquad (1)$$

Doing the same as we did before, but now in the radial direction yields

$$mg\cos\phi - N = ma_n$$

$$N = mg\cos\phi - ma_n. \qquad 2)$$

The sum of the moments is equal to the effective moment,

$$\overset{+}{\circlearrowright}\ \Sigma M_G = \Sigma M_{eff}$$

$$(f)\ r = I_G \alpha .$$

Now $I_G = \frac{2}{5}\ mr^2$ for a sphere

$$\therefore\ (f)\ r = \frac{2}{5}\ mr^2\alpha$$

$$f = \frac{2}{5}\ mr\alpha . \qquad (3)$$

Noting that the point of contact has a zero velocity and acceleration, $a_t = r\alpha$. Substituting this value of a_t into equation (1) and comparing that equation with equation (3) yields:

$$f = mg \sin\phi - mr\alpha = \frac{2}{5}\ mr\alpha$$

$$\alpha = \frac{5}{7}\ \frac{g \sin\phi}{r} .$$

Substituting this value of α into equation (3) yields:

$$f = \frac{2}{5}\ mr \left[\frac{5}{7}\ \frac{g \sin\phi}{r}\right]$$

$$f = \frac{2}{7}\ mg \sin\phi . \qquad (4)$$

Apply the principle of Conservation of Energy to the sphere. Let the potential energy be,

$$PE = mg\ (R+r) \cos\phi , \qquad (5)$$

and the kinetic energy, $KE = \frac{1}{2}\ I_C\omega^2$, (6)

where I_C is the centroidal moment of inertia about the instantaneous point of contact.

Now by the parallel-axis theorem

$$I_C = \frac{2}{5}\ mr^2 + mr^2$$

$$I_C = \frac{7}{5}\ mr^2 .$$

Substituting this value of I_C into equation (6) gives

$$KE = \frac{7}{10}\ mr^2\omega^2 .$$

Now, use the Conservation of Energy principle with the initial position $\phi = 0$, initial kinetic energy zero, to obtain:

$$\Delta KE = -\Delta PE$$

$$\frac{7}{10} mr^2\omega^2 = mg\ (R+r) - mg\ (R+r)\cos\phi,$$

for any final position at ϕ. Solving for ω^2 yields:

$$\omega^2 = \frac{10}{7} g \frac{(R+r)}{r^2} (1 - \cos\phi).$$

Now, $a_n = r\omega^2$. Substituting this value into equation (6) yields:

$$N = mg\cos\phi - mr\left[\frac{10}{7} g \frac{(R+r)}{r} (1 - \cos\phi)\right]$$

$$= mg\cos\phi - \frac{10}{7} mg \frac{R+r}{r} + \frac{10}{7} mg \frac{R+r}{r}\cos\phi$$

$$= \frac{mg}{7}\left\{\left[7 + 10\frac{(R+r)}{r}\right]\cos\phi - 10\frac{(R+r)}{r}\right\}.$$

When the sphere is about to leave the cylindrical surface the normal force is zero. For N = 0,

$$\left[7 + 10\frac{(R+r)}{r}\right]\cos\phi = 10\frac{(R+r)}{r}$$

$$\phi = \cos^{-1}\left\{\frac{10\frac{(R+r)}{r}}{7 + 10\frac{(R+r)}{r}}\right\}$$

$$= \cos^{-1}\left\{\frac{10R + 10r}{10R + 17r}\right\}.$$

• PROBLEM 18-18

A uniform cylinder of mass m and radius r rolls without slipping inside a spherically curved surface of radius R. Knowing that the cylinder is released from rest at the position shown, derive an expression for a) the linear velocity of the cylinder as it passes through B, and b) the magnitude of the vertical reaction at that instant.

Solution: (a) Since all the forces are conservative forces in the system, the principle of conservation of energy can be applied. For a conservative system,

$$T_1 + V_1 = T_2 + V_2$$

where T and V are kinetic and potential energies, respectively.

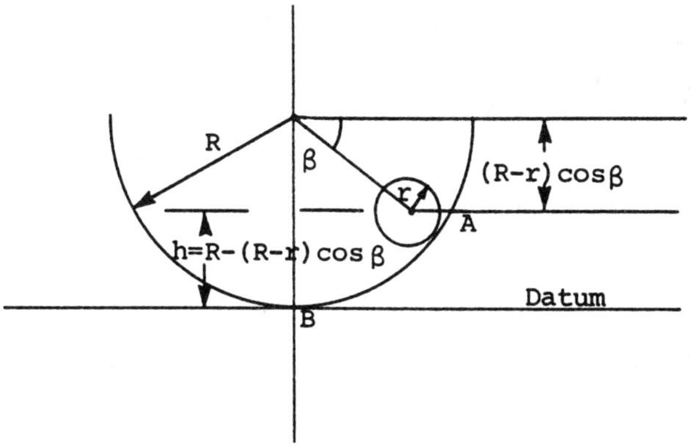

We know that $T_1 = 0$, since the cylinder starts from rest. Therefore, $V_1 = T_2 + V_2$. (1)

Choosing datum (zero potential energy) as shown in the figure, $V_1 = mgh$. Substituting for h yields:

$$V_1 = mg\,[R - (R-r)\cos\beta)].$$

The final kinetic energy of the cylinder is

$$T_2 = \frac{1}{2}mv^2 + \frac{1}{2}I\omega^2 \qquad (2)$$

where v is the linear velocity of the center of mass of the cylinder, ω is the angular velocity of the cylinder, and I is the centroidal moment of the inertia. The cylinder has an instantaneous center of zero velocity at the point of contact because it rolls without slipping. The velocity of any point on the cylinder is

$$v = r\omega \qquad \text{or} \qquad \omega = v/r.$$

The moment of inertia of a cylinder is $I = \frac{1}{2}mr^2$. Putting these values in eq. (2) gives

$$T_2 = \frac{1}{2}mv^2 + \frac{1}{2}\left(\frac{1}{2}mr^2\right)\frac{v^2}{r^2}$$

$$T_2 = \frac{1}{2}mv^2 + \frac{1}{4}mv^2$$

$$= \frac{3}{4}mv^2.$$

At point B the potential energy of the cylinder is $V_2 = mgr$.

Substituting the respective terms for V_1, T_2 and V_2 in equation (1) yields:

$$mg\,[R - (R-r)\cos\beta] = \frac{3}{4}mv^2 + mgr.$$

Dividing both sides by m and rearranging the terms we have

$$g\,[R - (R-r)\cos\beta] - gr = \frac{3}{4}v^2$$

or $$g\,[R - (R-r)\cos\beta - r] = \frac{3}{4}v^2.$$

Factoring yields:

$$g\,[(R-r)(1 - \cos\beta)] = \frac{3}{4}v^2. \qquad (3)$$

Multiplying both sides by $\frac{4}{3}$ and taking the square root we obtain:

$$v = \sqrt{\frac{4}{3}g\,(R-r)(1 - \cos\beta)}.$$

(b) By D'Alembert's principle, the resultant of forces acting on the body is equal to the inertia force on the body, hence

$$F_R - mg = \frac{mv^2}{(R-r)}. \qquad (4)$$

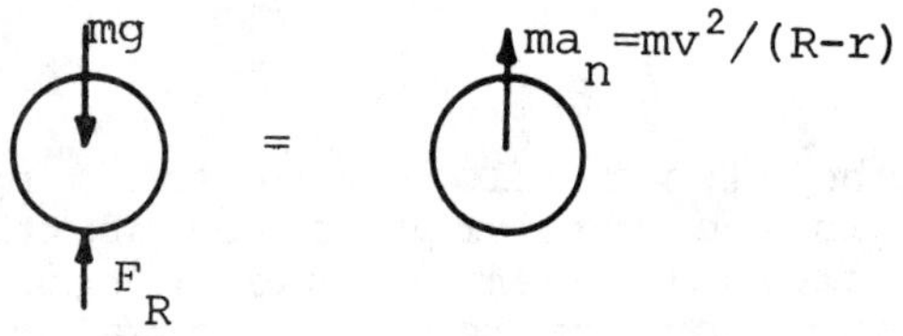

Substituting the value of v^2 from equation (3) into equation (4) we have

$$F_R - mg = \frac{m}{(R-r)}\left[\frac{4}{3}g\,(R-r)(1 - \cos\beta)\right]$$

$$F_R = mg\left(\frac{7}{3} - \frac{4}{3}\cos\beta\right).$$

From which we derive

$$F_R = \frac{mg\,(7 - 4\cos\beta)}{3}.$$

• PROBLEM 18-19

Two disks are connected by a cord. Each disk has a mass of 15 kg and a radius of 0.4m. If the initial angular velocity of the disk 2 is 20 rad/sec clockwise determine the height of disk 1 before the angular velocity of disk 2 is reduced to 5 rad/sec. Assume no loss of energy due to friction.

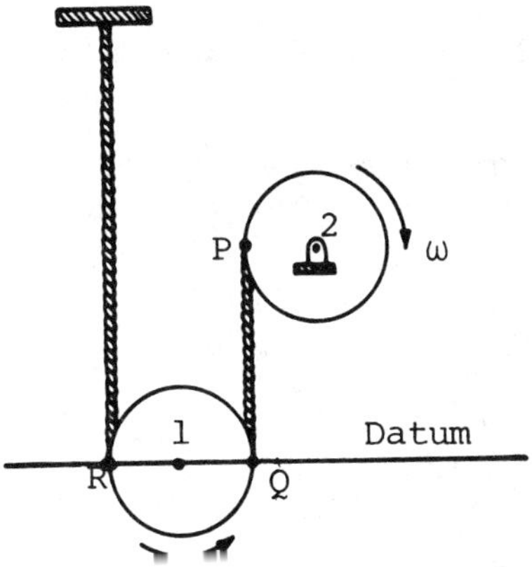

Solution: The system has only conservative forces, therefore the equation of conservation of energy can be applied. The equation is

$$T_1 + V_1 = T_2 + V_2 \tag{1}$$

where T and V are kinetic and potential energies, respectively. Choosing datum as shown, the energies of the two disks can be written as

$$\frac{1}{2} I_2\omega_{2_i}{}^2 + \frac{1}{2} I_1\omega_{1i}{}^2 = \frac{1}{2} I_2\omega_{2f}{}^2 + \frac{1}{2} I_1\omega_{1f}{}^2 + mgh \tag{2}$$

Note that disk A rotates simply about its center but also, at the same time, about point R in the figure.

Now the velocity of disk A at point Q is equal to the tangential velocity of disk B at point P. That is,

$$V_{Qi} = V_{Pi} = (\omega_{1i})(r_1) = (20)(.4) = 8 \text{ m/sec.}$$

Also, $\omega_{1i} = V_{Qi}/(\text{distance between R and Q})$

or $\omega_{1i} = \dfrac{8}{.8} = 10 \dfrac{\text{rad}}{\text{sec}}.$

Similarly, $V_{Q\text{final}} = V_{P\text{final}} = (\omega_{1f})(r_2) = (5)(.4) = 2.0$ m/sec

and $\omega_{1f} = V_{Qf}/(\overline{R\text{-}Q})$

or $\omega_{1f} = \dfrac{2.0}{.8} = 2.5 \dfrac{\text{rad}}{\text{sec}}.$

The mass moment of inertia of a disk is $\frac{1}{2}$ mr^2.

Therefore, $I_{1 \text{ center}} = I_2 = \frac{1}{2}(15)(.4)^2 = 1.2$ kg - m^2

and by the parallel-axis theorem

$$I_1 = 1.2 + mr^2 = 1.2 + (15)(.4)^2 = 3.6 \text{ kg - m}^2.$$

Substituting the above results in equation (2) and dropping the units for simplicity, we get

$$\frac{1}{2}(1.2)(20)^2 + \frac{1}{2}(3.6)(10)^2 = \frac{1}{2}(1.2)(5)^2 + \frac{1}{2}(3.6)(2.5)^2$$

$$+ (15)(9.81)\ h.$$

We can now solve this equation for h to obtain:

$$h = 2.67 \text{ m}.$$

• PROBLEM 18-20

A uniform cylinder rolls from rest down the side of a trough whose vertical dimension y is given by the equation $y = kx^2$. The cylinder does not slip from h_1 to the origin, but the surface of the trough is frictionless from the origin toward h_2. How far will the cylinder ascend toward h_2? Under the same conditions, will a uniform sphere of the same radius go farther or less far toward h_2 than the cylinder?

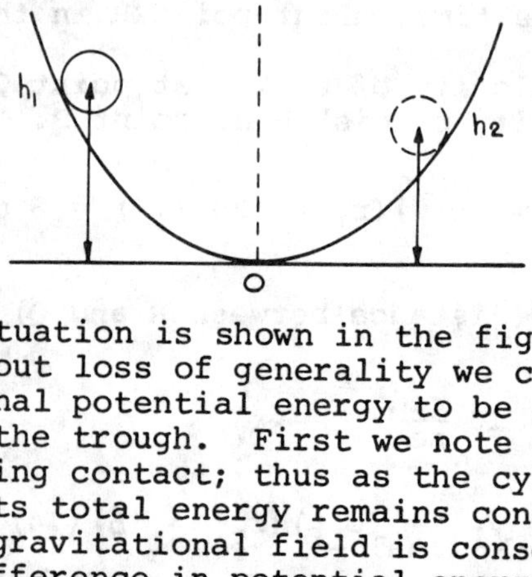

Solution: The situation is shown in the figure. For convenience and without loss of generality we choose the zero of the gravitational potential energy to be at the level of the bottom of the trough. First we note that no work is done at a rolling contact; thus as the cylinder rolls down the trough its total energy remains constant. Second, we note that the gravitational field is conservative and, therefore, the difference in potential energy between any two points in the field is independent of the path between those points. So, the shape of the trough does not matter; only the difference in elevation between the bottom and the positions at h_1 and h_2. Thus, by definition the gravitational potential energy at the origin is zero and at h_1 is given by mgh_1. Since the cylinder is rolling, its kinetic energy just before it reaches the origin is

$$K = \frac{1}{2} mv^2 + \frac{1}{2} I\omega^2 \qquad (1)$$

where v, ω are the velocity and angular velocity at the

origin and I is the moment of inertia of the cylinder. By the condition of rolling $v = \omega r$ and the kinetic energy becomes

$$K = \frac{1}{2} mv^2 \ (1 + I/mr^2)$$

where r is the radius of the cylinder. Since no work is done on the system, the conservation of energy says that the energy at h_1 is equal to the energy at the origin or

$$mgh_1 = \frac{1}{2} mv^2 \ (1 + I/mr^2). \tag{2}$$

Since the cylinder slides up the opposite side the linear kinetic energy at the bottom is converted into potential energy by h_2, that is

$$\frac{1}{2} mv^2 = mgh_2. \tag{3}$$

Substituting equation (3) into equation (2) gives

$$h_1 = (1 + I/mr^2)h_2. \tag{4}$$

Since the moment of inertia of a cylinder is $\frac{1}{2}mr^2$ equation (4) becomes

$$h_1 = \frac{3}{2}h_2 \qquad \text{or} \qquad h_2 = \frac{2}{3}h_1.$$

The cylinder rises to $\frac{2}{3}$ of its original height.

If the object is a sphere the analysis is the same except now $I = \frac{2}{5} mr^2$. Thus,

$$h_2' = \frac{h_1}{1 + I/mr^2} = \frac{h_1}{1 + 2/5} = \frac{5}{7} h_1.$$

Comparing with the result for the cylinder

$$h_2' = \frac{5}{7} h_1 = \frac{15}{21} h_1 \quad \text{or} \quad h_2' > h_2 = \frac{14}{21} h_1.$$

Therefore, the sphere will rise higher than the cylinder. Note that the result is independent of the mass or radius of sphere and cylinder and, therefore, implies that any sphere will rise higher than any cylinder.

• **PROBLEM** 18-21

A 5 ft. long 25-lb thin bar A' is pivoted about a point 0 which is 1 ft. from point A'. At point A' there is a spring with a constant of K = 1900 lb/in, compressed 1.5 in. Just before the bar is released, it is in a horizontal position. Determine the angular velocity of the bar and the reaction at the pivot 0 as the bar passes through a vertical position.

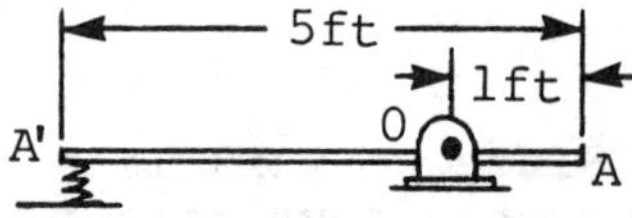

Fig. 1

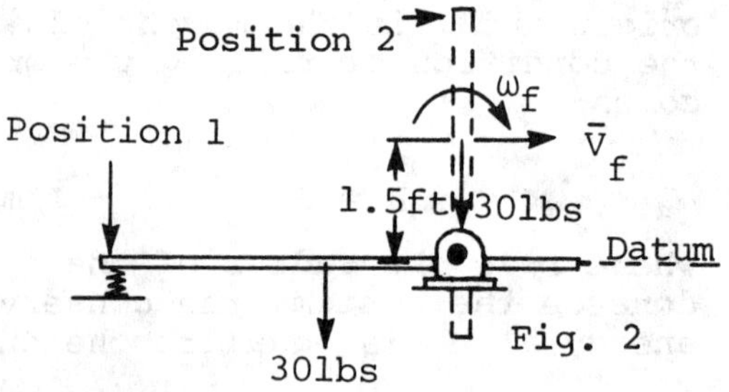

Fig. 2

<u>Solution:</u> **Since the forces involved, gravity and spring, are conservative, we may use the Principle of Conservation of Energy.**

Energy before = Energy after

$$T_i + V_i = T_f + V_f. \qquad (1)$$

The rod starts from rest, so

$$T_i = 0 \qquad (2)$$

V_i is composed of 2 terms, the gravitational and spring potential energies. If the datum of gravitational energy is chosen as the initial horizontal position, the potential due to gravity is 0.

The spring potential is

$$V_i = \frac{1}{2}kx^2 = \frac{1}{2}\left(1900\frac{\text{lb.}}{\text{in.}}\right)\left(1.5\,\text{in.}\right)^2 = 2{,}137.5 \text{ in. lb.} = 178.1 \text{ ft. lb.} \quad (3)$$

The T_f term is composed of 2 terms also. They are the translational and rotational energies.

$$T_f = T_{trans} + T_{rot}$$

$$T_f = \frac{1}{2}mv_f^2 + \frac{1}{2}I\omega_f^2.$$

Because one point of the rod remains stationary (due to the pivot),

$$v_f = \omega_f r = \omega_f\,(1.5 \text{ ft.}).$$

Also, $I = \frac{1}{12}m\ell^2 = \frac{1}{12}\frac{25 \text{ lb.}}{32.2 \text{ ft/s}^2}(5 \text{ ft.})^2$

$$= 1.617 \text{ ft.lb.s}^2$$

Therefore, $T_f = \frac{1}{2}mv_f^2 + \frac{1}{2}I\omega^2$

$$T_f = \frac{1}{2}\left(\frac{25 \text{ lb.}}{32.2 \text{ ft/s}^2}\right)\left[(1.5 \text{ ft})\omega_f\right]^2 + \frac{1}{2}(1.617 \text{ ft.lb.s}^2)\omega_f^2$$

$T_f = (1.682 \text{ ft.lb.s}^2)\omega_f^2.$ (4)

The V_f term has Z components, as did V_i. They are gravitational and spring potential energies.

The spring potential energy is 0 because the spring is unattached to the rod. So it is unstretched.

In the final position, the rod is vertical. Its center of mass is directly over the pivot

$V_f = (mg)\ h = (25 \text{ lb})(1.5 \text{ ft.}) = 37.5 \text{ ft. lb.}$ (5)

Substituting (2), (3), (4), and (5) into (1) yields

$$0 + 178.1 \text{ ft. lb} = (1.682 \text{ ft.lb.s}^2)\omega_f^2 + 37.5 \text{ ft. lb.}$$

$$\vec{\omega}_f = 9.14 \frac{\text{rad}}{\text{s}} \circlearrowright$$

The two components of the acceleration of G are a_N and a_T, the respective radial and tangential accelerations

$$a_N = \omega_f^2 r = \left(9.14 \frac{\text{rad}}{\text{s}}\right)^2 (1.5 \text{ ft.}) = 125.3 \text{ ft/s}^2$$

$$\vec{a}_N = 125.3 \frac{\text{ft}}{\text{s}^2} \downarrow$$

$$a_T = \alpha r$$

$$\vec{a}_T = \alpha r \rightarrow$$

The system of external forces is equivalent to the system of effective forces. The latter can be represented by the vector of components $m\vec{a}_T$ and $m\vec{a}_N$ attached at G, and the couple $I\vec{\alpha}$.

(See fig. 4).

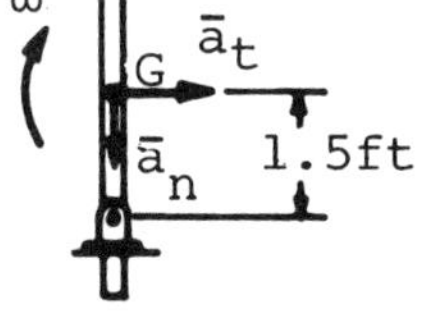

Fig. 3

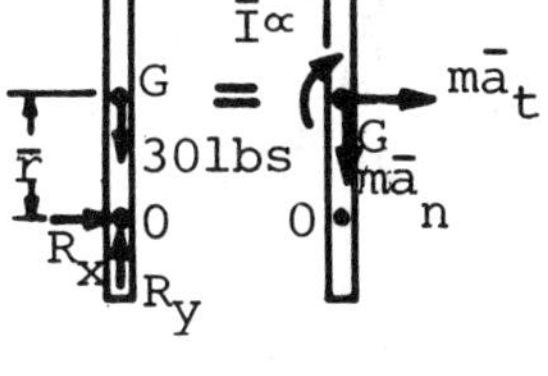

Fig. 4

Taking moments, cw positive

$$\Sigma M_0 = \Sigma (M_0)_{eff}$$

$$0 = I\alpha + m\ (r\alpha) r$$

$$\alpha = 0$$

Summing forces in the x direction gives

$$\Sigma F_x = \Sigma (F_x)_{eff}$$

$$R_x = m\,(r\alpha)$$

$$R_x = 0$$

Finally, equating the external forces to the effective forces give

$$\Sigma F_y = \Sigma (F_y)_{eff}$$

$$R_y - 25 \text{ lb} = -\,ma_N$$

$$R_y - 25 \text{ lb} = -\left(\frac{25 \text{ lb}}{32.2 \text{ ft/s}^2}\right)(125.3 \text{ ft/s}^2)$$

$$R_y = 122.3 \text{ lbs}$$

$$\vec{R}_y = -122.3, \text{ downwards.}$$

• PROBLEM 18-22

A slender homogeneous bar, shown in Figure 1, is 4 ft long and weighs 1.2 lb. A spring is attached to it as shown, 3 ft from the lower end which is hinged. When the bar is in a vertical position the spring is compressed 0.12 ft. The bar is released from rest. What should the spring constant be if the bar comes to rest when it just reaches the horizontal position, Figure 2?

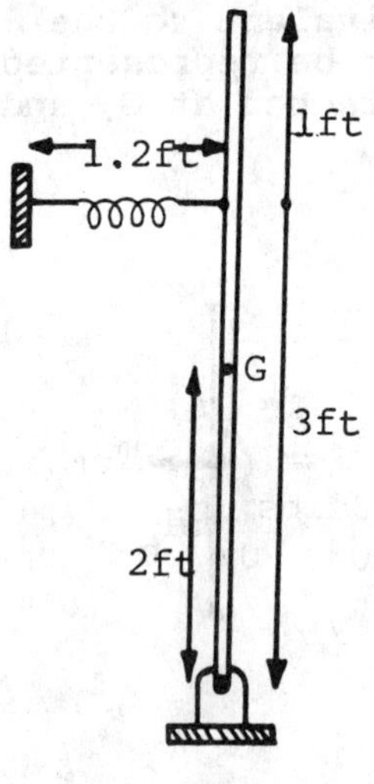

Fig. 1

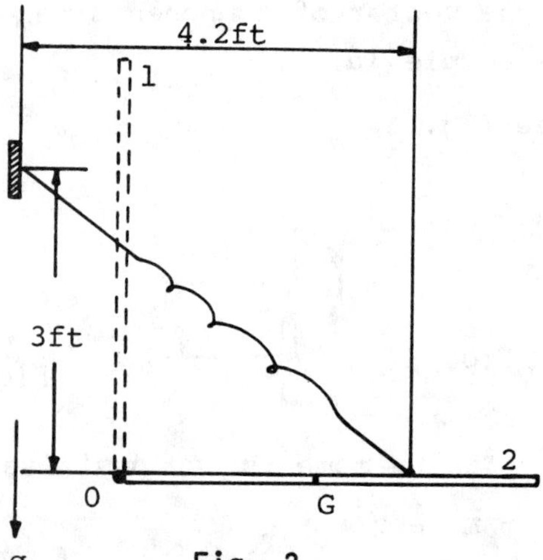

Fig. 2

Solution: The system is conservative since only spring and gravity forces are acting, and if the hinge is assumed to be frictionless. We therefore use the principle of conservation of mechanical energy, i.e.,

$$(KE)_1 + (PE)_1 = (KE)_2 + (PE)_2$$

(Bar in vertical position, 1) (Bar in horizontal position, 2.)

Further, since the bar is at rest in both positions, the above equation reduces to

$$(PE)_1 = (PE)_2 \qquad (1)$$

The potential energy of the system is the sum of that of the spring and the gravitational potential energy. Potential energy of a spring is $\frac{1}{2} K\delta^2$ where K is the spring constant and δ its extension or compression. In position 1, $\delta_1 = 0.12$ ft. Thus the free length of the spring is 1.2 + 0.12 = 1.32 ft. In position 2, the length of the spring is $(4.2^2 + 3^2)^{1/2} = 5.16$ ft. Thus $\delta_2 = 5.16 - 1.32 = 3.84$ ft. The gravitational potential energy of a mass m is mgh where h is the height of its center of gravity above the datum. Let us assume the datum to be the horizontal plane passing through the hinge at O. Thus the total potential energies in the two positions are:

$$(PE)_1 = \frac{1}{2} K (0.12)^2 + (1.2)(2)$$

$$= 0.0072 K + 2.4 \text{ ft-lb}$$

$$(PE)_2 = \frac{1}{2} K (3.84)^2 + (1.2)(0)$$

$$= 7.378 K \text{ ft-lb.}$$

Substituting these values into equation (1) yields

$$0.0072 K + 2.4 = 7.378 K$$

or

$$K = 0.3256 \text{ lb/ft.}$$

Note that this is not the equilibrium position. The bar will oscillate. The bottom position will be an amplitude extreme, where the bar is momentarily at rest. To consider the equilibrium position we need to use force equilibrium.

• PROBLEM 18-23

A child of mass m sits in a swing of negligible mass suspended by a rope of length ℓ. Assume that the dimensions of the child are negligible compared with ℓ. His father pulls the child back until the rope makes an angle of one radian with the vertical, then pushes with a force F = mg along the arc of the circle, releasing at the vertical.

a) How high up will the swing go?

b) How long did the father push?

Compare this with the time needed for the swing to reach the vertical position with no push.

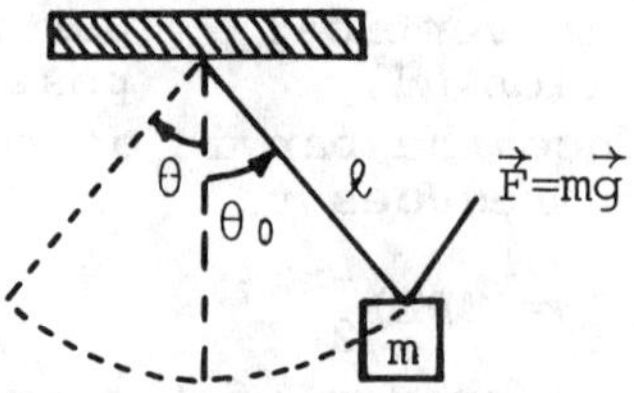

Solution: The father does work in pushing the swing. The work done increases the total energy of the swing and child as the swing moves from its initial position, $\theta_0 = 1$ radian, to the vertical position where the father stops pushing. From that point on, total energy of the swing and child remains constant.

Work done by the father is found from

$$W = \int \vec{F} \cdot d\vec{r} = \int F\, dr \cos\theta \qquad (1)$$

where r is the distance along the path.

$$W = \int_0^{\ell\theta_0} mg \cos(0°)\, dr = mg\ell\theta_0 . \qquad (2)$$

(a) Now we can use conservation of energy to find how high the swing goes. (Let the potential of the swing and child be 0 at the vertical position where the father lets go.)

$$(\text{PE at } \theta_0) + W = \text{PE at } \theta = E_{total} = \text{constant} \quad (3)$$

$$mg(\ell - \ell\cos\theta_0) + mg\ell\theta_0 = mg(\ell - \ell\cos\theta)$$

$$mg\ell(1 - \cos\theta_0 + \theta_0) = mg\ell(1 - \cos\theta)$$

$$1 - 0.54 + 1 = 1 - \cos\theta$$

$$\cos\theta = 0.46$$

$\theta = 63°$ above the horizon.

b) We need to find the tangential acceleration of the velocity of the swing at the bottom of its arc, which can be found by a second application of the conservation of energy principle as follows:

$$PE_{at\ \theta} + W = KE_{at\ bottom} = E_{total} \qquad (4)$$

$$mg\ell(1 - \cos\theta) + mg\ell\theta_0 = \frac{1}{2}mv^2$$

$$mg\ell(1.46) + mg\ell = \frac{1}{2}mv^2$$

$$v^2 = 2\ g\ell\ (2.46) = 4.92\ g\ell \qquad (5)$$

Application of Newton's Law allows us to find the tangential acceleration, a_T.

$$\Sigma F_T = ma_T$$

$$mg + mg \sin \theta_0 = ma_T \qquad (6)$$

But mg sin θ_0, the component of gravity in the tangential direction, does not stay constant. It gets smaller as the angle decreases. Thus, the tangential acceleration is not constant. We can obtain an approximate solution by integrating the gravity force between O and θ_0, and using that equivalent force.

$$\int_0^{\theta_0} mg \sin \theta \, d\theta = - mg \cos \theta \Big|_0^{\theta_0 = 1 \text{ rad}}$$

$$= [- 0.54 + 1]\ mg$$

$$= 0.46\ mg \qquad (7)$$

Substituting this value for mg sin θ_0 in Equation (6) yields

$$mg + 0.46\ mg = ma_T$$

$$a_T = 1.46\ g.$$

Now to find how long the father pushed write

$$a_T = \frac{dv}{dt} = 1.46\ g$$

$$\int_0^v dV = 1.46\ g \int_0^t dt$$

$$t = \frac{v}{1.46\ g}.$$

Using the value of v from Equation (5) yields

$$t = \frac{\sqrt{4.92\ g\ell}}{1.46\ g}$$

$$t = 3.37 \sqrt{\frac{\ell}{g}}\ .$$

The force decelerating the child will be the gravity force. This force will be of the magnitude given in equation (7). Writing F = ma,

$$0.46 \text{ mg} = \text{ma}$$

$$a = 0.46 \text{ g}.$$

The time will again be given by

$$t = \frac{v}{a} = \frac{v}{0.46 \text{ g}}. \tag{8}$$

If there is no push, conservation of energy yields a velocity of $\sqrt{2 \text{ g}\ell (1 - \cos \theta_0)}$. Substituting this in equation (8) yields

$$t = \frac{\sqrt{2 \text{ g}\ell (0.46)}}{0.46 \text{ g}}$$

$$t = 2.09 \sqrt{\frac{\ell}{g}}.$$

• PROBLEM 18-24

Consider the motion of a uniform cylinder of mass M and radius R rolling down a rough, inclined plane whose length is ℓ and whose angle of inclination is θ. Find the linear and the angular acceleration of the cylinder, its energy at the bottom of the plane, and the time taken for it to roll down the plane. Assume that on the rough plane the force of friction between the cylinder and the plane is large enough to cause the cylinder to roll without sliding.

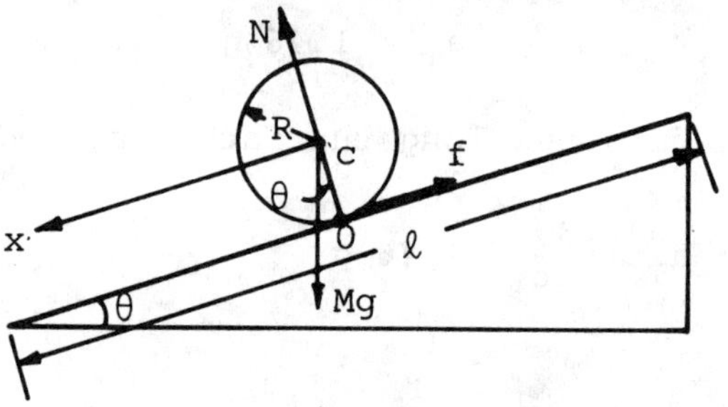

Fig. 1

Solution: This force of friction, shown in Fig. 1 as f, is exerted by the plane on the cylinder and causes the line of contact of cylinder and plane through O to be instantaneously at rest.

If the angular acceleration of the cylinder about the point of contact is α, then the linear acceleration $\ddot{x}$ of the center of mass down the plane is

$$\ddot{x} = \alpha R.$$

If the cylinder is rolling and not sliding, the angular velocity and the acceleration about the axis through O must be respectively equal to the angular velocity and acceleration about the axis through the center of mass C. In order to satisfy yourself on this point, consider the displacement of C and O when the cylinder rolls a short distance down the plane.

To obtain the angular acceleration, use the relationship $\tau_0 = I_0\alpha$ for the instantaneous motion of the cylinder about the axis through O. Thus

$$MgR \sin\theta = I_0\alpha. \qquad (1)$$

The moment of inertia about a horizontal axis through C is the same as that for a circular disk.

$$I_{C.M.} = \frac{MR^2}{2},$$

and by the theorem of parallel axes

$$I_0 = \frac{MR^2}{2} + MR^2 = \frac{3}{2}MR^2.$$

Substituting in Eq. 1 gives the angular acceleration α as

$$\alpha = \frac{2g \sin\theta}{3R}$$

and the acceleration of the center of mass parallel to the plane as

$$\ddot{x} = \alpha R = \frac{2}{3} g \sin\theta. \qquad (2)$$

This result may also be obtained by considering the motion of the center of mass. From the equations of motion for the translational and rotational acceleration of the center of mass, we have

$$Mg \sin\theta - f = M\ddot{x} \qquad (3)$$

and

$$fR = I_{CM}\alpha. \qquad (4)$$

Substituting for α and I_{CM} in Equation (4) yields

$$fR = \left(\frac{MR^2}{2}\right)\left(\frac{\ddot{x}}{R}\right).$$

Hence

$$f = \frac{M\ddot{x}}{2}.$$

Substituting this value for f in the Equation (3), we have

$$M\ddot{x} = Mg \sin\theta - \frac{M\ddot{x}}{2}$$

or $$\ddot{x} = \frac{2}{3} g \sin \theta$$

which agrees with Eq. 2.

Consider now the kinetic energy of the cylinder as it rolls down the plane. Let us assume that the cylinder starts from rest at the top of the plane and, after a time t_f, is at the bottom of the plane where it has an angular velocity of ω_f and a linear velocity of the center of mass equal to v. Thus at $t = 0$, $x = 0$, $\dot{x} = 0$, and $\omega = 0$; at $t = t_f$, $x = \ell$, $\dot{x} = v$, and $\omega = \omega_f$. The velocity v at the bottom of the plane can be obtained from Eq. 2. Multiplying this equation by the identity $\dot{x}\, dt = dx$ gives

$$\frac{1}{2} d\,(\dot{x}^2) = \frac{2}{3} g \sin \theta\, dx.$$

Integrating and applying the boundary conditions gives

$$\frac{v^2}{2} = \frac{2}{3} g\ell \sin \theta. \qquad (5)$$

This result can also be obtained from the conservation of energy equation, which, applied to this problem, gives for the work done and the change in kinetic energy

$$Mg\ell \sin \theta = \frac{1}{2} I_0 \omega_f^{\,2} = \frac{1}{2} I_{C.M.} \omega_f^{\,2} + \frac{1}{2} Mv^2.$$

Here we have taken the work done by the external force $\int \vec{F} \cdot d\vec{r}$ as equal to the loss in potential energy $Mg\ell \sin \theta$. Substituting for I_0 and $I_{C.M.}$ and setting $\omega_f^{\,2} = v^2/R^2$, we have

$$Mg\ell \sin \theta = \frac{3\ MR^2 v^2}{4\ R^2}$$

or $$g\ell \sin \theta = \frac{3}{4} v^2.$$

Thus from the expression for the energy we have

$$v^2 = \frac{4}{3} g\ell \sin \theta$$

which agrees with Eq. 5.

In obtaining the value of the velocity v at the bottom of the plane from the energy equation we assumed that the work done by the external force was $Mg\ell \sin \theta$ and that no work was done by or against the frictional force f. The force of friction f brings the cylinder instantaneously to rest along the line of contact through O. There is no work done by this force of friction since there is no displacement of the cylinder in the direction of the force. If work were done the cylinder would be sliding and not rolling.

The fact that from the two methods shown we obtained equal expressions for v^2 indicates that the friction force in fact does no work.

The time taken for the cylinder to roll down the plane is obtained by integrating Eq. 2 with respect to time. Integrating $\ddot{x} = \frac{2}{3} g \sin \theta$ gives

$$\dot{x} = \frac{2}{3} tg \sin \theta + C_1$$

where C_1 is zero from the boundary conditions. Integrating a second time gives

$$\bar{x} = \frac{t^2}{3} g \sin \theta + C_2$$

where the boundary conditions give $C_2 = 0$ and $\ell = (t_f^{\,2}/3) g \sin \theta$ or

$$t_f = \sqrt{\frac{3\ell}{g \sin \theta}}.$$

CONSERVATION OF MOMENTUM

• PROBLEM 18-25

The large flywheel shown in Fig. 1 weighs 1000 lb and is rotating in a horizontal plane. Its angular velocity is controlled by a rocket attached to the wheel. At t = 0, the rocket is fired and the thrust force exerted by the engine is a function of time, as shown in Fig. 2. What will be the angular velocity (rpm) of the wheel at t = 10 sec if the angular velocity of the wheel at t = 0 is 40 rpm?

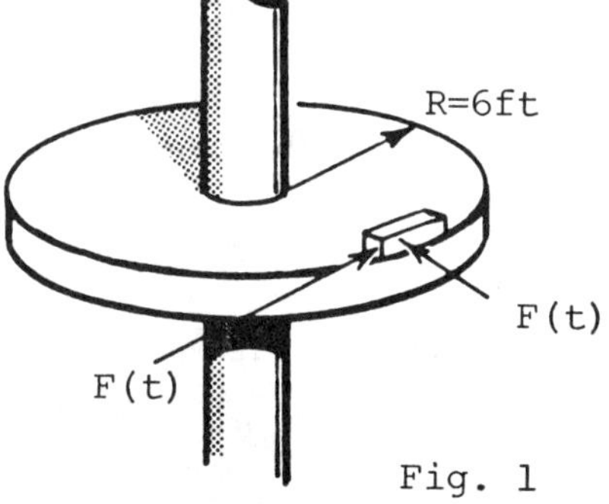

Fig. 1

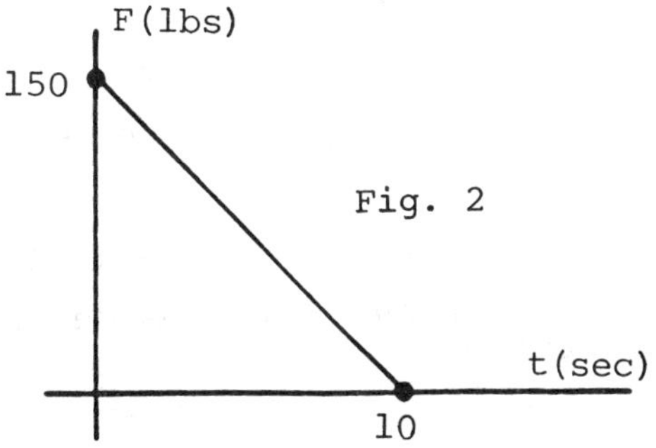

Fig. 2

Solution: Since the force (and thus the moment) is given as a function of time, and since it is desired to determine the change in the angular velocity of a rigid body, it appears that the impulse-momentum equation may be the best approach:

$$\int_A^B M_0 dt = I_0\omega_f - I_0\omega_i .$$

The equation is referred to the fixed point in the system, that is, the axis of rotation, which, incidentally, coincides with the center of mass of the disk in this case.

First of all, the moment of inertia of the body must be evaluated. The moment of inertia of the flywheel about its axis is equal to $\frac{1}{2} mR^2$. Thus,

$$I_0 = \frac{1}{2}\left(\frac{1000}{32.2}\right)(6)^2$$

$$= 559 \text{ lb-sec}^2\text{-ft.}$$

The impulse moment may be written as

$$\int_0^{10} M\, dt = \int F(t) R\, dt \qquad (1)$$

From Figure (2), the thrust force is a linear function of time. Therefore,

$$F(t) = mt + c$$

where m = slope of line

c = intercept on the y axis

thus, $F(t) = -15t + 150$

or $F(t) = 15(10 - t)$

substituting values of F(t) and R into (1), yields,

$$\int_0^{10} M\, dt = 6\int_0^{10} 15(10 - t)\, dt = 90\int_0^{10}(10 - t)\, dt$$

$$= 90\left(10t - \frac{t^2}{2}\right)\Bigg|_0^{10} = 9000 - 4500$$

$$= 4500 \text{ lb-ft-sec.}$$

Hence the impulse-momentum equation for the moment equation becomes

$$4500 = (559)\omega_f - (559)\left(\frac{2\pi}{60}\right)(40).$$

Solving for the angular velocity of the body after 10 sec,

$$\omega_f = 12.24 \text{ rad/sec}$$

or:

$$\omega_f = 117 \text{ rpm.}$$

• PROBLEM 18-26

A single blade of a DC-6 propeller is 2m long and has a mass of 50 kg. Treat this propeller blade as a uniform rigid rod. If this blade is rotated about one end at 2200 rpm, what outward force does it exert on the socket that holds it to the axis of rotation?

Solution: The outward force exerted on the socket is the centrifugal force. This force is equal and opposite to the centripetal force exerted by the socket on the blade to produce the normal acceleration of the blade. The magnitude of this force is given by

$$F = m\,a_n$$

where F is the centrifugal force, m is the mass of the propeller, and a_n is the normal acceleration of the center of the mass (G).

Now, $a_n = r\,\omega^2$

therefore $F = m\,r\,\omega^2.$

Substituting the values given we obtain:

$$F = (50 \text{ kg})\ (1 \text{ m}) \left(2200 \frac{\text{rev}}{\text{min}} \times \frac{1 \text{ min}}{60 \text{ sec}} \times 2\pi \frac{\text{rad}}{1 \text{ rev}}\right)^2 .$$

Simplifying yields:

$$F = 2{,}653.826 \frac{\text{kg - m}}{\text{sec}^2} \quad \text{or}$$

$$= 2.65 \times 10^6 \text{ Newtons.}$$

• PROBLEM 18-27

The 20-lb. ball shown in Fig. 1 is rolled with a velocity of 5.0 fps into the 30-lb. bar that is pinned at its upper end. The rebound velocity of the ball is observed to be 4.0 fps. Through what angle will the bar swing after the impact? The bar is four feet long.

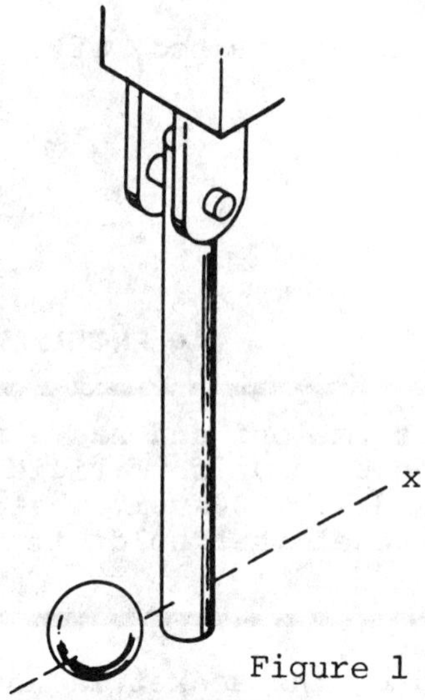

Figure 1

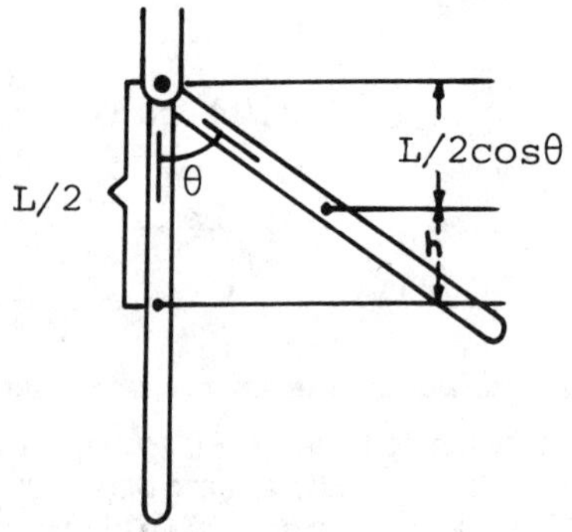

Fig. 2

Solution: Using the impulse-momentum equation for a single particle, the impulse of the bar on the ball may be evaluated:

$$\int_A^B F\,dt = m\dot{x}_B - m\dot{x}_A$$

$$= \frac{20}{32.2}[-5.0 - (4.0)]$$

$$= -5.58 \text{ lb-sec.}$$

The minus sign indicates that the impulse of the bar on the ball is in the negative x direction. By Newton's third law, it follows that the impulse of the ball on the bar is equal to 5.58 lb-sec. By using the impulse-momentum equation for the moment equation, the angular velocity of the bar directly after the impact may be determined as

$$\int_B^C M\,dt = I_o\omega_C - I_o\omega_B.$$

This impulse-momentum equation is referred to the fixed point O. Since the bar is at rest before the impulse, the angular velocity ω_B will be zero. The impulse moment is equal to

$$\int M\,dt = 5.58\ (4)$$
$$= 22.3 \text{ lb-sec-ft.}$$

The moment of inertia of a long slender bar about a point at the end of the bar is equal to $\frac{1}{3}mL^2$. Thus,

$$I_o = \frac{1}{3}\left(\frac{30}{32.2}\right)(4)^2$$

$$= 4.97 \text{ lb-sec}^2\text{-ft},$$

and it follows that $22.3 = 4.97\omega_c$.

Solving for the angular velocity of the bar after the impact,

$$\omega = 4.48 \text{ radians/sec.}$$

The angle through which the bar will rotate may be determined by use of the law of conservation of energy.

$$\begin{matrix}\text{Kinetic Energy} \\ \text{After Impact}\end{matrix} = \begin{matrix}\text{Gravitational Potential} \\ \text{Energy Gained}\end{matrix}$$

$$\frac{1}{2} I_o \omega^2 = \Delta\text{P.E.}$$

The only force on the bar as the bar swings upward is the gravitational force acting through the center of mass of the bar. The motion of the center of mass of the bar is described in Fig. 2. The work done by the gravitational force as the bar moves from its vertical position to some angle θ is

$$\Delta PE = \vec{W}\cdot\vec{h} = -W\left(\frac{1}{2}\right)(1 - \cos\theta)$$

so that the work-energy equation becomes

$$\frac{-30(4)}{2}(1 - \cos\theta) = \frac{-1}{2}(4.97)(4.48)^2$$

and, solving for the cosine of θ,

$$\cos\theta = 0.165.$$

Then solving for θ, it is seen that θ is equal to roughly 80 deg.

• PROBLEM 18-28

A spaceship of cylindrical symmetry is rotating about its cylindrical axis. In order to slow the rate of rotation, two strings of negligible mass with particles of equal mass on the ends are let out slowly and symmetrically on opposite sides of the spaceship in such a way that the strings stay nearly radial. After the strings are let out to full length, the particle on the end of each string rotates about the axis of the ship at the same angular rate as the ship rotates. If each particle has a mass of 10 kg and the moment of inertia of the spaceship is 1000 kg - m^2, what must be the length of each string in order that the final angular velocity of the spaceship be 1 percent of its initial angular velocity? Assume that the particles were initially a perpendicular distance of 2 m from the axis of the spaceship. Is angular momentum conserved if both strings are cut simultaneously?

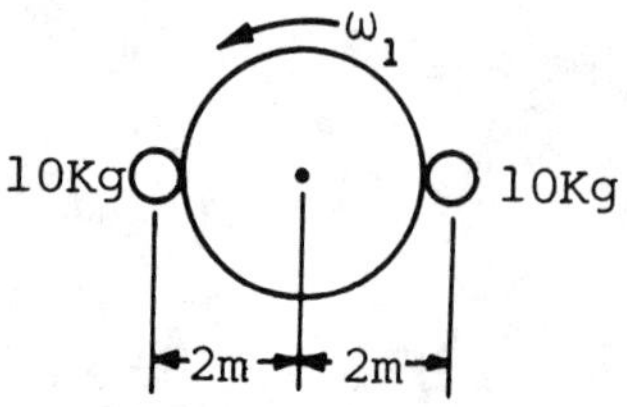

Fig. 1 Position 1

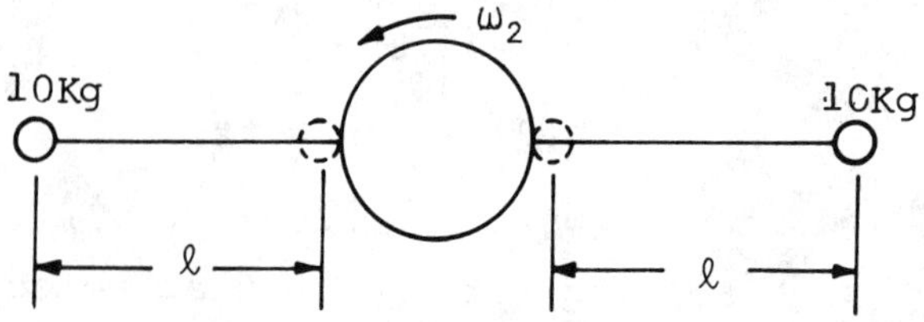

Fig. 2 Position 2

Solution: The angular momentum of the system is constant because no external moments are applied. All applied forces are internal to the system. The angular momentum of the spacecraft alone or of the 10 kg masses alone is not, of course, conserved.

Applying conservation of angular momentum,

$$I_{total_1} \omega_1 = I_{total_2} \omega_2. \tag{1}$$

Now, the total mass moments of inertia of the system in Fig. 1 is

$$I_{total_1} = I_s + 2\ mr^2$$

where I_s is the moment of inertia of the spacecraft alone, m is a particle mass of 10 kg and r is 2m. So, from the given data

$$I_{total_1} = 1000(kg - m^2) + 2\ (10\ kg)(2m)^2$$

$$= 1080\ kg - m^2. \tag{2}$$

For Fig. 2 we have a total moment of inertia of

$$I_{total_2} = I_s + 2m\ (r + \ell)^2$$

where I_s, m and r are as described above and ℓ is the length of string released on each side

$$I_{total_2} = 1000\ kg - m^2 + 2\ (10\ kg)(2m + \ell)^2. \tag{3}$$

Now, in Equation (1) it is desired that

$$\omega_2 = \frac{\omega_1}{100}.$$

Making this substitution in (1) and canceling ω_1, we get

$$I_{total_1} = I_{total_2}/100 \tag{4}$$

Substituting (2) and (3) into (4) results in

$$1080 = \frac{1000 + 20\ (2 + \ell)^2}{100}$$

$$107000 = 20\ (2 + \ell)^2$$

$$2 + \ell = 73.14$$

or, $\ell = 71.14$ m.

In other words, when 71.14 m of string have been released on each side and the masses are "straight out" on radial lines, then the final angular velocity of the spacecraft will be 1% of the initial velocity. When the strings are cut the masses still have angular momentum so the total angular momentum is conserved. The angular momentum of just the spaceship alone is reduced.

• PROBLEM 18-29

A meter stick whose mass is 150 gm can rotate about an axis through one end, O. Suppose that the stick starts from a vertical position and falls to the horizontal position where it strikes a fixed inelastic rod at a distance of 60 cm from the pivot. Find the impulse of the blow on the fixed inelastic rod and the impulse of the reaction at the support O.

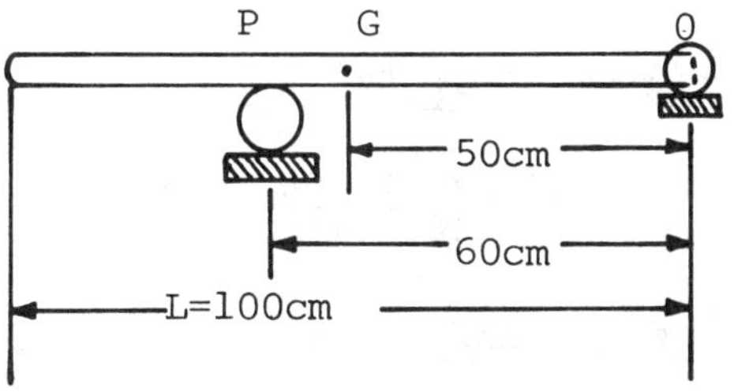

Fig. 1

Solution: As the stick falls it gains momentum. When it hits, all the momentum is transferred from the stick. The principles of conservation of energy and conservation of momentum will be used to solve this problem. The energy of a system minus any work done is constant and the momentum of a system minus any impulse is constant. We apply the principle of impulse and momentum to the stick, knowing that the final momentum of the stick is zero since it strikes a fixed inelastic rod. The equation of impulse and momentum is:

$$\text{momentum}_1 - \text{impulse} = \text{momentum}_2 \quad (\text{momentum}_2 = 0)$$

Diagramatically,

Take moments about point O

$$\circlearrowleft + \Sigma M_0: \quad I\omega_1 + mv_1\left(\frac{L}{2}\right) - F\Delta t\ (.6L) = 0 \qquad (1)$$

Now, $V_1 = \frac{L}{2}\,\omega_1$, $I = \frac{1}{3}mL^2$. Substitute these values into equation (1):

$$\frac{1}{3}mL^2\omega_1 + m\frac{L}{2}\omega_1\frac{L}{2} - F\Delta t\,(.6)L = 0.$$

Solve for impulse $F\Delta t$:

$$F\Delta t = \frac{7}{12}mL^2\omega_1 \times \frac{1}{.6L} \qquad (2)$$

Now we apply the Principle of Conservation of Energy to calculate the angular velocity of the rod at the impact. We write

$$\Delta PE = \Delta KE$$

$$mg\left(\frac{1}{2}L\right) = \frac{1}{2}I_0\omega_1^{\ 2} + \frac{1}{2}mv^2 \qquad (3)$$

where $I_0 = \frac{1}{3}mL^2$. Substituting I_0, $g = 9.81\ \frac{m}{s^2}$, and the given data into eq. (3) we get

$$\frac{mgL}{2} = \frac{1}{2} \times \frac{1}{3}mL^2\omega^2 + \frac{1}{2}m\omega^2\frac{L^2}{4}$$

$$g = \frac{1}{3}L\omega^2 + \frac{1}{4}L\omega^2 = \frac{7}{12}L\omega^2$$

$$\omega^2 = \frac{12g}{7L} = \frac{12\ (9.81\ m/s^2)(100\ cm/m)}{7\ (100\ cm)} = 16.817/sec^2$$

$\omega = 4.1$ rad/sec. And, from equation (2),

$$\text{Impulse} = F\Delta t = \frac{1}{.6} \times \frac{7}{12}(150)(100)(4.1) = 59800 \text{ gm cm/sec.}$$

Write the linear impulse and momentum equation:

$$+\!\downarrow \qquad mv_1 - F\Delta t - O_y\Delta t = 0.$$

Substituting $v_1 = \frac{L}{2}\omega_1$ and $F\Delta t = 59800$, and rewriting the above equation, we have

$$(150)\left(\frac{100}{2} \times 4.1\right) - 59800 = O_y\Delta t$$

$$\therefore\ O_y\Delta t = -\ 29050\ \frac{gm - cm}{sec}.$$

The negative sign indicates that the reaction O_y acts in the opposite direction from that drawn in figure (2).

Since there is no momentum in the horizontal direction, $O_x = 0$, hence $O_x\Delta t = 0$.

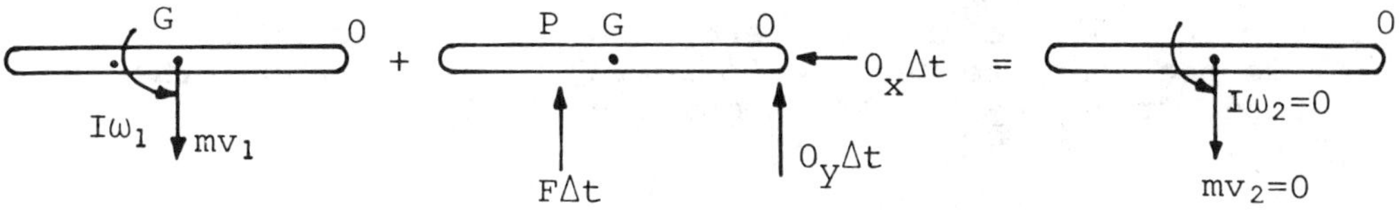

Fig. 2

Therefore, the impulse at O is

$$O\Delta t = 29050 \ \frac{\text{gm - cm}}{\text{sec}} \downarrow .$$

• **PROBLEM 18-30**

Consider the earth to be a uniform solid sphere. If all of the mass of the earth were suddenly moved radially into a thin-walled hollow spherical shell of the same radius, how long would it take the new earth to rotate on its axis? What fraction of the initial kinetic energy of rotation would be lost in this process?

Solution: Since there are no external forces or torques acting on the system, angular momentum is conserved. The angular momentum, L, of the earth is

$$L = I\omega \qquad (1)$$

where I is the moment of inertia and ω the angular velocity of the earth. By the conservation of angular momentum

$$L_1 = L_2 \qquad (2)$$

where L_1 is the angular momentum before the change to the shell and L_2 the angular momentum afterwards. Substitution of equation (1) into equation (2) gives

$$I_1 \omega_1 = I_2 \omega_2 .$$

Solving for ω_2 gives the angular velocity after the change:

$$\omega_2 = \omega_1 \frac{I_1}{I_2} . \qquad (3)$$

The moment of inertia of a uniform sphere of mass M and radius R is $I_1 = \frac{2}{5} mR^2$ while that of a shperical shell of the same mass and radius is $I_2 = \frac{2}{3} mR^2$. Thus equation (3) becomes

$$\omega_2 = \omega_1 \frac{\frac{2}{5} mR^2}{\frac{2}{3} mR^2} = \frac{3}{5} \omega_1 . \qquad (4)$$

To find the time of revolution, i.e., the period T, it is necessary to use the following relations among the period, frequency, ν , and angular velocity ω.

$$\omega = 2\pi\nu, \quad T = \frac{1}{\nu}$$

Using these in equation (4) gives

$$\frac{1}{T_2} = \frac{3}{5}\frac{1}{T_1} \quad \text{or} \quad T_2 = \frac{5}{3} T_1.$$

Since the original period $T_1 = 24$ hr, the final period is 40 hours. The second part of the problem is to determine the kinetic energy lost in the process. The kinetic energy of a rigid body is $K = \frac{1}{2} I\omega^2$. Letting K_1 be the initial kinetic energy and K_2 the final, the fraction of kinetic energy lost in the process , f_1, is

$$f = \frac{K_1 - K_2}{K_1} = 1 - \frac{K_2}{K_1} = 1 - \frac{I_2\,\omega_2^{\;2}}{I_1\,\omega_1^{\;2}}.$$

From equation (4) and the definitions of the moments of inertia given above one obtains

$$f = 1 - \frac{\frac{2}{3}}{\frac{2}{5}}\left(\frac{3}{5}\right)^2 = 1 - \frac{3}{5} = \frac{2}{5}.$$

Thus $\frac{2}{5}$ of the initial kinetic energy is lost in the process.

• PROBLEM 18-31

The vertical tail propeller on a helicopter prevents the cab from rotating when the speed of the main propeller changes. Knowing that the cab has a centroidal moment of inertia of 700 lb · ft · sec^2 and that the four main blades are 14 ft slender rods weighing 55 lbs, determine the final angular momentum of the cab after the speed of the main blades changes from 200 to 280 rpm. Assume the tail propeller of the helicopter to be inoperational.

Solution: If we consider the helicopter to be one system, there is no external momentum applied to it. Therefore, the angular momentum is conserved. The equation for conservation of angular momentum is

$$I_P\omega_{P1} + I_C\omega_{C1} = I_P\omega_{P2} + I_C\omega_{C2} \qquad (1)$$

where I_P, I_C are moments of inertia of the propeller and the cab, respectively. And ω_P, ω_C are the angular velocities of the propeller and the cab, respectively. Since the initial angular velocity of the cab is zero, equation (1) can be written as

$$I_P\omega_{P1} = I_P\omega_{P2} + I_C\omega_{C2}.$$

Solving for ω_{C2}, yields:

$$\omega_{C2} = \frac{I_P\omega_{P1} - I_P\omega_{P2}}{I_C}$$

or,
$$\omega_{C2} = \frac{I_P\ (\omega_{P1} - \omega_{P2})}{I_C}. \qquad (2)$$

Now, the absolute angular velocity of the propeller can be written in terms of the angular velocity of the cab as follows:

$$\omega_{P2} = \omega_{C2} + \omega_{P2/C2}$$

where $\omega_{P2/C2}$ is the angular velocity of propeller with respect to the cab. Note, $\omega_{C1} = 0$ therefore $\omega_{P1} = \omega_{P1/C2}$

Substitute the above expression in equation (2) yeilds

$$\omega_{C2} = \frac{I_P\ [\omega_{P1} - (\omega_{C2} + \omega_{P2/C2})]}{I_C}.$$

Multiply both sides by I_C and simplify:

$$\omega_{C2}I_C = I_P\omega_{P1} - I_P\omega_{C2} - I_P\omega_{P2/C2}$$

Rearrange terms and factor:

$$\omega_{C2}\ (I_C + I_P) = I_P\ (\omega_{P1} - \omega_{P2/C2})$$

$$\therefore\ \omega_{C2} = \frac{I_P\ (\omega_{P1} - \omega_{P2/C2})}{I_C + I_P}. \qquad (3)$$

Now, the moment of inertia of the propeller about the axis of rotation is

$$I_P = \frac{1}{12}\ m\ell^2 + m\ (\ell/2)^2$$

substituting the values, yields:

$$I_P = \frac{1}{12}\frac{55 \text{ lb}}{32.2 \text{ ft/sec}^2}(14 \text{ ft})^2 + \frac{55 \text{ lb}}{32.2 \text{ ft/sec}^2}(14/2 \text{ ft})^2$$

$$= 27.89 \text{ lb·ft·sec}^2 + 83.69 \text{ lb·ft·sec}^2$$

$$= 111.58 \text{ lb·ft·sec}^2.$$

For 4 propellers, the total moment of inertia is

$$I_P = 4 \times 11.58 \text{ lb·ft·sec}^2$$

$$= 446.3 \text{ lb·ft·sec}^2.$$

substituting this and the other values given, equation (3) yields:

$$\omega_{C2} = \frac{446.3 \text{ lb·ft·sec}^2 \ (200 \text{ rpm} - 280 \text{ rpm})}{(700 + 446.3) \text{ lb·ft·sec}^2}$$

$$= -31.15 \text{ rpm}. \quad \text{(Opposite to the blade's rotation.)}$$

Now, the angular momentum of the cab can be found as

$$I_C\omega_{C2} = (700 \text{ lbs·ft·sec}^2)\left(31.15 \text{ rpm} \times \frac{2\pi \text{ rad}}{60 \text{ sec}}\right)$$

$$=2283.2 \text{ lbs.ft.sec}$$

• PROBLEM 18-32

Two gears 1 and 2 have a mass of 15 Kg and 5 Kg and a radius of gyration of 250 mm and 65 mm respectively. A couple M of magnitude 8 N · M is applied to gear 2 when the system is at rest. Neglecting external friction, determine (a) the time required for the angular velocity of gear 2 to reach 550 rpm, (b) the tangential force which gear 2 exerts on gear 1.

This 'Problem' has been previously considered in example 18-9. In this case the principle of impulse and momentum is used as the method of solution.

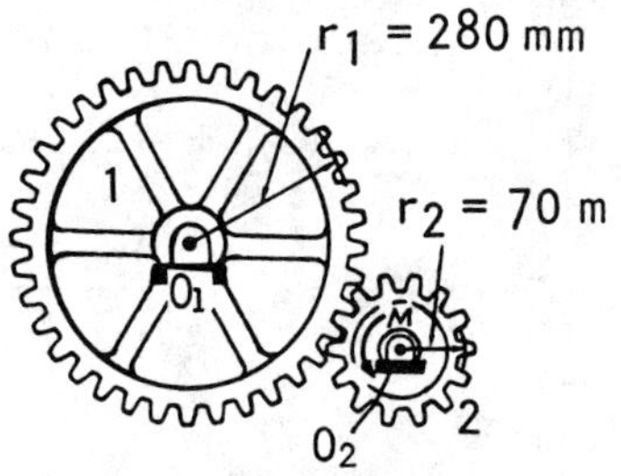

Solution: We apply the principle of impulse and momentum to each gear separately. Since all forces and the couple are constant, their impulses are obtained by multiplying them by the unknown time t. The centroidal moments of inertia are given by $I = m\,k^2$ where k is the radius of gyration:

$$I_1 = 15\ \text{kg}\ (.25\ \text{m})^2 \qquad I_2 = 5\ \text{kg}\ (.065\ \text{m})^2$$

$$I_1 = .937\ \text{kg} - \text{m}^2 \qquad I_2 = .021\ \text{kg} - \text{m}^2$$

The final angular velocities are:

$$(\omega_1)_f = \frac{r_2}{r_1}\ (\omega_1)_f \qquad (\omega_2)_f = 550\ \text{rpm} \times \frac{.1047\ \text{rad/sec}}{\text{rpm}} = 57.58$$

$$(\omega_1)_f = \frac{70\ \text{mm}}{280\ \text{mm}}\ (57.58)\ \text{rad/sec}\ (\omega_2)_f$$

$$(\omega_1)_f = 14.39\ \frac{\text{rad}}{\text{sec.}}$$

Principle of Impulse and Momentum for Gear 1: The systems of initial momenta, impulses, and final momenta are shown in three separate sketches.

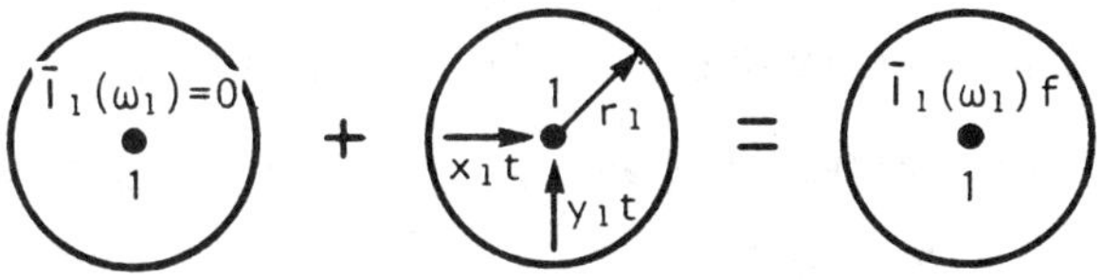

Using the angular momenta form of the Impulse-momentum equation we have

(Angular Momentum)$_1$ + External Impulsive Torque

= (Angular Momentum)$_2$

$$\circlearrowleft+\ 0 + T = -\ I_1 (\omega_{1f})_2$$

The impulsive torque is given by

$$T = -\ Ftr_1$$

where again, the positive direction is counterclockwise. Substituting values into the impulse momentum equation yields

$$Ft = (0.937\ \text{kg}\cdot\text{m}^2)(14.4\ \text{rad/s})$$

$$Ft(.280\ \text{m}) = 48.189\ \text{N}\cdot\text{s}$$

Principle of Impulse and Momentum for Gear 2:

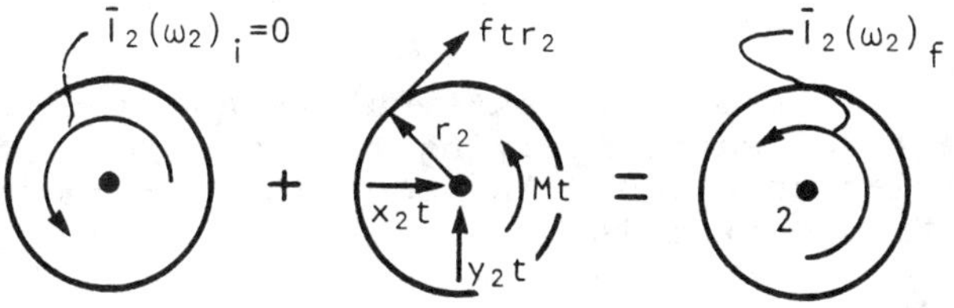

Next, apply the impulse-momentum equation to gear 2. Gear 2 has two impulsive torques, one from the force F, and one from the applied couple M. The sum of these impulsive torques is

$$Mt - Ftr_2$$

where again we use a counterclockwise torque as positive. We can substitute this in the impulse-momentum equation and use the value of Ft we obtained for Gear A, since this force is equal in magnitude and opposite in direction to the other force by Newton's Third Law.

$$\text{Syst Momenta}_i + \text{Syst Ext Imp}_{i-f} = \text{Syst Momenta}_f$$

$\circlearrowleft +$ moments about 2. $\quad 0 + Mt - Ftr_2 - \bar{I}_2\omega_{2f}$

$$+(8\ \text{N}\cdot\text{m})t - (48.19\ \text{N}\cdot\text{s})(0.070\ \text{m}) = (.021\ \text{kg}\cdot\text{m}^2)(57.6\ \text{rad/s})$$

$$t = .5728\ \text{sec.}$$

Recalling that $Ft = 48.19\ \text{N}\cdot\text{s}$, we write

$$F(.5728\,\text{s}) = 48.19\,\text{N}\cdot\text{s} \qquad F = +84.13\,\text{N}$$

Thus, the force exerted by gear 2 on gear 1 is $\vec{F} = 84.13\,\text{N}$

• PROBLEM 18-33

Two cylinder disks rotate on frictionless bearings whose axes lie along the same line perpendicular to the faces, as shown in the Figure. Initially the disks rotate with different angular velocities. The disks are now brought together so that their opposing faces come in contact and eventually they rotate together with the same angular velocity. If no external torques are applied during this process, what is the final common angular velocity of the two disks? What fraction of the initial kinetic energy of rotation is lost to friction during this process?

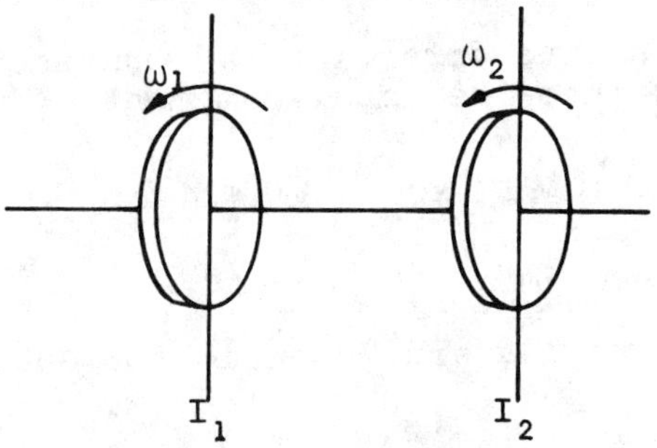

Solution: Since there is no external torque applied, this problem is best solved by applying the principle of conservation of angular momentum. The equation for the conservation of angular momentum is

$$I_1\omega_1 + I_2\omega_2 = I_1\omega'_1 + I_2\omega'_2. \qquad (1)$$

The left-hand side of the equation represents the initial angular momentum and the right-hand side is the final angular momentum.

Now from the given data, it is known that $\omega'_1 = \omega'_2 = \omega_f$. Therefore equation (1) can be written as

$$I_1\omega_1 + I_2\omega_2 = I_1\omega_f + I_2\omega_f \quad \text{or}$$

$$I_1\omega_1 + I_2\omega_2 = (I_1 + I_2)\omega_f.$$

Dividing both sides by $(I_1 + I_2)$, we get

$$\omega_f = \frac{I_1\omega_1 + I_2\omega_2}{I_1 + I_2}. \qquad (2)$$

The loss in kinetic energy (T) due to the friction between the two disks is the difference between the initial T and the final T. Thus, loss in T,

$$\Delta T = \left(\frac{1}{2} I_1\omega_1^2 + \frac{1}{2} I_2\omega_2^2\right) - \left(\frac{1}{2} I_1\omega_f^2 + \frac{1}{2} I_2\omega_f^2\right) \quad \text{or}$$

$$\Delta T = \left(\frac{1}{2} I_1\omega_1^2 + \frac{1}{2} I_2\omega_2^2\right) - \frac{1}{2}\left(I_1 + I_2\right)\omega_f^2$$

Substituting the expression for ω_f from equation (2) we get

$$\Delta T = \left(\frac{1}{2} I_1\omega_1^2 + \frac{1}{2} I_2\omega_2^2\right) - \frac{1}{2}\left(I_1 + I_2\right)\left(\frac{I_1\omega_1 + I_2\omega_2}{I_1 + I_2}\right)^2.$$

Expanding the above expression yields:

$$\Delta T = \frac{1}{2}\left[I_1\omega_1^2 + I_2\omega_2^2 - \frac{(I_1 + I_2)(I_1^2\omega_1^2 + I_2^2\omega_2^2 + 2I_1I_2\omega_1\omega_2)}{(I_1 + I_2)(I_1 + I_2)}\right]$$

$$\Delta T = \frac{1}{2}\frac{\left[(I_1 + I_2)(I_1\omega_1^2 + I_2\omega_2^2) - (I_1^2\omega_1^2 + I_2^2\omega_2^2 + 2I_1I_2\omega_1\omega_2)\right]}{I_1 + I_2}$$

Expanding the terms, and cancelling yields

$$\Delta T = \frac{1}{2}\frac{I_1I_2\omega_2^2 + I_1I_2\omega_1^2 - 2I_1I_2\omega_1\omega_2}{I_1 + I_2}.$$

Now factor:

$$\Delta T = \frac{\frac{1}{2} I_1 I_2 (\omega_1^2 - 2\omega_1\omega_2 + \omega_2^2)}{I_1 + I_2}$$

$$\Delta T = \frac{\frac{1}{2} I_1 I_2 (\omega_1 - \omega_2)^2}{I_1 + I_2}.$$

Now the fraction of the initial KE loss is

$$\frac{\Delta T}{T_{initial}} = \frac{\frac{1}{2} I_1 I_2 (\omega_1 - \omega_2)^2}{I_1 + I_2} \times \frac{1}{\frac{1}{2} I_1 \omega_1^2 + \frac{1}{2} I_2 \omega_2^2}$$

which can be written as

$$\frac{\Delta T}{T_{initial}} = \frac{I_1 I_2 (\omega_1 - \omega_2)^2}{(I_1 + I_2)(I_1\omega_1^2 + I_2\omega_2^2)}.$$

• PROBLEM 18-34

A circular hoop with a piece of lead glued inside of its rim, rolls down a straight incline at all times. Does the rate of change of angular momentum of the system about the center of the hoop equal the net external torque about this center? Does the rate of change of angular momentum of the system about its center of mass equal the net external torque about the center of mass?

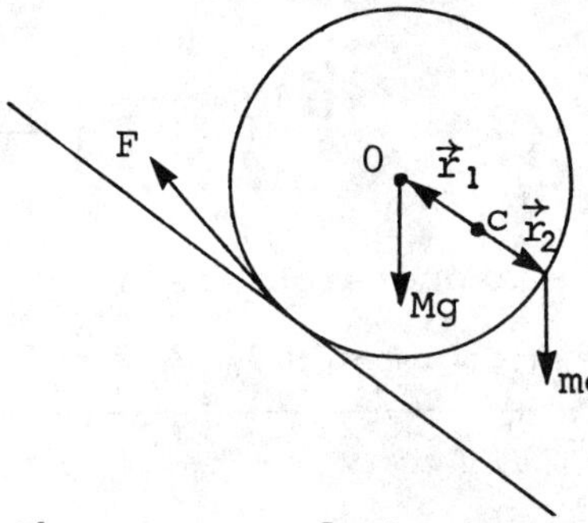

Solution: If the masses of the hoop and the piece of lead are M and m respectively, the gravitational forces at any instant are given by $M\vec{g}$ and $m\vec{g}$ as shown in the figure. For the hoop to roll down the incline there must be an external force F acting where the hoop and plane touch, as shown.

The equation of rotational motion of a rigid body about any axis is given by

$$\dot{\vec{L}} = \vec{\Gamma} \qquad (1)$$

where $\vec{\dot{L}}$ is the angular momentum and $\vec{\Gamma}$ is the total torque of all the forces about the axis.

If we consider the rotation about O, the center of the hoop, our equation of motion will be

$$\dot{L}_0 = \Gamma_0 \tag{2}$$

with
$$\Gamma_0 = \Gamma_0^{F} + \Gamma_0^{mg}, \tag{3}$$

Γ_0^{F} and Γ_0^{mg} being the torque due to the two forces $\vec{F}$ and and $m\vec{g}$ about O.

Eq. (2) then tells us that the rate of change of angular momentum about the center of the hoop is equal to the sum of the external torque and the torque due to the weight of the lead piece.

Now consider the rotational motion about the center of mass C of the system. The equation of motion in this case is

$$\dot{L}_C = \Gamma_C \tag{4}$$

with
$$\Gamma_C = \Gamma_C^{F} + \Gamma_C^{Mg} + \Gamma_C^{mg}, \tag{5}$$

Γ_C^{F}, Γ_C^{Mg} and Γ_C^{mg} being the torque about C due to the forces $\vec{F}$, $M\vec{g}$ and $m\vec{g}$, respectively.

Consider torques due to Mg and mg

$$\Gamma_C^{Mg} + \Gamma_C^{mg} = \vec{r}_1 \times M\vec{g} + \vec{r}_2 \times m\vec{g}$$

$$= (M\vec{r}_1 + m\vec{r}_2) \times \vec{g}$$

where $\vec{r}_1$ and $\vec{r}_2$ are the position vectors of M and m from C. The quantity within the parentheses is equal to zero since it defines the position vector of the center of mass of the system from C which is the center of mass itself.

From Eqs. (5) and (4) we then have

$$\dot{L}_C = \Gamma_C^{F}. \tag{6}$$

Thus, the rate of change of angular momentum of the system about its center of mass is equal to the net external torque about the center of mass.

A turntable of mass M and radius $\vec{R}$ is rotating with angular velocity ω_a on frictionless bearings. A spider of mass $\vec{m}$ falls vertically on to the rim of the turntable. What is the new angular velocity $\vec{\omega}_b$? The spider then slowly walks in toward the center of the turntable. What is the angular velocity $\vec{\omega}_c$ when the spider is at a distance $\vec{r}$ from the center? Assume that, apart from a negligibly small inward velocity along the radius, the spider has no velocity relative to the turntable.

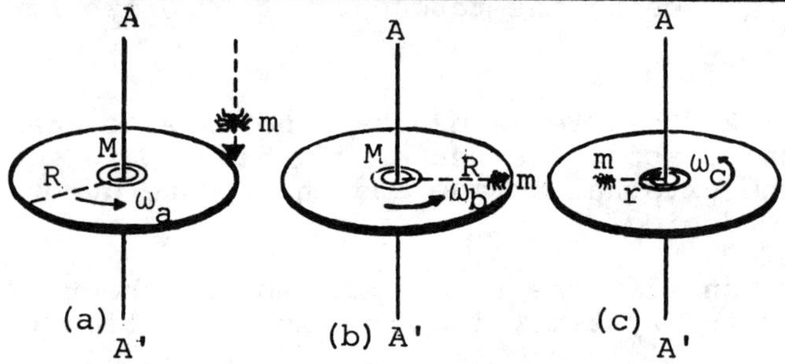

Solution: Consider the system which includes the turntable and the spider. Since the bearing is frictionless and the resistance of the air is to be ignored, no external couple acts on this system and its angular momentum must always remain the same. Just before the spider lands on the rim (figure a) the spider has no angular motion about the axis $\vec{A}A'$ and the angular momentum is contained entirely in the turntable. The turntable is a disc whose moment of inertia about its axis of symmetry is:

$$I_t = \tfrac{1}{2}MR^2$$

The angular momentum is therefore

$$L = I_t\omega_a$$

$$= \tfrac{1}{2} MR^2\omega_a$$

When the spider is standing on the rim (figure b) he takes up the motion of the turntable and both have an angular velocity $\vec{\omega}_b$. The moment of inertia of the spider is

$$I_{sb} = mR^2$$

since all of the spider's mass is at a distance $\vec{R}$ from the center of the turntable. The total moment of inertia of the system is

$$I_b = I_t + I_{sb}$$

$$= \tfrac{1}{2}MR^2 + mR^2$$

$$= \tfrac{1}{2}(M + 2m)R^2$$

The angular momentum is

$$L = I_b\omega_b$$

$$= \tfrac{1}{2}(M + 2m)R^2\omega_b.$$

Applying the law of conservation of angular momentum and equating the angular momenta before and after the spider lands,

$$\tfrac{1}{2}(M + 2m)R^2\omega_b = \tfrac{1}{2}MR^2\omega_a$$

$$\omega_b = \frac{M}{M + 2m}\,\omega_a$$

When the spider is at a distance $\vec{r}$ from the center (figure c), the angular velocity of both the spider and the turntable is $\vec{\omega}_c$. The moment of inertia of the spider is then

$$I_{sc} = mr^2$$

The total moment of inertia is

$$I_c = I_t + I_{sc}$$

$$= \tfrac{1}{2}MR^2 + mr^2$$

The angular momentum is

$$L = I_c\omega_c$$

$$= \left(\tfrac{1}{2}MR^2 + mr^2\right)\omega_c$$

Applying the law of conservation of angular momentum

$$\left(\tfrac{1}{2}MR^2 + mr^2\right)\omega_c = \tfrac{1}{2}MR^2\omega_a$$

$$\omega_c = \frac{\tfrac{1}{2}MR^2}{\left(\tfrac{1}{2}MR^2 + mr^2\right)}\,\omega_a$$

$$= \frac{\omega_a}{\left(1 + \dfrac{2mr^2}{MR^2}\right)}$$

Check that this agrees with the equation for $\vec{\omega}_b$

when $\vec{r} = \vec{R}$. As the spider walks inward and $\vec{r}$ decreases, the angular velocity increases since angular momentum must remain constant. When the spider reaches the center and $\vec{r} = 0$

$$\omega_c = \omega_a \quad \text{when } \vec{r} = 0.$$

• PROBLEM 18-36

Two men, each of whom weighs 150 lb, stand opposite each other on the rim of a small uniform circular platform which weighs 900 lb. Each man simultaneously walks clockwise and at a fixed speed once around the rim. The platform is free to rotate about a vertical axis through its center. Find the angle in space through which each man has turned. (The platform's moment of inertia is $I = \frac{1}{2}MR^2$, where M is its mass).

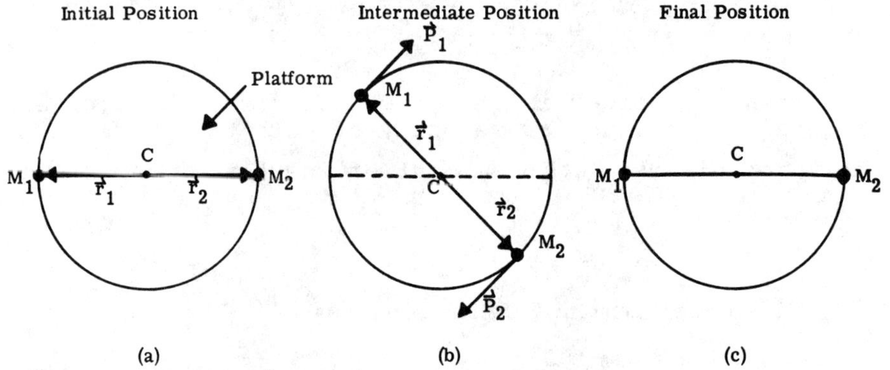

Solution: The figure (parts (a) through (c)) illustrates various positions of the 2 men, and the platform, relative to an observer on the ground. Our ultimate goal is to find an equation, for each man, describing his angular position, θ, as a function of time, relative to his initial position (figure (a)). It is equally acceptable to find an equation for the angular velocity $\vec{\omega}$ of each man, since this may be integrated to find θ. Rather than jumping right into a dynamic analysis of this problem, it might be worth our while to see if we can use any conservation relations in solving this problem.

Note that the angular momentum of the system consisting of the 2 men and platform is constant in time, since no external torques (i.e., friction) act on the system. Furthermore, using angular momentum as our conserved quantity will give us a relation between the angular kinematical variables (α,ω,θ) at 2 times during the motion of the system. We take these 2 times to be as illustrated in figures (a) and (b). The initial angular momentum of the system, L_0, is zero, since figure (a) shows the system at rest. The final angular momentum, $\vec{L}_f$, is due to the angular momentum of each man $(\vec{L}_1, \vec{L}_2)$ plus the platform's angular momentum, L_p. Hence,

$$\vec{L}_f = \vec{L}_1 + \vec{L}_2 + \vec{L}_p \qquad (1)$$

But, by the definition of the angular momentum of a particle about point C,

$$\vec{L}_1 = \vec{r}_1 \times \vec{p}_1$$

$$\vec{L}_2 = \vec{r}_2 \times \vec{p}_2$$

where $\vec{p}_1$ and $\vec{p}_2$ are the linear momentum of each man, and $\vec{r}_1$ and $\vec{r}_2$ are as illustrated in figure (a). Also, from figure (b)

$$\vec{r}_1 = -\vec{r}_2$$

$$\vec{p}_1 = -\vec{p}_2$$

since each man walks with the same velocity. Then

$$\vec{L}_1 = \vec{r}_1 \times \vec{p}_1$$

and

$$\vec{L}_2 = \vec{r}_1 \times \vec{p}_1 \tag{2}$$

Furthermore, the angular momentum of the platform is

$$\vec{L}_p = I\,\vec{\omega} \tag{3}$$

where $\vec{\omega}$ and I are its angular velocity and moment of inertia relative to C. (See figure (a)). Using (2) and (3) in (1)

$$\vec{L}_f = 2\vec{r}_1 \times \vec{p}_1 + I\,\vec{\omega} \tag{4}$$

But, the momentum of particle $1\left(\vec{p}_1\right)$ is related to its angular velocity $\left(\vec{\omega}_1\right)$ relative to an external observer by

$$\vec{p}_1 = m_1\vec{\omega}_1 \times \vec{r}_1$$

Therefore,

$$\vec{r}_1 \times \vec{p}_1 = m_1\vec{r}_1 \times \left(\vec{\omega}_1 \times \vec{r}_1\right) \tag{5}$$

Since, for any vectors $\vec{A}$, $\vec{B}$ and $\vec{C}$

$$\vec{A} \times (\vec{B} \times \vec{C}) = (\vec{A}\cdot\vec{C})\vec{B} - (\vec{A}\cdot\vec{B})\vec{C}$$

we obtain, using (5)

$$\vec{r}_1 \times \vec{p}_1 = m_1\vec{r}_1 \times \left(\vec{\omega}_1 \times \vec{r}_1\right) = m_1 r_1^2\left(\vec{\omega}_1\right) - m_1\left(\vec{r}_1\cdot\vec{\omega}_1\right)\vec{r}_1 \tag{6}$$

Now $\vec{\omega}_1$ is perpendicular to the plane of the platform, which contains $\vec{r}_1$. Then

$$\vec{\omega}_1 \cdot \vec{r}_1 = 0$$

and (6) becomes

$$\vec{r}_1 \times \vec{p}_1 = m_1\vec{r}_1 \times \left(\vec{\omega}_1 \times \vec{r}_1\right) = m_1 r_1^2\,\vec{\omega}_1 \tag{7}$$

Using (7) in (4)

$$\vec{L}_f = 2m_1 r_1^2\,\vec{\omega}_1 + I\,\vec{\omega} \tag{8}$$

By the principle of conservation of angular momentum

$$\vec{L}_f = \vec{L}_0 = 0$$

and (8) yields

$$2m_1 r_1^2\,\vec{\omega}_1 + I\,\vec{\omega} = 0$$

or

$$\vec{\omega}_1 = \left(\frac{-I}{2m_1 r_1^2}\right)\vec{\omega} \tag{9}$$

Since each man travels with the same speed,

$$\vec{\omega}_1 = \vec{\omega}_2 = \vec{\omega}\,' \tag{10}$$

Also, the masses of the 2 men are equal, and

$$m_1 = m_2 = m \,. \tag{11}$$

Noting that r_1 equals r_2, and that each equals the platform's radius, R, we obtain

$$r_1 = r_2 = R \,. \tag{12}$$

Using (10) through (12) in (9)

$$\vec{\omega}' = \left(\frac{-I}{2mR^2}\right)\vec{\omega} \tag{11}$$

Now

$$\omega = \frac{d\theta}{dt} \quad \text{and} \quad \omega' = \frac{d\theta'}{dt}$$

by definition, where θ and θ' are the angular positions of the platform and either man relative to an outside observer. Then

$$\frac{d\theta'}{dt} = \frac{-I}{2mR^2}\frac{d\theta}{dt}$$

or

$$\int_{\theta_0'}^{\theta'} d\theta' = \frac{-I}{2mR^2}\int_{\theta_0}^{\theta} d\theta$$

where $\theta' = \theta_0'$ and $\theta = \theta_0$ at $t = 0$. (figure (a)). Finally

$$\theta' - \theta_0' = \frac{-I}{2mR^2}\left(\theta - \theta_0\right)$$

This relates the net angle traversed by each man (relative to an observer on the ground) to the net angle which the platform rotates through. Using the given data

$$\theta' - \theta_0' = -\frac{\frac{1}{2}MR^2}{2mR^2}\left(\theta - \theta_0\right) = \frac{-M}{4m}\left(\theta - \theta_0\right)$$

$$\theta' - \theta_0' = \frac{-Mg}{4mg}\left(\theta - \theta_0\right) = \frac{-900 \text{ lb}}{600 \text{ lb}}\left(\theta - \theta_0\right)$$

$$\theta' - \theta_0' = -\frac{3}{2}\left(\theta - \theta_0\right) \tag{12}$$

Each man makes 1 revolution (or traverses 2π radians) relative to the disc. (See figure (c)). Then, relative to an outside observer, the man traverses an angle $\left(\theta' - \theta_0'\right)$ equal to 2π, plus the angle the disc turns through relative to him $\left(\theta - \theta_0\right)$. Hence,

$$\left(\theta' - \theta_0'\right) = \left(\theta - \theta_0\right) + 2\pi \tag{13}$$

Using (12) in (13)

$$\left(\theta' - \theta_0'\right) = -\frac{2}{3}\left(\theta' - \theta_0'\right) + 2\pi$$

$$\frac{5}{3}\left(\theta' - \theta_0'\right) = 2\pi$$

$$\theta' - \theta_0' = \frac{6\pi}{5} = 216^\circ$$

Then

$$\theta - \theta_0 = -\frac{2}{3}\left(\theta' - \theta_0'\right) = -144^\circ$$

The negative sign indicates that the men move in a direction opposite to the direction of motion of the disc.

IMPACT

• PROBLEM 18-37

The pendulum in figure 1 consists of a homogeneous prismatic bar of mass m and length ℓ, which hangs at rest on its frictionless suspension O. A spherical pellet of mass m travels with a horizontal velocity v_1 and strikes it in the middle. Immediately after the impact, the pellet moves with velocity v_2 to the right and the pendulum begins to rotate about O with initial angular velocity ω. Solve for v_2 and ω in terms of v_1 for the cases in which the collision is: 1.) completely elastic, and 2.) completely inelastic.

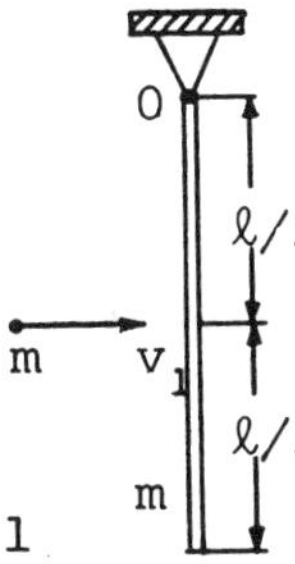

Fig. 1

Solution: Since not only the normal force, but also the reaction at O, becomes large during the impact, the principle of impulsive forces is useless for the determination of v_2 and ω. However, here we can apply the principle of impulsive moments about O.

The angular momentum about O of the system of two bodies before the impulse originates from the pellet alone. It is normal to the plane of motion and has the magnitude

$$H_{O1} = m\frac{\ell}{2}v_1. \tag{1}$$

The total angular momentum immediately after the impulse has the same direction and the magnitude

$$H_{O2} = m\frac{\ell}{2}v_2 + \frac{m\ell^2}{3}\omega. \tag{2}$$

The second term in equation (2) is the angular momentum of the bar, $I\omega$, where I is the moment of inertia of the bar about O, which is equal to $m\ell^2/3$.

Since no external forces that have a static moment about O become large during the impulse, we have $H_{O1} = H_{O2}$, or substituting the values from (1) and (2),

$$m\frac{\ell}{2}v_2 + \frac{m\ell^2}{3}\omega = m\frac{\ell}{2}v_1. \tag{3}$$

If the impulse is elastic the kinetic energy of the system is conserved. Thus we have

$$\frac{m}{2} v_2^2 + \frac{1}{2} \frac{m\ell^2}{3} \omega^2 = \frac{m}{2} v_1^2. \tag{4}$$

Hence, by (3) and (4) we have the following equations for the determination of v_2 and ω:

$$v_2 + \frac{2}{3} \ell \omega = v_1, \qquad v_2^2 + \frac{1}{3} \ell^2 \omega^2 = v_1^2.$$

In addition, we have the requirement that $(\ell/2)\omega \geq v_2$. The solution yields

$$\omega = \frac{12}{7} \frac{v_1}{\ell} \text{ and } v_2 = -\frac{1}{7} v_1.$$

Thus, both bodies move after the impulse, and in such a way that the speed of the pellet relative to the pendulum if $v_2' = \ell/2\omega - v_2 = v_1$ to the left. The velocity of the pellet relative to the pendulum is reversed.

If the pellet remains imbedded in the pendulum, the impulse is completely inelastic. Then, instead of (4), we have the condition

$$v_2 = \frac{\ell}{2}\omega. \tag{5}$$

and, from (3) and (4) it follows that

$$\omega = \frac{6}{7} \frac{v_1}{\ell} \text{ and } v_2 = \frac{3}{7} v_1.$$

• PROBLEM 18-38

A horizontal rod A'B' rotates freely about the vertical with a counterclockwise angular velocity of 8 rad/sec. Two solid spheres of radius 5 in., weighing 3 lbs. each are held in place at A and B by a cord which is suddenly cut. knowing that the centroïdal moment of inertia of the rod and pivot is $\bar{I}_r = 0.25 \text{ lb} \cdot \text{ft} \cdot \text{s}^2$, determine (a) the angular velocity of the rod after the spheres have moved to positions A' and B', (b) the energy lost due to the plastic impact of the sphere and the stops at A' and B'.

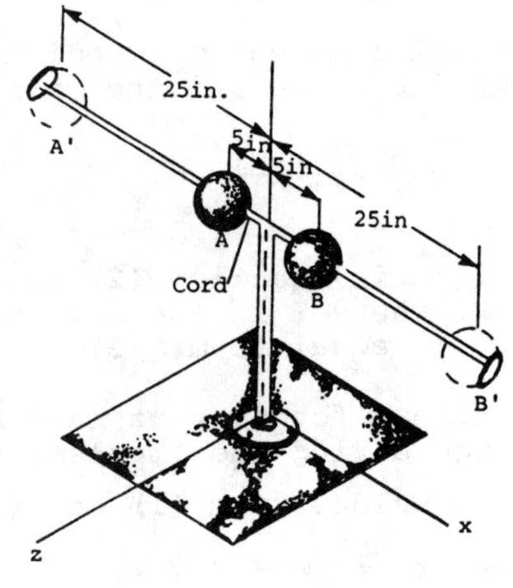

Solution: It is clear that part (a) can be solved by using conservation of total angular momentum since there are no external forces involved. The total angular momentum is

$$L = I_{rod}\, \omega + 2\, I_{sphere}\, \omega + 2\, (m_{sphere}\, r^2)\omega.$$

The last two terms follow from the parallel axis theorem. Setting $L_i = L_f$,

$$I_r\omega_i + 2I_s\omega_i + 2\, m_s\, r_i^{\,2}\omega_i = I_r\omega_f + 2I_s\omega_f + 2\, m_s r_f^{\,2}\omega_f$$

$$(I_r + 2I_s + 2\, m_s r_i^{\,2})\omega_i = (I_r + 2I_s + 2\, m_s r_f^{\,2})\omega_f. \qquad (1)$$

We are given $I_r = 0.25$ lb·ft·sec², $m_s = 3$ lb, $r_i = \frac{5}{12}$ ft and $r_f = \frac{25}{12}$ ft. We compute

$$I_s = \frac{2}{5}\, m_s a^2 = \frac{2}{5}\left(\frac{3\text{ lb.}}{(32.2\text{ ft/sec}^2)}\right)\left(\frac{5}{12}\text{ ft}\right)^2$$

$$= 0.00645\ \text{lb·ft·sec}^2,$$

and $m_s r_i^{\,2} = \left(\frac{3}{32.2}\right)\left(\frac{5}{12}\right)^2 = 0.01617\ \text{lb·ft·sec}^2$

$$m_s r_f^{\,2} = \left(\frac{3}{32.2}\right)\left(\frac{25}{12}\right)^2 = 0.4043\ \text{lb·ft·sec}^2.$$

Substituting these values into equation (1) yields:

$$[0.25 + 2\,(.00645) + 2\,(.01617)]\omega_i = [(0.25 + 2\,(.00645) + 2\,(.4043)]\omega_f$$

$$\omega_f = \frac{0.2963}{1.0715}\,\omega_i = .2765(8) = 2.21\ \text{rad/sec}$$

The kinetic energy lost due to the collision of the balls with the ends of the rod is equal to the kinetic energy loss of the system. The kinetic energy of the system for any arbitrary instant is

$$T = \frac{1}{2}\, I_r\omega^2 + 2\left(\frac{1}{2}\, m_s v^2 + \frac{1}{2}\, I_s\omega^2\right).$$

The velocity of the balls is $v = r\omega$. Therefore,

$$T = \frac{1}{2}\,(I_r + 2\, I_s + 2\, m_s r^2)\omega^2. \qquad (2)$$

Substituting known numerical values into equation (2) we get

$$T_i = \frac{1}{2}(0.2765\ (8)^2 = 9.48 \text{ ft·lb}$$

$$T_f = \frac{1}{2}(1.0715)(2.21)^2 = 2.62 \text{ ft·lb.}$$

Therefore, $\Delta T = T_f - T_i = -6.86$ ft·lb.

The energy of the system decreases by 6.86 ft·lb.

• PROBLEM 18-39

A rectangular package of height a and length 2a moves with a constant velocity V_1 down a conveyor belt A. At the end of the conveyor belt, the corner of the package strikes a rigid support at B. Assuming that the impact at B is perfectly elastic, derive an expression for the smallest magnitude of the velocity V_1 for which the package will rotate about B and reach conveyor belt C.

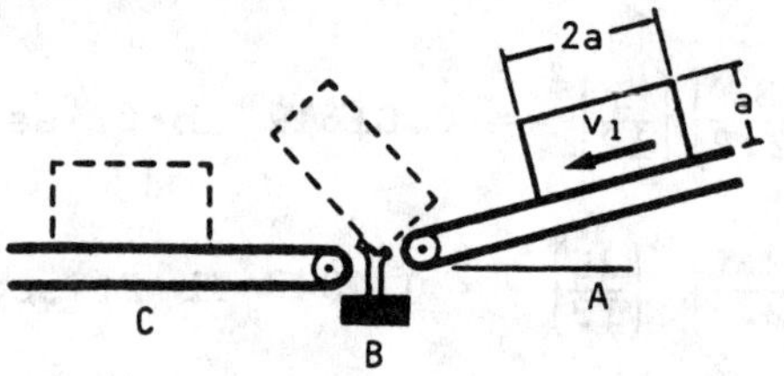

Fig. 1

Solution: The impact between the package and the support is perfectly elastic. The package rotates about B during the impact. Applying the principle of impulse and momentum, we observe that the only external impulsive force on the package is the impulsive reaction force at B.

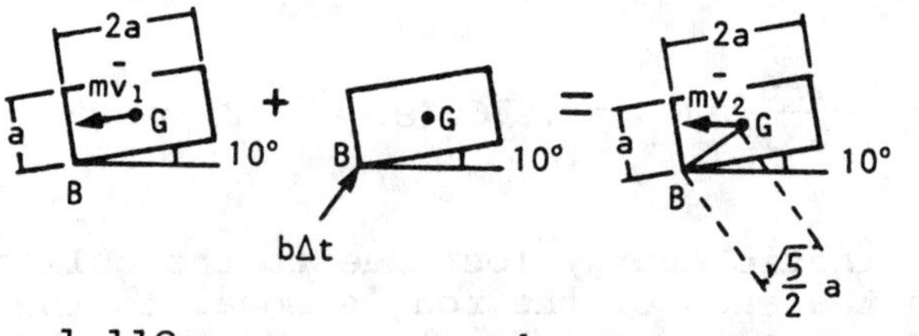

Fig. 2

$$GB = \tfrac{1}{2}\sqrt{5}a = 1.118a \qquad h_3 = GB = 1.118a$$

$$h_2 = GB \sin(26.5 + 10°)$$

$$= 0.665a$$

Position 2. $V_2 = Wh_2$. Recalling that $\bar{V}_2 = \frac{1}{2}\sqrt{5}\ a\omega_2$, we write

$$T_2 = \tfrac{1}{2}m\bar{v}_2^2 + \tfrac{1}{2}\bar{I}\omega_2^2 = \tfrac{1}{2}M[\tfrac{1}{2}\sqrt{5}\ a\omega_2]^2 + \tfrac{1}{2}[\tfrac{5}{12}ma^2]\omega_2^2 = \tfrac{5}{6}ma^2\omega_2^2$$

Position 3. The package must pass through position 3 to reach conveyor belt B. To determine the smallest velocity to reach this position, we choose $\bar{V}_3 = \omega_3 = 0$. Therefore $T_3 = 0$ and $V_3 = Wh_3$.

Conservation of Energy

$$T_2 + V_2 = T_3 + V_3$$

$$\frac{5}{6}ma^2\omega_2^2 + Wh_2 = 0 + Wh_3$$

$$\omega_2^2 = \frac{6W}{5ma^2}(h_3 - h_2) = \frac{6g}{5a^2}(h_3 - h_2) \qquad (3)$$

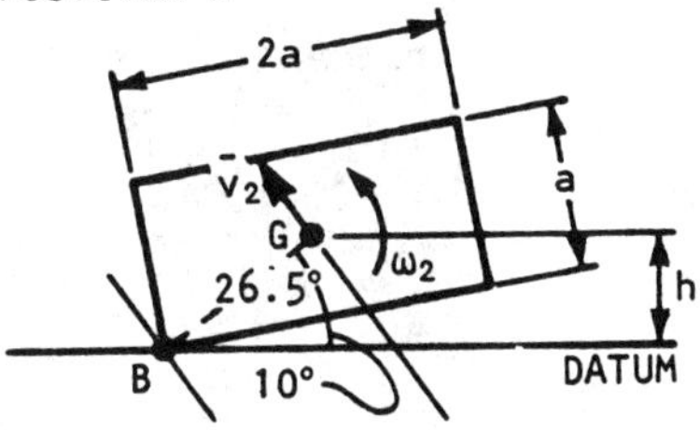

Fig. 3

GB = $1/2\sqrt{5}a$ = 1.118a
h_3 = GB sin(26.5° + 10°)
= 0.665a

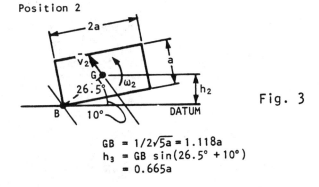

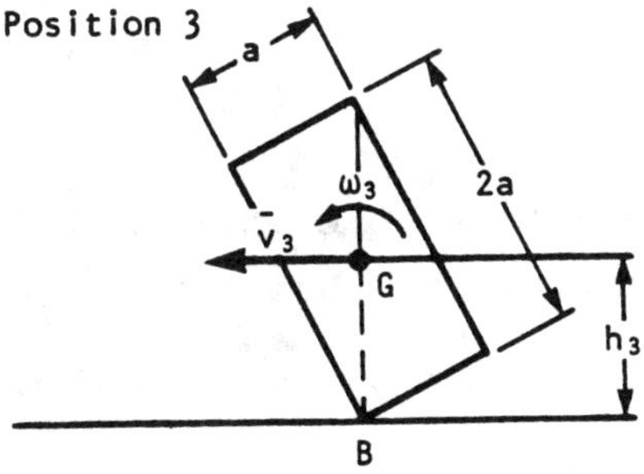

Fig. 4

h_3 = GB = 1.118a

Substituting in the values for h_2 and h_3 who Eq. (3), yields

$$\omega_2^2 = \frac{6g}{5a^2}(1.118a - 0.665a) = \frac{6g}{5a^2}(.453a), \ \omega_2 = 0.543g/a$$

$$\bar{V}_1 = \frac{5}{3}a\omega_2 = \frac{5}{3}a\sqrt{0.543g/a} \qquad \bar{V}_1 = 1.228\sqrt{ga}$$

A homogeneous bar of length ℓ, mass m with its axis inclined at an angle of 45°, falls vertically. Its lower end strikes a smooth horizontal plane. Let the velocity just before impact be v_1. Determine the motion immediately after the impulse for both elastic and completely inelastic impulse, with the lower end slipping along the horizontal plane.

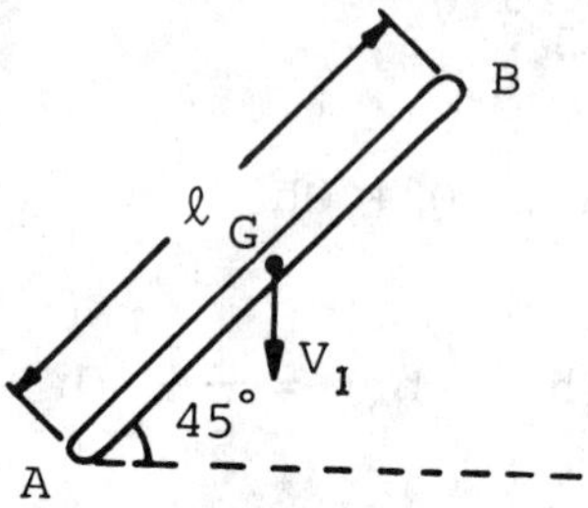

Solution: The impulse and momentum diagram is

In the x-direction the momentum is conserved. Hence

$$mv_x = 0 + 0 = 0$$

$$\therefore \quad v_x = 0.$$

Taking moments about point A,

$$mv_1 \frac{\ell}{2} \cos 45° + 0 = mv_y \left(\frac{\ell}{2} \sin 45°\right) + I_G\omega \qquad (1)$$

where I_G is the moment of inertia about G, the center of gravity of the bar,

$$I_G = \frac{m\ell^2}{12}. \qquad (2)$$

After the impact in an elastic collision $v_A = v_1$ and is in the upward direction. Using the relative velocity equation

$$\vec{v}_A - \vec{v}_G = \vec{v}_{A/G}$$

$$+ v_1 + v_y = + \omega \frac{\ell}{2} \cos 45°$$

$$v_y = - v_1 + \omega \frac{\ell}{2} \cos 45°. \qquad (3)$$

Substituting (3) in (1) and using (2)

$$\not{m} v_1 \frac{\ell}{2\sqrt{2}} = \frac{\not{m}\,\ell}{2\sqrt{2}} \left(- v_1 + \frac{\omega\ell}{2\sqrt{2}}\right) + \frac{\not{m}\ell^2}{12} \omega$$

$$v_1 = -v_1 + \frac{\omega \ell}{2\sqrt{2}} + \frac{\ell 2\sqrt{2}\,\omega}{12}$$

$$\omega = \frac{12\sqrt{2}\,v_1}{5\ell} \qquad (4)$$

Therefore from equations (3) and (4)

$$v_2 = v_y = -v_1 + \frac{12\sqrt{2}}{5\ell} v_1 \frac{\ell}{2\sqrt{2}}$$

$$= v_1/5.$$

For an inelastic impact $v_A = 0$ after the impact. Hence from equation (3)

$$v_y = \omega \frac{\ell}{2} \cos 45°. \qquad (5)$$

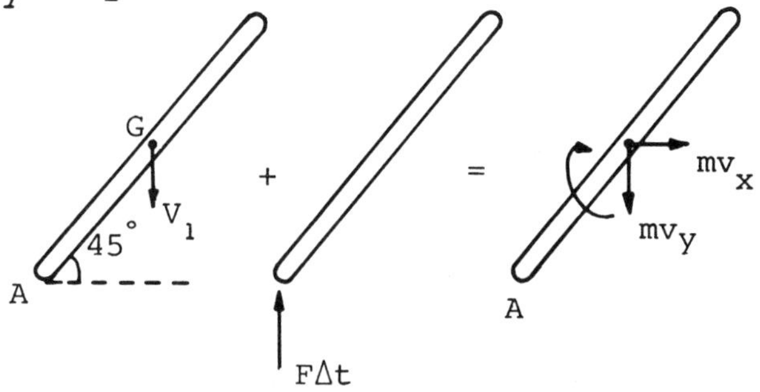

Substituting equation (5) in (1) yields:

$$mv_1 \frac{\ell}{2}\left(\frac{v_2}{2}\right) = m\omega \frac{\ell}{2}\left(\frac{v_2}{2}\right)\left(\frac{\ell}{2}\right)\left(\frac{v_2}{2}\right) + \frac{m\ell^2\omega}{12}$$

$$\frac{v_1 v_2}{4} = \omega\ell\left(\frac{1}{8} + \frac{1}{12}\right)$$

$$\omega = \frac{6\sqrt{2}\,v_1}{5\ell} \qquad (6)$$

Substituting in (6) in (5) yields:

$$v_y = \frac{\omega\ell}{2\sqrt{2}} = \frac{6\sqrt{2}\,v_1}{5\ell}\,\frac{\ell}{2\sqrt{2}} = \frac{3}{5} v_1 .$$

• PROBLEM 18-41

A homogeneous bar of mass m, length ℓ in a vertical position falls. Its lower end strikes a smooth inclined plane that has an angle of inclination of 30°. Assume that the bar has a velocity v_1 just before impact and that the lower end slips along the plane. Find the motion immediately after the impact for elastic and completely inelastic impulse.

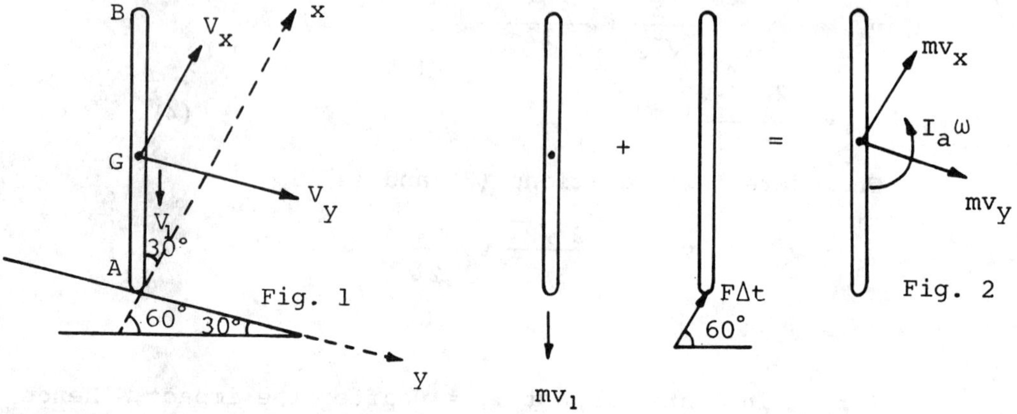

Fig. 1 Fig. 2

Solution: Let the coordinate system be chosen as shown in figure 1. The velocity of point A before impact is then

$$v_{Ax} = - v_1 \cos 30°$$

$$v_{Ay} = + v_1 \sin 30°.$$

The impulse diagrams are as shown.

Taking moments about A, we have

$$0 = I_G\omega - mv_x \sin 30° \left(\frac{\ell}{2}\right) - mv_y \cos 30° \left(\frac{\ell}{2}\right) . \qquad (1)$$

Where I_G, the moment of inertia is

$$I_G = \frac{m\ell^2}{12}. \qquad (2)$$

The momentum equation parallel to the plane yields:

$$mv_y = mv_1 \sin 30°$$

$$v_y = v_1 \sin 30° = \frac{v_1}{2} \qquad (3)$$

We have from equations (1), (2) and (3):

$$\frac{m\ell^2}{12}\omega - mv_x \frac{\ell}{4} - \frac{mv_1}{4} \ell \cos 30° = 0$$

$$\frac{\ell\omega}{12} - \frac{v_x}{4} - \frac{v_1}{4} \cos 30° = 0. \qquad (4)$$

We write the kinematic relationship between points A and G of the bar as

$$\vec{v}_A = \vec{v}_G + \vec{v}_{A/G'} \qquad (5)$$

where $\vec{V}_{A/G}$ is the velocity of A with respect to G. After the collision

$$\vec{V}_A = v'_{Ax}\,\hat{i} + v'_{Ay}\,\hat{j}$$

$$\vec{V}_G = v_x\,\hat{i} + v_y\,\hat{j}$$

$$\vec{V}_{A/G} = \frac{\ell\omega}{2}\ \sin 30°\ \hat{i} + \frac{\ell}{2}\ \omega\ \cos 30°\ \hat{j}.$$

Compare the components of the x-direction in Equation (5):

$$v_x = v'_{Ax} - \frac{\ell\omega}{2}\ \sin 30°. \qquad (6)$$

Now, for an elastic collision the speed of A must be the same before and after the collision, however, with a hard wall the velocity will be oppositely directed. Hence,

$$v'_{Ax} = -\ v_{Ax} = +\ v_1\ \cos 30°,$$

and equation (6) is

$$v_x = v_1\ \cos 30° - \frac{\ell\omega}{2}\ \sin 30°. \qquad (7)$$

Substitute equation (7) into equation (4):

$$\frac{\ell\omega}{12} = \frac{v_1\ \cos 30°}{4} + \frac{\ell\omega}{8}\ \sin 30° - \frac{v_1}{4}\ \cos 30° = 0$$

$$\omega = v_1\ \left(\frac{12\ \sqrt{3}}{7\ell}\right).$$

Now using equation (5) we may find the linear velocities of the center of mass G. From equation (7) we write directly,

$$v_x = v_1\ \cos 30° - \frac{\ell}{2}\left(v_1\ \frac{12\ \sqrt{3}}{7\ell}\right)\ \sin 30°$$

$$= \frac{v_1\ \sqrt{3}}{2} - v_1\ \frac{3\ \sqrt{3}}{7}$$

$$= v_1\ \frac{\sqrt{3}}{14}\ .$$

And, from equation (3), $v_y = v_1/2$. This completes the description of the bar's motion after the elastic impact.

For the inelastic impact $v_y = v'_{Ay} = v_1\ \sin 30° = \frac{v_1}{2}$

again. But now $v'_{Ax} = 0$.

Referring to equation (6), now

$$v_x = -\frac{\ell\omega}{2} \sin 30°. \qquad (8)$$

Substitute equation (8) into equation (4):

$$\frac{\ell\omega}{12} + \frac{\ell\omega}{16} - \frac{v_1\sqrt{3}}{8} = 0$$

$$\omega = v_1\left(\frac{6\sqrt{3}}{7}\right).$$

Substituting back into equation (8) we find

$$v_x = v_1\left(\frac{3\sqrt{3}}{14}\right).$$

• PROBLEM 18-42

A square plate of mass m suspended from two wires at A and B is hit at D in a direction perpendicular to the plate. Denoting by $\vec{F}\Delta t$ the impulse applied at D, determine immediately after impact (a) the velocity of the mass center G, (b) the angular velocity of the plate.

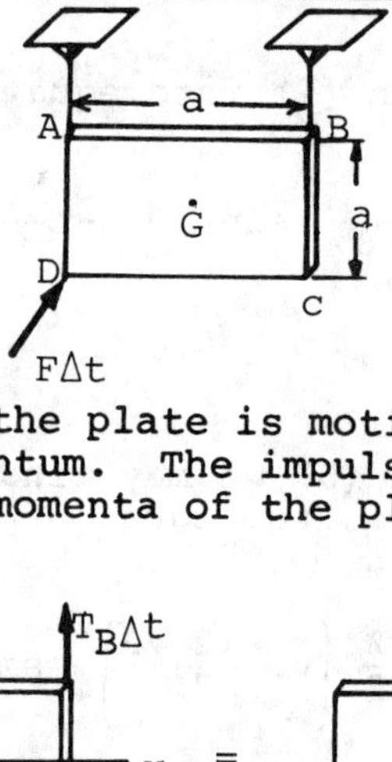

Fig. 1

Solution: Initially, the plate is motionless; it has no linear or angular momentum. The impulse applied, then, must be equal to the final momenta of the plate.

Fig. 2

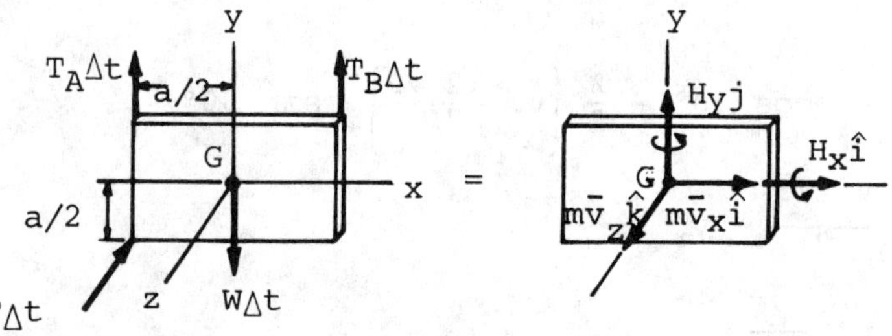

We shall assume that the wires remain taut and thus the components v_y of $\vec{v}$ and ω_z of $\vec{\omega}$ are zero. Thus yielding

$$\bar{v} = \bar{v}_x\,\hat{i} + \bar{v}_z\hat{k} \qquad (1)$$

$$\omega = \omega_x\,\hat{i} + \omega_y\,\hat{j} \qquad (2)$$

The impulse condition is shown in figure 2. Because the weight is not impulsive $\omega\Delta t \sim 0$ for a sudden impulse and $T_A\Delta t$ and $T_D\Delta t$ can also be ignored. Let the angular momentum be

$$L_G = I_x\omega_x \hat{i} + I_y\omega_y \hat{j}$$

$$= \frac{1}{6} ma^2\omega_x \hat{i} + \frac{1}{6} ma^2\omega_y \hat{j}$$

Since x, y, z axes are the principal axes of inertia, we obtain the following impulse-momentum equations:

$$0 = mv_x \qquad (3)$$

$$-F\Delta t = mv_z \qquad (4)$$

$$\frac{1}{2} a F\Delta t = L_x = \frac{1}{6} ma^2\omega_x \qquad (5)$$

$$-\frac{1}{2} a F\Delta t = L_y = \frac{1}{6} ma^2\omega_y \qquad (6)$$

From (1),(3) and (4):

$$\vec{v}_G = -\frac{F\Delta t}{m}\hat{k}.$$

From (2),(5) and (6):

$$\vec{\omega}_G = \frac{3\ F\Delta t}{ma}\hat{i} - \frac{3\ F\Delta t}{ma}\hat{j}\ .$$

• **PROBLEM** 18-43

A small elastic ball with a velocity V_1 of magnitude V_1 and a backspin ω_1 of magnitude ω_1, is thrown against a rough floor. The ball is observed to repeatly bounce back and forth between points P_1 and P_2 and always rising at an angle of 45° from the floor. Assuming perfectly elastic impact, determine a) the required magnitude, ω_1 of the backspin in terms of V_1 and r, and b) the minimum required value of the coefficient of friction.

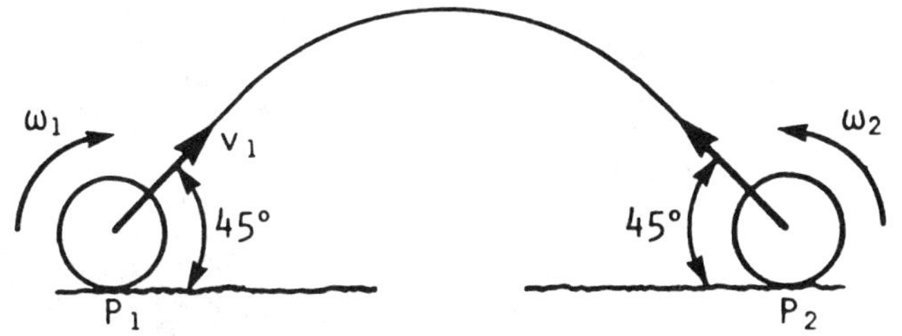

Fig. 1

Solution: The problem is best solved by using the principle of impulse and momentum. First draw the impulse and momentum diagrams for the ball at A.

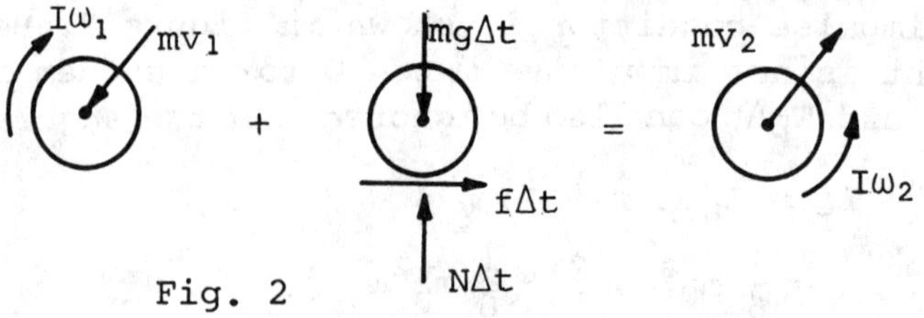

Fig. 2

Neglecting the impulse due to the weight of the ball, the impulse and momentum equations are,

$$\overset{+}{\leftarrow} \quad mv_1 \cos 45° - \int f\Delta t = -\, mv_1 \cos 45°$$

$$\therefore \int f\Delta t = 2\, mv_1 \cos 45°. \qquad (1)$$

$$+\uparrow \quad -\, mv_1 \sin 45° + \int N\Delta t = mv_1 \sin 45°$$

$$\int N\Delta t = 2\, mv_1 \sin 45° \qquad (2)$$

Angular impulse at the center of gravity of the ball is

$$\overset{\curvearrowleft}{+} \quad I\omega_1 - \left(\left|\int f\Delta t\right|\right) r = -\, I\omega_1$$

$$\left(\left|\int f\Delta t\right|\right) r = 2\, I\omega_1.$$

For a sphere, $I = \frac{2}{5} mr^2$, therefore, the above equation can be written as

$$\left(\left|\int f\Delta t\right|\right) r = 2\, \frac{2}{5}\, mr^2\omega_1.$$

or $\int f\Delta t = \frac{4}{5} mr\omega_1$ (3)

Equating equations (1) and (3) gives

$$2\, mv_1 \cos 45° = \frac{4}{5} mr\omega_1$$

Solving for ω_1, yields

$$\omega_1 = 1.77 \frac{v_1}{r}$$

To find the minimum value of the coefficient of friction, μ, set $f = \mu N$ and using equations (1) and (2), solve for μ.

$$\int fr\Delta t = \mu \int N\Delta t = 2\ mv_1 \cos 45°$$

$$= \mu\ (2\ mv_1 \sin 45°) = 2\ mv_1 \cos 45°$$

$$\therefore \quad \mu = \frac{2\ mv_1 \cos 45°}{2\ mv_1 \sin 45°} = \cot 45° = 1$$

• PROBLEM 18-44

> A Sphere with a velocity of 8 m/s moves to the right on a horizontal trajectory. It hits a rigid bar BB'. The bar is suspended from a rotating support. The mass of the sphere is 3 kgs. The mass of the bar is 9 kgs. and the coefficient of restitution between the bar and sphere is 0.95. Determine the angular velocity of the bar and the velocity of the sphere immediately after impact.

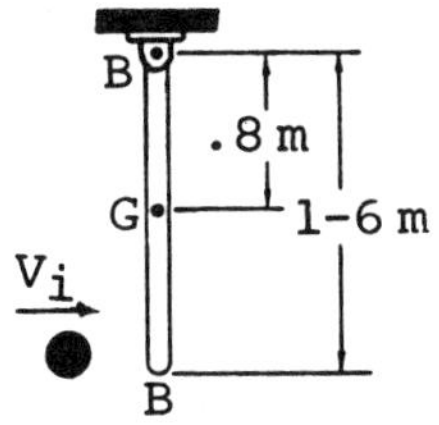

Solution: The only external force acting on this system is gravity. The rod is acted on by gravity but it only feels a torque when it is displaced from its downward position. Then, in the impulsive approximation that the collision does not cause any immediate displacements, we must have that the angular momentum before and immediately after the collision is conserved. The same approximation does allow sudden changes in momentum and, hence, velocity.

The initial angular momentum is:

$$L = m_s v_s r = (3\ \text{kg})\ (8\ \text{m/sec})\ (1.6\ \text{m}) = 38.4\ \text{kg-m}^2/\text{sec}.$$

The final angular momentum is:

$$L' = m_s v_s' r + I_r \omega' + m_r v_G' \left(\frac{r}{2}\right)$$

$$L' = m_s v_s' r + I_r \omega' + m_r \left(\frac{r}{2}\right)^2 \omega'$$

since $v_G' = \left(\frac{r}{2}\right)\omega'$

$$L' = (3\ \text{kg})\ v_s'\ (1.6\ \text{m}) + \left[\frac{1}{12}\ (9\ \text{kg})\ (1.6\text{m})^2\right]\omega'$$

$$+ \frac{(9\ \text{kg})\ (1.6)^2\ \omega'}{(2)^2}$$

$$L' = 4.8\ v_s' + 7.68\,\omega'. \qquad (1)$$

By conservation of angular momentum,

$$L = 38.4\ \frac{kg - m^2}{sec} = L'$$

$$38.4\ \frac{kg - m^2}{sec} = 4.8\ V_s' + 7.68\,\omega'$$

Given the coefficient of restitution we know that

$$v_B' - v_s' = e\ (v_s - v_B)$$

$$v_B' - v_s' = (0.95)\ (8\ m/sec. \qquad (2)$$

Finally, noting that the rod rotates rigidly about A, we write

$$v_B' = (1.6\,m)\,\omega'. \qquad (3)$$

Solving equations (1), (2) and (3) simultaneously gives

$$\omega' = +\ 4.87\ rad/sec$$

$$v_s' = -0.2\quad m/sec$$

or $\vec{\omega}' = 4.87$ rad/sec, counterclockwise

and $\vec{v}_s' = 0.2$ m/sec, to the right.

CHAPTER 19

THREE-DIMENSIONAL RIGID BODY DYNAMICS

INERTIA TENSOR

• PROBLEM 19-1

Diagonalize the inertia tensor

$$T = \begin{pmatrix} 7 & \sqrt{6} & -\sqrt{3} \\ \sqrt{6} & 2 & -5\sqrt{2} \\ -\sqrt{3} & -5\sqrt{2} & -3 \end{pmatrix}$$

and find the principle axes.

Solution: Let $\hat{e}_1$, $\hat{e}_2$, $\hat{e}_3$ be the unit vectors of the original coordinate system, and

$$\hat{e}_i' = x_i\hat{e}_1 + y_i\hat{e}_2 + z_i\hat{e}_3 , \quad i = 1, 2, 3$$

be the unit vectors of a new coordinate system. Also, let ψ_i be a column matrix (or 3 × 1 matrix) with x_i, y_i, z_i as its elements. Finding the new coordinate system in which the tensor is diagonal is equivalent to solving the eigenvalue problem,

$$T\psi_i = \lambda_i\psi_i, \tag{1}$$

where the λ's are the three diagonal elements of the tensor in the new coordinate system.

By carryingout the matrix multiplication of Eq. (1) and omitting the subscript i for simplicity,

$$\left.\begin{aligned} (7 - \lambda)x + \sqrt{6}y - \sqrt{3}z &= 0 \\ \sqrt{6}x + (2 - \lambda)y - 5\sqrt{2}z &= 0 \\ -\sqrt{3}x - 5\sqrt{2}y - (3 + \lambda)z &= 0. \end{aligned}\right\} \tag{2}$$

For nontrivial solutions λ must satisfy the equation,

$$\begin{vmatrix} 7-\lambda & \sqrt{6} & -\sqrt{3} \\ \sqrt{6} & 2-\lambda & -5\sqrt{2} \\ -\sqrt{3} & -5\sqrt{2} & -3-\lambda \end{vmatrix} = 0. \tag{3}$$

The expansion of the determinant is straightforward, resulting in the cubic equation,

$$\lambda^3 = 6\lambda^2 - 72 \quad + 320 = 0.$$

By inspection, $\lambda = 10$ is a solution. After factoring out $\lambda - 10$, the remaining quadratic equation gives $\lambda = 4, -8$. The three diagonal elements of tensor T' are

$$\lambda_1 = 4, \quad \lambda_2 = 10, \quad \lambda_3 = -8.$$

To find the principal axes, first substitute $\lambda = \lambda_1 = 4$ into Eqs. (2) and solve for x_1, y_1 and z_1. The result is:

$$y_1 = -\sqrt{2/3}x_1 \quad , \quad z_1 = x_1/\sqrt{3} \, ,$$

or $\hat{e}_i' = x_1(\hat{e}_1 - \sqrt{2/3}\hat{e}_2 + \sqrt{1/3}\hat{e}_3)$.

The fact that $\hat{e}_1'$ is a unit vector, i.e.,

$$1 = \hat{e}_1' \cdot \hat{e}_1' = x_1^2[1 + (2/3) + 1/3],$$

determines the value $x_1 = 1/\sqrt{2}$.

$$\hat{e}_1' = (\sqrt{3}\hat{e}_1 - \sqrt{2}\hat{e}_2 + \hat{e}_3)/\sqrt{6}.$$

By repeating the above procedure for the other two eigenvalues $\lambda_2 = 10$ and $\lambda_3 = -8$, we find the other two principal axes,

$$\hat{e}_2' = (\sqrt{3}\hat{e}_1 + \sqrt{2}\hat{e}_2 - \hat{e}_3)/\sqrt{6}$$

$$\hat{e}_3' = (\hat{e}_2 + \sqrt{2}\hat{e}_3)/\sqrt{3}.$$

• PROBLEM 19-2

Find the inertia tensor of a uniform rectangular block of mass M, dimensions a × b × c, about a set of axes through its center with the z-axis parallel to side c and the y axis parallel to a diagonal of rectangle a × b.

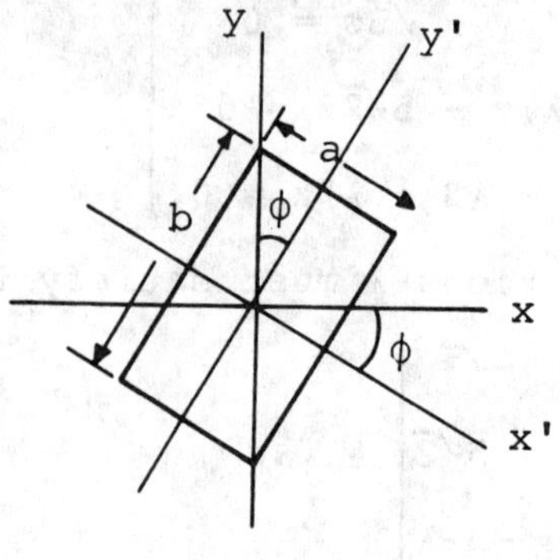

Solution: The block is shown in the diagram for a cross-seciton passing through the block's center 0 and perpendicular to the z-axis. The (xyz) coordinate and the principal axes coordinate (x'y'z') are related by

$$x = \cos\phi\, x' + \sin\phi\, y'$$

$$y = -\sin\phi\, x' + \cos\phi\, y'$$

$$z = z'$$

where $\cos\phi = b/\sqrt{a^2 + b^2}$, $\sin\phi = a/\sqrt{a^2 + b^2}$.

The inertia tensor of the block can be expressed in terms of integrals in the principal axes coordinates,

$$I_{xy} = I_{yx} = \int xy\,dm = (\cos^2\phi - \sin^2\phi)\int x'y'dm + \sin\phi\cos\phi\left[\int y'^2dm - \int x'^2dm\right]$$

$$I_{xz} = I_{zx} = \int xz\,dm = \cos\phi\int x'z'dm + \sin\phi\int y'z'dm$$

$$I_{yz} = I_{zy} = \int yz\,dm = \cos\phi\int y'z'dm - \sin\phi\int x'z'dm$$

$$I_{xx} = \int (y^2 + z^2)dm = \cos^2\phi\int y'^2dm + \sin^2\phi\int x'^2dm + \int z'^2dm + 2\sin\phi\cos\phi\int x'y'dm$$

$$I_{yy} = \int (x^2 + z^2)dm = \cos^2\phi\int x'^2dm + \sin^2\phi\int y'^2dm + \int z'^2dm + 2\sin\phi\cos\phi\int x'y'dm$$

$$I_{zz} = \int (x^2 + y^2)dm = \int x'^2dm + \int y'^2dm.$$

The evaluation of the above integrals are straightforward:

$$\int x'^2dm = \int x'^2 \frac{M}{abc}dx'dy'dz'$$

$$= \int_{-b/2}^{b/2} dy' \int_{-c/2}^{c/2} dz' \int_{-a/2}^{a/2} x^{12}\, dx' \frac{m}{abc}$$

$$= \frac{M}{abc}\cdot b\cdot c\cdot \frac{1}{3}\left[\left(\frac{a}{2}\right)^3 - \left(-\frac{a}{2}\right)^3\right] = Ma^2/12$$

$$\int x'y'dm = (M/abc)\int_{-a/2}^{a/2} x'dx' \int_{-b/2}^{b/2} y'dy' \int_{-c/2}^{c/2} dz' = 0.$$

Similarly

$$\int y'^2 \, dm = Mb^2/12 \qquad \int z'^2 \, dm = Mc^2/12$$

$$\int y'z' \, dm = \int z'x' \, dm = 0$$

Substituting these integrals obtains for the components of the inertia tensor,

$$I_{xx} = \frac{b^2}{a^2+b^2}\frac{Mb^2}{12} + \frac{a^2}{a^2+b^2}\frac{Ma^2}{12} + \frac{Mc^2}{12} = \frac{M}{12}\left(\frac{a^4+b^4}{a^2+b^2} + c^2\right)$$

$$I_{yy} = \frac{b^2}{a^2+b^2}\frac{Ma^2}{12} + \frac{a^2}{a^2+b^2}\frac{Mb^2}{12} + \frac{Mc^2}{12} = \frac{M}{12}\left(\frac{2a^2b^2}{a^2+b^2} + c^2\right)$$

$$I_{zz} = \frac{Ma^2}{12} + \frac{Mb^2}{12} = \frac{M}{12}(a^2 + b^2)$$

$$I_{xy} = I_{yx} = \frac{a}{\sqrt{a^2+b^2}}\frac{b}{\sqrt{a^2+b^2}}\left[\frac{Mb^2}{12} - \frac{Ma^2}{12}\right] = \frac{M}{12}\frac{ab}{a^2+b^2}(b^2 - a^2)$$

$$I_{xz} = I_{zx} = 0, \quad I_{yz} = I_{zy} = 0.$$

• PROBLEM 19-3

Find the components of the tensor corresponding to a rotation by an angle θ about the z-axis, followed by a rotation by an angle ψ about the y axis. Find its eigenvalues.

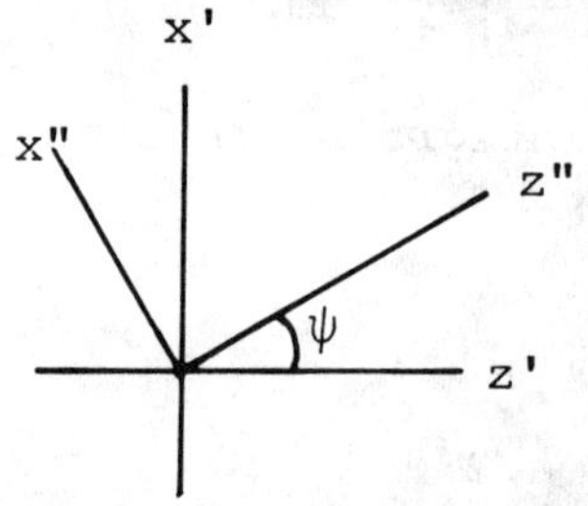

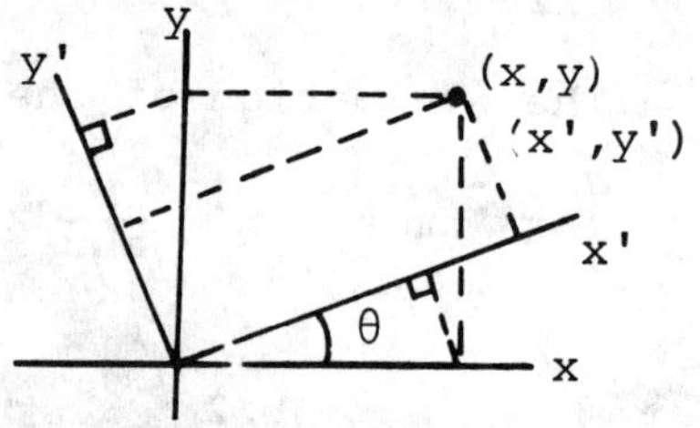

Solution: The coordinate transformation corresponding to a rotation by an angle about the z-axis is given by

$$x' = \cos\theta\, x + \sin\theta\, y$$

$$y' = -\sin\theta\, x + \cos\theta\, y$$

$$z' = z$$

or, in matrix notation,

$$R' = AR$$

where

$$R' = \begin{pmatrix} x' \\ y' \\ z' \end{pmatrix}, \quad R = \begin{pmatrix} x \\ y \\ z \end{pmatrix}, \quad A = \begin{pmatrix} \cos\theta & \sin\theta & 0 \\ -\sin\theta & \cos\theta & 0 \\ 0 & 0 & 1 \end{pmatrix}$$

and A is called the transformation tensor. Similarly, the transformation corresponding to a rotation by an angle about the y' axis is given by

$$R'' = BR'$$

where

$$R'' = \begin{pmatrix} x'' \\ y'' \\ z'' \end{pmatrix}, \quad B = \begin{pmatrix} \cos\psi & 0 & -\sin\psi \\ 0 & 1 & 0 \\ -\sin\psi & 0 & \cos\psi \end{pmatrix}.$$

It follows that

$$R'' = BR' = BAR = CR.$$

That is, the tensor for the overall transformation is

$$C = BA = \begin{pmatrix} \cos\psi\cos\theta & \cos\psi\sin\theta & -\sin\psi \\ -\sin\theta & \cos\theta & 0 \\ \sin\psi\cos\theta & \sin\psi\sin\theta & \cos\psi \end{pmatrix}.$$

The eigenvalues of the tensor are determined by the equation

$$\begin{vmatrix} \cos\psi\cos\theta-\lambda & \cos\psi\sin\theta & -\sin\psi \\ -\sin\theta & \cos\theta-\lambda & 0 \\ \sin\psi\cos\theta & \sin\psi\sin\theta & \cos\psi-\lambda \end{vmatrix} = 0.$$

Working out this lengthy expansion,

$$(\cos\psi\cos\theta-\lambda)[(\cos\theta-\lambda)(\cos\psi-\lambda)]$$
$$-\cos\psi\sin\theta[-\sin\theta(\cos\psi-\lambda)]$$
$$-\sin\psi[-\sin\psi\sin^2\theta - \sin\psi\cos\theta(\cos\theta-\lambda)] = 0.$$

Multiplying out all terms,

$$\cos^2\psi\cos^2\theta - \lambda\cos\theta\cos\psi - (\cos\psi\cos^2\theta + \cos^2\psi\cos\theta)\lambda$$
$$+ (\cos\theta + \cos\psi)\lambda^2 + \cos\psi\cos\theta\lambda^2 - \lambda^3$$
$$+ \cos^2\psi\sin^2\theta - \cos\psi\sin^2\theta\lambda + \sin^2\psi\sin^2\theta$$
$$+ \sin^2\psi\cos^2\theta - \sin^2\psi\cos\theta\lambda = 0.$$

Collecting equal powers of λ,

$$-\lambda^3 + (\cos\theta + \cos\psi + \cos\psi\cos\theta)\lambda^2$$
$$- [\cos\psi(\sin^2\theta + \cos^2\theta) + \cos\theta(\cos^2\psi + \sin^2\psi) + \cos\theta\cos\psi]\lambda$$
$$+ (\cos^2\psi\cos^2\theta + \cos^2\psi\sin^2\theta)$$
$$+ (\sin^2\psi\sin^2\theta + \sin^2\psi\cos^2\theta) = 0.$$

Letting $a = \cos\theta + \cos\psi + \cos\psi\cos\theta$, and using $\sin^2\theta + \cos^2\theta = \sin^2\psi + \cos^2\psi = 1$, the result is

$$-\lambda^3 + a\lambda^2 - a\lambda + \cos^2\psi + \sin^2\psi = 0;$$

that is,

$$-\lambda^3 + a\lambda^2 - a\lambda + 1 = 0.$$

This equation may be factored into

$$-(\lambda - 1)[\lambda^2 + (1 - a)\lambda + 1] = 0$$

by adding and subtracting $(\lambda^2 + \lambda)$.

Clearly, one root is $\lambda = 1$. The others can be obtained by the standard solution of the remaining quadratic expression:

$$\lambda = \frac{-(1 - a) \pm \sqrt{(1 - a)^2 - 4}}{2}.$$

Examine the discriminant. It will be negative for $-2 < (1 - a) < 2$, or $-3 < -a < 1$, or $3 > a > -1$. What are the possible values for a? The range of the cosines are -1 to +1, giveing the following extremum for a:

$$a = (-1) + (-1) + (-1)(-1) = -1$$

$$a = (+1) + (+1) + (+1)(+1) = 3.$$

The discriminant is always less than or equal to zero so we may rewrite λ as

$$\lambda = \frac{(a - 1)}{2} \pm i\,\frac{\sqrt{4 - (a - 1)^2}}{2}.$$

Now, using the following triangle, we can rewrite $\lambda = \cos\alpha \pm i\sin\alpha$, or

$$\lambda = e^{\pm i\alpha}.$$

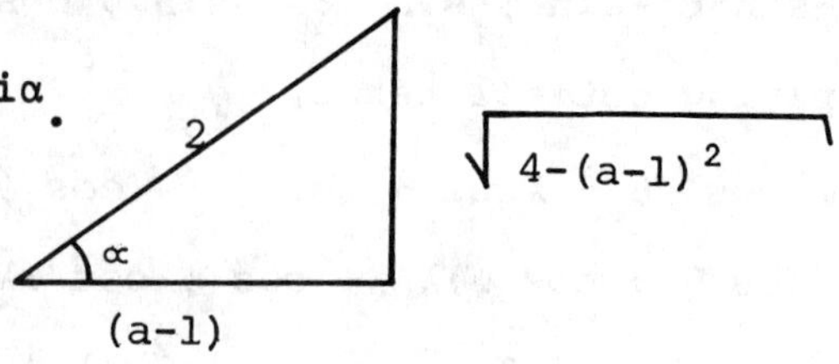

Summarizing, the three eigenvalues are

$$\lambda = 1, \quad e^{\pm i}$$

where $\cos\alpha = \frac{(a - 1)}{2} = (\cos\psi\cos\theta + \cos\psi + \cos\theta - 1)/2.$

• PROBLEM 19-4

Find the inertial tensor of a straight rod of length l, mass m, about its center. Use this result to find the inertia tensor about the centroid of an equilateral pyramid constructed of six uniform rods.

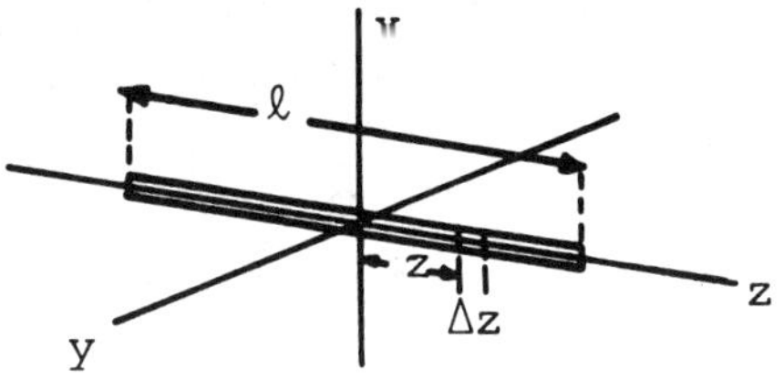

Fig. 1

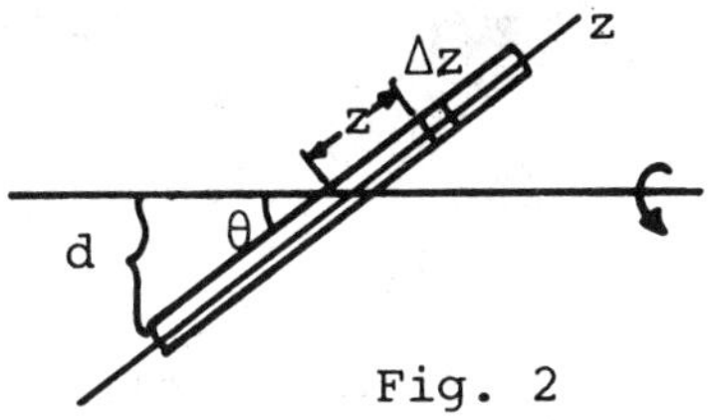

Fig. 2

Solution: First consider the single rod. Lay out coordinate axes as in figure 1. From the symmetry the x and y-axes are equivalent and will have the same moment of inertia. Clearly $I_z = 0$. No part of the rod has any distance to the z-axis so $\sum \Delta mr^2 = 0$. To find I_x or I_y, split the rod into short segments of length Δz; each piece is z from the axis. Thus, $I = \sum \Delta mr^2$ becomes

$$\sum \frac{m\Delta z}{\ell} z^2 = \frac{m}{\ell}\int_{-\ell/2}^{\ell/2} z^2 dz = \frac{m}{\ell}\left(\frac{z^3}{3}\right)_{-\ell/2}^{\ell/2} = \frac{m\ell^2}{12}.$$

Putting the three moments of inertia around these principal axes into a matrix form gives

$$I = \frac{m\ell^2}{12}\begin{pmatrix} 1 & 0 & 0 \\ 0 & 1 & 0 \\ 0 & 0 & 0 \end{pmatrix}.$$

A useful result in working towards the inertia tensor of the rod-pyramid is the following: If a rod is swung around an axis through its center inclined by the angle θ from the rod, we have

$$I_{axis} = \sum \Delta mr^2 = \sum\left(\frac{m\Delta z}{\ell}\right)(z\sin\theta)^2 = \frac{m}{\ell}\sin^2\theta \int_{-\ell/2}^{\ell/2} z^2 dz$$

$$= \frac{m}{\ell}\sin^2\theta\left(\frac{z^3}{3}\right)_{-\ell/2}^{\ell/2} = \frac{m\ell^2}{12}\sin^2\theta. \quad \text{See figure 2.}$$

If, instead, the rod is rotated around an axis through its end, inclined by the angle θ, the parallel-axis theorem may be used to obtain:

$$I = \frac{m\ell^2}{12}\sin^2\theta + md^2 = \frac{m\ell^2 \sin^2\theta}{12} + m\left(\frac{\ell \sin\theta}{2}\right)^2 = \frac{m\ell^2 \sin^2\theta}{3}.$$

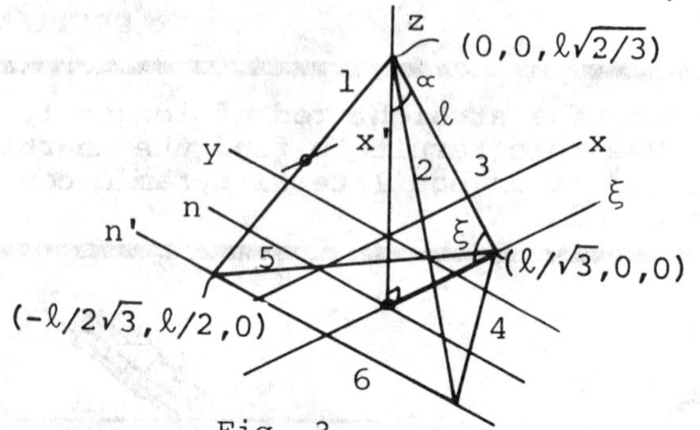

Fig. 3

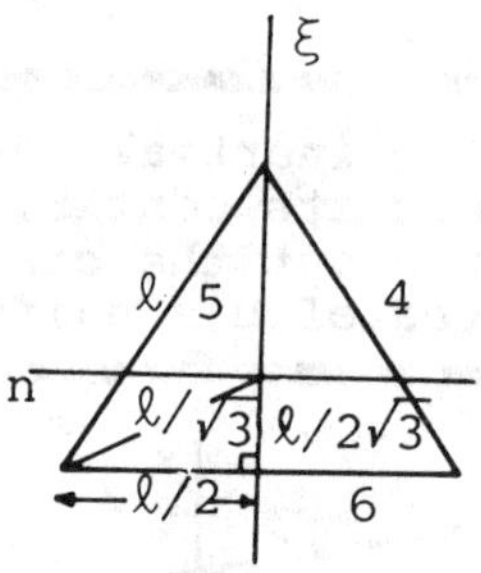

Fig. 4

Start on the rod-pyramid by establishing some facts about its geometry. The rods form the edges of a regular tetrahedron, its sides are equilateral triangles. Lay out a coordinate system as in figure 3.
The base triangle in the $\xi-\eta$ plane is shown in figure 4. Some lengths which are readily obtained from Pythagoras' theorem are indicated. By using the distance from one base tip to the center of the base triangle in another right triangle shown in figure 3, we find the height to be $\ell\sqrt{2/3}$, or the top has ξ-η-z-coordinates $(0,0,\ell\sqrt{2/3})$. The coordinates which will be used most often are these ξ-η-z-coordinates. To find the principal axes, note that the z-axis is a threefold symmetry axis, and, thus, should be a good choice. The z-ξ plane is a reflection plane, so one principal axis ought to be in this plane, parallel to the ξ-axis. Perpendicular to these two is the third, a y-axis, above the η-axis through the center of mass. Thus, as soon as the elevation of the center of mass is found, we can construct the the set of principal axes x,y,z. To find the center of mass, note the upper three rods (numbered 1, 2, and 3 in figure 3) have their upper end at elevation $\ell\sqrt{2/3}$; thus, their midpoints and individual centers of mass are at elevation $z = \ell/\sqrt{6}$. The lower three rods (4, 5, and 6) are at elevation 0, so averaging the z-values gives us an elevation of the center of mass of $\frac{1}{2}\ell/\sqrt{6} = 1/\sqrt{24} \sim 0.204\ell$. The x and y-axes which can be constructed by this process are also indicated on figure 3.

The parallel-axis theorem will be used repeatedly: If I_{cm} is the moment of inertia around an axis through the center of mass of an object, then the moment of inertia around a parallel axis passing a distance d from the center of mass is $I = I_{cm} + md^2$. Abbreviate this theorem P.A.T. First, find I_z, the moment of inertia of the figure if it is rotating around the z-axis. Each of the upper three rods has $I = m\ell^2 \sin^2 \alpha/3$ according to the result obtained earlier; they are rotating around an axis through their ends. From the right triangle in figure 3 notice that $\sin\alpha = 1/\sqrt{3}$, so $I = (m\ell^2/3)\cdot(1/3) = m\ell^2/9$. Each of the lower three rods is a distance d from the z-axis; from figure 4 observe that $d = \ell/(2\sqrt{3})$, so P.A.T. gives $I = m\ell^2/12 + m\ell^2/(2\sqrt{3})^2 = m\ell^2/6$. Add all these moments of inertia to find I_z:

$$I_z = 3\frac{m\ell^2}{9} + 3\frac{m\ell^2}{6} = m\ell^2 \cdot \frac{5}{6}.$$

Next, find I_x. Do so by first calculating I_ξ around the parallel ξ-axis, and use P.A.T.

I_ξ of rod 6 is $m\ell^2/12$; it is spinning around a perpendicular axis through its center.

I_ξ of rods 4 or 5 is $m\ell^2 \sin^2 30°/3 = m\ell^2/12$; these are spinning around an inclined axis through their ends.

Similarly, I_ξ of rod 3 is $m\ell^2 \sin^2 \zeta/3$. In the marked right triangle in figure 3 we find $\sin\zeta = \sqrt{2/3}$, so I_ξ here becomes $m\ ^2\ \frac{2}{9}$. To find I of rods 1 or 2, first find I around each midpoint and use P.A.T. By averaging the coordiantes of the top and bottom of rod 1, its midpoint is found at $\left(-\frac{\ell}{4\sqrt{3}}, \frac{\ell}{4}, \frac{\ell}{\sqrt{6}}\right)$. The distance d from a x'-axis to the ξ-axis (found in the η-z plane) is, thus, $\sqrt{\left(\frac{\ell}{4}\right)^2 + \left(\frac{\ell}{\sqrt{6}}\right)^2} = \ell\sqrt{\frac{11}{48}}$. The angle this rod makes with the x-direction (and the x'-axis) is given by $\hat{x}\cdot\vec{\ell}_1 = 1\cdot\ell\cdot\cos\delta$, or, since the vector $\vec{\ell}_1$ in the direction of rod 1 is $\left(\frac{\ell}{2\sqrt{3}}, -\frac{\ell}{2}, \ell\sqrt{\frac{2}{3}}\right)$, $\cos\delta = \frac{\ell}{2\sqrt{3}}$ and $\sin^2\delta = 1 - \cos^2\delta = 1 - \frac{1}{4\cdot 3} = 11/12$.

Thus, $I_{x'}$ of rod 1 is $m\ell^2 \sin^2\delta/12 = m\ell^2\cdot 11/144$. P.A.T. now gives $I_\xi = I_{x'} + md^2 = m\ell^2\left(\frac{11}{144} + \frac{11}{48}\right) = m\ell^2\cdot\frac{11}{36}$. Adding all the I's gives total $I_\xi = I_1 + I_2 + I_3 + I_4 + I_5 + I_6$

$$= m\ell^2\left(\frac{11}{36} + \frac{11}{36} + \frac{2}{9} + \frac{1}{12} + \frac{1}{12} + \frac{1}{12}\right) = m\ell^2\cdot\frac{39}{36}.$$

$I_x = I_\xi\ Md^2$ by the parallel axis theorem; here $d = \frac{\ell}{2\sqrt{6}}$:

$I_x = m\ell^2\cdot\frac{39}{36} - 6m\left(\frac{\ell}{2\sqrt{6}}\right)^2 = m\ell^2\cdot\frac{5}{6}$. Now go to I around the the y-axis. First, find the moment of inertia around a parallel axis, the 6-rod. Call this axis η'. Afterwards, use P.A.T. again. Rods 1, 2, 4, and 5 form 60° angles with η', so each has $I_{\eta'} = m\ell^2 \sin^2 60°/3 = m\ell^2/4$. Rod 3 has I around its own center of $m\ell^2/12$; its center is at $(\ell/(2\sqrt{3}), 0, \ell/\sqrt{6})$ by averaging the coordinates of its ends. The distance of this point from η', found in the ξ-z plane, is $\left[\left(\frac{\ell}{2\sqrt{3}} + \frac{\ell}{2\sqrt{3}}\right)^2 + (\Pi/\sqrt{6})^2\right]^{\frac{1}{2}} = \ell/\sqrt{2}$. Thus, $I_{\eta'} = m\ell^2/12 + m(\ell/\sqrt{2})^2 = m\ell^2 \cdot \frac{7}{12}$, by P.A.T. If we add all these $I_{\eta'}$'s then $I_{\eta'} = 4\, m\ell^2/4 + m\ell^2 \frac{7}{12} = m\ell^2 \cdot \frac{19}{12}$. (Rod 6 does not contribute to this sum, since it is on the axis.) Let d be the distance of η' from the center of mass; we find in the ξ-z plane that $d = \left[\left(\frac{\ell}{2\sqrt{3}}\right)^2 + \left(\frac{\ell}{2\sqrt{6}}\right)^2\right]^{\frac{1}{2}} = \ell/(2\sqrt{2})$. By P.A.T.

$I_y = I_{\eta'} - Md^2 = m\ell^2 \cdot \frac{19}{12} - 6m\left(\frac{\ell}{2\sqrt{2}}\right)^2 = m\ell^2 \cdot \frac{5}{6}$. Thus, in the set of perpendicular axes xyz through the center,

$$I = \begin{pmatrix} I_x & 0 & 0 \\ 0 & I_y & 0 \\ 0 & 0 & I_z \end{pmatrix} = \frac{5}{6}m\ell^2 \begin{pmatrix} 1 & 0 & 0 \\ 0 & 1 & 0 \\ 0 & 0 & 1 \end{pmatrix} = \frac{5}{36}M\ell^2 \begin{pmatrix} 1 & 0 & 0 \\ 0 & 1 & 0 \\ 0 & 0 & 1 \end{pmatrix} (M = 6m).$$

Since the I's are all equal, there is no way to tell the directions of the figure by spinning it; all directions are rotationally equivalent. We would find the same I for any axis through the center of mass.

• PROBLEM 19-5

i) Find the inertia tensor of a plane rectangle of mass M, dimensions a × b (a < b). ii) Use this result to find the inertia tensor about the center of mass of the house of cards shown. Use principal axes.

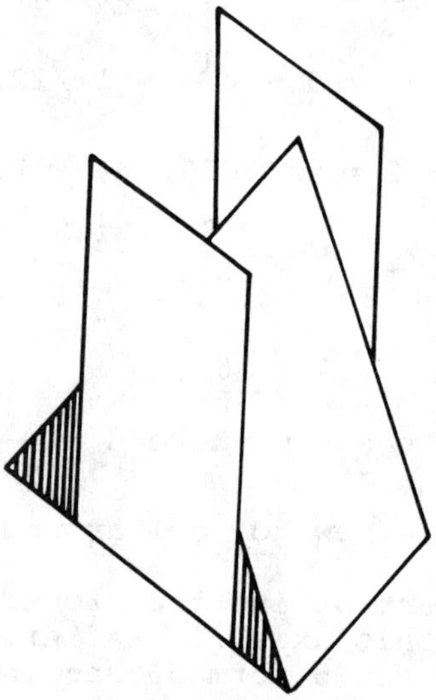

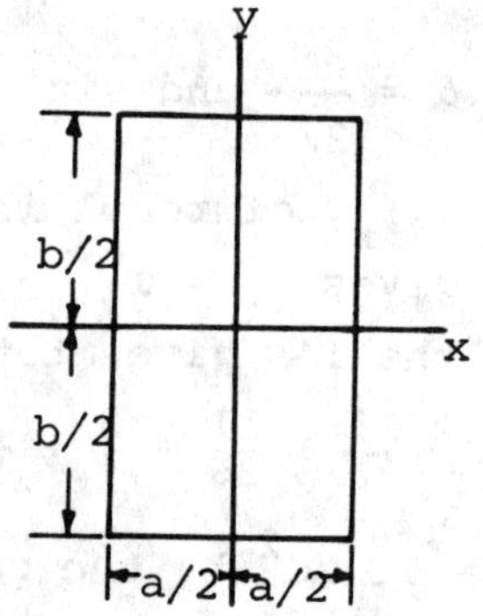

Solution: i) The rectangular plate and its principal axes are shown in the diagram. The calculation of the inertia tensor is straightforward:

$$I_1 = \int (y^2 + z^2)dm = (M/ab) \int y^2 \, dxdy$$

$$= (M/ab) \int_{-a/2}^{a/2} dx \int_{-b/2}^{b/2} y^2 dy = Mb^2/12$$

$$I_2 = \int (x^2 + z^2)dm = (M/ab) \int x^2 \, dxdy = Ma^2/12$$

$$I_3 = \int (x^2 + y^2)dm = M(a^2 + b^2)/12$$

$$I_{12} = (M/ab) \int_{-a/2}^{a/2} xdx \int_{-b/2}^{b/2} ydy = 0$$

$I_{ij} = 0$ for $i \neq j$.

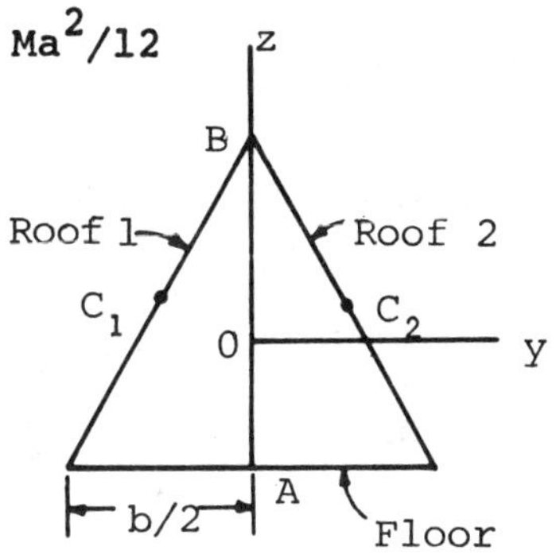

ii) Imagine a plane which is parallel to the two walls of the house and is midway between them. The cut this plane makes with the roofs and the floor is sketched in the diagram. Points A, C_1 and C_2 are, respectively, the center of the floor, roof 1, and roof 2. Wall 1 is a distance (a/2) below and wall 2 is a distance (a/2) above the plane of drawing.

From symmetry the center of mass of the house must be somewhere on the line AB, say point 0, where $\overline{A0} = h$. To determine h, we know that the center of mass of the roofs and the floor is at a point $b/2\sqrt{3}$ above point A, and the center of mass of the two walls is at a point b/2 above point A. Therefore,

$$h = [3M \times (b/2\sqrt{3}) + 2M \times (b/2)]/5M = \frac{1 + \sqrt{3}/2}{5} b.$$

The center of mass 0 is chosen as the origin of the principal axes coordinate system. The z-axis and the y-axis are as shown. The x-axis is perpendicular to and pointing out of the drawing.

Note here the parallel-axis theorem which will be used repeatedly,

$$I = I_{CM} + Md^2,$$

where I is the moment of inertia of a rigid body about a given axis, I_{CM} is the moment of inertia about an axis parallel to the given axis but passing through the center of mass and d is the perependicular distance between these two axes.

(a) Calculation of I_1

From Roof 1 $\quad I_1 = \frac{Mb^2}{12} + M\overline{0C_1}^2 = \frac{Mb^2}{12} + M\left[\left(\frac{\sqrt{3}}{4}b-h^2\right) + \left(\frac{b}{4}\right)^2\right]$

$$= Mb^2\left(\frac{19}{75} - \frac{3\sqrt{3}}{50}\right)$$

From Roof 2 $\quad I_1 = Mb^2\left(\frac{19}{75} - \frac{3\sqrt{3}}{50}\right)$

From Floor $\quad I_1 = \frac{Mb^2}{12} + Mh^2 = Mb^2\left(\frac{23}{150} + \frac{\sqrt{3}}{25}\right)$

From Wall 1 $\quad I_1 = \frac{M(a^2 + b^2)}{12} + M\left(\frac{b}{2} - h\right)^2 = \frac{Ma^2}{12} + Mb^2\left(\frac{61}{300} - \frac{3\sqrt{3}}{50}\right)$

From Wall 2 $\quad I_1 = \frac{Ma^2}{12} + Mb^2\left(\frac{61}{300} - \frac{3\sqrt{3}}{50}\right).$

Totalling up,

$$I_1 = \frac{Ma^2}{6} + Mb^2\left\{2 \times \frac{19}{75} - \frac{3\sqrt{3}}{25} + \frac{23}{150} + \frac{\sqrt{3}}{25} + \frac{61}{150} - \frac{3\sqrt{3}}{25}\right\}$$

$$= \frac{Ma^2}{6} + Mb^2\left(\frac{16}{15} - \frac{\sqrt{3}}{5}\right).$$

(b) Calculation of I_2.

First, the general formula

$$I = \ell^2 K_x + m^2 I_y + n^2 I_z - 2nmI_{yz} - 2\ell nI_{zx} - 2m\ell I_{xy}$$

is used to find the moment of inertia of a rigid body about any line in terms of the directional cosines (ℓ, m, n) of that line and the inertia tensor of the body. Apply the formula to find $I_{y''}$ and $I_{z''}$ of roof 2.

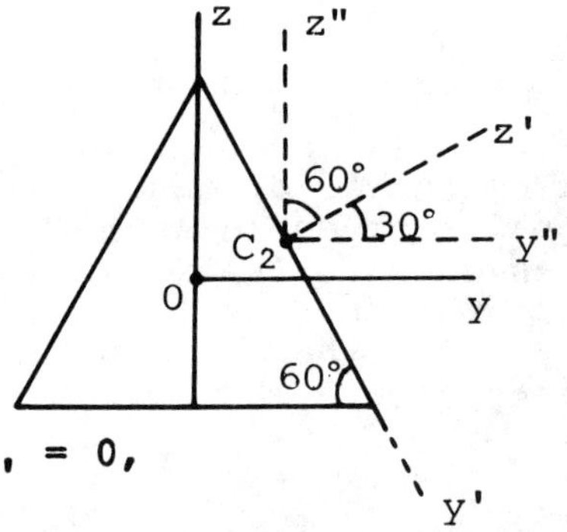

Since

$$I_{x'} = Mb^2/12$$

$$I_{y'} = Ma^2/12$$

$$I_{z'} = M(a^2 + b^2)/12$$

$$I_{x'y'} = I_{y'z'} = I_{z'x'} = 0,$$

and the directional cosines of the y" axis and the z" axis are, respectively,

$$(0,\ 1/2,\ \sqrt{3}/2) \quad \text{and} \quad (0,\ \sqrt{3}/2,\ 1/2),$$

we find

$$I_{y''} = \frac{Ma^2}{12} + \frac{Mb^2}{16}, \quad I_{z''} = \frac{Ma^2}{12} + \frac{Mb^2}{48}.$$

Now calculate I_2.

From both roofs $\quad I_2 = 2\left\{\frac{Ma^2}{12} + \frac{Mb^2}{16} + M\left(\frac{\sqrt{3}}{4}b - h\right)^2\right\}$

$$= \frac{Ma^2}{6} + Mb^2\left(\frac{17}{50} - \frac{3\sqrt{3}}{25}\right)$$

From floor $\quad I_2 = \frac{Ma^2}{12} + Mh^2 = \frac{Ma^2}{12} + Mb^2\left(\frac{7}{100} + \frac{\sqrt{3}}{25}\right)$

From both walls $\quad I_2 = 2\left\{\frac{Mb^2}{12} + M\left[\left(\frac{b}{2} - h\right)^2 + \left(\frac{a}{2}\right)^2\right]\right\}$

$$= \frac{Ma^2}{2} + Mb^2\left(\frac{61}{150} - \frac{3\sqrt{3}}{25}\right).$$

Summing,

$$I_2 = \frac{3}{4}Ma^2 + Mb^2\left(\frac{49}{60} - \frac{\sqrt{3}}{5}\right).$$

(c) Calculation of I_3

From both roofs $\quad I_3 = 2\left\{\frac{Ma^2}{12} + \frac{Mb^2}{48} + M\left(\frac{b}{4}\right)^2\right\}$

$$\frac{Ma^2}{6} + \frac{Mb^2}{6}$$

From floor $\quad I_3 = \frac{M}{12}(a^2 + b^2)$

From both walls $\quad I_3 = 2\left\{\frac{Ma^2}{12} + M\left(\frac{a}{2}\right)^2\right\} = \frac{2}{3}Ma^2.$

Adding,

$$I_3 = \frac{11}{12}Ma^2 + \frac{1}{4}Mb^2.$$

(d) Finally, because the house is symmetrical about the plane $y = 0$,

$$I_{12} = \int xy\,dm = 0 \text{ , and } I_{23} = 0.$$

The house is also symmetric about the plane $x = 0$, therefore,

$$I_{13} = 0.$$

That is, all the non-diagonal elements of hhe inertia tensor vanish.

Find the equation for the ellipsoid of inertia of an object in the shape of an ellipsoid whose equation is

$$\frac{x^2}{\ell^2} + \frac{y^2}{\omega^2} + \frac{z^2}{h^2} = 1.$$

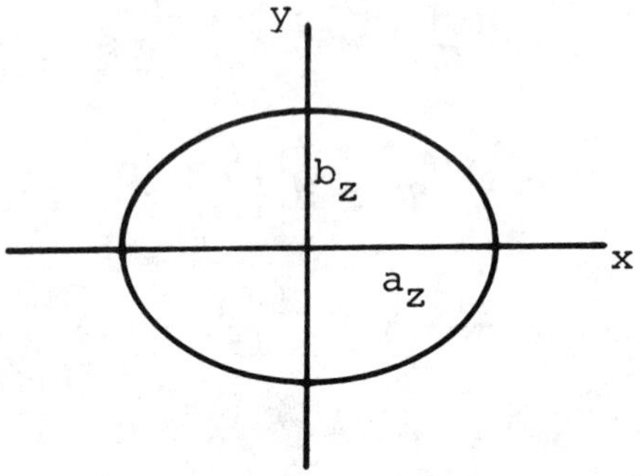

Solution: The equation for the ellipsoid of inertia (or momental ellipsoid) of an object is given by

$$x^2 I_x + y^2 I_y + z^2 I_z - 2xyI_{xy} - 2yzI_{yz} - 2zxI_{zx} = 1$$

wheje the I's are the components of the inertia tensor of the object.

The object in the problem is itself an ellipsoid given by

$$\frac{x^2}{\ell^2} + \frac{y^2}{\omega^2} + \frac{z^2}{h^2} = 1. \tag{1}$$

This implies that the coordinate axes x,y,z are the principal axes. Therefore, the products of inertia vanish, $I_{xy} = I_{yz} = I_{zx} = 0$, and then momental ellipsoid reduces to

$$x^2 I_x + y^2 I_y + z^2 I_z = 1. \tag{2}$$

Before calculating the moments of inertia, show that the volume of the ellipsoid is $V = (4\pi/3)\ell\omega h$.

Consider the cut made by the plane z = z and the ellipsoid as shown in the diagram. It is an ellipse with axes a_z and b_z. For constant z, equation (1) can be written as

$$\frac{x^2}{\ell^2} + \frac{y^2}{\omega^2} = 1 - \frac{z^2}{h^2}, \quad \text{or} \quad \frac{x^2}{\ell^2\left(1 - \frac{z^2}{h^2}\right)} + \frac{y^2}{\omega^2\left(1 - \frac{z^2}{h^2}\right)} = 1,$$

or $$\frac{x^2}{a_z^2} + \frac{y^2}{b_z^2} = 1$$

where $$a_z = \ell(1 - z^2/h^2)^{1/2}, \tag{3}$$

$$b_z = \omega(1 - z^2/h^2)^{1/2}.$$

The area of this ellipse is $\pi a_z b_z$. The volume of the slice is $\pi a_z b_z d_z$. The volume of the ellipsoid is

$$V = \int_{-h}^{h} \pi a_z b_z d_z = \int_{-h}^{h} \pi\ell\omega(1 - z^2/h^2)dz = \frac{4\pi}{3}\ell\omega h. \tag{4}$$

Next, show that the moment of inertia of a laminar ellipse $x^2/a^2 + y^2/b^2 = 1$ of uniform density ρ about the z-axis is $(\rho\pi ab/4)(a^2 + b^2)$.

$$I_z' = \iint (x^2 + y^2)\rho dxdy = \rho \int_{-a}^{a} dx \int_{-b\sqrt{1-x^2/a^2}}^{b\sqrt{1-x^2/a^2}} (x^2 + y^2)dy$$

$$= \rho \int_{-a}^{a} dx\left[2bx^2(1 - x^2/a^2)^{1/2} + \frac{2}{3}b^3(1 - x^2/a^2)^{3/2}\right]$$

$$= \rho\left[2b \cdot \frac{\pi a^3}{8} + \frac{2}{3}b^3 \cdot \frac{3\pi a}{8}\right]$$

$$= (\rho\pi ab/4)(a^2 + b^2). \tag{5}$$

Now calculate the moments of inertia of the ellipsoid.

$$I_z = \iiint (x^2 + y^2)dm = \int_{-h}^{h} dz I_z'$$

$$= \int_{-h}^{h} dz(\rho\pi a_z b_z/4)(a_z^2 + b_z^2)$$

$$I_z = (\rho\pi/4)\int_{-h}^{h} dz\ell\omega(1 - z^2/h^2)(\ell^2 + \omega^2)(1 - z^2/h^2)$$

$$= (\rho\pi\ell\omega(\ell^2 + \omega^2)/4)\int_{-h}^{h} (1 - z^2/h^2)^2 dz$$

$$= (\rho\pi\ell\omega(\ell^2 + \omega^2)/4)\cdot(16/15)h$$

$$= \left(\frac{4}{3}\pi \ell \omega h \cdot \rho\right) \cdot \frac{1}{5}(\ell^2 + \omega^2)$$

$$= \frac{M}{5}(\ell^2 + \omega^2).$$

We have used equations (3), (4),and (5). M is the total mass of the ellipsoid. Due to symmetry, it can be shown in a similar manner that

$$I_x = \frac{M}{5}(\omega^2 + h^2), \quad I_y = \frac{MM}{55}(h^2 + \ell^2).$$

After substituting these moments of inertia into equation (2), the result is the equation of momental ellipsoid,

$$(\omega^2 + h^2)x^2 + (h^2 + \ell^2)y^2 + (\ell^2 + \omega^2)z^2 = 5/M.$$

GENERAL SPATIAL MOTION

• PROBLEM 19-7

A rod AB with diameter d = 3 in. length L = 12 ft. and weight w = 60 lb. is connected to a rotating shaft, DE, by a pin. The shaft rotates with a constant angular velocity w = 10 rad/s. The rod is held at an angle of 45° to the shaft by a wire BC. Find the tension in the wire and the reaction at A.

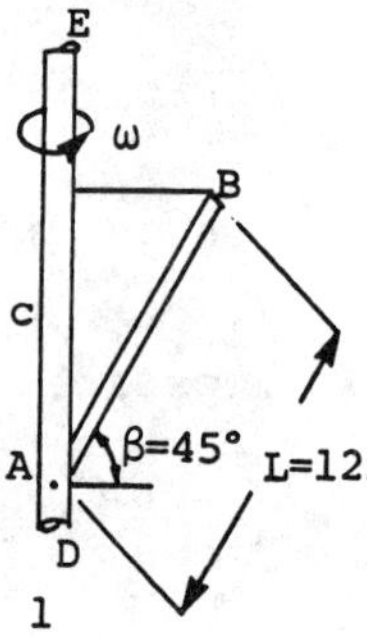

Fig. 1

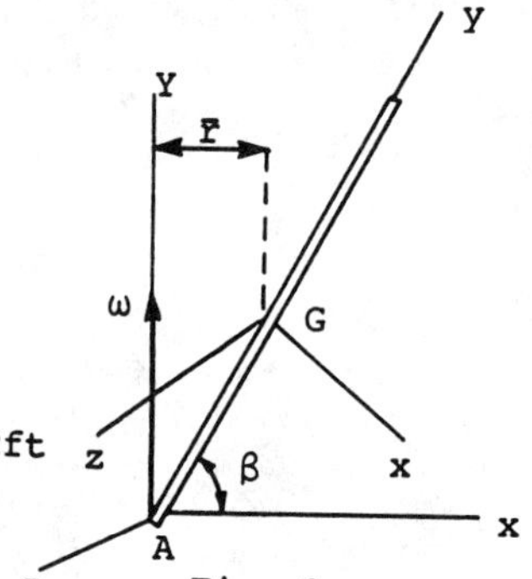

Fig. 2

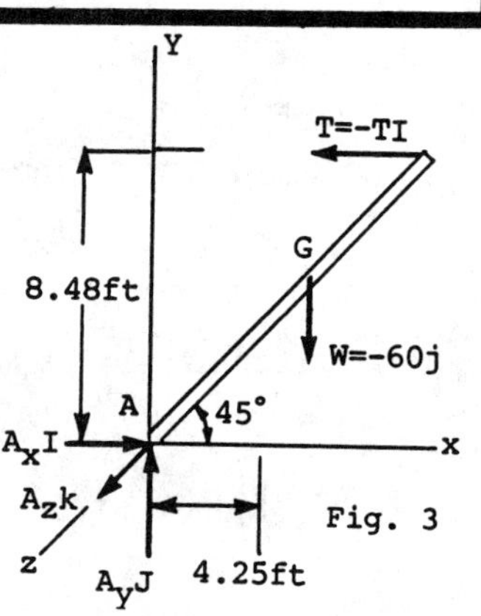

Fig. 3

Solution: The external forces acting on the rod, the weight $\vec{W}$, tension $\vec{T}$, and the reaction force $\vec{A}$ create a net force and torque which are equal to, respectively, $m\vec{a}_G$ and $\dot{\vec{H}}_G$ where G is the center of mass as seen in fig. 2.

Since G describes a horizontal circle of radius $r = \frac{1}{2}L \cos \beta$ at the constant rate ω, this can be expressed as

$$\vec{a}_G = -r\omega^2\hat{I} = -(\tfrac{1}{2}L \cos 45°)(\omega^2)\hat{I}$$

$$\vec{a}_G = -424 \text{ ft/sec}^2\hat{I} \qquad (1)$$

where I is a unit vector in the X direction of the X, Y, Z

frame, and the rod is in the position shown at the instant chosen. $\vec{H}_G$, the angular momentum, will be determined before finding $\dot{\vec{H}}_G$. Using the principal axes of the rod, x, y, and z,

$$I_x = \frac{1}{12}m\left(3\left(\frac{d}{2}\right)^2 + L^2\right),\ I_y = \frac{1}{2}m\left(\frac{d}{2}\right)^2,\ I_z = \frac{1}{12}m\left(3\left(\frac{d}{2}\right)^2 + L^2\right)$$

Expressing $\vec{\omega}$ in this coordinate system,

$$\omega_x = -\omega \cos \beta\ ,\quad \omega_y = \omega \sin \beta\ ,\quad \omega_z = 0.$$

Combining these for $\vec{H}_G = I_x\omega_x\hat{i} + I_y\omega_y\hat{j} + I_z\omega_z\hat{k}$,

$$\vec{H}_G = -\frac{1}{12}m\left(3\left(\frac{d}{2}\right)^2 + L^2\right)\omega \cos \beta\ \hat{i} + \frac{1}{2}m\left(\frac{d}{2}\right)^2 \omega \sin \beta\ \hat{j}$$

Since $\hat{i} = \cos(90° - \beta)\hat{I} - \sin(90° - \beta)\hat{J}$, $\hat{j} = \cos \beta\ \hat{I} + \sin \beta\ \hat{J}$,
$\hat{i} = \sin \beta\ \hat{I} - \cos \beta\ \hat{J}$, $\hat{j} = \cos \beta\ \hat{I} + \sin \beta\ \hat{J}$

$$\vec{H}_G = \left[-\frac{m}{12}\left(3\left(\frac{d}{2}\right)^2 + L^2\right) + \frac{m}{2}\left(\frac{d}{2}\right)^2 \omega \cos \beta \sin \beta\right]\hat{I}$$

$$+ \left[\frac{m}{12}\left(3\left(\frac{d}{2}\right)^2 + L^2\right)\omega \cos^2 \beta + \frac{m}{2}\left(\frac{d}{2}\right)^2 \omega \sin^2 \beta\right]\hat{J}$$

The magnitude of $\vec{H}_G$ does not change, so

$$\dot{\vec{H}}_G = \omega \times \vec{H}_G \text{ only.} \qquad \dot{\vec{H}}_G = (\omega\hat{J}) \times \vec{H}_G$$

$$\dot{\vec{H}}_G = -\left[-\frac{m}{12}\left(3\left(\frac{d}{2}\right)^2 + L^2\right) + \frac{m}{2}\left(\frac{d}{2}\right)^2\right]\omega^2 \cos \beta \sin \beta\ \hat{K}$$

$$\dot{\vec{H}}_G = (2271 \text{ lb ft})\hat{K} \tag{2}$$

In expressing that the sum of the torques is equal to the change in angular momentum, the change in angular momentum of the center of mass about the axis chosen must be added. Using point A and the Y axis,

$$\dot{\vec{H}}_{AG} = \frac{d}{dt}(m\vec{r} \times \vec{v}) = m(\vec{r} \times \vec{a}). \tag{3}$$

The equations of motion, using the system of forces shown in figure 3 and equations (1), (2) and (3), are

$$\sum\vec{F} = m\vec{a}$$

$$Ax\hat{I} + Ay\hat{J} + Az\hat{K} - W\hat{J} - T\hat{I} = \left(\frac{60}{9}\right)(-424)\hat{I} \tag{4}$$

and $$\sum M_A = \dot{\vec{H}}_G + \dot{\vec{H}}_{AB}$$

$$(12\sin 45°)\hat{J} \times (-T\hat{I}) + (6 \cos 45°)\hat{I} \times (-W)\hat{J} = 2271\hat{K}$$

$$+ \left(\frac{60}{9}\right)(6 \sin 45°)\hat{J} \times (-424)\hat{I}. \tag{5}$$

After substitution, equation (5) becomes

$$(8.49T - 255)\hat{K} = (2271 + 3352)\hat{K}$$

so $$T = 692 \text{ lb.}$$

Using this value in equation (4),

$$Ax\hat{I} + Ay\hat{J} + Az\hat{K} = 692\hat{I} + 60\hat{J} - 790\hat{I},$$

so $$\vec{A} = -(98 \text{ lb})\hat{I} + (60 \text{ lb})\hat{J}.$$

Now, it is clear what A and T are when the rod is rotating. The vectors rotate with the rod keeping the same orientation relative to the rod.

• PROBLEM 19-8

A part from a machine consists of a homogeneous disk connected to a rod. The rod is connected to a pivot at point 0. The disk rolls on a horizontal plate. The disk has a radius r = 10 cm and a mass m = 0.5 kg, and rotates counter clockwise at a constant angular velocity ω_1 = 10 rad/s about the rod. The rod has a length L = 50cm. Determine (a) the force (assuming vertical) exerted by the floor on the disk, (b) the reaction at the pivot 0.

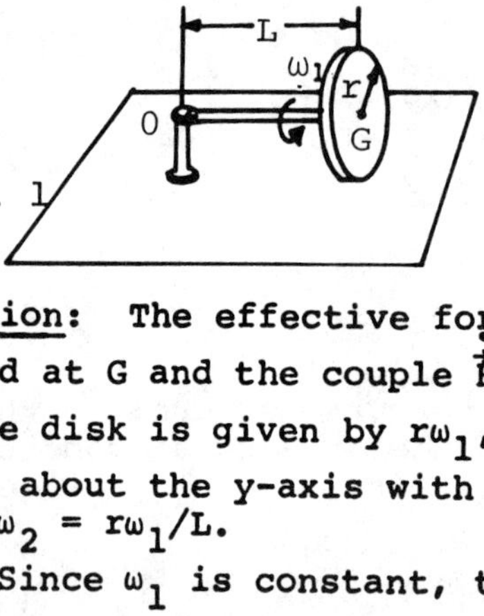

Fig. 1

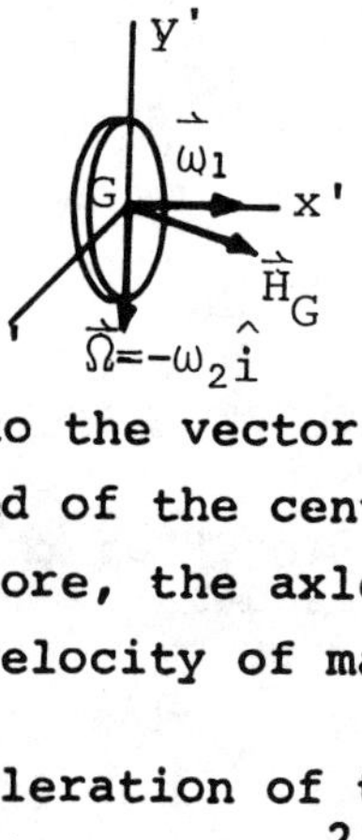

Fig. 2

Solution: The effective forces reduce to the vector $m\bar{a}$ attached at G and the couple $\dot{\vec{H}}_G$. The speed of the center, G, of the disk is given by $r\omega_1$, and, therefore, the axle rotates about the y-axis with an angular velocity of magnitude $\omega_2 = r\omega_1/L$.

Since ω_1 is constant, the only acceleration of the disk is centripetal acceleration of magnitude $L(\omega_2)^2$ directed in the negative x-direction. Therefore, we can write

$$m\bar{a} = -mL\omega_2^2 i = -mL(r\omega_1/L)^2 i = -(mr^2\omega_1^2/L)i. \tag{1}$$

The angular momentum of the disk about G maybe found from

$$\vec{H}_G = I_{x'}\omega_{x'}\hat{i} + I_{y'}\omega_{y'}\hat{j} + I_{z'}\omega_{z'}\hat{k}.$$

From $\omega_{x'} = \omega_1$, $\omega_{y'} = -\omega_2 = -\frac{r\omega_1}{L}$, $I_{x'} = \frac{1}{2}mr^2$ and

$I_{y'} = \frac{1}{4}mr^2$ and, since $\omega_{z'} = 0$, the result is

$$\vec{H}_g = \frac{1}{2}mr^2\omega_1(\hat{i} - \frac{r}{2L}\hat{j}),$$

where H_G is resolved into components along the rotating axes x', y', z', with x' along OG and y' vertical . See figure 2. Noting that the rate of change $(\dot{\vec{H}}_G)_{Gx'y'z'}$ of $\vec{H}_G$ with respect to the rotating frame is zero, and that the angular velocity $\vec{\Omega}$ of that frame is

$$\vec{\Omega} = -\omega_2\hat{j} = -(r\omega_1/L)\hat{j},$$

we have

$$\dot{\vec{H}}_G = (\dot{\vec{H}}_G)_{Gx'y'z'} + \vec{\Omega} \times \vec{H}_G = 0 - \frac{r\omega_1}{L}\hat{j} \times \frac{1}{2}mr^2\omega_1\left(\hat{i} - \frac{r}{2L}\hat{j}\right)$$

$$= \frac{1}{2}mr^2(r/L)\omega_1^2\hat{k}. \qquad (2)$$

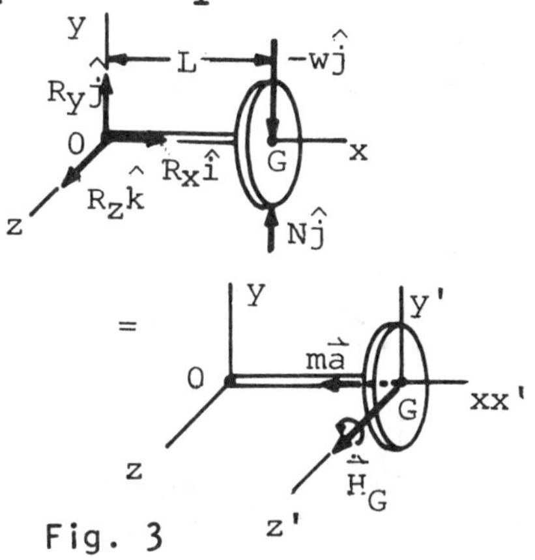

Fig. 3

In figure 3 a free body diagram of the system and the effective force and moment are shown. Expressing that the system of the external forces is equivalent to the system of the effective forces, write

$$\sum\vec{M}_0 = \sum(\vec{M}_0)_{eff}: \quad L\hat{i} \times (N\hat{j} - W\hat{j}) = \dot{\vec{H}}_G$$

$$(N - W)L\hat{k} = \frac{1}{2}mr^2(r/L)\omega_1^2\hat{k}$$

$$\vec{N} = [W + \frac{1}{2}mr(r/L)^2\omega_1^2]\hat{j}, \qquad (3)$$

$$= [(0.5)(9.81) + \frac{1}{2}(0.5)(0.05)(\frac{0.10}{0.50})^2(10)^2]\hat{j}$$

$$N = 4.955 \text{ N}\hat{j}$$

$$\sum\vec{F} = \sum\vec{F}_{eff}: \quad \vec{R} + N\hat{j} - W\hat{j} = m\vec{a}.$$

Substituting for N from (3), for ma from (1), and solving for R:

$$\vec{R} = -(mr^2\omega_1^2/L)\hat{i} - \tfrac{1}{2}mr(r/L)^2\omega_1^2\hat{j}$$

$$\vec{R} = -\frac{mr^2\omega_1^2}{L}\left(\hat{i} + \frac{r}{2L}\hat{j}\right)$$

$$= -(0.5 \times 0.1^2 \times 10^2/0.5)\hat{i} - (\tfrac{1}{2} \times 0.5 \times 0.1 \times (\tfrac{0.1}{0.5})^2 \times 10^2)\hat{j}$$

$$R = -1\,\hat{i} - 0.1\,\hat{j}\ \text{N}$$

• PROBLEM 19-9

> Consider the machine part in the preceding problem. All conditions remain the same, except that the disk rotates with an angular velocity of $\omega_1 = 15$ rad/sec. Determine (a) the angular velocity of the disk (b) its angular momentum about pivot 0 (c) its kinetic energy.

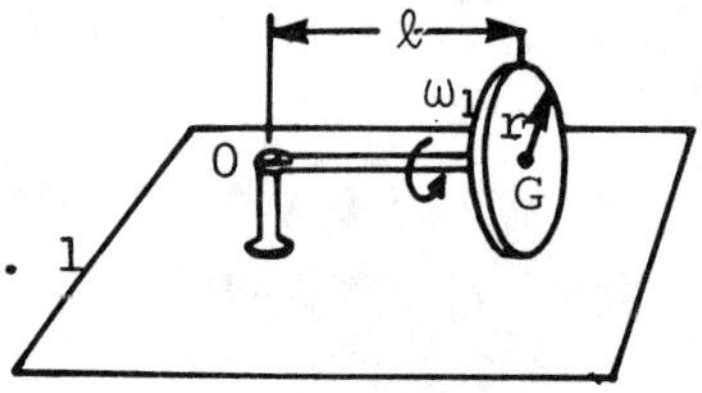

Fig. 1

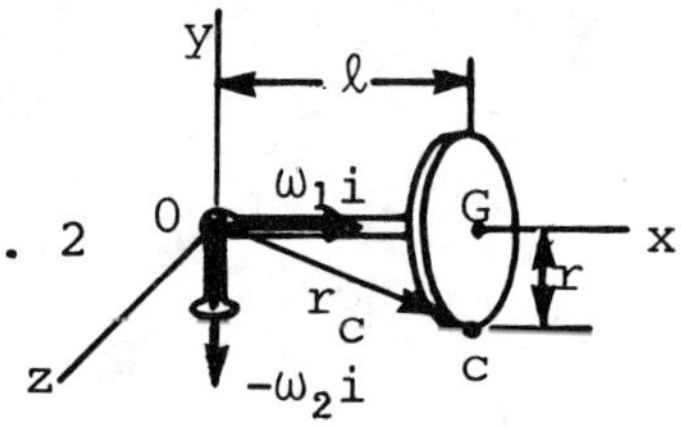

Fig. 2

Solution: a) The total angular velocity is the sum of the given rotation plus the resulting rotation about the pivot O; that is,

$$\vec{\omega} = \omega_1\hat{i} - \omega_2\hat{j}. \qquad (1)$$

The relevant quantities are shown in figure 2. To determine ω_2, write that the velocity of a point C, in contact with the floor, is instantaneously zero:

$$\vec{v}_C = \vec{\omega} \times \vec{v}_C = (\omega_1\hat{i} - \omega_2\hat{j}) \times (\ell\hat{i} - r\hat{j}) = 0$$

$$= (\ell\omega_2 - r\omega_1)\hat{k} = 0$$

so

$$\omega_2 = r\omega_1/\ell. \qquad (2)$$

Substituting into (1) for ω_2 yields

$$\vec{\omega} = \omega_1\hat{i} - \left(\frac{r\omega_1}{\ell}\right)\hat{j}.$$

$$= 15\,\hat{i} - (0.1 \times 15/0.5)\hat{j}$$

$$= 15\,\hat{i} - 3\,\hat{j}\ \text{rad/sec}$$

The reader who dislikes vector techniques may find equation (2) by geometrical reasoning.

b) In this problem the principal axes of the road and disk are aligned, so we may calculate the angular momentum components easily.

$$L_x = I_x\omega_x = \left(\frac{1}{2}mr^2\right)\omega_1$$

$$L_y = I_y\omega_y = \left(m\ell^2 + \frac{1}{4}mr^2\right)\left(-\frac{r\omega_1}{\ell}\right)$$

$$L_z = I_z\omega_z = 0 \quad \text{since} \quad \omega_z = 0.$$

Thus,

$$\vec{L} = \frac{1}{2}mr^2\omega_1\hat{i} - m\left(\ell^2 + \frac{1}{4}r^2\right)\frac{r\omega_1}{\ell}\hat{j}.$$

$$= (\frac{1}{2} \times 0.5 \times 0.1^2 \times 15)\hat{i}$$

$$- 0.5(0.5^2 + \frac{1}{4} \times 0.1^2)\frac{0.1 \times 15}{0.5}\hat{j}$$

$$L = 0.0375\ \hat{i} - 0.3788\ \hat{j}\ \text{Kg} \cdot \text{m}^2/\text{sec}$$

c) Calculate the kinetic energy using the formula

$$T = \frac{1}{2}(I_x\omega_x^2 + I_y\omega_y^2 + I_z\omega_z^2).$$

$$T = \frac{1}{2}\left[\frac{1}{2}mr^2\omega_1^2 + m(\ell^2 + \frac{1}{4}r^2)\left(-\frac{r\omega_1}{\ell}\right)^2\right] = \frac{1}{8}mr^2\left(6 + \frac{r^2}{\ell^2}\right)\omega_1^2.$$

$$= \frac{1}{8} \times 0.5 \times 0.1^2 \times (6 + \frac{0.1^2}{0.5^2})15^2$$

$$= 0.0849 \text{ joule}$$

• PROBLEM 19-10

For a uniform sphere with moments of inertia $I_1 = I_2 = I_3$, use the Euler equations to find the equation of motion of the sphere.

Solution: The Euler Equations are

$$I_1\frac{d\omega_1}{dt} + \left(I_3 - I_2\right)\omega_3\omega_2 = N_1$$

$$I_2\frac{d\omega_2}{dt} + \left(I_1 - I_3\right)\omega_1\omega_3 = N_2$$

$$I_3\frac{d\omega_3}{dt} + \left(I_2 - I_1\right)\omega_2\omega_1 = N_3$$

where the subscript 1 refers to the first principal axis of the sphere. N and ω are the net external torque and angular velocity of the sphere. Noting that $I_1 = I_2 = I_3$, we obtain

$$I_1\frac{d\omega_1}{dt} = N_1; \qquad I_2\frac{d\omega_2}{dt} = N_2; \qquad I_3\frac{d\omega_3}{dt} = N_3$$

Defining $I_1 = I_2 = I_3 = I$, we may write

$$\frac{d\omega_1}{dt} = \frac{N_1}{I}, \quad \frac{d\omega_2}{dt} = \frac{N_2}{I}, \quad \frac{d\omega_3}{dt} = \frac{N_3}{I} \tag{1}$$

In free motion $N = 0$, and (1) tells us that $\omega = \text{const}$. The result $\omega = \text{const}$ is a special feature of the free rotating sphere.

• PROBLEM 19-11

Shown in the figure is a schematic diagram of an armature that has been misaligned; that is, the axis of symmetry of the armature does not coincide with the axis of rotation of the body. The angle between the axis of symmetry and the axis of rotation is given as ϕ, as shown in the figure. If the armature is rotating with an angular velocity of $\dot{\theta}$, what are the reactions at the bearings A and B?

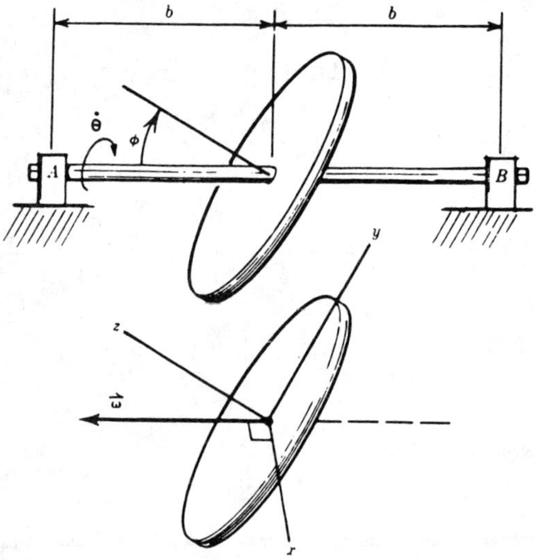

Solution: The reference axes for the body are chosen as defined in the figures. The axes shown are the principal axes for the body. First, write the moment of momentum (angular momentum) for the body:

$$\vec{H} = I_{xx}\omega_x\hat{i} + I_{yy}\omega_y\hat{j} + I_{zz}\omega_z\hat{k}.$$

The angular velocity of the body is

$$\vec{\omega} = -\dot{\theta}\sin\phi\hat{j} + \dot{\theta}\cos\phi\hat{k}.$$

The moments of inertia of a body are

$$I_{xx} = \int_V \rho(\gamma)\ (\gamma^2 - x^2)dv$$

$$I_{yy} = \int_V \rho(\gamma)\ (\gamma^2 - y^2)dv$$

and $$I_{zz} = \int_V \rho(\gamma)\ (\gamma^2 - z^2)dv$$

where $\rho(\gamma)$ is the density and v is the volume of the body under consideration. For the disc $\rho(\gamma) = \frac{m}{\pi R^2 h}$ where R is the radius of the disc and h is the width considered much smaller than R. Therefore,

$$I_{zz} = \frac{m}{\pi R^2 h} \int_0^R u\, du \int_0^{2\pi} d\theta \int_0^h dz\, u^2 = \frac{mR^2}{2}$$

using the cylindrical coordinates.

Also, using the perpendicular axis theorem,

$$I_{xx} + I_{yy} = I_{zz} \quad \text{and} \quad I_{xx} = I_{yy};$$

hence, $I_{xx} = I_{yy} = \frac{mR^2}{4}$. Thus, the momentum may be written as

$$H = -\frac{mR^2\dot{\theta}}{4} \sin \phi \hat{j} + \frac{mR^2\dot{\theta}}{2} \cos \phi \hat{k}.$$

Sinc Since the angular velocity of rotation of the body about the axis of rotation is a constant, and since the angle ϕ does not change, the time derivative of the moment of momentum becomes

$$\dot{\vec{H}} = -\frac{mR^2\dot{\theta}}{4} \sin \phi \dot{\hat{j}} + \frac{mR^2\dot{\theta}}{2} \cos \phi \dot{\hat{k}}.$$

Evaluating the time derivatives of the unit vectors $\hat{j}$ and $\hat{k}$,

$$\dot{\hat{j}} = \vec{\omega} \times \hat{j} = \begin{vmatrix} \hat{i} & \hat{j} & \hat{k} \\ 0 & -\dot{\theta} \sin \phi & \dot{\theta} \cos \phi \\ 0 & 1 & 0 \end{vmatrix} = -\dot{\theta} \cos \phi \hat{i}$$

$$\dot{\hat{k}} = \vec{\omega} \times \hat{k} = \begin{vmatrix} \hat{i} & \hat{j} & \hat{k} \\ 0 & -\dot{\theta} \sin \phi & \dot{\theta} \cos \phi \\ 0 & 0 & 1 \end{vmatrix} = -\dot{\theta} \sin \phi \hat{i}$$

The time derivative of the moment of momentum becomes

$$\dot{\vec{H}} = -\frac{mR^2\dot{\theta}^2}{8} \sin 2\phi \hat{i}$$

and then from the equation of motion, $\vec{M} = \dot{\vec{H}}$, it follows that

$$M_x = -\frac{mR^2\dot{\theta}^2}{8} \sin 2\theta$$

$$M_y = 0$$

$$M_z = 0.$$

This says that there will be a moment only about the x axis which will come from a force couple at the bearings. From the minus sign for M_x, the couple will tend to cause a clockwise turning about the x-axis. Therefore, in the position shown the reaction is down at B and up at A. But note that the x axis is fixed to the wheel. Thus, the moment of the body and the forces at the reactions depend upon the position of the x axis. When the x axis is in the position shown, the vertical reactions at A and B will be a maximum; when the x axis is vertical, the vertical reactions at A and B will be equal to zero and the horizontal reactions will be a maximum. Thus, the vertical reactions are harmonic functions of time.

The magnitude of the reactions may be evaluated by noting that the moment about the x axis in the position shown will be equal to $2A_v b$. (The forces A_v and B_v are equal in magnitude and opposite in direction. The two must add to zero since there is no motion of the center of mass of the sytem in that direction.) The magnitude of the vertical reaction at A is

$$A_v = \frac{mR^2\dot{\theta}^2}{16b} \sin 2\phi$$

and since the line of action of the reactive forces rotates with the body, it follows that

$$A_v(t) = \left[\frac{mR^2\dot{\theta}^2}{16b} \sin 2\phi\right] \cos \dot{\theta}t.$$

and

$$A_H(t) = \left[\frac{mR^2\dot{\theta}^2}{16b} \sin 2\phi\right] \sin \dot{\theta}t,$$

taking the positive horizontal direction to be into the paper. Of course,

$$B_v(t) = -A_v(t)$$

and

$$B_H(t) = -A_H(t).$$

• PROBLEM 19-12

The "gyro-pendulum." A gyro-pendulum consists of a bar in the form of a T supported by bearings at the points B and C as shown in Fig. 1. At point A, a distance R from CB, the pendulum carries a thin, circular disc (mass m, radius r) which spins with an angular velocity ω about OA, as indicated. Using the dimensions shown in the figure, determine the forces on the bearings when the pendulum passes its lowest point after having been released from the angle θ = 90°. Neglect the mass of the bar OA.

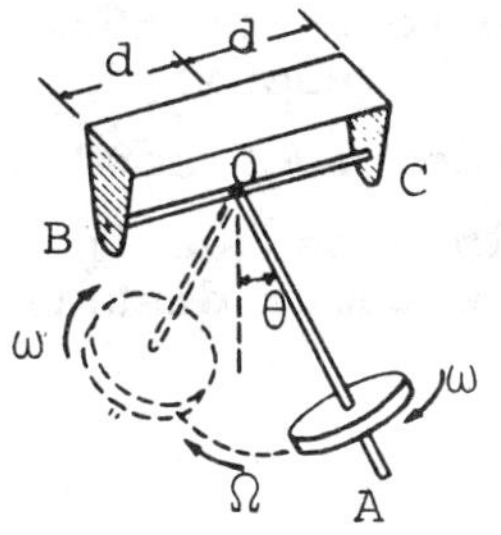

Fig. 1

Pendulum with spinning disc (gyro-pendulum).

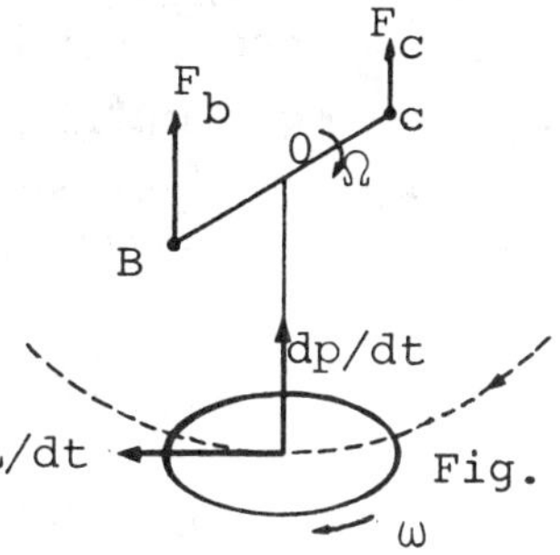

Fig. 2

Solution: If the angular velocity of the pendulum is denoted at its lowest point by Ω, the forces at B and C can be obtained as follows. The acceleration of the center of mass of the disc at the lowest point is $\Omega^2 R$, directed upward. At the lowest point, the spin angular momentum $\vec{L}$ of the disc points downward, and the magnitude of its rate of change is ΩL. The direction of $d\vec{L}/dt$ is the same as the direction of motion of A, i.e., horizontally toward the left, as shown in Fig. 2.

The remaining contribution to the angular momentum of the pendulum with respect to the point O, due to the motion of the center of mass, has no rate of change at the lowest point. Thus, if the vertical components of the forces on the axis BC from the bearings are denoted by F_b and F_c, the result is

$$F_b + F_c - mg = m\Omega^2 R,$$

$$(F_b - F_c)d = \Omega L = I_0 \omega \Omega.$$

From these equations, we find

$$F_b = \frac{mg}{2} + \frac{m\Omega^2 R}{2} + \frac{I_0 \omega \Omega}{2d}$$

$$F_c = \frac{mg}{2} + \frac{m\Omega^2 R}{2} - \frac{I_0 \omega \Omega}{2d}$$

Since there are no other components of $d\vec{p}/dt$ and $d\vec{L}/dt$, we realize that there are no horizontal force components at B and C.

The angular velocity Ω can be found directly from the work-energy principle. The moment of inertia with respect to the axis of rotation, BC, of the disc is obtained from the parallel-axis theorem. The moment of inertia with respect to the center-of-mass axis parallel to CB is $I_0 = mr^2/4$, and, therefore, $I = I_0 + mR^2 = mr^2/4 + mR^2$. The angular velocity is then obtained from $\frac{1}{2}I\Omega^2 = mgR$, and the result is

$$\Omega = \sqrt{\frac{8gR}{r^2 + 4R^2}}.$$

Insertion of this value of Ω into the equations for F_b and F_c gives the final answer.

It is interesting to note that the force on the bearing at C can become zero when the angular velocity of the disc is $\omega = (mg + m\Omega^2 R)d/I_0\Omega$. Thus, under such conditions, when the pendulum passes its lowest point the bearing at C can actually be removed for an instant without disturbing the motion.

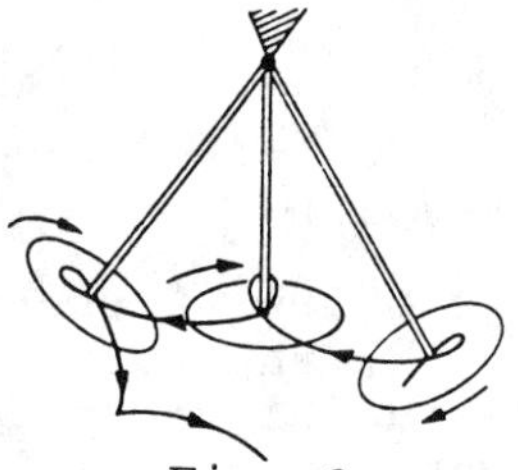

Fig. 3

If instead of a fixed axis of rotation we have a pivot suspension, it is impossible to produce any torque on the pendulum from the suspension to keep the pendulum oscillating in a vertical plane. To make the torque zero at the suspension point, the pendulum has to move out of the vertical plane and describe a precessional motion, usually with strong superimposed nutations as illustrated schematically in Fig. 3.

• PROBLEM 19-13

A symmetrical top is described by the fact that its moments of inertia about 2 of its principal axes are equal (i.e., $I_1 = I_2 \neq I_3$). Assuming that no external torques act, derive and solve the equations of motion of this body.

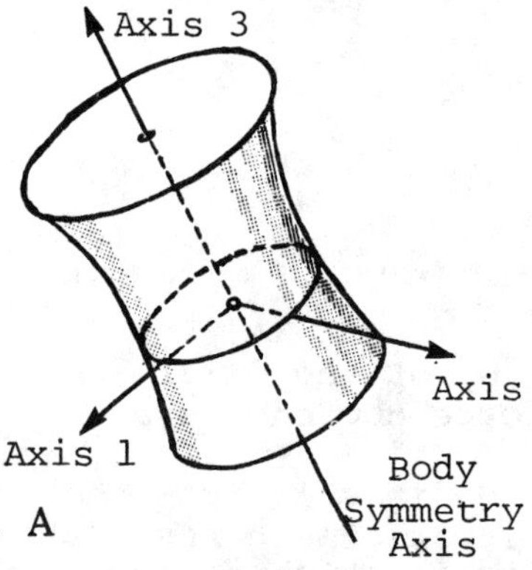

Fig. A

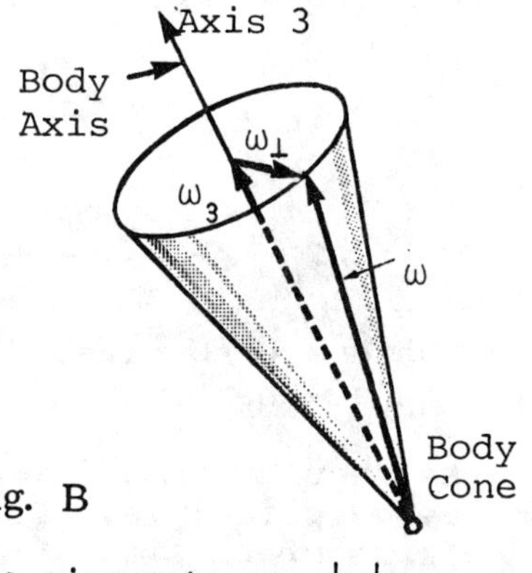

Fig. B

Consider the "free" rotation (N=0) of a body with axial symmetry (e.g. z axis). Then $T_1=T_2$, so ω_3=const.

In these circumstances $|\omega|$=const. The vector precesses at a constant rate around the body axis of symmetry.

Solution: The general motion of a rigid body is very complex. For cases in which the object doesn't rotate about a fixed axis, we cannot relate the angular momentum (L) to the angular acceleration by

$$L = I\alpha$$

This relation only holds for rotations about a fixed axis. We must use

$$\vec{N} = \frac{d\vec{L}}{dt}$$

where $\vec{N}$ is the net external torque on the body for these rotations. Another alternative is to use the Euler Equations

$$I_1 \frac{d\omega_1}{dt} + \left(I_3 - I_2\right)\omega_3\omega_2 = N_1$$

$$I_2 \frac{d\omega_2}{dt} + \left(I_1 - I_3\right)\omega_1\omega_3 = N_2 \qquad (1)$$

$$I_3 \frac{d\omega_3}{dt} + \left(I_2 - I_1\right)\omega_2\omega_1 = N_3$$

where the subscript 1 refers to the first principal axis of the rigid body (see fig. (A)), and similarly for the subscripts 2 and 3. Furthermore, $\vec{\omega}$ is the angular velocity of rotation of the body with reference to an inertial frame.

In this example, $I_1 = I_2 \neq I_3$ and $\vec{N} = 0$, whence

$$I_1 \frac{d\omega_1}{dt} + \left(I_3 - I_2\right)\omega_3\omega_2 = 0$$

$$I_2 \frac{d\omega_2}{dt} + \left(I_1 - I_3\right)\omega_1\omega_3 = 0 \qquad (2)$$

$$I_3 \frac{d\omega_3}{dt} = 0$$

Letting $I_1 = I_2 = I$, we obtain

$$\frac{d\omega_1}{dt} + \frac{\left(I_3 - I\right)}{I}\omega_3\omega_2 = 0$$

$$\frac{d\omega_2}{dt} + \frac{\left(I - I_3\right)}{I}\omega_1\omega_3 = 0$$

$$\frac{d\omega_3}{dt} = 0$$

Defining $\Omega \equiv \frac{I_3 - I}{I}\omega_3$, we may write

$$\frac{d\omega_1}{dt} + \Omega\omega_2 = 0$$

$$\frac{d\omega_2}{dt} - \Omega\omega_1 = 0 \qquad (3)$$

$$\frac{d\omega_3}{dt} = 0$$

A solution of (3) is given by

$$\omega_1 = A \cos \Omega t; \qquad \omega_2 = A \sin \Omega t ,$$

where A is a constant. We see that the component of the angular velocity perpendicular to the figure axis (axis 3) (see figure (A)) of the top rotates with a constant angular velocity Ω. The component ω_3 of the angular velocity along the figure axis is constant. Therefore the vector ω rotates uniformly with angular velocity Ω about the figure axis of the top. In other words, a top which spins about its figure axis with angular velocity ω_3 in force-free space will wobble with the frequency Ω.

For the earth I_3 is not exactly equal to I_1 because the earth is not exactly a sphere. The wobble is actually very well observed, giving rise to what is called the variation of latitude. The wobble is so interesting that the International Latitude Service maintains a number of observatories just for the purpose of measuring it.

• PROBLEM 19-14

> The large turntable shown in the figure is rotating with a constant angular velocity of $\dot{\psi}$ in the direction shown. In turn the small disk is spinning on its axis, AB, with a constatnt angular velocity of $\dot{\phi}$. If the system is moving in this diection, what is the dynamic force that the bearings A and B must exert on the spinning disk?

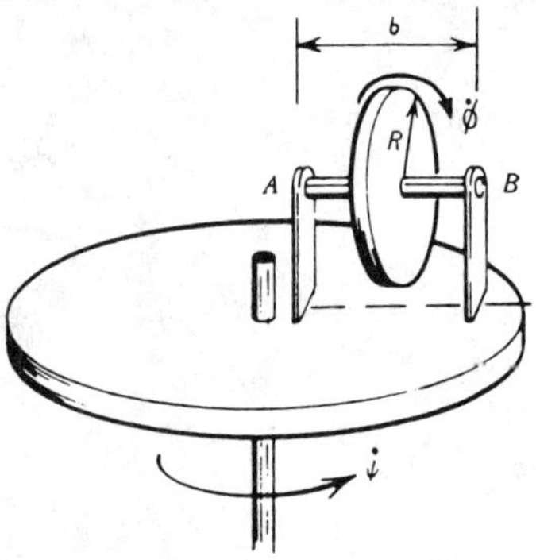

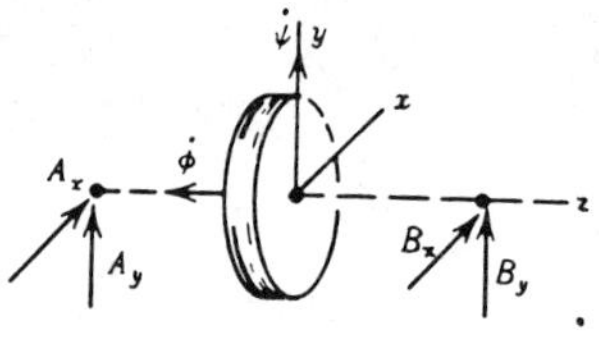

Figure

Solution: A free-body diagram of the spinning disk is shown in the figure. The axes shown are fixed in the body and have the orientation shown at this instant of time. They are the principal axes. The total angular velocity of the disk is a combination of the angular velocity due to the spin of the disk and the angular velocity due to the spin of the turntable. The moment of momentum (angular momentum) of the body is

$$\vec{H} = I_{xx}\omega_x\hat{i} + I_{yy}\omega_y\hat{j} + I_{zz}\omega_z\hat{k},$$

At this instant, the angular velocity of spin about the x, y, and z axes are zero, $\dot{\psi}$ and $-\dot{\phi}$, respectively. Thus, the moment of momentum of the body becomes

$$\vec{H} = I_{yy}\dot{\psi}\hat{j} - I_{zz}\dot{\phi}\hat{k}.$$

The time derivative of the moment of momentum is

$$\dot{\vec{H}} = I_{yy}\dot{\psi}\dot{\hat{j}} - I_{zz}\dot{\phi}\dot{\hat{k}}.$$

The angular velocities $\dot{\psi}$ and $\dot{\phi}$ are constant.

The time derivatives of the unit vectors $\hat{j}$ and $\hat{k}$ may be determined by taking the following cross-products:

$$\dot{\hat{j}} = \vec{\omega} \times \hat{j}$$

$$\dot{\hat{k}} = \vec{\omega} \times \hat{k}.$$

The angular velocity of the body is

$$\vec{\omega} = \dot{\psi}\hat{j} - \dot{\phi}\hat{k}$$

so that

$$\dot{\hat{j}} = \begin{vmatrix} \hat{i} & \hat{j} & \hat{k} \\ 0 & \dot{\psi} & -\dot{\phi} \\ 0 & 1 & 0 \end{vmatrix} = \dot{\phi}\hat{i}$$

$$\dot{\hat{k}} = \begin{vmatrix} \hat{i} & \hat{j} & \hat{k} \\ 0 & \dot{\psi} & -\dot{\phi} \\ 0 & 0 & 1 \end{vmatrix} = \dot{\psi}\hat{i}.$$

Finally, the time derivative of the moment of momentum becomes

$$\dot{\vec{H}} = (I_{yy}\dot{\psi}\dot{\phi} - I_{zz}\dot{\phi}\dot{\psi})\hat{i}$$

and, since $\vec{M} = \dot{\vec{H}}$, it follows that

$$M_x = (I_{yy} - I_{zz})\dot{\phi}\dot{\psi}.$$

Now, since the body is not accelerating in the direction of the motion of the center of mass (the center of mass does have a radial acceleration, since it is moving in a circular path), it must be that A_y and B_y are equal in magnitude and opposite in direction (ignoring the weight). Thus, they form a couple of magnitude $A_y b$. Then it follows that

$$A_y = \frac{I_{yy} - I_{zz}}{b}\dot{\phi}\dot{\psi}.$$

But, since $I_{zz} = \frac{1}{2}mR^2$ and $I_{yy} = \frac{1}{4}mR^2$, it follows that

$$A_y \quad \frac{-mR^2\dot{\phi}\dot{\psi}}{4b}.$$

Since the moment of the forces about the y axis is equal to zero, the forces A_x and B_x are equal to zero. Finally, the centripital acceleration alluded to before must come from force in the z-direction which may be assumed to be coming from bearing B only. It is

$$B_z = -mr\dot{\psi}^2.$$

We can use Euler's equations as an alternate method of determining the moment on the body. The moment of the bearing forces on the body may be determined directly from Euler's equations. It is noted from the figure that

$$\omega_z = 0 \qquad \dot{\omega}_z = 0$$

$$\omega_y = \psi \qquad \dot{\omega}_y = 0$$

$$\omega_z = -\phi \qquad \dot{\omega}_z = 0$$

and that

$$I_{xx} = I_{yy}.$$

Therefore, from the Euler's equations,

$$M_x = (I_{yy} - I_{zz})\dot{\psi}\dot{\phi}$$

$$M_y = 0$$

$$M_z = 0$$

which is the same result that was obtained by writing $\vec{M} = \dot{\vec{H}}$ directly. It goes without saying that it is much easier to obtain the moment equations by writing Euler's equations in this case, but the previous method is basic and does not involve the memorization of a set of complicated equations.

• PROBLEM 19-15

A thin, homogeneous disk of mass 10 lb. and radius 12 in. rotates at a constant rate, ω_2 = 30 rad./sec., with respect to the arm ABC which itself rotates at a constant rate, ω_1 = 5 rad./sec., about the x-axis. For the position shown, determine the dynamic reactions at the bearings D and E.

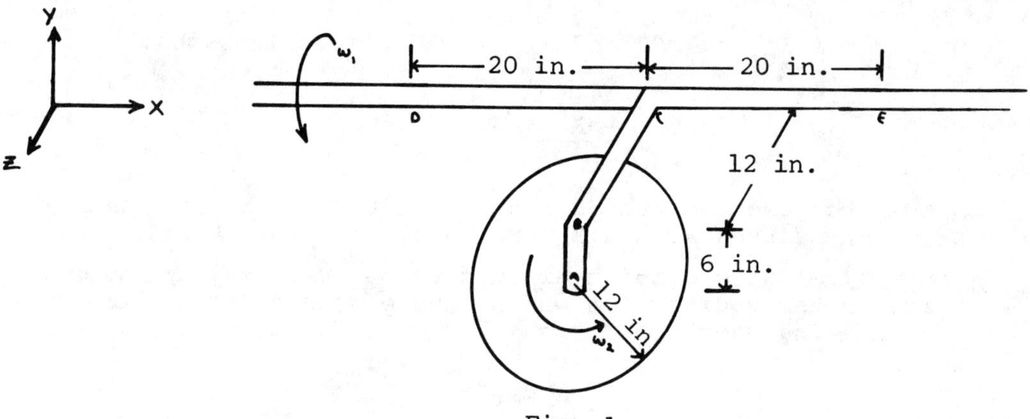

Fig. 1

Solution: The important forces in this problem are due to gyroscopic inertia and centripetal acceleration. The free body diagram of the system looks like this:

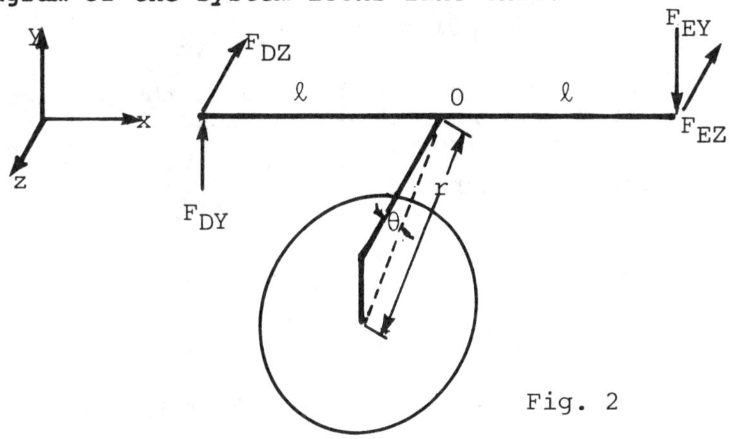

Fig. 2

The problem is governed by Newton's Second Law: The sum of the forces equals the mass times its acceleration, and the sum of the moments equals the rate of change of angular momentum.

Rate of change of angular momentum is governed by the familiar gyroscopic equation $H = I\vec{\omega}_p \times \vec{\omega}_s$. ω_s is the gyroscope's spin velocity, $\vec{\omega}_p$ is the gyroscope's angular velocity about the central axis and I is the gyroscope's moment of inertia about its own spin axis.

Combining Newton's Law and the above equation yields $\Sigma M = I\vec{\omega}_p \times \vec{\omega}_s$.

Summing moments about O yields

$$\sum M = [-(F_{EY})\ell - F_{DY}\ell]\hat{k} - (F_{DZ}\ell)\hat{j} + F_{EZ}\ell\hat{j},$$

or
$$-\ell\{[F_{EY} + F_{DY}]\hat{k} + (F_{DZ} - F_{EZ})\hat{j}\} = I\vec{\omega}_p \times \vec{\omega}_s.$$

From the problem statement,

$$\vec{\omega}_s = \omega_s\hat{j} \qquad \vec{\omega}_p = \omega_p\hat{i}.$$

It follows, then, that

$$-\ell\{[F_{EY} + F_{DY}]\hat{k} + (F_{DZ} - F_{EZ})\hat{j}\} = I\omega_s\omega_p(i \times j),$$

and, since $\hat{i} \times \hat{j} = k$,

$$-\ell\{[F_{EY} + F_{DY}]\hat{k} + (F_{DZ} - F_{EZ})\hat{j}\} = I\omega_s\omega_p\hat{k}.$$

Notice that there are no j terms on the right hand side. This means that

$$F_{DZ} - F_{EZ} = 0,$$

or $$F_{DZ} = F_{EZ} \qquad \text{i)}$$

and $$\ell\,[F_{EY} + F_{DY}] = -I\omega_s\omega_p. \qquad \text{ii)}$$

More equations are needed to solve the problem. These shall come from Newton's Force Equations. Since the rotating body moves in a circle, F_{EY}, F_{EZ}, F_{DZ} and F_{DY} provide the centripetal acceleration.

$$\sum\vec{F} = \frac{mv^2}{r}\left(\frac{-\vec{r}}{r}\right)$$

$$-[F_{DZ} + F_{EZ}]\hat{k} + [F_{DY} - F_{EY}]\hat{j} = \frac{mv^2}{r}(\sin\theta j - \cos\theta k).$$

Here, $v = r\omega_p$ and r is the distance from the shaft to the disc's center, m is the disc's mass, and θ is the angle between r and the horizontal.

Separating the components,

$$F_{DZ} + F_{EZ} = \frac{mv^2}{r}\cos\theta \qquad \text{iii)}$$

and $$F_{DY} - F_{EY} = \frac{mv^2}{r}\sin\theta \qquad \text{iv)}$$

i) through iv) form a solvable system of four equations in four unknowns.

Combining i) + iii) yields

$$2F_{DZ} = \frac{mv^2}{r}\cos\theta$$

$$F_{DZ} = F_{EZ} = \frac{mv^2}{2r}\cos\theta. \qquad \text{v)}$$

Combining ii) + iv) yields

$$\ell[F_{EY} + F_{EY} + \frac{mv^2}{r}\sin\theta\,] = -I\omega_s\omega_p$$

$$F_{EY} = \frac{\left(-I\omega_s\omega_p - \dfrac{mv^2\,\ell\sin\theta}{r}\right)}{2\ell} \qquad \text{vi)}$$

From the system geometry,

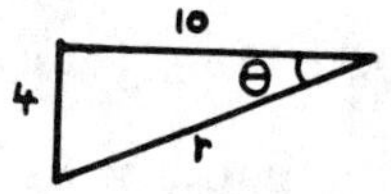

Fig. 3

$$\theta = \tan^{-1} 6/12$$

$$= 26.6°$$

$$r = \sqrt{(12^2 + 6^2)cm^2} = 13.42 \text{ in.}$$

$$= 1.12 \text{ ft.}$$

$$I = 1/2\ mr_D^2 \qquad r_D = \text{disc radius}$$

$$= 1/2(0.311\ \text{slug})(1.12\ \text{ft})^2 = 0.390\ \text{slug ft.}^2$$

Substituting values into the equations yields the final results.

From v)

$$F_{DZ} = F_{EZ} = \frac{0.311\ \text{slug}(1.12\ \text{ft} \times 5\text{r/s})^2}{2(1.12)} \cos 26.6°$$

$F_{DZ} = F_{EZ} = 3.89$ lbs. (The positive sign indicates that the direction on the free body diagram is correct.)

From vi)

$$F_{EY} = \frac{(0.390\ \text{slug ft}^2)(30\ \text{r/s})(5\ \text{r/s})}{2(1.67\ \text{ft})} - \frac{(0.311\ \text{slug})(1.12\ \text{ft} \times 5\ \text{r/s})^2 \sin(26.6°)}{2(1.12\ \text{ft})}$$

$= -19.47$ lbs. (Negative sign indicates direction opposite that assumed.)

From iv)

$$F_{DY} = \frac{(1.12\ \text{ft} \times 5\ \text{r/s})^2(0.311\ \text{slug})}{1.12\ \text{ft}} \sin(26.6°) - 19.47\ \text{lb}$$

$= -15.57$ lbs. (Negative sign indicates direction opposite that assumed.)

In vector form the reactions at D and E become:

at D, $(-15.57\hat{j} - 3.89\hat{k})$ lb.;

at E, $(-19.47\hat{j} - 3.89\hat{k})$ lb.

• PROBLEM 19-16

A symmetric rigid body moving freely in space is powered with jet engines symmetrically placed with respect to the 3-axis of the body which supply a constant torque N_3 about the symmetry axis. Find the general solution for the angular velocity vector as a function of time, relative to body axes, and describe how the angular velocity vector moves relative to the body.

Solution: Since we have principal axes, the 3-axis is one of symmetry; the inertia tensor may be written as

$$I = \begin{pmatrix} I_1 & 0 & 0 \\ 0 & I_1 & 0 \\ 0 & 0 & I_3 \end{pmatrix}$$

and the angular momentum will be $\vec{L} = (I_1\omega_1, I_1\omega_2, I_3\omega_3)$. The basic equation of motion in the (moving) body axes is

$$\frac{d\vec{L}}{dt} + \vec{\omega}\times\vec{L} = \vec{N}.$$

In components this becomes

$$dL_1/dt + (\omega_2 L_3 - \omega_3 L_2) = 0 \qquad (1)$$

$$dL_2/dt + (\omega_3 L_1 - \omega_1 L_3) = 0 \qquad (2)$$

$$dL_3/dt + (\omega_1 L_2 - \omega_2 L_1) = N_3. \qquad (3)$$

Using $L_i = I_i\omega_i$, (3) becomes $I_3 d\omega_3/dt + (\omega_1\omega_2 I_1 - \omega_2\omega_1 I_1) = N_3$, or $d\;_3/dt = N_3/I_3$, which has the solution $\omega_3 = (N_3/I_3)t + C$, by direct integration. Set

$C = -\frac{N_3}{I_3} t_0 + \omega_{30}$, where ω_{30} is the value of ω_3 at $t = t_0$.

Thus,

$$\omega_3 = \frac{N_3}{I_3}(t - t_0) + \omega_{30}. \qquad (4)$$

Substitute equation (4) into (1) and (2) along with $L_i = I_i\omega_i$ to yield

$$I_1 \frac{d\omega_1}{dt} + \omega_2\omega_3 I_3 - \omega_3\omega_2 I_1 = 0$$

$$\frac{d\omega_1}{dt} + \left[\frac{N_3}{I_3}(t - t_0) + \omega_{30}\right](I_3 - I_1)\omega_2 = 0. \qquad (5)$$

Likewise,

$$\frac{d\omega_2}{dt} - \left[\frac{N_3}{I_3}(t - t_0) + \omega_{30}\right](I_3 - I_1)\omega_1 = 0. \qquad (6)$$

It is necessary to solve the coupled equations (5) and (6). Add (5) + i(6) and let $z = \omega_1 + i\omega_2$. The result is

$$\frac{dz}{dt} + \left[\frac{N_3}{I_3}(t - t_0) + \omega_{30}\right](I_3 - I_1)(\omega_2 - i\omega_1) = 0, \qquad \text{or}$$

$$\frac{dZ}{dt} - i\left[\frac{N_3}{I_3}(t - t_0) + \omega_{30}\right](I_3 - I_1)Z = 0. \tag{7}$$

Trying the substitution,

$$Z = Z_0 e^{if(t)},$$

$$\frac{dZ}{dt} = Zif'(t).$$

In order for $Z_0 e^{if(t)}$ to be a solution of equation (7), notice that

$$f'(t) = \left[\frac{N_3}{I_3}(t - t_0) + \omega_{30}\right](I_3 - I_1),$$

so

$$f(t) = \frac{1}{2}\frac{N_3}{I_3}(t - t_0)^2(I_3 - I_1) + \frac{N_3}{I_3}\omega_{30}t(I_3 - I_1) + K.$$

Choosing $Z = Z_0$ at $t = t_0$, the $K = -\frac{N_3}{I_3}\omega_{30}t_0(I_3 - I_1)$ and

$$f(t) = \frac{1}{2}\frac{N_3}{I_3}\left[(t - t_0)^2 + 2\omega_{30}(t - t_0)\right](I_3 - I_1). \tag{8}$$

Write z_0 in its real and imaginary parts: $z_0 = a + ib$. Do the same for $\exp(f(t))$. Then $z = \omega_1 + i\omega_2 = (a + ib)\{\cos[f(t)] + i \sin[f(t)]\}$. Separate real and imaginary parts. This results in

$$\omega_1 = a \cos[f(t)] - b \sin[f(t)]$$

$$\omega_2 = b \cos[f(t)] + a \sin[f(t)].$$

At the time $t = t_0$ when $f(t = t_0) = 0$, we clearly have $\omega_{10} = a$, $\omega_{20} = b$, as the initial conditions.

Summarizing, write the complete solution as

$$\omega_3 = \frac{N_3}{I_3}(t - t_0) + \omega_{30}$$

$$\omega_2 = \omega_{20} \cos[f(t)] + \omega_{10} \sin[f(t)]$$

$$\omega_1 = \omega_{10} \cos[f(t)] - \omega_{20} \sin[f(t)],$$

where $f(t)$ is given in equation 8.

These equations describe a rotation of the ω-vector around the 3-axis which proceeds ever faster while the 3-

component increases linearly with time. Thus, the total ω-vector increases in magnitude. Its components perpendicular to the 3-axis describe an ellipse of constant size.

GYROSCOPIC MOTION

• PROBLEM 19-17

A gyroscope consists of a uniform circular disk of mass $M = 1$ kg and radius $R = 0.2$ m. The disk spins with an angular velocity $\omega = 400\ \text{sec}^{-1}$. The gyroscope precesses with the axes making an angle of 30° with the horizontal. The gyroscope wheel is attached to its axis at a point a distance $\ell = 0.3$ m from the pivot which supports the whole structure. What is the precessional angular velocity?

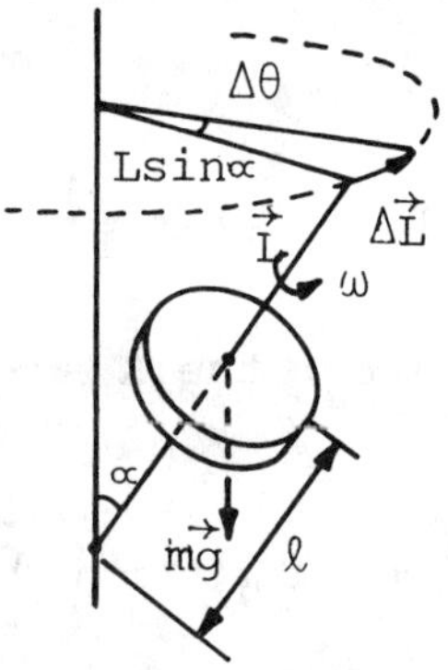

Solution: First, establish the general formula for regular precession. Refer to the figure. $|L| = I\omega$ for a symmetric object spinning around its symmetry axis. The torque τ from the weight is $mg\ell \sin\alpha$; this is directed perpendicular to the plane containing $\vec{L}$ and the vertical axis, according to the rule for a vector cross product. In a time Δt the torque effects a change $\Delta L = \tau\Delta t = mg\ell \sin\alpha\Delta t$ in the angular momentum, also perpendicular to $\vec{L}$. Thus, after Δt, the angular momentum will have changed its direction; the horizontal component will have moved by an angle $\Delta\theta = \Omega\Delta t$. ($\Omega$ is the precessional angular velocity.) There is no change in the magnitude of $\vec{L}$ since its increment $\Delta\vec{L}$ is perpendicular. Comparing the arc length to radius in the circle described by the tip of $\vec{L}$, the result is $\Delta\theta = \Delta L/(L \sin\alpha)$; $L \sin\alpha$ is the radius of this circle. Thus, $\Omega = \Delta\theta/\Delta t = (1/L \sin\alpha)(\Delta L/\Delta t)$. Substituting the above expressions for L and ΔL gives

$$\Omega = 1/(I\omega \sin\alpha)(mg\ell \sin\alpha\Delta t/\Delta t) = mg\ell/(I\omega).$$

For the uniform disk we have $I = \frac{1}{2}mR^2$, so $\Omega = 2g\ell/(R^2\omega)$.

Use $g = 9.8\ \text{m/s}^2$, $R = 0.2$ m, $\omega = 400\ \text{s}^{-1}$, $\ell = 0.3$ m. All these are mks units. Then,

$$\Omega = 2 \cdot 9.8 \cdot 0.3/[(0.2)^2 \cdot 400] \sim 0.368\ \text{s}^{-1}.$$

Note that the precessional angular velocity does not depend on the mass of the disk nor on the angle that its axes makes with the horizontal.

• PROBLEM 19-18

The flywheel of an automobile engine, mounted on the crank-shaft, is equivalent to a 16-in diameter steel plate of $\frac{15}{16}$-in thickness. At a time when the flywheel is rotating at 4000 rpm, the automobile is traveling around a curve of 600 ft. radius at a speed of 55 mi/hr. Determine, at that time, the magnitude of the couple exerted by the flywheel on the horizontal crankshaft. (Specific weight of steel = 490 lb/ft^3.)

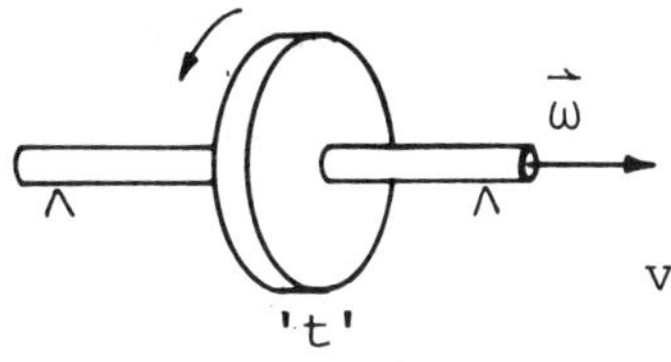

Fig. 1

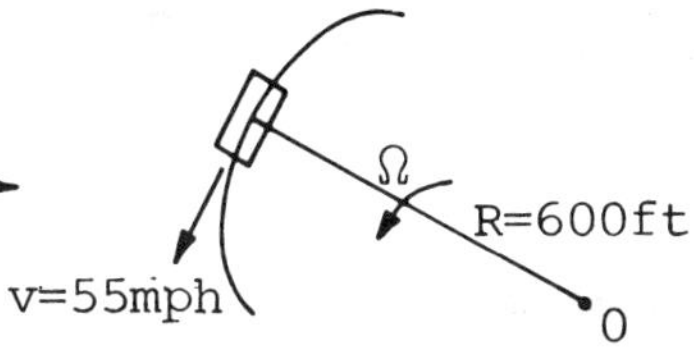

Fig. 2

Fig. 3

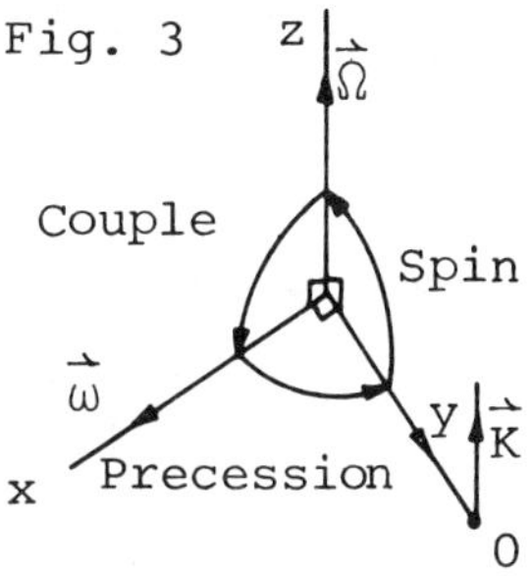

Solution: Whenever a spinning object also rotates about another axis, it is functioning as a gyroscope. Whenever the automobile goes around a curve, a couple is required to change the direction of the angular momentum. In this case the flywheel is precessing about the vertical axis through 0 with angular velocity Ω. The relationship among precession Ω, spin ω, and couple T for gyroscopic motion is shown in Figure 3. In this case Figure 3 shows that the couple tries to push the front of the flywheel down and the back up. Consider the x, y, z coordinate system which is fixed at the center of the flywheel with the x-axis along the axis of the flywheel. In this system $\vec{\omega} = \omega_1\vec{i}$. In the system fixed at 0, $\vec{\Omega} = \Omega\vec{K} = \Omega\vec{k}$. In general the angular momentum is $\vec{L} = A\omega_1\vec{i} + B\omega_2\vec{j} + C\omega_3\vec{k}$ where A, B and C are the moments of inertia about the respective axes. In this case $\vec{L} = A\omega\vec{i}$.

The rate of change of a vector which is rotating with angular velocity $\vec{\Omega}$ with respect to a fixed coordinate system is $\frac{d\vec{L}}{dt} = \frac{\partial\vec{L}}{\partial t} + \vec{\Omega} \times \vec{L}$ where $\frac{\partial\vec{L}}{\partial t}$ is the rate of change in the rotating system. In this case $\frac{\partial\vec{L}}{\partial t} = 0$. Also, from Newton's Second Law, the torque, T, is equal to the rate of change of the angular momentum. So,

$$\vec{T} = \frac{d\vec{L}}{dt} = \vec{\Omega} \times \vec{L} = \Omega\vec{k} \times A\omega\vec{i} = A\omega\Omega\vec{j}.$$

Thus, the couple is about the y-axis as shown in Figure 3.

$$\omega = 4000\frac{\text{rev}}{\text{min}} \cdot \frac{\text{min}}{60 \text{ sec}} \quad \frac{2\pi \text{ rad}}{\text{rev}} = 418.9 \text{ rad/sec}$$

$$\Omega = \frac{v}{R} = \frac{55 \text{ mph}}{600 \text{ ft}} = \frac{81 \text{ ft/sec}}{600 \text{ ft}} = .134 \text{ rad/sec.}$$

The moment of inertia A is about the symmetry axis of the disk so $A = \frac{1}{2}mr^2$. The weight of the disk is $W = mg = \rho\pi r^2 t$, so

$$m = \frac{\rho\pi r^2 t}{g} = \frac{490 \text{ lb/ft}^3 \pi (8 \text{ in})^2 (15/16 \text{ in}) \text{ft}^3}{32 \text{ ft/sec}^2 \quad (12 \text{ in})^3} = 1.67 \text{ lb}\cdot\text{sec}^2/\text{ft.}$$

Combining the above results gives the magnitude of the couple

$$T_y = \frac{1}{2}mr^2\omega\Omega = \frac{1}{2}\,\frac{1.67 \text{ lb sec}^2}{\text{ft}}\left(\frac{2}{3} \text{ ft}\right)^2 418.9\frac{\text{rad}}{\text{sec}} \; .134 \text{ rad/sec}$$

$$= 20.8 \text{ lb ft.}$$

• PROBLEM 19-19

The flywheel in a delivery truck is mounted with its axis vertical, and thus acts as a stabilizing gyroscope for the truck. Calculate the torque that would have to be applied to it when it is rotating at full speed to make it precess in a vertical plane.

Figure (A)

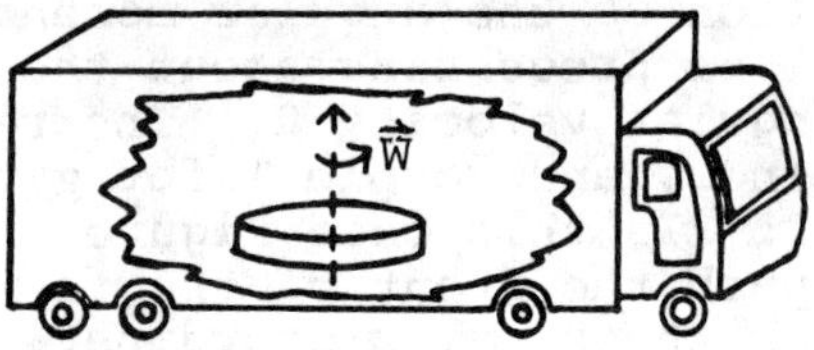

Figure (B): Blowup of Flywheel

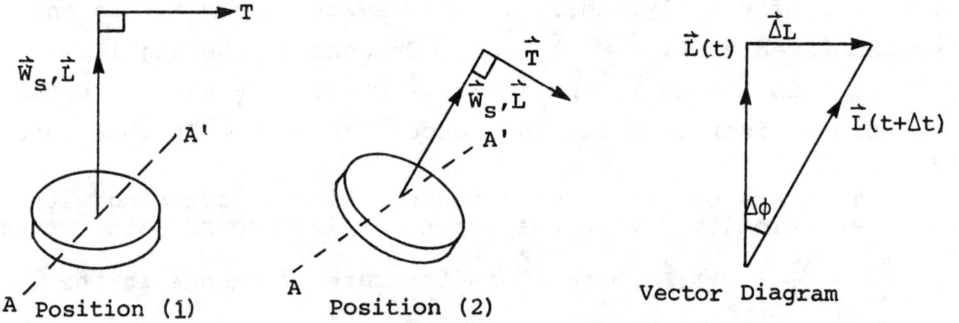

Position (1) Position (2) Vector Diagram

Solution: Figure (a) shows the situation. The flywheel is to precess about axis AA' as shown in figure (b). Originally, the angular momentum vector $\vec{L}$ is as shown in figure (b), position 1. After a time Δt, $\vec{L}$ has the new value $L(t + \Delta t)$, and the angular momentum vector has gone through an angle $\Delta\phi$, as shown in the vector diagram. Note that we have neglected the fact that there is a component of $\vec{L}$ along AA' due to the precession of the flywheel.

This approximation is valid if the rate of precession, w_p, is small. By the relation

$$\frac{d\vec{L}}{dt} = \vec{\tau} \tag{1}$$

where $\vec{\tau}$ is the net torque on the flywheel, we see that if we are to change $\vec{L}$, and thereby cause precession to occur, we must exert a torque $\vec{\tau}$. Now, the most efficient way for the torque to cause precession is if it acts in a direction perpendicular to $\vec{L}$, as shown in figure (b). As a result of (1), $d\vec{L}$ is also perpendicular to $\vec{L}$, and the length of $\vec{L}$ doesn't change. Hence

$$\frac{\Delta L}{\Delta t} \approx \tau$$

If $\Delta\vec{L}$ is small, and $\vec{\tau}$ is always perpendicular to $\vec{L}$,

$$\left|\frac{\Delta\vec{L}}{\vec{L}}\right| = \frac{\Delta L}{L} = \Delta\phi$$

or $\quad \tau \approx L \dfrac{\Delta\phi}{\Delta t} = L\, w_p$

$$\tau \approx L\, w_p \tag{2}$$

But $\quad L = I\, w_s \qquad (3)$

where I is the moment of inertia of the flywheel about its symmetry axis and w_s is the spin angular velocity of the disc. Using (3) in (2)

$$\tau \approx I\, w_s\, w_p$$

Also $\quad I = \frac{1}{2} Mr^2$

where M and r are the mass and radius of the disc. Finally

$$\tau \approx \frac{1}{2} Mr^2\, w_s\, w_p$$

• PROBLEM 19-20

Illustrated in figure 1 is a heavy symmetrical top comprised of a solid cone mounted on a rod that can be assumed to be weightless. The weight of the cone is 64 lb, and angular velocity of spin of 8000 rpm in the direction shown in the figure is imparted to the body. If the body is released in the horizontal plane, discuss the subsequent motion of the body.

Solution: This top, by virtue of its large angular velocity and the fact that it is constrained to spin about the symmetry axis shown, is equivalent to a gyroscope. It is known that for a gyroscope the angular momentum $\vec{H} = I\dot{\phi}$, the applied moment $\vec{M}$, and the angular velocity of precession ($\dot{\psi}$) are mutually orthogonal. Furthermore, since

$\vec{M} = \dot{\vec{H}}$ and $\dot{\vec{H}} = \dot{\vec{\psi}} \times \vec{H}$ and since the vectors are orthogonal,

$$M = \dot{\psi}H = \dot{\psi}I\dot{\phi}. \tag{1}$$

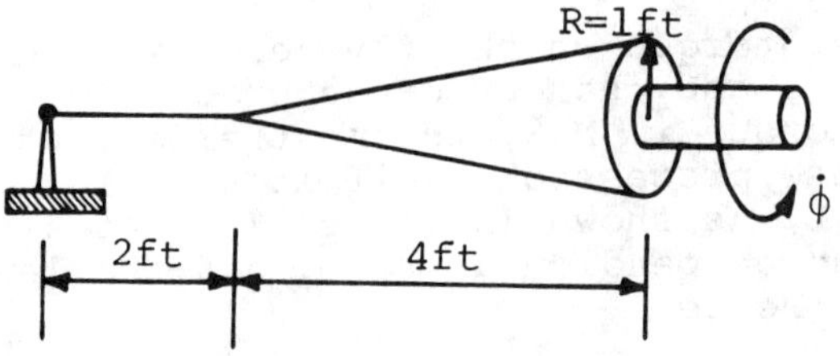

Fig. 1

So, the first thing to do would be to establish the spin and moment vectors. The spin, represented by the vector $\dot{\phi}$, is given, and the moment vector may be determined from the freebody diagram in Fig. 2.

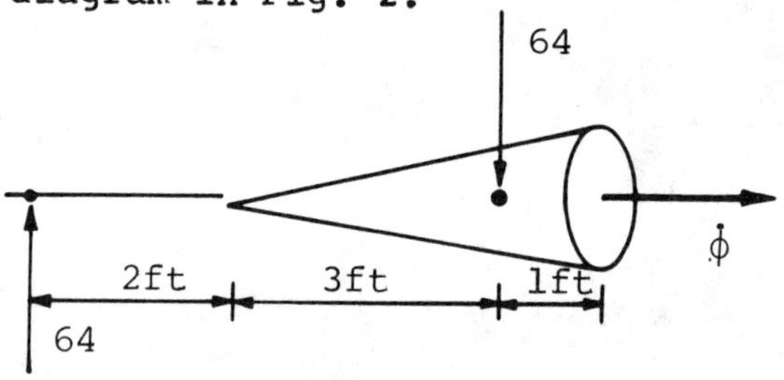

Fig. 2

At the time that the body is released, the reaction at the support will be equal to the weight of the body, and the two forces will create a couple; the moment of the couple is equal to

$$M = 64 \times 5$$

$$= 320 \text{ ft-lb}$$

and the vector representing the moment is in a direction shown in Fig. 3. Of course, that implies that the vector representing the angular velocity of precession ($\dot{\psi}$) would be in the direction shown in Fig. 3.

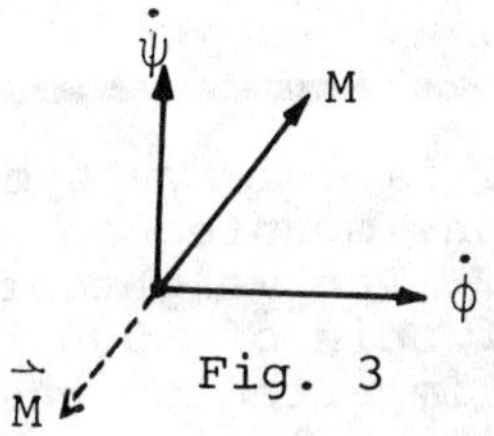

Fig. 3

The angular velocity of precession may be determined from Eq. (1).

$$\dot{\psi} = \frac{M}{I\dot{\phi}} .$$

The moment of inertia of a cone about its longitudinal axis

is $\frac{3}{10}mR^2$, so that

$$I = \frac{3}{10}\left(\frac{64}{32.2}\right)(1)^2$$

$$= 0.6 \text{ lb-sec}^2\text{-ft.}$$

The angular velocity of precession becomes

$$\dot{\psi} = \frac{320}{0.6[2\pi(8000)/60]}$$

$$\dot{\psi} = 0.64 \text{ radian/sec.}$$

Therefore, this top will rotate about the vertical axis with an angular velocity of 0.64 radian/sec.

• PROBLEM 19-21

The rate of steady precession $\dot{\phi}$ of the cone shown about the vertical is observed to be 20 rpm. Knowing that r = 100 mm and h = 200 mm, determine the rate of spin $\dot{\psi}$ of the cone about its axis of symmetry if $\beta = 135°$.

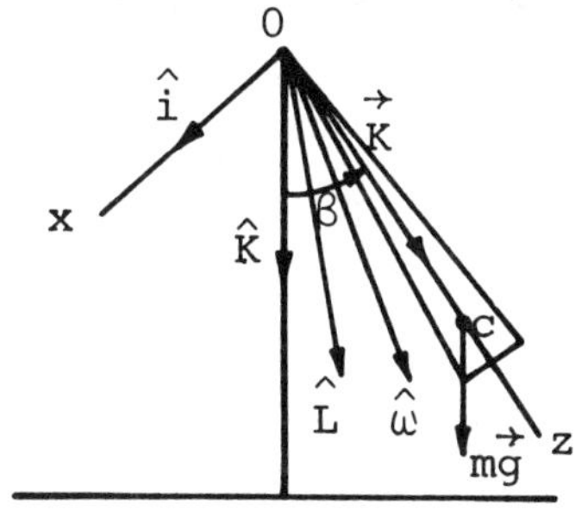

Fig. 1

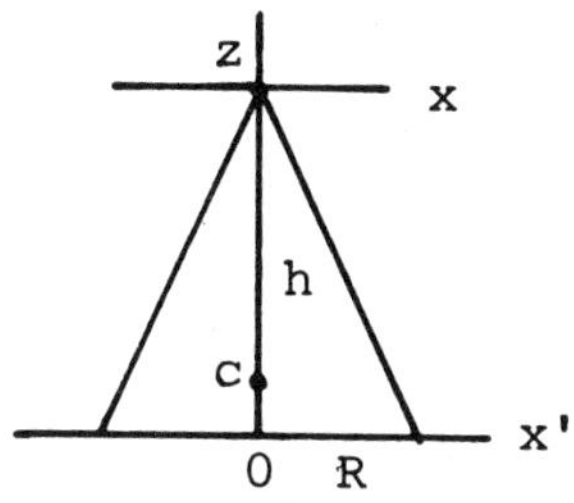

Fig. 2

Solution: Choose a coordinate system $\hat{i}, \hat{j}, \hat{k}$ which is fixed in the cone with $\hat{k}$ along the axis of symmetry. $\hat{K}$ is a unit vector fixed in space along the vertical. The cone has radius r, height h, and its center of mass is located a distance $\vec{a} = a\hat{k}$ from 0 on the symmetry axis. The cone precesses about the vertical with a constant angular velocity $p = \dot{\phi}$. The spin about the symmetry axis is $s = \dot{\psi}$.

From figure 1, $\hat{K} = \sin\beta\hat{i} + \cos\beta\hat{k}$. The angular momentum vector lies in the x, z plane.

$$\vec{\omega} = \omega_1\hat{i} + s\hat{k}.$$

The center of mass, C, rotates about $\vec{\omega}$ with a velocity

$$\vec{v}_c = \vec{\omega} \times \vec{a} = (\omega_1\hat{i} + s\hat{k}) \times a\hat{k} = -a\omega_1\hat{j}.$$

In the fixed system, C rotates around the vertical with velocity

$$\vec{v}_c = p\hat{k} \times \vec{a} = (p \sin \beta\hat{i} + p \cos \beta\hat{k}) \times a\hat{k}$$

$$= -ap \sin \beta\hat{j}.$$

Since both velocities are for the same point, a comparison gives

$$\omega_1 = p \sin \beta.$$

The only force acting on the top is the weight $\vec{W} = mg\hat{k}$. It produces a torque, $\vec{\tau}$, about 0 of

$$\vec{\tau} = \vec{a} \times \vec{W} = a\hat{k} \times mg\hat{k} = a\hat{k} \times mg(\sin \beta\hat{i} + \cos \beta\hat{k})$$

$$= amg \sin \beta\hat{j}. \tag{1}$$

The angular momentum is given by

$$\vec{L} = I_x\omega_1\hat{i} + I_z s\hat{k}.$$

$\vec{L}$ also lies in the x, z plane and rotates with it about the vertical. From Newton's Second Law $\vec{\tau} = \frac{d\vec{L}}{dt}$; the rate of change of a fixed vector in a rotating coordinate system is $\frac{d\vec{A}}{dt} = \vec{\Omega} \times \vec{A}$. In this case

$$\vec{\tau} = p\hat{k} \times \vec{L} = p(\sin \beta\hat{i} + \cos \beta\hat{k}) \times (I_x\omega_1\hat{i} + I_z s\hat{k})$$

$$= (p^2 I_x \cos \beta \sin \beta - pI_z s \sin \beta)\hat{j}.$$

Comparing this with equation (1) and substituting in $\omega_1 = p \sin \beta$ gives

$$amg \sin \beta = p^2 I_x \cos \beta \sin \beta - psI_z \sin \beta.$$

Solving for the spin gives

$$s = \frac{pI_x \cos \beta}{I_z} - \frac{amg}{pI_z}. \tag{2}$$

Next, determine the moments of inertia. For the right circular cone shown in Figure 2 the center of mass is at $\vec{z} = h/4$ and the moments of inertia are

$$I_z = \frac{3}{10}mR^2 \quad \text{and} \quad I_{x'} = \frac{m}{20}(3R^2 + 2h^2).$$

For this problem we need I_x, the moment of inertia about the x-axis through the vertex. To obtain this use the parallel axis Theorem, $I_0 = I_c + mr^2$ where r is the perpendicular distance between the parallel lines through 0 and C.

$$I_C = I_{x'} - mr_1^2 = I_{x'} - m\frac{h^2}{16}$$

$$I_x = I_C + mr_2^2 = I_C + \frac{9}{16}mh^2 = I_{x'} - \frac{1}{16}mh^2 + \frac{9}{16}mh^2$$

$$= I_{x'} + \frac{1}{2}mh^2 = \frac{3}{20}mR^2 + \frac{1}{10}mh^2 + \frac{1}{2}mh^2 = \frac{3}{20}m(R^2 + 4h^2).$$

So, $$I_x/I_z = \frac{3/20m(R^2 + 4h^2)}{3/10\ mR^2} = \frac{1}{2}\left[1 + 4\left(\frac{h}{R}\right)^2\right].$$

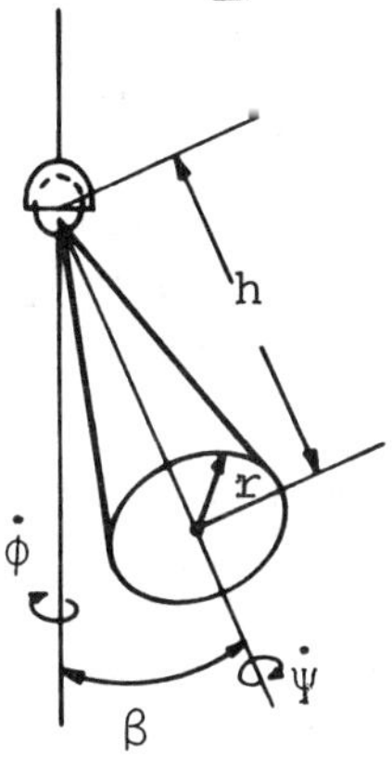

Fig. 3

For this problem p = 20 rpm=0.33rad/sec, R = 10 cm, h = 20 cm, $\beta = 135°$ and $a = h - \frac{1}{4}h = \frac{3}{4}h = 15$ cm. Substitution of these values into equation (2) gives

$$s = \frac{\pi}{2}\left[1 + 4\left(\frac{20}{10}\right)^2\right] \cos 135° - \frac{15\ \text{cm}\ \ 980\ \text{cm/sec}^2}{\pi\ \text{rad/sec}\ \frac{3}{10}\ (10\ \text{cm})^2}$$

$$= -18.88\ \text{rad/sec} - 155.97\ \text{rad/sec}$$

$$= -174.85\ \text{rad/sec} = -1670\ \text{rpm}.$$

This is the angular velocity of the cone about its symmetry axis. The negative sign means $\vec{s}$ points in the negative $\hat{k}$ direction.

• PROBLEM 19-22

A spinning target in a shooting gallery is dynamically equivalent to two thin disks of equal mass, m, connected by a light bar. The radius of the disks is a = 400 mm and the bar has a length 2a = 800 mm. The target is initially rotating about its' axis of symmetry with a rate ω_0 = 40 rpm. A bullet of mass $m_0 = m/100$ and traveling with a velocity v_0 = 100 m/s relative to the target, hits the target and is embedded in point C. Determine (a) the angular velocity of the target immediately after collision, (b) the precession of the ensuing motion, (c) the rates of precession and spin of the ensuing motion.

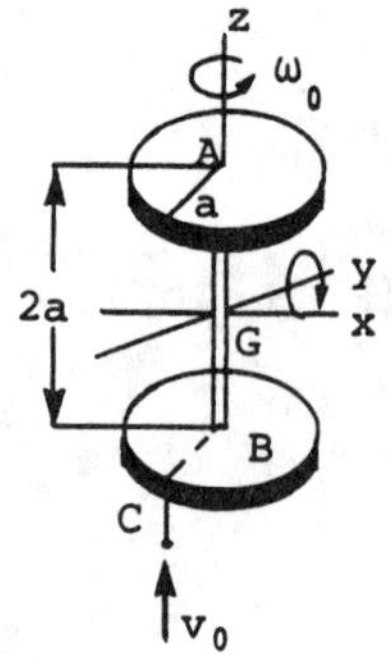

Fig. 1

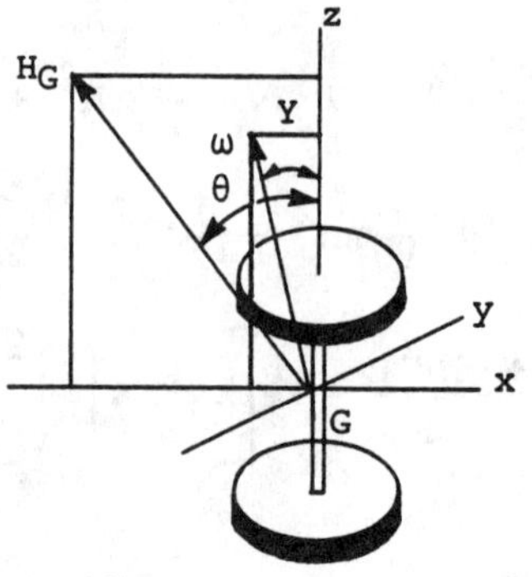

Fig. 2

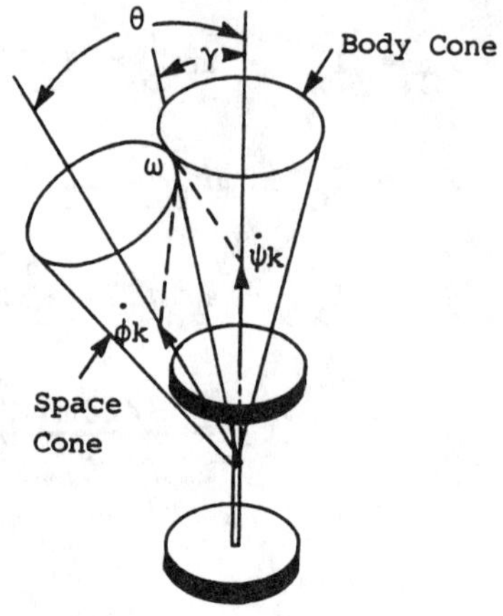

Fig. 3

Solution: We need the principal axes of the satellite to solve this problem. With the simplifying assumption that the meteorite does not change the principal axes, note that the x,y,z axes shown are the principal axes.

$$I = I_z = \frac{1}{2}ma^2.$$

Find $I' = I_x = I_y$ by use of the parallel-axis theorem. The moment of inertia of the disk with respect to an axis through it, parallel to the x or y axis in the figure, is $\frac{1}{4}mr^2$. Since the mass of each disk is $(\frac{1}{2}m)$ and the distance from the parallel axis to the axis in the figure is a, the moment of inertia of one disk is

$$I' = I_x = I_y = \frac{1}{4}(\frac{1}{2}m)a^2 + (\frac{1}{2}m)a^2.$$

Due to symmetry, it is possible to multiply by two to find the Moment of Inertia of the total system.

$$I' = I_x = I_y = 2\left[\frac{1}{4}(\frac{1}{2}m)a^2 + (\frac{1}{2}m)a^2\right] = \frac{5}{4}ma^2$$

The total angular momentum of the satellite-meteorite system is conserved during collision and remains constant during the ensuing motion. Taking moments about the coordinate origin G yields

$$\vec{H}_G = -a\hat{j} \times m_0v_0\hat{k} + I\omega_0\hat{k}$$

$$\vec{H}_G = -m_0v_0a\hat{i} + I\omega_0\hat{k}. \qquad (1)$$

Now determine the angular velocity after impact by invoking the conservation of $\bar{H}_G$. Using

$$\vec{H} = K_x\omega_x\hat{i} + I_y\omega_y\hat{j} + I_z\omega_z\hat{k},$$

the result is

$$-m_0v_0a = I'\omega_x = \frac{5}{4}ma^2\omega_x$$

$$\omega_x = \frac{4}{5}\frac{m_0 v_0}{ma};$$

$$0 = I'\omega_y\ ,\ \omega_y = 0;$$

$$I\omega_0 = I\omega_z$$

$$\omega_z = \omega_0.$$

$$\vec{\omega} = -\frac{4}{5}\frac{m_0 v_0}{ma}\hat{i} + \omega_0\hat{k}. \qquad (2)$$

While this equation is correct it is not very helpful since the body axes x,y,z are moving in space in an unknown way. For the satellite considered,

$$\omega_0 = 40 \text{ rpm} = 4.19 \text{ rad/sec}, \quad m_0/m = 1/100\ ,$$

$$a = 0.400 \text{ m}, \quad v_0 = 100 \text{ m/sec};$$

we find $\omega_x = -2$ rad/sec, $\omega_y = 0$, $\omega_z = 4.19$ rad/sec.

This gives

$$\omega = \sqrt{\omega_x^2 + \omega_z^2} = 4.64 \text{ rad/sec} = 44.3 \text{ rpm}$$

with angle $\gamma = \arctan \dfrac{-\omega_x}{\omega_z} = 25.5°$ to the z-axis. (Figure 2).

The angle θ shown in fig. 2 is between the precession axis and the z-axis. (H_G is fixed in space; the satellite wobbles around it.)

$$\theta = \arctan\left(\frac{m_0 v_0 a}{I\omega_0}\right) = \arctan\left(\frac{2m_0 v_0}{ma\omega_0}\right)$$

$$\theta = 50.0°.$$

Figure 3 shows $\vec{\omega}$, the space and body cones and $\dot{\phi}$ and $\dot{\psi}$ which, added vectorially, must equal $\vec{\omega}$. Using the law of sines,

$$\left|\frac{\omega}{\sin(\pi-\theta)}\right| = \left|\frac{\dot{\phi}}{\sin\gamma}\right| = \left|\frac{\dot{\psi}}{\sin(\theta-\gamma)}\right|;$$

$$\frac{\omega}{\sin\theta} = \frac{\dot{\phi}}{\sin\gamma} = \frac{\dot{\psi}}{\sin(\theta-\gamma)}\ .$$

This gives

$$\dot{\phi} = 24.9 \text{ rpm}\ ,\ \dot{\psi} = 24.0 \text{ rpm}.$$

A spacecraft, with a mass $m = 2000kg$, rotates with an angular velocity of $\omega = (0.05\ rad/s)\hat{i} + (0.15\ rad/s)\hat{j}$. Two small jets, located at points A and B, are turned on in a direction parallel to the z-axis. Each jet has a thrust of 25N. The radii of gyration of the spacecraft are $k_x = k_z = 1.5m$ and $k_y = 1.75m$. Determine the required time of operation for each jet, so that the angular velocity of the spacecraft reduces to zero.

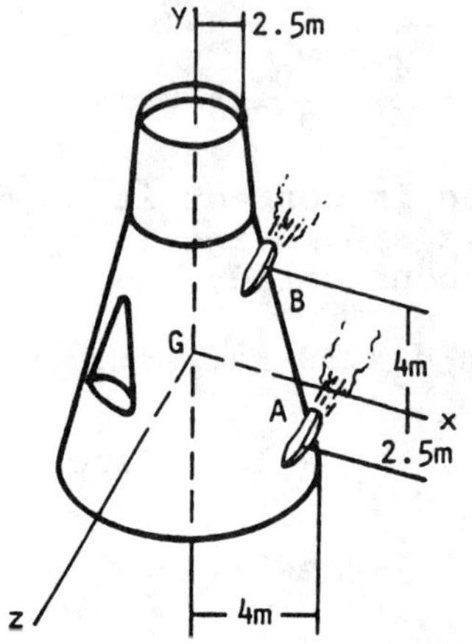

Fig. 1

Solution: Choose a coordinate system fixed in the body such that the rockets at A and B lie in the x, y plane and the origin is at the center of mass. The positions of A and B are $\vec{r}_1$ and $\vec{r}_2$ respectively. Qualitatively, what happens is both the rockets oppose the original motion about the y axis but B will increase the original motion about the x-axis while A will oppose it. Thus, A will have to be fired longer in order to compensate for the increase about the x-axis due to B. The torque, τ, produced by A and B is

$$\vec{\tau}_B = \vec{r}_1 \times \vec{F}_1 = y_1F_1\vec{i} - x_1F_1\vec{y}$$
$$\vec{\tau}_A = \vec{r}_2 \times \vec{F}_2 = -y_2F_2\vec{i} - x_2F_2\vec{j}. \quad (1)$$

From Newton's Second Law, $\vec{\tau} = I\vec{\alpha}$ where I is the moment of inertia and α is the angular acceleration. In components this becomes

$$\vec{\tau} = I_x\alpha_x\vec{i} + I_y\alpha_y\vec{j} + I_z\alpha_z\vec{k} \quad (2)$$

where I_x, I_y and I_z are the moments of inertia about the x, y, z axes. Comparing equations (2) and (3) gives

$$\alpha_x^A = \frac{d\omega_x^A}{dt} = -\frac{y_2F_2}{I_x} = -C_1\ ; \quad \frac{d\omega_y^A}{dt} = -\frac{x_2F_2}{I_y} = -C_2;; \quad \frac{d\omega_z^A}{dt} = 0 \quad (3)$$

$$\frac{d\omega_x^B}{dt} = \frac{y_1F_1}{I_x} = C_3 \;; \quad \frac{d\omega_y^B}{dt} = -\frac{x_1F_1}{I_y} = -C_4 \;; \quad \frac{d\omega_z^B}{dt} = 0.$$

These equations can be integrated directly to give

$$\begin{aligned} \Delta\omega_x^A &= -C_1t_A \;; \quad \Delta\omega_y^A = -C_2t_A \;; \quad \omega_z^A = \text{constant} \\ \Delta\omega_x^B &= C_3t_B \;; \quad \Delta\omega_y^B = -C_4t_B \;; \quad \omega_z^B = \text{constant} \end{aligned} \tag{4}$$

where $\Delta\omega = \omega(t) - \omega(0)$ is the change in angular velocity about each axis due to firing the rockets.

Since initially $\vec{\omega} = 0.2$ rad/sec $\vec{i}$ + 0.1 rad/sec $\vec{j}$; $\omega_z(0) = 0$, the constants in equation (4) must also be zero. Since we want the final angular velocity to be zero, the sum of the initial angular momentum plus the additional increments about each axis must be zero. So,

$$\omega_{x0} + \Delta\omega_x^A + \Delta\omega_x^B = 0$$

$$\omega_{y0} + \omega_y^A + \Delta\omega_y^B = 0.$$

Substitution from equation (3) gives

$$C_1t_A - C_3t_B = \omega_{x0}$$

$$C_2t_A + C_4t_B = \omega_{y0}$$

These are two simultaneous equations in the unknowns t_A and t_B. From the theory of equations, the solutions are

$$\begin{aligned} t_A &= \frac{1}{\Delta}\begin{vmatrix} \omega_{x0} & -C_3 \\ \omega_{y0} & C_4 \end{vmatrix} = \frac{1}{\Delta}(C_4\omega_{x0} + C_3\omega_{y0}) \\ t_B &= \frac{1}{\Delta}\begin{vmatrix} C_1 & \omega_{x0} \\ C_2 & \omega_{y0} \end{vmatrix} = \frac{1}{\Delta}(C_1\omega_{y0} - C_2\omega_{x0}) \end{aligned} \tag{5}$$

where

$$\Delta = \begin{vmatrix} C_1 & -C_3 \\ C_2 & C_4 \end{vmatrix} = C_1C_4 + C_2C_3.$$

For this problem $x_1 = 2.5$ m, $y_1 = 4$ m, $x_2 = 4$ m, $y_2 = 2.5$ m $m = 2 \times 10^3$ kg, $F_1 = F_2 = 25$ n, $k_x = k_z = 1.5$ m and $k_y = 1.75$ m.

From equation (3) the C values are

$$C_1 = y_2F_2/I_x = \frac{2.5\text{ m}\quad 25\text{ n}}{2\times 10^3\text{kg}(1.5\text{m})^2} = 0.0139\ \frac{\text{rad}}{\text{sec}^2}$$

$$C_2 = x_2F_2/I_y = \frac{4\text{ m}\quad 25\text{ n}}{2\times 10^3\text{kg}(1.75\text{m})^2} = 0.0163\ \frac{\text{rad}}{\text{sec}^2}$$

$$C_3 = y_1F_1/I_x = \frac{4\text{ m}\quad 25\text{ n}}{2\times 10^3\text{kg}(1.5\text{m})^2} = 0.0222\ \frac{\text{rad}}{\text{sec}^2}$$

$$C_4 = x_1F_1/I_y = \frac{2.5\text{ m}\quad 25\text{ n}}{2\times 10^3\text{kg}(1.75\text{m})^2} = 0.0102\ \frac{\text{rad}}{\text{sec}^2}$$

$$\Delta = C_1C_4 + C_2C_3 = 0.000503\text{ rad/s}^4.$$

Substitution in eq. (5) gives the times

$$t_A = \frac{C_4\omega_{x0} + C_3\omega_{y0}}{\Delta} = \frac{0.0102\times 0.05 + 0.0222\times 0.15}{0.000503}$$

$$= 7.634\text{ sec.}$$

$$t_B = \frac{C_1\omega_{y0} - C_2\omega_{x0}}{\Delta} = \frac{0.0139\times 0.15 - 0.0163\times 0.05}{0.000503}$$

$$= 2.525\text{ sec.}$$

Thus, it is necessary to fire A approximately three times as long as B in order to stop the capsule's rotation. This agrees with our initial qualitative remarks regarding the motion.

• PROBLEM 19-24

Consider the spacecraft found in the preceding problem. The craft has no initial angular velocity and only jet A is turned on for 1 sec. with a thrust of 25 N. Determine the axis of precession and the rates of precession and the spin after the jet has stopped.

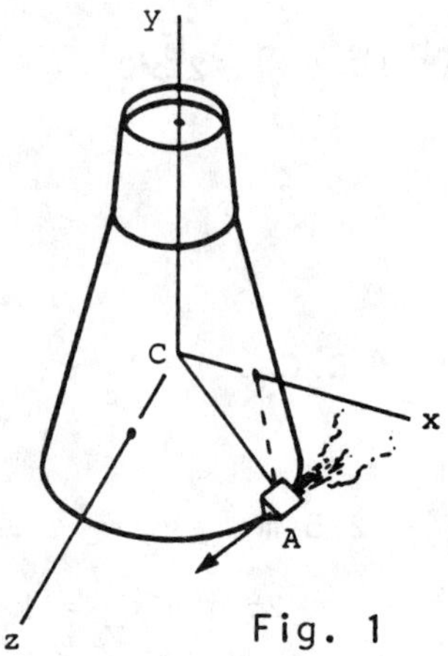

Fig. 1

Solution: The problem can be considered in two parts: The motion during the firing of the rocket and the motion after.

For the first part let t = 0 be the time when the rocket is turned on. The rocket produces a torque, $\vec{\tau}$, which, by Newton's Second Law, produces an angular acceleration, $\vec{\alpha}$.

$$\vec{\tau} = I\vec{\alpha} = I_x\alpha_x\vec{i} + I_y\alpha_y\vec{j} + I_z\alpha_z\vec{k} \tag{1}$$

where I is the moment of inertia and I_x, I_y, I_z are the principal moments about the x, y, z axes, respectively. In this problem the torque produced by the rocket is

$$\vec{\tau} = \vec{r} \times \vec{F} = (x\vec{i} - y\vec{j}) \times F\vec{k} = -yF\vec{i} - xF\vec{j}. \tag{2}$$

Equating equations (1) and (2) gives three first order differential equations for the components of the angular velocity, ω.

$$\alpha_x = \frac{d\omega_x}{dt} = -\frac{yF}{I_x} = -C_1 \ ; \quad \frac{d\omega_y}{dt} = -\frac{xF}{I_y} = -C_2 \ ; \quad \frac{d\omega_z}{dt} = 0.$$

The solutions to these equations are

$$\omega_x(t) = \omega_x(0) - C_1t \ ; \quad \omega_y(t) = \omega_y(0) - C_2t \ ; \quad \omega_z(t) = \omega_3(0).$$

However, at t = 0 the capsule is not rotating, so all ω(0) = 0. Thus,

$$\omega_x(t) = -C_1t \ , \quad \omega_y(t) = -C_2t \ , \quad \omega_z(t) = 0. \tag{3}$$

From equation (3) we see that $\vec{\omega}$ lies in the x, y plane while the rocket is being fired.

For this problem x = 4 m, y = 2.5 m, $k_x = k_z = 1.5$ m, $k_y = 1.75$ m, m = 2000 kg, F = 25 n, and the rocket is fired for 1 second. k_x, k_y and k_z are the radii of gyration of the capsule. The angular velocity at the end of 1 second is from equation (3).

$$\omega_x(1) = -\frac{2.5 \text{ m } 25 \text{ n}}{2000 \text{ kg}(1.5 \text{ m})^2} 1 \text{ sec.} = -0.0139 \text{ rad/sec} \tag{4}$$

$$\omega_y(1) = -\frac{4 \text{ m } 25 \text{ n}}{2000 \text{ kg}(1.75 \text{ m})^2} 1 \text{ sec.} = -0.0163 \text{ rad/sec} \tag{5}$$

$$\omega^2 = \omega_x^2(1) + \omega_y^2(1) = (0.0139)^2 + (0.0163)^2$$

$$\omega(1) = 0.02142 \text{ rad/sec.} \tag{6}$$

Figure 2 shows the directions of $\vec{\omega}$ and the angular momentum $\vec{L}$.

$$\tan\theta = \frac{\omega_x}{\omega_y} = \frac{0.0139}{0.0163} = 0.8528$$

$$\theta = 44.46°.$$

The angular momentum of the capsule is

$$\vec{L} = I_x \omega_x \vec{i} + I_y \omega_y \vec{j} + I_z \omega_z \vec{k}.$$

At the end of one second

$$\vec{L} = -2000(1.5)^2(0.0139)\vec{i} - 2000(1.75)^2(0.0163)\vec{j}$$

$$= -62.6\vec{i} - 99.8\vec{j} \qquad (7)$$

$$\tan \phi = \frac{L_x}{L_y} = \frac{62.6}{99.8} = 0.6273$$

$$\phi = 32.1°.$$

Next, consider the second part of the motion after the rocket has stopped firing. For simplicity, refer to this time as t = 0; i.e., the values of $\vec{\omega}$ and $\vec{L}$ given in equations (4), (5) and (7) become the initial values for the subsequent motion.

In general the angular form of Newton's Second Law can also be written as

$\vec{\tau} = \frac{d\vec{L}}{dt}$. Now $\vec{\tau} = 0$ and $\vec{L}$ = constant. Thus, $\vec{L}$ does not change after t = 0. It is a constant, fixed in space and remains at the angle ϕ with respect to the y-axis.

For a vector in a rotating coordinate system its rate of change with respect to a fixed system is

$$\frac{d\vec{A}}{dt} = \frac{\partial \vec{A}}{\partial t} + \vec{\omega} \times \vec{A},$$

where $\frac{\partial \vec{A}}{\partial t}$ is the rate of change in the rotating system. If we let $\vec{A} = \vec{L}$, then

$$\vec{\tau} = \frac{d\vec{L}}{dt} = \vec{\omega} \times \vec{L} + \frac{\partial \vec{L}}{\partial t}. \qquad (8)$$

This is known as Euler's equation. In this case $\vec{\tau} = 0$ and in terms of components, (8) becomes

$$I_x \omega_x + \omega_y \omega_z (I_z - I_y) = 0$$

$$I_y \omega_y + \omega_x \omega_y (I_x - I_z) = 0$$

$$I_z \omega_z + \omega_x \omega_y (I_y - I_x) = 0. \qquad (9)$$

Since the y axis is the axis of symmetry of the capsule, $I_x = I_z$. Therefore, the second equation above gives

$$\dot{\omega}_y = 0, \quad \text{or} \quad \omega_y(t) = \omega_y(0).$$

Thus, the y-component of ω is a constant equal to the value when the rocket stopped; i.e., from equation (5)

$\omega_y(t) = -0.0163$rad/sec. Let $\alpha = \frac{I_y - I_x}{I_x}$ and $\beta = \omega_y(0)$, then the other two equations can be written as

$$\dot{\omega}_x - \alpha\beta\omega_z = 0 \quad \text{and} \quad \dot{\omega}_z + \alpha\beta\omega_x = 0. \qquad (10)$$

These are two coupled, first order differential equations in ω_x and ω_z. To find a solution, let $\omega_x = Ae^{kt}$ and $\omega_z = Be^{kt}$ substitute in equation (10).

$$Ak - \alpha\beta B = 0 \quad \text{and} \quad Bk + \alpha\beta A = 0$$

$$k = \alpha\beta B/A \quad \text{and} \quad \alpha\beta B^2/A + \alpha\beta A = 0 \quad \text{or} \quad B^2 + A^2 = 0.$$

Thus, there is a solution if $B = \pm iA$ and $k = \pm i\alpha\beta \equiv \pm iP$

$$\omega_x = Ae^{ipt}, \quad \omega_z = \pm iAe^{-ipt}.$$

Since the original equations are linear, a linear combination of solutions is also a solution. Therefore,

$$\omega_x(t) = C\cos(pt + \theta) \quad \text{and} \quad \omega_z(t) = C\sin(pt + \theta) \qquad (8)$$

are also solutions as can be verified by direct substitution. The constants can be evaluated by use of the initial conditions. At $t = 0$,

$$\omega_x(0) = C\cos\theta \quad \text{and} \quad \omega_z(0) = C\sin\theta,$$

but $\omega_z(0) = 0$ from equation (3); therefore, in order to satisfy the second equation we must choose $\theta = 0$. This leaves $C = \omega_x(0) = -0.0139$rad/sec. from equation (4).

Since $\omega_x^2(t) + \omega_z^2(t) = C^2$, the tip of the ω vector describes a circle of radius C about the symmetry axis (y-axis) in the x, z plane. $\vec{\omega}$ also maintains its initial angular displacement, θ, with respect to the symmetry axis. The cone that traces out is called the body cone since its motion is with respect to the x, y, z axes fixed in the body.

From equation (8) the frequency of rotation of $\vec{\omega}$ about the symmetry axis is

$$P = \alpha\beta = \omega_y(0)\frac{I_y - I_x}{I_x} = -0.0163\frac{(1.75)^2 - (1.5)^2}{(1.5)^2}$$

$$= -0.0059 \text{ rad/sec.}$$

It has been shown above that the direction of $\vec{L}$ is fixed in space. So an observer in the fixed system also sees $\vec{\omega}$ rotate about $\vec{L}$ with a frequency of rotation Ω. This motion also generates a cone called the space cone. Both the body cone and the space cone are shown in Figure 3. Since $\vec{\omega}$ is common to both, the cones always have a common line of contact. Thus, $\vec{\omega}$ acts as an instantaneous axis and the general motion can be characterized by noting that the body cone rolls on the fixed space cone.

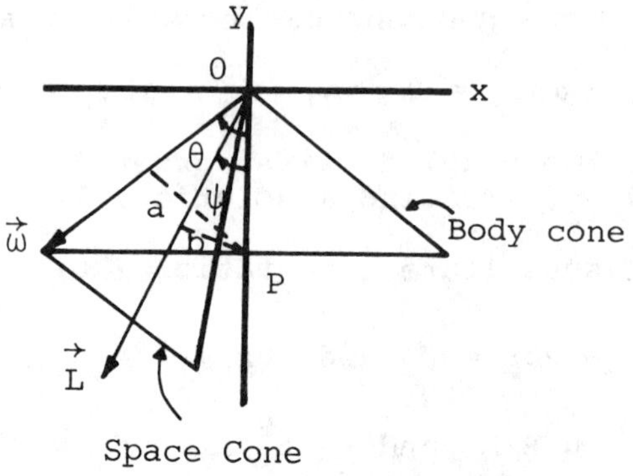

Fig. 3

We can find the rotation frequency Ω by noting that the point P in Figure 3 in a time dt will turn through a distance ds = aωdt about $\vec{\omega}$. It will also turn through a distance ds = bΩ dt with respect to L. Since it is the same point, the result is aω = bΩ. However, from Figure 3 a = OP sin θ and b = OP sin θ. Thus,

$$\Omega = \omega\frac{\sin\theta}{\sin\phi} = \omega\frac{\sin 44.46°}{\sin 32.1°} = 0.02142\ \frac{0.6489}{0.5314}$$

$$= 0.02616 \text{ rad/sec.}$$

• PROBLEM 19-25

Analyze the motion of a heavy symmetrical top.

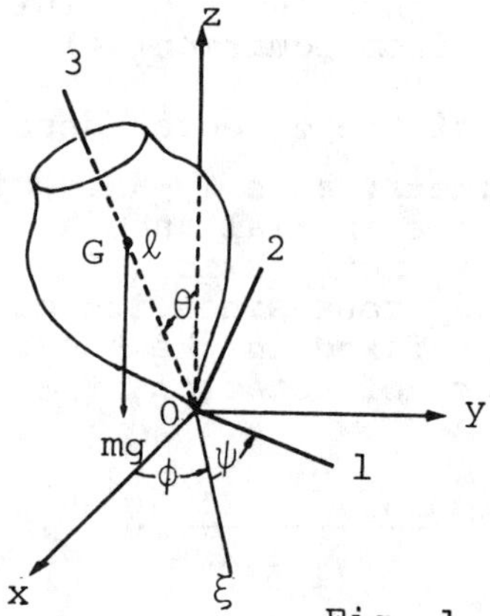

Fig. 1

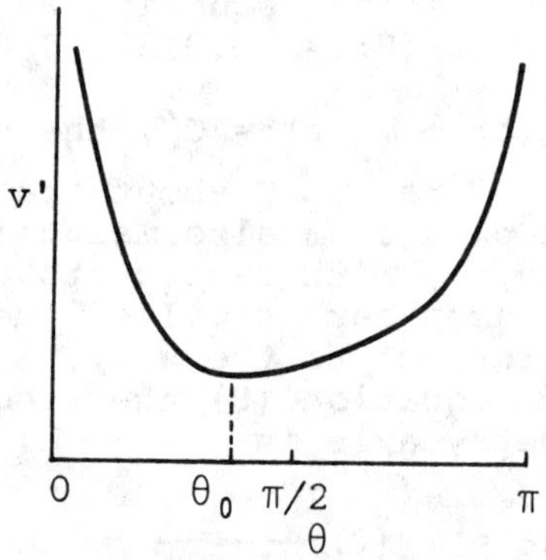

Fig. 2
Effective potential energy
For the symmetrical top $P_\psi \neq 0$

Solution: The symmetrical top is a rigid body pivoting around a fixed point in a gravitational field as shown in figure 1. The motion of the top can be best analyzed by considering two sets of axes. One is a space fixed coordinate system denoted by x, y, z and the other, a set of axes fixed in the body denoted by 1, 2, 3. The origin shall be the fixed point 'O' that lies on the axis of symmetry at a distance ℓ from the center of mass G. Due to symmetry, the moment of inertia $I_1 = I_2$. The orientation of the body frame with respect to the space frame is specified conveniently by the three Euler Angles θ, ϕ, and ψ as shown in the figure. When the body is in motion, these Euler Angles change in time. If θ alone changes while ϕ and ψ are fixed, the body rotates around the line of nodes with angular velocity $\dot{\theta}\hat{e}_\xi$ where $\hat{e}_\xi$ is the unit vector along ξ axes This is called the line of nodes. If ϕ alone changes, the body rotates around the z-axis with angular velocity $\dot{\phi}\hat{k}$ where k is the two unit vector along z-direction. If ψ alone changes, the body rotates around its 3-axis with angular velocity $\psi\hat{e}_3$ where e_3 is the unit vector along 3-axis. Then the angular velocity of the body is

$$\underset{\sim}{\omega} = \dot{\theta}\hat{e}_\xi + \dot{\phi}\hat{k} + \dot{\psi}\hat{e}_3.$$

Expressing the angular velocity in terms of the principal axes obtains

$$\omega_1 = \dot{\theta} \cos \psi + \dot{\phi} \sin \psi \sin \theta$$

$$\omega_2 = -\dot{\theta} \sin \psi + \dot{\phi} \sin \theta \cos \psi$$

$$\omega_3 = \dot{\psi} + \dot{\phi} \cos \theta.$$

Hence, the kinetic energy,

$$T = \frac{1}{2}I_1\omega_1^2 + \frac{1}{2}I_2\omega_2^2 + \frac{1}{2}I_3\omega_3^2,$$

is, with $I_1 = I_2$,

$$T = \frac{1}{2}I_1\dot{\theta}^2 + \frac{1}{2}I_1\dot{\phi}^2 \sin^2 \theta + \frac{1}{2}I_3(\dot{\psi} + \dot{\phi} \cos \theta)^2.$$

The potential energy is

$$'V' = mg\ell \cos \theta.$$

The Lagrangian function is given by

$$L = T - 'V'$$

$$= \frac{1}{2}I_1\dot{\theta}^2 + \frac{1}{2}I_1\dot{\phi}^2 \sin^2 \theta + \frac{1}{2}I_3(\dot{\psi} + \dot{\phi} \cos \theta)^2$$

$$- mg\ell \cos \theta.$$

The coordinates ψ and ϕ are absent in the Lagrangian function ((they may be disregarded), and, therefore, there are three integrals of motion:

$$\frac{\partial p_\psi}{\partial t} = \frac{\partial L}{\partial \psi} = 0$$

$$\frac{\partial p_\phi}{\partial t} = \frac{\partial L}{\partial \phi} = 0$$

$$\frac{\partial E}{\partial t} = -\frac{\partial L}{\partial t} = 0$$

where

$$p_\psi = I_3(\mathring{\psi} + \mathring{\phi}\cos\theta) \qquad \text{(a)}$$

$$p_\phi = I_1\mathring{\phi}\sin^2\theta + I_3\cos\theta(\mathring{\psi} + \mathring{\phi}\cos\theta) \qquad \text{(b)}$$

$$E = \tfrac{1}{2}I_1\mathring{\theta}^2 + \tfrac{1}{2}I_1\mathring{\phi}^2\sin^2\theta + \tfrac{1}{2}I_3(\mathring{\psi} + \mathring{\phi}\cos\theta)^2$$
$$+ mg\ell\cos\theta. \qquad \text{(c)}$$

Using (a) and (b) to eliminate $\mathring{\phi}$ and $\mathring{\psi}$ from (c),

$$E = \tfrac{1}{2}I_1\theta^2 + \frac{(p_\phi - p_\psi\cos\theta)^2}{2I_1\sin^2\theta} + \frac{p_\psi^2}{2I_3} + mg\ell\cos\theta. \qquad \text{(d)}$$

The problem now can be solved by the energy method. Setting

$$E' = E - \frac{p_\psi^2}{2I_3} \qquad \text{(e)}$$

$$'V' = \frac{(p_\phi - p_\psi\cos\theta)^2}{2I_1\sin^2\theta} + mg\ell\cos\theta, \qquad \text{(f)}$$

then

$$\theta = \left\{\frac{2}{I_1}[E' - 'V'(\theta)]\right\}^{1/2} \qquad \text{(g)}$$

and θ is given by

$$\left(\frac{I_1}{2}\right)^{1/2} t = \int_{\theta_0}^{\theta} \frac{d\theta}{[E' - 'V'(\theta)]^{1/2}}. \qquad \text{(h)}$$

The constant θ_0 is the initial value of θ. Using $\theta(t)$, equations (a) and (b) can be solved for $\mathring{\psi}$ and $\mathring{\phi}$ and integrated to give $\psi(t)$ and $\phi(t)$. Using the expression for ω_3, we have from equation (a), $p = I_3\omega_3$, so that ω_3 is a constant of motion. (If $\omega_3 = 0$, then 'V(θ)' corresponds to the energy of a spherical pendulum.) The torque is given by

$$'N' = -\frac{\partial 'V'}{\partial \theta} = mg\ell \sin\theta - \frac{(p_\phi - p_\psi \cos\theta)(p_\psi - p_\phi \cos\theta)}{I_1 \sin^3}$$

This equation shows that 'N' is positive in general (if $p_\phi \neq p_\psi$) for $\theta = 0$ and negative for $\theta = \pi$ and has one zero between 0 and π. Hence, the potential 'V' has a minimum at θ_0 as shown in figure 2 given by

$$mg\ell I_1 \sin^4\theta_0 - (p_\phi - p_\psi \cos\theta_0)(p_\psi - p_\phi \cos\theta_0) = 0. \tag{j}$$

If $E' = 'V'(\theta_0)$ and $\theta_i = \theta_0$, then $\mathring{\theta} = 0$ so that the axis of the top precesses uniformly at an angle θ_0 with veoocity

$$\mathring{\phi} = \frac{p_\phi - p_\psi \cos\theta_0}{I_1 \sin^2\theta_0}$$

determined from equations (a) and (b). From equation (j),

$$(p_\phi - p_\psi \cos\theta_0)\left(\frac{p_\psi}{\cos\theta_0} - p_\phi\right) = mg\ell\, I_1 \frac{\sin^4\theta_0}{\cos\theta_0}$$

$$(p_\phi - p_\psi \cos\theta_0)\left(p_\phi - p_\psi \cos\theta_0 - p_\psi \frac{\sin^2\theta_0}{\cos\theta_0}\right)$$

$$= -mg\ell\, I_1 \frac{\sin^4\theta_0}{\cos\theta_0}$$

$$(p_\phi - p_\psi \cos\theta_0)^2 - p_\psi(p_\phi - p_\psi \cos\theta_0)\frac{\sin^2\theta_0}{\cos\theta_0} + \frac{1}{4}p_\psi^2 \frac{\sin^4\theta_0}{\cos^2\theta_0}$$

$$= \frac{1}{4}p_\psi^2 \frac{\sin^4\theta_0}{\cos^2\theta_0} - mg\ell\, I_1 \frac{\sin^4\theta_0}{\cos\theta_0}$$

$$\left[(p_\phi - p_\psi \cos\theta_0) - \frac{1}{2}p_\psi \frac{\sin^2\theta_0}{\cos\theta_0}\right]^2$$

$$= \frac{1}{4}p_\psi^2 \frac{\sin^4\theta_0}{\cos^2\theta_0}\left(1 - \frac{4mg\ell\, I_1 \cos\theta_0}{p_\psi^2}\right)$$

$$(p_\phi - p_\psi \cos\theta_0) - \frac{1}{2}p_\psi \frac{\sin^2\theta_0}{\cos\theta_0}$$

$$= \pm \frac{1}{2}p_\psi \frac{\sin^2\theta_0}{\cos\theta_0}\left(1 - \frac{4mg\ell\, I_1 \cos\theta_0}{p_\psi^2}\right)^{1/2}.$$

Using $p_\psi = I_3\omega_3$, the result is

$$(p_\phi - p_\psi \cos\theta_0) = \frac{1}{2}I_3\omega_3 \frac{\sin^2\theta_0}{\cos\theta_0}\left[1 \pm \left(1 - \frac{4mg\ell\, I_1 \cos\theta_0}{I_3^2\omega_3^2}\right)^{1/2}\right]. \tag{k}$$

It can be seen that to preserve the reality of the root if $\theta_0 < \pi/2$, there is a minimum spin angular velocity below which the top cannot precess uniformly at the nagle θ_0:

$$\omega_{min} = \left(\frac{4mg\ell\, I_1}{I_3^2} \cos\theta_0\right)^{1/2}. \tag{l}$$

For $\omega_3 > \omega_{min}$ there are two roots for the equation (k). Hence, there are two possible values of $\dot{\phi}_0$, a slow and fast precession, both in the direction of ω_3 for $\omega_3 >> \omega_{min}$. The fast precession occurs with

$$p_\phi - p_\psi \cos\theta_0 \approx \frac{1}{2}I_3\omega_3 \frac{\sin^2\theta_0}{\cos\theta_0}[1 + 1] \text{ so that}$$

$$\dot{\phi} = \frac{p_\phi - p_\psi \cos\theta_0}{I_1 \sin^2\theta_0} \text{ yields } \dot{\phi}_0 = \frac{I_3}{I_1}\frac{\omega_3}{\cos\theta_0}.$$ For the slow

precession, $(1 - \varepsilon)^{1/2} \simeq 1 - \frac{1}{2}\varepsilon$ gives

$$p_\phi - p_\psi \cos\theta_0 \simeq \frac{1}{2}I_3\omega_3 \frac{\sin^2\theta_0}{\cos\theta_0}\left[1 - \left(1 - \frac{4mg\ell\, I_1 \cos\theta_0}{2p^2}\right)\right]$$

$$p_\phi - p_\psi \cos\theta_0 \simeq \frac{I_3\omega_3\, mg\ell\, I_1 \sin^2\theta_0}{p_\psi^2}$$

$$\simeq \frac{mg\ell\, I_1 \sin^2\theta_0}{I_3\omega_3}$$

so that $\dot{\phi}_0 = \frac{mg\ell}{I_3\omega_3}$. It is the slow precession that is ordinarily observed with a rapidly spinning top. For $\theta_0 > \pi/2$ (top hanging with the axis below the horizontal) there is one positive and one negative value of θ_0.

Also, there is a more general motion called 'nutation' or oscillation of the axis of the top in the θ-direction as it precesses. The axis oscillates between θ_1 and θ_2 which satisfies the equation.

$$E' = \frac{(p_\phi - p_\psi \cos\theta)^2}{2I_1 \sin^2\theta} + mg\ell \cos\theta \qquad \text{(m)}$$

where p_ϕ, p_ψ and E' are determined from the initial conditions. Multiplying (m) by $\sin^2\theta$ obtains a cubic equation in $\cos\theta$ which has two real roots between -1 and +1 with angles θ_1 and θ_2 are one non-physical root greater than 1. It is possible to obtain uniform precession if the two roots coincide at $\cos\theta_0$. If initially $\dot{\theta} = 0$, then the initial value $\cos\theta_1$ of $\cos\theta$ satisfies equation (m). Knowing one root, it is possible to factor the cubic equation to obtain the other roots.

During nutation, the precession velocity varies as

$$\dot{\phi} = \frac{p_\phi - p_\psi \cos\theta}{I_1 \sin^2\theta}.$$

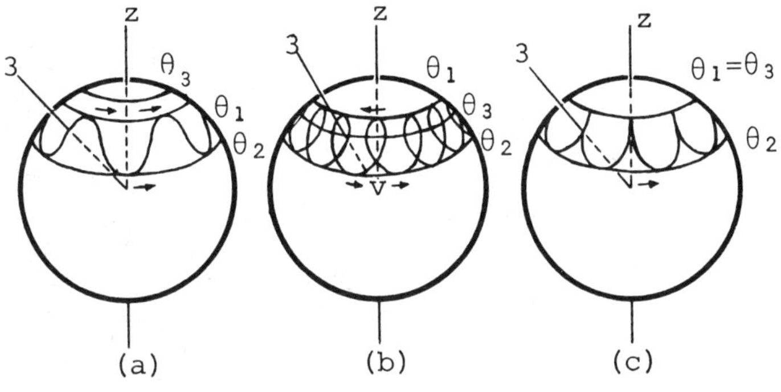

Fig. 3

Locus of top axis (3) on unit sphere.

Depending on the relationship of the precession and nutation velocity, several possible motions may occur as shown in figure 3. If $|p_\phi| < |p_\psi|$, we define

$$\cos\theta_3 = \frac{p_\phi}{p_\psi}$$

which is well defined value determined by the constants of motion. If $\theta_3 < \theta_1$ then $\dot{\phi}$ has the same sign as ω_3, and, throughout the nutation, the axis of the top traces the curve shown in figure 3(a). If $\theta_3 > \theta_1$, $\dot{\phi}$ changes sign during the nutation and the top axis traces the curve shown in figure 3b. It is also clear that if the top is set in motion initially above the horizontal plane with $\dot{\phi}$ opposing ω_3, the motion is like that shown in figure 3b.

Finally, consider the top axis which initially conincides with the z-axis. The line of nodes is indeterminate and the angle between 1 and the x-axis is then $(\psi + \phi)$.

Initially from equations (a) and (b),

$$p_\psi = I_3(\mathring{\psi} + \mathring{\phi}) = I_3\omega_3$$

$$p_\phi = I_3(\mathring{\psi} + \mathring{\phi}) = p_\psi.$$

Then

$$'V' = \frac{I_3^2\omega_3^2}{2I_1}\left[\frac{(1 - \cos\theta)^2}{\sin^2\theta} + \alpha\cos\theta\right]$$

where

$$\alpha = \frac{2I_1\,mg\ell}{I_3^2\,\omega_3^2}.$$

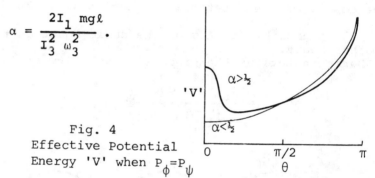

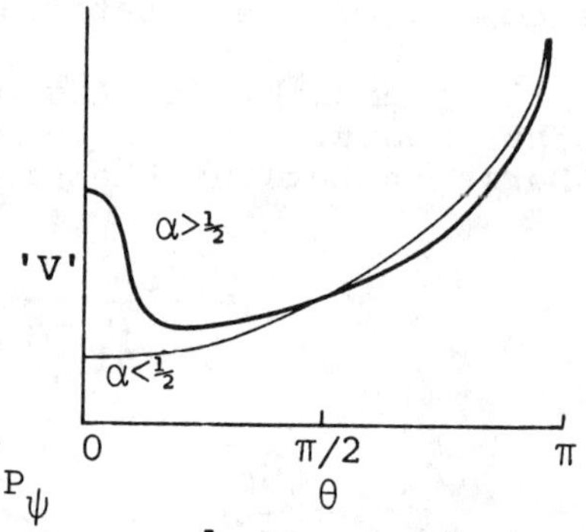

Fig. 4
Effective Potential
Energy 'V' when $P_\phi = P_\psi$

In figure 4 plot 'V' vs θ for both $\alpha < \frac{1}{2}$ and $\alpha > \frac{1}{2}$. Note that a rapidly spinning top for which $\alpha < \frac{1}{2}$ can spin stably about the vertical axis because the potential has a minimum at $\theta = 0$.

For the slowly spinning top and the potential has a maximum at $\theta = 0$, therefore, it is unstable. The minimum spin angular velocity below which the top cannot spin stably about the vertical axis occurs when $\alpha = \frac{1}{2}$, or

$$\omega_{min} = \left[\frac{4mg\ell\,I_1}{I_3^2}\right]^{1/2}.$$

All these considerations can be very easily verified experimentally with a top or gyroscope.

CHAPTER 20

MOVING COORDINATE FRAMES

MOTION OBSERVED FROM AN ACCELERATED FRAME

• PROBLEM 20-1

Suppose observer C in accelerating frame shines a narrow beam of light across his box perpendicular to the direction of acceleration of the box. Demonstrate qualitatively that observer C will observe this beam of light to be bent downward toward the floor. Use the principle of equivalence to demonstrate that light is bent when it passes across a gravitational field.

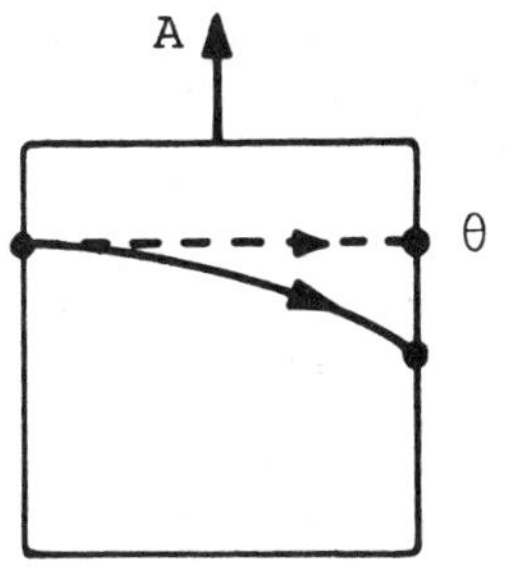

Solution: During the time of flight of the light beam across the box, the box will move in the direction of $\bar{a}$ by a distance: $\frac{1}{2}at^2$. Thus, the light beam will strike a distance $\frac{1}{2}at^2$ below the target point θ.

The same effect of deflection would be produced if the light beam were gravitationally attracted with an acceleration $-\bar{a}$ opposite to $\bar{a}$ (toward the floor of the box).

Since any experiment done inside the box cannot distinguish between an accelerated box or a gravitational force (principle of equivalence), and the acceleration of the box would clearly cause the deflection of the beam, it follows that a gravitational field would also produce a similar deflection.

• PROBLEM 20-2

Discuss the meaning of the word "real" as applied to physical observables. Are pseudo-forces "real"? If not, why do non-inertial observers experience them? If so, why do different observers disagree about their presence or absence?

Solution: Physicists classify forces as "real forces" or "pseudo-forces". Included in the list of real forces are those arising from gravity, electricity, magnetism and forces within the nuclei of atoms. Also included are the general forms of contact forces such as friction, Hooke's Law forces, and forces occuring during the collision of masses. All observers, regardless of what coordinate system they use, would agree about the existence, magnitude and direction of real forces. Newton's Laws, as applied in inertial reference frames, include only real forces.

Pseudo-forces arise only when the coordinate system being used is accelerated in any way relative to an inertial frame. If the student uses only inertial frames, pseudo-forces never arise. However, as the problems ahead will amply show, many problems are much easier to solve in a non-inertial reference frame.

The pseudo-forces are not real in the sense of a force arising due to interaction of some sort between two bodies. However, the existence of force-like properties in a non-inertial frame is not to be doubted. The key to this distinction is 1) the agent of the force; 2) the point of view provided by different observers.

Real forces as noted are apparent to any observer in any frame and are due to interactions, the existence of which does not depend on inertial properties.

Pseudo-forces are only apparent to observers within the accelerating frame and arise solely as an inertial reaction to the acceleration of the frame.

They arise due to an inertial mass's natural tendency to continue its motion unchanged as described by Newton's First Law. The accelerated frame is one in which the velocity at the entire frame is changing. Any objects within this frame that are free to do so will continue along with their previous motion. This means that they have a tendency to move in relation to the accelerated frame. But always in physics, the tendency of a body to change its motion in a frame is taken as the result of an applied force. Thus, the observer in the accelerated frame who sees the frame as static but observes a particular mass travel in its own path attributes this new motion to a force. However, the observer in another inertial frame, relative to which the non-inertial frame is accelerating, observes it as moving in a straight line; hence, no force is being applied to it while the frame it is within changes its motion.

• PROBLEM 20-3

A boy who is holding a suitcase stands in an elevator. The elevator has an acceleration $\vec{A}$, which may be directed upward ($\vec{A}$ positive) or downward ($\vec{A}$ negative). What force does the suitcase exert on the boy's hand?

Solution: This problem involves two different inertial coordinate systems. One system, S, is attached to the ground. The other system, S', is attached to the elevator which is constantly accelerated at a rate, A.

According to S: With respect to the coordinate system S attached to the ground, the suitcase, as well as the boy and the elevator, is accelerated and so the net interaction force $F - mg$ (contact and gravity) must equal the observed rate of change of momentum mA; that is,

$$F - mg = mA \quad \text{or} \quad F = m(g + A).$$

If A is positive (elevator accelerated upward) the contact force on the suitcase from the boy's hand is larger than the weight of the suitcase, and so the suitcase feels heavier. On the other hand, when A is negative (elevator accelerated downward) the suitcase feels lighter and, in fact, when A is -g, the suitcase exerts no force on the boy's hand. The elevator, boy, and suitcase are all in free-fall motion.

According to S': With respect to a coordinate system on the elevator, the suitcase is at rest relative to that system, and so it is concluded that the suitcase is in equilibrium. But it is in equilibrium under the influence of the two interaction forces, the force of the boy's hand $\bar{F}$, and the weight $m\vec{g}$, and the inertial force $\bar{F}_i = -m\vec{A}$. Since the acceleration according to S' is zero, we have

$$F - mg - mA = 0 \quad \text{or} \quad F = m(g + A).$$

In other words, the apparent weight of an object in an accelerated elevator is $m(g + A)$, and to an observer in the coordinate system on the elevator, there is no way of distinguishing whether the apparent weight of objects arises from gravitational force or from a uniform acceleration of the coordinate system.

• PROBLEM 20-4

A wedge of wood rests on a horizontal surface. A uniform sphere is placed on the upper surface of the wedge, which is accelerated along the horizontal surface with a constant acceleration. What must the acceleration of the wedge be in order that the sphere will remain at rest with respect to the wedge?

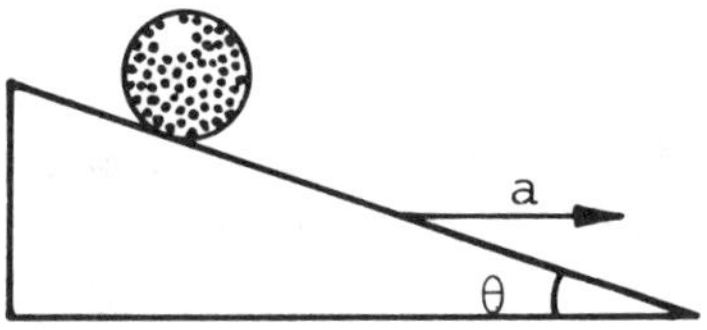

Fig. 1

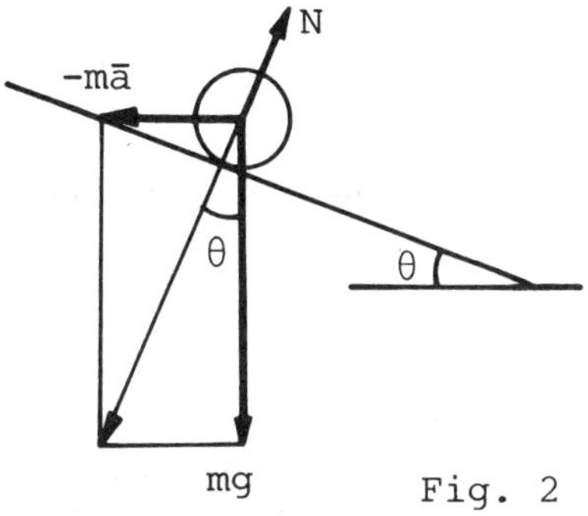

Fig. 2

Solution: To prevent the sphere from rolling either up or down the plane, the forces and pseudo-forces must cancel parallel to the plane's inclined surface. These forces and pseudo-forces are illustrated in the free-body diagram in figure (2). $\vec{N}$ is always perpendicular to the plane, but $-m\vec{a}$ and $m\vec{g}$ are not. However, the sum of $-m\vec{a}$ and $m\vec{g}$ are perpendicular to the plane if $-m\vec{a}$ is such that $\tan\theta = ma/mg$, the situation shown in the figure. Thus,

$$a = g \tan\theta .$$

• PROBLEM 20-5

A mass m hangs at the end of a rope which is attached to a support fixed on a trolley (as shown in the figures). Find the angle α it makes with the vertical, and its tension T when the trolley 1) moves with a uniform speed on horizontal tracks, 2) moves with a constant acceleration on horizontal tracks.

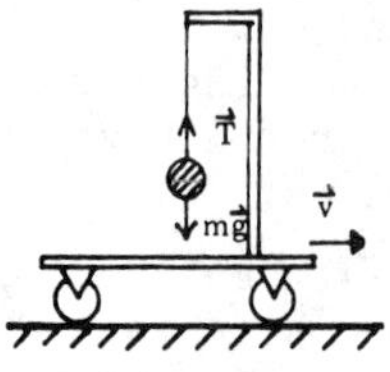

FIGURE A

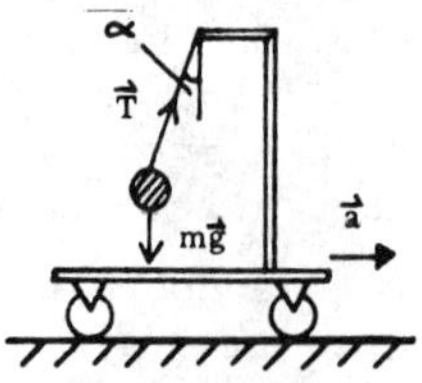

FIGURE B

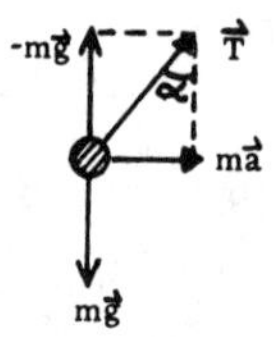

FIGURE C

Solution: 1) When the trolley moves with a uniform speed, the only external force acting on m is the gravitational force $m\vec{g}$. It is balanced by the tension $\vec{T}$ in the rope since m is in equilibrium (Fig.a). Hence, $\alpha = 0$. 2) When the trolley has acceleration a, the effect of the acceleration is transmitted to the mass through the rope. It is seen in (Fig. b) that the magnitudes of the gravitational and horizontal accelerations determine the magnitude of the angle α. The tension in the rope provides the upward force -mg to hold the mass (since m is in vertical equilibrium) as well as the external force ma acting on the mass as a result of the motion of the trolley, (Fig. c). Hence,

$$T \sin \alpha = ma \tag{1}$$

$$T \cos \alpha = mg \tag{2}$$

Dividing (1) by (2), the result is

$$\frac{T \sin \alpha}{T \cos \alpha} = \frac{ma}{mg}$$

$$\tan \alpha = \frac{a}{g}$$

Therefore, α is given by

$$\alpha = \tan^{-1} \frac{a}{g} \quad .$$

• PROBLEM 20-6

A light rod of length 2r is mounted on a vertical shaft which, in turn, is mounted on a car that has a constant acceleration $\vec{A}$ to the right, as shown in Fig. 1. The rod has a pivot at its center that enables it to rotate freely in a horizontal plane. On the ends of the rod are attached masses m and M, as shown. What is the motion of the rod with respect to a moving coordinate system attached to the car?

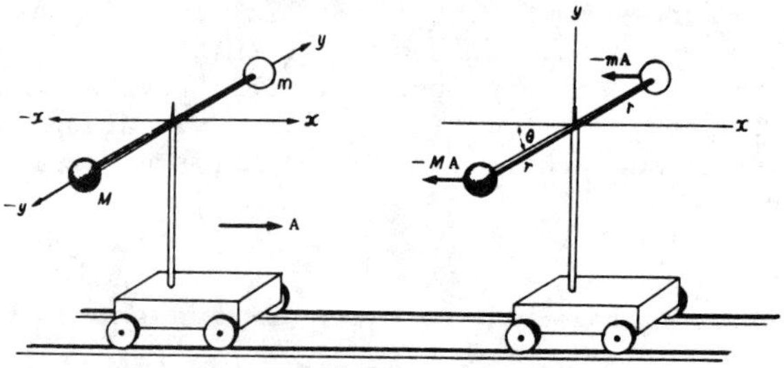

FIGURE 1

Solution: Gravitational forces are of no consequence so far as motion in the horizontal plane is concerned. Both masses, however, are subject to inertial forces mA and MA, directed toward the left. If the mass of M and that of m are equal, there is no tendency for the rod to rotate. If the masses are unequal, the larger mass will move toward the left (toward the negative x-axis).

More generally, the rod swings as a pendulum about the negative x-axis. When the rod makes an angle θ with the negative x-axis, there is a torque $\tau = -(M - m)Ar \sin\theta$, or $-(M - m)Ar\theta$ for small θ. This torque must be equal to the rate of change of angular momentum; that is, $I_0 d^2\theta/dt^2 = -(M - m)Ar\theta$, and since $I_0 = Mr^2 + mr^2$, the rod swings back and forth about $\theta = 0$ with a period of

$$T = 2\pi\sqrt{\frac{Mr^2 + mr^2}{(M - m)rA}} .$$

The system has a potential energy $(M - m)Ar(1 - \cos\theta)$, and if it is released from θ_0, the kinetic energy at $\theta = 0$ will be

$$E = (M - m)Ar(1 - \cos\theta_0) .$$

• PROBLEM 20-7

Discuss the laws of motion in an accelerating box (see figure 1). Show that an observer C in the box will measure the same acceleration for unequal masses such as A and B in his own coordinate system. Finally, show that the force, $\vec{F}$, required to provide the acceleration $\vec{a}$ to the box and its contents is less during the time that masses like A and B are in free fall.

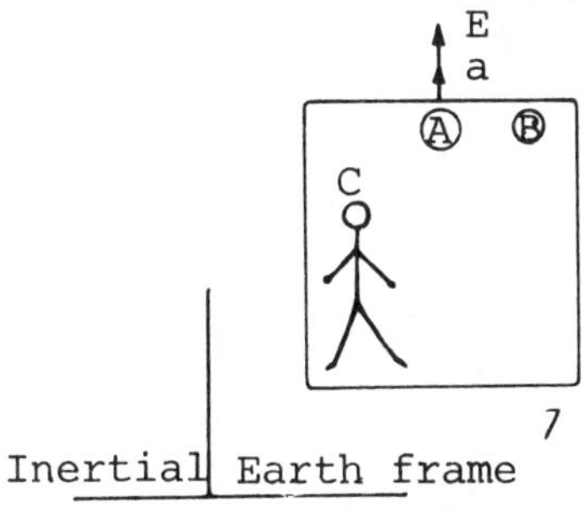

Fig. 1

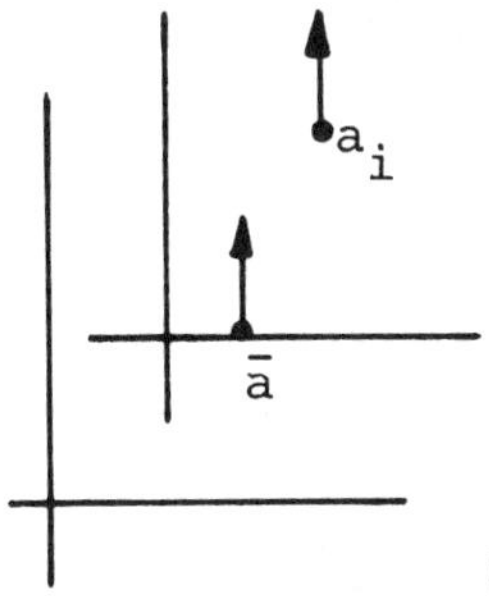

Fig. 2

Solution: Newton's Second Law applies to the box and all its contents

$$\Sigma \vec{f}_i = m_i\vec{A}_i \tag{1}$$

where f_i are all "real" forces and the accelerations are measured relative to an inertial frame of reference (the earth frame, but not C's frame of reference). Accelerations in the earth frame, $\vec{A}_i$'s, are related to $\vec{a}$, the acceleration of the frame and the accelerations measured by C (symbolized by $\vec{a}$) as follows (from figure 2):

$$\bar{A}_i = \bar{a} + \bar{a}_i \tag{2}$$

Putting this $\bar{A}_i$ into Newton's Second Law, results in

$$\Sigma \bar{f}_i = m_i\bar{a} + m_i\bar{a}_i . \tag{3}$$

Rearranging so that $m_i\bar{a}_i$ is by itself,

$$\Sigma \bar{f}_i - m_i\bar{a} = m_i\bar{a}_i . \tag{4}$$

This equation, derived from Newton's Second Law, states that $m\bar{a}_i$, the mass multiplied by the acceleration measured in the frame of C and the box, equals the sum of all the usual "real" forces plus the negative of $m\bar{a}$.

This last term is called a "pseudo-force". It appears in Newton's Second Law only because an accelerated reference frame is being used. Both the direction and magnitude depend on the coordinate system used, unlike "real" forces such as gravity and the forces exerted by springs, friction, electricity and magnetism.

Now, consider unequal masses A and B dropped within the box. In the earth frame both A and B have acceleration $-\bar{g}$. In C's system

$$\bar{a}_A = -\bar{g} - \bar{a}$$

$$\bar{a}_B = -\bar{g} - \bar{a}$$

and both masses have the same freefall acceleration, though not the familiar 9.8 m/sec^2.

Finally, consider the force $\bar{F}$ needed to accelerate the box and its contents relative to the earth. Applying $\Sigma f_i = m_i A_i$ to the box of mass M and all its contents,

$$\bar{F} - (m_A + m_B + m_C + M)g = (m_A + m_B + m_C + M)\bar{A}.$$

We find for F,

$$\bar{F} = (m_A + m_B + m_C + M)(\bar{A} + \bar{g}).$$

But, if one of the masses, say A, is in free fall, its acceleration is $\bar{A}_A = -\bar{g}$. The Newton's Second Law equation is now:

$$F' - (m_A + m_B + m_C + M)g = -m_A g + (m_B + m_C + M)\bar{A}.$$

Giving a smaller value for F,

$$\bar{F}' = (m_B + m_C + M)(\bar{A} + \bar{g})$$

by the amount

$$\bar{F}' - \bar{F} = -m_A(\bar{A} + \bar{g}).$$

The new value of F is what would be obtained if the value of m_A were neglected while it is in the condition called free fall.

ROTATING COORDINATE SYSTEMS AND CENTRIFUGAL FORCES

• PROBLEM 20-8

A rod is fixed on the turntable in a radial direction from the origin. A rider on the turntable starts at the center and climbs slowly outward along the rod. If the rider can hold onto the rod with a force equal to his own weight on earth, how far out along the rod can he move before he loses his grip?

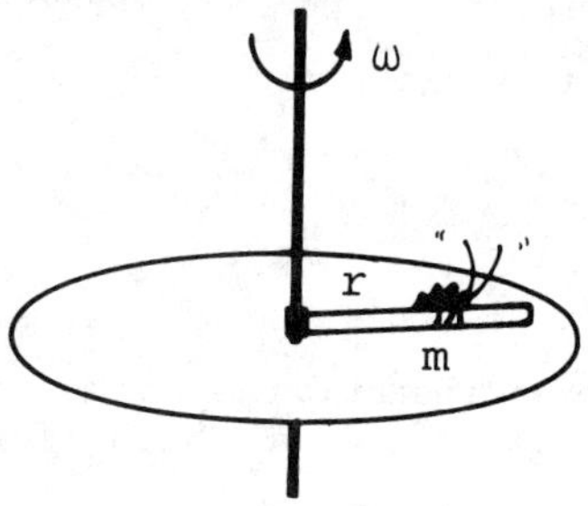

Solution: From the point of view of the rider there exists a centrifugal force, $f = mv^2/r$. As long as this does not exceed the maximum inward force he is able to exert, he can safely negotiate the rod. The maximum distance from the radius occurs when $mg = mv^2/r$.

Now, putting in $v = r\omega$, the result is

$$mg = m(r^2\omega^2)/r .$$

Solving for r yields

$$r = \frac{g}{\omega^2} .$$

• **PROBLEM 20-9**

If a turntable is accelerating with an angular acceleration $\dot{\omega}$ radians per second squared, show that a particle at rest with respect to the turntable will experience an additional pseudo-force given by the expression $-mr\dot{\omega}\hat{\theta}$, where $\hat{\theta}$ is unit vector lying in the plane of the turntable and pointing perpendicular to $\vec{r}$ in the direction of increasing θ .

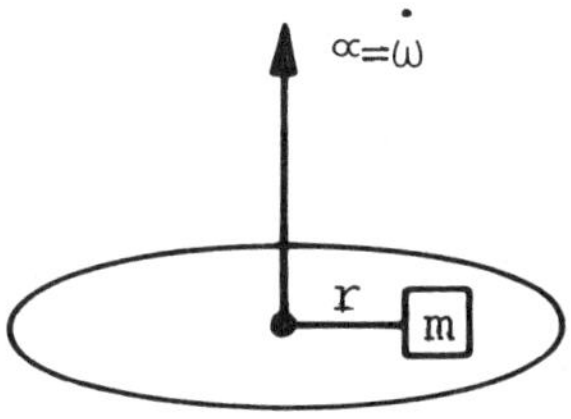

Solution: A particle at rest on the turntable undergoes a centripetal acceleration $a_c = v^2/R$ radially inward. This gives the effect in the rest system of the turntable of the familiar centrifugal force mv^2/r outward. But, in addition, there is also a tangential acceleration, $a_T = r\dot{\omega}$, at radius r. Connected with this is a tangentially directed pseudo-force, $f_t = -ma_t = -mr\dot{\omega}$. The direction is opposite to that of the angular acceleration $\vec{\dot{\omega}} = \dot{\omega}\hat{\theta}$, or $\vec{f}_i = -mr\dot{\omega}\hat{\theta}$.

• **PROBLEM 20-10**

A spring scale that measures the weight of a body does not register exactly the force of attraction exerted by the earth on the body (except when the body is being weighed at one of the two poles). The spin of the earth causes the body to move outward from the earth, the net result being that the weight of the body as measured by the scale will be less than the actual gravitational force. Determine the effect of the spin of the earth on the weight of bodies when measured by a spring scale.

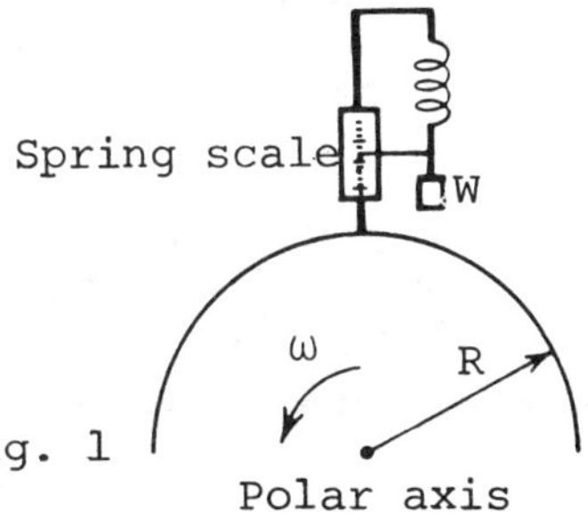

Fig. 1

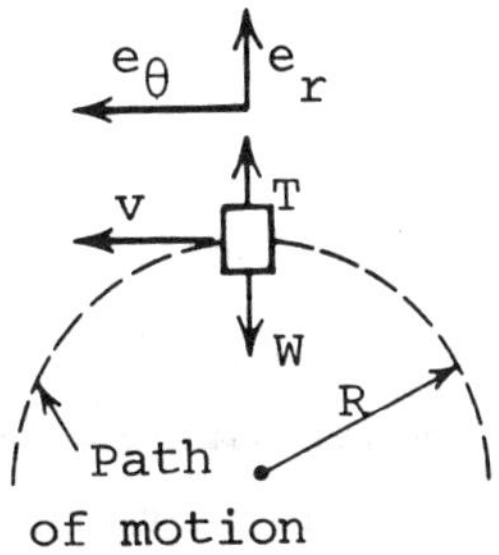

Fig. 2

Solution: Since the effect of the spin of the earth would be greatest when the body is measured at the equator, determine the apparent weight of the body at that point. A free-body diagram of the mass that is being weighed is shown in Fig. 2. The body moves along the path of motion because it is fixed to the moving earth. The force T in the spring is the measured weight of the body, and the force W is the actual weight (or gravitational attractive force) on the body. The unit vectors $\hat{e}_r$, $\hat{e}_\theta$ are also defined in the figure. The equation of motion for the particle in the $\hat{e}_r$ direction is

$$F_r = m(\ddot{r} - r\dot{\theta}^2) .$$

The body is moving in a circular path; the radius of the circular path is the radius of the earth (4000 miles) and the constant angular velocity of the particle is the angular velocity of the earth (one revolution every 24 hrs). Thus,

$$\ddot{r} = 0$$

$$r = 4000 \text{ miles} = 4000 \times 5280 = 2.11 \times 10^7 \text{ ft.}$$

$$\dot{\theta} = \frac{1 \text{ rev}}{24 \text{ hr}} = \frac{2\pi}{24(3600)} = 7.27 \times 10^{-5} \text{ radian/sec} .$$

From the free-body diagram of the mass, the force in the $\hat{e}_r$ direction is

$$F_r = T - W$$

and since the mass of the body is

$$m = \frac{W}{g} = \frac{W}{32.3} \text{ slugs}$$

the equation of motion of the body is

$$T - W = \frac{W}{32.2}\{-(2.11 \times 10^7)(7.27 \times 10^{-5})^2\} .$$

Solving for the force in the spring T, the result is

$$T = W(1.0000 - 0.00357)$$
$$= 0.996W .$$

Thus, it is seen that, even at the equator, the effect of the spin of the earth has little effect on the measured weight of the body. if, for instance, the actual weight of the body were 1000 lb, the measured weight would be 996 lb. Again it should be pointed out that the effect of the spin on the weight of the body will be less at all other points on the surface of the earth.

• PROBLEM 20-11

The apparent acceleration due to gravity g is the vector sum of the centrifugal acceleration away from the axis of the earth and the gravitational attraction of the earth, g_0, which is directed toward the center of the earth. Show that the resultant accleration g at any latitude θ is approximately

$$g = g_0\left(1 - \frac{\omega^2 R\cos^2\theta}{g_0}\right)$$

where R is the radius and ω is the angular velocity of the earth. (Note that ω is a relatively small quantity so that the higher order terms such as $\omega^2 R^2/g_0^2$ may be neglected as compared to 1).

Solution: In vector notation

$$\vec{g} = \vec{g}_0 + \vec{g}' \tag{1}$$

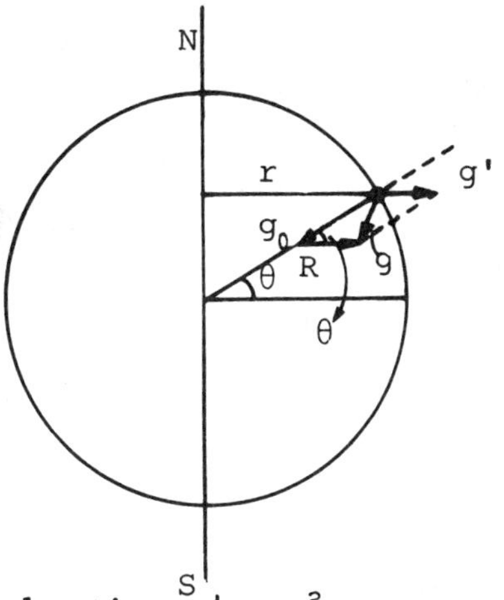

With the centrifugal acceleration $g' = r\omega^2$. (2)

Using the law of cosines to solve for the magnitude of $\bar{g}$,

$$g^2 = g_0^2 + (r\omega^2)^2 - 2g_0 r\omega^2 \cos\theta \,. \tag{3}$$

Of the three terms on the right,

$$g_0^2 \gg 2g_0 r\omega^2 \cos\theta \gg r^2\omega^4 \,, \tag{4}$$

so the smallest term may be dropped. This leaves

$$g^2 = g_0^2 - 2g_0 r\omega^2 \cos\theta \,. \tag{5}$$

Using

$$r = R\cos\theta \tag{6}$$

and factoring out g_0^2 results in

$$g^2 = g_0^2\left(1 - \frac{2R\omega^2 \cos^2\theta}{g_0}\right) . \tag{7}$$

Taking the square root,

$$g = g_0\left(1 - \frac{2R\omega^2 \cos^2\theta}{g_0}\right)^{\frac{1}{2}} . \tag{8}$$

Equation (8) is expanded using the binomial expansion equation:

$$(1 + \epsilon)^{\frac{1}{2}} = 1 + \frac{\epsilon}{2} - \frac{\epsilon^2}{8} + \frac{3\epsilon^3}{48} - \ldots \,. \tag{9}$$

In the present case,

$$\epsilon = \frac{-2R\omega^2 \cos^2\theta}{g_0} \,, \tag{10}$$

and only the first two terms of the expansion need to be retained since $\epsilon \ll 1$. This gives:

$$g \approx g_0\left(1 - \frac{R\omega^2 \cos^2\theta}{g_0}\right) . \tag{11}$$

• PROBLEM 20-12

Assuming that the sun is fixed in space and an inertial coordinate system could be fixed to it, one is interested in knowing how much error would be introduced if it were assumed that the surface of the earth could be considered for the origin of an inertial coordinate system. Specifically, we should like to know what point on the earth's surface has the maximum acceleration with respect to the sun and what is the magnitude.

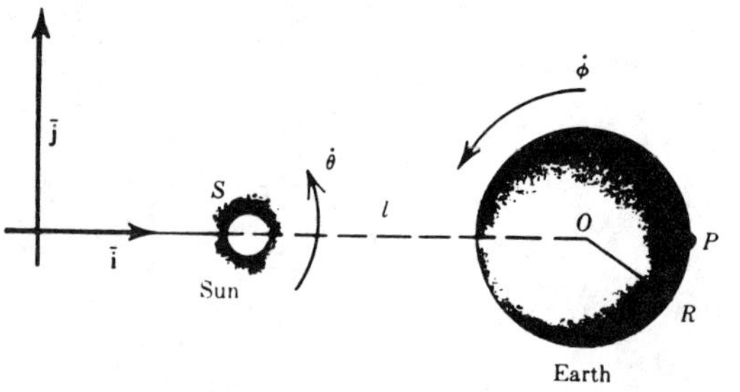

Fig. 1 Top view of the ecliptic plane

Solution: The plane defined by the earth as it moves in orbit about the sun is called the ecliptic plane. The plane defined by the equator of the earth is the equatorial plane. There is an inclination of roughly 23 deg. between the equatorial plane and the plane of the ecliptic, which implies that there is a similar inclination between the polar axis of the earth and a normal to the ecliptic plane. Assuming, however, that the ecliptic and the equatorial planes are coplanar, we are able to make a rough calculation of the acceleration of a point of the surface of the earth (assuming that the sun is inertially fixed). See Fig. 1.

The earth has a radius R of 4000 miles and an angular velocity φ of one revolution per day. The center of the earth, point O, is moving in a circular path about the sun. The distance ℓ is equal to 93 million miles, and the angular velocity of the line OS is equal to one revolution per 365 days. Since there is no angular acceleration involved, note that there will be only radial acceleration. And it follows that the maximum acceleration of the point on the earth's surface will be the point P shown in Fig. 1. Thus,

$$\vec{a}_P = \vec{a}_O + \vec{a}_{P/O} .$$

First of all, evaluate the acceleration of the center of the earth, point O:

$$\vec{a}_O = -\ell\theta^2\hat{i}$$

$$= -[93(5280) \times 10^6]\left(\frac{2\pi}{365(24)(3600)}\right)^2$$

$$\vec{a}_O = -0.0197\hat{i} \text{ ft/sec}^2 .$$

The acceleration of P with respect to O is

$$\vec{a}_{P/O} = -R\varphi^2\hat{i}$$

$$= -[4000(5280)]\left(\frac{2\pi}{(24)(3600)}\right)^2 \hat{i}$$

$$= -0.111\hat{i} \text{ ft/sec}^2$$

so that, finally,

$$\vec{a}_P = -[0.0197 + 0.111]\hat{i}$$

$$= -0.131\hat{i} \text{ fps}$$

From the results we may conclude, first of all, that the effect on the acceleration of the earth moving about the sun is somewhat smaller than the effect on the acceleration of the rotation of the earth on its own axis. Secondly, note that the total acceleration, though measurable, is very small in comparison with the acceleration due to gravity. Therefore, we conclude that this acceleration may often be reasonably neglected when calculating the motion of most terrestrial bodies.

• PROBLEM 20-13

A small car moves on a straight track on a horizontal table which can rotate about a vertical axis. The car is attached to the mid-point of a spring which is held stretched between the endpoints of the track, as shown in Fig. 1. The equilibrium position of the car is at the axis of rotation. When the table does not rotate, the period of oscillation of the car is T_0. What is the period of oscillation when the table rotates with an angular velocity ω ?

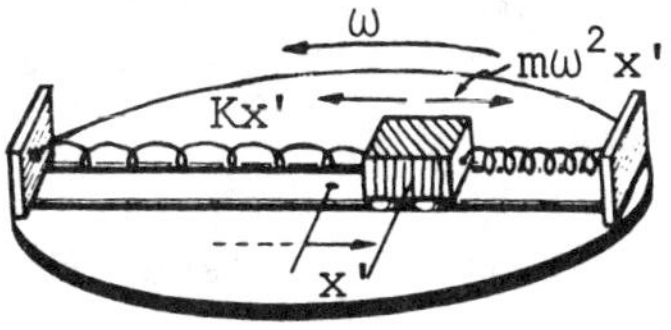

Fig. 1

Solution: Study the motion with respect to a coordinate system S' attached to the rotating table. At a distance x' from the equilibrium position we then have two forces acting on the car, the spring interaction force -Kx' and the inertial (centrifugal) force $m\omega^2 x'$, where K is the spring constant and m is the mass of the car. The equation of motion is then

$$m \frac{d^2 x'}{dt^2} = -Kx + m\omega^2 x' = -(K - m\omega^2)x' .$$

In other words, the centrifugal force reduces the restoring force, and the effective spring constant is $K - m\omega^2$. The period of oscillation is then

$$T = 2\pi \sqrt{\frac{m}{K - m\omega^2}} = \frac{\omega_0 T_0}{\sqrt{\omega_0^2 - \omega^2}} ,$$

where $\omega_0 = \sqrt{K/m}$. It is clear that when $\omega > \omega_0$, the car does not oscillate but has a positive acceleration at all times and moves outward from the axis of rotation.

• PROBLEM 20-14

A particle has the position vector $\vec{r} = x\hat{i} + y\hat{j} + z\hat{k}$ and is rotating around the z-axis with an angular velocity $\vec{\omega} = \omega\hat{k}$. Show that the direction of the centripetal acceleration, $\vec{\omega} \times (\vec{\omega} \times \vec{r})$, is directed inward toward the z-axis and perpendicular to this axis. Show that this acceleration is equal to zero if the particle lies anywhere on the z-axis.

Solution: Careful inspection of the figure may be sufficiently convincing, however, an algebraic argument follows.

Given $\vec{\omega} = \omega\hat{k}$ and $(\vec{r}) = x\hat{i} + y\hat{j} + z\hat{k}$. Therefore, the cross-product in parentheses is

$$\vec{\omega} \times \vec{r} = \omega\hat{k} \times (x\hat{i} + y\hat{j} + z\hat{k})$$

$$\vec{\omega} \times \vec{r} = (-\omega y\hat{i} + \omega x\hat{j} + 0\hat{k})$$

The vector $\vec{\omega}$ is now crossed with the line above.

$$\vec{\omega} \times (\vec{\omega} \times \vec{r}) = \omega\hat{k} \times (-\omega y\hat{i} + \omega x\hat{j})$$
$$= (-\omega^2 x\hat{i} - \omega^2 y\hat{j}) .$$

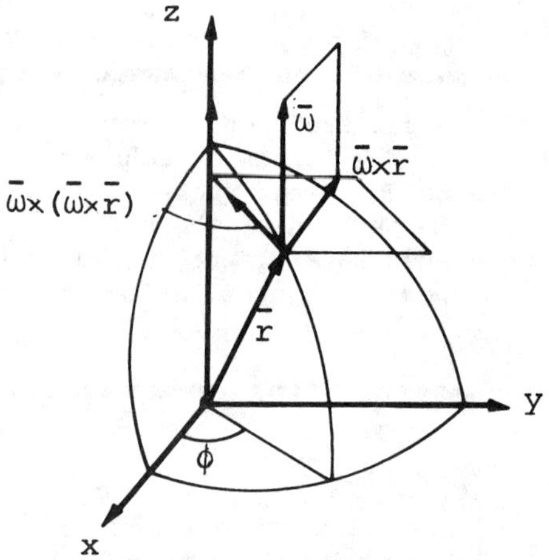

When this is compared to the X and Y parts of the vector $\vec{r}$, namely

$$x\hat{i} + y\hat{j} \ ,$$

one sees that $\vec{\omega} \times (\vec{\omega} \times \vec{r})$ is a negative multiple of $x\hat{i} + y\hat{j}$; that is,

$$\vec{\omega} \times (\vec{\omega} \times \vec{r}) = -\omega^2 (x\hat{i} + y\hat{j}) \ .$$

Therefore, $\vec{\omega} \times (\vec{\omega} \times \vec{r})$, besides being perpendicular to $\vec{\omega}$, is directed inward toward the z-axis just opposite to the X and Y parts of $\bar{r}$.

If $\vec{r}$ lies on the z-axis then $x = y = 0$. This immediately gives $\vec{\omega} \times (\vec{\omega} \times \vec{r}) = 0$.

• PROBLEM 20-15

A uniform cube of wood rests on a horizontal turntable with its inner edge in a tangential direction. The turntable is rotated from rest with a very small angular acceleration. What must be the conditions on the coefficient of static friction between the block and the turntable so that the block will slide rather than tip over?

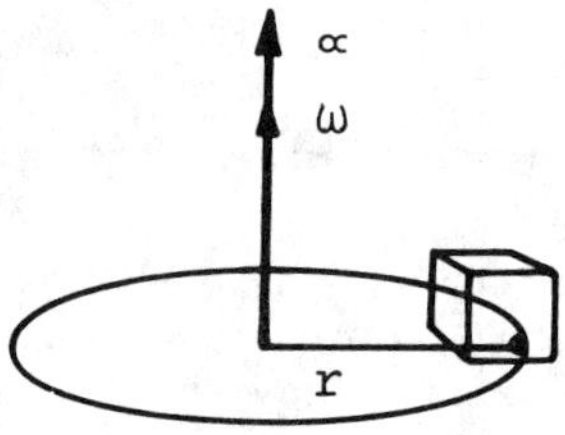

Fig. 1

Solution: Since the α is very small, there is a very slow increase in the angular velocity ω and the tangential velocity $v = r\omega$. There is almost no tangential acceleration $a_T \mp r\alpha$ so its effect will be negligible. First calculate the velocity at which the block slips and then the velocity at which the block tips outwards. See which occurs at the lowest velocity.

The block slides whenever the horizontal pseudo-force mv^2/r equals or exceeds the force of static friction, or

$$mv^2/r = f \le \mu_s N = \mu_s mg \ .$$

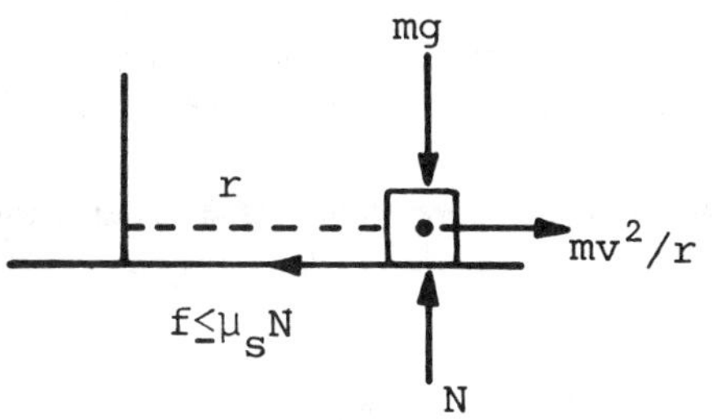

Fig. 2

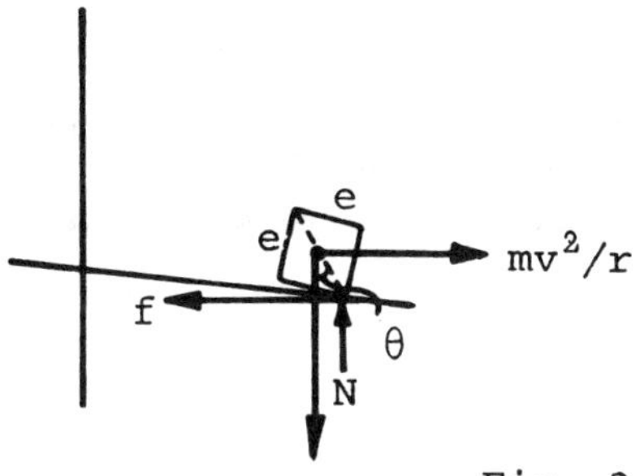

Fig. 3

The maximum velocity v is:

$$v_{max} = \sqrt{\mu_S \, gr} = \omega r .$$

The maximum ω is:

$$\omega_{max} = \sqrt{\frac{\mu_S g}{r}} .$$

Tipping occurs when the clockwise torques in figure 3 equal or exceed the counterclockwise torques. Taking torques about the outer edge as tipping begins eliminates the need to include torques from $\bar{f}$ and $\bar{N}$. The imbalance of torques is expressed by

$$\frac{mv^2}{r}\left(\frac{\ell\sqrt{2}}{2} \cos\theta\right) \geq mg\left(\frac{\ell\sqrt{2}}{2} \sin\theta\right) .$$

The term for the semidiagonal of the cube, $\frac{\ell\sqrt{2}}{2}$, cancels from both sides. The maximum CW torque and minimum CCW torque occur at the initial angle, $\theta = 45^\circ$. Once tipping starts it will continue to completion ($\theta \to 0^\circ$) unless v and ω are drastically and quickly lessened. At any rate, with a steadily rising ω, tipping begins when:

$$\frac{v^2_{max}}{r} \cos 45^\circ = g \sin 45^\circ$$

or, since $\cos 45^\circ = \sin 45^\circ$,

$$v^2_{max} = rg$$

$$v_{max} = \sqrt{rg} .$$

The v_{max} of this equation exceeds the previous v_{max} equation if $\mu < 1$.

In conclusion, if $\mu < 1$, the first expression for v_{max} is smaller, and sliding occurs first as ω is slowly built up.

But, if $\mu > 1$, the second expression for v_{max} is less, and tipping occurs first.

If $\mu = 1$ exactly, both slipping and tipping start at the same ω. Of course in reality, $\mu < 1$, so that sliding will always occur first.

• PROBLEM 20-16

A particle of mass m which is at rest on a frictionless horizontal surface is pushed by a stick of length L, pivoted at one end. The stick rotates about the pivot at a constant angular velocity ω, and the particle is initially just slightly to one side of the axis. What is the contact force F that is exerted on the particle by the stick, and what is the kinetic energy of the particle when it leaves the end of the stick? (Assume sliding without friction).

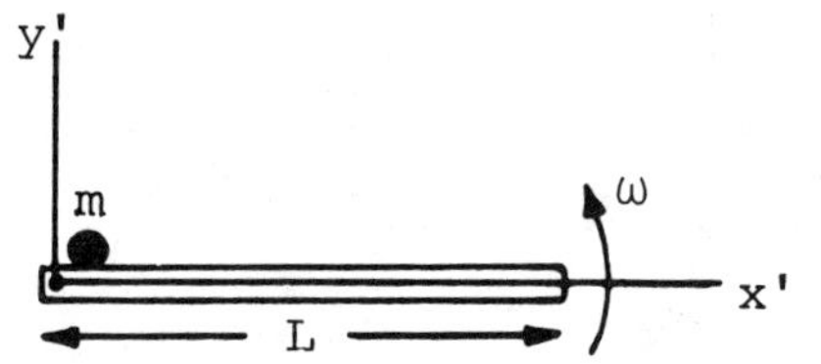

Fig. 1

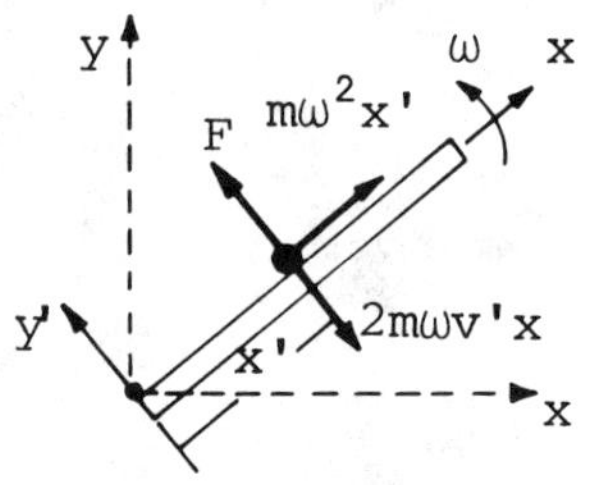

Fig. 2

Solution: In figure (2), we show the stationary reference frame, S with coordinate axes x and y, and the reference frames S' which moves with the stick and has coordinate axis x' and y'. Let us take the x'-axis to be along the length of the stick, as shown in Fig. 1. Then, according to the coordinate system S' which moves with the stick, the centrifugal force is $m\omega^2x'$ directed outward, and the coriolis force is $2m\omega v'_x$ directed perpendicular to the stick as shown in Figure (2), where x', v'_x are the position and velocity of the mass. According to S', the particle is in equilibrium in the y'-direction so that the net force in the y'-direction must be zero. Therefore, if the contact force on the particle is F, the result is $F - 2m\omega v'_x = 0$ or $F = 2m\omega v'_x$.

In the analysis according to the coordinate system S at rest, one reasons as follows. The distance x' measures the radial distance of the mass from the axis. The angular momentum is therefore $\ell = mx'^2\omega$. The torque, or rate of change of angular momentum, is

$$\tau = \frac{dp}{dt} = 2mx'\omega v'_x ,$$

but since the torque must be $\tau = F'_x$ the force F must be $2m\omega v'_x$. The interaction forces in S and S' agree.

The centrifugal force is $m\omega^2x'$ directed outward, and so as the particle moves from $x' = 0$ to $x' = L$, the work done on it is

$$W = \int_0^L m\omega^2 x'dx' = \frac{mL^2}{2}\omega^2 .$$

Thus, the kinetic energy according to S' is $mL^2\omega^2/2$. Since the kinetic energy is also $mv'^2_x/2$ the result is $v'_x = \omega L$ at $x' = L$.

According to $S, v'_x = \omega L$ is a radial velocity and, in addition, the mass has a tangential velocity equal to the velocity of the end of the stick. This tangential velocity is also ωL, so that the speed of the mass with respect to S is $\omega L\sqrt{2}$, and the kinetic energy according to S is $m\omega^2L^2$. In problems of this kind the motion in S' is often simpler than in S.

In the present example, the trajectory of the particle is simply a straight line in S', but a rather complicated spiral in S.

CORIOLIS FORCES

• PROBLEM 20-17

A river flows due north. Which side of the bank should be the most worn?

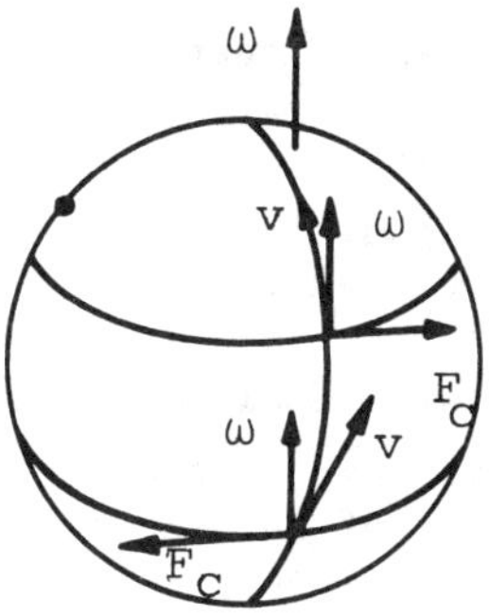

Solution: This will depend on the hemisphere in which the river exists. The force which presses the river against the bank is the Coriolis force:

$$\vec{F}_C = 2m\vec{v} \times \vec{\omega} .$$

In the northern hemisphere, it can be seen from the diagram that $\vec{v} \times \vec{\omega}$ is to the east, whereas in the southern hemisphere $\vec{v} \times \vec{\omega}$ is to the west. Thus, the river will tend to wear down its right bank in the northern hemisphere and left bank in the southern hemisphere.

• PROBLEM 20-18

A body is at rest in coordinate system S. In S', which is a rotating coordinate system, the same body rotates about the origin with angular velocity ω and has the speed $\omega r'$. Show that the expression for centrifugal plus Coriolis force in a moving system provides the fictitious centripetal force needed to explain the motion of the body from the point of view of coordinate system S'.

Solution: The fictitious force in a rotating coordinate system is given by:

$$\vec{F}_{fict.} = -2m\vec{\Omega} \times \vec{v}_{rot.} - m\vec{\Omega} \times (\vec{\Omega} \times \vec{r}) \qquad (1)$$

where $\vec{\Omega}$ is the vector angular velocity of the rotating system, $\vec{r}$ is the position vector of the body of mass m, and $\vec{v}_{rot.}$ is the velocity vector of the body as observed in the rotating system. The velocity of the body in the non-rotating inertial coordinate system, $\vec{v}_{in}$, is related to $\vec{v}_{rot.}$ by

$$\vec{v}_{in} = \vec{v}_{rot} + \vec{\Omega} \times \vec{r} . \qquad (2)$$

It is assumed that the origins of both coordinate systems coincide at all times. If we let $|\vec{\Omega}| = \omega$ and $\vec{v}_{in} = 0$, then

$$\vec{v}_{rot} = -\vec{\Omega} \times \vec{r} \qquad (3)$$

and

$$\vec{F}_{fict.} = 2m\,\vec{\Omega} \times (\vec{\Omega} \times \vec{r}) - m\,\vec{\Omega} \times (\vec{\Omega} \times \vec{r})$$

or

$$\vec{F}_{fict.} = m\,\vec{\Omega} \times (\vec{\Omega} \times \vec{r}) \,. \tag{4}$$

Now the triple vector product can be expanded as:

$$\vec{A} \times (\vec{B} \times \vec{C}) = (\vec{A}\cdot\vec{C})\vec{B} - (\vec{A}\cdot\vec{B})\vec{C} \tag{5}$$

Thus, (4) becomes:

$$\vec{F}_{fict.} = -m\,\omega^2\,\vec{r} \tag{6}$$

since $\vec{\Omega}\cdot\vec{r} = 0$ and $\vec{\Omega}\cdot\vec{\Omega} = \omega^2$. Note that the condition that the body rotates about the origin with speed $\omega\underline{r}'$ required that $\vec{r}'$ must be perpendicular to $\vec{\Omega}$. Thus, the force $\vec{F}_{fict}$ is directed toward the origin and has the magnitude $mr\omega^2$. Thus, in the rotating coordinate system, this force looks like a centripetal force.

• PROBLEM 20-19

Westerly winds blow from west to east in the northern hemisphere with an average speed v. If the density of the air is ρ , what pressure gradient is required to maintain the steady flow at air from west to east with this speed?

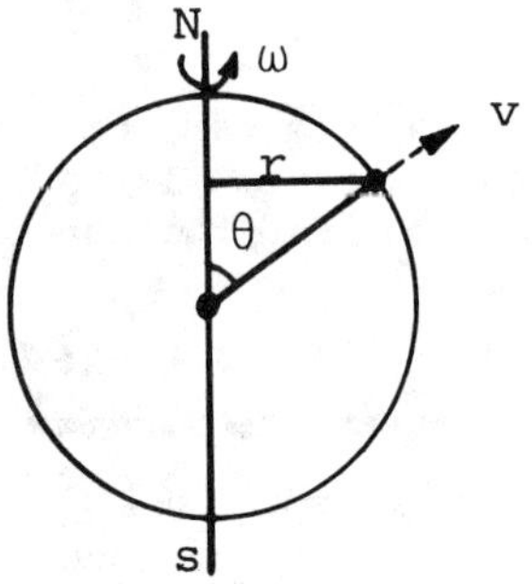

Fig. 1

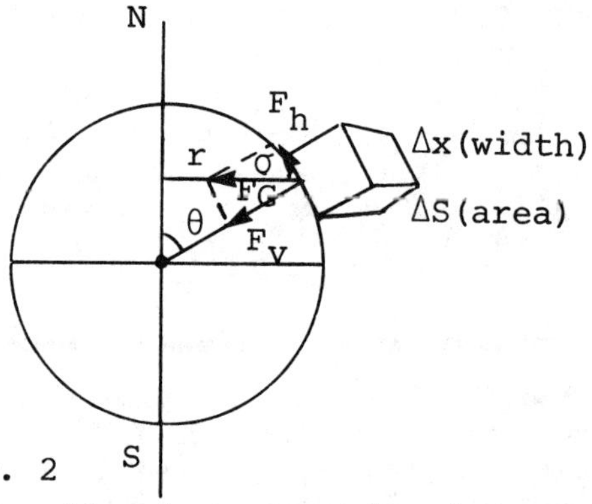

Fig. 2

Solution: The first step in solving the problem is to determine the acceleration of a mass of air moving over the earth as described. Due to the earth's daily rotation there is a tangential velocity term of $r\omega$. Added to this is the velocity over the earth's surface, v, in the same direction. The total tangential velocity is $(r\omega + v)$. Write the centripetal acceleration of the mass of air as:

$$a_c = \frac{v^2_{total}}{r} = \frac{(r\omega + v)^2}{r} \,. \tag{1}$$

The required inward (centripetal) force to produce the motion described is given by Newton's Second Law.

$$F_c = ma_c = \frac{m(r^2\omega^2 + 2r\omega v + v^2)}{v} \tag{2}$$

This force directed toward the axis of the earth is shown acting on a small cubical volume of air in Figure 2. Force F_c has vertical and horizontal components F_v and F_h respectively. Our problem calls for a study of the horizontal component which is

$$F_h = F_c \cos\theta \,. \tag{3}$$

This component is provided by having an atmospheric pressure which steadily increases from north to south with a gradient $d\rho/dx$. The gradient will produce a net northward force on the mass of

$$F_h = \Delta P \; \Delta s = \left(\frac{dp}{dx}\,\Delta x\right) \Delta s \; . \tag{4}$$

Putting equation (2) and (4) in equation (3),

$$\frac{m(r^2\omega^2 + 2r\omega v + v^2)}{r} \cos\theta = \left(\frac{dp}{dx}\;\Delta x\right) \Delta s \; . \tag{5}$$

The mass contained in the volume $V = \Delta s\,\Delta x$ is:

$$m = \rho V = \rho\;\Delta s\;\Delta x \; . \tag{6}$$

The mass m is now substituted into (5) to give

$$\frac{\rho\;\Delta S\;\Delta x}{r}\,(r^2\omega^2 + 2r\omega v + v^2)\cos\theta = \frac{dp}{dx}\,\Delta x\;\Delta S \; . \tag{7}$$

The terms ΔS and Δx cancel from both sides giving an expression for the gradient

$$\frac{dp}{dx} = \frac{\rho\cos\theta}{r}\,(r^2\omega^2 + 2r\omega v + v^2) \; . \tag{8}$$

The second and third terms on the right depend on the velocity of the wind, v, relative to the earth and are called the Coriolis effect. The third term is much smaller than the second since $r\omega$ approaches 1000 miles/hour at the equator but v is always much smaller. The additional gradient to maintain a steady west to east velocity above what is needed when $v = 0$ is given by retaining just the second term on the right.

$$\frac{dp}{dx} = 2\omega v\rho\,\cos\theta \; . \tag{9}$$

• PROBLEM 20-20

A particle of mass m located at latitude λ north moves with velocity $\bar{V}$ having components

$$\bar{V} = \bar{V}_e + \bar{V}_s + \bar{V}_v \; .$$

The subscripts e,s and v designate east, south, and vertical upward from the earth's center at the particular locality of the particle. See figure 1.

Show that the components of the Coriolis pseudo-force are:

$$f_s = 2\omega m v_e \sin\lambda$$

$$f_e = -2m\omega v_s \sin\lambda - 2m\omega v_v \cos\lambda$$

$$f_v = 2m\omega v_e \cos\lambda \; ,$$

and that for a velocity due south $(\bar{V} = \bar{V}_s)$, the pseudo-force is due west.

Solution: In general, $\vec{F} = -2m\vec{\omega} \times \vec{v}$ or

$$\vec{F} = -2m\omega\hat{i} \times (\vec{v}_s + \vec{v}_e + \vec{v}_v) \; .$$

Since the answers are in the local coordinate system, convert $\vec{\omega}$ to this system first.

$$\vec{\omega} = -\omega\cos\lambda\,\hat{s} + 0\hat{E} + \omega\sin\lambda\,\hat{v}$$

The $\bar{s},\bar{e}$ and $\bar{v}$ axis makes up an orthogonal right handed coordinate system, and so the cross products are formed just the same as if the labels had been X,Y and Z.

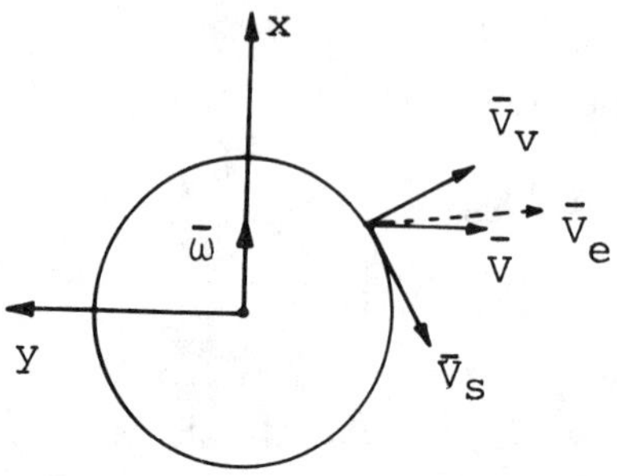

Fig. 1

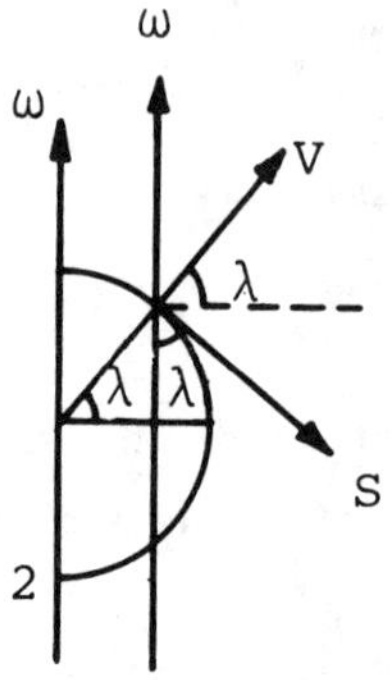

Fig. 2

$$\vec{F} = -2m(-\omega\cos\lambda\,\hat{s} + 0\hat{e} + \omega\sin\lambda\,\hat{v}) \times (V_s\hat{s} + V_e\hat{e} + V_v\hat{v})$$

leading to

$$f_s = -2m(0v_v - \omega\sin\lambda\, v_e) = +\,2m\omega v_e \sin\lambda$$

$$f_e = -2m(\omega\sin\lambda\, v_s + \omega\cos\lambda\, v_v)$$

$$= -2m\omega v_s \sin\lambda - 2m\omega v_v \cos\lambda$$

and

$$f_v = -2m(-\omega\cos\lambda\, v_e + 0N_s) = 2m\,\omega v_e \cos\lambda$$

In case $\vec{V}$ is due south, V_e and V_v both equal zero and then the only term remaining in $\vec{F}$ is

$$\vec{f}_e = -2m\,\omega\sin\lambda\,\hat{e}$$

and $\vec{F}$ is due west.

• PROBLEM 20-21

A particle at rest is dropped from an altitude of 1000 ft. Determine the drift due to the earth's rotation at the north pole, at a 45° latitude, and at the equator.

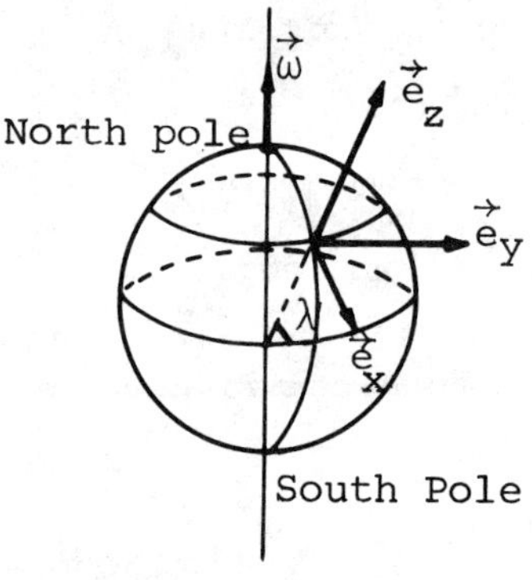

Fig. 1

Solution: The drift, or lateral deflection, of a falling body near the surface of the rotating earth is due to the effect known as the Coriolis force. The acceleration of the falling body is given by

$$\vec{a}_r = \vec{g} - 2\vec{\omega} \times \vec{v}_r \tag{1}$$

where $\vec{g}$ is the acceleration due to gravity, $\vec{\omega}$ is the angular velocity of rotation of the earth, and $\vec{v}_r$ is the radial component of velocity of the falling body.

Choose a coordinate system as shown in Figure 1, where $\vec{e}_x$, $\vec{e}_y$, and $\vec{e}_z$ are unit vectors in the X,Y,Z directions of a rotating local frame of reference at the surface of the earth with x southwards, y easterly, z upwards.

Also assume that the distance of fall is small enough so that g, the acceleration due to gravity, remains constant. Air resistance will be neglected. Then, the components of the earth's angular velocity with respect to the rotating coordinate system become:

$$\omega_x = -\omega \cos \lambda \tag{2}$$

$$\omega_y = 0 \tag{3}$$

$$\omega_z = \omega \sin \lambda \tag{4}$$

where λ is the latitude of the origin of the rotating coordinate system.

The components of velocity of the falling body may be written:

$$v_x \simeq 0 \tag{5}$$

$$v_y \simeq 0 \tag{6}$$

$$v_z \simeq -gt \tag{7}$$

where we have neglected the velocity components in the x and y directions as being small when compared with the velocity in the z direction. Then,

$$\vec{\omega} \times \vec{v}_r \simeq \begin{vmatrix} \vec{e}_x & \vec{e}_y & \vec{e}_z \\ -\omega \cos \lambda & 0 & \omega \sin \lambda \\ 0 & 0 & -gt \end{vmatrix} \tag{8}$$

$$\simeq -(\omega gt \cos \lambda)\vec{e}_y \quad . \tag{9}$$

Since the components of $\vec{g}$ are

$$g_x = 0 \tag{10}$$

$$g_y = 0 \tag{11}$$

$$g_z = -g \ , \tag{12}$$

The components of acceleration by combining (1), (9) and (12) are

$$a_x = 0 \tag{13}$$

$$a_y \simeq 2\omega gt \cos \lambda \tag{14}$$

$$a_z \simeq -g \ . \tag{15}$$

Thus, the Coriolis force produces a component of acceleration in the y direction, that is, toward the east. Integrating (14) twice, the result is:

$$v_y = \int a_y dt \simeq \int 2\omega gt \cos \lambda \, dt$$

or

$$v_y \simeq \omega g t^2 \cos \lambda \tag{16}$$

and

$$y = \int v_y dt \simeq \int \omega g t^2 \cos \lambda \, dt$$

or

$$y \simeq \tfrac{1}{3} \omega g t^3 \cos \lambda \ . \tag{17}$$

Integration of (15) twice gives the familiar result

$$z = -\tfrac{1}{2} gt^2 \tag{18}$$

from which the time of fall is found as

$$t = \sqrt{2z/g} \tag{19}$$

where z now represents the initial altitude of the body above the surface of the earth. Combining (17) and (19) gives an expression for the deflection of a body falling from rest from an altitude z at latitude λ:

$$y \simeq \tfrac{1}{3}\,\omega \cos \lambda \sqrt{(8z^3)/g}\ . \tag{20}$$

The value of ω for the rotation of the earth is:

$$\omega = \frac{2\pi \text{ rad/day}}{86400 \text{ sec/day}} = 7.29 \times 10^{-5} \text{ rad/sec.} \tag{21}$$

At the north pole $\lambda = 90^\circ$; hence, $y = 0$ and the body falls without deflection along the axis of the earth's rotation. At the equator $\lambda = 0$ and

$$y = \tfrac{1}{3}(7.29 \times 10^{-5}) \sqrt{\frac{8(1000)^3}{32.2}} = 0.38 \text{ ft.}$$

At a latitude of $\lambda = 45^\circ$,

$$y = \tfrac{1}{3}(7.29 \times 10^{-5})(.707) \sqrt{\frac{8(1000)^3}{32.2}} = 0.27 \text{ ft.}$$

• PROBLEM 20-22

A ballistic projectile is launched with an initial speed v_1 equal to 700 m/sec southward at an angle θ of sixty degrees from the horizontal from a point at $\lambda = 40$ degrees north latitude. Predict the point of impact. Neglect air resistance.

Solution: Assume that the effective acceleration g' due to the gravitational force and the centrifugal pseudo-force have been measured and are the same for every point on the path of the projectile.

First find the range of the projectile, ignoring the rotating frame of the earth. Its original vertical velocity is $v_1 \sin \theta$ (positive being upwards). When it hits the ground, it will have the same magnitude velocity, though now in a different direction. Therefore, its velocity will be $-v_1 \sin \theta$. The acceleration of gravity is given by $-g'$. We can then write

$$-v_1 \sin \theta = v_1 \sin \theta - g't_f$$

where t_f is the time of flight. Solving for t_f yields

$$t_f = \frac{2v_1 \sin \theta}{g'}\ . \tag{1}$$

Now the original horizontal velocity is given by $v_1 \cos \theta$. Since this velocity is constant, the range is given by

$$R = v_1 \cos \theta\, t_f\ .$$

Substituting for t_f yields

$$R = v_1 \cos \theta \left(\frac{2v_1 \sin \theta}{g'}\right)$$

$$R = \frac{v_1^2}{g'}\,(2 \cos \theta \sin \theta)\ .$$

But $2 \cos \theta \sin \theta = \sin 2\theta$ by a trigonometric identity. Therefore, the range can be written as

$$R = \frac{v_1^2 \sin 2\theta}{g'} \quad .$$

If $g' \approx g = 9.80 \text{ m/sec}^2$, the range is equal to 43.3 km. We assume that the curvature of the earth can be neglected in this distance.

The Coriolis force is given by

$$\vec{F}_c = 2m\,\vec{v} \times \vec{\omega} \tag{2}$$

In components,

$$F_c = 2m\,\omega\,v_V \sin(90^\circ - \lambda)$$
$$= 2m\,\omega\,v_V \cos \lambda \tag{3}$$

The force will be towards the west for v_V up (or positive) and towards the east for v_V down (or negative). Also,

$$F_c = 2m\,\omega\,v_{n-s} \sin \lambda$$
$$= 2m\,\omega\,v_s \sin \lambda \quad \text{(west)} \tag{4}$$
$$= 2m\,\omega\,v_n \sin \lambda \quad \text{(east)} \tag{5}$$

Only velocity components, north, south and vertically, contribute to the Coriolis force; east and west motion is perpendicular to ω. The equations (3), (4), (5) (with equation (3) broken down for the two vertical directions) give the (4) possible deflections in any Coriolis force problem.

If it were not for the Coriolis pseudo-forces, the answer would be 43.3 km south of the point of launch. According to equation (3) and (4), there will be Coriolis pseudo-force components due to both the vertical and the southward components of velocity. We can find the deviation from the previously predicted point of impact due to each of these pseudo-forces separately and then add these deviations to find the resultant deviation due to both of these pseudo-forces acting together. Let us analyze first the deflection corresponding to the vertical component of velocity v_V.

The vertical component of velocity as a function of time after launch will be given by the equation

$$v_V = v_1 \sin \theta - g't \,. \tag{6}$$

From Eqs. (3) the Coriolis pseudo-force due to this component of velocity will point due west during the ascent of the projectile and will have a magnitude $2m\omega v_V \cos \lambda = 2m\omega(v_1 \sin \theta - g't)\cos \lambda$. The westward acceleration a_{W1} due to this force will be given by

$$a_{W1} = 2\omega(v_1 \sin \theta - g't)\cos \lambda \,. \tag{7}$$

The westward component of velocity v_{W1} resulting from this acceleration can be found by integrating this equation with respect to time and setting the initial westward velocity equal to zero.

$$v_{W1} = 2\omega v_1 \sin \theta (\cos \lambda) t - \omega g' (\cos \lambda) t^2 \,,$$

The westward displacement y_{W1} of the particle resulting from this component of velocity can be found by integrating this equation with respect to time and setting the initial westward displacement equal to zero.

$$y_{W1} = \omega v_1 \sin \theta (\cos \lambda) t^2 - \tfrac{1}{3}\,\omega g' (\cos \lambda) t^3 \,. \tag{8}$$

Substituting the value for t_f from equation (1) yields

$$y_{W1} = \frac{4\omega v_1^3 \sin^3\theta \cos\lambda}{g'^2} - \frac{8}{3}\frac{\omega v_1^3 \sin^3\theta \cos\lambda}{g'^2}$$

$$y_{W1} = \frac{4}{3}\frac{\omega v_1^3 \sin^3\theta \cos\lambda}{g'^2} \qquad (9)$$

This is the westward deflection of the projectile due to the Coriolis pseudo-force corresponding to the vertical component of velocity. Note that the sign convention implied in equation (3) takes care of the direction of vertical motion. If equation (1) had come out negative, then the deflection would have been eastward.

It may be argued that there should be no net displacement of the projectile due to its vertical component of velocity. This is because the westward acceleration due to the vertical component of velocity of the projectile at any altitude on the way up will be equal in magnitude to the eastward acceleration due to the vertical component of velocity of the projectile at the same altitude on the way down. In fact, all this tells us is that the final westward acceleration, equation (7), will be zero. However, the net effect on the displacement, found after integrating, is non-zero.

The westward Coriolis pseudo-force on the projectile corresponding to the southward component of velocity v_S can be found from Equation (4). It has the value $2m\omega v_S \sin\lambda$. Since $v_S = v_1 \cos\theta$ remains constant during the flight, the westward acceleration a_{W2} due to this pseudo-force will be constant

$$a_{W2} = 2\omega v_1 \cos\theta \sin\lambda .$$

The westward displacement y_{W2} to this acceleration during the time of flight will be given by

$$y_{W2} = \tfrac{1}{2} a_{W2} t_f^2 .$$

From equation (1) this becomes

$$y_{W2} = \tfrac{1}{2}(2\omega v_1 \cos\theta \sin\lambda)\left(\frac{4v_1^2 \sin^2\theta}{g'^2}\right)$$

$$= \frac{4\omega v_1^3 \sin^2\theta}{g'^2} \cos\theta \sin\lambda . \qquad (10)$$

The total westward displacement of the projectile during its flight will be found by adding the right sides of equations (9) and (10).

$$y_W = y_{W1} + y_{W2} = \frac{4\omega v_1^3 \sin^2\theta}{g'^2}\left(\frac{\sin\theta \cos\lambda}{3} + \cos\theta \sin\lambda\right) .$$

Substituting the values $\omega = 0.73 \times 10^{-4}$ radian/sec, $v_1 = 700$ m/sec, $g' \approx g = 9.80$ m/sec^2, $\theta = 60$ degrees, and $\lambda = 40$ degrees, results in an answer of about 424 m for the total westward displacement. This could be a significant error in point of impact.

The complete answer to this example is that the projectile will land about 43.3 km south of the point of launch and about 424 m to the west.

• PROBLEM 20-23

Derive the general form of Newton's Second Law relative to an inertial coordinate system O for a mass particle moving in a rotating reference frame O'. Discuss the terms in the equation and then simplify it for the special cases of a) a fixed rotating frame, b) a fixed rotating frame with constant angular velocity and c) a turntable. Finally, for the turntable case, show that the Coriolis acceleration is perpendicular to, and acts towards the left of the particle's velocity relative to the center of the turntable.

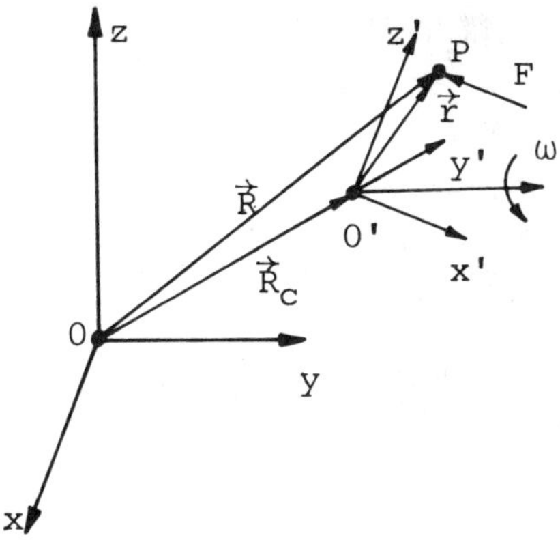

Solution: Starting with Newton's Second Law, $F = \frac{d(mv)}{dt}$, it is possible to derive the general equations of motion for a particle in a rotating reference frame O', relative to a fixed reference frame O .

$\frac{d(m\vec{v})}{dt}$ is evaluated as follows: First, note that $\vec{v} = \frac{d\vec{R}}{dt}$. This is just the definition of velocity relative to O, when $\vec{R}$ is from O to the particle. Next, since $\vec{R} = \vec{R}_c + \vec{r}$, $\vec{R}_c$ is the vector from O to O' and $\vec{r}$ is the vector from O' to the particle.

$$\frac{d\vec{R}}{dt} \text{ is then } \frac{d(\vec{R}_c + \vec{r})}{dt} = \frac{d\vec{R}_c}{dt} + \frac{d\vec{r}}{dt}$$

$\frac{d\vec{R}_c}{dt} = \dot{\vec{R}}_c$. $\vec{r}$, however, is rotating with respect to O . Therefore, in addition to its velocity with respect to O', $\vec{r}_{rel}$, the particle has a rotational velocity component $\vec{\omega} \times \vec{r}$. $\vec{\omega}$ is the rotation vector of O' relative to O .

Mathematically, this is derived as follows:

$\vec{r} = (x'\hat{i} + y'\hat{j} + z'\hat{k})$ where $\hat{i}$, $\hat{j}$ and $\hat{k}$ are unit vectors in the O' coordinate system.

$$\frac{d\vec{r}}{dt} = (\dot{x}'\hat{i} + \dot{y}'\hat{j} + \dot{z}'\hat{k}) + (x'\dot{\hat{i}} + y'\dot{\hat{j}} + z'\dot{\hat{k}})$$

Relative to the O' system, $\dot{\hat{i}}$, $\dot{\hat{j}}$ and $\dot{\hat{k}}$ would be zero. However, relative to O, $\hat{i}$, $\hat{j}$, and $\hat{k}$ change as the O' system rotates. In general, relative to O , $\dot{\hat{i}} = \vec{\omega} \times \hat{i}$, $\dot{\hat{j}} = \vec{\omega} \times \hat{j}$ and $\dot{\hat{k}} = \vec{\omega} \times \hat{k}$.

$\frac{d\vec{r}}{dt}$ then becomes

$$\frac{d\vec{r}}{dt} = (\dot{x}'\hat{i} + \dot{y}'\hat{j} + \dot{z}'\hat{k}) + \vec{\omega} \times (x'\hat{i} + y'\hat{j} + z'\hat{k})$$

$$\vec{\omega} \times (x'\hat{i} + y'\hat{j} + z'\hat{k}) = \vec{\omega} \times (\vec{r}) .$$

$(\dot{x}'\hat{i} + \dot{y}'\hat{j} + \dot{z}'\hat{k})$ is just the particles velocity relative to

O' . Therefore,

$$\vec{v} = \frac{d\vec{R}}{dt} = \dot{\vec{R}}_c + \vec{\omega} \times \vec{r} + \dot{\vec{r}}_{rel\ to\ 0'} \ .$$

$m\vec{v}$ then equals $m(\dot{\vec{R}}_c + \vec{\omega} \times \vec{r} + \dot{\vec{r}}_{rel})$.

$\frac{d(m\vec{v})}{dt} = \frac{md\vec{v}}{dt}$ when m is constant. The evaluation of $\frac{d\vec{v}}{dt}$ proceeds in the same way as the evaluation of $\frac{d\vec{R}}{dt}$.

$$\frac{d(\vec{v})}{dt} = \frac{d(\dot{\vec{R}}_c + (\vec{\omega} \times \vec{r}) + \dot{\vec{r}}_{rel})}{dt}$$

$$\frac{d(\vec{v})}{dt} = \frac{d(\dot{\vec{R}}_c)}{dt} + \frac{d(\vec{\omega} \times \vec{r})}{dt} + \frac{d(\dot{\vec{r}}_{rel})}{dt}$$

$$\frac{d(\dot{\vec{R}}_c)}{dt} = \ddot{\vec{R}}_c$$

$\frac{d(\vec{\omega} \times \vec{r})}{dt} = [\frac{d(\vec{\omega})}{dt} \times \vec{r}] + \vec{\omega} \times [\frac{d\vec{r}}{dt}]$ by the product rule.

From a previous result, $\frac{d\vec{r}}{dt}$ is equal to $(\vec{\omega} \times \vec{r}) + \dot{\vec{r}}_{rel}$.

$$\frac{d(\vec{\omega} \times \vec{r})}{dt} = \dot{\vec{\omega}} \times r + (\vec{\omega} \times \dot{\vec{r}}_{rel}) + \vec{\omega} \times (\vec{\omega} \times \vec{r})$$

$$\frac{d(\dot{\vec{r}}_{rel})}{dt} = \frac{d(\dot{x}'\hat{i} + \dot{y}'\hat{j} + \dot{z}'\hat{k})}{dt} = (\ddot{x}'\hat{i} + \ddot{y}'\hat{j} + \ddot{z}'\hat{k}) + (\dot{x}'\dot{\hat{i}} + \dot{y}'\dot{\hat{j}} + \dot{z}'\dot{\hat{k}}).$$

As before, $\dot{\hat{i}} = \vec{\omega} \times \hat{i}$, etc.

$$\frac{d(\dot{\vec{r}}_{rel})}{dt} = (\ddot{x}'\hat{i} + \ddot{y}'\hat{j} + \ddot{z}'\hat{k}) + \vec{\omega} \times (\dot{x}'\hat{i} + \dot{y}'\hat{j} + \dot{z}'\hat{k})$$

$$(\ddot{x}'\hat{i} + \ddot{y}'\hat{j} + \ddot{z}'\hat{k}) = \ddot{\vec{r}}_{rel\ to\ 0'}$$

$$\frac{d(\dot{\vec{r}}_{rel})}{dt} = \ddot{\vec{r}}_{rel} + \vec{\omega} \times \dot{\vec{r}}_{rel}$$

Summing the terms,

$$\vec{a}_{(relative\ to\ 0)} = \frac{d(\dot{\vec{R}}_c)}{dt} + \frac{d(\vec{\omega} \times \vec{r})}{dt} + \frac{d(\dot{\vec{r}}_{rel})}{dt}$$

$$= \ddot{\vec{R}}_c + (\dot{\vec{\omega}} \times \vec{r}) + (\vec{\omega} \times \dot{\vec{r}}_{rel}) + (\vec{\omega} \times (\vec{\omega} \times \vec{r}))$$

$$+ \ddot{\vec{r}}_{rel} + \vec{\omega} \times \dot{\vec{r}}_{rel}$$

$$\vec{a}_{(relative\ to\ 0)} = \ddot{\vec{R}}_c + 2(\vec{\omega} \times \dot{\vec{r}}_{rel}) + (\vec{\omega} \times \vec{\omega} \times \vec{r}) + (\dot{\vec{\omega}} \times \vec{r}) + \ddot{\vec{r}}_{rel}$$

and

$$F = m[\ddot{\vec{R}}_c + \ddot{\vec{r}}_{rel} + 2(\vec{\omega} \times \dot{\vec{r}}_{rel}) + (\vec{\omega} \times \vec{\omega} \times \vec{r}) + (\dot{\vec{\omega}} \times \vec{r})] \ .$$

F is the force applied to the particle; m is the particle's mass. $\ddot{\vec{R}}_c$ is merely the acceleration of point 0' with respect to 0 and $\ddot{\vec{r}}_{rel}$ is the acceleration of the particle relative to 0' caused by a change of distance with respect to 0'. The terms with $\vec{\omega}$ in them are terms resulting from the rotation of 0' relative to 0. $(\vec{\omega} \times (\vec{\omega} \times r))$

is the familiar centripetal acceleration term and $\dot{\vec{\omega}} \times \vec{r}$ is the familiar tangential acceleration term. The term $2(\vec{\omega} \times \dot{\vec{r}}_{rel})$ is called the Coriolis acceleration.

Suppose that $\dot{\vec{R}}_c$ and $\ddot{\vec{R}}_c$ are zero. Then O' is a fixed rotating coordinate system. When $\ddot{\vec{R}}_c = 0$, the force equation reduces to

$$\vec{F} = \vec{m}(\ddot{\vec{r}}_{rel} + (\vec{\omega} \times (\vec{\omega} \times \vec{r})) + (\dot{\vec{\omega}} \times \vec{r}) + 2(\vec{\omega} \times \dot{\vec{r}}_{rel}))$$

which is the form for a fixed rotating reference frame. If $\vec{\omega}$ is constant, then $\dot{\vec{\omega}} = 0$ and the equation becomes

$$\vec{F} = \vec{m}(\ddot{\vec{r}}_{rel} + (\vec{\omega} \times (\vec{\omega} \times \vec{r})) + 2(\vec{\omega} \times \dot{\vec{r}}_{rel})) .$$

Once again $\ddot{\vec{r}}_{rel}$ is translational acceleration relative to O', $\vec{\omega} \times (\vec{\omega} \times \vec{r})$ is centripetal acceleration and $2(\vec{\omega} \times \dot{\vec{r}}_{rel})$ is the Coriolis acceleration. These are the general accelerations present in a fixed rotating reference frame rotating with a constant angular velocity. A force must exist to cause each of them.

Now in order to solve the last piece of the problem, allow the general system with origin at O' to become a turntable and a particle.

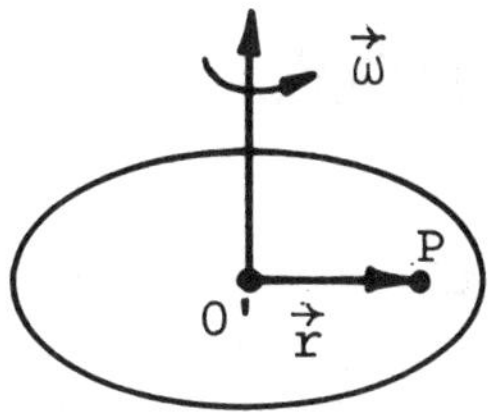

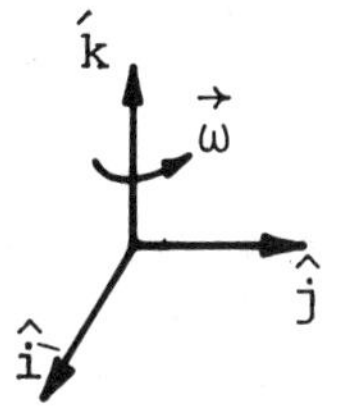

$\vec{\omega}$ is now equal to $\omega(\hat{k})$, and the force equation becomes

$$\vec{F} = m(\ddot{\vec{r}}_{rel} + \omega\hat{k} \times (\omega\hat{k} \times \vec{r}) + 2(\omega\hat{k} \times \dot{\vec{r}}_{rel}))$$

since a turntable is a fixed rotating coordinate system rotating at a constant angular velocity. The Coriolis acceleration is $2(\omega\hat{k} \times \dot{\vec{r}}_{rel})$. If $\dot{\vec{r}}_{rel}$ is in a tangential direction but opposite $\vec{\omega} \times \vec{r}$, then it is equal to $\dot{r}(A\hat{i} + B\hat{j})$ (the speed times an arbitrary combination of $\hat{i}$ and $\hat{k}$ which describe tangency).

When $\dot{\vec{r}}_{rel}$ is in a radial direction, $\dot{\vec{r}}_{rel} = \dot{r}(C\hat{i} + D\hat{j})$ once again, where C and D are constants describing radial motion. (A,B,C and D all depend on the position of the particle on the turntable and would involve sines or cosines of some angle θ between $\vec{r}$ and either the x or y axis.)

Evaluating the Coriolis acceleration yields

$$2(\omega\hat{k} \times \dot{\vec{r}}_{rel}) = 2(\omega\hat{k} \times \dot{r}(E\hat{i} + F\hat{j}))$$

(E stands for A or C, F stands for B or D), and, since $\hat{k} \times \hat{j} = -\hat{i}$ and $\hat{k} \times \hat{i} = \hat{j}$, the Coriolis acceleration is

$$\vec{C} = 2\,\omega\dot{r}(E\hat{j} - F\hat{i}) .$$

Graphically, the velocity $\dot{\vec{r}}_{rel}$ and Coriolis acceleration look like this:

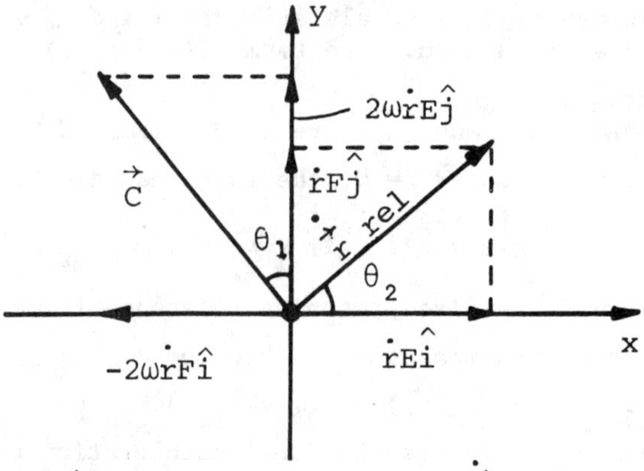

As one can see, $\vec{C}$ points to the left of $\dot{\vec{r}}_{rel}$. The angle between $\vec{C}$ and the Y axis (θ_1) equals the inverse tangent of

$$\left(\frac{2\omega\dot{r}F}{2\omega\dot{r}E}\right).$$

$\theta_1 = \tan^{-1}\left(\frac{F}{E}\right)$ as does θ_2 $\left(\theta_2 = \tan^{-1}\left(\frac{\dot{r}F}{\dot{r}E}\right) = \tan^{-1}\frac{F}{E}\right)$, the angle between $\dot{\vec{r}}_{rel}$ and the x-axis. $\vec{C}$ and $\dot{\vec{r}}_{rel}$ are perpendicular.

PROBLEMS WITH SEVERAL DIFFERENT KINDS OF FORCES

• PROBLEM 20-24

Analyze the motion of a bead of mass m free to slide along a smooth straight rod P shown in figure 1. P is rotating with angular velocity ω around the vertical axis μ to which it is inclined at an angle α. Regard the earth as an inertial system.

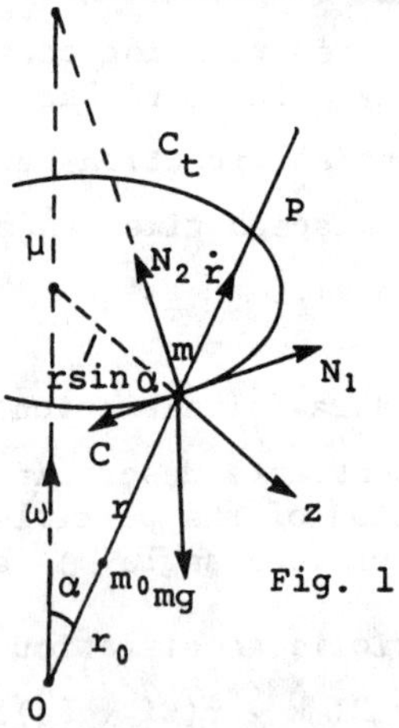

Fig. 1

Solution: The motion of m relative to P is rectilinear. The distance r from O can be used as position coordinate. Then, the relative velocity $\dot{r}$ is coaxial with the line P and is positive when outward. The actual forces are (a) the weight mg, and (b) the normal pressure which lies in the normal plane to P. The normal pressure can be resolved into the component N_1 tangential to the circle C_t of the motion of transport and the component N_2 lying in the vertical plane through P and μ.

The auxiliary forces are (a) the centrifugal force

$$Z = mr\omega^2 \sin \alpha$$

directed radially outward, and (b) the azimuthal Coriolis force

$$C = 2m\dot{r}\omega \sin \alpha .$$

Newton's law, resolved in the directions defined by r, N_1, and N_2, yields relations

$$m\ddot{r} = -mg \cos \alpha + mr\omega^2 \sin^2 \alpha ,$$

$$0 = N_1 - 2m\dot{r}\omega \sin \alpha ,$$

$$0 = N_2 - mg \sin \alpha - mr\omega^2 \sin \alpha \cos \alpha ,$$

of which the first embodies the actual equation of motion, whereas the other two furnish the normal pressures. With the notation

$$K = \omega \sin \alpha , \qquad (1)$$

the actual equation of motion assumes the form

$$\ddot{r} - K^2 r = -g \cos \alpha .$$

It is nonhomogeneous and has the general integral

$$r = \alpha \cosh Kt + b \sinh Kt + \frac{g}{K^2} \cos \alpha .$$

If the particle is released from rest at $t = 0$ at the position r_0, the integration constants found from the initial conditions $r(t = 0) = r_0$, $\dot{r}(t = 0) = 0$ are

$$a = r_0 - \frac{g}{K^2} \cos \alpha , \quad b = 0 ,$$

so that the motion is finally given by

$$r = \left(r_0 - \frac{g \cos \alpha}{\omega^2 \sin^2 \alpha}\right) \cosh Kt + \frac{g \cos \alpha}{\omega^2 \sin^2 \alpha}$$

with the notation of (1). Depending on whether r_0 is chosen larger or smaller than

$$r_0^* = \frac{g \cos \alpha}{\omega^2 \sin^2 \alpha} ,$$

the particle will slide with increasing speed upward or downward. Since no movement at all ensues when $r_0 = r_0^*$, this represents the only position of equilibrium, which is, indeed, unstable.

• PROBLEM 20-25

A particle moves along the straight line path from 0 to P on the disk shown in Fig. 1. The disk lies in the XY plane. For the position shown the particle is at P and is moving 10 ft/sec relative to the disk. At that same time the disk is rotating with an angular velocity about the Z axis of 5 rad/sec ounterclockwise. The radius of the disk is 5 ft.

a) Determine the absolute velocity of the particle at P.

b) Assume the particle at P is decelerating 6 ft/sec² and the disk has an angular acceleration of 9 rad/sec² counterclockwise. Determine the absolute acceleration of the particle P.

Solution: a) Choose the rotating set of x,y,z axes as shown in Fig. 1. Let $\hat{i}_0$, $\hat{j}_0$, $\hat{k}_0$ be the unit vectors in the X,Y,Z system and $\hat{i},\hat{j},\hat{k}$ be the unit vectors in the x,y,z system. Note that $\hat{k}$ and $\hat{k}_0$ are col-

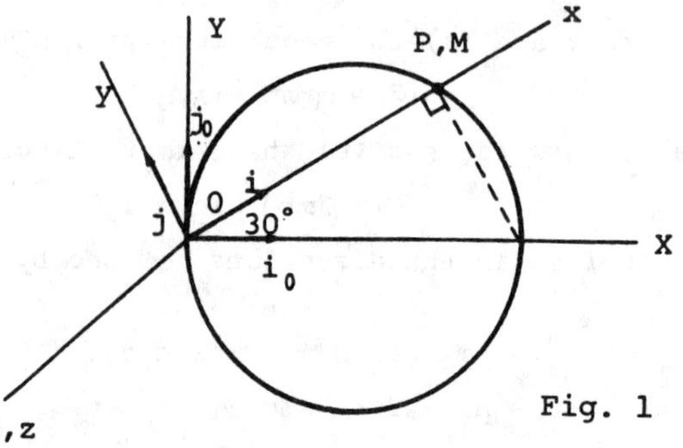

Fig. 1

linear for this choice of axes. Consider M as the point fixed in the rotating frame and coincident with the particle P at the instant considered. For relative velocities,

$$\vec{v}_P = \vec{v}_{P/M} + \vec{v}_M \,. \tag{1}$$

The velocity $\vec{v}_M$ of M is $\vec{\Omega} \times \overrightarrow{OM}$ where $\vec{\Omega} = 5\hat{k}_0$. By taking projections it can be seen that $OP = 10 \cos 30^\circ = 8.66$ and that OM is equal to $8.66 \cos 30^\circ \hat{i}_0 + 8.66 \sin 30^\circ \hat{j}_0 = 7.5\hat{i}_0 + 4.33\hat{j}_0$. Hence,

$$\vec{v}_M = (5\hat{k}_0) \times (7.5\hat{i}_0 + 4.33\hat{j}_0) = 37.5\hat{j}_0 - 21.7\hat{i}_0 \,.$$

The velocity $\vec{v}_{P/M} = 10\hat{i} = 10 \cos 30^\circ \,\hat{i}_0 + 10 \sin 30^\circ \,\hat{j}_0$. Substitute these values into Eq. (1) to obtain

$$\vec{v}_P = (8.66\hat{i}_0 + 5\hat{j}_0) + (37.5\hat{j}_0 - 21.7\hat{i}_0) = -13\hat{i}_0 + 42.5\hat{j}_0 \text{ ft/sec.}$$

b) For relative accelerations in a moving frame,

$$\vec{a}_P = \vec{a}_{P/M} + \vec{a}_M + 2\vec{\Omega} \times \vec{v}_{P/M} \tag{2}$$

Because M is a point fixed in the x,y,z axes and thus moves in a circular path of radius 8.66 ft about 0, its acceleration is

$$\vec{a}_M = (\vec{a}_M)_n + (\vec{a}_M)_t \,,$$

where the magnitude of $(\vec{a}_M)_n$ is $(OM)\Omega^2$ and of $(\vec{a}_M)_t$ is $(OM)\alpha$. Hence,

$$(\vec{a}_M)_n = -8.66(5)^2 \cos 30^\circ \hat{i}_0 - 8.66(5)^2 \sin 30^\circ \hat{j}_0$$

$$= -187.5\hat{i}_0 - 108.3\hat{j}_0$$

$$(\vec{a}_M)_t = -8.66(9)\sin 30^\circ \hat{i}_0 + 8.66(9)\cos 30^\circ \hat{j}_0$$

$$= -39\hat{i}_0 + 67.5\hat{j}_0$$

or

$$\vec{a}_M = -227\hat{i}_0 - 40.8\hat{j}_0 \,.$$

The relative acceleration with respect to the rotating x,y,z axes in

$$\vec{a}_{P/M} = -6\hat{i} = -6 \cos 30^\circ \hat{i}_0 - 6 \sin 30^\circ \hat{j}_0 = -5.2\hat{i}_0 - 3\hat{j}_0 \,.$$

The Coriolis acceleration is

$$2\vec{\Omega} \times \vec{v}_{P/M} = 2(5\hat{k}_0) \times (8.66\hat{i}_0 + 5\hat{j}_0)$$

$$= 86.6\hat{j}_0 - 50\hat{i}_0 \,.$$

Substitute these values into Eq. (2) to obtain

$$\vec{a}_P = (-5.2\hat{i}_0 - 3\hat{j}_0) + (-227\hat{i}_0 - 40.8\hat{j}_0) + (86.6\hat{j}_0 - 50\hat{i}_0)$$

$$= -282\hat{i}_0 + 42.8\hat{j}_0 \text{ ft/sec}^2 \,.$$

• **PROBLEM 20-26**

A helicopter traveling with a constant forward velocity of 50 ft/sec hoists an object at a speed of 2 ft/sec. Determine the velocity and acceleration of the object with respect to the ground at the instant the hoist line makes an angle of 30 degrees with the vertical and experiences an angular velocity and angular acceleration of 1 rad/sec clockwise and 5 rad/sec^2 counterclock-wise. At the instant in question, 20 ft of line are out.

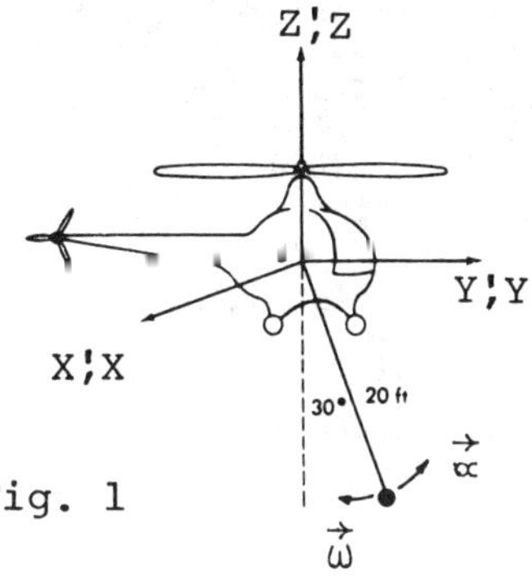

Fig. 1

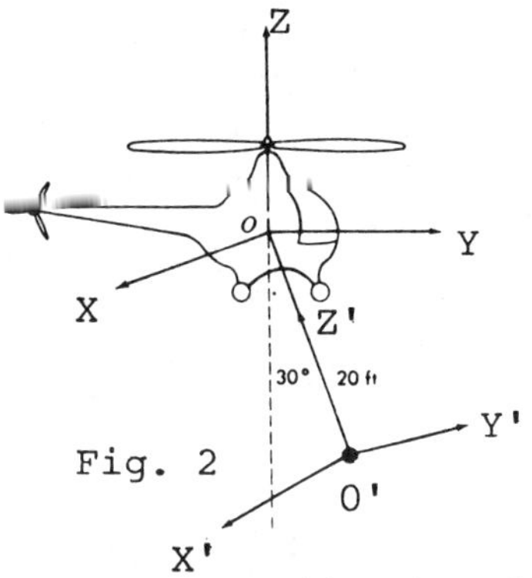

Fig. 2

Solution: The equations relating the kinematic variables in a fixed (unprimed) coordinate axes to the variables in the accelerating (primed) system are

$$\vec{v} = \dot{\vec{R}} + \dot{\vec{\rho}}' + \vec{\Omega} \times \vec{\rho}' \quad (1)$$

and

$$\vec{a} = \ddot{\vec{R}} + \ddot{\vec{\rho}}' + 2\vec{\Omega} \times \dot{\vec{\rho}}' + \vec{\Omega} \times (\vec{\Omega} \times \vec{\rho}') + \dot{\vec{\Omega}} \times \vec{\rho}' \quad (2)$$

where $\vec{R}$ is the vector from the origin of the fixed axes to the moving axes and $\vec{\Omega}$ is the rotation of the moving axes relative to the fixed set. In this problem two sets of coordinate axes are used, illustrating the difference in the mathematics and the constancy of the physics, i.e., final results.

I) The XYZ axes are coincident with the helicopter axes at the moment under consideration although fixed with respect to the earth. The X'Y'Z' axes are fixed to the helicopter (figure 1).

In this case $\vec{R} = 0$ since the axes are coincident, and $\vec{\Omega} = 0$ since the helicopter is not rotating. $\dot{\vec{R}} = 50$ ft/sec $\hat{j}$, the velocity of the helicopter, $\ddot{\vec{R}} = 0$. Finally, $\vec{\rho}'$ measured to the object is

$$\vec{\rho}' = 20(0.500\,\hat{j}' - 0.866\,\hat{k}')$$
$$= 10\,\hat{j}' - 17.32\,\hat{k}' \text{ ft.}$$

Since the axes are aligned coincidently, $\hat{i} = \hat{i}'$, $\hat{j} = \hat{j}'$, $\hat{k} = \hat{k}'$ and

$$\vec{\rho}' = 10\,\hat{j} - 17.32\,\hat{k} \text{ ft.} \quad (3)$$

$$\dot{\vec{\rho}}' = \frac{d'\vec{\rho}'}{dt'} + \vec{\omega}' \times \vec{\rho}' \quad (4)$$

where ω' is the angular velocity at the object in the ω' system; $\vec{\omega} = -1\,\hat{i}$ rad/sec.

$\left|\frac{d'\vec{\rho}'}{dt'}\right|$ is given as -2 ft/sec (the object is being drawn in), so

$$\dot{\vec{\rho}}' = 2(0.500\,\hat{j} - 0.866\,\hat{k}) + (-1\,\hat{i}) \times (10\,\hat{j} - 17.32\,\hat{k}) \text{ ft/sec.}$$

$$\dot{\vec{\rho}}' = -18.32\,\hat{j} - 8.72\,\hat{k} \text{ ft/sec.} \quad (5)$$

so $$\ddot{\vec{\rho}}' = \frac{d'\dot{\vec{\rho}}'}{dt'} + 2\vec{\omega} \times \frac{d'\vec{\rho}'}{dt'} + \vec{\omega}' \times (\vec{\omega}' \times \vec{\rho}') + (\dot{\vec{\omega}}' \times \vec{\rho}') .$$

Here, $\frac{d'\dot{\vec{\rho}}'}{dt'} = 0$; $\dot{\vec{\omega}}' = 5\,\hat{i}$ rad/sec²,

$$\ddot{\vec{\rho}}' = 2(-1\,\hat{i}) \times -2(0.500\,\hat{j} - 0.866\,\hat{k})$$
$$+ (-1\,\hat{i}) \times [(-1\hat{i}) \times (10\,\hat{j} - 17.32\,\hat{k})]$$
$$+ (5\,\hat{i}) \times (10\,\hat{j} - 17.32\,\hat{k}) \text{ ft/sec}^2 .$$

$$\ddot{\vec{\rho}}' = 80.06\,\hat{j} + 69.32\,\hat{k} \text{ ft/sec}^2 . \qquad (6)$$

Note that the equations used to find $\dot{\vec{\rho}}$ (v relative) and $\ddot{\vec{\rho}}$ (a relative), equations (4) and (5), are analogous to equations (1) and (2). This provides the connection to solution II.

Finally, since the helicopter is in translation only, $\vec{\Omega} = \dot{\vec{\Omega}} = 0$, so

$$2\vec{\Omega} \times \dot{\rho}' = \vec{\Omega} \times (\vec{\Omega} \times \vec{\rho}') = \dot{\vec{\Omega}}' \times \vec{\rho}' = 0 .$$

Now applying equations (1) and (2) and using the values obtained in equations (3), (5) and (6),

$$\vec{v} = 50\,\hat{j} + (-18.32\,\hat{j} - 8.72\,\hat{k})$$
$$= 31.68\,\hat{j} - 8.27\,\hat{k} \text{ ft/sec.} \qquad (8)$$

and

$$\vec{a} = 80.06\,\hat{j} + 69.32\,\hat{k} \text{ ft/sec}^2 . \qquad (9)$$

II) Now the X'Y'Z' axes are attached to the object (figure 2). Immediately we know that

$$\frac{d'\vec{\rho}'}{dt} = v_{rel} = 0 \quad \text{and} \quad \frac{d^2{}'\vec{\rho}'}{dt^2} = a_{rel} = 0 ,$$

and, since the object is at the origin of X'Y'Z', $\vec{\rho}' = 0$.

The angular motion of the object is now the angular motion at the XYZ frame, so

$$\vec{\Omega} = -1\,\hat{i} \text{ rad/sec} .$$
$$\dot{\vec{\Omega}} = 5\,\hat{i} \text{ rad/sec}^2 .$$
$$\vec{R} = 20(.500\,\hat{j} - 0.866\,\hat{k}) = 10\,\hat{j} - 17.32\,\hat{k} \text{ ft.}$$
$$\dot{\vec{R}} = 50\,\hat{j} + (-2)(0.500\,\hat{j} - 0.866\,\hat{k})$$
$$+ (-1\,\hat{i}) \times (10\,\hat{j} - 17.32\,\hat{k})$$
$$= 31.68\,\hat{j} - 8.27\,\hat{k} \text{ ft/sec.} \qquad (10)$$

$\dot{\vec{R}}$ is the sum of the velocity of the helicopter relative to the unprimed axes plus the velocity at O' relative to the helicopter (from equation (4)).

$$\ddot{\vec{R}} = 2(-1\,\hat{i}) \times (-2)(0.500\,\hat{j} - 0.866\,\hat{k})$$
$$+ (-1\,\hat{i}) \times [(-1\,\hat{i}) \times (10\,\hat{j} - 17.32\,\hat{k})]$$
$$+ (5\,\hat{i}) \times (10\,\hat{j} - 17.32\,\hat{k})$$
$$= 80.06\,\hat{j} + 69.32\,\hat{k} \text{ ft/sec}^2 . \qquad (11)$$

$\ddot{\vec{R}}$ is the sum of the acceleration components of O' with respect to O, the centripetal, tangental, and Coriorlis contributions plus the acceleration of the helicopter relative to O (zero in this case).

$$\vec{\rho} = 0 , \text{ so } \vec{\Omega} \times (\vec{\Omega} \times \vec{\rho}') = \dot{\vec{\Omega}} \times \rho' = 0$$

and

$$\dot{\vec{\rho}} = 0 , \text{ thus, } 2\vec{\Omega} \times \dot{\vec{\rho}}' = 0 .$$

Applying equation (1) and (2) yield, using (10) and (11),

$$v = 31.68\,\hat{j} - 8.72\,\hat{k} \text{ ft/sec},$$

and

$$\vec{a} = 80.06\,\hat{j} + 69.32\,\hat{k} \text{ ft/sec}^2 ,$$

identical with equations (8) and (9).

CHAPTER 21

SPECIAL RELATIVITY

RELATIVE VELOCITIES AND THE LORENTZ CONTRACTION

• PROBLEM 21-1

A spaceship moving away from the Earth at a velocity $v_1 = 0.75$ c with respect to the Earth, launches a rocket (in the direction away from the Earth) that attains a velocity $v_2 = 0.75$ c with respect to the spaceship. What is the velocity of the rocket with respect to the Earth?

Solution: Newtonian relativity states that the laws of mechanics are the same in all inertial reference frames but that the laws of electrodynamics are not. Since such a theory was not acceptable, the theory of relativity was developed and resolved this problem. One of the statements of this new theory was that the velocity of light is independent of any relative motion between the light source and the observer. Experiments have shown this to be true. In support of this statement Einstein showed that ordinary mechanical velocities do not add algebraically. Instead, for the addition of two velocities v_1 and v_2, the sum is

$$v = \frac{v_1 + v_2}{1 + \frac{v_1 v_2}{c^2}}$$

For small velocities ($v >> c$), this reduces to

$$V = v_1 + v_2,$$

as in Newtonian mechanics, since the term $v_1 v_2/c^2$ is very small compared to one and can be ignored.

Since the velocities in this problem are large, the velocity of the rocket with respect to the Earth is

$$V = \frac{v_1 + v_2}{1 + \frac{v_1 v_2}{c^2}} = \frac{0.75c + 0.75c}{1 + \frac{(0.75c)\ (0.75c)}{c^2}} = \frac{1.5c}{1 + 0.5625} = 0.96c.$$

Therefore, in spite of the fact that the simple sum of the two velocities exceeds c, the actual velocity relative to the Earth is slightly less than c.

• PROBLEM 21-2

A rocket has a length L = 600 m measured at rest on the earth. It moves directly away from the earth at constant velocity, and a pulse of radar is sent from earth and reflected back by devices at the nose and tail of the rocket. The reflected signal from the tail is detected on earth 200 sec after emission and the signal from the nose 17.4×10^{-6} sec later. Calculate the distance of the rocket from the earth and the velocity of the rocket with respect to the earth.

Solution: First we will make a non-relativistic calculation. The speed of the pulse is 3×10^8 m/sec (equal to c, the speed of light). If the rocket is a distance R from the earth then the pulse travels 2R and

$$(3 \times 10^8 \text{ m/sec})(200 \text{ sec}) = 2R, \quad R = 3 \times 10^{10} \text{m}$$

To calculate the speed v of the rocket, we note that the pulse from the front end arrived 17.4×10^{-6} sec after that from the back. Thus this pulse was sent into space a distance

$$\frac{1}{2}(17.4 \times 10^{-6} \text{ sec})(3 \times 10^8 \text{ m/sec}) = 2.61 \times 10^3 \text{ m}$$

farther. This distance is equal to L + vt where

$$t = \frac{1}{2}(17.4 \times 10^{-6} \text{ sec}) = 8.7 \times 10^{-6} \text{ sec}$$

Thus: $L + vt = 2.61 \times 10^3$ m

$$v = \frac{2.61 \times 10^3\text{m} - L}{t} = \frac{2.61 \times 10^3\text{m} - 0.6 \times 10^3\text{m}}{8.7 \times 10^{-6}\text{sec}} = 2.31 \times 10^8\text{m/sec}$$

The factor of $\frac{1}{2}$ arises because we are interested only in the time of the rocket to ground and not in the total time of flight from ground to ground.

According to the above calculation (which we might suspect from the start to be incorrect), the ratio v/c = 0.77. Thus it is very probable that relativistic effects will be important. To the observer on the ground the Lorentz length contraction makes the rocket's length appear to be

$L\sqrt{1 - \beta^2}$ where $\beta = v/c$. Then the pulse travels a distance

$L\sqrt{1 - \beta^2} + vt$ farther, which is equal to ct where $t = \frac{1}{2} \times 17.4 \times 10^{-6}$sec.

Thus $$L\sqrt{1 - \beta^2} + vt = ct$$

which can be solved to obtain

$$\beta = \frac{(ct/L)^2 - 1}{(ct/L)^2 + 1}$$

$$= 0.9$$

The distance R will be the same as in the previous calculation.

• PROBLEM 21-3

What is the Lorentz contraction of an automobile traveling at 60 mph? (60 mph is equivalent to 2682 cm/sec.)

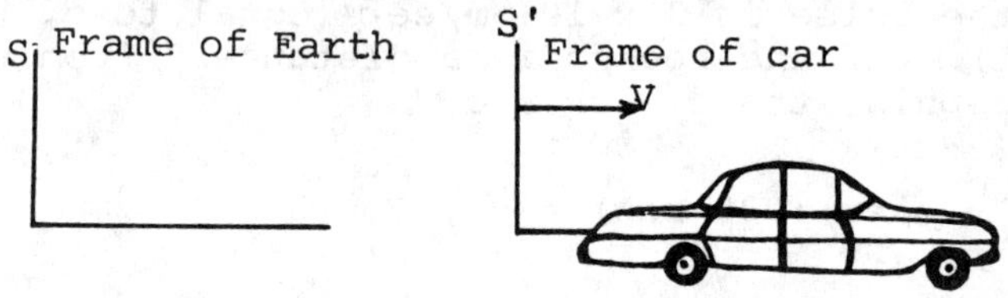

Solution: Suppose we are given two frames of reference moving relative to one another with a velocity v (see figure). If we are dealing with classical physics and want to relate the coordinates of an event occurring in the S-frame (x, y, z, t) to the coordinates of an event occurring in the S'-frame (x', y', z', t'), we use the Galilean transformation, or

$$x' = x - vt$$

$$y' = y \text{ (If v is in the x-direction only).}$$

$$z' = z$$

$$t' = t.$$

In relativistic physics, this transformation is invalid, and must be replaced by the Lorentz transformation, or

$$x' = \frac{x - vt}{\sqrt{1 - v^2/c^2}}$$

$$y' = y$$

$$z' = z$$

$$t' = \frac{t - vx/c^2}{\sqrt{1 - v^2/c^2}}.$$

Now, we may relate distances measured in S' to distances as measured in S. Let us imagine the measurement of a distance parallel to the x'-axis in the S' frame. In order to measure the length of a rod in S, we must locate both ends of the rod (x_1, x_2) at the same time ($t_1 = t_2$) in S. Hence, the length in S' is

$$x_2' - x_1' = \frac{(x_2 - x_1) - v(t_2 - t_1)}{\sqrt{1 - v^2/c^2}}.$$

But $t_1 = t_2$, therefore,

$$x_2' - x_1' = \frac{x_2 - x_1}{\sqrt{1 - v^2/c^2}}.$$

Hence $(x_2 - x_1) = (x_2' - x_1')\sqrt{1 - v^2/c^2}$. (1)

Since $\sqrt{1 - \frac{v^2}{c^2}} < 1$ we then have $x_2 - x_1 < x_2' - x_1'$. The observer in S measures a smaller rod length (contracted) than the observer in the rod's rest frame, S'. Now, we calculate the length of the car in S, $(x_2 - x_1)$. If V_r is 2682 cm/sec.

$$\frac{V_r}{c} = \frac{2682}{3 \times 10^{10}}$$

$$= 8.94 \times 10^{-8}$$

$$\left(\frac{V_r}{c}\right)^2 = 8.0 \times 10^{-15}.$$

When x is very much less than 1,

$$\sqrt{1 - x} = 1 - \frac{1}{2}x \text{ approximately.}$$

Therefore, $$\sqrt{1 - \left(\frac{V_r}{c}\right)^2} \approx [1 - (4.0 \times 10^{-15})].$$

Substituting in (1)

$$x_2 - x_1 \approx (x_2' - x_1')(1 - 4.0 \times 10^{-15}).$$

This means that the change in length of a meter rule is only 4.0×10^{-15} meters, or 4.0×10^{-13} cm. Since the diameter of an atom is about 10^{-8} cm, the diameter of a nucleus is about 10^{-12} cm and the size of the electron is about 10^{-13} cm, this contraction is clearly negligible. Again we see that the difference between relativistic and classical physics is not important for the velocities we are normally concerned with.

REST MASS AND RELATIVISTIC MASS

• PROBLEM 21-4

If two 1 gram masses with equal and opposite velocities of 10^5 cm/sec collide and stick together, what is the additional rest mass of the joined pair?

FIGURE 1 FIGURE 2

Solution: In figure 1, both particles are moving towards each other with the same velocity in our frame of reference. In figure 2 they have collided and stuck together. The velocity of the joined body would seem to be zero, but we must prove that this is so. By the law of conservation of momentum, the total momentum before and after collision must be conserved. We are told that the velocities of the two particles are equal but opposite and we assume that they are collinear so that the problem is one dimensional making the velocities v and - v respectively. Therefore, the momentum before the collision is mv + m (- v). Since this is equal to 0, the momentum after the collision must also equal 0, or $m_{final} v_{final} = 0$. Since no energy is produced in the collision, no mass can be lost (it can only be gained) so that m_{final} is not 0, meaning that v_{final} is 0.

Further, we note that the two particles had total kinetic energy $2 \cdot \frac{1}{2} mv^2$ before the collision, and 0 kinetic energy after the collision. Therefore, if we assume that all of the kinetic energy was converted into rest mass, we can calculate the additional rest mass of the joined pair. Since $E = mc^2$ for the system, the kinetic energy lost has an equivalent mass given by $\Delta E = \Delta mc^2$. Hence,

$$\Delta m = \frac{mv^2 - 0}{c^2}$$

$$\Delta m = \frac{1 \text{ gm } (10^5 \text{ cm/sec})^2}{(3 \times 10^{10} \text{ cm/sec})^2} = 1.1 \times 10^{-11} \text{ gm}$$

• PROBLEM 21-5

The Berkeley synchrocyclotron was designed to accelerate protons to a kinetic energy of 5.4×10^{-11} J, which corresponds to a particle speed of about 2×10^8 m/s. At this speed the relativistic mass increase is important and must be taken into account in the design of the accelerator. Calculate the percentage increase in the proton mass encountered in this instrument.

Solution: The relation between relative mass, m, and rest mass, m_0, is given by

$$m = \frac{m_0}{\sqrt{1 - v^2/c^2}}, = \frac{m_0}{\sqrt{1 - (2 \times 10^8/3 \times 10^8)^2}}$$

$$= \frac{m_0}{\sqrt{1 - \frac{4}{9}}} = \frac{m_0}{\sqrt{\frac{5}{9}}} = \frac{3m_0}{\sqrt{5}} = 1.34\ m_0 .$$

The percentage increase in mass is the fractional increase in mass times 100, or

$$\frac{m - m_0}{m_0} \times 100 = \frac{1.34m_0 - m_0}{m_0} \times 100 = 34 \text{ per cent.}$$

The proton mass at a speed of 2×10^8 m/s is 34 per cent greater than its rest mass.

• **PROBLEM 21-6**

What is the fractional increase of mass for a 600-mi/hr jetliner?

Solution: Fractional increase of mass is defined as change in mass divided by the original mass or $\Delta m/m_0$. The equation for the variation of mass with velocity is

$$m = \frac{m_0}{\sqrt{1 - \beta^2}}, \qquad \beta^2 = \frac{v^2}{c^2}$$

Therefore the change in mass is

$$\Delta m = m - m_0 = m_0 \left(\frac{1}{\sqrt{1 - \beta^2}} - 1 \right)$$

and $$\frac{\Delta m}{m_0} = \frac{1}{\sqrt{1 - \beta^2}} - 1$$

For velocities much less than light, $\beta = v/c$ is very small and $1/\sqrt{1 - \beta^2}$ can be approximated by $1 + \beta^2/2$. Since 600 mi/hr is small compared to c, we can say that for this problem the fractional mass is

$$\frac{\Delta m}{m_0} \cong (1 + \beta^2/2) - 1 = \beta^2/2$$

$$v = 600 \text{ mi/hr} \cong 2.7 \times 10^4 \text{ cm/sec}$$

$$\beta = \frac{v}{c} = \frac{2.7 \times 10^4 \text{ cm/sec}}{3 \times 10^{10} \text{ cm/sec}} \cong 10^{-6}$$

Therefore, $\frac{\Delta m}{m_0} \cong \frac{1}{2}\beta^2 \cong .5 \times 10^{-12}$

so that the mass is increased by only a trivial amount.

• PROBLEM 21-7

At what velocity is the mass of a particle twice its rest mass?

Solution: When the mass of an object that is travelling at a velocity approaching the speed of light, c, is measured, it is found to be larger than the mass measured when the object is at rest. The mass associated with an object travelling at any velocity $\vec{v}$ is called the particle's relativistic mass, and is given by the formula

$$m(v) = \frac{m_0}{\sqrt{1 - v^2/c^2}} \qquad (1)$$

where m_0 is the rest mass of the particle.

We are asked to find the velocity at which the mass of a particle (meaning its relativistic mass) is equal to twice the particle's rest mass. Writing this as an equation,

$$m(v) = 2m_0$$

Using (1), this may be written as

$$\frac{m_0}{\sqrt{1 - \frac{v^2}{c^2}}} = 2m_0$$

$$\frac{1}{\sqrt{1 - \frac{v^2}{c^2}}} = 2$$

Multiplying both sides by $\frac{1}{2}\sqrt{1 - \frac{v^2}{c^2}}$, we obtain

$$\sqrt{1 - \frac{v^2}{c^2}} = \frac{1}{2}$$

$$1 - \frac{v^2}{c^2} = \frac{1}{4}$$

$$\frac{v^2}{c^2} = \frac{3}{4}$$

$$\frac{v}{c} = \frac{\sqrt{3}}{2}$$

$$\frac{v}{c} = 0.866$$

$$v = 0.866 \times 3 \times 10^{10} \text{ cm/sec}$$

$$v = 2.664 \times 10^{10} \text{ cm/sec}$$

• PROBLEM 21-8

A photon rocket is being propelled through space. The propellant consists of photons. The mass of the rocket is not constant throughout the motion, for it continually loses mass in the form of photons. Use the principles of conservation of momentum and conservation of energy to show that only about 5 per cent of the mass of the photon rocket remains after the rocket has reached a speed of 99.5 per cent of the speed of light.

Solution: Before takeoff, the total momentum of the photon-rocket system is zero as the rocket is at rest. Since no external force F acts on the system, then

$$F = 0 = \frac{dP}{dt}$$

where P is the total momentum of the system. Hence P remains constant in time. Consequently, after attaining final speed, the total momentum must still be zero; therefore, the momentum of the photons must equal the final spaceship momentum. (These momentum vectors are oppositely directed to give zero total momentum.) When the rocket has reached its final speed, its rest mass will be a fraction f of the original total rest mass m_0.

The fraction (1 - f) of the original rest mass has been ejected in the form of photons.

total ejected photon momentum = p

$$\text{final rocket momentum} = \frac{(fm_0)v}{\sqrt{1 - v^2/c^2}} = \gamma\,(fm_0)v \text{ where}$$

$$\gamma = \frac{1}{\sqrt{1 - v^2/c^2}}$$ is the relativistic correction of the rest mass due to the rocket velocity.

Therefore, by conservation of momentum,

$$p = \gamma f m_0 v$$

Since the rocket speed si 99.5 per cent of the speed of light, v may be replaced by c with negligible error.

$$p = \gamma f m_0 c$$

The total mass-energy of the rocket system must remain constant in time, for no energy enters or leaves the closed ejected photons and rocket system. Then

$$\text{total initial energy} = m_0 c^2$$

$$\text{final rocket energy} = \gamma\,(fm_0)c^2$$

$$\text{total ejected photon energy} = pc$$

since initial energy = final energy

$$m_0 c^2 = \gamma f m_0 c^2 + pc$$

Substituting for p from the previous equation gives

$$m_0 c^2 = \gamma f m_0 c^2 + \gamma f m_0 c^2 = 2\gamma f m_0 c^2$$

Solving for f,

$$f = \frac{1}{2\gamma}$$

At a speed of 99.5 per cent c, the value of γ is

$$\frac{1}{\sqrt{1 - \left(\frac{.995\ c}{c}\right)^2}} = 1.0 \times 10^1$$

$$\text{Therefore, } f = \frac{1}{(2)\ (1 \times 10^1)} = 0.05 \text{ per cent}$$

Therefore, only 5 per cent of the original mass of the rocket remains when the rocket has achieved a speed of 99.5 per cent c.

THE LORENTZ TRANSFORMATION AND LORENTZ INVARIANCE

• PROBLEM 21-9

Show that in the limit as $v/c \to 0$, the equation of Lorentz transformation will reduce to the form of classical mechanics.

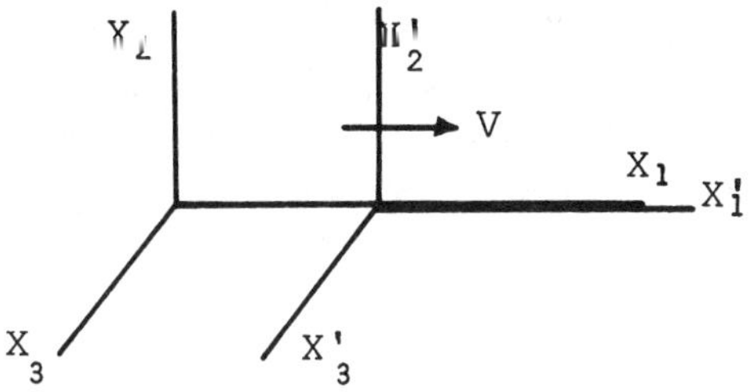

Solution: Suppose the "unprimed" and "primed" reference frames travel with a relative velocity v. Choose Cartesian coordinates within the two systems such that the x_1 and x'_1 axes coincide with the direction of relative motion, with the positive direction along the velocity of the "primed" frame, as seen in the "unprimed" one. See the figure, x_2, x'_2, and x_3, x'_3 axes are perpendicular to this unique direction.

If we set the clocks in the two reference frames such that they both read 0 as the two origins coincide, the Lorentz transformation takes the form

$$\begin{cases} x'_1 = \gamma(x_1 - vt) & (1) \\ x'_2 = x_2 & (2) \\ x'_3 = x_3 & (3) \\ t' = \gamma(t - x_1 v/c^2) & (4) \end{cases}$$

γ stands for $(1 - v^2/c^2)^{-\frac{1}{2}}$. We want to show that as $v/c \to 0$, this set of equations reduce to the appropriate transformation equations of classical mechanics, the Galilean transformation. With the same choice of coordinates as above, the Galilean transformation is:

$$\begin{cases} x'_1 = x_1 - vt & (5) \\ x'_2 = x_2 & (6) \\ x'_3 = x_3 & (7) \\ t' = t & (8) \end{cases}$$

Equations (2) and (3) are identical to (6) and (7), so these need no further work. As $v/c \to 0$, the factor $\gamma \to 1$, and equation (1) takes the same form as (5). Let us rewrite equation (4): $t' = \gamma[t - (v/c)(x/c)]$. From the form of the second term in the bracket, we see that as long as x/c stays finite, this term will become negligible, and the right side reduces to just t (as $\gamma \to 1$, again.) Thus, each of the Lorentz transformation equations has the same form as the corresponding Galilean one in the limit $v/c \to 0$.

• PROBLEM 21-10

Show that two simultaneous events $t_1 = t_2$, taking place at different places (i.e., x_1 not equal to x_2) in the S coordinate system, are not simultaneous in system S'.

Solution: We shall solve this problem by using the Lorentz transformation for time

$$t' = \gamma \left(t - \frac{\beta}{c} x\right) \quad (1)$$

where $\gamma = 1/\sqrt{1 - \left(\frac{v}{c}\right)^2}$ and $\beta = \frac{v}{c}$.

First, let us write down the Lorentz transformations for the two events.

$$t'_1 = \gamma \left(t_1 - \frac{\beta}{c} x_1\right) \quad (2)$$

and

$$t'_2 = \gamma \left(t_2 - \frac{\beta}{c} x_2\right). \quad (3)$$

Next, subtract eq. (3) from eq. (2). This yields

$$t'_1 - t'_2 = \gamma \left(t_1 - \frac{\beta}{c} x_1\right) - \gamma \left(t_2 - \frac{\beta}{c} x_2\right). \quad (4)$$

Simplifying eq. (4), we obtain

$$t'_1 - t'_2 = \gamma \left[(t_1 - t_2) - \frac{\beta}{c} (x_1 - x_2)\right]. \quad (5)$$

Next, note that the term $t_1 - t_2 = 0$, that is, the two events are simultaneous in the S coordinate system. Thus eq. (5) reduces to:

$$t'_1 - t'_2 = -\gamma \frac{\beta}{c} (x_1 - x_2). \quad (6)$$

We also are told that the two events in S take place at two different positions, that is $x_1 \neq x_2$. Then from eq. (6), we get

$$t'_1 \neq t'_2.$$

Thus, the two events are not simultaneous in the S' reference frame.

• PROBLEM 21-11

Show from the transformation equations that if two Lorentz transformations, with relative velocities given by β_1 and β_2 respectively, are carried out consecutively, the result is the same as that of a single Lorentz transformation with relative velocity β given by

$$\beta = (\beta_1 + \beta_2)/1 + \beta_1\beta_2$$

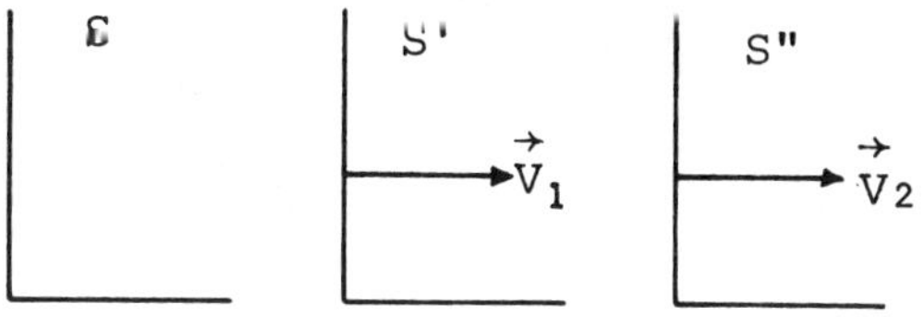

Fig. 1

Solution: There are two methods we could use to answer this question. Method one would involve using the Lorentz transformation equations for the coordinates and time, while the second method involves using the Lorentz transformation equation for velocity. Since the second method is algebraically simpler, we shall use that method.

In Fig. 1, reference frame S" has a relative velocity $\vec{V}_2$ with respect to S' and reference frame S' has a relative velocity $\vec{V}_1$ with respect to S. We want to find the relative velocity $\vec{V}$ of frame S" with respect to S. The velocity transformation equation for the x-component is given by

$$v_x = \frac{v'_x + v}{1 + \frac{v'_x v}{c^2}} \tag{1}$$

In eq. (1), v is the relative velocity of frame S' relative to S. In this case this is equal to V_1. Also, note that v'_x is the x-component of velocity of an object with respect to S'. Here, this is equal to V_2. Substituting these values into eq. (1) we obtain:

$$V = \frac{V_2 + V_1}{1 + \frac{V_2 V_1}{c^2}}, \tag{2}$$

where V is the relative velocity in the x-direction of S" relative to S. Dividing eq. (2) by the speed of light c, we obtain the desired result,

$$\beta = \frac{\beta_2 \quad \beta_1}{1 + \beta_1 \beta_2}, \tag{3}$$

where $\beta = \frac{V}{c}$, $\beta_1 = \frac{V_1}{c}$ and $\beta_2 = \frac{V_2}{c}$.

We have shown that the relative velocity of frame S" with respect to S is given by eq. (3). This means that the Lorentz transformation equations relating an individual point event in S" and S are given by:

$$x'' = \frac{1}{\sqrt{1 - \beta^2}} (x - Vt)$$

$$y'' = y$$

$$z'' = z$$

$$t'' = \frac{1}{\sqrt{1 - \beta^2}} (t - \frac{\beta}{c} x),$$

where V and β are given by eqs. (2) and (3) respectively.

• PROBLEM 21-12

Two events occur at the same place in an inertial reference frame, S, and are separated by a time interval of 3 sec. What is the spatial separation between these two events in an inertial reference frame S' in which the two events are separated by a time interval of 5 sec.?

Solution: The simplest way to solve this problem is to use the fact that the space-time interval is an invariant. This relationship is given by:

$$(c\Delta t)^2 - (\Delta x)^2 - (\Delta y^2) - (\Delta z)^2 = (c\Delta t')^2 - (\Delta x')^2 - (\Delta y')^2 - (\Delta z')^2 \tag{1}$$

This means that the space-time interval has the same value for all inertial reference frames

We are given that in frame S, the two events occur at the same place; that is $\Delta x = 0$, $\Delta y = 0$ and $\Delta z = 0$. Substituting these values into eq. (1) we obtain,

$$(c\Delta t)^2 = (c\Delta t')^2 - (\Delta x')^2 - (\Delta y')^2 - (\Delta z')^2. \tag{2}$$

Furthermore, we also know the time intervals in the two reference frames, $\Delta t = 3$ sec and $\Delta t' = 5$ sec. Substituting these values into eq. (2) and solving for the spatial interval yields,

$$(\Delta x')^2 + (\Delta y')^2 + (\Delta z')^2 = c^2 (\Delta t'^2 - \Delta t^2) \text{ or}$$

$$(\Delta x')^2 + (\Delta y')^2 + (\Delta z')^2 = 16c^2 \qquad (3)$$

In eq. (3) the left side is equal to the spatial distance squared. Thus the spatial distance d' between the two events is

$$d' = 4c$$

where $c = 3 \times 10^8$ m/sec.

or $d' = 12 \times 10^8$ m.

• PROBLEM 21-13

Using the Lorentz Transformation, give some examples of the addition of velocity from one frame of reference to another moving with respect to the first.

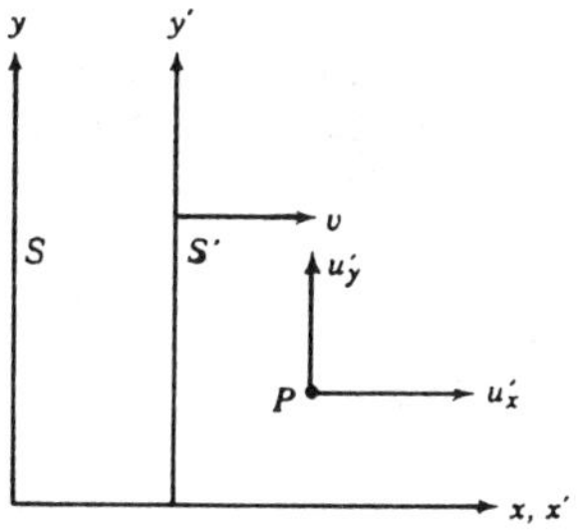

Fig 1

Solution: First, the most important example is the one which demonstrates that the Lorentz Transformation satisfies Einstein's postulate that the speed of light is constant in all inertial frames of reference.

Consider a light signal traveling along the x'direction with a speed c. If the S' system is moving in the x direction with a speed $c/2 = v$ relative to the S system and $u_x' = c$, then according to the Galilean transformation equation the speed of light signal as measured by an observer in the S system would be 3c/2. This is not the result given by the Einstein equation for the addition of velocities,

$$u_x = \frac{u_x' + v}{1 + \frac{u_x'v}{c^2}}.$$

This equation gives for the speed of light in the S system

$$u_x = \frac{c + c/2}{1 + c^2/2c} = \frac{1.5c}{1.5} = c.$$

That is, the speed of light in both the S and S' systems is c.

As a further example of the addition relationship, consider an object P to be moving in the y' direction with a speed $u_y' = dy'/dt'$, and also moving in the x' direction with a speed u_x' (Fig. 1) with system S' moving in the x, x' direction relative to S.

Now we can not simply apply the formula used before for u_x because it is used only when the motion is only along one direction, the x axis. However, using the definition of velocity and the differential form of the Lorentz Transformation,

$$dx = \gamma\,(dx' + v dt'),$$

$$dy = dy',\ dz = dz',$$

$$dt = \gamma\,(dt' + v dx'/c^2),$$

$$\gamma = 1/\sqrt{1 - v^2/c^2}$$

we may find u_y.

The speed of object P as measured in the S system in the y direction is $u_y = dy/dt$. Now $y = y'$ and $dy = dy'$, hence

$$u_y = \frac{dy}{dt} = \frac{dy'}{\gamma\,(dt' + v\,dx'/c^2)} = \frac{u_y'}{\gamma\,(1 + vu_x'/c^2)}$$

$$= \frac{u_y'\sqrt{1 - v^2/c^2}}{(1 + vu_x'/c^2)}.$$

If $u_x' = 0$, that is, the object P is moving only in the y' direction in system S' with speed u_y', then the speed of P measured in the S system is

$$u_y = u_y'\sqrt{1 - v^2/c^2}.$$

• PROBLEM 21-14

In a particular inertial reference frame S, two events occur a distance 6×10^8 m apart and are separated by a time interval of (a) 2 sec (b) 1 sec. What is the proper time interval between these two events?

Solution: Let us solve this problem by using the invariance of the space-time interval. This relationship is defined by the equation

$$(c\Delta t)^2 - (\Delta x)^2 - (\Delta y)^2 - (\Delta z)^2 = (c\Delta t')^2 - (\Delta x')^2 - (\Delta y')^2 - (\Delta z')^2. \qquad (1)$$

Next let us define proper time. The proper time is the time interval recorded in a reference frame in which the two events occur at the same place. Thus, to obtain the proper time interval in reference frame S', set $\Delta x' = \Delta y' = \Delta z' = 0$ in eq. (1) and solve the resultant equation for $(\Delta t')^2$

$$(\Delta t')^2 = (\Delta t)^2 - \frac{1}{c^2}\left[(\Delta x)^2 + (\Delta y)^2 + (\Delta z)^2\right]. \quad (2)$$

It is given that in S the two events are separate by a distance of 6×10^8 m, i.e.,

$$(\Delta x)^2 + (\Delta y)^2 + (\Delta z)^2 = (6 \times 10^8)^2 = 36 \times 10^{16}\ m^2,$$

and $$c^2 = (3 \times 10^8)^2 = 9 \times 10^{16}\ m^2.$$

Substituting these values into eq. (2) we obtain,

$$(\Delta t')^2 = (\Delta t)^2 - 4. \quad (3)$$

(a) In this part it is given that in frame S the time interval between the two events is given by $\Delta t = 2$ sec. Substituting this value into eq. (3) we can solve for the proper time interval $\Delta t'$,

$$\Delta t' = 0.$$

Note that in S', these two events are simultaneous. Furthermore the relative speed of the two frames is

$$\frac{\Delta x}{\Delta t} = \frac{6 \times 10^8}{2} = 3 \times 10^8\ m/sec.$$

We recognize this value as being equal to the speed of light.

(b) Here we are given that $\Delta t = 1$ sec. When we substitute this value into eq. (3) we obtain

$$(\Delta t')^2 = -3.$$

This is a spacelike interval and no reference frame S' can be found in which the two events occur at the same place.

• PROBLEM 21-15

Show that the inverse transformation of the Lorentz Equation has the form

$$x = (x' + vt')/\sqrt{1 - \frac{v^2}{c^2}}$$

$$y = y'$$

$$z = z'$$

$$t = \left(t' + x' \frac{v}{c^2}\right) / \sqrt{1 - \frac{v^2}{c^2}}$$

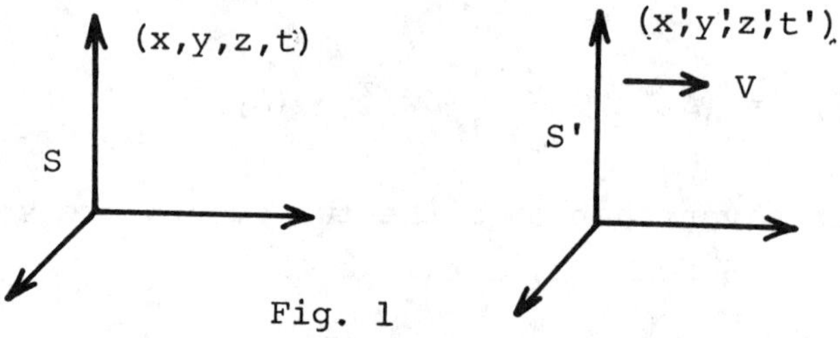

Fig. 1

Solution: In the above diagram, x, y, z, t and x', y', z', t', are coordinates in the system S and S', respectively.

An observer at rest in system S sees system S' moving at a velocity v away from him. The equations that relate the coordinates of S' to those of S are the Lorentz Transformation Equations. They are:

$$x' = \frac{x - vt}{\sqrt{1 - \frac{v^2}{c^2}}} \tag{1}$$

$$y' = y \tag{2}$$

$$z' = z \tag{3}$$

$$t' = \frac{t - x\left(\frac{v}{c^2}\right)}{\sqrt{1 - \frac{v^2}{c^2}}} \tag{4}$$

Another way of looking at the above situation is to imagine an observer at rest relative to system S'. He sees system S moving away from him. So, the velocity of S relative to S' is -v. (See figure 2.)

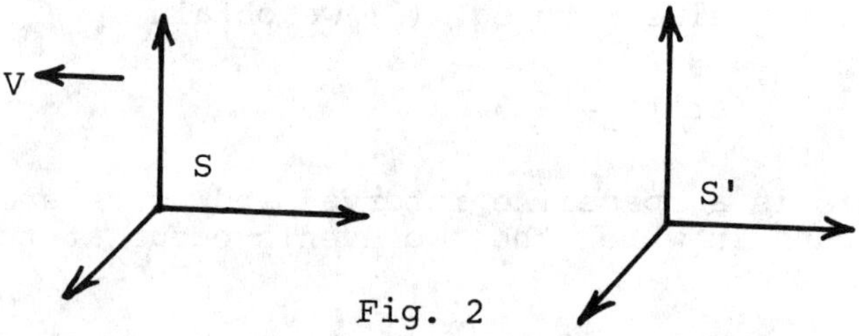

Fig. 2

The inverse Lorentz Transformation Equations allow the coordinates of system S to be expressed in terms of those in system S'.

To do this, replace v by -v, change all primed coordinates to unprimed coordinates and all unprimed coordinates to primed coordinates in equations (1), (2), (3), (4).

Doing this gives

$$x = \frac{x' + vt'}{\sqrt{1 - \frac{v^2}{c^2}}} \qquad (5)$$

$$y = y' \qquad (6)$$

$$z = z' \qquad (7)$$

$$t = \frac{t' + x'\left(\frac{v}{c^2}\right)}{\sqrt{1 - \frac{v^2}{c^2}}} \qquad (8)$$

Equations (5), (6), (7), (8) are the inverse Lorentz Transformation Equations.

• PROBLEM 21-16

Show that no reference frame can be found by consecutive Lorentz transformations in which the speed of a given particle is greater than the speed of light if the speed of the particle is less than the speed of light in any such frame.

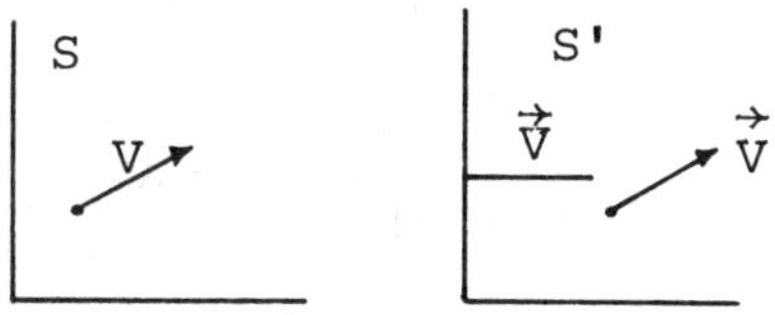

Fig. 1

Solution: Since we are interested in the speed of the particle, we shall use the velocity addition equations to obtain the solution. In Fig. 1, the velocity of the particle relative to reference frames S' and S is given respectively by $\vec{V}'$ and $\vec{V}$. Also note that reference frame S' is moving at a speed v relative to frame S.

In component form, the transformation equations for velocity are given by:

$$V'_x = \frac{V_x - v}{1 - \frac{vV_x}{c^2}}, \qquad (1)$$

$$V'_y = \frac{V_y/\gamma}{1 - \frac{vV_x}{c^2}}, \qquad (2)$$

and

$$V'_z = \frac{V_z/\gamma}{1 - \frac{vV_x}{c^2}} \qquad (3)$$

where $\gamma = 1/\sqrt{1 - \left(\frac{v}{c}\right)^2}$.

Since we are interested in the speed of the particle, we want to calculate the quantity

$$(V')^2 = \left(V'_x\right)^2 + \left(V'_y\right)^2 + \left(V'_z\right)^2 .$$

Using eqs. (1), (2) and (3) we obtain

$$(V')^2 = \left[\left(V_x - v\right)^2 + \left(V_y/\gamma\right)^2 + \left(V_z/\gamma\right)^2\right] \Big/ \left(1 - \frac{vV_x}{c^2}\right)^2 . \quad (4)$$

The problem states that $V' < c$ or $(V')^2 < c^2$ and we are supposed to show that $V < c$ or $V^2 < c^2$. Notice that the expression $c^2 - (V')^2$ is positive, i.e.,

$$c^2 - (V')^2 > 0 \quad (5)$$

Using eq. (5), we can rewrite eq. (4) in the following way:

$$c^2 - (V')^2 = \left\{ c^2 \left(1 - \frac{vV_x}{c^2}\right)^2 - \left[\left(V_x - v\right)^2 + \left(\frac{V_y}{\gamma}\right)^2 + \left(\frac{V_z}{\gamma}\right)^2\right]\right\} \Big/ \left(1 - \frac{vV_x}{c^2}\right)^2 > 0 \quad (6)$$

In eq. (6) the denominator $\left(1 - \frac{vV_x}{c^2}\right)^2$ is always positive. Therefore, in order for eq. (6) to be valid, the numerator must also be positive. Thus we have

$$c^2 \left(1 - \frac{vV_x}{c^2}\right)^2 - \left[\left(V_x - v\right)^2 + \left(\frac{V_y}{\gamma}\right)^2 + \left(\frac{V_z}{\gamma}\right)^2\right] > 0. \quad (7)$$

Next simplify eq. (7). First, multiply by γ^2, this yields,

$$c^2\gamma^2 \left(1 - \frac{vV_x}{c^2}\right)^2 - \left[\gamma^2 \left(V_x - v\right)^2 + V_y^{\,2} + V_z^{\,2}\right] > 0. \quad (8)$$

To simplify eq. (8) further, expand all terms in parentheses. The resultant equation is:

$$c^2\gamma^2 - 2vV_x\gamma^2 + \frac{v^2V_x^{\,2}\gamma^2}{c^2} - V_x^{\,2}\gamma^2 + 2vV_x\gamma^2 - v^2\gamma^2 - V_y^{\,2} - V_z^{\,2} > 0 \quad (9)$$

In eq. (9) notice that the second term will cancel with the fifth term. We can also combine the first with the sixth term, i.e., $c^2\gamma^2 - v^2\gamma^2 = c^2$. Making these simplifications we obtain:

$$c^2 + \frac{v^2 V_x^{\,2}\gamma^2}{c^2} - V_x^{\,2}\gamma^2 - V_y^{\,2} - V_z^{\,2} > 0 \qquad (10)$$

Eq. (10) can be simplified by combining the second and third terms,

$$\frac{v^2 V_x^{\,2}\gamma^2}{c^2} - V_x^{\,2}\gamma^2 = V_x^{\,2}\gamma^2\left(\frac{v^2}{c^2} - 1\right) = - V_x^{\,2}. \qquad (11)$$

Substituting eq. (11) into eq. (10) we obtain,

$$c^2 - V_x^{\,2} - V_y^{\,2} - V_z^{\,2} > 0 \quad \text{or}$$

$$c^2 - V^2 > 0 \qquad (12)$$

where $V^2 = V_x^{\,2} + V_y^{\,2} + V_z^{\,2}$ is the speed squared of the particle relative to frame S. Thus from eq. (12) we can deduce that $V > c$ and the speed of the particle is less than the speed of light. This derivation can be repeated for any number of consecutive Lorentz transformations and the conclusion is that the speed of the particle is less than the speed of light in any such frame.

CHAPTER 22

PARTICLE VIBRATIONS

• PROBLEM 22-1

A body of 0.5 kgm mass is connected to three equal springs each of 1-m relaxed length. The spring constant for each of the springs is K = 2 n/m, and in their relaxed state the springs are fixed to three corners of an equilateral triangle, as shown in Fig. 1. The mass m is displaced to point A, then released. What will be the kinetic energy of m if it returns to point B?

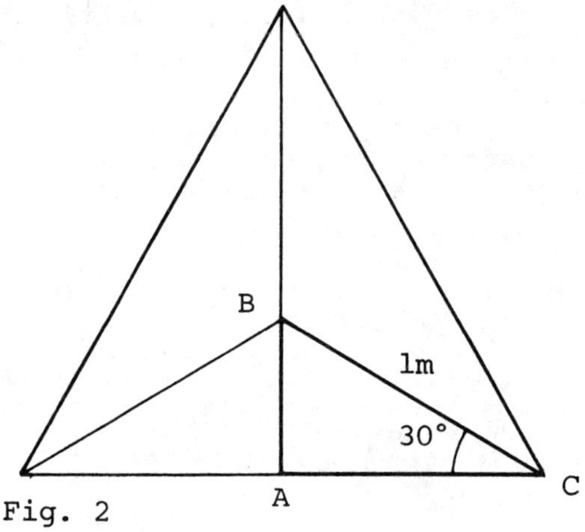

Fig. 2

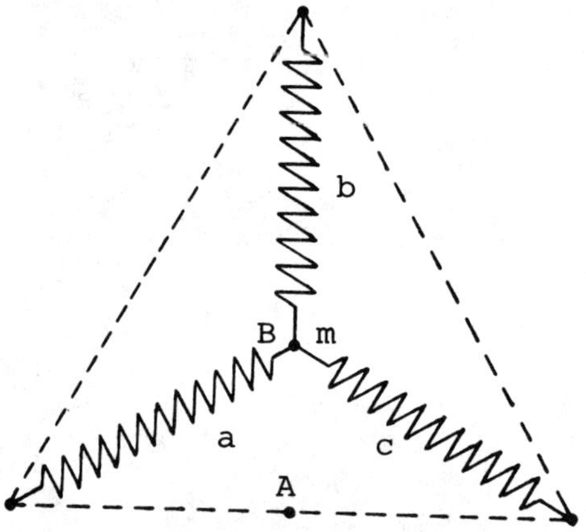

Fig. 1

Solution: The springs are relaxed in the position shown, and the potential energy is zero at B. The kinetic energy, when the body returns to B, will be equal to the potential energy gained by the body at A.

The lengths of the springs when the body is at point A are required. This calculation involves some simple geometry as illustrated in figure 2.

The triangle ABC contains the information needed. The angle ACB is equal to one-half the vertex angle of the equilateral triangle (all of which are 60°). Therefore, that angle is 30°. The length $\overline{AC}$ = 1 cos 30° = 0.87 m and the length $\overline{BA}$ = 1 sin 30° = 0.50 m.

But, $\overline{AC}$ is the compressed length of springs a and c while $\overline{BA}$ is the expanded length of spring b. The potential energy at A is, then

$$V_A = \frac{1}{2} Kx_a^2 + \frac{1}{2} Kx_b^2 + \frac{1}{2} Kx_c^2$$

$$= 0.017 + 0.017 + 0.25 = 0.28 \text{ joule.}$$

This, as the opening remark noted, is also the kinetic energy at point B and the speed is $v \simeq 1$ m/sec.

• PROBLEM 22-2

Show that the period of the simple pendulum of small amplitude is given by the expression $T = 2\pi\sqrt{\frac{L}{g}}$. Is this equation true at the surface of the moon as well as at the surface of the earth? Compare the period of the pendulum at the surface of the earth to that at the surface of the moon.

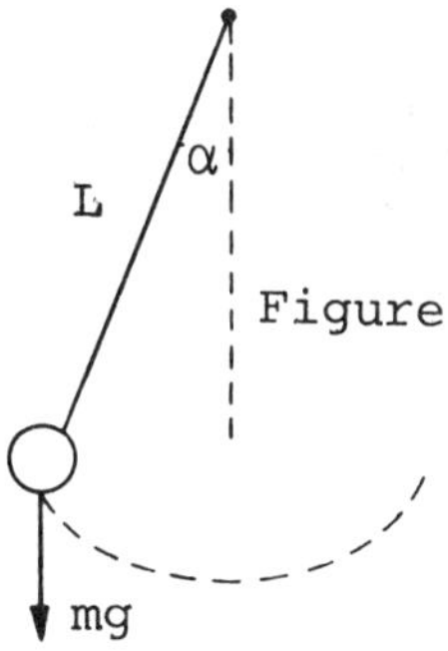

Figure 1

Solution: For small angles the differential equation can be shown to be

$$\ddot{\alpha} = -\frac{g}{L}\alpha \qquad (1)$$

where α = angle of swing in radians

g = gravitational constant

L = length of string.

This is the equation for simple harmonic motion, the solution for which is

$$\alpha = D \sin(\omega t + \phi) \qquad (2)$$

where D and ϕ are the amplitude and phase, respectively, and depend on the initial conditions. ω is not arbitrary but is such that

$$\omega = \sqrt{\frac{g}{L}} \qquad (3)$$

where ω is the circular frequency in radians.

Since T, the period of motion is

$$T = \frac{2\pi}{\omega},$$

$$\omega = \frac{2\pi}{T}. \qquad (4)$$

Substituting (4) in (3)

$$\frac{2\pi}{T} = \sqrt{\frac{g}{L}}$$

Inverting both sides

$$\frac{T}{2\pi} = \sqrt{\frac{L}{g}}$$

$$T = 2\pi\sqrt{\frac{L}{g}}. \qquad (5)$$

In Eq. (5), g is the local value of gravity. Thus the equation would apply on the moon as long as the appropriate value of $g = g_{\text{(}} (g_{moon})$ is used.

Evluating Eq. (5) on the earth and on the moon and taking the quotient of the two equations yields:

$$\frac{T_e}{T_m} = \frac{2\pi\sqrt{\frac{L}{g_e}}}{2\pi\sqrt{\frac{L}{g_m}}}$$

Canceling and simplifying

$$\frac{T_e}{T_m} = \sqrt{\frac{\frac{1}{g_e}}{\frac{1}{g_m}}} = \sqrt{\frac{g_m}{g_e}}.$$

Since $g_m \cong \frac{1}{6} g_e$ we get

$$\frac{T_e}{T_m} = \sqrt{\frac{\frac{1}{6} g_e}{g_e}}$$

$$= \sqrt{\frac{1}{6}} = .408 .$$

i.e., the period of oscillation on earth is shorter or the oscillation faster (higher frequency).

Write the differential equation of motion and determine the natural frequency for small oscillations of the pendulum shown below.

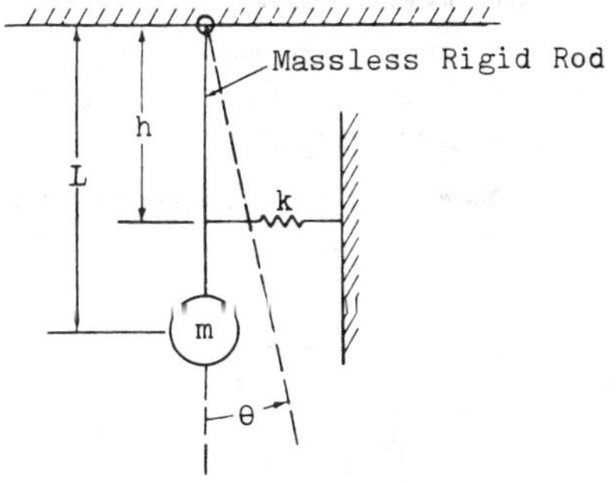

Figure 1

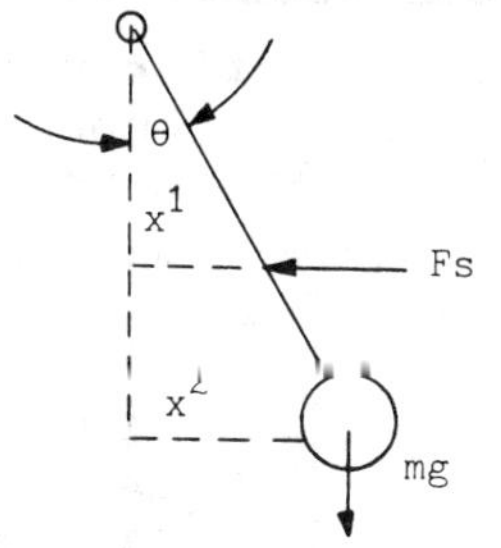

Figure 2

Solution: First, draw a free body diagram of the deflected pendulum:

Here, F_s is the restoring force of the spring and is given as

$$F_s = kx_1$$

where $x_1 = h \sin \theta$. Therefore,

$$F_s = k\ h \sin \theta.$$

Now apply the rotational form of Newton's Law; i.e.,

$$\Sigma T = I\ddot{\theta} \qquad (1)$$

where $\ddot{\theta}$ is the angular acceleration,

ΣT is the sum of the external torques and I is the mass moment of inertia. For a rod of length L with mass m concentrated at its end

$$I = mL^2.$$

Taking the sum of the torques about the pivot we obtain, upon substitution into Eq. (1),

$$-(k\ h \sin \theta)h - mg\ x_2 = mL^2\ \ddot{\theta}.$$

But, since $x_2 = L \sin \theta$,

$$mL^2\ \ddot{\theta} + (kh^2 + mg\ L)\sin \theta = 0.$$

This equation can be simplified if we restrict ourselves to small oscillations θ. In that case, $\sin \theta \approx \theta$ and

$$\ddot{\theta} + \left(\frac{kh^2 + mg\ L}{mL^2}\right) \theta = 0.$$

Now, observe that this is the governing equation of a simple harmonic oscillator with natural frequency

$$\omega_0 = \left(\frac{kh^2 + mg\,L}{mL^2}\right)^{\frac{1}{2}}.$$

• PROBLEM 22-4

Determine the differential equation and the natural frequency for liquid in a U tube, as shown in figure 1. Solve this by energy conservation and by Newton's Second Law.

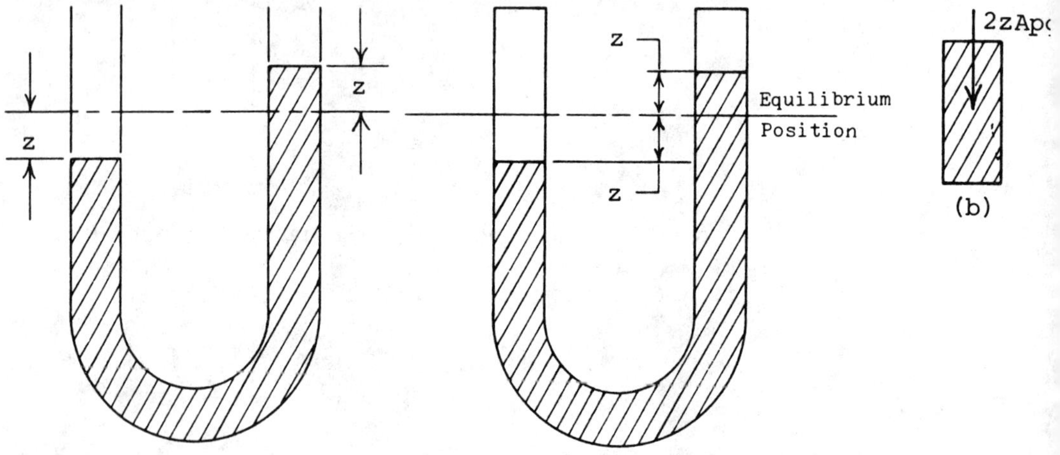

Fig. 1

Fig. 2 (a) (b)

Solution by Energy Method: Since each particle of liquid possesses the same average speed at any instant, the kinetic energy of the liquid

$$T = \tfrac{1}{2}mv^2 = \tfrac{1}{2}\rho AL\dot{z}^2$$

where ρ is the liquid density, L the total liquid column length, and A the cross-sectional area of the manometer tube. For an arbitrary liquid displacement as shown, the potential-energy gain of the liquid may be determined by considering the work done against the gravitational field in moving a column of liquid of length z from the left side to the right side of the tube. This gives for

$$V = mgz = (\rho Az)(g)(z)$$

or $$V = \rho Agz^2.$$

Now, set the sum of the expression for the kinetic and potential energy equal to a constant:

$$\tfrac{1}{2}\rho AL\dot{z}^2 + \rho gAz^2 = E = \text{constant}.$$

Since the time rate of change of energy is zero,

$$\frac{d}{dt}(T + V) = 0 = \rho AL\dot{z}\ddot{z} + 2\rho gA\dot{z}z$$

or $\ddot{z} + \left(\frac{2g}{L}\right) z = 0.$ (1)

This differential equation should easily be recognized as that governing the simple harmonic motion. The base form is:

$$m\ddot{x} = -kx$$

or $\ddot{x} + \left(\frac{k}{m}\right) x = 0.$ (2)

This equation applies in spring, pendulum, and other small oscillation problems. The natural frequency is given by $\omega = \sqrt{k/m}$. For our equation (a) we have

$$\omega = \left(\frac{2g}{L}\right)^{\frac{1}{2}} .$$

Solution by Force Determination: Examine Figure 2. The portion of the liquid in the right-hand column whose length is 2z is unbalanced and subjected to a restoring force equal to its own weight, $w = mg = 2\rho gAz$. Taking the upward direction to be positive in the right tube, the weight force becomes negative.

According to Newton's Law, $F = m\ddot{z}$, with m the mass of the entire liquid system;

$$-2\rho gAz = \rho AL\ddot{z}$$

or $\ddot{z} + \left(\frac{2g}{L}\right) z = 0$ as before.

• PROBLEM 22-5

Consider a bob attached to an inextensible string of length L, the other end of which is attached to a fixed point. If the bob is released from a position other than the equilibrium position, show that its equation of motion will be given by

$$L\ddot{\alpha} = -g \sin \alpha$$

which is independent of the mass of the bob. If α is small enough, show that the result can be rewritten in the approximate form

$$\ddot{\alpha} = -\frac{g}{L}\alpha .$$

Then find the general solution of the equation.

Solution: Since this is a problem involving rotation, use the torque equation $\tau = I\ddot{\alpha}$. Choose a coordinate system with an origin at the support point.

$$\tau = I\ddot{\alpha} \qquad (1)$$

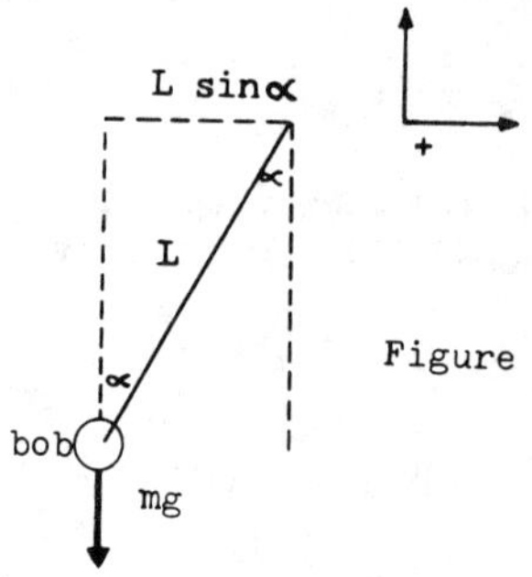

Figure 1

Now $\mathcal{T}$ = Force × perpendicular lever arm

$$= (-\ mg)\ L\ \sin\ \alpha .$$

The negative sign on the force indicates that it is directed downward.

I = moment of inertia

$$= mL^2.$$

Substituting into Eq. (1)

$$-\ mg\ L\ \sin\ \alpha = m\ L^2\ \ddot{\alpha}\ .$$

Didiving through by mL and rearranging yields

$$L\ddot{\alpha} = -\ g\ \sin\ \alpha. \qquad (2)$$

Note that the mass of the bob cancels out. Since L and g are positive constants, the negative sign indicates that the angular force ($\sim \ddot{\alpha}$) is opposite to sin α, i.e., the force is a restoring force.

This equation can be simplified and solved for small angles by using a Taylor expansion of sin α

$$\sin\ \alpha = \alpha - \frac{\alpha^3}{3!} + \frac{\alpha^5}{5!} - \frac{\alpha^7}{7!} + \ldots \cong \alpha \quad \underline{\text{if } \alpha \text{ small}}\ .$$

Substituting into (2)

$$L\ddot{\alpha} = -\ g\alpha$$

or

$$\ddot{\alpha} = -\ \frac{g}{L}\ \alpha. \qquad (3)$$

This is the basic equation for simple harmonic motion (SHM), sometimes, called the harmonic oscillator (HO).

We will now try to find a general solution of the form

$$\alpha = D\ \sin\ (\omega t + \phi) \qquad (4)$$

where D is an arbitrary amplitude and ϕ an arbitrary phase factor, both of which must be determined by initial conditions. ω is the circular frequency and must be measured in radians per time unit.

Differentiating (4) once and twice:

$$\dot{\alpha} = -\omega D \cos(\omega t + \phi)$$

$$\ddot{\alpha} = -\omega^2 D \sin(\omega t + \phi). \qquad (5)$$

Comparing Eq. (4) with Eq. (5) yields:

$$\ddot{\alpha} = -\omega^2 \alpha. \qquad (6)$$

We notice the similarity with Eq. (3). Thus we make the association

$$\omega^2 = \frac{g}{L}$$

$$\omega = \sqrt{\frac{g}{L}}. \qquad (7)$$

Substituting into Eq. (4)

$$\alpha = D \sin\left(\sqrt{\frac{g}{L}}\, t + \phi\right). \qquad (8)$$

Note that ω is independent of amplitude D or phase ϕ. This is basic reason swinging pendula can be used as the governors in clocks.

• PROBLEM 22-6

> A ball of mass m is supported by a frictionless horizontal surface and is restrained by two wires of length ℓ_1 and ℓ_2, as shown. It is displaced slightly from equilibrium and released. Determine the natural frequency for small oscillations. At equilibrium, the tension in each wire is T.

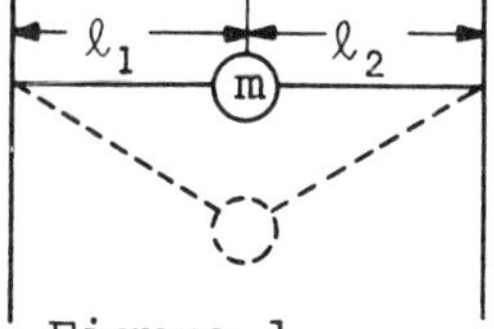

Figure 1

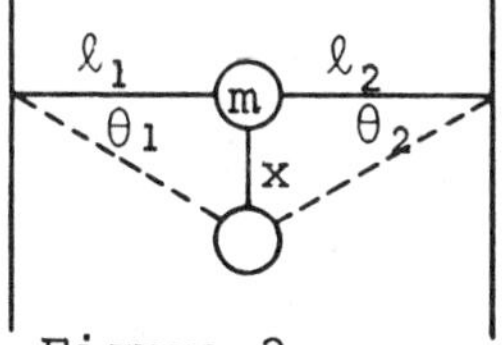

Figure 2
Space Diagram

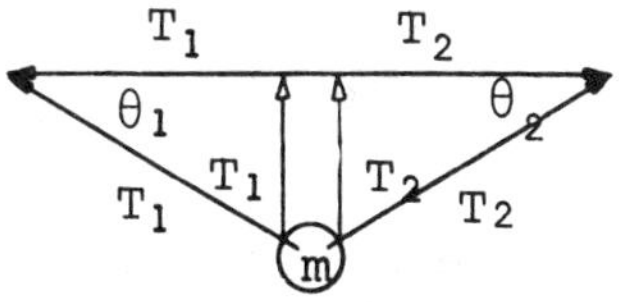

Figure 3
Force Diagram

Solution: The key to this problem is realizing that forces in wires must lie along the wires; there can be no shearing forces. This means that the force diagram (Fig. 3) is similar to the space diagram (Fig. 2).

From Fig. 3 the net restoring force is

$$F = T_1 + T_2 \ . \qquad (1)$$

By trigonometry

$$\frac{T_2}{T'} = \sin\theta_2 \qquad T_2 = T' \sin\theta_2 \qquad (2)$$

$$\frac{T_1}{T'} = \sin\theta_1 \qquad T_1 = T' \sin\theta, \qquad (3)$$

Substituting Eqs. (2) and (3) into (1) and factoring out T'

$$F = T' (\sin\theta_1 + \sin\theta_2).$$

For small displacements $T' \stackrel{\sim}{\sim} T$ so

$$F = T (\sin\theta_1 + \sin\theta_2). \qquad (4)$$

Also for small θ, $\sin\theta \simeq \tan\theta$

$$\sin\theta_1 \simeq \tan\theta_1 \qquad \sin\theta_2 \simeq \tan\theta_2$$

$$F = T (\tan\theta_1 + \tan\theta_2).$$

From Fig. 2: $\tan\theta_1 = \frac{x}{\ell_1}$

$$\tan\theta_2 = \frac{x}{\ell_2} \ .$$

Substituting in Eq. (4)

$$F = T\left(\frac{x}{\ell_1} + \frac{x}{\ell_2}\right)$$

Combining over a common denominator

$$F = Tx\left(\frac{\ell_1 + \ell_2}{\ell_1 \ell_2}\right) .$$

Substituting the force into Newton's Second Law results in

$$- Tx\left(\frac{\ell_1 + \ell_2}{\ell_1 \ell_2}\right) = m\ddot{x} \ . \qquad (5)$$

A minus sign had to be added because this is a restoring force, i.e., the direction of the force is oppposite the displacement.

We recognize Eq. (5) as a case of SIMPLE HARMONIC MOTION because it can be written as

$$\ddot{x} = -\frac{T}{m}\left(\frac{\ell_1 + \ell_2}{\ell_1 \ell_2}\right) x \qquad (6)$$

i.e. $\ddot{x} = -\frac{k}{m} x$ $\qquad$ k = constant.

One solution of this equation is $x = A \sin(\omega t + \phi)$ where A is an arbitrary amplitude, ϕ is an arbitrary phase factor, ω is the circular frequency, and equals $= \sqrt{k/m}$.

Thus by examination of Eq. (6)

$$\omega = \sqrt{\frac{T(\ell_1 + \ell_2)}{m \ \ell_1 \ \ell_2}} .$$

• PROBLEM 22-7

Obtain an approximate solution to the problem of the swinging pendulum.

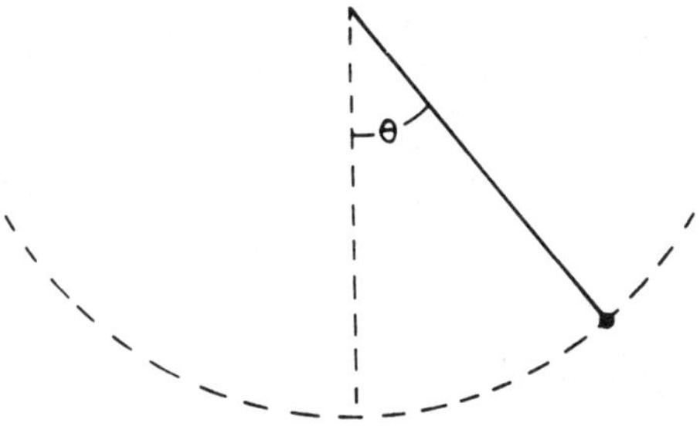

FIG.1

Solution: In the simple pendulum problem a mass m is suspended from a point P by a rigid connection, and is thus constrained to move on the arc of a circle. The two forces acting on m are the constraint force of the connection and gravity. From Newton's Second Law

$$my'' + g \sin y = 0 \qquad (a)$$

where y measures the angle from the vertical and g is the acceleration due to gravity. Equation (a) is second order and requires two initial conditions which we take to be

$$y(0) = 0 \qquad \text{and} \qquad y'(0) = v. \qquad (b)$$

Because equation (a) is non-linear (due to the sin y term) we resort to an approximation.

To begin, let us assume m = g, and the initial velocity is v = 1/2. Then (a) becomes

$$y'' + \sin y = 0. \qquad (c)$$

If y is small, $\sin y \approx y$ and the approximation to (c) is

$$y'' + y = 0. \qquad (d)$$

This is the harmonic oscillator equation whose solution is

$$y = A \sin x + B \cos x$$

which may be verified by direct substitution in (d). Since

$y(0) = 0$, we must have $B = 0$, and since $y'(0) = 1/2$. we must have $A = 1/2$. Thus the solution to our approximate equation (d) is

$$y = \frac{1}{2} \sin x . \tag{e}$$

We note that in solution (e) $y_{max} = 1/2$ when $x = \pi/2$ and that the period of oscillation is 2π.

Let us now investigate further. Since

$$\sin x = x - \frac{x^3}{3!} + \frac{x^5}{5!} - \ldots$$

we may write (e) as

$$y = \frac{1}{2}\left(x - \frac{x^3}{6} + \frac{x^5}{120} - \ldots\right) .$$

Keeping the first three terms

$$y \approx \frac{1}{2}\left(x - \frac{x^3}{6} + \frac{x^5}{120}\right) . \tag{f}$$

We find y_{max} by letting $y' = 0$.

$$y' = \frac{1}{2}\left(1 - \frac{x^2}{2} + \frac{x^4}{24}\right)$$

$$0 = \frac{1}{2}\left(1 - \frac{x^2}{2} + \frac{x^4}{24}\right) .$$

This is a quadratic equation in $p = x^2$,

$$p^2 - 12p + 24 = 0$$

or, by the quadratic formula

$$p = x^2 = 6 \pm 2\sqrt{3}.$$

Since $\sqrt{3} = 1.732$

$$x^2 = 2.536$$

$$x = 1.592.$$

As we noted earlier, y_{max} occurs when $x = \pi/2 = 1.571$. Our series approximation (f) compares quite well with the expected value: 1.592 versus 1.571.

Let us now return to our original equation

$$y'' + \sin y = 0.$$

Assume

$$y = x + a_3x^3 + a_5x^5 + \ldots . \tag{g}$$

then $y'' = 6a_3x + 20a_5x^5 + \ldots$ (h)

if we approximate

$$\sin y \approx y - \frac{y^3}{6}$$

then substituting from (g)

$$\sin y \approx (x + a_3x^3 + a_5x^5) - \frac{(x + a_3x^3 + a_5x^5)^3}{6} + \ldots$$

or $\sin y \approx x + \left(a_3 - \frac{1}{6}\right) x^3 + \ldots$ (i)

Substituting from (h) and (i) into $y'' + \sin y = 0$,

$$(6a_3)x + (20a_5)x^3 + x + \left(a_3 - \frac{1}{6}\right) x^3 = 0.$$

Thus $6a_3 + 1 = 0$

and $20a_5 + \left(a_3 - \frac{1}{6}\right) = 0$.

or $a_3 = -\frac{1}{6}$,

$$a_5 = \frac{1}{60} .$$

Our power series approximate solution is thus

$$y = x - \frac{x^3}{6} + \frac{x^5}{60} .$$

• PROBLEM 22-8

In the configuration shown, there are two spring systems. A 100 kg block is suspended from each system. The spring constants are K_a = 5kN/m and K_b = 10kN/m. If the block is constrained to move in the vertical direction only, and is displaced .05m down from its equilibrium position determine for each spring system: (1) period of vibration, (2) the maximum velocity of the block, (3) and the maximum acceleration of the block.

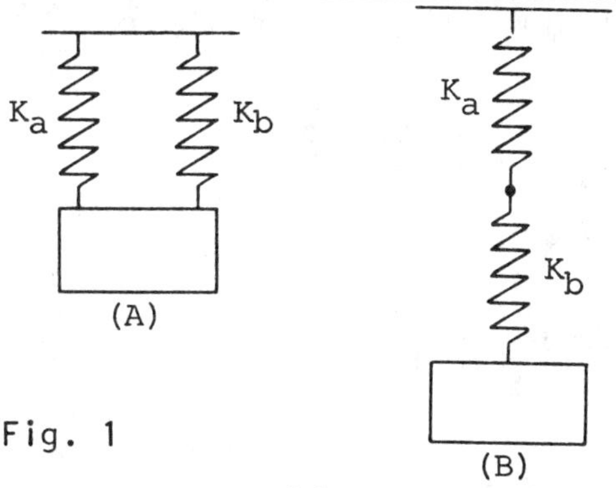

Fig. 1

Solution: Parallel arrangement (a): To determine the behaviour of the spring system, we must first find a single spring arrangement equivalent to the parallel arrangement. The spring constant for the equivalent system can be found by ignoring the block and calculating the magnitude of force $\vec{F}$, necessary to change the system by the amount ε (figure 2). For the parallel spring system, both springs are changed by an amount ε during deformation. The subsequent magnitude of the forces exerted by the two springs are $K_a\varepsilon$ and $K_b\varepsilon$.

For equilibrium conditions we write:

$$F = k_a\varepsilon + k_b\varepsilon = (k_a + k_b)\varepsilon.$$

The constant k of the single equivalent spring is

$$k = \frac{F}{\varepsilon} = k_a + k_b = \frac{5\text{kN}}{\text{m}} + \frac{10\text{kN}}{\text{m}} = \frac{15\text{kN}}{\text{m}}$$

$$k = 15{,}000 \text{ N/m}.$$

Using this spring constant, we have, for the period of system

(1) $$T = 2\pi\sqrt{\frac{m}{k}} = 2\pi\sqrt{\frac{100}{15000}} = .513 \text{ sec.}$$

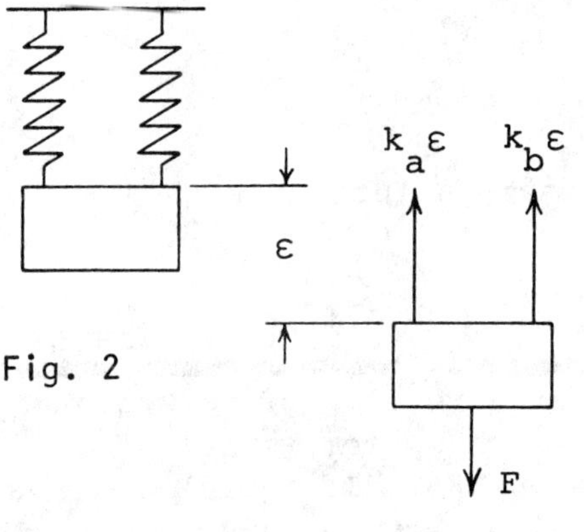

Fig. 2

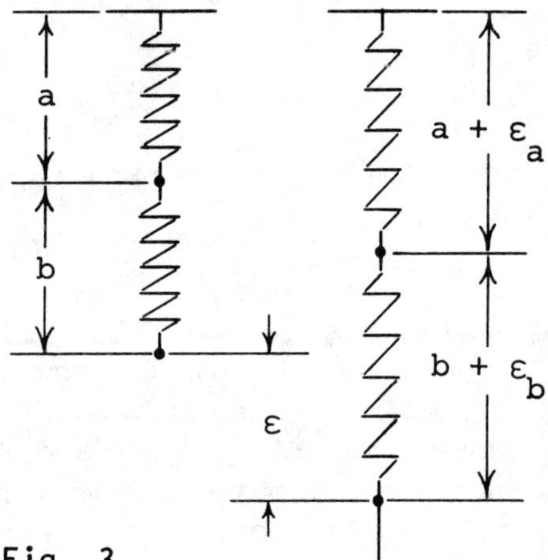

Fig. 3

Using the formulas for maximum velocity and maximum acceleration $v_m = x_m\omega$ and $a_m = x_m\omega^2$ where $\omega\sqrt{k/m}$, we have for system A:

(2) $$v_m = (0.050 \text{ m})\left(\sqrt{\frac{15000}{100}}\right)$$

$$= 0.612 \text{ m/sec},$$

(3) $$a_m = (0.050)\left(\frac{15000}{100}\right)$$

$$= 7.50 \text{ m/sec}^2.$$

Series Arrangement (b): We may use the same method used for case (a). Again, not considering the mass, we find the ratio between the magnitude of the force $\vec{F}$ and the deflection ε it causes (figure 3). In this case, according to Newton's Third Law, the force on both springs is $\vec{F}$. We may write:

$$\varepsilon = \varepsilon_a + \varepsilon_b = \frac{F}{k_a} + \frac{F}{k_b} = F\left(\frac{1}{k_a} + \frac{1}{k_b}\right),$$

Yielding $$k = \frac{F}{\varepsilon} = \left(\frac{1}{k_a} + \frac{1}{k_b}\right)^{-1} = \left(\frac{1}{5\frac{kN}{m}} + \frac{1}{10\frac{kN}{m}}\right)^{-1}$$

$$= 3.333 \frac{kN}{m}.$$

For system (B) we have

(1) $$T = 2\pi\sqrt{\frac{m}{k}} = 2\pi\sqrt{\frac{100}{3333}} = 1.088 \text{ sec}$$

(2) $$v_m = x_m\sqrt{\frac{k}{m}} = (0.050)\left(\sqrt{\frac{3333}{100}}\right) = 0.289 \frac{m}{sec}$$

(3) $$a_m = x_m\left(\frac{k}{m}\right) = (0.050)\left(\frac{3333}{100}\right) = 1.67 \frac{m}{sec^2}.$$

• PROBLEM 22-9

What is the period of a small oscillation of an ideal pendulum of length ℓ, if it oscillates in a truck moving in a horizontal direction with acceleration a? (See figure a).

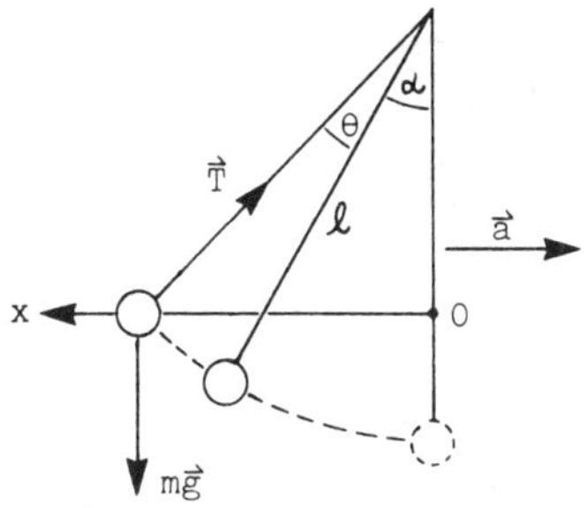

Fig. A

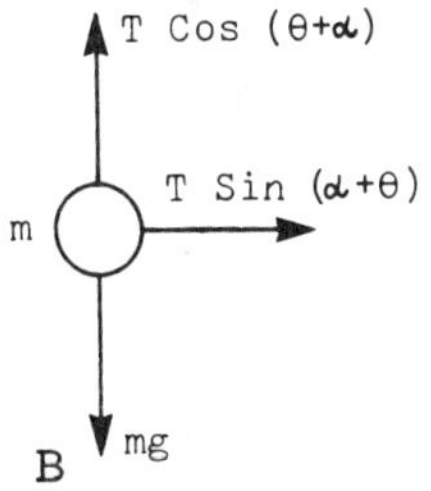

Fig. B

Solution: Let the equlibrium position be given by the angle α. In this position, the net force on m is along the horizontal axis and equals ma. The angle α is determined by the equations

$$T \sin \alpha = ma \qquad (1)$$

$$T \cos \alpha = mg \qquad (2)$$

where g is the gravitational acceleration. When the pendulum is displaced by a small amount θ, it will perform a simple harmonic motion around the equilibrium position. The force on m along the horizontal x-axis is $T \sin(\theta + \alpha)$, as shown in Fig. b. Newton's Second Law for the motion along this axis is

$$m \frac{d^2x}{dt^2} = - T \sin (\theta + \alpha) \tag{3}$$

where x is the distance from the vertical.

For small θ, we can expand sin (θ + α) as

$$\sin (\theta + \alpha) \underset{-}{\sim} \sin \alpha + \theta \cos \alpha.$$

Substituting in (3)

$$m \frac{d^2x}{dt^2} \underset{-}{\sim} - T \sin \alpha - \theta T \cos \alpha. \tag{4}$$

Combining (1), (2) and (4), we get

$$m \frac{d^2x}{dt^2} \underset{-}{\sim} - ma - \theta mg$$

$$\frac{d^2x}{dt^2} \underset{-}{\sim} - a - \theta g. \tag{5}$$

θ and ℓ are related geometrically as

$$x = \ell \sin (\theta + \alpha) = \ell \sin \alpha + \theta \ell \cos \alpha$$

therefore $\frac{d^2x}{dt^2} \underset{-}{\sim} \ell \cos \alpha \frac{d^2\theta}{dt^2}$.

Substituting in (5),

$$\ell \cos \alpha \frac{d^2\theta}{dt^2} \underset{-}{\sim} - \alpha - \theta g$$

$$\frac{d^2\theta}{dt^2} \underset{-}{\sim} - \frac{g}{\ell \cos \alpha} \left[\theta + \frac{a}{g} \right]. \tag{6}$$

If we make the following substitution in (6)

$$\phi = \theta + \frac{a}{g}$$

we get $\frac{d^2\phi}{dt^2} = - \frac{g}{\ell \cos \alpha} \phi$.

This is the differential equation for a simple harmonic motion. Therefore its solution is

$$\phi = A \sin (\omega t + B)$$

where A and B are the constant of integration and

$$\omega = \sqrt{\frac{g}{\ell \cos \alpha}} .$$

The expression for θ is

$$\theta = A \sin (\omega t + B) - \frac{a}{g} .$$

The boundary conditions are $\theta = 0$ at $t = 0$ and $\theta = \theta_{max}$ at $t = \frac{\tau}{2}$ (where τ is the period).

Determine the constants A, B:

i) $t = 0$: $A \sin B = \frac{a}{g}$

ii) $t = \frac{\tau}{2}$: $A \sin\left(\frac{\omega\tau}{2} + B\right) - \frac{a}{g} = \theta_{max}$.

The period of the motion is

$$\tau = \frac{2\pi}{\omega} = 2\pi\sqrt{\frac{\ell \cos \alpha}{g}} .$$

• PROBLEM 22-10

A mass m is connected by means of a linear spring k to a wall which experiences velocity shock. At t = 0, the wall shown in Fig. 1 instantaneously attains a constant velocity u to the right. If the mass is at rest at t = 0, and if the spring is initially undeformed, determine the response of the system and the spring force as a function of time. Assume frictional influences to be negligible.

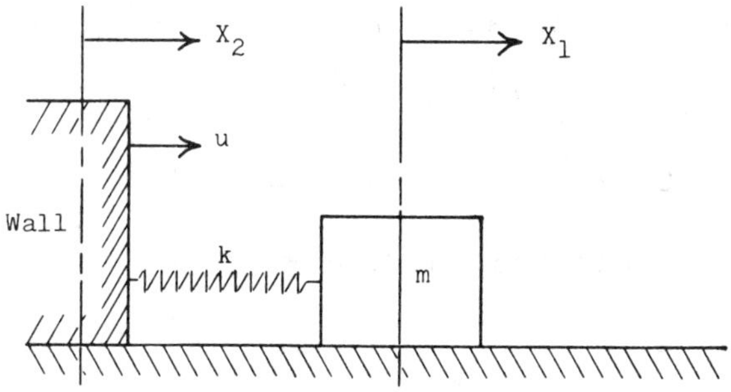

fig 1

Solution: The only force acting parallel to the direction of motion is associated with the deformation $x_1 - x_2$ of the spring. The equation of motion is therefore

$$-k(x_1 - x_2) = m\ddot{x}_1 .$$

Since the displacement of the wall may be written

$$x_2 = ut$$

the equation of motion becomes

$$m\ddot{x}_1 + kx_1 = kut$$

or $$\ddot{x}_1 + \omega_0^2 x_1 = \omega_0^2 ut$$

where $\omega_0^2 = k/m$.

The transient solution $x_{1,t}$ applies to the homogeneous

equation $\ddot{x}_1 + \omega_0^2 x_1 = 0$:

$$x_{1,t} = C_1 \sin \omega_0 t + C_2 \cos \omega_0 t.$$

Because the forcing function is linear in time, the assumed form of $x_{1,ss}$, the steady-state solution is

$$x_{1,ss} = At + B$$

where A and B are constants to be evaluated by substitution into the differential equation of motion. Since $\dot{x}_{1,ss} = A$ and $\ddot{x}_{1,ss} = 0$,

$$\omega_0^2 (At + B) = \omega_0^2 ut$$

or $A = u$, $B = 0$.

The complete solution is therefore

$$x_1(t) = C_1 \sin \omega_0 t + C_2 \cos \omega_0 t + ut$$

to which the conditions $x_1 = 0$, $\dot{x}_1 = 0$ at $t = 0$ must be applied. From the condition on initial position, $C_2 = 0$. Applying the condition on velocity,

$$\dot{x}_1 = \omega_0 C_1 \cos \omega_0 t + u.$$

At $t = 0$,

$$0 = \omega_0 C_1 + u \qquad C_1 = -\frac{u}{\omega_0} .$$

The response, plotted in Figure 2, is

$$x_1(t) = u \left(t - \frac{1}{\omega_0} \sin \omega_0 t\right).$$

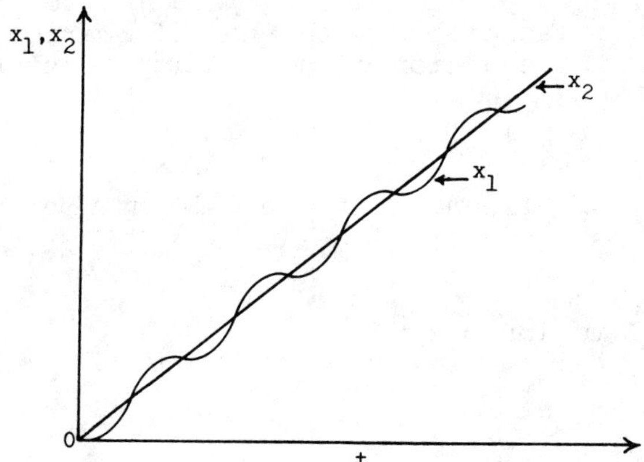

fig 2

The spring force

$$F = -k(x_1 - x_2) = \frac{ku}{\omega_0} \sin \omega_0 t .$$

• PROBLEM 22-11

A body of mass m, on a horizontal frictionless plane, is attached to two horizontal springs of spring constant k_1 and k_2 and equal relaxed lengths L. The free ends of the springs are pulled apart and fastened to two fixed walls a distance 3L apart. Determine the equlibrium position of the body and the frequency of oscillation about the equilibrium position.

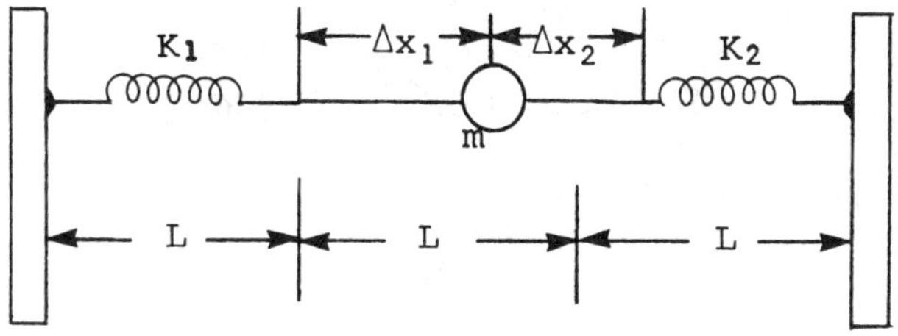

Fig. 1

Fig. 2

Solution :

(a) At equlibrium the net force on the mass is zero. The only important forces come from the springs, and since they pull in opposite directions, the force from spring one is of the same magnitude as the force from spring two.

$$\Sigma F_x = 0$$

$$-F_1 + F_2 = 0 \quad \text{(motion to the right is positive).}$$

$$F_1 = F_2 . \tag{1}$$

From Hooke's law, $F = -k\Delta x$, and (1)

$$F_1 = -k_1\Delta x_1 \qquad F_2 = -k_2\Delta x_2$$

$$-k_1\Delta x_1 = -k_2\Delta x_2$$

$$\frac{k_1}{\Delta x_2} = \frac{k_2}{\Delta x_1} \tag{2}$$

(where Δx_1 and Δx_2 are the distances each spring must be stretched in order to attach it.)

To find the equlibrium position, remember that the total change in displacement was $(3L - 2L) = L$.

$$\therefore \quad L = \Delta x_1 + \Delta x_2$$

$$\Delta x_1 = L - \Delta x_2 . \tag{3}$$

Combining (2) and (3) yields,

$$\frac{k_1}{\Delta x_2} = \frac{k_2}{L - \Delta x_2}$$

$$k_1(L - \Delta x_2) = K_2(\Delta x_2)$$

$$(R_1 + k_2)\Delta x_2 = k_1 L$$

$$\Delta x_2 = \frac{k_1 L}{(k_1 + k_2)} \,. \qquad (4)$$

Fig. 1 shows that the equilibrium position is $L + \Delta x_2$ or $\left(L + \frac{k_1 L}{k_1 + k_2}\right)$ units away from the wall where k_2 is attached.

(b) When the mass is displaced from equlibrium an amount Δx, $F_1 + F_2$ change by an amount equal to $+ (k_1 \Delta x)$ and $- (k_2 \Delta x)$ respectively.

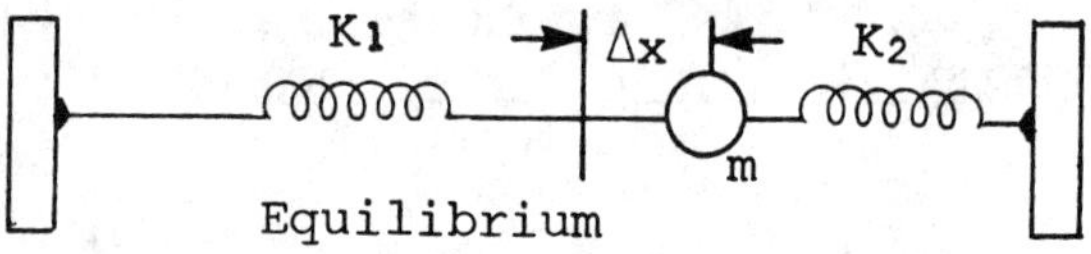

Fig. 3

The net force on the mass is then $- (F_1 + k_1 \Delta x) + (F_2 - k_2 \Delta x)$ or $F_2 - F_1 - (k_1 + k_2)\Delta x$. This is also equal to the mass times its acceleration. Since $F_2 = F_1$,

$$- (k_1 + k_2)\Delta x = m \ (\ddot{x})$$

and if equilibrium is defined as $x = 0$, then

$$m\ddot{x} + (k_1 + k_2)x = 0$$

which is the familiar equation of simple harmonic motion. Here however

$$k_{total} = k_1 + k_2 \rightarrow m\ddot{x} + (k_{tot})x = 0$$

since $\omega = \sqrt{\frac{k_{tot}}{m}}$, $\omega = \sqrt{\frac{(k_1 + k_2)}{m}}$

• PROBLEM 22-12

A vertical spring of length 2L and spring constant K is suspended at one end. A body of mass m is attached to the other end of the spring. The spring is compressed to half its length and then released. Determine the kinetic energy of the body, and its maximum value, in the ensuing motion.

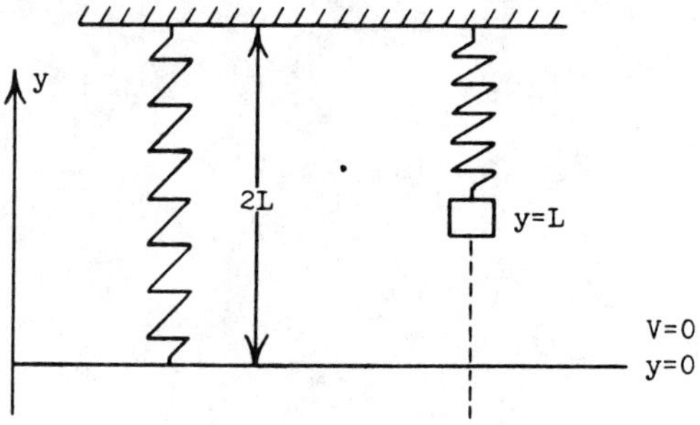

fig 1

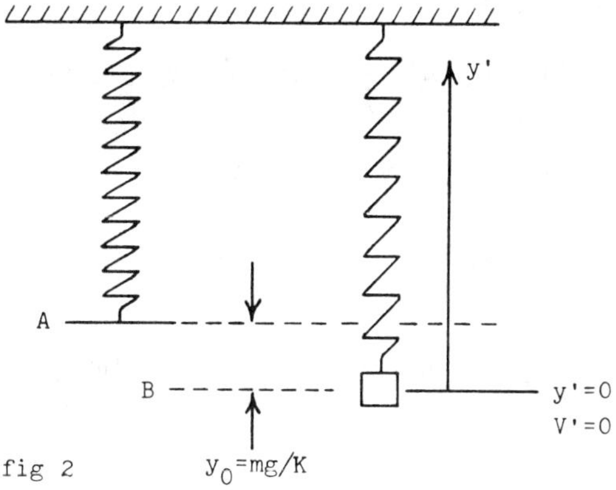

fig 2

Solution: This problem can be done using the relaxed position of the spring or the equilibrium position as a reference. We shall do both and then contrast the two methods.

The body is acted on by two conservative forces, the spring force and gravity. If the position of the body is measured from the relaxed position of the spring by the coordinate y (positive upward), and if the potential energy is set equal to zero at y = 0, see figure 1, the potential energies that correspond to the two forces are $Ky^2/2$ and mgy, respectively. The total potential energy is then

$$V = \tfrac{1}{2}Ky^2 + mgy. \tag{1}$$

The law of conservation of energy, KE + V = H = const, then gives the following expression for the calculation of the kinetic energy KE:

$$KE + \tfrac{1}{2}Ky^2 + mgy = \text{const} = H. \tag{2}$$

The constant value of H is determined from the fact that KE = 0 when y = L (the initial conditions). Therefore we get $H = KL^2/2 + mgL$, and from eq. (2) we find that the kinetic energy is then

$$KE = \frac{K}{2}(L^2 - y^2) + mg(L - y). \tag{3}$$

The kinetic energy will have a maximum and the potential energy a minimum value where dE/dy = 0 or dV/dy = 0, that is, using the latter condition on equation (1), at the point y_0 given by Ky_0

$$y_0 = -\frac{mg}{K}. \tag{4}$$

This is the static equilibrium point of the body. The maximum kinetic energy then is

$$KE_{max} = \frac{K}{2}(L^2 - y_0^2) + mg(L - y_0)$$

$$= \frac{K}{2}\left(L^2 - \frac{m^2g^2}{K^2}\right) + mg\left(L + \frac{mg}{K}\right)$$

$$= \frac{K}{2}\left[\left(L^2 - \frac{m^2g^2}{K^2}\right) + \frac{2\ mg}{K}\left(L + \frac{mg}{K}\right)\right]$$

$$= \frac{K}{2}\left[L^2 + \frac{2\ mg}{K} L + \frac{m^2g^2}{K^2}\right]$$

$$= \frac{K}{2}\left(L + \frac{mg}{K}\right)^2 \qquad (5)$$

and the corresponding minimum value of the potential energy is

$$V_{min} = -\frac{1}{2} Ky_0^2 = \frac{1}{2}\frac{m^2g^2}{K} .$$

We have found the potential energy formula (equation (1)) and the equilibrium point (y_0). We may create a new potential function $V' = V - V_1(y_0)$, that is

$$V' = V - V_{min}$$

$$= \frac{1}{2} Ky^2 + mgy + \frac{1}{2} Ky_0^2$$

$$= \frac{1}{2} Ky^2 - Ky_0y + \frac{1}{2} Ky_0^2$$

$$= \frac{1}{2} K (y - y_0)^2.$$

And defining $y' = y - y_0$, that is, using the equilibrium point as a reference, we have reduced the problem to a simple spring potential

$$V' = \frac{1}{2} Ky'^2.$$

Use the reference system shown in figure 2. The new initial condition is that the body is released from rest at

$$y' = y - y_0 = L + \frac{mg}{K} .$$

The total energy H is given by

$$H = KE + \frac{1}{2} Ky'^2$$

so that $H = 0 + \frac{1}{2} K \left(L + \frac{mg}{K}\right)^2$

using the appropriate initial condition. This accounts for the kinetic energy during the motion

$$KE = \frac{1}{2} K \left[\left(L + \frac{mg}{K}\right)^2 - y'^2\right] . \qquad (6)$$

For a simple spring potential, the maximum kinetic energy occurs when y' = 0. At that point, the kinetic energy is equal to the total energy;

$$KE_{max} = H = \frac{1}{2} K \left(L + \frac{mg}{K}\right)^2 \quad . \tag{7}$$

We see by comparing equations (e) and (g) that both methods give the same result for KE_{max}. It can be easily checked that the substitution of y' in equation (6) will result in equation (3).

In fact, both approaches to this problem give correct results. Once the equilibrium is determined, the transformation of coordinates is recommended for studies of motion of a body under the influence of a spring force and of a constant force (such as gravity).

• PROBLEM 22-13

A particle of mass m is subject to the influence of the potential energy,

$$V(x) = (1 - \alpha x)e^{-\alpha x}$$

in the region $x \geqq 0$, where α represents a positive constant. Determine the location of the equilibrium point(s), the nature of the equilibrium, and the natural frequency of oscillation for motion restricted to small displacements in the neighborhood of equilibrium.

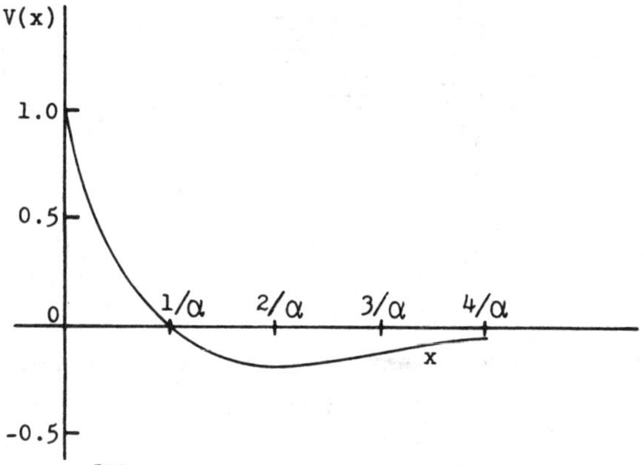

Solution: Since $F = -\frac{dV}{dx}$ the condition for equilibrium is $\frac{dV}{dx} = 0$. For the potential energy given,

$$\frac{dV}{dx} = e^{-\alpha x} (\alpha^2 x - 2\alpha)$$

and there is one equilibrium point at $x = 2/\alpha$. The second derivative evaluated at equilibrium provides a test for the nature of equilibrium.

If $\frac{dF}{dx} < 0$ then, as x is increased, F decreases; therefore, it must become negative and as x decreases, F increases, therefore, it must become positive. This means F points towards the equilibrium point for displacements in either direction. That is the nature at a stable equilibrium. $\frac{dF}{dx} = -\frac{d^2V}{dx^2} < 0$ or $\frac{d^2V}{dx^2} > 0$ is the condition for stable equilibrium.

$$\frac{d^2V}{dx^2} = e^{-\alpha x}(3\alpha^2 - \alpha^3 x).$$

At $x = 2/\alpha$, $d^2V/dx^2 = \alpha^2 e^{-2}$. Since $d^2V/dx^2 > 0$, equilibrium is stable.

Expansion of the force in a Taylor series gives:

$$F(x) = F(x_0) + \left.\frac{dF}{dx}\right|_{x=x_0}(x - x_0) + \left.\frac{d^2F}{dx^2}\right|_{n=n_0}\frac{(x - x_0)^2}{2!} + \ldots$$

The value x_0, expanded about, is the equilibrium point, so $F(x_0) = 0$. If we assume small displacements, then $(x - x_0)^2$ and higher terms may be ignored, so

$$F(x) = \left.\frac{dF}{dx}\right|_{x=x_0}(x - x_0)$$

which is a form of Hooke's law with $F' = -k$, the spring constant. Here $\frac{dF}{dx} = -\frac{d^2V}{dx^2}$ so

$$k = \left.\frac{d^2V}{dx^2}\right|_{x=x_0}.$$

The natural circular frequency is, therefore,

$$\omega_0 = \left(\frac{k}{m}\right)^{\frac{1}{2}} = \left(\frac{\alpha^2 e^{-2}}{m}\right)^{\frac{1}{2}} = \frac{\alpha}{e}(m)^{\frac{1}{2}}.$$

A plot of V(x) as a function of x, indicates that for large displacements about $x = 2/\alpha$, the motion is anharmonic.

• PROBLEM 22-14

A 16-lb. weight is attached to the end of a spring suspended from a ceiling. It comes to rest in its equilibrium position, thereby stretching the spring 1 foot. The weight is then pulled down 6 inches below its equilibrium position and released at t = 0 with an initial velocity of 1 ft/sec, directed downward. Neglecting the resistance of the medium and assuming that no external forces are present, determine the amplitude, period and frequency of the resulting motion.

Solution: The general differential equation which must be solved for this type of problem is of the form

$$m\frac{d^2x}{dt^2} + kx = 0 \qquad (1)$$

where m = mass, k = spring constant, x = x(t) is the displacement of the mass and is a function of t(the time elapsed).

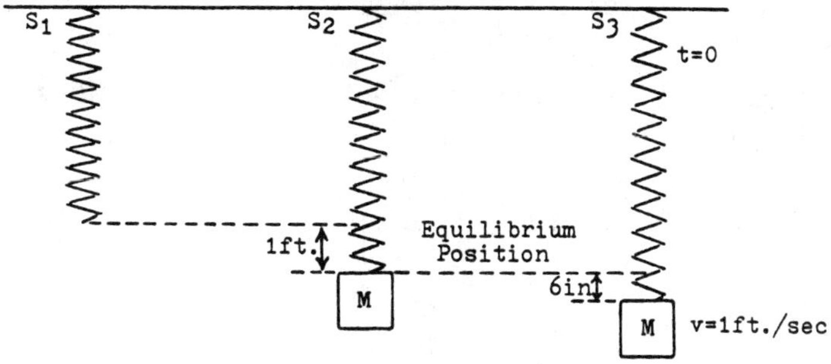

S_1: Spring, with no mass attatched to it, at rest

S_2: Downward displacement is taken to be positive, upward to be negative

S_3: Mass is displaced 6 inches (note; +6inches) and given a positive initial velocity of 1ft./sec.

To solve for m and k, use F = mg. Thus the mass = $\frac{F}{g} = \frac{16 \text{ lb}}{32 \text{ ft/sec}^2} = \frac{1}{2} \text{ lb/ft/sec}^2 = \frac{1}{2}$ (slugs) = m.

Hooke's law is given by F = - ks where s is the amount of elongation of the spring from its natural length. So here we have

$$16 \text{ lbs} = k \text{ (1 ft)} \qquad \text{or} \qquad k = 16 \text{ lbs/ft.}$$

So we have $\frac{1}{2} \frac{d\,x}{dt} + 16x = 0$ or $\frac{d^2x}{dt^2} + 32x = 0$. (2)

We also have initial conditions, namely: $x(0) = \frac{1}{2}$ ft

$$\left.\frac{dx}{dt}\right|_0 = 1 \text{ ft/sec.}$$

The auxiliary equation corresponding to equation (1) is: $m^2 + 32 = 0$. Using the quadratic formula to solve, we get

$$\frac{-0 \pm \sqrt{0^2 - 4(1)(32)}}{2} = \frac{\pm i\sqrt{128}}{2} = \pm 4\sqrt{2}\, i = m$$

so that the solution is of the form

$$x(t) = C_1 \sin 4\sqrt{2}t + C_2 \cos 4\sqrt{2}t. \qquad (3)$$

Now we can solve for C_1 and C_2 by applying the initial conditions, namely

$$x(0) = \frac{1}{2} = C_1 \sin 0 + C_2 \cos 0 = C_2$$

and $\frac{dx}{dt} = 4\sqrt{2}\, C_1 \cos 4\sqrt{2}t - 4\sqrt{2}\, C_2 \sin 4\sqrt{2}t$

$$\left.\frac{dx}{dt}\right|_0 = 1 = 4\sqrt{2}\, C_1 \text{ or } C_1 = \frac{1}{4\sqrt{2}} = \frac{\sqrt{2}}{8}.$$

Thus the solution of the differential equation, (2) is

$$x(t) = \frac{\sqrt{2}}{8} \sin 4\sqrt{2}t + \frac{1}{2} \cos 4\sqrt{2}t \ . \tag{4}$$

Now construct the right triangle ABC where $AB = \frac{1}{2}$, $BC = -\frac{\sqrt{2}}{8}$.

$$\overline{AC} = \sqrt{\left(\frac{1}{2}\right)^2 + \left(-\frac{\sqrt{2}}{8}\right)^2}$$

$$\overline{AC} = \frac{3}{8}\sqrt{2} \ .$$

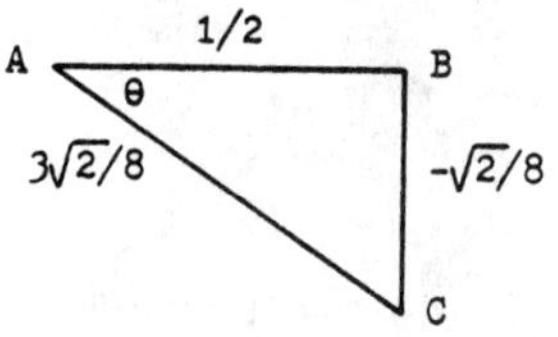

Note $$\sin\theta = \frac{\text{Opp.}}{\text{Hyp}} = \frac{-\frac{\sqrt{2}}{8}}{\frac{3}{8}\sqrt{2}} = -\frac{1}{3}$$

$$\cos\theta = \frac{\text{Adj}}{\text{Hyp}} = \frac{\frac{1}{2}}{\frac{3}{8}\sqrt{2}} = \frac{4}{3\sqrt{2}} = \frac{2\sqrt{2}}{3} \ . \tag{5}$$

Rewriting equation (4) (note we are just multiplying by 1)

$$x(t) = \frac{3\sqrt{2}}{8}\left[\frac{1}{3}\sin 4\sqrt{2}t + \frac{2}{3}\sqrt{2}\cos 4\sqrt{2}t\right]$$

and noting $\cos(\alpha + \beta) = \cos\alpha\cos\beta - \sin\alpha\sin\beta$

we get $$x(t) = \frac{3\sqrt{2}}{8}(-\sin\theta\sin 4\sqrt{2}t + \cos\theta\cos 4\sqrt{2}t)$$

$$= \frac{3\sqrt{2}}{8}\cos(4\sqrt{2}t + \theta) \ . \tag{6}$$

Here θ is determined by (5); namely

$$\tan\theta = \frac{-1}{2\sqrt{2}} \quad \text{or} \quad \theta = \tan^{-1}\left(\frac{-1}{2\sqrt{2}}\right) \approx -.35 \text{ radians} \ .$$

Thus equation (6) can be rewritten

$$x(t) = \frac{3\sqrt{2}}{8}\cos(4\sqrt{2}t - .35) \ .$$

In general the solution to equation 1 is

$$x = C\cos\left(\sqrt{\frac{k}{m}}\,t + \theta\right) \ . \tag{7}$$

The constant C is defined to be the amplitude of the motion and it gives the maximum positive displacement of the mass from its equilibrium position. The motion is periodic, oscillating from $x = C$ to $x = -C$. If $x = C$ then by equation (7)

$$\sqrt{\frac{k}{m}}\, t + \theta = \pm\, 2n\pi \qquad \text{where } n = 0, 1, 2, \ldots \quad t > 0 .$$

Transposing we get $t = \sqrt{\frac{m}{k}}\,(\pm\, 2n\pi - \theta)$. (8)

Now the time interval between two successive maxima is called the period of the motion. Thus, if

$$t_1 = \sqrt{\frac{m}{k}}\,(\pm\, 2n\pi - \theta) \quad \text{and } t_2 = \sqrt{\frac{m}{k}}\,(\pm\, 2(n+1) - \theta)$$

$$t_2 - t_1 = \frac{2\pi}{\sqrt{\frac{k}{m}}} = \text{Period} .$$

The reciprocal of the period which gives the number of oscillations per second is called frequency of motion. Now looking at our problem, namely

$$x(t) = \frac{3\sqrt{2}}{8} \cos\,(4\sqrt{2}t - .35)$$

we see that the amplitude is $\frac{3\sqrt{2}}{8}$, the period is

$$\frac{2\pi}{\sqrt{\frac{k}{m}}} = \frac{2\pi}{\sqrt{\frac{16}{1/2}}} = \frac{\pi}{2\sqrt{2}} = \frac{\sqrt{2}\,\pi}{4}$$

and the frequency is $\frac{4}{\sqrt{2}\pi} = \frac{2\sqrt{2}}{\pi}$.

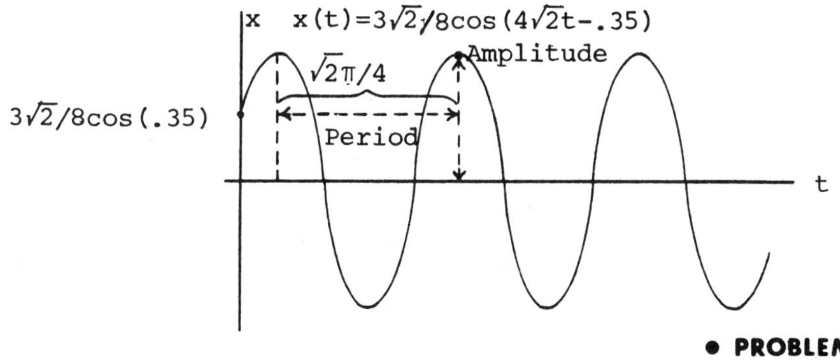

• PROBLEM 22-15

Consider a pendulum which is oscillating with an amplitude so large that we may not neglect the θ^3 term in the expansion of sin θ, as we do for a harmonic oscillator. What is the effect on the motion of the pendulum of the term in θ^3? This is an elementary example of an anharmonic oscillator. Anharmonic or nonlinear problems are usually difficult to solve exactly (except by computers), but approximate solutions are often adequate to give us a good idea of what is happening.

Solution: The expansion of sin θ (for small θ) to terms of order θ^3, commonly expressed as "expansion to $O(\theta^3)$," is

$$\sin\theta = \theta - \frac{1}{6}\theta^3 + \ldots ,$$

so that the equation of motion of a simple oscillator becomes, to this order,

$$\frac{d^2\theta}{dt^2} + \omega_0{}^2\theta - \frac{\omega_0{}^2}{6}\theta^3 = 0, \tag{1}$$

where $\omega_0{}^2$ denotes the quantity g/L. This is the equation of motion of an anharmonic oscillator.

We see if we can find an approximate solution to (1) of the form

$$\theta = \theta_0 \sin\omega t + \varepsilon\theta_0 \sin 3\omega t. \tag{2}$$

It is now evident that (2) can only be an approximate solution at best. It remains for us to determine ε, and also ω; while ω must reduce to ω_0 at small amplitudes, it may differ at large amplitudes. For simplicity we suppose that θ = 0 at t = 0.

An approximate solution of this type to a differential equation is called a perturbation solution, because one term in the differential equation perturbs the motion which would occur without that term. As you have seen, we arrived at the form of (2) by guided guesswork. It is easy enough to try several guesses and to reject the ones which do not work.

We have from (2)

$$\ddot{\theta} = -\omega^2\theta_0 \sin\omega t - 9\omega^2\varepsilon\theta_0 \sin 3\omega t;$$

$$\theta^3 = \theta_0{}^3 (\sin^3\omega t + 3\varepsilon \sin^2\omega t \sin 3\omega t + \ldots),$$

where we discarded the terms of order ε^2 and ε^3 because of our assumption that we can find a solution with $\varepsilon \ll 1$. Then the terms of (1) become

$$\omega_0{}^2\theta = \frac{\omega_0{}^2}{6}\theta^3 - \ddot{\theta}$$

$$\omega_0{}^2\,\theta_0 \sin\omega_0 t + \varepsilon\,\omega_0{}^2\theta_0 \sin 3\omega t \simeq \frac{\omega_0{}^2}{6}\theta_0{}^3 \times$$

$$(\sin^3\omega t + 3\varepsilon \sin^2\omega t \sin 3\omega t) +$$

$$\omega^2\theta_0 \sin\omega t + 9\,\omega^2\,\varepsilon\,\theta_0 \sin 3\omega t.$$

The coefficient of the term $\sin^2\omega t \sin 3\omega t$ in the previous equation is small by $O(\varepsilon)$ or by $O(\theta_0{}^2)$, compared with the other terms in the equation. Since ε and θ_0 are small quantities, we neglect this term.

Using the trigonometric identity

$$\sin^3 \omega t = \frac{3 \sin \omega t - \sin 3\omega t}{4} \qquad \text{we get}$$

$$\omega_0{}^2\theta_0 \sin \omega t + \varepsilon\, \omega_0{}^2\theta_0 \sin 3\omega t \simeq \left(\frac{1}{8}\omega_0{}^2\theta_0 + \omega^2\theta_0\right) \times$$

$$\sin \omega t + \left(9\,\omega^2\varepsilon\theta_0 - \frac{1}{24}\omega_0{}^2\theta_0^3\right) \sin 3\omega t. \qquad (3)$$

In order for this equation to hold, the corresponding coefficients of sin ωt and sin $3\omega t$ on each side of the equation must be equal:

$$\omega_0{}^2 = \frac{\omega_0{}^2\theta_0{}^2}{8} + \omega^2 \qquad (4)$$

$$\varepsilon\omega_0{}^2 = -\frac{1}{24}\omega_0{}^2\theta_0{}^2 + 9\varepsilon\omega^2 \qquad (5)$$

From (4), $\omega^2 \simeq \omega_0{}^2 \left(1 - \frac{1}{8}\theta_0{}^2\right)$

or $\omega \simeq \omega_0 \left(1 - \frac{1}{8}\theta_0{}^2\right)^{\frac{1}{2}} \simeq \omega_0 \left(1 - \frac{1}{16}\theta_0{}^2\right)$

using the binomial equation for the square root. We see that as $\theta_0 \to 0$, $\omega \to \omega_0$. The frequency shift $\Delta\omega$ is

$$\Delta\omega = \omega - \omega_0 = \frac{\omega_0}{16}\theta_0{}^2.$$

In (5), we make the substitution $\omega^2 \simeq \omega_0{}^2$ and obtain an expression for ε.

$$\varepsilon\,\omega_0{}^2 \simeq -\frac{1}{24}\omega_0{}^2\theta_0{}^2 + 9\varepsilon\omega_0{}^2$$

$$\varepsilon \simeq \frac{\theta_0{}^2}{8 \times 24} = \frac{1}{16}\theta_0{}^2.$$

We think of ε as giving the fractional admixture of the sin $3\omega t$ term in a solution for θ dominated by the sin ωt term.

Why did we not include in (2) a term in sin $2\omega t$? Try for yourself a solution of the form

$$\theta = \theta_0 \sin \omega t + \eta\theta_0 \sin 2\omega t,$$

and see what happens. You will find $\eta = 0$. The pendulum generates chiefly third harmonics, i.e., terms in sin $3\omega t$, and not second harmonics. The situation would be different for a device for which the equation of motion included a term in θ^2.

What is the frequency of the pendulum at large amplitudes? There is no single frequency in the motion. We have seen that the most important term (the largest component) is in sin ωt, and we say that ω is the fundamental frequency of the pendulum. To our approximation ω is given by (4). The term in sin $3\omega t$ is called the third harmonic of the fundamental frequency. Our argument following (2) suggests that an infinite number of harmonics are present in the exact motion, but that most of these are very small. The amplitude in (2) of the fundamental component of the motion is θ_0; the amplitude of the third harmonic component is $\varepsilon\theta_0$.

• PROBLEM 22-16

Determine the root-mean-square (rms) values of displacement, velocity, and acceleration for a damped forced harmonic oscillator operating at steady state.

Solution: The general expression for the displacement is

$$x = x_{max} \cos(\omega t + \theta), \tag{a}$$

for the velocity

$$v = \dot{x} = -\omega x_{max} \sin(\omega t + \theta), \tag{b}$$

and for the amplitude

$$a = \dot{v} = -\omega^2 x_{max} \cos(\omega t + \theta). \tag{c}$$

ω is the circular frequency, and the period $T = 2\pi/\omega$.

The defining expression for rms values is

$$g_{rms} = \left[\frac{\int_0^T g^2\, dt}{\int_0^T dt} \right]^{\frac{1}{2}} \tag{d}$$

where g is an arbitrary periodic function of time, with period T.

Applying equation (d) to function (a) yields

$$x_{rms} = \left[\frac{\int_0^{2\pi/\omega} [x_{max} \cos(\omega t + \theta)]^2\, dt}{\int_0^{2\pi/\omega} dt} \right]^{\frac{1}{2}}$$

$$= \frac{x_{max}}{(2\pi/\omega)^{\frac{1}{2}}} \left[\int_0^{2\pi/\omega} \cos^2(\omega t + \theta)\, dt \right]^{\frac{1}{2}}.$$

Shown here are some general trigonometric identities for reference:

$$\cos^2\alpha = \frac{\alpha}{2} + \frac{1}{2}\cos 2\alpha \tag{e}$$

$$\sin^2 \alpha = \frac{\alpha}{2} - \frac{1}{2} \cos 2\alpha . \qquad \text{(f)}$$

Applying (a) to our expression for x_{rms},

$$x_{rms} = \frac{x_{max}}{(2\pi/\omega)^{\frac{1}{2}}} \left[\int_0^{2\pi/\omega} [\tfrac{1}{2} + \tfrac{1}{2} \cos^2 (\omega t + \theta)] \, dt \right]^{\frac{1}{2}}$$

$$= \frac{x_{max}}{\sqrt{2}} .$$

Following this procedure for function (b) yields

$$v_{rms} = \left[\frac{\int_0^{2\pi/\omega} [- \omega x_{max} \sin (\omega t + \theta)]^2 \, dt}{\int_0^{2\pi/\omega} dt} \right]^{\frac{1}{2}}$$

$$= \frac{\omega x_{max}}{(2\pi/\omega)^{\frac{1}{2}}} \left[\int_0^{2\pi/\omega} \sin^2 (\omega t + \theta) dt \right]^{\frac{1}{2}}$$

Using identity (f)

$$= \frac{\omega x_{max}}{(2\pi/\omega)^{\frac{1}{2}}} \left[\int_0^{2\pi/\omega} [\tfrac{1}{2} - \tfrac{1}{2} \cos^2 (\omega t + \theta)] \, dt \right]$$

$$= \frac{\omega x_{max}}{\sqrt{2}} .$$

Finally, using function (c),

$$a_{rms} = \frac{\int_0^{2\pi/\omega} [- \omega^2 x_{max} \cos (\omega t + \theta)]^2 \, dt}{\int_0^{2\pi/\omega} dt}$$

we can write down the answer immediately as follows:

$$a_{rms} = \frac{\omega^2 x_{max}}{\sqrt{2}} .$$

• PROBLEM 22-17

Solve for the period and the tension of a simple pendulum with and without the small angle approximation.

Solution: A simple pendulum consists of a small mass suspended by a light inextensible cord from a fixed support. We shall assume that there are no damping forces so that theoretically, once the pendulum is oscillating it continues indefinitely with the same amplitude. Let O in Fig. 1 be the fixed point of support, OP the length of the pendulum, and m the mass of the particle at P. The point O is taken as the origin of the radial vector $\vec{r}$ which at time t is along OP and makes an angle θ with the vertical through O. We shall now set up the equations of motion of the particle m along and perpendicular to the radius vector OP.
In polar coordinates the position of a particle

$\vec{r} = r\hat{r}$. Differentiating

$$\dot{\vec{r}} = \vec{v} = \dot{r}\hat{r} + r\dot{\hat{r}}$$

$$\vec{v} = \dot{r}\hat{r} + r\dot{\theta}\hat{\theta} \qquad \text{(a)}$$

and $$\dot{\vec{v}} = \vec{a} = \ddot{r}\hat{r} + \dot{r}\dot{\hat{r}} + \dot{r}\dot{\theta}\hat{\theta} + r\ddot{\theta}\hat{\theta} + r\dot{\theta}\dot{\hat{\theta}}$$

$$\vec{a} = (\ddot{r} - r\dot{\theta}^2)\hat{r} + (r\ddot{\theta} + 2\dot{r}\dot{\theta})\hat{\theta}. \qquad \text{(b)}$$

This gives, for the force along the radial vector, F_r

$$F_r = m(\ddot{r} - r\dot{\theta}^2)$$

and this is equal to the components of the applied forces acting on mass m in the direction OP. Thus

$$m(\ddot{r} - r\dot{\theta}^2) = mg \cos \theta - T$$

where T is the tension in the cord at the angle θ. Similarly, for the transverse force F_θ in the direction of increasing θ, we have from Eq. (b)

$$F_\theta = m(2\dot{r}\dot{\theta} + r\ddot{\theta}) = - mg \sin \theta.$$

Since the string is assumed to be inextensible, the vector $\vec{r}$ has a constant magnitude equal to l, and it follows that

$$\dot{r} = 0 \qquad \text{and} \qquad \ddot{r} = 0.$$

The equations of motion become

$$- ml\dot{\theta}^2 = mg \cos \theta - T \qquad \text{(c)}$$

$$ml\ddot{\theta} = - mg \sin \theta. \qquad \text{(d)}$$

To make Eq. (d) linear, let us assume that θ is so small that sin θ may be set equal to θ in radian measure. This really assumes that θ^3 and higher powers of θ are negligibly small compared to θ, since the expansion of sin θ in radian measure is

$$\sin \theta = \theta - \frac{\theta^3}{3!} + \frac{\theta^5}{5!} \cdots .$$

Thus for θ = 3°, sin 3° = 0.05234 and 3° = 0.05236 radian.

If therefore the maximum angular amplitude of the pendulum is a few degrees, we may write Eq. (d) approximately as

$$\ddot{\theta} = -\frac{g}{l}\theta . \tag{e}$$

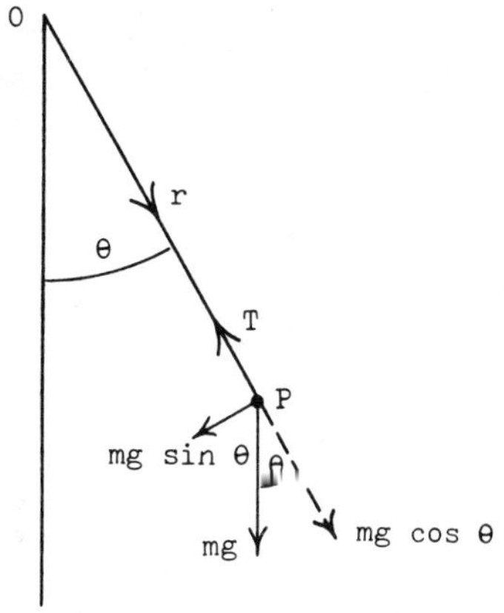

This is the general form of a differential equation for simple harmonic motion whose period is:

$$P = 2\pi\sqrt{\frac{l}{g}} . \tag{f}$$

The tension T exerted by the cord on the particle depends on the angular displacement as given by Eq. (c), namely

$$T = mg \cos\theta + ml\dot{\theta}^2$$

where $ml\dot{\theta}^2$ is the centrifugal or fictitious force.

In the small angle approximation $\cos\theta \simeq 1$ so

$$T = mg + ml\dot{\theta}^2 .$$

We shall now drop the assumption of a small angular amplitude and investigate the period for relatively large amplitudes. To do this, we must integrate Eq. (d) or

$$\ddot{\theta} = -\frac{g}{l}\sin\theta$$

A useful identity, which we shall use here, is

$$\dot{\theta}dt = \frac{d\theta}{dt}dt = d\theta .$$

Multiplying both sides of equation (d) with this relationship yields:

$$\ddot{\theta}\dot{\theta}dt = -\frac{g}{l}\sin\theta\, d\theta$$

or $\quad \frac{1}{2}d(\dot{\theta}^2) = \frac{g}{l}d(\cos\theta)$

Integrating and using the initial conditions that the pendulum has a maximum angular amplitude of α or $\dot{\theta} = 0$ when $\theta = \alpha$

$$\dot{\theta}^2 = 2\,\frac{g}{l}\,(\cos\theta - \cos\alpha) \qquad (g)$$

Thus $$\frac{d\theta}{\sqrt{\cos\theta - \cos\alpha}} = \sqrt{\frac{2g}{l}}\,dt.$$

The time taken by the pendulum to swing from $+\alpha$ to $-\alpha$ is half of the period of oscillation P_α for the angular amplitude α or

$$\int_{-\alpha}^{+\alpha} \frac{d\theta}{\sqrt{\cos\theta - \cos\alpha}} = \sqrt{\frac{2g}{l}} \int_0^{(P\alpha)/2} dt = \sqrt{\frac{2g}{l}}\,\frac{P_\alpha}{2} \; .$$

Using the transformation

$$\cos\theta = 1 - 2\sin^2\frac{\theta}{2}$$

the integral becomes

$$\int_{-\alpha}^{+\alpha} \frac{d\theta}{\sqrt{\sin^2\alpha/2 - \sin^2\theta/2}} = \sqrt{\frac{g}{l}}\,P_\alpha \; . \qquad (h)$$

Change variables using the substitution

$$\sin\frac{\theta}{2} = \sin\frac{\alpha}{2}\sin\phi. \qquad (g)$$

The limits on the integral $+\alpha$, $-\alpha$ apply to $\pi/2$, $-\pi/2$. Differentiating equation (g), we have:

$$\frac{1}{2}\cos\frac{\theta}{2}\,d\theta = \sin\frac{\alpha}{2}\cos\phi\,d\phi$$

or $$d\theta = \frac{2\sin(\alpha/2)\cos\phi\,d\phi}{\sqrt{1 - \sin^2(\alpha/2)\sin^2\phi}} \; .$$

Substituting in Eq. (h) for the period, it follows that

$$\sqrt{\frac{g}{l}}\,P_\alpha = \int_{-\pi/2}^{+\pi/2} \frac{2d\phi}{\sqrt{1 - \sin^2(\alpha/2)\sin^2\phi}}$$

$$= \int_{-\pi/2}^{+\pi/2} 2\,d\phi \left(1 - \sin^2\frac{\alpha}{2}\sin^2\phi\right)^{-\frac{1}{2}} .$$

Using the binomial expansion,

$$(1-x)^{-n} = 1+nx+ \frac{n\ (n+1)}{2!}\ x^2+ \frac{n\ (n+1)(n+2)}{3!}\ x^3+ \ .....\ ;$$

$$= \int_{-\pi/2}^{+\pi/2} 2\ d\phi \left(1 + \frac{1}{2} \sin^2 \frac{\alpha}{2} \sin^2 \phi + \frac{3}{8} \sin^4 \frac{\alpha}{2} \sin^4 \phi ...\right).$$

Using, for example

$$\int_{-\pi/2}^{\pi/2} \sin^2 \phi\ d\phi = \int_{-\pi/2}^{\pi/2} \frac{(1 - \cos 2\phi)d\phi}{2} = \frac{\pi}{2}\ ;$$

$$= 2 \left[\pi + \frac{1}{2} \sin^2 \left(\frac{\alpha}{2}\right) \frac{\pi}{2} + \frac{3}{8} \sin^4 \left(\frac{\alpha}{2}\right) \frac{3\pi}{8} ... \right] .$$

Hence, $P_\alpha = 2\pi \sqrt{\frac{l}{g}} \left(1 + \frac{1}{4} \sin^2 \frac{\alpha}{2} + \frac{9}{64} \sin^4 \frac{\alpha}{2} ...\right)$.

If the period for an angular amplitude of approximately zero is called P_0, where

$$P_0 = 2\pi \sqrt{\frac{l}{g}}$$

then the period P_α for an angular amplitude α is

$$P_\alpha = P_0 \left(1 + \frac{1}{4} \sin^2 \frac{\alpha}{2} + \frac{9}{64} \sin^4 \frac{\alpha}{2} ... \right) \quad . \qquad \text{(i)}$$

From the three terms in the expansion given in Eq. (i) we obtain for an angular amplitude of 90° a period of

$$P_{90°} = P_0 \left(1 + \frac{1}{8} + \frac{9}{256}\right)$$

$$= 1.16\ P_0 \quad .$$

Even for this amplitude another term in the expansion should be added ir real accuracy is desired. However, in most experiments, amplitudes as large as 90° are not encountered. For an amplitude of 20° the contribution of the $\sin^2 \alpha/2$ term is 0.0075 and of the $\sin^4\alpha/2$ term is 0.00013. If the amplitude is smaller than 20°, then the $\sin^4\alpha/2$ term may be neglected without making an appreciable error. A further simplification can be made by replacing $\sin^2\alpha/2$ by $\alpha^2/4$. Thus for α = 20° the value of $\sin^2 10°$ is approximately 0.0308 and $\alpha^2/4$ is approximately 0.0305 in radian measure. With these limitations, we may write the period

$$P_\alpha = P_0\left(1 + \frac{\alpha^2}{16}\right) .$$

If the angular amplitude is 4° or 0.0698 radian, then

$$P_{4°} = P_0\ (1 + 0.0003) = 1.0003\ P_0$$

The periods differ by three parts in 10,000. The tension in the cord can now be evaluated for any angular displacement. From Eq. (c) the tension in the cord is

$$T = mg \cos \theta + ml\dot{\theta}^2$$

and substituting for $\dot{\theta}^2$ from Eq. (g), we have

$$T = mg (3 \cos \theta - 2 \cos \alpha).$$

• PROBLEM 22-18

Consider free, damped motion of a mass m attached to a spring of force constant k. Let the coefficient of damping be a. Then show that the resulting motion of the system is completely determined by the quantity $a^2 - 4km$.

Solution: The differential equation of the system is derived from Newton's Second Law:

$$m \frac{d^2x}{dt^2} = - a \frac{dx}{dt} - kx$$

or $$m \frac{d^2x}{dt^2} + a \frac{dx}{dt} + kx = 0. \qquad (a)$$

The characteristic equation is obtained by assuming a solution of the form $x = e^{\lambda t}$. Substituting this in (a) and dividing by $e^{\lambda t}$ we obtain the characteristic equation:

$$m\lambda^2 + a\lambda + k = 0 .$$

By the quadratic formula

$$\lambda = \frac{- a \pm \sqrt{a^2 - 4km}}{2m} . \qquad (b)$$

We call the quantity which determines the resulting motion by

$$\Delta = a^2 - 4km.$$

Then $$\lambda_1 = \frac{- a + \sqrt{\Delta}}{2m}, \qquad \lambda_2 = \frac{- a - \sqrt{\Delta}}{2m} . \qquad (c)$$

Case I: $\Delta > 0$ or $a^2 > 4km$

In this case λ_1 and λ_2 are real numbers. Our solution is

$$x(t) = c_1e^{\lambda_1 t} + c_2e^{\lambda_2 t} \qquad (d)$$

where c_1 and c_2 are arbitrary constants. We note that λ_1 and λ_2 are negative. To show this, assume

$$\lambda_1 > 0 ;$$

then $\frac{-a + \sqrt{a^2 - 4km}}{2m} > 0,$

$$-a + \sqrt{a^2 - 4km} > 0$$

$$\sqrt{a^2 - 4km} > a > 0 .$$

Squaring, $a^2 - 4km > a^2$

$$-4km > 0 .$$

This is a contradiction since km is positive and $-4km$ must therefore be negative. Consequently, our assumption that $\lambda_1 > 0$, is false, and $\lambda_I < 0$. We may similarly show that $\lambda_2 < 0$. Thus, in solution (d)

$$\lim_{t\to\infty} c_1 e^{\lambda_1 t} + c_2 e^{\lambda_2 t} \to 0.$$

This is called overdamped motion.

Case II: $\Delta = 0, \quad a^2 = 4km$

In this case λ_1 and λ_2 are real and equal. Our general solution is

$$x(t) = c_1 e^{\lambda_1 t} + c_2 t\, e^{\lambda_2 t} ,$$

$$x(t) = e^{\lambda_1 t}(c_1 + c_2 t) \qquad \text{(e)}$$

where c_1 and c_2 are arbitrary constants.

Since $\lambda_1 = \frac{-a}{2m} < 0,$

$$\lim_{t\to\infty} e^{\lambda_1 t}(c_1 + c_2 t) = 0$$

and the system is said to be critically damped.

Case III: $\Delta < 0, \quad a^2 < 4km$

In this case the roots λ_1 and λ_2 are complex:

$$\lambda_1 = \frac{-a + i\sqrt{-\Delta}}{2m}, \qquad \lambda_2 = \frac{-a - i\sqrt{-\Delta}}{2m}$$

where i is the complex number such that $i^2 = -1$.

The solution is then

$$x(t) = c_1 e^{\lambda_1 t} + c_2 e^{\lambda_2 t} , \qquad \text{or}$$

$$x(t) = e^{\frac{-a}{2m}}\left(c_1 e^{i\frac{\sqrt{-\Delta}}{2m}} + c_2 e^{-i\frac{\sqrt{-\Delta}}{2m}}\right) , \qquad \text{(f)}$$

where c_1 and c_2 are arbitrary constants. By Euler's formula

$$e^{i\theta} = \cos\theta + i\sin\theta,$$

or $$e^{-i\theta} = \cos\theta - i\sin\theta\,.$$

If we let $\theta = \frac{\sqrt{-\Delta}}{2m}t$, then (f) becomes

$$x(t) = e^{-\frac{a}{2m}}\left[(c_1 + c_2)\cos\theta + i\,(c_1 - c_2)\sin\theta\right]. \quad (g)$$

If c_1 and c_2 are complex numbers they must be complex conjugates of each other. To see this, let $c_1 = \alpha + i\beta$. Then c_1 contains two arbitrary constants, α and β. Two arbitrary real constants are all that are required in the solution of a second order differential equation. Thus we may let $c_2 = \alpha - i\beta$ and still have two arbitrary constants. Hence

$$c_1 + c_2 = 2\alpha$$

and $$c_1 - c_2 = 2i\beta\,.$$

Substituting into (g)

$$x(t) = e^{-\frac{a}{2m}}\,(2\alpha\cos\theta - 2\beta\cos\theta)\cdot$$

Let $2\alpha = A$ and $-2\beta = B$ to obtain our final form of the solution

$$x(t) = e^{-\frac{a}{2m}}\left(A\cos\left(\frac{\sqrt{-\Delta}}{2m}\right)t + B\cos\left(\frac{\sqrt{-\Delta}}{2m}\right)t\right).$$

This is damped oscillatory motion. We summarize in a table.

If $a^2 - 4mk$ is	The Motion is
POSITIVE	OVER damped
ZERO	CRITICALLY damped
NEGATIVE	OSCILLATORY damped

• PROBLEM 22-19

An oscillatory system with viscous friction can have its motion divided into three classes; overdamped, critically damped, and underdamped (damped oscillatory) motion. Find the condition which determines the type of motion for a given system and describe the various motions.

Solution: The general equation for damped motion is

$$m\ddot{y} + b\dot{y} + ky = 0.$$

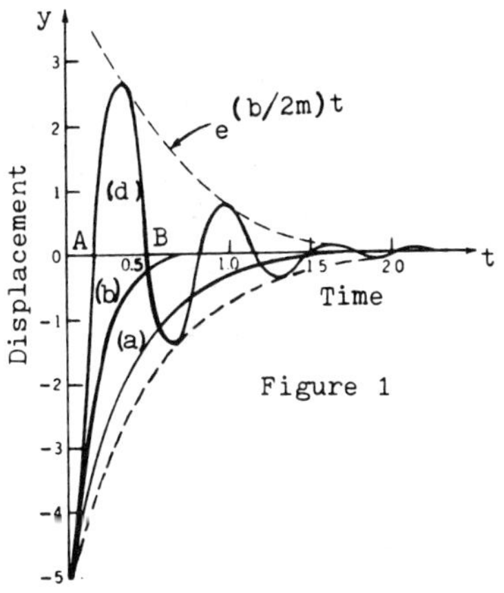

Graph of displacement of mass m on the end of the spring plotted against time. (a) Overdamped motion. (b) Critically damped motion. (c) Damped oscillatory motion.

If we solve this equation with the substitution $y = e^{st}$, for the characteristic equation we have

$$ms^2 + bs + k = 0$$

so that

$$s = \frac{-b \pm \sqrt{b^2 - 4km}}{2m} .$$

This can be rewritten as

$$s = -\frac{b}{2m} \pm \sqrt{\left(\frac{b}{2m}\right)^2 - \left(\frac{k}{m}\right)} .$$

The type of solution obtained depends on the value of the discriminant

$$\left[\left(\frac{b}{2m}\right)^2 - \frac{k}{m}\right].$$

For oscillatory motion, the argument of the exponential must have an imaginary part, thus

$$\left(\frac{b}{2m}\right)^2 - \frac{k}{m} < 0$$

is the condition for underdamped or oscillatory motion. Entirely real arguments will lead to non-oscillatory damped behavior. When the roots are equal $(b/2m)^2 = k/m$ we get exponential behavior and this is referred to as critically damped. When $(b/2m)^2 > k/m$ there is overdamped motion.

Note that even in the case of oscillatory motion, we have the real part of the exponential $-b/2m$ so that the oscillations are damped by the amount $e^{-(b/2m)t}$ (unless, of course, $b = 0$!). All of these behaviors are displaced in Figure 1.

To continue, assume values for m, b and k leading to

the different types of motion and then solve for the equations of motion.

Overdamped Motion:

Let k = 2500 dynes/cm, b =1000 dyne-sec/cm, and m = 25 g. Also assume for initial conditions that y(t=0) = - 5 cm, $\dot{y}(0) = 0$. First we check the discriminant's value,

$$b^2 - 4km = \left(1{,}000\ \frac{\text{dyne - sec}}{\text{cm}}\right)^2 - 4\left(2500\ \frac{\text{dynes}}{\text{cm}}\right)(25\ \text{gm})$$

$$= 750{,}000\ \frac{\text{dyne}^2 - \text{sec}^2}{\text{cm}^2}\ .$$

$$s_1 = \frac{-\ b - \sqrt{b^2 - 4km}}{2m}$$

$$= \frac{-\ 1{,}000\ \frac{\text{dyne - sec}}{\text{cm}} - \sqrt{\frac{750{,}000\ \text{dyne}^2 - \text{sec}^2}{\text{cm}^2}}}{2\ (25\ \text{gm})}$$

$$s_1 = \frac{-\ 37.32}{\text{sec}}$$

$$s_2 = \frac{-\ 1{,}000\ \frac{\text{dyne - sec}}{\text{cm}} + \sqrt{\frac{750{,}000\ \text{dyne}^2 - \text{sec}^2}{\text{cm}^2}}}{2\ (25\ \text{gm})}$$

$$s_2 = \frac{-\ 2.7}{\text{sec}}$$

Therefore $Y = A\ e^{-37.3t} + B\ e^{-2.7t}$.

This is correct for overdamped motion. For the two constants A and B, apply the initial conditions;

$y(0) = -\ 5 = A + B$

$\dot{y}(0) = 0 = -\ 37.3A - 2.7B.$

Solving these equations we have

A = 0.390 cm, B = - 5.39 cm.

Thus the displacement at any time t is

$$y = +\ 0.390\ e^{-37.3t} - 5.39\ e^{-2.7t} \qquad \text{(a)}$$

This is curve (a) in figure 1.

Critically Damped Motion:

Keep the same m and k and solve for the viscous coefficient that will give critical damping (a realistic exercise for design engineers).

$$\left(\frac{b_c}{2m}\right)^2 = \frac{k}{m}$$

$$b_c^2 = 4\ km = 4\ (2500)(25)$$

$$b_c = 500\ \frac{dyne - sec}{cm}\ .$$

The roots $s_1 = s_2 = -\ 500/2.25 = -\ 10\ sec^{-1}$. Thus the displacement at any time t is

$$y = (A + Bt)e^{-10t}\ .$$

Using the same initial conditions,

$-\ 5\ cm = A$ and

$$\dot{y} = B\ e^{-10t} - 10\ (A + Bt)\ e^{-10t}$$

so $0 = B - 10A.$

So $A = -\ 5$ cm, $B = -\ 50$ cm, and the equation of motion is

$$y = -\ 5e^{-10t}\ (1 + 10t). \qquad (b)$$

This is curve (b) in figure 1. This curve (and solution) also give the shortest time in which the mass may reach its equilibrium position. This is very desirable in many practical situations to reduce vibrations effectively.

Underdamped Motion:

For any damping constant less than 500 dyne - sec/cm, oscillations will occur before the motion is damped out. Let b = 100 dyne - sec/cm to illustrate. The discriminant is

$$b^2 - 4\ km = \left(100\ \frac{dyne - sec}{cm}\right)^2 - 4\ \left(2500\ \frac{dyne}{cm}\ 25\ gm\right)$$

$$= -\ 240{,}000\ \frac{dynes^2 - gm^2}{cm^2}\ .$$

$$\frac{\sqrt{b^2 - 4km}}{2m} = i\ \left(\frac{\sqrt{240{,}000\ \frac{dyne^2 - gm^2}{cm^2}}}{2\ (25\ gm)}\right)$$

$$= \frac{96}{sec}\ i$$

and $\frac{b}{2m} = 2\ sec^{-1}$. The roots are

$$s_1 = (-\ 2 + i\ \sqrt{96})\ sec^{-1}$$

$$s_2 = (-\ 2 - i\ \sqrt{96})\ sec^{-1}.$$

The solution, then, is

$$y = e^{-2t} \left[A\ e^{i\sqrt{96}\ t} + B\ e^{-i\sqrt{96}\ t}\right]. \qquad (c)$$

One can rationalize this formula, the result of which is, without proving it here,

$$y = k\ e^{-2t} \cos\ (\sqrt{96}\ t - \phi).$$

No proof is really needed, both complex exponentials have the same root ($\sqrt{96} = 9.8$) therefore, the complex term $\left[A\ e^{i\sqrt{96}t} + B\ e^{-i\sqrt{96}t}\right]$ in equation (c) will equal some sum of sine and cosine terms. However, $k \cos\ (9.8\ t - \phi)$ also represents such a possibility, with two arbitrary constants k and ϕ. How you proceed from equation (c) is somewhat arbitrary.

Proceeding with the problem, apply the initial conditions to get

$$-5 = k \cos \phi.$$

Also,
$$\dot{y} = -2\ k\ e^{-2t} \cos\ (9.8\ t - \phi) - 9.8\ k\ e^{-2t} \sin\ (9.8\ t - \phi)$$

at $t = 0$, $\dot{y} = 0$

so, $0 = -2\ k \cos \phi + 9.8\ k \sin \phi.$

Thus $\tan \phi = \dfrac{2}{9.8} = 0.204$

and $\phi = 11.53°$ or $191.53°$.

The difference of 180° leads to absorbing a minus sign into the cosine function, so we will work with $\phi = 11.53°$ only.

The value of k is derived from

$$k = -\frac{5\ \text{cm}}{\cos \phi} = -5.10\ \text{cm}.$$

The displacement, then, is given by

$$y = -5.10\ e^{-2t} \cos\ (9.80\ t - 11.53°). \qquad (d)$$

This is curve (d) in figure 1. Note also the amplitude envelope $e^{-b/2m} = e^{-2t}$ illustrated.

Incidentally equation (d) has a mixed cosine argument and it should be changed 11.53° to radians. The conversion gives

$$11.53° \cdot \frac{2\pi\ \text{radians}}{360°} = 0.20\ \text{rad}.$$

So $y = -5.10\, e^{-2t} \cos(9.80\, t - 0.20)$.

The frequency of the system is not the natural frequency of the spring-mass system $\omega_0 = (k/m)^{\frac{1}{2}}$ instead

$$\omega = \left[\left(\frac{b}{2m}\right)^2 - \frac{k}{m} \right]^{\frac{1}{2}}.$$

Of course, as $b \to 0$, $\omega \to \omega_0$.

• PROBLEM 22-20

Explain the mathematics of analyzing oscillatory motion using complex exponentials and rotating phasors.

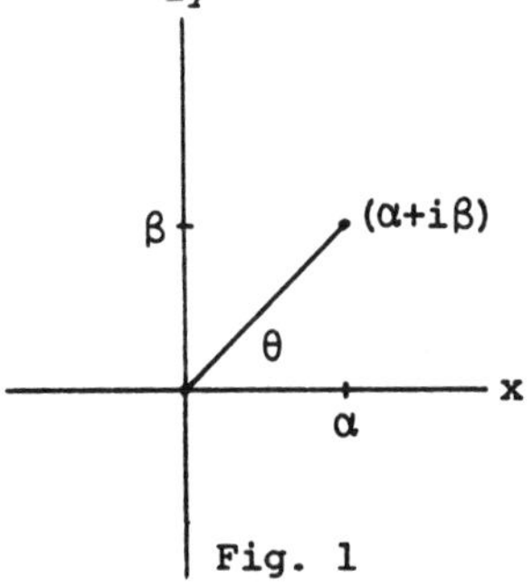

Fig. 1

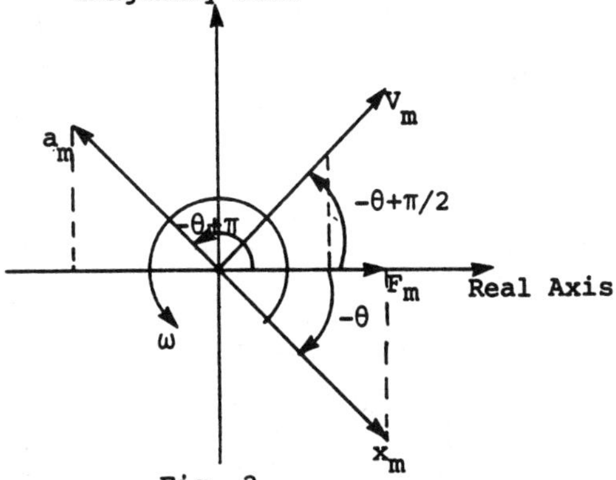

Fig. 2

Solution: We will work with most general equation possible

$$m\ddot{x} + b\dot{x} + kx = F_m \cos \omega t \qquad \text{(a)}$$

defining $\omega_0 = \sqrt{k/m}$.

Complex exponentials are often preferred because they are easier to differentiate and integrate. However, they do require care in use since they consist of real and imaginary terms. Therefore, in a simple problem like this, their advantage may not be obvious. However, an understanding of these techniques will enable the reader to perform more difficult problems and calculations with greater ease.

Using Euler's identity,

$$e^{i\omega t} = \cos \omega t + i \sin \omega t. \qquad \text{(b)}$$

We have

$$\cos \omega t = \text{real part } e^{i\omega t} . \qquad \text{(c)}$$

$$= R\left(e^{i\omega t}\right) .$$

This gives us for equation (a),

$$m\ddot{x} + b\dot{x} + kx = F_m R\left(e^{i\omega t}\right) .$$

One of the methods in using complex notation is to drop the explicit real part stipulation, treat the entire complex numbers throughout, and take the real parts of any quantity obtained to find the answer desired. Thus write (c) as

$$m\ddot{x} + b\dot{x} + kx = F_m e^{i\omega t} . \qquad (d)$$

Now we must solve this differential equation to get the motion of the general harmonic oscillator. We will start by assuming a solution of the form $x = x_m e^{st}$ and substituting into (d),

$$mx_m s^2 e^{st} + bx_m s e^{st} + kx_m e^{st} = F_m e^{i\omega t}$$

$$x_m e^{st} (ms^2 + bs + k) = F_m e^{i\omega t} . \qquad (e)$$

The time dependent exponentials must be equal, therefore $s = i\omega$. Then cancelling the exponentials in equation (e) yields

$$x_m (ms^2 + bs + k) = F_m$$

$$x_m = \frac{F_m}{ms^2 + bs + k} = \frac{F_m}{-m\omega^2 + i\omega b + k} .$$

Dividing the numerator and denominator of the right hand side by k and using $\omega_0^2 = k/m$ we get

$$x_m = \frac{F_m/k}{1 - (\omega/\omega_0)^2 + (i\omega b/k)} . \qquad (f)$$

The denominator here is complex, we can rationalize it by some manipulation.

A complex number is defined as coordinates on a complex plane (fig. 1).

The point shown is $(\alpha + i\beta)$ however, using the angle $\theta = \tan^{-1}(\beta/\alpha)$ shown we can redefine α and β. The new relations are

$$\alpha = (\alpha^2 + \beta^2)^{1/2} \cos\theta$$

$$\beta = (\alpha^2 + \beta^2)^{1/2} \sin\theta .$$

Using these

$$\alpha + i\beta = (\alpha^2 + \beta^2)^{1/2} (\cos\theta + i \sin\theta)$$

with Euler's identity yields

$$\alpha + i\beta = (\alpha^2 + \beta^2)^{1/2} e^{i\theta}$$

$$\theta = \tan^{-1} (\beta/\alpha). \qquad (g)$$

The identity (g) is completely general. Applying it to equation (f), let

$$\alpha = 1 - (\omega/\omega_0)^2$$

$$\beta = \frac{\omega b}{k}$$

and the denominator of (f) is

$$\left\{\left[\left[1 - \left(\frac{\omega}{\omega_0}\right)^2\right]^2 + \left(\frac{\omega b}{k}\right)^2\right]^{\frac{1}{2}} e^{i\theta}\right\}$$

with $\theta = \tan^{-1} \dfrac{\omega b/k}{[1 - (\omega/\omega_0)^2]}$.

Now for x_m,

$$x_m = \frac{F_m/k\ e^{-i\theta}}{\{[1 - (\omega/\omega_0)^2]^2 + (\omega b/k)^2\}^{\frac{1}{2}}} \qquad (g)$$

and for $x = x_m e^{i\omega t}$, the solution to the forced, damped harmonic oscillator equation (d) is

$$x = \frac{(F_m/k)\ e^{i(\omega t - \theta)}}{\{[1 - (\omega/\omega_0)^2]^2 + (\omega b/k)^2\}^{\frac{1}{2}}} \qquad (h)$$

With this result the real part is obtained as

$$x = \frac{(F_m/k) \cos (\omega t - \theta)}{\{[1 - (\omega/\omega_0)^2]^2 + (\omega b/k)^2\}^{\frac{1}{2}}} = x_m \cos (\omega t - \theta). (i)$$

This indicates that x follows the behavior of the driving force $F = F_m \cos \omega t$ but out of phase, lagging by the phase angle θ. These concepts will be illustrated with a phasor diagram at the end of this problem.

Another quantity of interest is the velocity $v = \dot{x}$. We could, now, differentiate equation (i) but to complete the heuristic intentions of this solution we will work with equation (h). However, to save space we will use x_m rather than the explicit expression equation (g) from here on.

$$\dot{x} = x_m\ i\omega\ e^{i(\omega t - \theta)}. \qquad (j)$$

but $i = e^{i\pi/2}$ (Euler's identity) so

$$\dot{x} = x_m \omega\ e^{i(\omega t - \theta + \pi/2)} \qquad (k)$$

or $\dot{x} = v_m \cos(\omega t - \theta + \pi/2)$

$= - v_m \sin(\omega t - \theta)$, $v_m = x_m\omega$.

Differentiating equation j for the acceleration,

$a = - x_m \omega^2 e^{i(\omega t - \theta)}$

$= - a_m \cos(\omega t - \theta)$, $a_m = x_m\omega^2$.

Using $e^{i\pi} = -1$ (Euler's identity) we can also write

$$a = x_m \omega^2 e^{i(\omega t - \theta + \pi)} . \qquad (\ell)$$

We will now display different representations containing the information of equations (h), (k), (ℓ) using rotating phasors. These are quantities of constant magnitude (equal to the amplitudes of the respective equations) but varying direction. We will represent them as vectors rotating with angular velocity ω. By drawing these vectors, called phasors, with the proper phase angles measured counterclockwise we can represent the complex relationships, and the projections of these vectors onto the real axis supply instantaneous values of physical interest. The phasor diagram is in figure 2.

If this entire picture is rotated at the velocity ω, all the quantities in their proper relationships can be read off the real axis as indicated above.

CHAPTER 23

RIGID BODY VIBRATIONS

• PROBLEM 23-1

Explain how Kater's pendulum, a specially constructed compound pendulum, enables precise measurements of the acceleration of gravity.

Solution: Kater's pendulum is constructed so that there are two knife edges, one on each side of the center of gravity, from which the pendulum may be suspended. In addition, there are movable weights which slide along the length of the pendulum to change the location of the center of mass. The figure shows one such pendulum with knife edges at A and B. It is the object of the experiment to arrange for the periods of oscillation about each of these two knife edges to be almost equal. This is done by adjusting the position of the weights.

Let us assume that the period about one knife edge at a distance of L_1 from the center of mass is P_1, and that the period about the other at a distance of L_2 is P_2. The adjustment is made so that P_1 is nearly equal to P_2. The period P of a compound pendulum is given by

$$P = 2\pi\sqrt{\frac{k^2 + L^2}{Lg}},$$

where k is the radius of gyration of the pendulum about the center of mass. Squaring this equation, yields P_1^2 for the length L_1

$$P_1^2 L_1 g = 4\pi^2(k^2 + L_1^2)$$

and P_2^2 for the length L_2

$$P_2^2 L_2 g = 4\pi^2(k^2 + L_2^2).$$

By subtraction

$$g(P_1^2 L_1 - P_2^2 L_2) = 4\pi^2(L_1^2 - L_2^2) \quad \text{or}$$

$$\frac{4\pi^2}{g} = \frac{P_1^2 L_1 - P_2^2 L_2}{L_1^2 - L_2^2} = \frac{1}{2}\left(\frac{P_1^2 + P_2^2}{L_1 + L_2} + \frac{P_1^2 - P_2^2}{L_1 - L_2}\right). \quad (a)$$

If considerable time and care are taken, then P_1 may be made so nearly equal to P_2 that the second term is negligible. Assuming that P_1 may be taken as equal to P_2, which we shall set equal to P, then

$$\frac{4\pi^2}{g} = \frac{P^2}{L_1 + L_2}$$

or $$g = \frac{4\pi^2}{P^2}(L_1 + L_2).$$

The distance between the knife edges, $L_1 + L_2$, can be measured very accurately. If the period is accurately known, then a really accurate determination of the value of g can be made. If the period P_1 is not exactly equal to P_2, then the first term on the right in equation (a) can be evaluated accurately. The second term, which is very small, need not be evaluated so accurately. By balancing the pendulum on a knife edge at its center of mass, the distances L_1 and L_2 can be measured with sufficient accuracy.

• PROBLEM 23-2

As shown in the figure, a ball having a mass m is attached to the lower end of a rigid bar whose length is ℓ and whose weight is negligible. The upper end of the bar is pinned to a slider which is constrained to move in a horizontal slot. The slide is subjected to a harmonic motion such that its displacement from the position shown is $y = Y \sin \omega t$. Derive the differential equation for small oscillations ($y << \ell$) of the mass m.

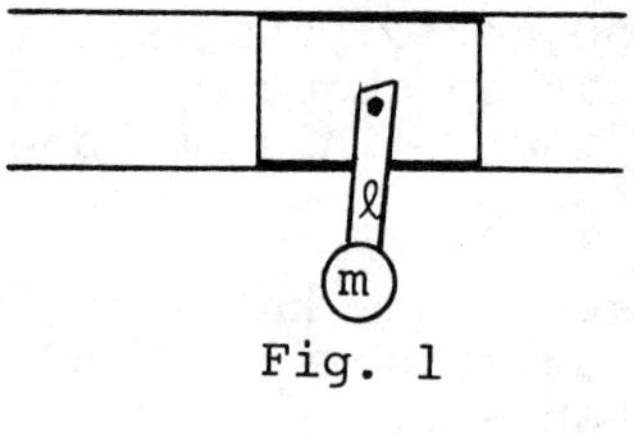

Fig. 1

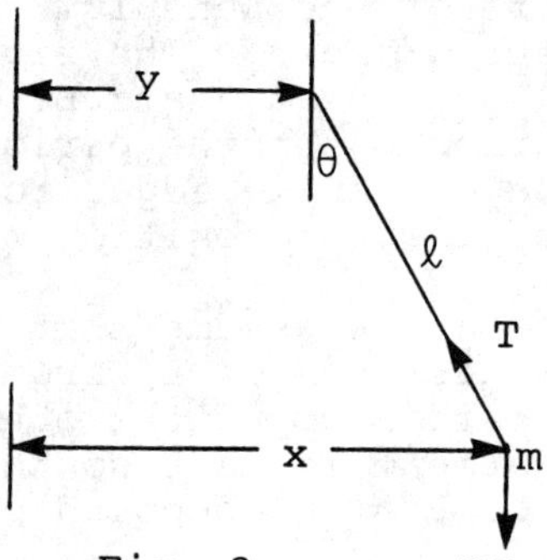

Fig. 2

Solution: The schematic, figure 2, of the apparatus includes the free body forces on the mass m. In assuming small oscillations, we will take $x << \ell$ too.

We will also use the small angle approximations $\cos\theta = 1$ and $\sin\theta = \theta$ for small angles. In that case,

$$T = mg\cos\theta \approx mg,$$

and the transverse force, $T\sin\theta$, on the mass is

$$F = mg\sin\theta \approx mg\theta.$$

In this case, with two degrees of freedom for the pendulum, $\sin\theta = \theta \approx \frac{x-y}{\ell}$, which can be seen from the diagram.

Thus, Newton's Second Law yields

$$m\ddot{x} = -mg\frac{(x-y)}{\ell}.$$

the minus sign is inserted since the traverse component of the tension is restorative, much like a spring force. Rearranging,

$$m\ddot{x} + \frac{mgx}{\ell} = \frac{mgy}{\ell}$$

or, $$\ddot{x} + \frac{g}{\ell}x = \frac{g}{\ell}Y\sin\omega t.$$

• PROBLEM 23-3

A homogeneous slender beam having weight W and length 2L is pinned at the left-hand end by a smooth pin O and is held from falling by a spring having a constant k and located as shown. Assume the equilibrium position occurs when the beam is horizontal. Determine the natural frequency and the period of the system. Assume the angular displacements are small.

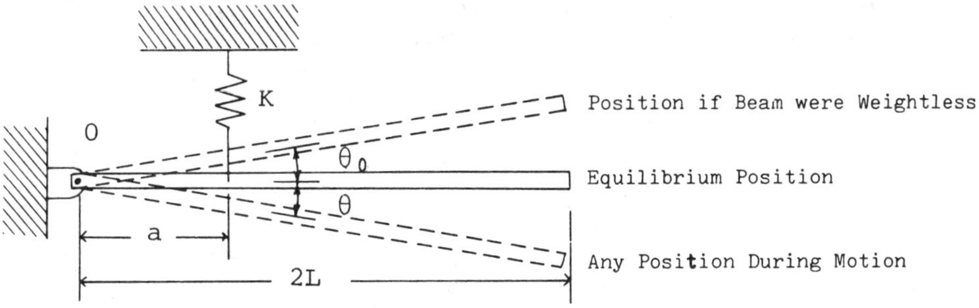

Solution: The figure illustrates the position the beam would assume if it was weightless. That is defined by the spring's unstretched length. We name the angle between this position and equilibrium θ_0. Assuming θ_0 is small, the extension of the spring at equilibrium is $a\theta_0$. Remember, that in finding moments we can consider the mass of an extended body to be concentrated at its center of gravity, in this case, the center of the beam. We can sum moments about 0 at equilibrium to give

$$\Sigma M_0 = -(ka\theta_0)a + WL = 0. \qquad \text{(a)}$$

Away from equilibrium, the sum of the moments is equal to the product of the moment of inertia of the beam

$$I = \frac{1}{3} m(2L)^2 = \frac{4}{3}\frac{W}{g} L^2$$

and the angular acceleration $\ddot{\theta}$. Thus,

$$\Sigma M = -ka\ (\theta + \theta_0)a + WL = \frac{4}{3}\frac{W}{g} L^2\ \ddot{\theta}.$$

Expanding and using equation (a) yields

$$-ka^2\theta = \frac{4}{3}\frac{W}{g} L^2\ \ddot{\theta},$$

$$\ddot{\theta} + \frac{3}{4}\frac{kg}{W}\frac{a^2}{L^2}\ \theta = 0.$$

This equation is in the form governing simple harmonic motion. The frequency and the period, by inspection, are

$$f = \frac{1}{2\pi}\sqrt{\frac{3}{4}\frac{kg}{W}\frac{a^2}{L^2}}$$

$$T = \frac{1}{f} = 2\pi\frac{L}{a}\sqrt{\frac{4}{3}\frac{W}{kg}}\ .$$

• PROBLEM 23-4

The slender bar in the figure, length 5a, is suspended by light rigid rods in two modes as shown. Determine which mode of oscillation will have the greater frequency, and derive the ratio f_1/f_2.

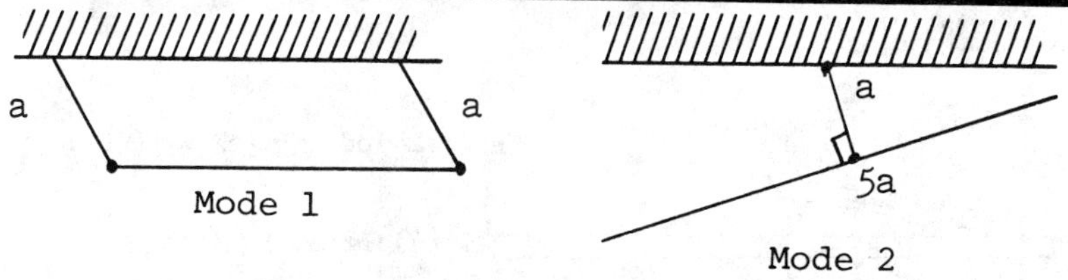

Fig. 1

Solution: Mode 1. The bar undergoes translation only with the center of mass moving through an arc. The free body diagram (Figure 2), when the rods are displaced an angle θ from the vertical, provides us with the equation of motion.

For small angles of θ, $T \simeq \frac{1}{2}mg$ and the equation of motion in the horizontal direction is

$$\Sigma F_x = Ma_x$$

$$\Sigma F_x = -2\left(\tfrac{1}{2} Mg\right) \sin\theta \simeq -Mg\theta = Ma_x .$$

But, since $a_x \simeq a\ddot{\theta}$,

$$-Mg\theta = Ma\ddot{\theta}, \quad \text{and}$$

$$a\ddot{\theta} + g\theta = 0,$$

so $\omega_1 = \sqrt{g/a}$.

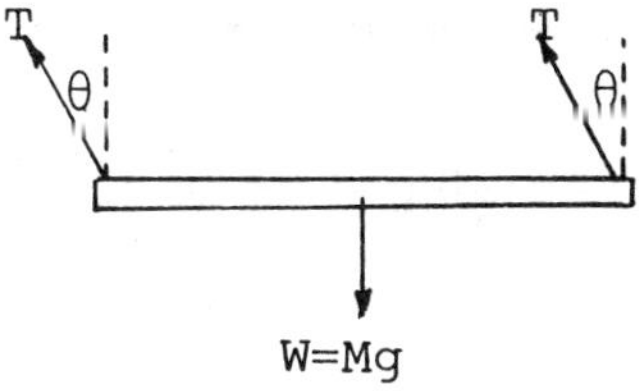

Fig. 2

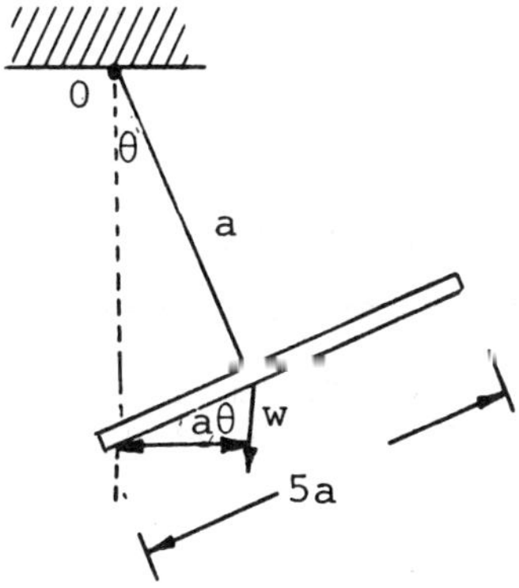

Fig. 3

Mode 2. This bar undergoes rotation as well as translation. Taking both motions as a rotation about point O, Figure 3 leads to the equation of motion.

$$M_0 = I_0 \alpha \tag{1}$$

The torque, M_0, about O is $-mga\theta$.

The moment of inertia, I_0, about O is, by the parallel axis theorem,

$$I_0 = I_G + Mr_G^2$$

$$I_0 = \frac{1}{12} M (5a)^2 + Ma^2$$

$$I_0 = \frac{37}{12} Ma^2 . \tag{2}$$

From equations (1) and (2),

$$-mga\,\theta = \frac{37}{12} ma^2 \ddot{\theta}$$

or $$\left(\frac{37}{12} a\right) \ddot{\theta} + g\theta = 0,$$

and $$\omega_2 = \sqrt{\frac{g}{\frac{37}{12} a}} = \omega_1 \sqrt{\frac{12}{37}\frac{a}{g}} = \sqrt{\frac{12}{37}}\, \omega_1 .$$

Therefore, $\omega_1 > \omega_2$

and $$\frac{\omega_1}{\omega_2} = \frac{2\pi f_1}{2\pi f_2} = \frac{f_1}{f_2} = \sqrt{\frac{37}{12}} = 1.756.$$

A Borda pendulum is made up of a spherical bob of radius R and a long wire of negligible weight. If the length of the wire from the point of suspension to the center of the bob is ℓ, show that the period of the pendulum for small oscillations is $\rho = 2\pi\sqrt{\frac{\ell}{g}\left(1 + \frac{2}{5}\frac{R^2}{\ell^2}\right)}$. If this pendulum is considered as a simple pendulum, show that the period calculated on this basis is in error by about 2% when $R = \frac{1}{5}\ell$.

Solution: A pendulum is considered simple if the weight at the end of the wire is a point mass. If the shape or size of the weight is of importance then the pendulum is considered to be a compound pendulum. Thus, Borda's pendulum is analyzed as a compound pendulum.

The theory of the compound pendulum indicates that the analysis developed for the simple pendulum may be used if ℓ is replaced by k_o^2/h where k_o is the radius of gyration about the support of the compound pendulum and h is the distance from the support to the center of gravity.

The radius of gyration can be obtained by using its defining equation,

$$mk_o^2 = I. \tag{1}$$

The moment of inertia in this problem is obtained by using the parallel axis theorem. The center of mass is the center of the bob, a distance ℓ from the point of suspension. Thus,

$$I = m\ell^2 + \frac{2}{5}mR^2 \tag{2}$$

since the moment of inertia of a sphere of radius R through its mass center is $\frac{2}{5}mR^2$.

Combining equations (1) and (2) yields

$$k_o^2 = \ell^2 + \frac{2}{5}R^2.$$

As defined above, h is simply equal to ℓ given in the problem. Thus, the simple pendulum equivalent to the given pendulum has length ℓ' where

$$\ell' = \frac{k_o^2}{\ell} = \frac{\ell^2 + 2/5R^2}{\ell} = \ell\left(1 + \frac{2}{5}\frac{R^2}{\ell^2}\right). \tag{3}$$

For a simple pendulum, the period is

$$P = 2\pi\sqrt{\frac{\ell}{g}}. \tag{4}$$

Thus, the given pendulum has period

$$P = 2\pi \sqrt{\frac{\ell}{g}\left(1 + \frac{2}{5}\frac{R^2}{\ell^2}\right)}. \qquad (5)$$

The period calculated if the pendulum is considered simple is given by equation (4). The error in equation (4) may be calculated using

$$\text{error} = \frac{P_C - P_s}{P_C} \times 100\%$$

where P_C = Period of the compound pendulum and P_s = Period of the simple pendulum. Let $P_s = 2\pi\sqrt{\ell/g} = 1$ to simplify the calculation. Then,

$$= \frac{\sqrt{1 + \frac{2}{5}\frac{R^2}{\ell^2}} - 1}{\sqrt{1 + \frac{2}{5}\frac{R^2}{\ell^2}}}$$

Substituting $\frac{R}{\ell} = \frac{1}{5}$ yields

$$\text{error} = 1.6\%.$$

• PROBLEM 23-6

A uniform rod of mass m and length 2h is suspended from the same horizontal level by two vertical cords of length ℓ which are attached to its end. Show that the period of this bifilar suspension for small oscillations about a vertical axis through the center of the rod is

$$P = 2\pi\sqrt{\frac{I\ell}{mgh^2}},$$

where I is the moment of inertia of the rod about a vertical axis through its center of mass.

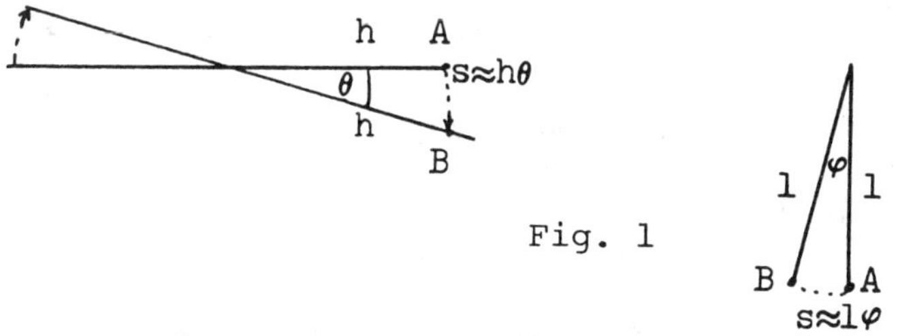

Fig. 1

Solution: As the tip of the rod rotates from A to B, the rod turns through angle θ and the supporting string turns through angle φ. The arc lengths are the same, and for small angles,

$$s = h\theta \simeq \ell\phi. \qquad (1)$$

Each string must support half the weight of the rod. Hence, the tension in either string is equal to mg/2.

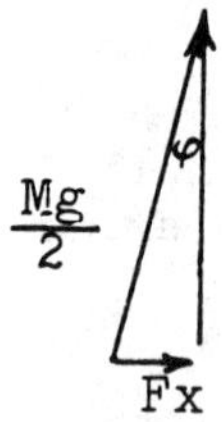

Fig. 2

Figure 2 shows a vector triangle for the force at either end of the rod. F_x is essentially horizontal if the angle ϕ is small.

Comparing this triangle with the side view in Figure 1, we may use the similarity of the triangles to obtain the proportion

$$\frac{F_x}{mg/2} = \frac{s}{\ell} \quad . \tag{2}$$

Substituting for s from equation (1),

$$\frac{F_x}{mg/2} = \frac{h\theta}{\ell}$$

and multiplying by mg/2 gives

$$F_x = \frac{h\theta}{\ell} \cdot \frac{mg}{2} \; .$$

Since this force acts at either end of the rod, and is essentially perpendicular to the rod, the total torque relative to the center of the rod from these two forces is

$$L = -\,2\left(\frac{h\theta}{\ell} \cdot \frac{mg}{2}\right) \cdot h$$

where the negative sign reflects the fact that the force opposes an increase in the angle θ. Then

$$L = -\,\frac{mgh^2}{\ell}\;\theta \; . \tag{3}$$

Using $L = I\,\frac{d^2\theta}{d^2t}$ for a pure rotation,

$$I\,\frac{d^2\theta}{dt^2} = -\,\frac{mgh^2}{\ell}\;\theta \; .$$

Dividing by I, and transposing the term on the right gives the differential equation in standard form

$$\frac{d^2\theta}{dt^2} + \frac{mgh^2}{\ell I}\;\theta = 0. \tag{4}$$

From which we may identify

$$\omega^2 = \frac{mgh^2}{\ell I} \qquad \text{or} \qquad \omega = \frac{2\pi}{P} = \sqrt{\frac{mgh^2}{\ell I}}$$

which gives $$P = 2\pi\sqrt{\frac{\ell I}{mgh^2}} .$$

• PROBLEM 23-7

Find the value of the damping constant c which produces critical damping in the system shown. The moment of inertia of the disk is I, its radius is R. The disk is attached by a rigid rod to the dashpot shown. Assume small oscillations so that the spring remains horizontal and the rod remains vertical.

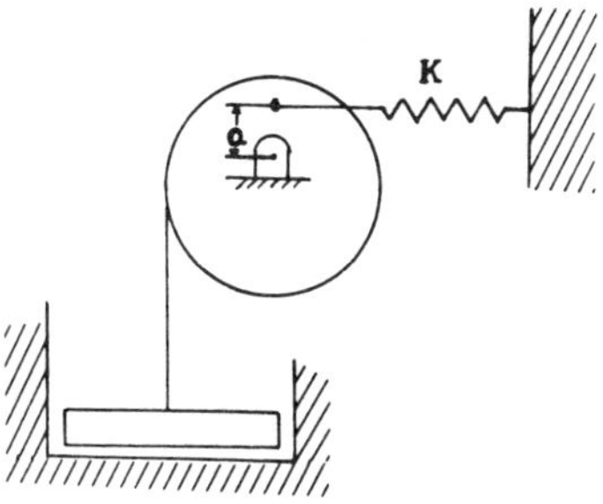

Solution: The figure shows two forces on the disk which will cause two torques, the sum of which

$$T = I\ddot{\theta} . \tag{1}$$

A displacement of θ will stretch or compress the spring aθ. Taking θ positive for counterclockwise rotation, the force from the spring is $F_s = ka\theta$, positive to the left. The force from the dashpot is $F_d = cR\dot{\theta}$, which, for c positive, provides torque with the same sense as the spring.

The torque from both is clockwise for ccw displacements, therefore

$$T = -\ a^2k\theta - R^2c\dot{\theta} . \tag{2}$$

Substituting equation (2) into (1) and rearranging,

$$I\ddot{\theta} + R^2c\dot{\theta} + a^2k\theta = 0. \tag{3}$$

This equation has the form of the generic damped oscillator equation:

$$m\ddot{x} + b\dot{x} + kx = 0 \tag{4}$$

for which $b_c = 2\sqrt{km}$. This comes from setting the discriminant equal to zero in the characteristic equation when solving equation (4). Comparing Eqs. (3) and (4), remembering $b = 2\sqrt{km}$, gives

$$R^2c = 2\sqrt{a^2kI}$$

$$c = \frac{2a}{R^2}\sqrt{kI}.$$

• PROBLEM 23-8

A disk, whose moment of inertia is I, is attached to a stepped shaft as shown. The torsional constants for the two pieces of the shaft are K_1 and K_2. Determine the torsional frequency of vibration.

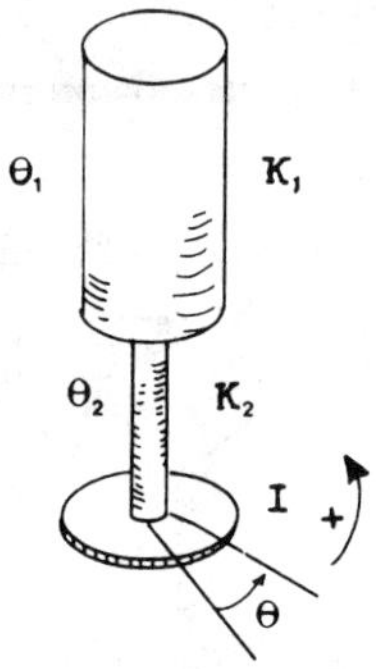

Solution: When the disk is in a disturbed position from equilibrium, the torque vs θ relationship will be:

$$\theta = \theta_1 + \theta_2 = \frac{T}{K_1} + \frac{T}{K_2} = T\left[\frac{K_1 + K_2}{K_1K_2}\right].$$

The equation of motion is then obtained from

$$\Sigma M_{AR} = I_{AR}\,\alpha = -\theta\left[\frac{K_1K_2}{K_1 + K_2}\right] = I\ddot{\theta}.$$

Consequently, the fundamental equation of motion is

$$I\ddot{\theta} + \frac{K_1K_2}{K_1 + K_2}\theta = 0.$$

Now, recognizing that this falls into the basic differential equation form that is analogous to $m\ddot{x} + kx = 0$, $\omega = \sqrt{\frac{K}{m}}$ with $f = \frac{1}{2\pi}\omega$. Then

$$f = \frac{1}{2\pi}\sqrt{\frac{K_1K_2}{I(K_1 + K_2)}}\ \text{HZ}.$$

• PROBLEM 23-9

A weightless beam supports a weight W and is itself supported by a spring, the spring constant is k, and a hinge O. The weight is a distance L from the hinge, and the spring is attached at a distance a from the hinge. The spring-weight equilibrium position is the beam in a horizontal state. Determine the natural frequency of the system, assuming small displacements.

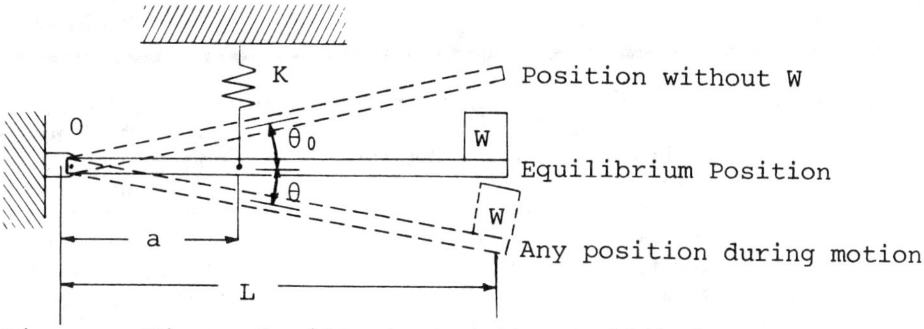

Solution: Figure 1, illustrates the equilibrium position of the weightless beam with no weight on it. This defines the relaxed position of the spring. The most convenient point to take moments about is point O, in equilibrium position.

$$\Sigma M_0 = (-k\ a\ \theta_0)\ a \cos\theta_0 + WL = 0.$$

Using the small angle approximation for the stretch of the spring, we have,

$$\Sigma M_0 = -(k\ a\ \theta_0)\ a + WL = 0. \qquad \text{(a)}$$

Away from equlibrium, the sum of the moments is equal to the product of the moment of inertia of the weight (remember the rod is assumed to be massless)

$$I = ML^2 = \frac{W}{g} L^2$$

and the angular acceleration $\ddot{\theta}$. Thus,

$$\Sigma M = -k\ a\ (\theta + \theta_0)\ a + WL = \frac{W}{g} L^2 \ddot{\theta}.$$

Expanding,

$$-k\ a^2\theta - k\ a^2\theta_0 + WL = \frac{W}{g} L^2\ \ddot{\theta}.$$

Equation (a) establishes that $-ka^2\theta_0 + WL = 0$ so

$$-ka^2\theta = \frac{W}{g} L^2\ddot{\theta}. \qquad \text{(b)}$$

Note that the gravity term WL and θ_0 drop out from the equation of motion. This is a common feature, and an advantage of using the equilibrium position as a reference for the spring-gravity system.

Rearranging equation (b) yields,

$$\ddot{\theta} + \frac{kg\ a^2}{W\ L^2}\theta = 0.$$

This equation is in the basic form governing simple harmonic motion, and the frequency is given by

$$f = \frac{1}{2\pi}\omega = \frac{1}{2\pi}\frac{a}{L}\sqrt{\frac{kg}{W}}.$$

The thin homogeneous plate in the figure oscillates in the plane of the paper in a viscous medium **about the equilibrium position shown.** The weight of the plate is w, and the viscous resistance to the motion of any element of the surface whose area is dA is given by $dF = \nu\, v\, dA$, where ν is the coefficient of viscosity and v is the velocity of the element. Determine the frequency of the damped and undamped oscillations, and derive the expression for the coefficient of viscosity that will provide critical damping.

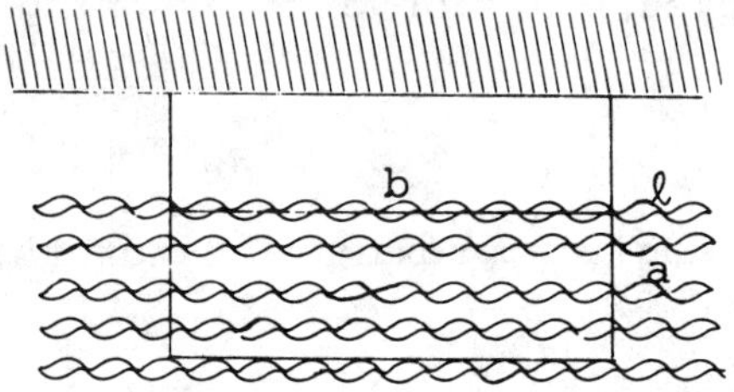

Solution: In this problem, we must realize that the viscous drag force is caused by both sides of the plate. Because the plate is supported by two wires, as shown in Figure 1, it has translational motion only. All points on the plate move through the same arc simultaneously, hence, all of the plate move at the same velocity at any instant. This means that the total drag force

$$F_d = 2\,\nu\, v(ab).$$

If we assume small oscillations only, then the horizontal restoring force from the support wires is approximately $w \sin\theta \simeq w\theta$. The velocity is $v = \ell\dot{\theta}$, the acceleration $a = \ell\ddot{\theta}$ so

$$\Sigma F_x = ma_x$$

$$- w\theta - \nu\ell\dot{\theta}\, 2ab = \frac{w}{g}\,\ell\ddot{\theta}$$

$$\frac{w}{g}\,\ell\ddot{\theta} + \nu\ell(2ab)\dot{\theta} + w\theta = 0.$$

This takes the form of a damped harmonic oscillator ($m\ddot{x} + b\dot{x} + kx = 0$) for which the critical damping $b_c = 2\sqrt{km}$. Thus ν_c is found from .

$$\nu_c\, \ell(2ab) = 2\sqrt{\left(\frac{w}{g}\,\ell\right)(w)}$$

$$\nu_c = \frac{w}{ab\ell}\sqrt{\frac{\ell}{g}} = \frac{w}{abg}\sqrt{\frac{g}{\ell}} \ .$$

Determine the natural frequency for the system shown for small oscillations so that $\sin\theta \simeq \theta$. Assume the pendulum is attached to the hinge by a thin, massless, rigid rod and that the mass of the springs is negligible.

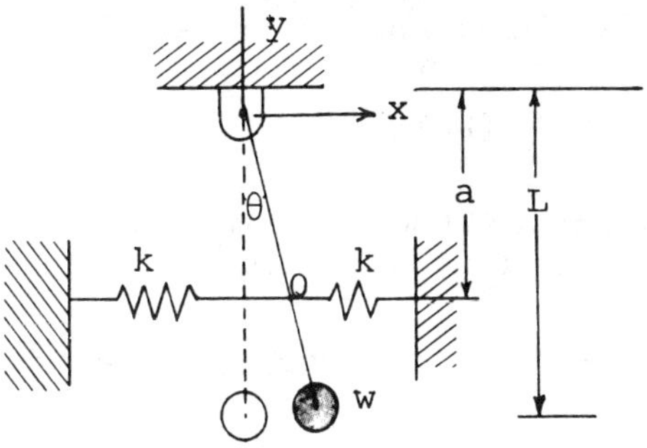

Fig. 1

Solution: Since there are only conservative forces acting, we shall use the energy method to obtain the equation of motion for the system. Fig. 1 illustrates the system where the pendulum is displaced by a small angle θ. Note that the motion is constrained to take place in the x-y plane.

First, calculate the potential energy of the system. There are three potential energy terms to consider; the gravitational potential energy of the weight w, and the elastic potential energy due to the two springs. Let us evaluate the gravitational potential energy of the weight w. When the pendulum is at an angle θ, then the weight is raised by the distance $h = L - L\cos\theta$. Thus, the potential energy is given by:

$$(PE)_w = mgh = w(L - L\cos\theta). \tag{1}$$

In Eq. (1) the zero level of the potential energy has been chosen with the pendulum hanging straight down.

Next, calculate the elastic potential energy stored in the two springs. In Fig. 2, note that when the pendulum is at the angle θ, one of the springs is stretched by the amount $a\theta$ while the other spring is compressed by that same amount. Therefore, the elastic potential energy value is:

$$(PE)_s = \tfrac{1}{2}k(a\theta)^2 + \tfrac{1}{2}k(a\theta)^2 = ka^2\theta^2. \tag{2}$$

Again, the zero position of the potential energy is the pendulum hanging straight down.

The sum of Eqs. (1) and (2) is the total potential energy of the system, this yields:

$$(PE) = w(L - L\cos\theta) + ka^2\theta^2. \tag{3}$$

Next, we have to determine the kinetic energy of the system. Since the rod and the springs have no mass, the only kinetic energy is associated with the motion of the weight w. In Fig. 2, observe that the speed of the weight is given by:

$$v = a\dot{\theta},$$

and therefore the kinetic energy is equal to:

$$(KE) = \tfrac{1}{2}mv^2 = \frac{1}{2}\frac{w}{g}(L\dot{\theta})^2. \tag{4}$$

The total energy of the system is equal to the sum of Eqs. (3) and (4) and is given by:

$$E = \frac{1}{2}\frac{w\ L^2}{g}\dot{\theta}^2 + w\ (L - L\cos\theta) + ka^2\theta^2. \tag{5}$$

Differentiate Eq. (5) with respect to time. This yields:

$$\dot{E} = \frac{w\ L^2}{g}\dot{\theta}\ddot{\theta} + w\ L\sin\theta\ \dot{\theta} + 2\ ka^2\theta\dot{\theta}. \tag{6}$$

In Eq. (6) $\dot{E} = 0$. Because the energy of the system is constant since there are no dissipative forces. Furthermore, one can simplify Eq. (6) by dividing by $w\ L^2\dot{\theta}/g$. This gives,

$$\ddot{\theta} + \frac{g}{L}\sin\theta + \frac{2ka^2g}{w\ L^2}\theta = 0. \tag{7}$$

Eq. (7) can be further simplified if we assume small oscillations. For small oscillations, the angle θ is small and we can set $\sin\theta \simeq \theta$. We then obtain

$$\ddot{\theta} + \left(\frac{g}{L} + \frac{2ka^2g}{w\ L^2}\right)\theta = 0. \tag{8}$$

Eq. (8) is the desired differential equation of motion. We could proceed to find the general solution of Eq. (8) but this is not necessary because we recognize it as the equation of motion of the simple harmonic oscillator. The angular frequency in radians per second is given by:

$$\omega = \sqrt{\frac{g}{L} + \frac{2ka^2g}{wL^2}}\quad \text{rad/sec}.$$

Expressing the previous equation in terms of cycles per minute yields:

$$f = \frac{60}{2\pi}\sqrt{\frac{g}{L} + \frac{2ka^2g}{w\ L^2}}\quad \text{cpm}. \tag{9}$$

We can check our answer by letting k = 0. For this case Eq. (9) should reduce to the answer for a simple pendulum; and it does.

Analyze the motion of a compound pendulum. Find (a) the length of the equivalent simple pendulum, (b) the significance of this length in relation to other properties of the pendulum, (c) the relationship of period to the distance between its point of suspension and its center of mass.

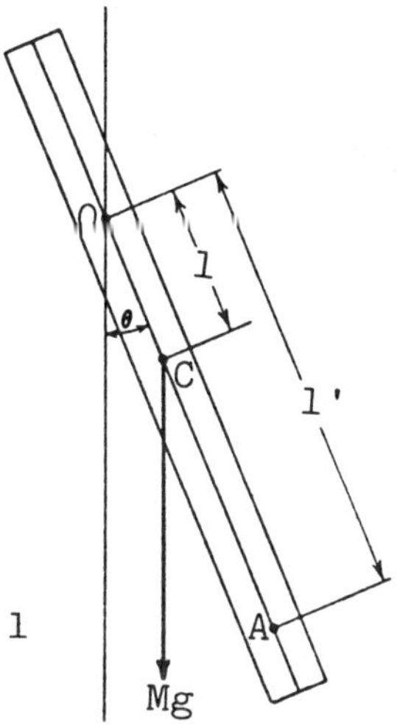

Fig. 1

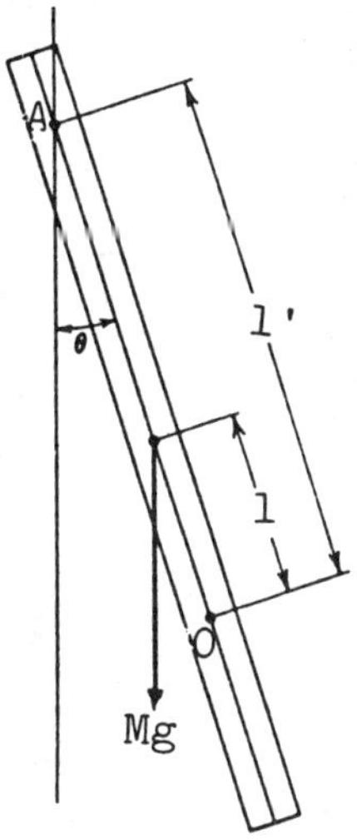

Fig. 2

Solution: The compound pendulum is any extended rigid body suspended from a point and free to swing to and fro in a plane. This is to be distinguished from the simple pendulum, which is a small mass suspended from a point by a weightless string or rod so that all the force and inertia acts on the small mass.

Analyze the idealized compound pendulum shown in Figure 1. This meter stick idealization will make the calculations easier leaving out no essential details.

Let the rod have a mass M and be suspended at O from a frictionless support a distance ℓ from the center of mass C. Suppose that at some instant of time the rod makes an angle θ with the vertical. At this time the torque τ_0 exerted by the rod about the axis through O is

$$\tau_0 = Mg\ell \sin\theta .$$

This torque is related to the angular acceleration $\ddot{\theta}$, which in this case gives

$$Mg\ell \sin\theta = -I_0\ddot{\theta} \qquad \text{(a)}$$

where I_0 is the moment of inertia of the rod about O. The negative sign is used since the torque tends to decrease θ whereas $\ddot{\theta}$ is positive in the direction of increasing θ. We are assuming that there are no torques due to friction or air resistance, which is, of course, only approximately correct in practice.

We shall further assume that the maximum angle of oscillation is so small that sin θ may be replaced by the angle θ. If this is not done, then the expression for the period becomes relatively complicated but may be evaluated by the same method as for the simple pendulum. With this assumption, Eq. (a) becomes

$$\ddot{\theta} = -\frac{Mg\ell\theta}{I_0}$$

or $$\ddot{\theta} + \frac{Mg\ell}{I_0} = 0.$$

This is of the general form for simple harmonic motion $\ddot{\theta} + \omega^2\theta = 0$. By comparison

$$\omega = \sqrt{\frac{Mg\ell}{I_0}} .$$

The period $P_0 = \frac{1}{f} = \frac{2\pi}{\omega}$

so $$P_0 = 2\pi\sqrt{\frac{I_0}{Mg\ell}} . \qquad \text{(b)}$$

The moment of inertia of the rod about O may be written by the theorem of parallel axes in terms of the moment of inertia about the center of mass as

$$I_0 = I_{C.M.} + M\ell^2 .$$

If k is the radius of gyration of the rod about the center of mass, then

$$I_{C.M.} = Mk^2 \qquad \text{and} \qquad I_0 = M(k^2 + \ell^2).$$

Thus the period about O from Eq. (b) is

$$P_0 = 2\pi\sqrt{\frac{k^2 + \ell^2}{\ell g}} . \qquad \text{(c)}$$

An ideal simple pendulum having this same period P_0 would have a length ℓ' given by

$$P_0 = 2\pi\sqrt{\frac{\ell'}{g}} .$$

This length ℓ' is related to the length ℓ between the point of support O and the center of mass C of the physical pendulum by

$$\ell' = \frac{k^2 + \ell^2}{\ell} = \frac{k^2}{\ell} + \ell . \qquad \text{(d)}$$

We shall see that this length ℓ', called the length of the equivalent simple pendulum, has physical importance. Suppose that the pendulum is now suspended from an axis through A a distance ℓ' from O, the former axis of suspension. It is then readily shown that the period of oscillation about A is the same as that about O. The moment of inertia I_A of the rod about A in Fig. 2 is

$$I_A = Mk^2 + M(\ell' - \ell)^2$$

using the parallel axis theorem. By equation (d), this becomes

$$I_A = Mk^2 + M\left(\frac{k^2}{\ell}\right)^2 = \frac{Mk^2}{\ell^2}(\ell^2 + k^2).$$

The torque is now

$$\iota_A = Mg(\ell' - \ell) = -I_A\ddot{\theta},$$

so the period, by analogy with equation (b), is given by

$$P_A = 2\pi\sqrt{\frac{I_A}{Mg(\ell' - \ell)}} = 2\pi\sqrt{\frac{Mk^2(\ell^2 + k^2)\ell}{\ell^2 Mgk^2}}$$

$$= 2\pi\sqrt{\frac{(k^2 + \ell^2)}{\ell g}} = P_0.$$

The two points O and A are called the center of suspension and the center of oscillation respectively. It follows that, if either point is the center of oscillation, the other is the center of suspension.

We shall next prove that there are two points of suspension at different distances from the center of mass and on the same side of the center of mass at which the periods are equal. From Eq. (d)

$$\ell^2 - \ell\ell' + k^2 = 0.$$

The solution of this quadratic equation is

$$\ell = \frac{\ell' \pm \sqrt{\ell'^2 - 4k^2}}{2}$$

so there are two length, ℓ_1 and ℓ_2 which will give periods equal to P_0 of the equivalent simple pendulum. They are

$$\ell_1 = \frac{\ell' + \sqrt{\ell'^2 - 4k^2}}{2}, \ell_2 = \frac{\ell' - \sqrt{\ell'^2 - 4k^2}}{2}. \quad \text{(e)}$$

The period of the compound pendulum of length ℓ_1 is

$$P = 2\pi\sqrt{\frac{M(k^2 + \ell_1^2)}{Mg\ell_1}}$$

which may readily be shown by substitution to be equal to

$$P = 2\pi\sqrt{\frac{\ell'}{g}}.$$

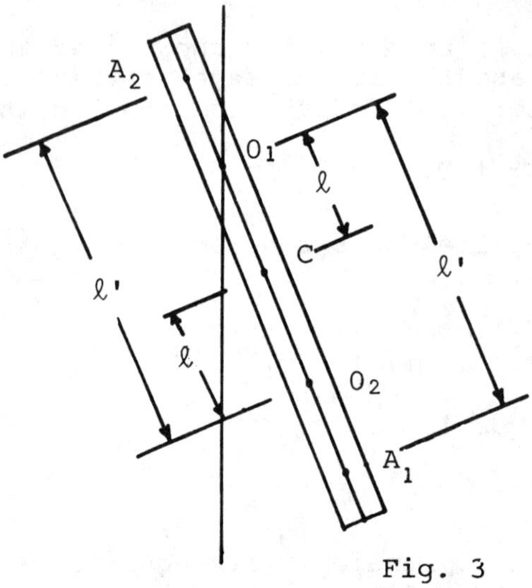

Fig. 3

This is a general proof of the existence of these two equivalent lengths. A simple argument, based on the symmetry of the chosen rod may clarify this behavior although it is not generally applicable. Examine Figure 1 and 2, although we said, and labeled as such, that the pendulum was turned around in going from one figure to another, it is possible, due to the rod's symmetry, to consider the rod to have simply been moved downwards from O in Figure 1 to A in Figure 2. Upon examining Figure 3 where A and O from Figure 1 have been relabeled as A_1 and O_1 and from Figure 2 as A_2 and O_2, this may be clearer. So on either side of C, there are two points which give the same period of oscillation. We can also see that when $\ell = \frac{1}{2}\ell'$, A_2 and O_1 will coincide as well as O_2 and A_1. From the mathematical proof above, the condition for ℓ_1 to equal ℓ_2 is for

$$\ell'^2 - 4k^2 = 0.$$

Putting $\ell = \frac{1}{2}\ell'$ into equation (d) yields

$$\ell' = \frac{2k^2}{\ell'} + \frac{\ell'}{2}$$

$$2\ell'^2 = 4k^2 + \ell'$$

$$\ell'^2 = 4k^2 \quad \text{or} \quad \ell'^2 - 4k^2 = 0.$$

Shown in Figure 4 is a plot of the period of the pendulum vs. the distance of the suspended point from the center of mass. The plot follows directly from equation (c). When the pendulum is suspended at its center of mass, no oscillations occur and the period is infinite.

The distance of the point of the suspension from the center of mass for which the period is a minimum may be calculated by setting $dP/d\ell$ equal to zero in the expression

$$P = 2\pi\sqrt{\frac{k^2 + \ell^2}{\ell g}}\,.$$

When the differentiation is carried out, it is found that $dP/d\ell = 0$ when $\ell = k$.

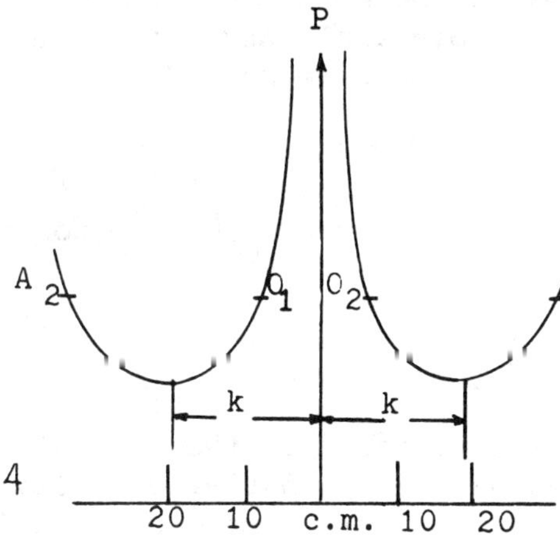

Fig.4

This occurs again when the points giving equal periods coincide, that is when $\ell'^2 = 4k^2$. Since $\ell' = 2\ell$, we also know $4\ell^2 = 4k^2$ or $\ell = k$. Note also that the trough of the curves in Figure 4 occurs where $dP/d\ell = 0$ and where the labeled points will coincide. The minimum period of oscillation is therefore

$$P_{min} = 2\pi \sqrt{\frac{2k}{g}} \, .$$

If this minimum period is experimentally determined, then the radius of gyration k of the physical pendulum can be found. This radius of gyration k may also be found from the product of any of the two distances ℓ_1, ℓ_2, for from Eq. (e) it follows that

$$\ell_1 \ell_2 = k^2.$$

• PROBLEM 23-13

Derive an expression for the period of oscillation for a disk of radius r and weight w rolling on a track with constant radius of curvature (see Figure 1). Assume the disk rolls without slipping and that the oscillations are small.

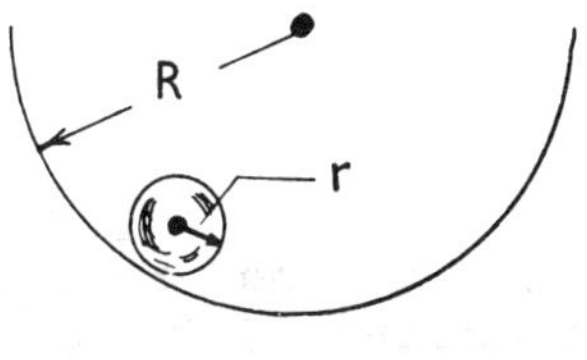

Fig. 1

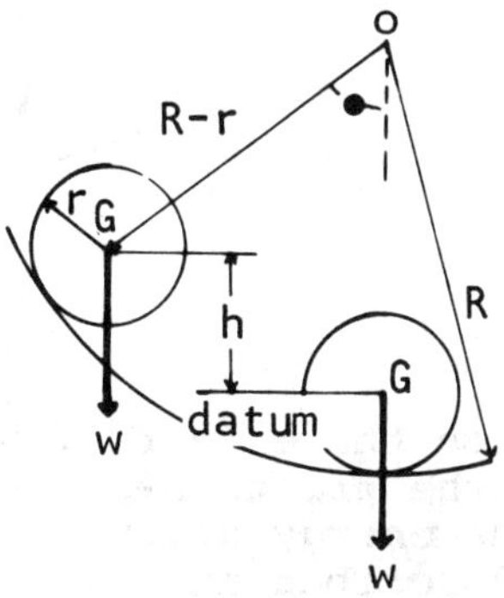

Fig. 2

Solution: A treatment of the potential energy formula in the small angle approximation will yield a spring-like potential, as shall be shown here. Choosing the position of the center of the disk at the bottom of the curve for the potential energy datum, i.e., potential energy is 0, and h for the height above this reference we have (see Figure 2)

$$V = Wh = W(R - r)(1 - \cos\theta).$$

However, $(1 - \cos\theta) = 2\sin^2\left(\frac{\theta}{2}\right) \simeq \frac{\theta^2}{2}$ for small oscillations, so

$$V = W(R - r)\,\frac{\theta^2_{max}}{2}\ . \qquad \text{(a)}$$

If the angular velocity of the disk's center of mass about 0 is denoted by $\dot{\theta}$, the angular velocity of the disk's rotation about point of contact by ω and the instantaneous linear velocity by v we have,

$$v = (R - r)\dot{\theta}\ , \qquad \text{(b)}$$

$$\omega = \frac{v}{r} = \frac{R - r}{r}\dot{\theta}\ . \qquad \text{(c)}$$

This yields for the kinetic energy

$$T = \tfrac{1}{2}mv^2 + \tfrac{1}{2}I\omega^2$$

$$= \tfrac{1}{2}m(R - r)^2\dot{\theta}^2 + \tfrac{1}{2}(\tfrac{1}{2}mr^2)\left(\frac{R - r}{r}\right)^2\dot{\theta}^2$$

$$= \frac{3}{4}\,m\,(R - r)^2\,\dot{\theta}^2. \qquad \text{(d)}$$

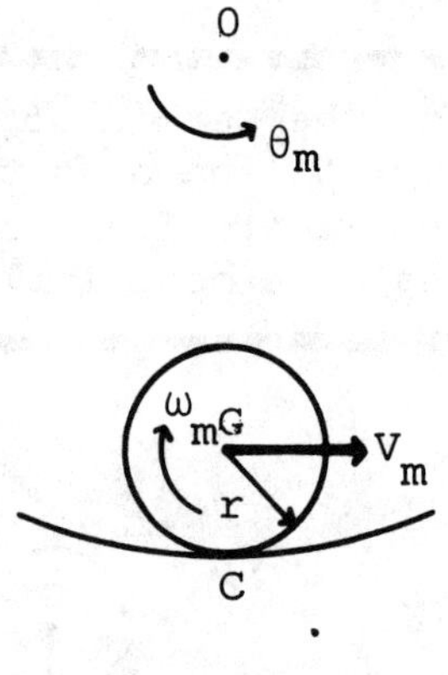

Fig. 3

Since there is rolling without slipping, no work is done by the frictional force and equations (a) and (d) complete the energy analysis of this problem. These equations can be described in a more familiar form, i.e., we define

$$K' = W(R - r) = mg(R - r)$$

$$m' = \frac{3}{2} m(R - r)^2 .$$

Then, $V = \frac{1}{2}K'\theta^2$, $\qquad T = \frac{1}{2}m'\dot{\theta}^2$.

With these expressions the period of small oscillations can be described as:

$$T = 2\pi\sqrt{\frac{m'}{K'}} = 2\pi\sqrt{\frac{3}{2}\frac{(R - r)}{g}} .$$

• PROBLEM 23-14

A disk with moment of inertia I = 10.6 lb-sec²-in., hangs from a wire attached to its center. It is observed to have a period of torsional vibration equal to 1.4 sec. A disk of unknown moment of inertia is substituted for the first disk and the new period of oscillation is 2.2 sec. What is the unknown moment of inertia?

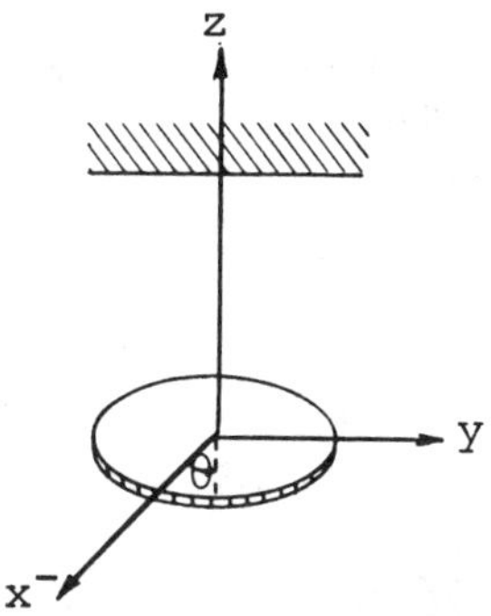

Solution: To determine the unknown moment of inertia we have to find an equation for the period of oscillation for the system. This equation can be obtained from the equation of motion for the system. Since the disk is in pure rotation about the z-axis, we shall use Euler's equation to obtain the equation of motion for the disk,

$$I_{zz}\ddot{\theta} = T_z . \qquad (1)$$

In Eq. (1) T_z is the restoring torque exerted by the wire on the disk when the disk has rotated by an angle θ. According to Hooke's law, this torque is proportional to the angle of rotation if the angle of rotation is small. Thus we have

$$T_z = - K\theta \qquad (2)$$

where K is the torsional spring constant. Substituting Eq. (1) into Eq. (2) and rearranging the equation yields:

$$\ddot{\theta} + \frac{K}{I_{zz}}\theta = 0 . \qquad (3)$$

Eq. (3) is a second order linear differential equation with constant coefficients. The solution to this equation is:

$$\theta = C_1 \sin \sqrt{\frac{K}{I_{zz}}}\, t + C_2 \cos \sqrt{\frac{K}{I_{zz}}}\, t, \tag{4}$$

where the constants C_1 and C_2 are determined from the initial conditions. In this problem we are not interested in these values. In Eq. (4) the trigonometric functions repeat themselves whenever the phase change **is** 2π. This observation allows us to determine the period of the motion, that is,

$$\sqrt{\frac{K}{I_{zz}}}\, T = 2\pi,$$

or the period of the motion is

$$T = 2\pi \sqrt{\frac{I_{zz}}{K}} \ . \tag{5}$$

We shall now use Eq. (5) to calculate the unknown moment of inertia. This is accomplished by solving Eq. (5) for I_{zz} yielding,

$$I_{zz} = \frac{KT^2}{4\pi^2} \ . \tag{6}$$

Let us designate the unknown moment of inertia by I'_{zz} and the corresponding period by T'. This yields,

$$I'_{zz} = \frac{KT'^2}{4\pi^2} \ . \tag{7}$$

Now divide Eq. (7) by Eq. (6) and rearrange this equation to obtain:

$$I'_{zz} = I_{zz} \frac{T'^2}{T^2} \ . \tag{8}$$

Substituting all known values into Eq. (8) we can evaluate I'_{zz}. Thus we find

$$I'_{zz} = 10.6 \frac{(2.2)^2}{(1.4)^2} = 26.2 \text{ lb sec}^2 \text{ in.}$$

Note that this method of determining an unknown moment of inertia will also work for objects having more complex shapes.

• PROBLEM 23-15

An electric motor with a mass of 100 kg rests on 2 rubber pads with an equivalent spring rate of 100 N/mm. Inside the motor the rotor has an imbalance equivalent to a 25 g mass with an eccentricity of 150 mm. The motor is limited to vertical motion. Determine

(a) the resonance speed for this system

(b) the amplitude of the vibration when the motor is operating at a speed of 1000 rpm.

Fig. 1

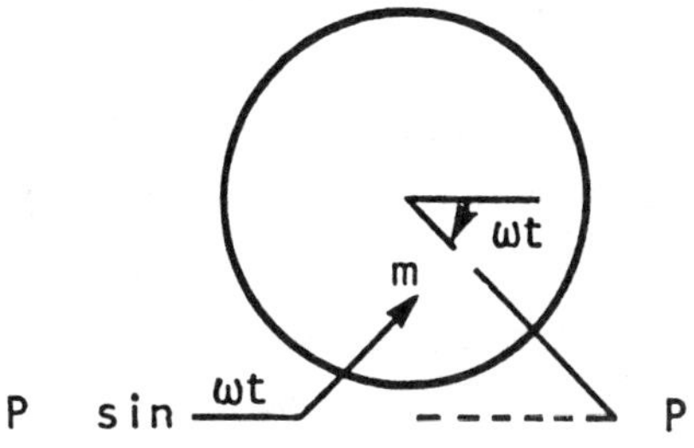

Fig. 2

Solution: This problem is basically the forced harmonic oscillator. The driving force comes from the motor's reaction to the rotor's unbalanced motion. The frequency of this reaction is equal to the frequency of the rotor's revolutions. The spring rate for this problem is equal to the net spring rate from the two rubber pads connected in parallel.

$$k_{eg} = k_1 + k_2 = 2k \qquad \text{so}$$

$$k_{eg} = 2(100 \text{ N/mm}) = 200 \text{ N/mm} = 200{,}000 \text{ N/m}$$

The mass of the motor is 100 kg.

The natural frequency of the motor is, then,

$$\omega_o = \frac{k}{m} = \frac{200{,}000}{100} = 44.72 \text{ rad/sec} = 427 \text{ rpm}$$

Resonance occurs when the driving force frequency is equal to the natural frequency. Therefore, the resonance speed is 427 rpm.

Knowing the magnitude of the driving force will give us the amplitude of the vibrations. The force is a result of Newton's Second and Third Laws.

The maximum force on the 25 g equivalent weight is $F_{max} = ma_{max}$. This same force is experienced by the motor (in the opposite direction). Given circular motion, the component of the 25 g mass's location in the direction on vibration is given by $y = y_{max} \sin \omega t$, and, differentiating twice, $a = -\omega^2 y_{max} \sin \omega t$. Therefore,

$$F_{max} = -my_{max}\,\omega^2.$$

$= 25\ g = 0.025\ kg$

$\omega = 1000\ rpm = 104.7\ rad/sec$

and $y_{max} = 150\ mm = 0.15\ m$

This yields,

$$F_{max} = -\ my_{max}\ \omega^2 = -(0.025)(0.15)(104.7)^2 = -41.11\ N$$

The amplitude for forced vibrations without damping is given by

$$y_f = \frac{F_{max}/k}{1 - (\omega_f/\omega_0)^2}$$

$$= \frac{-41.11\ N/200\ N/mm}{1 - (104.7/44.72)^2}$$

$$= 0.0459\ mm$$

• PROBLEM 23-16

A steel sphere of radius r executes simple harmonic oscillations of small amplitude on a concave surface of radius R. Show that the period of the oscillations ρ is given by

$$\rho = 2\pi\sqrt{\frac{7}{5}(R-r)/g}.$$

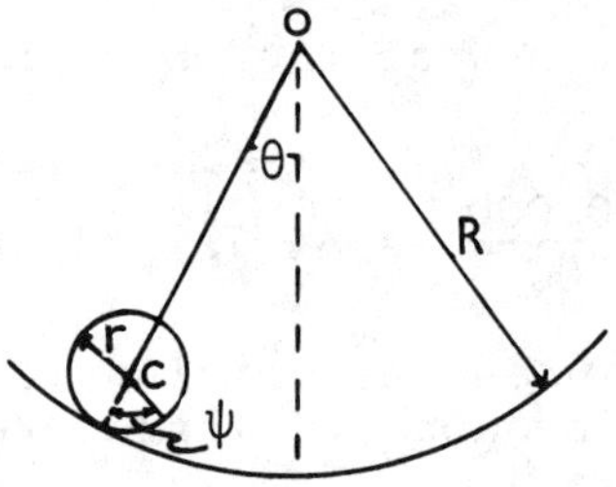

Fig. 1

Solution: The easiest way to do this problem is using the energy method. By assuming that this is a conservative, harmonic system, one can say that the maximum kinetic energy is equal to the maximum potential energy, and that the ball's motion is of the form $\theta = \theta_{max} \sin \omega t$.

The maximum potential energy is equal to mg (R-r) (1-cosθ) where m is the ball's mass, g is the gravitational constant and θ is the angle traveled along the arc. The maximum kinetic energy is equal to $1/2\ I_c\ \dot{\psi}^2 + 1/2\ I_o\dot{\theta}^2$ where I_o and I_C are the moments of inertia about 0 and C, respectively, and ψ is the rotation angle of the ball. Since the ball rolls without slipping, $r\dot{\psi} = (R-r)\dot{\theta}$.

I_o, by the parallel axis theorem, is:

$$I_o = m\,(R-r)^2 + \frac{2}{5}\,mr^2;$$

$I_C = \frac{2}{5}\,mr^2$ since the moment of inertia of a sphere is equal to $\frac{2}{5}\,mR^2$, where m is the mass of the ball.

If $PE_{max} = KE_{max}$,

$$mg\,(R-r)\,(1-\cos\theta) = \frac{1}{2}\,m\left[(R-r)^2 + \frac{2}{5}\,r^2\right]\dot{\theta}^2 + \frac{1}{2}\,m\left[\frac{2}{5}\,r^2\right]\dot{\psi}^2.$$

Substituting $\frac{(R-r)}{r}\,\dot{\theta}$ for $\dot{\psi}$ in the above relation yields:

$$mg\,(R-r)\,(1-\cos\theta) = \frac{1}{2}\,m\left[(R-r)^2 + \frac{2}{5}\,r^2\right]\dot{\theta}^2 + \frac{1}{2}\,m\left[\frac{2}{5}\,r^2\right]\frac{(R-r)^2}{r^2}\,\dot{\theta}^2,$$

and the r^2 terms in the second K.E. term cancel.

Now, if $\theta = \theta_{max}\sin\omega t$,

$$\dot{\theta} = \omega\theta_{max}\sin\omega t$$

and

$$\dot{\theta} = \omega\theta.$$

Substituting this into the previous equation,

$$mg\,(R-r)\,(1-\cos\theta) = \frac{1}{2}\,m\left[(R-r)^2 + \frac{2}{5}\,r^2\right]\omega^2\theta^2 + \frac{1}{2}\,m\left[\frac{2}{5}(R-r)^2\right]\omega^2\theta^2.$$

Now, by a trig identity,

$$(1 - \cos\theta) = 2\sin^2\left(\frac{\theta}{2}\right).$$

But, if θ is small then $\sin\left(\frac{\theta}{2}\right) = \frac{\theta}{2}$. Then,

$$(1 - \cos\theta) - 2\left(\frac{\theta}{2}\right)^2 = \frac{\theta^2}{2}.$$

Substituting this value for $1-\cos\theta$ in the equation yields

$$mg\,(R-r)\,\frac{\theta^2}{2} = \frac{1}{2}\,m\left[(R-r)^2 + \frac{2}{5}\,r^2\right]\omega^2\,\theta^2 + \frac{1}{2}\,m\left[\frac{2}{5}\,(R-r)^2\right]\omega^2\,\theta^2.$$

Dividing through by m, θ^2, and $\frac{1}{2}$ and solving for ω^2 yields the following:

$$\omega^2 = \frac{g\,(R-r)}{\left[(R-r)^2 + \frac{2}{5}\,r^2 + \frac{2}{5}\,(R-r)^2\right]}.$$

Finally, if R > > r then the $\frac{2}{5}\,r^2$ term will be small and

$$\omega^2 \approx \frac{g\ (R-r)}{\frac{7}{5}\ (R-r)^2}$$

$$\omega^2 \approx \frac{5g}{7\ (R-r)}$$

By definition, $P = 2\pi\ \sqrt{1/\omega^2}$

therefore, $P = 2\pi\sqrt{\dfrac{7\ (R-r)}{5g}}$.

• PROBLEM 23-17

Develope an expression for the period of vibration for a cylinder of weight W and radius r suspended in a spring loaded sling (see figure 1)

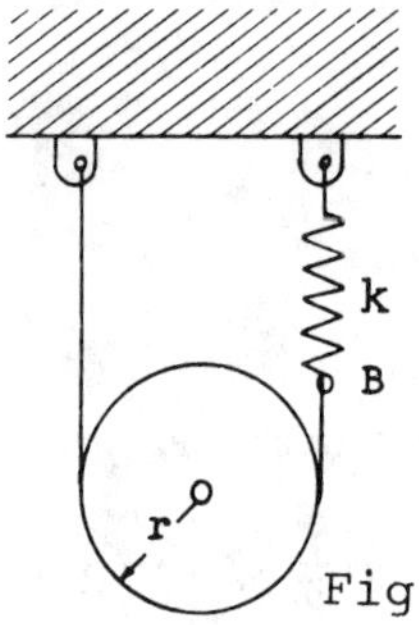

Fig. 1

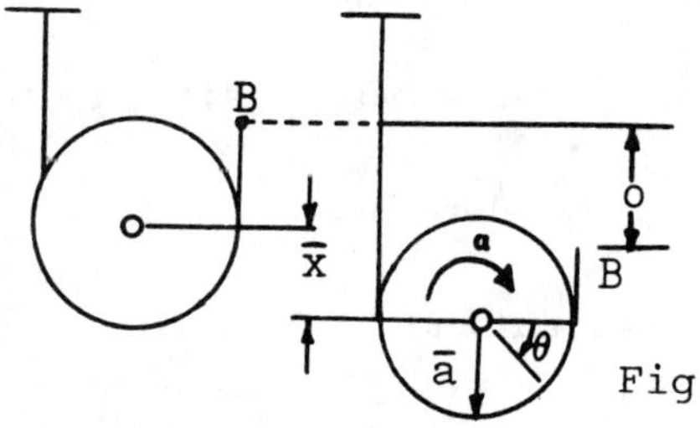

Fig. 2

Solution: Examine Figure 2 to understand the nature of the motion of the cylinder. When the spring is extended, that is, the point of attachment B is lowered a distance δ, the cylinder drops down and rotates. Since the string wraps around the cylinder, displacement is taken up evenly by the right and left supports, that is, a displacement δ causes the cylinder to move downward $x = \delta/2$. Also, the cylinder rolls with no slippage against the cord. In general, for rolling the linear displacement of the center is equal to the product of the radius and the angle subtended, $x = r\theta$. We have then

$$\delta = 2x = 2r\theta \tag{a}$$

and $\vec{a} = \ddot{x} = r\ddot{\theta} = r\vec{\alpha}$. (b)

$\vec{a}$ and $\vec{\alpha}$ are shown in Figure 2.

So far we have only made kinematical considerations, the relevant forces on the cylinder are shown in Figure 3. This system of tensions and weight can be expressed as a torque $T = (T_1 - T_2)r$ and a linear force $F_x = W - (T_1 + T_2)$.

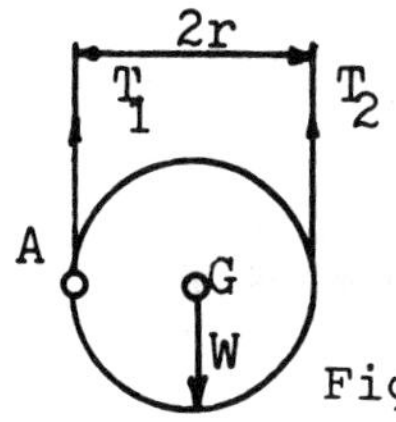

Fig. 3

These yield the equations of motion

$$(T_1 - T_2)r = I\ddot{\theta} \qquad (c)$$

and $$W - (T_1 + T_2) = m\ddot{x} . \qquad (d)$$

When the cylinder is in equilibrium,

$$T_1 = T_2 = \tfrac{1}{2}W.$$

Given a displacement δ, T becomes

$$T_2 = \tfrac{1}{2}W + k\delta = \tfrac{1}{2}W + k(2r\theta).$$

We can eliminate T_1 from the equations (c) and (d) to yield

$$[(W - T_2 - m\ddot{x}) - T_2]r = I\ddot{\theta}$$

$$Wr - 2\,T_2 r = I\ddot{\theta} + m\ddot{x}r.$$

Using the expression for T_2 and equation (b) and $I = \tfrac{1}{2}mr^2$ we get

$$Wr - 2r(\tfrac{1}{2}W + 2kr\theta) = \tfrac{1}{2}mr^2\ddot{\theta} + mr^2\ddot{\theta}$$

$$- 4kr^2\theta = \frac{3}{2}\, mr^2\ddot{\theta}$$

or $$\ddot{\theta} + \frac{8}{3}\frac{k}{m}\,\theta = 0.$$

This establishes that the motion is simple harmonic. The equation for general simple harmonic motion is

$$\ddot{\theta} + \omega^2\theta = 0 .$$

By comparison

$$\omega^2 = \frac{8}{3}\frac{k}{m}$$

$$\omega = \sqrt{\frac{8}{3}\frac{k}{m}} .$$

The frequency f is given by the equation $\omega = 2\pi f$, so

$$f = \frac{\omega}{2\pi} = \frac{1}{2\pi}\left(\frac{8k}{3m}\right)^{\frac{1}{2}} .$$

The period $T = \frac{1}{f} = \frac{2\pi}{\sqrt{\frac{8k}{3m}}} = 2\pi \left(\frac{3m}{8k}\right)^{\frac{1}{2}}$.

• PROBLEM 23-18

Consider the systems shown in the figures below, all of which perform oscillatory motion.

a) The first system is a cylinder of mass m and radius r attached to a pivot by a very light bar. Calculate the characteristic frequency f_a of this system.

b) A cylinder of mass m in a frictionless track slides back and forth. Calculate f_b for this system. Compare this frequency to f_a.

c) The cylinder, mass m, radius r, now rolls back and forth without slipping. Calculate f_c for this system. Compare this frequency to f_a.

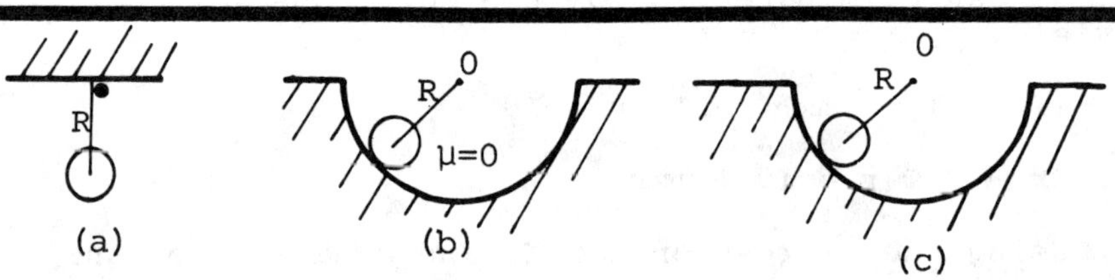

Fig. 1

Solution: a) The first case is that of a compound pendulum to the extent that the cylinder does not act like a point mass. It is known that a compound pendulum is equivalent to a simple pendulum of length $\ell = k_o^2/h$, where k_o is the radius of gyration and h is the distance from the pivot to the center of mass; in this case, h = R. Using the parallel axis theorem,

$$I = mR^2 + \frac{1}{2} mr^2 .$$

Setting $I = mk_o^2$ yields $k_o^2 = R^2 + \frac{1}{2} r^2$.
For a simple pendulum,

$$f = \sqrt{g/\ell}.$$

Hence, $f_a = \sqrt{\frac{g}{k_o^2/h}} = \sqrt{\frac{g}{(R^2 + \frac{1}{2}r^2)/R}}$

$$= \frac{\sqrt{gR}}{\sqrt{R^2 + \frac{1}{2}r^2}} = \frac{\sqrt{g/R}}{\sqrt{1 + (r^2/2R^2)}} .$$

Note that for r << R, $f_a \sim \sqrt{g/R}$, the simple pendulum expression.

b) This back and forth motion involves no rolling, that is, no turning of the cylinder. Therefore, despite its bulk, it acts like a point particle. A free body diagram at an arbitrary position is shown in figure 2.

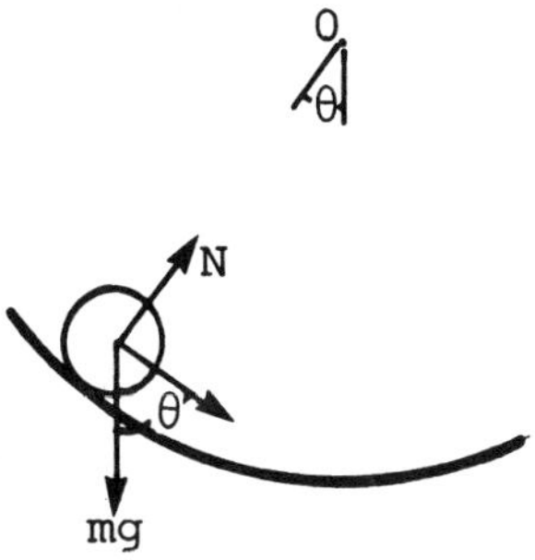

Fig. 2

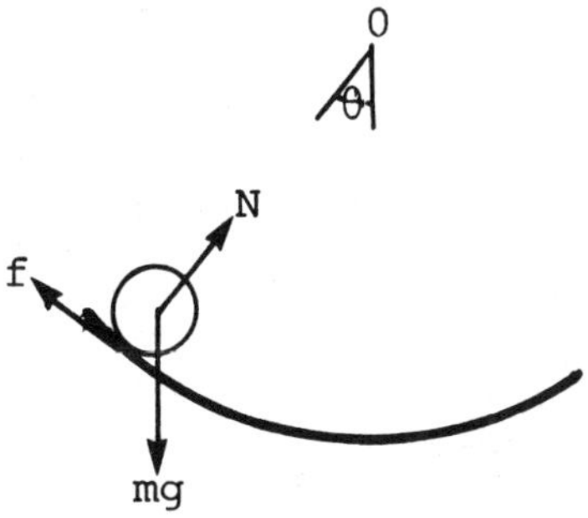

Fig. 3

The oscillitory motion is produced by the mg sin θ force component. The equation of motion, assuming sin θ ≈ θ yields:

$$m\,(R\ddot{\theta}) + mg\,\theta = 0$$

or, $$\ddot{\theta} + (g/R)\,\theta = 0.$$

This is in the standard form for oscillations so that

$$f_b = \sqrt{g/R}.$$

This (and figure 2) is equivalent to the simple pendulum and differs with f_a by the factor $1/\sqrt{1 + r^2/2R^2}$.

c) The free body diagram, assuming rolling only, is shown in figure 3. Assuming small θ again, the equation of motion is

$$m\,(R\,\ddot{\theta}) + (mg\theta - f) = 0.$$

If α is the angular velocity of the cylinder, the following equations also apply:

$$\text{torgue} = I\alpha = r\,f$$

and $$\alpha r = -\,R\,\ddot{\theta}.$$

The minus sign in the last equation is assigned because acceleration of the cylinder in a given direction results in $\ddot{\theta}$ and α being in opposite senses.

Solving the last two equations yields

$$f = -\frac{I}{r^2}R\ddot{\theta} = -\frac{1/2\ mr^2}{r^2}R\ddot{\theta} = -\frac{1}{2}mR\ddot{\theta},$$ since the moment of inertia of a cylinder through its central axis is $\frac{1}{2}mr^2$.

Now, substituting in the first equation gives

$$mR\ddot{\theta} + mg\theta - \left(-\frac{1}{2}mR\ddot{\theta}\right) = 0$$

$$m\frac{3}{2}R\ddot{\theta} + mg\theta = 0$$

$$\ddot{\theta} + \frac{2}{3}g/R\theta = 0$$

from which $f_c = \sqrt{2g/3R}$.

Note that for the reasonable value range $0 < r < R$,

$$f_b > f_a > f_c.$$

The hierachy derives from the "degree of rotation" in the system. That is, the less rotating the cylinder does, the faster the oscillations are approaching the limit $\sqrt{g/R}$.

• PROBLEM 23-19

The system illustrated consists of two homogeneous rods connected by hinge B. Hinge O is fixed and hinge A is constrained to move up and down the vertical line. Use energy conservation to formulate the differential equation of motion. Assume all hinges are frictionless. Also determine the differential equation of motion for small oscillations and solve this equation. What is the equivalent length ℓ_0 of a simple pendulum with the same natural frequency?

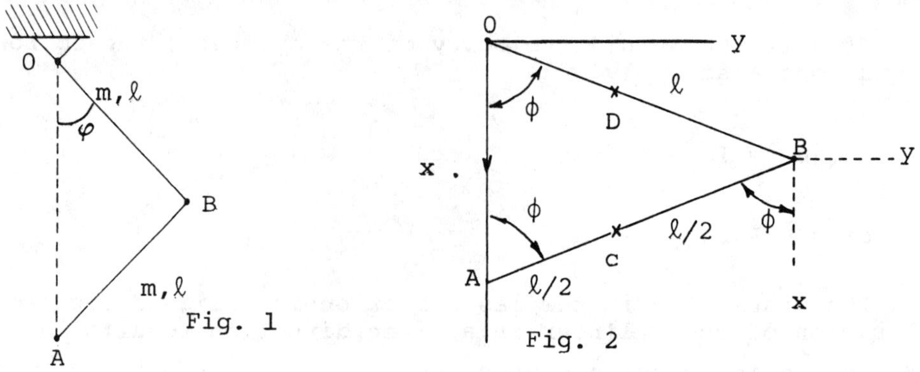

Fig. 1 Fig. 2

Solution: In this problem there are only conservative forces acting. Therefore, the total mechanical energy is constant. To obtain the equation of motion for this system we will calculate the total mechanical energy and differentiate this equation with respect to time.

First, calculate the kinetic energy of the system. From Fig. 2 we see that the motion takes place in the x-y plane and furthermore, we will choose as our origin the fixed point O. Since the rod OB is in pure rotation about the z-axis, we have that

$$(KE)_{OB} = \tfrac{1}{2}(I_{ZZ})_O\ \omega^2. \tag{1}$$

In Eq. (1) $(I_{ZZ})_O = \frac{1}{3} m\ell^2$ and $\omega = \dot{\psi}$. Substituting these values into Eq. (1) yields,

$$(KE)_{OB} = \frac{1}{6} m\ell^2\dot{\phi}^2 \quad . \tag{2}$$

Next, calculate the kinetic energy of rod AB. The kinetic energy of rod AB is equal to the kinetic energy of the center of mass plus the kinetic energy relative to the center of mass or

$$(KE)_{AB} = \tfrac{1}{2}\ mv_C^2 + \tfrac{1}{2}(I_{ZZ})_C\ \omega^2 \quad . \tag{3}$$

Since both rods are of equal length, we see from Fig. 2 that they will have the same angular speed $\dot{\phi}$ and the second term in Eq. (3) is equal to

$$\frac{1}{2}(I_{ZZ})_C \quad \omega^2 = \frac{1}{2}\left(\frac{1}{12}\ m\ell^2\right)\dot{\phi}^2 \ . \tag{4}$$

To evaluate the first term in Eq. (3) we have to determine the velocity of point C relative to point O. This is accomplished by noting that the velocity of point C is equal to the velocity of point B relative to O plus the velocity of point C relative to B

$$\vec{v}_C = \vec{v}_B + \vec{v}_{C/B} \ . \tag{5}$$

Now point C is in pure rotation relative to point B, i.e., $\vec{v}_C = \vec{\omega} \times \vec{r}_{BC}$. Also note that the angular velocity $\vec{\omega}$ of rod AB relative to point B is equal in magnitude in direction to the angular velocity of rod OB relative to point O. Thus,

$$\vec{v}_C = -\ \vec{\dot{\phi}} \times \vec{r}_{BC}. \tag{6}$$

$$\vec{v}_B = \vec{\dot{\phi}} \times \vec{r}_{OB} \ .$$

Substituting this and equation (6) into equation (5) yields

$$\vec{v}_C = \dot{\vec{\phi}} \times \vec{r}_{OB} - \dot{\vec{\phi}} \times \vec{r}_{BC} \,. \qquad (7)$$

Let us evaluate Eq. (7) using Cartesian coordinates. From Fig. 2 we see that

$$\vec{v}_C = \dot{\phi}\, k \times (\ell \cos\phi\, \hat{i} + \ell \sin\phi\, \hat{j})$$

$$- \dot{\phi}\, k \times \left(\frac{\ell}{2} \cos\phi\, \hat{i} - \frac{\ell}{2} \sin\phi\, \hat{j}\right).$$

Evaluating the cross products yields

$$\vec{v}_C = \dot{\phi}\, \ell \cos\phi\, \hat{j} - \dot{\phi}\, \ell \sin\phi\, \hat{i} - \dot{\phi}\, \frac{\ell}{2} \cos\phi\, \hat{j} - \dot{\phi} \sin\phi\, \hat{i},$$

or, $$\vec{v}_C = -\frac{3}{2}\dot{\phi}\, \ell \sin\phi\, \hat{i} + \frac{1}{2}\dot{\phi}\, \ell \cos\phi\, \hat{j}. \qquad (8)$$

Using Eq. (8) $v_C^{\,2}$ can be evaluated.

$$v_C^{\,2} = \vec{v}_C \times \vec{v}_C = \frac{9}{4}\dot{\phi}^2 \ell^2 \sin^2\phi + \frac{1}{4}\dot{\phi}^2 \ell^2 \cos^2\phi;$$

and substituting the trigonometric identity

$\cos^2\phi = 1 - \sin^2\phi$ we obtain,

$$v_C^{\,2} = \frac{1}{4}\dot{\phi}^2 \ell^2 + 2\dot{\phi}^2 \ell^2 \sin^2\phi \,. \qquad (9)$$

We now can obtain an expression for the kinetic energy of the rod AB. Substituting Eqs. (4) and (9) into Eq. (3) gives

$$\frac{1}{2} m v_C^2 + \frac{1}{2}(I_{ZZ})_C\, \dot{\phi}^2 = \frac{1}{2} m \left[\frac{1}{4}\dot{\phi}^2 \ell^2 + 2\dot{\phi}^2 \ell^2 \sin^2\phi\right]$$

$$+ \frac{1}{24} m\ell^2 \dot{\phi}^2,$$

and simplifying, we get

$$(KE)_{AB} = \frac{1}{6} m\ell^2 \dot{\phi}^2 + m\ell^2 \dot{\phi}^2 \sin^2\phi \,. \qquad (10)$$

The sum of Eqs. (2) and (10) is the total kinetic energy of the system. Thus,

$$(KE)_{TOT} = \frac{1}{3} m\ell^2 \dot{\phi}^2 + m\ell^2 \dot{\phi}^2 \sin^2\phi \,. \qquad (11)$$

Next, calculate the potential energy of the system. In this problem, there is only gravitational potential energy. We shall choose the zero of potential energy at $x = 0$. Since the rods are homogeneous, the center of mass of each rod is located at the center of each rod.

In Fig. 2 these points are labelled D and C. Thus the potential energy of rod OB is given by,

$$(PE)_{OB} = - mg \left(\frac{\ell}{2} \cos \phi\right) = - \frac{1}{2} mg\ell \cos \phi \qquad (12)$$

and the potential energy of rod AB is equal to,

$$(PE)_{AB} = - mg \left(\ell \cos \phi + \frac{\ell}{2} \cos \phi\right) = - \frac{3}{2} mg\ell \cos \phi \qquad (13)$$

The total potential energy of the system is equal to the sum of Eqs. (12) and (13)

$$(PE)_{TOT} = - 2 mg \ell \cos \phi. \qquad (14)$$

We now have an expression for the total mechanical energy of the system, that is, the total energy is equal to the sum of Eqs. (11) and (14),

$$E_{TOT} = \frac{1}{3} m \ell^2 \dot{\phi}^2 + m\ell^2 \dot{\phi}^2 \sin^2 \phi - 2mg \ell \cos \phi . \qquad (15)$$

Next, differentiate Eq. (15) with respect to time. Performing this operation yields,

$$\dot{E} = \frac{1}{3} m\ell^2 (2\dot{\phi})\ddot{\phi} + m\ell^2 (2\dot{\phi})\ddot{\phi} \sin^2 \phi$$

$$+ m\ell^2 \dot{\phi}^2 (2 \sin \phi)\cos \phi(\dot{\phi}) - 2 mg \ell (- \sin \phi)\dot{\phi},$$

and simplifying,

$$\dot{E} = \frac{2}{3} m\ell^2 \dot{\phi}\ddot{\phi} + m\ell^2 (2\dot{\phi}\ddot{\phi} \sin^2 \phi + 2\dot{\phi}^3 \sin \phi \cos \phi)$$

$$+ 2mg \ell \dot{\phi} \sin \phi .$$

Since there are no dissipating forces, the total energy is constant, so $\dot{E} = 0$. Thus the right hand side of the above equation is equal to zero.

Dividing through by $2m\ell^2\dot{\phi}$ gives,

$$\left(\frac{1}{3} + \sin^2 \phi\right)\ddot{\phi} + (\sin \phi \cos \phi)\dot{\phi}^2 + \left(\frac{g}{\ell}\right) \sin \phi = 0. \qquad (16)$$

Eq. (16) is the equation of motion for the system. Next, obtain the equation of motion for small angles of oscillations.

Using the small angle approximations that $\sin \phi = \phi$, and $\cos \phi = 1$, we get by substituting into Eq. (16):

$$\left(\frac{1}{3} + \phi^2\right)\ddot{\phi} + [\phi(1)]\dot{\phi}^2 + \left(\frac{g}{\ell}\right)\phi = 0 .$$

Ignoring all ϕ^2 or $\dot{\phi}^2$ terms since they will be very small, yields,

$$\frac{1}{3}\ddot{\phi} + \left(\frac{g}{\ell}\right)\phi = 0 \qquad \text{or}$$

$$\ddot{\phi} + \left(\frac{3g}{\ell}\right)\phi = 0 \quad . \tag{18}$$

Eq. (18) is the desired equation of motion. We recognize that Eq. (18) has the same form as the equation of motion for a simple harmonic oscillator and the general solution is,

$$\phi = C_1 \sin\sqrt{\frac{3g}{\ell}}\,t + C_2 \cos\sqrt{\frac{3g}{\ell}}\,t. \tag{19}$$

In Eq. (19) the constants C_1 and C_2 are determined from the known conditions at some time t. Next, write down the equation of motion for small oscillations of a simple pendulum of length ℓ_0. This equation is:

$$\ddot{\phi} + \left(\frac{g}{\ell_0}\right)\phi = 0. \tag{20}$$

Comparing Eqs. (18) and (20), note that they are identical if the length of the simple pendulum is

$$\ell_0 = \frac{\ell}{3}\,.$$

• PROBLEM 23-20

The system shown in Fig. 1 consists of a block of mass m, a uniform disk of mass m, a spring having a spring constant C and a weightless cord. The disk is free to rotate about its central axis and the cord does not slip on the disk. Assuming the block moves only up and down, determine the motion of the block if at t = 0 it passes the equilibrium position with a downward velocity v_0. Also find all external and internal forces acting on the system and give the condition that v_0 has to satisfy so that the cord always remains taut.

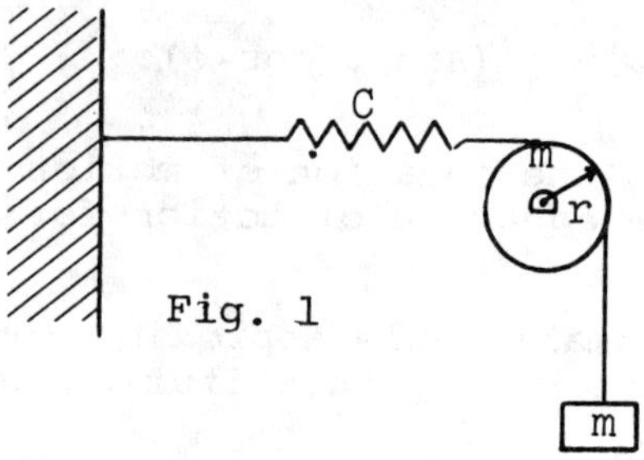

Fig. 1

Solution: This problem involves both translational and rotational motion. We will use Newton's Second Law to obtain the equation of motion for the block of mass m and Euler's equation to obtain the equation of motion for the disk.

Refer to the free body diagrams (Figs. 2 and 3). First, use Newton's Second Law applied to the block

$$\Sigma F = m\ddot{x} \qquad \text{or}$$

$$mg - s_2 = m\ddot{x} \; . \qquad (1)$$

Next, use Euler's equation about the axis of rotation of the block

$$T = I\ddot{\theta} \qquad \text{or,}$$

$$s_2 r - s_1 r = I\ddot{\theta} \; . \qquad (2)$$

In Eqs. (1) and (2), both x and θ are measured from their respective equilibrium positions.

Now, relate the coordinates x and θ. Since the cord remains taut, we have

$$x = r\theta \; .$$

The second derivative with respect to time of this equation is given by

$$\ddot{x} = r\,\ddot{\theta}. \qquad (3)$$

Also note that the moment of inertia of the disk of uniform thickness is

$$I = \tfrac{1}{2}mr^2 \; . \qquad (4)$$

Substituting Eqs. (3) and (4) into Eq. (2), yields

$$s_2 r - s_1 r = \tfrac{1}{2}\,mr^2\left(\frac{\ddot{x}}{r}\right) \; .$$

Simplifying this equation yields

$$2s_2 - 2s_1 = m\ddot{x} \; . \qquad (5)$$

We now have two equations of motion, Eqs. (1) and (5). Eliminate the force s_2 between these two equations. Do this by solving Eq. (1) for s_2

$$s_2 = mg - m\ddot{x}$$

and substituting this expression into Eq. (5)

$$2(mg - m\ddot{x}) - 2s_1 = m\ddot{x} \; .$$

Rearranging this equation, we obtain

$$3m\ddot{x} = 2mg - 2s_1 \quad . \qquad (6)$$

To solve equation (6) we need an expression for the force s_1. Figs. 1 and 2 illustrate that this force is caused by the spring. When the block of mass m is in its equilibrium position, then the force s_1 with which the spring pulls on the cord, is equal to mg. But if the block moves a distance x downwards, the spring is stretched an additional distance x and the total force with which the spring pulls on the cord is equal to

$$s_1 = mg + cx \qquad (7)$$

where c is the spring constant.

Substituting Eq. (7) into Eq. (6) and simplifying yields

$$\ddot{x} + k^2x = 0 \qquad (8)$$

where $k^2 = \frac{2c}{3m}$.

Eq. (8) is a second order linear differential equation with constant coefficients. The general solution of this equation is

$$x = A \sin kt + B \cos kt. \qquad (9)$$

In Eq. (9) the constants A and B are determined from the initial conditions. When t = 0, x = 0, and $\dot{x} = v_0$. Substitute the initial conditions into Eq. (9) to get

$$0 = A \sin k(0) + B \cos k(0) = 0 + B$$

so $B = 0$.

The derivative of Eq. (9) is

$$\dot{x} = Ak \cos kt - Bk \sin kt . \qquad (10)$$

Substituting the initial conditions,

$$\dot{x} = Ak \cos 0 - Bk \sin kt = Ak - 0$$

so $A = \frac{\dot{x}}{k}$.

Substituting these values into Eq. (9) we then have the desired solution,

$$x = \frac{v_0}{k} \sin kt. \qquad (11)$$

Next, determine all the external and internal forces (see Figs. 2 and 3). From Eq. (7) we have that

$$s_1 = mg + cx . \qquad (7)$$

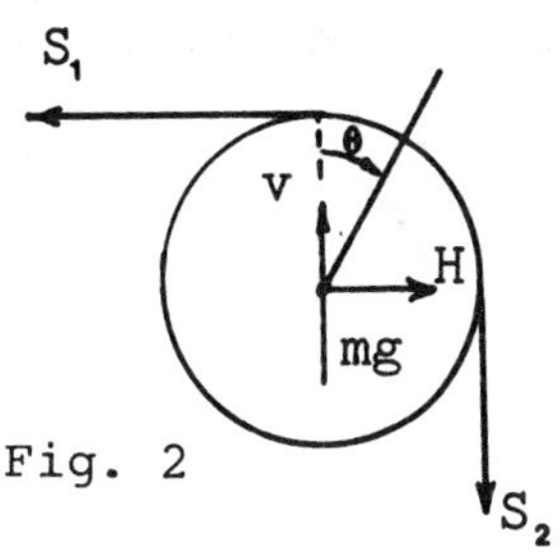

Fig. 2

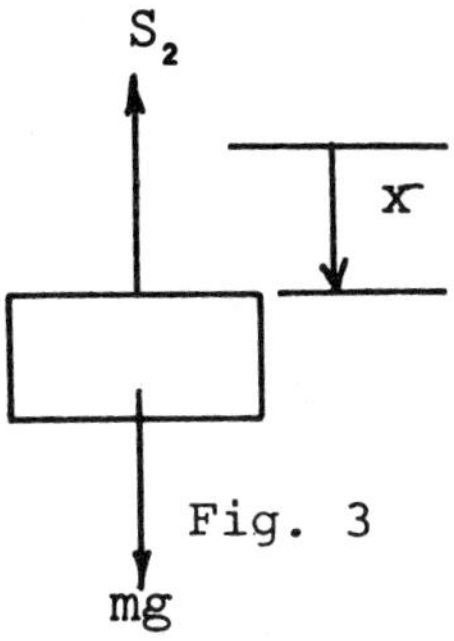

Fig. 3

Substituting Eq. (8) into Eq. (1) and simplifying gives us the force s_2.

$$s_2 = mg - m\ddot{x} = mg + mk^2x. \qquad (12)$$

Now, find the hinge support forces V and H. Since the center of mass of the disk is stationary, the vector sum of all of the force acting on the disk is equal to zero. Thus for the horizontal direction we get,

$$- H + s_1 = 0 \quad , \qquad (13)$$

and for the vertical direction,

$$- V + mg + s_2 = 0. \qquad (14)$$

Using Eqs. (7) and (13) solve for H,

$$H = s_1 = mg + cx,$$

and using Eqs. (12) and (14) solve for V,

$$V = mg + s_2 = 2mg + mk^2x.$$

Finally, determine the condition on v_0 so that the cord will always remain taut.

The condition necessary so that the string remains taut is that the string remain under tension, or $s_1 > 0$, $s_2 > 0$. s_1, or s_2 less than zero would refer to a change in direction of the forces or a compression,which in the case of non-compressible strings refers to a loss of tautness.

That is, from equations (7) and (12)

$$mg + cx > 0$$

$$mg + mk^2x > 0$$

or

$$x > - mg/c \qquad (15)$$

$$x > - g/k^2. \qquad (16)$$

Substituting $k^2 = 2c/3m$ into equation (16) gives

$$x > - 3mg/2c \,,$$

so we see that $x > -\, mg/c$ satisfies both conditions.

The minimum value for x from equation (11) is when $\sin\theta = 1$, or $x = -\, v_0/k$, so we have

$$-\frac{v_0}{k} > \frac{mg}{c}$$

or $$v_0 < \frac{kmg}{c} \; .$$

Substituting for k

$$v_0 < g\sqrt{\frac{2c}{3m}}\;\frac{m}{c}$$

$$v_0 < g\sqrt{\frac{2}{3}\frac{m}{c}} \; .$$

• PROBLEM 23-21

Determine the natural frequency of oscillation of the spring-mass system shown in Fig. 1 by two methods, force and energy. Assume pure rolling between the cylinder and the plane, and small angular displacement.

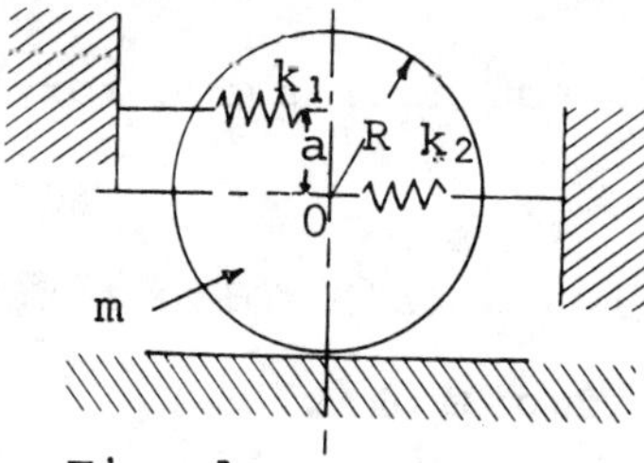

Fig. 1

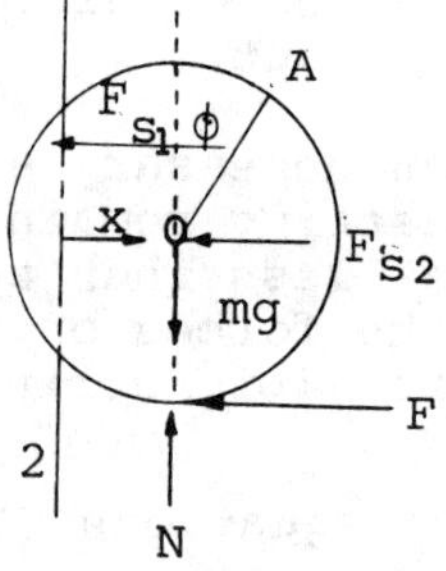

Fig. 2

Solution: Referring to the free-body diagram, Fig. 2, in which the cylinder is displaced a small amount to the right, Newton's second law written for the mass center (O) is

$$-\, k_1(x + a\phi) - k_2 x - F = m\ddot{x}$$

where F_{s1}, the spring force due to spring 1, is based upon a total deformation $(x = a\phi)$, and F represents the force of the plane upon the cylinder.

Summing moments of force about the mass center and equating to $I\ddot{\phi}$, we have, in the small angle approximation,

$$FR - k_1(x + a\phi)a = I\ddot{\phi}.$$

Multiplying the first equation by R and adding to the second, F is eliminated:

$$-k_1(x + a\phi)(R + a) - k_2xR = m\ddot{x}R + I\ddot{\phi}.$$

Since we have pure rolling, $x = R\phi$ and $\ddot{x} = R\ddot{\phi}$, and these may be substituted in the above expression to yield an equation in x or ϕ alone. Thus x and ϕ are not independent and the system possesses only one degree of freedom. In terms of ϕ,

$$-k_1(R\phi + a\phi)(R + a) - k_2R^2\phi = mR^2\ddot{\phi} + I\ddot{\phi}$$

or, $$\ddot{\phi} + \frac{[k_1(R + a)^2 + k_2R^2]\phi}{mR^2 + I} = 0$$

which has the familiar form of an undamped single degree of freedom oscillator. The natural circular frequency is, by inspection,

$$\omega_0 = \left[\frac{k_1(R + a)^2 + k_2R^2}{mR^2 + I}\right]^{\frac{1}{2}}.$$

Employing an energy approach, the kinetic- and potential-energy terms are

$$T = \tfrac{1}{2}m\dot{x}^2 + \tfrac{1}{2}I\dot{\phi}^2 = \tfrac{1}{2}mR^2\dot{\phi}^2 + \tfrac{1}{2}I\dot{\phi}^2$$

$$V = \tfrac{1}{2}k_1(x + a\phi)^2 + \tfrac{1}{2}k_2x^2 = \tfrac{1}{2}k_1(R\phi + a\phi)^2 + \tfrac{1}{2}k_2R^2\phi^2.$$

The force of friction does no work (because there is zero slip) and consequently the energy is constant:

$$\frac{d}{dt}[\tfrac{1}{2}mR^2\dot{\phi}^2 + \tfrac{1}{2}I\dot{\phi}^2 + \tfrac{1}{2}k_1(R\phi + a\phi)^2 + \tfrac{1}{2}k_2R^2\phi^2] = 0$$

This yields,

$$mR^2\dot{\phi}\ddot{\phi} + I\dot{\phi}\ddot{\phi} + k_1(R\phi + a\phi)(R\dot{\phi} + a\dot{\phi}) + k_2R^2\phi\dot{\phi} = 0,$$

$$\ddot{\phi}(mR^2 + I) + k_1(R + a)^2\phi + k_2R^2\phi = 0,$$

$$\ddot{\phi} + \frac{[k_1(R + a)^2 + k_2R^2]\phi}{mR^2 + I}$$

as before.

• PROBLEM 23-22

Shown in the Figure 1 is a schematic diagram of an electrical machine that has an unbalanced rotor. The unbalance is represented schematically by a mass m, removed at a distance e from the axis of rotation. The machine is operating at a frequency of p radians per second, that is, the angular velocity of the rotor is equal to $2\pi p$ revolutions/sec. Write the equation of motion of the mass m of the engine block. Neglect weight in the free-body diagrams.

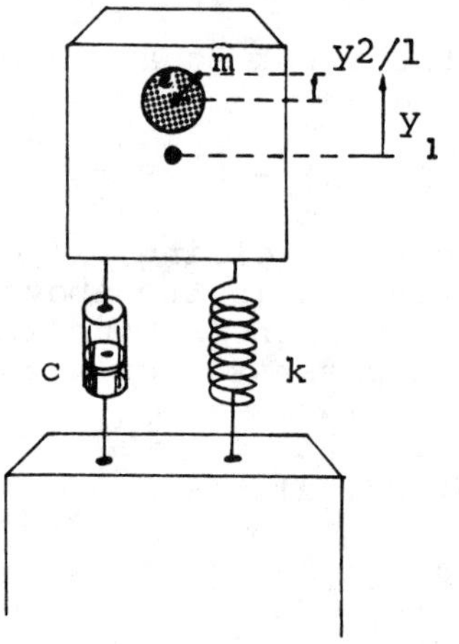

Fig. 1

Solution: Figure 1 shows the coordinates to be used for this problem. y_1 locates the position of the motor M, y_2 (not shown) locates the position of the rotor imbalance m. Both of these are with respect to a fixed axis. $y_{2/1}$ locates the position of m relative to the center of the rotor M.

The force that the engine block is exerting on the rotor in the y direction is given by $F_y(t)$. By Newton's second law, this force may be determined if the acceleration of the mass m is known. Thus,

$$F_y = m\ddot{y}_2 .$$

But, employing the relative acceleration equation for motion in a plane,

$$\ddot{y}_2 = \ddot{y}_1 + \ddot{y}_{2/1} . \qquad (a)$$

The acceleration in the y direction of the second mass with respect to the first is equal:

$$\ddot{y}_{2/1} = - ep^2 \sin pt$$

so that the force on the unbalance in the y direction is equal to

$$R_y = m(\ddot{y}_1 - ep^2 \sin pt) . \qquad (b)$$

Now, by Newton's third law, the force that the engine block exerts on the rotor must be equal in magnitude and opposite in direction to the force that the rotor exerts on the engine block. This was assumed in drawing R_y in the free-body diagram of Fig. 2. Therefore the equation of motion of the engine block is

$$F_y = M\ddot{y}_1$$

$$-R_y - c\dot{y}_1 - ky_1 = M\ddot{y}_1$$

or, by using the determined value of R_y, equation (b),

$$(M + m)\ddot{y}_1 + c\dot{y}_1 + ky_1 = mep^2 \sin pt .$$

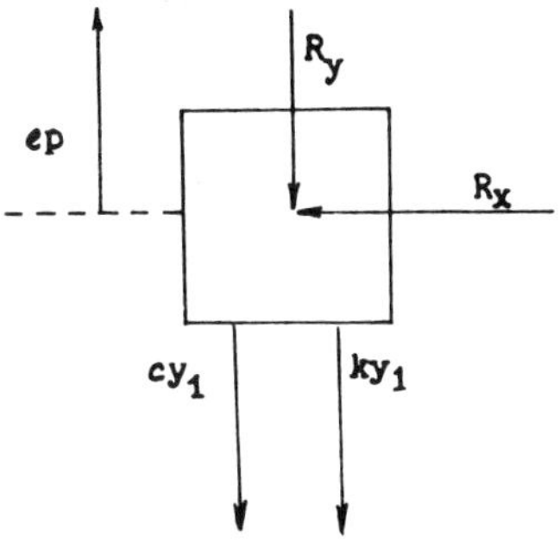

Fig. 2(a)

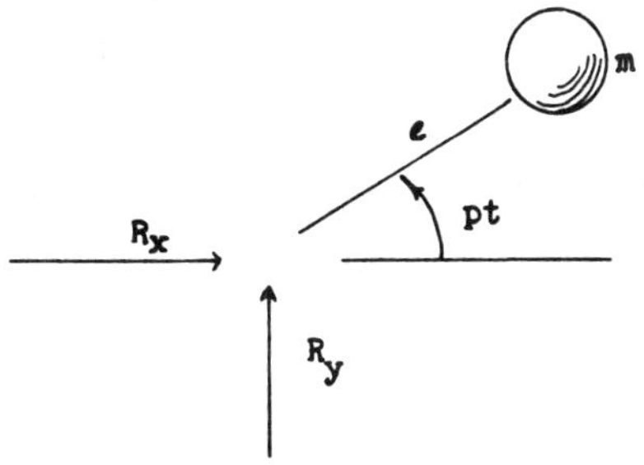

Fig. 2(b)

If the unbalance is fairly small, the magnitude of m will be small in comparison to the magnitude of M, so that the equation may be written as

$$M\ddot{y}_1 + c\dot{y}_1 + ky_1 = mep^2 \sin pt \quad . \qquad (c)$$

There is a standard solution for this equation in its most general form

$$m\ddot{x} + b\dot{x} + kx = F_m \sin \omega t .$$

It is, $x = \dfrac{F_m/k}{\sqrt{[1 - (\omega/\omega_0)^2]^2 + 4m\omega}} \sin (\omega t - \phi)$

where $\omega_0 = \sqrt{\dfrac{k}{m}}$, and

$$\tan \phi = \frac{4m\omega}{1 - (\omega/\omega_0)^2} \quad .$$

The equation of motion for the block is then,

$$y_1(t) = \frac{mep^2/k}{\sqrt{(1 - (p/p_0)^2)^2 + 4mp}} \sin (pt - \phi)$$

$$= \frac{er^2}{(1 - r^2) + 4mp} \sin (pt - \phi)$$

where $r^2 = \frac{p^2}{p_0^2}$ and $p_0{}^2 = \frac{k}{m}$. And

$$\tan \phi = \frac{4mp}{1 - r^2} .$$

• PROBLEM 23-23

An electric dipole, consisting of a pair of oppositely charged electrons attached to a massless rod of length ℓ, is placed in a uniform electric field of magnitude E. Find the natural frequency of oscillation if the dipole, shown in Fig. 1, is slightly rotated from its equilibrium position and released. If a damping torque proportional to angular velocity of the dipole is present, determine the frequency of oscillation.

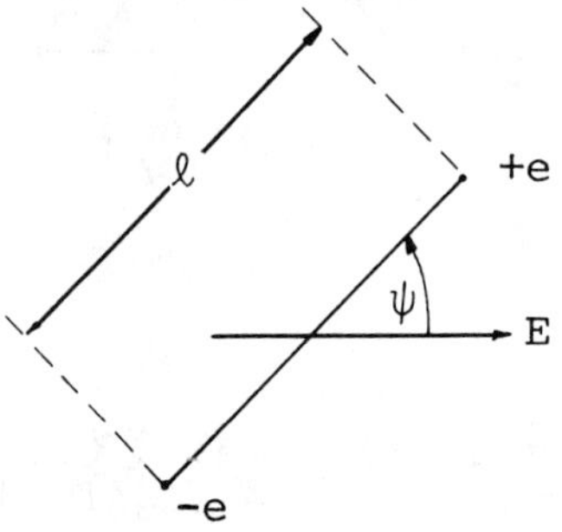

Fig. 1

Solution: Although the net force on the dipole is zero (eE - eE), the net torque is $- eE\ell \sin \psi$. The torque is equated to the product of system moment of inertia (about the center of mass) and the angular acceleration. For the undamped case,

$$- eE\ell \sin \psi = I\ddot{\psi} = \tfrac{1}{2}m\ell^2\ddot{\psi}$$

where $I = 2m(\ell/2)^2 = \tfrac{1}{2}m\ell^2$.

For small angles, the equation of motion is,

$$\ddot{\psi} + \frac{2eE}{m\ell}\psi = 0.$$

The undamped natural circular frequency is, therefore,

$$\omega_0 = \left(\frac{2eE}{m\ell}\right)^{\frac{1}{2}} .$$

With damping present,

$$- eE\ell\psi - C\dot{\psi} = \tfrac{1}{2}m\ell^2\ddot{\psi}$$

where C represents the torsional damping coefficient (torque per unit angular velocity). The above equation may be written

$$\ddot{\psi} + \frac{2C}{m\ell^2}\dot{\psi} + \frac{2eE}{m\ell}\psi = 0 .$$

Assuming a solution of the form $\psi = e^{st}$, taking the appropriate derivatives and substituting into the equation of motion,

$$s^2 + \left(\frac{2C}{m\ell^2}\right) s + \frac{2eE}{m\ell} = 0$$

and by using the quadratic formula

$$s = -\frac{C}{m\ell^2} \pm \left[\left(\frac{C}{m\ell^2}\right)^2 - \frac{2eE}{m\ell}\right]^{\frac{1}{2}}.$$

The condition for oscillation of the dipole is

$$\frac{2eE}{m\ell} > \left(\frac{C}{m\ell^2}\right)$$

and the circular frequency of oscillation is

$$\Omega = \left[\frac{2eE}{m\ell} - \left(\frac{C}{m\ell^2}\right)^2\right]^{\frac{1}{2}} .$$

As the damping coefficient increases, Ω diminishes, until at $C = (2eEm\ell^3)^{\frac{1}{2}}$, oscillation ceases.

• PROBLEM 23-24

A model is used to find the resistance of water to the motion of a ship at low speeds. The model is placed in a tank and the bow and stern are attached to the ends of the tank by identical springs with spring constant k. When the model is displaced lengthwise from its position of equilibrium, it is observed that the amplitudes of the oscillations decrease such that the ratio of each maximum displacement to the preceeding one in the same direction is 0.85. The frequency of the oscillation is 0.75 vibrations per second. Determine the drag constant of proportionality if it is assumed that the resistance varies in proportion to the first power of the speed.

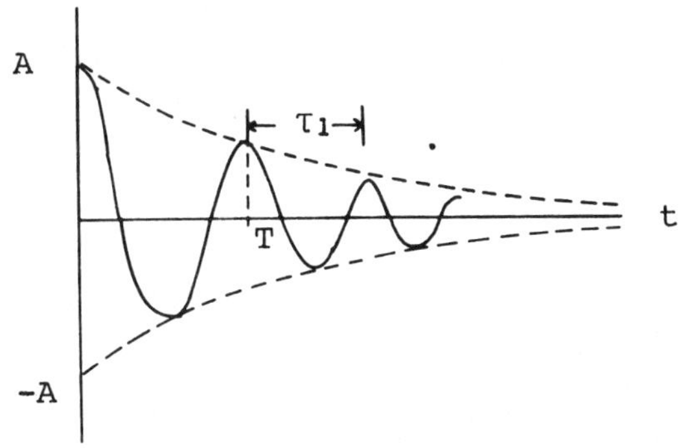

Solution: The differential equation describing the damped oscillatory motion is

$$m\ddot{x} + b\dot{x} + 2kx = 0 \tag{1}$$

where m is the mass of the ship model, b is the constant of proportionality of the resistance of the water, and 2k is the combined effective spring constant of the two springs which are in effect connected in parallel with each other. Equation (1) may be written in the form

$$\ddot{x} + 2\beta\dot{x} + \omega_0^2 x = 0 \tag{2}$$

where $\beta = \frac{b}{2m}$ is called the damping parameter and $\omega_0 = \sqrt{\frac{2k}{m}}$ is the undamped, or natural angular frequency of oscillation of the spring-mass system. The solution of equation (2) for the underdamped case $(\omega_0^2 > \beta^2)$ is found by standard techniques:

$$x(t) = A\, e^{-\beta t} \cos(\omega_1 t - \delta) \tag{3}$$

where A is the initial displacement from the equilibrium position, $\omega_1 = \sqrt{\omega_0^2 - \beta^2}$ is the frequency of oscillation of the damped oscillatory system, and δ is a phase angle which depends on the initial velocity. The displacement x(t) is shown in the graph below as a function of time.

The dashed line represents the envelope of the oscillation curve and is given as $Ae^{-\beta t}$. The solid line represents the oscillation of the ship model, and is given by equation (3) for the case $\delta = 0$. The maximum displacement at time T is $x_1 = Ae^{-\beta T}$, and at time $T + \tau_1$, one vibration later, is $x_2 = Ae^{-\beta(T+\tau_1)}$.

Thus, the ratio of the two displacements is

$$\frac{x_2}{x_1} = \frac{Ae^{-\beta(T+\tau_1)}}{Ae^{-\beta T}} = e^{-\beta\tau_1}. \tag{4}$$

τ_1 is the period of oscillation and is related to the the angular frequency by $\tau_1 = \frac{2\pi}{\omega_1}$. Taking natural logarithms of both sides of equation (4), we get:

$$\ln\left(\frac{x_2}{x_1}\right) = -\beta\tau_1 . \tag{5}$$

The quantity $\beta\tau_1$ is called the logarithmic decrement. From equation (2),

$$\beta\tau_1 = \left(\frac{b}{2m}\right)\left(\frac{2\pi}{\omega_1}\right) = \left(\frac{b}{2m}\right)\left(\frac{1}{f}\right) \tag{6}$$

where $f = \frac{\omega_1}{2\pi}$ is the frequency of oscillation. Solving for b using equations (5) and (6),

$$b = 2mf\ (\beta\tau_1) = -\ 2mf\ \ln\left(\frac{x_2}{x_1}\right) .$$

Substituting numerical data, the constant of proportionality is

$$b = 2\ m\ (0.75)\ \ln\ (0.85) = 0.24\ m.$$

• PROBLEM 23-25

The slender rod shown is 12 ft long and weighs 96 lb. (a) Derive the differential equation for small oscillations of the rod. (b) Solve the differential equation and determine the value of the spring constant k required for critical damping if c_c = 20 lb-sec/ft. (c) What will be the angular displacement for the critically damped system at t = 1 sec if the initial conditions at t = 0 are $\dot{\theta}$ = 4 rad/sec and θ = 0.

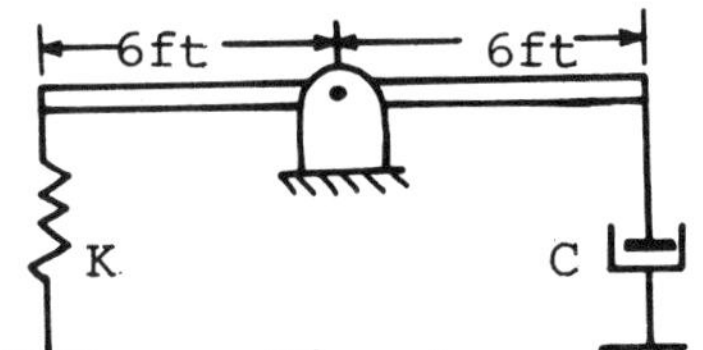

Fig. 1

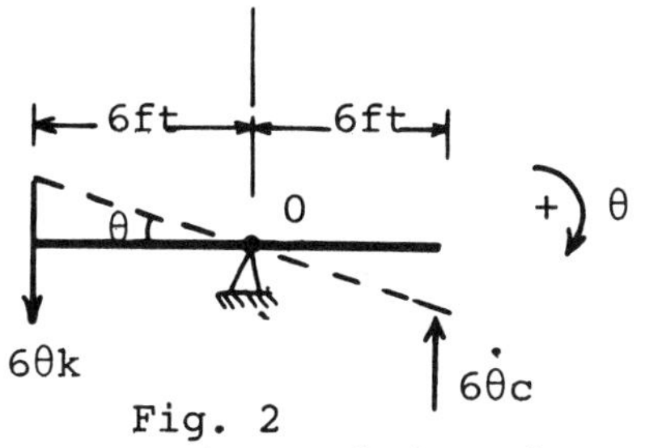

Fig. 2

Solution: Euler's equation $\Sigma M_0 = I\alpha$ is needed. The moment about O due to the dashpot is the force it exerts times the moment arm.

$$F_c = c\dot{x} = c(\dot{\theta}r) = 6c\dot{\theta}$$

So its moment is $M_c = -\ 6(6c\dot{\theta})$, since + θ is assumed clockwise.

The moment due to the spring is

$$M_s = F_s r = (-\ ks)6 = -\ k(6\theta)(6) .$$

Therefore, the sum of all the moments is

$$\Sigma M_0 = M_s + M_c = -\ 36k\theta - 36c\dot{\theta}, \qquad (1)$$

which is equal to $I\ddot{\theta}$ from Euler's equation.

For a rod suspended from its center, its moment of inertia is

$$I = \frac{1}{12} m\ell^2 = \frac{1}{12}\frac{W}{g}\ell^2 = \frac{1}{12}\frac{W}{g}(12)^2 . \qquad (2)$$

So $\Sigma M_0 = I\ddot{\theta}$ becomes, from Eq. (1) and (2)

$$-36k\theta - 36c\dot{\theta} = \frac{w}{g}(12)\ddot{\theta} \quad , \quad \text{or}$$

$$\frac{w}{g}(12)\ddot{\theta} + 36c\dot{\theta} + 36k\theta = 0 \; .$$

Simplifying, $\ddot{\theta} + \frac{3gc}{w}\dot{\theta} + \frac{3gk}{w}\theta = 0.$ (3)

The general damped oscillator equation is

$$m\ddot{x} + b\dot{x} + k'x = 0. \qquad (4)$$

b_c is found by setting the discriminant in the characteristic equation equal to 0, or $b_c = 2\sqrt{k'm}$.

By comparison of equations (3) and (4),

$$b = \frac{3gc}{w} \qquad \text{and } k' = \frac{3gk}{w} = \omega^2, \qquad (4(a))$$

so $b_c = 2\sqrt{k'm} = \dfrac{3gc_c}{w}$

and $c_c = \dfrac{2w}{3g}\sqrt{k'm}$.

Substituting for k',

$$c_c = \frac{2w}{3g}\sqrt{\frac{3gk}{w}m} = 2\sqrt{\frac{w^2}{9g^2}}\sqrt{\frac{3gk}{w}m} = 2\sqrt{\frac{wk}{3g}} \qquad (5)$$

Solving Eq. (5) for k, which is what the question asks for, gives

$$k = \frac{3}{4}\frac{c_c{}^2 g}{w} \; .$$

Putting in given values,

$$k = \left(\frac{3}{4}\right)\frac{(20)^2\,(32.2)}{96} = 100.6 \text{ lb/ft.}$$

The solution to Eq. (3) is most easily found by referring again to the generic equation. Its solution is

$$\theta = Ae^{p+t} + Be^{p-t}$$

where $p_{\pm} = \dfrac{-b \pm \sqrt{b^2 - 4mk'}}{2m}$.

However, with critical damping, $p_+ = p_- = \dfrac{-b}{2m}$ and $b_c = 2\sqrt{k'm}$.

Therefore $p_+ = p_- = \sqrt{\frac{k'}{m}}$.

When the roots of the characteristic equation are equal, the solution is

$$\theta = (A + Bt)e^{-\sqrt{\frac{k'}{m}}\,t} = (A + Bt)e^{-\omega t} \qquad (6)$$

where $\omega = \sqrt{\frac{k'}{m}}$.

Differentiating Eq. (6), we get

$$\dot{\theta} = Be^{-\omega t} + (A + Bt)(-\omega)e^{-\omega t} . \qquad (7)$$

The initial conditions are at t = 0, $\dot{\theta}$ = 4 rad/s, and $\theta = 0$.

Substituting these values into Eq. (6) to find the constant A,

$$0 = [A + B(0)]e^{-\omega(0)}$$

$$0 = A(1)$$

$$A = 1 .$$

Putting the initial conditions on Eq. (7),

$$4 = Be^0 + (0 + B(0))(-\omega)e^0$$

$$4 = B + 0$$

$$B = 4 .$$

Equation (6) becomes

$$\theta = 4te^{-\omega t} .$$

Now, $\omega = \sqrt{\frac{3gk}{W}} = \sqrt{\frac{(3)(32.2)(100.6)}{96}} = 10.06$, from Eq. (4(a)), so θ at t = 1 is

$$\theta(t = 1) = 4e^{-10.06} = .000171 \text{ rad},$$

indicating that the critical damping has virtually halted the motion already.

• PROBLEM 23-26

An electric motor with a mass of 50 kg is supported by four springs having a spring rate of 400 N/mm. The rotor having an imbalance equivalent to a mass of 40 g with an eccentricity of 150 mm. The motor is restrained to move vertically only. The operating speed of the motor is 1500 rpm. Determine the steady-state amplitude of vibration (a) neglecting damping (b) for a damping factor $c/c_c = 0.15$.

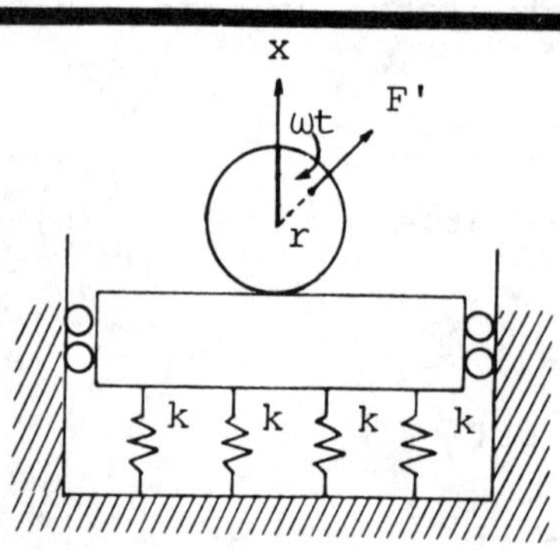

Fig. 1

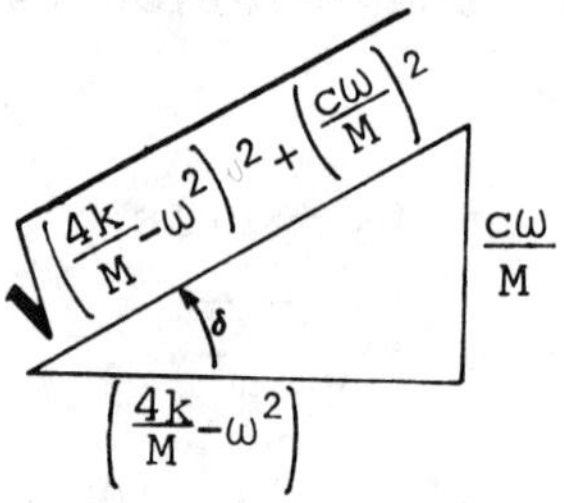

Fig. 2

Solution: In this problem we shall obtain the differential equation of motion by applying Newton's Second Law of Motion. We shall then discuss the general solution of this equation and in particular, the steady state motion of the system.

Apply Newton's second law in the vertical direction,

$$\Sigma F = m\ddot{x} \quad . \tag{1}$$

In Eq. (1), ΣF represents all the vertical forces acting on the motor when it is rotating and when the motor is displaced from its equilibrium position by a distance x. For the case of the damped motion there are three forces acting. These are as follows:

$$F_1 = -\ 4kx. \tag{2}$$

This is the linear restoring force due to the four springs. Next there is the frictional force or damping force. This force is proportional to the velocity and in the opposite direction, that is,

$$F_2 = -\ c\dot{x}\ , \tag{3}$$

where c is the coefficient of damping. Finally, there is a force acting on the motor due to the unbalance of the rotor. This force is indicated as F' in Fig. 1. The vertical component of this force is given by

$$F_3 = F' \cos \omega t, \tag{4}$$

where F' is the centrifugal force or,

$$F' = m\frac{v^2}{r} = m\ \omega^2 r. \tag{5}$$

In Eq. (5), $m = 40\ g = 40 \times 10^{-3}$ kg, $\omega = 1500 \left(\frac{2\pi}{60}\right) \frac{\text{rad.}}{\text{sec.}}$, and $r = 150$ mm $= 0.15$ m. The resultant force acting on the motor is the sum of Eqs. (2),(3) and (4), thus we get:

$$F = F_1 + F_2 + F_3 = -4kx - c\dot{x} + F' \cos \omega t. \quad (6)$$

Substituting Eq. (5) into Eq. (6) and substituting the resulting equation into Eq. (1) yields:

$$-4kx - c\dot{x} + m\omega^2 r \cos \omega t = M\ddot{x}. \quad (7)$$

Dividing Eq.(7) by M and rearranging the equation we obtain the desired differential equation of motion, which is:

$$\ddot{x} + \frac{c}{M}\dot{x} + \frac{4k}{M}x = \frac{m\omega^2 r}{M} \cos \omega t. \quad (8)$$

We recognize Eq. (8) as the differential equation of motion for the forced, damped harmonic oscillator.

Next, find the general solution of Eq. (8). The solution will be the general solution of the homogeneous equation plus any particular solution of the inhomogeneous equation.

Consider the homogeneous equation,

$$\ddot{x} + \frac{c}{M}\dot{x} + \frac{4k}{M}x = 0. \quad (9)$$

Assume the solution is of the form,

$$x = Ae^{pt}. \quad (10)$$

So $\dot{x} = Ape^{pt}$, and $\ddot{x} = Ap^2e^{pt}$.

Substituting Eq. (10) into Eq. (9) and cancelling common terms, yields the characteristic equation,

$$p^2 + \frac{c}{M}p + \frac{4k}{M} = 0. \quad (11)$$

Solving Eq. (11) for p gives,

$$p = -\frac{c}{2M} \pm \sqrt{\left(\frac{c}{2M}\right)^2 - \frac{4k}{M}}. \quad (12)$$

There are three different cases to consider:

Case 1. $\sqrt{\left(\frac{c}{2M}\right)^2 - \frac{4k}{M}} = 0$ or

$$c = 4\sqrt{kM} \quad (13)$$

This is the case where the motion is critically damped. Let us call this constant c_c, $c_c = 4\sqrt{kM}$.

Case 2. $\left(\frac{c}{2M}\right)^2 - \frac{4k}{M} > 0$

This is the case where p is a real number and the motion is overdamped.

Case 3. $\left(\frac{c}{2M}\right)^2 - \frac{4k}{M} < 0$

For this situation, p is a complex number and the motion is underdamped.

No matter which case we consider, each of the solutions is multiplied by the exponential term $e^{-\frac{c}{2M}t}$ since we assumed the solution is of the form e^{pt}. Thus the solution to the homogeneous equation dies out exponentially with time. This is called the transient solution. We are not concerned with this solution but are interested in the steady state motion.

Now we must determine a particular solution of the imhomogeneous differential equation, that is, we have to find a particular solution of Eq. (8). We will assume the solution is of the form

$$x = A \cos(\omega t - \delta). \tag{14}$$

The constants A and δ are to be determined so that Eq. (14) satisfies Eq. (8). Let us substitute Eq. (14) into Eq. (8), this yields:

$$-A\omega^2 \cos(\omega t - \delta) - A\omega \frac{c}{M} \sin(\omega t - \delta)$$
$$+ A \frac{4k}{M} \cos(\omega t - \delta) = \frac{m\omega^2 r}{M} \cos \omega t. \tag{15}$$

This equation must be satisfied for all t. The simplest method to determine A and δ is to choose the two values $t = \frac{\pi}{2\omega}$ and $t = \frac{\pi}{2\omega} + \delta$. Substituting $t = \frac{\pi}{2\omega}$ into Eq. (15); we get,

$$-A\omega^2 \cos\left(\frac{\pi}{2} - \delta\right) - A\omega \frac{c}{M} \sin\left(\frac{\pi}{2} - \delta\right)$$
$$+ A \frac{4k}{M} \cos\left(\frac{\pi}{2} - \delta\right) = 0.$$

We can simplify this equation by dividing by A and using the identities $\cos\left(\frac{\pi}{2} - \delta\right) = \sin \delta$ and $\sin\left(\frac{\pi}{2} - \delta\right) = \cos \delta$. Thus we get,

$$-\omega^2 \sin \delta - \frac{c\omega}{M} \cos \delta + \frac{4k}{M} \sin \delta = 0, \quad \text{and}$$

$$\left(\frac{4k}{M} - \omega^2\right) \sin \theta = \frac{c\omega}{M} \cos \delta \qquad \text{or}$$

$$\tan\delta = \frac{\frac{c\omega}{M}}{\left(\frac{4k}{M} - \omega^2\right)} \qquad (16)$$

See Figure 2.

Substituting $t = \frac{\pi}{2\omega} + \delta$ into Eq. (15), yields

$$- A\omega^2 \cos\frac{\pi}{2} - A\omega\frac{c}{M}\sin\frac{\pi}{2} + A\frac{4k}{M}\cos\frac{\pi}{2}$$

$$= \frac{m\omega^2 r}{M}\cos\left(\frac{\pi}{2} + \delta\right), \text{ and}$$

$$- A\omega\frac{c}{M} = \frac{m\omega^2 r}{M}\cos\left(\frac{\pi}{2} + \delta\right).$$

The above equation can be simplified by using the identity $\cos\left(\frac{\pi}{2} + \delta\right) = -\sin\delta$ and rearranging the equation. Then,

$$-\sin\delta = \frac{-A\frac{c\omega}{M}}{\frac{m\omega^2 r}{M}} \quad . \qquad (17)$$

Using Fig. 2 (which was obtained from Eq. (16))we obtain a second expression for sin δ, namely

$$\sin\delta = \frac{\frac{c\omega}{M}}{\sqrt{\left(\frac{4k}{M} - \omega^2\right)^2 + \left(\frac{c\omega}{M}\right)^2}} \quad . \qquad (18)$$

Equating Eqs. (17) and (18) and solving for A, yields

$$A = \frac{\frac{m\omega^2 r}{M}}{\sqrt{\left(\frac{4k}{M} - \omega^2\right)^2 + \left(\frac{c\omega}{M}\right)^2}} \quad . \qquad (19)$$

We have now obtained the particular solution of the inhomogeneous equation. Collect all the pertinent equations so that it is easier to discuss them. We have that

$$x = A\cos(\omega t - \delta) \qquad (14)$$

where $$A = \frac{\frac{m\omega^2 r}{M}}{\sqrt{\left(\frac{4k}{M} - \omega^2\right)^2 + \left(\frac{c\omega}{M}\right)^2}} \qquad (19)$$

and $$\tan \delta = \frac{\frac{c\omega}{M}}{\left(\frac{4k}{M} - \omega^2\right)} . \tag{16}$$

In order to assure ourselves that this is a solution for all times t, we should substitute the solution into Eq. (8) and convince ourselves that Eq. (8) is always satisfied. We shall not do this here.

Next, evaluate the amplitude of the steady state vibration, i.e. evaluate Eq. (19).

First, consider the case where the damping is negligible. This means that we want to set the damping constant c = 0. Eq. (19) then reduces to:

$$A = \frac{\frac{m\omega^2 r}{M}}{\left|\left(\frac{4k}{M} - \omega^2\right)\right|} . \tag{20}$$

Substituting all known values into Eq. (20), yields:

$$A = \frac{\dfrac{(4 \times 10^{-2}\ \text{kg})(50\ \pi\ \text{rad/sec})^2(1.5 \times 10^{-1}\ \text{m})}{5.0 \times 10\ \text{kg}}}{\left|\left[\dfrac{4(4 \times 10^5)\text{N/m}}{5.0 \times 10\ \text{kg}} - (50\ \pi)^2\right]\right|}$$

$$= \frac{2.961}{7.326 \times 10^3}\ \text{m}.$$

Evaluating the above expression we find,

$$A = 4.042 \times 10^{-4}\ \text{m} = 0.4042\ \text{mm}.$$

Notice that we used the absolute value sign in the denominator of Eq. (20). This is to assure that the amplitude A is positive.

Next, consider the case where the damping factor $\frac{c}{c_c}$ is equal to 0.15. In order to evaluate Eq. (19), first determine the coefficient of damping, c. Using Eq. (13) we can evaluate the critical damping constant c_c, that is

$$c_c = 4\sqrt{kM} = 4\sqrt{4 \times 10^5(50)} = 1.789 \times 10^4\ \text{kg/sec},$$

and using this result we can find the coefficient of damping c,

$$c = 0.15\ c_c = 2.683 \times 10^3\ \text{kg/sec}.$$

Now substitute all known values into Eq. (19) and evaluate the amplitude A,

$$A = \frac{\dfrac{(4 \times 10^{-2}\ \text{kg})(50\pi\ \text{rad/sec})^2(1.5 \times 10^{-1}\ \text{m})}{5.0 \times 10\ \text{kg}}}{\sqrt{\left[\dfrac{4\times4\times10^5\ \text{N/m}}{5.0 \times 10\ \text{kg}} - (50\pi)^2\right]^2 + \left[\dfrac{\left(2.683\times10^3\ \frac{\text{kg}}{\text{sec}}\right)\left(50\pi\frac{\text{rad}}{\text{sec}}\right)}{5.0 \times 10\ \text{kg}}\right]^2}}$$

Evaluating A we get,

$$A = \frac{2.961}{\sqrt{5.367 \times 10^7 + 1.324 \times 10^6}}$$

or, $A = 0.399 \times 10^{-3}\text{m} = 0.399$ mm.

It is also interesting to evaluate the amplitude for the two limiting cases when ω equals zero and when ω is very large. For these cases we get:

$A = 0 \qquad$ as $\omega \to 0$,

and $A = \frac{mr}{M} \qquad$ as $\omega \to \infty$.

These results are due to the fact that the amplitude of the driving force depends on the frequency of rotation of the rotor.

• PROBLEM 23-27

A slender bar, connected to a string at one end and roller at the other end, is allowed to move horizontally. Determine the natural frequency of oscillation for a small initial horizontal displacement of the bar. Neglect friction.

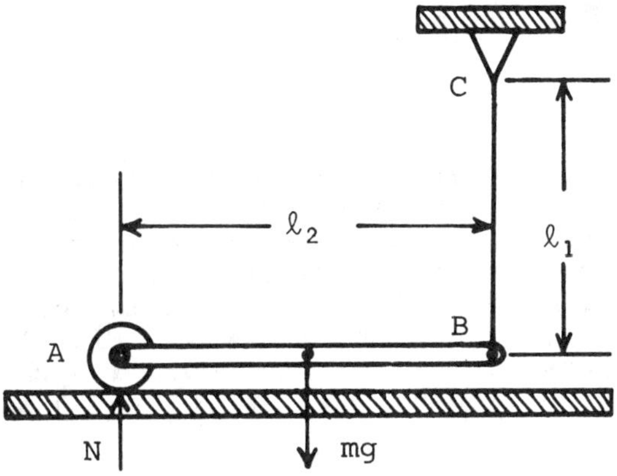

Solution: The method used to solve this problem is very general and can be used in many instances. First, the torque about a fixed point, here C, is found. Then this is set equal to the product of the moment of inertia of the system (taken about the same point as the torque) and the angular acceleration, $\ddot{\theta}$. Next, the definition of simple harmonic motion (SHM) is used in conjunction with the above equation to obtain the frequency. The definition of SHM used here is: angular acceleration is proportional to angular displacement and oppositely directed to the latter. The constant of proportionality is always ω^2 where ω, the angular velocity, equals 2π times the frequency, f, of oscillation. Mathematically, the above becomes:

$$T = I\ddot{\theta}, \tag{1}$$

where T must be written in the form:

$$T = -k\theta, \tag{2}$$

and $$\ddot{\theta} = -\omega^2\theta. \quad \text{(SHM)} \tag{3}$$

Eqs. (2) and (3) are substituted into Eq. (1) which is then solved for the frequency via

$$\omega = 2\pi f. \tag{4}$$

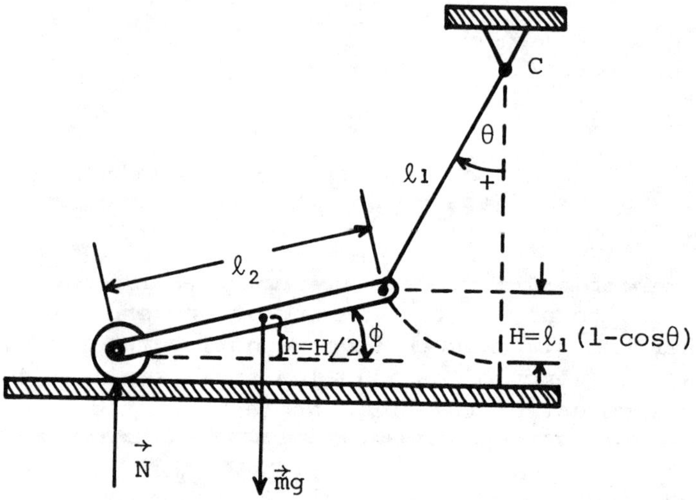

Fig. 1

The positive direction assigned to torque must be the same as the positive increase in the angel θ. Hence, in Fig. (1), since the angle increases clockwise, the direction of positive torque must be clockwise. If the angle were shown to the right of vertical, the positive direction of torque would have to be counterclockwise. Fig. 1 has been drawn making the assumption that the cord length ℓ is much greater than the rod length, ℓ_2. Also, ϕ and θ are assumed to be small.

The sum of torque about point C is:

$$T = +N(\ell_2\cos\phi + \ell_1\sin\theta) - mg\left(\frac{\ell_2\cos\phi}{2} + \ell_1\sin\theta\right) \tag{5}$$

The rod's moment of inertia about its supported end is $mL^2/3$ and hence its moment of inertia about the point C is, by Steiner's Theorem:

$$I = m\left(\frac{\ell_2^{\,2}}{3} + \ell_1^{\,2}\right) \qquad (6)$$

For simplicity, assume $\ell >> L$ so that $L^2/3$ may be dropped in Eq. (6). Also, assume that angles ϕ and θ are small enough to allow $\cos\phi \cong 1$, $\sin\theta \cong \theta$, and $N = mg/2$. Eq. (5) then reduces to:

$$T = + N\ell_2 + N\ell_1\theta - mg\ell_2/2 - mg\ell_1\theta$$

Since $N = mg/2$ this reduces to:

$$T = -\frac{mg\ell_1\theta}{2}. \qquad (7)$$

Eq. (6) becomes, under the assumptions made:

$$I = m\ell_1^2. \qquad (8)$$

Substitute Eq. (7) and (8) into Eq. (1):

$$-\frac{mg\ell_1\theta}{2} = m\ell_1^2\ddot{\theta}$$

or

$$\ddot{\theta} = -\frac{g}{2\ell_1}\theta. \qquad (9)$$

Since Eq. (9) is the same form as Eq. (3), it follows that

$$\omega^2 = \frac{g}{2\ell_1}.$$

Substitute $\omega = 2\pi f$ in this and solve for f to get:

$$f = \frac{1}{2\pi}\sqrt{\frac{g}{2\ell_1}} \qquad (10)$$

• PROBLEM 23-28

The slender rod in the figure is 12 ft long and weighs 96 lbs.

a) Derive the differential equation for small oscillations of the rod.

b) Find the value of the spring constant k required for critical damping if the damping constant of the dash-pot is c = 20 lb - sec/ft.

c) What will be the angular displacement for the critically damped system at t = 1 sec if the initial conditions at t = 0 are $\dot{\theta}$ = 4 rad/sec and θ = 0?

Solution: In Fig. 2, the rod is shown displaced from its equilibrium position by an angle θ, and moving in the counter-clockwise direction with an angular velocity $\dot{\theta}$.

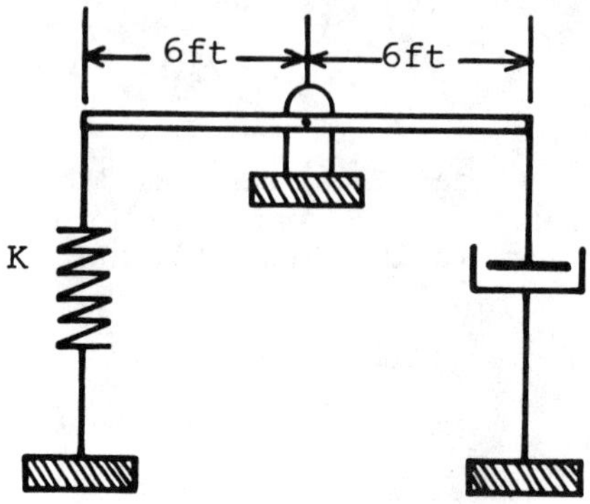

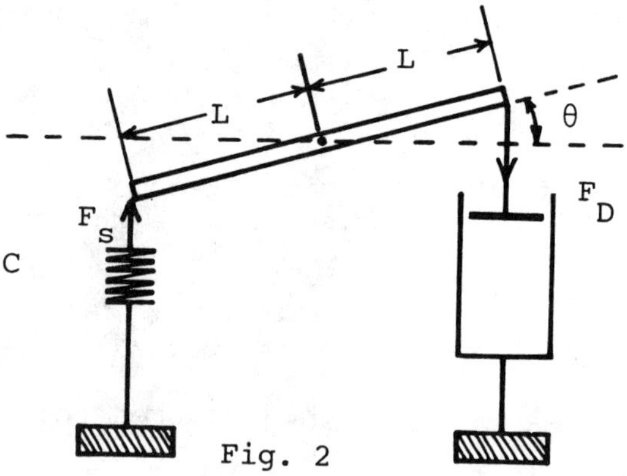

Fig. 2

Fig. 1

The left-hand end of the rod is displaced a distance, x, thereby compressing the spring by an amount, x. The compressed spring exerts a force, F_s, on the end of the rod in a direction toward the equilibrium position. Using the spring constant, k, the force can be written:

$$F_s = - kx = - k (L\theta) \tag{1}$$

where, for small displacements from equilibrium we have written $x = L\theta$.

The force exerted by the dashpot on the right-hand end of the rod is proportional to the velocity of the end of the rod and opposite in direction to the velocity. Thus, the damping force, F_D, can be written:

$$F_D = - cv = - c (L\dot{\theta}) \tag{2}$$

where the velocity of the end has been written as $v = L\dot{\theta}$.

Now, using Newton's Second Law for rotating bodies,

$$\Sigma \Gamma = I\ddot{\theta} \tag{3}$$

where $\Sigma \Gamma$ is the sum of the moments, or torques, of the forces acting on the rod, and $I = \frac{1}{12} m (2L)^2$ is the moment of inertia of the rod about an axis through its midpoint. Thus:

$$F_s L + F_D L = I\ddot{\theta} \tag{4}$$

or

$$- kL^2\theta - cL^2\dot{\theta} = \frac{1}{3} mL^2\ddot{\theta}. \tag{5}$$

Rearranging terms yields the differential equation for the motion of the rod:

$$\ddot{\theta} + \frac{3c}{m} \dot{\theta} + \frac{3k}{m} \theta = 0. \tag{6}$$

The mass, m, of the rod may be expressed in terms of the weight, ω, by $m = \frac{\omega}{g}$, where g is the acceleration due to gravity.

Equation (6) may be put in a more convenient form as:

$$\ddot{\theta} + 2\beta\dot{\theta} + \omega_0^2\theta = 0 \qquad (7)$$

where $\beta = \frac{3c}{2m}$ is called the damping parameter, and $\omega_0 = \sqrt{\frac{3k}{m}}$ is called the characteristic angular frequency of oscillation in the absence of damping. The general solution for Eq. (7) can be shown, using standard techniques for solving differential equations, to be:

$$\theta(t) = e^{-\beta t}\left[Be^{\sqrt{\beta^2 - \omega_0^2}\,t} + Ce^{-\sqrt{\beta^2 - \omega_0^2}\,t}\right] \qquad (8)$$

The constants B and C must be determined from initial conditions. Three cases of general interest arise, and can be identified as follows:

Case 1, Under damped: $\omega_0^2 > \beta^2$

Case 2, Critically damped: $\omega_0^2 = \beta^2$

Case 3, Over damped: $\omega_0^2 < \beta^2$

For case 2, the critically damped motion of the rod, we find:

$$\omega_0^2 = \beta^2$$

or

$$\frac{3k_c}{m} = \left(\frac{3c}{2m}\right)^2. \qquad (9)$$

Thus, the spring constant required for critical damping is

$$k_c = \left(\frac{m}{3}\right)\left(\frac{3c}{2m}\right)^2 = \frac{3c^2}{4m} = \frac{3gc^2}{4W}. \qquad (10)$$

Substituting numerical values:

$$k_c = \frac{3\ (32.2\ \text{ft/s}^2)(20\ \text{lb-s/ft})^2}{4\ (96\ \text{lb})} = 101\ \text{lb/ft}. \qquad (11)$$

For the critically damped motion, the general solution to the differential equation must be rewritten as:

$$\theta(t) = (B + Ct)\,e^{-\beta t} \qquad (12)$$

To evaluate the constants of integration, B and C, substitute the initial conditions into Eq. (12). First, at $t = 0$, we have $\theta = 0$. Thus, Eq. (12) becomes:

$$\theta(0) = 0 = (B + C(0))\, e^{0} \qquad (13)$$

or $$B = 0$$

Differentiating (12) yields:

$$\dot{\theta}(t) = Ce^{-\beta t} - \beta Cte^{-\beta t} - \beta Be^{-\beta t} \qquad (14)$$

Next, at $t = 0$, $\dot{\theta} = 4$ rad/sec, and we know $B = 0$. Thus

$$\dot{\theta}(0) = 4 = Ce^{0} - \beta C(0)e^{0}$$

or $$C = 4 \text{ rad/sec.}$$

Therefore, the equation of motion for the critically damped system is:

$$\theta(t) = 4t\, e^{-\beta t} \qquad (15)$$

where $$\beta = \frac{3c}{2m} = \frac{3gc}{2W}.$$

To find the angular displacement of the rod at $t = 1$ sec, subject to the initial conditions given, let $t = 1$ in Eq. (15):

$$\theta(1) = 4(1)\, e^{-\beta} = 4e^{-\frac{3gc}{2W}}$$

$$\theta(1) = 4e^{-\frac{3(32.2)(20)}{2(96)}} = 1.71 \times 10^{-4} \text{ rad.}$$

CHAPTER 24

SYSTEMS HAVING MULTI-DEGREES OF FREEDOM

COUPLED HARMONIC OSCILLATORS

• PROBLEM 24-1

Two masses, m_1 and m_2, initially at rest on a frictionless surface, are connected by a linear spring, as shown in Fig. 1-a. The spring is compressed an amount δ and released. Derive an expression for the velocity of each mass when the spring returns to its undeformed length L.

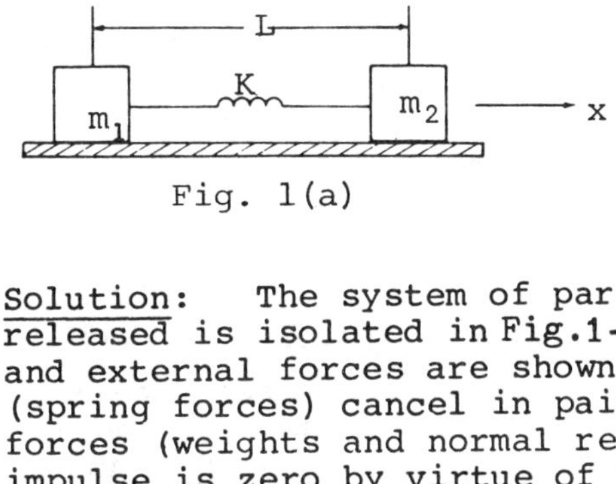

Fig. 1(a)

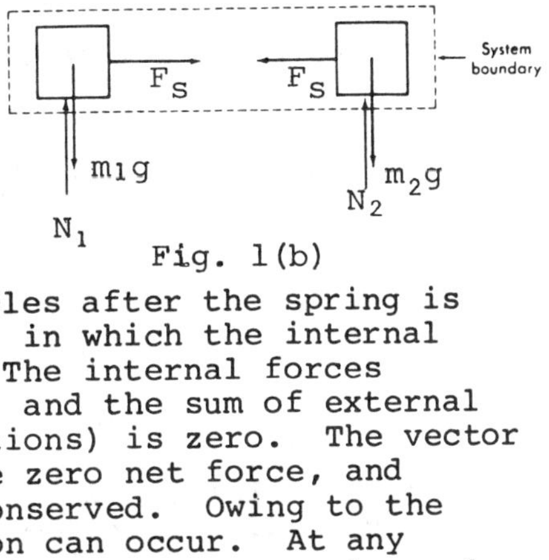

Fig. 1(b)

Solution: The system of particles after the spring is released is isolated in Fig.1-b, in which the internal and external forces are shown. The internal forces (spring forces) cancel in pairs, and the sum of external forces (weights and normal reactions) is zero. The vector impulse is zero by virtue of the zero net force, and therefore, linear momentum is conserved. Owing to the system constraints, only X-motion can occur. At any arbitrary time, the sum of the linear momenta must equal a constant, which in this case is zero, because the initial system momentum is zero:

$$m_1\dot{x}_1 + m_2\dot{x}_2 = 0 \qquad \dot{x}_1 = -\left(\frac{m_2}{m_1}\right)\dot{x}_2 .$$

Another equation is required since there are two unknowns, $\dot{x}_1$ and $\dot{x}_2$. In the absence of dissipative forces, mechanical energy is conserved and the system exchanges kinetic and elastic potential energy. The total system energy at any time must equal the maximum elastic potential energy (obtained by initially compressing the spring, that is, doing work on the system).

$$E = \frac{1}{2}k\delta^2 = \frac{1}{2}m_1\dot{x}_1^2 + \frac{1}{2}m_2\dot{x}_2^2$$

where $\dot{x}_1$ and $\dot{x}_2$ represent the velocities corresponding to the undeformed state of the spring (maximum velocities). Simultaneous solution of the equations gives

$$\dot{x}_2 = \left[\frac{k\delta^2}{(m_2^2 + m_1 m_2)/m_1}\right]^{1/2} \quad \text{and}$$

$$\dot{x}_1 = -\left(\frac{m_2}{m_1}\right)\left[\frac{k\delta^2}{(m_2^2 + m_1 m_2)/m_1}\right]^{1/2} = -\left[\frac{k\delta^2}{(m_1^2 + m_1 m_2)/m_2}\right]^{1/2}$$

• PROBLEM 24-2

In the study of, for example, the vibrations of a diatomic molecule, we deal with the oscillations of two interacting bodies, as illustrated in Fig. 1. Here, two bodies of masses m_1 and m_2 are connected with a spring of force constant K. If the spring is compressed and then released, the bodies will oscillate to and fro with the same frequencies. Sometimes the center of mass is at rest, but, in general, there is a translational motion superimposed on the oscillations. In any event, the oscillatory motion is the motion of the bodies with respect to the center of mass. We wish to determine the characteristic frequency of the oscillation and the ratio between the oscillation amplitudes and energies of the two bodies.

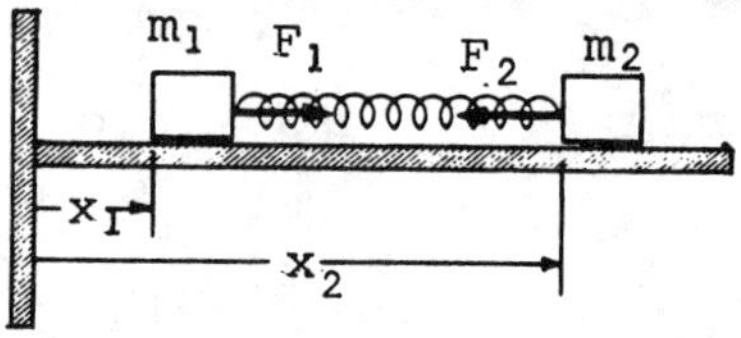

Fig. 1 Two-body oscillator

Solution: If the position coordinates of the two bodies are x_1 and x_2 with respect to a fixed coordinate system as shown in Fig. 1, the length of the spring is $x_2 - x_1$, and if the relaxed length is d, the extension of the spring is $x = x_2 - x_1 - d$. The force on m_2 is then $F_2 = -Kx$ and the force on m_1 is $F_1 = Kx$. Thus, the equations of motion for the two bodies become

$$m_2 \frac{d^2x_2}{dt^2} = -Kx \tag{1}$$

and $$m_1 \frac{d^2x_1}{dt^2} = Kx. \tag{2}$$

If we multiply Eq. (1) by m_1, and Eq. (2) by m_2, and then subtract the equations, we obtain

$$\frac{m_1 m_2}{m_1 + m_2} \frac{d^2x}{dt^2} = -Kx, \tag{3}$$

where $x = x_2 - x_1 - d$. In other words, the extension x of the spring satisfies an equation which is the same as that for a single-mass oscillator where one end of the spring is held fixed and the other carries the mass μ:

$$\mu = \frac{m_1 m_2}{m_1 + m_2} . \tag{4}$$

The quantity μ is called the reduced mass of the oscillator.

The characteristic angular frequency of the oscillation, which is the same for both m_1 and m_2, is then

$$\omega_0 = \sqrt{\frac{K}{\mu}} . \tag{5}$$

The energy of oscillation can be expressed either as the maximum potential energy $V_m = Kx_m^2/2$, where x_m is the maximum extension of the spring or as the sum of the maximum kinetic energies of the two bodies at the moment when the spring is relaxed. These kinetic energies can be expressed as $E_1 = p_1^2/2m_1$ and $E_2 = p_2^2/2m_2$, where p_1 and p_2 are the momenta of the bodies as measured with respect to the center of mass. Thus, $p_1 = p_2 = p$, and the total mechanical energy of oscillation is

$$H = \frac{1}{2} Kx_m^2 = \frac{1}{2} p_m^2 \left(\frac{1}{m_1} + \frac{1}{m_2}\right) = \frac{1}{2} \frac{p_m^2}{\mu} . \tag{6}$$

Clearly, the ratio between the energies of oscillation of the two bodies is

$$\frac{E_1}{E_2} = \frac{m_2}{m_1} .$$

If the magnitudes of the displacement amplitudes of the two bodies are X_1 and X_2, we have $X_1/X_2 = m_2/m_1$ and $x_m = X_1 + X_2$. From these relations we can determine X_1 and X_2, in terms of x_m.

If the center of mass is not at rest, but has a velocity V, the total energy of the ststem (mass $M = m_1 + m_2$)is the sum of the center-of-mass energy $MV^2/2$ and the oscillatory energy. Thus, by subtracting the center-of-mass energy $MV^2/2$ from the total energy (which often is known), we obtain the mechanical energy of oscillation from which the properties of the oscillatory motion can be found as shown above.

• PROBLEM 24-3

Two bodies of masses m_1 and m_2 are free to move along a horizontal straight track. They are connected by a spring with spring constant K. The system is initially at rest. An instantaneous impulse J is given to m_1 along the direction of the track. Determine the motion of the system and, in particular, find the energy of oscillation of the bodies.

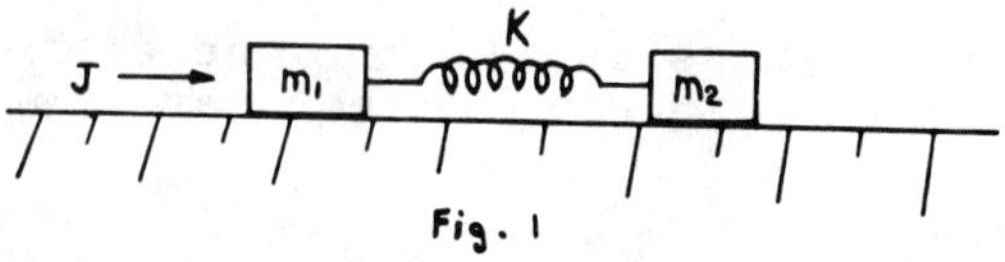

Fig. 1

Solution: Impulse represents the total change in momentum of a body. Since m is at rest at first, it acquires a momentum J when it is given an instantaneous impulse J. That is, $J = m_1v_1$, where v_1 is the velocity acquired by m_1. However, this momentum is eventually shared by both bodies when their center of mass starts moving with velocity V. Since the momentum is conserved,

$$J = m_1v_1 = (m_1 + m_2)V$$

or $$V = \frac{J}{m_1 + m_2} = \frac{J}{M}, \text{ where } M = m_1 + m_2.$$

The above equation gives the velocity of center of mass. The kinetic energy due to translation of the two bodies E_t, is simply given by the formula $KE = \frac{1}{2} mv^2$. That is, $E_t = \frac{1}{2} MV^2$

or $$E_t = \frac{1}{2} M \cdot \left(\frac{J}{M}\right)^2 = \frac{J^2}{2M}$$

The energy used up in compressing the spring is the difference between the initial kinetic energy of mass m_1 $\left(\frac{1}{2} m_1v_1^2\right)$ and the final translational kinetic energy of the system $\left(= \frac{1}{2}(m_1 + m_2)V^2\right)$. Representing this energy by H_0 yields

$$H_0 = \frac{1}{2} m_1v_1^2 - \frac{1}{2}(m_1 + m_2)V^2$$

$$= \frac{1}{2} m_1 \left(\frac{J}{m_1}\right)^2 - \frac{1}{2}(m_1 + m_2)\left(\frac{J}{m_1 + m_2}\right)^2$$

$$= \frac{J^2}{2m_1} - \frac{J^2}{2(m_1 + m_2)} = \frac{J^2}{2}\left[\frac{1}{m_1} - \frac{1}{m_1 + m_2}\right]$$

$$= \frac{J^2}{2}\left[\frac{m_1 + m_2 - m_1}{(m_1 + m_2)m_1}\right] = \frac{J^2 m_2}{2(m_1 + m_2)m_1}.$$

The maximum compression x_{max} is obtained from the formula for the potential energy of a compressed spring, namely, $PE = \frac{1}{2} kx^2$.

Thus, $$H_0 = \frac{K x_{max}^2}{2}$$

or $$x_{max} = \sqrt{\frac{2H_0}{K}} = \sqrt{\frac{2 J^2 m_2}{2m_1(m_1 + m_2)}}$$

$$x_{max} = J\sqrt{\frac{m_2}{m_1(m_1 + m_2)}}$$

or $$x_{max} = J\sqrt{\left(\frac{m_2}{m_1}\right)\frac{1}{M}}$$

where $M = m_1 + m_2$.

The motion of the bodies after impact will include an oscillation superimposed on the major translation. From a previous problem one may remember that

$$\omega = \sqrt{\frac{K}{M}}$$

for this system where M is the reduced mass,

$$M = \frac{m_1 m_2}{m_1 + m_2}$$

$$\omega = \sqrt{\frac{K(m_1 + m_2)}{m_1 m_2}}$$

The motion of each mass is then:

$$x_1 = Vt + \bar{x}_1 \cos\sqrt{\frac{K(m_1 + m_2)}{m_1 m_2}}\ t$$

$$x_2 = Vt + \bar{x}_2 \cos\sqrt{\frac{K(m_1 + m_2)}{m_1 m_2}}\ t$$

V is the velocity of the mass center, $\bar{x}_1 + \bar{x}_2$ are the maximum relative motions of the two masses.

• PROBLEM 24-4

Two masses, m_1 and m_2, are attached by rigid rods of length ℓ_1 and ℓ_2 to a fixed support as shown in fig. 1. This is the double pendulum. T_1 is the tension in rod ℓ_1 and T_2 is the tension in ℓ_2. The angular displacement of the pendulums is given by θ_1 and θ_2. Finally, let x_1 denote the x-coordinate of m_1 and x_2 the x-coordinate of m_2. Find $x_1(t)$ and $x_2(t)$.

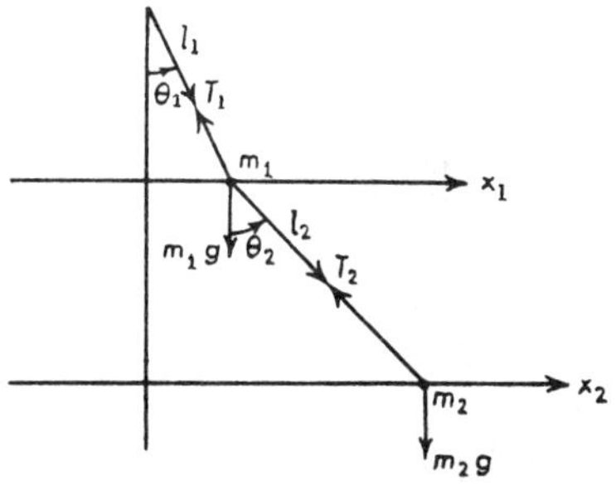

Solution: From Newton's second law for the horizontal motion and forces,

$$m_1\ddot{x}_1 = T_2 \sin \theta_2 - T_1 \sin \theta_1 \quad (a)$$

$$m_2\ddot{x}_2 = - T_2 \sin \theta_2 \quad (b)$$

where dots denote time derivatives. If θ_1 and θ_2 are small, we approximate (a) and (b) as

$$m_1\ddot{x}_1 = T_2\theta_2 - T_1\theta_1 \quad (c)$$

$$m_2\ddot{x}_2 = - T_2\theta_2. \quad (d)$$

Now, since the tension in the rods is due to the weight of the masses,

$$T_1 = m_1 g \cos \theta_1 + m_2 g \cos \theta_2 \cos (\theta_2 - \theta_1)$$

and $T_2 = m_2 g \cos \theta_2$.

Again making the small angle approximation,

$$T_1 = (m_1 + m_2)g, \quad (e)$$

$$T_2 = m_2 g. \quad (f)$$

Finally,

$$x_1 = \ell_1 \sin \theta_1 \cong \ell_1\theta_1 \quad (g)$$

$$x_2 = \ell_1 \sin \theta_1 + \ell_2 \sin \theta_2$$

$$x_2 \cong \ell_1\theta_1 + \ell_2\theta_2 \quad (h)$$

Thus,

$$\theta_1 = \frac{x_1}{\ell_1} \quad (i)$$

$$\theta_2 = \frac{x_2 - x_1}{\ell_2} \quad (j)$$

Substituting (e), (f), (i) and (j) into (c) and (d),

$$m_1\ddot{x}_1 = m_2 g \frac{x_2 - x_1}{\ell_2} - (m_1 + m_2)g \frac{x_1}{\ell_1}$$

and $m_2\ddot{x}_2 = - m_2 g \dfrac{x_2 - x_1}{\ell_2}$.

To save writing, let

$$u = \frac{m_2}{m_1} \quad (k)$$

$$p = \frac{g}{\ell_1} + gu \left(\frac{1}{\ell_2} - \frac{1}{\ell_1}\right) \quad (\ell)$$

$$\sigma = \frac{g}{\ell_2} \quad (m)$$

$$\nu = 1 + u. \quad (n)$$

Then our differential equations in operator notation are

$$(D^2 + p)x_1 - u\sigma x_2 = 0 \qquad (p)$$

$$(-\sigma)x_1 + (D^2 + \sigma)x_2 = 0. \qquad (q)$$

To eliminate x_1 we operate on (q) by $(D^2 + p)$ and multiply (p) by σ, then add:

$$[(D^2 + p)(D^2 + \sigma) - \mu\sigma^2]x_2 = 0$$

or $$[D^4 + \nu(w_1{}^2 + w_2{}^2)D^2 + \nu w_1{}^2 w_2{}^2]x_2 = 0 \qquad (r)$$

where $$w_1{}^2 = \frac{g}{\ell_1}$$

$$w_2{}^2 = \frac{g}{\ell_2} = \sigma.$$

Equation (r) is 4th order, homogeneous with constant coefficients. Thus, the auxiliary equation is

$$Z^4 + \nu(w_1{}^2 + w_2{}^2)Z^2 + \nu w_1{}^2 w_2{}^2 = 0$$

This is a quadratic equation in Z^2; thus,

$$Z^2 = \frac{\nu}{2}(w_1{}^2 + w_2{}^2) \pm \frac{1}{2}\sqrt{\nu^2(w_1{}^2 + w_2{}^2) - 4\nu w_1{}^2 w_2{}^2}\ .$$

For simplicity we denote the two values of Z^2 above by

$-\lambda_+{}^2$ and $-\lambda_-{}^2$. Hence,

$$x_2 = A\sin(\lambda_+ t + \gamma_1) + B\sin(\lambda_- t + \gamma_2)$$

where A, B, γ_1, γ_2 are arbitrary constants.

To find x_1, we note from (q) that

$$x_1 = \frac{1}{\sigma}(D^2 + \sigma)x_2. \qquad \text{Thus,}$$

$$x_1 = A\left(1 - \frac{\lambda_1{}^2}{w_2{}^2}\right)\sin(\lambda_+ t + \gamma_1) + B\left(1 - \frac{\lambda_2{}^2}{w_2{}^2}\right)\sin(\lambda_- t + \gamma_2).$$

• **PROBLEM 24-5**

Two masses, m_1 and m_2, are connected by springs of spring constants k_1 and k_2 and natural lengths L_1 and L_2 as shown in fig. 1.

Point P is fixed and O_1 and O_2 mark the equilibrium positions of the springs. Suppose at $t = 0$, m_1 is displaced a distance a_1 from O_1 and m_2 is displaced a distance a_2

from O_2, and then both masses are released. Find the position $x_1(t)$ of m_1 and position $x_2(t)$ of m_2 for all $t > 0$. Assume no friction or external forces.

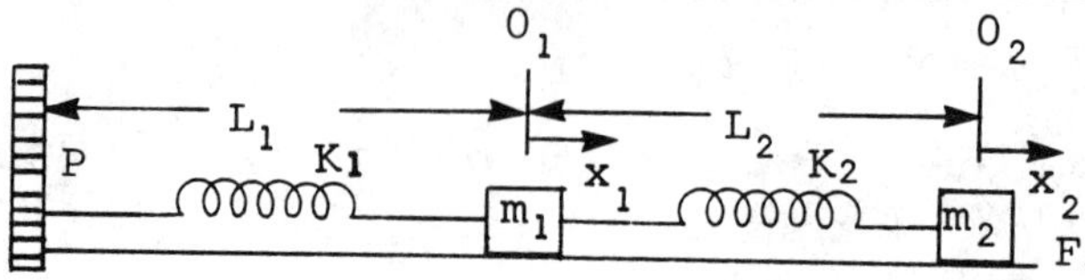

Fig. 1

Solution: Assume x_1 and x_2 are positive when measured to the right. Consider the forces acting on m_1. There will be force F_1 due to spring k_1, and F_2 due to spring k_2. By Hooke's law $F_1 = -k_1x_1$ and $F_2 = -k_2|s|$ where $|s|$ is the total elongation or compression of spring k_2. From our definition of x_1 and x_2, $F_2 = -k_2(x_1 - x_2)$. Applying Newton's second law to m_1 we have

$$m_1 \frac{d^2x_1}{dt^2} = -k_1x_1 + k_2(x_2 - x_1). \tag{a}$$

Mass m_2 experiences only a force from spring k_2 which is

$$m_2 \frac{d^2x_2}{dt^2} = -k_2(x_2 - x_1). \tag{b}$$

The initial conditions are

$$x_1(0) = a_1, \qquad \left.\frac{dx_1}{dt}\right|_{t=0} = 0 \tag{c}$$

$$x_2(0) = a_2, \qquad \left.\frac{dx_2}{dt}\right|_{t=0} = 0 \tag{d}$$

Equations (a) and (b) represent two coupled simultaneous equations which must be solved.

Let us now give numerical values to simplify the solution process:

$$m_1 = m_2 = 1,$$

$$k_1 = 3, \qquad k_2 = 2$$

$$a_1 = -1, \qquad a_2 = 2.$$

Substituting into (a) and (b)

$$\frac{d^2x_1}{dt^2} = -3x_1 + 2(x_2 - x_1)$$

or $$\frac{d^2x_1}{dt^2} + 5x_1 - 2x_2 = 0 \tag{e}$$

and $$\frac{d^2x_2}{dt^2} = -2(x_2 - x_1)$$

or $$\frac{d^2x_2}{dt^2} - 2x_1 + 2x_2 = 0. \tag{f}$$

Writing (e) and (f) in operator notation,

$$(D^2 + 5)x_1 - 2x_2 = 0 \qquad (g)$$

$$-2x_1 + (D^2 + 2)x_2 = 0. \qquad (h)$$

Apply the operator $D^2 + 2$ to (g), multiply (h) by 2 and then add the resulting equations to eliminate x_2. Thus,

$$[(D^2 + 2)(D^2 + 5) - 4]x_1 = 0$$

$$(D^4 + 7D^2 + 6)x_1 = 0. \qquad (i)$$

Equation (i) has constant coefficients and is homogeneous.

We thus assume a solution of the form $x_1 = e^{pt}$ and substitute in (i) to produce the auxiliary equation

$$p^4 + 7p^2 + 6 = 0 .$$

Let $p^2 = q$, then

$$q^2 + 7q + 6 = 0$$

$$(q + 6)(q + 1) = 0$$

Thus, our solutions are

$$p^2 = -6$$

and $p^2 = -1$

which give a solution set of $\langle i\sqrt{6}, -i\sqrt{6}, i, -i \rangle$ where i is the imarginary $\sqrt{-1}$. Our general solution is, thus,

$$x_1 = c_1 e^{\sqrt{6}it} + c_2 e^{-\sqrt{6}it} + c_3 e^{it} + c_4 e^{-it}$$

where c_1, c_2, c_3 and c_4 are arbitrary constants. Using Euler's identity

$$e^{i\theta} = \cos\theta + i\sin\theta,$$

and the fact that c_1 and c_2 as well as c_3 and c_4 may be chosen as complex conjugates, we write the general solution as

$$x_1 = c_5 \sin t + c_6 \cos t + c_7 \sin\sqrt{6}t + c_8 \cos\sqrt{6}t \qquad (j)$$

where no c_5, c_6, c_7, c_8 are arbitrary real constants.

In order to find x_2 we multiply (g) by 2 and apply the operator $D^2 + 5$ to (h), and then add to obtain

$$(D^4 + 7D^2 + 6)x_2 = 0$$

which is the same operator that was applied to x_1. By analogy the solution is

$$x_2 = k_1 \sin t + k_2 \cos t + k_3 \sin \sqrt{6}t + k_4 \cos \sqrt{6}t \qquad (k)$$

where k_1, k_2, k_3 and k_4 are arbitrary real constants.

Since the system is 4th order, we can only have a total of 4 independent arbitrary constants. Thus, we must find the relationship between c_5, c_6, c_7, c_8 and k_1, k_2, k_3, k_4. This is accomplished by substituting (j) and (k) into (g) and (h). We then use the linear independence of sin t, cos t, sin $\sqrt{6}t$, and cos $\sqrt{6}t$ to obtain, after much algebra,

$$k_1 = 2c_5, \quad k_2 = 2c_6, \quad k_3 = -\frac{1}{2}c_7 \text{ and } k_4 = -\frac{1}{2}c_8.$$

Hence,

$$x_1 = c_5 \sin t + c_6 \cos t + c_7 \sin \sqrt{6}t + c_8 \cos \sqrt{6}t$$

$$x_2 = 2c_5 \sin t + 2c_6 \cos t - \frac{1}{2}c_7 \sin \sqrt{6}t - \frac{1}{2}c_8 \cos \sqrt{6}t$$

To evaluate c_5, c_6, c_7, c_8 we must apply the initial conditions

$$x_1(0) = -1 \qquad \text{and} \qquad \left.\frac{dx_1}{dt}\right|_{t=0} = 0;$$

thus,

$$-1 = c_6 + c_8$$

$$0 = c_5 + \sqrt{6}\,c_7.$$

Next, $x_2(0) = 2$ and $\left.\dfrac{dx_2}{dt}\right|_{t=0} = 0.$

Thus

$$2 = 2c_6 - \frac{1}{2}c_8$$

$$0 = 2c_5 - \frac{\sqrt{6}}{2}c_7 .$$

Solving these four equations in four unknowns:

$$c_5 = 0 \quad , \qquad c_6 = \frac{3}{5}$$

$$c_7 = 0 \quad , \qquad c_8 = -\frac{8}{5}$$

The final solution is

$$x_1 = \frac{3}{5}\cos t - \frac{8}{5}\cos \sqrt{6}t$$

$$x_2 = \frac{6}{5}\cos t + \frac{4}{5}\cos \sqrt{6}t \quad .$$

FORCED VIBRATION

• PROBLEM 24-6

> The system of coupled oscillators shown is subjected to the applied force, $F = F_0 \cos \omega t$, applied to mass M_1. Find the steady state solution. Sketch for the amplitude and phase of each oscillator as functions of ω.

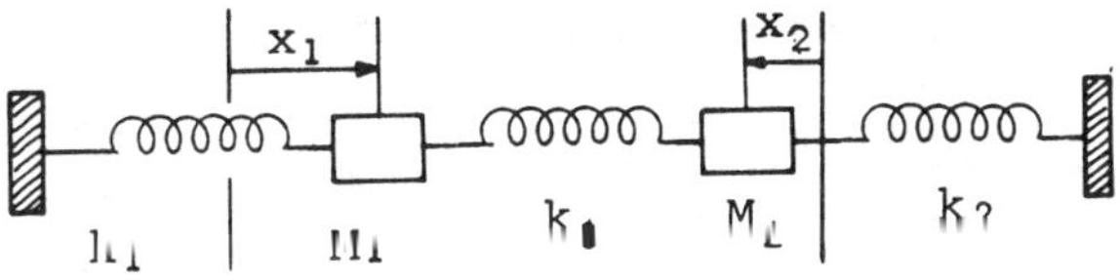

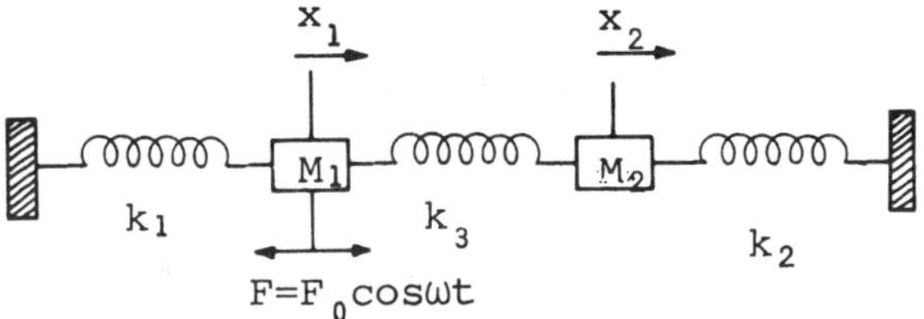

Solution: The basic procedure used to solve a problem like this is to: 1) Write the equations of motion. 2) Assume form $X_n = A_n \cos \omega t + B_n \sin \omega t$, where ω is the frequency of the driving function, and A_n and B_n are arbitrary amplitudes. 3) Differentiate X_n and use the expressions in the equations of motion to solve for the constants A_n and B_n.

The key assumption in this approach is that $X_n = A_n \cos \omega t + B_n \sin \omega t$ for the steady state. Experience and results using different methods of analysis suggest that this type of steady state motion is usually the type found in such an undamped system.

1) Equations of motion

Use Newtons Second Law, $\Sigma F = M\ddot{x}$, to find the equations of motion for this system. Assume a positive displacement for each mass with $x_1 > x_2$, then sum the spring forces which may be found using Hooke's Law. Forces to the right are positive.

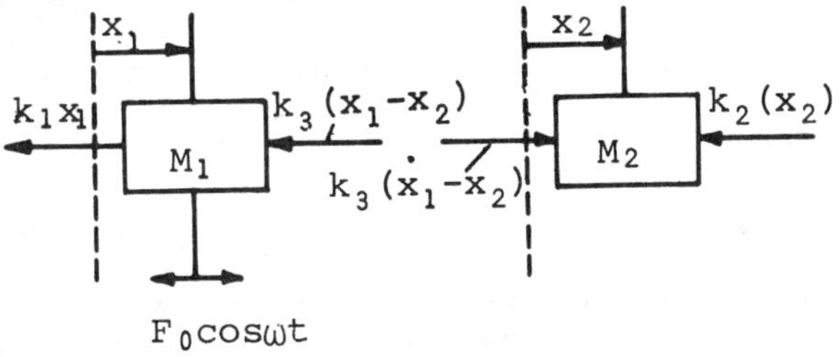

(Remember that the force in spring 3 is due to the relative displacement of the blocks, $(x_1 - x_2)$.)

For block 1 the equation of motion is

$$M_1\ddot{x}_1 = - K_1x_1 - K_3 (x_1 - x_2) + F_0 \cos \omega t$$

$$M_1\ddot{x}_1 + (K_1 + K_3)x_1 - K_3x_2 = F_0 \cos \omega t$$

For block 2

$$M_2\ddot{x}_2 = - K_2(x_2) + K_3(x_1 - x_2)$$

$$M_2\ddot{x}_2 + (K_2 + K_3)x_2 - K_3x_1 = 0$$

2) Now let

$$x_1 = A_1 \cos \omega t + B_1 \sin \omega t$$

$$x_2 = A_2 \cos \omega t + B_2 \sin \omega t.$$

Differentiating, $\dot{x}_1 = - A_1 \omega \sin \omega t + B_1 \omega \cos \omega t$

$$\ddot{x}_1 = - A_1 \omega^2 \sin \omega t + B_1 \omega^2 \sin \omega t$$

$$\dot{x}_2 = - A_2 \omega \sin \omega t + B_2 \omega \cos \omega t$$

$$\ddot{x}_2 = - A_2 \omega^2 \cos \omega t + B_2 \omega^2 \sin \omega t.$$

Note that: $\ddot{x}_1 = - \omega^2x_1$

$$\ddot{x}_2 = - \omega^2x_2$$

3) Substituting $- \omega^2x_1$ for $\ddot{x}_1$ in the equation of motion yields

$$- M_1\omega^2 x_1 + (K_1 + K_3)x_1 - K_3x_2 = F_0 \cos \omega t \qquad \text{(a)}$$

$$- M_2\omega^2 x_2 + (K_2 + K_3)x_2 - K_3x_1 = 0 \qquad \text{(b)}$$

Substituting $A_1 \cos \omega t + B_1 \sin \omega t$ for x_1 in (a),

$$[- M\omega^2 + (K_1 + K_3)](A_1 \cos \omega t + B_1 \sin \omega t)$$

$$- K_3 (A_2 \cos \omega t + B_2 \sin \omega t) = F_0 \cos \omega t$$

Next, set the coefficients of the sin ωt terms on the left equal to the coefficients of the sin ωt terms on the right.

i) $[- M_1\omega^2 + (K_1 + K_3)]B_1 - K_3B_2 = 0$

Similarly, for the cos ωt terms

ii) $[- M_1\omega^2 + (K_1 + K_3)]A_1 - K_3A_2 = F_0$

Following the same procedure for (b),

$$[-M\ \omega^2 + (K_2 + K_3)](A_2 \cos \omega t + B_2 \sin \omega t)$$

$$- K_3(A_1 \cos \omega t + B_1 \sin \omega t) = 0$$

iii) $[- M_2\omega^2 + (K_2 + K_3)]A_2 - K_3A_1 = 0$ (cos ωt terms).

iv) $[- M_2\omega^2 + (K_2 + K_3)]B_2 - K_3B_1 = 0$ (sin ωt terms).

Equations i) - iv) make up a solvable system of four equations in four unknowns.

From iii) $$A_1 = \frac{[-M_2\omega^2 + (K_2 + K_3)]A_2}{K_3} \qquad \text{(v)}$$

From iv) $$B_1 = \frac{[- M_2\omega^2 + (K_2 + K_3)]B_2}{K_3} \qquad \text{(vi)}$$

Substituting these relationships into i) + ii) yields

$$\frac{[- M_1\omega^2 + (K_1 + K_3)][- M_2\omega^2 + (K_2 + K_3)]A_2}{K_3}$$

$$- K_3A_2 = F_0 \qquad \text{(vii)}$$

and $$\frac{[- M_1\omega^2 + (K_1 + K_3)][- M_2\omega^2 + (K_2 + K_3)]B_2}{K_3} - K_3B_2 = 0.$$

In general,

$$\frac{[- M_1\omega^2 + (K_1 + K_3)][- M_2\omega^2 + (K_2 + K_3)]}{K_3} \neq K_3;$$

therefore, $B_2 = 0$ and from (vi), $B_1 = 0$.

A_2 from vii) is

$$\frac{F_0\ (K_3)}{[- M_1\omega^2 + (K_1 + K_3)][- M_2\omega^2 + (K_2 + K_3)] - K_3^2}.$$

From v)

$$A_1 = \frac{F_0}{\frac{[- M_2\omega^2 + (K_1 + K_3)] - K_3^2}{[- M_2\omega^2 + (K_2 + K_3)]}}$$

$X_1(t) = A_1 \cos \omega t$

$$= \frac{F_0\ [- M_2\omega^2 + K_2 + K_3] \cos \omega t}{[- M_2\omega^2 + K_1 + K_3][- M_2\omega^2 + K_2 + K_3] - K_3^2}$$

$$= \frac{- F_0\ [M_2\omega^2 - (K_2 + K_3)] \cos \omega t}{[M_2\omega^2 - (K_1 + K_3)][M_2\omega^2 - (K_2 + K_3)] - K_3^2}$$

and $X_2(t) = A_2 \cos \omega t$

$$x_2(t) = \frac{F_0\ K_3 \cos \omega t}{[M_1\ \omega^2 - (K_1 + K_3)][M_2\omega^2 - (K_2 + K_3)] - K_3^2}$$

Remember that these are steady state solutions. To find the transient motions when the system is first driven, one must use another type of analysis. Laplace transforms would be particularly useful. Also note that the steady state motions here are either in phase with or 180° out of phase with the driving force.

The amplitude of $x_1(t)$ is simply

$$A_n = \frac{x_n(t)}{\cos \omega t} = x_n \max$$

$$x_1 \max = \frac{-F_0 [M_2\omega^2 - (K_2 + K_3)]}{[M_2\omega^2 - (K_1 + K_3)][M_2\omega^2 - (K_2 + K_3)] - K_3^2}$$

Similarly,

$$x_2 \max = \frac{F_0 K_3}{[M_1\omega^2 - (K_1 + K_3)][M_2\omega^2 - (K_2 + K_3)] - K_3^2}$$

For arbitrary K_1, K_2, K_3, M_1 and M_2, x_1 and x_2 will have resonant points when their denominator equals zero; that is, when ω^2 is a root of

$$[M_1\omega^2 - (K_1 + K_3)][M_2\omega^2 - (K_2 + K_3)] - K_3^2 = 0,$$

x_1 and x_2 will go to infinity. ω_1^2 + ω_2^2 are these roots.
Also, x_1 will be zero when

$$\omega^2 = \omega_0^2 = \frac{K_2 + K_3}{M_2}$$

(its numerator is 0), or F_0 is zero. x_2 will be zero only if F_0 is 0; however, when F_0 is zero, the system is not being driven and the problem becomes trivial. Therefore, x_1 will become zero at one value of ω^2, but x_2 will never be zero in a problem of interest.

The signs of x_1 and x_2 switch after each resonant point or zero. Also note that x_1 and x_2 start as positive values when $\omega^2 = 0$. The graphs of x_1 + x_2 vs. ω^2 look like this:

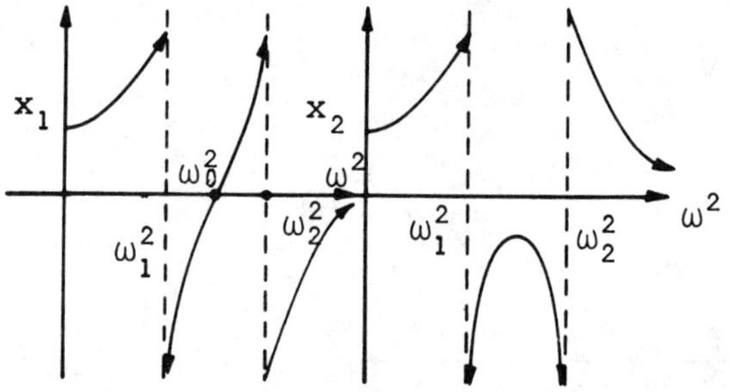

When x is positive, the motion is in phase with the force; when x is negative, the motion is 180° out of phase.

This type of system is often used to stop vibrations of a driven mass. By selecting the proper K_2, K_3 and M_2, one can set ω_0 equal to the driving frequency, thus making $x_1 = 0$. When used for this purpose, the system is called a vibration absorber.

NORMAL MODES OF VIBRATION AND NATURAL FREQUENCIES

• PROBLEM 24-7

Two pendulums, each of mass m and length ℓ, are suspended from the same horizontal line. The angular displacement is $\theta(t)$ and $\psi(t)$. The masses are coupled by a spring of spring constant K. The spring is unextended when both pendulums are in the vertical position ($\theta = \psi = 0$). Find the equation of motion of these coupled pendulums if $\theta(0) = \theta_0$ and $\psi(0) = \psi_0$; also $\theta(0) = 0$ and $\psi(0) = 0$.

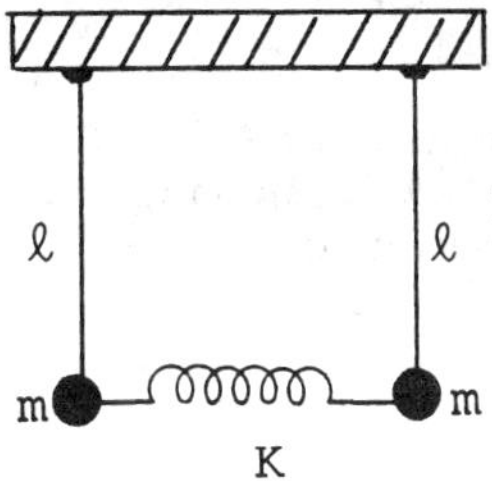

Fig. 1

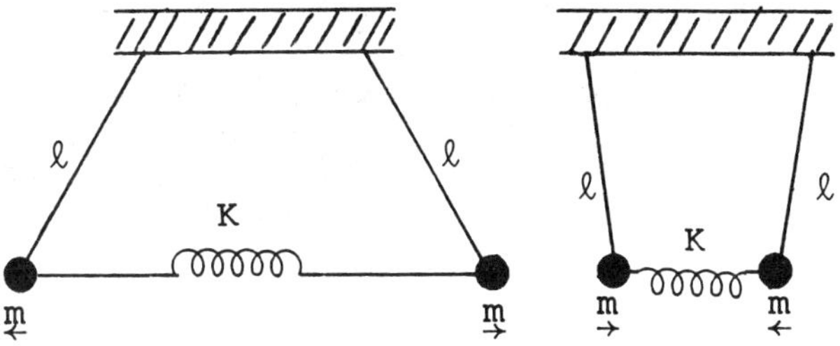

Fig. 2

Solution: If we assume small angular displacements then $\sin\theta \approx \theta$ and $\sin\psi \approx \psi$. In such a case the application of Newton's second law to each pendulum yields

$$m\ddot{\theta} + g\theta = -\ell K(\theta - \psi)$$

and $$m\ddot{\psi} + g\psi = \ell K(\theta - \psi).$$

If we let $w = \sqrt{\frac{g}{\ell}}$ and

$k = \frac{\ell K}{m}$, the coupled equations of this system are

$$\ddot{\theta} + w^2\theta = -k(\theta - \psi) \qquad \text{(a)}$$

$$\ddot{\psi} + w^2\psi = k(\theta - \psi). \qquad \text{(b)}$$

We observe that if we add (a) and (b) we obtain

$$\ddot{\theta} + \ddot{\psi} + w^2(\theta + \psi) = 0 \qquad \text{(c)}$$

and if we subtract (b) from (a) we obtain

$$\ddot{\theta} - \ddot{\psi} + w^2(\theta - \psi) = -2k(\theta - \psi). \qquad \text{(d)}$$

In (c) and (d) our dependent variables appear as $\theta + \psi$ or $\theta - \psi$. Thus, it is natural to define

$$y = \theta + \psi,$$

$$z = \theta - \psi.$$

Then (c) and (d) are

$$\ddot{y} + w^2y = 0 \qquad \text{(e)}$$

$$\ddot{z} + (w^2 + 2k)z = 0 \qquad \text{(f)}$$

since $$\ddot{\theta} + \ddot{\psi} = \frac{d^2(\theta + \psi)}{dt^2}$$

and $$\ddot{\theta} - \ddot{\psi} = \frac{d^2(\theta - \psi)}{dt^2}\ .$$

Equations (e) and (f) are uncoupled in the coordinates y and z.

The process of uncoupling a system of equations by suitable choice of a change of variable is called going to normal coordinates. Thus y and z are the normal coordinates of this problem. If we let $p^2 = w^2 + 2k$ the equations to be solved, (e) and (f), are

$$\ddot{y} + w^2y = 0$$

$$\ddot{z} + p^2z = 0$$

These are the simple harmonic oscillator equations whose solutions is

$$y = c_1 \cos wt + c_2 \sin wt \qquad \text{(g)}$$

and $$z = c_3 \cos pt + c_4 \sin pt. \qquad \text{(h)}$$

where c_1, c_2, c_3 and c_4 are constants to be determined from the initial conditions. We now reverse the change of coordinates to

$$\theta = \frac{1}{2}(y + z)$$

$$\psi = \frac{1}{2}(y - z).$$

Substituting from (g) and (h),

$$\theta = \frac{1}{2}(c_1 \cos wt + c_2 \sin wt + c_3 \cos pt + c_4 \sin pt)$$

$$\psi = \frac{1}{2}(c_1 \cos wt + c_2 \sin wt - c_3 \cos pt - c_4 \sin pt).$$

Applying the initial conditions,

$$\theta(0) = \frac{1}{2}(c_1 + c_3) = \theta_0$$

$$\psi(0) = \frac{1}{2}(c_1 - c_3) = \psi_0$$

which yields

$$c_1 = \theta_0 + \psi_0$$

$$c_2 = \theta_0 - \psi_0.$$

Also $\dot{\theta}(0) = \frac{1}{2}(wc_2 + pc_4) = 0$

$$\dot{\psi}(0) = \frac{1}{2}(wc_2 - pc_4) = 0$$

which yields

$$c_2 = c_4 = 0.$$

Thus our final solution is

$$\theta = \frac{1}{2}[(\theta_0 + \psi_0)\cos wt + (\theta_0 - \psi_0)\cos pt] \qquad (i)$$

$$\psi = \frac{1}{2}[(\theta_0 + \psi_0)\cos wt - (\theta_0 - \psi_0)\cos pt]. \qquad (j)$$

We note two cases of special initial conditions:

(I) $\theta_0 = \psi_0$, then

$$\theta = \psi = \frac{1}{2}(\theta_0 + \psi_0)\cos wt$$

$$= \theta_0 \cos wt$$

and the pendulums oscillate in unison (phase). The spring remains unstretched.

(II) $\theta_0 = -\psi_0$, then

$$\theta = -\psi = \theta_0 \cos pt$$

and the pendulums oscillate exactly out of phase (fig. 2).

These two special cases are called the normal modes of vibration.

• PROBLEM 24-8

Two masses m_1, m_2 connected with springs of negligible mass and force constant k_1, k, k_2 move along a frictionless horizontal plane. If m_1 has a positive displacement of x_1 and m_2 a positive displacement of x_2 from their positions of equilibrium, show that the equations of motion of m_1, m_2 are

$$m_1\ddot{x}_1 = -k_1x_1 - k(x_1 - x_2)$$

$$m_2\ddot{x}_2 = -k_2x_2 + k(x_1 - x_2).$$

If $y_1 = x_1\sqrt{m_1}$, $y_2 = x_2\sqrt{m_2}$, $k_1 + k = m_1\omega_1^2$, $k_2 + k = m_2\omega_2^2$ and $k = c\sqrt{m_1m_2}$, show that the equations of motion can be reduced to

$$\ddot{y}_1 + \omega_1^2 y_1 - cy_2 = 0$$

$$\ddot{y}_2 + \omega_2^2 y_2 - cy_1 = 0$$

Solve these equations by setting $y_1 = A \sin \omega t$ and $y_2 = B \sin \omega t$, and show that

$$\omega^2_h = \frac{\omega_1^2 + \omega_2^2}{2} + \frac{1}{2}\sqrt{(\omega_1^2 - \omega_2^2)^2 + 4c^2}$$

$$\omega^2_\ell = \frac{\omega_1^2 + \omega_2^2}{2} - \frac{1}{2}\sqrt{(\omega_1^2 - \omega_2^2)^2 + 4c^2}$$

Show that $\frac{A}{B}$ is positive for $\omega = \omega_\ell$ or for the in-phase component, while $\frac{A}{B}$ is negative for $\omega = \omega_h$ or for the out-phase component. Also show that $\omega^2_h + \omega^2_\ell = \omega_1^2 + \omega_2^2$ and that ω_h and ω_ℓ are the angular frequencies of the normal modes of vibration.

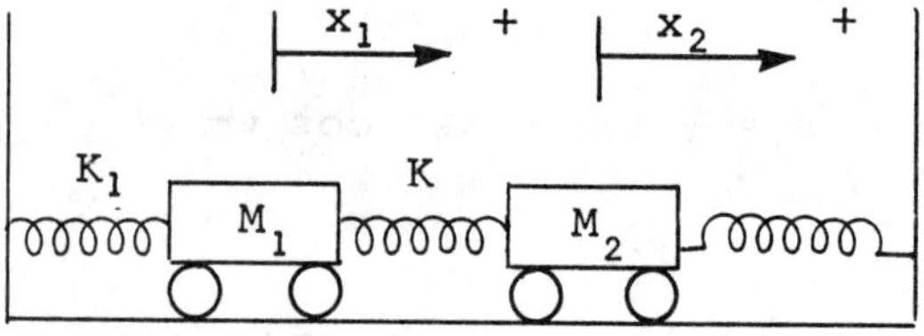

Fig. 1

<u>Solution</u>: Let m_1 be displaced x_1 units to the right and m_2 displaced x_2 units to the right. For purposes of visualization assume $x_1 > x_2$. Spring k_1 is stretched x_1 while k is compressed a distance x_1 on its left and allowed to expand a smaller amount, x_2, on its right for a total compression of $(x_1 - x_2)$. The springs k_1 and k exert a force on m_1 opposite to its displacement. Applying F = ma to m_1 gives

$$-k_1x_1 - k(x_1 - x_2) = m_1\ddot{x}_1 \qquad (1)$$

Spring k_2 exerts a force on m_2 to its left while k exerts a force on m_2 to its right, the positive direction. Applying F = ma to m_2 gives

$$-k_2x_2 + k(x_1 - x_2) = m_2\ddot{x}_2 \qquad (2)$$

These equations are called coupled linear differential equations of the second order. They are coupled in the sense that Eq.(1) contains x_2 and Eg. (2) contains an x_1. This relation arises from the coupling spring, k.

Combine like terms in each equation to obtain

$$\ddot{x}_1 = -\frac{(k_1 + k)}{m_1} x_1 + \frac{k}{m_1} x_2 \qquad (3)$$

$$\ddot{x}_2 = \frac{k}{m_2} x_1 - \frac{(k_2 + k)}{m_2} x_2 \qquad (4)$$

Let $\omega_1{}^2 = \dfrac{k_1 + k}{m_1}$ and $\omega_2{}^2 = \dfrac{k_2 + k}{m_2}$ (5)

Upon substitution of (5) into (3) and (4), the result is

$$\ddot{x}_1 = -\omega_1{}^2 x_1 + \frac{k}{m_1} x_2 \qquad (6)$$

$$\ddot{x}_2 = \frac{k}{m_2} x_1 - \omega_2{}^2 x_2 \qquad (7)$$

Next substitute $x_1 = y_1/\sqrt{m_1}$, $x_2 = y_2/\sqrt{m_2}$ and their corresponding second derivatives into Egs. (6) and (7) to obtain

$$\frac{\ddot{y}_1}{\sqrt{m_1}} = -\omega_1{}^2 \frac{y_1}{\sqrt{m_1}} + \frac{k}{m_1} \frac{y_2}{\sqrt{m_1}} \qquad (8)$$

$$\frac{\ddot{y}_2}{\sqrt{m_2}} = \frac{k}{m_2} \frac{y_1}{\sqrt{m_2}} - \omega_2{}^2 \frac{y_2}{\sqrt{m_2}} \qquad (9)$$

A clever observation shows that if we substitute $k = c\sqrt{m_1 m_2}$ into the above equations they become

$$\ddot{y}_1 = -\omega_1^2 y_1 + c\, y_2 \qquad (10)$$

$$\ddot{y}_2 = c\, y_1 - \omega_2{}^2 y_2 \qquad (11)$$

These equations reduce to the simple case of a mass oscillating on the end of a spring when $c = 0$. This means $k = 0$; i.e., the coupling spring is removed from the system.

A normal mode of oscillation is defined as one in which the masses of the system vibrate at the same frequency, but not necessarily the same amplitude. In order to find the normal modes of oscillation of the coupled Egs. (10) and (11), assume the respective solutions of common frequency:

$y_1 = A \sin \omega t$ and $y_2 = B \sin \omega t$.

Substitute these and their second derivatives into Egs. (10) and (11) to obtain

$$-\omega^2 A \sin \omega t = -\omega_1{}^2 A \sin \omega t + cB \sin \omega t \qquad (12)$$

$$-\omega^2 B \sin \omega t = cA \sin \omega t - \omega_2{}^2 B \sin \omega t. \qquad (13)$$

Divide both sides of each equation by $\sin \omega t$ and rearrange as

$$(\omega_1{}^2 - \omega^2)A - c\,B = 0$$

$$-c\,A + (\omega_2{}^2 - \omega^2)B = 0$$

These equations are homogeneous, linear, simultaneous equations in the unknowns A and B. (ω is also unknown, but for the moment assume it is given.) Homogeneous refers to the fact that the right side is zero. These equations have a solution other than A = B = 0 only if the determinant of the system is zero. This requirement determines the values of ω. So,

$$\begin{vmatrix} (\omega_1^2 - \omega^2) & -c \\ -c & (\omega_2^2 - \omega^2) \end{vmatrix} = 0$$

The determinant becomes

$$(\omega_1^2 - \omega^2)(\omega_2^2 - \omega^2) - c^2 = 0$$

or

$$\omega_1^2\omega_2^2 - \omega_1^2\omega^2 + \omega_2^2\omega^2 + \omega^4 - c^2 = 0$$

$$\omega^4 - (\omega_1^2 + \omega_2^2)\omega^2 + (\omega_1^2\omega_2^2 - c^2) = 0$$

Since this is a quadratic equation in ω^2, the quadratic formula gives

$$\omega^2 = \frac{(\omega_1^2 + \omega_2^2) \pm \sqrt{(\omega_1^2 + \omega_2^2)^2 - 4(\omega_1^2\omega_2^2 - c^2)}}{2} .$$

Expansion and rearrangement of the radicand results in

$$\omega^2 = \frac{(\omega_1^2 + \omega_2^2)}{2} \pm \frac{\sqrt{(\omega_1^2 - \omega_2^2)^2 + 4c^2}}{2} \qquad (16)$$

This tells us that there are two different frequencies that result in normal mode oscillations; the one when the positive sign is used in Eq. (16) will be called ω_h^2 and the other ω_ℓ^2. So,

$$\omega_h^2 = \frac{\omega_1^2 + \omega_2^2}{2} + \frac{\sqrt{(\omega_1^2 - \omega_2^2)^2 + 4c^2}}{2} \qquad (17)$$

$$\omega_\ell^2 = \frac{\omega_1^2 + \omega_2^2}{2} - \frac{\sqrt{(\omega_1^2 - \omega_2^2)^2 + 4c^2}}{2} \qquad (18)$$

Inspection of Eqs. (17) and (18) shows that

$$\omega_h^2 + \omega_\ell^2 = \omega_1^2 + \omega_2^2 . \qquad (19)$$

If c = 0, (i.e., no coupling), then Eq. (17) shows $\omega_h = \omega_1$. Therefore, when there is coupling

$$\omega_h > \omega_1 . \qquad (20)$$

Similarly, Eq. (18) shows that

$$\omega_\ell < \omega_2 . \qquad (21)$$

Rearranging Eqs. (14) and (15) gives

$$\frac{A}{B} = \frac{c}{\omega_1^{\,2} - \omega^2} \tag{22}$$

$$\frac{A}{B} = \frac{\omega_2^{\,2} - \omega^2}{c} \tag{23}$$

It follows from Eqs. (22) and (20) that when $\omega = \omega_h$, A and B have opposite displacements, i.e., A/B is negative. Physically, this means that one mass is initially given a positive displacement and the other a negative one, then they will oscillate at the higher frequency. Similarly, from Eqs. (23) and (21), it follows that when $\omega = \omega_\ell$, then A/B is positive and the masses oscillate in phase with each other. This mode of oscillation will occur when each mass is displaced in the same direction before being released.

• PROBLEM 24-9

Find the two normal modes of vibration for a pair of identical damped coupled harmonic oscillators.

$$m\ddot{x}_1 + b\dot{x}_1 + (k + k_c)x_1 + k_c x_2 = 0$$

$$m\ddot{x}_2 + b\dot{x}_2 + (k + k_c)x_2 + k_c x_1 = 0.$$

Let $k' = k + k_c$.

Solution: The system given has two degrees of freedom and, therefore, it also has two characteristic frequencies of vibration. In order to solve the problem assume a solution of the form

$$x_1(t) = Ae^{j\omega t}$$

$$x_2(t) = Be^{j\omega t}$$

where $j = \sqrt{-1}$.

This assumption is based on previous experience with simple, one degree of freedom, harmonic oscillators.

Now,

$$\dot{x}_1 = Aj\omega\, e^{j\omega t} \qquad \dot{x}_2 = Bj\omega\, e^{j\omega t}$$

$$\ddot{x}_1 = -A\omega^2\, e^{j\omega t} \qquad \ddot{x}_2 = -B\omega^2\, e^{j\omega t}$$

Substituting these results into the original equations yields

$$-A\omega^2 m\, e^{j\omega t} + bAj\omega e^{j\omega t} + (k + k_c)Ae^{j\omega t} + k_c Be^{j\omega t} = 0$$

$$-B\omega^2 me^{j\omega t} + bBj\omega e^{j\omega t} + (k + k_c)Be^{j\omega t} + k_c Ae^{j\omega t} = 0$$

Since $e^{j\omega t} \neq 0$, divide both equations by $e^{j\omega t}$.

$$- A\omega^2 m + bAj\omega + (k + k_c)A + k_c B = 0 \qquad (1)$$

$$- B\omega^2 m + bBj\omega + (k + k_c)B + k_c A = 0$$

Putting these equations in matrix form,

$$\begin{bmatrix} k + k_c + bj\omega - m\omega^2 & k_c \\ k_c & k + k_c + bj\omega - m\omega^2 \end{bmatrix} \begin{bmatrix} A \\ B \end{bmatrix} = 0$$

$$\text{Now, } A = \frac{\begin{vmatrix} 0 & k_c \\ 0 & k + k_c + bj\omega - m\omega^2 \end{vmatrix}}{\begin{vmatrix} k + k_c + bj\omega - m\omega^2 & k_c \\ k_c & k + k_c + bj\omega - m\omega^2 \end{vmatrix}}$$

by Kramer's Rule.

Similarly for B,

$$B = \frac{\begin{vmatrix} k + k_c + bj\omega - m\omega^2 & 0 \\ k_c & 0 \end{vmatrix}}{\begin{vmatrix} k + k_c + bj\omega - m\omega^2 & k_c \\ k_c & k + k_c + bj\omega - m\omega^2 \end{vmatrix}}$$

In both cases the determinant in the denominator must equal zero for A or B to have a value other than zero.

Therefore,

$$\begin{vmatrix} k + k_c + bj\omega - m\omega^2 & k_c \\ k_c & k + k_c + bj\omega - m\omega^2 \end{vmatrix} = 0$$

This is the characteristic frequency equation for the system. In expanded form the characteristic equation becomes

$$(k + k_c + bj\omega - m\omega^2)^2 - k_c^2 = 0.$$

This is the difference of two squares and can be expanded as shown below:

$$[(k + k_c + bj\omega - m\omega^2) - k_c]$$

$$[k + k_c + bj\omega - m\omega^2) + k_c] = 0$$

or $$[m\omega^2 - k - bj\omega][m\omega^2 - k - 2k_c - bj\omega] = 0.$$

Either term may equal zero for the equation to be zero; therefore,

$$m\omega^2 - k - bj\omega = 0$$

will supply one root and

$$m\omega^2 - k - 2k_c - bj\omega = 0$$

will yield the second characteristic frequency.

Solving these equations using the quadratic formula,

$$\omega_1 = \frac{bj \pm \sqrt{-b^2 - 4km}}{2m}$$

$$\omega_2 = \frac{bj \pm \sqrt{-b^2 + 4(k + 2k_c)m}}{2m}.$$

$x_1(t)$ then is

$$Ae^{j\frac{(bj \pm \sqrt{-b^2 + 4km})t}{2m}}$$ and by Euler's theorem,

$$x_1(t) = Ae^{-bt/2m} \cos\left(\frac{\sqrt{-b^2 + 4km}}{2m}t + \theta\right).$$

Here θ is a phase shift angle determined by initial conditions.

Similarly,

$$x_2(t) = Be^{-bt/2m} \cos\left(\frac{\sqrt{-b^2 + 4km}}{2m} + \theta\right)$$

in the first mode (first characteristic frequency).

In the second mode,

$$x_1(t) = Ce^{-bt/2m} \cos\left[\frac{\sqrt{-b^2 + 4(k + k_c)m}}{2m} t + \theta\right]$$

$$x_2(t) = De^{-bt/2m} \cos\left[\frac{\sqrt{-b^2 + 4(k + k_c)m}}{2m} t + \theta\right]$$

To find the relative amplitudes of $x_1 + x_2$ (the mode shape), divide $x_1(t)$ by $x_2(t)$.

$$\frac{x_1(t)}{x_2(t)} = \frac{A}{B} \quad \text{for both frequencies.}$$

From the simplified system equations (1),

$$\frac{A}{B} = \frac{k_c}{m\omega^2 - bj\omega - (k + k_c)}$$

$\frac{A}{B}$ for ω_1 is

$$\frac{A}{B} = \frac{k_c}{m\ (\omega_1^2) - bj\omega_1 - (k + k_c)}$$

$$\frac{A}{B} = \frac{k_c}{m\ \frac{(bj \pm \sqrt{-b^2 + 4km})^2}{(2m)^2} - bj\ \frac{(bj \pm \sqrt{-b^2 + 4km})}{2m} - (k+k_c)}$$

$$= \frac{k_c}{\frac{1}{4}\frac{(-b^2 \pm 2bj\sqrt{-b^2+4km} + -b^2+4km)}{m} + \frac{b^2}{2m}\ \frac{\pm bj\sqrt{-b^2+4km}}{2m} - (k+k_c)}$$

$$= \frac{k_c}{\frac{-b^2}{2m} + k - k - k_c}$$

$$= \frac{k_c}{\frac{-b^2}{2m} - k_c}$$

In most physical systems b is small and $\frac{A}{B} \approx -1$.

$\frac{A}{B}$ for ω_2 is:

$$\frac{A}{B} = \frac{k_c}{m\ \frac{(bj \pm \sqrt{-b^2+4(k+2k_c)m})^2}{2m} - bj\ \frac{(bj \pm \sqrt{-b^2+4(k+2k_c)m})}{2m} - k - k_c}$$

$$\frac{A}{B} = \frac{k_c}{\frac{1}{4m}\ (-b^2 \pm 2bj\sqrt{b^2+4(k+2k_c)m} - b^2 + 4(k+2k_c)m) + \frac{b^2}{2m} \pm \frac{bj\sqrt{b^2+4(k+2k_c m)}}{2m} - k - k_c}$$

$$\frac{A}{B} = \frac{k_c}{(k + 2k_c) - \frac{b^2}{2m} - (k + k_c)} = \frac{k_c}{\frac{-b^2}{2m} + k_c}\ .$$

Again, $\frac{b^2}{2m}$ is small for most physical systems and $\frac{A}{B} \approx 1$.

Therefore, when the system vibrates at $\omega = \omega_1$

$A \approx - B$ and $x_1(t)x - x_2(t) \approx x(t)$

$$x(t) \approx Ae^{-b/2m\ t} \cos\left(\sqrt{-b^2 + 4km}\ t + \theta\right),$$

and when the system vibrates at its other natural frequency,

$A \approx B$ and $x_1(t) \approx x_2(t) \times x(t)$

$$x(t) = Ae^{-b/2m\ t} \cos\left(\sqrt{-b^2 + 4(k + 2k_c)m}\ t + \theta\right).$$

In actuality, the system might vibrate at either of these frequencies or possibly some combination of the two, depending on the conditions which caused the vibration. That is, the vibratory mode is fixed by the initial conditions.

• PROBLEM 24-10

Given the double pendulum illustrated below, set up the equations of motion and find the approximate natural frequencies. Assume that the oscillations are small.

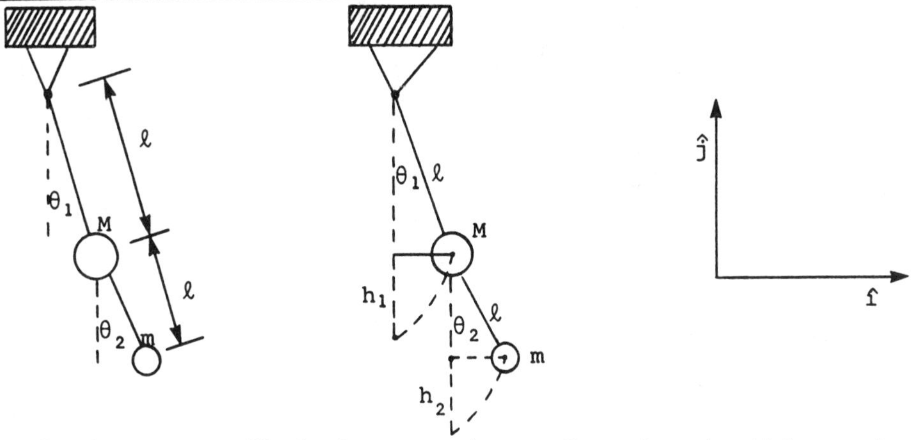

Solution: To find the equations of motion in this problem, it may be easiest to use Langranges method. Langranges equation states that:

$$\frac{d}{dt}\frac{\partial T}{\partial \dot{q}_i} - \frac{\partial T}{\partial q_i} + \frac{\partial V}{\partial q_i} = Q_i ,$$

where V is potential energy, T is kinetic energy, Q_i is a generalized force, q_i is a general coordinate and $\dot{q}_i$ is a generalized velocity.

In this problem, $q_1 = \theta_1$ and $q_2 = \theta_2$. Similarly, $\dot{q}_1 = \dot{\theta}_1$ and $\dot{q}_2 = \dot{\theta}_2$.

V is due to gravitational forces only in this problem.

If one defines V = 0 when θ_1 and $\theta_2 = 0$, then

$$V = Mgh_1 + mgh_1 + mgh_2$$

where $h_1 = \ell(1 - \cos\theta_1)$; $h_2 = \ell(1 - \cos\theta_2)$.

Combining all of the above yields

$$V = g\ell[(M + m)(1 - \cos\theta_1) + m(1 - \cos\theta_2)]$$

Assume that the pendulum balls can be represented as particles in this case.

$$T = \frac{MV_1{}^2}{2} + \frac{mV_2{}^2}{2}$$

V_1 is the velocity of M, V_2 is the velocity of m.

$$\vec{V}_1 = (\ell\dot{\theta}_1(\cos\theta_1\ \hat{i} + \sin\theta_1\ \hat{j})$$

$$\vec{V}_2 = \vec{V}_1 + (\ell\dot{\theta}_2(\cos\theta_2\ \hat{i} + \sin\theta_2\ \hat{j})$$

If $\theta_1 + \theta_2$ are small, then

$$V_1^2 \approx (\ell\dot{\theta}_1)^2 \quad \text{and } V_2^2 \approx (\ell\dot{\theta}_1 + \ell\dot{\theta}_2)^2$$

$$T = \frac{M\ell^2}{2}(\dot{\theta}_1)^2 + \frac{m\ \ell^2(\dot{\theta}_1 + \dot{\theta}_2)^2}{2} .$$

To find the equations of motion, one differentiates in accordance with Langrange's relation, with Q_1 and $Q_2 = 0$, since this is a free vibration.

$$\frac{d}{dt}\frac{\partial T}{\partial\dot{\theta}_1} - \frac{\partial T}{\partial\theta_1} + \frac{\partial V}{\partial\theta_1} = Q_1$$

$$\frac{\partial T}{\partial\dot{\theta}_1} = \frac{\partial\left[\frac{M\ell^2}{2}(\dot{\theta}_1)^2 + \frac{m\ell^2}{2}(\dot{\theta}_1 + \dot{\theta}_2)^2\right]}{\partial\dot{\theta}_1}$$

$$\frac{\partial T}{\partial\dot{\theta}_1} = M\ell^2\dot{\theta}_1 + m\ell^2(\dot{\theta}_1 + \dot{\theta}_2)$$

$$\frac{d}{dt}\left(\frac{\partial T}{\partial\dot{\theta}_1}\right) = M2^2\ddot{\theta}_1 + m\ell^2\ddot{\theta}_1 + m\ell_2\ddot{\theta}_2$$

$$= (M + m)\ell^2\ddot{\theta}_1 + m\ell^2\ddot{\theta}_2$$

$$\frac{\partial T}{\partial\theta_1} = \frac{\partial\left[\frac{M\ell^2}{2}(\dot{\theta}_1)^2 + \frac{m\ell^2}{2}(\dot{\theta}_1{}^2 + \dot{\theta}_2{}^2)\right]}{\partial\theta_1} = 0$$

$$\frac{\partial V}{\partial \theta_1} = \frac{\partial\, q\ell[(M + m)(1 - \cos\theta_1) + m(1 - \cos\theta_2)]}{\partial \theta_1}$$

$$= q\ell(M + m)\sin\theta_1 \ .$$

Combining the above yields one equation of motion,

$$(M + m)\ell^2\ddot{\theta}_1 + m\ell^2\ddot{\theta}_2 + q\ell(M + m)\sin\theta_1 = 0$$

Following the same procedure for θ_2 yields

$$\frac{d}{dt}\left(\frac{\partial T}{\partial \dot{\theta}_2}\right) = m\,\ell^2\ddot{\theta}_1 + m\ell^2\ddot{\theta}_2$$

$$\frac{\partial T}{\partial \theta_2} = 0$$

$$\frac{\partial V}{\partial \theta_2} = mq\ell \sin\theta_2$$

$$m\ell^2\ddot{\theta}_1 + m\ell^2\ddot{\theta}_2 + mq\ell \sin\theta_2 = 0$$

It has already been assumed that the angles are small, therefore, $\sin\theta \simeq \theta$, and it follows that

$$(M + m)\ell^2\ddot{\theta}_1 + m\ell^2\ddot{\theta}_2 + q\ell(M + m)\theta_1 = 0$$

$$m\ell^2\ddot{\theta}_1 + m\ell^2\ddot{\theta}_2 + mq\ell\,\theta_2 = 0$$

These are the system's equations of motion. In matrix form they look like this:

$$\begin{bmatrix} (M + m)\ell^2 & m\ell^2 \\ m\ell^2 & m\ell^2 \end{bmatrix}\begin{bmatrix} \ddot{\theta}_1 \\ \ddot{\theta}_2 \end{bmatrix} + \begin{bmatrix} (M + m)q\ell & 0 \\ 0 & mq\ell \end{bmatrix}\begin{bmatrix} \theta_1 \\ \theta_2 \end{bmatrix} = 0$$

Assuming that the solution is simple harmonic motion, $\theta_n = a_n \sin\omega t$

$$\ddot{\theta}_n = -\omega^2 a_n \sin\omega t = -\omega^2\theta_n \ .$$

Using this important result,

$$-\omega^2\begin{bmatrix} (M + m)\ell^2 & m\ell^2 \\ m\ell^2 & m\ell^2 \end{bmatrix}\begin{bmatrix} \theta_1 \\ \theta_2 \end{bmatrix} + \begin{bmatrix} (M + m)g\ell & 0 \\ 0 & mq\ell \end{bmatrix}\begin{bmatrix} \theta_1 \\ \theta_2 \end{bmatrix} = 0$$

The equation can be simplified by multiplying by

$$\begin{bmatrix} \frac{1}{(M + m)q\ell} & 0 \\ 0 & \frac{1}{mq\ell} \end{bmatrix}$$

$$-\omega^2 \begin{bmatrix} \frac{\ell^2}{q\ell} & \frac{m\ell^2}{(M+m)q\ell} \\ \frac{m\ell^2}{mq\ell} & \frac{m\ell^2}{mq\ell} \end{bmatrix} \begin{bmatrix} \theta_1 \\ \theta_2 \end{bmatrix} + \begin{bmatrix} 1 & 0 \\ 0 & 1 \end{bmatrix} \begin{bmatrix} \theta_1 \\ \theta_2 \end{bmatrix} = 0$$

If θ_1 and $\theta_2 \neq 0$, then for the equation to be satisfied, the following determinant must be zero.

$$\begin{vmatrix} 1 - \omega^2 \left(\frac{\ell}{q}\right) & \frac{-\omega^2 m\ell}{(M+m)q} \\ -\omega^2 \left(\frac{\ell}{q}\right) & 1 - \omega^2 \frac{\ell}{q} \end{vmatrix} = 0$$

Evaluating the determinant and simplifying,

$$\left(1 - \omega^2 \left(\frac{\ell}{q}\right)\right) \left(1 - \omega^2 \frac{\ell}{q}\right) - \frac{\omega^4 \quad m\ell^2}{(M+m)q^2} = 0$$

$$1 - 2\omega^2 \frac{\ell}{q} + \omega^4 \frac{\ell^2}{q^2} - \frac{\omega^4 \, m\ell^2}{(M+m)q^2} = 0$$

$$\omega^4 \left(\frac{\ell^2}{q^2} - \frac{m \;\ell^2}{(M+m)q^2}\right) - \omega^2 \left(2 \frac{\ell}{q}\right) + 1 = 0$$

From the quadratic equation,

$$\omega^2 = \frac{2 \frac{\ell}{q} \pm \sqrt{\frac{4\ell^2}{q^2} - \frac{4\ell^2}{q^2}\left(1 - \frac{m}{M+m}\right)}}{\frac{2\ell^2}{q^2}\left(1 - \frac{m}{M+m}\right)}$$

Simplifying,

$$\omega^2 = \frac{\frac{\ell}{q}\left[2 \pm \sqrt{4 - 4\left(1 - \frac{m}{M+m}\right)}\right]}{\frac{2\ell^2}{q^2}\left(1 - \frac{m}{M+m}\right)}$$

$$\omega^2 = \frac{q}{\ell}\left[\frac{2 \pm \sqrt{4 - 4 + \frac{4m}{M+m}}}{2\left(1 - \frac{m}{M+m}\right)}\right]$$

$$\omega^2 = \frac{q}{\ell}\left[\frac{2 \pm 2\sqrt{\frac{m}{M+m}}}{2\left(1 - \frac{m}{M+m}\right)}\right]$$

Finally,

$$\omega^2 = \frac{g}{\ell}\left[\frac{1 \pm \sqrt{\dfrac{m}{M + m}}}{1 - \dfrac{m}{M + m}}\right]$$

The approximate natural frequencies are

$$\omega_1{}^2 = \frac{g}{\ell}\left[\frac{1 + \sqrt{\dfrac{m}{M + m}}}{1 - \dfrac{m}{M + m}}\right]$$

$$\omega_2{}^2 = \frac{g}{\ell}\left[\frac{1 - \sqrt{\dfrac{m}{M + m}}}{1 - \dfrac{m}{M + m}}\right]$$

CHAPTER 25

CONTINUOUS AND DEFORMABLE MEDIA

FLUID FLOW PROBLEMS

• PROBLEM 25-1

Determine the flow velocity of an incompressible fluid emanating from a small hole in the side of a container open to the atmosphere.

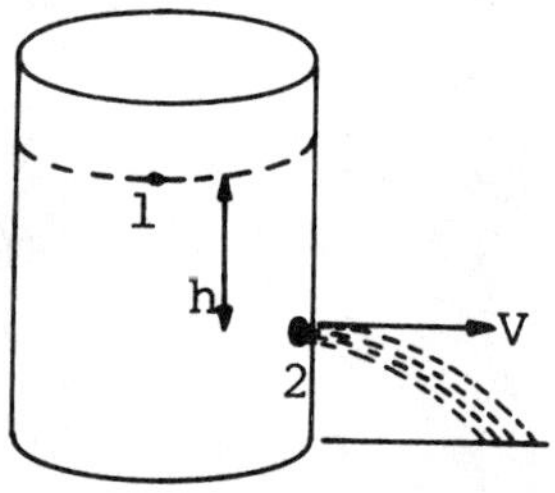

Solution: As shown in the Figure, h is the height of the surface above the hole. At any two points in an incompressible fluid, Bernoulli's Theorem holds:

$$p_1 + \frac{1}{2}\rho v_1^2 + \rho g h_1 = p_2 + \frac{1}{2}\rho v_2^2 + \rho g h_2. \quad (1)$$

Where p is the pressure, ρ the density, v the velocity and h the height above the reference level.

In this case we take 1 to be a point at the upper surface and 2 a point at the center of the hole. If the area of surface is large compared to that of the hole, then the surface drops very slowly and $v_1 \approx 0$. If the reference level is chosen to be the center of the hole, then $h_2 = 0$. Notice also that, since both the surface and the hole are exposed to the atmosphere, they are at atmospheric pressure, p_0, so $p_1 = p_2 = p_0$. With these conditions equation (1) becomes

$$\rho g h = \frac{1}{2}\rho v_2^2$$

$$v_2 = \sqrt{2gh}.$$

This result is known as Torricelli's Theorem. Notice that the magnitude of the velocity is the same as if the water had free fallen through a distance h.

• PROBLEM 25-2

1) Evaluate the potential energy u per unit mass as a function of pressure p for an ideal gas of molecular weight M at temperature T. 2) For the steady isothermal flow of this gas through a pipe of varying cross section and varying height above the earth, find expressions for the pressure, density, and velocity of the gas as functions of: s, the cross sectional area, and h, the height of the pipe. Assume pressure p_0, velocity v_0, at a point in the pipe at height $h = 0$, cross section s_0.

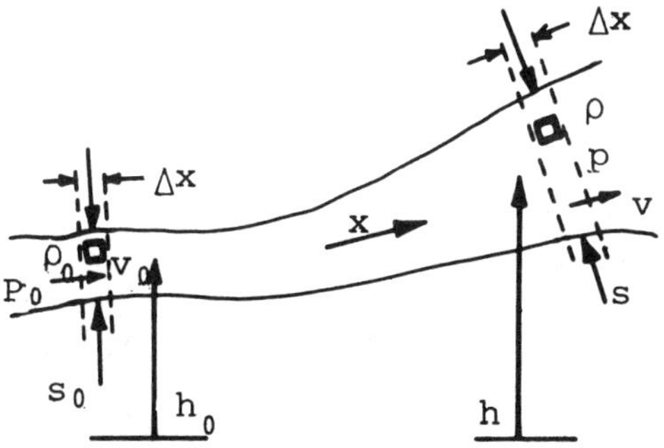

Solution: 1) When a perfect gas is compressed, energy is stored in the form of pressure. The amount of energy stored is equal to the work done when the system is pressurized. If the initial pressure is p_0 and the final pressure is p, then the change in potential energy is:

$$\Delta\mu = \mu - \mu_0 = \int_{p_0}^{p} v\, dp$$

or
$$\mu = \int_{p_0}^{p} v\, dp + \mu_0 \qquad (1)$$

Here μ_0 is the potential energy at p_0, and v is the specific volume of the gas.

According to ideal gas law, $v = \frac{RT}{Mp}$ where R is the universal gas constant, M is the molecular weight of gas, and T is the temperature (in °k).

Substituting the volume expression into the equation (1), yields,

$$\mu = \int_{p_0}^{p} \frac{RT}{Mp}\, dp + \mu_0$$

Since, $\frac{RT}{M}$ is constant, it can be taken outside the integral. Then integration gives,

$$\mu = \frac{RT}{M} \ln \frac{p}{p_0} + \mu_0$$

2) Consider the system in the figure.

where: ρ is the density of the gas, h_0 and h are elevations above the earth, (measured to the center of the pipe) Δx is an infinitesimal distance along the pipe, and s_0 and s are cross sectional areas.

It is necessary that we simplify the actual situation by assuming that the flow proceeds only in the x direction. According to this model, the energy and momentum in a volume $s\Delta x$ will always be the same. By making Δx very small, it is also possible to assume that the gas's properties are constant across Δx.

Now we can find an expression for pressure in terms of the velocities and cross sectional areas by using the principle of conservation of momentum. The total momentum at the initial station is the same as the total momentum at the final station:

$$m_0 = m \qquad \text{(m is total momentum)} \tag{2}$$

We know that m_0 = (mass)(velocity) - (a)

and, mass = (volume)(density)

mass = $(s\Delta x)$ (ρ),

So $m_0 = (s_0\Delta x\ \rho)v_0$.

Similarly, $m = (s\Delta x\ \rho)v$.

Substituting these values of m_0 and m into (2), yields

$$(s_0\ \Delta x\ \rho_0)\ v_0 = (s\Delta x\ \rho)v$$

After simplification, this equation may be written as

$$\frac{\rho_0}{\rho} = \frac{vs}{v_0 s_0} \tag{3}$$

From the ideal gas law, $pv = \frac{RT}{M}$, but $v = \frac{1}{\rho}$ so

$$p = \frac{\rho\ RT}{M}.$$

Since the flow is isothermal, the temperature of the gas remains constant. In other words, $T_0 = T$. Therefore the pressures on the initial station and the final stations are

$$p_0 = \frac{\rho_0 RT}{M}, \text{ and } p = \frac{\rho \ RT}{M}, \text{ respectively.}$$

Using both of these relations, we may write

$$\frac{p}{p_0} = \frac{\rho \ RT/M}{\rho_0 \ RT/M}$$

or

$$\frac{p}{p_0} = \frac{\rho}{\rho_0} \qquad (4)$$

From equation (3),

$$\frac{\rho}{\rho_0} = \frac{v_0 s_0}{vs}$$

So equation (4) can be written as

$$\frac{p}{p_0} = \frac{v_0 s_0}{vs} \qquad (4')$$

or

$$p = p_0 \frac{v_0 s_0}{vs}$$

Now use the principle of energy conservation to find an expression which relates all the variables. Assume that energy lost as heat due to friction between the gas and the pipe is minimal; then the total energy at the initial station is the same as the total energy at the final station.

The total energy at the first station is the potential and kinetic energy of each particle integrated over the volume $s_0 \Delta x$.

$$E_0 = \int_{volume} \left(\rho_0 gh_0 + \frac{1}{2} \rho_0 v_0{}^2 + \mu_0\right) dv$$

Since all of these quantities are assumed to be constant over the volume,

$$E_0 = s_0 \Delta x \left(\rho_0 \ gh_0 + \frac{1}{2} \rho_0 \ v_0{}^2 + \mu_0\right).$$

Let us choose h_0 and μ_0 as reference points for the two potential functions. Then,

$$E_0 = s_0 \Delta x \left(\frac{1}{2} \rho_0 \ v_0{}^2\right)$$

Similarly the total energy at the second station is:

$$E = \int_{volume} \left[\rho g(h - h_0) + \frac{1}{2} \rho v^2 + \frac{RT}{M} \ln \frac{p}{P_0} + \mu_0\right] dv$$

$$E = s\Delta x \left[\rho gz + \frac{1}{2}\rho v^2 + \frac{RT}{M} \ln \frac{p}{p_o}\right]$$

where $z = h - h_0$.

Since $E_0 = E$

$$s_0\Delta x \left(\frac{1}{2} \rho_o v_o^2\right) = s\Delta x \left[\rho gz + \frac{1}{2}\rho v^2 + \frac{RT}{M} \ln \frac{p}{p_o}\right]$$

or $$\frac{1}{2} s_o \rho_o v_o^2 = s \left[\rho gz + \frac{1}{2}\rho v^2 + \frac{RT}{M} \ln \frac{p}{p_o}\right]$$

Then, $$\frac{s_0 \rho_0 v_0^2}{s \rho} = 2\ gz + v^2 + \frac{2}{\rho} \left(\frac{RT}{M} \ln \frac{p}{p_0}\right)$$

$$- v^2 + \frac{s_0 \rho_0 v_0^2}{s\rho} = 2\ gz + \frac{2}{\rho}\ \frac{RT}{M} \ln \frac{p}{p_o}$$

This can be solved for v.

Now substitute the values of $\frac{\rho_0}{\rho}$ (from Eq. (3)) and $\frac{p_0}{p}$ (from Eq. (4)). Then:

$$- v^2 + \frac{s_0}{s}\ \frac{v}{v_0}\ \frac{s}{s_0}\ v_0^2 = 2gz + \frac{2vs}{\rho_0 v_0 s_0} \left(\frac{RT}{m} \ln \frac{v_0 s_0}{vs}\right)$$

or $$- v^2 + vv_0 = 2gz + \frac{2v\ s\ R\ T}{\rho_0 v_0 s_0 m} \ln \frac{v_0 s_0}{vs}$$

• PROBLEM 25-3

A fluid of viscosity η flows steadily between two infinite parallel plane walls a distance of ℓ apart. The velocity of the fluid depends only on the distance from the walls. The total fluid current between the walls in any unit length measured along the walls perpendicular to the direction of the flow is I. Find the velocity distribution and the pressure gradient parallel to the walls, assuming the pressure varies only in the direction of flow.

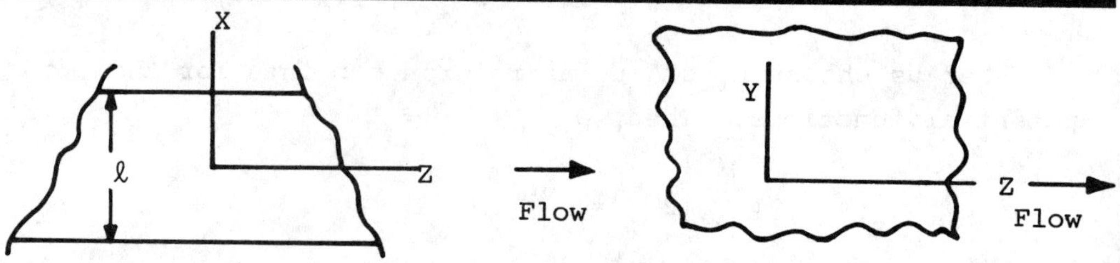

Solution: This problem can be solved by applying and integrating the governing differential equations for a viscous incompressible fluid in steady flow. These (Navier-Stokes) equations, assuming negligible body forces are

$$u \frac{\partial u}{\partial x} + v \frac{\partial u}{\partial y} + w \frac{\partial u}{\partial z} = - \frac{1}{\rho} \frac{\partial p}{\partial x} + \nu \left(\frac{\partial^2 u}{\partial x^2} + \frac{\partial^2 u}{\partial y^2} + \frac{\partial^2 u}{\partial z^2} \right) \tag{1}$$

$$u \frac{\partial v}{\partial x} + v \frac{\partial v}{\partial y} + w \frac{\partial v}{\partial z} = - \frac{1}{\rho} \frac{\partial p}{\partial y} + \nu \left(\frac{\partial^2 v}{\partial x^2} + \frac{\partial^2 v}{\partial y^2} + \frac{\partial^2 v}{\partial z^2} \right) \tag{2}$$

$$u \frac{\partial w}{\partial x} + v \frac{\partial w}{\partial y} + w \frac{\partial w}{\partial z} = - \frac{1}{\rho} \frac{\partial p}{\partial z} + \nu \left(\frac{\partial^2 w}{\partial x^2} + \frac{\partial^2 w}{\partial y^2} + \frac{\partial^2 w}{\partial z^2} \right) \tag{3}$$

where u, v and w are the components of velocity in the x, y and z directions respectively; ρ is the mass density; p is the pressure, and ν is the kinematic viscosity.

There are no velocity components in the x or y directions, so

$$u = v = 0.$$

Also, the velocity, w, varies only in the x direction:

$$w = w(x)$$

$$\frac{\partial w}{\partial y} = \frac{\partial w}{\partial z} = 0; \text{ hence } \frac{\partial^2 w}{\partial y^2} = \frac{\partial^2 w}{\partial z^2} = 0.$$

The equations (1), (2) and (3), therefore, reduce to

$$\frac{\partial p}{\partial x} = 0 \tag{4}$$

$$\frac{\partial p}{\partial y} = 0 \tag{5}$$

$$\frac{1}{\rho} \frac{\partial p}{\partial z} = \nu \frac{\partial^2 w}{\partial x^2}. \tag{6}$$

Equations (4) and (5) clearly imply that p is only a function of z (i.e., p = p(z)).

Therefore, $\frac{\partial p}{\partial z}$ is also a function of z only. But w = w(x); thus, $\frac{\partial^2 w}{\partial x^2}$ is also a function of x only. The sole way, then, that Equation (6) can be satisfied is for both sides to equal a constant. Let

$$\frac{\partial^2 w}{\partial x^2} \equiv \frac{d^2 w}{dx^2} = c_1. \tag{7}$$

Substituting Equation (7) into Equation (6) results in:

$$\frac{\partial p}{\partial z} \equiv \frac{dp}{dz} = \eta\, c_1. \tag{8}$$

where η is the dynamic viscosity given by

$$\eta = \rho \nu.$$

To obtain the velocity equation, integrate Equation (7) twice, obtaining:

$$w = c_1 \frac{x^2}{z} + c_2 x + c_3 \tag{9}$$

Now, substituting the boundary conditions, $w = 0$ at $x = \pm \frac{\ell}{z}$ (i.e., velocity of fluid at the wall is zero) we have,

$$0 = c_1 \left(\frac{\ell^2}{8}\right) + c_2 \left(\frac{\ell}{2}\right) + c_3 \tag{10}$$

and

$$0 = c_1 \left(\frac{\ell^2}{8}\right) - c_2 \left(\frac{\ell}{2}\right) + c_3 \tag{11}$$

Subtracting Eq. (11) from Eq. (10) produces the result

$$c_2 = 0 \tag{12}$$

Adding Eq. (10) and Eq. (11) with $c_2 = 0$ results in

$$c_3 = - c_1 \left(\frac{\ell^2}{8}\right) \tag{13}$$

With Eq. (12) and Eq. (13), Eq. (9) simplifies to

$$w = c_1 \left(\frac{x^2}{2} - \frac{\ell^2}{8}\right) \tag{14}$$

Now, the mass of fluid moved in time, T, is

$$\int_0^T I \, dt = IT.$$

But this mass of fluid is also given by

$$IT = \rho \; T \int_{-\ell/2}^{\ell/2} w \, dx,$$ (since mass = density x velocity x area)

Canceling T and substituting Eq. (14), this becomes

$$I = \rho \, c_1 \int_{-\ell/2}^{\ell/2} \left(\frac{x^2}{2} - \frac{\ell^2}{8}\right) dx.$$

Performing the integration, yields

$$I = \rho \; c_1 \left[\frac{x^3}{6} - \frac{\ell^2 x}{8}\right] \Bigg|_{-\ell/2}^{\ell/2}$$

$$= \rho \; c_1 \left[\frac{1}{6}\left(\frac{\ell^3}{8} + \frac{\ell^3}{8}\right) - \frac{\ell^2}{8}\left(\frac{\ell}{2} + \frac{\ell}{2}\right)\right]$$

$$= \rho \; c_1 \left[\frac{\ell^3}{24} - \frac{\ell^3}{8}\right]$$

$$I = - \frac{\rho \, c_1 \, \ell^3}{12}.$$

Solving for c_1 yields

$$c_1 = - \frac{12 \; I}{\rho \quad \ell^3}. \tag{15}$$

When Eq. (15) is substituted into Eq. (14) the result is

$$w = \frac{-\;12\;I}{\rho\,\ell^3}\left(\frac{x^2}{2} - \frac{\ell^2}{8}\right)$$

or,

$$w = \frac{3\;I}{2\;\rho\ell^3}\left(\ell^2 - 4x^2\right). \tag{16}$$

Eq.(16) describes the velocity distribution as a function of x.

The pressure gradient is obtained now by substituting Eq.(15) into Eq.(8):

$$\frac{dp}{dz} = - \frac{12\;\eta\;I}{\rho\,\ell^3}$$

It is interesting to note that when x = 0, the velocity w is maximum. In other words, the maximum point velocity occurs midway between the planes at x = 0. Thus, equation in this case, becomes

$$w_{max} = \frac{3\;I}{2\;\rho\;\ell^3}\;(\ell^2)$$

$$= \frac{3\;I}{2\rho\,\ell}$$

STRING PROBLEMS

• PROBLEM 25-4

A string of length ℓ is tied at $x = \ell$. The end at x = 0 if forced to move sinusoidally so that

$$u(0,t) = A \sin \omega t.$$

Find the motion of the string when all points on the string vibrate with the same angular frequency ω.

Solution: A common procedure in problems of this type is to assume a solution to the governing differential equation that can meet the boundary conditions and the imposed motion.

The one-dimensional wave equation which governs the motion u(x,t) is

$$\frac{\partial^2 u}{\partial t^2} = c^2 \frac{\partial^2 u}{\partial x^2}. \tag{1}$$

Because of the prescribed motion we assume a solution of form

$$u = X(x) \sin (\omega t). \tag{2}$$

Substituting this solution into the wave equation,

$$-\omega^2 X \sin (\omega t) = c^2 X'' \sin (\omega t)$$

Cancelling sin (ωt) and rearranging,

$$X'' + \left(\frac{\omega}{c}\right)^2 X = 0. \tag{3}$$

The general solution to this second-order ordinary differential equation is

$$X = a \sin \left(\frac{\omega x}{c}\right) + b \cos \left(\frac{\omega x}{c}\right). \tag{4}$$

where a and b are constants.

Now the boundary conditions on u are

$$u(0,t) = A \sin (\omega t)$$

$$u(\ell,t) = 0.$$

when these are substituted into Eq. 2, we have the transformed boundary conditions on X

$$X(0) = A \tag{5}$$

$$X(\ell) = 0. \tag{6}$$

From Equations (4) and (5),

$$b = A. \tag{7}$$

From Equations (4), (6) and (7),

$$0 = a \sin\left(\frac{\omega \ell}{c}\right) + A \cos\left(\frac{\omega \ell}{c}\right)$$

Solving for a yields

$$a = -A \cot\left(\frac{\omega \ell}{c}\right). \qquad (8)$$

Substituting the values of constants into (4), and letting $k = \omega/c$ gives

$$X = -A \cot(k\ell) \sin(kx) + A \cos(kx)$$

or $\quad X = A[\cos(kx) - \cot(k\ell) \sin(kx)]$

Substituting this into equation (2) the solution becomes,

$$u = A \sin \omega t \ [\cos(kx) - \cot(k\ell) \sin(kx)] .$$

• **PROBLEM** 25-5

Calculate the velocity of a transverse pulse in a string under a tension of 20 lb if the string weighs 0.003 lb/ft.

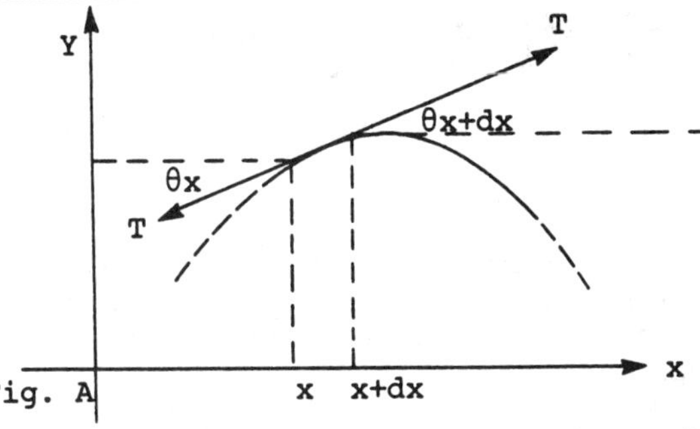

Fig. A

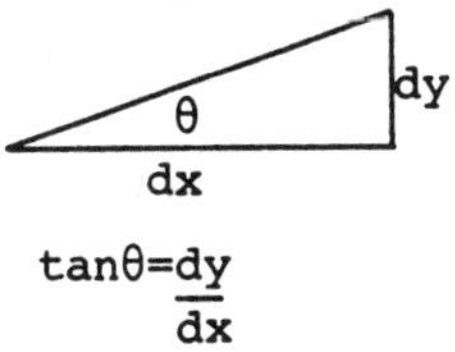

Fig. B

Solution: Consider the section of string shown in figure A, which is under the tension T. This segment can be interpreted as being part of the pulse being transmitted along the string. The net force acting in the y direction is

$$F_y = T \sin \theta_{x+dx} - T \sin \theta_x$$

Since θ_x and θ_{x+dx} are small, we have approximately,

$$\sin \theta_x \approx \theta_x \approx \tan \theta_x$$

$$\sin \theta_{x+dx} \approx \theta_{x+dx} \approx \tan \theta_{x+dx}$$

Therefore, $F_y = T(\tan \theta_{x+dx} - \tan \theta_x)$

but $\tan \theta_{x+dx} = \left(\frac{\partial y}{\partial x}\right)_{x+dx}$

$$\tan \theta_x = \left(\frac{\partial y}{\partial x}\right)_x$$ (See fig. B)

∂y and ∂x have the same meaning as dy and dx, they are small increments of y and of x, respectively. The ∂ symbol indicates that more than one variable is under consideration.

Hence, $$F_y = T\left[\left(\frac{\partial y}{\partial x}\right)_{x+dx} - \left(\frac{\partial y}{\partial x}\right)_x\right] \qquad (1)$$

Now, by Taylor's theorem, given a continuous and differentiable function G(x),

$$G(x + dx) = G(x) + \frac{\partial G}{\partial x}\,dx + \ldots$$

$$G(x + dx) - G(x) \approx \frac{\partial G}{\partial x}\,dx \qquad (2)$$

Higher order terms have been neglected because dx is so small that terms involving $(dx)^2$ or higher must necessarily be negligible. If $G(x) = \partial y/\partial x$, then we have from (2)

$$\left(\frac{\partial y}{\partial x}\right)_{x+dx} - \left(\frac{\partial y}{\partial x}\right)_x = \frac{\partial}{\partial x}\left(\frac{\partial y}{\partial x}\right) dx = \frac{\partial^2 y}{\partial x^2}\,dx$$

Hence, (1) becomes $F_y = T\,\frac{\partial^2 y}{\partial x^2}\,dx$

Denote the mass per unit length of the string by μ (i.e. $\mu = dm/dx$). The mass of the segment dx is then $m = \mu dx$. By Newton's second law, the vertical acceleration of the string segment, a_y, is

$$F_y = ma_y$$

$$T\,\frac{\partial^2 y}{\partial x^2}\,dx = (\mu dx)\,\frac{\partial^2 y}{\partial t^2}$$

$$\frac{\partial^2 y}{\partial x^2} = \frac{\mu}{T}\,\frac{\partial^2 y}{\partial t^2} \qquad (3)$$

Equation (3) is of the form of the wave equation

$$\frac{\partial^2 y}{\partial x^2} = \frac{1}{v^2}\,\frac{\partial^2 y}{\partial t^2}$$

where y(x, t) is the displacement of the wave at position x and at time t, and v is the velocity of the wave. For the transverse wave, then

$$v = \sqrt{\frac{T}{\mu}}$$

Returning to our original problem,

$T = 20$ lb,

$$\mu = \frac{0.003}{32} \frac{\text{lb/ft}}{\text{ft/sec}^2},$$

$$v = \sqrt{\frac{20 \text{ lb} \times 32 \text{ ft/sec}^2}{0.003 \text{ lb/ft}}} = 461 \frac{\text{ft}}{\text{sec}}.$$

• PROBLEM 25-6

A particular case of the vibrating string problem is the plucked string problem. Assume the string is such that the constant $\alpha^2 = \left(\frac{T}{\rho}\right) = 2500$ and that the ends of the string are fixed at $x = 0$ and $x = 1$. Let the mid-point of the string be displaced in the xy plane a distance of 0.01 in the direction of the positive y axis. Find an expression for the displacement, $y(x,t)$.

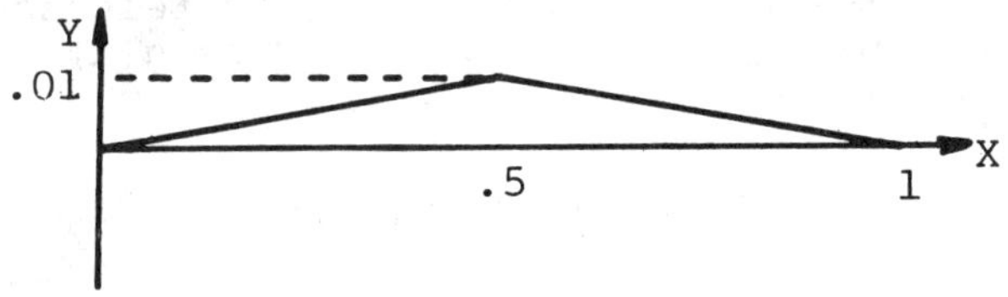

Solution: First sketch a figure representing the given problem:

The constant $\alpha^2 = \left(\frac{T}{\rho}\right)$ represents the ratio of the tension of the string to its linear density, ρ. The displacement, $y(x,t)$, satisfied the partial differential equation

$$\alpha^2 \frac{\partial^2 y}{\partial x^2} = \frac{\partial^2 y}{\partial t^2}$$

where $\alpha^2 = \left(\frac{T}{\rho}\right)$. It also satisfied the boundary conditions:

$$y(0,t) = 0, \qquad 0 \le t < \infty;$$

$$y(L,t) = 0, \qquad 0 \le t < \infty;$$

and the initial conditions

$$y(x,0) = f(x), \qquad 0 \le x \le L;$$

$$\frac{\partial y(x,0)}{\partial t} = g(x), \qquad 0 \le x \le L.$$

By the method of separation of variables, the solution to the above problem is found to be the infinite series

$$y(x,t) = \sum_{n=1}^{\infty} \left[\sin \frac{n\pi x}{L}\right] \left[a_n \sin \frac{n\pi \alpha t}{L} + b_n \cos \frac{n\pi \alpha t}{L}\right] \quad (1)$$

$$\text{where } a_n = \frac{2}{n\pi\alpha} \int_0^L g(x) \sin \frac{n\pi x}{L}\, dx \quad (n = 1, 2, 3, \ldots) \quad (2)$$

$$b_n = \frac{2}{L} \int_0^L f(x) \sin \frac{n\pi x}{L}\, dx \quad (n = 1, \ldots) \quad (3)$$

To obtain a solution of form (1) for the plucked string problem we must first find expressions for $f(x)$ and $g(x)$, the initial velocity. Examining the figure,

$$f(x) = \begin{cases} \dfrac{x}{50}, & 0 \le x \le \dfrac{1}{2} \\ -\dfrac{x}{50} + \dfrac{1}{50}, & \dfrac{1}{2} \le x \le 1 \end{cases} \quad (4)$$

Since we assume that the string is released from rest, the initial velocity must be zero. Thus $g(x) = 0$, $0 \le x \le 1$ and from (2), $a_n = 0$ $(n = 1, \ldots)$.

Substituting (4) into (3) yields

$$b_n = 2\int_0^{\frac{1}{2}} \frac{x}{50} \sin n\pi x\, dx + 2\int_{\frac{1}{2}}^{1} \left(-\frac{x}{50} + \frac{1}{50}\right) \sin n\pi x\, dx. \quad (5)$$

Using integration by parts, (5) is found to be

$$b_n = \frac{2}{25n^2\pi^2} \sin \frac{n\pi}{2} \quad (n = 1, 2, \ldots), \quad (6)$$

When n is even, $\sin \frac{n\pi}{2} = 0$ and $b_n = 0$. When n is odd the coefficients b_n are given by

$$b_{2n-1} = \frac{(-1)^{n-1}\, 2}{25\pi^2\, (2n-1)^2} \quad (n = 1, 2, \ldots).$$

Thus, the displacement of the plucked string for the given data is:

$$y(x,t) = \frac{2}{25\pi^2} \sum_{n=1}^{\infty} \frac{(-1)^{n-1}}{(2n-1)^2} \sin\,[(2n-1)\pi x]$$

$$\cos\,[50\,(2n-1)\pi t].$$

• PROBLEM 25-7

> A string of mass m, length L, and tension $\mathcal{T}$, fixed at both ends is distorted into the shape of the parabola, y = c(L - x)x and released from rest. Determine the motion.

Solution: The differential equation for the motion of the string is

$$\frac{\partial^2 y}{\partial x^2} = \frac{m}{L\mathcal{T}} \frac{\partial^2 y}{\partial t^2} = \frac{1}{v^2} \frac{\partial^2 y}{\partial t^2}$$

The boundary conditions are $y(0,t) = y(L,t) = 0$; the initial conditions are $y(x,0) = c(L - x)x$ and $\partial y/\partial t$ at $(x,0) = 0$.

The solution to this differential equation can be written as an infinite series

$$Y(x,t) = \sum_{n=1}^{\infty} \sin \frac{n\pi x}{L} \left(a_n \sin \frac{n\pi vt}{L} + b_n \cos \frac{n\pi vt}{L} \right) \quad (1)$$

This can be verified by substitution in the differential equation. The coefficients a_n and b_n must be determined from the initial conditions. The condition $y(x,0) = c(L - x)x$ gives

$$\sum_{n=1}^{\infty} b_n \sin \frac{n\pi x}{L} = c(L - x)x \qquad 0 \le x \le L$$

And by inverting the Fourier Series we obtain

$$b_n = \frac{2}{L} \int_0^L c(L - x')x' \sin \frac{n\pi x'}{L}\, dx' \qquad n = 1, 2, \ldots$$

The coefficients a_n are determined from the condition $\frac{\partial y}{\partial t}(x,0) = 0$, which gives

$$\sum_{n=1}^{\infty} a_n \sin \frac{n\pi x}{L} = 0 \qquad 0 \le x \le L$$

Inverting the Fourier Series again, we have

$$a_n = \frac{2L}{n\pi v} \int_0^L (0) \sin \frac{n\pi x'}{L}\, dx'$$

Hence

$$y(x,t) = \sum_{n=1}^{\infty} \frac{2}{L} \left[\int_0^L \left(cLx' \sin \frac{n\pi x'}{L} \right.\right.$$

$$\left. - cx'^2 \sin \frac{n\pi x'}{L} \right) dx' \cos \frac{n\pi vt}{L} \sin \frac{n\pi x}{L}$$

$$\left. + \frac{L}{n\pi v} \int_0^L (0) \sin \frac{n\pi x'}{L} dx' \sin \frac{n\pi vt}{L} \sin \frac{n\pi x}{L} \right]$$

There are two integrals to evaluate:

$$\int_0^L x' \sin \left(\frac{n\pi x'}{L} \right) dx' = - \frac{L^2}{n\pi} \cos n\pi = - \frac{L^2}{n\pi} (-1)^n$$

and
$$\int_0^L x'^2 \sin \left(\frac{n\pi x'}{L} \right) dx' = \frac{L^3}{n\pi} \left[\frac{2(\cos n\pi - 1)}{n^2\pi^2} - \cos n\pi \right]$$

Thus
$$y(x,t) = \sum_{n=1}^{\infty} \frac{2c}{L} \left\{ - \frac{L^3}{n\pi} (-1)^n - \frac{L^3}{n\pi} \left[\frac{2(\cos n\pi - 1)}{n^2\pi^2} \right.\right.$$

$$\left.\left. - (-1)^n \right] \right\} x \cos \left(\frac{n\pi vt}{L} \right) \sin \left(\frac{n\pi x}{L} \right)$$

or
$$y(x,t) = 4cL^2 \sum_{n=1}^{\infty} \frac{1 - (-1)^n}{n^3\pi^3} \cos \left(\frac{n\pi vt}{L} \right) \sin \left(\frac{n\pi x}{L} \right)$$

By setting $n = 2\ell - 1$, only terms involving odd n appear, and we have

$$y(x,t) = 8cL^2 \sum_{\ell=1}^{\infty} \frac{1}{(2\ell - 1)^3\pi^3} \cos \left[\frac{(2\ell - 1)\pi vt}{L} \right]$$

$$\sin \left[\frac{(2\ell - 1)\pi x}{L} \right]$$

There are two nodes, $x = 0$ and $x = L$. The maximum displacement of y, at any time, occurs at $x = L/2$.

• PROBLEM 25-8

Develop and solve the problem of the vibrating string by considering an elastic, flexible string stretched tightly and clamped at the two ends. Use separation of variables in determining the solution.

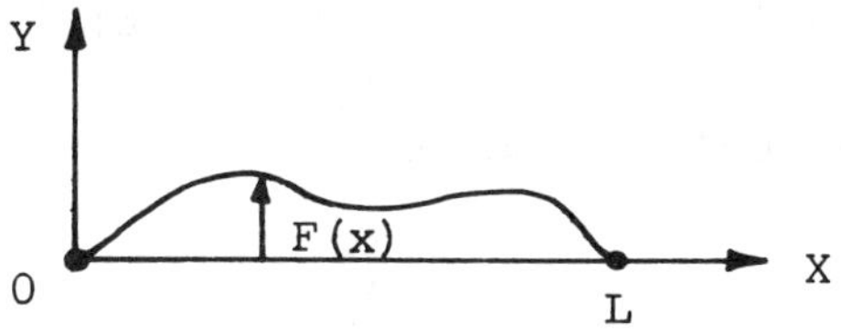

Fig. 1

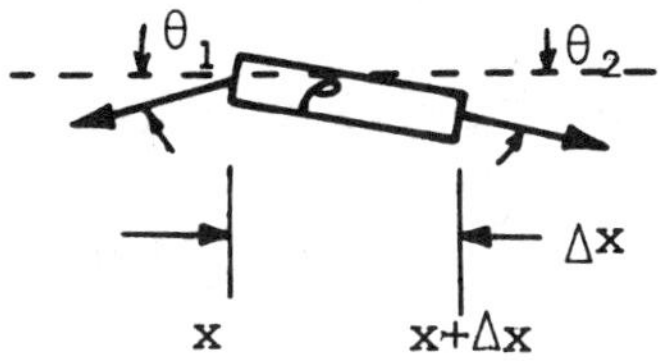

Fig. 2

Solution: Consider a perfectly elastic flexible string of length L stretched tightly and clamped at the two ends. For convenience, we fix the ends on the x axis at x = 0 and x = L. Suppose that for each x in the interval $0 < x < L$, the string is displaced into the xy plane and that for each such x, the displacement from the x axis is given by f(x), a known function of x. This is illustrated in Figure 1.

Suppose that at t = 0 the string is released from the initial position defined by f(x), with an initial velocity given at each point in the interval $0 \leq x \leq L$ by g(x), where g(x) is a known function of x. The string vibrates, and its displacement in the y direction at any point x at any time t will be a function of both x and t. We wish to determine this displacement as a function of x and t; we denote it by y(x,t).

This assumption, that y = y(x,t), is valid provided our system, as given in Figure 1, fulfills certain conditions.

Concerning the vibrations, assume that all motion is confined to the xy plane and that each point on the string moves on a straight line perpendicular to the x axis as the string vibrates. Also, assume that the displacement y at each point of the string is small compared with the length L. (This is equivalent to assuming that the length of any segment of the string is considered constant.) In addition, we require that at times during motion, the angle between the string and the x axis at each point be small.

As for the string itself, assume that its linear density ρ, and its tension T, are both constant at all times. We require that the tension T of the string be great enough to make the string act as though it were perfectly flexible. See Figure (2).

Let Δs represent an element of arc of the string. Since the tension is assumed to be constant, the net upward vertical force acting on Δs is given by:

$$T \sin \theta_2 - T \sin \theta_1. \qquad \text{(a)}$$

Since it is assumed that the angle between the string and the x axis is small:

$$\sin\theta \cong \theta \cong \tan\theta.$$

Then the force given in (a) becomes

$$T\tan\theta_2 - T\tan\theta_1. \tag{b}$$

Since, in Figure (2), the slope is given by

$$\frac{\partial y}{\partial x} = \tan\theta,$$

(b) can be rewritten as:

$$T\left.\frac{\partial y}{\partial x}\right|_{x+\Delta x} - T\left.\frac{\partial y}{\partial x}\right|_{x} \tag{c}$$

Newton's Law requires that this net force (c) be equal to the mass of the string times the acceleration. For a segment of string of arc length Δs, the mass is given by $\rho\Delta s$, where ρ is the linear density, which we have assumed is constant. The acceleration of our arc element Δs is given by $\left(\frac{\partial^2 y}{\partial t^2} + \varepsilon\right)$, where the term $(\rho\Delta s)\varepsilon$ can be interpreted as the force difference at the points (x,y) and $(x + \Delta x, y + \Delta y)$, such that

$$\lim_{\Delta s \to 0} \varepsilon = 0. \tag{d}$$

For this problem, Newton's Law assumes the form

$$\text{Force} = \text{mass x acceleration}$$

$$= (\rho\Delta s)\left(\frac{\partial^2 y}{\partial t^2} + \varepsilon\right)$$

Equating this expression with (c) yields, for the net force acting on an element of arc Δs;

$$T\left[\left.\frac{\partial y}{\partial x}\right|_{x+\Delta x} - \left.\frac{\partial y}{\partial x}\right|_{x}\right]$$

$$= (\rho\Delta s)\left(\frac{\partial^2 y}{\partial t^2} + \varepsilon\right). \tag{e}$$

By assuming small vibrations, we have imposed the condition that

$$\Delta x \cong \Delta s.$$

Making this substitution, (e) becomes

$$T\left[\left.\frac{\partial y}{\partial x}\right|_{x+\Delta x} - \left.\frac{\partial y}{\partial x}\right|_{x}\right]$$

$$= (\rho\Delta x)\left(\frac{\partial^2 y}{\partial t^2} + \varepsilon\right).$$

Dividing by $\rho\Delta x$ gives:

$$\frac{T}{\rho}\ \frac{\left[\left.\frac{\partial y}{\partial x}\right|_{x+\Delta x} - \left.\frac{\partial y}{\partial x}\right|_{x}\right]}{\Delta x} = \left(\frac{\partial^2 y}{\partial t^2} + \varepsilon\right).$$

Taking the limit as $\Delta x \to 0$, in which case $\varepsilon \to 0$, by (d) gives:

$$\frac{T}{\rho}\frac{\partial}{\partial x}\left(\frac{\partial y}{\partial x}\right) = \frac{\partial^2 y}{\partial t^2},$$

or $$\frac{\partial^2 y}{\partial t^2} = a^2\frac{\partial^2 y}{\partial x^2}, \qquad \text{(f)}$$

where $$a^2 = \frac{T}{\rho}.$$

This is the one dimensional wave equation, a second order partial differential equation that can be solved by using separation of variables. Before determining the solution, however, we must establish our boundary and initial conditions.

Since we imposed the condition that the ends of the string are fixed at $x = 0$ and $x = L$ for all time t, the displacement $y(x,t)$ must satisfy the two boundary conditions

$$y(0,t) = 0, \qquad 0 \le t < \infty;$$

$$y(L,t) = 0, \qquad 0 \le t < \infty. \qquad \text{(g)}$$

To determine the initial conditions that $y(x,t)$ must satisfy, recall that at $t = 0$ the string was released from the initial position defined by $f(x)$ $(0 \le x \le L)$, with initial velocity $g(x)$ $(0 \le x \le L)$. Thus, $y(x,t)$ must fulfill the two initial conditions:

$$y(x,0) = f(x) \qquad 0 \le x \le L;$$

$$\frac{\partial Y(x,0)}{\partial t} = g(x), \qquad 0 \le x \le L. \qquad \text{(h)}$$

We may now state the problem: Find a function $y(x,t)$ which simultaneously satisfies the partial differential equation (f), the boundary conditions (g), and the initial conditions (h).

In order to apply the method of separation of variables, assume that the partial differential equation (f) has product solutions of the form XT, where X is a function of x alone, and T is a function of t alone:

$$Y(x,t) = X(x)T(t). \qquad (i)$$

Differentiating and substituting (i) into the differential equation (f) yields:

$$a^2 T \frac{d^2X}{dx^2} = X \frac{d^2T}{dt^2},$$

or

$$\frac{a^2}{X} \frac{d^2X}{dx^2} = \frac{1}{T} \frac{d^2T}{dt^2}$$

Since the right side of this equation is a function only of t, and the left side a function only of X, for the equation to be true both sides must equal a constant, given by k:

$$\frac{a^2}{X} \frac{d^2X}{dx^2} = k, \qquad (j)$$

$$\frac{1}{T} \frac{d^2T}{dt^2} = k, \qquad (k)$$

Rewriting (j) and (k), we obtain, respectively, the two ordinary differential equations,

$$\frac{d^2X}{dx^2} - \frac{k}{a^2} X = 0, \qquad (\ell)$$

and

$$\frac{d^2T}{dt^2} - kT = 0. \qquad (m)$$

We must now use the boundary conditions (g). Since $Y(x,t) = X(x)T(t)$, it follows that $y(0,t) = X(0)T(t)$ and that $y(L,t) = X(L)T(t)$. Thus our boundary conditions become

$$X(0)T(t) = 0, \qquad 0 \leq t < \infty;$$

$$X(L)T(t) = 0, \qquad 0 \leq t < \infty.$$

In order for this system to have a nontrivial solution $T(t)$, it is necessary that

$$X(0) = 0 \quad \text{and} \quad X(L) = 0. \qquad (n)$$

If $k = 0$, the general solution of (ℓ) is

$$X(x) = c_1 + c_2 x, \qquad (o)$$

where c_1 and c_2 are arbitrary constants. Applying the boundary conditions (n) to the solution (o) yields

$$X(0) = 0 = c_1,$$

$$X(L) = c_1 + c_2 L = 0 = c_2 L.$$

Since this requires that both c_1 and c_2 be zero, dismiss the case $k = 0$, for it only yields the trivial solution.

If $k > 0$, the general solution of (ℓ) is

$$X(x) = c_1 \exp\left(\frac{\sqrt{k}x}{a}\right) + c_2 \exp\left(-\frac{\sqrt{k}x}{a}\right). \quad \text{(p)}$$

Applying the boundary conditions (n) to (p):

$$c_1 + c_2 = 0$$

$$c_1 \exp\left(\frac{\sqrt{k}L}{a}\right) + c_2 \exp -\left(\frac{\sqrt{k}L}{a}\right) = 0.$$

Using Cramer's Rule, solve this system for c_1 and c_2:

$$c_1 = \frac{\begin{vmatrix} 0 & 1 \\ 0 & \exp\left(-\frac{\sqrt{k}L}{a}\right) \end{vmatrix}}{\begin{vmatrix} 1 & 1 \\ \exp\left(\frac{\sqrt{k}L}{a}\right) & \exp\left(-\frac{\sqrt{k}L}{a}\right) \end{vmatrix}} = 0,$$

$$c_2 = \frac{\begin{vmatrix} 1 & 0 \\ \exp\left(\frac{\sqrt{k}L}{a}\right) & 0 \end{vmatrix}}{\begin{vmatrix} 1 & 1 \\ \exp\left(\frac{\sqrt{k}L}{a}\right) & \exp -\left(\frac{\sqrt{k}L}{a}\right) \end{vmatrix}} = 0.$$

Since $c_1 = c_2 - 0$, we have only the trivial solution for $k > 0$.

For $k < 0$, equation (ℓ) has the general solution

$$X(x) = c_1 \sin \frac{\sqrt{-k}}{a} x + c_2 \cos \frac{\sqrt{-k}}{a} x. \quad \text{(q)}$$

In order for (q) to satisfy the boundary conditions (n), it is necessary that

$$X(0) = c_2 = 0; \text{ i.e., } \quad c_2 = 0$$

and

$$X(L) = c_1 \sin \frac{\sqrt{-k}L}{a} + c_2 \cos \frac{\sqrt{-k}L}{a} = 0.$$

Therefore, $c_1 \sin \dfrac{\sqrt{-k}L}{a} = 0.$ (r)

Since we want the solution (q) to be nontrivial, assume that $c_1 \neq 0$. In order to satisfy equation (r), we have to conclude that

$$\sin \frac{\sqrt{-k}L}{a} = 0.$$

This implies

$$\frac{\sqrt{-k}L}{a} = n\pi, \qquad (n = 1, 2, 3, \ldots).$$

Therefore, determine k by restricting it to those values for which

$$k = -\frac{n^2 \pi^2 a^2}{L^2}, \quad (n = 1, 2, 3, \ldots). \qquad \text{(s)}$$

Say that (s) yields the characteristic values k_n (n = 1, 2. 3, ...) of the system under consideration, noting that (s) can generate only negative numbers; this is in agreement with our assumption that $k < 0$. Substituting (s) into (q), yields the corresponding nontrivial solutions

$$X_n(x) = c_n \sin \frac{n\pi x}{L}, \quad (n = 1, 2, 3, \ldots), \qquad \text{(t)}$$

where c_n (n = 1, 2, 3, ...) are arbitrary constants. We may refer to (t) as the characteristic functions of the system, since the function X(x) in the assumed solution (i) must be of the form given in (t). More generally, it can be concluded that for each value of n (n = 1, 2, 3, ...), we obtain functions X_n of the form (t) which serve as the function X in the product solution of our system, which is given by (i).

Now turn to the differential equation that T, the other member of our product solution, must satisfy. Substituting k from (s) into (m), gives

$$\frac{d^2T}{dt^2} + \frac{n^2 \pi^2 a^2}{L^2} T = 0,$$

where n = 1, 2, 3, ... Since the coefficient of T in this equation is always positive, a solution is obtained that is valid for each n = 1, 2. 3, ...

$$T_n = c_{n,1} \sin \frac{n \pi at}{L} + c_{n,2} \cos \frac{n \pi at}{L}, \quad \text{(u)}$$

where $c_{n,1}$ and $c_{n,2}$ are arbitrary constants. As in the case of X_n given in (t), the function T_n given in (u) yields, for each value of n, (n = 1, 2, 3, ...) an apppropriate function to serve in the product solution (i).

Therefore, for each positive integral value of n, we obtain corresponding solutions as the product

$$X_nT_n = \left[c_n \sin \frac{n\pi x}{L}\right] \left[c_{n,1} \sin \frac{n\pi at}{L} + c_{n,2} \cos \frac{n\pi at}{L}\right].$$

If we set $a_n = c_nc_{n,1}$ and $b_n = c_nc_{n,2}$ (n = 1, 2, 3, ...) the product solution may be written as:

$$y_n(x,t) = \left[\sin \frac{n\pi x}{L}\right] \left[a_n \sin \frac{n\pi at}{L} + b_n \cos \frac{n\pi at}{L}\right],$$

$$(n = 1, 2, 3, \ldots), \quad \text{(v)}$$

and it is guaranteed that each of these solutions (v) satisfies both the partial differential equation (f) and the two boundary conditions (g) for all values of the constants a_n and b_n.

Now try to satisfy the two initial conditions (h). In general no single one of the solutions (v) will satisfy these conditions. For example, if the first initial condition (h) is applied to a solution of the form (v) we must have

$$b_n \sin \frac{n\pi x}{L} = f(x), \qquad 0 \leq x \leq L$$

where n is some positive integer; and this is impossible unless f happens to be a sine function of the form $A \sin \left(\frac{n\pi x}{L}\right)$ for some positive integer n.

Approach this problem from a theoretical standpoint. We know, from the Principle of Superposition, that every finite linear combination of solutions of (f) is also a solution of (f); furthermore, assuming appropriate convergence, an infinite series of solutions of (f) is also a solution of (f). This suggests that we should form either a finite linear combination or an infinite series of the solutions (v) and attempt to apply the initial conditions (h) to the "more general" solutions thus obtained. In general no finite linear combination will satisfy these conditions, and we must resort to an infinite series.

Therefore, form an infinite series

$$\sum_{n=1}^{\infty} Y_n(x,t) = \sum_{n=1}^{\infty} \left[\sin \frac{n\pi x}{L}\right] \left[a_n \sin \frac{n\pi at}{L} + b_n \cos \frac{n\pi at}{L}\right]$$

of the solutions (v). Assuming appropriate convergence, the sum of this series is also a solution of the differential equation (f). Denoting this sum by $y(x,t)$, write

$$y(x,t) = \sum_{n=1}^{\infty} \left[\sin \frac{n\pi x}{L}\right] \left[a_n \sin \frac{n\pi at}{L} + b_n \cos \frac{n\pi at}{L}\right], \qquad \text{(w)}$$

and note that $y(0,t) = 0$ and $y(L,t) = 0$. Thus, assuming appropriate convergence, the function y given by (w) satisfies both the differential equation (f) and the two boundary conditions (g).

Now apply the initial conditions (h) to the series solution (w). The first condition $y(x,0) = f(x)$, $0 \le x \le L$, reduces to

$$\sum_{n=1}^{\infty} b_n \sin \frac{n\pi x}{L} = f(x), \quad 0 \le x \le L. \qquad \text{(x)}$$

Thus to satisfy the first initial condition (h), it is necessary to determine the coefficients b_n so that (x) is satisfied. This is a problem in Fourier sine series. Using the coefficient formula yields,

$$b_n = \frac{2}{L} \int_0^L f(x) \sin \frac{n\pi x}{L}\, dx \quad (n = 1, 2, 3, \ldots). \qquad \text{(y)}$$

Thus in order for the series solution (w) to satisfy the initial condition $y(x,0) = f(x)$, $0 \le x \le L$, the coefficients b_n in the series must be given by formula (y).

The only condition which remains to be satisfied is the second initial condition (h), which is

$$\frac{\partial y(x,0)}{\partial t} = g(x), \qquad 0 \le x \le L.$$

From (w) we find that

$$\frac{\partial y}{\partial t}(x,t) = \sum_{n=1}^{\infty} \left[\frac{n\pi a}{L}\right] \left[\sin \frac{n\pi x}{L}\right] \left[a_n \cos \frac{n\pi at}{L} - b_n \sin \frac{n\pi at}{L}\right].$$

The second initial condition reduces this to

$$\sum_{n=1}^{\infty} \frac{a_n n\pi a}{L} \sin \frac{n\pi x}{L} = g(x), \qquad 0 \le x \le L.$$

Letting $A_n = \frac{a_n n\pi a}{L}$ $(n = 1, 2, 3, \ldots)$, this takes the form

$$\sum_{n=1}^{\infty} A_n \sin \frac{n\pi x}{L} = g(x), \qquad 0 \le x \le L. \tag{z}$$

Thus to satisfy the second initial condition (h), it is necessary to determine the coefficients A_n so that (z) is satisfied. This is another problem in Fourier sine series, so use the coefficient formula to obtain

$$A_n = \frac{2}{L} \int_0^L g(x) \sin \frac{n\pi x}{L}\, dx \qquad (n = 1, 2, 3, \ldots).$$

Since $A_n = \frac{a_n n\pi a}{L}$ $(n = 1, 2, 3, \ldots)$, we find that

$$a_n = \frac{L}{n\pi a} A_n = \frac{2}{n\pi a} \int_0^L g(x) \sin \frac{n\pi x}{L}\, dx \qquad (n = 1, 2, 3, \ldots). \tag{aa}$$

Thus in order for the series solution (w) to satisfy the second initial condition (h), the coefficients a_n in the series must be given by (aa).

Therefore, the formal solution of the problem consisting of the partial differential equation (j), the two boundary conditions (g), and the two initial conditions (h) is:

$$y(x,t) = \sum_{n=1}^{\infty} \left[\sin \frac{n\pi x}{L}\right] \left[a_n \sin \frac{n\pi at}{L} + b_n \cos \frac{n\pi at}{L}\right],$$

where $a_n = \frac{2}{n\pi a} \int_0^L g(x) \sin\left(\frac{n\pi x}{L}\right) dx \quad (n = 1, 2, 3, \ldots),$

and $b_n = \frac{2}{L} \int_0^L f(x) \sin\left(\frac{n\pi x}{L}\right) dx \quad (n = 1, 2, 3, \ldots).$

In summary, the principal step in the solution of this problem was first to assume the product solution XT given by (i). This led to the ordinary differential equation (j) for the function X and the ordinary differential equation (k) for the function T. We then considered the boundary conditions (g) and found they reduced to the boundary conditions (n) on the function X. Thus the function X(x) had to be a nontrivial solution of the problem consisting of (j) and (n) in order to be part of the product solution (i) of the partial differential equation (f). Solving this problem, yielded, for solutions, the functions $X_n(x)$ given by (t). We then returned to the differential equation (k) for the function T(t) and obtained the solutions $T_n(t)$ given by (u). Thus for each positive integral value of n, the product solutions X_nT_n were denoted by y_n and given by (v). Each of these solutions y_n satisfied both the partial differential equation (f) and the boundary conditions (g), but no one of them satisfied the initial conditions (h). In order to satisfy these initial conditions, we formed an infinite series of the solutions Y_n. We thus obtained the formal solution y given by (w), in which the coefficients a_n and b_n were arbitrary. We applied the initial conditions to this series solution and thereby determined the coefficients a_n and b_n. We thus obtained the formal solution Y given by (w) in which the coefficients a_n and b_n are given by (aa) and (y), respectively. We emphasize that this solution is a formal one, for in the process of obtaining it we made assumptions of convergence which we did not justify.

• PROBLEM 25-9

The midpoint of a stretched string of length ℓ is pulled a distance of u = ℓ/10 from its equilibrium position. The string is then released. Find an expression for its motion by the Fourier series method.

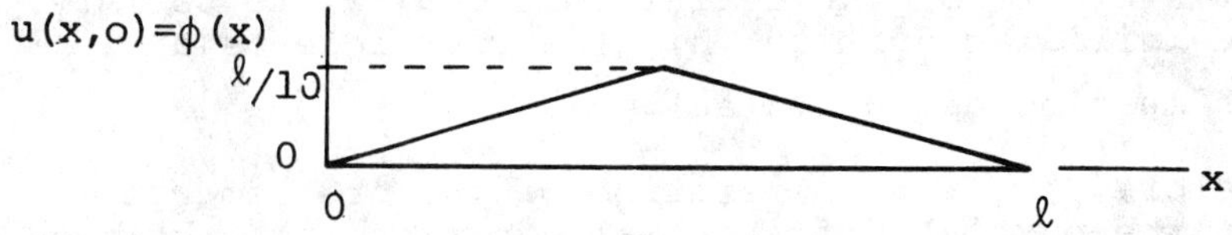

Solution: Starting with the one-dimensional wave equation

$$\frac{\partial^2 u}{\partial t^2} = c^2 \frac{\partial^2 u}{\partial x^2} \tag{1}$$

we assume a solution for the displacement u of form

$$u(x,t) = X(x)\ T(t). \tag{2}$$

Inserting the boundary conditions will lead to an infinite number of solutions. But because the wave equation is linear, the sum of these solutions is also a solution. When the initial string deflection is inserted into the "summed" solution, Fourier series result. The theory of Fourier series is used to evaluate the unknown coefficients in the displacement equation.

To begin the solution, differentiate Eq.(2) twice with respect to x. Then

$$\frac{d^2u}{dx^2} = T\,\frac{d^2x}{dx^2} \tag{i}$$

Now differentiating equation (2), with respect to t, gives,

$$\frac{d^2u}{dt^2} = x\,\frac{d^2T}{dt^2} \tag{ii}$$

Substituting (i) and (ii) into equation (1) gives

$$X\,\frac{d^2T}{dt^2} = c^2\ T\,\frac{d^2X}{dx^2}.$$

Separating the variables,

$$\frac{1}{T}\,\frac{d^2T}{dt^2} = \frac{c^2}{X}\,\frac{d^2X}{dx^2}. \tag{3}$$

Now, since the left side is a function of t alone and the right side is a function of x alone, the equation can be satisfied only if both sides equal a constant, say λ. λ might be 0, positive or negative. We have then two ordinary differential equations

$$\frac{d^2T}{dt^2} - \lambda T = 0 \tag{4}$$

$$\frac{d^2X}{dx^2} - \frac{\lambda}{c^2}\,X = 0. \tag{5}$$

If $\lambda = 0$ in these two equations, then T and X must be linear functions of t and x, respectively. But this would produce a solution in Eq.(2) that is no oscillatory and, hence, not acceptable. So, $\lambda \neq 0$. The solutions to Eq.(4) and Eq.(5) are then

$$T = c_1\ e^{\sqrt{\lambda}\ t} + c_2\ e^{-\sqrt{\lambda}\ t}$$

$$X = c_3\, e^{\sqrt{\lambda}\, x/c} + c_4\, e^{-\sqrt{\lambda}\, x/c}.$$

Substitution of these expressions in Eq.(2) gives

$$u(x,t) = \left(c_1\, e^{\sqrt{\lambda}t} + c_2\, e^{-\sqrt{\lambda}t}\right)\left(c_3\, e^{\sqrt{\lambda}x/c} + c_4\, e^{-\sqrt{\lambda}x/c}\right).$$

To produce an oscillatory solution, it must be true that λ is negative, say

$$\lambda = -\beta^2.$$

The solution for u is then

$$u(x,t) = \left(c_1\, e^{i\beta t} + c_2\, e^{-i\beta t}\right)\left(c_3\, e^{i\beta x/c} + c_4\, e^{-i\beta x/c}\right)$$

or, alternatively,

$$u(x,t) = (a \sin \beta t + b \cos \beta t)\left(d \sin \frac{\beta x}{c} + e \cos \frac{\beta x}{c}\right). \quad (6)$$

The two boundary conditions are

$$u(0,t) = 0$$

$$u(\ell,t) = 0$$

Substitution of the first boundary condition into Eq. (6) produces

$$0 = e\,(a \sin \beta t + b \cos \beta t)$$

or

$$e = 0 \quad (7)$$

The second boundary condition gives with $e = 0$

$$0 = (a \sin \beta t + b \cos \beta t)\, d \sin \frac{\beta \ell}{c}$$

We cannot have both d and e zero or else $u(x,t) = 0$. This implies that

$$\sin \frac{\beta \ell}{c} = 0. \quad (8)$$

We seek those values of β which satisfy Eq. (8). There are an infinite number of those values given by

$$\frac{\beta \ell}{c} = n\pi, \qquad n = 1, 2, 3, \ldots$$

or,

$$\beta_n = \frac{n\pi c}{\ell} \qquad n = 1, 2, 3, \ldots \quad (9)$$

The solution associated with the nth value of β has the form (from Eq. (6), (7) and (9))

$$u_n(x,t) = \sin\frac{n\pi x}{\ell}\left(A_n \sin\frac{n\pi ct}{\ell} + B_n \cos\frac{n\pi ct}{\ell}\right). \qquad (10)$$

We now seek a solution to the given problem from the infinite series of all the u_n's,

$$u(x,t) = \sum_{n=1}^{\infty} u_n(x,t)$$

$$= \sum_{n=1}^{\infty} \sin\frac{n\pi x}{\ell}\left(A_n \sin\frac{n\pi ct}{\ell} + B_n \cos\frac{n\pi ct}{\ell}\right). \qquad (11)$$

To satisfy the initial displacement, we set $t = 0$,

$$u(x,0) = \phi(x) = \sum_{n=1}^{\infty} B_n \sin\frac{n\pi x}{\ell} \qquad (12)$$

where $\phi(x)$ is depicted in the figure. The problem of determining the B_n's to satisfy Eq. (12) is nothing but the problem of expanding a given function $\phi(x)$ in a half-range Fourier sine series over the interval $(0,\ell)$. Thus, we have

$$B_n = \frac{2}{\ell}\int_0^{\ell} \phi(x) \sin\frac{n\pi x}{\ell}\, dx$$

Since the equation for $\phi(x)$ changes at $x = \ell/2$, we break the integral into two parts

$$B_n = \frac{2}{\ell}\left\{\int_0^{\ell/2} \frac{x}{5} \sin\frac{n\pi x}{\ell}\, dx + \int_{\ell/2}^{\ell} \left(-\frac{x}{5} + \frac{\ell}{5}\right) \sin\frac{n\pi x}{\ell}\, dx\right\}.$$

To facilitate integration, let $v = \frac{n\pi x}{\ell}$ and $dv = \frac{n\pi}{\ell}\, dx$.

$$\text{Then } B_n = \frac{2}{\ell}\left\{\frac{\ell}{n\pi}\int_0^{\frac{n\pi}{2}} \frac{v\ell}{5n\pi} \sin v\, dv + \frac{\ell}{n\pi}\int_{\frac{n\pi}{2}}^{n\pi}\left(-\frac{v\ell}{5n\pi} + \frac{\ell}{5}\right) \sin v dv\right\}$$

Integrating,

$$B_n = \frac{2}{\ell}\left\{\left(\frac{\ell}{n\pi}\right)^2 \left[\frac{\sin v - v\cos v}{5}\right]\Bigg|_0^{\frac{n\pi}{2}} + \left(\frac{\ell}{n\pi}\right)\left[\frac{-\ell}{5n\pi}(\sin v - v\cos v) - \frac{\ell}{5}(\cos v)\right]\Bigg|_{\frac{n\pi}{2}}^{n\pi}\right\}$$

and, inserting the limits of integration,

$$B_n = \frac{2}{\ell}\left\{\left(\frac{\ell}{n\pi}\right)^2\left(\frac{1}{5}\right)\left(\sin\frac{n\pi}{2} - \frac{n\pi}{2}\cos\frac{n\pi}{2}\right) + \left(\frac{\ell}{n\pi}\right)\left[\frac{-\ell}{5n\pi}\left(-n\pi \not{\cos}\, n\pi\right.\right.\right.$$

$$\left.\left.\left. - \sin\frac{n\pi}{2} + \frac{n\pi}{2}\cos\frac{n\pi}{2}\right) - \frac{\ell}{5}\left(\not{\cos}\, n\pi - \cos\frac{n\pi}{2}\right)\right]\right\}$$

$$= \frac{2\ell}{(n\pi)^2}\left\{\frac{2}{5}\left(\sin\frac{n\pi}{2} - \frac{n\pi}{2}\not{\cos}\frac{n\pi}{2}\right) + \frac{n\pi}{5}\left(\not{\cos}\frac{n\pi}{2}\right)\right\}$$

$$B_n = \frac{4\ell}{5(n\pi)^2}\sin\frac{n\pi}{2}. \qquad (13)$$

Since the string is initially at rest,

$$\left.\frac{\partial u}{\partial t}\right|_{x,0} = 0$$

Taking this derivative of Eq. (11) and setting t = 0,

$$0 = \sum_{n=1}^{\infty}\left(\sin\frac{n\pi x}{\ell}\right)\left(\frac{n\pi c}{\ell}\right)A_n$$

which implies

$$A_n = 0, \text{ for all } n. \qquad (14)$$

Substituting Eq. (13) and (14) into solution (11), we have finally

$$u(x,t) = \sum_{n=1}^{\infty}\frac{4\ell}{5(n\pi)^2}\sin\frac{n\pi}{2}\sin\frac{n\pi x}{\ell}\cos\frac{n\pi ct}{\ell}.$$

We note that in the series when n is even $\sin\frac{n\pi}{2} = 0$. The summation then needs to be performed only over odd values of n. It may be further simplified by letting

$$n = 2m + 1$$

or

$$m = (n - 1)/2$$

Then we have an alternative expression

$$u(x,t) = \frac{4\ell}{5\pi^2}\sum_{m=0}^{\infty}\frac{(-1)^m}{(2m+1)^2}\sin\frac{(2m+1)\pi x}{\ell}\cos\frac{(2m+1)\pi ct}{\ell}.$$

CHAPTER 26

VARIATIONAL METHODS

METHOD OF VIRTUAL WORK

• **PROBLEM 26-1**

Using the method of virtual work, determine the moment of the couple M required to maintain the equilibrium of the mechanism shown.

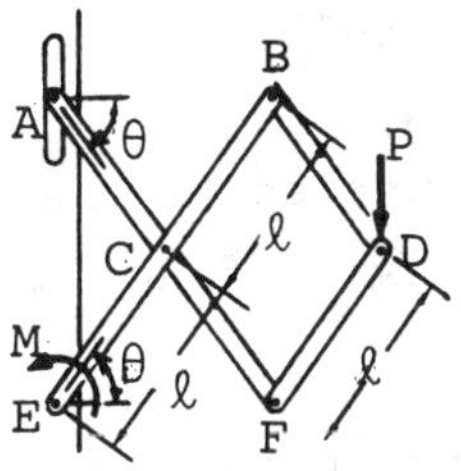

Fig. 1

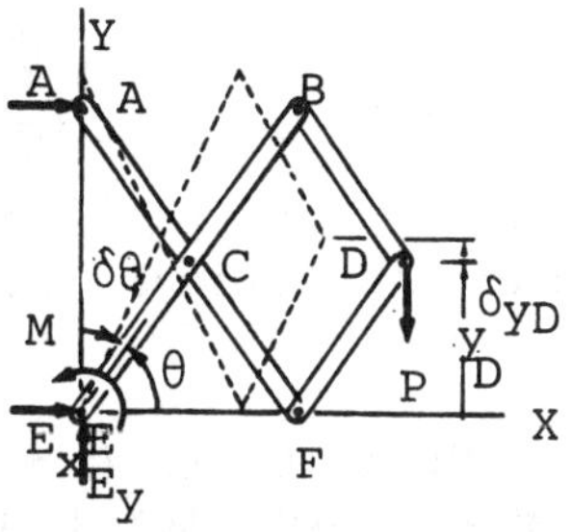

Fig. 2

Solution: If the mechanism moves slightly outwards as shown in Figure 2 we can find the virtual displacement of point D. Segment ECB and FD are always parallel, so $y_D = \ell \sin\theta$ and

$$\delta y_D = \ell \cos\theta \, \delta\theta.$$

In making this virtual displacement, the point E does not move, so there is no work from E_x and E_y. Also there is no work at A since point A moves vertically, but the slot allows only horizontal force.

We can now set the sum of the virtual work that does exist equal to zero to get

$$+ M\delta\theta - P\delta y_D = 0$$

$$+ M\delta\theta - P(\ell \cos\theta\delta\theta) = 0$$

So $\quad M = Pl \cos\theta.$

Note: positive signs are assigned if the displacement and

force (or couple) are in the same direction; negative, if they are opposite.

• PROBLEM 26-2

A hydraulic-lift table is used to raise a 1000-kg crate. It consists of a platform and of two identical linkages on which hydraulic cylinders exert equal forces. (Only one linkage and one cylinder are shown.) Members EDB and CG are each of length 2a, and member AD is pinned to the midpoint of EDB. If the crate is placed on the table, so that half of its weight is supported by the system shown, determine the force exerted by each cylinder in raising the crate for $\theta = 60°$, $a = 0.70$m, and $L = 3.20$m.

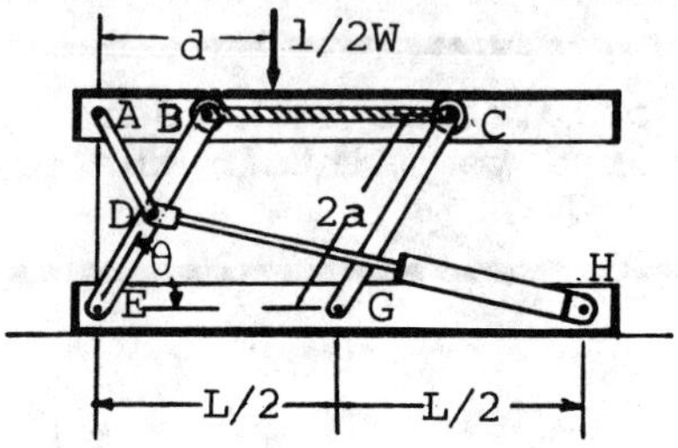

Fig. 1

Solution: Let $\vec{F}_{DH}$ be the force exerted by the hydraulic cylinder at point D. Let the downward force of the load be $\frac{1}{2}\vec{W}$.

Since E and G are stationary points no work is done by the reaction force at these locations. Let y be the elevation of the platform, and s the extension of the hydraulic piston. From the principle of virtual work

$$-\frac{1}{2}W\delta y + F_{DH}\delta s = 0. \qquad (1)$$

If we write

$$y = (ED)\sin\theta$$

$$y = 2a\sin\theta$$

then

$$\delta y = 2a(\cos\theta)\delta\theta. \qquad (2)$$

By the law of cosines

$$s^2 = a^2 + L^2 - 2aL\cos\theta$$

so

$$2s\delta s = -2aL(-\sin\theta)\delta\theta$$

$$\delta s = \frac{aL\sin\theta}{s}\delta\theta. \qquad (3)$$

Substituting δy and δs in (1) we obtain

$$\left(-\frac{1}{2}W\right)2a\cos\theta\,\delta\theta + F_{DH}\frac{aL\sin\theta}{s}\,\delta\theta = 0$$

$$F_{DH} = W\frac{s}{L}\cot\theta.$$

Using the given numerical data:

$$W = mg = (1000)(9.81) = 9.81 \text{ kN}.$$

and $\quad s^2 = (.70)^2 + (3.20)^2 - 2(.70)(3.2)\cos 60°$

$$s^2 = 8.40$$

thus $\quad s = 2.91$ m.

Finally,

$$F_{DH} = (9.81)\frac{(2.91)}{(3,20)}\cot 60°$$

$$F_{DH} = 5.15 \text{ kN}.$$

• PROBLEM 26-3

A 400-lb weight is attached to the lever AO as shown. The constant of the spring BC is k = 250 lb/in., and the spring is unstretched when $\theta = 0$. Determine the position or positions of equilibrium, and state in each case whether the equilibrium is stable, unstable, or neutral. See Fig. 1.

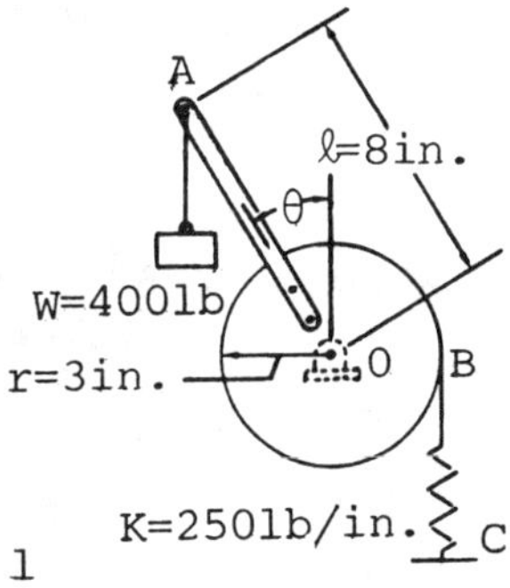

Fig. 1

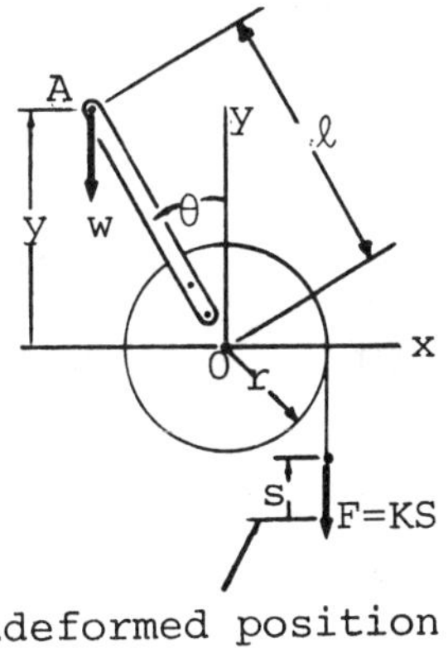

Fig. 2

Solution: Let s denote the amount the spring is stretched from equilibrium. (See Fig. 2 for notation.)

The potential of the spring and the weight is, respectively,

$$V_s = \frac{1}{2}ks^2$$

$$V_g = Wy.$$

If we use the relation $s = r\theta$ and $y = \ell \cos\theta$, the total potential energy can be written

$$V = V_e + V_g = \frac{1}{2} k\, r^2\theta^2 + W\ell \cos\theta.$$

To find the position of equilibrium we seek the minimum of V. Thus we set

$$\frac{dV}{d\theta} = 0.$$

$$kr^2\theta - W\ell \sin\theta = 0,$$

$$\sin\theta = \frac{kr^2}{W\ell}\ \theta.$$

Substituting the given values,

$$\sin\theta = \frac{(250)(3)^2}{(400)(8)}\ \theta$$

$$\sin\theta = 0.703\ \theta.$$

Expanding $\sin\theta$ in a power series, truncating, and solving algebraically for θ we obtain

$$\theta = 0 \quad \text{and}$$

$$\theta = 80.4°.$$

The minimum value will occur when $d^2V/d\theta^2 > 0$.

Since $\dfrac{d^2V}{d\theta^2} = kr^2 - W \cos\theta$

we observe that

$$\left.\frac{d^2V}{d\theta^2}\right|_{\theta=0} = -950 < 0$$

and

$$\left.\frac{d^2V}{d\theta^2}\right|_{\theta=80.4°} = 1716 > 0.$$

Hence we have stable equilibrium at $\theta = 80.4°$.

• PROBLEM 26-4

Derive Lagrange's equation of motion for a uniform thin disk that rolls without slipping on a horizontal plane under the action of a horizontal force F applied at its center.

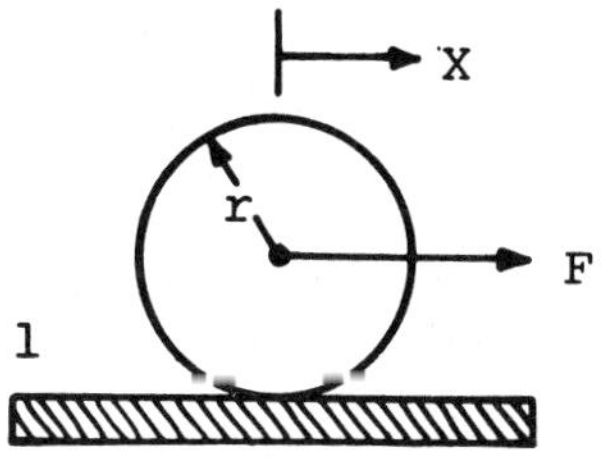

Fig. 1

Solution: Let the disk be of radius r, rolling in the x-direction which measures the position of the center of the disk.

The kinetic energy is

(1) $$T = \frac{1}{2} mv^2 + \frac{1}{2} I\omega^2,$$

where $v = dx/dt$, $I = \frac{1}{2} mr^2$ is the moment of inertia, and $\omega = \frac{v}{r}$ is the angular velocity. Substituting appropriate expressions we write (1) as

$$T = \frac{1}{2} mv^2 + \frac{1}{2} \left(\frac{1}{2} mr^2\right)(v/r)^2$$

(2) $$T = \left(3/4\right) mv^2$$

The generalized force is given

$$Q_i = \frac{d}{dt}\left(\frac{\partial T}{\partial \dot{q}_i}\right) - \frac{\partial T}{\partial q_i} .$$

In our 1-dimensional case

$$Q = \frac{d}{dt}\left(\frac{\partial T}{\partial v}\right) - \frac{\partial T}{\partial x}$$

$$Q = \frac{d}{dt}\left(\frac{3}{2} mv\right)$$

$$Q = \frac{3}{2} ma$$

where $a = dv/dt$ is the acceleration. This is clearly Newton's second law.

Derive the equations of motion for the systen shown. The masses move on a smooth horizontal plane.

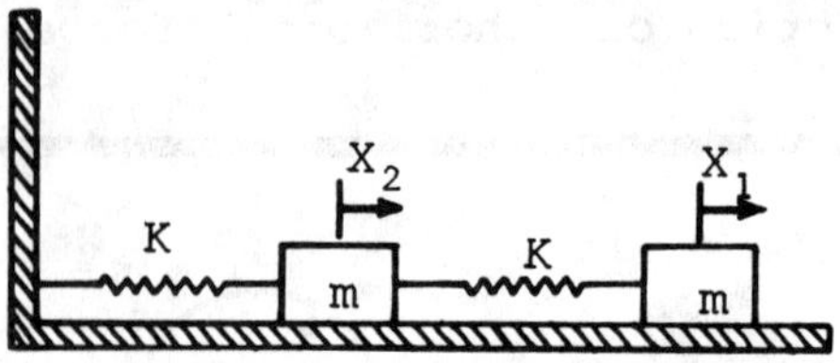

Fig. 1

Solution: Let x_1 and x_2 be the displacement of the blocks from the equilibrium position. Then the kinetic energy is given by

$$T = \frac{1}{2} m\dot{x}_1^2 + \frac{1}{2} m\dot{x}_2^2 .$$

The potential energy is proportional to the amount the spring is stretched, squared so

$$V = \frac{1}{2} kx_2^2 + \frac{1}{2} k(x_1 - x_2)^2.$$

The Lagrangian for the system is

$$L = T - V$$

The Euler-Lagrange equations of motion are

$$\frac{d}{dt}\frac{\partial L}{\partial \dot{x}_i} - \frac{\partial L}{\partial x_i} = 0.$$

Hence,

$$\frac{d}{dt}(m\dot{x}_i) + k(x_1 - x_2) = 0$$

and

$$\frac{d}{dt}(m\dot{x}_2) - k(x_1 - x_2) + kx_2 = 0$$

which reduce to

$$m\ddot{x}_1 = -k(x_1 - x_2)$$

$$m\ddot{x}_2 = -kx_2 + k(x_1 - x_2)$$

which are clearly Newton's Second Law equations.

• PROBLEM 26-6

Apply Lagrange's equations of the second kind to the problem of the one-dimensional, undamped, unforced linear oscillator.

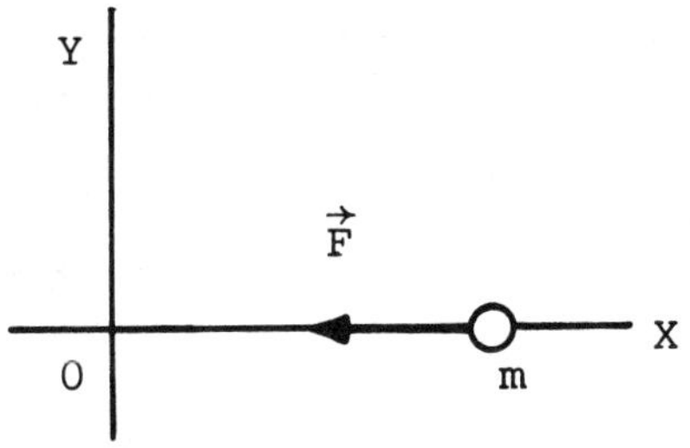

Solution: For a one dimensional oscillator (x-direction) the generalized force is the same as the real force:

$$Q_\ell = \sum_{j=1}^{3n} F_j \frac{\partial x_j}{\partial q_\ell}$$

$$Q_1 = F_x \frac{\partial x}{\partial x} = F_x = -kx,$$

there is only one dimension and one Q.

The kinetic energy is $T = \frac{1}{2} m\dot{x}^2$. So Lagrange's equation of the second kind:

$$Q_\ell - \frac{d}{dt}\frac{\partial T}{\partial \dot{q}_\ell} + \frac{\partial T}{\partial q_\ell} = 0 \qquad (\ell = 1 \ldots 3n)$$

is,

$$-kx - \frac{d}{dt}\frac{\partial}{\partial \dot{x}}\left(\frac{1}{2} m\dot{x}^2\right) + \frac{\partial}{\partial x}\left(\frac{1}{2} m\dot{x}^2\right) = 0$$

or

$$-kx - m\ddot{x} = 0.$$

So the equation of motion is

$$m\ddot{x} + kx = 0.$$

Which is the usual form of the equation of motion for a harmonic oscillator with solution:

$$x = A \cos \omega t + B \sin \omega t$$

where A,B are constants

and $\omega^2 = k/m$.

A uniform rigid bar of weight W and length ℓ is maintained in a horizontal equilibrium position by a vertical spring at its end whose modulus is given as k. The other end of the bar is pinned to the wall. If the bar is depressed slightly and released, determine the equation of motion of the bar by use of Lagrange's equation.

Solution: The kinetic energy of the rotating bar is given by

$$T = \frac{1}{2} I\dot{\theta}^2 = \frac{1}{2}\cdot\left(\frac{1}{3}\frac{W}{g}\ell^2\right)\dot{\theta}^2$$

The potential energy of the system is composed of the sum of the potential of the gravity force and the spring force. Both forces are conservative.

The spring is stretched in the horizontal position to hold the horizontal equilibrium position of the bar. The amount of stretch may be found by considering the equilibrium of the moments:

$$\left(\frac{1}{2}\ell\right) W = \ell\, k\, \delta_{st}$$

$$\delta_{st} = \frac{1}{2}\frac{W}{k} .$$

The potential energy for any angle θ is

$$V = \frac{1}{2} k\, [\ell\, \theta + \delta_{st}]^2 - W \frac{\ell}{2}\theta.$$

Putting in the value for δ_{st} we get

$$V = \frac{1}{2} k\, [\ell\, \theta + \frac{1}{2}\frac{W}{k}]^2 - \frac{1}{2} W\ell\theta$$

$$= \frac{1}{2} k \left(\ell^2\theta^2 + \frac{\ell W\theta}{k} + \frac{1}{4}\frac{W^2}{k^2}\right) - \frac{1}{2} W\ell\theta$$

$$= \frac{1}{2} k\, \ell^2\theta^2 + \frac{1}{8}\frac{W^2}{k} .$$

Since Lagrange's equations involve differentials throughout we may take

$$L = T - V = \frac{1}{6}\frac{W}{g}\ell^2\dot{\theta}^2 - \frac{1}{2} k\ell^2\theta^2.$$

Applying Lagrange's equation with $g_i = \theta$ we have

$$\frac{d}{dt}\left(\frac{\partial L}{\partial\dot{\theta}}\right) - \frac{\partial L}{\partial\theta} = 0$$

$$\frac{d}{dt}\left(\frac{1}{3}\frac{W}{g}\ell^2\dot{\theta}\right) + k\ell^2\theta = 0$$

$$\ddot{\theta} + \frac{3gk}{W}\theta = 0.$$

• PROBLEM 26-8

A uniform rod of length L and mass m is pivoted at one end and can swing in a vertical plane. A homogeneous disk of mass m, radius r, of uniform thickness, is attached by a pivot at its center to the free end of the rod. See Figure 1. Ignoring friction, formulate the equations of motion for the system using Lagrange's equations and determine the motion.

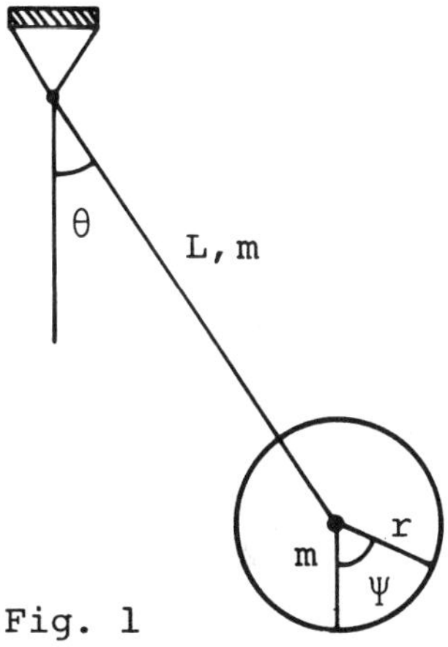

Fig. 1

Solution: For a multi-degree of freedom conservative system, the Lagrange's equations of motion are

$$\frac{d}{dt}\left(\frac{\partial T}{\partial \dot{q}_i}\right) - \frac{\partial T}{\partial q_i} + \frac{\partial V}{\partial q_i} = 0 \qquad (1)$$

$$i = 1,2,\ldots,n,$$

where $T = T(q_i,\dot{q}_i)$ and $V = V(q_i)$ are the kinetic and potential energies of the system, and q_i are the n independent generalized coordinates. Let the angle ϕ define the position of the rod relative to vertical, and angle ψ the position of the disk. These are the coordinates for the problem,

$$q_i = \phi \ ; \ q_2 = \psi$$

We first determine the expressions for kinetic and potential energy. The kinetic energy consists of three parts as follows:

(a) Rod swinging about the pivot

$$T_1 = \frac{1}{2} I_r \dot{\phi}^2$$

where $I_r = mL^2/3$, the moment of inertia of the rod about the fixed pivot.

(b) Disk swinging (without rotation) at end of the rod.

$$T_2 = \frac{1}{2} I_p \dot{\phi}^2$$

without rotation the disk can be considered a point mass with $I_p = 0 + mL^2$.

(c) Disk rotating about the pivot at its center

$$T_3 = \frac{1}{2} I_m \dot{\psi}^2$$

where $I_m = \frac{1}{2} m r^2$, the moment of inertia of disk about the pivot.

Thus the total kinetic energy of the system is

$$T = \frac{1}{6} m L^2 \dot{\phi}^2 + \frac{1}{2} m L^2 \dot{\phi}^2 + \frac{1}{4} m r^2 \dot{\psi}^2$$

$$= \frac{2}{3} m L^2 \dot{\phi}^2 + \frac{1}{4} m r^2 \dot{\psi}^2 \qquad (2)$$

The potential energy of the system consists of two parts as follows:

(a) Rod. When in position ϕ, its c.g. is $L(1-\cos\phi)/2$ above the datum at $\phi = 0$. The potential energy is

$$V_1 = mg\, L(1-\cos\phi)/2.$$

(b) Disk. When the rod is in position ϕ, the c.g. of the disk is $L(1-\cos\phi)$ above the datum at $\phi = 0$. The potential energy is

$$V_2 = mg\, L(1-\cos\phi)$$

Thus the total potential energy of the system is

$$V = \frac{3}{2} mg\, L(1-\cos\phi). \qquad (3)$$

Now we substitute these expressions for T and V in equation (1) to obtain the equations of motion.

First, for the coordinate ϕ

$$\frac{\partial T}{\partial \dot{\phi}} = \frac{4}{3} m L^2 \dot{\phi} \; ; \; \frac{d}{dt}\left(\frac{T}{\partial \dot{\phi}}\right) = \frac{4}{3} m L^2 \ddot{\phi}$$

$$\frac{\partial T}{\partial \phi} = 0 \qquad ; \frac{\partial V}{\partial \phi} = \frac{3}{2} mg\, L \sin\phi .$$

Therefore the equation of motion is

$$\frac{4}{3} m\, L^2 \ddot{\phi} + \frac{3}{2} mg\, L \sin\phi = 0$$

or

$$\ddot{\phi} + (9g/8L) \sin\phi = 0. \tag{4}$$

For small oscillations $\sin\phi \approx \phi$, equation (4) becomes

$$\ddot{\phi} + (9g/8L)\phi = 0. \tag{5}$$

Next we consider coordinate ψ, for which

$$\frac{\partial T}{\partial \dot{\psi}} = \frac{1}{2} mr^2 \dot{\psi}; \ \frac{d}{dt}\left(\frac{\partial T}{\partial \dot{\psi}}\right) = \frac{1}{2} mr^2 \ddot{\psi}$$

$$\frac{\partial T}{\partial \psi} = 0 \qquad ; \frac{\partial V}{\partial \psi} = 0$$

The equation of motion is

$$\frac{1}{2} mr^2 \ddot{\psi} = 0 \tag{6}$$

The expressions for motions of the system are obtained by solving equations (5) and (6). As each of these equations contains only one coordinate, they are independent and can be solved separately. The general solution of equation (5) is

$$\phi = A \sin \omega_n t + B \cos \omega_n t \tag{7}$$

where $\omega_n = (9g/8L)^{1/2}$ is the angular frequency of the simple harmonic motion. The constants A and B can be determined from initial conditions.

From equation (6) we see that

$$\ddot{\psi} = 0$$

therefore $\quad \dot{\psi} = C$

and $\quad \psi = Ct + D \qquad (8)$

where C and D are constants which can be determined from initial conditions on the disk. From the solution we see that the disk can move at a constant (or zero) angular velocity.

• **PROBLEM 26-9**

For the following problems, derive the equations of motion by use of Lagrange's equation.

(a) A particle of mass m is shot vertically upwards in a uniform gravitational field. Neglect air resistance (Figure 1).

(b) A mass m is attached to the lower end of a vertical spring of stiffness k. The upper end of the spring is fixed. The mass is constrained to move in a vertical straight line (Figure 2).

(c) A particle of mass m is attached to the lower end of a light cord of length L, the upper end of which is fixed. The mass and string can move in a vertical plane in the gravitational field (Figure 3).

(d) A uniform rod of mass m is suspended by a frictionless horizontal peg at a distance h from its center of mass. The rod can swing in a vertical plane under the action of gravity (Figure 4).

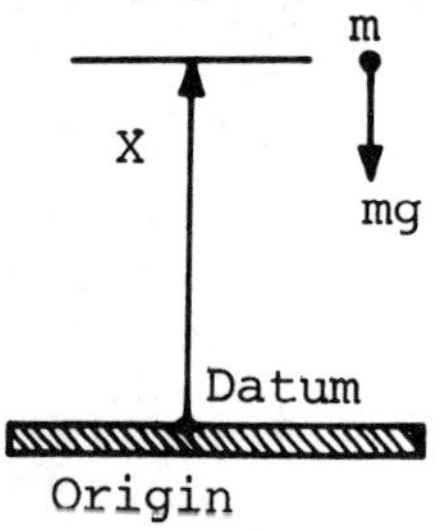

Fig. 1

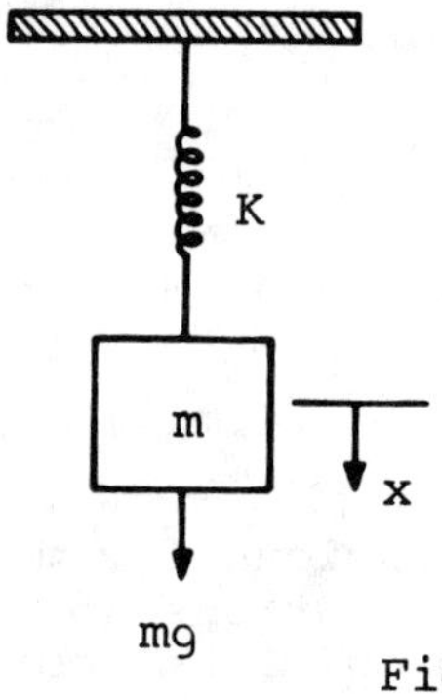

Fig. 2

Solution: Lagrange's equations for conservative mechanical systems are given below. A conservative system is one in which the total energy of the system is constant, and there is no addition or dissipation of energy. For single degree of freedom systems, as the ones in this problem, let q be the coordinate describing the position of the system. Then the kinetic energy T and potential energy V are

$$T = T(q,\dot{q}) \quad ; \quad V = V(q).$$

The Lagrange's equation of motion for such a system is

$$\frac{d}{dt}\left(\frac{\partial T}{\partial \dot{q}}\right) - \frac{\partial T}{\partial q} + \frac{\partial V}{\partial q} = 0. \qquad (1)$$

An alternate form of Lagrange's equation can be written if we define the Lagrangian of the system as L = T-V. Noting that V is only a function of q and not of $\dot{q}$, equation (1) can be written as

$$\frac{d}{dt}\left(\frac{\partial L}{\partial \dot{q}}\right) - \frac{\partial L}{\partial q} = 0. \qquad (2)$$

(a) As shown in Figure 1, we define the position of the particle by x, measured from a given origin. The velocity of the mass is $\dot{x}$. Therefore the kinetic energy is

$$T = \frac{1}{2} m\dot{x}^2.$$

The potential energy is

$$V = mgx$$

if we assume that V = 0 at the datum. Now substituting these expressions in equation (1), and noting that $q = x$ and $\dot{q} = \dot{x}$, and since

$$\frac{\partial T}{\partial \dot{x}} = m\dot{x} \;;\; \frac{d}{dt}\left(\frac{\partial T}{\partial \dot{x}}\right) = m\ddot{x}$$

$$\frac{\partial T}{\partial x} = 0 \;;\; \frac{\partial V}{\partial x} = mg$$

the equation of motion becomes

$$m\ddot{x} + mg = 0$$

or $$\ddot{x} = -g. \tag{3}$$

(b) As shown in Figure 2, let x be the position of the mass, with x = 0 when the spring is unstretched. The kinetic energy is

$$T = \frac{1}{2} m\dot{x}^2.$$

Potential energy of the spring is

$$V_s = \frac{1}{2} kx^2.$$

Potential energy of the mass due to its position in the gravitational field is

$$V_g = -mgx$$

where we use x = 0 as the datum for measuring V_g. Thus the total potential energy is

$$V = \frac{1}{2} kx^2 - mgx$$

We substitute these expressions in equation (1), and note that

$$\frac{\partial T}{\partial \dot{x}} = m\dot{x} \;;\; \frac{d}{dt}\left(\frac{\partial T}{\partial \dot{x}}\right) = m\ddot{x}$$

$$\frac{\partial T}{\partial x} = 0 \;;\; \frac{\partial V}{\partial X} = kx - mg.$$

The equation of motion for the mass is

$$m\ddot{x} + kx - mg = 0$$

or $\ddot{x} + (k/m)x = -g.$ (4)

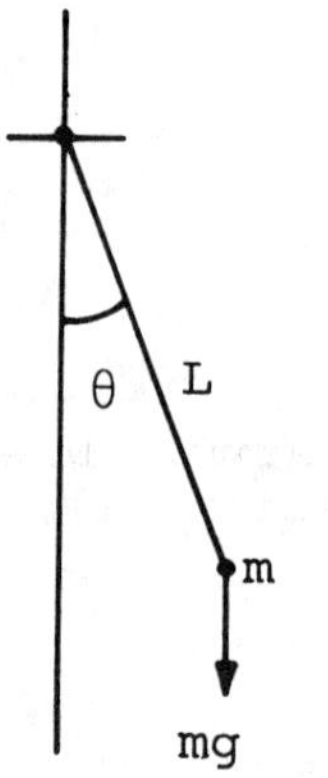

Fig. 3

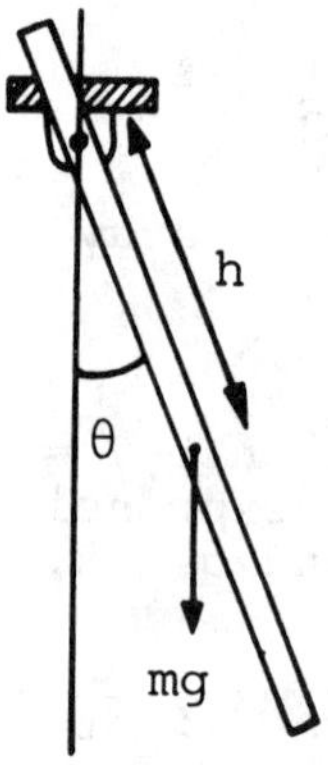

Fig. 4

(c) The simple pendulum is shown in Figure 3. The particle can move along a circular path of radius L. Let θ be the angle from a vertical line to the string. The velocity of the mass therefore is $(L\dot{\theta})$. The kinetic energy is

$$T = \frac{1}{2} m (L\dot{\theta})^2$$

For the potential energy expression we assume the datum to be the lowest position of the mass, i.e., when $\theta = 0$. When the pendulum is at some angle θ, the position of the mass above the datum is $L(1-\cos\theta)$. Therefore the potential energy is

$$V = mg\, L(1-\cos\theta)$$

Now we substitute these expressions in equation (1), noting that $q = \theta$ and $\dot{q} = \dot{\theta}$, and

$$\frac{\partial T}{\partial \dot{\theta}} = m\, L^2 \dot{\theta} \;; \quad \frac{d}{dt}\left(\frac{\partial T}{\partial \dot{\theta}}\right) = m\, L^2 \ddot{\theta}$$

$$\frac{\partial T}{\partial \theta} = 0 \quad ; \quad \frac{\partial V}{\partial \theta} = mg\, L \sin\theta \, .$$

The equation of motion for the simple pendulum is

$$m\, L^2 \ddot{\theta} + mg\, L \sin\theta = 0$$

or $\ddot{\theta} + (g/L)\sin\theta = 0.$ (5)

If the oscillations are of small magnitude, then $\sin\theta \approx \theta$ and the equation of motion becomes

$\ddot{\theta} + (g/L)\theta = 0.$ (6)

(d) Let the rod pivot about point O, Figure 4, and let I be its mass moment of inertia about the axis through O. Let θ define the position of the rod. The kinetic energy of the rod is

$$T = \frac{1}{2} I \dot{\theta}^2.$$

The potential energy of the rod at any instant is

$$V = mgh\ (1-\cos\phi)$$

where we assume the potential energy to be measured from the datum at $\theta = 0$.

We substitute these expressions in equation (1), noting that

$$\frac{\partial T}{\partial \dot{\theta}} = I\dot{\theta} \quad . \quad \frac{d}{dt}\left(\frac{\partial T}{\partial \dot{\theta}}\right) \quad I\ddot{\theta}$$

$$\frac{\partial T}{\partial \theta} = 0 \quad ; \quad \frac{\partial V}{\partial \theta} = mgh\ \sin\theta.$$

The equation of motion becomes

$$I\ddot{\theta} + mgh\ \sin\theta = 0. \tag{7}$$

If θ is small, then $\sin\theta \approx \theta$, and the equation becomes

$$I\ddot{\theta} + mgh\theta = 0. \tag{8}$$

We note that in three of the above problems, parts (b), (c) and (d), the equations of motion, (4), (6) and (8) are of the form

$$\ddot{q} + \omega_n^2 q = F \tag{9}$$

where F is a constant, or zero. The motion described by equation (9) is of the simple harmonic type, of frequency ω_n. The general solution of equation (9) is

$$q(t) = A \sin \omega_n t + B \cos \omega_n t + F/\omega_n^2. \tag{10}$$

• PROBLEM 26-10

A heavy particle slides down a curve in the vertical plane starting from the point P = (a, A). Find the curve such that the particle reaches the vertical line, x = b, in the shortest time.

Solution: This is a variant of the brachistochrone problem proposed by Jakob Bernoulli. It involves variable end points. Thus, we visualize a competition between curves whose end points lie on two given vertical lines, x = a and x = b. In the given problem one of the ends is fixed while the other is variable.

Consider **first** the problem of variable end points.

That is, we wish to find the curve for which the functional

$$J[y] = \int_a^b F(x, y, y')dx \tag{1}$$

has an extremum.

The variation, δJ of the functional (1) is

$$\delta J = \int_a^b (F_y \delta y + F_{y'} \delta y')\ dx. \tag{2}$$

Note that in (2) if $\delta y(a) = \delta y(b) = 0$, we would obtain the fixed end point problem, $y(a) = A$; $y(b) = B$. Here, however, this need not be the case. Hence, setting $\delta J = 0$:

$$\delta J = \int_a^b (F_y \delta y + F_{y'} \delta y')\ dx$$

$$= \int_a^b F_y \delta y\ dx + \int_a^b F_{y'} \delta y'\ dx = 0. \tag{3}$$

Let $u = F_{y'}$; $dv = \delta y'\ dx$. Then, using integration by parts,

$$\int_a^b F_{y'} \delta y'\ dx = F_{y'} \delta y(x) \Big|_{x=a}^{x=b} - \int_a^b \frac{d}{dx} F_{y'} \delta y(x) dx. \tag{4}$$

Substituting (4) into (3),

$$\delta J = \int_a^b \left(F_y - \frac{d}{dx} F_{y'} \right) \delta y\ dx + F_{y'} \Big|_{x=b} \delta y(b)$$

$$- F_{y'} \Big|_{x=a} \delta y(a). \tag{5}$$

If, now, $\delta y(a) = \delta y(b) = 0$, we obtain the second order differential equation

$$F_y - \frac{d}{dx} F_{y'} = 0. \tag{6}$$

But, if $h(x)$ is a solution of Euler's equation, (6), then $\delta J = 0$ implies:

$$\int_a^b \left(F_y - \frac{d}{dx} F_{y'} \right) \delta y\ dx = 0.$$

Thus we obtain from (5) the natural boundary conditions:

$$F_{y'}\Big|_{x=a} = F_{y'}\Big|_{x=b} = 0, \tag{7}$$

since h(x) is arbitrary.

We must now apply the above theoretical machinery to the given problem.

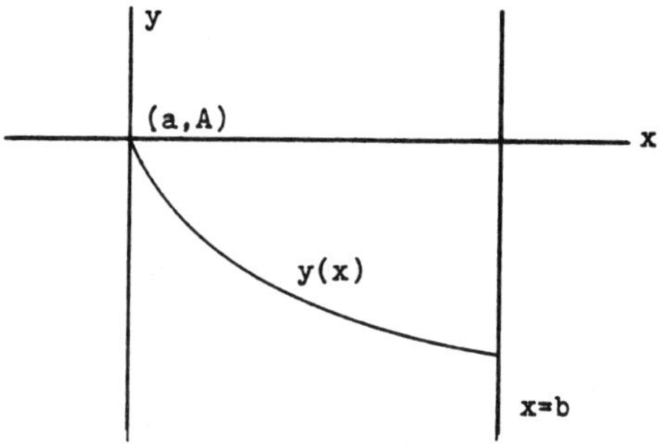

The velocity of the particle along the curve is given by:

$$v = \frac{ds}{dt} = \sqrt{1 + y'^2}\ \frac{dx}{dt}\ . \tag{8}$$

Equation (8) is a first order, separable equation. Separating the variables,

$$dt = \frac{\sqrt{1 + y'^2}}{v}\ dx. \tag{9}$$

But from physics, the velocity of a particle falling under the influence of gravity is given by:

$$v = \sqrt{2gy}\ .$$

Hence (9) becomes:

$$dt = \frac{\sqrt{1 + y'^2}}{\sqrt{2gy}}\ dx. \tag{10}$$

Integrating (10) we obtain the time of travel, i.e.

$$T = \int \frac{\sqrt{1 + y'^2}}{\sqrt{2gy}}\ dx. \tag{11}$$

The equation of motion (the Euler equation) is

$$\frac{d}{dx}\,(F_{y'}) - F_y = 0, \tag{12}$$

where $F_{y'} = \frac{\partial}{\partial y'}\left[\frac{\sqrt{1 + y'^2}}{\sqrt{2gy}}\right] = \frac{1}{\sqrt{2gy}} \cdot \frac{y'}{(1 + y'^2)^{1/2}}$,

$$F_y = \frac{\partial}{\partial y}\left[\frac{\sqrt{1 + y'^2}}{\sqrt{2gy}}\right] = \frac{-\sqrt{1 + y'^2}}{2\sqrt{2g}\,(y)^{3/2}} .$$

Equation (12) becomes:

$$\frac{d}{dx}\left[\frac{1}{\sqrt{2gy}} \cdot \frac{y'}{(1 + y'^2)^{1/2}}\right] + \frac{\sqrt{1 + y'^2}}{2\sqrt{2g}\,(y)^{3/2}} = 0,$$

or, $$\frac{d}{dx}\left[\frac{1}{y^{1/2}} \frac{y'}{(1 + y'^2)^{1/2}}\right] + \frac{(1 + y'^2)^{1/2}}{2y^{3/2}} = 0,$$

or, $$-\frac{1}{2y^{3/2}} \frac{y'}{(1 + y'^2)^{1/2}} + \frac{y''}{y^{1/2}(1 + y'^2)^{3/2}}$$

$$+ \frac{(1 + y'^2)^{1/2}}{2y^{3/2}} = 0.$$

After simplification, we obtain:

$$2yy'' + (y'^4 - y'^3 + 2y'^2 - y') + 1 = 0$$

which is a non-linear, second order, differential equation.

The general solution of the Euler equation consists of a family of cycloids:

$$x = r(\theta - \sin\theta) + c,\ y = r(1 - \cos\theta).$$

But, by our original assumption the curve must pass through the origin. Hence, $c = 0$. To find the other constant, r, we use the natural boundary condition:

$$F_{y'} = \frac{y'}{\sqrt{2gy}\,\sqrt{1 + y'^2}} = 0 \quad \text{for } x = b. \tag{13}$$

Equation (13) implies $y' = 0$ and hence,

$$r = \frac{b}{\pi} .$$

Thus, the required curve is given by:

$$x = \frac{b}{\pi}(\theta - \sin\theta),\ y = \frac{b}{\pi}(1 - \cos\theta).$$

(a) A particle of mass m slides down a curved frictionless path in the vertical plane. Using variational calculus show that the path must be a cycloid if the particle is to move from one point to another in minimum time.

(b) Show that the time for the particle starting from rest at any initial point on the cycloid to the lowest point is $\pi(b/2g)^{1/2}$ where b is the diameter of the generating circle.

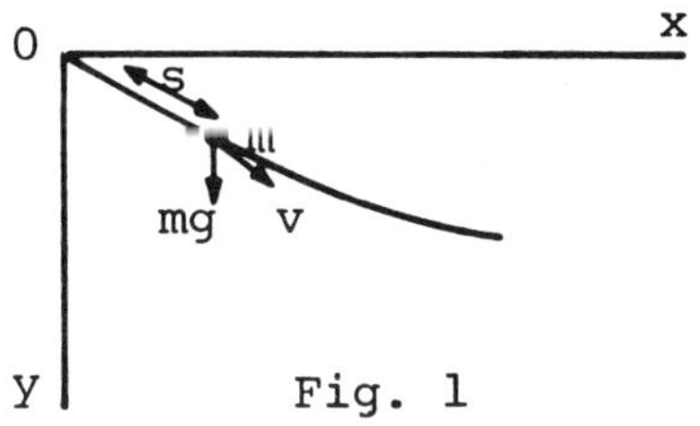

Fig. 1

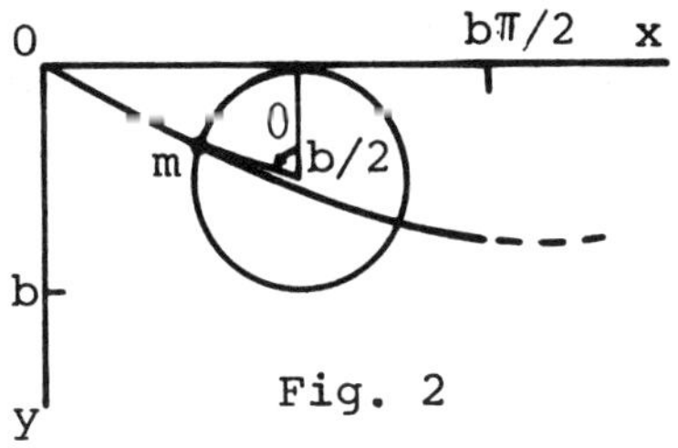

Fig. 2

Solution: The basic problem solved by using variational calculus is as follows:

To find a function y(x) between (x_1, y_1) and (x_2, y_2) which makes the integral

$$I = \int_{x_1}^{x_2} f(x,y,y')dx \tag{1}$$

an extremum, i.e. a minimum or a maximum. In equation (1), f is a given function of x, y and y' (=dy/dx).

The function y(x) is found by solving the Euler-Lagrange equation

$$\frac{\partial f}{\partial y} - \frac{d}{dx}\left(\frac{\partial f}{\partial y'}\right) = 0. \tag{2}$$

An alternate form of the Euler-Lagrange equation is

$$\frac{d}{dx}\left(f - y'\frac{\partial f}{\partial y'}\right) = \frac{\partial f}{\partial x} \tag{3}$$

If f does not contain x explicitly, $\frac{\partial f}{\partial x} = 0$, and equation (3) reduces to

$$f - y'\frac{\partial f}{\partial y'} = \text{constant}. \tag{4}$$

(a) Figure 1 shows a curved path in the vertical plane along which the mass m slides. Assume that the x axis is the datum for determining the potential energy of the particle. Thus in any position (x,y) the potential energy is (-mgy). Let v be its velocity. Then its kinetic energy is $mv^2/2$. Assume that the particle starts from rest at the origin O. Using the law of conservation of energy gives us

$$mv^2/2 - mgy = 0$$

or $$v = (2gy)^{1/2} . \tag{5}$$

If s is the distance measured along the path, then, by definition

$$v = \frac{ds}{dt} . \tag{6}$$

Also, we can write an infinitesimal change in s, ds, in terms of dy and dx as

$$ds = (dx^2+dy^2)^{1/2} = dx(1+y'^2)^{1/2} \tag{7}$$

or $$\frac{ds}{dt} = \frac{dx}{dt} (1+y'^2)^{1/2} . \tag{8}$$

Now, combining equations (5), (6) and (8) we get

$$\frac{dx}{dt} (1+y'^2)^{1/2} = (2gy)^{1/2}$$

or $$dt = \left(\frac{1+y'^2}{2gy}\right)^{1/2} dx. \tag{9}$$

Upon integrating both sides of equation (9) we obtain the expression for the time required for the particle to move for x = 0 to some arbitrary x.

$$t = \int_0^t dt = (1/2g)^{1/2} \int_{x_o}^{x} \left(\frac{1+y'^2}{y}\right)^{1/2} dx \tag{10}$$

The time required is to be minimized. Now compare equation (10) with (1). Ignoring to coefficient $(1/2g)^{1/2}$, we see that

$$f = \left(\frac{1+y'^2}{y}\right)^{1/2} \tag{11}$$

Now we can write the Euler-Lagrange equation which must be solved to determine y(x). We note that since f is not an explicit function of x, we can use equation (4). For the given f

$$\frac{\partial f}{\partial y'} = y' \left[\frac{1}{y(1+y'^2)}\right]^{1/2} . \tag{12}$$

Substituting this and f from equation (11) into equation (4) we get after simplification

$$\frac{1}{[y(1+y'^2)]^{1/2}} = \text{constant} = \frac{1}{b^{1/2}} \quad \text{(Say)}$$

or $$y(1+y'^2) = b. \tag{13}$$

Now, solving for $y'(=dy/dx)$ from equation (13) we have

$$\frac{dy}{dx} = \left(\frac{b-y}{y}\right)^{1/2} . \tag{14}$$

Equation (14) is the required differential equation to be solved for $y(x)$, which is the required path of the particle. It cannot be integrated directly. We introduce a new variable θ such that

$$y = b(1-\cos\theta)/2. \tag{15}$$

From (15)

$$dy = b \sin\theta \; d\theta/2. \tag{16}$$

Now we solve equation (14) for dx, and substitute for y and dy from (15) and (16)

$$dx = \left(\frac{y}{b-y}\right)^{1/2} dy$$

$$= \left[\frac{b(1-\cos\theta)/2}{b-b(1-\cos\theta)/2}\right]^{1/2} b \sin\theta \; d\theta/2$$

$$= \left[\frac{1-\cos\theta}{1+\cos\theta}\right]^{1/2} b \sin\theta \; d\theta/2$$

$$= \left[\frac{(1-\cos\theta)\sin^2\theta}{1+\cos\theta}\right]^{1/2} b \; d\theta/2$$

$$= b(1-\cos\theta)d\theta/2. \tag{17}$$

Integrating both sides of equation (17) we obtain x as function of θ

$$x = \int_0^x dx = \frac{b}{2}\int_0^\theta (1-\cos\theta) \; d\theta$$

$$= b(\theta-\sin\theta)/2. \tag{18}$$

Equations (15) and (18) define the path of the particle. These are the equations of a cycloid, which is generated by a circle of diameter b, rolling on the x axis. This is illustrated in Figure 2. At the apex of the cycloid

$$\theta = 0 \text{ and } x = y = 0$$

At the lowest point of the curve

$$\theta = \pi \; ; \; x = b\pi/2 \quad \text{and } y = b.$$

(b) To find the time required for the particle to travel from an initial position (x_o, y_o) and $\theta = \theta_o$, to the lowest point, $y = b$, we use equation (10). Now, since $v = [2g(y-y_o)]^{1/2}$, equation (10) becomes

$$t = (1/2g)^{1/2} \int_{x_o}^{x} \left(\frac{1+y'^2}{y-y_o} \right)^{1/2} dx \tag{19}$$

Using equations (13), (16) and (17), the above equation becomes

$$t = (b/2g)^{1/2} \int_{y_o}^{b} \frac{dy}{[(y-y_o)(b-y)]^{1/2}}$$

$$= (b/2g)^{1/2} \; - \sin^{-1} \frac{-2y+y_o+b}{b-y_o} \Bigg|_{y_o}^{b}$$

$$= (b/2g)^{1/2} \; [\sin^{-1}(1) - \sin^{-1}(-1)]$$

$$= (b/2g)^{1/2} \; [\pi/2 - (-\pi/2)]$$

$$= \pi(b/2g)^{1/2} \tag{20}$$

This is the required value of elapsed time.

And, since the value of y_o drops out of the expression when the limits of integration are computed, the elapsed time does not depend on the initial position.

• PROBLEM 26-12

Find the amplitude and frequency of small oscillations of the planar double pendulum.

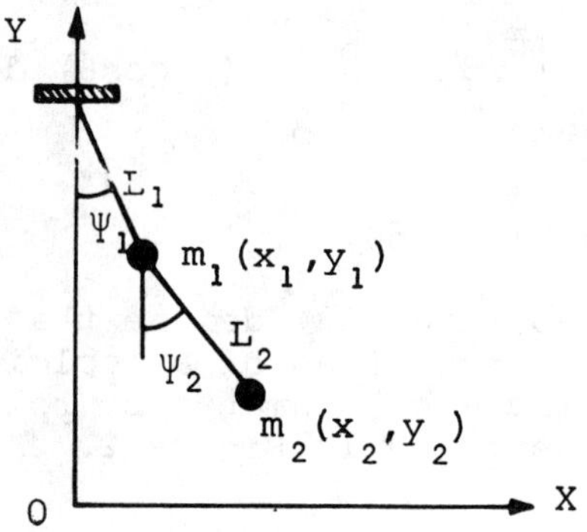

Fig. 1

Solution: In Fig. 1 note the planar double pendulum. The kinetic energy is

$$T = \frac{1}{2} m_1(\dot{x}_1^2 + \dot{y}_1^2) + \frac{1}{2} m_2(\dot{x}_2^2 + \dot{y}_2^2)$$

and the gravitational potential energy is

$$V = m_1 g y_1 + m_2 g y_2 .$$

It is most convenient to use Ψ_1 and Ψ_2 as generalized coordinates of the problem.

Changing to these coordinates we have

$$x_1 = L_1 \sin \Psi_1$$

$$x_2 = L_1 \sin \Psi_1 + L_2 \sin \Psi_2$$

$$y_1 = (L_1+L_2) - L_1 \cos \Psi_1$$

$$y_2 = (L_1+L_2) - L_1 \cos \Psi_1 - L_2 \cos \Psi_2 .$$

Differentiating with respect to time yields

$$\dot{x}_1 = L_1(\cos \Psi_1)\dot{\Psi}_1$$

$$\dot{x}_2 = L_1(\cos \Psi_1)\dot{\Psi}_1 + L_2(\cos \Psi_2)\dot{\Psi}_2$$

$$\dot{y}_1 = L_1(\sin \Psi_1)\dot{\Psi}_1$$

$$\dot{y}_2 = L_1(\sin \Psi_1)\dot{\Psi}_1 + L_2(\sin \Psi_2)\dot{\Psi}_2 .$$

Squaring these and substituting into the expressions for T and V yields

$$(1) \quad T = \frac{1}{2} m_1(L_1^2\dot{\Psi}_1^2) + \frac{1}{2} m_2\left[L_1^2\dot{\Psi}_1^2 + L_2\dot{\Psi}_2^2 + 2L_1L_2\dot{\Psi}_1\dot{\Psi}_2 (\cos \Psi_1 \cos \Psi_2 + \sin \Psi_1 \sin \Psi_2)\right]$$

$$(2) \quad V = m_1 g L_1(1-\cos \Psi_1) + m_2 g L_1(1-\cos \Psi_1) + m_2 g L_2(1-\cos \Psi_2).$$

The Lagrangian of the system is then given by

$$L = T - V.$$

Since we are asked for small oscillations $\Psi_1 = \Psi_2 = 0$. Note that $\Psi_1 = \Psi_2 = 0$ is the point of minimum potential energy, and hence a point of stable equilibrium. Using the standard approximations of sin and cos for small angles, ($\sin \Psi \doteq \Psi$ and $\cos \Psi = 1-\Psi^2/2$) we obtain

(3) $T = \frac{1}{2} m_1 L_1^2 \dot{\Psi}_1^2 + \frac{1}{2} m_2 (L_1^2 \dot{\Psi}_1^2 + L_2^2 \dot{\Psi}_2^2 + 2L_1 L_2 \dot{\Psi}_1 \dot{\Psi}_2)$

(4) $V = m_1 g \left[L_2 + \frac{L_1 \Psi_1^2}{2} \right] + m_2 g \left[\frac{L_1 \Psi_1^2}{2} + \frac{L_2 \Psi_2^2}{2} \right].$

By Euler-Lagrange equations

$$\frac{d}{dt} \frac{\partial L}{\partial \dot{\Psi}_i} - \frac{\partial L}{\partial \Psi_i} = 0, \qquad i = 1,2$$

we have

(5) $m_1 L_1^2 \ddot{\Psi}_1 + m_2 L_1^2 \ddot{\Psi}_1 + m_2 L_1 L_2 \ddot{\Psi}_2 + m_1 g L_1 \Psi_1 + m_2 g L_1 \Psi_1 = 0$

(6) $m_2 L_2^2 \ddot{\Psi}_2 + m_2 L_1 L_2 \ddot{\Psi}_1 + m_2 g L_2 \Psi_2 = 0.$

Equations (5) and (6) are coupled.

We select the trial solution

$$\Psi_1 = A_1 \cos(\omega t + \alpha).$$

$$\Psi_2 = A_2 \cos(\omega t + \alpha).$$

Substituting into (5) and (6) we obtain

(7) $\left[(m_1+m_2)(g-L_1\omega^2)\right] A_1 - \left[m_2 L_2 \omega^2\right] A_2 = 0$

(8) $\left[-L_1\omega^2\right] A_1 + \left[g-L_2\omega^2\right] A_2 = 0$

For the system (7) and (8) to have a solution the determinant of the coefficients of A_1 and A_2 must vanish. Thus we require

$$0 = \det \begin{bmatrix} (m_1+m_2)(g-L_1\omega^2) & -m_2 L_2 \omega^2 \\ -L_1\omega^2 & g-L_2\omega^2 \end{bmatrix}$$

or, after simplifying,

(9) $m_1 L_1 L_2 \omega^4 - (m_1+m_2)(L_1+L_2) g\omega^2 + (m_1+m_2) g^2 = 0.$

This is a quadratic in ω^2. To simplify the algebra we take

$$L_1 = L_2 = L$$

$$m_1 = m_2 = m$$

Then (9) becomes

$$mL^2\omega^4 - 4Lmg\omega^2 + 2mg = 0.$$

The roots are

$$\omega_1^2 = (2 + \sqrt{2})(g/L)$$

$$\omega_2^2 = (2 - \sqrt{2})(g/L)$$

or, since ω_1 and ω_2 must be positive

$$\omega_1 = \sqrt{(2+\sqrt{2})(g/L)}$$

$$\omega_2 = \sqrt{(2-\sqrt{2})(g/L)}$$

Using these values we have the general solution

$$\Psi_1 = A_1 \cos(\omega_1 t + \alpha) + B_1 \cos(\omega_2 t + \beta)$$

$$\Psi_2 = A_2 \cos(\omega_1 t + \alpha) + B_2 \cos(\omega_2 t + \beta)$$

However, solving (7) and (8) for A_1 and A_2 yields

$$\frac{A_1}{A_2} = \frac{-\sqrt{2}}{2} \qquad (\omega = \omega_1)$$

and

$$\frac{B_1}{B_2} = \frac{\sqrt{2}}{2} \qquad (\omega = \omega_2).$$

Thus the final form of the solution is

$$\Psi_1 = A_1 \cos(\omega_1 t + \alpha) + B_1 \cos(\omega_2 t + \beta)$$

$$\Psi_2 = \frac{\sqrt{2}}{2}\left[-A_1 \cos(\omega_1 t + \alpha) + B_1 \cos(\omega_2 t + \beta)\right]$$

where A_1, B_1, α, β are to be determined by initial conditions.

• PROBLEM 26-13

Analyze the motion of the spherical pendulum by Lagrangian formalism.

Solution: A spherical pendulum is a bob of mass m and is free to move in space at the end of a rod of length R. Spherical coordinates are natural to the problem (see Fig. 1).

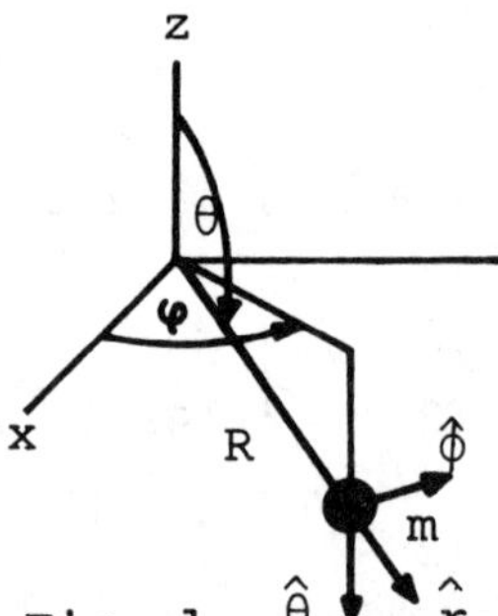

Fig. 1

In these coordinates the kinetic energy is

$$T = \frac{1}{2} mv^2$$

$$T = \frac{1}{2} m(R^2\dot{\theta}^2 + R^2\sin^2\theta\dot{\phi}^2) \tag{1}$$

The potential energy due to gravity is

$$V = mg\ R\cos\theta \tag{2}$$

where the plane Z = 0 is taken as the zero level.

The L-grangian is thus

$$L = T - V$$

$$L = \frac{1}{2} mR^2\dot{\theta}^2 + \frac{1}{2} mR^2\sin^2\theta\dot{\phi}^2 - mgR\cos\theta \tag{3}$$

and the Euler-Lagrange equations for θ and ϕ are

$$\frac{d}{dt}(mR^2\dot{\theta}) - mR^2\dot{\phi}^2 \sin\theta\cos\theta - mgR\sin\theta = 0 \tag{4}$$

and

$$\frac{d}{dt}(mR^2\sin^2\theta\dot{\phi}) = 0 \tag{5}$$

Equation (5) is immediate:

$$mR^2\sin^2\theta\dot{\phi} = \text{constant} = P_\phi\ . \tag{6}$$

Also we note that since $\frac{\partial L}{\partial t} = 0$, the Euler-Lagrange equations of the second kind yield

$$\dot{\theta}\frac{\partial L}{\partial\dot{\theta}} + \dot{\phi}\frac{\partial L}{\partial\dot{\phi}} = \text{constant}$$

or

$$\frac{1}{2} mR^2\dot{\theta}^2 + \frac{1}{2} mR^2\sin^2\theta\dot{\phi}^2 + mgR\cos\theta = E \tag{7}$$

where the constant E is, of course, the energy.

Solving (6) for $\dot{\phi}$ and substituting in (7) yields

$$\frac{1}{2} mR^2\dot{\theta}^2 + \frac{P_\phi^{\ 2}}{2mR^2\sin^2\theta} + mgR\cos\theta = E.$$

We call

$$'V'(\theta) = mgR\cos\theta + \frac{P_\phi^{\ 2}}{2mR^2\sin^2\theta} \tag{8}$$

the effective potential. Thus

$$\frac{1}{2} mR^2\dot{\theta}^2 = E - 'V'(\theta). \tag{9}$$

Since LHS of (9) must be positive we require

$$E > 'V'(\theta)$$

(See Fig. 2.)

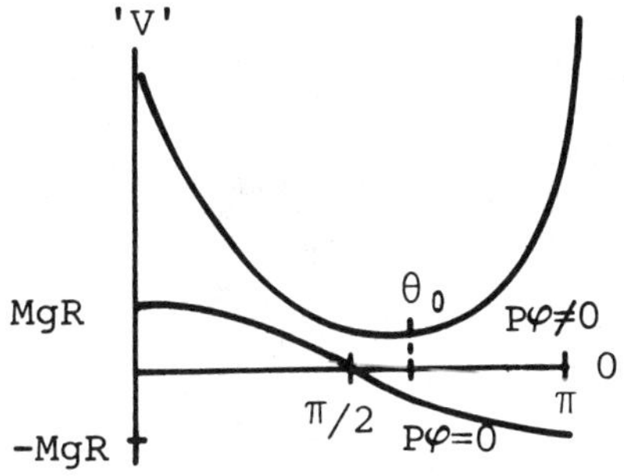

Fig. 2

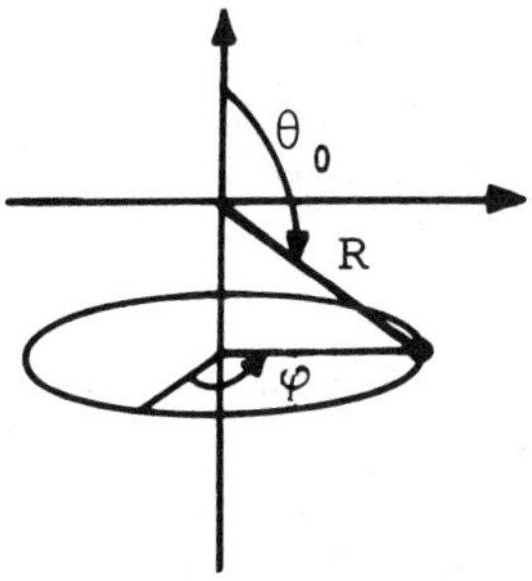

Fig. 3

Consider the minimum of 'V' at $\theta = \theta_o$, which is obtained via

$$\left.\frac{d'V'}{d\theta}\right|_{\theta=\theta_o} = 0$$

so

$$-mgR\sin\theta_o - \frac{P_\phi^{\ 2}\cos\theta_o}{mR^2\sin^3\theta_o} = 0. \tag{10}$$

Substituting for P_ϕ from (6) we find

$$\dot{\phi}^2 = \frac{g}{R}\frac{1}{(-\cos\theta_o)}$$

which represents uniform circular motion at $\theta = \theta_o$.

The energy of this circular motion is

$$E_o = \frac{mgR}{2}\left[\frac{2-3\sin^2\theta_o}{\cos\theta_o}\right] \tag{11}$$

which is obtained from (7).

If we perturb the energy slightly from E_o, the bob will execute simple harmonic motion about the circle $\theta = \theta_o$. To see this evaluate

$$\left.\frac{d^2(V)}{d\theta^2}\right|_{\theta=\theta_o} = k$$

$$\frac{mgR}{-\cos\theta_o}(1+3\cos^2\theta_o) = k \qquad (12)$$

Taking a Taylor expansion of 'V' about $\theta = \theta_o$ we find

$$'V'(\theta) = E_o + \frac{1}{2}k(\theta-\theta_o)^2 \qquad (13)$$

The energy equation is now

$$\frac{1}{2}mR^2\dot{\theta}^2 + \frac{1}{2}k(\theta-\theta_o)^2 = E-E_o$$

which is the harmonic oscillator of energy $E-E_o$, coordinate $\theta-\theta_o$, mass mR^2, spring constant k. The frequency of oscillation is therefore

$$\omega^2 = \frac{k}{mR^2} = \frac{g}{R}\left[\frac{1+3\cos^2\theta_o}{-\cos\theta_o}\right]$$

and $\qquad \theta = \theta_o \cos \omega t.$

• PROBLEM 26-14

A charge e is moving in a region of free space, in which an electric field $\vec{E}$ and a magnetic field $\vec{B}$ are present. Using Maxwell's equations, show that the force acting on the charge may be derived from a velocity dependent potential U and determine U.

Solution: The charge e moving in free space in which both electric and magnetic fields are present, experiences a force which depends not only on the strength of the electric and magnetic fields but also on its velocity.

Thus the Lorentz force on the charge e travelling with a velocity $\vec{v}$ is

$$\vec{F} = e\vec{E} + e\vec{v} \times \vec{B} \qquad (a)$$

The electric and magnetic fields obey the Maxwell's equations in free space

$$\vec{\nabla} \times \vec{E} = -\frac{\partial \vec{B}}{\partial t} \quad \text{(b)}$$

$$\vec{\nabla} \cdot \vec{B} = 0 \quad \text{(c)}$$

$$\vec{\nabla} \cdot \vec{E} = 0 \quad \text{(d)}$$

and $$\vec{\nabla} \times \vec{B} = \mu_o \varepsilon_o \frac{\partial \vec{E}}{\partial t} \quad \text{(e)}$$

where ε_o and μ_o are the permittivity and permeability of free space.

In the case of forces depending on the velocities, it may still be possible to find a potential function U which depends on the coordinates and velocities such that the generalized force Q_j is given by

$$Q_j = -\frac{\partial U}{\partial q_j} + \frac{d}{dt}\left[\frac{\partial U}{\partial \dot{q}_j}\right]$$

A Lagrangian, L, then can be defined as L = T - U so that the equation of motion is given by

$$\frac{d}{dt}\left[\frac{\partial L}{\partial \dot{q}_k}\right] - \frac{\partial L}{\partial q_k} = 0 \qquad k = 1 \ldots f$$

From equation (a), $\vec{\nabla} \times \vec{E} \neq 0$ and hence we cannot write $\vec{E}$ as a gradiant of a scalar function. But from equation (c) it follows that $\vec{B}$ can be represented as a curl of a vector

$$\vec{B} = \vec{\nabla} \times \vec{A} \quad \text{(f)}$$

where $\vec{A}$ is called the magnetic vector potential. Then substituting equation (f) in equation (b) we obtain

$$\vec{\nabla} \times \vec{E} + \frac{\partial}{\partial t}(\nabla x A) = \nabla \times (\vec{E} + \frac{\partial \vec{A}}{\partial t}) = 0$$

Since the curl of a gradient is always zero we can get

$$\vec{E} + \frac{\partial \vec{A}}{\partial t} = -\vec{\nabla}\phi$$

where ϕ is a scalar potential function.

i.e. $$\vec{E} = -\vec{\nabla}\phi - \frac{\partial \vec{A}}{\partial t} \quad \text{(g)}$$

Writing the Lorentz force $\vec{F}$ given by equation (a) in terms of the vector and scalar potentials $\vec{A}$ and ϕ we obtain

$$\vec{F} = e\, E + e\, \vec{v} \times \vec{B}$$

$$\vec{F} = e(-\vec{\nabla}\phi - \frac{\partial \vec{A}}{\partial t}) + e\, \vec{v} \times (\vec{\nabla} x \vec{A})$$

Let us consider the x-component of the force F_x, explicitly. It consists of two parts: $(\vec{v} \times \vec{\nabla} \times \vec{A})_x$ and $\left(-\frac{\partial Q}{\partial x} - \frac{\partial A_x}{\partial t}\right)$.

$$[\vec{v} \times (\vec{\nabla} \times \vec{A})]_x = v_y \left(\frac{\partial A_y}{\partial x} - \frac{\partial A_x}{\partial y}\right) - v_z \left(\frac{\partial A_x}{\partial z} - \frac{\partial A_z}{\partial x}\right)$$

$$= v_y \frac{\partial A_y}{\partial x} + v_z \frac{\partial A_z}{\partial x} + v_x \frac{\partial A_x}{\partial x} - v_y \frac{\partial A_x}{\partial y}$$

$$- v_z \frac{\partial A_x}{\partial z} - v_x \frac{\partial A_x}{\partial x} \qquad \text{(h)}$$

where a term $v_x \frac{\partial A_x}{\partial x}$ is added and subtracted.

Also the time derivative of $\vec{A}$ is

$$\frac{d\vec{A}}{dt} = \frac{\partial \vec{A}}{\partial x}\frac{dx}{dt} + \frac{\partial \vec{A}}{\partial y}\frac{dy}{dt} + \frac{\partial \vec{A}}{\partial z}\frac{dz}{dt} + \frac{\partial \vec{A}}{\partial t}$$

$$= \dot{x}\frac{\partial \vec{A}}{\partial x} + \dot{y}\frac{\partial \vec{A}}{\partial y} + \dot{z}\frac{\partial \vec{A}}{\partial z} + \frac{\partial \vec{A}}{\partial t}$$

$$= (\vec{v} \cdot \vec{\nabla})\vec{A} + \frac{\partial \vec{A}}{\partial t} \qquad \text{(i)}$$

where $\vec{v} = \dot{x}\,\hat{i} + \dot{y}\,\hat{j} + \dot{z}\,\hat{k}$

From equation (i)

$$\frac{\partial A_x}{dt} = v_x \frac{\partial A_x}{\partial x} + v_y \frac{\partial A_x}{\partial y} + v_z \frac{\partial A_x}{\partial z} + \frac{\partial A}{\partial t} \qquad \text{(j)}$$

Using equation (j)

$$(\vec{v} \times \vec{\nabla} \times \vec{A})_x = \frac{\partial}{\partial_x}(\vec{v} \cdot \vec{A}) - \frac{dA_x}{dt} + \frac{\partial A_x}{\partial t} \qquad \text{(k)}$$

Using equation (k) we can write for the x-component of the Lorentz force immediately

$$F_x = e\left[-\frac{\partial}{\partial_x}(\phi - \vec{v}\cdot\vec{A}) - \frac{d}{dt}\frac{\partial}{\partial v_x}(\vec{v}\cdot\vec{A})\right] \qquad \text{(l)}$$

where we have used the identity

$$\frac{d}{dt}\frac{\partial}{\partial v_x}(\vec{v}\cdot\vec{A}) = \frac{d}{dt}\left[\frac{\partial}{\partial v_x}(v_x A_x + v_y A_y + v_z A_z)\right] = \frac{dA_x}{dt}$$

Similarly

$$F_y = e\left[-\frac{\partial}{\partial y}(\phi-\vec{v}\cdot\vec{A}) - \frac{d}{dt}\frac{\partial}{\partial v_y}(\vec{v}\cdot\vec{A})\right]$$

and $$F_z = e\left[-\frac{\partial}{\partial z}(\phi-\vec{v}\cdot\vec{A}) - \frac{d}{dt}\frac{\partial}{\partial v_z}(\vec{v}\cdot\vec{A})\right]$$

If we now define

$$U = e\,(\phi - \vec{v}\cdot\vec{A})$$

as the generalized potential function which depends on the coordinates and the velocities, then equation (1) is

$$F_x = -\frac{\partial U}{\partial x} + \frac{d}{dt}\frac{\partial U}{\partial \dot{x}}$$

which is the generalized Lagrangian equation of motion.

• PROBLEM 26-15

A flyball governor for a steam engine is shown in a simplified form in the figure. Two balls, each of mass m are attached by means of four hinged arms of negligible mass and each of length L, to sleeves which are located on a thin vertical rod. The upper sleeve is fastened to the rod. The lower sleeve has mass M and negligible moment of inertia, and is free to slide up and down the rod as the balls move in and out from the rod. The whole assembly rotates with constant angular velocity ω.

(a) Ignoring friction, set up the equation of motion by Lagranges method. Discuss the motion.

(b) Determine the value of the height z of the lower sleeve above its lowest position as function of ω for a steady rotation of the system.

(c) Find the frequency of small oscillations of the system about the steady state.

Solution: (a) As generalized coordinate we choose the angle θ shown in the figure. The Lagrange's equation of motion then is

$$\frac{d}{dt}\frac{\partial T}{\partial \dot{\theta}} - \frac{\partial T}{\partial \theta} + \frac{\partial V}{\partial \theta} = 0. \qquad (1)$$

First we find the expressions for the kinetic and potential energies. The velocity of each mass m has components in two mutually perpendicular directions: $\omega L \sin\theta$ tangential to the circle in which the masses rotate, and $L\dot{\theta}$ perpendicular to the upper arms. The kinetic energy of each mass m is

$$T_m = \frac{1}{2}m[(\omega L\sin\theta)^2 + (L\dot{\theta})^2],$$

therefore for both masses

$$T_{2m} = m[(\omega L \sin\theta)^2 + (L\dot{\theta})^2].$$

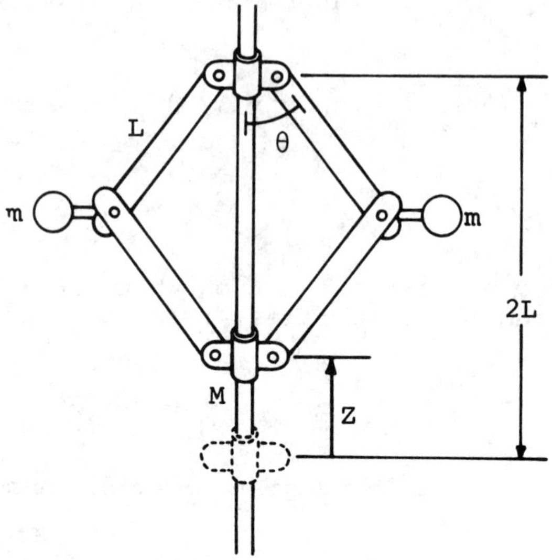

The position of the lower sleeve is

$$z = 2L(1-\cos\theta)$$

therefore its velocity is $2L\dot{\theta}\sin\theta$ and the kinetic energy

$$T_M = \frac{1}{2} M(2L\dot{\theta}\sin\theta)^2.$$

Thus the total kinetic energy of the system is

$$T = m[(\omega L \sin\theta)^2 + (L\dot{\theta})^2] + 2M(L\dot{\theta} \sin\theta)^2. \quad (2)$$

The potential energy, measured from the lowest position of the masses is

$$V = 2mgL(1-\cos\theta) + Mg[2L(1-\cos\theta)]$$

$$= 2(m+M)gL(1-\cos\theta). \quad (3)$$

In order to substitute these expressions in equation (1) we note that

$$\frac{\partial T}{\partial \dot{\theta}} = 2mL^2\dot{\theta} + 4ML^2\sin^2\theta\ \dot{\theta}$$

$$\frac{d}{dt}\left(\frac{\partial T}{\partial \dot{\theta}}\right) = 2mL^2\ddot{\theta} + 4ML^2\sin^2\theta\ \ddot{\theta} + 8ML^2\sin\theta\cos\theta\ \dot{\theta}^2$$

$$\frac{\partial T}{\partial \theta} = 2mL^2\omega^2\sin\theta\cos\theta + 4ML^2\sin\theta\cos\theta\ \dot{\theta}^2$$

$$\frac{\partial V}{\partial \theta} = 2(m+M)gL\sin\theta$$

Thus the equation of motion for the system is

$$(2m+4M\sin^2\theta)L^2\ddot{\theta} + 4ML^2\sin\theta\cos\theta\ \dot{\theta}^2$$

$$- 2mL^2\omega^2\sin\theta\cos\theta + 2(m+M)gL\sin\theta = 0 \quad (4)$$

Motion of the system consists of the rotation at constant speed ω, on which is superimposed an oscillatory motion in the plane of the linkage, i.e., a swinging of the masses m and an up and down motion of sleeve M.

(b) In steady state the whole system rotates at speed ω, with no oscillatory motion. Thus in steady state

$$\ddot{\theta} = \dot{\theta} = 0, \quad \text{and let } \theta = \theta_o .$$

Substituting these values in equation (4) we obtain

$$2mL^2\omega^2\sin\theta_o\cos\theta_o = 2(m+M)gL\sin\theta_o$$

or
$$\cos\theta_o = \frac{(m+M)g}{mL\omega^2} \quad (5)$$

which yields the steady state value θ_o of the angle θ. Substituting this in the expression for z given above, we obtain

$$z_o = 2L\left[1 - \frac{(m+M)g}{mL\omega^2}\right] \quad (6)$$

(c) In order to study the motion of the system about the steady state, we write

$$\theta(t) = \theta_o + \delta\theta(t)$$

where $\delta\theta$ is a small change in θ, superimposed on the steady state value θ_o. Thus $\dot{\theta} = \delta\dot{\theta}$ and $\ddot{\theta} = \delta\ddot{\theta}$. For the purpose of brevity we use the notation

$$C_o = \cos\theta_o \quad \text{and} \quad S_o = \sin\theta_o .$$

As $\delta\theta$ is small we use the following approximations

$$\sin\delta\theta = \delta\theta \ ; \quad \cos\delta\theta = 1$$

$$\delta\dot{\theta}^2 = \delta\theta^2 = 0$$

$$\sin\theta = \sin(\theta_o+\delta\theta) = S_o + C_o\delta\theta$$

$$\sin\theta\cos\theta = S_oC_o + (C_o^2 - S_o^2)\delta\theta.$$

We now substitute the above expressions in equation (4) in order to obtain the equation of motion in terms of $\delta\theta$ which would describe the perturbed motion about the steady state. We get

$$2(m+2MS_o^2)L^2\delta\ddot{\theta} - 2mL^2\omega^2C_oS_o$$
$$- 2mL^2\omega^2(C_o^2-S_o^2)\delta\theta + 2(m+M)gLS_o$$
$$+ 2(m+M)gLC_o\delta\theta = 0. \qquad (7)$$

With the help of relation (5), the above equation simplifies, after cancelling terms, to

$$\delta\ddot{\theta} + \omega_n^2\delta\theta = 0 \qquad (8)$$

where

$$\omega_n^2 = \frac{m\omega^2S_o^2}{m+2MS_o^2} = \frac{(m+M)gS_o^2}{(m+2MS_o^2)C_oL}. \qquad (9)$$

Equation (8) is the equation of motion for the system for small motions about the steady state. It describes a simple harmonic motion of frequency ω_n, which we note from equation (9) is proportional to the speed ω of the governor. The general solution of equation (8) is

$$\delta\theta = A\sin\omega_n t + B\cos\omega_n t. \qquad (10)$$

• PROBLEM 26-16

Analyze the Betatron oscillations in a Particle accelerator, by considering the Lagrangian for velocity dependent potentials.

Solution: In a betatron (or any circular particle accelerator like Cyclotron or Synchrotron), the charged particles move in circular vacuum chambers. Their orbits are controlled by the magnetic fields. Since the particles revolve many times while they are being accelerated, it is essential that the orbits be stable. We shall consider the stability of the orbit at constant energy neglecting the acceleration process.

Let us use Cylindrical Polar Coordinates and assume that the magnetic field is symmetrical about the z-axis

$$\vec{B}(\rho,\phi,z) = B_z(\rho,z)\hat{z} + B_\rho(\rho,z)\hat{\rho} \qquad (1)$$

Since the total energy E is constant, we assume B as constant. Let the field be entirely vertical in the median plane z = 0

$$\vec{B}(\rho,\phi,0) = B_{z_o}(\rho)\hat{z} \qquad (2)$$

The equilibrium orbit depends on the energy E and is given by $\rho = a(E)$. We are interested in the stability of this orbit and would like to know whether the charged particles execute small oscillations about this orbit.

The Lagrangian for the case of a charged particle moving in a magnetic field alone and no electric field is given by

$$L = T - U$$

where $$U = -e\vec{v}\cdot\vec{A} \tag{3}$$

where $\vec{v}$ is the velocity of the particle and $\vec{A}$ is the vector potential. The vector potential is related to the magnetic field by

$$\vec{B} = \vec{\nabla} \times \vec{A} \tag{4}$$

Where the magnetic field is symmetric about the z-axis, the vector potential is entirely in the ϕ direction

$$\vec{A} = A_\phi(\rho,z)\hat{\phi} \tag{5}$$

From equation (4) we have

$$B_z(\rho,z)\hat{z} = \frac{1}{\rho}\left(\frac{\partial}{\partial\rho}\rho A_\phi\right)\hat{z} \tag{6}$$

and $$B\rho(\rho,z)\hat{\rho} = -\frac{\partial A_\phi}{\partial z}\hat{\rho} \tag{7}$$

Equation (6) can be solved for A_ϕ in terms of the magnetic field. Since the particle is moving in a circular orbit

$$\vec{v} = \rho\dot{\phi}\hat{\phi} \tag{8}$$

and $$U = -e\rho\dot{\phi}A_\phi(\rho,z) \tag{9}$$

Hence $$L = \frac{1}{2}m(\dot{\rho}^2+\rho^2\dot{\phi}^2+\dot{z}^2) + e\rho\dot{\phi}A_\phi(\rho,z) \tag{10}$$

This expression for the Lanrangian is non-relativistic and is useful only for particles whose velocities are much smaller than the velocity of light. (In the real Betatron problems relativistic Lagrangian must be employed.) The generalized momenta are given by

$$p_\rho = \frac{\partial L}{\partial\dot{\rho}} = m\dot{\rho} \tag{11}$$

$$p_\phi = m\rho^2\dot{\phi} + e\rho A_\phi \tag{12}$$

$$p_z = m\dot{z} \tag{13}$$

The Hamiltonian is given by

$$H = \sum_i p_i\dot{q}_i - L = \frac{p_e^2}{2m} + \frac{p_z^2}{2m} + \frac{[p_\phi - \frac{e}{c}\rho A_\phi(\rho,z)]^2}{2m\rho^2} \tag{14}$$

It is easily seen that ϕ is an ignorable coordinate (not explicitly present in H), so p_ϕ may be taken as a given constant. This Hamiltonian has the form

$$H = 'T' + 'V'$$

with
$$'T' = \frac{1}{2} m(\dot{\rho}^2 + \dot{z}^2) \tag{15}$$

and
$$'V' = \frac{(p_\phi - \frac{e}{c} \rho A_\phi)^2}{2m\rho^2} \tag{16}$$

where we have used $\dot{\phi} = \dfrac{(p_\phi - \frac{e}{c} \rho A_\phi)}{m\rho^2}$ from (12). Here 'T' can be considered to be an effective kinetic energy and 'V' an effective potential energy. The steady state solution is obtained by

$$\frac{\partial 'V'}{\partial z} = \frac{e}{c} \rho\dot{\phi}B_\rho = 0 \qquad \text{using (7) and (8)}$$

and
$$\frac{\partial 'V'}{\partial \rho} = - \rho\dot{\phi}(m\dot{\phi} + \frac{e}{c} B_z) = 0 \quad \text{using (6) and (8)}$$

The later equation gives

$$\dot{\phi} = - \frac{e}{cm} B_{z_o} (\rho) \tag{17}$$

We may solve this equation for ρ given $\dot{\phi}$, or alternatively, for $\dot{\phi}$ with ρ = a the radius of the equilibrium orbit.

To study the small deviations from the equilibrium orbit, we set

$$\rho = a + n$$

where a is the equilibrium orbit radius. We also define

$$\omega = \dot{\phi} = - \frac{\rho B_{z,o}(a)}{cm} \tag{18}$$

$$n = - \left[\frac{a}{B_z} \frac{\partial B_z}{\partial \rho}\right]_{z=0,\rho=a} \tag{19}$$

Expanding 'V' and 'T' in powers of n, z $\dot{n}$ and $\dot{z}$, and using $\frac{\partial V}{\partial \rho} = 0$ and $\frac{\partial V}{\partial z} = 0$ at equilibrium, it is not difficult to show that

$$T = \frac{1}{2} m(\dot{n}^2 + \dot{z}^2) \tag{20}$$

$$'V' = \frac{1}{2} m\omega^2(1-n)n^2 + \frac{1}{2} m\omega^2 nz^2 \tag{21}$$

then deriving (21) it is necessary to use

$$\frac{\partial B_z}{\partial \rho} = \frac{\partial B_\rho}{\partial z} \quad \text{which follows from}$$

$$\vec{\nabla} \times \vec{B} = 0.$$

The quantity n is called the field index. We see immediately that the motion is stable only if

$$0 < n < 1.$$

In a cyclotron, the field is nearly constant at the center, so that $n \ll 1$, and then falls rapidly near the outside edge of the magnet. In a betatron or synchrotron, the magnetic field has a constant value of n and B increases in magnitude as the particles are accelerated so as to keep a constant. We see from equation (19) that the value of n does not change as B_z is increased provided the shape of the magnetic field as a function of radius does not change, that is, provided $\partial B_z/\partial z$ increases in proportion to B_z.

Since the variables x,z are separated in 'T' and 'V', we can immediately write down the betatron oscillation frequencies. It is convenient to express them in terms of the numbers of betatron oscillations per revolution, v_x and v_z:

$$v_x = \frac{\omega_x}{\omega} = (1-n)^{1/2}$$

$$v_z = \frac{\omega_z}{\omega} = n^{1/2} .$$

If there are imperfections in the accelerator, so that B_z is not independent of ϕ, the difference between B_z and its average value gives rise to a periodic force acting on the coordinate x. The resulting perturbation of the orbit can be treated by solving the corresponding forced harmonic oscillator equations. If B_ρ is not zero everywhere in the median plane (z = 0), vertical forces act which drive the vertical betatron oscillations. In general, such imperfections also lead to variations in the field index n, so that $n = n(\phi)$, and n for the steady motion becomes a periodic function of time. In alternating gradient accelerators, the field index n is deliberately made to vary periodically in azimuth ϕ. The solution of this problem is too complex for inclusion here.

HAMILTON'S PRINCIPLE

• PROBLEM 26-17

Describe the variational formulation of mechanics.

Solution: In the variational formulation of mechanics a system is assumed to be governed by a principle which says that the system will move or evolve in such a way that some quantity will be extremized.

Hamilton's Principle states that the integral of the difference of kinetic and potential energy must be extremized

$$\delta \int_{t_1}^{t_2} (T-V)dt = 0 \tag{1}$$

where δ is the functional variation operator.

We define T-V as the Lagrangian function

$$L = T-V.$$

By applying the calculus of variation we obtain from (1) the Euler-Lagrange equations. If

$$L = L(q_i, \dot{q}_i, t), \quad i = 1,2,\ldots,N$$

then

$$\frac{\partial L}{\partial q_i} - \frac{d}{dt}\frac{\partial L}{\partial \dot{q}_i} = 0, \quad i = 1,2,\ldots,N \tag{2}$$

where the q_i are the generalized coordinates of the problem.

Equations (2) are sometimes known as the Euler-Lagrange equations of the first kind; the Euler-Lagrange equations of the second kind which also came from the calculus of variation are

$$\frac{\partial L}{\partial t} = \frac{d}{dt}\left[L - \dot{q}_i \frac{\partial L}{\partial \dot{q}_i}\right] = 0 \tag{3}$$

Finally, suppose some constraint conditions are imposed on the coordinates such as

$$f_j(q_i) = 0, \quad j=1,2,\ldots,M,$$

then the Euler-Lagrange equations with undetermined multipliers are

$$\frac{\partial L}{\partial q_i} - \frac{d}{dt}\frac{\partial L}{\partial \dot{q}_i} + \sum_{\ell=1}^{M} \lambda_\ell \frac{\partial f_\ell}{\partial q_i} = 0$$

where the λ_ℓ are the undetermined multipliers which act as additional generalized coordinates.

• PROBLEM 26-18

Derive Hamilton's principle from Newton's Laws.

Solution: Consider a single particle of mass m acted on by a force $\vec{f}$. Then Newton's second law is:

$$m \frac{d^2\vec{r}}{dt^2} - \vec{f} = 0 \tag{1}$$

where $\vec{r}$ is the position vector of the particle. The solution of Eq.(1):

$$\vec{r} = \vec{r}\,(t) \tag{2}$$

is the true path of the particle. Let

$$\vec{r} = \vec{r} + \delta\vec{r} \tag{3}$$

be a varied path with the restriction that

$$\delta\vec{r}\Big|_{t_1} = \delta\vec{r}\Big|_{t_2} = 0 \tag{4}$$

where $t_1 < t < t_2$ is the time interval of interest.

We begin the derivation by taking the scalar product of $\delta\vec{r}$ and (1), and then integrating from t_1 to t_2:

$$\int_{t_1}^{t_2} \left(m \frac{d^2\vec{r}}{dt^2} \cdot \delta\vec{r} - \vec{f} \cdot \delta\vec{r} \right) dt = 0. \tag{5}$$

Now, $$m \int_{t_1}^{t_2} \frac{d^2\vec{r}}{dt} \cdot \delta\vec{r}\, dt = m \frac{d\vec{r}}{dt} \cdot r \Big|_{t_1}^{t_2}$$

$$- m \int_{t_1}^{t_2} \frac{d\vec{r}}{dt} \cdot \frac{\delta d\vec{r}}{dt}\, dt$$

when integrated by parts.

Since by (4) $\delta\vec{r}(t_1) = \delta\vec{r}(t_2) = 0$ the first term vanishes. Also,

$$\frac{d\vec{r}}{dt} \cdot \delta \frac{d\vec{r}}{dt} = \frac{1}{2} \delta \frac{d\vec{r}}{dt} .$$

Thus, $$m \int_{t_1}^{t_2} \frac{d^2\vec{r}}{dt^2} \cdot \delta\vec{r}\, dt = -\frac{m}{2} \int_{t_1}^{t_2} \delta \left(\frac{d\vec{r}}{dt} \right)^2 dt.$$

We note, however, that:

$$\frac{1}{2} m \left(\frac{d\vec{r}}{dt}\right)^2 \equiv T,$$

is the kinetic energy of the particle.

Thus (5) becomes:

$$\int_{t_1}^{t_2} (\delta T + \vec{f} \cdot \delta\vec{r})\, dt = 0. \qquad (6)$$

If the force field, $\vec{f}$, is conservative, there exists a function, $v(x, y, z)$, such that

$$\vec{f} \cdot d\vec{r} = -dv.$$

This means:

$$\vec{f} = -\operatorname{grad} v,$$

where grad v is the vector with components:

$$\left(\frac{\partial v}{\partial x}, \frac{\partial v}{\partial y}, \frac{\partial v}{\partial z}\right).$$

It follows that if $\vec{f}$ is conservative there is a single valued function, $\Phi(x, y, z)$, such that

$$\vec{f} \cdot \delta\vec{r} = -\delta\Phi.$$

Equation (6) may be written as

$$\int_{t_1}^{t_2} (\delta T - \delta\Phi)\, dt = 0.$$

Φ is called the potential energy of the system, and

$$L \equiv T - \Phi$$

is called the Lagrangian of the system. Hamilton's principle is thus:

$$\delta \int_{t_1}^{t_2} L\, dt = 0.$$

• PROBLEM 26-19

Derive the Hamiltonian, Hamilton's equations of motion, and conservation of energy for a generalized, dynamical system starting from the Lagrangian formulation.

Solution: A system is assumed to be described by n generalized coordinates $q_1, q_2, \ldots, q_n$ and their generalized velocities $\dot{q}_1, \dot{q}_2, \ldots, \dot{q}_n$. The Lagrangian is:

$$L = T - V \tag{1}$$

where $T = \sum_i^n \sum_j^n a_{ij} \dot{q}_i \dot{q}_j$ is a symmetric, homogeneous function called the kinetic energy. $V = V(q_1, \ldots, q_n)$ is the potential energy. The equations of motion are given by the Euler-Lagrange equations:

$$\frac{d}{dt} \frac{\partial L}{\partial \dot{q}_i} = \frac{\partial L}{\partial q_i} . \tag{2}$$

We now define the generalized momentum by:

$$p_i \equiv \frac{\partial L}{\partial \dot{q}_i} . \tag{3}$$

The Hamiltonian is defined as the total energy of the system. Thus:

$$H = T + V. \tag{4}$$

We wish to show first that

$$H = \sum_{k=1}^{n} p_k \dot{q}_k - L. \tag{5}$$

To derive this we note that T is a symmetric quadratic form:

$$T = \sum_i \sum_j a_{ij} \dot{q}_i \dot{q}_j .$$

Then,

$$\frac{\partial T}{\partial \dot{q}_k} = \sum_i \sum_j a_{ij} \left[\frac{\partial \dot{q}_i}{\partial \dot{q}_k} \dot{q}_j + \dot{q}_i \frac{\partial \dot{q}_j}{\partial \dot{q}_k} \right]$$

$$= \sum_i \sum_j a_{ij} \left[\delta_{ik} \dot{q}_j + \dot{q}_i \delta_{jk} \right]$$

$$= \sum_j a_{kj} \dot{q}_j + \sum_i a_{ij} \dot{q}_i ;$$

but, since i is a dummy variable and $a_{ij} = a_{ji}$, then,

$$\frac{\partial T}{\partial \dot{q}_k} = 2 \sum_i a_{ik} \dot{q}_i .$$

Multiplying by $\dot{q}_k$ and summing over k:

$$\sum_k \dot{q}_k \frac{\partial T}{\partial \dot{q}_k} = 2 \sum_k \sum_i a_{ik} \dot{q}_i \dot{q}_k = 2T.$$

Thus, $$T = \frac{1}{2} \sum_k \dot{q}_i \frac{\partial T}{\partial \dot{q}_k} . \qquad (6)$$

Since V is independent of $\dot{q}_k$:

$$\frac{\partial T}{\partial \dot{q}_k} = \frac{\partial}{\partial \dot{q}_k} (T - V) = \frac{\partial}{\partial \dot{q}_k} L = p_k .$$

Equation (6) becomes:

$$T = \frac{1}{2} \sum_k \dot{q}_k p_k . \qquad (7)$$

Now, $$L = T - V$$

$$H = T + V.$$

Adding:

$$L + H = 2T.$$

Substituting from (7):

$$H = \sum \dot{q}_k p_k - L,$$

which establishes (5).

Using (5) we obtain Hamilton's equations of motion,

$$\frac{\partial H}{\partial \dot{q}_i} = p_i - \frac{\partial L}{\partial \dot{q}_i} = p_i - p_i = 0,$$

which implies H is independent of $\dot{q}_i$.

$$\frac{\partial H}{\partial p_i} = \dot{q}_i - \frac{\partial L}{\partial p_i} = \dot{q}_i - 0.$$

Thus, $$\frac{\partial H}{\partial p_i} = \dot{q}_i . \qquad (8)$$

Next, $$\frac{\partial H}{\partial q_i} = - \frac{\partial L}{\partial q_i} = - \frac{d}{dt} \frac{\partial L}{\partial \dot{q}_i} = - \frac{d}{dt} p_i .$$

Thus, $$\frac{\partial H}{\partial q_i} = - \dot{p}_i . \qquad (9)$$

Equations (8) and (9) are the canonical equations of motion.

Finally, multiply (8) by $\dot{p}_i$ and (9) by $\dot{q}_i$. Sum over i and add to obtain

$$\frac{\partial H}{\partial p_i}\dot{p}_i + \frac{\partial H}{\partial q_i}\dot{q}_i = \dot{q}_i\dot{p}_i - \dot{q}_i\dot{p}_i = 0.$$

But the left hand side is just $\frac{dH}{dt}$. Thus,

$$\frac{dH}{dt} = 0$$

which is the statement of conservation of energy.

EXAMPLES OF SYSTEMS SUBJECT TO CONSTRAINTS

• PROBLEM 26-20

Consider a disk or radius R_o and infinitesimal thickness which rolls without slipping on a horizontal surface as the plane of the disk remains vertical. Determine the equations of constraint.

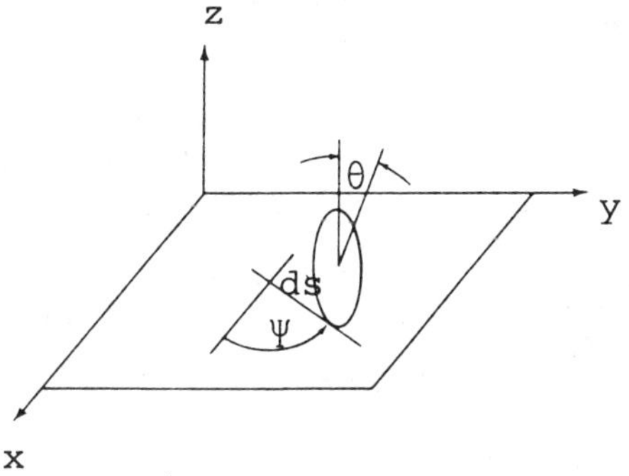

Fig. 1

Solution: The condition that the disk rolls without slipping is that $ds = R_o d\theta$ where

$$ds = dx \cos \Psi + dy \sin \Psi$$

(see Fig. 1). Thus

$$dx \cos \Psi + dy \sin \Psi = R_o d\theta \qquad (1)$$

which is a non-holonomic constraint (a non-integrable differential relation).

Another constraint is obviously

$$z = R_o. \qquad (2)$$

Finally, since the disk is not to "twirl," the line normal to the disk must stay perpendicular to ds.

Thus

$$- dx \sin \Psi + dy \cos \Psi = 0. \tag{3}$$

Equations (1), (2) and (3) represent the equations of constraint.

• PROBLEM 26-21

A particle of mass m is constrained to move on the surface of a sphere of radius R_o by means of its attachment to a rigid massless rod as shown in Fig. 1. Determine the position of the particle when in equilibrium.

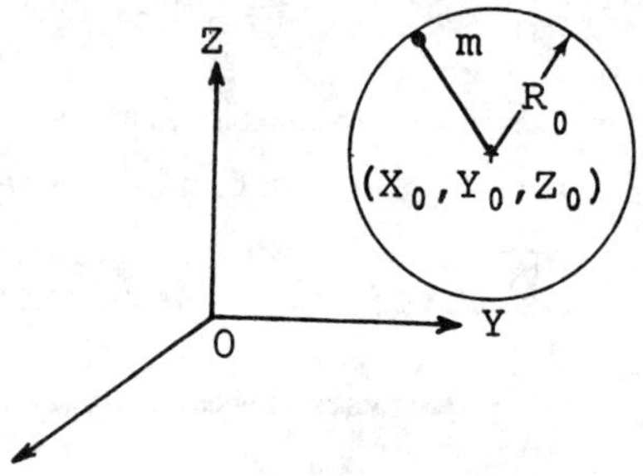

Fig. 1

Solution: The usual approach of considering virtual work is to say $\vec{F}_{total} = 0$, however we may separate the total force into an external part $\vec{F}$ and a constraint part R so that $\vec{F} + \vec{R} = 0$ signifies equilibrium. We can take the virtual work, resulting from virtual displacement $\delta\vec{r}$,

$$\delta W = (\vec{F} + \vec{R}) \cdot \delta\vec{r} = 0 \tag{a}$$

since $\qquad \vec{F} + \vec{R} = 0.$

The method involves developing a general expression for $\vec{R}$ to go into the equations above. If we find a constraint equation of the form

$$\phi(x,y,z,t) = 0 \tag{b}$$

we can find $\delta\phi$, a virtual displacement, as

$$\delta\phi = \frac{\partial\phi}{\partial x}\delta x + \frac{\partial\phi}{\partial y}\delta y + \frac{\partial\phi}{\partial z}\delta z$$

$(\delta t = 0)$. Now we assume that some multiplier λ, in general not constant, times $\delta\phi$ will equal $\vec{R} \cdot \delta\vec{r}$. Then we may write

$$F_x\delta_x + F_y\delta_y + F_z\delta_z + \lambda\frac{\partial\phi}{\partial x}\delta x + \lambda\frac{\partial\phi}{\partial y}\delta y + \lambda\frac{\partial\phi}{\partial z}\delta z = 0$$

or

$$(F_x + \lambda\frac{\partial\phi}{\partial x})\delta x + (F_y + \lambda\frac{\partial\phi}{\partial y})\delta y + (F_x + \lambda\frac{\partial\phi}{\partial z})\delta z = 0.$$

The components of the constraint force are given by

$$R_x = \lambda \frac{\partial\phi}{\partial x}, \quad R_y = \lambda \frac{\partial\phi}{\partial y}, \quad R_z = \lambda \frac{\partial\phi}{\partial z}.$$

For equilibrium we set

$$\left[F_x + \lambda \frac{\partial\phi}{\partial x}\right] = 0, \quad \left[F_y + \lambda \frac{\partial\phi}{\partial y}\right] = 0 \text{ and } \left[F_z = \lambda \frac{\partial\phi}{\partial z}\right] = 0$$

independently since δx, δy, and δz are arbitrary. And these three equations along with equation (b), the equation of constraint gives us the four equations we need to solve for the equilibrium points x_o, y_o, z_o and for λ which gives us the constraint forces.

Now for the problem given us the constraint equation is

$$\phi(x,y,z,t) = (x - x_o)^2 + (y - y_o)^2 + (z - z_o)^2 - R_o^2 = 0.$$

The external force, which is due only to gravity, is simply

$$F = 0\hat{i} + 0\hat{j} - mg\hat{k}.$$

Using the first of our equilibrium conditions.

$$F_x + \lambda \frac{\partial\phi}{\partial x} = \lambda 2(x - x_o) = 0.$$

Since λ is arbitrary and not necessarily zero, $x - x_o = 0$. Next

$$F_y + \lambda \frac{\partial\phi}{\partial y} = 0 + \lambda 2(y - y_o) = 0$$

and therefore $y - y_o = 0$. Finally,

$$F_z + \lambda \frac{\partial\phi}{\partial z} = -mg + \lambda 2(z - z_o) = 0$$

We already found $(x - x_o) = (y - y_o) = 0$. Therefore the constraint equation gives us $(z - z_o)^2 = R_o^2$, $(z - z_o) = \pm R_o$. Using this value we can solve for λ

$$\lambda = \pm \frac{mg}{2R_o}$$

Thus there are two points of equilibrium, one at the top, the other at the bottom of the sphere: $x = x_o$, $y = y_o$, $z = z_o + R_o$; $x = x_o$, $y = y_o$, $z = z_o - R_o$.

The reaction forces are now readily determined:

$$R_x = \lambda \left.\frac{\partial\phi}{\partial x}\right|_{x=x_o,\, y=y_o,\, z=z_o \pm R_o} = 0$$

$$R_y = \lambda \left.\frac{\partial\phi}{\partial y}\right|_{x=x_o, y=y_o, z=z_o \pm R_o} = 0$$

$$R_z = \lambda \left.\frac{\partial\phi}{\partial z}\right|_{x=x_o, y=y_o, z=z_o \pm R_o}$$

$$= \pm \frac{mg}{2R_o}\, 2(z - z_o) = \pm \frac{mg}{2R_o}\, 2(\pm R_o) = mg.$$

• PROBLEM 26-22

A particle moves on the outer surface of a hoop of radius R_o as shown. Determine its position as a function of time. At what point does the particle leave the hoop?

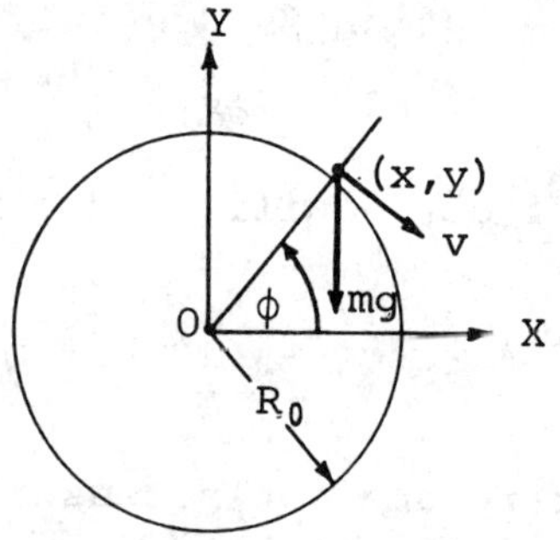

Solution: Solve this problem using Lagrange undetermined multipliers. To do so express the constraint in differential form and temporarily relax the constraints to allow determination of the constraint force.

The two generalized coordinates can be chosen to be x and y or ϕ and r (note we are not yet applying $r = R_o$). Either will be good for determining the motion while the particle is on the hoop. Since the contact force between the hoop and particle is along the radius, the radius will be a convenient generalized coordinate. Hence, choose ϕ and r as the two coordinates.

Find the equation of constraint in differential form. The constraint equation is $r = R_o$. Therefore we find

$$dr = 0. \tag{1}$$

Lagrange's equations take the form

$$\frac{d}{dt}\frac{\partial L}{\partial \dot{q}_i} - \frac{\partial L}{\partial q_i} = \sum_j a_{ij}\lambda_j \tag{2}$$

where $\lambda_j \sum_j a_{ij}\, dq_i = 0.$

In this problem j = 1 and our constraint equation (1) is

$$\lambda dr = 0. \tag{3}$$

This gives us $a_r = 1$, $a_\phi = 0$.

The Lagrangian

$$L = T - V = \frac{1}{2} m(\dot{r}^2 + (r\dot{\phi})^2) - mgr \sin\phi.$$

Again notice that we are taking r to be variable. We apply the constraint $r = R_o$ only at the end of the problem.

Lagrange's equation (2) is

$$\frac{d}{dt}\left(\frac{\partial L}{\partial \dot{r}}\right) - \frac{\partial L}{\partial r} = \lambda$$

$$m\ddot{r} - mr\dot{\phi}^2 + mg \sin\phi = \lambda \tag{4}$$

and

$$\frac{d}{dt}\left(\frac{\partial L}{\partial \dot{\phi}}\right) - \frac{\partial L}{\partial \phi} = 0$$

$$mr^2\ddot{\phi} + 2m\dot{r}\dot{\phi} + mgr \cos\phi = 0. \tag{5}$$

Only now after completing the differentials, use the constraint. Since $r = R_o$, $\dot{r} = \ddot{r} = 0$, and the equation of motion is

$$\lambda = mg \sin\phi - mR_o\dot{\phi}^2 \tag{6}$$

$$mR_o^2\ddot{\phi} + mgR_o \cos\phi = 0. \tag{7}$$

Equation (7) gives the equation of motion

$$\ddot{\phi} = -\frac{g}{R_o} \cos\phi.$$

Using $\ddot{\phi} = \frac{d}{dt}\left(\frac{d\phi}{dt}\right) = \frac{d\phi}{dt}\frac{d}{d\phi}\left(\frac{d\phi}{dt}\right) = \dot{\phi}\frac{d\dot{\phi}}{d\phi}$,

yielding $\dot{\phi}d\dot{\phi} = -\frac{g}{R_o} \cos\phi d\phi$

$$\frac{\dot{\phi}^2}{2} = -\frac{g}{R_o} \sin\phi + C. \tag{8}$$

Since $r = R_o$,

$$\dot{\phi} = \frac{v}{R_o}, \text{ so}$$

$$\frac{1}{2} v^2 = -gR_o \sin\phi + R_o^2 C.$$

If we take as initial conditions $\phi(o) = \frac{\pi}{2}$, $v(o) = 0$ (particle starts from rest at the top of the hoop) then

$$R_o^2 C = gR_o \quad \text{or}$$

$$\frac{1}{2} v^2 = gR_o(1-\sin\phi).$$

Note that we clearly went the long route for this equation which (if multiplied through by m) is simply the conservation of energy equation with the top of the hoop as datum.

The advantage of Lagrange Multipliers arises in answering constraint related questions. For example, when does the particle leave the hoop? The condition is that the constraint force, λ, goes to zero. Setting $\lambda = 0$ in equation (6),

$$0 = mg \sin\phi - mR_o\dot{\phi}^2.$$

Since $C = \frac{g}{R_o}$, $\dot{\phi}^2$ from equation (8) is

$$\dot{\phi}^2 = 2\frac{g}{R_o}(1-\sin\phi).$$

This gives us

$$mg \sin\phi = 2mg(1-\sin\phi)$$

$$3mg \sin\phi = 2mg$$

$$\phi = \sin^{-1}(2/3) = 41.8°$$

So the particle leaves the hoop at $(r,\phi) = (R_o, 41.8°)$.

• PROBLEM 26-23

Determine the motion of a bead subject to no external applied force, constrained to move on a straight wire rotating at constant angular velocity about an axis at one end, the axis being perpendicular to the wire. Solve this problem in cartesian and polar coordinates using Lagrange's equations of the second kind and find the constraint force.

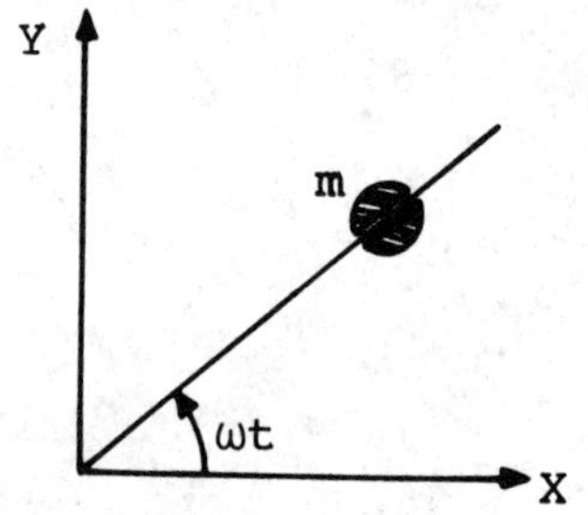

Fig. 1

Solution: Cartesian Coordinates: The constraint equation (Figure 1) is

$$y \cos \omega t - x \sin \omega t = 0 \qquad \text{(a)}$$

In differential form it is

$$\cos \omega t\, dy - \sin \omega t\, dx = 0. \qquad \text{(b)}$$

The Lagrangian is simply equal to the kinetic energy so Lagrange's equations are

$$m\ddot{x} = -\lambda \sin \omega t$$
$$m\ddot{y} = \lambda \cos \omega t \qquad \text{(c)}$$

Eliminating λ,

$$m\ddot{x} + m\ddot{y} \tan \omega t = 0$$

or

$$m\ddot{x} \cos \omega t + m\ddot{y} \sin \omega t = 0 \qquad \text{(d)}$$

Now that we have found the equation of motion in cartesian coordinates we will make the obvious substitution for $\ddot{x}$ and $\ddot{y}$ by introducing polar coordinates, $x = R \cos \omega t$ and $y = R \sin \omega t$ (note this also solves equation (a)):

$$\dot{x} = \dot{R} \cos \omega t - R\omega \sin \omega t$$
$$\dot{y} = \dot{R} \sin \omega t + R\omega \cos \omega t$$
$$\ddot{x} = \ddot{R} \cos \omega t - R\omega^2 \cos \omega t - 2\dot{R}\omega \sin \omega t$$
$$\ddot{y} = \ddot{R} \sin \omega t - R\omega^2 \sin \omega t + 2\dot{R}\omega \cos \omega t.$$

Substitution for the acceleration components in the equation of motion (d) yields

$$m(\ddot{R} \cos \omega t - R\omega^2 \cos \omega t - 2\dot{R}\omega \sin \omega t) \cos \omega t$$
$$+ m(\ddot{R} \sin \omega t - R\omega^2 \sin \omega t + 2\dot{R}\omega \cos \omega t) \sin \omega t = 0$$

or

$$\ddot{R} = \omega^2 R = 0.$$

The general solution of the above equation is

$$R = A \cosh \omega t + B \sinh \omega t.$$

If at $t = 0$, $R = R_o$ and $\dot{R} = 0$, $A = R_o$ and $B = 0$. Therefore,

$$R = R_o \cosh \omega t$$

and

$$x = R_o \cosh \omega t \cos \omega t$$
$$y = R_o \cosh \omega t \sin \omega t.$$

From equation (c)

$$\lambda = \frac{m\,\ddot{y}}{\cos \omega t} = 2mR_o\omega^2 \sinh \omega t. \qquad \text{(e)}$$

This clearly is the normal force which acts perpendicular to R and in the direction of increasing ωt.

Polar Coordinates:

The kinetic energy in polar coordinates is

$$T = L = \frac{1}{2} m (\dot{R}^2 + R^2\dot{\phi}^2).$$

The constraint equation is

$$\phi - \omega t = 0 \tag{f}$$

and, in differential form,

$$\lambda d\phi = 0.$$

Thus Lagrange's equations are

$$m\ddot{R} = mR\dot{\phi}^2 = 0$$

$$mR^2\ddot{\phi} + 2mR\dot{R}\dot{\phi} = \lambda. \tag{g}$$

From (f) we know that $\dot{\phi} = \omega$, so equations (g) are

$$\ddot{R} - R\omega^2 = 0$$

$$2mR\dot{R}\omega = \lambda.$$

We get as the solution to the first equation

$$R = A \cosh \omega t + B \sinh \omega t.$$

Using the same initial conditions, $R(o) = R_o$ and $\dot{R}(o) = 0$ we get

$$R = R_o \cosh \omega t.$$

Putting this into the equation for λ we get

$$\lambda = 2mR_o^2\omega^2 \cosh \omega t \sinh \omega t.$$

In this case λ is the torque exerted by the wire on the bead. And we see that $\lambda = NR$ where N is the normal force we determined earlier in equation (e) and R is the solution

$$R = R_o \cosh \omega t.$$

Certainly the solution using polar coordinates throughout, has been easier although care is needed in interpreting the quantity λ. In these situations a dimensional analysis can help prevent confusion.

• PROBLEM 26-24

Use the Lagrange formulation to find the tension in the rope of an Atwood's machine.

Solution: Atwood's machine is shown in Fig. 1, where the length of the rope supporting masses m_1 and m_2 is ℓ. The vertical portion of the rope suspending m_1 is x. Let τ be the tension in the rope. Assume ℓ is constrained to be constant. To find the corresponding force of constraint, imagine the constraint as being violated and see what force would restore the system.

Thus, take x and ℓ as generalized coordinates. The only forces acting on the masses are gravity and rope tension.

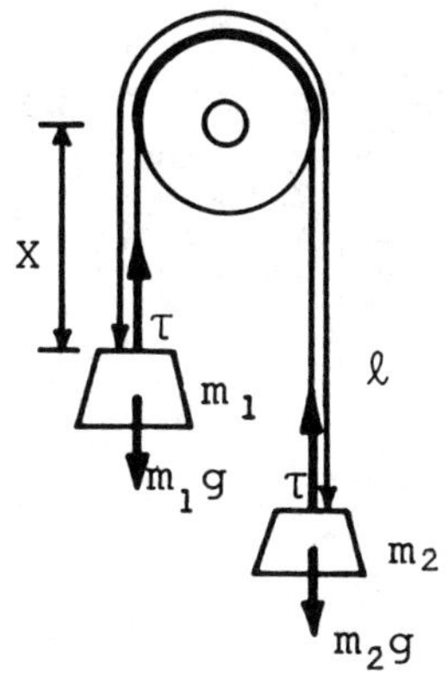

Fig. 1 Atwood's Machine

Obtain the generalized forces by variational procedure. Suppose x is varied by δx, while ℓ is held constant. Then the variational work will correspond to the system's change of potential energy

$$\delta W = (m_1 - m_2) g \delta x$$

But by definition the generalized force Q_x is given by $\delta W = Q_x \delta x$. Thus

$$Q_x = (m_1 - m_2) g.$$

Now let ℓ vary by $\delta\ell$ and hold x fixed. Then

$$\delta W = (m_2 g - \tau)\delta\ell = Q_\ell \delta\ell$$

so

$$Q_\ell = m_2 g - \tau.$$

The kinetic energy of the system is

$$T = \frac{1}{2} m_1 \dot{x}^2 + \frac{1}{2} m_2 (\dot{\ell} - \dot{x})^2.$$

(Where $\dot{\ell} = \ddot{\ell} = 0.$)

Using the Lagrange equations

$$\frac{d}{dt} \frac{\partial T}{\partial \dot{q}} - \frac{\partial T}{\partial q} = Qq$$

where q is x or ℓ we obtain

$$(m_1 + m_2)\ddot{x} = (m_1 - m_2) g$$

$$-m_2\ddot{x} = m_2 g - \tau .$$

Solving these equations of motion we obtain

$$x = x_o + v_o t + \frac{1}{2}\left(\frac{m_1 - m_2}{m_1 + m_2}\right) g t^2$$

$$\tau = m_2(g+\ddot{x}) = \left(\frac{2m_1 m_2}{m_1 + m_2}\right) g .$$

• PROBLEM 26-25

Solve the problem of one cylinder rolling over another. Assume there is static friction $f \leq \mu F$ to cause rolling until the normal force F becomes small enough at which point the cylinder slips without friction until, at another point, the rolling cylinder falls off the bottom cylinder. The bottom cylinder does not move. Determine the point where sliding first begins and the equations of motion up until that point.

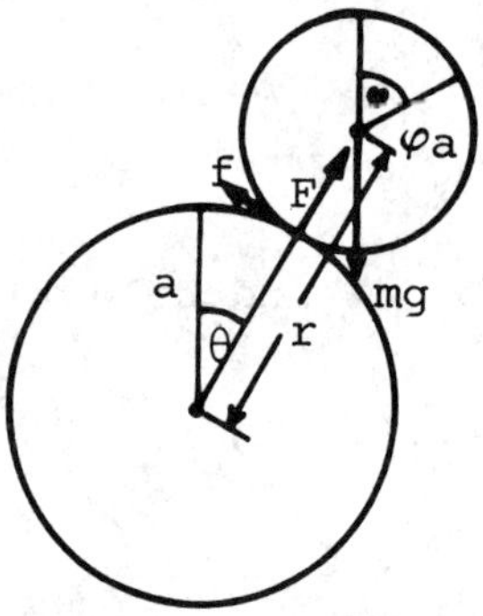

Solution: Before attempting to solve the problem let us describe the expected motion. First the smaller cylinder, of radius αa, and mass m, starts from rest with its center directly above the center of the other cylinder, of radius a. Anticipate some mathematical difficulties with this part of the motion since it is in unstable equilibrium. Physically, there is no difficulty, since the slightest disturbance will cause the cylinder to roll down, but mathematically there may be no expression for the needed slight disturbance.

The moving cylinder will fly off the static one at some point. We can state this condition by saying that when the interaction force goes to zero the cylinders may separate. That is, when the normal force F = 0.

The cylinder can not continue to roll right up until this happens. Some minimum friction, say f_r, must exist to cause rolling. The static friction $f \leq \mu F$ will equal this f_r as long as F is big enough to allow it. However when

$$f_r = \mu F \tag{a}$$

rolling will cease and slipping will start. This is actually a simplifying assumption we are making. In reality some simultaneous rolling and slipping will occur as the cylinder goes from only rolling to only slipping.

Our problem then is to analyze rolling motion from $\theta = 0$ to θ_1, then analyze slipping motion from θ_1 to θ_2 when the cylinder flies off and to find θ_1 and θ_2 by considering the values of f and F and the conditions of equation (a) and F = 0 as mentioned above.

Let us find that part of the motion when the moving cylinder rolls without slipping. There is then only one degree of freedom, and we shall specify the position of the cylinder by the angle θ between the vertical and the line connecting the centers of the two cylinders. In order to compute the kinetic energy, we introduce the auxiliary angle ϕ through which the moving cylinder has rotated about its axis. The condition that the cylinder roll without slipping leads to the equation of constraint:

$$a\dot{\theta} = \alpha a(\dot{\phi} - \dot{\theta}), \tag{b}$$

which can be integrated in the form

$$a(1 + \alpha)\theta = a\alpha\phi. \tag{c}$$

If we were concerned only with the rolling motion, we could now proceed to set up the Lagrange equation for θ, but inasmuch as we need to know the forces of constraint F and f, it is necessary to introduce additional coordinates which are maintained constant by these constraining forces. The frictional force f_r maintains the constraint (c), and an appropriate coordinate is

$$\gamma = \theta - \frac{\alpha\phi}{1 + \alpha} . \tag{d}$$

So long as the cylinder rolls without slipping, $\gamma = \theta$; γ measures the angle of slip around the fixed cylinder. The normal force F maintains the distance r between the centers of the cylinders:

$$r = a + \alpha a = (1 + \alpha)a. \tag{e}$$

Now we can apply Lagrange multipliers to solve this part of the motion. The coordinates we will use are r, θ, and γ. As noted r and γ are involved only to allow us to find F and f_r. The equations we need are

$$\lambda_j \sum_i a_{ij}\, d_{qi} = 0$$

and

$$\frac{d}{dt}\left(\frac{\partial L}{\partial \dot{q}_i}\right) - \frac{\partial L}{\partial q_i} = \sum_j \lambda_j a_{ij} .$$

The differential equation of constraint has to be developed from the equations of constraint which was already stated. From equation (e),

$$dr = 0.$$

For γ we must combine (d) and (e). From (d) we have

$$(1+\alpha)\gamma = (1+\alpha)\theta - \alpha\phi$$

From (c) we have $a(1+\alpha)\theta - a\alpha\phi = 0$

so $\quad a(1+\alpha)\gamma = 0$

and $\quad a(1+\alpha)d\gamma = 0.$

We have then

$$\lambda_1 dr = 0 \quad , \quad a_{r1} = 1$$

and $\quad \lambda_2 a(1+\alpha)d\gamma = 0, \quad a_{\gamma 2} = a(1+\alpha)$

when we solve the equations $\lambda_1 = F$ and $\lambda_2 = f_r$. The Lagrangian is

$$L = T - V.$$

$$T = \frac{1}{2} m(\dot{r}^2 + r^2\dot{\theta}^2) + \frac{1}{2} I\dot{\phi}^2.$$

After substituting for ϕ from equation (d) and using $I = \frac{1}{2} m\alpha^2 a^2$ we get

$$T(r,\theta,\gamma) =$$

$$\frac{1}{2} m\dot{r}^2 + \frac{1}{2} mr^2\dot{\theta}^2 + \frac{1}{4} m(1+\alpha)^2 a^2 (\dot{\theta}^2 - 2\dot{\gamma}\dot{\theta} + \dot{\gamma}^2)$$

The potential energy function is

$$V = mgr \cos\theta.$$

The Lagrange equations for θ, r and γ are now

$$mr^2\ddot{\theta} + 2mr\dot{r}\dot{\theta} + \frac{1}{2} m(1+\alpha)^2 a^2\ddot{\theta} - \frac{1}{2} m(1+\alpha)^2 a^2\ddot{\gamma} - mgr \sin\theta = 0$$

$$m\ddot{r} - mr\dot{\theta}^2 + mg \cos\theta = \lambda_1 - \frac{1}{2} m(1+\alpha)^2 a^2\ddot{\theta} + \frac{1}{2} m(1+\alpha)^2 a^2\ddot{\gamma}$$

$$= a(1+\alpha)\lambda_2.$$

Inserting the constraints $\gamma = 0$, $r = (1+\alpha)a$, these equations become

$$\frac{3}{2}\, m(1+\alpha)^2 a^2 \ddot{\theta} = + mga(1+\alpha)\sin\theta \tag{f}$$

$$\lambda_1 = -m(1+\alpha)a\dot{\theta}^2 + mg\cos\theta \tag{g}$$

$$\lambda_2 = -\frac{1}{2}\, m(1+\alpha)a\ddot{\theta}. \tag{h}$$

Equation (f) can be solved by the energy method. The total energy, so long as the cylinder rolls without slipping, is $T + V$. Therefore,

$$\frac{3}{4}\,(1+\alpha)^2 ma\dot{\theta}^2 + (1+\alpha)mga\cos\theta = E, \tag{i}$$

is a constant. The friction force does no work here. Since the cylinder starts from rest at $\theta = 0$

$$E = (1+\alpha)mga.$$

Using this to solve for $\dot{\theta}$ in (i) we get

$$\frac{3}{4}\,(1+\alpha)^2 ma^2\dot{\theta}^2 + (1+\alpha)mga(\cos\theta - 1) = 0,$$

$$\dot{\theta}^2 = -\frac{4}{3}\,\frac{g}{(1+\alpha)a}\,(\cos\theta - 1) = 2\left[\frac{4}{3}\,\frac{g}{(1+\alpha)a}\right]\sin^2\frac{\theta}{2}$$

$$\dot{\theta} = 2\left[\frac{\beta g}{a}\right]^{1/2} \sin\frac{\theta}{2} \tag{j}$$

with $$\beta = \frac{2}{3}\,\frac{1}{(1+\alpha)}\,.$$

We can integrate (j) to solve for θ

$$\int_o^\theta \frac{\frac{1}{2}\,d\theta}{\sin\frac{\theta}{2}} = \left[\frac{\beta g}{a}\right]^{1/2} \int_o^t dt$$

$$\ln\left[\tan\frac{\theta}{2}\right]\Bigg|_o^\theta = \left[\frac{\beta g}{a}\right]^{1/2} t.$$

Now we see the difficulty we warned against earlier. Substituting the lower limit $\theta = 0$ in the L.H.S. leads to $\ln 0 = -\infty$. This corresponds to the infinite amount of time the cylinder might stay in its equilibrium position. In reality, though, some slight disturbance is bound to displace the upper cylinder some amount, say θ_o, so that we get

$$\tan\frac{\theta}{4} = \left(\tan\frac{\theta_o}{4}\right)\exp\left[\left(\frac{\beta g}{a}\right)^{1/2} t\right].$$

This motion applies until

$$f_r = \mu F \text{ (equation (a))}.$$

In our multipliers notation this is

$$\lambda_2 = \mu\lambda_1.$$

Using equations (g) and (h) we have

$$-\frac{1}{2} m(1+\alpha)a\ddot{\theta} = -\mu m(1+\alpha)a\dot{\theta}^2 + \mu mg \cos\theta.$$

Using equation (f) for $\ddot{\theta}$ and (j) for $\dot{\theta}$ we get:

$$+\frac{1}{2} m(1+\alpha)a \frac{mga(1+\alpha)\sin\theta}{\frac{3}{2} m(1+\alpha)^2 a^2} = -\mu m(1+\alpha)a\ 4 \frac{\beta g}{a} \sin^2 \frac{\theta}{2} + \mu mg \cos\theta$$

which reduces to:

$$\frac{2}{3}\frac{1}{2} mg \sin\theta = -\mu m(1+\alpha)4\beta g \sin^2 \frac{\theta}{2} + \mu mg \cos\theta$$

Substituting for β:

$$\frac{1}{3} mg \sin\theta = -\mu m(1+\alpha)4g \frac{2}{3} \frac{1}{(1+\alpha)} \sin^2 \frac{\theta}{2} + \mu mg \cos\theta$$

Making the trigonometric substitution for $\sin^2 \frac{\theta}{2}$:

$$\frac{1}{3} mg \sin\theta = -\mu m4g \frac{2}{3} \left(\frac{1-\cos\theta}{2}\right) + \mu mg \cos\theta$$

Cancelling:

$$\frac{1}{3} \sin\theta = -\mu \frac{4}{3} + \mu \frac{4}{3} \cos\theta + \mu\cos\theta$$

Combining:

$$\frac{1}{3} \sin\theta = -\frac{4}{3} \mu + \frac{7}{3} \mu\cos\theta$$

$$\sin\theta = -4\mu + 7 \mu\cos\theta$$

INDEX

Numbers on this page refer to PROBLEM NUMBERS, not page numbers

Numbers on this page refer to PROBLEM NUMBERS, not page numbers

Numbers on this page refer to PROBLEM NUMBERS, not page numbers

Numbers on this page refer to PROBLEM NUMBERS, not page numbers

Numbers on this page refer to PROBLEM NUMBERS, not page numbers

Numbers on this page refer to PROBLEM NUMBERS, not page numbers

Numbers on this page refer to <u>PROBLEM NUMBERS</u>, not page numbers

Numbers on this page refer to PROBLEM NUMBERS, not page numbers

A particularly useful reference for those in math, science, engineering and other technical fields. Includes the most-often used formulas, tables, transforms, functions, and graphs which are needed as tools in solving problems. The entire field of special functions is also covered. A large amount of scientific data which is often of interest to scientists and engineers has been included.

Available at your local bookstore or order directly from us by sending in coupon below.

RESEARCH & EDUCATION ASSOCIATION
61 Ethel Road W., Piscataway, New Jersey 08854
Phone: (732) 819-8880 **website: www.rea.com**

VISA MasterCard

Charge Card Number

☐ Payment enclosed
☐ Visa ☐ MasterCard

Expiration Date: ________ / ________
Mo Yr

Please ship the **"Math Handbook"** @ $34.95 plus $4.00 for shipping.

Name ____________________

Address ____________________

City ____________________ State ______ Zip ______

Available at your local bookstore or order directly from us by sending in coupon below.

RESEARCH & EDUCATION ASSOCIATION
61 Ethel Road W., Piscataway, New Jersey 08854
Phone: (732) 819-8880 **website: www.rea.com**

VISA MasterCard

Charge Card Number

☐ Payment enclosed
☐ Visa ☐ MasterCard

Expiration Date: ______ / ______
Mo Yr

Please ship **"The Handbook of Electrical Engineering"** @ $38.95 plus $4.00 for shipping.

Name ______________________________

Address ______________________________

City ______________________ State ______ Zip ______

The HANDBOOK of CHEMICAL ENGINEERING

Staff of Research and Education Association

Available at your local bookstore or order directly from us by sending in coupon below.

RESEARCH & EDUCATION ASSOCIATION
61 Ethel Road W., Piscataway, New Jersey 08854
Phone: (732) 819-8880 website: www.rea.com

VISA *MasterCard*

Charge Card Number
☐ Payment enclosed
☐ Visa ☐ MasterCard

[ ][ ][ ][ ][ ][ ][ ][ ][ ][ ][ ][ ][ ][ ][ ][ ]

Expiration Date: _____ / _____
Mo   Yr

Please ship the **"The Handbook of Chemical Engineeing"** @ $38.95 plus $4.00 for shipping.

Name _____

Address _____

City _____ State _____ Zip _____

Available at your local bookstore or order directly from us by sending in coupon below.

RESEARCH & EDUCATION ASSOCIATION
61 Ethel Road W., Piscataway, New Jersey 08854
Phone: (732) 819-8880 **website: www.rea.com**

VISA MasterCard

Charge Card Number

☐ Payment enclosed
☐ Visa ☐ MasterCard

Expiration Date: ________ / ________
Mo Yr

Please ship **"GRE General with CAT Software"** @ $34.95 plus $4.00 for shipping.

Name ______________________________

Address ______________________________

City ______________________ State ________ Zip ____________

Available at your local bookstore or order directly from us by sending in coupon below.

RESEARCH & EDUCATION ASSOCIATION
61 Ethel Road W., Piscataway, New Jersey 08854
Phone: (732) 819-8880 **website: www.rea.com**

VISA MasterCard

Charge Card Number

☐ Payment enclosed
☐ Visa ☐ MasterCard

Expiration Date: ______ / ______
Mo Yr

Please ship **"GRE Computer Science"** @ $22.95 plus $4.00 for shipping.

Name ____________________

Address ____________________

City ____________________ State ______ Zip ______

Available at your local bookstore or order directly from us by sending in coupon below.

RESEARCH & EDUCATION ASSOCIATION
61 Ethel Road W., Piscataway, New Jersey 08854
Phone: (732) 819-8880 **website: www.rea.com**

Charge Card Number

☐ Payment enclosed
☐ Visa ☐ MasterCard

Expiration Date: ______ / ______
Mo Yr

Please ship **"GRE Physics"** @ $27.95 plus $4.00 for shipping.

Name ________________________________

Address ______________________________

City ____________________ State ______ Zip ________

Available at your local bookstore or order directly from us by sending in coupon below.

RESEARCH & EDUCATION ASSOCIATION
61 Ethel Road W., Piscataway, New Jersey 08854
Phone: (732) 819-8880 **website: www.rea.com**

VISA MasterCard

Charge Card Number

☐ Payment enclosed
☐ Visa ☐ MasterCard

Expiration Date: ______ / ______
Mo Yr

Please ship REA's **"FE/EIT AM Exam"** @ $49.95 plus $4.00 for shipping.

Name ______________________

Address ______________________

City ______________ State ______ Zip ______

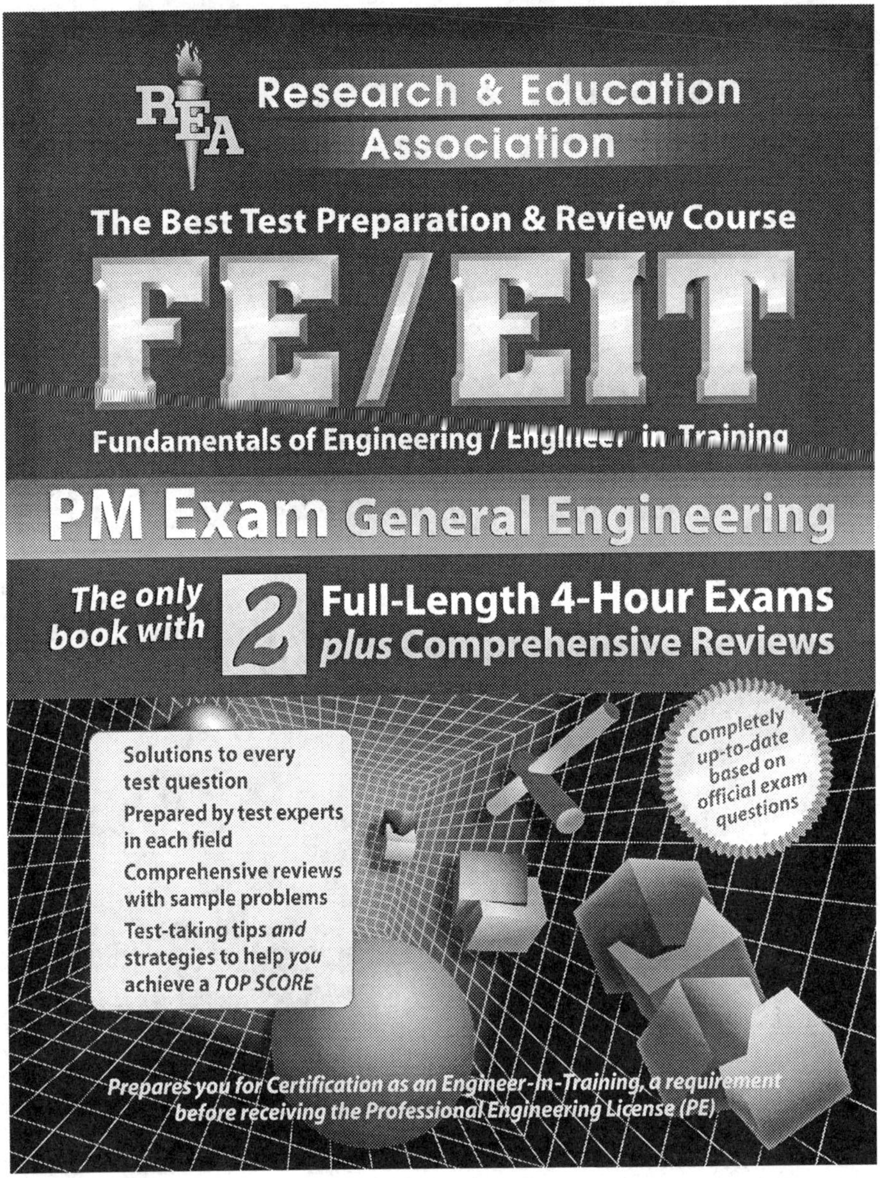

Available at your local bookstore or order directly from us by sending in coupon below.

RESEARCH & EDUCATION ASSOCIATION
61 Ethel Road W., Piscataway, New Jersey 08854
Phone: (732) 819-8880 **website: www.rea.com**

VISA MasterCard

Charge Card Number

☐ Payment enclosed
☐ Visa ☐ MasterCard

Expiration Date: ________ / ________
Mo Yr

Please ship REA's **"FE/EIT PM Exam"** @ $39.95 plus $4.00 for shipping.

Name ______________________________

Address ______________________________

City ____________________ State ______ Zip ______

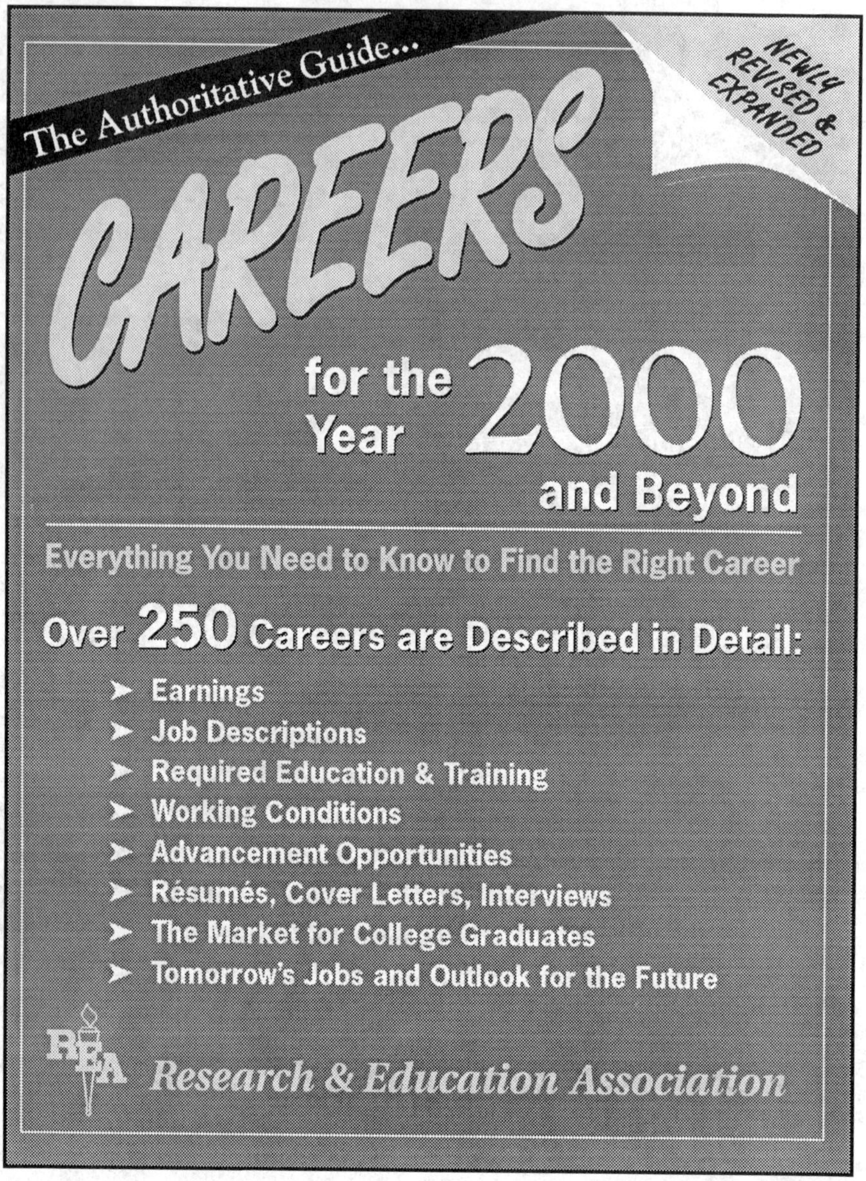

Available at your local bookstore or order directly from us by sending in coupon below.

RESEARCH & EDUCATION ASSOCIATION
61 Ethel Road W., Piscataway, New Jersey 08854
Phone: (732) 819-8880 **website: www.rea.com**

VISA MasterCard

☐ Payment enclosed
☐ Visa ☐ MasterCard

Charge Card Number ____________________

Expiration Date: ______ / ______
Mo Yr

Please ship **"Careers for the Year 2000 and Beyond"** @ $21.95 plus $4.00 for shipping.

Name ____________________

Address ____________________

City ____________ State ______ Zip ______

REA's Problem Solvers

The "PROBLEM SOLVERS" are comprehensive supplemental textbooks designed to save time in finding solutions to problems. Each "PROBLEM SOLVER" is the first of its kind ever produced in its field. It is the product of a massive effort to illustrate almost any imaginable problem in exceptional depth, detail, and clarity. Each problem is worked out in detail with a step-by-step solution, and the problems are arranged in order of complexity from elementary to advanced. Each book is fully indexed for locating problems rapidly.

ACCOUNTING
ADVANCED CALCULUS
ALGEBRA & TRIGONOMETRY
AUTOMATIC CONTROL SYSTEMS/ROBOTICS
BIOLOGY
BUSINESS, ACCOUNTING, & FINANCE
CALCULUS
CHEMISTRY
COMPLEX VARIABLES
DIFFERENTIAL EQUATIONS
ECONOMICS
ELECTRICAL MACHINES
ELECTRIC CIRCUITS
ELECTROMAGNETICS
ELECTRONIC COMMUNICATIONS
ELECTRONICS
FINITE & DISCRETE MATH
FLUID MECHANICS/DYNAMICS
GENETICS
GEOMETRY
HEAT TRANSFER
LINEAR ALGEBRA
MACHINE DESIGN
MATHEMATICS for ENGINEERS
MECHANICS
NUMERICAL ANALYSIS
OPERATIONS RESEARCH
OPTICS
ORGANIC CHEMISTRY
PHYSICAL CHEMISTRY
PHYSICS
PRE-CALCULUS
PROBABILITY
PSYCHOLOGY
STATISTICS
STRENGTH OF MATERIALS & MECHANICS OF SOLIDS
TECHNICAL DESIGN GRAPHICS
THERMODYNAMICS
TOPOLOGY
TRANSPORT PHENOMENA
VECTOR ANALYSIS

If you would like more information about any of these books, complete the coupon below and return it to us or visit your local bookstore.

RESEARCH & EDUCATION ASSOCIATION
61 Ethel Road W. • Piscataway, New Jersey 08854
Phone: (732) 819-8880 **website: www.rea.com**

Please send me more information about your Problem Solver books

Name ______________________________

Address ______________________________

City ____________________ State ______ Zip ____________

REA's Test Preps

The Best in Test Preparation

- REA "Test Preps" are **far more** comprehensive than any other test preparation series
- Each book contains up to **eight** full-length practice tests based on the most recent exams
- **Every** type of question likely to be given on the exams is included
- Answers are accompanied by **full** and **detailed** explanations

REA has published over 60 Test Preparation volumes in several series. They include:

Advanced Placement Exams (APs)
Biology
Calculus AB & Calculus BC
Chemistry
Computer Science
English Language & Composition
English Literature & Composition
European History
Government & Politics
Physics
Psychology
Spanish Language
Statistics
United States History

College-Level Examination Program (CLEP)
Analyzing and Interpreting Literature
College Algebra
Freshman College Composition
General Examinations
General Examinations Review
History of the United States I
Human Growth and Development
Introductory Sociology
Principles of Marketing
Spanish

SAT II: Subject Tests
American History
Biology E/M
Chemistry
English Language Proficiency Test
French
German

SAT II: Subject Tests (cont'd)
Literature
Mathematics Level IC, IIC
Physics
Spanish
Writing

Graduate Record Exams (GREs)
Biology
Chemistry
Computer Science
Economics
Engineering
General
History
Literature in English
Mathematics
Physics
Psychology
Sociology

ACT - ACT Assessment

ASVAB - Armed Services Vocational Aptitude Battery

CBEST - California Basic Educational Skills Test

CDL - Commercial Driver License Exam

CLAST - College-Level Academic Skills Test

ELM - Entry Level Mathematics

ExCET - Exam for the Certification of Educators in Texas

FE (EIT) - Fundamentals of Engineering Exam

FE Review - Fundamentals of Engineering Review

GED - High School Equivalency Diploma Exam (U.S. & Canadian editions)

GMAT - Graduate Management Admission Test

LSAT - Law School Admission Test

MAT - Miller Analogies Test

MCAT - Medical College Admission Test

MSAT - Multiple Subjects Assessment for Teachers

NJ HSPT- New Jersey High School Proficiency Test

PPST - Pre-Professional Skills Tests

PSAT - Preliminary Scholastic Assessment Test

SAT I - Reasoning Test

SAT I - Quick Study & Review

TASP - Texas Academic Skills Program

TOEFL - Test of English as a Foreign Language

TOEIC - Test of English for International Communication

RESEARCH & EDUCATION ASSOCIATION
61 Ethel Road W. • Piscataway, New Jersey 08854
Phone: (732) 819-8880 **website: www.rea.com**

Please send me more information about your Test Prep books

Name ______________________________

Address ______________________________

City ____________________ State ______ Zip __________